TABLE OF ATOMIC WEIGHTS AND NUMBERS

Based on the 1981 Report of the Commission on Atomic Weights of the International Union of Pure and Applied Chemistry. Scaled to the relative atomic mass of carbon-12, and precise to ±1 unit in the last place, except where noted by footnote p. A value in parentheses is the atomic mass number of the isotope of longest known half-life.

Element	Symbol	Atomic Number	Atomic Weight		Element	Symbol	Atomic Number	Atomic Weight	
Actinium	Ac	89	227.0278	(L)	Molybdenum				(g)
Aluminum	Al	13	26.98154		Neodymium				(p, g)
Americium	Am	95	(243)		Neon				(g, m)
Antimony	Sb	51	121.75	(p)	Neptunium	Np	93	237.0482	(L)
Argon	Ar	18	39.948	(g, r)	Nickel	Ni	28	58.69	
Arsenic	As	33	74.9216		Niobium	Nb	41	92.9064	
Astatine	At	85	(210)		Nitrogen	N	7	14.0067	
Barium	Ba	56	137.33	(g)	Nobelium	No	102	(259)	
Berkelium	Bk	97	(247)		Osmium	Os	76	190.2	(g)
Beryllium	Be	4	9.01218		Oxygen	O	8	15.9994	(p, g, r)
Bismuth	Bi	83	208.9804		Palladium	Pd	46	106.42	(g)
Boron	B	5	10.81	(m, r)	Phosphorus	P	15	30.97376	
Bromine	Br	35	79.904		Platinum	Pt	78	195.08	(p)
Cadmium	Cd	48	112.41	(g)	Plutonium	Pu	94	(244)	
Calcium	Ca	20	40.08	(g)	Polonium	Po	84	(209)	
Californium	Cf	98	(251)		Potassium	K	19	39.0983	
Carbon	C	6	12.011	(r)	Praseodymium	Pr	59	140.9077	
Cerium	Ce	58	140.12	(g)	Promethium	Pm	61	(145)	
Cesium	Cs	55	132.9054		Protactinium	Pa	91	231.0359	(L)
Chlorine	Cl	17	35.453		Radium	Ra	88	226.0254	(g, L)
Chromium	Cr	24	51.996		Radon	Rn	86	(222)	
Cobalt	Co	27	58.9332		Rhenium	Re	75	186.207	
Copper	Cu	29	63.546	(p, r)	Rhodium	Rh	45	102.9055	
Curium	Cm	96	(247)		Rubidium	Rb	37	85.4678	(p, g)
Dysprosium	Dy	66	162.50	(p)	Ruthenium	Ru	44	101.07	(p, g)
Einsteinium	Es	99	(252)		Samarium	Sm	62	150.36	(p, g)
Erbium	Er	68	167.26	(p)	Scandium	Sc	21	44.9559	
Europium	Eu	63	151.96	(g)	Selenium	Se	34	78.96	(p)
Fermium	Fm	100	(257)		Silicon	Si	14	28.0855	(p)
Fluorine	F	9	18.998403		Silver	Ag	47	107.8682	(p, g)
Francium	Fr	87	(223)		Sodium	Na	11	22.98977	
Gadolinium	Gd	64	157.25	(p, g)	Strontium	Sr	38	87.62	(g)
Gallium	Ga	31	69.72		Sulfur	S	16	32.06	(r)
Germanium	Ge	32	72.59	(p)	Tantalum	Ta	73	180.9479	
Gold	Au	79	196.9665		Technetium	Tc	43	(98)	
Hafnium	Hf	72	178.49	(p)	Tellurium	Te	52	127.60	(p, g)
Helium	He	2	4.00260	(g)	Terbium	Tb	65	158.9254	
Holmium	Ho	67	164.9304		Thallium	Tl	81	204.383	
Hydrogen	H	1	1.00794	(p, g, m, r)	Thorium	Th	90	232.0381	(g, L)
Indium	In	49	114.82	(g)	Thulium	Tm	69	168.9342	
Iodine	I	53	126.9045		Tin	Sn	50	118.69	(p)
Iridium	Ir	77	192.22	(p)	Titanium	Ti	22	47.88	(p)
Iron	Fe	26	55.847	(p)	Tungsten	W	74	183.85	(p)
Krypton	Kr	36	83.80	(g, m)	(Unnilhexium)	(Unh)	106	(263)	(n)
Lanthanum	La	57	138.9055	(p, g)	(Unnilpentium)	(Unp)	105	(262)	(n)
Lawrencium	Lr	103	(260)		(Unnilquadium)	(Unq)	104	(261)	(n)
Lead	Pb	82	207.2	(g, r)	Uranium	U	92	238.0289	(g, m)
Lithium	Li	3	6.941	(p, g, m, r)	Vanadium	V	23	50.9415	
Lutetium	Lu	71	174.967		Xenon	Xe	54	131.29	(p, g, m)
Magnesium	Mg	12	24.305	(g)	Ytterbium	Yb	70	173.04	(p)
Manganese	Mn	25	54.9380		Yttrium	Y	39	88.9059	
Mendelevium	Md	101	(258)		Zinc	Zn	30	65.38	
Mercury	Hg	80	200.59	(p)	Zirconium	Zr	40	91.22	(g)

(g) Geologically unusual samples of this element are known that have different isotopic compositions. For such samples, the atomic weight given here may not apply as precisely as indicated.

(L) This atomic weight is for the relative mass of the isotope of longest half-life.

(m) Modified isotopic compositions can occur in commercially available materials that have been processed in undisclosed ways, and the atomic weight given here might be quite different for such samples.

(n) Name is temporary until scientists agree to a more traditional name.

(p) Precise to ±3 units in the last place, except for hydrogen for which the atomic weight is precise to ±7 units in the last place.

(r) Ranges in isotopic compositions of normal samples obtained on earth do not permit a more precise atomic weight for this element.

Fundamentals of General, Organic, and Biological Chemistry

About the Author

JOHN HOLUM is on the faculty of Augsburg College, Minneapolis, MN. He did his undergraduate work at St. Olaf College and earned the Ph.D. (organic chemistry) at the University of Minnesota. Additional studies were taken as sabbatical leaves at California Institute of Technology and Harvard University. In 1974 he was given the Distinguished Teaching Award of the Minnesota Section of the American Chemical Society. He is a member of Phi Beta Kappa, Phi Lambda Upsilon, Sigma Xi, and Sigma Pi Sigma. The National Science Foundation has awarded him several research grants and a Science Faculty Fellowship. He is the author or co-author of several texts in chemistry, all published by John Wiley and Sons, and a reference work, *Topics and Terms in Environmental Problems* (Wiley-Interscience). He has also authored papers for the *Journal of the American Chemical Society,* the *Journal of Organic Chemistry,* and the *Journal of Chemical Education.* He has been active on the Examinations Committee and the Committee on Chemistry for Professional Health Care Students of the Division of Chemical Education of the ACS, and has spoken often at Divisional and Regional meetings, as well as at conferences of the Two-Year College Chemistry Association. His textbooks in chemistry for professional health care students have been widely used in America and abroad for 25 years.

JOHN R. HOLUM
Augsburg College

Fundamentals of General, Organic, and Biological Chemistry

Third Edition

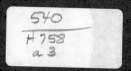

John Wiley & Sons

New York   Chichester   Brisbane   Toronto   Singapore

Cover art by Karl Gerstner
"Color Form Yellow"
Diversion
1970–1975, 1977
Nitrocellulose on phenolic resin plates
800 × 800 mm unframed

Production Supervised by Linda R. Indig
Photo researched by John Schultz and Elyse Rieder
Photos edited by Stella Kupferberg
Designed by Kevin Murphy
Illustrations by John Balbalis with the
assistance of the Wiley Illustration Department.
Manuscript edited by Pam Landau under
the supervision of Bruce Safford.

Library of Congress Cataloging in Publication Data:

Holum, John R.
    Fundamentals of general, organic and biological
chemistry.

    Includes index.
    1. Chemistry. I. Title.
QD31.2.H62   1986   540   85-16912
ISBN 0-471-81517-9
Printed in the United States of America
10 9 8 7 6 5 4 3 2 1

Preface

This text is intended primarily for students who need to know about the processes of life at the molecular level and whose programs make them limit the scope of this study to two college terms. These students include those who are preparing for careers in the professional health sciences such as nurses and nurse practitioners, medical technologists, histotechnologists, dietitians, nutritionists, inhalation therapists, home economists, physical therapists, teachers of health and physical education, other teachers, and students in the humanities who desire a broader survey of chemistry than is provided by the traditional freshman chemistry course. The text does not assume that chemistry was studied recently at the high school level.

In our use of earlier editions, we found that we can fit the first 10 to 11 chapters into one semester. Thus the first semester provides the introduction to chemical principles and types of inorganic substances — particularly acids, bases, salts, buffers, and their associated equilibria — that must be provided before the study of organic functional groups and biochemistry should begin. We have also found that the remainder of the book can be studied in the second semester where many chapters are shorter than usual, and where not all topics are treated each year in as much depth as the text provides.

For many years the **theme** of our course and the theme of this book and its earlier editions has been the molecular basis of life. We believe that no other text so consciously remains faithful to a consistent theme, and we know from years of experience that students respond with gratitude to the careful effort to make the chemistry they study as relevant as possible to their professional needs. Our students are overwhelmingly career-oriented, and we capitalize on this rather than discourage it. For example, we know that it motivates students to high efforts, as they patiently work their way through atomic and molecular structure — topics that initially seem to be so remote from their careers — to be gently reminded that life does have a molecular basis and we can't talk about it until we know what it means to be a molecule. And we give many hints early in the course of the relevance of this or that topic to the study of life at the molecular level. When acids, bases, and buffers are studied not as things in themselves but as substances whose interactions profoundly affect health, even those students who have most resisted chemistry begin to see and to accept its importance to their professions.

We are seeing an increasing number of professional nurses with an RN certificate but without a B.S. degree who are stepping back into the degree program. To do this at Augsburg College, they must have completed a one-year course in chemistry that includes roughly one term devoted to organic chemistry and biochemistry. With these students, without exception, there is no motivational problem about learning chemistry. As practicing nurses, they have seen how much they have to know about the molecular basis of life and disease. Moreover, for their own professional satisfaction, they *want* to know more chemistry, and some are considerably surprised by this development in their lives.

One major change between this edition and the previous edition occurs in the first 10 chapters. All stoichiometry, including problems associated with molar concentrations, are now consolidated into one chapter, new Chapter 5 ("Quantitative Relationships in Chemical Reactions"). Another important change has been to incorporate old Chapter 9 ("Important Metals and Nonmetals") into other chapters.

When the study of organic chemistry starts, frequent mention is made of the kinds of biological chemicals that happen to have the particular functional group currently being studied, because students appreciate the reminders that the theme of the course continues throughout all of the study of organic chemistry. However, we believe that the most sound pedagogical approach is to introduce these groups, one after the other, as they occur among the *simplest,* monofunctional compounds. Large, complex structures such as glucose or hemoglobin or DNA can be sources of terror rather than wonder when they are introduced too early.

Only the barest minimum organic chemistry is included, because the available time has to be very carefully allotted. Thus alkyl halides are barely mentioned because this system occurs nowhere among the biochemicals to be studied later. Very little is done with aromatic chemistry, because the details of aromatic electrophilic substitution reactions will not be exploited later. (What it means to be *aromatic,* and some of the characteristic reactions of the benzene ring, are studied, because this ring does occur among some amino acids and proteins.) In the first edition we tried to teach the aldol and the Claisen condensations because this chemistry can be applied when we study certain metabolic pathways. We no longer do so, however, because it simply took too much time for the benefit, so these reactions are not presented in this edition.

The emphasis in the chapters on organic chemistry is on the chemical properties of functional groups rather than on strategies for making organic compounds. We emphasize three types of reactants — water, oxidizing agents, and reducing agents — types of reactants that abound in cells. We do look briefly at certain mechanisms of reactions so that students can see that they are rational, that the reactions don't just happen by the operation of "lassos." However, the elaborate paraphernalia of mechanistic terminology that students who go on in chemistry must learn is here kept to the barest of levels. Thus the chapters on organic chemistry do not constitute a survey of organic chemistry such as one might obtain even in just a one-term course. These chapters are wholly devoted to preparing the students using this text for the biochemistry and molecular biology that follows — no more, no less.

Chapters 19 to 29 focus everything that has gone before onto a study of the principal organic substances found in cells — carbohydrates, lipids, proteins, and nucleic acids — and their chief functions as chemicals when they are in cells. Among these chapters are found most of the changes from the preceding edition. Significant portions of what was Chapter 9 in the second edition ("Important Metals and Nonmetals") are now in Chapter 25 ("Extracellular Fluids of the Body") where they work better, pedagogically.

Some changes in the biochemistry chapters make the level of the treatment more realistic for the freshman level. They reflect our own experiences in the classroom. Thus in Chapter 26 ("Biochemical Energetics") we have found that the treatment flows much better when we do not try to relate phosphate group transfer potential to the concept of free energy. Also in Chapter 26, we have reduced the level of the discussion of the chemiosmotic theory by omitting a detailed discussion of the electron flow that follows the involvement of the iron-sulfur protein.

Recent developments in the area of lipoprotein complexes and the transport of cholesterol in the blood prompted a new opening section to Chapter 28 ("Metabolism of Lipids"). Although this material is new, it is not difficult, and future health professionals ought to have a background in this area.

There are innumerable small changes in the text. The preparation of this edition involved a total rewriting of the entire book, sentence by sentence, in an effort to improve its clarity and readability for students. Of course, not every sentence was changed, but every sentence was scrutinized to see if a worthwhile change were warranted.

The design of this edition is similar to that of the previous edition. There are **margin comments,** but there is now more assurance that they are truly marginal and not integrally important to the paragraphs by which they are placed. Some are reminders. Some restate a point. Some are small, illustrative tables to which the neighboring paragraph refers. Some are structures that need not be memorized.

There are **Special Topics** on matters of current interest, and a list is provided following the Contents. One new Special Topic is NMR imaging as a diagnostic tool. Another goes into the probable connection between the regulation of fatty acid catabolism in brown adipose tissue and our ability to eat without gaining excessive weight.

Our overall goal has been to make this text the most modern, up-to-date, and readable college freshman level text in general, organic, and biological chemistry that is available anywhere. We are well aware that perfection is an ever-receding horizon, but we owe our students our finest efforts. We have gone to much greater pains than ever before to produce a zero mechanical defect product. Extra professional help in proofreading and in checking the answers to Review Exercises and Review Problems has been engaged. As always, the author stands responsible for any errors that remain.

A comprehensive package of instructional materials, described in detail on page xi, is available to help the students. The accompanying laboratory manual has been revised and updated with special attention given to improving the clarity of the experimental directions.

Two packages of computer-aided instructional disks have been prepared to accompany this text. Four-color, overhead transparencies can be obtained. (These are the actual transparencies, not masters that would be used to make them.) A Teachers' Manual includes the answers to all of the Practice Exercises and Review Exercises.

OTHER DESIGN FEATURES THAT AID STUDENTS

Chemistry is one of the disciplines in which important scientific terms can be sharply defined. We have tried to do so at the first occasion of using the term or as soon thereafter as possible, at or near the place where the **key term** is highlighted by a boldface color treatment. Then our aim has been to use these terms as carefully and consistently as possible. As an aid in reviewing, these terms are listed at the ends of the chapters. Then, at the end of the book, there is a **Glossary** where each of the key terms is defined.

Each chapter has a **Summary** that uses the key terms in a narrative survey. Each main section of each chapter also begins with a **summary statement** that announces what is coming and that serves during test review periods to highlight the major topics.

A new design feature is the use of labels to identify sets of **Review Exercises** that are about a common topic. (These labels replace the Index of Exercises, Questions, and Problems of the previous edition.) Within most chapters are several **Practice Exercises,** and most of these immediately follow a **worked example** that provides a step-by-step description of how to solve a certain kind of problem. The **factor-label** method is used for nearly all computations. The **answers** to all Practice Exercises are found at the back of the book together with the answers to selected Review Exercises. The *Study Guide* that accompanies this book has answers to other Review Exercises.

The **Appendix on Mathematical Concepts** has been enlarged to provide more review of what exponentials are and how to manipulate them. New to this Appendix are discussions of how to use pocket calculators to handle exponentials and to carry out chain operations.

Continuing a long tradition, we have tried to make the **Index** the most thorough, most cross-referenced index in any text of this type.

JOHN R. HOLUM
Augsburg College

Acknowledgments

Over the many years of writing instructional materials, my family — Mary, my wife, and our daughters, Liz, Ann, and Kathryn — have been Gibralters of support. They, rather than teaching or writing, are my career and so such teaching and writing is seen by us as one of the ways by which our family has tried to be of help to others. I am pleased to say "thank you" to them for being such nice people.

Here, at Augsburg College, I have enjoyed many years of support from Dr. Earl Alton, Chemistry Department Chair, and Dr. Charles S. Anderson, President. My freedom to write stems in no small measure from the freedom that these caring people have accorded me.

Nice people abound at John Wiley & Sons, too. They do good work. I think of the special support of my Chemistry Editor, Dennis Sawicki, and of that of the Executive Editor, Clifford Mills. Chief illustrator John Balbalis has been skillful, artistic, and faithful in handling artwork for many years. Picture editor Stella Kupferberg solves problems and makes this facet of production worry-free. The designer, Kevin Murphy, stands in the long Wiley tradition of artistry and imagination. Senior copy editor Bruce Safford smoothly managed the overall conversion of the manuscript into a final book. The copy editor Pam Landau, has been superb in smoothing out stylistic and grammatical problems. And when it comes to error-free competence and an utterly nonabrasive but yet unyielding manager of deadlines, no one surpasses Linda Indig as a supervisor of production. All in all, it's an impressive team, and I count myself to be fortunate indeed for having become associated with John Wiley & Sons in the first place.

Part of the process of preparing a manuscript involves the professional critiques of teachers. I am pleased to acknowledge and to thank the following additional people for their work: Hugh Akers, Associate Professor, Lamar University; Charles E. Bell, Jr., Professor of Chemical Sciences, Old Dominion University; Lois Dalla-Riva, Professor of Chemistry, Golden West College; Dr. Estelle Gearon, Professor of Chemistry, Montgomery College; Dr. Arlin Gyberg, Augsburg College; Dr. Robert G. Martinek, Manager, Chicago Laboratory, Illinois Department of Health; Sandra Olmstead, Professor, Augsburg College; Dr. James R. Paulson, Chemistry Department, University of Wisconsin-Oshkosh; Salvatore Profeta, Jr., Adjunct Professor of Chemistry, Louisiana State University; and Dr. Neal Thorpe, Augsburg College.

J.R.H.

Supplementary Materials for Students and Teachers

The complete package of supplements available to help students study and teachers plan the course and operate the associated laboratory work includes the following:

Laboratory Manual for Fundamentals of General, Organic, and Biological Chemistry, 3rd edition. This has been prepared by Professor Sandra Olmsted. An instructor's manual is a section in the general Teachers' Manual described below.

Study Guide for Fundamentals of General, Organic, and Biological Chemistry, 3rd edition. This softcover book contains chapter objectives, chapter glossaries, additional worked examples and exercises, sample examinations for each chapter, and the answers to those Review Exercises for which answers are not in the text.

Teachers' Manual for Fundamentals of General, Organic, and Biological Chemistry, 3rd edition. This softcover supplement is available to teachers, and it contains all of the usual services for *both the text and the laboratory manual.* Those who adopt this book may also request from John Wiley & Sons a set of questions that can be used to prepare examinations. (The address is given in the next paragraph.)

Transparencies. Instructors who adopt this book can receive from John Wiley & Sons, without charge, a set of slides (actual slides, not slide masters), many in color, of several figures and tables in this book. Write to: Chemistry Editor, John Wiley & Sons, Inc., 605 Third Ave., New York, NY 10158.

Computer-Aided Instructional Packages. Two sets of instructional software are available as supplements, one for the use of students when they are reviewing and the other for their use as pre-lab study material. Professor Richard Cornelius of Wichita State University wrote the lecture-oriented software, and Sandra and Richard Olmsted (Augsburg College) prepared the lab-oriented materials. The latter package has several units that work exceptionally well in classrooms equipped with electronic blackboards. The animated sequences that explain difficult concepts such as mutarotation, DNA-directed polypeptide synthesis, and osmosis are outstanding teaching devices. For further information write to: Chemistry Editor, John Wiley & Sons, Inc., 605 Third Ave., New York, NY 10158 or to Prof. Sandra Olmsted, Augsburg College, Minneapolis, MN 55454.

J.R.H.

Contents

Index to Special Topics

Chapter 1
Goals, Methods, and Measurements

A goal, an ability to concentrate, a willingness to build bit by bit, and the support of many people all go into education in any field. A sound understanding of the molecular basis of life, the overarching topic of this book, is a building block in many careers.

1.1   CONCERNING THE MOLECULAR BASIS OF LIFE

The theme of this book is the molecular basis of life.

A molecule is an extremely tiny particle, and the molecular level of life is the level at which molecules interact.

Centuries before anyone believed in atoms or molecules, people could not help but notice that many different kinds of animals drank at the same water holes, breathed the same air, ate the same kinds of food, and delighted in the same salt licks. Ancient farmers knew that the droppings of animals could nourish plants, and that animals could eat plants and prosper. Some animals could eat weaker animals and grow. Evidently, at some deep level of existence, living things could exchange parts. The curious who looked at nature carefully found an astonishing unity to it, and one of our goals is to discover this unity, too. In this course we will go deep into the cell to its molecular level to learn how events there make great differences in retaining health, preventing illness and pain, and curing injury and disease. The selection and the organization of every topic in this book have been made with this general theme in mind — the molecular basis of life.

Our Strategy.   Life at the molecular level involves molecules and chemical changes or reactions that are often complicated. The symbols we use for them, however, are much less complex than many things you have mastered — a highway map, for example. You learned how to read and understand dozens of maps after you learned the meanings of barely a quarter page of map symbols. The most complicated molecules of living systems similarly carry "map signs," so before we tackle them in Chapters 19 through 29, we will learn the molecular map signs common among simpler substances, organic compounds (Chapters 11–18), concentrating on what is essential to our goal.

To understand the molecules and the reactions of life, we need a general background in the properties of water, of acids and bases, and other things. This study occurs mainly in the first third of the book. Because life has a molecular basis, we obviously must deal with molecules and the atoms they are made of. Before starting with these (in Chapters 2–10), we will need to learn something about the nature of scientific facts in general and the ways they are measured and recorded.

1.2   SCIENTIFIC FACTS, PHYSICAL QUANTITIES, AND MEASUREMENTS

A scientific fact usually involves a physical quantity that is related to an official standard of measurement.

In science, the most reliable facts are those that can be observed in repeated observations or measurements.

Some facts are reproducible and some facts aren't. A reproducible fact is one that can be observed over and over again by independent observers. It might be an event — a chemical reaction, for example, that can be carried out repeatedly — or it might be some property of an object or substance that anyone can check over and over.

Scientists regard reproducible facts as the best of all kinds for developing scientific hypotheses and theories, but they also know that nonreproducible facts are often essential. The past medical history of a patient, for example, can be important information in making a diagnosis today, but the patient can't be put through this past again just to check some details. This makes the many *measured* and confirmed facts of the medical history much more useful than vague statements about feelings. Thus scientists, including physicians, nurses, and others who try to figure out what's wrong with an ailing person use facts, and they know there are different kinds having different reliabilities. They also know that the chief value of a fact is usually its use in building a hypothesis.

Hypotheses and Theories in Science.   A **hypothesis** is a conjecture, subject to being disproved, that explains a set of facts in terms of a common cause and that serves as the basis for designing additional tests or experiments. Every preliminary diagnosis of some ailment is a hypothesis based on initially observed or reported facts. A preliminary diagnosis always suggests what new information should be sought — new lab tests, or X rays, or possibly obtaining an in-depth medical history. Often reported facts and lab data will be double-checked. The results of these could make the initial hypothesis untrue. If so, the new facts, *combined with the old facts,* would be used to construct a new hypothesis — a revised diagnosis of what ails the patient.

The aim in testing a hypothesis is not to prove the hypothesis but to discover the truth about it.

Making, testing, discarding, and rethinking hypotheses are everyday occurrences to scientists in all fields, including medicine. Sometimes, one or more scientists will begin to hypothesize on a grander scale than suggested by our example of a medical diagnosis. These broad speculations in the past led to theories about diseases that have stood the test of time so well that they now are accepted as facts themselves — for example, the germ theory of infectious diseases, or the vitamin theory of nutrition-related diseases. You can see from these examples that a theory differs in scope from a hypothesis. A **theory** is an explanation for a large number of facts, observations, and hypotheses in terms of one or a few fundamental assumptions of what the world or a small part of the world is like. Later in our study we will encounter Dalton's atomic theory and its revisions, the kinetic theory of gases, and other theories.

Scientific Method. The progression from isolated facts to hypotheses to theories occurs often in the work of scientists, and anyone who uses the same approach to a problem is applying the **scientific method.** This method is as much an attitude of mind as it is a procedure. Scientists, like all people, have feelings, but when they are doing scientific work, they insist on letting facts control their reasoning. They try very hard not to let their reasoning either ignore facts or invent those that don't exist. If they fail at this, they sooner or later will be found out and, quite possibly, be professionally ruined. (If they're not found out, they're probably not working on anything that anyone else cares about, and eventually they will run out of research support!)

One realization made long ago which contributed much to scientific progress was that in studying nature, "How?" questions are much more useful in building theories than "Why?" questions. For example, if we ask, "Why do we get sick?" instead of "How do we get sick?," we will be bogged down in speculations that humans have never resolved to the satisfaction of everyone. The question, "Why do we get sick?" has been answered in the past by such responses as "Because of evil spirits," or "Because of sin," or "Because an enemy put a hex on me." The question "*How* do we get sick?" allows all speculation on the "Why?" to continue, but it gets at *mechanism.* Knowing the physical or chemical *mechanisms* of the ways various illnesses develop doesn't answer all questions, but this knowledge has certainly helped to reduce pain and suffering in the world. Science doesn't claim to address all questions, but it has been extremely successful with those that ask "How?" Scientists use the language of "Why?" After all, both "Why?" and "How?" are ways of asking "What causes . . . ?" We will sometimes encounter examples of such usage in this book. But almost always "Why?" means "How?" in science.

Properties and Physical Quantities. A **property** is any characteristic of something that we can use to identify and recognize it when we see it again. For example, some properties of liquid water are that it is colorless, odorless, and tasteless; that it dissolves sugar and table salt but not butter; that it makes a thermometer read 100 °C (212 °F) when it boils (at sea level); and that if it is mixed with gasoline it will sink, not float. If you were handed a glass containing a liquid having these properties your initial hypothesis undoubtedly would be that it is water. (Before you drank much of it, however, no doubt you would do some more tests, especially tests to see if it contained traces of some poison or harmful bacteria.)

Notice how much a description of water's properties depends on human senses — our abilities to see, taste, feel, and to sense hotness or coldness. Our senses, however, are quite limited, so inventors have developed many devices and instruments for extending the senses and thus have made possible finer and sharper observations. These devices are equipped with scales or readout panels, and from them we observe and record physical quantities.

A **physical quantity** is a property to which we can assign both a numerical value and a unit. One of your own properties to which we can do this is your height. The numerical value and its unit together tell at a glance how much greater your height is than some agreed-upon reference of height, such as the foot, or the inch, or the meter. *The unit in a physical quantity is just as important as the number.* If you said that your height is "two," people would ask, "Two what?" If you said "two yards," they would know what you meant, provided they knew what a yard is. (But they might ask, "*Exactly* two yards?") This example shows that we can't describe a property by a physical quantity without using both a number and a unit.

2 = a number
2 yards = a physical quantity

Physical quantity = number × unit

A physical quantity is obtained by a **measurement,** an operation by which we compare an unknown physical quantity with one that is known. Maybe as you were growing up, someone in your family measured your height by comparing it with how many sticks (they might have been one-foot rulers) it took to equal your height. Usually the number of sticks did not match your height exactly, so fractions of sticks called inches (each with their own fractions) were also used.

The *properties* of something, as we have said, are characteristics that can be used to identify it, and the properties of interest to scientists are those that can be described by one or more physical quantities. However, the measurements of some properties change the object or substance into something else. We can measure, for example, how much gasoline it takes to drive a car 100 miles, but this measurement uses up the gasoline, which changes as it burns into carbon dioxide (the fizz in soda pop) and water. Properties such as height or weight that can be measured without changing the object into something different are called **physical properties,** and those that necessarily involve a change into something new are called **chemical properties.** Charcoal, for example, is black, and this color is one of the physical properties of charcoal. We can take note of it without changing charcoal into something that isn't charcoal. Charcoal's ability to burn in air to give carbon dioxide and heat is one of its chemical properties. This can't be observed without letting a sample of charcoal change into something else. We will only be concerned with physical properties in this chapter together with the physical quantities that we use to describe them.

Large inertia goes with large mass.

Base Units of Measurement.   The most common measurements in chemistry are those of mass, volume, temperature, time, and quantity of chemical substance.

Mass is the measure of the inertia of an object. Anything said to have a lot of inertia such as a train engine, a massive boulder, or an ocean liner is very hard to get into motion, or if it is in motion, it is difficult to slow it down or make it change course. It is this inherent resistance to any kind of change in motion that we call **inertia,** and *mass* is our way of describing inertia quantitatively. A large inertia means a large *mass*. (A large mass doesn't always mean a large *weight*. For the difference between mass and weight, see Special Topic 1.1.)

Quantitative describes something expressible by a number and a unit.

Volume of cube = $(l)^3$.

The **volume** of an object is the space it occupies, and space is described by means of a more basic physical quantity, length. The volume of a cube, for example, is the product of (length) × (length) × (length), or (length)^3. **Length** is a physical quantity that describes how far an object extends in some direction, or it is the distance between two points.

A fundamental quantity such as mass or length is called a **base quantity** and any other quantity such as volume that can be described in terms of a base quantity is called a **derived quantity.** Another base quantity in science is **time** — our measure of how long events last. We need this quantity to describe how rapidly the heart beats, for example, or how fast some chemical reaction occurs. Still another important physical quantity is **temperature degree,** which we use to describe the hotness or coldness of an object.

The other two SI base quantities are electric current and luminous intensity. Their base units are called the ampere and the candela, respectively.

All of these base quantities are necessary to all sciences, but chemistry has a special base quantity called the *mole* that describes a certain amount of a chemical substance. It consists of a particular (and very large) *number* of tiny particles without regard to their masses or volumes. We will not study this base quantity further until we know more about these particles.

Mass and *weight* are two terms that are often confused and used as if they were identical. They are, however, quite different concepts. Your own *mass* doesn't depend on where you are, whether here on earth or on the moon, but your *weight* does. You would weigh on the earth about six times what you would on the moon, because **weight** is the measure of how strongly the gravitational force pulls on you. This force is six times stronger on the earth than on the moon.

The confusion between these two concepts arises because of the way we measure an unknown mass. We usually do this by comparing how hard the gravitational force acts on this unknown mass with how hard it acts on reference masses *at the same location.* When we use an equal-arm balance—pictured on the left, below—we place the object whose mass we want to know on one pan and then place objects (weights) whose masses we know on the other pan until the pans balance each other. When they do, the gravitational force acting on the unknown mass and the total known mass is the same. The operation is properly called *weighing,* but the result is the *mass* of the object in question.

Although few laboratories today use equal-arm balances, the balances that are used operate on essentially the same principle. These include single-pan balances, such as the one pictured below on the right, or the familiar triple-beam balances.

When we use a spring-loaded scale such as pictured below in the middle or when we use a typical bathroom scale (the word *balance* wouldn't be proper) the reading for a given mass taken here on earth would be six times what it would be on the moon. A spring-loaded scale has a pointer that is attached to the spring and moves as the spring stretches in response to some load. What makes the load stretch the spring is the gravitational force acting on the load. The scale of numbers along which the pointer moves was made here on earth by hanging known masses on the spring and marking appropriate numbers on the scale. The mark on the scale denoting say, 6 kg, was put on the scale because a mass known to be 6 kg streches the spring as far as this mark. If such a scale that has been given its marks here on earth were carried to the moon and used there, a 6-kg mass would pull the pointer only to the 1-kg mark. We'd have to renumber the scale for use anywhere that the gravitational force differs from that on earth.

Although we can use the word *weighing* for the procedure used in a laboratory for measuring a mass, we are actually determining a mass, not a weight. Whenever possible in this book, we will preserve this distinction, because it contributes to clearer thinking.

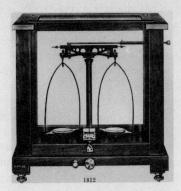

Equal-arm balance

Spring scale

Single-pan balance

Standards of Measurements.   To measure and report an object's mass, its temperature, or any of its other base or derived physical quantities we obviously need some units and some references. By international treaties among the nations of the world, the reference units and standards are decided by a diplomatic organization called the General Conference of Weights and Measures, headquartered in Sèvres, a suburb of Paris, France. The General Conference has defined a unit called a **base unit** for each of seven base quantities, but we need units only for the five that we have already mentioned — mass, length, time, temperature, and mole. We also need some units for the derived quantities — for example, for volume, density, pressure, and heat. The standards and definitions of base and derived quantities and units together make up what is now known as the **International System of Units** or the **SI** (after the French name, *Système Internationale d'Unités*).

Each base unit has a reference **standard,** which is some physical description or embodiment of the base unit. Long ago, the reference (such as it was) for the *inch* was "three

FIGURE 1.1
The SI standard mass and the former standard meter. Shown here are the U.S. copies kept at the National Bureau of Standards in Washington, D.C. (Courtesy of the National Bureau of Standards, Washington, D.C.)

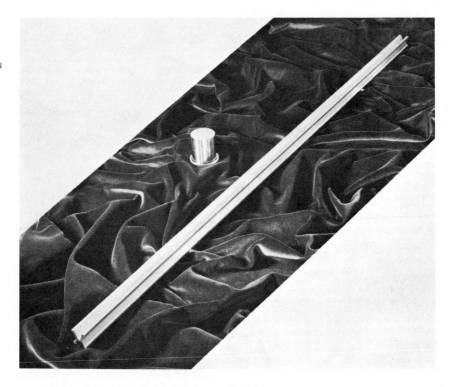

barleycorns, round and dry, laid end to end." Obviously, which three barleycorns were picked had a bearing on values of length under this "system." And if the barleycorns got wet, they sprouted. You can see that a reference standard ought to have certain properties if it is to serve the needs of all countries. It should be entirely free of such risks as corrosion, fire, war, theft, or plain skulduggery, and it should be accessible at any time to scientists in any country. The improvement that the SI represents over its predecessor, the *metric system,* is not so much in base units as in their reference standards. Before we survey what these are, we have to take note of an important mathematical feature of the SI.

Considering the large differences in what is studied among the sciences, it would have been unreasonable to expect the General Conference to devise base units having sizes convenient to all. Astronomers must describe huge distances, for example, but chemists usually measure small distances. To make it convenient for any scientist to adapt any base unit to a particular use, the SI, like its predecessor the metric system, devised multiples and submultiples of base units. What makes the SI (or the older metric system) vastly superior to any earlier systems of measurement is that there is a decimal or base-10 relationship among these multiples and submultiples. Multiplying or dividing by 10 is much easier than by 2, or 3, or any non-10 number because all we have to do is move the decimal point. (Compare this with converting, say, a distance in miles into yards, or feet, or inches. And even before doing this, you'd have to find out which kind of "mile" was meant. There are at least five!)

The SI base unit of length is called the **meter,** abbreviated **m,** and its reference is the *standard meter.* Until 1960, the standard meter was the distance separating two thin scratches on a bar of platinum-iridium alloy stored in an underground vault in Sèvres (Figure 1.1). This bar, of course, could have been lost or stolen, so the new reference for the meter is based on a property of light, something available everywhere, in all countries, and that obviously can't be lost or damaged. This change in reference didn't change the actual length of the meter, it only changed its official reference.[1]

An alloy is a mixture of two or more metals made by stirring them together in their molten states.

[1] Unfortunately for those without good backgrounds in physics, the SI reference for the meter involves a number of unfamiliar concepts. Officially, the standard meter is 1,650,763.73 wavelengths in vacuum of the orange – red line of the spectrum of krypton-86. It takes this many wavelengths to go from one scratch to the other on the earlier reference, the meter bar.

FIGURE 1.2
Relative sizes of common SI and U.S. Customary units of length, volume, and mass.

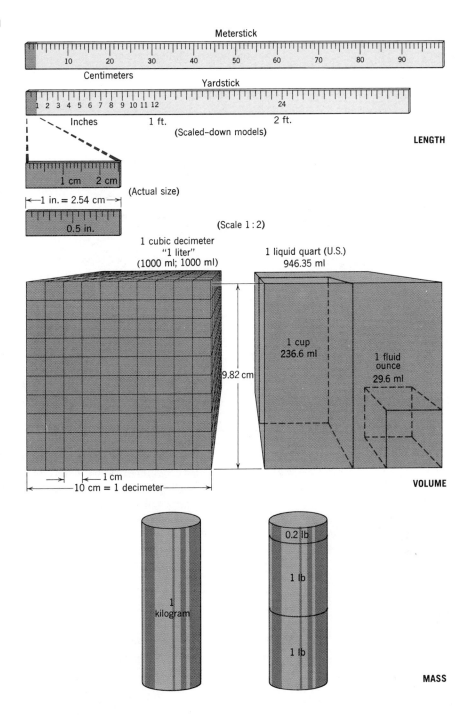

Meterstick

Centimeters

Yardstick

Inches   1 ft.   2 ft.
(Scaled–down models)   **LENGTH**

(Actual size)
1 cm   2 cm

├── 1 in. = 2.54 cm ──┤
0.5 in.   (Scale 1:2)

1 cubic decimeter
"1 liter"
(1000 ml; 1000 ml)

1 liquid quart (U.S.)
946.35 ml

9.82 cm

1 cup
236.6 ml

1 fluid ounce
29.6 ml

1 cm
10 cm = 1 decimeter   **VOLUME**

1 kilogram

0.2 lb

1 lb

1 lb   **MASS**

In the United States, older units are now legally defined in terms of the meter. For example the yard (yd), roughly nine-tenths of a meter (Figure 1.2), is defined as

$$1 \text{ yd} = 0.9144 \text{ m} \quad \text{(exactly)}$$

And the foot (ft), roughly three-tenths of a meter is defined as

$$1 \text{ ft} = 0.3048 \text{ m} \quad \text{(exactly)}$$

TABLE 1.1
Some Common Measures of Length[a]

SI		U.S. Customary	
1 kilometer (km) = **1000** meters (m)		1 mile (mi)	= **5280** feet (ft)
1 meter = **100** centimeters (cm)			= **1760** yards (yd)
1 centimeter = **10** millimeters (mm)		1 yard	= **3** feet (ft)
		1 foot	= **12** inches (in.)
1 meter = 39.37 inches		1 foot = 30.48 centimeters	
= 3.280 feet		= 0.3048 meter	
= 1.093 yard		1 inch = **2.54** centimeters	

[a] Numbers in boldface are exact.

1 cm   2 cm

In chemistry, the meter is usually too long for convenience, and submultiples are often used, particularly the **centimeter, or cm,** and the **millimeter, or mm.** The centimeter is 1/100 m, or there are 100 cm in a meter. The millimeter is 1/1000 m, or there are 1000 mm in a meter. The millimeter is 1/10 cm. Expressed mathematically, these definitions become

$$1 \text{ m} = 100 \text{ cm} \quad \text{or} \quad 1 \text{ cm} = 0.01 \text{ m}$$
$$1 \text{ m} = 1000 \text{ mm} \quad \text{or} \quad 1 \text{ mm} = 0.001 \text{ m}$$
$$1 \text{ cm} = 10 \text{ mm} \quad \text{or} \quad 1 \text{ mm} = 0.1 \text{ cm}$$

The inch (in.) is about $2\frac{1}{2}$ centimeters; more exactly,

$$1 \text{ in.} = 2.54 \text{ cm (exactly)}$$

Table 1.1 gives several important relationships between various units of length.

The SI base unit of mass is named the **kilogram,** abbreviated **kg,** and its reference is the *standard kilogram mass,* a cylindrical block of platinum-iridium alloy housed at Sèvres under the most noncorrosive conditions possible (Figure 1.1) This is the only SI reference that could still be lost or stolen, but no alternative has yet been devised. Duplicates made as much like the original as possible are stored in other countries. One kilogram has a mass roughly equal to 2.2 pounds (Figure 1.2), and Table 1.2 gives a number of useful relationships among various units of mass, including some old apothecaries' units. The most often used units of mass in chemistry and medicine are the kilogram, the **gram (g),** the **milligram (mg),** and the **microgram (μg).** These are related as follows.

$$1 \text{ kg} = 1000 \text{ g} \quad \text{or} \quad 1 \text{ g} = 0.001 \text{ kg}$$
$$1 \text{ g} = 1000 \text{ mg} \quad \text{or} \quad 1 \text{ mg} = 0.001 \text{ g}$$
$$1 \text{ mg} = 1000 \text{ μg} \quad \text{or} \quad 1 \text{ μg} = 0.001 \text{ mg}$$

The body weight of an individual is usually expressed in kilograms, and the smaller units appear in expressions of concentration or as dosages of medications. Lab experiments in chemistry usually involve grams or milligrams of substances.

The SI unit of volume, one of the important derived units, is the cubic meter, m^3, but this is much too large for convenience in chemistry. An older unit, the **liter,** abbreviated **L,** is accepted as a *unit of convenience* for use with the SI, but it isn't yet an official SI unit. The liter occupies a volume of 0.001 m^3 (exactly), and one liter is almost the same as one liquid quart (Figure 1.2):

$$1 \text{ quart (qt)} = 0.946 \text{ L}$$

Even the liter is often too large for convenience in chemistry, and two submultiples are commonly used, the **milliliter (mL)** and, particularly in some clinical situations, an ex-

The SI kilogram mass is the only SI standard that is a manufactured object.

1 kg of butter.

One cubic meter holds a little over 250 gallons.

TABLE 1.2
Some Common Measures of Mass[a]

SI	
1 kilogram (kg)	= **1000** grams (g)
1 gram	= **1000** milligrams (mg)
1 milligram	= 1000 micrograms (μg, γ, or mcg)[b]

U.S. Customary (avoirdupois)[c]	
1 short ton	= **2000** pounds (lb avdp)
1 pound	= **16** ounces (oz avdp)
1 ounce	= **16** drams (dr avdp)
1 dram	= 437.5 grains (grain)[d]

Apothecaries'	
1 pound (lb ap)	= **12** ounces (oz ap, or ℥) = **5760** grains
1 ounce	= **8** drams (dr ap, or ℈) = **480** grains
1 dram	= **60** grains

Other Relationships

1 kilogram = 2.205 lb avdp = 2.679 lb ap = 15,432 grains	
= 35.27 oz avdp = 32.15 oz ap	
1 lb avdp = 453.6 grams	1 lb ap = 373.2 grams
1 oz avdp = 28.35 grams	1 oz ap = 31.10 grams
1 grain = 0.0648 gram = 64.8 milligrams	
1 gram = 15.43 grains	1 dr ap = 3.887 grams

[a] Numbers in boldface are exact.

[b] The microgram is sometimes called a *gamma* in medicine and biology.

[c] These are the common units in the United States, not the apothecaries' units.

[d] The National Bureau of Standards has adopted no symbol for *grain*. Pharmacists usually symbolize it by *gr*. There is another apothecaries' weight called the *scruple* (20 grains), but it is no longer listed in the *U.S. Pharmacopoeia*.

One drop of water is about 60 mg.

tremely small unit called the **microliter (μL).** These are related as follows:

$$1 \text{ L} = 1000 \text{ mL} \quad \text{or} \quad 1 \text{ mL} = 0.001 \text{ L}$$
$$1 \text{ mL} = 1000 \ \mu\text{L} \quad \text{or} \quad 1 \ \mu\text{L} = 0.001 \text{ mL}$$

Table 1.3 gives several other relationships among units of volume. Figure 1.3 shows apparatus used to measure volumes in the lab.

The SI unit of time is called the **second,** abbreviated **s.** The SI *definition,* however, involves complexities of atomic physics that are entirely beyond our needs. Fortunately, the

FIGURE 1.3
Apparatus for measuring liquid volumes. (*a*) Graduated cylinder — a "graduate". (*b*) Graduated pipet. (*c*) Volumetric flask. A line is etched part way up the narrow neck, and when the liquid level is at this level, its volume has the value also etched on the flask.

(*a*)

(*b*)

(*c*)

TABLE 1.3
Some Common Measures of Liquid Volume[a]

SI	
1 cubic meter (m³)	= **1000** liters (L)
1 liter	= **1000** milliliters (mL)
1 milliliter	= **1000** microliters (µL, λ, or lambda)
U.S. Customary and Apothecaries'	
1 gallon (gal)	= **4** liquid quarts (liq qt)
1 liquid quart	= **2** liquid pints (liq pt)
1 liquid pint	= **16** liquid ounces (liq oz) (fluidounce, f℥, fl oz, in the apothecaries' system)
1 liquid ounce	= **8** fluidrams (f℈) = **480** minims (℔)
1 fluidram	= **60** minims (℔)
Other Relationships	
1 cubic meter	= 264.2 gallons
1 liter	= 1.057 liquid quarts = 2.113 liquid pints
	= 33.81 liquid ounces
1 milliliter	= 16.23 minims
1 liquid ounce	= 29.57 milliliters
1 liquid quart	= 946.4 milliliters
1 fluidram	= 3.696 milliliters
Miscellaneous Approximate Equivalents (unofficial)	
1 liquid pint	= 2 cups = 4 gills
1 cup	= 16 tablespoonfuls = 250 mL
1 tablespoon	= 3 teaspoonfuls = 15 mL
1 teaspoonful	= 5 mL

[a] The apothecaries' system uses the U.S. Customary system for liquid measures. Numbers in boldface are exact.

duration of the SI second is the same as before, for essentially all purposes. The second is 1/18,400 of a mean solar day. Decimal-based multiples and submultiples of the second are used in science, but so are such deeply entrenched old units as minute, hour, day, week, month, and year.

The SI unit for degree of temperature is called the **kelvin, K.** (Be sure to notice that the abbreviation is K, not °K.) This degree used to be called the **degree Celsius** (°C), and even earlier the degree centigrade (also °C). Then it was defined as 1/100 the interval between the freezing point of water (named 0 °C) and the boiling point of water (named 100 °C). The most extreme coldness possible is −273.15 °C, and this is named 0 K on the Kelvin scale. The kelvin is the name of the degree on this scale, and it is identical with the Celsius degree. Only the *numbers* assigned to points on the scale differ. (See Figure 1.4, where the scales are compared.) Because 0 K corresponds to −273.15 °C, we have the following simple relationships between kelvins and degrees Celsius (where we follow common practice of rounding 273.15 to 273):

$$°C = K - 273$$
or
$$K = °C + 273$$

The *kelvin* is named after William Thomson, Baron Kelvin of Largs (1842–1907), a British scientist.

PRACTICE EXERCISE 1   Normal body temperature is 37 °C. What is this in kelvins?

The Kelvin scale isn't used much in biology and medicine. In chemistry it is used mostly in describing temperatures of gases. The Celsius scale is more popular and it is rapidly supplanting the old, familiar Fahrenheit scale in medicine. The **degree Fahrenheit** (°F) is $\frac{5}{9}$ the size of

FIGURE 1.4
Relationships between the Kelvin, Celsius, and Fahrenheit scales of temperature.

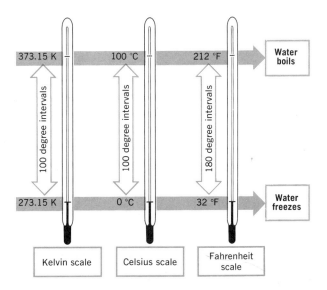

the degree Celsius. To convert a Celsius temperature, t_C, to a Fahrenheit temperature, t_F, we can use either of the following equations:

$$t_C = \frac{5\ °C}{9\ °F}\,(t_F - 32\ °F)$$

$$t_F = \frac{9\ °F}{5\ °C} \times t_C + 32\ °F$$

Table 1.4 gives some common temperatures in both °C and °F.

Scientific Notation and Physical Quantities. The typical human red blood cell has a diameter of 0.000008 m. Whether we want to write it, say it, or remember it, 0.000008 m is an awkward number, and to make life easier scientists have developed a method called **scientific notation** for recording very small (or very large) numbers. In scientific notation (sometimes called exponential notation), a number is written as the product of two numbers. The first is a decimal number with a value usually between 1 and 10, although sometimes a wider range is used. Following this number is a times (×) sign and then the number 10 with an exponent or power. For example, we can write the number 4000 as follows:

Appendix A has a review of exponential numbers.

$$4000 = 4 \times 1000 = 4 \times 10 \times 10 \times 10$$
$$= 4 \times 10^3$$

When the decimal point is omitted, we assume that it is after the last digit in the number.

Notice that the exponent 3 is the number of places to the left that we have to move the decimal point in 4000 to get to 4, which is a number in the desirable range:

$$4\underset{3\ \ \ 2\ \ \ 1}{0\,0\,0}$$

TABLE 1.4
Some Common Temperature Readings in °C and °F

	°F	°C
Room temperature	68	20
Very cold day	−20	−29
Very hot day	100	38
Normal body temperature	98.6	37
Hottest temperature the hands can stand	120	49

If our large number is 42,195, the number of meters in a marathon distance, we can rewrite it as follows after figuring out that we have to move the decimal point four places to the left to get a decimal number between 1 and 10:

$$42,195 \text{ m} = 4.2195 \times 10^4 \text{ m}$$

In rewriting numbers smaller than 1 in scientific notation, we have to move the decimal point to the *right* to get a number in the acceptable range of 1 to 10. This number of moves is the value of the *negative* exponent of 10. For example, we can rewrite the number 0.000008 as

$$0.000008 = 8 \times 10^{-6}$$

You should not continue until you are satisfied you can change large or small numbers into scientific notation. For practice, do the following exercises.

PRACTICE EXERCISE 2   Express each number in scientific notation. Let the decimal part be a number between 1 and 10.

(a) 545,000,000   (b) 5,670,000,000,000   (c) 6454

(d) 25   (e) 0.0000398   (f) 0.00426

(g) 0.168   (h) 0.00000000000987 (see footnote 2)

Prefixes to SI Base Units.   If we rewrite 3000 m as 3×10^3 m and try to pronounce the result, we have to say "three times ten to the third meters." There's nothing basically wrong with this, but it's clumsy. This is why the SI has names for several exponential expressions — not independent names but prefixes that can be attached to the name of any unit. For example, 10^3 has been assigned a prefix name of *kilo-*, abbreviated *k-*. Thus 1000 or 10^3 meters can be called 1 kilometer. Abbreviated, this becomes 10^3 m = 1 km.

With just a few exceptions, the prefixes set by the SI go with exponentials in which the power is 3, 6, 9, 12, 15, and 18 or with $-3, -6, -9, -12, -15$, and -18. These are all divisible by 3. Table 1.5 has a list of the SI prefixes and their symbols. Those given in boldface are so often encountered in chemistry, biology, and medicine they should be learned now. Notice there are four prefixes that do not go with powers divisible by 3. The SI hopes their usage will gradually fade away, but this hasn't happened yet. The two in boldface have to be learned. However, *centi* is used almost entirely in just one physical quantity, the centimeter, and *deci* is limited almost completely to another physical quantity, the deciliter (100 mL or $\frac{1}{10}$ L). The *deciliter* is particularly common on clinic report sheets. For example, the clinical lab might report that a patient's blood contains 145 milligrams of glucose per 100 milliliters. The report sheet would show this as 145 mg/dL instead of 145 mg/100 mL (thus saving space on a crowded sheet). A typical report sheet is shown in Figure 1.5. Notice that dL is written as dl and that mL is written as ml. Don't let things like this bother you. Even in the sciences life has its inconsistencies.

To take advantage of these prefixes, we sometimes have to modify a rule used in converting a large or small number into scientific notation. The goal in this conversion will now

1 dL = 1 $\times$ 10^{-1} L = 1/10th liter
But 1/10th liter = 100 mL
Therefore,

1 dL = 100 mL

---

[2] Some of the numbers in this exercise illustrate a small problem that the SI is trying to get all scientists to handle in a uniform way. In part (h), for example, you might have gotten a bit dizzy trying to count closely spaced zeros. The SI recommends (and most European scientists have accepted the suggestion) that in numbers having four or more digits the digits be grouped in threes separated by thin spaces. For large numbers, just omit the commas. Thus 545,000,000 would be written as 545 000 000. The number 0.00000000000987 becomes 0.000 000 000 009 87. It will be a while before you see this usage very often in the United States, but when you do you'll now know what it means. Incidentally, European scientists use a comma instead of a period to locate the decimal point. You might see this yourself soon when you first weigh anything in the lab. If the weighing balance was made in Europe, a reading such as 1,045 g means 1.045 g.

TABLE 1.5
SI Prefixes for Multiples and Submultiples of Base Units[a]

	Prefix	Symbol
1 000 000 000 000 000 000 = 10^{18}	exa	E
1 000 000 000 000 000 = 10^{15}	peta	P
1 000 000 000 000 = 10^{12}	tera	T
1 000 000 000 = 10^{9}	giga	G
1 000 000 = 10^{6}	**mega**	**M**
1 000 = 10^{3}	**kilo**	**k**
100 = 10^{2}	hecto	h
10 = 10^{1}	deka	da
0.1 = 10^{-1}	**deci**	**d**
0.01 = 10^{-2}	**centi**	**c**
0.001 = 10^{-3}	**milli**	**m**
0.000 001 = 10^{-6}	**micro**	μ
0.000 000 001 = 10^{-9}	nano	n
0.000 000 000 001 = 10^{-12}	pico	p
0.000 000 000 000 001 = 10^{-15}	femto	f
0.000 000 000 000 000 001 = 10^{-18}	atto	a

[a] The most commonly used prefixes and their symbols are in boldface. Thin spaces instead of commas are used to separate groups of three zeros to illustrate the format being urged by the SI (but not yet widely adopted in the United States).

be to get the exponential part of the number to match one with an SI prefix even if the decimal part of the number isn't between 1 and 10. For example, we know that the number 545,000,000 can be rewritten as 5.45×10^{8}, but 8 isn't divisible by 3, and there isn't an SI prefix to go with 10^{8}. If we counted 9 spaces to the left, however, we could use 10^{9} as the exponential part:

$$5\,4\,5\,0\,0\,0\,0\,0\,0 = 0.545 \times 10^{9}$$
$$9\;8\;7\;6\;5\;4\;3\;2\;1$$

We usually put a zero in front of a decimal point in numbers that are less than 1, such as in 0.545. This zero just helps us remember the decimal point, and it doesn't count as a significant figure.

Now we could rewrite 545,000,000 m as 0.545×10^{9} m or 0.545 Gm (gigameter),[3] because the prefix *giga*, abbreviated G, goes with 10^{9}. We also could have rewritten 545,000,000 as 545×10^{6}, and then 545,000,000 m could have been written as 545 Mm (megameters) because *mega* goes with 10^{6}.

EXAMPLE 1.1   REWRITING PHYSICAL QUANTITIES USING SI PREFIXES

Problem: Bacteria that cause pneumonia have diameters roughly equal to 0.0000009 m. Rewrite this using the SI prefix that goes with 10^{-6}.

Solution: In straight exponential notation, 0.0000009 m is 9×10^{-7} m, but -7 is not divisible by 3 and no SI prefix goes with 10^{-7}. If we move the decimal six places instead of seven to the right, however, we get 0.9×10^{-6} m. The prefix for 10^{-6} is *micro* with the symbol μ, so

$$0.0000009 \text{ m} = 0.9 \times 10^{-6} \text{ m} = 0.9 \ \mu\text{m}$$

The diameter of one of these a bacteria is 0.9 micrometers (0.9 μm).

---

[3] We will never again see the gigameter, or Gm, used as a unit of length in this book. It is fortunate that we don't need it because nurses, physicians, and pharmacists often write "Gm" instead of "g" as the symbol for "gram" in order to avoid any confusion with an archaic but still-used unit, the "grain." If you ever see "Gm" in your professional career, it will almost certainly mean "gram."

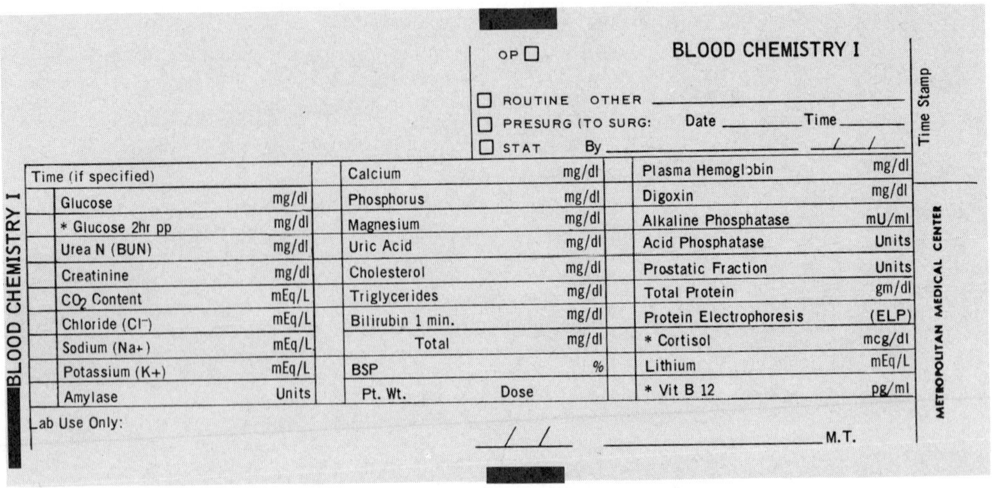

FIGURE 1.5

A typical clinical report form for blood chemistry analysis. The units are mg/dl = milligrams per deciliter; mEq/l = milliequivalents per liter; mU/ml = microunits per milliliter; gm/dl = grams per deciliter; mcg/dl = micrograms per deciliter; and pg/ml = picograms per milliliter. (Courtesy Clinical Laboratory, Metropolitan Medical Center, Minneapolis, MN.)

BLOOD CHEMISTRY I						
Time (if specified)		Calcium	mg/dl	Plasma Hemoglobin	mg/dl	
Glucose	mg/dl	Phosphorus	mg/dl	Digoxin	mg/dl	
* Glucose 2hr pp	mg/dl	Magnesium	mg/dl	Alkaline Phosphatase	mU/ml	
Urea N (BUN)	mg/dl	Uric Acid	mg/dl	Acid Phosphatase	Units	
Creatinine	mg/dl	Cholesterol	mg/dl	Prostatic Fraction	Units	
CO₂ Content	mEq/L	Triglycerides	mg/dl	Total Protein	gm/dl	
Chloride (Cl⁻)	mEq/L	Bilirubin 1 min.	mg/dl	Protein Electrophoresis	(ELP)	
Sodium (Na+)	mEq/L	Total	mg/dl	* Cortisol	mcg/dl	
Potassium (K+)	mEq/L	BSP	%	Lithium	mEq/L	
Amylase	Units	Pt. Wt. Dose		* Vit B 12	pg/ml	

OP ☐   BLOOD CHEMISTRY I

☐ ROUTINE   OTHER _____
☐ PRESURG (TO SURG: Date _____ Time _____
☐ STAT   By _____   / /

Lab Use Only:

_____ / / _____ M.T.

Time Stamp

METROPOLITAN MEDICAL CENTER

PRACTICE EXERCISE 3

Complete the following conversions to exponential notation by supplying the exponential parts of the numbers.

(a) $0.0000398 = 39.8 \times$ ____   (b) $0.000000798 = 798 \times$ ____

(c) $0.000000798 = 0.798 \times$ ____   (d) $16500 = 16.5 \times$ ____

PRACTICE EXERCISE 4

Write the abbreviation of each of the following. All are common in chemistry and medicine and should be learned.

(a) Milliliter   (b) Microliter   (c) Deciliter

(d) Millimeter   (e) Centimeter   (f) Kilogram

(g) Microgram   (h) Milligram

PRACTICE EXERCISE 5

Write the full name that goes with each of the following abbreviations.

(a) kg   (b) cm   (c) dL   (d) μg

(e) mL   (f) mg   (g) mm   (h) μL

PRACTICE EXERCISE 6

Rewrite the following physical quantities using the standard SI abbreviated forms to incorporate the exponential parts of the numbers.

(a) 1.5×10^6 g   (b) 3.45×10^{-6} L   (c) 3.6×10^{-3} g

(d) 6.2×10^{-3} L   (e) 1.68×10^3 g   (f) 5.4×10^{-1} m

PRACTICE EXERCISE 7

Express each of the following physical quantities in a way that uses an SI prefix.

(a) 275,000 g   (b) 0.0000625 L   (c) 0.000000082 m

1.3   ACCURACY AND PRECISION IN PHYSICAL QUANTITIES

The way in which the number part of a physical quantity is expressed discloses something about the precision of the measurement but nothing about its accuracy.

If you were taking a medication and knew that small overdoses caused bad side effects and small underdoses did no good, you would care about both the *accuracy* and the *precision* used in measuring each dose. Most people use these terms interchangeably, but they are quite different. **Accuracy** means the closeness of a measurement or the average of several measurements to the true value. It means freedom from error by the person taking the measurement and freedom from any malfunction of the instrument. **Precision** means the degree to which successive measurements agree with each other. It also means the fineness of the measurement when only one is made.

Figure 1.6 illustrates the difference between accuracy and precision in the measurement of someone's height. Each dot represents one measurement. In the first set of results, the dots are tightly clustered close to or exactly at the true value, and obviously a skilled person was at work with a carefully manufactured meter stick. This set illustrates both high precision and great accuracy. In the second set, a skilled person, without realizing it, evidently used a faulty meter stick, one that was mislabeled by a few centimeters. The precision is as great as that shown by the first set, because the successive measurements agree well with each other. But they're all untrue, so the accuracy is poor. In the third set of measurements, someone with a good meter stick did careless work. Only by accident do the values average to the true value, so the accuracy, in terms of the average, turned out to be high, but the precision is terrible and no one would really trust the average. The last set displays no accuracy and no precision. No matter what physical quantities we use, we want to be able to judge how accurately and precisely they were measured.

When we read the value of some physical quantity, we have no way of telling from it alone if it emerged from an accurate measurement. Someone might write on a report sheet "4.5678 mg of antibiotic," but in spite of all its digits we can't tell if the balance was working or if the person using it knew how to handle it and read it correctly. A skilled and careful experimenter frequently checks the instruments against references of known accuracy. Thus the question of *accuracy* is a human problem. Employ trained people, give them good instruments, and reward consistently good results, and problems of accuracy are well-managed.

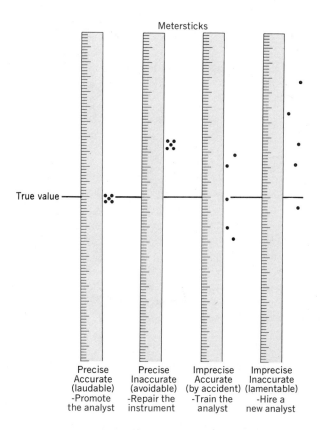

Metersticks

True value

| Precise Accurate (laudable) -Promote the analyst | Precise Inaccurate (avoidable) -Repair the instrument | Imprecise Accurate (by accident) -Train the analyst | Imprecise Inaccurate (lamentable) -Hire a new analyst |

FIGURE 1.6
Accuracy and precision.

Significant Figures.   We can indicate something about the precision of a measurement by the way in which we write its value. We do this by the way in which we round off the value to leave it with a certain number of *significant figures*. The number of **significant figures** in a physical quantity is the number of digits known with complete certainty to be correct plus one more. For example, the quantity "4.56 mg of antibiotic" has three significant figures. The first two, the 4 and the 5, are known to be correct, but the analyst is acknowledging a small uncertainty in the last digit. Unless otherwise stated, the uncertainty is assumed to be *one* unit of the last digit, so this report means that the mass of the sample is closer to 4.56 mg than to 4.55 mg or 4.57 mg. If the mass had been reported to be 4.560 mg, then it has four significant figures. The 4, 5, and 6 are certainly true, but there is some uncertainty in the last digit, 0. The true mass is closer to 4.560 mg than to 4.559 mg or 4.561 mg. Thus "4.560 mg" discloses a greater precision or fineness of measurement than 4.56 mg. Sometimes you'll see a report with a value such as 4.560 ± 0.001 mg or 4.560 ± 0.005. The symbol $\pm$ stands for "plus or minus," and what follows it in numbers such as these indicates how much uncertainty is carried in the last digit.

> The $\pm$ sign in 4.560 ± 0.001 mg means that the physical quantity is in the range of 4.559 to 4.561 mg.

Figuring out how many significant figures are in a number is easy, provided we have an agreement on how to treat zeros. Are all of the zeros counted as *significant* in such quantities as 4,500,000 people, or 0.0004500 L, or 400,005 m? We will use the following rules to decide.

1.   Zeros sandwiched between nonzero digits are always counted as significant.

Thus both 400,005 and 400.005 have six significant figures.

2.   Zeros that do no more than set off the decimal point on their *left* are never counted as significant figures.

Although such zeros are necessary to convey the general *size* of a quantity, they don't say anything about the *precision* of the measurement. Thus such quantities as 0.045 mL, 0.0045 mL, and 0.00045 mL all have only two significant figures.

> The quantities 4056 g and 4506 g both have four significant figures.

3.   Trailing zeros to the *right* of the decimal point are always significant.

Trailing zeros are any that come at the very end of the number, as in 4.56000. This has three trailing zeros, and because they are to the right of the decimal point, all are significant. This number has six significant figures and represents considerable precision or fineness of measurement.

> The other zeros are needed to locate the decimal points in these numbers, and they definitely are important in this sense. They just have nothing to do with precision.

4.   Trailing zeros that are to the *left* of the decimal point are counted as significant only if the author of the book or article has somewhere said so.

The zeros in 4,500,000 are trailing zeros, but are they significant? Suppose this number stands for the population of a city. A city's population changes constantly as people are born and die, and as they move in and out. No one could claim to know a population is *exactly* 4,500,000—not 4,499,999 and not 4,500,001, but 4,500,000. Most scientists handle this problem by restating the number in scientific notation, so that any desired trailing zeros can be placed *after* the decimal point. By doing this, as many or as few of such zeros can be given to convey the proper degree of precision. If the census bureau feels that the population is known to be closer to 4,500,000 than to 4,400,000 or to 4,600,000 people, and that no better precision than this is possible, then it should show only two significant figures in the result. It should report the population as 4.5×10^6 people. Giving the population as 4.50×10^6 people indicates greater precision—to three significant figures. There are four significant figures in 4.500×10^6.

Not everyone agrees with this way of handling trailing zeros that stand to the *left* of the decimal point, so you have to be careful. Some say they aren't significant unless the decimal point is actually given, as in 45,000. L. When the decimal point is the last item in a number, however, it is easily forgotten at the time of making the record. This problem is avoided by switching to scientific notation so that all trailing zeros come after the decimal point. This is the practice we will usually follow in this book, unless noted otherwise, or unless the context makes the intent very clear.

Rounding Off Calculated Physical Quantities.   When we mathematically combine the values of two or more measurements, we usually have to round the result so that it has no more significant figures than allowed by the original data. Normally, such rounding is done at the *end* of a calculation (unless specified otherwise) to minimize the errors introduced by rounding. There are four simple rules for rounding.

1.   When we multiply or divide quantities, the result can have no more significant figures than carried by the least precise quantity (the one with the fewest significant figures).

2.   When we add or subtract numbers, the result can have no more decimal places than are in the number having the fewest decimal places.

3.   When the first of the digits to be removed by rounding is 5 or higher, round the digit to its left *upward* by one unit. Otherwise, drop it and all others after it.

4.   Treat exact numbers as having an infinite number of significant figures.

An *exact number* is any that we define to be so, and we usually encounter exact numbers in statements relating units. For example, all of the numbers in the following expressions are exact and, for purposes of rounding calculated results, have an infinite number of significant figures:

We use the period in the abbreviation of inch (in.) to avoid any confusion with the preposition *in*, which has the same spelling.

$$1 \text{ in.} = 2.54 \text{ cm} \qquad \text{(exactly, as defined by law)}$$
$$1 \text{ L} = 1000 \text{ mL} \qquad \text{(exactly, by the definition of mL)}$$

The significance of having an infinite number of significant figures lies in our not letting such numbers affect how we round results. It would be silly to say that the "1" in "1 L" has just one significant figure when we intend, by definition, that it be an exact number.

EXAMPLE 1.2   ROUNDING THE RESULT OF A MULTIPLICATION OR A DIVISION

Problem: A floor is measured as 11.75 m long and 9.25 m wide. What is its area, correctly rounded?

Solution: Area = (length) × (width)
                                                      = 11.75 m × 9.25 m
                                                      = 108.6875 m²   (not rounded)

Resist the impulse that some owners of new calculators have of keeping all the digits they paid for.

But the measured width, 9.25 m, has only three significant figures, whereas the length, 11.75 m, has four. We have to round the calculated area to three significant figures:

                                        Area = 109 m²   (correctly rounded)

EXAMPLE 1.3   ROUNDING THE RESULT OF AN ADDITION (OR A SUBTRACTION)

Problem: Samples of a medication having masses of 1.12 g, 5.1 g, and 0.1657 g are mixed. How should the total mass of the resulting sample be reported?

Solution: The sum of the three values, obtained with a calculator, is 6.3857 g, which shows four places following the decimal point. However, one mass is precise only to the first decimal place, so we have to round to this place. The final mass should be reported as 6.4 g. Notice that the value of the second sample mixed, 5.1 g, says nothing about the third or fourth decimal places. We don't know if the mass is 5.101 g or 5.199 g, or what; the sample just wasn't measured precisely. This is why we can't know anything beyond the first decimal place in the sum.

The following numbers are the numerical parts of physical quantities. After the indicated mathematical operations are carried out, how must the results be expressed?

(a) 16.4×5.8 (b) $5.346 + 6.01$

(c) 0.00467×5.6324 (d) $2.3000 - 1.00003$

(e) $16.1 + 0.004$ (f) $(1.2 \times 10^2) \times 3.14$

(g) $9.31 - 0.00009$ (h) $\dfrac{1.0010}{0.0011}$

1.4 THE FACTOR – LABEL METHOD IN CALCULATIONS

In calculations involving physical quantities, the units are multiplied or canceled as if they were numbers.

Many people have developed a mental block about any subject that requires the use of mathematics. They know perfectly well how to multiply, divide, add, and subtract, but the problem is in knowing *when*, and no pocket calculator tells this. We said earlier that the inch is defined by the relationship, 1 in. = 2.54 cm. This fact has to be used when a problem asks for the number of centimeters in some given number of inches, but for some people the problem arises in knowing whether to divide or multiply.

Science teachers have worked out a method called the *factor – label* method for correctly setting up such a calculation and *knowing* that it is correct. The **factor – label method** takes a relationship between units stated as an equation (such as 1 in. = 2.54 cm), expresses the relationship in the form of a fraction, called a **conversion factor,** and then multiplies some given quantity by this conversion factor. In this multiplication, identical units (the "labels") are multiplied or canceled as if they were numbers. If the remaining units for the answer are right, then the calculation was set up correctly. We can learn how this works by doing an example, but first let's see how to construct conversion factors.

Some call the factor-label method the cancel-unit or the factor-unit method.

The relationship, 1 in. = 2.54 cm, can be restated in either of the following two ways and both are examples of conversion factors:

$$\frac{2.54 \text{ cm}}{1 \text{ in.}} \quad \text{or} \quad \frac{1 \text{ in.}}{2.54 \text{ cm}}$$

When we divide both sides of the equation 2.54 cm = 1 in. by 2.54 cm, we get:

$$\frac{2.54 \text{ cm}}{2.54 \text{ cm}} = \frac{1 \text{ in.}}{2.54 \text{ cm}}$$

This only restates the relationship of the centimeter and the inch, it doesn't change it. The use of a conversion factor just changes units, not actual quantities.

If we read the divisor line as "per," then the first conversion factor says "2.54 cm per 1 in." and the second says "1 in. per 2.54 cm." These are merely alternative ways of saying that "1 in. equals 2.54 cm." Any relationship between two units can be restated as two conversion factors. For example,

$$1 \text{ L} = 1000 \text{ mL} \qquad \frac{1000 \text{ mL}}{1 \text{ L}} \quad \text{or} \quad \frac{1 \text{ L}}{1000 \text{ mL}}$$

$$1 \text{ lb} = 453.6 \text{ g} \qquad \frac{453.6 \text{ g}}{1 \text{ lb}} \quad \text{or} \quad \frac{1 \text{ lb}}{453.6 \text{ g}}$$

Restate each of the following relationships in the forms of their two possible conversion factors:

(a) 1 g = 1000 mg   (b) 1 kg = 2.205 lb

Suppose we want to convert 5.65 in. into centimeters. The first step is to write down what has been given — 5.65 in. Then we multiply this by the one conversion factor relating inches to centimeters that lets us cancel the unit no longer wanted and leaves the unit we want.

$$5.65 \; \cancel{\text{in.}} \times \frac{2.54 \; \text{cm}}{1 \; \cancel{\text{in.}}} = 14.4 \; \text{cm} \qquad \text{(rounded correctly from 14.351 cm)}$$

Notice how the units of "in." cancel. Only "cm" remains, and it is on top in the numerator where it has to be. Suppose we had used the wrong conversion factor:

The arithmetic is correct, but the result is still all wrong.

$$5.65 \; \text{in.} \times \frac{1 \; \text{in.}}{2.54 \; \text{cm}} = 2.22 \; \frac{(\text{in.})^2}{\text{cm}} \qquad \text{(correctly rounded)}$$

That's right. We *must* do to the units exactly what the times sign and the divisor line tell us, and (in.) times (in.) equals (in.)^2 just as $2 \times 2 = 2^2$. Of course, the units in the answer, (in.)^2/cm, make no sense, so we know with certainty that we can't set up the solution this way. The reliability of the factor–label method lies in this use of the units (the "labels") as a guide to setting up the solution. Now let's work an example.

EXAMPLE 1.4   USING THE FACTOR–LABEL METHOD

Problem: How many grams are in 0.230 lb?

Solution: From Table 1.2, we find that 1 lb = 453.6 g, so we have our pick of the following conversion factors:

$$\frac{453.6 \; \text{g}}{1 \; \text{lb}} \quad \text{or} \quad \frac{1 \; \text{lb}}{453.6 \; \text{g}}$$

To change 0.230 lb into grams, we want "lb" to cancel and we want "g" in its place in the numerator. Therefore we pick the first conversion factor; it's the only one that can give this result.

$$0.230 \; \cancel{\text{lb}} \times \frac{453.6 \; \text{g}}{1 \; \cancel{\text{lb}}} = 104 \; \text{g} \qquad \text{(correctly rounded)}$$

There are 104 g in 0.230 lb. (We rounded from 104.328 g to 104 g because the given value, 0.230 lb, has only three significant figures. Remember that the "1" in "1 lb" has to be treated as an exact number because it's in a definition.)

PRACTICE EXERCISE 10   The *grain* is an old unit of mass still used by some pharmacists and physicians, and 1 grain = 0.0648 g. How many grams of aspirin are in an aspirin tablet containing 5.00 grain of aspirin?

Often there isn't one conversion factor that does the job, and two or more have to be used. For example, we might want to find out how many kilometers are in, say, 26.22 miles, but our tables don't have a direct relationship between kilometers and miles. However, if we can find in a table that 1 mile = 1609.3 m and that 1 km = 1000 m, we can still work the problem. We'll see in the next example how we can string two (or more) conversion factors together before doing the calculation that gives the final answer.

EXAMPLE 1.5   USING THE FACTOR–LABEL METHOD. STRINGING CONVERSION FACTORS

Problem: How many kilometers are there in 26.22 miles, the distance of a marathon race? Use the following relationships:

$$1 \; \text{mile} = 1609.3 \; \text{m}$$
$$1 \; \text{km} = 1000 \; \text{m}$$

Solution: The given relationships provide the following sets of conversion factors:

$$\frac{1 \text{ mile}}{1609.3 \text{ m}} \quad \text{or} \quad \frac{1609.3 \text{ m}}{1 \text{ mile}}$$

and

$$\frac{1 \text{ km}}{1000 \text{ m}} \quad \text{or} \quad \frac{1000 \text{ m}}{1 \text{ km}}$$

Now let's write down the given, 26.22 mile, and pick a conversion factor that lets us cancel "mile."

$$26.22 \text{ mile} \times \frac{1609.3 \text{ m}}{1 \text{ mile}}$$

If we paused to carry out this calculation, the answer would be in meters (m), not in kilometers (km). Therefore *before doing this calculation,* use another conversion factor that lets us cancel "m." In principle, we could keep on doing this—stringing out conversion factors—until we found the unit that we wanted for the answer.

$$26.22 \text{ mile} \times \frac{1609.3 \text{ m}}{1 \text{ mile}} \times \frac{1 \text{ km}}{1000 \text{ m}} = 42.20 \text{ km} \quad \text{(correctly rounded)}$$

The marathon distance is 42.20 km.

PRACTICE EXERCISE 11   Using the relationships between units given in tables in this chapter, carry out the following conversions. Be sure that you express the answers in the correct number of significant figures.

(a) How many milligrams are in 0.324 g (the aspirin in one normal tablet)?
(b) A long-distance run of 10.0×10^3 m is how long in feet? (This is the 10-km distance.)
(c) A prescription calls for 5.00 fluidrams of a liquid. What is this in milliliters?
(d) One drug formulation calls for a mass of 10.00 drams (10.00 dr ap). If only an SI balance is available, how many grams have to be weighed out?
(e) How many microliters are in 0.00478 L?

On page 11, equations were given relating degrees Celsius and degrees Fahrenheit. The use of these equations illustrates further examples of how units no longer wanted cancel, as you can demonstrate by using those equations to work the following Practice Exercises.

PRACTICE EXERCISE 12   A patient has a temperature of 104 °F. What is this in degrees Celsius?

PRACTICE EXERCISE 13   If the water at a beach is reported as 15 °C, what is this in degrees Fahrenheit? (Would you care to swim in it?)

1.5 DENSITY AND SPECIFIC GRAVITY

The specific gravity of a body fluid such as urine can be used to diagnose certain conditions.

Both the mass of some chemical sample and its volume are examples of **extensive properties,** those that are directly proportional to the size of the sample. Length is also an extensive property. An **intensive property** is independent of the sample's size. Temperature and color are such intensive properties, for example. Generally, intensive properties disclose some

TABLE 1.6
Densities of Some Common Substances

Substance	Density (g/cm³)
Aluminum	2.70
Bone	1.7 – 2.0
Butter	0.86 – 0.87
Cement, set	2.7 – 3.0
Cork	0.22 – 0.26
Diamond	3.01 – 3.52
Glass	2.4 – 2.8
Gold	18.88
Iron	7.87
Marble	2.6 – 2.8
Mercury	13.55
Milk	1.028 – 1.035
Wood, balsa	0.11 – 0.14
ebony	1.11 – 1.33
maple	0.62 – 0.75
teak	0.98

essential quality of a substance that is true for any sample size, and this is why scientists find intensive properties particularly useful.

Density. One useful intensive property of a substance, particularly if it is a fluid, is its density. **Density** is the mass per unit volume of a substance:

$$\text{Density} = \frac{\text{mass}}{\text{volume}}$$

The density of mercury, the silvery liquid used in most thermometers, is 13.60 g/mL, making mercury one of the most dense substances known. The density of liquid water is 1.0 g/mL. Table 1.6 gives the densities of several common substances.

Don't make the mistake of confusing *heaviness* with *denseness*. A pound of mercury is just as heavy as a pound of water or a pound of feathers — a pound is a pound. But a pound of mercury occupies only $\frac{1}{13.6}$ the volume of a pound of water.

The density of a substance varies with temperature, because for samples of most substances the volume but not the mass changes with temperature. Most substances expand in volume when warmed and contract when cooled. The *effect* isn't great if the substance is a liquid or a solid. For example, the density of mercury changes only from 13.60 g/mL to 13.35 g/mL when its temperature changes from 0 °C to 100 °C, a density change of only about 2%. Table 1.7 gives the density of water at several temperatures. Notice that to two significant figures the density is 1.0 g/mL in the range of temperatures from 0 °C to 30 °C, which covers room temperature and below. Water is unusual in that its density becomes *smaller* when it is cooled from 3.98 °C to the temperature at which it freezes.

One of the uses of density is in calculating what volume of a liquid to take when the problem or experiment specifies a certain mass. Often it is easier (and sometimes safer) to measure a volume than a mass, as we will note in the next example.

EXAMPLE 1.6   USING DENSITY TO CALCULATE VOLUME FROM MASS

Problem: Concentrated sulfuric acid is a thick, oily, and very corrosive liquid that no one would want to spill on the pan of an expensive balance, to say nothing of the skin. It is an example of a liquid that is usually measured by volume instead of by mass, but suppose an

TABLE 1.7
Density of Water at Various Temperatures

Temperature °C	Density (g/mL)
0	0.99987
3.98	1.00000
5	0.99999
10	0.99973
15	0.99913
20	0.99823
25	0.99707
30	0.99567
35	0.99406
40	0.99224
45	0.99025
50	0.98807
60	0.98324
80	0.97183
90	0.96534
100	0.95838

experiment called for 25.0 g of sulfuric acid. What volume should be taken to obtain this mass? The density of sulfuric acid is 1.84 g/mL.

Solution: The given value of density means that 1.84 g acid = 1.00 mL acid. This gives two possible conversion factors:

$$\frac{1.84 \text{ g acid}}{1 \text{ mL acid}} \quad \text{or} \quad \frac{1 \text{ mL acid}}{1.84 \text{ g acid}}$$

The "given" in our problem, 25.0 g acid, should be multiplied by the second of these conversion factors to get the unit we want, mL.

$$25.0 \text{ g acid} \times \frac{1 \text{ mL acid}}{1.84 \text{ g acid}} = 13.6 \text{ mL acid}$$

Thus if we measure 13.6 mL of acid, we will obtain 25.0 g of acid. (The pocket calculator result is 13.58695652, but we have to round to three significant figures.)

PRACTICE EXERCISE 14   An experiment calls for 16.8 g of methyl alcohol, the fuel for fondue burners, but it is easier to measure this by volume than by mass. The density of methyl alcohol is 0.810 g/mL, so how many milliliters have to be taken to obtain 16.8 g of methyl alcohol?

PRACTICE EXERCISE 15   After pouring out 35.0 mL of corn oil for an experiment, a student realized that the mass of the sample also had to be recorded. The density of the corn oil is 0.918 g/mL. How many grams are in the 35.0 mL?

Specific Gravity.   The **specific gravity** of a liquid is the ratio of the mass contained in a given volume to the mass of the identical volume of water at the same temperature. If we arbitrarily say that the "given volume" is 1.0 mL, then the water sample has a mass of 1.0 g (or extremely close to this over a wide temperature range). This means that dividing the mass of some liquid that occupies 1.0 mL by the mass of an equal volume of water is like dividing by

FIGURE 1.7
A hydrometer designed to serve as a urinometer.

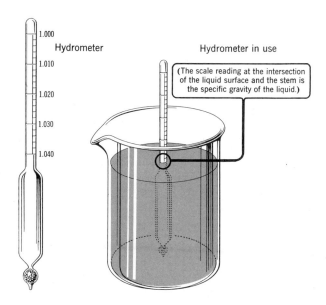

Hydrometer

Hydrometer in use

(The scale reading at the intersection of the liquid surface and the stem is the specific gravity of the liquid.)

1 — numerically — but all the units cancel. Specific gravity has no units, and a value of specific gravity is numerically so close to its density that we usually say they are the numerically the same. This fact has resulted in a rather limited use of the concept of specific gravity, but one use occurs in medicine.

In clinical work, the idea of a specific gravity surfaces most commonly in connection with urine specimens. Normal urine has a specific gravity in the range of 1.010 to 1.030. It's slightly higher than water because the addition of wastes to water usually increases its mass more rapidly than its volume. Thus the more wastes in 1 mL of urine the higher is its specific gravity. Figure 1.7 shows one of the traditional ways of quickly measuring the specific gravities of urine specimens — the urinometer. Its use, however, has largely been supplanted by a method — the use of a refractometer — that needs only one or two drops of urine for the measurement. The refractometer is an instrument that measures the ratio of the speed of light through air to its speed through the sample being tested, and this ratio can be correlated with the concentration of dissolved substances in the urine. (*How* the refractometer does this is beyond the scope of our study.)

The clinical significance of a change in the concentration of substances dissolved in the urine — however this is measured — is that it indicates a change in the activity of the kidneys, and this might indicate either kidney damage or some damage elsewhere that generates wastes the kidneys remove.

SUMMARY

Molecular basis of life One of the many ways of looking at life is to examine its molecular basis, the way in which well-being depends on chemicals and their properties. Down at the level of nature's tiniest particles, we find the "parts" that nature shuffles from organism to organism in the living world.

Scientific facts and physical quantities Scientific facts are usually capable of being checked by independent observers, and these facts often are physical quantities that pertain to physical properties, whether these are extensive or intensive properties. Physical properties — those that can be studied without changing the substance into something else — include mass, volume, time, temperature, color, and density. For our purposes, the important base quantities are (with the names of the SI base units given in parentheses)

mass (kilogram), length (meter), time (second), temperature degree (kelvin), and quantity of chemical substance (mole). The liter is an important derived unit for volume. Special prefixes can be attached to the names of the base units to express multiples or submultiples of these units. To select a prefix we have to be able to put very large or very small numbers into scientific notation.

Precision and accuracy Whether we obtain data from direct measurements or by calculations, we have to be careful not to imply too much precision by using the incorrect number of significant figures. When we add or subtract numbers, the decimal places in the result can be no more than the least number of decimal places among the original numbers. When we multiply or divide, we have to round the result to show the same number of significant figures as are in the least precise original number.

Factor-label method   The units of the physical quantities involved in a calculation are multiplied or canceled as if they were numbers. To convert a physical quantity into its equivalent in other units, we multiply the quantity by a conversion factor that permits the final units to be correct. The conversion factor is obtained from a defined relationship between the units.

Density and specific gravity   "How heavy?" isn't the same as "How dense?" Density is the ratio of mass to volume. When the density of something expressed in units of g/mL is divided by the density of water in the same units, the result is the specific gravity. Specific gravity can be used to estimate how much substances are dissolved in a given volume of a fluid, such as urine.

KEY TERMS

The following terms were introduced in this chapter, and their meanings should be mastered before you continue to the next chapter. There is a **Glossary** at the end of this book where concise definitions can be found. It's one thing to master new terms simply by repeatedly using them, but it's also perfectly acceptable to initiate this usage by plain, old-fashioned memorization. One of the major purposes of this study is to add a number of technical terms to your vocabulary. Not only will the following terms be used again in this book, you will often encounter many of them in the professional reading you will do to keep abreast of your field and remain up to date.

accuracy	factor–label method	mass	property
base quantity	gram (g)	measurement	scientific method
base unit	hypothesis	meter (m)	scientific notation
centimeter (cm)	inertia	microgram (μg)	second (s)
chemical property	intensive property	microliter (μL)	significant figures
conversion factor	International System of Units (SI)	milligram (mg)	specific gravity
degree Celsius	kelvin (K)	milliliter (mL)	standard
degree Fahrenheit	kilogram (kg)	millimeter (mm)	temperature degree
density	length	physical property	theory
derived quantity	liter (L)	physical quantity	time
extensive property		precision	volume

SELECTED REFERENCES

1 "Brief History of Measurement Systems with a Chart of the Modernized Metric System." *Special Publication 304A,* National Bureau of Standards, 1972.
2 W. G. Davies and J. W. Moore. "Adopting SI Units in Introductory Chemistry." *Journal of Chemical Education,* April 1980, page 303.
3 F. M. Pipkin and R. C. Ritter. "Precision Measurements and Fundamental Constants." *Science,* February 25, 1983, page 913.

REVIEW EXERCISES

The answers to Review Exercises that require a calculation and whose numbers are marked with an asterisk are given in an Appendix. The answers to all other Review Exercises are given in the *Study Guide to Fundamentals of General, Organic, and Biological Chemistry,* 3rd edition.

Molecular Basis of Life

1.1   Give some common experiences which illustrate that at some deep level of existence we can use the same "parts" as, say, a kitten.

1.2   Life, besides having a molecular basis, can be studied in terms of other bases. What are some? (One, for example, is the psychological basis.)

Scientific Method

1.3   You're driving down a dry highway (and sober), and the car starts to respond poorly to the steering wheel. You probably devise what — what is the better term — a theory or a hypothesis? Why?

1.4   What is it about experimental data — clinical lab results, for example — that make them more useful to someone trying to develop a medical diagnosis than descriptions of general feeling (important as these can be, too)?

1.5   What mental attitude goes along with the use of the scientific method?

1.6   You're studying at home, alone, at night when suddenly the lights in your room go out. Assuming that you instinctively use the scientific method in such a situation, what do you do next? And then after that?

Physical Quantities, Properties, and Measurements

1.7 What is meant by the word *property?*

1.8 How does a *physical property* differ from a *chemical property?*

1.9 How does a physical quantity differ from a number?

1.10 What is meant by the *inertia* of some object, and how is its inertia related to its mass?

1.11 Why is *volume* considered to be a less fundamental quantity than *length?*

1.12 What general name do we give to any fundamental quantity in terms of which less fundamental quantities are defined?

1.13 Name the five base quantities to be used in this book.

1.14 What is the name of the base unit for each of the following quantities?
   (a) time   (b) mass
   (c) temperature degree   (d) length
   (e) quantity of chemical substance

1.15 What relationship does a *reference standard* have to a *base quantity?*

1.16 What are some of the properties that a reference standard should have, ideally?

1.17 Which reference standard in the SI is least ideal in terms of its being secure from any kind of physical or chemical loss?

1.18 What are the names and symbols for the common submultiples that we will use in this book for each of the following base units?
   (a) meter   (b) liter
   (c) kilogram

1.19 How many centimeters make 1 m?

1.20 How many millimeters make 1 cm?

1.21 How many grams are in 1 kg?

1.22 How many milligrams are in 1 g?

1.23 Is the yard slightly shorter or slightly longer than the meter?

1.24 Is the liquid quart slightly smaller or slightly larger than the liter?

Temperature Degrees and Temperature Scales

1.25 The value -273.15 °C is a peculiar number to pick to be equivalent to 0 K. Why was it selected?

1.26 Suppose you took an unmarked, unetched thermometer and immersed it in a slush of ice and water until the mercury level stopped changing and then you marked the mercury level with a wax crayon. What is the name of this line in the Celsius system? In the Fahrenheit system? On the Kelvin scale?

1.27 Suppose that you repeated the experiment of Exercise 1.26 only this time immersed the unmarked thermometer in boiling water (at sea level), and you marked the mercury level after it stopped changing. What is the name given to this mark in each of the three temperature scales?

1.28 How many scale divisions are there between the two marks created by the experiments described in Exercises 1.26 and 1.27 when the distances are subdivided into (a) Fahrenheit degrees, (b) Celsius degrees, (c) kelvins?

1.29 How do the kelvin and the degree Celsius compare in size?

1.30 How do the degree Celsius and the degree Fahrenheit compare in size?

•1.31 Water freezes at 0 °C and it boils at 100 °C. What are these values in kelvins?

1.32 Many people would find a temperature of 294 K to be comfortable. What is this in degrees Celsius?

•1.33 A German recipe calls for baking batter at 210 °C. What Fahrenheit setting should you use?

1.34 A weather report from a station in Alaska gave the outside temperature as -40 °F. What is this in °C?

•1.35 The pool temperature at an Austrian health spa is 31 °C. What is this in °F?

1.36 An infant's temperature is 38.5 °C. Is this normal? (Do a calculation. Normal temperature is considered to be 98.6 °F.)

•1.37 Wishing to make an outside temperature of 109 °F seem less hot to a friend visiting from Europe, you tell him what it is in °C. What number do you report? (Will this gesture work?)

1.38 Anesthetic ether is a vapor above 94.3 °F. What is this in °C?

SI Prefixes and Scientific Notation

1.39 Rewrite the following physical quantities with their units abbreviated.
   (a) 26 micrograms of vitamin E
   (b) 28 millimeters wide
   (c) 5.0 deciliters of solution
   (d) 55 kilometers in a distance
   (e) 46 microliters of solution
   (f) 64 centimeters long

1.40 Rewrite the following physical quantities with their units written out in full.
   (a) 125 mg of water   (b) 25.5 mL of blood
   (c) 15 kg of salt   (d) 12 dL of sugar solution
   (e) 2.5 μg of insulin   (f) 16 μL of fluid

1.41 Rewrite the following data in scientific notation in which the decimal part of the number is between 1 and 10.
   (a) 0.013 L   (b) 0.000006 g
   (c) 0.0045 m   (d) 1455 s

1.42 Express each of these quantities in scientific notation and limit the decimal part of the number to a number between 1 and 10.
   (a) 24,605 m   (b) 654,115 g
   (c) 0.0000000095 L   (d) 0.00000568 s

1.43 Use a suitable SI prefix to express each of the quantities in Exercise 1.41.

1.44 Reformulate the numbers in the quantities of Exercise 1.42 so that they can be expressed in units that employ suitable SI prefixes.

Accuracy, Precision, and Significant Figures

1.45 According to one almanac, the population of the United States in 1790 was 3,939,214. Rewrite this figure in scientific notation retaining only three significant figures.

1.46 When a meter stick was used to take five successive measurements of a person's height, the following data were recorded: 172.7 cm, 172.9 cm, 172.6 cm, 172.6 cm, 172.8 cm. The meter stick had earlier been checked against an official reference standard and found to be good. The true value of the height was verified as 172.7 cm.
(a) Can the measurements be described as *accurate?* Why?
(b) Can they be described as *precise?* Why?

1.47 Examine the following numbers:
(A) 3.7200×10^3 (B) 3720 (C) 3.720
(D) 0.03720 (E) 37,200 (F) 0.00372
(G) 3.720×10^3 (H) 0.0372 (I) 3.72×10^9
(a) Which of these numbers has three significant figures? Identify them by their letters.
(b) Which has four significant figures?
(c) Which of them has five significant figures?

1.48 If we multiply $3.4462 \times 55.1 \times 10^8$, how many significant figures can we permit the answer to have?

1.49 If we add 0.00014 to 1.36, how many places after the decimal point can we permit to stand in the answer?

1.50 Rewrite the following number according to the number of significant figures specified by each part:

144,549.09

(a) Seven (b) Five (c) Four
(d) Three (e) One (f) Two

1.51 The relationship between the gram and the microgram is given by

$$1 \text{ g} = 100,000 \ \mu g$$

How many significant figures are considered to be in each number?

Converting Between Units

1.52 Write each of the following relationships between units for physical quantities in the forms of two conversion factors.
(a) 1 mile = 5280 ft
(b) 1 dram = 60 grains
(c) 1 lb = 453.6 g
(d) 1 ounce = 480 grains
(e) 1 m = 39.37 in.
(f) 1 liquid ounce = 480 minims

1.53 Given the relationships of Exercise 1.52, which of the two calculated answers is more likely to be correct in each part? You should be able to make this kind of judgment without actually doing a calculation.
(a) 250 minims = 0.520 liquid ounce or 1.20×10^4 liquid ounce
(b) 3.50 ounce = 0.00729 grain or 1.68×10^3 grain
(c) 0.350 lb = 159 g or 0.000772 g

1.54 What are the units in the result of the following calculation?

$$1.0 \text{ kg} \times 1 \ \frac{m}{sec} \times 1 \ \frac{m}{sec}$$

(The resulting units are the SI units for energy, a derived quantity.)

*1.55 Convert each of the following physical quantities into the units specified. Use tables in this chapter to find relationships between units.
(a) 75.5 in. into centimeters (the height of an adult male)
(b) 50.5 kg into pounds (the mass of an adult female)

1.56 Using relationships between units found in this chapter, convert each of the following quantities into the new units specified.
(a) 70.0 kg into pounds (the mass of an adult male)
(b) 64.0 in. into centimeters (the height of an adult female)

*1.57 A 500-mL bottle of soda contains how many liquid ounces (to three significant figures)?

1.58 If a gas tank holds 16.0 U.S. gallons, how many liters does it hold?

*1.59 Driving a car with a mass of 4.6×10^3 lb, you come to a bridge with a sign reading "Closed to all vehicles weighing more than 1.5×10^3 kg." Should you cross? (Do the calculation.)

1.60 A foreign car has a mass of 915 kg. What is this in pounds?

*1.61 If you can get 32 equal-sized butter pats from a quarter-pound (0.25 lb) stick of butter, what is the mass of each pat in grams?

1.62 The *carat* is a measurement jewelers use to describe the mass of a precious stone; 1 carat = 200 mg. What is the mass in milligrams of a diamond rated as 0.750 carat?

*1.63 Mt. Everest in Nepal is the highest mountain in the world— 29,028 ft. What is this in meters? In kilometers?

1.64 The highest mountain in the United States is Alaska's Mt. McKinley at 6194 m. How high is this in feet? In miles?

Density and Specific Gravity

1.65 Why is density called an intensive property?

1.66 If dissolving something in water increased the mass of the system by 0.5 g for each 0.5 mL increase in its volume, how would the density be affected?

1.67 Within limits, dissolving a solid in water increases the mass of the system more rapidly than the volume. What does this do to the density of the system: cause it to increase, decrease, or remain the same?

1.68 What is the difference between density and specific gravity?

1.69 What fact about water makes the density of some object and its specific gravity the same (or very nearly so, depending on the number of significant figures used)?

1.70 In which fluid does a urinometer float sink farther, one of low density or one of high density?

*1.71 Taking the density of lead to be 11.35 g/cm³, how many pounds of lead fill a milk container with a volume of 1.00 qt?

1.72 Aluminum has a density of 2.70 g/cm³. A milk carton with a volume of 1.00 qt could hold how many grams of aluminum? How many kilograms? How many pounds?

*1.73 To avoid spilling any liquid onto an expensive balance, a student obtained 30.0 g of acetic acid by measuring a corresponding volume. The density of acetic acid is 1.06 g/mL. What volume in milliliters was taken?

1.74 In an experiment to test how well methyl alcohol works as an antifreeze, a technician took 275 mL of this liquid. Its density is 0.810 g/mL. How many grams of methyl alcohol were taken?

Chapter 2
Matter and Energy

The unusual thermal properties of water, one of the topics in this chapter, help the body maintain its constant internal temperature throughout vigorous exercise in both cold and hot climates.

2.1 STATES AND KINDS OF MATTER

Elements and compounds always have definite compositions, but mixtures do not.

ICE

Solid state
Shape: definite
Volume: definite

WATER

Liquid state
Shape: indefinite; same
as the container
Volume: definite

STEAM

Gas state
Shape: indefinite; same
as entire container
Volume: indefinite; same
as entire container

Matter is anything that occupies space and has mass. This includes literally everything, and because there is such a huge variety of things, making some sense out of matter and its behavior might seem impossible. Throughout history, however, humans have had a powerful impulse to sort and classify things whenever confronted by what is very complex. Biologists, for example, have created kingdoms, phylla, species, subspecies, and groups for plants and animals. One reason for such activity is simply to be able to find things and recognize them again, but another reason is that by sorting and classifying we focus our minds on possible *causes* for the nature of things.

States of Matter.   Chemists have classified matter in a number of ways that make the study of matter easier and lead to basic explanations of the nature of matter. One such classification, for example, is according to the physical condition of aggregation—the *states* of matter. There are three **states of matter**—solid, liquid, and gas—and our many experiences with ice, liquid water, and steam are reminders of these states. The advantage in recognizing the states of matter is that each can be studied by itself with the hope of finding some features that all solids have, for example, or that all gases have. It simplifies a study to take it in parts. Gases, as we will see in Chapter Six, have many properties in common, and this fact served as a clue to the fundamental nature not only of gases but of solids and liquids as well. Each of the three states is characterized by an ability or an inability to hold a shape and have a definite volume, as the figures in the margin illustrate.

Substances.   The materials of which matter of any kind consists have the very broad name of **substances.** Most matter that you see consists of two or more substances each of which contributes something to the overall set of properties. Freshly squeezed orange juice, for example, consists of water, some vitamins, a little fruit sugar, citric acid, some of the pulp of the fruit, and other substances. If you remove any of these substances from the juice, you will notice the change (although some changes, such as removing the vitamins, might not be too apparent right away). You probably have already sensed that if for some reason we want to learn more about the fundamental nature of orange juice, we have to study its individual substances. The same reasoning applies to anything that is complex, and it makes the study of complex things much easier. We know that we can't learn everything we want to know about people by studying the substances that make them up, but we surely can answer *some* questions and make it easier to treat illness or preserve health. In short, the best way to begin a study of the physical and chemical properties of any material object is to study its constituent substances.

Elements, Compounds, and Mixtures.   There are just three kinds of substances—elements, compounds, and mixtures. An **element** is a substance that cannot be broken down into simpler substances. Elements include many familiar things—aluminum, copper, gold, iron, chromium, for example, or the oxygen and the nitrogen in the air we breathe. Water isn't an element because we can break it down into oxygen and hydrogen, which are elements, and we can make oxygen and hydrogen recombine to give water. Inside the front cover of this book is a list of the known elements. There are just a few over 100 of them, but we'll be concerned with only about a dozen. A little over 96% of the bulk mass of the human body consists of substances made from only four elements—carbon, nitrogen, hydrogen, and oxygen.

Some naturally occurring elements, such as uranium and radium, are radioactive, too.

Just 90 elements occur naturally; the rest have been made by physicists who used very expensive equipment. These elements in addition to a few that occur naturally have a special property called **radioactivity**—the ability to emit one kind of dangerous radiation or another. Several of the radioactive elements are useful in medicine, and we will return to their special properties in Chapter Ten.

At room temperature, all but 13 elements are solids, two are liquids, and 11 are gases. All but about 20 elements are metals. A **metal** is any substance that has a shiny surface when polished, can be hammered into sheets or drawn into wires, and — most important — is a good conductor of electricity. Sometimes two or more metals are melted, mixed together, and allowed to cool to give an **alloy** that has special properties. Steel, for example, is actually a family of alloys — chromium steel, nickel steel, and many others — each of which has particularly useful properties such as unusual resistance to corrosion, or to breaking under high stretching and breaking forces. A few alloys are used to replace bones or to strengthen them, and they must be unusually resistant to corrosion. **Nonmetals** — for example, carbon, sulfur, phosphorus, hydrogen, oxygen, and any of the gaseous elements — generally can't be given a shiny finish; they can't be worked into sheets or wires; and they do not conduct electricity. (Carbon in the form of graphite is an exception because it is a good electrical conductor.) For purposes of our study, the important features of metals and nonmetals are their somewhat opposite *chemical* properties, as we will study in the next chapter.

Compounds are substances made from two or more elements that have combined in such a way that they *always* occur in a proportion by mass which is both definite and unique for the compound. The third and last substance is the **mixture.** A mixture consists of two or more substances that are present in a proportion which can vary considerably.

To refine the distinctions between elements, compounds, and mixtures, we have to compare how compounds and mixtures are made. Compounds are made by chemical **reactions,** events in which substances called **reactants** change into different substances called **products.** Reactants and products almost always have at least some physical properties that are quite different. For example, consider as two reactants the elements iron and sulfur — the first a typical metal, and the second a brittle, yellow nonmetal. If you heat sulfur to about 120 °C and stir the molten material with a white-hot iron rod (not, incidentally, an average, everyday experiment, because it is dangerous), the iron and sulfur change or *react with* each other. They combine to give a blackish, decidedly nonmetal-like solid that, unlike iron, won't conduct electricity or be attracted to a magnet. The product isn't bright yellow, like sulfur, either, and it no longer melts at the same temperature at which sulfur melts. Some dramatic changes in physical properties have occurred as the result of this reaction.

It would take some careful experiments, but we could show that for every 1.000 g of iron that "disappears" from the rod, 0.574 g of sulfur combines with it to give 1.574 g of new material. We have a new substance, a *compound* called iron(II) sulfide, and the proportions of the elements in this compound are always 1.000 g of iron to 0.574 g of sulfur, regardless of how often we make it or from which part of the world we obtain our samples of iron and sulfur. (Don't worry yet about how to name compounds; we'll use names only as labels now.)

To dramatize further the difference between iron(II) sulfide and its original elements, consider that we could mix iron bits and sulfur powder by simply stirring them together at room temperature in just any ratio by mass. The overall appearance of this aggregation could be varied from dark and iron-like to light yellow and sulfur-like, according to the mass ratio. Regardless of the ratio, we could remove the iron from the sulfur in this *mixture* of the two simply by using a strong magnet — in other words, by using a physical operation, one that doesn't make new substances. (See Figure 2.1.) In contrast, to separate the iron from the sulfur in a sample of the *compound,* iron(II) sulfide, we would have to carry out chemical reactions.

A solution of sugar in water is a much more familiar example of a mixture than iron and sulfur. This solution can be made as rich in sugar as syrup, or it can be made with such a small amount of sugar that it would barely taste sweet. This mixture can also be prepared in almost any proportion we wish, but all that we would have to do to recover the sugar is let the water evaporate — a physical change, because it doesn't change the water or the sugar into something that isn't water or sugar. You can see by these examples that mixtures and compounds differ in two important ways, in the methods needed to isolate their constituents and in the

The current U.S. nickel coin is actually an alloy of copper (75%) and nickel (25%).

Carbon also occurs in a pure form in diamonds.

Obviously, this experiment has to be done very carefully by experienced people.

Lemonade is little more than lemon-flavored sugar-water, and its sweetness can be varied from nearly sour to syrupy sweet.

Sugar is a *compound*. Water is, too. Sugar dissolved in water constitutes a *mixture* of compounds — a *solution*.

FIGURE 2.1
Mixture versus compound. The magnet can pull the iron filings from the *mixture* of iron and sulfur on the left. The iron cannot be separated from the *compound* of iron and sulfur, iron(II) sulfide, on the right by this purely physical action.

flexibility of their mass ratios. The components of mixtures can be separated from each other by using physical methods, but the separation of the elements making up a compound requires chemical reactions.

The proportions of the constituents in a mixture can vary widely. The proportions by mass of the elements that make up a compound are so constant that this feature is an essential part of the very definition of a compound. Compounds obey the **law of definite proportions,** the first important scientific law in our study. Like all scientific laws, it wasn't just thought up. It was *discovered* by a great amount of experimental work carried out by many scientists who began to observe the particular natural pattern that could be expressed by the law.

> Law of Definite Proportions In a given chemical compound, the elements are always combined in the same proportion by mass.

In this chemical context, the opposite of *pure* is not *impure,* because *impure* carries an extra biological meaning, such as "dangerous to human health." Sugar-water isn't necessarily dangerous to health; it's just a mixture of two pure substances.

In a sense, elements also obey this law, because an element consists of 100% of itself. Chemists call elements and compounds **pure substances** because of the constancy of their composition. This implies that there are actually just two kinds of substances, pure substances and mixtures.

The law of definite proportions is one of the laws of chemical combination. Another such law was suggested earlier when we pointed out that in the formation of iron(II) sulfide, 1.000 g of iron always combines with 0.574 g of sulfur and 1.574 g of iron(II) sulfide forms. This mass relationship between reactants and products has always been observed for chemical reactions, and these observations are behind the **law of conservation of mass.**

> Law of Conservation of Mass In any chemical reaction, the sum of the masses of the reactants always equals the sum of the masses of the products.

Any widespread regularity in nature generally suggests some very basic truth about the world in which we live, and the laws of chemical combination were examples of regularity that demanded an answer to the question: What must be true about substances to explain these laws? John Dalton was the first to offer an answer, as we'll see next.

2.2 DALTON'S ATOMIC THEORY OF MATTER

John Dalton

John Dalton (1766–1844), an English scientist, was the first to find a reasonable explanation for both definite proportions in compounds and the conservation of mass in reactions. Dalton reasoned that matter must be made of very tiny individual particles that could undergo a variety of chemical reactions without breaking apart or losing any mass. In order to explain the *definite* compositions observed for compounds, he said that these tiny particles simply cannot exist as major fragments of themselves. Each particle is an unbreakable unit.

The idea of such particles had been around for centuries, because ancient Greek philosophers had proposed it. The Greek word for "not cut" is *atomos,* and from this term came our word *atom.* Dalton revived this ancient belief in "not cuttable" particles with the enormously important difference that he had solid evidence—laws of chemical combination.

The chief postulates of **Dalton's atomic theory** are the following.

1. Matter consists of definite particles called **atoms.**

2. Atoms are indestructible.

3. All atoms of one particular element are identical in mass.

4. Atoms of different elements have different masses.

5. By becoming stuck together in different ways, atoms form compounds in definite ratios *by atoms.*

The only way, said Dalton, that we can observe definite ratios *by mass* in compounds is that compounds possess definite ratios of atoms, each kind of atom having its own unique and definite mass.

Dalton would have explained the definite ratio by mass in iron(II) sulfide as illustrated in Figure 2.2 The large circles labeled "Fe" represent iron atoms and the smaller circles labeled "S" stand for sulfur atoms. If we assume that the *atoms* of these two elements combine in a ratio of 1 atom of iron to 1 atom of sulfur, each atom with its own mass, *then the mass ratio can't help but be a constant.*

Pyrite is often found as golden crystals embedded in rock samples. Many a novice gold miner has felt an increased heartbeat upon finding what experienced miners called fool's gold.

Law of Multiple Proportions.   Powerful evidence for Dalton's atomic theory came from the study of compounds that can be made from the same elements but with different mass ratios. The mineral pyrite, like iron(II) sulfide, can be broken down to iron and sulfur, but the ratio by mass in pyrite is 1.000 g of iron to 1.148 g of sulfur. The mass ratio in iron(II) sulfide is 1.000 g of iron to 0.574 g of sulfur. Notice that 1.148 is exactly twice the size of 0.574. In other words, there is a simple *whole-number* ratio between the grams of sulfur combined with 1.000 g of iron in the two compounds. If Dalton is right, there *must* be a simple whole-number ratio such as this because atoms combine as whole units, as whole particles. Figure 2.3 shows why this is so. It shows several possible combinations of intact iron and sulfur atoms. Three of the four are known compounds—the first two and the fourth.

Tin is the metal used to coat the inner surfaces of "tin" cans. Tin, unlike iron and less-expensive steels, won't rust in an environment of food juices.

Two compounds of tin and oxygen further illustrate the special relationship among compounds made from the same elements. In one compound of tin and oxygen, 1.000 g of oxygen is combined with 3.710 g of tin. In another compound, 1.000 g of oxygen is combined with 7.420 g of tin. Now compare 7.420 g of tin with 3.710 g of tin:

$$\frac{7.420}{3.710} = \frac{2}{1}$$

FIGURE 2.2
Mass ratio versus atom ratio in iron(II) sulfide.

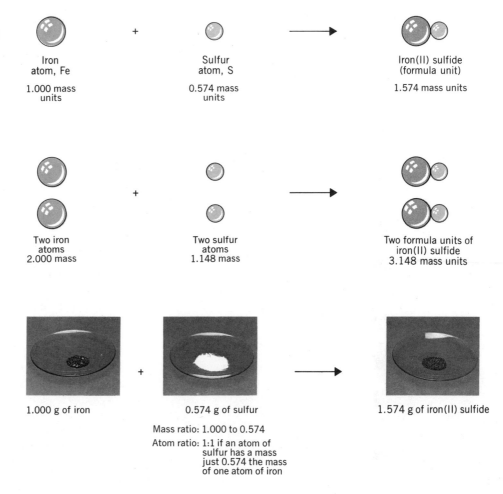

Iron
atom, Fe

1.000 mass
units

+

Sulfur
atom, S

0.574 mass
units

Iron(II) sulfide
(formula unit)

1.574 mass units

Two iron
atoms
2.000 mass

+

Two sulfur
atoms
1.148 mass

Two formula units of
iron(II) sulfide
3.148 mass units

1.000 g of iron

+

0.574 g of sulfur

1.574 g of iron(II) sulfide

Mass ratio: 1.000 to 0.574
Atom ratio: 1:1 if an atom of
sulfur has a mass
just 0.574 the mass
of one atom of iron

The ratio of the quantities of tin in the two compounds that combine with the same mass of oxygen is a simple, whole-number ratio — 2 to 1. These and several other examples led to the third law of chemical combination, the **law of multiple proportions.**

Law of Multiple Proportions Whenever two elements form more than one compound, the different masses of one that combine with the same mass of the other are in the ratio of small whole numbers.

The many examples that illustrated this law, combined with Dalton's astute interpretations, virtually compelled scientists to believe that atoms exist. Since Dalton's time, so much additional evidence has accumulated that the existence of atoms is taken as fact.

Chemical Formulas.   The name *pyrite* is not an informative name; it discloses nothing about the composition of this sulfide of iron. The name *iron(II) sulfide* is much more informative, at least after we have learned what the II stands for. The name iron(II) sulfide tells us the names of the elements in the compound, but it does not directly show their ratio either by atoms or by mass. Figure 2.3 introduced the use of some abbreviated symbols for elements, Fe and S, and these symbols together with those for the other elements are used to construct the most informative symbols for compounds that we have, chemical **formulas.**
Each element has been assigned a symbol consisting of one or two letters. Those with

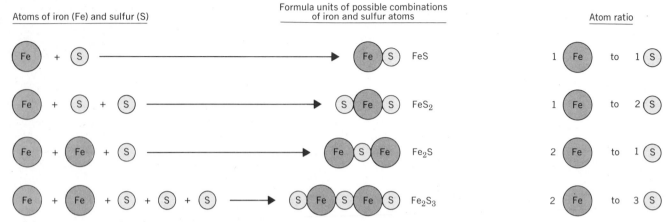

Atoms of iron (Fe) and sulfur (S)

Formula units of possible combinations
of iron and sulfur atoms

Atom ratio

FIGURE 2.3

Possible multiple proportions for combinations of iron atoms and sulfur atoms. If atoms remain essentially intact (and suffer no detectable loss in mass) when they form compounds, then they must assemble in *whole-number ratios by atoms* regardless of the masses of the individual atoms.

Among the trickier, but still common, pairs of chemical symbols are:

P = phosphorus
K = potassium

S = sulfur
Na = sodium

I = iodine
Fe = iron

which we will most often work are given in Table 2.1, and the complete list of atomic symbols appears in a table on the inside front cover of this book. Many elements, such as those in the first column of Table 2.1, have single-letter symbols, usually (but not invariably) the capitalized first letter. Of course there are more elements than letters in the alphabet, so several elements have names beginning with the same letter — for example, carbon, calcium, chlorine, chromium, cobalt, and copper. Therefore many atomic symbols consist of the first two letters with the first letter *always* capitalized and the second letter always lowercase. Examples appear in the second column of Table 2.1. The third column shows how the first letter and some letter in the name that stands beyond the second place are combined to make a symbol. Thus chlorine has the symbol Cl and chromium has the symbol Cr. The last column of Table 2.1 lists some elements named long ago when Latin was the almost universal language of educated people, and so the symbols of some elements were derived from Latin names, as shown.

Students often find the symbols for sodium (Na) and potassium (K) the trickiest, so be sure to take some extra time to fix these firmly in mind. They are important elements involved at the molecular level of life.

Chemists employ more than one kind of chemical formula, but we don't have to learn about them all in this chapter. All chemical formulas, however, disclose the kinds of atoms present and their numerical ratios. We have already represented iron(II) sulfide as FeS. Notice that the symbols of the elements are combined without spaces between them. The formula of pyrite is FeS_2. The 2 is called a *subscript,* and subscripts in formulas always *follow* the symbols to which they refer as a way of showing the ratios of the elements. The subscript 1 is always "understood." We write the formula of iron(II) sulfide as FeS, not Fe_1S_1, and the formula correctly shows that iron atoms and sulfur atoms have combined in a ratio of 1 to 1. In FeS_2, the ratio is 1 atom of Fe to 2 atoms of S; and in Fe_2S_3, the fourth possibility listed in Table 2.1 and which has the name iron(III) sulfide, the ratio is 2 atoms of Fe to 3 atoms of S. You can see that formulas give chemical information more clearly than names. It is too early in our study to discuss the rules for writing either formulas or names, so don't worry about this phase yet. For the present, just be sure you can spot the difference between say, FeS and fes, or that you can tell that CO can't possibly be the symbol for an element — the second letter is not in lowercase — but that Co might be and is — cobalt. (CO is the formula of carbon monoxide, a compound made of carbon, C, and oxygen, O.)

The formulas for the three sulfides of iron, FeS, FeS_2, and Fe_2S_3, illustrate a general practice. Whenever a compound is made from a metal and a nonmetal, the symbol for the metal is placed first. The symbols for carbon monoxide, CO, and methane, CH_4 (natural gas), illustrate another general practice or convention. Whenever a formula involves carbon and

TABLE 2.1
Names and Symbols of Some Common Elements[a]

C	Carbon	Al	Aluminum	Cl	Chlorine	Ag	Silver (*argentum*)
H	Hydrogen	Ba	Barium	Mg	Magnesium	Cu	Copper (*cuprum*)
O	Oxygen	Br	Bromine	Mn	Manganese	Fe	Iron (*ferrum*)
N	Nitrogen	Ca	Calcium	Pt	Platinum	Pb	Lead (*plumbum*)
S	Sulfur	Li	Lithium	Zn	Zinc	Hg	Mercury (*hydrargyrum*)
P	Phosphorus	Si	Silicon	As	Arsenic	K	Potassium (*kalium*)
I	Iodine	Co	Cobalt	Cs	Cesium	Na	Sodium (*natrium*)
F	Fluorine	Ra	Radium	Cr	Chromium	Au	Gold (*aurum*)

[a] The names in parentheses in the last column are the Latin names from which the atomic symbols were derived.

one other nonmetal, the symbol for carbon comes first. Let's work an example to illustrate how to use these rules and to become more familiar with chemical formulas and the use of subscripts.

EXAMPLE 2.1   WRITING CHEMICAL FORMULAS FROM ATOMIC COMPOSITIONS

Problem: Aluminum, a metal with the symbol Al, and sulfur, a nonmetal with the symbol S, form a compound in which the atom ratio is 2 atoms of Al to 3 atoms of S. Write the formula.

Solution: The symbol for aluminum has to come first because aluminum is a metal. The answer is Al_2S_3 in which the 2 goes with Al and the 3 with S.

EXAMPLE 2.2   WRITING CHEMICAL FORMULAS FROM ATOMIC COMPOSITION

Problem: Carbon, a nonmetal with the symbol C, and chlorine, another nonmetal but with the symbol Cl, form a compound in which the ratio of atoms is 1 of C to 4 of Cl. Write the formula.

Solution: By convention, the symbol of carbon comes first, so the answer is CCl_4. (The subscript, 1, of carbon is understood.)

PRACTICE EXERCISE 1   Write the formula of the compound between sodium, a metal, and sulfur in which the atom ratio is 2 atoms of sodium to 1 atom of sulfur.

PRACTICE EXERCISE 2   Give the names of the elements that are present and the ratio of their atoms in K_2CO_3.

These exercises don't take us very far into the subject of chemical formulas. We will do much more with this topic later, but you should now feel a bit more comfortable about the labels on bottles of chemicals that you will see in the laboratory. Realize, of course, that when you see a formula such as FeS on a bottle, this does not mean that only one atom of iron and one of sulfur are present. Always remember that formulas disclose *ratios*, not absolute numbers. The actual numbers in even a speck barely visible under a microscope are extremely large because atoms are exceedingly tiny. Atoms are so small that no one has ever clearly seen a single atom, not even with the most powerful microscope, although (as seen in Figure 2.4) spots of light caused by individual atoms have been obtained.

Returning to the formula of pyrite, FeS_2, some students ask why we can't write it as Fe_2S_4. A ratio of 4 to 2 is the same as 2 to 1. In fact, there is nothing fundamentally wrong with Fe_2S_4. It is only a matter of mutual consent among chemists to use the smallest whole

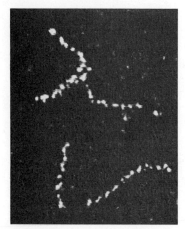

FIGURE 2.4
The large white spots are caused by individual thorium atoms in this high-magnification scanning electron microscope picture.

numbers that describe the ratio. Later in our study, we will find many situations in which this agreement is set aside in favor of another.

If Fe stands for an atom, what kind of particle, if any, does FeS stand for? The most general name we can give to the particle having the composition of the formula of a compound is **formula unit.** Other names such as *molecule* or *set of ions* or *ion group* will be used when we are farther along in our study. *Formula unit* includes all of these more special names. The formula unit of FeS consists of one iron atom and one sulfur atom. In fact, we can extend the term *formula unit* to include atoms. The chemical formula for sodium is given simply by its symbol, Na; hence, one formula unit of sodium is one atom of sodium.

Chemical Equations.   Now that we know something about chemical formulas we can learn how to use them in describing chemical reactions by means of chemical equations. A **chemical equation** is a special shorthand description of a reaction that groups the symbols of the reactants, separated by plus signs, on one side of an arrow and places the symbols of the products, also separated by plus signs, on the arrowhead side of the arrow. A very simple example is the formation of iron(II) sulfide from iron and sulfur.

$$Fe + S \longrightarrow FeS$$

Translation:

> iron reacts with sulfur
> in a ratio of one atom of
> Fe to one atom of S...........to give.......iron(II) sulfide

A more complicated example of an equation is the equation for the formation of aluminum sulfide, Al_2S_3, from aluminum and sulfur. This equation introduces **coefficients,** numbers that stand in front of formulas to show the proportions of formula units that are involved.

$$2Al + 3S \longrightarrow Al_2S_3$$

Translation:

> aluminum reacts with sulfur
> in a ratio of 2 aluminum
> atoms to 3 sulfur atoms..........to give......1 formula unit of aluminum sulfide

As with subscripts, whenever a coefficient is 1, and 1 isn't written; it is understood.

Our objective here is simply to recognize equations and translate them, not to write them. Notice, however, an essential feature of any chemical equation. It is a **balanced equation,** one in which all atoms present in the reactants occur somewhere among the products. For example, in the equation for the formation of Al_2S_3, there are 2 Al atoms on the left and 2 on the right in Al_2S_3. Similarly, there are 3 S atoms on the left and 3 on the right. The unbreakability of atoms and the conservation of mass in chemical reactions ensures this kind of balance. As Dalton said so long ago, when reactions occur the atoms of the reactants rearrange; they do not break up or disappear.

Some Features of Chemical Reactions.   It is one matter to notice that a change has occurred in some system but another matter to decide if the change was a physical change or a chemical reaction. The absolutely necessary and sufficient condition that determines whether the event was a chemical reaction is that substances change into other substances. However, to tell if this has happened, we rely on changes in physical appearances or physical properties. As a chemical reaction occurs, the physical properties associated with the reactants disappear and those of the products emerge. Such changes might be in color, in odor (but always be very cautious when checking odors), or in physical state—a gas might bubble out, or a new-appearing solid might separate out. If the event seems to be able to occur, once started, without any further intervention and it releases heat, this usually means it is a *chemical* change. Most chemical reactions that "go by themselves" release heat. If none of these tests decides the matter, then more sophisticated measures must be used, steps that positively identify different compounds.

2.3 FORMS OF ENERGY

Chemical energy is the energy stored in substances and that changes into other forms of energy such as heat, light, sound, and kinetic energy when the substances undergo particular reactions.

We have to be interested in energy in our study of the molecular basis of life, because almost all reactions either require energy in order to occur or they release energy. **Energy** is the ability to cause various kinds of changes. We say that things have energy when they have the ability to cause changes. Energy comes in a variety of forms having names that we associate with the kind of change — names such as light, sound, heat, and work.

Kinetic is from the Greek kinetikos, meaning ''of motion.''

Kinetic Energy. One kind of energy is called **kinetic energy,** the energy associated with motion. A moving car has kinetic energy; so does an avalanche, or a falling star, or a running child. The kinetic energy (K.E.) of a moving object can be calculated by the equation

$$K.E. = \tfrac{1}{2}mv^2$$

Velocity, a derived unit, is distance/time, and in the SI its units are meters per second (m/s).

where m = mass and v = velocity. Thus the kinetic energy of a moving object is directly proportional to its mass and to the *square* of its velocity. If you double the velocity of a moving car, its kinetic energy increases by a factor of four (because $4 = 2^2$). The energy quadruples, not doubles, and this is why a small increase in velocity can be much more dangerous than the numbers might indicate.

When the brakes are applied to a moving car, all of the kinetic energy of the car is converted into heat, and the brake shoes and brake drums become hotter. If the brakes squeal, a small amount of the kinetic energy changes into sound energy. Sound energy is associated with the ability to change the noise level. You can see by these simple examples that we can convert energy from one form into another. It is conceivable that the brake drums could get so hot that they would glow in the dark. Now some of the car's kinetic energy is being converted into light. Light energy is related to the ability to change the level of illumination.

A lot of play energy comes from the chemical energy in a sandwich.

Energy Conservation. Consider now a peanut butter sandwich and a small child. What an astonishing amount of activity is made possible by this sandwich! If the child receives energy for its activities from such a source, it must mean that substances can have a special kind of energy — energy in storage, so to speak. A simpler example showing that substances have energy in storage is a piece of paper. You can hold it in your hands comfortably, until someone puts a match to it. Now the paper burns, giving off heat and light (and maybe a little sound). If such energy can come from the paper, the paper must have possessed it in some way, because scientists have discovered something very fundamental about energy — the **law of conservation of energy.**

> Law of Conservation of Energy Energy can be neither created nor destroyed; it can only be transformed.

Chemical Energy. Burning a piece of paper doesn't create energy from nothing; it releases it from storage. Generally speaking, it is stored in the forces that hold the atoms together, forces called chemical bonds that undergo adjustments when the substances undergo a chemical reaction. The energy residing in the paper or the sandwich even when it is at room temperature and doing nothing is called **chemical energy.**

The chemical energy in paper becomes heat energy.

Chemical energy is one of the kinds of **potential energy.** This is energy that something possesses simply by virtue of its location or because of its chemical composition. A boulder poised at the edge of a high cliff is not in motion, but anyone immediately below it fears its potential for causing change — its potential energy, which it possesses not only because of its mass but also because of its location.

The substances in a peanut butter sandwich or in a piece of paper have chemical energy not because of their locations but because they have chemical properties. One of the chemical properties is that these substances can combine with oxygen from the air, undergo energy-releasing chemical reactions, and liberate their chemical or stored energy in other forms such as heat. If the reactions of the substances in the peanut butter sandwich occur under the very special conditions of a human body, then some of the energy — about half — is released not as heat but as the kinetic energy in the form of muscles in motion, or in other forms useful to the body.

2.4   HEAT ENERGY

The human body can absorb or release large quantities of heat with little change in temperature because its water content gives it a high heat capacity.

Heat is the energy that transfers from one object to another when the two are at different temperatures and in some kind of contact. We say that heat flows from the object with the higher temperature to the one with the lower temperature and, if left to itself, the flow continues until both objects reach the same intermediate temperature. Heat is thus a temperature-changing capacity possessed by any object. To get heat to flow, all we have to do is put the object next to one with a lower temperature.

Turning up a burner on a pan of boiling water won't cook the potatoes faster, because it doesn't raise the temperature. It just boils the water away faster. (*Then* the temperature soars as the potatoes turn blacker and blacker.)

Heat is also a physical-state-changing capacity. A block of ice at 0 °C in contact with a warm radiator will not itself undergo a change in temperature. It will simply melt — change its physical state from solid to liquid. As long as the freshly melted water is in contact with some ice, its temperature is the same as that of the ice. The temperature at which a solid changes into a liquid is called the **melting point** of the solid. If you put a pan of water at 100 °C on a hot burner, the flame's higher temperature won't raise the temperature of the water. It will cause the water to boil and change its state from liquid to gas (vapor). The temperature at which this occurs is called the **boiling point.** Thus an object at a higher temperature in contact with one at a lower temperature causes in the latter either a rise in temperature or a change in state.

In nearly all popular books on nutrition and diet, the word *calorie* actually means *kilocalorie.*

The Calorie.   It takes a certain amount of heat to make the temperature of 1 g of water change from 14.5 °C to 15.5 °C, and the name of this quantity is the **calorie,** abbreviated **cal.** (Although the specific Celsius degree, the one between 14.5 and 15.5 °C, is specified in this formal definition of the calorie, the transition between any of the Celsius degrees on the scale between the 0 °C and 100 °C marks requires essentially an identical quantity of heat.)

The calorie is an extremely small amount of heat, so often it is convenient to use a multiple called the **kilocalorie, kcal:**

$$1 \text{ kcal} = 1000 \text{ cal}$$

The symbol Δ is the Greek capital *delta.* Pronounce Δ*t* as "delta tee." When Δ is in front of any other symbol, it means a *change* in the value of whatever the other symbol represents. Thus Δ*E* refers to a *change* in energy; Δ*m* is a change in mass.

All other substances require specific quantities of heat in order for the temperature of a one-gram sample to change by one Celsius degree, and the general name for this thermal property is **specific heat.** We can define specific heat by the following equation.

$$\text{Specific heat} = \frac{\text{cal}}{\text{g } \Delta t} \qquad (2.1)$$

where cal = calories, g = mass in grams, and Δ*t* = the change in temperature in Celsius degrees. Translated into words, Equation 2.1 says that "the specific heat equals the calories

TABLE 2.2
Specific Heats of Some Substances

Substance	Specific Heat (cal/g °C)[a]
Ethyl alcohol	0.58
Gold	0.031
Granite	0.192
Iron	0.12
Olive oil	0.47
Water (liquid)	1.00

[a] These values are good in the temperature range of several degrees Celsius on either side of room temperature.

absorbed (or released) per gram per Celsius degrees of temperature change.'' The physical units for specific heat, taken from the defining equation (2.1), have to be cal/g °C.

Table 2.2 gives the specific heats of various substances. Notice that metals have very low values. To one significant figure, the specific heat of iron is 0.1 cal/g °C. In other words, it takes only one-tenth of a calorie to make the temperature of a 1-g sample of iron rise by 1 °C. Put another way, the 1 cal that raises the temperature of 1 g of water by only 1 °C can raise the temperature of the same mass of iron by 10 °C. Iron undergoes large changes in temperature from the gain or loss of a relatively small amount of heat. In the next section we will see how tremendously significant is the high specific heat of water at the molecular level of life. To get a better understanding of the concept of specific heat we will work an example of its use.

To two significant figures, water's specific heat is 1.0 cal/g °C over the entire range of 0 to 100 °C.

EXAMPLE 2.3   USING SPECIFIC HEAT DATA

A large iron nail has a mass of about 25.4 g.

Problem: The specific heat of iron is 0.106 cal/g °C in the temperature range pertaining to this problem. If a 25.4-g piece of iron at 20.0 °C received 115 cal of heat, what would its temperature change to?

Solution: What we need to find is the value of Δt, and we can use Equation 2.1. *It is essential that we carry along all of the units as we place data into this equation, because we have to be sure that they will cancel properly to leave the answer in the correct unit.* NEVER OMIT THE UNITS IN PHYSICAL QUANTITIES DURING CALCULATIONS UNTIL THEY CANCEL OR MULTIPLY PROPERLY TO GIVE THE CORRECT FINAL UNITS. THE UNITS ARE OUR CHIEF CHECK ON SETTING UP A SOLUTION CORRECTLY.

$$\text{Specific heat of iron} = \frac{0.106 \text{ cal}}{\text{g °C}} = \frac{115 \text{ cal}}{25.4 \text{ g} \times \Delta t}$$

To solve this for Δt we have to cross-multiply. If you think this mathematical procedure is something you can't do, turn to Appendix A where it is described using this particular problem. Cross-multiplication gives us

$$\Delta t = \frac{(115 \text{ cal}) \times (\text{g °C})}{(25.4 \text{ g}) \times (0.106 \text{ cal})}$$

Notice how we can cross-multiply units just like numbers and that now all of the units except °C cancel. After doing the arithmetic, we get

$$\Delta t = 42.7 \text{ °C} \qquad \text{(rounded from 42.71282127)}$$

In other words, 115 cal of heat will raise the temperature of a 25.4-g piece of iron by 42.7 °C. Therefore its new temperature is 20.0 °C + 42.7 °C = 62.7 °C.

PRACTICE EXERCISE 3   Suppose that the same amount of heat used in the example, 115 cal, was absorbed by 25.4 g of water instead of iron, with the initial temperature also 20.0 °C. What will be the final temperature of the water in degrees Celsius? (The specific heat of water in this range is 0.998 cal/g °C.) This exercise demonstrates the superior ability of water to absorb heat, compared to iron, without experiencing a large change in temperature.

Heat Capacity and Thermal "Cushions."   The high specific heat of water is a major factor in the body's ability to manage its heat budget without experiencing large swings in body temperature. Changes of even a few degrees from 37.0 °C—normal body temperature—result in death, unless corrected soon. Because the adult body is about 60% water, however, it has a substantial thermal "cushion" called its *heat capacity*.

Notice the difference between *specific heat* and *heat capacity*:

specific heat = cal/g °C

heat capacity = cal/°C

If we divide heat capacity by grams (g), we get specific heat. Thus specific heat is the heat capacity of a substance *per gram*.

All objects have this thermal property of heat capacity. Its value is proportional to the total mass of the object, so heat capacity is an extensive property. The **heat capacity** of an object is the quantity of heat that it can absorb (or release) per degree change in Celsius temperature. Its common units are calories per degree Celsius (cal/°C). For calculating purposes, we can define it by the following equation:

$$\text{Heat capacity} = \frac{\text{cal}}{\Delta t} \qquad (2.2)$$

The estimated heat capacity of a 70-kg adult male is 5×10^4 cal/°C. If such a person generated this much heat without being able to get rid of any, his temperature would rise by only 1 °C. In contrast, the temperature of the same mass of iron would rise by about 10 °C, a change that, if it occurred in a human, would cause death. You can see how the body's relatively high heat capacity helps the system protect itself against harmful fluctuations in temperature. The heat generated by the body each day, however, is so great that additional mechanisms are needed. To understand how they work, we have to study other thermal properties of substances, particularly their heats of vaporization.

The use of the word *vapor* is usually limited to talking about the gaseous form of something that at ordinary temperatures is a liquid or a solid. Thus we speak of "water vapor," but we don't refer to air as a vapor.

Heats of Vaporization and Fusion.   The change of a liquid to its gaseous or vapor state is called either **vaporization** or **evaporation,** and the associated verbs are *to vaporize* and *to evaporate*. The opposite change, the conversion of a vapor to its liquid form, is called **condensation,** and the verb is *to condense*. To vaporize a liquid requires a constant addition of heat, as you no doubt have experienced when you have boiled water on the stove. Of course, a liquid doesn't have to be at its boiling point to evaporate. Wet clothing does dry out, and if it is next to the skin, much of the heat needed for evaporation is taken from the body. You've also experienced this, no doubt, whenever you've noticed how cold wet jeans can feel. The heat needed to change 1 g of a substance from its liquid to its gaseous form is called its **heat of vaporization.** Its value varies somewhat with the temperature at which the liquid is evaporating. For water at its boiling point, the heat of vaporization is 539.6 cal/g; at body temperature (37 °C), water's heat of vaporization is about 580 cal/g. Table 2.3 gives the heats of vaporization for several substances at their boiling points. Notice that the value for water is considerably higher than for the others. As we will see in greater detail in the next section, this fact means that the body can get rid of a lot of heat by letting a little water evaporate from the skin.

EXAMPLE 2.4   USING HEAT OF VAPORIZATION DATA

Problem: How much heat in calories is needed to convert 10.0 g of liquid water to steam at 100 °C?

TABLE 2.3
Heats of Vaporization of Some Substances

Substance	Heat of Vaporization (at the boiling point) (cal/g)
Benzene	94.1
Chloroform	59.0
Ethyl alcohol	204
Ethyl chloride	93
Diethyl ether	84
Gasoline	76–80[a]
Water	539.6

[a] This is the range in values for the individual compounds present in gasoline.

Gaseous water, below 100 °C, can be called *water vapor,* but when the temperature of gaseous water is about 100 °C (water's boiling point), it's called *steam.*

Solution: We're given 10.0 g of water, and we know that the heat of vaporization is 539.6 cal/g. This value means that we have available the following two conversion factors:

$$\frac{539.6 \text{ cal}}{1 \text{ g}} \quad \text{or} \quad \frac{1 \text{ g}}{539.6 \text{ cal}}$$

(Treat the "1" in these conversion factors as an exact number.) Therefore, to get the answer in the right units, we have to multiply 10.0 g by the first factor:

$$10.0 \text{ g} \times \frac{539.6 \text{ cal}}{1 \text{ g}} = 5396 \text{ cal} \quad \text{(unrounded)}$$

Because 10.0 has only three significant figures, we have to round the answer and use scientific notation to express the result as 5.40×10^3 cal. This is the same as 5.40 kcal, because

$$5.40 \times 10^3 \text{ cal} \times \frac{1 \text{ kcal}}{10^3 \text{ cal}} = 5.40 \text{ kcal}$$

PRACTICE EXERCISE 4   How much heat in kilocalories is needed to evaporate 1.0 kg of water from the body at 37 °C? The heat of vaporization of water at this temperature is 5.8×10^2 cal/g.

Changing a solid to its liquid state also requires a characteristic quantity of heat called the heat of fusion. The **heat of fusion** of a substance is the heat needed to change 1 g of it to a liquid at the same temperature, the melting point. Table 2.4 gives heats of fusion for some common substances; notice again the unusually high value for water. These high values for water's heats of fusion and vaporization are an indication of how strongly water's formula units cling to each other.

Ice is much more effective in an ice pack than liquid water, even if the liquid is at essentially the same temperature, because of the high heat of fusion of water, 79.67 cal/g. The melting of a small amount of ice draws considerable heat without any change in temperature, but if the water is a liquid, its absorption of heat can occur only if its temperature rises. We have learned that the specific heat of liquid water is (rounded) 1.0 cal/g °C. Its heat of fusion is (also rounded) 80 cal/g. If *liquid* water at roughly 0 °C is to be used to absorb 80 cal and yet rise in temperature only 1 °C, then 80 g of liquid would be needed. But if *solid* water at roughly 0 °C can be used, then only 1 g is needed to absorb 80 cal because this much heat is needed to melt the ice.

TABLE 2.4
Heats of Fusion of Some Substances

Substance	Heat of Fusion (cal/g)
Benzene	30
Ethyl alcohol	24.9
Gold	15.0
Iron	65.7
Sulfur	10.5
Water	79.67

EXAMPLE 2.5   USING HEAT OF FUSION DATA

Problem: How much heat is needed to melt an ice cube with a mass of 30.0 g (about 1 ounce) if the ice temperature is 0 °C?

Solution: The value for the heat of fusion of ice, 79.67 cal/g, makes available to us two conversion factors:

$$\frac{79.67 \text{ cal}}{1 \text{ g}} \quad \text{or} \quad \frac{1 \text{ g}}{79.67 \text{ cal}}$$

If we multiply the given, 30.0 g, by the first factor, the remaining unit will be "cal."

$$30.0 \text{ g} \times \frac{79.67 \text{ cal}}{1 \text{ g}} = 2390 \text{ cal}$$

We have to express the answer as 2.39×10^3 cal to show the right number of significant figures, three. Thus one ice cube, by melting, can absorb considerable heat. If this heat were removed from 250 mL of liquid water—about one glassful—with an initial temperature of 25 °C (77 °F), the water temperature would drop by about 10 °C and become 15 °C (59 °F).

PRACTICE EXERCISE 5   An ice pack was prepared using 375 g of ice at 0 °C. As this ice melts, how much heat in calories and in kilocalories will be removed from the surroundings?

One important fact must be remembered about all of these thermal properties of water. If a certain quantity of heat has to be absorbed to cause a change in one direction, then exactly the same quantity has to be released to go in the opposite direction. For example, if 80 cal have to be absorbed by ice at 0 °C to *melt* one gram, then to *freeze* one gram of water at the same temperature requires that we remove 80 cal. Similarly, if 540 cal must be absorbed by liquid water at 100 °C to vaporize one gram, then when the same mass of steam at 100 °C condenses, it releases the identical quantity of heat. This is why steam is so much more dangerous in contact with the skin than very hot water—although both are life-threatening. When *steam* contacts the much cooler skin, it condenses and all of its heat of vaporization is released—some into the skin itself.

PRACTICE EXERCISE 6   If 16.4 g of steam change to the liquid state at 100 °C, how many kilocalories of heat are released?

Heat and Chemical Reactions.   In a very broad sense, chemical reactions are either spontaneous or they are not. Spontaneous events are those that, once started, continue with no further human intervention. Combustion (burning) is a common example. The flame from a tiny match can initiate a gigantic forest fire. A reaction such as combustion that continuously releases heat is called an **exothermic** reaction, and most (but not all) spontaneous reactions are exothermic.

Many chemical reactions can be made to take place if we continuously supply them with heat. The experiment described earlier in which a hot iron rod was used to stir molten sulfur to make iron(II) sulfide is an example. Reactions such as this that require a continuous input of heat are called **endothermic** reactions.

2.5 METABOLISM AND THE BODY'S HEAT BUDGET

Heat generated by metabolism is lost by radiation, conduction, convection, and the evaporation of water.

The minimum activities inside the body that must take place just to maintain muscle tone, control body temperature, circulate the blood, breathe, make certain substances, and otherwise operate tissues and glands during periods of rest are called the body's **basal activities.** The sum total of all of the chemical reactions that supply the energy for the basal activities is called the body's **basal metabolism.** The rate at which chemical energy is used for basal activities is called the **basal metabolic rate,** and it is customarily given in kcal/min or in the units of kcal/kg h (kilocalories per kilogram of body weight per hour).

Measurements of basal metabolic rates are taken when the person is lying down, has done no vigorous exercise for several hours, has eaten no food for at least 14 hours, and is otherwise awake but at complete rest. A 70-kg (154-lb) adult male has a basal metabolic rate of 1.0 to 1.2 kcal/min. The rate for a 58-kg (128 lb) woman is 0.9 to 1.1 kcal/min. Under other activities, the metabolic rate is higher, of course, as the data in Table 2.5 show.

EXAMPLE 2.6   COMPUTING DAILY CALORIE NEEDS

Problem: A woman student was found to have a basal metabolic rate of 1.1 kcal/min. During a typical day, she spent 6.0 hr sleeping, 6.0 hr in class and lab, 8.0 hr in seated activities such as eating, studying, and resting, 2.0 hr in walking—her rate was 3.0 mph, and 2.0 hr in moderately strenuous recreational activities. How many calories of energy were needed during these 24 hours to sustain these activities? Assume that the data for a 58-kg woman apply to this individual. Assume also that the values at the higher end of each range apply.

Solution: One way is to prepare a table using the data in Table 2.5 and remembering that each hour has 60 min. We will do necessary rounding after the last summing up.

Activity	Rate of Energy Expenditure	Minutes	Energy
Sleep, 6.0 hr	1.1 kcal/min	360 min	396 kcal
Class/lab, 6.0 hr	2.0 kcal/min	360 min	720 kcal
Seated, 8.0 hr	2.0 kcal/min	480 min	960 kcal
Walking, 2.0 hr	3.9 kcal/min	120 min	468 kcal
Exercise, 2.0 hr	5.9 kcal/min	120 min	708 kcal
Total			3252 kcal

The total has to be rounded to two significant figures. Thus the woman used 3.3×10^3 kcal in the day.

TABLE 2.5
Average Energy Expenditures by Individuals According to Daily Activities[a]

Categories of Activities	Rate of Energy Expenditure (kcal/min)	
	Man (70 kg)	Woman (58 kg)
Sleeping, reclining	1.0–1.2	0.9–1.1
Very light Seated and standing activities; painting trades, auto and truck driving; laboratory work; typing; playing musical instruments; sewing and ironing	1.2–2.5	1.1–2.0
Light Walking on level, 2.5–3 mph; tailoring; pressing; garage work; electrical trades; carpentry; restaurant trades; cannery workers; washing clothes; shopping with light load; golf; sailing; table tennis; volleyball	2.5–4.9	2.0–3.9
Moderate Walking 3.5–4 mph; plastering; weeding and hoeing; loading and stacking hay bales; scrubbing floors; shopping with heavy load; cycling; skiing; tennis; dancing	5.0–7.4	4.0–5.9
Heavy Walking with load uphill; tree felling; work with pick and shovel; basketball; climbing; football	7.5–12.0	6.0–10.0

[a] Data are for mature, adult men and women and are from *Recommended Dietary Allowances,* 8th ed. (Committee on Dietary Allowances, Committee on Interpretation of the Recommended Dietary Allowances, Food and Nutrition Board, National Research Council, National Academy of Sciences. Washington, D.C., 1974).

PRACTICE EXERCISE 7   A 70-kg adult male student carried out the same activities as those described for the female student in Example 2.6. Assuming that his basal metabolic rate was 1.2 kcal/min, and assuming that his other activities used energy at the rates given by the higher ends of the ranges, what were his energy needs for the day?

Mechanisms for Losing Heat.   The higher the metabolic rate, the more heat the body must release to preserve its temperature. Because of the relatively high heat of vaporization of water at body temperature, the evaporation of water becomes an important vehicle for this release. Table 2.6 shows the daily water budget of a typical adult male. Notice particularly how much water is lost by evaporation — 40% or 1.0 L. This means that 1.0 kg of water evaporates, and because the heat of vaporization at body temperature is 5.8×10^2 cal/g, then 5.8×10^2 kcal of heat are removed from the body each day by evaporation. Compare this with the 20×10^2 to 40×10^2 kcal per day of food energy taken in by eating from which roughly half does appear as heat. Very roughly, upwards of half of the heat lost by the body each day is released by means of the body water that evaporates. (Half of 20×10^2 kcal is 10×10^2 kcal, and the figure of 5.8×10^2 kcal is 58% of this.)

Evaporation that occurs without the activities of the sweat glands is called **insensible perspiration** because we do not notice that it happens. When the sweat glands work and beads of perspiration emerge, the evaporation is called **sensible perspiration.** Both forms help to cool the body, but in hot weather and during strenuous exercise, sensible perspiration accelerates. Either form, of course, cannot be sustained without the intake of ample fluids.

TABLE 2.6
Water Budget of the Human Body

Water Intake		Water Outgo	
As drink	1.2 L	Evaporation	
In food	1.0 L	from skin	0.5 L
Made by metabolism	0.3 L	from lungs	0.5 L
		Urine	1.4 L
		Feces	0.1 L
Total intake	2.5 L	Total outgo	2.5 L

Infrared radiation is sometimes referred to as heat rays.

The body has other mechanisms for releasing heat besides perspiration. One is by **radiation,** the same phenomenon that occurs with a radiator or a hot iron. Radiation from the body is like light radiation, except it isn't visible light but infrared radiation. The uncovered head in cold weather radiates as much as half of the body's heat production. This is why experienced mountaineers say, "If you feet are cold, put on your hat." The hat helps the entire body retain heat. It also reduces the work of another mechanism for the loss of heat — conduction. **Conduction** is the direct transfer of heat from a warmer body to a colder object. It happens, for example, when we place an ice pack on an inflamed area of the skin, or when we sit on a cold surface, or when we put bare hands onto cold machinery or tools.

Finally, another mechanism for losing heat from the body is by **convection.** This happens whenever we let the wind or a draft sweep away the warm, thin layer of air next to the skin. Both waffle-weave undergarments and heavy wool slacks or sweaters are filled with tiny pore spaces that trap this layer of warm air. Air is a very poor conductor of heat, so as long as the warm air layer is held close to the skin, little heat is lost by convection.

hyper- = over or above

therm = heat

hypo- = under or below

One reason why the body tries to maintain a steady temperature is that even small changes in temperature affect the rates of chemical reactions, including those of metabolism. If body temperature rises — a condition called **hyperthermia** — metabolic processes speed up. To sustain this, the body needs more oxygen — about 7% more for every 1 degree Fahrenheit increase. To deliver this oxygen, the heart must work harder, so a sustained condition of hyperthermia creates problems for the heart. Sometimes, a hyperthermic patient is almost fully immersed in ice water if this is considered necessary to lower the patient's body temperature quickly.

The opposite of hyperthermia is **hypothermia** — a condition of a lower than normal body temperature. Under this condition, the rates of metabolism slow down, including those reactions that keep vital functions working normally. Special Topic 2.1 describes the progression of events when someone is a victim of hypothermia.

SUMMARY

Matter Matter, anything with mass that occupies space, can exist in three physical states — solid, liquid, and gas. Broadly, the three kinds of matter are elements, compounds, and mixtures. Elements and compounds are designated as pure substances, and they obey the law of definite proportions. Mixtures, which do not obey this law, can be separated by operations that cause no chemical changes, but to separate the elements that make up a compound requires chemical reactions. A chemical reaction is an event in which substances change into different substances with different formulas. Elements can be classified as metals or nonmetals, or as radioactive or nonradioactive.

Dalton's atomic theory The law of definite proportions and the law of conservation of mass in chemical reactions led John Dalton to the theory — now regarded as well-established fact — that all matter consists of discrete, noncuttable particles called atoms. The atoms of the same element all have the same mass, and those of different elements have different masses. When atoms of different elements combine to form compounds, they combine as *whole* atoms; they do not break apart. Dalton realized that when different elements combine in different proportions by atoms, the resulting compounds must display a pattern now summarized by the law of multiple proportions.

The body responds to a fall in its temperature by trying to increase its rate of metabolism so that more heat is generated internally. Uncontrollable shivering is the outward sign of this response, and it sets in with a drop in temperature of only 2 to 3 °F (measured rectally). If the temperature continues to drop, the shivering will be violent for a period of time. Then loss of memory—amnesia—sets in at about 95 to 91 °F. The muscles become more rigid as the core temperature drops to the range of 90 to 86 °F, and the individual must have outside help immediately. The victim no longer has any ability to take life-saving steps. The heartbeat becomes erratic, unconsciousness sets in (87 to 78 °F), and below 78 °F death occurs by heart failure or pulmonary edema.

Death by hypothermia has often been called death by *exposure,* and it can happen even if the air temperature is above freezing. If you become soaked by perspiration or rain and the wind comes up, an outside temperature of 40 °F is dangerous. Those who fall into cold water (32 to 35 °F) seldom live longer than 15 to 30 min.

The legendary St. Bernard dogs who brought little casks of brandy to blizzard victims in the Swiss Alps were like agents of death to anyone who drank the brandy. A shot of brandy in a hypothermic individual worsens the situation. Alcohol *enlarges* blood capillaries, and when the capillaries near the skin's surface (which are loaded with the most chilled blood in the hypothermic body) suddenly enlarge, the chilled blood moves quickly to the body's core. This rapid drop in *core* temperature is particularly life-threatening.

If a victim of hypothermia is conscious and able to swallow food or drink, administer warm, nonalcoholic fluids and sweet foods. As quickly as possible, get the victim dry and out of the wind. Get into a dry sleeping bag with the victim, so that your own body warmth can be used. It is a genuine medical emergency and prompt aid is vital.

Symbols, formulas, and equations Every element is given a one or a two-letter symbol, and it can stand either for the element or for one atom of the element. The symbol for a compound—called a formula—consists of the symbols of the atoms that make up one of its formula units, and subscripts (following their associated atom symbols) are used to show the proportions of the different atoms present. To describe a chemical reaction, the symbols of the substances involved as reactants, separated by plus signs, are on one side of an arrow that points to the symbols of the products, also separated by plus signs. The equation is balanced when all atoms that occur among the reactants are present in like numbers among the products. Coefficients, numbers standing in front of formulas, are employed to show the correct balance. In both formulas and equations, the numbers used for subscripts and coefficients are generally the smallest whole numbers that show the correct proportions.

Forms of energy When something is able to cause a change in motion, position, illumination, sound, or chemical composition, it has energy of one form or another—kinetic, light, sound, chemical, and potential. Energy is not created nor does it disappear into nothing; it can only be transformed from one type into another or from one place to another. Spontaneous chemical reactions are usually exothermic —they release heat—but many reactions can be made to occur by continuously heating the reactants. These are endothermic reactions.

Heat energy Energy that transfers because of temperature differences—heat—either changes other temperatures or causes changes in physical state, such as melting or freezing, boiling or condensing. The heat that changes the temperature of one gram of a substance by one degree Celsius is called the substance's specific heat. When the substance is water, the quantity of heat is called the calorie. The heat capacity of an object—an extensive property—is the heat that the whole object will absorb or release as its temperature changes by one degree Celsius.

Heat and metabolism Because the body contains so much water, and because water has a relatively high specific heat, the body has a high heat capacity. Its temperature doesn't change much as heat is absorbed or released. The body loses heat by radiation, conduction, convection, and the evaporation of water. The body's loss of heat by the evaporation of water takes advantage of water's relatively high heat of vaporization. Because of water's relatively high heat of fusion, an ice pack is roughly 80 times more efficient than the same mass of ice-cold water in cooling an inflamed area of the body. Without proper management of the body's heat budget, a condition of hyperthermia—a temperature increase—or hypothermia—a temperature decrease—will develop. Hyperthermia results in an increased rate of metabolism, because heat accelerates reactions, and this eventually strains the heart. Hypothermia slows down body reactions, and this is also life-threatening.

KEY TERMS

The following important terms were introduced in this chapter, and their meanings should be mastered before continuing to the next chapter. Concise definitions can be found in the Glossary at the end of this book.

alloy	element	hypothermia	potential energy
atom	endothermic	kilocalorie (kcal)	product
basal activities	energy	kinetic energy	pure substance
basal metabolic rate	equation, balanced	law of conservation of energy	radiation, heat
basal metabolism	equation, chemical	law of conservation of mass	radioactivity
boiling point	evaporation	law of definite proportions	reactant
calorie (cal)	exothermic	law of multiple proportions	reaction
chemical energy	formula	matter	specific heat
coefficient	formula unit	melting point	states of matter
compound	heat	metal	substance
condensation	heat capacity	mixture	vaporization
conduction	heat of fusion	nonmetal	
convection	heat of vaporization	perspiration, insensible	
Dalton's atomic theory	hyperthermia	perspiration, sensible	

SELECTED REFERENCES

1 L. E. Strong. "Differentiating Physical and Chemical Changes." *Journal of Chemical Education,* October 1970, page 689.

2 T. B. Tripp. "The Definition of Heat." *Journal of Chemical Education,* December 1976, page 782.

3 H. A. Bent. "Energy and Exercise." *Journal of Chemical Education,* October 1978, page 659.

4 J. Silberner. "Hyperthermia. Hot Stuff in Cancer Treatment." *Science News,* August 30, 1980, page 141.

5 J. G. Neuwirth, Jr. "Principles of Cold Weather Adaptability." *Wilderness Camping,* November–December, 1973, page 26.

6 J. A. Miller. "Beyond the Ice Pack." *Science News,* October 7, 1978, page 250.

REVIEW EXERCISES

The answers to the Review Exercises that require a calculation and whose numbers are marked by an asterisk are given in an Appendix. The answers to all other Review Exercises appear in the *Study Guide* that accompanies this book.

Matter

2.1　If we think about a warm room, bathed in sunlight, we might imagine that energy itself occupies space. Why don't we call it matter?

2.2　What are the names of the three states of matter?

2.3　In terms of shapes and volumes, how do solids differ from gases?

2.4　We can push a hand through a basin of water, but we can't move our hand through a block of ice. Yet, the same formula units (molecules) are present in both. What can we say about the relative *mobilities* of the formula units of water in these two states?

2.5　Unless the wind is strongly against us, we can move through air with no noticeable difficulty. Air consists of a mixture of the formula units of nitrogen and oxygen, both gases. What can we say about the relative *mobilities* of these formula units as contrasted with the formula units in a liquid such as water?

2.6　Indium is a substance that can be given a shiny finish, and it conducts electricity well. Is indium more likely to be a metal or a nonmetal?

2.7　Phosphorous is a dark-reddish, powdery substance that doesn't conduct electricity. Is it more likely to be a metal or a nonmetal?

2.8　What is the essential difference between a chemical reaction and a physical change?

2.9 Which of the following events are physical changes?
(a) The splintering of rock into gravel
(b) The change in the color of leaves during the fall season
(c) The decay of plant remains on a forest floor
(d) The eruption of an oil well just "brought in" by drillers
(e) The seeming disappearance of sugar as it is stirred into coffee or tea
(f) The swing of a compass needle near another magnet
(g) The formation of a raindrop in a cloud

2.10 What fact is true about all elements and that distinguishes them from compounds and mixtures?

2.11 Roughly, how many elements are known: 80, 100, 200, or 2000?

2.12 Roughly, how many elements are not solids under ordinary room conditions, but are either liquids or gases: 2, 13, 44, 75, 180, or 1885?

2.13 What important facts about compounds distinguish them from either elements or mixtures?

2.14 What is meant in chemistry when something is described as a *pure substance?* What two kinds of substances are called *pure?*

2.15 What is the general name given to the event by which elements are changed into compounds?

2.16 Hydrogen reacts with oxygen to give water. What are the *reactant(s)* and the *product(s)* in this event?

2.17 What is an alloy?

2.18 What important mass relationship between reactants and products in a chemical reaction is always observed?

Dalton's Atomic Theory

2.19 If a sulfur atom had the same mass as an iron atom, what would be the *mass* ratio of iron to sulfur in iron(II) sulfide in which the atom ratio is 1 to 1?

2.20 In compounds consisting of just two elements, the mass ratios of the elements are never found to be simply 1 to 1. Which postulate of Dalton's atomic theory is based on this fact?

*2.21 Platinum forms two compounds with oxygen. In one, the combining ratio is 6.10 g of platinum to 1.00 g of oxygen. In the other, the combining ratio is 12.2 g of platinum to 1.00 g of oxygen. By means of a calculation, show how these two compounds illustrate the law of multiple proportions.

2.22 Two compounds of the elements tin and chlorine are known. When they are broken down into the separate elements, the following mass data are obtained from 100.0-g samples of each compound.

Compound A: 62.20 g of tin and 37.40 g of chlorine

Compound B: 45.56 g of tin and 54.44 g of chlorine

(a) From compound A, for every 1.000 g of tin, how many grams of chlorine are obtained?
(b) From compound B, for every 1.000 g of tin, how many grams of chlorine are obtained?

(c) What is the ratio of the mass of chlorine that combines with 1.000 g of tin in compound B to the mass that combines to make compound A?
(d) What law of chemical combination is illustrated by the results of part (c)?

2.23 What, fundamentally, must happen in order for some change to be described as chemical and not physical?

2.24 What are some observations or measurements that we can make to determine if a given change is chemical and not just physical?

Chemical Symbols, Formulas, and Equations

2.25 The symbol BN stands for a compound (boron nitride), not an element. How can we tell this *from the symbol itself?*

2.26 What are the symbols of the following elements?
(a) Iodine   (b) Lithium   (c) Zinc
(d) Lead   (e) Nitrogen   (f) Barium

2.27 Write the symbols for the following elements:
(a) Carbon   (b) Chlorine   (c) Copper
(d) Calcium   (e) Fluorine   (f) Iron

2.28 What are the symbols of the following elements?
(a) Hydrogen   (b) Aluminum   (c) Manganese
(d) Magnesium   (e) Mercury   (f) Sodium

2.29 Write the symbols for the following elements:
(a) Oxygen   (b) Bromine   (c) Potassium
(d) Silver   (e) Phosphorus   (f) Platinum

2.30 The symbol S represents the *element* sulfur. What else does this symbol stand for?

2.31 What are the names of the elements represented by the following symbols?
(a) P   (b) Pt   (c) Pb   (d) K
(e) Ca   (f) C   (g) Hg   (h) H
(i) Br   (j) Ba   (k) F   (l) Fe

2.32 Give the name of each element represented by the following symbols:
(a) S   (b) Na   (c) N   (d) Zn
(e) I   (f) Cu   (g) O   (h) Li
(i) Mn   (j) Mg   (k) Ag   (l) Cl

2.33 One formula unit of water is made from two hydrogen atoms and one oxygen atom. Which of the following formulas is the best for water?
(a) HO_2 (b) H_2O (c) $2HO$ (d) H_2O_4 (e) H_4O_2

2.34 Suppose that we wrote the formula of iron(II) sulfide as Fe_2S_2 instead of FeS. Would this violate a natural law — for example, the law of definite proportions — or a convention, or both?

2.35 What essential feature of FeS makes it a compound, not a mixture?

2.36 Write a balanced equation that symbolizes the following description of a chemical reaction: Iron combines with sulfur to give FeS_2.

Energy

2.37 Matter is a *thing*. What is energy?

2.38 Kinetic energy refers to what kind of energy? What is the equation that defines kinetic energy?

2.39 What is the law of conservation of energy?

2.40 If energy is conserved, what happens to the kinetic energy in a rock avalanche or slide?

2.41 A stick of dynamite lying on the ground has no potential energy that relates to its *location*, but it still has potential energy. What kind?

2.42 A lighted candle represents the conversion of chemical energy into what two other forms?

2.43 Is chemical energy one form of potential energy or kinetic energy?

*2.44 In the equation for kinetic energy, if m is in kilograms and v is in meters per second (m/s), the calculated value of the kinetic energy has the units of kg m^2/s^2, and these units define the SI unit of energy called the joule (J). The relationship between the joule and the calorie is 1 cal = 4.184 J (exactly).
  (a) How much kinetic energy, in joules, does a large station wagon with a mass of 1.97×10^3 kg (2.17 ton) have when it is traveling at a velocity of 24.4 m/s (55.0 mph)?
  (b) How much energy does this station wagon have in calories? In kilocalories?

2.45 Referring to the information about kinetic energy and the joule given in Exercise 2.44, what velocity in m/s does a compact car with a mass of 910 kg have if its kinetic energy is 5.86×10^5 J?

Heat Energy

2.46 *Heat* is the name we give to the form of energy that transfers from one object to another in what two kinds of situations?

2.47 What is the name of the temperature reading at which a substance changes from the solid to the liquid state?

2.48 The numerical value of the specific heat of granite is 0.192. What are the units (as studied in this book)?

2.49 Consider a body of water with a volume of 10^5 m^3. Which would be the larger value for this, its heat capacity or its specific heat? What relationship exists between heat capacity and specific heat?

2.50 The specific heat of gold is 0.031 cal/g °C. In order to calculate the heat capacity of a given bar of gold, what additional information is needed?

*2.51 What is the heat capacity of 1.00 g of water? Of 10.0 g of water?

2.52 The heat capacity of gold is 0.031 cal/g °C and that of olive oil is 0.471 cal/g °C. Suppose that you had 10.0 g samples of each at a temperature of 20.0 °C, and that each sample absorbed 25.0 cal of heat. First, judge which sample would experience a greater rise in temperature, and then calculate the final temperatures of each.

*2.53 Describe two circumstances in which water can absorb heat and not experience a change in temperature.

2.54 The numerical value for the heat of fusion of gold is 15. What are the units (as used in this book)?

2.55 Which substance would experience a greater change of its mass from the solid to the liquid form upon absorbing 25 cal of heat, gold (at its melting point) or ice (at 0 °C)? Explain.

2.56 If ice at 0 °C can absorb heat without any change in temperature, why doesn't this constitute a violation of the law of conservation of energy?

2.57 Why is ice at 0 °C far superior in an ice pack than the same mass of liquid water at 0.0005 °C?

2.58 Which could melt more ice, the heat in 100 g of water as steam at 100.0001 °C or as a liquid at 99.9999 °C? Explain.

2.59 What two terms can be used for the change of a liquid to its vapor state at any temperature at which this occurs?

2.60 What happens to the heat that changes water to steam at 100 °C when the steam condenses?

2.61 Water in the gaseous state is sometimes called *steam* and sometimes *water vapor*. Under what circumstance is each term used?

2.62 Would the physical change of steam to liquid water at just under 100 °C be properly described as an endothermic or an exothermic change?

2.63 What kind of change, exothermic or endothermic, is necessary to make ice at 0 °C change to liquid water?

2.64 Give an example of an exothermic chemical change.

*2.65 The specific heat of gold is 0.031 cal/g °C. If a bar of gold with a mass of 1.0 g at 25 °C is given the same quantity of heat as is needed to melt 0.10 g of ice at 0 °C, what will be the new temperature of the bar of gold?

2.66 The specific heat of iron is 0.119 cal/g °C. If a piece of iron with a mass of 1.00 g at 20.0 °C is given the same quantity of heat as is needed to vaporize 0.100 g of water at 100 °C, what will be the new temperature of the iron?

*2.67 The heat of vaporization of ethyl alcohol is 204 cal/g. How many grams of ethyl alcohol can be vaporized by the heat released when 100.0 g of steam (at 100 °C) condenses?

2.68 Ethyl chloride is sometimes used as a local anesthetic when boils are lanced. Because its boiling point is very low (12 °C), it quickly evaporates when sprayed on the skin at the site of the boil. The energy for this evaporation comes from the skin, which experiences a local drop in temperature — the desired response. At the lower temperature, the chemical reactions needed to send a pain signal don't occur rapidly enough to matter. The heat of vaporization of ethyl chloride is 93 cal/g. Suppose that a physician sprays 2.0 g of ethyl chloride on a boil, and that the affected site has a mass of 5.0 g. Further suppose that the specific heat of the exposed site is the same as that of water, 1.00 cal/g °C, and that the temperature at the site is initially 37 °C. If only the heat at this site and none from the surrounding air goes to make the ethyl chloride evaporate, what will be the final temperature of the exposed site, in °C and in °F? (A fall of skin temperature to 50 °F always causes it to feel numb and soon to lose all sense of touch and pain.)

Basal Activities and Metabolism

2.69 List six activities that are basal activities of the body.

2.70 What constitutes the body's *basal metabolism,* in general terms?

2.71 What does the term *basal metabolic rate* mean?

2.72 To one significant figure, what is the basal metabolic rate for adults in kcal/min?

2.73 What are the ways in which water becomes part of the body and, to two significant figures, how many liters of water are involved by each route?

2.74 What percent of the water lost by the body is lost by a change of state?

2.75 What name is given to the loss of body water by the evaporation that does not involve the sweat glands?

2.76 Name the body's three mechanisms for losing heat that do not involve evaporation directly.

2.77 What is the difference between radiation and conduction as means for losing heat from the body?

2.78 Wearing woolen clothing minimizes heat loss from the body by what mechanism?

2.79 What is hypothermia, and why is it dangerous to life?

2.80 What is hyperthermia, and how can it be life-threatening?

*2.81 Suppose that an adult woman with a weight of 58 kg carried out the following activities over a period of one day:

Sleeping, 8.0 h

Very light activities, 12.0 h

Light activities, 3.0 h

Heavy activities, 1.0 h

How many kilocalories are expended for these activities? Use the higher of the two rates (in kcal/min) given in Table 2.5 as the basis for your calculations. (Thus the metabolic rate for sleeping would be taken as 1.1 kcal/min. Remember, there are 60 min/h.)

2.82 Suppose that an adult male with a weight of 70 kg performed the following activities in a 24-hour day:

Sleeping, 8.0 h

Very light activities, 4.0 h

Moderate activities, 4.0 h

Heavy activities, 8.0 h

What is his energy requirement in kilocalories for this period? Use the higher of the two rates in the data of Table 2.5.

*2.83 The National Academy of Sciences uses the following conversion factors for the energy content of foods: proteins, 4.0 kcal/g; carbohydrates, 4.0 kcal/g; food fat, 9.0 kcal/g. If a 1-cup serving of milk (250 g) contains 8.4 g of protein, 12 g of carbohydrate, and 9.6 g of food fat, what is the energy content of this serving in kilocalories?

2.84 A 100-g portion (about ⅔ cup) of roasted, salted peanuts contains (besides some undigestible food fiber) the following: protein, 26 g; carbohydrate, 19 g; food fat, 50 g.

(a) What is the energy content of this portion in kilocalories? (Use the conversion factors given in Exercise 2.83.)

(b) If a man walking at 3.5 mph needs 5.0 kcal/min to maintain this activity, how many hours does he have to walk to "work off" the kilocalorie intake of the 100-g serving of peanuts?

(c) How far does he have to walk, in miles?

Chapter 3
Atomic Theory and the Periodic System of the Elements

The solar system, uncluttered by asteroids or comets, once served as a beautiful model for the structure of the atom. Atoms aren't flat, however, nor do electrons move in neat little orbits. We'll study a better model in this chapter.

3.1 THE NUCLEAR ATOM

The particles that make up atoms — nuclei and electrons — have opposite electrical charges, but the atom itself is uncharged.

Like most successful theories in science, Dalton's atomic theory opened many more areas for further study than it closed. Because the mass relationships of the elements in chemical compounds successfully pointed the way for Dalton, many scientists began to look for ways to measure the relative masses of the atoms of all of the elements. Others became increasingly interested in what atoms themselves might be made of.

Subatomic Particles.   Dalton had declared that atoms cannot be broken, but what he actually postulated was that they are not broken *in chemical reactions*. He couldn't have known about other ways of splitting atoms. Techniques for splitting atoms weren't developed until early in the 1930s when scientists subjected various elements to a number of high-energy conditions, and they found that atoms are made of smaller, subatomic particles. Several have been identified, but only three are needed to account for the masses and the chemical properties of atoms. These three **subatomic particles** are the **electron, proton,** and **neutron.** Their electrical conditions and their masses are summarized in Table 3.1.

The proton and the neutron appear to be made of still smaller particles that physicists have named *quarks*.

Two of the subatomic particles, the proton and the electron, carry electrical charge. You no doubt have had many experiences with the *fact* if not the vocabulary of such charges. For example, if you have ever received a shock after walking across a carpet (or touching a bare wire!), the spark that gave this shock was a discharge of an electrical charge. There are fundamentally two kinds of charge that give rise to two important phenomena, and you have probably experienced both. Have you ever tried to flick away a small piece of plastic, the kind used to wrap record albums, only to have it stick stubbornly to your fingers? Most annoying! You shake your hand harder, but it still sticks. This is because you are momentarily carrying one kind of electrical charge, and the plastic has picked up the opposite kind. *Opposite charges attract*. This is the first of two rules of behavior of electrical charges that are among the most important rules in all of chemistry. Maybe you have seen your hair stand on end after blow-drying it. The individual hairs act as if they repel each other, because each hair has picked up the *same* kind of charge. *Like charges repel* — this is the second of the two rules about electrical charges.

Like charges repel.

Attracting and repelling are opposites, so we designate one kind of charge as positive and the other as negative, and give them $+$ and $-$ signs. The proton has one unit of positive charge, $1+$. The electron has one unit of negative charge, $1-$. Electrons tend to repel each other, because they are like-charged; protons also tend to repel each other for the same reason. (Whether they actually succeed in pushing each other apart depends on how free they are to move.)

Electrons and protons tend to attract each other, because they are oppositely charged. The neutron has no charge — hence its name.

Atomic Mass Unit.   Table 3.1 gives the masses of the three subatomic particles both in grams and in a unit new to our study, the **atomic mass unit** or **amu.** This unit was invented

TABLE 3.1
Properties of Three Subatomic Particles

Name	Mass		Electrical Charge	Common Symbols
	in grams	in amu		
Electron	9.109534×10^{-28} g	0.0005486 amu	$1-$	e^-
Proton	1.672649×10^{-24} g	1.007277 amu	$1+$	p^+ or p
Neutron	1.674954×10^{-24} g	1.008665 amu	0	n

to let us express the masses of protons and neutrons in numbers that are simpler than those we have to use when we employ the gram unit:

$$1 \text{ amu} = 1.6606 \times 10^{-24} \text{ g}$$

Protons and neutrons have masses of 1.0 amu each, when we round to two significant figures. The electron has a mass only 1/1836 the mass of the proton. This is so small, relatively speaking, that we ignore it in working out relative masses of atoms. We say, for example, that the mass of an atom (in amu) is made up entirely of the sum of the masses of its neutrons and protons. This sum of neutrons and protons is called the atom's **mass number.**

No known atom has more than 108 electrons, and they contribute only about 0.02% to the mass of such an atom.

Mass number = number of protons + number of neutrons

Atomic Numbers.   We learned in the last chapter than an **element** is a *substance* that cannot be further broken down by chemical means into simpler substances and that it consists of *atoms*. At **atom** is the smallest sample of an element that has its chemical properties. We can now add that in an element *all of the atoms have the same number of protons*. This number is called the **atomic number** of the element.

An alphabetically arranged list of the elements with their atomic numbers appears inside the front cover.

Atomic number = number of protons in each atom of an element

Thus each element has its own unique atomic number, and to date the atomic numbers run to 108. Element number 6, for example, is carbon. Every atom of carbon has six protons.

Isotopes, Neutrons, and Mass Numbers.   Although every atom in any given element has the identical number of protons, these atoms do not always have the same number of neutrons. Among the atoms of almost any given element, the numbers of neutrons vary over a small range. These variations in numbers of neutrons are responsible for substances called isotopes. An **isotope** of an element is made up of those atoms of the element that not only have identical atomic numbers but also identical mass numbers. In other words, all atoms of any particular *isotope* have identical numbers of neutrons as well as identical numbers of protons.

Most elements consist of mixtures of a small number of isotopes, but the isotopes of the same element have essentially identical chemical properties. Naturally occurring hydrogen, the simplest element, consists of two isotopes. Both have the same number of protons per atom or the same atomic number, 1. They differ in their numbers of neutrons and, therefore, in mass numbers. One hydrogen isotope has a mass number of 1 and the other has a mass number of 2. The first isotope, occasionally called *protium,* has just one proton in each atom and no neutrons. The second isotope, usually called *deuterium,* has one proton per atom, but also one neutron. (This is why its mass number is 2, because 1 proton + 1 neutron = 2 particles with mass of 1 amu each.) (See Table 3.2.) As you can see from the last column in this table, deuterium occurs only in a trace amount relative to the chief isotope. For every 100,000 atoms of hydrogen in nature, 99,985 have a mass number of 1 and only 15 have a mass number of 2. There is a third isotope of hydrogen, called *tritium,* that has a mass number of 3 (1 proton + 2 neutrons), but it doesn't occur in nature except as a product of atomic reactors and their atomic wastes.

Hydrogen is the only element whose isotopes have unique names.

About 250 isotopes occur naturally. About 1100 more have been made using nuclear reactors. Many synthetic isotopes are used in medicine.

Isotopes, as we have said, share the same chemical properties. When an element undergoes *chemical* changes, the atoms of its various isotopes do not react differently nor do they separate from each other. Therefore we can use a common atomic symbol for an element

TABLE 3.2
Isotopes of Hydrogen

Name	Symbol[a]	Protons	Neutrons	Mass Number	Natural Abundance
Hydrogen	$_1^1H$	1	0	1	99.985%
Deuterium	$_1^2H$	1	1	2	0.015%
Tritium	$_1^3H$	1	2	3	trace[b]

[a] Sometimes deuterium is symbolized as $_1^2D$ and tritium as $_1^3T$.

[b] Tritium is radioactive and decays (which we will study in Chapter 10). Its only occurrence is as part of short-lived radioactive wastes from atomic power plants and in cosmic rays.

in almost all situations. When we need a symbol that specifies just one isotope, we use the atomic symbol but add the specific mass number and atomic number as a left superscript and a left subscript, as illustrated in the following:

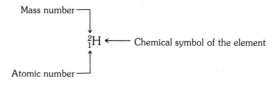

Deuterium

EXAMPLE 3.1   INTERPRETING SYMBOLS FOR ISOTOPES

Problem: The atoms of one of the isotopes of carbon have six protons and seven neutrons. Write the symbol for this isotope.

Solution: We need the chemical symbol, which is C, and we also need the atomic number, which is given by the number of protons, 6. The mass number is the sum of the protons and neutrons.

$$\text{Mass number} = \text{protons} + \text{neutrons}$$
$$= 6 + 7$$
$$= 13$$

You aren't expected to memorize the atomic numbers or mass numbers that go with the elements, although you're bound to learn a few through repeated usage.

The symbol, therefore, is $_6^{13}C$.

PRACTICE EXERCISE 1   The most abundant isotope of oxygen consists of atoms with eight protons and eight neutrons. Write its special symbol.

PRACTICE EXERCISE 2   How many neutrons and protons are in the atoms of each of the following isotopes?

(a) $_8^{17}O$ (b) $_7^{14}N$ (c) $_{17}^{37}Cl$ (d) $_{17}^{35}Cl$

PRACTICE EXERCISE 3   If $_6^{12}C$ is the correct symbol for one isotope of carbon what is incorrect about the symbol $_7^{14}C$ for another isotope of carbon?

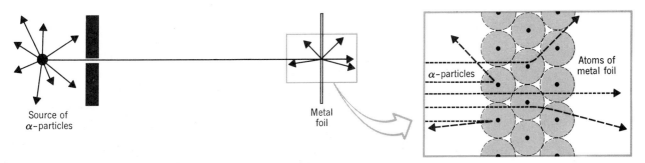

FIGURE 3.1
The discovery of the atomic nucleus. The α (alpha) particles consist of two protons and two neutrons, a "package" that some radioactive elements emit in high-energy streams called alpha rays. Most of the α particles passed through the foil, although some were deflected. A few particles bounded right back as if they had hit something far more massive than an electron.

Rutherford's discovery of the nucleus earned him the 1908 Nobel prize in chemistry and, in 1930, the title Baron Rutherford of Nelson.

Ernest Rutherford (1871 – 1937)

The Atomic Nucleus.   The impression of the atom that one might get from Dalton's theory is that it is a hard sphere, like a billiard ball (only very small). In 1911, British scientists working under Ernest Rutherford found evidence that pointed to a softer image. They found that when streams of certain subatomic particles from a radioactive element were allowed to strike a very thin metal foil, most of the particles sailed right through with no change in course. Only a few bounced back, and many went through with various angles of deflection. It was as if the metal foil were mostly empty space, like chain-link fencing, but that at some places there were particles massive enough to bounce the subatomic "bullets" back again. (See Figure 3.1.) By studying the angles at which many of the particles careened through the foil, Rutherford deduced that the massive particles in the foil were positively charged and contained virtually all of the mass of an atom. Thus, he concluded, an atom must be mostly empty space around a dense, massive inner core, to which he gave the name **nucleus.**

We now know that any given atom has only one nucleus and that all of the atom's protons and neutrons are located in it. The protons cause the positive charge on the nucleus and the *number* of protons equals the size of this charge. In other words, the atomic number of an element turns out to be equal to the size of the positive charge carried by the element's atomic nuclei.

Atomic number = + charge on nucleus = number of protons

Now we need to find out where the electrons are in atoms. This knowledge will help us understand many chemical properties of substances at the molecular level of life.

3.2 ELECTRON CONFIGURATIONS

The electrons in an atom are confined to particular regions of space near the nucleus, and they possess particular values of energy.

We may seem far from the molecular basis of life right now, but don't give up. Remember that an apple seed doesn't look much like an apple, and right now we're planting conceptual seeds for everything that follows.

In the next chapter, we will learn that the fundamental event that occurs when atoms combine to form compounds is that the electrons of the atoms become redistributed relative to the nuclei. Therefore to understand how atoms combine to give compounds, we have to learn the number of electrons a particular atom possesses and, more important, where they are. The question of how many electrons is easy, because we know that atoms are electrically neutral; the number of electrons must, therefore, equal the number of protons. And we know how many protons are present from the atomic number.

> Atomic number = number of protons = number of electrons

Thus an atom carrying six protons in its nucleus has a total charge of 6 +. To balance this and let the atom be electrically neutral, there must be a total of 6 − in negative charge. This requires six electrons.

At atom's electrons are not randomly distributed. They are constrained to particular patterns, and the specific description of the arrangement of electrons about a nucleus is called the atom's **electron configuration.**

The Bohr Atom — An Early Atomic Model.

In 1913, only two years after Rutherford's discovery of the nucleus, Niels Bohr (1885–1962), a Danish physicist, proposed that an atom's electrons are in very rapid motion around the nucleus, and that they follow paths, called *orbits,* much as planets follow orbits around the sun. Bohr's image of an atom is an example of a scientific **model,** a mental construction — often involving a picture or drawing — used to explain a number of facts. The **Bohr model** of the atom was quickly dubbed the ''solar system'' model, and it is still commonly used in the media to accompany almost any discussion of things atomic.

The Bohr model was soon discarded — it didn't explain enough facts — but two of the ideas or postulates behind the model are still firmly fixed in atomic theory. Bohr's first postulate was that electrons cannot be just anywhere, like buzzing mosquitoes around one's head; they are confined to what came to be called *allowed energy states.* (Bohr likened these to planetary orbits.)

Bohr's second postulate was that as long as the atom's electrons remain in allowed energy states, the atom neither radiates nor absorbs any energy associated with the electrons' movements. Not that it isn't possible to make an electron move from one allowed state to another. This is what happens, for example, when an iron fireplace poker is heated until it glows in the dark. Before being heated, the iron atoms are nearly all in their *ground state* arrangement of electrons — all their electrons are in the lowest energy states available. The heat causes electrons in iron atoms to shift to higher energy states called *excited states.* This is partly how the iron actually soaks up heat energy. As soon as atoms have become excited this way, the electrons begin to shift back to lower states *and the difference in energy between the excited state and the lower state is emitted as light.*

Scientists had known long before Bohr appeared that the emitted light did not possess every conceivable value of light energy, but had only certain values. This is what led Bohr to believe that atoms had only certain allowed energy states. The precise quantity of energy that is emitted when one electron changes from a higher to a lower energy state is called a **quantum** of energy, and sometimes a **photon** of energy. Sometimes electricity rather than heat is used to generate excited atoms that then emit light. Sodium vapor lamps or mercury vapor lamps along highways work this way. And you know that the sodium vapor lamps have a characteristic color. Excited sodium atoms don't emit all colors of light, just yellow light, and the pure yellow light from sodium lamps is characterized by photons of just two, very nearly equal values of energy.

Heisenberg's Uncertainty Principle.

The Bohr model worked well only for one element, hydrogen. The problem, as a German physicist, Werner Heisenberg (1901–1976), soon realized, is that the mathematical calculations based on the Bohr model assumed that both the location of an electron and its energy can be precisely known at any given instant. This is untrue. The electron is small enough that any act of measuring its location gives it a nudge and changes its energy, and any attempt to find its energy changes its location. Both the location and energy of an electron cannot be simultaneously known. There is always some uncertainty about either the location or the energy of an electron, and this statement came to be called the **Heisenberg uncertainty principle.**

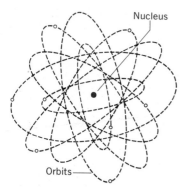

The Bohr ''solar system'' model

Niels Bohr

Light quantum emitted

Heisenberg won the 1932 Nobel prize in physics.

The idea that the very act of measuring something actually alters what is being measured has profound implications for the fields of pyschology, sociology, poll taking, and even television news.

As a consequence of the work of Heisenberg, scientists soon gave up the idea that an electron moves in a fixed orbit. Instead, the location of a moving electron came to be described in terms of probabilities. The question became: In what particular parts of the space surrounding a nucleus is it *likely* that an electron will be? It's like asking: If I go out a certain distance from the nucleus in one particular direction, what are the chances of an electron being there? The answer depends first and foremost on what energy state we're talking about.

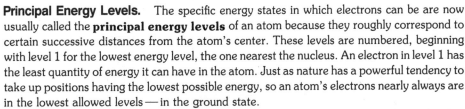

Main energy levels

Principal Energy Levels. The specific energy states in which electrons can be are now usually called the **principal energy levels** of an atom because they roughly correspond to certain successive distances from the atom's center. These levels are numbered, beginning with level 1 for the lowest energy level, the one nearest the nucleus. An electron in level 1 has the least quantity of energy it can have in the atom. Just as nature has a powerful tendency to take up positions having the lowest possible energy, so an atom's electrons nearly always are in the lowest allowed levels — in the ground state.

Not all electrons can crowd into level 1. Remember that electrons are like-charged and so they repel each other. Each principal energy level has a limit to its number of electrons, and the limit for level 1 is only two. At level 2, the limit is eight electrons — a larger number because there is more room farther from the nucleus. Table 3.3 summarizes what has been observed about all levels.

Although the principal energy levels are spaced at certain distances from the atom's nucleus, don't think of a level as some sort of shelf. If you've ever studied a whole onion, you know that it exists in sections, each one a hollow sphere with thick walls, and each successively larger. They share a common center, so they are called concentric spheres. Think of the principal energy levels as concentric spheres, too, and imagine that each one has a definite thickness. In fact, many scientists call the principal energy levels an atom's **electron shells.**

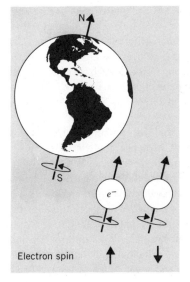

Electron spin

Atomic Orbitals. The thickness of an electron shell allows for some fine structure. All of the electron shells except the one corresponding to the lowest energy (level 1) have a small number of **sublevels.** It's as if an atom is an apartment house for electrons (with the nucleus in the basement) and each floor is a principal shell. Each floor can have one, two, three, four, or five apartments called sublevels. The complexity doesn't end here.

Each sublevel has a certain definite number of spaces where electrons can be. These are called **atomic orbitals,** and they are particularly shaped volumes of space that can hold up to two electrons apiece. It's as if each apartment has a certain number of rooms for electrons, but no room can hold more than two.

The final complexity is that electrons are *spinning* particles. Like the earth, electrons spin about an axis, and they can spin in one or the other of two opposite ways. When an atomic orbital holds two electrons, they can be present only if one electron spins in a direction opposite to that of the other. Part of the reason is that spinning electrons behave like tiny magnets, and by spinning oppositely, the two tiny magnets behave as if they have a magnetic attraction for each other. This helps to overcome the electrical repulsion experienced between two like-charge electrons that must be in the same small space.

Now let's look more closely at the kinds of rooms, the atomic orbitals, that are available in the apartments. Each atomic orbital has a particular shape. These shapes have been deduced from the mathematical operations used to calculate the probabilities of finding

TABLE 3.3
The Principal Energy Levels

Principal Level Number	1	2	3	4	5	6	7
Maximum number of electrons observed in nature[a]	2	8	18	32	32	18	8

[a] In theory, levels 5, 6, and 7 could accommodate 50, 72, and 98 electrons, respectively. Levels 1, 2, 3, and 4 are filled in theory as well as among certain elements by the numbers of electrons given.

electrons. The shape of an orbital is simply what you get when you wrap an imaginary envelope around enough of a particular space to enclose a region of high probability — say, 90% — of having an electron somewhere within it.

It turns out that at principal level 1, there is only one sublevel and it constitutes an entire atomic orbital. (It's as if the ground floor — the first sublevel — of the apartment house has only one apartment and it consists of just one room.) Figure 3.2 shows the shape of this orbital. It's called the $1s$ orbital — 1 for level 1 and s for a German word of no interest here but we can think of its as meaning "spherical." The $1s$ orbital looks like a sphere when viewed from the outside. In its center is the nucleus. The surface of the sphere encloses a space within which the probability of finding an electron belonging to level 1 is greater than 90%. An electron in a $1s$ orbital moves about, but we can't know *exactly* how or where. Fortunately, this doesn't matter. It will turn out to be important that we know only if an electron is in a $1s$ state, not precisely where in this state.

The fact that an electron does move about within an orbital — it moves about extremely rapidly — gives us another useful image, that of an **electron cloud.** An electron moves so rapidly within an orbital that it seems to be everywhere at once. At least the influence of its negative electrical charge is evened out within the orbital, somewhat as a cloud is an evened-out distribution of moisture in a volume of very humid air.

Principal level 2 has two sublevels. One sublevel consists of only one orbital. It's named the $2s$ orbital, and it corresponds to slightly less energy than that of the other sublevel. The $2s$ orbital looks like a sphere from the outside, which is why it is designated an s orbital — it resembles the s orbital at level 1. The other sublevel holds three orbitals. They all correspond to the same energy, but they are pointed at right angles to each other, as seen in Figure 3.3. Each looks like two small spheres pierced by one of the axes — x, y, and z. Orbitals that have this kind of shape are named p orbitals, and when they belong to the 2 level, they are named the $2p$ orbitals — the $2p_x$, the $2p_y$, and the $2p_z$. We say that each of them has two *lobes,* and where these lobes touch is where the three axes — x, y, and z — touch. This point between lobes is called a *node.*

Principal level 3 has three sublevels. (You may notice that the number of sublevels at a principal level is the same as the number of the principal level. At level 1, there is one sublevel; at level 2, there are two sublevels; and now at level 3, there are three sublevels.) One of the sublevels at level 3 is an orbital all by itself — the $3s$ orbital; it's of the s-type because it looks from the outside like the other s orbitals. At the second sublevel of level 3, there are p-type orbitals — the $3p_x$, the $3p_y$, and the $3p_z$. They resemble the p orbitals of level 2. Finally, the third sublevel at level 3 is made up of five orbitals called d orbitals. Having mentioned them, we'll not give them individual names nor deal with them further here. We won't be needing them for the elements that are most important in our study. At principal level 4, there are four sublevels, one $4s$ orbital, three $4p$ orbitals, five $4d$ orbitals, and seven of a new type, called f orbitals.

$\overline{2p_x}\ \overline{2p_y}\ \overline{2p_z}$

$\overline{2s}$

Sublevels at
main level 2

$3d$ _ _ _ _ _
$3p$ _ _ _
$3s$ _
Sublevels at
main level 3

$4f$ _ _ _ _ _ _ _
$4d$ _ _ _ _ _
$4p$ _ _ _
$4s$ _
Sublevels at
main level 4

FIGURE 3.2
The $1s$ orbital. *(a)* Imagine that the space around a nucleus is made up of layer upon layer of extremely thin, concentric shells. At each distance away from the nucleus there is a point on the curve that indicates the probability of finding a $1s$ electron at this distance. Notice that the probability is zero for a zero distance — the electron is not on the nucleus. The probability reaches a maximum at the radius marked a_0, which is 52.9 pm (1 pm = 10^{-12} m). *(b)* One of the thin spheres described in *(a)* encloses a space within which the total probability of finding an electron is large, say, 90%. This sphere is the "envelope" discussed in the text, and its shape is the shape of the $1s$ orbital.

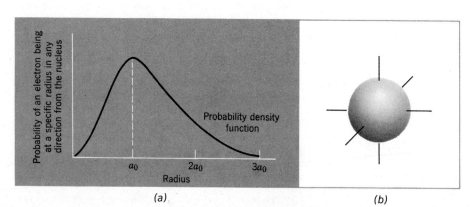

(a)   (b)

FIGURE 3.3
The orbitals at principal energy level 2. From the outside, the 2s orbital looks like a 1s orbital, seen in Figure 3.2. However, its radius is larger. Each of the 2p orbitals has the same shape and energy as the others, but they are oriented along the three different axes.

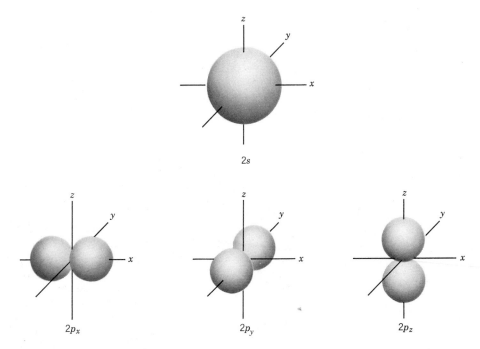

The circles in Figure 3.4 represent all of the types of atomic orbitals that we have just introduced. The thick, horizontal stripes correspond to the main levels. (Don't worry about the arrows in this figure yet.) Notice that the 3d orbitals actually correspond to slightly lower energies than the 4p orbitals. There are other peculiarities such as this that we have to leave to more advanced treatments.

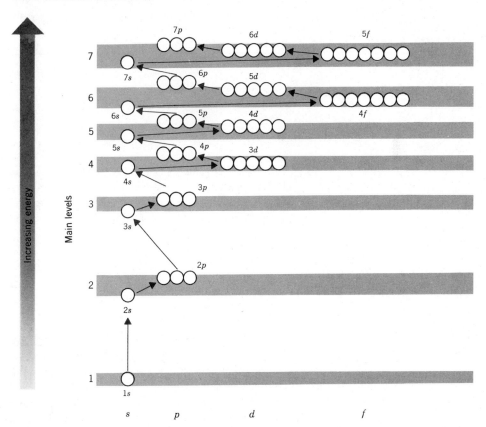

FIGURE 3.4
The atomic orbitals, showing their relative energies and the order in which the sublevels fill in accordance with the aufbau rules.

Think of the names of individual orbitals as addresses of individual electrons. An electron in, say, the $3p_y$ orbital is located in principal level 3, in a sublevel that consists of p orbitals, and it specifically is in the p orbital whose axis is the y axis.

We said that electrons can be in the same orbital only if their spins are in opposite directions. We also said that no orbital can hold more than two electrons. Wolfgang Pauli (1900–1958), an Austrian-born physicist, was the first to realize this, and we call the rule the **Pauli Exclusion Principle.**

Wolfgang Pauli won the 1945 Nobel prize in physics.

> Pauli Exclusion Principle An orbital can hold as many as two electrons, but only if they have opposite spin.

As a symbol for the direction of spin, we use an arrow, which can point either down ($\downarrow$) or up ($\uparrow$). However, because two electrons in the same orbital *must* have opposite spins, we seldom need such arrows. Usually, the symbol used for a pair of electrons in the same orbital is a superscript number as in $1s^2$.

This number designates the principal level ———

$1s^2$

——— This superscript says that there are two electrons in this orbital.

——— This letter specifies the kind of orbital.

Hund's Rule.   We're almost ready to describe electron configurations of atoms, but there is one more rule that governs them, called **Hund's rule.**

> Hund's Rule Electrons at the same sublevel spread out among the sublevel's orbitals as much as possible.

Hund's rule arises when we have to decide how to distribute two or more electrons among orbitals of identical energy, such as three p orbitals at the *same* principal energy level. Because electrons are like-charged, they tend to be as far from each other as possible. Therefore when it makes no difference in terms of the orbital energies, electrons spread out among the same set of orbitals. When they do, they have the same spins. In other words, the distribution on the left is more stable than the one on the right.

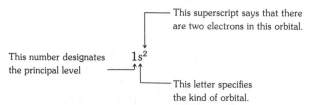

Now we're finally ready to study electron configurations of some representative elements.

Electron Configurations of Elements 1 through 20.   The question we raise here is what is the most stable distribution of the electrons among the orbitals of, say, element 13, aluminum? (Remember, we need to know this to *understand* why different elements have certain chemical properties. Instead of memorizing the electron configurations of every kind of atom, we use just a few rules — the **aufbau rules** — and figure out the electron configuration whenever we need it from the atomic number.)

Aufbau = "building up" in German.

1. The atomic number tells us how many electrons to distribute.

2. Electrons are placed into the orbitals of *lowest* energy that are available, provided that:

    (a) No more than two electrons go into the same orbital, and then only if they have opposite spin — Pauli exclusion principle.

    (b) Electrons are spread out as much as possible — retaining the same spins — when orbitals of the *same* sublevel are open — Hund's rule.

Let us apply these rules to a few elements, beginning with the simplest, hydrogen, atomic number 1.

A hydrogen atom has one electron. We have learned that the $1s$ orbital has the lowest associated energy, so this is where the electron resides.

H $1s^1$

meaning $1s$ ↑

Helium is a gas used to fill dirigibles such as the Goodyear blimp. It is much less dense than air, and it won't burn.

Helium has atomic number 2 and therefore its atoms have two electrons. Both can (and must) go into the $1s$ orbital.

He $1s^2$

meaning $1s$ ↑↓

Lithium, atomic number 3, has three electrons in each atom. The first two fill the $1s$ orbital. The third electron must go to the next lowest orbital, the $2s$. Here is where the arrows in Figure 3.4 help; they show the order of filling of the orbitals, and notice that the order goes from lowest energy successively upward to higher energy values.

Li $1s^2 2s^1$

meaning $2p$ __ __ __
$2s$ ↑
$1s$ ↑↓

Think of the three lines at the $2p$ sublevel as representing, in order, the $2p_x$, $2p_y$, and the $2p_z$ orbitals.

Beryllium, atomic number 4, has four electrons per atom. The first two fill the $1s$ orbital, and the last two fill the $2s$ orbital. None enters a $2p$ orbital, because these orbitals are at a higher energy level (Figure 3.4), and we must fill the lower energy orbitals first.

Be $1s^2 2s^2$

meaning $2p$ __ __ __
$2s$ ↑↓
$1s$ ↑↓

Borax, a cleansing agent, contains boron.

Boron, atomic number 5, has five electrons per atom. The first four fill the $1s$ and the $2s$ orbitals, exactly as in beryllium (number 4), and boron's fifth electron enters a $2p$ orbital. We don't know which of the three takes it, so we just arbitrarily assign it to the $2p_x$ orbital.

B $1s^2 2s^2 2p_x^1$

meaning $2p$ ↑ __ __
$2s$ ↑↓
$1s$ ↑↓

The next element, carbon (atomic number 6), is of central importance at the molecular

All but a handful of the roughly six million compounds of carbon are classified as *organic compounds*.

level of life, because its atoms make up most if not quite all of the "backbones" of molecules, other than water, that are in living cells.

$$\text{C} \quad 1s^2 2s^2 2p_x^1 2p_y^1$$
$$\text{meaning} \quad 2p \uparrow \uparrow \underline{}$$
$$2s \uparrow\downarrow$$
$$1s \uparrow\downarrow$$

Notice that carbon illustrates the application of Hund's rule. The last two electrons go into *separate* orbitals at the 2p sublevel.

Nitrogen, number 7, is another very important element among biological chemicals. It also illustrates Hund's rule.

Air is about 79% nitrogen.

$$\text{N} \quad 1s^2 2s^2 2p_x^1 2p_y^1 2p_z^1$$
$$\text{meaning} \quad 2p \uparrow \uparrow \uparrow$$
$$2s \uparrow\downarrow$$
$$1s \uparrow\downarrow$$

Air is about 21% oxygen.

Oxygen (atomic number 8) illustrates that we don't start with level 3 until level 2 is filled. Oxygen's eighth electron goes into a 2p orbital, not the 3s.

$$\text{O} \quad 1s^2 2s^2 2p_x^2 2p_y^1 2p_z^1$$
$$\text{meaning} \quad 2p \uparrow\downarrow \uparrow \uparrow$$
$$2s \uparrow\downarrow$$
$$1s \uparrow\downarrow$$

Fluorine is so reactive that it burns with water.

Fluorine (atomic number 9) has nine electrons, and we continue to fill the 2p orbitals.

$$\text{F} \quad 1s^2 2s^2 2p_x^2 2p_y^2 2p_z^1$$
$$\text{meaning} \quad 2p \uparrow\downarrow \uparrow\downarrow \uparrow$$
$$2s \uparrow\downarrow$$
$$1s \uparrow\downarrow$$

Neon is the gas in "neon" lights.

With the next element, neon (atomic number 10), we complete the filling of all of the atomic orbitals at level 2.

$$\text{Ne} \quad 1s^2 2s^2 2p_x^2 2p_y^2 2p_z^2$$
$$\text{meaning} \quad 2p \uparrow\downarrow \uparrow\downarrow \uparrow\downarrow$$
$$2s \uparrow\downarrow$$
$$1s \uparrow\downarrow$$

Level 2 now has its maximum of eight electrons, two at the 2s sublevel and six at the 2p sublevel.

With element 11, sodium, we start to fill the third main level.

$$\text{Na} \quad 1s^2 2s^2 2p_x^2 2p_y^2 2p_z^2 3s^1$$
$$\text{meaning} \quad 3s \uparrow \underline{}\ \underline{}$$
$$2p \uparrow\downarrow \uparrow\downarrow \uparrow\downarrow$$
$$2s \uparrow\downarrow$$
$$1s \uparrow\downarrow$$

The aufbau rules, as least as far as we have developed them, continue to apply through element 18, argon. We won't be concerned with the electron configurations of elements beyond element 20, calcium, because nearly all of the elements that are important at the molecular level of life are among the first 20. These include potassium (19) and calcium (20). If you look again at Figure 3.4, you will see that the 4s orbital accepts electrons before the 3p. Therefore the electron configuration of potassium is $1s^2 2s^2 2p^6 3s^2 3p^6 4s^1$. (Notice that when a p sublevel has all its orbitals filled, we can save a little space by writing p^6 instead of showing the individual orbitals with two electrons each.)

When we write the electron configuration of an element, we do so for any of its isotopes. Atoms of isotopes of the same element differ only in their numbers of neutrons, but these are in the nucleus and have nothing to do with electron configurations. Because isotopes have the same electron configurations, they have the same chemical properties.

The electron configurations of all of the elements through number 103 are in Appendix B.

EXAMPLE 3.2   **WRITING AN ELECTRON CONFIGURATION**

Problem: An atom has 15 electrons. Write its electron configuration.

Solution: With 15 electrons, we know that both levels 1 and 2 are filled; they take $2 + 8 = 10$ electrons. The remaining five must be in level 3.

$$1s^2 2s^2 2p^6 3s^2 3p_x^{\,1} 3p_y^{\,1} 3p_z^{\,1}$$

meaning $3p \uparrow \ \uparrow \ \uparrow$

$3s \uparrow\downarrow$

$2p \uparrow\downarrow \ \uparrow\downarrow \ \uparrow\downarrow$

$2s \uparrow\downarrow$

$1s \uparrow\downarrow$

PRACTICE EXERCISE 4   Using the one-line representation rather than the system with arrows, write the electron configuration of each of the following elements. Use the table inside the front cover to find out their atomic numbers.

(a) magnesium   (b) chlorine   (c) argon   (d) calcium

Relative Atomic Masses, Mass Numbers, and Atomic Weights.   Even though an atom's neutrons contribute nothing to the chemistry of an isotope, they do contribute to mass. And in the lab, we have to use measurements of mass to obtain particular numbers of atoms. We are interested in *numbers* of atoms, because atoms combine as whole units when they form compounds. However, we can't count atoms directly. Atoms are so small that even if we had a sample of the element carbon consisting of 6.02×10^{23} atoms, the sample would have a mass of only 12.0 g. Even if we could see, handle, and count atoms, it would take nearly 200 thousand billion centuries to assemble a pile of carbon atoms this size, counting at a rate of 2 atoms per second. Obviously, we need an indirect method for counting particles this small.

You probably know that it is possible to count such things as pennies by weighing them. All we need is to measure the mass of one penny or, if the pennies are of different ages and have not all worn down similarly, the average mass of several pennies. If a penny has an average mass of say, of 3.00 g (to pick a round number), and a pile of pennies has a mass of 300 g, you can tell that there are 100 pennies in the pile. We do something like this to count atoms by weighing them. What we need is something analogous to the average mass of all of the atoms of the various isotopes that make up a naturally-occurring sample of an element. We'll work with the element magnesium, which consists of three isotopes, $^{24}_{12}Mg$, $^{25}_{12}Mg$, and $^{26}_{12}Mg$. In naturally-occurring magnesium, 78.70% of all atoms are magnesium-24; 10.13% are magnesium-25; and 11.17% are magnesium-26. Because most of the atoms have mass numbers of 24, we'd naturally expect that the mass of the "average" atom is closer to 24 than to 25 or 26 amu. The accepted value for the average atomic mass of magnesium is 24.305 amu.

The strictly numerical part of the average atomic mass (in amu) of the atoms of an element, as they occur naturally, is called the **atomic weight** of the element. A Table of Atomic Weights and Numbers appears inside the front cover. Notice that the atomic weight of carbon is 12.001. We just saw that the atomic weight of magnesium is 24.305. In other words, an atom of magnesium is roughly twice as heavy as an atom of carbon. Remember that Dalton said that the atoms of different elements differ in mass. The atomic weights provide a way to describe how different. Notice also in this table that atomic weights are reported to varying numbers of significant figures. This only means that it hasn't yet been possible to measure atomic weights as precisely for some elements as for others.

Most scientists use the expression *atomic weight* instead of the technically better term, *atomic mass*.

3.3   THE PERIODIC LAW AND THE PERIODIC TABLE

In a given family of representative elements, the atoms of all members have the same outside-level electron configurations.

If each of the 108 elements were completely unlike any of the others, the study of chemistry would be far more difficult. Fortunately, chemists have found that the elements can be sorted into a small number of families whose members have much in common. Dimitri Mendeleev, a Russian scientist, was the first to notice this as he wrote a chemistry textbook for his students, published in 1869.

Dimitri Mendeleev (1834–1907)

Periodic Properties.   When he organized some of the physical and chemical properties of the elements known in his time, Mendeleev observed that the properties seemed to go through cycles from the elements of lowest atomic weight to the highest. The cycles weren't geometrically perfect, but there were definite rises and falls. One improvement, which came with the discovery of atomic numbers a few decades after Mendeleev's work, was to use atomic numbers instead of atomic weight as the basis of ranking the elements. As seen in Figure 3.5a, the temperatures at which the first 20 elements boil do not keep rising ever higher as the atomic numbers increase. Instead, the boiling points rise and fall. The ionization energies of the elements display a similar rising and falling in values, as seen in Figure 3.5b. An element's ionization energy is the energy needed to make one electron leave each atom in a sample whose mass equals the element's atomic weight in grams. Notice that in the plot of boiling points versus atomic numbers, helium, neon, and argon are at the bottoms of cycles. They are at the tops in Figure 3.5b. They seem to form a set of elements having similar properties.

The combining abilities of the atoms of one element for atoms of another also go through a cyclical rise, fall, rise, fall pattern. For example, most of the first 20 elements form binary compounds with hydrogen. A **binary compound** is one made of only two elements.

An atom of atomic number:	3	4	5	6	7	8	9	10
—can bind these many H atoms:	1	2	3	4	3	2	1	0
The formulas are	LiH	BeH$_2$	BH$_3$	CH$_4$	NH$_3$	H$_2$O	HF	—

This pattern repeats itself as we go to still higher atomic numbers.

An atom of atomic number:	11	12	13	14	15	16	17	18
—can bind these many H atoms:	1	2	3	4	3	2	1	0
The formulas are	NaH	MgH$_2$	AlH$_3$	SiH$_4$	PH$_3$	H$_2$S	HCl	—

In both series, we increase the number of hydrogens from 1 to 4 and then see the number fall back again. The number of hydrogens doesn't just keep rising to ever higher values as we go to elements of higher atomic numbers.

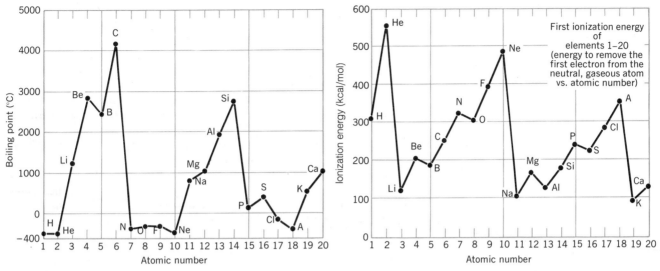

FIGURE 3.5
Periodicity and two properties of elements 1–20. (a) Boiling points versus atomic numbers. (b) Ionization energies versus atomic numbers.

Periodic means the reoccurrence of something in a regularly repeating way, such as lamp posts or sunsets.

Just as neon (10) and argon (18) seem to be similar with respect to boiling points and ionization energies, they also similarly form no compound with hydrogen; neither does helium. Elements 6 and 14 are similar in their ability to bind four hydrogen atoms, and they similarly share high boiling points (Figure 3.5a). Periodically, with increasing atomic number, physical and chemical properties reoccur — more or less. This is the essence of the **periodic law,** one of the important laws of nature.

Periodic Law The properties of the elements are a periodic function of their atomic numbers.

Periodic Table.   The heart of Mendeleev's discovery (using atomic weights as the basis, not atomic numbers) was that by making breaks in the list of elements and starting new rows in order to place similar elements, such as 3 and 11, in the same column, the rest of the columns *automatically* lined up vertically into sets of similar elements. The result of doing this is called the **periodic table,** shown inside the front cover of this book. (It incorporates elements discovered since the time of Mendeleev.) Each horizontal row in the periodic table is called a **period** and each vertical column is called a **group.**

Mendeleev had the boldness to leave blanks in the columns whenever this seemed necessary to get elements to line up vertically in families. He even went so far as to declare that these blanks represented elements that had not yet been discovered — and he was right. In order to achieve the best vertical sorting into families, he even switched some pairs of elements from their order of increasing atomic weights. He listed, for example, tellurium (atomic weight 127.6) *before* iodine (atomic weight 126.9), because iodine seemed to fit far better with fluorine, chlorine, and bromine in group VIIA than with oxygen, sulfur, and selenium in group VIA. When atomic numbers were discovered, it was gratifying to find that placing tellurium before iodine put these elements in the correct order according to increasing atomic numbers. Mendeleev's observation and his periodic law aroused considerable interest in finding underlying causes for the similarities of members of the same groups.

Hydrogen isn't a member of group IA in any chemical sense, but it is often located in the periodic table where it gives this appearance.

You have probably noticed that the periods in the periodic table are not all of the same length and that several are broken. This is necessary if the highest priority is to be the nature of the vertical columns or groups — the members of these groups must, as much as possible, have similar chemical properties. Thus period 1 is very short, containing only hydrogen and helium, and like the next two periods it is separated into two parts. The groups have both numbers and letters. One set of periods make up an A-series — IA, IIA, IIIA, and so forth up to VIIA. This set plus group 0 are called the **representative elements.** The groups in the B-series, clustered near the middle of the periodic table, are called the **transition elements.** The two rows of elements placed outside the table are the **inner transition elements.** (The table would not fit well on the page if the inner transition elements were not handled this way.) Elements 58 through 71 constitute the *lanthanide* series, named after element 57 which just precedes this series. The series of elements 90 to 103 is the *actinide series.*

Several of the groups of representative elements also have common family names. Those in group IA, for example, are called the **alkali metals,** because they all react with water to give an alkaline or caustic (skin-burning) solution. If we let M represent any alkali metal, the general equation for this reaction is

$$2M \;+\; 2H_2O \longrightarrow 2MOH \;+\; H_2$$

| Alkali metal | Water | Alkali metal hydroxide | Hydrogen |

NaOH flakes are an ingredient in one kind of drain cleaner. Use it very carefully and keep it out of reach of children.

For example, if M is sodium, Na, the alkali metal hydroxide that forms is sodium hydroxide, NaOH, commonly known as lye.

The elements in group IIA are called the **alkaline earth metals,** because they are commonly found in "earthy" substances such as limestone. For example, calcium carbonate or $CaCO_3$ is a compound of calcium, an alkaline earth metal of group IIA, and this compound is the chief substance in limestone.

The elements in group VIIA are called the **halogens** after a Greek word signifying salt-forming ability. Chlorine of group VIIA, for example, is present in a chemically combined form in table salt, NaCl (sodium chloride).

The elements in group 0, all gases, were discovered after Mendeleev's work. Except for a few compounds that xenon and krypton form with fluorine and oxygen, these elements chemically react with nothing. For this reason, they are called the **noble gases** (*noble* signifying limited activity).

Other groups of representative elements are named simply after the first member — for example, the **carbon family** (group IVA), the **nitrogen family** (group VA), and the **oxygen family** (VIA).

PRACTICE EXERCISE 5   Referring to the periodic table, pick out the symbols of the elements as specified.

(a)   A member of the carbon family: Sr, Sn, Sm, S
(b)   A member of the halogen family: C, Ca, Cl, Co
(c)   A member of the alkali metals: Rn, Ra, Ru, Rb
(d)   A member of the alkaline earth metals: Mg, Mn, Mo, Md
(e)   A member of the noble gas family: Ac, Al, Am, Ar

The Periodic Table and Electron Configurations.   The numbers of electrons housed in the principal energy levels of the atoms of several representative elements are given in Table 3.4, family by family. We will use the term **outside level** for the highest *occupied* main level of any given element. When we look at the outside levels, family by family, we see two very striking facts. The atoms of each family all have identical numbers of electrons in their outside levels, and the numbers in the outside levels differ from family to family. All atoms of group IA metals have one electron in the outside level. In group IIA, all atoms have two outside level electrons. In fact, *among the representative elements, the group number (I, II, and so forth through VII) corresponds to the number of electrons in the outside level.* The group number for the noble gases used to be VIII, so giving this group number 0 doesn't make it as much an exception to this rule as it might seem. Helium, the first element in this group, has just two electrons in its outside level, but this level (number 1) can't hold more than two anyway.

The electron configurations of the series B elements, such as the various transition elements, have their own fairly regular patterns, but our study will not require much knowledge of them. We can mention that as we go from element 21 (Sc, scandium) through element 30 (Zn, zinc) — the transition elements in period 4 — the $4s$ orbital usually holds one or two electrons, whereas the five $3d$ orbitals successively fill. You can see this by consulting Appendix B, where you will also be able to see how the transition elements in period 5 (elements 39 through 48) experience the successive filling of the $4d$ orbitals. As soon as the d orbitals fill among each of these transition elements, we reenter the region of the representative elements of the table. The progressive fillings of $4f$ and $5f$ orbitals take place among the lanthanide and actinide series, as Appendix B shows.

EXAMPLE 3.3   FINDING INFORMATION IN THE PERIODIC TABLE

Problem: How many electrons are in the outside level of an atom of iodine?

Solution: Until you become more familiar with the locations of certain elements in the table, you'll have to use the Table of Atomic Weights and Numbers (inside the front cover) to find the atomic number of a given element. Doing this, we find that the atomic number of iodine is 53. Now we use the periodic table, and find that iodine is in group VIIA. Being one of the A-type elements, we know that iodine is a *representative* element, which means that its group number is the same as the number of outside level electrons — 7.

PRACTICE EXERCISE 6   How many electrons are in the outside level of an atom of each of the following elements?

(a) Potassium   (b) Oxygen   (c) Phosphorus   (d) Chlorine

Metals and Nonmetals in the Periodic Table.   Another use of the table is to tell us at a glance which elements are metals and which are not. As seen in Figure 3.6, the nonmetals are all in the upper right-hand corner. All of the noble gases (group 0) and the halogens (group VIIA) are nonmetals. The great majority of all elements are metals. The few lying along the borderline between metals and nonmetals are sometimes called **metalloids,** and they have properties that are partly metallic and partly nonmetallic.

If you compare the locations of the elements in Figure 3.6 with the electron configurations given in Table 3.4, you will see that all of the nonmetals, except hydrogen and helium, have four to eight electrons in their outside levels. All of the elements that are metals, except for hydrogen and helium, have atoms with one, two, or three outside-level electrons. Only

FIGURE 3.6
Locations of metals, nonmetals, and metalloids in the periodic table.

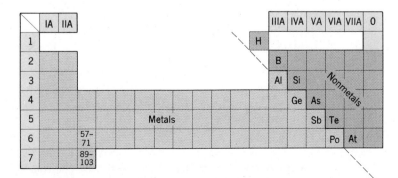

TABLE 3.4
Electron Configurations among Four Families of Representative Elements

Family	Element	Number	Principal Level Number						
			1	2	3	4	5	6	7
GROUP IA	Lithium	3	2	1					
The alkali metals	Sodium	11	2	8	1				
	Potassium	19	2	8	8	1			
	Rubidium	37	2	8	18	8	1		
	Cesium	55	2	8	18	18	8	1	
	Francium	87	2	8	18	32	18	8	1
GROUP IIA	Beryllium	4	2	2					
The alkaline earth metals	Magnesium	12	2	8	2				
	Calcium	20	2	8	8	2			
	Strontium	38	2	8	18	8	2		
	Barium	56	2	8	18	18	8	2	
	Radium	88	2	8	18	32	18	8	2
GROUP VIIA	Fluorine	9	2	7					
The halogens	Chlorine	17	2	8	7				
	Bromine	35	2	8	18	7			
	Iodine	53	2	8	18	18	7		
	Astatine	85	2	8	18	32	18	7	
GROUP O	Helium	2	2						
The noble gases	Neon	10	2	8					
	Argon	18	2	8	8				
	Krypton	36	2	8	18	8			
	Xenon	54	2	8	18	18	8		
	Radon	86	2	8	18	32	18	8	

among some elements with high atomic numbers do we find metals with more than three outside-level electrons. Tin and lead of group IVA are common examples.

Other Uses of the Periodic Table. Perhaps the most noteworthy consequences of having a set of elements in the same group is that they tend to form compounds of similar formulas, and that such compounds have at least some similar properties. We have already learned that the alkali metals form compounds called hydroxides that have the general formula MOH and that all of these are caustic substances. Thus if we had never handled rubidium hydroxide, RbOH, we would be very careful with it, because it is the hydroxide of an alkali metal. (And it actually is very caustic.)

All of the binary compounds involving the halogens (group VIIA) and hydrogen have the common formula HX (where X can be F, Cl, Br, or I). If we know that a solution in water of, say, HCl could give an acid burn to skin, we would be very careful when handling similar solutions of HF, HBr, and HI. These solutions are able to destroy or neutralize the caustic properties of the alkali metal hydroxides according to the following general equation:

$$HX + MOH \longrightarrow MX + H_2O$$

For example, HCl and NaOH react as follows:

$$HCl + NaOH \longrightarrow NaCl + H_2O$$

Water (H_2O) and sodium chloride (NaCl), or table salt, form in this reaction, and we know that a solution of salt in water can't give either an alkali burn or an acid burn to the skin. Because the reaction of an acid with an alkali destroys a characteristic property of any alkali and any acid, it is an important reaction and deserves a special name. We call it **neutralization.**

These examples are intended to show how useful the periodic table can be, especially when taken together with the sorting of the elements made possible by the periodic law. Make no mistake; one purpose of this study is to learn factual knowledge of the chemical properties of important substances. The periodic table greatly reduces the quantity of this work, however, because it helps us organize the information. We will often be able to write *general* equations that apply to several reactions, instead of having to learn each and every reaction.

SUMMARY

Atomic structure Atoms are electrically neutral particles that are the smallest representatives of an element that can display the element's chemical properties. Each atom has one nucleus—a hard inner core—surrounded by enough electrons to balance the positive charge on the nucleus. All of the atom's protons and neutrons are in the nucleus, and each of these subatomic particles has a mass of 1.0 amu. The mass of the electron is only 1/1836th the mass of a proton. The proton has a charge of $1+$, the electron's charge is $1-$, and the neutron is electrically neutral.

The atomic number of an element is the number of protons in each atom. It also equals the number of electrons in one atom. Nearly all elements occur as a mixture of a small number of isotopes. The isotopes of an element share the same atomic number, but their atoms have small differences in their numbers of neutrons. To fully characterize an isotope, we have to specify both the atomic number and the mass number—the number of protons plus neutrons. The atomic weight of an element is numerically equal to the average mass of the atoms of all its isotopes as they occur naturally together.

Electron configurations To write an electron configuration of an atom, we place its electrons one by one into the available orbitals, starting with the one of lowest energy, the 1s orbital. Each orbital can hold two electrons (if their spins are opposite), but where two or more orbitals are available at the same sublevel, electrons spread out.

Orbitals are spaces where probabilities are high of finding electrons that have a particular energy and geometrical orientation with respect to the x-y-z axes. Each occupied orbital can be viewed as an electron "cloud." It isn't possible to obtain precise information about

an electron's simultaneous location and energy, but the knowledge of where electrons are most likely to be—information obtained from electron configurations—is sufficient for understanding chemical properties.

As long as electrons remain in their orbitals, an atom neither absorbs nor radiates energy. However, an atom can absorb the energy that corresponds exactly to the difference in energy between two orbitals, provided the orbital of higher energy has a vacancy. The absorbed energy makes an electron move to the higher orbital. When it drops back down, this energy is radiated as a quantum or photon of light.

Periodic properties Because many properties of the elements are periodic functions of atomic numbers, the elements fall naturally into groups or families that we can organize into vertical columns in a periodic table. Atoms of the representative elements that belong to the same family have the same number of outside-level electrons—a number that corresponds to the group number itself. (Helium does not fit this generalization.)

The horizontal rows of the periodic table are called periods. The long periods that contain transition and inner transition elements involve the systematic filling of inner orbitals. The nonmetallic elements are in the upper right-hand corner of the periodic table, and the metals—the great majority of the elements—make up the rest of the table. (At the border between metals and nonmetals occur the elements—the metalloids—that have both metallic and nonmetallic properties.)

KEY TERMS

The following terms were emphasized in this chapter, and their meanings should be mastered before you continue to succeeding chapters.

alkali metals	atomic orbital	carbon family	element
alkaline earth metals	atomic weight	electron	group
atom	aufbau rules	electron cloud	halogens
atomic mass unit (amu)	binary compound	electron configuration	Heisenberg uncertainty principle
atomic number	Bohr model of the atom	electron shell	

Hund's rule	neutralization	oxygen family	principal energy level
inner transition element	neutron	Pauli exclusion principle	proton
isotope	nitrogen family	period	quantum
mass number	noble gases	periodic law	subatomic particle
metalloid	nucleus	periodic table	sublevel
model, scientific	outside level	photon	transition element
			representative element

SELECTED REFERENCES

1 T. F. Tremble. "Mendeleev's Discovery of the Periodic Law." *Journal of Chemical Education,* January 1981, page 28.

2 F. H. Firsching. "Anomalies in the Periodic Table." *Journal of Chemical Education,* June 1981, page 478.

3 W. C. Fernelius and W. H. Powell. "Confusion in the Periodic Table of the Elements." *Journal of Chemical Education,* June 1982, page 504.

4 G. T. Seaborg. "The New Elements." *American Scientist,* May–June 1980, page 279.

5 G. T. Seaborg. "The Periodic Table. Tortuous Path to Man-Made Elements." *Chemical and Engineering News,* April 16, 1979, page 46.

REVIEW EXERCISES

The answers to Review Exercises marked with asterisks are in an Appendix. The answers to the rest of the Review Exercises are in the *Study Guide to Fundamentals of General, Organic, and Biological Chemistry,* 3rd edition.

THE NUCLEAR ATOM

3.1 What are the names, electrical charges, and masses of the three subatomic particles? (Use the amu as the unit of mass, and omit the mass of the lightest of the three.)

3.2 Name two subatomic particles that attract each other.

3.3 What subatomic particles have forces of repulsion between them?

3.4 We will learn in the next chapter that it is possible for particles of an atomic size to have unequal numbers of electrons and protons. If particle M has 13 protons and 10 electrons, what is the electrical charge on M?

3.5 It costs energy to make an electron leave an atom, but which of the following changes would cost the *least* energy, and why?
(a) Removal of an electron from a particle having 12 protons and 12 electrons.
(b) Removal of an electron from a particle having 12 protons and 11 electrons.

3.6 Particle M has 13 protons and 10 electrons, and particle Y has 8 protons and 10 electrons. Do they attract each other, repel each other, or leave each other alone? Explain.

3.7 The atomic mass unit is roughly on the order of what size, 10^{-230} g, 10^{-23} g, 10^{-3} g, or 10^{23} g?

3.8 To five significant figures, calculate the mass in grams of a sample of 6.0220×10^{23} hydrogen atoms, each one having just one proton. Ignore the mass of the electron. How does the result compare with the atomic weight of hydrogen? (To two significant figures, are the results the same or different?)

*3.9 The mass of one proton, which we said was 1.0 amu, is actually 1.007277 amu, and the mass of one neutron is 1.008665 amu.
(a) What is the total mass in amu of a helium nucleus that consists of 2 protons and 2 neutrons? (Calculate to 7 significant figures.)
(b) The observed mass of one helium *atom* is 4.002604 amu. Given that the mass of one electron is 0.0005486 amu (and that a helium atom has two electrons), calculate the observed mass of one helium *nucleus.*
(c) Calculate the difference in amu between the observed mass of the helium nucleus and the mass you calculate by summing the masses of two protons and two neutrons. (Retain five significant figures.) You no doubt notice that the observed mass is less that the calculated mass. This difference is real. What we haven't discussed within the chapter is the Einstein relationship between mass and energy. Einstein found that mass can be converted to energy, and energy to mass. His famous equation, $E = mc^2$, gives the relationship, where E is the energy obtained when an amount of mass, m, "disappears" into energy; c in the Einstein equation is the velocity of light.
(d) For each amu loss in mass, 1.49454×10^{-10} joule of energy is released. Using the result of part (c), calculate the joules of energy released when two protons and two neutrons come together — fuse — to form *one* helium nucleus. The result doesn't seem to be very much, does it? However, go to the next part.
(e) Calculate the number of joules of energy that would be released if 6.0220×10^{23} helium nuclei form — about a total of 4 g of helium. This much energy could keep a 100-watt light bulb going for roughly 900 years. These calculations give an idea of the huge energy potentially available by nuclear fusion, the same kind of process that is believed to generate the energy of the sun.

ISOTOPES

3.10 Elements and isotopes are *substances,* not tiny particles—although they consist of such particles. What is the distinction between the terms *element* and *isotope of an element?*

3.11 What are the names and the compositions of the nuclei of the three isotopes of hydrogen? What do these nuclei have in common? In what specific way do they differ? In *atoms* having these nuclei, what else is the same?

3.12 Which of the following are isotopes? (Use the hypothetical symbols for your answer.)

M has 12 protons and 13 neutrons

Q has 13 protons and 13 neutrons

X has 12 protons and 12 neutrons

Z has 13 protons and 12 neutrons

3.13 Write the atomic symbol for the isotope of oxygen (atomic number 8) that has a mass number of 18.

3.14 Write the atomic symbol for the isotope of cobalt (atomic number 27) that has 33 neutrons. (This is a radioactive isotope used in cancer treatment.)

3.15 Carbon, atomic number 6, has three isotopes that are found in nature. The most abundant (98.89%) has a mass number of 12. The isotope with a mass number of 13 makes up 1.11% of naturally-occurring carbon. The third isotope—obviously present in the merest trace, because 98.89% + 1.11% = 100.00%—makes possible the dating of ancient artifacts.
(a) What is the same about these isotopes?
(b) In what specific feature of atomic structure do they differ?

3.16 Later, when we want to describe some *chemical* reaction of sulfur, which consists principally of two isotopes (mass numbers 32 and 34), we will use the symbol S. Why won't we have to use the special symbols $^{32}_{16}$S and $^{34}_{16}$S?

THE BOHR ATOM MODEL

3.17 Briefly describe the picture of atomic structure that emerged from Niels Bohr's theory.

3.18 What features of Bohr's theory are still valid?

3.19 What does the term *allowed energy state* mean?

3.20 What is meant by the term *ground state?*

3.21 If an electron drops from a higher energy state into a ground state, what else happens?

THE ORBITAL MODEL OF THE ATOM

3.22 Some kind of contact, if only by photons, is made between one making a measurement and what is being measured. If you sent energy into an atom to get precise information about the location of an electron, could you also get precise information about the electron's energy?

3.23 Heisenberg found that what physical quantities of an electron cannot be known with precision and accuracy at the same instant?

3.24 What can still be known (actually, calculated) about an electron's position if we cannot think of it as traveling in a definite orbit?

3.25 How many sublevels and how many orbitals are at principal level 1?

3.26 At principal level 2, how many sublevels are there? How many orbitals are in each? What symbols are used for these orbitals?

3.27 How many sublevels are at principal level 3? How many orbitals are in each?

3.28 When one electron is in an *s*-type sublevel at principal level 4, what symbol is used for this?

ELECTRON CONFIGURATIONS

3.29 What is the maximum number of electrons that can be in each?
(a) Principal level 2
(b) The *p*-type orbitals of principal level 3
(c) The 4*s* orbital
(d) Principal level 3

3.30 Write the one-line electron configuration of silicon (atomic number 14).

3.31 Using only the information in Figure 3.4, and the fact that krypton has atomic number 36, write the one-line electron configuration of a krypton atom.

3.32 The electron configuration of iron is $1s^22s^22p^63s^23p^63d^64s^2$. Using only this information, answer the following questions.
(a) What is the atomic number of iron?
(b) Is principal level 3 completely filled? If not, which sublevel is partially filled?
(c) Are there likely to be any electrons in this atom with unpaired spin? If not, why? If there are, where are they likely to be and why?

3.33 Suppose that we have the following sublevels and orbitals available.

$$2p \ \underline{} \ \underline{} \ \underline{}$$
$$2s \ \underline{}$$

(a) Name and state the aufbau rules that are illustrated if we have one, two, or three electrons to distribute.
(b) Besides the rules needed for part (a), what is the name and the statement of the aufbau rule that would be illustrated if six electrons have to be distributed?

3.34 Consider the two possible electron configurations:

$$1s^22s^22p_x^2\,2p_y^2\,2p_z^2\,3s^1 \quad \text{and} \quad 1s^22s^22p_x^2\,2p_y^2\,2p_z^2\,3p_x^1$$

$$\textbf{1} \qquad\qquad\qquad\qquad \textbf{2}$$

(a) Would atoms having these configurations be atoms of the *same* element of different elements? How can you tell?
(b) Which is more stable, **1** or **2**? Explain.
(c) What does it take to convert the atom in the more stable of these two states into the other state?

3.35 Write the one-line electron configurations of the elements that have the following atomic numbers, using only these numbers and the aufbau rules:
(a) 11 (b) 8 (c) 17 (d) 20

3.36 Write the one-line electron configurations of the elements that have the following atomic numbers:
(a) 12   (b) 19   (c) 9   (d) 3

ATOMIC WEIGHTS

3.37 Mass numbers are always whole numbers. Atomic weights almost never are. Explain.

3.38 Iodine, which is needed in the diet to have a healthy thyroid gland, occurs in nature as only one isotope, $^{127}_{53}I$. Using just this information, what is the atomic weight of iodine (to three significant figures)?

3.39 Bromine, in the same chemical family as iodine, occurs in nature as a nearly 1-to-1 mixture of two isotopes, $^{79}_{35}Br$ and $^{81}_{35}Br$. Using just this information, what is the atomic weight of bromine (to two significant figures)?

3.40 To two significant figures, the atomic weight of magnesium is 12 and the atomic weight of sulfur is 24.
(a) How much heavier are sulfur atoms than magnesium atoms?
(b) If we counted out in separate piles 10^{23} atoms of magnesium and the identical number of sulfur atoms, which pile of atoms would have the larger mass, and by what factor?
(c) If we weighed out 2.0 g of magnesium and we wanted a sample of sulfur that had the identical number of atoms, how many grams of sulfur should we weigh out?

•3.41 How much heavier are carbon atoms than hydrogen atoms (on the average)? Calculate the answer to two significant figures. Use data from the Table of Atomic Weights and Numbers inside the front cover.

•3.42 Carbon atoms and oxygen atoms can chemically combine to form particles consisting of these atoms in a ratio of 1 to 1. (The substance that forms is the poisonous gas, carbon monoxide.)
(a) How much heavier are oxygen atoms than carbon atoms (to three significant figures)?
(b) If all of the atoms in a sample of 12.0 g of carbon are to combine entirely with oxygen atoms, how many grams of oxygen are needed?

THE PERIODIC TABLE

3.43 What general fact about the chemical elements makes possible the stacking of these elements in the kind of array seen in the periodic table?

3.44 What do the terms *group* and *period* refer to in the periodic table?

3.45 Without consulting the periodic table, is an element in group VA a representative or a transition element?

3.46 Without consulting the periodic table, is an element in period 2 a representative or a transition element?

3.47 How does the modern form of the periodic law differ from Mendeleev's form?

3.48 How would period 4 of the periodic table be different if the elements were arranged in their order of increasing atomic weights?

3.49 Suppose that an atom has the following electron configuration:

$$1s^22s^22p^63s^23p^63d^54s^1$$

Without consulting the periodic table, answer the following questions.
(a) Is this element a representative or a transition element? How can you tell?
(b) What is the number of its outside level, and how many electrons are in it?
(c) Is the element a metal or a nonmetal? How can you tell?

3.50 The atoms of an element have the following electron configuration:

$$1s^22s^22p^63s^23p^63d^{10}4s^24p_x{}^14p_y{}^14p_z{}^1$$

Without using the periodic table, answer the following questions.
(a) What is the group number of this element? How can you tell?
(b) Is the element a metal or a nonmetal? How can you tell?

3.51 Give the group number and the chemical family name of the set of elements to which each of the following belongs:
(a) sodium   (b) bromine   (c) sulfur   (d) calcium

3.52 For each of the following elements give the group number and the name of family to which it belongs:
(a) iodine   (b) phosphorus   (c) magnesium   (d) lithium

3.53 Considering the pattern described on page 64 for the formulas of the binary compounds of the representative elements in periods 2 and 3 of the periodic table, what are the likely formulas for the binary compounds with hydrogen of the following elements?

(a) K   (b) Ca   (c) Ga   (d) Ge
(e) As   (f) Se   (g) Br

3.54 The accompanying diagram shows a section of the periodic table, except that hypothetical atomic symbols are used.

7	8	9	10
15	16	17	18
W	X	Y	Z
33	34	35	36

The numbers are atomic numbers, but not all of the symbols are given numbers. Without referring to an actual periodic table, answer the following questions.
(a) What elements, if any, are shown as belonging to the same family as Y? Write the number(s).
(b) What elements, if any, are in the same period as W? Write the number(s).

(c) Above the first horizontal row, write the group numbers for the elements of this row.

(d) If the outside level of an atom of number 33 has five electrons, how many electrons are in the outside level of an atom of number 15? Of number 16? Of number 36?

(e) If element 9 (let's give it the symbol M) has a compound with hydrogen with the formula HM, what is a good possi- bility for the formula of a compound between hydrogen and number 17?

(f) If number 10 is one of the noble gases, which is also a noble gas, 9, 11, or 36?

(g) If number 18 is a nonmetal, and number 33 is a metalloid, are the elements represented here metals or nonmetals?

Chapter 4
Chemical Compounds and Chemical Bonds

The sharp angles and smooth planes of crystals, seen here in pyrite are the results of highly ordered arrangements of the atoms, as we'll study in this chapter.

4.1 IONIC COMPOUNDS

Strong forces of attraction exist between oppositely charged ions in a large number of chemical compounds.

Whatever holds us together — skin and bones or the membranes of cells and blood vessels — must withstand a number of stresses and strains, both physical and chemical. In this chapter, we ask a fundamental question: What holds things together? What has to be true at the atomic level of small particles if we are to understand how some kinds of matter can stick together in bulk, and other kinds cannot?

We have already had a hint of a possible answer — the basic law of nature that unlike charges attract. At the heart of the "stickiness" of matter are electrical forces of attraction. Atoms, however, are electrically *neutral,* so how can they become stuck together strongly enough to account for the strengths of muscles, bones, blood vessels, and the like? The answer is that atoms are able to reorganize their electrons and nuclei into new particles in which there are net forces of attraction called **chemical bonds.**

Molecule is from a Greek term meaning *little mass.*

Two Kinds of Compounds.   There are two important ways by which electrons and nuclei can become reorganized relative to each other. One way leads to a new kind of small particle called a molecule. A **molecule** is an electrically neutral particle consisting of two or more atomic nuclei surrounded by a swarm of enough electrons to make the particle electrically neutral. Compounds whose smallest particles are molecules with two or more nuclei from *different* elements are called **molecular compounds.** Sugar, vitamin C, cholesterol, aspirin, and water are examples. Only rarely are molecules made from atoms of the *same* element. For example, both the nitrogen and the oxygen in the air we breathe consist not of separate, individual atoms but of molecules. The nitrogen molecule is made up of two nitrogen atoms, and its symbol is N_2. Similarly, oxygen occurs as molecules made up of two oxygen atoms and symbolized as O_2. We will see how these and other molecules are held together later in this chapter.

Molecules made from two atoms are called **diatomic molecules.**

The other way of reorganizing atoms into compounds produces tiny particles of opposite electrical charge called *ions,* and these strongly attract each other. Compounds consisting of oppositely charged ions are called *ionic compounds,* and we will study these next.

Electron Transfers and the Formation of Ions.   Sodium chloride (table salt) is a typical ionic compound. Its parent elements, sodium and chlorine, cannot be stored in each others' presence because they react together — violently, in fact. The following changes in electron configurations occur among billions upon billions of sodium and chlorine atoms. (These are the *overall* changes — the net results. To simplify this discussion, we're ignoring the fact that the element chlorine consists of molecules, Cl_2, instead of individual atoms.)

For purposes of illustration, we have picked the sodium-23 and chlorine-35 isotopes.

$$\left(\begin{array}{c} 11\ p^+ \\ 12\ n \end{array}\right) 1s^2 2s^2 2p^6 3s^1 + \left(\begin{array}{c} 17\ p^+ \\ 18\ n \end{array}\right) 1s^2 2s^2 2p^6 3s^2 3p_x^2 3p_y^2 3p_z^1 \longrightarrow$$

One sodium atom   One chlorine atom
Na   Cl

$$\left[\left(\begin{array}{c} 11\ p^+ \\ 12\ n \end{array}\right) 1s^2 2s^2 2p^6 \right]^+ + \left[\left(\begin{array}{c} 17\ p^+ \\ 18\ n \end{array}\right) 1s^2 2s^2 2p^6 3s^2 3p_x^2 3p_y^2 3p_z^2 \right]^-$$

Outer octet   Outer octet

One sodium ion   One chloride ion
Na^+ Cl^-

The new particle with the sodium nucleus is no longer an atom, because it no longer is electrically neutral. It bears a net charge of $1+$. (Its total positive charge from the 11 protons in the nucleus is $11+$, but it has only 10 electrons for a total negative charge of $10-$.) Similarly, the new particle with a chlorine nucleus isn't an atom either. It carries a net electrical charge of $1-$. (It has 17 protons for $17+$ but 18 electrons for $18-$, so the net is $1-$.)

Electrically charged particles at the atomic level of size are called **ions.** When an ion has just one nucleus, it is named after the parent element, either the same name or something close to it. All ions derived from metals have the same name as the element (plus the word *ion*). Ions derived from nonmetals have names that end in *-ide,* as in the *chloride ion,* whose parent element is chlorine. The names, symbols, and electrical charges of several common ions are given in Table 4.1, and they must be learned now. (A complete list of the ions mentioned in this book is in Appendix C.)

The word *ion* is from the Greek *ienai,* meaning to go or to move. Ions, unlike atoms, can move in response to an electrical attraction.

Ionic Compounds and Ionic Bonds.   The reaction between sodium and chlorine is typical of the reactions between any group IA element and any group VIIA element — electron transfers occur from the metal to the nonmetal and ions form. Of course, it isn't physically possible to arrange a chemical meeting between just one atom of sodium and one of chlorine. Any visible sample — even the tiniest speck — has upwards of at least 10^{18} atoms. Therefore when actual samples of these elements are mixed, a storm of electron transfers occurs, and countless numbers of oppositely charged ions form. Of course, because like charges repel, the new sodium ions repel each other. The chloride ions also repel each other. However, sodium ions and chloride ions attract each other — unlike charges attract. Out of all these attractions and repulsions, the storm subsides into firm, hard, and regularly shaped crystals of sodium chloride. Spontaneously, the unlike charged ions, Na^+ and Cl^-, nestle together as nearest

TABLE 4.1
Some Important Ions[a]

Group	Element	Symbol for Neutral Atom	Symbol for Its Common Ion	Name of Ion
IA	Lithium	Li	Li^+	Lithium ion
	Sodium	Na	Na^+	Sodium ion
	Potassium	K	K^+	Potassium ion
IIA	Magnesium	Mg	Mg^{2+}	Magnesium ion
	Calcium	Ca	Ca^{2+}	Calcium ion
	Barium	Ba	Ba^{2+}	Barium ion
IIIA	Aluminum	Al	Al^{3+}	Aluminum ion
VIA	Oxygen	O	O^{2-}	Oxide ion
	Sulfur	S	S^{2-}	Sulfide ion
VIIA	Fluorine	F	F^-	Fluoride ion
	Chlorine	Cl	Cl^-	Chloride ion
	Bromine	Br	Br^-	Bromide ion
	Iodine	I	I^-	Iodide ion
Transition Elements	Silver	Ag	Ag^+	Silver ion
	Zinc	Zn	Zn^{2+}	Zinc ion
	Copper	Cu	Cu^+	Copper(I) ion (cuprous ion)[b]
			Cu^{2+}	Copper(II) ion (cupric ion)
	Iron	Fe	Fe^{2+}	Iron(II) ion (ferrous ion)
			Fe^{3+}	Iron(III) ion (ferric ion)

[a] Other common ions are listed in Table 4.3

[b] The names in parentheses are older names, but still often used.

FIGURE 4.1
The structure of a sodium chloride crystal. *(a)* This schematic shows the alternating pattern of Na^+ and Cl^- ions. *(b)* Sodium ions have much smaller diameters than chloride ions. As seen here the sodium ions are surrounded by chloride ions as nearest neighbors, and the like-charged ions are just a little farther apart.

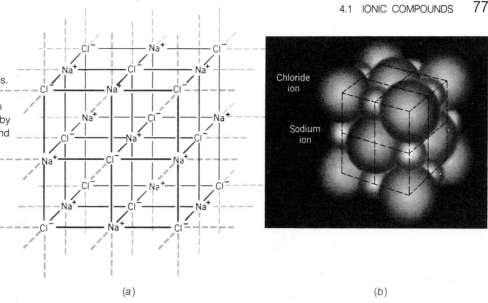

(a)   *(b)*

Crystal of sodium chloride

neighbors and like-charged ions stay just a little farther apart, as seen in Figure 4.1. Na^+ ions have Cl^- ions as closest neighbors, and Cl^- ions have Na^+ ions as closest neighbors. The forces of repulsion — Na^+ from Na^+ and Cl^- from Cl^- — are still there in the crystal, but because they act over slightly longer distances they cannot overcome the forces of attraction of Na^+ to Cl^-.

As you can see in Figure 4.1, the ions end up very regularly spaced within the crystal, and this explains why the crystal itself has a regular shape, as the photo in the margin illustrates. Such crystals can be broken, of course — shattered would better describe what happens. A hammer blow can make one layer of ions shift over so that suddenly like-charged ions momentarily become closest neighbors. Now the net force — at least along this layer — is one of repulsion, not attraction, and the crystal splits apart.

The force of attraction between oppositely charged ions is called an **ionic bond,** and substances made of ions are called **ionic compounds.** The ionic bond does not extend in any single, unique direction. The force of attraction from Na^+, for example, radiates equally in all directions, like light from a light bulb. One particular Na^+ ion doesn't belong to any particular Cl^- ion. There is no such thing as a separate, discrete particle consisting of just one Na^+ ion and one Cl^- ion that belong exclusively to each other. We mention this because the formula used for sodium chloride — NaCl — might be incorrectly interpreted this way. The formula for sodium chloride or for any ionic compound is meant only to disclose the ratio of the ions present. The ratio in sodium chloride is 1 to 1. It *must* be 1 to 1, because the net electrical charge of a compound is zero — always — and the charge of $1+$ on one sodium ion is balanced exactly by the charge of $1-$ on one chloride ion. As we said, ionic compounds are represented by formulas that give only the *ratios* of the particles present, and this kind of formula is called an **empirical formula.**

Many familiar substances besides sodium chloride are ionic compounds. Sodium bicarbonate in baking soda, barium sulfate in "barium X-ray cocktails," sodium hydroxide in lye and drain cleaners, and calcium sulfate in plaster and plaster of paris, are other examples of ionic substances. Every fluid in every living thing contains dissolved ions. Even among the huge molecules in living things — those of proteins, for example — there are often a few or several electrically charged sites. People in professional health care fields speak of the "electrolyte balance" of this or that body fluid. An **electrolyte** is any substance that can furnish ions in a solution in water. Thus at the molecular level of life, ions are everywhere.

Ions and the Periodic Table.   Not all elements readily form ions, but when we locate those that do in the periodic table, an interesting pattern emerges. Generally all metals can form ions, and all their ions have positive charges. If the metal is among the *representative* elements (those in the A series, such as group IA, IIA, and so forth), the positive charge on the ion is the same as the group number. For example, the ions of the group IA metals all have charges of $1+$, Li^+, Na^+, K^+, and Cs^+. Two of these ions, Na^+ and K^+, are particularly important in the fluids of the body, but we need more background in chemistry before we can discuss what they do. The ions of the group IIA metals have charges of $2+$, for example, Mg^{2+}, Ca^{2+}, and Ba^{2+}. The calcium ion and the magnesium ion are important in various ways in the body. We will be interested in only one element in group IIIA, aluminum, and its ion has a charge of $3+$, Al^{3+}.

The ions related to the transition metals are all positively charged, but there is no simple correlation between their group numbers and the size of their positive charges. Generally, the charges vary from $1+$ to $3+$, and several transition metals can exist as ions of more than one charge. Iron, for example, can exist as Fe^{2+} or Fe^{3+} ions. The first, Fe^{2+}, is present as part of the hemoglobin molecule in blood, the species that carries oxygen from the lungs to cells.

The nonmetallic elements of groups VIA and VIIA can form negatively charged ions — a charge of $2-$ for those in group VIA and a charge of $1-$ for the elements in group VIIA. (See also Table 4.1.) The chloride ion from group VIIA is the chief nonmetal ion in blood.

The nonmetal elements in groups IVA and VA form ions so rarely that we will assume they don't form ions at all. If we encounter an exception, we will be careful to note it.

The metal *elements in groups IVA and VA can form positively charged ions.*

4.2 THE OCTET RULE

Atoms and ions whose outside energy levels hold eight electrons are substantially more stable than those that do not.

Thus far we haven't asked *why* sodium atoms and chlorine atoms, in each other's presence, should engage in a flurry of electron transfers that lead to ions that then nestle into crystals. The plain truth is that no one knows — not any more than any scientist knows *why* there is a universe, for example. Earlier in this section we reported on *how* nature behaves, but we didn't explain why. We don't have to leave it at this, however, because we can draw some correlations that are a bit more satisfying.

G. N. Lewis (1875 – 1946), an American chemist, was chiefly responsible for the development of the octet rule.

The noble gases are

Helium	He
Neon	Ne
Argon	Ar
Krypton	Kr
Xenon	Xe
Radon	Rn

The Outer Octet. If you look back at Figure 3.6, you will see that the noble gas elements have the highest ionization energies. They require the greatest input of energy to suffer the loss of an electron. We also noted in Chapter 3 that the group 0 elements are called *noble gases* because they participate in very few reactions. Evidently, there is something quite special and stable about two kinds of electron configurations — the two kinds that occur among the atoms of the noble gases. One is the **outer octet,** the existence of eight electrons in whichever principal level happens to be the *outside* level — the highest, occupied principal energy level. The other is a *filled* level 1 when it is the outside level (as in helium). We can summarize these observations by one statement — *noble gas configurations are conditions of unusual chemical stability.* Let us see how all of this helps us to understand how the elements in the same chemical family, if they form ions at all, form ions of the same electrical charge.

A sodium atom, in group IA, has one electron in its outside level — level 3 — but after losing this electron, it has a new outside level — the former inner level number 2, which happens to have eight electrons. Now the particle has an outer octet, and it also has the electron configuration of a member of the noble gas family — neon. Whatever causes an outer octet to lend stability to the neon *atom* evidently is at work in the sodium *ion,* because the sodium ion is an unusually stable particle, too — provided that there is something such as an oppositely charged ion available, which lets the overall system be electrically neutral.

A chlorine atom, in group VIIA, has *seven* electrons in its highest occupied energy level — number 3. When it accepts one electron, its outside level acquires an outer octet. Thus the chloride *ion* has the same electron configuration as an atom of one of the noble gases — argon. The chloride ion, provided that an oppositely charged system is nearby, is also a particle of unusual chemical stability.

When we put sodium and chlorine together, the electron transfers between sodium and chlorine generate particles with noble gas configurations. In other words, these two chemicals — given the opportunity in terms of each other's actual presence — spontaneously change in a direction that leads to greater stability for both. They change from particles that don't have noble gas configurations to those that do. It does not matter that the new particles are charged. The atoms sacrifice neutrality for stability. Of course, once electrically (and oppositely) charged, they *must* attract each other. It's in the nature of unlike charges.

We haven't explained *why* a noble gas configuration is stable. We are only pointing out that for some reason it is. The pattern we noted for the formation of sodium chloride is so consistently observed among the reactions of the representative metallic elements with those that are nonmetals that it amounts virtually to a law of nature called the **octet rule.**

Octet Rule The atoms of the reactive representative elements tend to undergo those chemical reactions that most directly give them electron configurations of the nearest noble gas.

Table 4.2 shows how the atoms of some common, representative elements achieve noble gas electron configurations by giving up or acquiring electrons. The octet rule together with the aufbau rules enable us to figure out the electron configurations of the ions that are most likely to exist for any of the first 20 elements. When we can do this, we can also figure out the charges on these ions. We will work some examples to show how.

TABLE 4.2
Electron Configurations of Ions and Comparable Noble Gases

Group	Common Element	Atomic Number	ATOMS Electron Configurations (Main Levels Only)						Ion	IONS Electron Configurations (Main Levels Only)						Nearest Noble Gas
			1	2	3	4	5	6		1	2	3	4	5	6	
IA	Li	3	2	1					Li^+	2						Helium
Alkali metals	Na	11	2	8	1				Na^+	2	8					Neon
	K	19	2	8	8	1			K^+	2	8	8				Argon
IIA	Mg	12	2	8	2				Mg^{2+}	2	8					Neon
Alkaline earth	Ca	20	2	8	8	2			Ca^{2+}	2	8	8				Argon
metals	Ba	56	2	8	18	18	8	2	Ba^{2+}	2	8	18	18	8		Xenon
VIA	O	8	2	6					O^{2-}	2	8					Neon
Oxygen family	S	16	2	8	6				S^{2-}	2	8	8				Argon
VIIA	F	9	2	7					F^-	2	8					Neon
Halogens	Cl	17	2	8	7				Cl^-	2	8	8				Argon
	Br	35	2	8	18	7			Br^-	2	8	18	8			Krypton
	I	53	2	8	18	18	7		I^-	2	8	18	18	8		Xenon
0	He	2	2													
Noble gases	Ne	10	2	8					The noble gases do not form							
	Ar	18	2	8	8				stable ions							
	Kr	36	2	8	18	8										
	Xe	54	2	8	18	18	8									

EXAMPLE 4.1   USING THE OCTET RULE

Problem: When nutritionists speak of the calcium requirement of the body, they always mean the calcium *ion* requirement. Calcium has atomic number 20. What charge does the calcium ion have, and what is the symbol of this ion?

Solution: There are two methods for solving this kind of problem, and you should learn both. The first is to write the electron configuration of the atom and the second is to exploit the periodic table.

Using the aufbau rules, we write the electron configuration of element 20, remembering that 4*s* fills before electrons go into 3*d*:

$$1s^2 2s^2 2p^6 3s^2 3p^6 4s^2$$

The atom has two electrons in the outside level — number 4 — and eight electrons in the next level down — 2 3*s* electrons plus 6 3*p* electrons. Only by losing *both* of the 4*s* electrons can this atom get a new outside level that holds the octet. Losing one electron won't do. Neither will losing three or more. Therefore the only stable ion that calcium can form is one with the following electron configuration:

$$1s^2 2s^2 2p^6 3s^2 3p^6$$

$0 - (2-) = 2+$

By losing two electrons the net charge on the particle becomes $2+$, so the symbol for the calcium ion is Ca^{2+}.

The second way to arrive at this answer is to find the position of calcium in the periodic table — group IIA. Because all of the group IIA elements have 2 outside-level electrons and all can acquire configurations of the nearest noble gases by losing 2 electrons, all of the group IIA ions bear charges of $2+$. Therefore the group IIA ions are Be^{2+}, Mg^{2+}, Ca^{2+}, Sr^{2+}, Ba^{2+}, and Rn^{2+}.

EXAMPLE 4.2   USING THE OCTET RULE

Problem: Oxygen can exist as the oxide ion in such substances as calcium oxide, an ingredient in cement. What is the symbol for the oxide ion, including its electrical charge?

Solution: As in Example 4.1, we will solve this in two ways. First, the electron configuration of an oxygen atom is

$$1s^2 2s^2 2p_x^2 2p_y^1 2p_z^1$$

The outside level — number 2 — has a total of six electrons, just two short of an octet and a neon configuration. We could also say that the oxygen atom has six too many electrons to have the helium configuration. However, gaining two electrons to become like neon is much simpler than losing six, so oxygen achieves a noble gas configuration most directly by accepting two electrons from some metal atom donor. The new configuration, then, is

$$1s^2 2s^2 2p_x^2 2p_y^2 2p_z^2$$

The symbol for the oxide ion is O^{2-}.

In the approach using the periodic table, we note that oxygen is in group VIA, so this tells us that its outside level has six electrons. It must pick up two electrons — not just one and not more than two — to have a noble gas configuration. These two extra electrons give the particle a charge of $2-$, so we can write O^{2-} directly.

EXAMPLE 4.3   USING THE OCTET RULE

Problem: Hardly any element is involved in more compounds than carbon. (Roughly 6 million carbon compounds are known.) Can carbon atoms change to ions? If so, what is the symbol of the ion?

Solution: We will just use the shorter method of solving this. When we find carbon's place in the periodic table, we see that it is in group IVA. This tells us that carbon atoms have four outside-level electrons. To achieve a noble gas configuration, a carbon atom either must lose these four electrons (and become helium-like) or gain four electrons (and become neon-like). In one or two very rare situations, carbon can do the latter — become the C^{4-} ion. Because this is so rare we ignore it, and to make things simpler, we will stick to the rule that any *nonmetal* atoms in groups IVA and VA do not form ions.

PRACTICE EXERCISE 1   Write the electron configuration of an atom of each of the following elements, and from this deduce the charge on the corresponding ion. If the atom isn't expected to have a corresponding ion, state so. The numbers in parentheses are atomic numbers. Do not use the periodic table for this Practice Exercise.

(a) Potassium (19)   (b) sulfur (16)   (c) silicon (14)

PRACTICE EXERCISE 2   Write the electron configurations of the *ions* of the elements in Practice Exercise 1 that can form ions.

PRACTICE EXERCISE 3   Relying on their locations in the periodic table, write the symbols of the ions of each of the following elements. Always remember that no symbol of an ion is complete without its electrical charge. (The numbers in parentheses are atomic numbers.)

(a) Cesium (55)   (b) Fluorine (9)
(c) Phosphorus (15) (d) Strontium (38)

4.3 FORMULAS AND NAMES OF IONIC COMPOUNDS

The ratio of the ions in an ionic compound, which the formula gives, ensures that the compound is electrically neutral.

The formula of an ionic compound gives the ratio in which the ions assemble, and they must do so in a ratio that permits all of the opposite charges to neutralize each other. Compounds, as we have said, are electrically neutral. The formula of a compound that consists solely of calcium ions, Ca^{2+}, and oxide ions, O^{2-}, must show that these have assembled in a ratio of 1 to 1, because only in this ratio can the charge of $2+$ on the calcium ion cancel the charge of $2-$ on the oxide ion. A compound made of calcium ions and chloride ions must have *two* chloride ions for every calcium ion, because it takes two Cl^- ions to give enough negative charge to cancel the charge of $2+$ of a calcium ion.

Writing Formulas and Using Subscripts.   By common agreement among the world's chemists, the formula of an ionic compound always has the positive ion shown first. Also by common agreement, the charges are not shown in the formula. They are "understood" — which is one reason that the charges on the several most common ions have to be memorized.

The numbers that indicate the ratio of the constituent particles in the formula of a substance are called **subscripts.** These numbers follow and are placed half a line below the symbols for the particles with which they are associated. The subscript 1 is always omitted; it is understood because just writing a symbol means that you're taking at least one of what it represents. Thus we write the formula of the compoud of the calcium ion and the oxide ion as CaO, and the formula for the compound between the calcium ion and the chloride ion is written as $CaCl_2$ and not as Cl_2Ca nor as $Ca^{2+}Cl_2^-$.

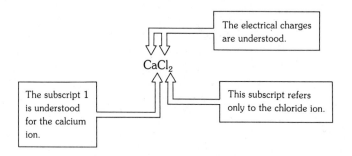

A formula that gives only the ratios of atoms is called an empirical formula after the Latin *empiricus,* something experienced.

Another convention for the formulas of ionic compounds is that the subscripts are normally those that express the ratio of ions in the smallest whole numbers. We might, for example, write Ca_2Cl_4 for the compound of calcium and chlorine, or $Ca_{0.5}Cl$. Either one, after all, does correctly show a ratio of calcium ion to chloride ion of 1 to 2. However, chemists normally write it $CaCl_2$, because $1:2$ is the correct ratio expressed in the smallest whole numbers.

Naming Ionic Compounds.   To name an ionic compound, we assemble the names of the ions in the order in which they appear in the formula, only the word *ion* is never used. For example,

NaCl   sodium chloride (sodium *ion* plus chloride *ion*)
CaO   calcium oxide (calcium *ion* plus oxide *ion*)
$CaCl_2$ calcium chloride (calcium *ion* plus two chloride *ions*)

In the older names, the *-ic* ending denotes the higher value of positive charge. The *-ous* ending refers to the lower value.

Four metal ions at the bottom of Table 4.1 are given two names; for example, copper(I) ion or cuprous ion. The first name is the modern name, but the second is still widely used. Both have to be learned. The roman numeral in "copper(I)" or "iron(III)" stands for the positive charge on the ion, Cu^+ or Fe^{3+}. (Notice there is no space between "copper" and "(I)" in "copper(I).")

There are two kinds of operations involving names and formulas that you must learn — how to write formulas from names and write names from formulas. We will limit the next examples and exercises to those that involve the ions in Table 4.1. (Appendix C has further information.)

EXAMPLE 4.4   WRITING FORMULAS FROM NAMES OF IONIC COMPOUNDS

Aluminum oxide is a buffing powder for polishing metals.

Problem: Write the formula of aluminum oxide.

Solution: We first use the name to identify the ions, including their electrical charges — Al^{3+} and O^{2-}. We know that $3+$ isn't canceled by $2-$, so we can't simply write AlO. The lowest common multiple of 2 and 3, however, is 6 ($2 \times 3 = 6$). Therefore if we can pick the smallest number of aluminum ions that give a total of $6+$ and the smallest number of oxide ions that give a total of $6-$, we will have the ratio. This requires $2Al^{3+}$, because $[2 \times (3+) = 6+]$, and it requires a $3O^{2-}$, because $[3 \times (2-) = 6-]$. The ratio in aluminum oxide, therefore, is $2Al^{3+}$ to $3O^{2-}$, and writing the aluminum first in the formula gives us the answer:

$$Al_2O_3$$

The strategy in Example 4.4 to find and use the lowest common multiple of the numbers in the charges on the ions always works. Try Practice Exercise 4 to develop experience in using this approach.

PRACTICE EXERCISE 4   Write the formulas of the following compounds:

(a)   Silver bromide (a light-sensitive chemical used in photographic film)
(b)   Sodium oxide (a very caustic substance that changes to lye in water)
(c)   Ferric oxide (the chief component in iron rust)
(d)   Copper(II) chloride (an ingredient in some laundry-marking inks)

EXAMPLE 4.5   WRITING NAMES FROM FORMULAS OF IONIC COMPOUNDS

$FeCl_3$ is used in photoengraving.   **Problem:** Write the name of $FeCl_3$, using both the modern and the older forms.

Solution: The symbol Fe stands for one of the two ions of iron, but which? What charge does it bear, in other words? This is a situation in which we have to use our knowledge of the charge on one of the ions, Cl^-, to figure out the charge on the other. The Cl_3 part of $FeCl_3$ tells us that there must be a total negative charge of $3-$, because $3 \times (1-) = 3-$. The lone Fe part of $FeCl_3$ must therefore provide $3+$ in charge, so we must be dealing with the Fe^{3+} ion—named either the iron(III) ion or the ferric ion. The compound is named either iron(III) chloride or ferric chloride.

PRACTICE EXERCISE 5   Write the names of each of the following compounds. When a name can be given in a modern form or an older form, write both.

(a) CuS (b) NaF (c) FeI_2 (d) $ZnBr_2$ (e) Cu_2O

The technique used in Example 4.5 of finding the electrical charge on one ion from the known charge on the other in an ionic compound can be very useful. It greatly reduces the number of ions and their charges that have to be memorized, because so many can be figured out this way. Work Practice Exercise 6; it involves ions that are not in Table 4.1.

PRACTICE EXERCISE 6   What are the charges on the *metal* ions in each of these substances?

(a) Cr_2O_3 (a green pigment in stained glass)
(b) HgS ("Chinese red," a bright, scarlet-red pigment)
(c) $CoCl_2$ (an ingredient in invisible ink)

4.4 OXIDATION AND REDUCTION

The loss of electrons is oxidation and the gain of electrons is reduction.

We have already seen the usefulness of classifying substances. We have begun to sort the millions of substances into elements, compounds, and mixtures, for example. We also sorted the elements, when we described some as metals and some as nonmetals. The value of recognizing some basis for creating classes of substances is that we can then make *general* statements about the classes that apply to all (or nearly all) of its members, and we reduce the quantity of information we might have to learn.

Redox Reactions.   Chemical reactions can also be usefully classified. We will introduce only one family of reactions now—the **redox reactions.** The common feature of *all* redox reactions is a change in the oxidation numbers of two of the elements involved among the reactants. For the kinds of elements that form the ions we have studied thus far, the **oxidation number** is simply the electrical charge—number and sign—on the ion. For

The earliest examples of oxidation involved oxygen itself as the oxidizing agent; hence, the name.

People used to call the conversion of some ore such as iron ore to the metal a *reduction* of the ore to the metal. This is where we got the general name, reduction.

example, the oxidation number of sodium in NaCl is $1+$. The oxidation number of chlorine is $1-$. In $FeCl_3$, the oxidation number of iron is $3+$. The oxidation numbers in MgS are $2+$ for magnesium and $2-$ for sulfur. *The oxidation number of an element is zero.* Now let's return to the reaction of sodium with chlorine and see how it illustrates a redox reaction.

When a sodium atom, Na, transfers an electron to a chlorine atom, Cl, the oxidation number of sodium changes from 0 to $1+$. We define any change that makes an oxidation number more positive as an **oxidation.** Of course, this can happen only if an electron leaves, so *oxidation means the loss of electrons.* We say that the sodium atom is oxidized to the sodium ion.

When a chlorine atom accepts an electron and changes from Cl to Cl^-, the oxidation number of chlorine changes from 0 to $1-$. We define any change that makes an oxidation number more negative as a **reduction.** This can't take place without the particle accepting an electron, so *reduction means the gain of electrons.*

A reduction cannot occur without an oxidation of something else. Electrons don't just leave from or go to outer space; they *transfer.* A reaction that involves a reduction also *must* involve an oxidation, and any such reaction is a redox reaction.

Redox reactions are at the heart of biological oxidations, the chain of chemical reactions whereby we use oxygen from the air to oxidize chemicals obtained by the partial breakdown of food molecules. These oxidations provide energy in forms the body can use. We will study these reactions much later, because we need considerably more background first, including some more of the vocabulary used in discussions of redox reactions.

The reactant that donates electrons and causes another reactant to be reduced is called the **reducing agent.** Sodium is the reducing agent in the reaction with chlorine. The reactant that accepts electrons and so oxidizes something else is called an **oxidizing agent.** Chlorine is the oxidizing agent in the reaction with sodium. It will help in this kind of analysis to break a redox reaction up into its two parts, one an oxidation and the other a reduction. We can write these parts as actual reactions if we temporarily treat an electron as a chemical species.

$$Na \longrightarrow Na^+ + e^-$$ oxidation—Na is oxidized to Na^+
$$Cl + e^- \longrightarrow Cl^-$$ reduction—Cl is reduced to Cl^-
$$\text{Sum:} \quad Na + Cl + e^- \longrightarrow Na^+ + Cl^- + e^-$$

We can add chemical equations just like algebraic equations, and we can cancel identical formulas on opposite sides of the arrow.

But we can cancel e^- from both sides, because equal numbers of e^- appear each side, and this leaves us with

$$Na \quad + \quad Cl \quad \longrightarrow Na^+ + Cl^-$$
Reducing   Oxidizing
agent   agent

Notice that an oxidizing agent is always itself reduced in a redox reaction, and that a reducing agent is always itself oxidized.

EXAMPLE 4.6   **ANALYZING A REDOX REACTION**

Problem: In the previous chapter we used the reaction of hot iron with molten sulfur to illustrate several principles.

$$Fe + S \longrightarrow FeS$$

Determine the oxidation numbers of iron in sulfur in FeS, and decide what is the oxidizing agent and what is the reducing agent in this redox reaction.

Solution: From Table 4.1 (and soon, it must be said, from memory) we know that the charge on Fe in FeS must be $2+$ and that the charge on S in FeS must be $2-$. Therefore iron has an oxidation number of $2+$ in FeS, and sulfur's oxidation number in this compound is $2-$. Because the reactants are both elements, they have oxidation numbers of 0. We can see that iron's oxidation number becomes more positive, so iron is oxidized. Sulfur's oxidation number becomes more negative, so it is reduced. Iron is the reducing agent and sulfur is the oxidizing agent.

PRACTICE EXERCISE 7   Identify by their chemical symbols what is oxidized and what is reduced in the following reactions. Also identify what is the oxidizing agent and what is the reducing agent.

(a) $Mg + S \rightarrow MgS$

(b) $CuCl_2 + Zn \rightarrow ZnCl_2 + Cu$

4.5 MOLECULAR COMPOUNDS

When electron transfers between the atoms of two reactive elements are very unfavorable in energy terms, these atoms can often achieve octets by sharing electrons.

We have noted that it is extremely rare for atoms of the nonmetal elements in groups IVA and VA to change into ions. Yet one of them — carbon — probably occurs in more different chemical compounds than any other single element. Evidently, nature has other ways of organizing chemical bonds than that of the ionic bond, and our purpose in this section is to learn about one of them. Usually, but not always, the bonds left for us to study occur between atoms of nonmetals; they do not often involve the metals.

The compounds made from nonmetals do not consist of oppositely charged ions. Instead, they are made of particles new to our study, molecules. A **molecule** is a small, electrically neutral particle consisting of at least two nuclei and enough electrons to make the whole system neutral. A compound that consists of molecules is called a **molecular compound.** Some elements are also molecular, as we will see next.

Molecular Orbitals.   Unlike the group 0 noble gas elements, the elements in group VIIA, the halogens, consist of diatomic molecules with the formulas F_2, Cl_2, Br_2, and I_2. The other diatomic elements are oxygen, O_2, hydrogen, H_2, and nitrogen, N_2. We cannot explain these diatomic particles as consisting of oppositely charged ions, because no halogen atom exists as a positively charged ion. For example, F^+ would have the electron configuration $1s^2 2s^2 2p_x^2 2p_y^2$, and the outside level (level 2) would have six, not eight electrons (an outer octet). We know that to fashion any kind of bond we must have an attraction between some kind of oppositely charged particles, so if we cannot use F^+ and F^- to make F_2, there must be another way to get a force of attraction between these two atoms.

If we had some way of concentrating extra electron density *between* the two fluorine nuclei, they would be attracted toward this region and be held a small distance apart by this attraction. In other words, *a bond could exist even though ions would not be involved.* Such a region of extra electron density can form between atoms of nonmetals, and Figure 4.2 shows how it happens to two fluorine atoms. On the left in the figure, we see two separated fluorine atoms, but only their half-filled p_z orbitals are pictured. (The nodes occur where their nuclei are.) Imagine that these two atoms move toward each other. Eventually, the spaces occupied by the p_z orbitals start to blend. The two atoms, of course, can't come *completely* together, because the nuclei repel each other more and more strongly as the atoms come closer to each other. Nevertheless, this blending gives the system an advantage it did not have as separated atoms — it gives the system the chance to be more stable than the separated atoms could be.

The technical term for this blending of atomic orbitals is the **overlapping of orbitals.** When this overlapping begins as the two atoms approach each other, the electron of one atom senses the presence of a second nucleus and starts to be attracted to it even as it retains its attraction to its original, parent nucleus. The other p_z electron has the identical experience. The result is that these two electrons find themselves attracted simultaneously to both nuclei,

The electron configuration of F is

$1s^2 2s^2 2p_x^2 2p_y^2 2p_z^1$

A region of *high electron density* is one with a thick or dense electron cloud.

The impulse to greater stability is always a major driving force in nature and is behind so much of what occurs naturally.

FIGURE 4.2
The covalent bond in the F–F molecule.

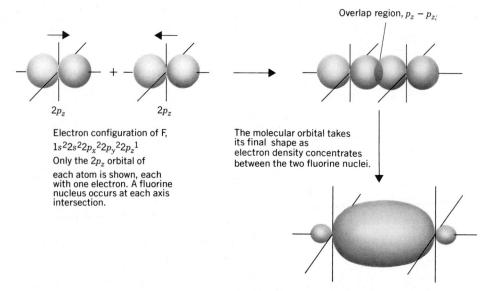

Overlap region, $p_z - p_{z;}$

Electron configuration of F,
$1s^2 2s^2 2p_x^2 2p_y^2 2p_z^1$
Only the $2p_z$ orbital of
each atom is shown, each
with one electron. A fluorine
nucleus occurs at each axis
intersection.

The molecular orbital takes
its final shape as
electron density concentrates
between the two fluorine nuclei.

and they tend to stay as much as possible in the region where they can enjoy this attraction to the fullest. Both electrons are now settled into a new region of space that encompasses both nuclei.

This overlapping of the two atomic orbitals has produced a **molecular orbital**. Like an atomic orbital, a molecular orbital can hold only two electrons and then only if their spins are in opposite directions. (This is why two electrons is the limit to the total number that can be present as two atomic orbitals overlap to form one molecular orbital.) Because orbital overlap permits electron density to concentrate between nuclei, the overlapping parts become larger and other parts, those away from the region between the two nuclei, shrink — as Figure 4.2 shows.

The force of attraction that is made possible by the overlapping of atomic orbitals and that holds atomic nuclei near each other is called a **covalent bond**. By the formation of this region populated by a pair of electrons, the two nuclei experience what is called **electron sharing**. Electron sharing and the formation of a covalent bond go hand in hand.

Co- from cooperative; *-valent* from the Latin *valere,* to be strong, signifying strong binding.

Electron-Dot Structures.   The new particle, F_2, is an example of a molecule, a neutral particle with more than one nucleus, as we said, and now we can add to the meaning of the term that a molecule is held together by covalent bonds. Like atoms and ions, molecules need symbols, and one approach is to represent a molecule by an electron-dot structure. We start with electron-dot symbols for atoms. In this kind of symbol we place dots representing the outer-level electrons around the atomic symbol, keeping the dots separated and at the corners of a square until their number exceeds four. The fifth, sixth, seventh, and eighth electrons are placed so as to create pairs of electrons. Thus the electron-dot structures of the atoms in the second period of the periodic table are as follows:

Remember, only the outside-level electrons are shown. The inside-level electrons are just understood to be there.

$$\cdot \text{Li} \quad \cdot \text{Be} \cdot \quad \cdot \overset{\cdot}{\text{B}} \cdot \quad \cdot \overset{\cdot}{\underset{\cdot}{\text{C}}} \cdot \quad \cdot \overset{\cdot}{\underset{\cdot}{\text{N}}} \cdot \quad \cdot \overset{\cdot\cdot}{\underset{\cdot\cdot}{\text{O}}} \cdot \quad : \overset{\cdot\cdot}{\underset{\cdot\cdot}{\text{F}}} \cdot \quad : \overset{\cdot\cdot}{\underset{\cdot\cdot}{\text{Ne}}} :$$

PRACTICE EXERCISE 8   Write the electron dot symbols for the atoms of the period 3 elements of the periodic table.

In electron-dot symbolism, we can represent the formation of F_2 as follows:

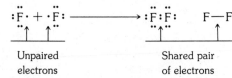

$$: \overset{\cdot\cdot}{\underset{\cdot\cdot}{\text{F}}} \cdot \; + \; \cdot \overset{\cdot\cdot}{\underset{\cdot\cdot}{\text{F}}} : \longrightarrow : \overset{\cdot\cdot}{\underset{\cdot\cdot}{\text{F}}} : \overset{\cdot\cdot}{\underset{\cdot\cdot}{\text{F}}} : \qquad \text{F——F}$$

Unpaired
electrons

Shared pair
of electrons

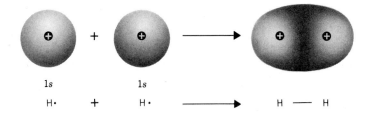

FIGURE 4.3
The covalent bond in H–H. The overlapping of the 1s orbitals of two separated hydrogen atoms, each with one electron, gives a molecular orbital with a pair of electrons. The dark shading indicates that electron density concentrates between the two nuclei, and the nuclei are attracted toward this region. This is what holds the nuclei near each other, and this attraction constitutes the covalent bond.

Structural Formulas. Because molecules that have several atoms would require us to write a large number of dots, chemists usually replace the shared electron pairs with straight lines and entirely omit the other outer-level electrons that are not involved in a covalent bond. An example was given above to the right of the electron-dot structure of F_2. The symbol of a molecule that uses lines to represent shared electron pairs — covalent bonds — to connect atomic symbols is called a **structural formula** or, simply, a **structure.**

Covalent Bonds and the Octet Rule. The fluorine molecule conforms to the octet rule if we consider that a shared pair counts for both atoms. Thus each fluorine atom in the molecule has an octet. The other members of the halogen family (as we would expect) form diatomic molecules in exactly the same way as fluorine. The only important difference is that the electrons involved in sharing are in outer levels with higher and higher level numbers.

Chlorine	$:\ddot{C}l\cdot + \cdot\ddot{C}l: \longrightarrow :\ddot{C}l:\ddot{C}l:$	or	Cl—Cl
Bromine	$:\ddot{B}r\cdot + \cdot\ddot{B}r: \longrightarrow :\ddot{B}r:\ddot{B}r:$	or	Br—Br
Iodine	$:\ddot{I}\cdot + \cdot\ddot{I}: \longrightarrow :\ddot{I}:\ddot{I}:$	or	I—I
	Electron-dot structures		Structural formulas

It is also possible for two s-type atomic orbitals to overlap and create a molecular orbital. Hydrogen, H_2, is an example, as seen in Figure 4.3. In electron-dot symbolism we can write

Hydrogen $H\cdot + \cdot H \longrightarrow H:H$ or   H—H

By sharing a pair of electrons, each hydrogen atom acquires the helium configuration of two outer-level electrons — another condition of stability.

An s orbital can overlap with a p orbital provided that just one lobe of the p orbital becomes involved, not both simultaneously. Figure 4.4 shows the formation of the molecular

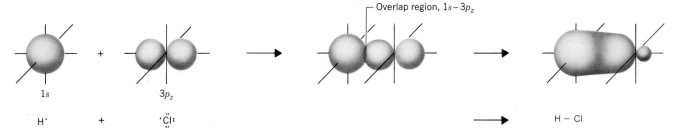

FIGURE 4.4
The covalent bond in H–Cl. The 3p_z orbital of a chlorine atom, which holds one electron, overlaps with the 1s orbital of a hydrogen atom, also with one electron. In the new molecular orbital, electron density concentrates between the two atomic nuclei.

The chlorine atom has its $3p_z$ orbital half-filled.

$$1s^2 2s^2 2p^6 3s^2 3p_x^2 3p_y^2 3p_z^1$$

orbital that provides the covalent bond in a molecule of hydrogen chloride. We can represent this as follows:

$$\text{Hydrogen chloride} \quad \text{H} \cdot + \cdot \ddot{\underset{..}{\text{Cl}}} : \longrightarrow \text{H} : \ddot{\underset{..}{\text{Cl}}} : \quad \text{or} \quad \text{H—Cl}$$

4.6 COVALENCE NUMBERS AND THE PERIODIC TABLE

When the group number of a nonmetal element is subtracted from 8, the result is the covalence number of the element.

All atoms in group VIIA have seven outer-level electrons and need one more electron for octets. When they acquire this one additional electron by *sharing,* they become involved in *one* covalent bond, as we just saw. Let's move back to group VIA and consider an atom of oxygen. It has six outer-level electrons. As we see in its electron-dot structure, $\cdot \ddot{\text{O}} \cdot$, the atom is two electrons short of an octet. If the octet is to be provided by electron *sharing,* the atom must get a share of two more electrons. Hydrogen atoms can meet this need, but we *need two:*

$$\text{H} \cdot + \cdot \ddot{\underset{..}{\text{O}}} \cdot + \cdot \text{H} \longrightarrow \text{H} : \ddot{\underset{..}{\text{O}}} : \text{H} \quad \text{or} \quad \text{H—O—H}$$
$$\text{Water}$$

Thus an oxygen atom has to form *two* covalent bonds to acquire an octet.

We are beginning to see that the number of covalent bonds a nonmetal can form is related to its location in the periodic table. If we subtract the element's group number in the periodic table from eight, we get the number of covalent bonds that atoms of this element can have:

$$8 - [\text{group number}] = \text{number of covalent bonds}$$

We can illustrate this further by considering nitrogen.

Nitrogen is in group VA of the periodic table, which tells us that a nitrogen atom has five outer-level electrons. Its electron-dot structure is $\cdot \dot{\text{N}} \cdot$, and it needs a share in three more electrons to have an octet. Thus $8 - [\text{group number}] = 8 - 5 = 3$. If hydrogen atoms are to provide these three shared electrons, we need three hydrogen atoms.

Ammonia is an important nitrogen fertilizer.

$$\text{H} \cdot + \cdot \ddot{\text{N}} \cdot + \cdot \text{H} \longrightarrow \text{H} : \ddot{\underset{..}{\text{N}}} : \text{H} \quad \text{or} \quad \text{H—N—H}$$
$$+ \qquad\qquad\qquad \text{H} \qquad\qquad\qquad\quad |$$
$$\cdot \qquad\qquad\qquad \text{Ammonia} \qquad\qquad\quad \text{H}$$
$$\text{H}$$

Carbon is in group IVA, so a carbon atom has four outer-level electrons. It needs a share of four more to have an octet. If these four electrons are to be provided by hydrogen atoms, we need four of them.

Methane is the chief constituent of natural gas.

$$\begin{array}{c} \text{H} \\ \cdot \\ + \\ \cdot \\ \text{H} \cdot + \cdot \text{C} \cdot + \cdot \text{H} \\ \cdot \\ + \\ \cdot \\ \text{H} \end{array} \longrightarrow \begin{array}{c} \text{H} \\ \ddot{} \\ \text{H} : \text{C} : \text{H} \\ \ddot{} \\ \text{H} \\ \text{Methane} \end{array} \quad \text{or} \quad \begin{array}{c} \text{H} \\ | \\ \text{H—C—H} \\ | \\ \text{H} \end{array}$$

PRACTICE EXERCISE 9 Locate sulfur in the periodic table, write its electron-dot structure, and then write the electron-dot structure and the conventional structural formula of its compound with hydrogen.

No algebraic sign goes with a covalence number because we're not involved with ions.

Covalence Numbers.   The number of covalent bonds that an atom can have in a molecule is called the **covalence number.** Thus the covalence number of chlorine is 1. The covalence number of oxygen is 2; that of nitrogen is 3; and that of carbon is 4. You can see that when we subtract the group number of a nonmetalic, representative element from 8, the result is the covalence number for all members of the group. (Metals and transition elements in general are able in some circumstances to form covalent bonds, but we will not encounter any examples and we will not require rules to govern these situations.)

PRACTICE EXERCISE 10   What is the expected covalence number of silicon?

Once we know the covalence number, we know how many lines must extend from the corresponding atomic symbol in a structural formula. Because the covalence number of carbon is 4, in all structures where the symbol C appears there must be four lines that extend from this symbol. Similarly, from each H in a structure, one line must extend.

EXAMPLE 4.7   DEDUCING STRUCTURES OF MOLECULES

Phosphine is present in the very unpleasant odor of decaying fish.

Problem: Phosphine is a compound of phosphorus and hydrogen. Deduce its structural formula.

Solution: First, we have to find phosphorus in the periodic table. When we do, we see that it is in group VA, and we can tell that it is both a representative element and a nonmetal. The rest is easy. Subtracting the group number, 5, from 8 leaves 3, the number of covalent bonds — the number of lines — that must extend from P in the structure. In other words, by forming three covalent bonds, the phosphorus atom achieves an outer octet. We need three hydrogens to meet this need, so we write the letter P, put three lines from it (more or less extending away from each other), and attach the letter H to each line.

Don't worry about the *angles* these lines (bonds) make to each other. We will take up the question of the geometries of molecules later.

$$H\!-\!P\!-\!H$$
$$\mid$$
$$H$$

It would be incorrect to write H—H—P—H or H—H—H—P as the structure of phosphine, because neither has enough lines from P and both have too many from at least one H.

PRACTICE EXERCISE 11   Deduce the structure of the compound between silicon and hydrogen.

The important covalence numbers are
H 1 O 2 N 3 C 4
F 1 S 2
Cl 1
Br 1
I 1

Molecular Formulas and Structural Formulas.   The covalence numbers of the elements that will be particularly important to our study are summarized in a table in the margin. Because we will need them so often, it's a good idea simply to memorize them now. They will help us to avoid errors in writing structural formulas. We should be able to look at a structure and tell if it's possible under the rules of covalence — the covalence numbers. They enable us to tell which of the possible structural formulas is acceptable for a given molecular formula. A **molecular formula** just gives the composition of one molecule, but a structural formula shows the arrangement of the atoms. For water, the molecular formula is H_2O, and the only correct structural formula is H—O—H, not H—H—O. The latter is incorrect because it shows only one bond from oxygen and two bonds from one of the hydrogen atoms.

EXAMPLE 4.8 USING COVALENCE NUMBERS TO JUDGE STRUCTURES

Problem: The molecular formula of hydrogen peroxide, a common bleach, is H_2O_2. Which of the following structures is correct, under the rules?

$$H—H—O—O \quad \text{or} \quad H—O—O—H$$

Solution: Each oxygen can have only two bonds (lines), but it *must* have two, not one. Each hydrogen *must* have one bond (line). The first structure violates both of these rules, but the second passes the tests and is the correct structure for hydrogen peroxide.

PRACTICE EXERCISE 12 Which is the correct structure for propane, C_3H_8, a common heating and cooking fuel?

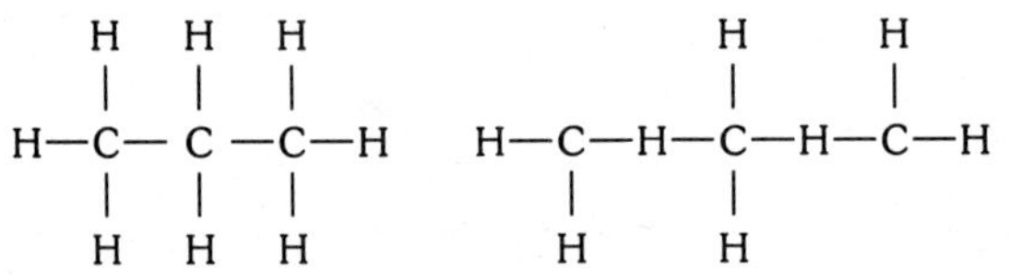

Multiple Bonds. In many molecules, two and sometimes even three electrons have to be shared between two atoms to satisfy the octet rule. The ethylene molecule, C_2H_4, is an example. Its structure is

Ethylene is a gas at room temperature, and it is the raw material for making polyethylene plastics.

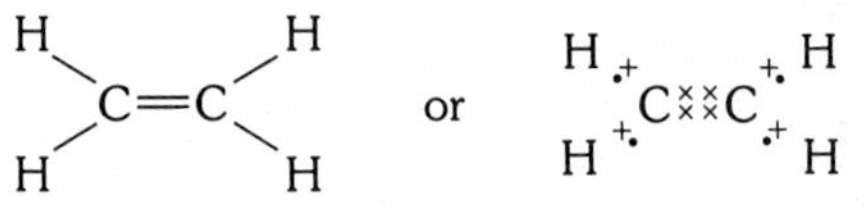

Notice that each carbon has four bonds, as it must. The electron-dot structure, which uses a tiny x for each outer shell electron of carbon, shows where the bonding electrons originate. Each carbon atom supplies four outer-level electrons. We can see that each atom has an octet provided that we count all of the shared electrons for each atom.

Some molecules have more than one double bond. Carbon dioxide, CO_2 is an example:

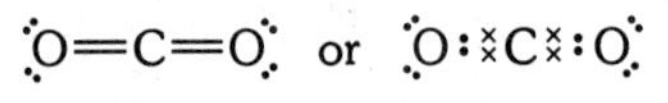

$$\ddot{O}{=}C{=}\ddot{O} \quad \text{or} \quad \ddot{O}{:}\overset{x}{\underset{x}{C}}{:}\ddot{O}$$

The nitrogen molecule, N_2, illustrates the triple bond. To get an octet for each nitrogen, three pairs of electrons must be shared:

$$:N{\equiv}N: \quad \text{or} \quad :N{:}{:}{:}N:$$

By sharing three pairs (and only this way), each atom has an octet.

The goal here is not to be able to construct electron-dot structures of complicated molecules, particularly when a double or a triple bond is involved. What must be mastered is the ability to look at a given structure — either one in a book or reference, or one that you have drawn as an answer to a question — and tell if it satisfies the rules of covalence and is therefore possible.

PRACTICE EXERCISE 13   Examine each of the following structures and decide if it represents a possible compound.

(a) H—C≡C—H   (b) H—O—N═O   (c) F—O—F   (d) H—S—H
                                                                      |
                                                                      H

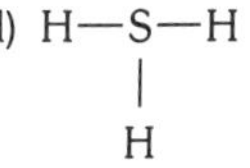

Multiple Covalence Numbers.   Once we get to the third row of the periodic table, we are dealing with atoms whose outer level is number 3. This level is not limited to eight electrons. Therefore, the atoms of nonmetals in the third and higher rows can sometimes become involved in more covalent bonds than our rules thus far allow. For example, sulfur, easily forms both sulfur dioxide, SO_2, and sulfur trioxide, SO_3. The covalence of sulfur can't be 2 in either of these substances, as it is in hydrogen sulfide, H_2S. We will not study the theory that explains multiple covalence, but you should be aware that the atoms of some of the elements which occur often among molecules at the molecular level of life — elements such as sulfur, phosphorus, and nitrogen — can have more than one covalence number. We will treat them on a case-by-case basis.

SO_2 and SO_3 are both air pollutants that contribute to acid rain. Coal and oil contain traces of sulfur, which is oxidized as the fuels burn.

4.7 POLYATOMIC IONS AND A FIRST LOOK AT ACIDS AND BASES

Many important ions are electrically charged clusters of atoms held together by covalent bonds.

Exceptions to the rules of covalence occur among several polyatomic ions. A **polyatomic ion** is a cluster of atoms held together by covalent bonds, but which has a net electrical charge. The ammonium ion, NH_4^+, is a particularly important example, because it and substances like it have functions at the molecular level of life. To understand how this ion can exist, we must expand our understanding of the covalent bond.

Coordinate Covalent Bonds.   We consider the ammonium ion as having ammonia as its electrically neutral "parent." The correct electron-dot structure of ammonia, NH_3, as we have just seen, is

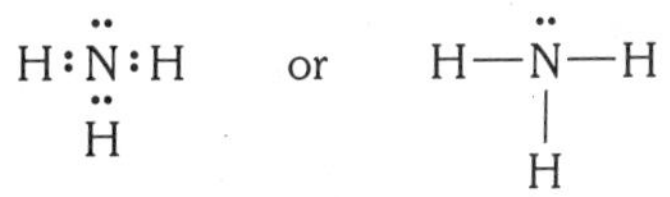

$$H:\overset{\cdot\cdot}{\underset{\cdot\cdot}{N}}:H \quad\quad \text{or} \quad\quad H—\overset{\cdot\cdot}{\underset{|}{N}}—H$$
$$H \phantom{\quad\quad \text{or} \quad\quad H—N—}H$$

The atomic orbitals that overlap to make a covalent bond, such as the three bonds in ammonia, can have no more than two electrons between them, because the new molecular orbital can hold only two electrons. However, *these two electrons do not initially have to have originated from separate atomic orbitals.* One atomic orbital can be empty and the other can provide *both* electrons. The result of the overlap of an empty orbital and a doubly occupied orbital is still a covalent bond because one pair of electrons is shared.

Now notice that the nitrogen atom in a molecule of ammonia has one pair of outer-level electrons that isn't part of a covalent bond. This unshared pair isn't doing any work. It's just there providing an outer octet (as well, of course, as contributing to the overall zero charge on the molecule). One of the important properties of the ammonia molecule is that the orbital that holds its unshared pair can overlap with an *empty* 1s orbital of a hydrogen atom. We can visualize this by means of electron-dot symbols as follows, where we have to represent a hydrogen atom with an *empty* 1s orbital, not as H· but as H^+. In other words, it's no longer a true atom, but an ion called the hydrogen ion.

$$
\text{H}-\overset{\cdot\cdot}{\text{N}}-\text{H} + \text{H}^+ \longrightarrow \left[\text{H}-\overset{\overset{\displaystyle \text{H}}{|}}{\underset{\underset{\displaystyle \text{H}}{|}}{\overset{\cdot\cdot}{\text{N}}}}-\text{H} \right]^+ \text{ or } \left[\text{H}-\overset{\overset{\displaystyle \text{H}}{|}}{\underset{\underset{\displaystyle \text{H}}{|}}{\text{N}}}-\text{H} \right]^+ \text{ or } \text{NH}_4{}^+
$$

Ammonia   Hydrogen   Ammonium
                      ion   ion

The new bond to the fourth hydrogen is a covalent bond, too, but sometimes it's useful to have a name which indicates it arose by the overlap of an empty orbital with a doubly occupied orbital. When we want to indicate this, we call the bond a **coordinate covalent bond.**

The ammonium ion bears a net charge of $1+$ because we have added the $1+$ charge of H^+ to a particle, NH_3, that has zero charge. The resulting cluster of atoms, all held together by covalent bonds, is therefore an example of a polyatomic ion as we have defined this term.

In the structure of the ammonium ion, the nitrogen still has an octet, only now all four electron pairs of the octet are involved in covalent bonds. All four of these bonds are equivalent. The molecule cannot remember which bond formed in which way, so you can see that a *coordinate covalent bond* and a *covalent bond* are not different *once they have formed.* This is why the distinctions between these two terms is seldom important.

Before considering other polyatomic ions we want to introduce two families of compounds that are not just sources of polyatomic ions, but are also involved either in supplying or in combining with an ion that we used to make $NH_4{}^+$, the hydrogen ion.

Acids and Bases — An Introduction.   If you're wondering where we got the hydrogen ion, H^+, there is a large family of substances that provide it. The hydrogen ion is an ion furnished by all common **acids** in water. For example, hydrochloric acid is actually a 1 to 1 mixture of hydrogen ions and chloride ions in water. Sulfuric acid, H_2SO_4, in water furnishes hydrogen ions and a polyatomic ion, the sulfate ion, $SO_4{}^{2-}$ (and some $HSO_4{}^-$ ions). Nitric acid provides, besides the hydrogen ion, the nitrate ion, $NO_3{}^-$. Dilute solutions of these acids all have very tart tastes (but don't experiment with them unless your instructor shows you what to do — some acids are poisons!). The tartness of lemon juice is caused by citric acid, and the tartness of vinegar is caused by acetic acid. Another property common to acids is that they corrode metals such as iron. The reason that acids have properties in common such as these is that acids, in common, furnish the hydrogen ion. The negative ions of acids are not responsible for the properties that all acids have in common.

The tartness of acidic solutions and their ability to corrode certain metals can be destroyed if we add to the solution the right amount of any member of another family of substances called **bases.** The reaction of an acid with a base is called **neutralization.** Bases are compounds that react with and remove hydrogen ions from a solution, and ammonia can do this. Ammonia is just one of several bases, and the source of H^+ when we constructed the ammonium ion earlier was an acid. The acid might have been hydrochloric acid. If we represent hydrochloric acid by its separated ions, H^+ and Cl^-, its reaction with ammonia can be written as follows:

$$
\text{NH}_3 + [\text{H}^+ + \text{Cl}^-] \longrightarrow [\text{NH}_4{}^+ + \text{Cl}^-]
$$

Ammonia   Hydrochloric acid   Ammonium chloride

When the ammonia molecule combines with the hydrogen ion — and they combine, as you

In water H—Cl molecules break up and both H^+ and Cl^- ions become available.

We briefly introduce acids and bases here. They are the subjects of two later chapters, but you'll probably be using them in the lab before then.

Antacid tablets contain mild bases that neutralize stomach acid.

Hydrochloric acid is present in gastric juice in the stomach.

can see, in a 1 to 1 ratio—the resulting solution contains only two ions, NH_4^+ and Cl^-. The solution no longer has either the H^+ ion or a species such as NH_3, which can combine with H^+. Therefore the solution doesn't have the characteristic properties either of the acid or the base. Both have been neutralized. The acid and the base have neutralized each other.

Other Polyatomic Ions.   Table 4.3 lists several of the important polyatomic ions. We won't take the time to do the analysis, but it can be shown that the atoms in each are held together by covalent bonds—coordinate or otherwise—but that the total number of electrons isn't quite equal to the total number of positive charges furnished by the nuclei.

Chemical Formulas Involving Polyatomic Ions.   In many of the reactions of compounds that provide polyatomic ions, the ions stay together as intact units. Therefore they are shown as units in chemical formulas, as in the following examples:

NH_4Cl ammonium chloride, an ingredient in smelling salts

$NaOH$ sodium hydroxide, a raw material for making soap

NH_4NO_3 ammonium nitrate, a fertilizer

$NaNO_2$ sodium nitrite, a preservative in bacon and bologna

Na_3PO_4 sodium phosphate, a powerful cleaning agent

Na_2CO_3 sodium carbonate, washing soda

$NaHCO_3$ sodium bicarbonate, baking soda (not baking *powder*)

TABLE 4.3
Some Important Polyatomic Ions

Name	Formula
Ammonium ion	NH_4^+
Hydronium ion[a]	H_3O^+
Hydroxide ion	OH^-
Acetate ion	$C_2H_3O_2^-$
Carbonate ion	CO_3^{2-}
Bicarbonate ion[b]	HCO_3^-
Sulfate ion	SO_4^{2-}
Hydrogen sulfate ion[c]	HSO_4^-
Phosphate ion	PO_4^{3-}
Monohydrogen phosphate ion	HPO_4^{2-}
Dihydrogen phosphate ion	$H_2PO_4^-$
Nitrate ion	NO_3^-
Nitrite ion	NO_2^-
Hydrogen sulfite ion[d]	HSO_3^-
Sulfite ion	SO_3^{2-}
Cyanide ion	CN^-
Permanganate ion	MnO_4^-
Chromate ion	CrO_4^{2-}
Dichromate ion	$Cr_2O_7^{2-}$

[a] This ion is known only in a water solution.

[b] Formal name: hydrogen carbonate ion.

[c] Common name: bisulfate ion.

[d] Common name: bisulfite ion.

Whenever a formula has to show more than one polyatomic ion, we have to place parentheses about this ion and put a subscript *outside* the closing parenthesis. One example is ammonium sulfate.

$(NH_4)_2SO_4$

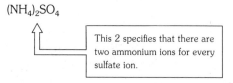

This 2 specifies that there are two ammonium ions for every sulfate ion.

Ammonium sulfate is a fertilizer that supplies both nitrogen and sulfur.

Other compounds whose formulas require parentheses are the following:

$Ca(NO_3)_2$ calcium nitrate
$Mg_3(PO_4)_2$ magnesium phosphate
$Al_2(SO_4)_3$ aluminum sulfate

Appendix C contains a summary of the rules for naming the kinds of compounds we have been using thus far, but notice from these examples that nothing new has been added to what we have already covered. We make names of ionic compounds from the names of their ions (omitting *ion*) in the usual way.

PRACTICE EXERCISE 14   Spend some time memorizing the names and formulas of the polyatomic ions that your instructor has assigned, and then drill yourself by writing the formulas of the following compounds.

(a)   Potassium bicarbonate
(b)   Sodium monohydrogen phosphate
(c)   Ammonium phosphate

PRACTICE EXERCISE 15   Write the name of each of the following compounds.
(a) NaCN   (b) KNO_3 (c) $NaHSO_3$ (d) $(NH_4)_2CO_3$ (e) $NaC_2H_3O_2$

4.8   POLAR MOLECULES

Even electrically neutral molecules can attract each other if they are polar.

We have just seen two ways in which electrical forces of attraction are generated to make atoms stick together in compounds. The first involves oppositely charged ions — the ionic bond — and the second is the attraction of two nuclei toward the same shared pair of electrons — the covalent bond. Molecules in which covalent bonds exist are themselves electrically neutral, so the question now is: What makes it possible for such neutral particles to attract each other? If molecules are neutral, how can they stick together? Sugar molecules, for example, stack together naturally to make beautiful crystals that are not easy to melt. What holds sugar molecules together in such crystals? We'll introduce the answer here and continue its development in succeeding chapters.

Electron Clouds and Partial Charges.   Shared electrons create a region between the two nuclei that has a high density of negative charge. Sometimes this region is said to have an **electron cloud,** because we can imagine that the electrons swarm like bugs in what is sometimes called a cloud of bugs.

When atoms that have different nuclear charges are joined by a covalent bond, the electron cloud of the shared pair usually becomes stabilized nearer one end of the bond than the other. For atoms of the same period in the periodic table, the nucleus with the *larger* positive charge is the one that pulls the electron cloud somewhat away from the nucleus of lower positive charge. The positive-charge effect of the nucleus of lower charge is therefore not entirely canceled *where it is*. The electron cloud is too thin in negative charge to do this. In other words, where the electron cloud is too thin, a fraction of a positive charge is still exerting whatever influences any charge can exert. Such a fractional charge is called a *partial charge,* and we use the Greek lowercase letter delta, δ, to stand for *partial*. Thus a partial positive charge has the symbol $\delta+$. The hydrogen fluoride molecule, for example, has a $\delta+$ charge at its hydrogen end. The hydrogen nucleus has only a charge of $1+$, but the fluorine nucleus has a charge of $9+$, nine times as great. Moreover, because fluorine's other electrons are not concentrated between the two nuclei, they cannot entirely shield the influence of fluorine's nuclear charge from the shared pair. This is why the electron cloud of the shared pair is pulled toward the fluorine end of the H—F molecule. This is why the hydrogen end does not retain enough electron density to neutralize fully the positive charge of the hydrogen nucleus where it is. It is also the reason why the hydrogen end has a $\delta+$ charge. Now let's look at the other end.

When the fluorine end of the H—F molecule pulls the electron density of the shared pair toward itself, the total electron density at the fluorine end of the molecule is more than enough to neutralize the positive charge that is there because of fluorine's nucleus. In other words, the electron cloud is thicker than necessary at this end, and this end of the molecule has a partial negative charge, $\delta-$. We can't say precisely what the sizes of these fractions are, but because the molecule as a whole is electrically neutral, the algebraic sum of the $\delta+$ and the $\delta-$ must be

$$\delta+ \; + \; \delta- \; = \; 0$$

zero.

Polar Bonds.   When a covalent bond has a $\delta+$ at one end and a $\delta-$ at the other, it is called a **polar bond.** For a good analogy, we can think of a magnet. A magnet has two poles, north and south, so it is polar in a magnetic sense. And you no doubt have played with toy magnets enough to know that two magnets can stick to each other *provided they are lined up properly*. In fact, if you have a lot of magnets, and line them up correctly, you can make all of them cling together. You just have to make sure that poles of opposite kind are nearest neighbors and that poles that are alike are as far apart as possible.

We can symbolize the polarity of the bond in hydrogen fluoride in either of two ways, as seen in structures **1** and **2.**

$$\overset{\delta+ \;\; \delta-}{\text{H—F}} \quad \overset{\longmapsto}{\text{H—F}}$$
$$\quad \textbf{1} \qquad\quad \textbf{2}$$

In **2,** the arrow points toward the end of the bond that is richer in electron density. At the other end, there is a hint of the positive character by the merger of the arrow with a plus sign. Because there are two partial, opposite charges in H—F, this molecule is sometimes said to have an **electrical dipole.**

Electronegativity.   The relative ability of an atom of an element to draw electron density toward itself from another atom that it holds by a covalent bond is called this element's **electronegativity.** Fluorine has the highest electronegativity of all of the elements, because its atoms have the highest nuclear positive charge while being shielded by only level-1 and level-2 electrons. Oxygen, which stands just to the left of fluorine in the periodic table, has the next highest electronegativity. Its atoms have one less charge than fluorine atoms, and also are shielded by only level 1 and 2 electrons. The element with the third highest electronegativity, you might now guess, lies just to left of oxygen. It is nitrogen with atoms that have one less charge on their nuclei than oxygen atoms.

Figure 4.5 shows the relative electronegativities of several elements, metals and non-metals, and their locations in the periodic table. Notice that carbon isn't the element with the fourth highest electronegativity; chlorine ranks fourth. The chlorine atom has a large positive

FIGURE 4.5
Relative electronegativities.

IA						
H 2.20	IIA	IIIA	IVA	VA	VIA	VIIA
Li 0.97	Be 1.47	B 2.01	C 2.50	N 3.07	O 3.50	F 4.10
Na 1.01	Mg 1.23	Al 1.47	Si 1.74	P 2.06	S 2.44	Cl 2.83
K 0.91	Ca 1.04				Se 2.48	Br 2.74
Rb 0.89	Sr 0.99				Te 2.01	I 2.21

charge — 17+ on its nucleus — but this doesn't make chlorine even more electronegative than fluorine because chlorine has level 3 electrons as well as those in levels 1 and 2. These extra electrons evidently provide enough shielding of a chlorine nucleus to make the chlorine atom less able than fluorine to be electronegative. Besides, the covalent bond in a molecule such as H—Cl is a longer bond than it is in H—F. The shared pair has its electron density farther from the chlorine nucleus to start with, and this also makes it harder for this nucleus to pull electron density toward itself. On balance, chlorine is less electronegative than fluorine, but more so than carbon.

The tendency of metal atoms is to *give up* electrons, not to attract them.

Notice in Figure 4.5 that metals have the lowest electronegativities. In fact, the general trend is that as you move to the right in the same period or as you move upward in the same group, the electronegativities become larger. The most electronegative element, as we said, is fluorine and the least electronegative is cesium, the last element in the group IA family (below rubidium, Rb, in Figure 4.5).

Although you're not asked to memorize any numerical values for electronegativities, you should learn what the trends are in the periodic table. Moreover, we'll be working so often with oxygen, nitrogen, carbon, and hydrogen, that you should memorize the order of their electronegativies — O > N > C > H. We'll see shortly how knowing this can be helpful.

Polar Molecules.   It's easy to tell if a bond is polar; it always is if the atoms that the bond joins have different electronegativities. But to decide if a bond is polar isn't quite the same as deciding if a *molecule* as a whole is polar. The question of the polarity of a *molecule* is what we're really after. For diatomic molecules such as H—F and H—Cl that have only one bond, when the bond is polar so is the molecule. Because such diatomics as H—H and F—F involve identical atoms and one bond, we can tell right away that these molecules can't be polar.

Whether larger molecules are polar in an overall sense depends not just on the presence of polar bonds but also on the *geometry* of the molecule. It's possible for the polarities of individual bonds to cancel each other. Consider, for example, the carbon dioxide molecule, **3**:

$$O=C=O \qquad \begin{matrix} H \\ \diagdown \\ O-H \end{matrix}$$

3   **4**

Because oxygen is more electronegative than carbon, each carbon-oxygen bond system must be polar. But these two dipoles point in exactly opposite directions, so in an overall sense they cancel each other. This leaves the molecule as a whole nonpolar. The water molecule, **4**, on the other hand, is known to be angular. In it, the individual bond polarities can't cancel, and the water molecule is polar — quite polar, in fact.

Another useful way of thinking about the polarity of a molecule involves the idea of a *center of density of charge,* something like a "balance point" for electrical charge. A **polar molecule** is one in which the center of density of positive charge is not at the same place as the center of density of negative charge. We don't have to be able to pinpoint these centers exactly to know if they are in the same place or not. The symmetry of the carbon dioxide molecule, **3,** tells us that all of the positive charges on the three nuclei balance around the center of the carbon nucleus. Similarly, all of the negative charges contributed by all of the

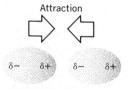

Attraction

Two polar molecules

Two polar molecules can attract each other.

FIGURE 4.6
Polar molecules attract each other in a crystal of a molecular compound.

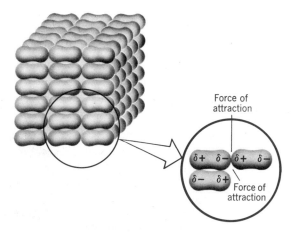

electrons, must also balance *around the identical point.* Because these two centers are in the same place, the molecule is nonpolar. Wherever these two centers are in the water molecule, **4,** we know that because of the angularity of the molecule they can't be at the same place. Hence, we know that this molecule must be polar.

Polar molecules can stick together (Figure 4.6) just as polar magnets can. The polarity of molecules is an essential concept if we are to understand how molecules that are electrically neutral can still attract each other, which they sometimes strongly do in a large number of important biochemicals.

EXAMPLE 4.9   PREDICTING MOLECULAR POLARITY

Problem: Place $\delta+$ and $\delta-$ signs at the correct ends of each of the bonds in the following structures (whose correct geometries are shown). Then decide if each molecule as a whole is polar or nonpolar. (Use information in Figure 4.5 as needed.) A three-dimensional view of the carbon tetrachloride molecule is given. This molecule is entirely symmetrical.

$$\text{I—Br}$$

Iodine bromide

105°
F   F
   O

Oxygen fluoride

Cl
  C   109.5°
Cl   Cl
  Cl

Carbon tetrachloride

Solution: Because bromine is more electronegative than iodine, we have to place the partial charges as follows in iodine bromide, and the molecule is polar.

$$\overset{\delta+ \quad \delta-}{\text{I—Br}}$$

Because fluorine is more electronegative than oxygen, we have to place the partial charges in oxygen fluoride as follows. Because the molecule is angular, the centers of density of positive and negative charge cannot be at the same location, so the molecule must be (and is) polar.

$$\overset{\delta-}{F} \quad \overset{\delta-}{F}$$
$$\underset{\delta+}{O}$$

Because chlorine is more electronegative than carbon, we have to place the partial charges by each bond in carbon tetrachloride as follows:

$$\overset{\delta-}{\underset{\underset{Cl}{|}}{\overset{Cl}{\underset{}{}}}}$$

Cl—C$^{\delta+}$—Cl

Remember, however, that this molecule is symmetrical. Therefore the balance point—the center of charge density—for all positive charge has to be at the center of the carbon nucleus. Similarly, the center of all negative charge density has to be in the identical place—the symmetrical disposition of the chlorine atoms about this center guarantees this result. Hence, the molecule as a whole is not polar.

PRACTICE EXERCISE 16   The structure of chloroform is just like that of carbon tetrachloride (in Example 4.9, above) except that one Cl has been replaced by H. Using the structure of carbon tetrachloride as a model, make the needed changes to draw a structure of chloroform and then place $\delta+$ and $\delta-$ signs by each atom. Finally, decide if the molecule as a whole is polar.

SUMMARY

Ionic bonds and ionic compounds   A reaction between a metal and a nonmetal usually goes by the transfer of an electron from the metal atom to the nonmetal atom. The metal atom changes into a positively charged ion and the nonmetal atom becomes a negatively charged ion. The oppositely charged ions aggregate in whatever whole-number ratio ensures that the product is electrically neutral. The electrical force of attraction between the ions is called an ionic bond, and compounds made of ions are called ionic compounds.

Octet rule   Among the representative elements of the periodic table, a useful guide for predicting how many electrons transfer is that the resulting ions must have electron configurations of the nearest noble gases. Metal atoms lose their outer-level electrons to achieve this, and nonmetal atoms gain enough new outer-level electrons to acquire such configurations.

Formulas   The formulas of ionic compounds begin with the symbol of the metal ion (without the sign indicating the charge). Subscripts are used to give the ratios of the ions. When two or more polyatomic ions have to be indicated in the formula, parentheses must enclose the symbols for these ions. The names of ionic compounds are based on the names of the ions (except that the word *ion* is omitted).

Redox reactions   Electron transfers mean changes in oxidation numbers and are called redox reactions. An atom that loses electrons and whose oxidation number becomes more positive is oxidized, and one that gains electrons and a more negative oxidation number is reduced. Elements have zero oxidation numbers. Simple, monatomic ions have oxidation numbers that equal their electrical charges. Anything that causes an oxidation is called an oxidizing agent; a reducing agent is anything that causes a reduction. Metals tend to be reducing agents and nonmetals are oxidizing agents.

Molecular compounds   Atoms of nonmetals can form molecules by sharing outer level electrons in pairs. Each shared pair constitutes one covalent bond, and each pair is counted as the joint property of both atoms when the system is checked to see if it follows the octet rule. The shared pair exists in a molecular orbital formed by the overlapping of atomic orbitals. The pair creates a region of relatively high electron density between the two atoms toward which the nuclei are electrically attracted, and this attraction is called a covalent bond.

Covalence numbers   The minimum number of covalent bonds a nonmetal atom can have equals the number of electrons it must get by sharing to have an octet. The covalence number equals [8 − group number]. Sometimes two or three pairs of electrons are shared, which means that double or triple bonds occur.

Polyatomic ions   In polyatomic ions, the atoms are joined by covalent bonds, but the overall numbers of electrons and protons do not balance. In these ions, coordinate covalent bonds can also occur.

Acids and bases   Acids are substances that can furnish H^+ ions, and bases are compounds that can combine with H^+ ions. Acids and bases react to neutralize each other.

Polar molecules   Even though molecules are electrically neutral, they can still be polar. If individual bond polarities, caused by electronegativity differences between the joined atoms, do not cancel, the molecule is polar and it can stick to adjacent polar molecules much as magnets can stick together.

KEY TERMS

The following terms were defined in this chapter, and their meanings should be mastered before going on to the next chapter.

acid	electron cloud	molecular orbital	polar bond
base	electronegativity	molecule	polar molecule
chemical bond	electron sharing	neutralization	polyatomic ion
coordinate covalent bond	empirical formula	octet rule	redox reaction
covalence number	ion	outer octet	reducing agent
covalent bond	ionic bond	oxidation	reduction
diatomic molecule	ionic compound	oxidation number	structural formula
electrical dipole	molecular compound	oxidizing agent	structure
electrolyte	molecular formula	overlapping of orbitals	subscripts

SELECTED REFERENCES

1 A. N. Stranges. "Reflections on the Electron Theory of the Chemical Bond: 1900–1925." *Journal of Chemical Education,* March 1984, page 185.

2 W. B. Jensen. "Abegg, Lewis, Langmuir, and the Octet Rule." *Journal of Chemical Education,* March 1984, page 191.

REVIEW EXERCISES

The *Study Guide* that accompanies this text contains answers to these Review Exercises.

Ions and Ionic Compounds

4.1   The term *chemical bond* is another name for what kind of force?

4.2   Lithium (group IA) and fluorine (group VIIA) can react to form ionic bonds. Use electron configurations (including the composition of the nuclei) as we did in Section 4.1 for the reaction of sodium and chlorine to show how a lithium atom and a fluorine atom can change to ions.

4.3   Show how magnesium atoms and fluorine atoms can cooperate to form ions that will aggregate in the correct ratio. (Follow the directions given in Review Exercise 4.2.)

4.4   Write diagrams to show how oxygen atoms and sodium atoms can cooperate to form ions that will aggregate in the correct ratio. (Follow the directions given in Review Exercise 4.2.)

4.5   If M is the symbol of some representative metal, and the symbol of its ion is M^{3+}, in what group in the periodic table is M?

4.6   An atom of the representative nonmetal X can accept two electrons and become an ion. In what group in the periodic table is X?

4.7   An element is in group IIA. What charges can its ions have?

4.8   A representative element forms an ion with a charge of $3-$. What group of the periodic table is this element in?

4.9   Two kinds of electron configurations are exceptionally stable, chemically. Describe them in your own words.

4.10   Study the following electron configuration. Is it likely that this atom can be changed to a stable ion? If so, what is the electrical charge on the ion?

$$1s^2 2s^2 2p^6 3s^2 3p^6 3d^{10} 4s^2 4p_x^2 4p_y^2 4p_z^1$$

4.11   Examine the following electron configuration of an atom. Can this atom be changed into a reasonably stable ion? If so, what is the electrical charge on the ion?

$$1s^2 2s^2 2p^6 3s^2 3p^6 3d^{10} 4s^2 4p^6 5s^2$$

4.12   Study the electron configuration given below. Can this atom change into a reasonably stable ion? If so, what electrical charge does the ion have?

$$1s^2 2s^2 2p^6 3s^2 3p^6 3d^{10} 4s^2 4p^6 4d^{10} 5s^2 5p^6$$

4.13   Element M is a transition element. Are its ions more likely to be positively or negatively charged? Explain.

4.14   Using only what can be deduced from electron configurations built from the atomic numbers using the aufbau rules, write electron configurations of each of the following *ions.* The atomic numbers are given in the parentheses.
(a) Calcium ion (20)   (b) Aluminum ion (13)

4.15   In certain compounds, the hydrogen atom can exist as a negatively charged ion called the *hydride ion.* What is likely to be the electron configuration of this ion? Write a symbol for the ion.

4.16   Some substances are described as *electrolytes.* What kinds of compounds are most likely to be electrolytes?

4.17 It is common in studies of heart conditions to hear scientists speak of the *sodium level* of the blood. *Level* refers to the concentration, the ratio of substance to volume. To what specifically does *sodium* refer?

Oxidation and Reduction

4.18 What is the oxidation number of the sodium ion?

4.19 What is the oxidation number of the Pb(IV) ion?

4.20 What is the oxidation number of the metal in each of the following compounds? Write also the chemical symbol of the metal *ion*.
(a) BiF_3 (b) CdI_2 (c) Cr_2S_3
(d) Gd_2O_3 (e) $SnCl_2$ (f) TiF_3

4.21 If the oxidation number of oxygen is always $2-$ (except in one compound with fluorine, F_2O), what are the oxidation numbers of the other elements in the following compounds?
(a) Mn_2O_7 (b) TiO_4 (c) W_2O_5 (d) Rb_2O

4.22 Consider the following reaction.

$$2Al + 3S \longrightarrow Al_2S_3 \quad \text{(an ionic compound)}$$

For the answer to each of the following parts, write a chemical symbol.
(a) What is reduced?
(b) What is the oxidizing agent?
(c) What is oxidized?
(d) What is the reducing agent?

4.23 Answer each of the parts of this question by writing a chemical symbol. Each part refers to the following reaction:

$$Mg + Br_2 \longrightarrow MgBr_2$$

(a) What is the reducing agent?
(b) What is the oxidizing agent?
(c) Which substance is oxidized?
(d) Which substance is reduced?

4.24 The action of oxygen on FeO can cause the following change:

$$4FeO + O_2 \longrightarrow 2Fe_2O_3$$

(a) What is the oxidation number of Fe in FeO?
(b) What is its oxidation number in Fe_2O_3?
(c) Is Fe oxidized in this reaction?
(d) If so, what is the oxidizing agent? (Write its formula.)
(e) What is reduced by this reaction? (Write its formula.)

Names and Formulas Involving Monatomic Ions

4.25 Give the correct symbols, including the charges, for the following ions:
(a) Potassium ion   (b) Aluminum ion
(c) Iodide ion   (d) Copper(I) ion
(e) Barium ion   (f) Sulfide ion
(g) Sodium ion   (h) Ferric ion
(i) Oxide ion   (j) Lithium ion
(k) Silver ion   (l) Magnesium ion
(m) Cupric ion   (n) Bromide ion

(o) Calcium ion   (p) Iron(II) ion
(q) Fluoride ion   (r) Chloride ion
(s) Zinc ion   (t) Barium ion

4.26 Give the names of the following ions. When an ion has more than one name, one older and the other a modern name, write both.
(a) Na^+ (b) Fe^{3+} (c) Li^+ (d) O^{2-}
(e) S^{2-} (f) Ba^{2+} (g) Cu^+ (h) I^-
(i) Al^{3+} (j) K^+ (k) Zn^{2+} (l) F^-
(m) Fe^{2+} (n) Cl^- (o) Ca^{2+} (p) Cu^{2+}
(q) Br^- (r) Mg^{2+} (s) Ag^+

4.27 If the older name of the Hg^{2+} ion is *mercuric ion,* what is the most likely name of Hg^+? (Actually, this ion exists doubled up as Hg_2^{2+}, but the charge is still $1+$ per Hg.)

4.28 An older name for the Sn^{4+} ion is *stannic ion.* Which is the more likely formula for the *stannous ion,* Sn^{2+} or Sn^{5+}?

4.29 The ion Pb^{2+} was once (and often still is) called the *plumbous ion.* What is its modern name?

4.30 The ion Au^{3+} has the older name of *auric ion.* What is its modern name?

4.31 Write the formula of each of the following compounds.
(a) Lithium chloride   (b) Barium oxide
(c) Aluminum sulfide   (d) Sodium bromide
(e) Cupric oxide   (f) Ferric chloride

4.32 What are the formulas of the following compounds?
(a) Cuprous sulfide   (b) Potassium fluoride
(c) Sodium sulfide   (d) Calcium iodide
(e) Magnesium chloride   (f) Ferrous bromide

4.33 Write the names of the following compounds. Wherever two names are possible, a modern name and an older name, write both.
(a) $FeBr_3$ (b) $MgCl_2$ (c) NaF
(d) ZnO   (e) $CuBr_2$ (f) Li_2O

4.34 What are the names of the following compounds? (If both a modern and an older name are possible, give both names.)
(a) KI   (b) CaS   (c) $BaCl_2$
(d) Al_2O_3 (e) $FeCl_2$ (f) AgI

Molecules and Molecular Compounds

4.35 In general terms, what is a *molecule,* and in what way or ways is a molecule different from an atom? From an ion?

4.36 If two atomic orbitals overlap and partially merge their spaces to form a new orbital, what is this new orbital called?

4.37 What provides the electrical force of attraction that is responsible for the covalent bond?

4.38 Describe what happens when two hydrogen atoms form a molecular orbital and a covalent bond?

4.39 Two helium atoms (atomic number 2) do not combine to give a diatomic molecule, He_2. In terms of what rule we have studied about the overlap of orbitals, why doesn't a covalent bond form between two helium atoms?

4.40 Draw figures to illustrate a brief discussion of how two chlorine atoms develop a covalent bond in Cl_2.

4.41 Discuss how the hydrogen atom and a chlorine atom interact to make a molecular orbital and the H—Cl molecule. Draw figures to illustrate your discussion.

4.42 Examine the following electron configuration and answer the questions about it:

$$1s^2 2s^2 2p^6 3s^2 3p^6 3d^{10} 4s^2 4p_x{}^2 4p_y{}^2 4p_z{}^1$$

(a) Can the atom that has this electron configuration participate in the formation of a covalent bond? If so, how many bonds can it have?

(b) If this atom has the symbol X, what is the structure of its molecular compound with hydrogen?

4.43 Study the following electron configuration:

$$1s^2 2s^2 2p^6 3s^2 3p_x{}^1 3p_y{}^1$$

(a) What covalence number does an atom with this electron configuration have?

(b) If its atomic symbol is Z, what is the structure of its molecular compound with hydrogen?

4.44 Hydrazine, N_2H_4, has been used as a rocket fuel. Which of the following structures is correct for hydrazine, A or B?

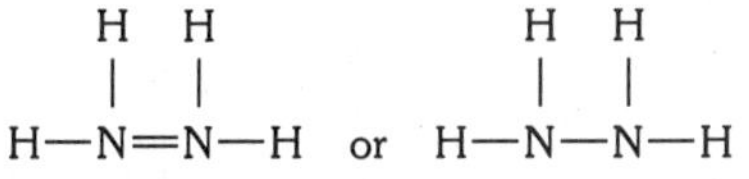

4.45 The raw material for polypropylene, widely used in making indoor-outdoor carpeting, is propylene, C_3H_6. Which of the following structures for propylene is correct, A or B?

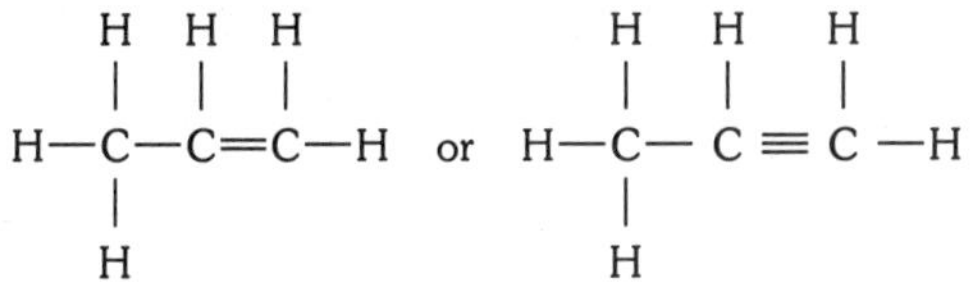

4.46 A molecular compound between germanium and hydrogen has the following structure:

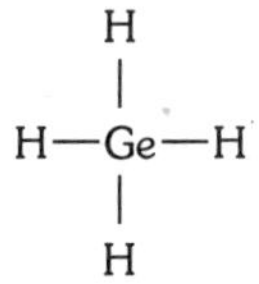

(a) What is the covalence number of Ge in this compound?

(b) Germanium is one of the representative elements. Without referring to the periodic table, using only the answer to part (a), to what group in the periodic table does germanium belong?

4.47 The molecular formula of phosphine is PH_3. Write a structural formula for this compound.

4.48 The molecular formula of carbon disulfide is CS_2. Write a structural formula that is consistent with the covalence numbers of carbon and sulfur.

4.49 Antimony, atomic number 51, is a representative element. It forms a compound with hydrogen called stibine. Using the location of antimony in the periodic table as a clue, write the molecular and structural formulas of stibine.

4.50 One of the components of a certain brand of chlorine bleach has the molecular formula HClO, hypochlorous acid. Write a structural formula of this compound that is consistent with the covalence numbers of H, Cl, and O, which were given in this chapter.

Polyatomic Ions and Formulas Involving Them

4.51 What are the names of the following ions?
(a) $HCO_3{}^-$ (b) $SO_4{}^{2-}$ (c) $NO_3{}^-$
(d) OH^- (e) $NH_4{}^+$ (f) CN^-
(g) $HPO_4{}^{2-}$ (h) $CrO_4{}^{2-}$ (i) $CO_3{}^{2-}$
(j) $MnO_4{}^-$ (k) $HSO_4{}^-$ (l) $HSO_3{}^-$
(m) $NO_2{}^-$ (n) $PO_4{}^{3-}$ (o) $Cr_2O_7{}^{2-}$
(p) $H_2PO_4{}^-$ (q) H_3O^+ (r) $C_2H_3O_2{}^-$

4.52 Write the formulas of the following ions:
(a) Carbonate ion (b) Nitrate ion
(c) Hydroxide ion (d) Ammonium ion
(e) Phosphate ion (f) Cyanide ion
(g) Hydronium ion (h) Monohydrogen phosphate ion
(i) Nitrite ion (j) Bicarbonate ion
(k) Bisulfate ion (l) Dihydrogen phosphate ion
(m) Dichromate ion (n) Bisulfite ion
(o) Chromate ion (p) Sulfate ion
(q) Sulfite ion (r) Acetate ion

4.53 Write the formulas of the following compounds:
(a) Ammonium phosphate
(b) Potassium monohydrogen phosphate
(c) Magnesium sulfate
(d) Calcium carbonate
(e) Lithium bicarbonate
(f) Potassium dichromate
(g) Ammonium bromide
(h) Iron(III) nitrate

4.54 What are the formulas of the following compounds?
(a) Sodium dihydrogen phosphate
(b) Copper(II) carbonate
(c) Silver nitrate
(d) Zinc bicarbonate
(e) Potassium bisulfate
(f) Ammonium chromate
(g) Calcium acetate
(h) Ferric sulfate

4.55 Write the names of the following compounds:
(a) $NaNO_3$ (b) $CaSO_4$
(c) KOH (d) Li_2CO_3
(e) NH_4CN (f) Na_3PO_4
(g) $KMnO_4$ (h) $Mg(H_2PO_4)_2$

4.56 What are the names of the following compounds?
(a) K_2HPO_4 (b) $NaHCO_3$

(c) NH_4NO_2 (d) $ZnCrO_4$
(e) $LiHSO_4$ (f) $Ca(C_2H_3O_2)_2$
(g) $K_2Cr_2O_7$ (h) $NaHSO_4$

4.57 What is the total number of atoms of all kinds in one formula unit of each of the following substances?
(a) $(NH_4)_2SO_4$ (b) $Al_2(HPO_4)_3$ (c) $Ca(HCO_3)_2$

4.58 One formula unit of each of the following substances has how many atoms of all kinds?
(a) $(NH_4)_3PO_4$ (b) $Fe_2(SO_4)_3$ (c) $Ca(C_2H_3O_2)_2$

4.59 Iron(III) glycerophosphate, $Fe_2[C_3H_5(OH)_2PO_4]_3$, has sometimes been used to treat iron deficiency anemia. How many atoms of all kinds are present in one formula unit of this substance?

Acids and Bases

4.60 What is the symbol for the ion supplied by any common *acid?*

4.61 What is the name of the family of compounds whose members can destroy the tartness of an acid and its ability to corrode metals?

4.62 What two kinds of substances react when neutralization occurs?

Polar Molecules

4.63 Suppose that X and Y can form a diatomic molecule, $X—Y$, and that Y is less electronegative than X.
(a) Is the molecule $X—Y$ polar?
(b) If so, where are the $\delta+$ and the $\delta-$ partial charges located?

4.64 Study the following molecules. For those that are polar, write in $\delta+$ and $\delta-$ where they are properly located.
(a) $H—H$ (b) $H—F$ (c) $F—F$
(d) $Cl—Cl$ (e) $N\equiv N$ (f) $H—I$

4.65 Suppose that X and Y are elements in the same group of nonmetals in the periodic table, but X stands above Y. Which of the two has the higher electronegativity? Explain.

4.66 Suppose that M and Z are in the same period in the periodic table, but M precedes Z. Which of these elements has the higher electronegativity? Explain.

Chapter 5
Quantitative Relationships in Chemical Reactions

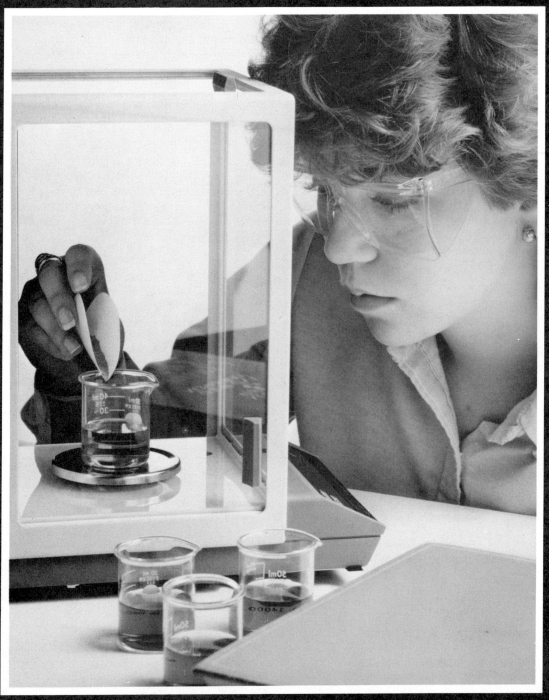

Atoms, ions, and molecules interact as particles which we can count only indirectly, by weighing. We'll see how it's done in this chapter.

5.1 BALANCED EQUATIONS—A SECOND LOOK

The coefficients of a balanced equation give the ratios, not by mass but by formula units, in which the chemicals interact.

In the preceding chapters we looked rather closely at the *structures* of substances—what particles make them up, and how these particles are organized and held together in elements, ionic compounds, and molecular compounds. There is more to learn about atomic, ionic, and molecular structure, but now is an appropriate time to get more into the *properties* of substances. We put the study of structure ahead of the study of properties because structure determines properties. Substances behave as they do because of their structures.

One of the most general and important properties of substances involves their weight relationships when they chemically react. We'll study this next, and we do so not just because it's an important topic to chemists but also because weight relationships in chemical change are at the quantitative heart of almost anything done in medicine that involves medications, diets, intravenously administered solutions, dialysis machines, solutions used to test for substances in blood and urine, respiratory therapy, clinical analyses, and many other areas.

Quantitative aspects of chemical changes begin with what we learned in Chapter Two—atoms and ions can combine only in definite, whole-number ratios to form compounds. It follows that compounds can participate in chemical reactions only in whole-number proportions by their formula units. The branch of chemistry that deals with the possible proportions of reacting chemicals is called **stoichiometry.** The starting place for its study is the balanced equation.

'Stoke-ee-ah-meh-tree' from the Greek *stoicheion,* meaning "element," and -*metron,* meaning "measure."

Balanced Equations.   A chemical equation is a **balanced equation** when all of the atoms given among the reactants appear in identical numbers among the products. Most chemical reactions have equations in which one or more of the formulas must be multiplied by two or some larger number in order to show the correct balance. These multipliers of formulas in chemical equations are called **coefficients.** For example, an equation might show $3H_2O$. The 3 is a coefficient and the 2 is a subscript. The 3 tells us that three molecules of water are involved.

We are never allowed to change subscripts, once we have the right formulas, just to get an equation to balance. For example, changing H_2O to H_2O_2 makes a change from the formula for water to the formula for hydrogen peroxide, an entirely different substance. We can adjust coefficients, however, in balancing an equation, as we will see in the next worked example.

EXAMPLE 5.1   BALANCING A CHEMICAL EQUATION

Problem: Sodium, Na, reacts with chlorine, Cl_2, to give sodium chloride, NaCl. Write the balanced equation for this reaction.

Remember, the formula unit of chlorine is a molecule, Cl_2, not an atom.

Solution: The first step is to set down all of the correct formulas in the format of an equation. (Never worry about the coefficients until the correct formulas are down—then, never change the formulas.)

$$Na + Cl_2 \longrightarrow NaCl \quad \text{(unbalanced)}$$

So far, we see two chlorine atoms on the left (in Cl_2) but only one on the right. We can't fix this by writing $NaCl_2$, because this isn't the correct formula for sodium chloride. The only way we are allowed to get two Cl atoms on the right is to put a coefficient of 2 in front of NaCl:

$NaCl_2$ doesn't even exist.

$$Na + Cl_2 \longrightarrow 2NaCl \quad \text{(unbalanced)}$$

Of course, writing 2 in front of NaCl makes it a multiplier for both Na and Cl, so now we have two Na atoms on the right and just one on the left. To fix this, we write a 2 before the Na:

$$2Na + Cl_2 \longrightarrow 2NaCl \quad \text{(balanced)}$$

Notice particularly how we used a subscript—the 2 in Cl_2—to suggest a coefficient for another formula on the other side of the arrow. This is standard strategy in balancing equations.

EXAMPLE 5.2 **BALANCING A CHEMICAL EQUATION**

Problem: Iron, Fe, can be made to react with oxygen, O_2, to form an oxide with the formula Fe_2O_3. Write the balanced equation for this reaction.

Solution: We first write down the correct formulas in the format of an equation

Oxygen, remember, exists as molecules, O_2, not as atoms.

$$Fe + O_2 \longrightarrow Fe_2O_3 \quad \text{(unbalanced)}$$

Next we exploit subscripts to suggest coefficients. Oxygen has a subscript of 2 in O_2 and a subscript of 3 in Fe_2O_3. To get a balance, we use the 3 as a coefficient for O_2 and the 2 as a coefficient for Fe_2O_3. This is cross-switching the numbers.

$$Fe + 3O_2 \longrightarrow 2Fe_2O_3 \quad \text{(unbalanced)}$$

Now there are six oxygen atoms on the left (in $3O_2$) and six on the right (in $2Fe_2O_3$). Of course, the coefficient of 2 in the formula on the right also means that there are 4 Fe atoms on the right. To fix this, we simply use a coefficient of 4 on the left, for Fe:

$$4Fe + 3O_2 \longrightarrow 2Fe_2O_3 \quad \text{(balanced)}$$

PRACTICE EXERCISE 1   In the presence of an electrical discharge, oxygen, O_2, can be changed into ozone, O_3. Write the balanced equation for this reaction.

PRACTICE EXERCISE 2   Aluminum, Al, reacts with oxygen to give aluminum oxide, Al_2O_3. Write the balanced equation for this change.

Sometimes, the adjustments to coefficients produce balanced equations whose coefficients can be divided by a common divisor. For example, we might have obtained the following equation for the reaction of sodium with chlorine in Example 5.1:

$$4Na + 2Cl_2 \longrightarrow 4NaCl$$

This is a balanced equation and all of its formulas are correct, so there is nothing fundamentally wrong with it. However, chemists generally (but not always) write balanced equations that use the set of smallest whole numbers possible for the coefficients. We will follow this rule unless a special situation warrants something else.

When the formulas in an equation include polyatomic ions, and when it is obvious that they do not themselves change, then treat them as whole units in balancing equations.

EXAMPLE 5.3   BALANCING EQUATIONS INVOLVING POLYATOMIC IONS

Problem: When water solutions of ammonium sulfate, $(NH_4)_2SO_4$, and lead nitrate, $Pb(NO_3)_2$, are mixed, a white solid separates that has the formula $PbSO_4$ (lead sulfate). Ammonium nitrate, NH_4NO_3, is the other product, but it remains dissolved. Represent this reaction by a balanced equation.

Solution: As usual, we start by simply writing the correct formulas in the format of an equation.

The symbol (*aq*) stands for **aqueous solution,** meaning a solution in which water is the solvent. The symbol (*s*) means a solid, one that forms directly and is not in solution.

$$(NH_4)_2SO_4(aq) + Pb(NO_3)_2(aq) \longrightarrow PbSO_4(s) + NH_4NO_3(aq)$$

Because polyatomic ions are involved, we next examine the formulas to see if any of them change or if they all appear to react as whole units — and we can see that they do remain as intact units. The subscript of 2 in $(NH_4)_2SO_4$ suggests that we use 2 as the coefficient in the formula on the right where (NH_4) occurs:

$$(NH_4)_2SO_4(aq) + Pb(NO_3)_2(aq) \longrightarrow PbSO_4(s) + 2NH_4NO_3(aq)$$

This automatically brought into balance the units of (NO_3) on each side of the arrow. The equation is now balanced.

PRACTICE EXERCISE 3   Balance each of the following equations:

(a) $Ca + O_2 \longrightarrow CaO$
(b) $KOH + H_2SO_4 \longrightarrow H_2O + K_2SO_4$
(c) $Cu(NO_3)_2 + Na_2S \longrightarrow CuS + NaNO_3$
(d) $AgNO_3 + CaCl_2 \longrightarrow AgCl + Ca(NO_3)_2$
(e) $Al + H_2SO_4 \longrightarrow Al_2(SO_4)_3 + H_2$
(f) $CH_4 + O_2 \longrightarrow H_2O + CO_2$

5.2 AVOGADRO'S NUMBER

The number of formula units needed for a sample of any pure substance to have a mass numerically equal to its formula weight in grams is 6.02×10^{23} — Avogadro's number.

The coefficients of a balanced equation tell us what proportions of the particles — atoms, molecules, or other kinds of formula units — are involved. For example, the equation for the reaction of sodium with chlorine,

$$2Na + Cl_2 \longrightarrow 2NaCl$$

can be interpreted as follows:

$$2 \text{ atoms of Na} + 1 \text{ molecule of } Cl_2 \longrightarrow 2 \text{ formula units of NaCl}$$

However, we can't carry out a reaction on such a small scale, as we noted in Chapter 3. We have to use much larger numbers of particles to obtain samples that can be manipulated in the lab. The question then is, what number should we use as a standard number of formula units? The number to which chemists throughout the world have long agreed is called

Amadeo Avogadro (1776–1856)

Avogadro's number, after Amadeo Avogadro, an Italian chemist who was interested in stoichiometry, and its value (to three significant figures) is 6.02×10^{23}:

$$6.02 \times 10^{23} = \text{Avogadro's number}$$

It's a pure number that has a special name, like the name *dozen* for 12. In fact, Avogadro's number is sometimes called the chemist's "dozen." Just as "dozen" can be used to signify 12 of anything, so *Avogadro's number* can be used to signify 6.02×10^{23} of anything— electrons, protons, atoms, virus particles, anything.

By now you're probably wondering why a chemist would choose such a complicated number as the standard number of formula units. The reason lies in its relationship to atomic weights. We have seen that 1 amu $= 1.6606 \times 10^{-24}$ g. We also know that the atomic weight of sodium—as one example—is 22.99, which means that one atom of sodium has a mass of 22.99 amu. Let's do a simple calculation to see how many grams a sample of sodium weighs that has Avogadro's number of sodium atoms:

$$6.022 \times 10^{23} \, \text{Na atoms} \times 22.99 \, \frac{\text{amu}}{1 \, \text{Na atom}} \times 1.6606 \times 10^{-24} \, \frac{\text{g}}{\text{amu}} = 22.99 \, \text{g Na}$$

Thus Avogadro's number of sodium atoms make up a sample of sodium that has a mass of 22.99 g, and this is numerically equal to the atomic weight of sodium. Thus the logic of Avogadro's number is that it delivers a mass that is numerically equal to something familiar about an element, namely its atomic weight. For example, Avogadro's number of chlorine atoms has a mass of 35.5 g, which numerically equals the atomic weight of chlorine.

EXAMPLE 5.4   UNDERSTANDING AVOGADRO'S NUMBER

Problem: How many carbon atoms are in 6.00 g of carbon?

Solution: We solve this by working with the basic meaning of Avogadro's number—it's the number of formula units in as many grams of a substance that numerically equal the formula weight. We first have to look up the atomic weight of carbon. Because carbon's atomic weight is 12.0, 12.0 g of carbon must consist of Avogadro's number of carbon atoms, namely 6.02×10^{23} carbon atoms. In other words, 12.0 g of carbon $= 6.02 \times 10^{23}$ atoms of carbon. This gives us two conversion factors:

$$\frac{6.02 \times 10^{23} \text{ atoms C}}{12.0 \text{ g C}} \quad \text{or} \quad \frac{12.0 \text{ g C}}{6.02 \times 10^{23} \text{ atoms C}}$$

If we multiply what is given, 6.00 g C, by the first conversion factor, the units "g C" will cancel, and our answer will be in atoms of carbon:

$$6.00 \, \text{g C} \times \frac{6.02 \times 10^{23} \text{ atoms C}}{12.0 \, \text{g C}} = 3.01 \times 10^{23} \text{ atoms C}$$

Thus 6.00 g of carbon contains 3.01×10^{23} atoms of carbon.

PRACTICE EXERCISE 4   How many atoms of gold are in 1.00 oz of gold? Note: 1.00 oz = 28.4 g.

5.3 FORMULA WEIGHTS AND MOLECULAR WEIGHTS

The sum of the atomic weights of all of the atoms that are included in a chemical formula is the formula weight of the chemical.

Because atoms lose no mass when they combine to form compounds, we can expand the idea of an atomic weight and have a formula weight for every compound. The **formula weight** of a compound is simply the sum of the atomic weights of all of the atoms present in one formula unit. For example, the formula weight of NaCl is calculated as follows, where we round atomic weights to the first decimal place before we use them in a calculation, and we follow a common practice that lets the unit of *amu* be left understood.

$$
\begin{array}{lr}
\text{1 atom of Na in NaCl gives} & 23.0 \\
\text{1 atom of Cl in NaCl gives} & \underline{35.5} \\
\text{Total} & 58.5
\end{array}
$$

The formula weight of NaCl is 58.5. We can invest this number with various meanings, according to our needs. We can say, for example, that one formula unit of NaCl has a mass of 58.5 amu. Or we can say that Avogadro's number of these formula units has a mass of 58.5 g.

The idea of a formula weight is general; it applies to anything with a definite formula, including elements as well as compounds. The formula of sodium, for example, is Na, so we can just as well say that its formula weight is 23.0 as to say that this is its atomic weight. The element chlorine occurs as a diatomic molecule, Cl_2, so its formula weight is twice its atomic weight or $2 \times 35.5 = 71.0$. You should learn, before we continue, that a synonym for formula weight, namely **molecular weight,** is used by many scientists. However, *formula weight* is a more general term, and it's the one that we will use.

EXAMPLE 5.5   CALCULATING A FORMULA WEIGHT

Problem: Some baking powders contain ammonium carbonate, $(NH_4)_2CO_3$. Calculate its formula weight.

Solution: First, look up and write down the atomic weights of all of the elements present, rounding each to the first decimal place:

$$N, 14.0 \quad H, 1.0 \quad C, 12.0 \quad O, 16.0$$

For the rest of this book, our policy is to round values of atomic weights to their first decimal point before starting any calculations.

Notice that in each formula unit of $(NH_4)_2CO_3$ N occurs 2 times, H occurs 8 times, C occurs 1 time, and O, 3 times. Therefore

$$
\begin{array}{llll}
2\,N & +8\,H & +1\,C & +3\,O & = (NH_4)_2CO_3 \\
\end{array}
$$
$$2 \times 14.0 + 8 \times 1.0 + 1 \times 12.0 + 3 \times 16.0 = 96.0 \quad \text{(correctly rounded)}$$

The formula weight of ammonium carbonate is 96.0. This means that one formula unit has a mass of 96.0 amu, and it means that Avogadro's number of these units has a total mass of 96.0 g.

PRACTICE EXERCISE 5   Calculate the formula weights of the following compounds:

(a) $C_9H_8O_4$ (aspirin)   (b) $Mg(OH)_2$ (milk of magnesia)
(c) $Fe_4[Fe(CN)_6]_3$ (ferric ferrocyanide or Prussian blue, an ink pigment)

5.4 THE MOLE

The standard reference unit for the *amount of each pure chemical substance* is the formula weight of the substance taken in grams and has the name *mole*.

The formula weight of any substance — element or compound — taken in grams is called one **mole** of the substance. The mole is the SI base unit for quantity of substance. Its standard abbreviation is **mol.** The actual mass that corresponds to one mole of a substance depends on

Mol stands for both the plural and the singular.

the formula weight, *so the actual mass that equals one mole varies from substance to substance.* However, the number of formula units does not vary. Regardless of the substance, if you have 1 mol of it, you have Avogadro's number of its formula units. Thus 1 mol of Na has a mass of 23.0 g and it consists of 6.02×10^{23} Na atoms; 1 mol of Cl_2 has a mass of 71.0 g, and it consists of 6.02×10^{23} Cl_2 molecules:, and 1 mol of NaCl has a mass of 58.5 g and consists of 6.02×10^{23} formula units of NaCl.

What is *constant* about one mole of any substance is not the mass but the number of formula units.

Think of the *mole* as the lab-sized unit of a substance, a quantity of the substance that can be manipulated experimentally and that can be taken in fractions or in multiples. For example, the formula weight of H_2O is 18.0, so 1 mol of H_2O weighs 18.0 g. If we wished, we could weigh out a smaller sample, say 1.80 g, and then we would have 0.100 mol of water, because 1.80 is 1/10th of 18.0. Or, we could take 36.0 g of H_2O, and then have 2.00 mol, because 36.0 is 2 times 18.0.

With the concept of a mole, we can now think about the coefficients in a balanced equation at two levels at the same time. For example, to return to an equation we used in connection with the laws of chemical combination, the reaction of Fe with S, notice that each formula has a coefficient of 1. Beneath each formula in the equation we can see various ways of interpreting these coefficients.

Fe + S $\longrightarrow$ FeS
1 atom of Fe + 1 atom of S $\longrightarrow$ 1 formula unit of FeS
1 dozen atoms of Fe + 1 dozen atoms of S $\longrightarrow$ 1 dozen formula units of FeS
6.02×10^{23} atoms Fe + 6.02×10^{23} atoms S $\longrightarrow$ 6.02×10^{23} formula units of FeS
1 mol of Fe + 1 mol of S $\longrightarrow$ 1 mol of FeS

Notice that the proportions all remain the same, provided we work with *formula units* of one kind or another. All that changes is the *scale* of the reaction—the actual numbers, not their proportions in relationship to each other. *The coefficients in a balanced equation give us the proportions of substances in moles.* We will use this interpretation almost exclusively from here on. Thus to use once more an equation that we have employed before:

$$2Na + Cl_2 \longrightarrow 2NaCl$$

we can now interpret this to mean that for every *2 mol* of Na that reacts, 1 mol of Cl_2 also reacts and 2 mol of NaCl forms.

There are three kinds of calculations involving moles that have to be learned. One is using an equation's coefficients to determine how many moles of one substance must be involved in a given reaction if a certain number of moles of another are involved. The key step in such a problem is to use the equation's coefficients to set up conversion factors, as we will see in the next example.

EXAMPLE 5.6 USING THE MOLE CONCEPT

Problem: How many moles of oxygen are needed to combine with 0.500 mol of hydrogen in the reaction that produces water by the following equation?

$$2H_2 + O_2 \longrightarrow 2H_2O$$

Solution: The coefficients tell us that 2 mol of H_2 combines with 1 mol of O_2. This relation, which pertains just to this particular reaction, lets us select between the following conversion factors:

$$\frac{2 \text{ mol } H_2}{1 \text{ mol } O_2} \quad \text{or} \quad \frac{1 \text{ mol } O_2}{2 \text{ mol } H_2}$$

What these ratios tell us is that *for this reaction* 2 mol of H_2 is chemically equivalent to 1 mol of O_2. We have to choose one of these ratios to multiply by the given quantity, 0.500 mol of H_2, to find out how much O_2 is needed. The correct ratio is the second one:

$$0.500 \; \cancel{\text{mol } H_2} \times \frac{1 \text{ mol } O_2}{2 \; \cancel{\text{mol } H_2}} = 0.250 \text{ mol } O_2$$

In other words, 0.500 mol of H_2 requires 0.250 mol of O_2 for this reaction.

PRACTICE EXERCISE 6   How many moles of H_2O are made from the 0.250 mol of O_2 in Example 5.6?

PRACTICE EXERCISE 7   Nitrogen and oxygen combine at high temperature in an automobile engine to produce nitrogen monoxide, NO, an air pollutant. The equation is $N_2 + O_2 \rightarrow 2NO$. To make 8.40 mol of NO, how many moles of N_2 are needed? How many moles of O_2 are also needed?

PRACTICE EXERCISE 8   Ammonia, an important nitrogen fertilizer, is made by the following reaction: $3H_2 + N_2 \rightarrow 2NH_3$. In order to make 300 mol of NH_3, how many moles of H_2 and how many moles of N_2 are needed?

Molar Mass.   The next kind of calculation we have to learn is the conversion of moles to grams. This is necessary because laboratory balances do not read in moles. If they did, we'd have to have a separate balance, marked for moles, for each and every possible formula weight! Instead, the balances read in grams, so after we have done the calculations to find out how many moles of a substance we must use, we have to translate this into the grams of the substance before we can weigh it out. For this, we use a formula weight in still another way. We write the units of g/mol after a formula weight, and when we do we have given the _____ of the substance — the number of grams per mole. Thus the molar mass of sodium, which has an atomic weight of 23.0, is 23.0 g Na/mol Na. The units of g/mol represent a ratio, so these units give us two possible conversion factors, as we will see in the next worked example.

EXAMPLE 5.7   CONVERTING MOLES TO GRAMS

Problem: In Example 5.6 we found that 0.250 mol of O_2 was needed. How many grams of O_2 are in 0.250 mol of O_2?

Solution: We first must calculate the formula weight of oxygen, which is two times its atomic weight (16.0) or $2 \times 16.0 = 32.0$. Therefore the molar mass of oxygen is 32.0 g O_2/mol O_2. This fact gives us the following conversion factors that relate mass of O_2 to moles of O_2:

$$\frac{32.0 \text{ g } O_2}{1 \text{ mol } O_2} \quad \text{or} \quad \frac{1 \text{ mol } O_2}{32.0 \text{ g } O_2}$$

We next multiply what was given, 0.250 mol of O_2, by whichever conversion factor lets us cancel "mol O_2" and leaves us with "g O_2," the desired final unit. This means that we use the first conversion factor:

$$0.250 \; \cancel{\text{mol } O_2} \times \frac{32.0 \text{ g } O_2}{1 \; \cancel{\text{mol } O_2}} = 8.00 \text{ g } O_2$$

Thus 0.250 mol of O_2 has a mass of 8.00 g of O_2.

PRACTICE EXERCISE 9   An experiment calls for 24.0 mol of NH_3. How many grams is this?

Sometimes an experiment starts with a specification that a given number of grams of some substance be prepared, perhaps for some application as a potential pharmaceutical. Now we have to be able to convert grams to moles in order to plan the experiment according to some given balanced equation. The next worked example shows how to convert grams to moles, and you will see that this is similar to the moles to grams conversion. We will again use a molar mass to find the right conversion factor.

EXAMPLE 5.8   CONVERTING GRAMS TO MOLES

Problem: A student was asked to prepare 12.5 g of NaCl. How many moles is this?

Solution: The relationship between grams and moles of NaCl is given by the molar mass of NaCl. We have already calculated that the formula weight of NaCl is 58.5, so its molar mass is simply 58.5 g NaCl/mol NaCl. Therefore we have these two possible conversion factors:

$$\frac{58.5 \text{ g NaCl}}{1 \text{ mol NaCl}} \quad \text{or} \quad \frac{1 \text{ mol NaCl}}{58.5 \text{ g NaCl}}$$

If we multiply the second ratio by the given, 12.5 g NaCl, the units will cancel properly and the result will be the moles of NaCl in this sample:

$$12.5 \text{ g NaCl} \times \frac{1 \text{ mol NaCl}}{58.5 \text{ g NaCl}} = 0.214 \text{ mol NaCl} \quad \text{(rounded from 0.2136752137)}$$

Thus 12.5 g of NaCl consists of 0.214 mol of NaCl.

PRACTICE EXERCISE 10   A student was asked to prepare 6.84 g of aspirin, $C_9H_8O_4$. How many moles is this?

With the ability to make these kinds of calculations, we can put them together in the context of a very common laboratory situation — how many grams of one substance are needed to make a given amount of another according to some equation? The next worked example illustrates how this is handled.

EXAMPLE 5.9   MOLE CALCULATIONS USING BALANCED EQUATIONS

Problem: Aluminum oxide can be used as a white filler for paints. How many grams of aluminum are needed to make 24.4 g of Al_2O_3 by the following equation?

$$4Al + 3O_2 \longrightarrow 2Al_2O_3$$

Solution: It's usually a good idea right at the start of a problem such as this to compute any needed formula weights and write them down for reference. When we do this, we have

Al, 27.0 Al_2O_3, 102.0

Because the equation's coefficients refer to *moles,* we must first find out how many moles are in 24.4 g of Al_2O_3. We learned how to do this in the previous example. The formula weight of Al_2O_3, 102.0, gives us its molar mass, 102.0 g Al_2O_3/mol Al_2O_3. We use this fact in the following calculation:

$$24.4 \text{ g } Al_2O_3 \times \frac{1 \text{ mol } Al_2O_3}{102.0 \text{ g } Al_2O_3} = 0.239 \text{ mol of } Al_2O_3$$

Now we can use the coefficients of the equation, where we see that 4 mol of Al gives 2 mol of Al_2O_3. In other words, we can select one of the following conversion factors:

$$\frac{4 \text{ mol Al}}{2 \text{ mol Al}_2O_3} \quad \text{or} \quad \frac{2 \text{ mol Al}_2O_3}{4 \text{ mol Al}}$$

Now we multiply 0.239 mol of Al_2O_3 by the first factor.

$$0.239 \text{ mol Al}_2O_3 \times \frac{4 \text{ mol Al}}{2 \text{ mol Al}_2O_3} = 0.478 \text{ mol Al}$$

The problem called for the answer in grams of Al, not moles of, so we next have to convert 0.478 mol of Al into grams of Al. We studied how to do this in Example 5.7. We use the formula weight of Al to devise the correct conversion factor.

$$0.478 \text{ mol Al} \times \frac{27.0 \text{ g Al}}{1 \text{ mol Al}} = 12.9 \text{ g Al}$$

This is the answer: it takes 12.9 g of Al to prepare 24.4 g of Al_2O_3 according to the given equation.

Before working the next practice exercise, think about the overall strategy that we used in this example. As diagrammed in Figure 5.1, we moved the calculations from the grams-level to the moles-level first. We had to do this because otherwise the equation's coefficients are useless to the solution. Then we used these coefficients to relate the moles of one substance to the moles of another. Finally, we moved back to the grams level so that we could use laboratory balances marked in grams, not moles.

PRACTICE EXERCISE 11   How many grams of oxygen are needed for the experiment described in Example 5.9? Use a diagram of the solution in the style of Figure 5.1 as you work out the answer.

PRACTICE EXERCISE 12   If 28.4 g of Cl_2 are used up in the following reaction, how many grams of Na are also used up, and how many grams of NaCl form?

$$2Na + Cl_2 \longrightarrow 2NaCl$$

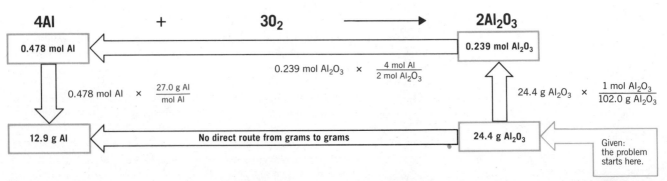

FIGURE 5.1
All calculations involving masses of reactants and products that participate in a chemical reaction must be worked out at the mole level. There is no direct route from grams of one substance to grams of another.

5.5 REACTIONS IN SOLUTION

Virtually all of the chemical reactions studied in the lab and that occur in living systems take place in an aqueous solution.

If the particles of one substance are to react with those of another, they must have enough freedom to move about to find and hit each other. This kind of freedom exists when substances are gases (as we'll learn more about in the next chapter) or are liquids, but not when they are solids. To get one solid to react with another, chemists can melt them together, but it is far more common to dissolve them in something so they are then not actually solids anymore, but are in the liquid state as they fundamentally are in a solution. Then their particles — ions or molecules — can move about. In order to learn in the next section about mass-relationships when the reactants are in solution, we first have to learn some of the common terms used in describing solutions.

A **solution** is a uniform mixture of particles that are of atomic, ionic, or molecular size. A minimum of two substances is needed to have a solution. One is called the solvent and all of the others are called the solutes. The **solvent** is the medium into which the other substances are mixed or dissolved. The solvent is usually a liquid such as water, and unless we state otherwise, in our studies we will always be dealing with **aqueous solutions,** to use the term employed when water is the solvent. A **solute** is anything that is dissolved by the solvent. In an aqueous solution of sugar, the solute is sugar and the solvent is water. The solute can be a gas. Club soda is a solution of carbon dioxide in water. The solute can be a liquid. For example, antifreeze is mostly a solution of a liquid such as propylene glycol — the pure antifreeze — in water, the solvent. (When we mix two liquids to make a solution, which one we call the solvent is a matter of arbitrary choice. If water is one of the liquids, it's generally called the solvent, but it isn't an important point.)

Several terms are used to describe a solution. For example, a **dilute solution** is one in which the ratio of solute to solvent is very small — for example, a few crystals of sugar dissolved in a glass of water. Most of the aqueous solutions in living systems have more than two solutes and they are dilute in each of the solutes. In a **concentrated solution,** the ratio of solute to solvent is large. Syrup, for example, is a concentrated solution of sugar in water. Some solutions are **saturated solutions,** which means that it isn't possible to dissolve more of the solute in them (assuming that the temperature of the solution is kept constant). If more solute is added to a solution that is saturated in this solute, the extra solute will just stay separate. If the solute is a solid, it will generally sink to the bottom and remain there. An **unsaturated solution** is one in which the ratio of solute to solvent is lower than that of the corresponding saturated solution. If more solute is added to an unsaturated solution, at least some of it will dissolve.

It isn't easy, but sometimes a **supersaturated solution** can be made. This is an unstable system in which the ratio of dissolved solute to solvent is actually higher than that of a saturated solution. We can sometimes make a supersaturated solution by carefully cooling a saturated solution. The ability of most solutes to dissolve in water decreases with temperature, so when a saturated solution is cooled, some of the now excess solute should separate. But this doesn't always happen, especially if all dust is excluded and the system is not disturbed in any way as it is cooled. If the excess solute does not separate, then we have a supersaturated solution. If we now scratch the inner wall of the container with a glass rod, or if we add a crystal — a "seed" crystal — of the pure solute to the system, the excess solute will usually separate immediately. This event can be dramatic and pretty to observe (see Figure 5.2). The separation of a solid from a solution is called **precipitation,** and the solid is referred to as the **precipitate.**

The amount of solute needed to give a saturated solution in a given quantity of solvent at a specific temperature is called the **solubility** of the solute in the given solvent. Table 5.1 gives some examples that show how widely solubilities can vary. Notice particularly that a saturated solution can still be quite dilute. For example, only a very small amount of lead sulfate is dissolved in 100 g of a saturated aqueous solution. Notice also that the solubilities of

(a)    *(b)*    *(c)*    *(d)*

FIGURE 5.2
Supersaturation. *(a)* This solution is supersaturated. *(b)* Some seed crystals are added to it. *(c)* and *(d)* Excess solute rapidly separates from the solution.

TABLE 5.1
Solubilities of Some Substances in Water

	Solubilities (g/100 g water)			
Solute	0 °C	20 °C	50 °C	100 °C
Solids				
Sodium chloride, NaCl	35.7	36.0	37.0	39.8
Sodium hydroxide, NaOH	42	109	145	347
Barium sulfate, $BaSO_4$	0.000115	0.00024	0.00034	0.00041
Calcium hydroxide, $Ca(OH)_2$	0.185	0.165	0.128	0.077
Gases				
Oxygen, O_2	0.0069	0.0043	0.0027	0
Carbon dioxide, CO_2	0.335	0.169	0.076	0
Nitrogen, N_2	0.0029	0.0019	0.0012	0
Sulfur dioxide, SO_2	22.8	10.6	4.3	1.8 (at 90 °C)
Ammonia, NH_3	89.9	51.8	28.4	7.4 (at 96 °C)

solids generally increase with temperature. (Calcium hydroxide is an exception that we will understand when we make a more detailed study in a later chapter of solutions and how things dissolve.) Gases become less and less soluble in water as the temperature rises, assuming that the measurements are made under the same pressure.

5.6 MOLAR CONCENTRATION

The unit of *moles per liter,* mol/L, is one of the most useful units for describing concentrations.

As we said earlier, most of the chemical reactions we will study occur in an aqueous solution. This is true both of reactions carried out in the lab and of reactions that occur in living systems. The question, now, is: How are mass relationships in chemical reactions handled when the reactants are in solution? To deal with this problem, chemists rely on a piece of quantitative information about the solution known as its concentration. The **concentration** of a solution is the ratio of the quantity of solute to some given unit of the solution. The units can be

FIGURE 5.3
The preparation of 1 liter of a 1 *M* solution.

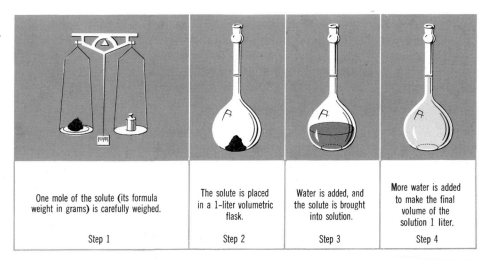

One mole of the solute (its formula weight in grams) is carefully weighed.	The solute is placed in a 1-liter volumetric flask.	Water is added, and the solute is brought into solution.	More water is added to make the final volume of the solution 1 liter.
Step 1	Step 2	Step 3	Step 4

anything we wish, but for the calculations we are discussing the best units are those of moles of solute per liter of solution. The special name for this ratio is called the solution's **molar concentration,** or **molarity,** abbreviated *M*. The molar concentration of a solution — its molarity — is the number of moles of solute per liter of solution:

M = moles/liter

mol = moles

$$M = \frac{\text{mol solute}}{\text{L solution}} = \frac{\text{mol solute}}{1000 \text{ mL solution}}$$

A bottle might, for example, have the label "0.10 *M* NaCl." If so, we know that the solution in this container has a concentration of 0.10 mol of NaCl per liter of solution (or per 1000 mL of solution). A particularly important fact concerning calculations is that the molarity on the label can be interpreted as giving two conversion factors we can use in calculations. In our example, they are:

$$\frac{0.10 \text{ mol NaCl}}{1000 \text{ mL NaCl solution}} \quad \text{and} \quad \frac{1000 \text{ mL NaCl solution}}{0.10 \text{ mol NaCl}}$$

To prepare a solution of known molarity, we use a special piece of laboratory glassware called a *volumetric flask.* Figure 5.3 shows how we could make one liter of a 1 *M* solution of some solute. The concentration tells us that the solution has a concentration of 1 mol/L, and the mass in grams that corresponds to 1 mol is first weighed out. Then it's placed in the volumetric flask and enough water is added to bring the solute into solution. Finally, enough additional water is carefully added until the level of the well-mixed solution reaches the mark on the neck of the flask that tells us that the flask now holds 1 L. Volumetric flasks of a number of capacities are available, ranging from 1 mL to 5 L and several intermediate sizes.

The concept of molarity will become clearer by studying how to do some of the calculations associated with it. In the next worked example we'll see what kinds of calculations have to be done in order to prepare a certain volume of a solution that has a given molar concentration.

EXAMPLE 5.10   PREPARING A SOLUTION OF KNOWN MOLAR CONCENTRATION

Problem: How much sodium bicarbonate, $NaHCO_3$, is needed to prepare 500 mL of 0.125 *M* $NaHCO_3$?

Solution: The label indirectly refers to *moles,* but the question asks for the answer in grams. Before we can calculate the grams needed, we have to find out how many moles of $NaHCO_3$ are required. Here is where the given concentration provides what we need most — a conversion factor for calculating moles of $NaHCO_3$ in the given volume, 500 mL of 0.125 M $NaHCO_3$. Actually, the molarity gives us the following conversion factors of which we'll need to pick one.

$$\frac{0.125 \text{ mol } NaHCO_3}{1000 \text{ mL } NaHCO_3 \text{ solution}} \quad \text{or} \quad \frac{1000 \text{ mL } NaHCO_3 \text{ solution}}{0.125 \text{ mol } NaHCO_3}$$

If we multiply the given volume, 500 mL of $NaHCO_3$ solution, by the first conversion factor, the volume units will cancel and we will learn how many moles that we need.

$$500 \text{ mL } NaHCO_3 \text{ solution} \times \frac{0.125 \text{ mol } NaHCO_3}{1000 \text{ mL } NaHCO_3 \text{ solution}} = 0.0625 \text{ mol } NaHCO_3$$

In other words, the solution we have to prepare has to contain 0.0625 mol of $NaHCO_3$. Now we have to use this fact to calculate the number of grams of $NaHCO_3$ required, and for this we have to use a conversion factor that the formula weight of $NaHCO_3$, 84.0, makes available. This formula weight gives us the following conversion factors:

$$\frac{84.0 \text{ g } NaHCO_3}{1 \text{ mol } NaHCO_3} \quad \text{or} \quad \frac{1 \text{ mol } NaHCO_3}{84.0 \text{ g } NaHCO_3}$$

If we multiply 0.0625 mol of $NaHCO_3$ by the first of these factors, then the units of mol $NaHCO_3$ will cancel and our answer will be in what we want, grams:

$$0.0625 \text{ mol } NaHCO_3 \times \frac{84.0 \text{ g } NaHCO_3}{1 \text{ mol } NaHCO_3} = 5.25 \text{ g } NaHCO_3$$

Thus to prepare 500 mL of 0.125 M $NaHCO_3$, we have to weigh out 5.25 g of $NaHCO_3$, dissolve it in some water in a 500 mL volumetric flask, and then carefully add water until its level reaches the mark, making sure that the contents become well mixed.

PRACTICE EXERCISE 13   How many grams of each solute are needed to prepare the following solutions?

(a) 250 mL of 0.100 M H_2SO_4 (b) 100 mL of 0.500 M glucose ($C_6H_{12}O_6$)

Another calculation that must sometimes be made when a reactant is available only in a solution of known molar concentration is to find out what volume of the solution must be taken in order to obtain a certain quantity of its solute. The next worked example shows how this is done.

EXAMPLE 5.11   USING SOLUTIONS OF KNOWN MOLAR CONCENTRATION

Problem: In an experiment to see if mouth bacteria can live on mannitol ($C_6H_{14}O_6$), the sweetening agent used in some sugarless gums, a student needed 0.100 mol of mannitol. It was available as a 0.750 M solution. How many milliliters of this solution must be used in order to obtain 0.100 mol of mannitol?

Solution: The two conversion factors that are provided by the given concentration are:

$$\frac{0.750 \text{ mol mannitol}}{1000 \text{ mL mannitol solution}} \quad \text{and} \quad \frac{1000 \text{ mL mannitol solution}}{0.750 \text{ mol mannitol}}$$

Therefore

$$0.100 \text{ mol mannitol} \times \frac{1000 \text{ mL mannitol solution}}{0.750 \text{ mol mannitol}} = 133 \text{ mL mannitol solution}$$

Thus, 133 mL of 0.750 M mannitol solution holds 0.100 mol of mannitol.

PRACTICE EXERCISE 14   To test sodium carbonate, Na_2CO_3, as an antacid, a scientist needed 0.125 mol of Na_2CO_3. It was available as 0.800 M Na_2CO_3. How many milliliters of this solution are needed for 0.125 mol of Na_2CO_3?

Once solutions of known molar concentration have been prepared, then the most common kind of calculation involves the stoichiometry of some reaction when at least one reactant is in solution. We'll study this in connection with a problem in acid-base neutralization. You may recall that we briefly introduced the nature of an acid and a base in the previous chapter (on page 92). We learned that an acid is a substance that can furnish H^+ ions and a base is something that can react with H^+ ions in a reaction called an acid-base neutralization. In the next worked example, we'll see how we can do stoichiometric calculations for such a reaction.

EXAMPLE 5.12   STOICHIOMETRIC CALCULATIONS THAT INVOLVE MOLAR CONCENTRATIONS

Problem: Potassium hydroxide, KOH, is a common base. It reacts with hydrochloric acid as follows:

$$HCl(aq) \ + \ KOH(aq) \longrightarrow KCl(aq) \ + H_2O$$

Hydrochloric   Potassium   Potassium
acid   hydroxide   chloride

How many milliliters of 0.100 M KOH are needed to neutralize the acid in 25.0 mL of 0.0800 M HCl?

Solution: We first calculate the moles of acid present. The molarity of the acid gives us these two conversion factors:

$$\frac{0.0800 \text{ mol HCl}}{1000 \text{ mL HCl solution}} \quad \text{or} \quad \frac{1000 \text{ mL HCl solution}}{0.0800 \text{ mol HCl}}$$

Therefore we multiply the given, 25.0 mL of HCl solution, by the first factor:

$$25.0 \text{ mL HCl solution} \times \frac{0.0800 \text{ mol HCl}}{1000 \text{ mL HCl solution}} = 0.00200 \text{ mol HCl}$$

The next step is to find out how many moles of base are needed to neutralize 0.00200 mol of HCl. Here is where we use the coefficients of the balanced equation. They tell us that the ratio is 1 : 1, which means 1 mol of HCl is equivalent chemically (in this equation) to 1 mol of KOH. Thus 0.00200 mol of HCl requires 0.00200 mol of KOH.

Finally, we have to calculate the volume (in mL) of the KOH solution that contains 0.00200 mol of KOH. The molarity of the KOH solution gives us the option of the following conversion factors:

$$\frac{0.100 \text{ mol KOH}}{1000 \text{ mL KOH solution}} \quad \text{or} \quad \frac{1000 \text{ mL KOH solution}}{0.100 \text{ mol KOH}}$$

We can see that if we multiply 0.00200 mol of KOH by the second conversion factor, we'll have the right units:

$$0.00200 \text{ mol KOH} \times \frac{1000 \text{ mL KOH solution}}{0.100 \text{ mol KOH}} = 20.0 \text{ mL KOH solution}$$

Thus 20.0 mL of 0.100 M KOH solution exactly neutralizes the acid in 25.0 mL of 0.0800 M HCl.

The next worked example shows how to solve a problem in which the mole-to-mole ratio of acid to base in the equation is not $1:1$.

EXAMPLE 5.13   STOICHIOMETRIC CALCULATIONS THAT INVOLVE MOLAR CONCENTRATIONS

Problem: Sodium hydroxide, NaOH, another common base, can react with sulfuric acid by the following equation:

$$H_2SO_4(aq) + 2NaOH(aq) \longrightarrow Na_2SO_4(aq) + 2H_2O$$

Sulfuric   Sodium   Sodium
acid   hydroxide   sulfate

How many milliliters of 0.125 M NaOH provide enough NaOH to react completely with the sulfuric acid in 16.8 mL of 0.118 M H_2SO_4 by the given equation?

Solution: Because of the coefficients in the equation, we know that we have to match the moles of sodium hydroxide to the moles of sulfuric acid on a 2-to-1 basis. Hence, we start by asking how many moles of sulfuric acid are in 16.8 mL of 0.118 M H_2SO_4. Then, we will relate this number of moles to the moles of NaOH that match it according to the coefficients. Finally, we will find out how many milliliters of the NaOH solution hold this calculated number of moles of NaOH.

First, then, the moles of H_2SO_4 that are neutralized:

The standard abbreviation for solution is *soln*.

$$16.8 \text{ mL } H_2SO_4 \text{ soln} \times \frac{0.118 \text{ mol } H_2SO_4}{1000 \text{ mL } H_2SO_4 \text{ soln}} = 0.00198 \text{ mol } H_2SO_4$$

Next, the moles of NaOH that chemically match 0.00198 mol of H_2SO_4 according to the coefficients of the equation:

$$0.00198 \text{ mol } H_2SO_4 \times \frac{2 \text{ mol NaOH}}{1 \text{ mol } H_2SO_4} = 0.00396 \text{ mol NaOH}$$

Finally, the volume of 0.125 M NaOH solution that holds 0.00396 mol NaOH:

$$0.00396 \text{ mol NaOH} \times \frac{1000 \text{ mL NaOH soln}}{0.125 \text{ mol NaOH}} = 31.7 \text{ mL NaOH soln}$$

Thus 31.7 mL of 0.125 M NaOH solution are needed to neutralize all of the sulfuric acid in 16.8 mL of 0.118 M H_2SO_4 solution.

Figure 5.4 provides a pictorial summary — a calculation flow chart — of the steps used to solve the problem of Example 5.13.

PRACTICE EXERCISE 15   Blood isn't supposed to be acidic, but in some medical emergencies it tends to become so. To stop and reverse this trend, the emergency-care specialist might administer a dilute solution of sodium bicarbonate intravenously. Sodium bicarbonate neutralizes acids. For example, it reacts with sulfuric acid (which is *not* present in blood) as follows:

$$2NaHCO_3(aq) + H_2SO_4(aq) \longrightarrow Na_2SO_4(aq) + 2CO_2(g) + 2H_2O$$

How many milliliters of 0.112 M H_2SO_4 will react with 21.6 mL of 0.102 M NaHCO₃ *according to this equation?*

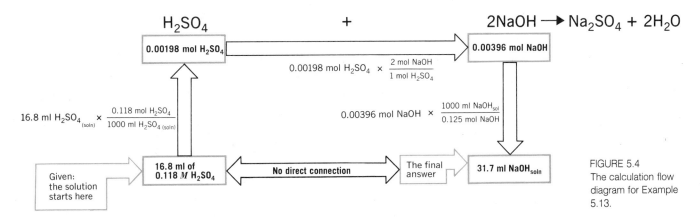

FIGURE 5.4
The calculation flow diagram for Example 5.13.

5.7   PREPARING DILUTE SOLUTIONS FROM CONCENTRATED SOLUTIONS

The amount of solute in a dilute solution that is prepared by adding solvent to a concentrated solution is identical to the amount of solute present before the solvent is added.

It costs less to ship concentrated solutions than dilute solutions because less mass of solution is needed per mole of solute shipped.

Chemicals are often purchased as concentrated reagents which then must be diluted to the concentration desired for some purpose. Sometimes the specific directions for making these dilutions are printed on the label of the concentrated solution, but such directions won't cover all possible needs (and sometimes no directions are given). In this section we'll study how to do the calculations needed if we have to prepare a dilute solution of some desired concentration from a more concentrated solution.

One basic (and somewhat obvious) idea works here. The actual quantity of solute in the particular volume of concentrated solution that we take is identical with the amount of solute in the dilute solution. After all, we only add *solvent*, not solute. Suppose we have to prepare a certain volume of a dilute solution that will have some desired molar concentration. The moles of solute that must be in this dilute solution can be calculated in the usual way from the data on volume and molarity:

dil = dilute

concd = concentrated

$$\text{mol solute} = \text{liters}_{\text{dil soln}} \times \frac{\text{mol solute}}{\text{liter}_{\text{dil soln}}} = \text{liters}_{\text{dil soln}} \times M_{\text{dil soln}}$$

This same number of moles of solute must be obtained from the volume of concentrated solution that we have to take and dilute. (And the purpose of this calculation is to find out this volume.) Thus we can also calculate the moles of solute by the following equation:

$$\text{mol solute} = \text{liter}_{\text{concd soln}} \times \frac{\text{mole solute}}{\text{liter}_{\text{concd soln}}} = \text{liters}_{\text{concd soln}} \times M_{\text{concd soln}}$$

We now have two expressions for moles of solute, and they must equal each other, as we have said. Therefore

$$\text{liters}_{\text{dil soln}} \times M_{\text{dil soln}} = \text{liters}_{\text{concd soln}} \times M_{\text{concd soln}}$$

We need not use the unit of liters in this equation. We can use any volume unit that we please provided that it's the same unit on both sides of the equation. Because we normally use the mL unit, our equation can be changed to give the following standard working equation for doing dilution problems.

$$\text{mL}_{\text{dil soln}} \times M_{\text{dil soln}} = \text{mL}_{\text{concd soln}} \times M_{\text{concd soln}}$$

Now let's work an example to show how this equation is used in a practical, lab situation.

EXAMPLE 5.14 DOING THE CALCULATIONS FOR MAKING DILUTIONS

Problem: Hydrochloric acid can be purchased at a concentration of 1.00 M HCl. How can we prepare 500 mL of 0.100 M HCl?

Solution: What the question really asks is how many milliliters of 1.00 M HCl would have to be diluted to a final volume of 500 mL to make a solution with a concentration of 0.100 M HCl. We first assemble the known data:

$$mL_{dil\,soln} = 500\ mL \qquad mL_{concd\,soln} = ?$$
$$M_{dil\,soln} = 0.100\ M \qquad M_{concd\,soln} = 1.00\ M$$

Now we use the equation:

$$mL_{dil\,soln} \times M_{dil\,soln} = mL_{concd\,soln} \times M_{concd\,soln}$$
$$500\ mL \times 0.100\ M = mL_{concd\,soln} \times 1.00\ M$$

Rearranging terms to solve for $mL_{concd\,soln}$ gives us

$$mL_{concd\,soln} = \frac{500\ mL \times 0.100\ M}{1.00\ M}$$
$$= 50.0\ mL$$

In other words, if we take 50.0 mL of 1.00 M HCl, place this in a 500-mL volumetric flask, and add water to the mark, we will have 500 mL of 0.10 M HCl.

Figure 5.5 shows the steps for doing a dilution of 0.50 M NaCl to give 100 mL of a solution with a concentration of 0.10 M NaCl.

PRACTICE EXERCISE 16 The concentrated sulfuric acid that can be purchased from chemical supply houses is 18 M H_2SO_4. How could we use this to prepare 250 mL of 1.0 M H_2SO_4?

A Word of Caution. When certain very concentrated solutions are diluted with water, a great deal of heat can be generated, enough to make the system boil and to spatter hazardous materials onto fabric, skin, or into the eyes. Concentrated sulfuric acid, in particular, must be handled very carefully. *The only safe course is to pour the concentrated solution slowly, with stirring, into water.* In this way the concentrated and more dense solution will slowly liberate heat as it sinks and dissolves. This lets the heat dissipate, and the mixture will not boil. If you mistakenly added water *to* a concentrated solution, the first water added would tend to float, and it would interact where the surfaces meet in such a way that a relatively large quantity of heat would quickly develop in a small volume. There could be enough heat to cause the system to boil. Remember, if you ever have to make a dilute aqueous solution from any concentrated acid or alkali, *add the concentrated solution slowly to the water as you stir the mixture and, as always, wear protective eye wear.*

SUMMARY

Stoichiometry The coefficients in a balanced equation give the proportions of the chemicals involved either in formula units or in moles. A quantity of a substance equal to its formula weight taken in grams is one mole of the substance, so to calculate a molar mass just find the formula weight and attach the units g/mol. One mole of any pure substance — element or compound — consists of 6.02×10^{23} of its formula units. This number is named Avogadro's number. In working problems involving balanced equations and quantities of

FIGURE 5.5
Preparing a dilute solution by dilution.
(a) The calculated volume of the more
concentrated solution is withdrawn and
(b) placed in a volumetric flask that
already contains some of the water that
will be needed. *(c)* Now additional
water is added slowly as the new solution is swirled to promote mixing until
the final volume is reached. Then the
new solution is transferred *(d)* to a dry
bottle.

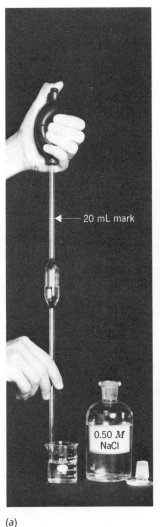

(a)

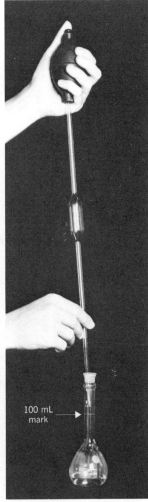

(b)

(c)

(d)

substances, be sure to solve them at the mole level where the coefficients can be used. Then, as needed, convert moles to grams.

Solutions A solution is made up of a solvent and one or more
solutes, and the ratio of quantity of solute to some unit quantity of
solvent or of solution is called the concentration of the solution. A
solution can be described as dilute or concentrated according to its
ratio of solute to solvent being small or large. A solution can also be
described as unsaturated, saturated, or supersaturated according to
whether it can be made to dissolve any more solute (at the same
temperature). Each substance has a particular solubility in a given

solvent at a specified temperature, and this is often expressed as the
grams of solute that can be dissolved in 100 g of the solvent.

Molar Concentration The most popular quantitative description
of the concentration of a solution is the ratio of the moles of solute per
liter (or 1000 mL) of solution. This is the molar concentration or the
molarity of the solution. When we have to prepare one solution by
diluting a more concentrated solution, the equation we use is

$$\text{mL}_{\text{dil soln}} \times M_{\text{dil soln}} = \text{mL}_{\text{concd soln}} \times M_{\text{concd soln}}$$

KEY TERMS

The following terms were emphasized in this chapter, and their
meanings should be mastered before you go to the next chapter.

Avogadro's number	concentration	formula weight	molarity
coefficient	equation, balanced	molar concentration	molar mass

mole (mol)	solubility	solution, concentrated	solution, unsaturated
molecular weight	solute	solution, dilute	stoichiometry
precipitate	solution	solution, saturated	solvent
precipitation	solution, aqueous	solution, supersaturated	

SELECTED REFERENCES

1 D. Kolb. "The Chemical Equation Part 1: Simple Reactions." *Journal of Chemical Education,* March 1978, page 184.

2 L. E. Strong. "Balancing Chemical Equations." *Chemistry.* January 1974, page 13.

3 D. Kolb. "The Mole." *Journal of Chemical Education,* November 1978, page 728.

4 H. A. Bent. "Should the Mole Concept be X-Rated?" *Journal of Chemical Education,* January 1985, page 59.

5 J. N. Lazonby, J. E. Morris, and D. J. Waddington. "The Mole: Questioning Format Can Make a Difference." *Journal of Chemical Education,* January 1985, page 60.

6 D. Todd. "Five Avogadro's Number Problems." *Journal of Chemical Education,* January 1985, page 76.

REVIEW EXERCISES

Answers to the Review Exercises whose numbers are marked by an asterisk are in Appendix D. The *Study Guide* that accompanies this text contains answers to the other Review Exercises. Remember to round atomic weights to their first decimal points *before* using them in any calculations.

Balanced Equations

5.1   Write in your own words what the following equation says.

$$2C + O_2 \longrightarrow 2CO \qquad \text{(carbon monoxide)}$$

5.2   What does the following equation state? Write it in your own words.

$$N_2 + 3H_2 \longrightarrow 2NH_3 \qquad \text{(ammonia)}$$

5.3   What would be a more acceptable way of writing the following balanced equation?

$$8H_3PO_4 + 16NaOH \longrightarrow 8Na_2HPO_4 + 16H_2O$$

5.4   Balance the following equations:
(a) $SO_2 + O_2 \rightarrow SO_3$
(b) $CaO + HNO_3 \rightarrow Ca(NO_3)_2 + H_2O$
(c) $AgNO_3 + MgCl_2 \rightarrow AgCl + Mg(NO_3)_2$
(d) $HCl + Ca(OH)_2 \rightarrow CaCl_2 + H_2O$
(e) $C_2H_6 + O_2 \rightarrow CO_2 + H_2O$

5.5   Balance each of the following equations:
(a) $NaHCO_3 + H_2SO_4 \rightarrow Na_2SO_4 + H_2O + CO_2$
(b) $Fe_2O_3 + H_2 \rightarrow Fe + H_2O$
(c) $Ca(OH)_2 + HNO_3 \rightarrow Ca(NO_3)_2 + H_2O$
(d) $NO + O_2 \rightarrow NO_2$
(e) $Al_2O_3 + H_2SO_4 \rightarrow Al_2(SO_4)_3 + H_2O$

Avogadro's Number

5.6   What is Avogadro's number?

5.7   Why did scientists select Avogadro's number and not some less complicated number to use in defining chemical units for substances?

*5.8   How many molecules are in 6.00 g of H_2O?

5.9   A sample of aspirin ($C_9H_8O_4$) with a mass of 0.180 g (roughly the amount of aspirin in a typical 5-grain tablet) consists of how many molecules of aspirin?

*5.10   One dose of a particular medication contains 6.02×10^{20} molecules. How many grams are in this dose? The formula weight of the medication is 150.

5.11   A sample of impure water contains 3.01×10^{18} molecules of the impurity. How many milligrams of the impurity are present if its formula weight is 240?

Formula Weights

5.12   What law of nature makes it possible to compute formula weights simply by adding all of the atomic weights of the atoms given in a formula?

*5.13   Calculate the formula weight of each of the following substances:
(a) NaOH   (b) $CaCO_3$ (c) H_2SO_4
(d) Na_2CO_3 (e) $KMnO_4$ (f) Na_3BO_3

5.14   Calculate the formula weight of each of the following substances:
(a) $Mg(HCO_3)_2$ (b) Na_2HPO_4 (c) HNO_3
(d) $KC_2H_3O_2$ (e) $(NH_4)_2SO_4$ (f) $Ca_3(PO_4)_2$

Moles

5.15   How do we calculate the quantity of mass present in 1 mol of any substance?

5.16   What is the relationship between Avogadro's number and 1 mol of any substance?

5.17   What is the practical difficulty in the lab of having balances read in moles instead of in mass units?

5.18   In converting grams to moles or moles to grams, what units are given to the formula weight of, say, H_2O, a molecular compound with a formula weight of 18.0?

5.19 The formula weight of bromine, Br_2, is 159.8. What are the two conversion factors that we can write that use this information?

*5.20 How many grams are in 2.50 mol of each of the compounds listed in Review Exercise 5.13?

5.21 How many grams are in 0.575 mol of each of the compounds given in Review Exercise 5.14?

*5.22 Calculate the number of moles that are in 75.0 g of each of the compounds listed in Review Exercise 5.13.

5.23 How many moles are in 28.6 g of each of the compounds given in Review Exercise 5.14?

Stoichiometry Involving Balanced Equations

*5.24 In the equation for the reaction of iron with oxygen,

$$4Fe + 3O_2 \longrightarrow 2Fe_2O_3$$

what two conversion factors express the *mole* relationship between each of the following pairs of substances?
(a) Fe and O_2 (b) Fe and Fe_2O_3 (c) O_2 and Fe_2O_3

5.25 Ethane, C_2H_6, burns according to the following equation:

$$2C_2H_6 + 7O_2 \longrightarrow 4CO_2 + 6H_2O$$

What two conversion factors express the *mole* relationship between each of the following pairs of compounds?
(a) C_2H_6 and O_2 (b) CO_2 and C_2H_6
(c) H_2O and O_2 (d) C_2H_6 and H_2O

*5.26 Butane, C_4H_{10}, is the fluid used in cigarette lighters. It burns according to the following equation:

$$2C_4H_{10} + 13O_2 \longrightarrow 8CO_2 + 10H_2O$$

(a) How many moles of oxygen are needed to react completely with 4.0 mol of butane?
(b) How many moles of water form from the burning of 10 moles of butane?
(c) To make 16 moles of carbon dioxide by this reaction, how many moles of oxygen are needed?

5.27 In the last step of the most commonly used method to convert iron ore into iron, the following reaction occurs:

$$Fe_2O_3 + 3CO \longrightarrow 2Fe + 3CO_2$$

(a) To make 350 mol of Fe, how many moles of CO are needed?
(b) If one batch began with 35 mol of Fe_2O_3, how many moles of iron are made?
(c) How many moles of CO are required to react with 125 mol of Fe_2O_3?

*5.28 The Synthane process for making methane, CH_4, from coal is as follows, where we use carbon, C, to represent coal.

$$C + 2H_2 \longrightarrow CH_4$$

(a) How many moles of hydrogen are needed to combine with 37.5 mol of C?
(b) How many moles of methane can be made from 86 mol of H_2?

5.29 Gasohol contains ethyl alcohol, C_2H_6O, which burns according to the following equation:

$$C_2H_6O + 3O_2 \longrightarrow 2CO_2 + 3H_2O$$

(a) If 475 g of ethyl alcohol burn this way, how many grams of oxygen are needed?
(b) If 326 g of CO_2 form in one test involving this reaction, how many grams of ethyl alcohol burned?
(c) If 92.6 g of O_2 are consumed by this reaction, how many grams of water form?

*5.30 An industrial synthesis of chlorine is carried out by passing an electric current through a solution of NaCl in water. Other commercially valuable products are sodium hydroxide and hydrogen.

$$2NaCl + 2H_2O \xrightarrow[\text{current}]{\text{electric}} 2NaOH + H_2 + Cl_2$$

(a) How many grams of NaCl are needed to make 775 g of Cl_2?
(b) How many grams of NaOH are also produced?
(c) How many grams of hydrogen are made as well?

5.31 Phosphoric acid, H_3PO_4, is needed to convert phosphate rock into a fertilizer called *triple phosphate*. One way to make phosphoric acid is by the following reaction:

$$P_4O_{10} + 6H_2O \longrightarrow 4H_3PO_4$$

(a) To make 1.00×10^3 kg of phosphoric acid (one metric ton), how many kilograms of P_4O_{10} are needed?
(b) How many kilograms of water are also required?

*5.32 One method that can be used to neutralize an acid spill in the lab is to sprinkle it with powdered sodium carbonate. For example, sulfuric acid, H_2SO_4, can be neutralized by the following reaction:

$$H_2SO_4 + Na_2CO_3 \longrightarrow Na_2SO_4 + CO_2 + H_2O$$

If 45.0 g of sulfuric acid are spilled, what is the minimum number of grams of sodium carbonate that have to be added to it to complete this reaction?

SOLUTIONS

5.33 A solution of sodium chloride at 50 °C was found to contain 36.5 g NaCl/100 g water. Using information in a table in this chapter, determine if this solution was saturated, unsaturated, or supersaturated.

5.34 A solution of barium sulfate at 20 °C contains 0.00024 g $BaSO_4$ per 100 g water. Is it saturated, unsaturated, or supersaturated? Would it be described as dilute or concentrated?

5.35 Suppose you do not know and do not have access to a reference in which to look up the solubility of potassium chloride, KCl, in water at room temperature. Yet you need a solution that you know beyond doubt is saturated. How could such a solution be made?

5.36 Suppose you have a saturated solution of vitamin C in water at 10 °C. Without changing either the quantity of solvent or the quantity of solute, what could you do to make this solution unsaturated? What could you do to find out if this solution could be made supersaturated?

Molar Concentration

5.37 What are the differences between *molecule, mole,* and *molarity?*

5.38 What is another term for *molarity?*

5.39 When a volumetric flask is used to prepare a solution having some specified molarity, does the one who makes this solution know precisely how much *solvent* is used? Why is this information unnecessary for the uses to which the solution might be put?

•5.40 Calculate the number of grams of solute that would be needed to make each of the following solutions:
(a) 500 mL of 0.200 M NaCl
(b) 250 mL of 0.125 M $C_6H_{12}O_6$ (glucose)
(c) 100 mL of 0.100 M H_2SO_4
(d) 500 mL of 0.400 M KOH

5.41 How many grams of solute are needed to make each of the following solutions?
(a) 250 mL of 0.120 M Na_2CO_3
(b) 500 mL of 0.100 M NaOH
(c) 100 mL of 0.750 M $KHCO_3$
(d) 250 mL of 0.100 M $C_{12}H_{22}O_{11}$ (sucrose)

•5.42 How many milliliters of 0.10 M HCl contain 0.025 mol of HCl?

5.43 How many milliliters of 1.0 M H_2SO_4 would have to be taken to obtain 0.0025 mol of H_2SO_4?

•5.44 If you need 0.0010 mol of $NaHCO_3$ for an experiment, how many milliliters of 0.010 M $NaHCO_3$ would you have to measure out?

5.45 The stockroom has a supply of 0.10 M H_2SO_4. If you have to have 0.075 mol of H_2SO_4, how many milliliters of this stock solution do you have to take?

•5.46 There is a supply of 1.00 M NaOH in the lab. How many milliliters of this solution have to be taken in order to obtain 10.0 g of NaOH?

5.47 The stock supply of sulfuric acid is 0.50 M H_2SO_4. How many milliliters of this solution contain 5.0 g of H_2SO_4?

Stoichiometry of Reactions in Solution

•5.48 Barium sulfate, $BaSO_4$, is very insoluble in water, and a slurry of this compound is the "barium cocktail" given to patients prior to taking X-rays of the intestinal tract. It can be made by the following reaction:

$$Ba(NO_3)_2(aq) + Na_2SO_4(aq) \longrightarrow BaSO_4(s) + 2NaNO_3(aq)$$

In one use of this reaction to make barium sulfate (it is collected by filtering the final solution) a chemist used 250 mL of 0.100 M $Ba(NO_3)_2$. How many milliliters of 0.150 M Na_2SO_4 were needed to supply enough solute for this reaction?

5.49 How many milliliters of 0.100 M HCl are required to react completely with (and be neutralized by) 25.4 mL of 0.158 M Na_2CO_3 if they react according to the following equation?

$$2HCl(aq) \quad + Na_2CO_3(aq) \longrightarrow 2NaCl(aq) + CO_2(g) + H_2O$$
Hydrochloric   Sodium
acid   carbonate

•5.50 Calcium hydroxide, $Ca(OH)_2$, is an ingredient in one brand of antacid tablets. It reacts with and neutralizes the acid in gastric juice, hydrochloric acid, as follows:

$$Ca(OH)_2(s) + 2HCl(aq) \longrightarrow CaCl_2(aq) + 2H_2O$$

A tablet that contains 2.00 g of $Ca(OH)_2$ can neutralize how many milliliters of 0.100 M HCl?

5.51 A nitric acid spill can be neutralized by sprinkling solid sodium carbonate on it. The reaction is

$$Na_2CO_3(s) + 2HNO_3(aq)$$
$$\longrightarrow 2NaNO_3(aq) + CO_2(g) + H_2O$$

In one accident, 25.0 mL of 16.0 M HNO_3 (concentrated nitric acid, a dangerous chemical) spilled onto a stone desk top. Will 40.0 g of Na_2CO_3 be enough to neutralize this acid by the given equation? How many grams are needed?

Preparing Dilute Solutions from Concentrated Solutions

•5.52 Concentrated acetic acid is 17 M $HC_2H_3O_2$. How would you prepare 100 mL of 2.0 M $HC_2H_3O_2$?

5.53 Concentrated nitric acid is 16 M HNO_3. How would you prepare 500 mL of 1.0 M HNO_3?

Chapter 6
States of Matter, Kinetic Theory, and Equilibria

Far below the snow and ice is rock hot enough to turn water to steam under great pressure and create this beautiful geyser action in Yellowstone National Park.

6.1 THE GASEOUS STATE

The four properties that completely define the physical state of any gas are pressure, temperature, volume, and moles.

Thus far we have been studying the *kinds* of matter — elements and compounds — and their fundamental particles — atoms, ions, and molecules. We will now study the *states of matter* — gases, liquids, and solids — the kinetic theory of these states, and equilibria among them. We begin with gases, because they are the easiest, and because our understanding of why gases behave as they do will help us understand the other states.

The Physical Quantities Used to Describe Gases.   Gases are unique. Unlike liquids and solids, all gases obey the same set of physical laws. It doesn't matter that different gases have different *chemical* properties. They share the same laws concerning their physical properties, and these laws involve just four physical quantities — volume, temperature, pressure, and the number of moles.

As you know, the container of a gas has to be entirely closed, or the gas will eventually spread out into the surrounding atmosphere. A gas always spreads out to occupy the entire volume of its container. This ability is called **diffusion,** and we will see later in this chapter how it works. Volume, symbolized as *V*, is thus one physical quantity we must give in order to describe a specific gas sample. It's usually stated in liters (L) or milliliters (mL).

Liquids and solids, of course, do not spread out like this.

Temperature, *T*, is another property that must be recorded if a gas sample is to be fully characterized. This is because a change in the temperature of an enclosed gas can affect its volume if the container's walls are flexible or one of them can move (as in an automobile cylinder). Any temperature scale could be used, but a gas temperature has to be given in kelvins (K) to make the calculations simple.

$$K = °C + 273$$

Pressure, *P*, is still another physical quantity that must be recorded when a specific sample of a gas is described. You're aware of air pressure in bicycle and automobile tires. You've also seen warnings printed on aerosol spray cans — "Do Not Incinerate." When an enclosed gas is heated, its pressure rises, and it might go high enough to burst the can. Picnickers who put unopened cans of beans on a campfire soon learn about the power of gas pressure as they pick beans from their hair.

The fourth physical quantity needed to describe a given gas sample is its number of moles, *n*. One mole of a gas — any gas — at some particular pressure and temperature occupies a certain volume, the same volume (more or less) for any gas. If we want two moles of the gas at the same pressure and temperature, we need twice the volume.

Volume, temperature, pressure, and moles are the four physical quantities we need to study gases. One of the remarkable facts about all gases is that if we know any three of these properties, the fourth can have only one value, a value that we can calculate with acceptable precision.

Pressure.   If we divide the force acting on a given area by the area itself, the result is called the **pressure.** Pressure is force per unit area:

$$pressure = \frac{force}{area}$$

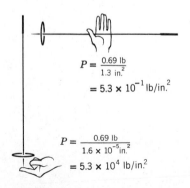

$$P = \frac{0.69 \text{ lb}}{1.3 \text{ in.}^2}$$
$$= 5.3 \times 10^{-1} \text{ lb/in.}^2$$

$$P = \frac{0.69 \text{ lb}}{1.6 \times 10^{-5} \text{in.}^2}$$
$$= 5.3 \times 10^4 \text{ lb/in.}^2$$

The weight of an object is a measure of the force it can exert when it is subject only to the gravitational attraction of the earth. If you've ever done something such as balancing a ski pole or similar object sideways on the palm of one hand, its weight is distributed over a fairly large area of the hand and you feel little force anywhere. (The simple calculations in the margin give a rough idea.) However, when you try to balance the pole by letting its tip rest on your finger, the entire weight is concentrated on a very small area, and the pressure you feel is very strong and painful. As the margin calculations show, the pressure could be as much as 100,000 times larger. Thus the distinction between force (or weight) and pressure is very important.

Units of Pressure. The unit that has served as the parent for other pressure units is the pressure exerted by air on the earth's surface. Because air consists of matter, it has weight; which means that the air is pulled toward the earth by the earth's gravitational attraction. Imagine a perfect cylinder of still air with invisible walls. Its area at the bottom is 1 square inch, and the cylinder reaches to outer space. This column of air has a weight at sea level of about 14.7 lb when the temperature is 0 °C. Because this weight rests on an area of 1 in.², the pressure exerted by the column of air is the ratio 14.7 lb/in.²

The distance to outer space is less from a high mountain top than it is at sea level, so our column of air has less air in it when its bottom rests at a high altitude. This is why air pressure decreases with altitude, and on top of Mt. Everest, the highest mountain on earth, the pressure is about one-third what it is at sea level.

The pushing ability of air caused by its weight is the basis of one way to measure pressure, the mercury or Torricelli **barometer,** named after Evangelista Torricelli (1608 – 1647), an Italian scientist. (See Figure 6.1.) It consists of a glass tube of a little less than a meter in length that has been sealed at one end, filled with mercury, and then inverted into a container of mercury. Some mercury immediately runs out of the tube. However, no air can get in to fill the gap that this creates at the top of the tube. There is no air in this gap, and any space where no air or other gas is found is called a **vacuum.** (The gap does not involve a *perfect* vacuum because this space contains a very small trace of vaporized mercury.) In other words, virtually nothing inside the tube at its top pushes down on the entrapped mercury. Thus the outside air pressure is unopposed, and it forces the remaining mercury to stay in the tube. The exact height of the column fluctuates somewhat with temperature and weather, but at sea level the height is about 760 mm. Because 25.4 mm = 1 in., this height is equivalent to 29.9 in. of mercury, the pressure unit commonly used in weather reports.

From the behavior of a mercury barometer at sea level came one of the standards for measuring pressure, the standard atmosphere. One **standard atmosphere,** symbolized **atm,** is the pressure exerted by a column of mercury that is 760 mm high at a temperature of 0 °C:

$$1 \text{ atm} = 760 \text{ mm Hg} \qquad (0 °C)$$

A smaller unit, one that is probably still the most widely used unit of pressure in medicine and clinical work in the United States, is called the **millimeter of mercury,** or **mm Hg** (pronounced *em em aitch gee*).

$$1 \text{ mm Hg} = \frac{1}{760} \text{ atm} \qquad (6.1)$$

The scientific and medical worlds are now in a period of transition concerning pressure units. Most scientists are uncomfortable about using a measurement of length — the millimeter — to express a measurement of pressure — force per unit area — so they have adopted a new name for the mm Hg, the **torr.**

$$1 \text{ torr} = 1 \text{ mm Hg} \qquad (6.2)$$

However, it is our understanding that the *mm Hg* is still most often used on clinical report sheets as well as in the professional journals that nurses, physicians, and other health care professionals read, and we will use this unit.

By *outer space* we mean space beyond the earth's atmosphere where the atmosphere is so thin as to be almost nonexistent.

$$760 \text{ mm} \times \frac{1 \text{ in.}}{25.4 \text{ mm}} = 29.9 \text{ in.}$$

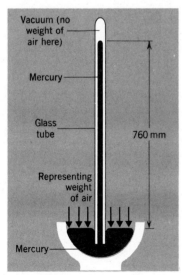

FIGURE 6.1
The Torricelli barometer.

In medicine, pressure units are needed in some discussions of anesthetic gases and the respiratory gases (O_2 and CO_2).

The SI has also entered the picture. The SI wants all derived units such as those of pressure to be based on SI base units, such as meter, kilogram, and so on. The SI unit for force is called the *newton,* and the SI unit for area is the meter square (m^2). Therefore the SI unit for pressure is the ratio of these — newtons/m^2 — and is called the *pascal,* and symbolized as *Pa.* The unit we have called the *atmosphere,* in fact, is now defined in relationship to the pascal:

kPa = kilopascal

$$1 \text{ atm} = 101{,}325 \text{ Pa} \quad \text{(exactly)}$$
$$1 \text{ atm} = 101.325 \text{ kPa} \quad \text{(exactly)}$$

$$101{,}325 \, \frac{Pa}{atm} \times \frac{1 \text{ atm}}{760 \text{ mm Hg}}$$
$$= 133.322 \text{ Pa/mm Hg}$$

During your professional career you will probably see the pascal or an SI multiple such as the kilopascal (kPa) used more and more, especially in formal scientific articles. The pascal is an extremely small unit of pressure, because it takes over 100,000 such units to equal just 1 atm. As shown in the margin, 1 mm Hg (1 torr) = 133.322 Pa.

6.2   THE PRESSURE–VOLUME RELATIONSHIP AT CONSTANT TEMPERATURE

When a fixed number of moles of gas is kept at constant temperature, the volume is inversely proportional to the pressure.

Anyone who has used a bicycle pump knows that when the pressure on a gas is increased, the volume decreases. Robert Boyle (1627–1691), an English scientist, discovered that the relationship between *P* and *V* is particularly simple if the amount of gas (the sample size in mass) is fixed and the temperature is held constant. Using a device similar to that shown in Figure 6.2, he found that if he doubled the pressure under these conditions, he cut the volume in half. If he decreased the pressure by, say, 50%, the volume increased by 50%.

Whenever one quantity decreases in proportion to an increase in the other we say that they are *inversely proportional* to each other. Boyle discovered that the pressure and volume are inversely proportional (at constant mass and temperature). In time, other scientists found that all the gases they studied had this property, so now we have a law of nature called the **pressure–volume law,** or **Boyle's law.**

> Pressure–Volume Law (Boyle's Law) The volume of a fixed quantity of gas is inversely proportional to its pressure at a constant temperature.

Robert Boyle (1627–1691)

FIGURE 6.2
J-tube apparatus for pressure–volume data. On the left, the pressures in the two arms of the tubes are the same and equal, say, 754 mm Hg. A volume of gas, V_1, has been trapped in the shorter arm with the sealed end. On the right, enough mercury has been added to make the mercury column in the longer arm extend 754 mm above the top of the shorter mercury column. In other words, the pressure on the entrapped gas is now 754 + 754 mm Hg or twice the initial value. This has squeezed the volume of the gas to half its original size. Higher pressures can be applied to obtain further Boyle's law data. All that is needed is a longer tube, made of strong material, into which more mercury can be poured.

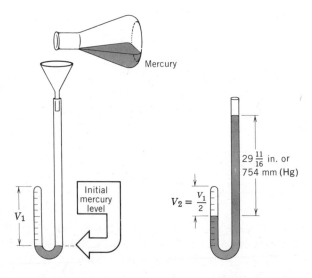

Put in mathematical form, this law states that

The symbol $\propto$ stands for *is proportional to*.

$$V \propto \frac{1}{P} \qquad (T \text{ and mass constant}) \qquad (6.3)$$

Introducing a proportionality constant, C, we can change Equation 6.3 into the following equation:

$$V = \frac{1}{P} \times C$$

or, rearranging terms,

$$PV = C \qquad (6.4)$$

The proportionality constant is usually not determined or calculated because Equation 6.4 is easily transformed into a more useful form. Its constant, whatever its value, has to be just that—a constant—provided that we use the same mass of gas and the same temperature. Thus if P_1 and V_1 are the values of pressure and volume initially, and we change one of them (at the same T and mass) so that we have new values, P_2 and V_2, then the pressure–volume law tells us that both of the following equations are true:

The same constant, C, applies. This is at the heart of Boyle's law.

$$P_1 V_1 = C$$
$$P_2 V_2 = C$$

The two quantities that C equals must equal each other, so we can write the following equation, the equation that is the most useful in doing pressure–volume law calculations.

$$P_1 V_1 = P_2 V_2 \qquad (6.5)$$

Boyle's law is like all scientific laws that can be put into a mathematical form. It can be used to predict the future. It's only an extremely restricted part of the future, of course, but suppose we have a question about the future such as: "If I start with 750 mL of a gas at 25 °C and a pressure of 760 mm Hg, what will be the future volume if, at the same temperature, I make the pressure 720 mm Hg (and don't let any gas escape)?" Equation 6.5 lets us make the calculation. We'll do two worked examples. The first solves the problem by using a "plugging the numbers into the equation" approach. The second approach will rely more on your *understanding* the pressure–volume law. You should be able to use either method.

EXAMPLE 6.1   DOING PRESSURE–VOLUME LAW CALCULATIONS

Problem: A given mass of oxygen occupies 500 mL at 760 mm Hg at 20 °C. At what pressure will it occupy 450 mL at the same temperature?

Solution: We have the following data:

$$P_1 = 760 \text{ mm Hg} \qquad P_2 = ?$$
$$V_1 = 500 \text{ mL} \qquad V_2 = 450 \text{ mL}$$

Therefore using Equation 6.5 directly:

$$(760 \text{ mm Hg})(500 \text{ mL}) = (P_2)(450 \text{ mL})$$

To solve for P_2 (that is, to have it stand alone on one side of the equals sign), we divide both sides by 450 mL:

$$\frac{(760 \text{ mm Hg})(500 \text{ mL})}{(450 \text{ mL})} = \frac{(P_2)(450 \text{ mL})}{(450 \text{ mL})}$$

$$P_2 = \frac{(760 \text{ mm Hg})(500)}{(450)}$$

$$P_2 = 844 \text{ mm Hg} \quad \text{(correctly rounded)}$$

Thus the new pressure must be 844 mm Hg.

PRACTICE EXERCISE 1   If 660 mL of helium at 20 °C is under a pressure of 745 mm Hg, what volume will the sample occupy (at the same temperature) if the pressure is changed to 375 mm Hg?

In the next worked example, we'll see how a little reasoning that uses the fundamental pressure – volume relationship can contribute a great deal to your basic understanding of what is happening. Before we discuss it, notice two variations of Equation 6.5. We can solve this equation for P_2 before we enter any numbers into it by dividing both sides by V_2:

$$P_2 = P_1 \times \left(\frac{V_1}{V_2} \right) \tag{6.6}$$

$\qquad\qquad\qquad$ ⌐ A ratio of volumes

In other words, *to find the second value of pressure we multiply the given value by a ratio of volumes.* If the problem gives *two* values for volume, both V_1 and V_2, then two ratios are possible, V_1/V_2 or V_2/V_1. You don't have to remember Equation 6.6 to select the correct ratio to multiply by the given pressure. If you remember the basic fact given by the pressure – volume law — P and V vary inversely — you choose the ratio that produces the right kind of change, a pressure-raising or a pressure-lowering change. In Example 6.1, the volume was going to be reduced, so this required that the pressure be increased; when you can make this analysis, you have caught the essence of the pressure – volume law. The volumes given were 500 mL and 450 mL, and only the ratio of 500 mL/450 mL could, when multiplied by the given pressure, yield a *larger* value of pressure. This ratio, 500 mL/450 mL, is a number *larger* than 1. The other ratio, 450 mL/500 mL, is less than 1, so if you used this to multiply by the given pressure, the result would have been a *smaller* pressure. And this would have violated the pressure – volume law's requirement as applied to this problem.

The second variation of Equation 6.5 is like the first. If the problem supplies two values of pressure, P_1 and P_2, and you have to find what a given volume, V_1, changes to, the need is for the appropriate ratio of pressures. By rearranging Equation 6.5, we get

$$V_2 = V_1 \times \left(\frac{P_1}{P_2} \right) \tag{6.7}$$

$\qquad\qquad\qquad$ ⌐ A ratio of pressures

In the next worked example, we will see how picking the right ratio of pressures, based on the basic requirement of the pressure – volume law, helps us to solve the problem.

EXAMPLE 6.2   DOING PRESSURE – VOLUME LAW CALCULATIONS

Problem: A sample of nitrogen at 25 °C occupies a volume of 5.65 L at a pressure of 740 mm Hg. If the pressure, at the same temperature, is changed to 760 mm Hg, what is the final volume?

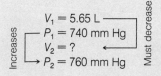

Solution: The pressure is being increased, so the volume will have to decrease—this is what we can predict from the pressure–volume law. Therefore we need a ratio of pressures that is less than 1 to multiply by the given volume. Of the two possible ratios, only the ratio 740 mm Hg/760 mm Hg meets this requirement. So, we multiply it by the given volume.

$$V_2 = 5.65 \text{ L} \times \frac{740 \text{ mm Hg}}{760 \text{ mm Hg}}$$

$$= 5.50 \text{ L} \quad \text{(correctly rounded)}$$

Thus increasing the pressure on 5.65 L of nitrogen at 25 °C from 740 mm Hg to 760 mm Hg decreases the volume to 5.50 L.

Notice that the pressure units cancel. Therefore they could be in any units as long as they are the same.

PRACTICE EXERCISE 2   If 2.5 L of a gaseous anesthetic is at a pressure of 760 mm Hg and the pressure changes to 730 mm Hg, what is the new volume, assuming that the temperature stays the same and no gas is lost?

6.3 LAW OF PARTIAL PRESSURES

The total pressure of a mixture of gases is the sum of the partial pressures of the individual gases.

Both the air we inhale and the air we exhale are mixtures of gases, and it is sometimes useful to be able to predict what happens to the total pressure when we mix two or more gases together (gases that do not chemically react). John Dalton discovered the relevant relationship, and we now know it either as the **law of partial pressures** or as **Dalton's law.**

> Law of Partial Pressures (Dalton's Law) The total pressure of a mixture of gases is the sum of their individual partial pressures.
>
> $$P_{Total} = P_a + P_b + P_c + \cdots \qquad (6.8)$$

The **partial pressure** of a gas in a mixture of gases is the pressure that this gas would have if it were all alone in the same container. It's the pressure that this gas would exert if all of the other gases disappeared. The subscripts in Equation 6.8,—a, b, and c—are identifiers for the individual gases, but in a real situation the formulas of the gases are generally used. For example, the partial pressure of O_2 might be symbolized as P_{O_2} or as PO_2. You'll see both kinds of symbols in various references and professional journals.

EXAMPLE 6.3   USING THE LAW OF PARTIAL PRESSURES

Problem: At sea level and 0 °C, the partial pressure of the nitrogen in clean, dry air is 601 mm Hg. That is, $P_{N_2} = 601$ mm Hg when the total pressure is 760 mm Hg. If oxygen is the only other constituent—which is virtually true—what is the partial pressure of the oxygen?

Solution: Apply Equation 6.8, we can write

$$P_{Total} = P_{N_2} + P_{O_2}$$

Therefore

$$760 \text{ mm Hg} = 601 \text{ mm Hg} + P_{O_2}$$

Or,

$$P_{O_2} = 760 \text{ mm Hg} - 601 \text{ mm Hg}$$
$$= 159 \text{ mm Hg}$$

The partial pressure of oxygen is 159 mm Hg (159 torr).

PRACTICE EXERCISE 3   At the top of Mt. Everest (29,000 ft, 8.8 km) the total atmospheric pressure is only 250 mm Hg. Under this total pressure of the air, the partial pressure of the nitrogen is 198 mm Hg. What is the partial pressure of oxygen (assuming no other gas is present)?

The value of the partial pressure of oxygen you calculated in Practice Exercise 3 is, incidentally, too low to force oxygen from the lungs into the bloodstream at a rate fast enough to sustain normal activity for nearly all humans. Virtually all climbers who intend to do much activity above 15,000 ft need to breathe oxygen-enriched air conducted through a face mask and supplied from a pressurized tank.

Vapor Pressure.   We learned in Chapter 2 that liquids can change to their vapor states by evaporation. Therefore the air space in contact with any liquid includes not only air but the vapor or gas generated by this evaporation. The partial pressure exerted by the vapor that originates by evaporation is called the **vapor pressure** of the liquid.

It's helpful to think of a liquid's vapor pressure as its *escaping tendency*.

To measure the vapor pressure of a liquid, we have to make sure that the vapor doesn't escape, as it would if the liquid were in an open container. The apparatus described in Figure 6.3 shows how this can be managed. Pressure-measuring instruments are called **manometers,** and the U-shaped glass tube (see Figure 6.3) that supports a meter stick is an example of an *open-end* manometer (meaning that one end is open to the atmosphere). Its two arms contain mercury, the same dense, liquid element found in most thermometers. If the pressure increases on the surface of the mercury in one arm but not in the other, the level with the higher pressure sinks down, and the other level rises. The difference in the heights of these

FIGURE 6.3
Measuring the vapor pressure of a liquid. The liquid whose vapor pressure is to be measured is added to the glass bulb that is connected to an open-end manometer. On the left, just before the liquid is added, the manometer shows that both mercury levels are the same, so the pressure in the bulb equals the atmospheric pressure outside the bulb. At the instant the liquid is added, the second valve is shut, which seals the bulb from the atmosphere. Now the pressure exerted by the developing vapor builds up, which forces one mercury level down and causes the other to rise. The difference in the heights of the two mercury levels equals the vapor pressure of the liquid. (From J. E. Brady and J. R. Holum, *Fundamentals of Chemistry,* 2nd ed., John Wiley and Sons, Inc., New York, 1984. Used by permission.)

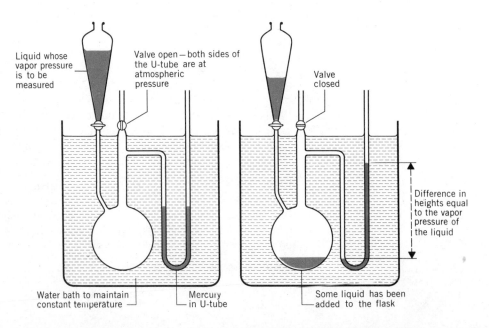

Liquid whose vapor pressure is to be measured

Valve open—both sides of the U-tube are at atmospheric pressure

Valve closed

Difference in heights equal to the vapor pressure of the liquid

Water bath to maintain constant temperature

Mercury in U-tube

Some liquid has been added to the flask

TABLE 6.1
Vapor Pressure of Water

Temperature (°C)	Vapor Pressure (mm Hg)	Temperature (°C)	Vapor Pressure (mm Hg)
18	15.5	32	35.7
20	17.5	34	39.9
22	19.8	36	44.6
24	22.4	37	47.1
26	25.2	38	49.7
28	28.3	40	55.3
30	31.8		

two levels is a measure of the difference in pressure experienced by the two mercury levels.

Initially, before putting any liquid into the bulb of the apparatus in Figure 6.3, the mercury level is the same in both of the arms of the manometer. This tells us that the pressure inside the bulb is initially the same as outside — atmospheric pressure. As long as any of the liquid form is present, the vapor pressure that the liquid can generate depends only on the identity of the liquid *and the temperature.* Different liquids have different vapor pressures at a given temperature, and the vapor pressures of all liquids increase with increasing temperature. Table 6.1, for example, shows how the vapor pressure of water changes with temperature.

Humid means having a high concentration of water vapor.

Water Vapor and the Respiratory Gases. Air that is in contact with water always contains water vapor. The air we inhale is usually somewhat humid, and as soon as it makes contact with the warm (37 °C), moist surfaces of lung tissue it picks up even more water vapor. Therefore the air we exhale is richer in water vapor than the air we inhale; stated technically, the partial pressure of water vapor in exhaled air is greater than that in inhaled air. The data in Table 6.2 give actual values. Inhaled air, as you can see, is mostly nitrogen and oxygen, and the partial pressure of water in inhaled air has been arbitrarily selected to correspond to a relative humidity of 20%. The very low value of the partial pressure of carbon dioxide means, of course, that inhaled air has very little carbon dioxide. The metabolism of the body, however, generates carbon dioxide, which is given up by the bloodstream as the blood flows through capillaries in the lungs. Water vapor will also be carried out — as you've no doubt noticed when you've breathed on eyeglasses before cleaning them or have been outside when it's cold enough to "see your breath." Thus the exhaled air has relatively large values of

TABLE 6.2
The Composition of Air During Breathing

Gas	Partial Pressure (in mm Hg) Inhaled Air	Exhaled Air	Alveolar Air[a]
Nitrogen	594.70	569	570
Oxygen	160.00	116	103
Carbon dioxide	0.30	28	40
Water vapor	5.00[b]	47	47
Totals	760.00	760	760

[a] Alveolar air is air within the alveoli, thin-walled air sacs enmeshed in beds of fine blood capillaries. Little more than bubbles of tissue, these sacs are the terminals of the successively branching tubes that make up the lungs. We have about 300 million alveoli in our lungs.

[b] A partial pressure of water vapor of 5.00 mm Hg corresponds to air with a relative humidity of about 20%, a familiar weather-report term that we will not define further here.

FIGURE 6.4
The collection of a gas by the displacement of water produces a gas sample that contains water vapor. On the left, the gas-collecting bottle, which was previously filled with water and inverted over a basin of water, is filling as the gas bubbles up through the water. During this time, water vapor effectively saturates the gas. The result is the sample of wet gas indicated on the right. (Adapted by permission from J. E. Brady and J. R. Holum, *Fundamentals of Chemistry,* 2nd ed., John Wiley and Sons, Inc., 1984.)

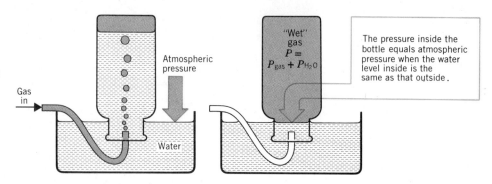

P_{H_2O} and P_{CO_2}. Notice, however, that regardless of how the individual partial pressures change during breathing, the total pressure remains a constant. The air we breathe obeys the law of partial pressures.

Notice also that if you compare the data in Tables 6.1 and 6.2, the partial pressure of the water vapor in exhaled air is the same as the vapor pressure of water at body temperature (37 °C), 47 mm Hg. This means that exhaled air holds as much water vapor as it can at 37 °C; we say that the exhaled air is *saturated* in water vapor. When the body exports warm water vapor this way, it also exports heat, as we discussed in Section 2.5.

Collecting Gases over Water.   Another situation in which we need to know about the vapor pressure of water and the law of partial pressures is when we collect a gas over water for some experiment. Figure 6.4 shows how this can be done in the lab.

The measured volume of the gas in the collecting bottle of Figure 6.4 contains water vapor, but to know how much gas we have prepared, we want to know the volume of the gas if it were dry. To know what volume the gas would occupy if it were dry, we do a calculation. Remember that the measured pressure is the sum of the partial pressures of the gas and the water vapor. Therefore the first step in this calculation is to correct the measured pressure for the partial pressure that the water has in the experiment. Then we do a Boyle's law calculation using this corrected (and lower) value of pressure and the volume that the wet gas occupied in the bottle. This will give us the correct volume for the gas if it were collected dry. We'll do an example to show how.

EXAMPLE 6.4   CALCULATING A DRY VOLUME FROM A WET VOLUME OF A GAS

Problem: On a day when the atmospheric pressure was 744 mm Hg, a student collected a sample of oxygen over water at 20 °C by the apparatus shown in Figure 6.4. By making the water level inside the bottle the same as its level outside, the inside pressure was known to be 744 mm Hg, too. The volume of the gas was 325 mL. Calculate the partial pressure of the oxygen in the bottle, and then calculate what volume the oxygen would have if all of the water vapor were removed and the gas pressure were to become 760 mm Hg.

Solution: To find the value of P_{O_2}, we use the law of partial pressures. From Table 6.1, we find that the vapor pressure of water at 20 °C is 17.5 mm Hg. This is the value of P_{H_2O}.

$$P_{Total} = P_{O_2} + P_{H_2O}$$
$$744 \text{ mm Hg} = P_{O_2} + 17.5 \text{ mm Hg}$$
$$P_{O_2} = 727 \text{ mm Hg} \quad \text{(correctly rounded)}$$

In other words, we have, in effect, *dry* oxygen in a volume of 325 mL at a pressure of 727 mm Hg and a temperature of 20 °C.

The remainder of the problem is to calculate what the volume of this dry oxygen would be at the specified pressure, 760 mm Hg (and 20 °C). This is a straightforward Boyle's law calculation in which we need a volume-*decreasing* ratio of pressures, because we are increasing the pressure from 727 mm Hg to 760 mm Hg:

We could also calculate V_2 by using Equation 6.5.

$$V_2 = 325 \text{ mL} \times \frac{727 \text{ mm Hg}}{760 \text{ mm Hg}}$$

$$= 311 \text{ mL} \quad \text{(correctly rounded)}$$

Thus the collected oxygen, when dry at 20 °C and 760 mm Hg, would have a volume of 311 mL.

PRACTICE EXERCISE 4 A student prepared a sample of nitrogen, N_2, by collecting it at 740 mm Hg in a 325 mL glass-collecting bottle over water at 22 °C. Calculate the partial pressure of the nitrogen, and then find what the dry volume of this nitrogen would be at 22 °C and 760 mm Hg.

6.4 THE TEMPERATURE–VOLUME RELATIONSHIP AT CONSTANT PRESSURE

If the mass and pressure of a gas are kept constant, the volume is directly proportional to the Kelvin temperature.

Because of the excitement over hot-air balloons in his time, the French physicist, Jacques Alexander Cesar Charles (1746–1823), became interested in the effect of temperature on the volume of a gas. In order to study only the temperature effect, he devised a way to maintain a fixed mass of gas at a constant pressure. Today, we could carry out such experiments in an apparatus such as illustrated in Figure 6.5.

In Figure 6.6 we see several plots of volumes versus temperatures from different experiments involving different samples of a gas. When the dots are connected, they form several straight lines. Each line shows how a certain quantity of gas changes in volume as the temperature changes. The different lines correspond to different masses of the gas. This

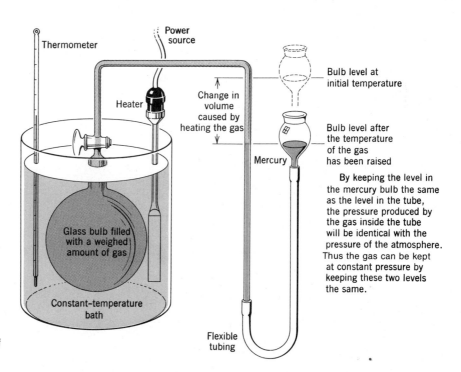

FIGURE 6.5
Obtaining temperature–volume data for a gas at constant pressure. The temperature of the gas can be changed by varying the temperature of the water bath.

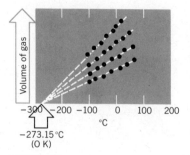

FIGURE 6.6
Plots of temperature–volume data obtained at constant pressure. For different masses of the same gas, the volume is directly proportional to the temperature.

particular gas happens to change to its liquid form at -100 °C, so no actual points can be plotted below this temperature. However, down to -100 °C the points are in straight lines, so we can extend the lines to see with considerable confidence how the gas volume would change if it could not liquify when further cooled. All the lines converge to one point, and this point corresponds to each volume being reduced to a hypothetical value of zero. We say "hypothetical" because it's physically impossible for any matter to have a zero volume — to say nothing of having a *negative* volume! When many gases were studied and when the measurements were further refined, the results were the same. The temperature at which these plots all converged, regardless of the gas, was -273.15 °C. Evidently, this is the coldest temperature possible in nature. The impossibility of having a negative volume for a gas, even one that never liquifies, assures us that -273.15 °C is the lowest temperature we can hope to reach.

The realization that -273.15 °C is something of an absolute low prompted scientists to invent a new scale of temperature that defines the lowest degree of coldness as zero, but otherwise keeps the same intervals of degrees as the Celsius scale. We've already introduced it as the Kelvin scale whose degree divisions are called kelvins, and we know that we can convert a reading in degrees Celsius to kelvins by the following equation, where we round 273.15 to 273 for all but the most precise work and for all of our own needs:

$$K = °C + 273$$

When we express the temperature of a gas in kelvins, the direct proportionality between volume and temperature (at constant pressure) can be expressed very simply in what is called the **temperature-volume law** or **Charles' law.**

Temperature-Volume Law (Charles' Law) The volume of a fixed mass of any gas is directly proportional to its Kelvin temperature, if the gas pressure is kept constant.

$$V \propto T \qquad \text{(constant } P \text{ and mass)} \qquad (6.9)$$

Expression 6.9 can be changed into an equation in the usual way by introducing a proportionality constant. We'll call this constant C' to distinguish it from the symbol we used in connection with Boyle's law:

$$V = T \times C' \qquad (6.10)$$

Or,

$$\frac{V}{T} = C'$$

$\dfrac{V_1}{T_1} = C'$ and

$\dfrac{V_2}{T_2} = C'$. Hence

$\dfrac{V_1}{T_1} = \dfrac{V_2}{T_2}$

Using subscripts as we did in connection with Boyle's law, we can restate this relationship as follows:

$$\frac{V_1}{T_1} = \frac{V_2}{T_2} \qquad \text{(constant } P \text{ and mass)} \qquad (6.11)$$

Notice that we can transform this equation to calculate a new volume, V_2, by multiplying the old volume, V_1, by a ratio of kelvins:

Remember, the temperatures must be in kelvins.

$$V_2 = V_1 \times \left(\frac{T_2}{T_1}\right)$$

A ratio of kelvins

Similarly, a new temperature can be found by multiplying the given temperature by a ratio of volumes.

The volumes can be in any units as long as they are the same units for both V_1 and V_2.

$$T_2 = T_1 \times \left(\frac{V_2}{V_1}\right)$$

└── A ratio of volumes

EXAMPLE 6.5   DOING TEMPERATURE–VOLUME LAW CALCULATIONS

Problem: Some of the total anesthetics used in surgery are gases at 37 °C, body temperature. If 1.50 L of a gas is used at 20 °C, to what volume does the gas change when the temperature becomes 37 °C at the same pressure?

Solution: The conditions given are those that let us use the temperature–volume law, but we *must* change degrees Celsius into kelvins in order to use it.

For 20°C, $T = 20 + 273 = 293$ K (T_1)

For 37 °C, $T = 37 + 273 = 310$ K (T_2)

We also know V_1 to be 1.50 L, so we could plug these given values into Equation 6.11 to find V_2. Or we could note that the temperature is increasing so we need a ratio of temperatures that exceeds 1 to multiply by the given volume in order to ensure that the result is a larger volume than before. This ratio is (310 K)/(293 K). Therefore

$$V_2 = 1.50 \text{ L} \times \frac{310 \text{ K}}{293 \text{ K}}$$

$$= 1.59 \text{ L} \quad \text{(correctly rounded)}$$

Thus the warming of the anesthetic as it enters the patient's lungs causes a rather small increase in volume. Still, anesthesiologists and inhalation therapists must know about it.

PRACTICE EXERCISE 5   A sample of cyclopropane, an anesthetic, with a volume of 575 mL at a temperature of 30 °C, was cooled to 15 °C at the same pressure. What was the new volume?

6.5 THE UNIVERSAL GAS LAW

If we know any three of the four physical properties of a gas — P, V, T, and n — we can use the universal gas law to calculate what the fourth *must* be.

The gas laws, as we have studied them so far, assume the size of the sample in moles is a constant. What happens if we increase the number of moles? The answer depends on what else we let change and on what else we decide to keep constant. For example, if we decide to maintain a constant pressure and temperature for the sample, then more moles of the gas will naturally require more room — a larger volume. In fact, at constant pressure and temperature, the volume of a gas is directly proportional to the number of moles:

$$V \propto n \quad \text{(at constant } P \text{ and } T) \quad (6.12)$$

Avogadro's Law. The relationship given by the proportionality of 6.12 can be converted into an equation simply by inserting a proportionality constant, which we can designate as C^*.

$$V = C^*n \quad \text{(at constant } P \text{ and } T) \quad (6.13)$$

Amadeo Avogadro is credited with discovering an important fact about Equation 6.13 — the proportionality constant, C^*, is the same for all gases provided that the comparisons are made at the same temperature and pressure. Thus if we have, say, 0.50 mol of oxygen at 25 °C and 760 mm Hg, its volume is identical with the volume of 0.50 mol of nitrogen at 25 °C and 760 mm Hg. Or, to turn it around, if we have, say, 45 L of oxygen at 25 °C and 760 mm Hg, it involves the identical number of moles as 45 L of nitrogen or any other gas under the same conditions. This relationship is now expressed in another gas law, the volume–mole relationship or **Avogadro's law.**

> Volume–Mole Relationship (Avogadro's Law) Equal volumes of gases have equal numbers of moles when compared at the same pressure and temperature.

Instead of learning how to solve problems involving just this law, we now have a basis for learning about a much more general law—actually, it's an equation—that combines the three gas laws we have studied, those of Boyle, Charles, and Avogadro.

The Universal Gas Law.   For any gas, the result of multiplying P and V and dividing by T is, experimentally, proportional to the number of moles of the gas, n:

$$\frac{PV}{T} \propto n$$

The proportionality constant for this relationship is symbolized by R. Therefore

$$\frac{PV}{T} = nR$$

The usual form in which this equation is written is

> $$PV = nRT \tag{6.14}$$

This equation is called the **universal gas law** and the constant, R, is called the **universal gas constant.**

The value of R, a constant for all gases, can be calculated once we know what 1 mol of any gas occupies at a given temperature and pressure. To provide a common reference, scientists throughout the world have agreed to designate 760 mm Hg and a temperature of 273 K (0 °C) as the reference conditions for gases, and these values are called the **standard conditions of temperature and pressure,** or **STP** for short.

At STP, the volume of 1 mol of a gas—its **molar volume**—is 22.4×10^3 mL (22.4 L), as the experimental data in Table 6.3 show. If we use this value for V, 273 K for T,

TABLE 6.3
Molar Volumes of Some Gases at STP

Gas	Formula	Molar Volume (liters)	Mass (g)
Helium	He	22.398	4
Argon	Ar	22.401	20
Hydrogen	H_2	22.410	2
Nitrogen	N_2	22.413	28
Oxygen	O_2	22.414	32
Carbon dioxide	CO_2	22.414	44

The numerical value of R depends on the units used. Had we used

$P = 1.00$ atm

$V = 22.4$ L

$n = 1.00$ mol

$T = 273$ K

then $R = 0.0821$ L atm/mol K

760 mm Hg for P, and 1.00 mol for n in Equation 6.14, then we can solve for R:

$$R = \frac{PV}{nT}$$

$$= \frac{(760 \text{ mm Hg})(22.4 \times 10^3 \text{ mL})}{(1.00 \text{ mol})(273 \text{ K})}$$

$$= 6.24 \times 10^4 \ \frac{\text{mm Hg mL}}{\text{mol K}}$$

It is very important that this numerical value of R be associated with the particular units of P, V, T, and n that are used here. Whenever the universal gas law (Equation 6.14) is used together with this value of R, the pressure must be in mm Hg, the volume must be in mL, and the temperature (as in *all* gas law calculations) must be in K.

EXAMPLE 6.6   USING THE UNIVERSAL GAS LAW

Problem: A sample of oxygen with a volume of 250 mL at a pressure of 7.45×10^4 mm Hg and a temperature of 20 °C is allowed to expand as its pressure falls to 750 mm Hg and the temperature changes to 37 °C (essentially atmospheric pressure and body temperature). How many moles of oxygen are in this sample and what volume will the gas sample have after the expansion?

Solution: The units given here for P and V match the units for these quantities required by the value of the gas constant, R, that we want to use. The temperature value, however, must be translated into kelvins:

$$T = {}^\circ C + 273$$
$$= 20 + 273 = 293 \text{ K}$$

Now we can calculate the number of moles, using $PV = nRT$

$$(7.45 \times 10^4 \text{ mm Hg})(250 \text{ mL}) = n \times \left(6.24 \times 10^4 \ \frac{\text{mm Hg mL}}{\text{mol K}} \right) (293 \text{ K})$$

$$n = \frac{7.45 \times 10^4 \times 250 \times \text{mol}}{6.24 \times 10^4 \times 293}$$

$$= 1.02 \text{ mol}$$

Of course, this value of n is the same as under the new conditions:

$$V = ? \quad P = 750 \text{ mm Hg} \quad T = 310 \text{ K} \quad n = 1.02 \text{ mol}$$

Therefore

$$(750 \text{ mm Hg})(V) = (1.02 \text{ mol}) \left(6.24 \times 10^4 \ \frac{\text{mm Hg mL}}{\text{mol K}} \right) (310 \text{ K})$$

$$V = \frac{1.02 \times 6.24 \times 10^4 \times 310 \text{ mL}}{750}$$

$$V = 2.63 \times 10^4 \text{ mL}$$

Thus a huge expansion in volume accompanies the other changes; the new volume is 26,300 mL.

PRACTICE EXERCISE 6   In Example 6.4 we found that the student had collected 311 mL of dry oxygen at 760 mm Hg and 20 °C. How many moles of oxygen were in this sample?

PRACTICE EXERCISE 7   If we use the subscripts 1 and 2 in the usual way to express initial and final conditions, show that the universal gas law can be rewritten as follows whenever we deal with a fixed quantity of gas (that is, when n, the number of moles, is a constant):

$$\frac{P_1 V_1}{T_1} = \frac{P_2 V_2}{T_2} \qquad \text{(at constant } n\text{)}$$

PRACTICE EXERCISE 8   Use the equation derived in Practice Exercise 7 to calculate the final volume of the gas sample that was described in Example 6.6.

6.6   THE KINETIC THEORY OF GASES

All of the laws of the physical behavior of gases can be explained in terms of the model of a gas described by the kinetic theory.

As the gas laws we have studied unfolded over the decades, more and more scientists asked the question: "What must gases really be like if these laws are true?" What is particularly remarkable is that these laws hold for any and all gases, particularly when a gas isn't close to condensing to a liquid (that is, isn't under high pressure or isn't very cold). There aren't *general* laws like the gas laws for liquids and solids.

The Ideal Gas.   When the accuracy and the precision of measurements are sufficiently high, no gas obeys any of the gas laws *exactly* over all ranges of pressures and temperatures. Many come quite close to this ideal equation-fitting behavior, however, so it was natural for scientists to theorize about a hypothetical gas that would fit the gas law equations exactly under all circumstances. Such a gas is called an **ideal gas,** and scientists used the behavior of real gases to postulate the following characteristics of the ideal gas.

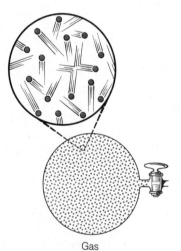

Gas
Mostly empty space
Random motions

Model of an Ideal Gas

1. The ideal gas consists of a large number of extremely tiny particles in a state of chaotic, utterly random motion.

2. The particles are perfectly hard, and when they collide they lose no energy because of friction.

3. The particles neither attract nor repel each other.

4. The particles move in accordance with the known laws of motion.

The laws of motion are part of the science of physics, the science of motion and energy.

Because this model postulated that the fundamental truth about gases is they consist of tiny particles *in motion,* the model and the equations based on it came to be called the **kinetic theory of gases.**

Theorists used these postulates and the mathematical equations for the laws of motion to see if they could derive the gas laws theoretically, and they were splendidly successful. In fact, some historians of science have called the kinetic theory one of the greatest triumphs of the human mind in all of history! Because this book has not assumed its users have the needed background in physics and mathematics to go through the kinetic theory in mathematical detail, we will settle for less rigorous but still very useful descriptions of how the kinetic model explains the gas laws.

FIGURE 6.7

The kinetic theory and the pressure–volume relationship (Boyle's law). The pressure of the gas is proportional to the frequency of collisions per unit area. When the gas volume is made smaller in going from (a) to (b), the frequency of the collisions per unit area of the container's walls increases. This is how the pressure increase occurs.

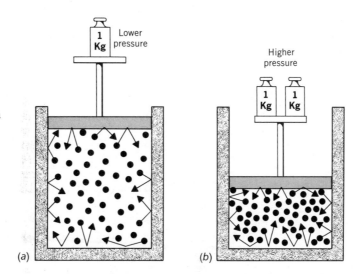

Theorists could show that the pressure exerted by a gas arises from the innumerable collisions per second that the particles make with each unit of area of the walls of the container. If we imagine, then, that the volume of the container is made less, the wall area against which the gas particles collide is also less. Moreover, the particles don't have to travel as far to hit a wall, and the change in volume cannot have affected the average speed at which the particles are traveling. Thus reducing the volume of the gas increases the frequency with which its particles hit the walls, so the gas pressure must increase (see Figure 6.7). When the theoretical calculations on which the preceding description is based were carried out, the result was identical with Boyle's pressure–volume law for real gases. This kind of agreement and others like it are the chief support for our confidence that the model of an ideal gas closely describes real gases, too.

Theorists also were able to use the model of an ideal gas to show that the temperature of a gas is directly proportional to the average kinetic energy of the gas particles. In other words, when we heat a gas and raise its temperature, we cause its particles to move around with greater average energy. When we cool a gas, we cause its particles to move with lower average energy. In fact, it should in theory be possible to cool a gas so much that the average kinetic energy of its particles drops to zero. Because this average kinetic energy could not become less than zero — there's no such thing as *negative* kinetic energy — there ought to be a lower limit to which something can be cooled. This limit is -273.15 °C, and scientists used this to define the Kelvin scale of temperature. As we have already learned, this temperature (which we usually round to -273 °C) is called absolute zero.

We can also understand the temperature–volume law (Charles' law) in terms of the kinetic theory. If we make the gas particles move with more energy (heat the gas), they will hit the container's walls more frequently and with greater energy. If we want to prevent the pressure from rising — and constant pressure is a condition of this law — we have to let the volume expand. Thus gas volume at constant pressure is proportional to the gas temperature.

If we don't let the volume expand, then the pressure of the gas must rise. This relationship between pressure and temperature at constant volume is actually another gas law that we haven't mentioned yet — the **pressure–temperature law** discovered by Joseph Gay-Lussac.

Pressure–Temperature Law (Gay-Lussac's Law) The pressure exerted by a fixed quantity of gas is directly proportional to its Kelvin temperature when the gas volume is kept constant.

Because this law adds nothing to the universal gas law, we will not go into separate calculations involving it. However, as you may know, you would see this law in action if you ever tossed an unopened can of baked beans into a campfire. The rising steam pressure inside the can eventually becomes large enough to rupture the can and distribute beans at random over the bystanders.

Another property of gases that the kinetic theory explains is the ability of a gas to diffuse or spread out throughout its entire container. Particles in random motion eventually find their way into all parts of the container. Almost everyone has experienced the diffusion of perfume, cologne, or after-shave fragrances throughout a room.

6.7 THE LIQUID AND SOLID STATES AND DYNAMIC EQUILIBRIA

The kinetic picture of molecules in rapid, random motion helps to explain some of the physical properties of liquids and solids.

Liquid
Densely packed
Random motions

Molecules of a liquid, like those of a gas, are in a state of constant, chaotic motion, but they are always in touch with one set of neighbors or another. In a liquid we are closer than in a gas to a balance between the forces that attract molecules toward each other and the forces, which arise from collisions, that keep the molecules moving around. The molecules in a liquid are quite tightly packed, but like marbles or beads in a small box, they can shift around a lot. The packing is so tight that the volume of a liquid is barely changed even if subjected to a pressure hundreds of times that of the atmosphere. There is no general pressure–volume law for liquids. There is no general temperature–volume law for liquids either, although liquids do expand slightly when they are heated. The lack of the empty spaces between the molecules of a liquid — a condition sharply in contrast to the empty spaces in a gas — explains why there are no laws similar to the gas laws that apply to all liquids.

Vapor Pressure and Dynamic Equilibrium. The kinetic idea of molecules in random motion helps us to understand how a liquid possesses a vapor pressure, and why this vapor pressure depends only on the liquid's temperature, not on its mass (provided at least *some* of the liquid is present).

The vapor pressure of several liquids are plotted against temperature in Figure 6.8. Notice that ether develops high vapor pressures even at low temperatures, and any liquid that behaves this way is called a **volatile liquid.** Such liquids readily evaporate from open containers at room temperature. In contrast, propylene glycol (in Figure 6.8) has a very low vapor pressure until its temperature is well above the boiling point of water. Such a liquid is

FIGURE 6.8
Equilibrium vapor pressure versus temperature. (Ether was once widely used as an anesthetic. Acetic acid is the sour component in vinegar. Propylene glycol is in several brands of antifreeze mixtures.)

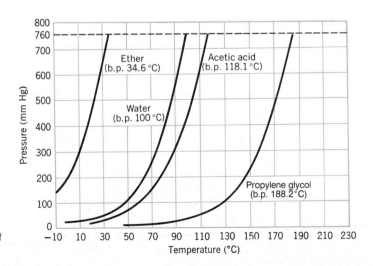

You can leave a nonvolatile liquid such as salad oil uncovered indefinitely with no noticeable loss of the liquid.

termed a **nonvolatile liquid,** because it evaporates very slowly at room temperature.

Suppose that we place a sample of ether in the glass bulb of the device that we pictured in Figure 6.3, page 132. Initially the pressure inside the bulb is the same as outside — atmospheric pressure. For a very brief moment we can imagine that no ether molecules are in the air space above the sample. Ether is volatile, however, so we also know that this situation quickly changes. Ether molecules escape (evaporate) into the space above, and as more and more do this, they force the mercury level to change. This change, after all, is the only way that the system can make room for the ether vapor. In other words, the ether is exerting its vapor pressure.

Eventually, more and more ether molecules, return and rejoin the liquid ether. This happens because their motions and collisions occur at random and some escaped molecules inevitably get turned around. Now we have ether molecules coming and going — some leaving the liquid for the vapor and others returning. The two processes are said to *oppose* each other because one exactly reverses the effect of the other. Eventually, the rates of the two opposing processes become equal. The rate at which liquid ether is changing to vapor exactly equals the rate at which ether vapor changes to liquid ether. The situation is one simple example of an extremely important class of phenomena in nature — a **dynamic equilibrium.** We say *dynamic* because there is considerable coming and going; and by *equilibrium* we mean that there is no *net* change.

Some Dynamic Equilibria in Nature.   We will cite just a few familiar illustrations of dynamic equilibria in nature to show the importance of this class of phenomena. For example, the earth receives energy from the sun; this is a temperature-raising process. The earth also radiates energy to outer space much as a warm pressing-iron radiates energy; this is a temperature-lowering process. Globally, there is essentially an almost perfect match between these opposing processes — the coming and going of thermal energy — and the earth is in a state of thermal equilibrium. You can imagine what it would be like if it weren't. If, for example, the earth radiated, on the average, *less* energy than it received, the planet would eventually become too hot for life. On the other hand, if the earth radiated *more* energy than it received, it would eventually become too cold for life. Just a few degrees change in the average overall global temperature would be a disaster. Special Topic 6.1 describes a relatively modest effect of just a slight overall warming of the earth's atmosphere.

Another example of dynamic equilibrium in nature involves the oxygen content of the atmosphere. Oxygen gas is consumed as a chemical by several processes in nature — the breathing of all animals (respiration), the burning of things (combustion), the decay of dead plants and animals, and reactions such as rusting. However, essentially as fast as oxygen is consumed, it is replaced by a process called **photosynthesis.** This is the manufacture by plants of complex compounds and oxygen from carbon dioxide, water, and soil minerals with the aid of solar energy. Photosynthesis not only makes more plant materials, it also synthesizes oxygen. The opposing processes of oxygen-use and oxygen-regeneration are evidently in an essentially perfect balance on earth, because the concentration of oxygen in air hasn't changed since measurements were first made decades ago.

In living things there are many situations that we can analyze in terms of opposing processes even though no permanent equilibrium exists. (Nothing would be born or would die if all equilibria were permanent.) For example, most adults want their weight-gaining activities — eating — to be in exact, dynamic equilibrium with their weight-losing activities. The *net* result depends on which of the two opposing changes occurs more rapidly or if their rates exactly balance.

The Vocabulary of Dynamic Equilibria.   The whole phenomenon of dynamic equilibria and how they can be upset is at the center of much of our ability to understand wellness and illness at molecular levels. We therefore need to introduce some terms associated with equilibria, and for the sake of simplicity, we will continue to employ a very simple example of an equilibrium, that between a liquid and its vapor.

SPECIAL TOPIC 6.1   EQUILIBRIUM IN THE EARTH'S HEAT BUDGET

About 2% of all of the water on the planet or beneath its surface is found frozen in the ice caps of the Antarctic and Greenland. This may seem to be a small amount of water, but keeping it as ice, in place, is of considerable importance, particularly to people who live in the heavily populated areas along the coastlines of the earth's several oceans. If all of the polar ice melted and were released as water to the world's oceans, many large cities would be flooded. This would occur if a global warming trend raised the mean global temperature by only one or two degrees. The impact on the lower 48 states of the United States is seen in Figure 6.9. Clearly, we don't want the earth to radiate less heat than it receives, because this would cause such a global warming trend.

A warming trend could occur if the concentration of carbon dioxide in our atmosphere were to continue to rise steadily. It has risen throughout this century as we have burned coal, oil, and natural gas at rates never before seen on our planet. Carbon dioxide is one product of combustion, and its molecules in the atmosphere have the ability to capture some of the energy that is being radiated from the earth and to reradiate it back to the earth. Thus as the carbon dioxide concentration of the atmosphere increases, less and less energy leaves per unit of time, and this causes an upward drift in the global temperature.

If the opposite were to happen, if the earth radiated heat faster than it absorbed heat, the earth's average temperature would drop and we might see a new ice age. If a new ice age were to *add* ice to the ice caps, taking this water from the oceans by means of the cycle of evaporation and precipitation, the ocean levels would fall. Figure 6.9 also shows how a fall of 300 to 400 feet would extend the land area of the continental United States.

A cooling of the earth could be caused by an increase in the concentration of atmospheric dusts and particles such as those from major volcanic eruptions. These particles reflect incoming sunlight back to outer space, so the earth receives solar energy at a slower rate. The particles, however, do not have a compensating effect on the energy that radiates from the earth. Thus particles in the atmosphere cause a global cooling trend.

Any dust hurled into the atmosphere by volcanoes eventually settles, so the cooling effect does not last indefinitely. Nevertheless, even a cooling that lasts only one or two years deeply affects human affairs. For example, what is thought to be the mightiest volcanic eruption in history occurred in 1815 when Tambora, a volcano on the small island of Sumbawa in Indonesia, exploded and ejected pumice and ash in a volume estimated to be a quarter to half the volume of Lake Erie, one of our Great Lakes. The small global cooling that this caused affected the weather in the eastern United States so much that 1816 was called the year without a summer. Crops were killed by frosts occurring in every summer month, and general famine threatened the population in the following winter. Historians believe that this triggered in 1817 the first large-scale migration of people from the Northeastern United States to the Middle West.

Just what might be happening now, gradual global warming or cooling, is a controversial question among atmospheric scientists. Such changes occur very slowly, but you can see how important it is that there be equilibrium in the earth's heat budget.

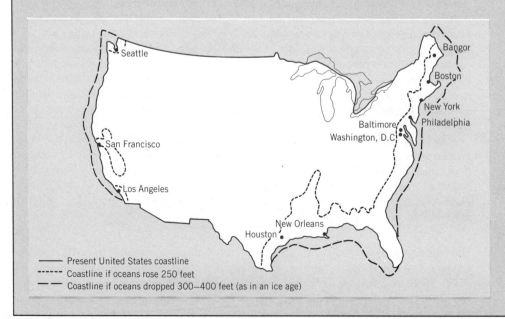

Present United States coastline
Coastline if oceans rose 250 feet
Coastline if oceans dropped 300—400 feet (as in an ice age)

FIGURE 6.9
(From *River of Life. Water: The Environmental Challenge*, 1970, U.S. Department of the Interior, Conservation Yearbook, Vol. 6. U.S. Government Printing Office, Washington, D.C.)

To represent a dynamic equilibrium, we use an equation with double arrows, which we can illustrate as follows, using our example:

$$\text{liquid} + \text{heat} \rightleftharpoons \text{vapor} \tag{6.15}$$

Anytime we use two oppositely pointing arrows in an equation, we signify that we have an equilibrium between the materials on the left and those on the right sides of arrows. An equation is called an *equilibrium expression* or an *equilibrium equation* whenever it has two oppositely pointing arrows. The change from left to right is called the *forward reaction,* and the opposing change from right to left is called the *reverse reaction.*

Often but not always, an equilibrium expression includes an indication of how any energy associated with the opposing changes is involved. For example, the *heat* term in equilibrium expression 6.15 is the heat of vaporization that we discussed on page 40. It appears on the left side of the arrows because the forward change consumes heat, so heat is like a reactant.

Once an equilibrium has become established no net change occurs spontaneously. But this doesn't mean that we can no longer force a change. For example, the liquid–vapor equilibrium of 6.15, above, is affected by heat, as we have already noted. The forward reaction is endothermic, so if we add heat to a liquid that is in equilibrium with its vapor, some of the liquid has to go into the vapor state. At least for a time, the rate at which the forward change occurs will be faster than the rate of the reverse change. For a time, the system will not be in equilibrium. We say that the addition of heat *upsets the equilibrium.* However, once we stop adding heat and maintain the temperature of the system at a constant but higher value, the rate at which vapor molecules return to the liquid will catch up, and the opposing rates will become the same again. Both will be faster than they were at the lower temperature, but when both are the same there is no further net change. Once again, we have dynamic equilibrium. At this higher temperature, of course, the vapor pressure is higher because we have more moles of the gas (the vapor) in a space that isn't much larger. As Figure 6.8 showed, the vapor pressure of any liquid increases with temperature. The vapor pressures plotted in this figure are called the *equilibrium* vapor pressures, which means that each value is the vapor pressure when the liquid and its vapor are in equilibrium at the particular temperature.

Some students wonder what happens to the vapor pressure if we put more liquid into a larger bulb, or if we remove all but a few drops of the liquid. In other words, they are asking if the vapor pressure of a liquid is somehow dependent on the size of the liquid sample or at least on its surface area. The answer is no. Vapor pressure is independent of sample size, provided at least some of the liquid is present. (If it isn't present, we're dealing with a gas and we have no equilibrium situation. Now the gas laws apply.) To be sure, if the surface area is made larger, liquid molecules will be able to escape into the vapor state at a faster rate. This stands to reason because there are more places where escape is possible. But there are now also more places at which molecules can return. The opposing rates of leaving and returning are *both* larger, but they still are equal to each other; there is no change to the net vapor pressure. If we reduce the surface area, perhaps by having a very small liquid sample present, the opposing rates of both leaving and returning between liquid and vapor states are now lower. However, as long as some liquid is present (and we have not changed the temperature), the concentration of vapor in particles per unit volume will still be the same as with any larger-sized sample. Hence, the liquid will still exert whatever value of vapor pressure that goes with the particular temperature being used. The size of the liquid sample and its surface area do not matter.

Le Chatelier's Principle and Equilibria.   The effect of a higher temperature on a liquid's vapor pressure illustrates an important principle that applies to all equilibria. This principle is called **Le Chatelier's principle,** after Henri Louis Le Chatelier (1850–1936), a French chemist.

> Le Chatelier's Principle If a system is in equilibrium and something occurs that can upset the equilibrium, the system will shift in whatever way that will help to restore the equilibrium.

The "something that can upset the equilibrium" is often called a *stress* on the equilibrium. In the liquid–vapor equilibrium, one stress that can be applied is the addition of heat. As we just saw, we can predict what happens to the equilibrium under this stress. In fact, the great value of Le Chatelier's principle is that we can use it to predict the way a system at equilibrium *must* change under a given stress. The stress of raising the temperature—of adding heat—accelerates whatever change absorbs this stress. Equilibria *always* change in whichever way absorbs an applied stress. It is as if the system "rolls with any punches."

Whenever a system at equilibrium changes in response to a stress, we say that the equilibrium *shifts*. This means that one of the two opposing changes becomes faster than the other. Whichever becomes faster is said to be *favored*. If the forward reaction becomes faster, we say that the equilibrium shifts to the right. And it shifts to the left if the reverse reaction becomes favored for any reason. If the applied stress is not too great, the rate of the initially unfavored (slower) reaction will catch up to the other reaction. Both then become equal again, and once more we have equilibrium. However, it isn't the identical equilibrium that existed before the intrusion of the stress. It is a new equilibrium because the actual quantities of materials represented as reactants and products have changed. If the forward reaction has become favored by the stress, then some of the reactants have changed into products. If 20 g of ether exist in equilibrium as, say, 18 g of liquid and 2 g of vapor at one temperature, at a higher temperature we might see 17 g of liquid and 3 g of vapor in equilibrium.

Normal Boiling Points.   Each liquid has a particular temperature at which its equilibrium vapor pressure exactly equals 760 mm Hg. What is now different in the liquid is that its molecules can enter the vapor state not just at the surface but *everywhere* in the liquid. Bubbles of the vapor can now form *beneath* the surface, and they cause quite a commotion as they rise everywhere. This, of course, is the action that we call **boiling.**

The temperature at which a liquid's equilibrium vapor pressure equals 760 mm Hg—1 atm—is called the **normal boiling point** of the liquid. If a liquid is heated in a location where the atmospheric pressure is not 760 mm Hg, boiling, of course, can still occur. It happens at whatever temperature the vapor pressure equals the new pressure, but the associated temperature is not called the *normal* boiling point. For example, in Denver, Colorado, at an elevation of 1 mile where the atmospheric pressure is usually lower than it is at sea level, water boils at about 95 °C instead of 100 °C.

As anyone who has tried to cook at a higher altitude where the pressure is lower knows, it takes longer than at sea level. The chemical reactions that occur during cooking are all endothermic. Therefore they do not occur as rapidly at a temperature of 95 °C as they do at 100 °C. No matter how high you turn up the stove setting as you prepare a soft boiled egg using boiling water, you cannot raise the temperature of boiling water above its value at your altitude. Turning up the stove boils the water away *faster,* but it does not raise its temperature. Of course, you could use a special pan with a tight lid, such as a pressure cooker. Now the steam cannot escape and its pressure can build up until the safety valve is activated. This higher pressure means that the temperature of the boiling water in the pressure cooker is higher than in the open vessel, so the chemical reactions of cooking occur more rapidly.

These same principles are at work in steam sterilization equipment. To ensure that bacteria and viruses on a surgical instrument are both quickly and completely destroyed, the instrument is placed in the equivalent of a pressure cooker where the water and steam temperature can be raised well above the normal boiling point of water.

Just a casual glance at the water in this beaker would tell you that it is boiling. Why?

Pressure cooker.

Solid (ionic)
—Densely and orderly packed
—Vibrations about fixed points

The Solid State and Kinetic Theory.   Not only are atoms, molecules, or ions tightly packed in a solid, they also have fixed positions and fixed neighbor particles. This does not mean that they are completely at rest. They jiggle about their fixed positions, but in the solid the forces of attraction between particles are just too strong to permit any of the movement that occurs in liquids (or gases).

As the temperature of a solid is raised, the vibrations of the individual particles become more and more intense. Eventually neighboring particles strike each other strongly enough to overcome the forces of attraction. Now the solid passes over into the liquid state — it *melts*. If the temperature of the system is carefully controlled, the rate at which particles leave the solid state and move around as a liquid can be made equal to the rate at which they return and take up fixed positions in the solid again. In other words, at the right temperature, the following equilibrium will exist:

$$\text{solid} + \text{heat} \rightleftharpoons \text{liquid}$$

The *heat* in this equilibrium is the *heat of fusion* that we described on page 41. The temperature at which an equilibrium exists between the solid and the liquid states of a substance is called its **melting point.** The forward change in this equilibrium is endothermic, so if we put a stress on the equilibrium by the addition of heat — by raising the temperature — the equilibrium will shift to the right. This shift is in the direction of using up the stress — namely, the added heat — and Le Chatelier's principle tells us that equilibria shift to absorb stresses. Of course, this means that the solid melts. If we remove heat — if we lower the temperature of the system — Le Chatelier's principle tells us that the equilibrium must now shift to the left to release heat. It is as if the attempt to lower the temperature by removing heat demands that the equilibrium do whatever is necessary to supply the heat. It must shift to the left, and any of the substance remaining in the liquid state now changes to the solid state.

London Forces and the States of Matter.   From time to time we have referred to forces of attraction between ions or molecules. Oppositely charged ions bearing full charges generally have the strongest forces of attraction between them, and ionic compounds are all solids at room temperature. It is similarly easy to understand how the molecules of polar, covalent substances such as water can have forces of attraction between them. They have $\delta+$ charges and $\delta-$ charges that can attract each other. However, even the most completely nonpolar substances, such as the noble gases, can be changed into liquids and solids by lowering the temperature sufficiently. This fact raises the question: How can forces of attraction develop between nonpolar particles?

Atoms and molecules have electron clouds that surround their nuclei. As one particle approaches another on collision course, these somewhat "soft" electron clouds tend to repel each other — at least temporarily — as the collision or near miss develops. (See Figure 6.10.) This distortion gives the approaching particles a *temporary* polarity, and the distortion is called **polarization.** We say that the particles become temporarily polarized.

In a sample that contains billions and billions of particles, it is easy to imagine that there will be innumerable molecules that are temporarily polarized. Therefore they can experience a real (although weak) force of attraction that is named the **London force** (after a physicist, Fritz London). At sufficiently low temperatures, London forces can cause even nonpolar substances to change from a gas to the liquid state or from the liquid to the solid state.

Because London forces are related to electron clouds, the larger the overall electron cloud per molecule the more the molecule can be polarized. Therefore substances with large molecules or atoms generally can experience larger London forces than those with small molecules or atoms. Because a large electron cloud implies a large formula weight, the rule for the boiling points of nonpolar substances is that *the higher the formula weight, the higher is the boiling point.* The boiling points of the noble gases, given in the margin, illustrate this rule.

—Electron cloud

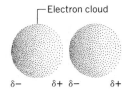

$\delta-$ $\delta+$ $\delta-$ $\delta+$

FIGURE 6.10
London forces. Molecules can be temporarily polarized by coming close to each other, and the force of attraction between such temporary dipoles is called a London force.

Noble Gas	Atomic Weight	Boiling Point (°C)
He	4.00	−269
Ne	20.2	−246
Ar	39.9	−186
Kr	84.8	−152
Xe	131	−107
Rn	222	−62

Increasing b.p.

6.8 THE KINETIC THEORY AND RATES OF CHEMICAL REACTIONS

The rate of a chemical reaction is dependent on the temperature, the concentration of reactants, and the presence or absence of a catalyst that lowers the energy of activation.

In chemical reactions, electrons and nuclei become reorganized. The electron configurations in the particles of the reactants switch over to the electron configurations in the products. If all it ever took for such an event to occur were a gentle coming together of the reactant particles, no reactant would be stable in the presence of anything else. Yet, many substances are stable and can be stored in the presence of air, moisture, glass, people, and other potential reactants.

The kinetic theory helps us understand why some combinations of reactants do nothing to each other, why others can stand each other until the temperature rises too much, and why still other combinations can't be stored under any circumstances. The field of chemistry that deals with the rates of chemical reactions is called **kinetics.**

Energy of Activation.   For the particles of two reactants to change each other chemically, the particles have to collide. Only by a collision of some sort can the electrons of the reactant particles be induced to become rearranged in their locations relative to atomic nuclei. Generally, very light tap-like collisions will not work.

To accomplish the rearrangement of electrons relative to nuclei that occurs in a reaction, the electron – nuclei systems of the reactants have to be energized by some mechanism, and a collision is the usual mechanism. (Some reactions can be launched solely by the action of light.) We have to remember that nature operates under a law of conservation of energy. When two moving particles are about to collide, each has a certain kinetic energy of motion. But imagine a collision in which both particles stop as a result. (This happens all the time in highway accidents.) If they stop, their kinetic energy goes to zero, because K.E. $= (1/2)mv^2$ and the value of v is now zero.

What happens to the kinetic energy? Is it lost? If so, what about the law of conservation of energy? Actually, the energy that existed as kinetic energy is not lost; it's transformed into potential energy. There are different ways by which this can happen, but in the realm of ultrasmall particles, such as atoms, ions, and molecules, the kinetic energy usually causes stable electron – nuclei arrangements to change temporarily to less stable arrangements. Now things can happen. One is simply to revert back to the original electron – nuclei arrangements of the reactants. If this were to happen — and it often does — the particles that collide may possibly bounce away from each other. The potential energy in the temporary and unstable arrangement at the instant of collision reconverts to kinetic energy of motion much as a bouncing ball can hit a sidewalk, momentarily stop, then bounce away — still as a ball and not as some other substance. In other words, some collisions lead to nothing.

The *total* energy — kinetic plus potential — stays constant throughout the change, but it becomes apportioned differently.

In other situations, however, reactant particles that achieve less stable configurations by collision might, during the brief moment, experience an electron – nuclei reorganization that gives new chemical species. Thus the conversion of some or all of the kinetic energy of collision into some potential energy of unstable configurations makes a chemical reaction possible. Some collisions, if violent enough, lead to particles of the products. We dub such events successful collisions, because they lead to products.

Each combination of chemicals and each chemical reaction is different, but for each one there is a certain minimum energy that must be involved during a collision before electrons will shift into new arrangements that result in new chemical bonds. This minimum value of energy is called the reaction's **energy of activation.** Figure 6.11, part *a*, illustrates what this means, and you can see why the energy of activation is sometimes called the "energy hill" or the energy barrier of a reaction.

The vertical axis represents changes in the *fraction* of all of the collisions that are occurring. The horizontal axis corresponds to values that the collision energy can have, ranging from zero on the left to very large values — values approaching infinity — on the right of this axis. Using the imagery we obtained from the kinetic theory of gases and liquids, it isn't

FIGURE 6.11
Energy of activation. (a) Only a small fraction of all of the collisions, represented by the ratio of areas, $A/(A + B)$, has enough energy for reaction. (b) This fraction greatly increases when the temperature of the reacting mixture is increased.

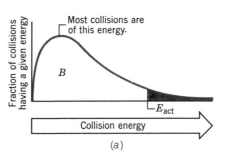

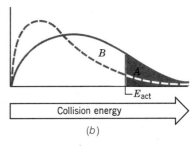

hard to imagine that the particles are moving over a large range of speeds, ranging from very low values (even a zero value for some, for a moment) to very high speeds. Therefore some collisions will be mere taps, whereas others will be extremely violent.

Figure 6.11a shows that in a sample of reactant particles engaged in colliding, some collisions will be such slight taps that virtually no kinetic energy changes into potential energy. However, the fraction of all collisions that have zero collision energy is essentially zero. Then, as we follow the curve from the intersection of the axes, as the value of the collision energy increases, the fraction of the collisions having a particular collision energy also increases until we reach a maximum value for this fraction. Then, continuing, collisions that have increasingly higher energies become less and less likely, so the fractions having such high energies decline and the curve begins to move back down. Eventually we reach a value of collision energy that provides the exact amount of energy to enable the electron–nuclei rearrangement—the chemical reaction—to occur. This is the energy of activation, symbolized as E_{act} in the figure.

When collisions have this much energy, or any higher value, the reaction involves enough energy for it to take place. Colliding particles that do not achieve the energy of activation just bounce away unchanged, chemically. Of course, it isn't enough to have sufficient energy. The colliding particles must hit each other just right, much as the runners in a relay race have to pass the baton correctly regardless of how fast or slowly they are moving at this critical moment in the race.

Only a small, total fraction of all of the collisions have enough energy to be successful— that is, to result in a reaction. This fraction is represented by the ratio of the shaded area marked A in the figure to the total area under the curve, $(A + B)$. Thus this ratio can be expressed as $A/(A + B)$.

You can see from the figure that if the energy of activation were very high—meaning a high energy barrier—the shaded area on the right in the figure would be even smaller, so the fraction $A/(A + B)$ would be much smaller. On the other hand, if the energy of activation were very low—a small energy barrier—this fraction would be large. In the extreme, the fraction might equal 1, which would mean that every collision would be successful no matter how low the energy of the collision were. In practical terms, such a reaction would be extremely rapid—an explosion, essentially—because it would mean that simply mixing the reactants causes instantaneous change.

At the other extreme, the fraction might be virtually zero, which would mean that no reaction ever occurs, and the "reactants" are eternally stable in each other's presence.

This analysis of the extremes tells us that the rate of a reaction depends on its energy of activation. We can consider that the **rate of a reaction** is the number of successful collisions that occur each second in each unit of volume of the reacting mixture. A high energy of activation means a slow rate, and a low energy of activation means a fast rate.

Effect of Temperature on Reaction Rate. If the temperature of the mixture of reactants is raised, the kinetic theory tells us that the particles experience a higher average kinetic energy. Of course, this means that the average energy of the collisions is raised, too, as shown in part (b) of Figure 6.11. Raising the temperature of a mixture of reactants makes the curve flatten out, and it shifts the maximum of the curve to the right, toward a higher collision energy. However, the value of the energy of activation for this reaction stays essentially unchanged when the temperature is changed. Therefore when the curve flattens and shifts to the right, a larger part of the area under the curve moves to the right of E_{act}. Thus the fraction of collisions represented by $A/(A + B)$ increases, which means that the reaction rate increases. This is how we explain a fact you probably already know — raising the temperature of a mixture of reacting chemicals increases the rate of the reaction.

Implications at the Molecular Level of Life.   The effect of temperature on reaction rate is very great. As a rule of thumb, an increase of only 10 °C doubles or triples the rates of most reactions. This is a huge rate acceleration for such a small temperature change, and it has serious implications for health. Health care scientists tell us that a rise in body temperature of only 1 °F increases the rate of the reactions of metabolism so much that the oxygen requirement of tissue increases by 7%. This places an extra strain on the heart, because it has to speed up the delivery of oxygen from the lungs.

All reactions in a healthy human body occur at a constant temperature of 37 °C, normal body temperature. Chemicals that do not react outside the body at this temperature, such as sugar and oxygen, do react inside the body. The next question we face is Why? What more do we need to know to understand this particular aspect of metabolism? Most of what we have said about the kinetic theory and reaction rates is background leading up to the answer. But before we can give the answer, we have some more background to obtain.

Heat of Reaction.   We must make an important distinction between a reaction's energy of activation and the heat of reaction. Figure 6.12 will help us do this. It shows a kind of plot that is new to our study — a *progress of reaction diagram*. In such a diagram, the vertical axis gives *relative* values of the potential energies of the substances — either the reactants or the products, depending on which part of the plot we focus our attention. The horizontal axis is used simply to help us visualize the direction of the chemical change. The symbols of the species involved are included on the diagram to keep the nature of the reaction in view.

To follow the progress of the reaction diagrammed in Figure 6.12, begin on the left at site A with the reactants — carbon and oxygen in our example. We know that these two substances are quite stable in each other's presence at or near room temperature. Coal (mostly carbon), after all, can be stored with no trouble in air (with is 21% oxygen). As you know, to induce carbon and oxygen to react, we have to heat them; we ignite them, in other words. Heating their particles gives them higher kinetic energies, and as the temperature rises, more and more collisions become closer to being successful. We are now moving in the diagram both upward toward higher potential energies of the reactants and rightward in the direction of the products. We are climbing the energy hill.

Eventually we provide sufficient energy of activation, and we are at the top of the energy barrier at location B. The electrons and nuclei of the reactants can now rearrange to give molecules of carbon dioxide. A great deal of the potential energy in the complex of electrons and nuclei at the top of the barrier now transforms into the kinetic energy of the newly-forming molecules of carbon dioxide. There is quite a drop in potential energy as the reaction progresses, finally, to products at site C in the diagram. Some of this potential energy goes to repay the cost of climbing the energy hill, but there is a net excess that appears to be liberated from the reaction mixture. This net energy difference between the reactants and the products is called the **heat of reaction.**

As we know, once the reaction of carbon and oxygen starts, it continues spontaneously. The reaction is exothermic, and some of the energy represented in Figure 6.12 in the change from B to C activates still unchanged particles of the reactants. The heat of reaction in this case represents the conversion of some of the chemical energy in carbon and oxygen into the kinetic energy of the molecules of CO_2.

Not all reactions liberate energy as in our example. Many chemical changes won't occur unless there is a continuous input of energy. One example is the conversion of potassium chlorate into potassium chloride and oxygen, as shown in the progress of reaction diagram of Figure 6.13. In this example, a good share of the energy of activation (A to B in this figure) is permanently retained by the product molecules as internal or potential energy. The net energy retained is represented in Figure 6.13 by the vertical distance between A and C. This reaction

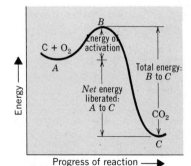

FIGURE 6.12
Progress of reaction diagram for the exothermic reaction of carbon with oxygen that produces carbon dioxide.

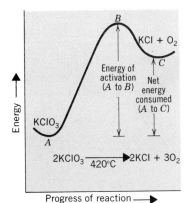

FIGURE 6.13
Progress of reaction diagram for the endothermic conversion of potassium chlorate into potassium chloride and oxygen.

is endothermic, and there is a net conversion of kinetic energy (supplied by the steady input of heat) into the potential or chemical energy of the products. Thus you can see that both exothermic and endothermic reactions have energies of activation, but in the exothermic reaction there is still a net release of energy, whereas in the endothermic reaction there is a net absorption of energy.

Effect of Concentration of Reaction Rate.   We have described the rate of a reaction in terms of the frequency of successful collisions, and have noted that this frequency is some fraction of all of the collisions. And we have seen how this frequency can be increased by increasing the average violence of all of the collisions by raising the temperature.

Another way to increase the frequency of successful collisions is simply to increase the frequency of all collisions. Even if we don't increase their average violence, by making collisions of *all* kinds occur more often we will make successful collisions more probable. The way to accomplish this is to increase the concentrations of the reactant particles. It's like going from a stroll down a lonely country lane to an aisle of a very crowded store. An increase in the concentration of people in motion causes an increase in the "excuse-me" kind of bumps and collisions. If the molar concentration of one reactant is doubled, the frequency of all collisions must double because there are twice as many of its particles *in the same volume.*

One of the spectacular results of increasing the concentration of a reactant can be observed by comparing the rate at which something burns in air with the rate of the same reaction in pure oxygen. Air is about 21% oxygen, which means that in every 100 liters of air there are 21 liters of oxygen (and virtually all of the rest is nitrogen). You can make steel wool glow and give off sparks if you direct a bunsen burner flame at it when it is in air; if glowing steel wool is thrust into pure oxygen, however, it bursts into flame. Obviously, if you ever work near oxygen tents or visit a patient being given oxygen or oxygen-enriched air, you will want all flames and cigarettes excluded. Someone has estimated that if the air we breathe were 30% oxygen instead of 21%, no forest fire could ever be put out, and eventually all of the world's forests would disappear. The effect of concentration on reaction rate is clearly a very important matter in survival and in health.

Catalysts.   Now we can deal with the question of why chemicals that do not react at body temperature when they are outside of the body react readily inside. One of the interesting and most important phenomena in all of nature is the acceleration of a reaction rate by a trace amount of some chemical that does not permanently change, chemically, as the reaction proceeds. This phenomenon is called **catalysis,** the chemical responsible for it is called a **catalyst,** and the verb is *to catalyze.*

The catalysts in living systems are called **enzymes,** and a special enzyme is involved in virtually every single reaction in any living system. Enzymes are but one family of substances that belong to a very large family of biochemicals called the proteins. You can easily observe the action of an enzyme if you have access to a dilute solution of hydrogen peroxide, which is sold in drugstores as a bleach and disinfectant, and a slice of raw liver. Hydrogen peroxide decomposes as follows:

$$2H_2O_2 \longrightarrow 2H_2O + O_2$$

This is a very slow reaction at room temperature or body temperature, and if you look at a sample of hydrogen peroxide, you won't notice any bubbling action. However, if you add a tiny slice of liver to the hydrogen peroxide solution, an enzyme in liver called *catalase* catalyzes this decomposition, and you'll see a vigorous evolution of oxygen. The frothing that you see if you ever use hydrogen peroxide to disinfect a wound is the same reaction. Hydrogen peroxide is toxic, and it can form in certain reactions of the metabolism of oxygen. Therefore in the liver (and also in the kidneys), catalase acts to detoxify hydrogen peroxide. The enzyme is not itself permanently changed by this work.

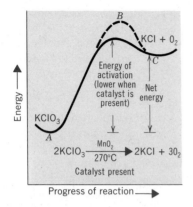

FIGURE 6.14
Progress of reaction diagram for the endothermic, catalyzed conversion of potassium chlorate into potassium chloride and oxygen. The dashed-line curve shows where the energy barrier went in the uncatalyzed reaction sketched in Figure 6.13. Notice that the net energy consumed, the heat of reaction, is identical to that of the uncatalyzed reaction, but the energy of activation is lower.

Catalase makes a reaction occur much faster than it would at the same temperature in the absence of a catalyst. A catalyst can also cause a reaction to take place at a much lower temperature than otherwise. A classic example of this effect is the decomposition of potassium chlorate into potassium chloride and oxygen that we mentioned earlier. Notice in the following equations how the temperature at which the reaction can occur varies according to the presence or absence of manganese dioxide, a catalyst for the decomposition:

Without MnO_2, the temperature has to be 420 °C

$$2KClO_3 + heat \xrightarrow[420\ °C]{} 2KCl + 3O_2$$

With MnO_2, the temperature can be only 270 °C

$$2KClO_3 + heat \xrightarrow[270\ °C]{MnO_2} 2KCl + 3O_2$$

The rates of the evolution of oxygen are equal under the sets of conditions given here. The rate observed at 420 °C in the absence of MnO_2, is observed in the presence of MnO_2 at a substantially lower temperature.

With or without the catalyst, the decomposition of potassium chlorate is endothermic, as we saw in its progress of reaction diagram of Figure 6.13. Figure 6.14 shows the progress of reaction diagram for the same reaction except that the catalyst is present. It illustrates some of the major facts about the entire phenomenon of catalysis—a catalyst does not change the heat of reaction; it lowers the energy of activation. This is why the reaction happens faster. The energy barrier is reduced, so the fraction of all collisions that have enough energy is larger with the catalyst than without. Figure 6.15 uses drawings similar to those in Figure 6.11 to illustrate what a lowering of the energy of activation can do.

In summary, a catalyst either makes a reaction go faster at the same temperature or it permits the reaction to occur at the same rate at a lower temperature. The catalyst accomplishes this by lowering the energy of activation of a reaction. Catalysts do not affect the heat of reaction, the net energy released or absorbed by the system as the reaction occurs. Thus we add the effect of a *catalyst* to the *temperature* of the reaction and to the *concentrations* of the reactants as the important factors that govern the rates of reactions.

None of the preceding discussion explained *how* a catalyst works. Each works by a different mechanism, but all we have discussed thus far is what catalysts do. When you study the subject of enzymes later, you will learn something about their mechanism of action.

FIGURE 6.15
The effect of lowering the energy of activation on the frequency of successful collisions—the rate of the reaction. (a) When the energy of activation is high, the ratio of areas in the drawing that represents the fraction of successful collisions, $A/(A + B)$, is small and the rate of reaction is slow. (b) The ratio is much larger—the rate is much faster—when the energy of activation is less. Catalysts lower energies of activation and so increase reaction rates.

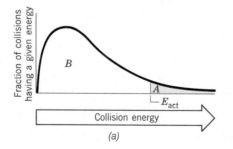

(a)

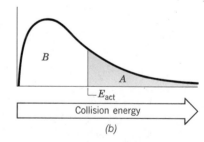

(b)

SUMMARY

Gas properties The four important variables for describing the physical properties of gases are moles (n), temperature T (in kelvins), volume (V), and pressure (P). We express pressure—force per unit area—in atmospheres (atm), mm Hg, or torr. Other important physical quantities in the study of gases are partial pressures, standard pressure and temperature (STP = 273 K and 760 mm Hg), and the molar volume at STP (22.4 L).

Gas laws All real gases obey, more or less, some important laws. Gas pressure is inversely proportional to volume (when n and T are fixed)—the pressure–volume law (Boyle's law). Gas volume is directly proportional to the Kelvin temperature (when n and P are fixed)—the temperature–volume law (Charles' law). Gas volume is also directly proportional to the number of moles (when T and P are fixed)—Avogadro's law. In fact, 1 mol of any gas under the same conditions of T and P has as many particles as 1 mol of any other gas. Another gas law is that the total pressure of a mixture of gases equals the sum of their partial pressures (Dalton's law). Boyle's, Charles', and Avogadro's law combine in the universal gas law, $PV = nRT$, where R, the universal gas constant, holds for all gases. This law also includes the pressure–temperature law (Gay-Lussac's law), which says that the pressure of a gas is directly proportional to its Kelvin temperature (at fixed n and V).

Kinetic theory   If we imagine that an ideal gas consists of a huge number of very tiny, very hard particles in random, chaotic motion without attracting or repelling each other, the gas laws can be derived from the laws of motion (with some help from statistics). Out of this kinetic theory of gases came the insight that the Kelvin temperature of a gas is directly proportional to the average kinetic energy of the gas particles.

Liquid state   Liquids do not follow common laws like gases, because essentially no space separates liquid particles from each other. In liquids, there are forces of attraction between the particles caused either by permanent dipoles or temporary dipoles (as in the case of London forces). Liquids can evaporate, and this "escaping tendency" gives rise to vapor pressure. Each liquid has a particular equilibrium vapor pressure that is a constant at each temperature, provided that care is taken to ensure a true dynamic equilibrium exists between the liquid and the vapor state. This pressure is not a function of the quantity of the liquid. When the liquid's vapor pressure equals the pressure of the atmosphere, the liquid boils. The normal boiling point

is the temperature at which boiling occurs when the pressure of the atmosphere is 760 mm Hg (1 atm).

Solid state   The particles in a solid vibrate about fixed equilibrium points, but if the solid is heated, these vibrations eventually become so violent that the particles enter the liquid state.

Dynamic equilibria   When the rates of two opposing changes, whether physical or chemical, are equal, the system is in dynamic equilibrium. A liquid and its solid form are in dynamic equilibrium at the melting point, for example. If some stress, such as the addition or removal of heat, upsets an equilibrium, the equilibrium shifts in whichever direction tends to absorb the stress. If a change is exothermic, the reverse reaction is favored by adding heat. If a change is endothermic, the forward reaction is favored by adding heat.

Kinetic theory and chemical reactions   Virtually all chemical reactions have an energy of activation. Reactant particles more frequently surmount this barrier — the reaction happens faster — the more concentrated they are and the more readily their collisions have the proper combined total collision energy. Raising the temperature of a reacting mixture increases the frequency of successful collisions (makes the rate of reaction faster). A catalyst, such as any enzyme, lowers the energy of activation without affecting the overall heat of reaction, so a catalyst also increases the rate of a reaction.

KEY TERMS

The following terms were introduced in this chapter, and their meanings should be mastered before you continue to the next chapter.

atmosphere, standard (atm)	energy of activation (E_{act})	melting point	rate of reaction
Avogadro's law	enzyme	millimeter of mercury (mm Hg)	standard conditions of temperature and pressure (STP)
barometer	equilibrium, dynamic	molar volume	
boiling	heat of reaction	nonvolatile liquid	temperature–volume law
boiling point, normal	ideal gas	partial pressure	torr
Boyle's law	kinetic theory of gases	photosynthesis	universal gas constant (R)
catalysis	kinetics	polarization	universal gas law
catalyst	law of partial pressures	pressure	vacuum
Charles' law	Le Chatelier's principle	pressure–temperature law	vapor pressure
Dalton's law of partial pressures	London force	pressure–volume law	volatile liquid
diffusion	manometer		

SELECTED REFERENCES

1   R. A. Hermens. "Boyle's Law Experiment," *Journal of Chemical Education*, September 1983, page 764.

2   J. A. Barker and D. Henderson. "The Fluid Phases of Matter," *Scientific American*, November 1981, page 130.

3   R. R. Festa. "Gay-Lussac: Chemist Extraordinary," *Journal of Chemical Education*, October 1981, page 789.

4   R. S. Treptow. "Le Chatelier's Principle," *Journal of Chemical Education*, June 1980, page 417.

REVIEW EXERCISES

The answers to Review Exercises that require a calculation and whose numbers are marked with an asterisk are given in an Appendix. The answers to the other Review Exercises are given in the *Study Guide* that accompanies this book.

Pressure and Other Variables

6.1 What are the four principal physical quantities used in describing the physical state of any gas?

6.2 What name do we give to the property of a gas that ensures that the gas occupies its entire container, whatever its shape or volume?

6.3 What is the difference between pressure and force?

6.4 What causes the atmosphere to exert a force on the earth?

6.5 What is one standard atmosphere of pressure?

6.6 Why is the high density of mercury an advantage over water for use in a Torricelli barometer?

6.7 How many mm Hg are in 1 atm?

6.8 How many mm Hg are in 1 torr?

6.9 Why doesn't all of the mercury run out of a Torricelli barometer?

*6.10 If a force of 300 lb acts on an area of 1 ft², what is the pressure in pounds per square inch?

6.11 Which exerts the higher pressure, a force of 150 lb acting on 25 in.² or a force of 50 lb acting on 5 in.²? (Do the calculations.)

*6.12 At the summit of Mt. McKinley (20,320 ft, 6194 m), the atmospheric pressure is 0.460 atm. What is this pressure in mm Hg? In torr?

6.13 The highest altitude at which a pilot or plane passenger could survive without a pressurized cabin but while breathing oxygen-enriched air is about 40,000 ft (8 miles, 13 km). The air pressure there is about 150 mm Hg. What is this in atm?

*6.14 If 29.92 in. Hg corresponds to 760 mm Hg, what is the atmospheric pressure in mm Hg on a day when it is reported as 28.95 in. Hg?

6.15 For a measured boiling point to be scientifically useful, the pressure at which it is measured must also be recorded and reported. In one old report the pressure was given as 29.82 in. Hg. What is this in mm Hg? (1 in. Hg = 25.40 mm Hg)

Specific Gas Laws

6.16 What fact about gases did Boyle discover?

6.17 When we say that volume is inversely proportional to pressure, what does *inversely proportional* mean?

6.18 In order for P_1V_1 to equal P_2V_2 what conditions must be true about the measurements at the moments designated by the subscripts 1 and 2?

6.19 What is the law of partial pressures?

6.20 What four gases are in exhaled air?

6.21 Of the four important variables in the study of the physical properties of gases, which are assumed to be held at constant values in each of the following gas laws?
(a) Boyle's law
(b) Charles' law
(c) Law of partial pressures
(d) Avogadro's law
(e) Gay-Lussac's law

6.22 Describe in your own words how we can figure out the dry volume of a gas after we have measured its volume when the gas is saturated with water vapor.

6.23 Describe in your own words how the method of ratios works for finding the volume of a gas when we know both its initial volume and pressure and its final pressure.

6.24 Why was a value of -273.15 °C selected as 0 K?

6.25 State the volume–temperature law.

6.26 Describe in your own words how the method of ratios works for finding a new value of the volume of a gas when we know its initial volume and both its initial and final temperatures. (Assume that n and P are constant.)

6.27 What is Avogadro's law?

*6.28 A sample of oxygen with a volume of 525 mL and a pressure of 750 mm Hg has to be given a volume of 475 mL. What pressure is needed if the temperature is to be kept constant?

6.29 What is the new pressure on a sample of helium initially with a volume of 1.50 L and a pressure of 745 mm Hg if its volume becomes 2.10 L at the same temperature?

*6.30 The value of P_{N_2} at the summit of Mt. McKinley (Review Exercise 6.12) is 277 mm Hg on a day when the atmospheric pressure there is 350 mm Hg. Assuming that the air is made up only of nitrogen and oxygen, what is the partial pressure of oxygen in mm Hg up there?

6.31 At an elevation of 40,000 ft (Review Exercise 6.13), the partial pressure of nitrogen is 119 mm Hg on a day when the air pressure at this elevation is 150 mm Hg. What is the value of P_{O_2}?

6.32 When nitrogen is prepared and collected over water at 30 °C and a total pressure of 739 mm Hg, what is its partial pressure in mm Hg?

6.33 If you were to prepare oxygen and collect it over water at a temperature of 26 °C and a total pressure of 752 mm Hg, what would be its partial pressure?

*6.34 Suppose that you needed 250 mL of dry oxygen at 760 mm Hg and 22 °C. How many milliliters of wet oxygen would you have to collect at this same total pressure and this temperature to have this much dry oxygen?

6.35 Assuming that bunsen burner gas is pure methane (CH_4), if you collected 325 mL of methane over water in a gas collecting bottle at 740 mm Hg and 26 °C, how many milliliters of dry methane are present?

*6.36 A sample of oxygen was warmed from 15 °C to 30 °C at constant pressure. Its initial volume was 1.75 L. What is its final volume in liters?

6.37 In order to change a 400-mL sample of nitrogen at 25 °C to a 200 mL sample with the same pressure, what must become of the temperature? (Give your answer in degrees Celsius.)

Universal Gas Law

6.38 What are the standard conditions of temperature and pressure?

6.39 What is meant by a *molar volume?*

6.40 Under what circumstances is the molar volume equal to 22.4 L?

6.41 What is the equation for the universal gas law?

6.42 To use the value of R that we calculated in this chapter, in what specific units must P, V, n, and T be?

6.43 At STP how many molecules of hydrogen are in 22.4 L?

*6.44 When an electric current is passed through water under suitable conditions, the water breaks down into hydrogen and oxygen according to the following equation:

$$2H_2O \longrightarrow 2H_2 + O_2$$

In one experiment, a dry sample of one of these gases was collected. Its volume at 748 mm Hg and 23.0 °C was 875 mL.
(a) How many moles of this gas were obtained?
(b) This sample of gas had a mass of 1.136 g. What is the formula weight of this gas? (Remember that a formula weight is numerically equal to the *ratio* of grams to moles.)
(c) Which gas was it, oxygen or hydrogen?

6.45 Consider the experiment described in Review Exercise 6.44.
(a) How many moles of hydrogen were obtained?
(b) What volume was occupied by this sample of hydrogen at 748 mm Hg and 23.0 °C?

*6.46 One source of industrial hydrogen is methane. At a high temperature, methane (CH_4) decomposes ("cracks") as follows into carbon and hydrogen:

$$CH_4 \underset{heat}{\longrightarrow} C + 2H_2$$

(a) If 1.00 mol of CH_4 is used, how many moles of hydrogen are produced?
(b) If 100 L of methane gas, initially at 740 mm Hg and 20 °C, are used, how many liters of hydrogen will be obtained when they are measured under the same conditions of temperature and pressure?
(c) If a sample of methane with a mass of 50.0 g is cracked, what volume of hydrogen is produced when measured at 750 mm Hg and 25 °C?

6.47 Carbon dioxide can be removed from exhaled air by making the air pass through granulated sodium hydroxide. The reaction is

$$NaOH + CO_2 \longrightarrow NaHCO_3$$

Sodium   Carbon   Sodium
hydroxide   dioxide   bicarbonate

After this system had operated for some time, it was found that 12.4 g of NaOH had been used up.
(a) How many moles of CO_2 were responsible for this amount of change?
(b) If 11.6 g of NaOH were used up in a separate operation, how many milliliters of CO_2 gas caused this change if the gas volume were measured at 740 mm Hg and 25 °C?

Kinetic Theory of Gases

6.48 Scientists asked, "What must gases be like for the gas laws to be true?" What was their answer?

6.49 What is true about an ideal gas that is not strictly true about any real gas?

6.50 Dalton's law of partial pressures implies that gas molecules from different gases actually leave each other alone in the mixture, both physically and chemically (except at moments of collisions, when they push each other around). Which one of the three postulates in the model of an ideal gas is based on Dalton's law?

6.51 How does the kinetic theory of gases explain the phenomenon of gas pressure?

6.52 How does the kinetic theory of gases account for Boyle's law (in general terms)?

6.53 For 1 mol of an ideal gas, those working out the kinetic theory found that the product of pressure and volume is proportional to the average kinetic energy of the ideal gas particles.
(a) Using the universal gas law (which makes no mention of kinetic energy), to which of the four physical quantities used to describe a gas is the product of pressure and volume for 1 mol of a gas also proportional?
(b) If the product of P and V is proportional both to the average kinetic energy of the ideal gas particles and to the Kelvin temperature, what does this say about the relationship between the average kinetic energy and this temperature?

6.54 What happens to the motions of gaseous molecules at 0 K?

6.55 How does the kinetic theory explain (in general terms) the volume–temperature law?

6.56 The pressure–temperature law (Gay-Lussac's law) can be explained in terms of the kinetic theory in what way (in general terms)?

The Liquid State and Vapor Pressure

6.57 Why aren't there universal laws for the physical behavior of liquids (or solids) as there are for gases?

6.58 How does the kinetic theory explain
(a) How vapor pressure arises?
(b) Why vapor pressure rises with increasing liquid temperature?

6.59 Dimethylsulfoxide (DMSO) is a controversial pain-killing drug permitted by only a few states. Its boiling point is 189 °C. Is it more volatile or less volatile than water? Explain.

Dynamic Equilibria and Changes of State

6.60 Compare the following expressions.

A   water + heat $\longrightarrow$ water vapor

B   water + heat $\longleftarrow$ water vapor

C   water + heat $\rightleftharpoons$ water vapor

Answer the following questions by using the letter A, B, or C to indicate which expression best represents the answer.

(a) Which expression describes heat being liberated from the system?

(b) Which expression describes the net formation of liquid water from water vapor?

(c) Which expression describes a net endothermic change?

(d) Which expression shows opposing changes?

(e) Which expression, A or B, represents the forward change in expression C? Which represents the reverse change?

(f) Expression C can be the correct description for the water – water vapor system at 1 atm only if the temperature is what?

(g) What is true about the opposing changes in expression C?

(h) What special term applies to expression C? (It does not represent a *reaction* or a permanent *change,* but what?)

6.61 When a liquid and its vapor are in dynamic equilibrium at a given temperature, the rates of what two changes are equal?

6.62 Some solids, like dry ice (solid carbon dioxide), pass directly from their solid state to their vapor state when they absorb heat. This process is called *sublimation.* Write an equilibrium expression for sublimation.

6.63 Explain the fact that the equilibrium vapor pressure of, say, water vapor over liquid water is the same if we have a liter of water or just a few milliliters.

6.64 Why does a liquid's boiling point decrease as the atmospheric pressure decreases?

6.65 Cooking an egg involves heat-induced chemical reactions as well as some physical changes. Why does it take longer to prepare a soft-boiled egg in Denver than in New York City?

6.66 Explain in your own words how a substance such as nitrogen, which consists of nonpolar molecules, can be converted into a liquid by cooling it. How do the forces of attraction that are needed if molecules are to be in the liquid state develop?

6.67 Why do nonpolar substances of high formula weight generally have higher boiling points than those of low formula weight?

6.68 What is Le Chatelier's principle?

6.69 Consider the following equilibrium in which all of the substances are gases:

$$N_2 + O_2 + heat \rightleftharpoons 2NO$$

Nitric oxide

(a) Which reaction, the forward or the reverse, is endothermic?

(b) Which reaction, the forward or the reverse, will be favored by adding heat?

6.70 What do we call the temperature at which a solid is in equilibrium with its liquid form?

6.71 The following are some common observations. Using the kinetic theory, explain how each occurs in terms of what molecules are doing.

(a) Moisture evaporates faster in a breeze than in still air.

(b) Ice melts much faster if it is crushed than if it is left in one large block.

(c) Even if hung out to dry in below-freezing weather, wet clothes will become completely dry even though they freeze first.

6.72 At room temperature, nitrogen is a gas, water is a liquid, and sodium chloride is a solid. What do these facts tell us about the relative strengths of electrical forces of attraction in these substances?

Kinetic Theory and Rates of Reactions

6.73 In terms of what we visualize as happening when two molecules interact to form products, how do we explain the existence of an energy barrier to the reaction—an energy of activation?

6.74 Study the accompanying progress of reaction diagram for the conversion of carbon monoxide and oxygen to carbon dioxide, and then answer the questions. The equation for the reaction is

$$2CO(g) + O_2(g) \longrightarrow 2CO_2(g)$$

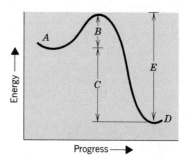

(a) What substance or substances occur at position A?

(b) What substance or substances occur at position D?

(c) Which letter labels the arrow that represents the heat of reaction?

(d) Which letter labels the arrow that stands for the energy of activation?

(e) Is this reaction endothermic or exothermic? How can you tell?

(f) Which letter labels the arrow that would correspond to the energy of activation if the reaction could go in reverse?

6.75 Suppose that the following hypothetical reaction occurs:

$$A + B \longrightarrow C + D$$

Suppose further that this reaction is endothermic and that the energy of activation is numerically twice as large as the heat of reaction. Draw a progress of reaction diagram for this reaction, and draw and label arrows that correspond to the energy of activation and the heat of reaction.

6.76 The reaction of *X* and *Y* to form *Z* is exothermic. For every mole of *Z* produced, 10 kcal of heat are generated. The energy of activation is 3 kcal. Sketch the energy relationships on a progress of reaction diagram.

6.77 How do we explain the rate-increasing effect of a rise in temperature?

6.78 As a rule of thumb, how much of a temperature increase doubles or triples the rates of most reactions?

6.79 Explain how a rise in body temperature can lead to a strain on the heart.

6.80 How can we increase the frequency of all collisions in a reacting mixture without raising the temperature?

6.81 When an increase in the concentration of one or more reactants causes an increase in the rate of a reaction, how do we explain this?

6.82 When an increase in the rate of a reaction has been caused by an increase in the concentration of one of the reactants, which of the following factors has been changed? (Identify them by letter.)

    A   the energy of activation

    B   the heat of reaction

    C   the frequency of collisions

    D   the frequency of successful collisions

6.83 In what way, if any, does a catalyst affect the following factors of a chemical reaction?
(a) the heat of reaction
(b) the energy of activation
(c) the frequency of collisions
(d) the frequency of successful collisions

6.84 What is the general name for the catalysts found in living systems?

6.85 Once we have selected a particular reaction, we have to accept whatever energy of activation and heat of reaction that goes with it. However, there are three things that we might try to help speed up the reaction. What are they?

Chapter 7
Water, Solutions, and Colloids

The geometry of the molecules of water, ammonia, and methane can be explained by a model as uncomplicated as four balloons tied together at a common point.

7.1 WATER

Many important properties of water are explained by the polarity of its molecules and the hydrogen bonds that exist between them.

We take in more water than all other materials combined. We use it as the fluid in all cells, as a heat-exchange agent, and as the carrier in the bloodstream for distributing oxygen and all molecules from food, all hormones, minerals, and vitamins, and all disease-fighting agents.

Water is a superb solvent. It can dissolve at least trace amounts of almost anything, including rock. It is particularly good at dissolving ionic substances and the more polar molecular compounds.

In this chapter we will focus on water and some of the physical properties of aqueous solutions. To understand many aspects of life at the molecular level, we need to know why water dissolves some things well but not others, and to achieve this goal we must study in greater detail how a water molecule is put together.

The Polarity of Hydrogen Fluoride, Ammonia, and Water. In the last chapter we learned an important rule of thumb — boiling points of similar substances rise with formula weights. However, three simple substances with low formula weights — water, ammonia, and hydrogen fluoride — are striking exceptions. Figure 7.1 displays plots of the boiling points versus the formula weights of the hydrides of the elements in Groups IVA through VIIA. (Hydrides are compounds of hydrogen with some other element.)

Look first at the plot that is lowest in the figure and which concerns the hydrides of carbon (CH_4), silicon (SiH_4), and germanium (GeH_4). Their boiling points follow the rule of thumb we mentioned. They increase regularly with increasing formula weight. The next plot, however, doesn't follow this rule as well. Ammonia, NH_3, is badly off the straight line on which the other Group VA hydrides fall — the hydrides of phosphorus, PH_3, arsenic, AsH_3, and antimony, SbH_3. Ammonia boils far higher than "it should." Similarly, hydrogen fluoride, HF, does not fit well with the plot of the boiling points of the hydrides of the halogens. And water is the worst of all. This hydride of oxygen, a Group VIA element, boils far higher than we would expect according to the rule. Water "should" boil at about $-100\ °C$, as the dashed-line extension of the plot for the Group VIA hydrides indicates, but it actually boils at $+100\ °C$.

The plots of Figure 7.1 suggest that forces of attraction between molecules in samples of HF, H_2O, and NH_3, but not methane (CH_4), are considerably higher than we could account for just on the basis of London forces (Section 6.7). The molecules of these three hydrides, but not methane, must have permanent dipoles and those of methane do not. We will look next at a theory that will help us to understand how water has such a high polarity.

These are *binary* hydrides because they are made of just two elements.

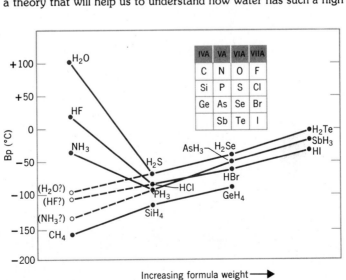

FIGURE 7.1
Boiling points versus formula weights for the binary, nonmetal hydrides of the elements in Groups IVA, VA, VIA, and VIIA.

VSEPR Theory and the Geometry of a Water Molecule. To explain the permanent dipole in the water molecule, we have to go a step beyond what we have already learned about molecular structure. We have to learn about molecular geometry and how it arises. There is no problem when the molecule has only two atoms, as in the H—F molecule, because a diatomic molecule can't be anything but linear. The water molecule, however, has three atoms. If they were all in a straight line, H—O—H, the molecule would have no permanent dipole. The dipole of one H—O bond would exactly cancel the dipole of the other, just as two equally matched and exactly opposed tug-of-war teams cancel each other's efforts. The three atoms in a water molecule, however, form an angle of 104.5°.

Linear means in a straight line.

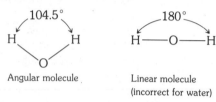

Angular molecule

Linear molecule
(incorrect for water)

The easiest way to understand why there is an angle in the water molecule is by a simple and successful theory with a long but very descriptive name — the **valence-shell electron-pair repulsion theory,** or the **VSEPR** theory, for short. *Valence-shell* refers to the outside energy level of an atom. *Electron-pair* refers to the tendency of electrons in the valence shell to occur in pairs, particularly when some are involved in covalent bonds. Remember, when an atom in a molecule has an octet, and when we count shared pairs, there are four *pairs* of electrons in the valence shell. *Repulsion* refers to the effect that an electron cloud of one pair of electrons has on the electron cloud of another pair in the valence shell.

VSEPR theory says that the directions of the bonds at a central atom are determined largely by the strong tendency of the electron pairs of the valence-shell to stay as far apart from each other as possible. A good analogy is to imagine that each pair's electron cloud resembles the shape of a balloon and that there is an imaginary line or axis that extends through each balloon from the point where it's tied off directly over to the middle of the opposite surface of the balloon. Now try to imagine how four identical balloons must arrange themselves most comfortably if we tie them all close together at one point. The chapter-opening photograph shows the result. The imaginary axes would not lie in a plane and point to the corners of a square, making angles of 90° to each other, because the balloons can be less crowded if their axes point at wider angles. The least crowded arrangement — and, hence, the most stable — occurs when the axes make angles of 109.5° with each other. This array is called **tetrahedral** because the axes point to the corners of a regular tetrahedron.

The tetrahedral angle of 109.5° is quite close to the known angle in the water molecule, 104.5°. As diagrammed in Figure 7.2, two pairs of electrons in the valence shell of oxygen in

The central atom in H_2O is O. In NH_3, it's N.

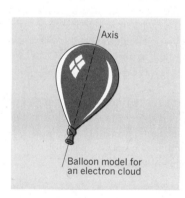

Axis

Balloon model for an electron cloud

FIGURE 7.2
The geometry of the water molecule. The electron clouds associated with the electron pairs of the valence shell of the oxygen atom have their axes pointed approximately to the corners of a regular tetrahedron (a geometric figure shown on the right).

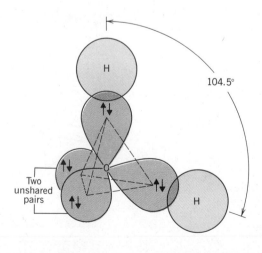

104.5°

Two unshared pairs

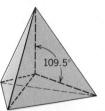

109.5°

A regular tetrahedron is a four-sided space bounded by identical equilateral triangles. Any two lines from the corners to the midpoint of the space make an angle of 109.5°.

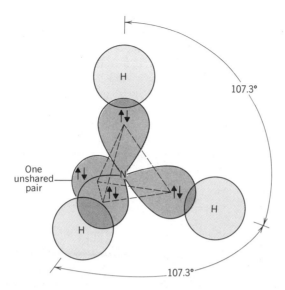

One
unshared
pair

107.3°

107.3°

the water molecule are not shared. Hence, their electron clouds tend to repel each other a bit more than the clouds of the two shared pairs. As a result, the bond angle is squeezed from 109.5° to a slightly lower value.

VSEPR Theory and the Ammonia Molecule.   When nitrogen is the central atom, as it is in ammonia, the VSEPR theory again predicts a tetrahedral bond angle of 109.5°, because there are four valence-shell pairs of electrons. The actual bond angle in ammonia is 107.3°, also quite close. (See Figure 7.3.)

VSEPR Theory and the Methane Molecule.   When carbon is the central atom, as in methane, CH_4, VSEPR theory also predicts bond angles of 109.5° for each of the H—C—H bonds. As seen in Figure 7.4, theory and fact coincide exactly. The bond angles in methane are all 109.5°.

A very simple idea — electron clouds repel each other — accounts very well for the bond angles in three molecules of great importance in our study — water, ammonia, and methane. What VSEPR theory does not do, however, is to explain the molecular orbitals in these hydrides, but we don't need this explanation until a later chapter. We want to get back to the substance of central importance in this chapter — water.

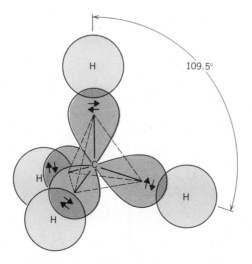

109.5°

FIGURE 7.4
The geometry of the methane
molecule. VSEPR theory predicts bond
angles for H—C—H to be 109.5°,
and these are observed.

The direction of the polarity of the H—O bond is

$$\overrightarrow{\text{H—O}}$$

In a chain, the important link is the *weakest* link, not the strongest one.

The Hydrogen Bond in Water.   Because oxygen is much more electronegative than hydrogen, each of the two H—O bonds is very polar. Because the water molecule is angular, these individual polarities do not cancel each other. Thus the water molecule as a whole is a polar molecule. It is so polar, in fact, that between water molecules there is a strong enough force of attraction to be called a bond. It's not a covalent bond or an ionic bond, so it has its own name — **hydrogen bond.** A hydrogen bond is the force of attraction between the $\delta+$ on H, when the H is bonded to oxygen, nitrogen, or fluorine (the three most electronegative elements) and the $\delta-$ on some other oxygen, nitrogen, or fluorine atom.

Here are most of the several possibilities where hydrogen bonds can exist. Only partial structures are shown, and a dotted line is used to represent the force of attraction that we have now named the hydrogen bond. (The solid lines, of course, are covalent bonds.)

$$\underset{\text{H—F}}{\overset{\delta+\quad\delta-}{}}\cdots\underset{\text{H—F}}{\overset{\delta+\quad\delta-}{}} \qquad \underset{\text{H—O}}{\overset{\delta+\quad\delta-}{}}\cdots\underset{\text{H—O}}{\overset{\delta+\quad\delta-}{}} \qquad \underset{\text{H—O}}{\overset{\delta+\quad\delta-}{}}\cdots\underset{\text{H—N}}{\overset{\delta+\quad\delta-}{}}$$

$$\underset{\text{H—N}}{\overset{\delta+\quad\delta-}{}}\cdots\underset{\text{H—O}}{\overset{\delta+\quad\delta-}{}} \qquad \underset{\text{H—N}}{\overset{\delta+\quad\delta-}{}}\cdots\underset{\text{H—N}}{\overset{\delta+\quad\delta-}{}}$$

The hydrogen bond is a bridging bond between molecules, as illustrated for water in Figure 7.5. Like all bonds, the hydrogen bond is a force of attraction. However, it is by no means as strong as a covalent bond, being roughly only 5% as strong. But this is strong enough to make a difference not just in water but in such important yet disparate substances as muscle proteins, cotton fibers, and the chemicals of genes, DNA. In fact, among those biochemicals where hydrogen bonds occur, these weak bonds are more important structurally than any other bond precisely because they are weak, not strong.

Largely because of the hydrogen bond, water has relatively high heats of fusion and vaporization, as we learned in Section 2.4. An extra large input of heat per gram is necessary to melt or boil water because energy is needed to overcome the force of attraction called the hydrogen bond.

Nitrogen is less electronegative than oxygen, and the H—N bonds in ammonia aren't as polar as the H—O bonds in water. Therefore the hydrogen bonds between molecules in ammonia aren't as strong as those that exist in water. This is chiefly why ammonia doesn't have nearly as high a boiling point as that of water. Despite this lessened polarity of the H—N bond, and as suggested by the plots in Figure 7.1, hydrogen bonding does occur in ammonia. It also occurs in protein molecules and in gene molecules, where H—N bonds abound. This is principally why we study the hydrogen bond.

Surface Tension.   All liquids possess a surface tension, but that of water is unusually high. **Surface tension** is a phenomenon in which the surface acts as if it were a thin, invisible,

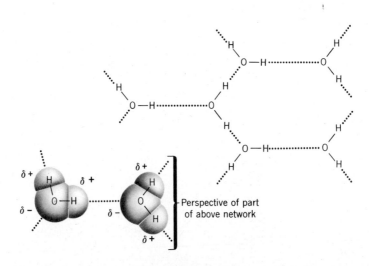

FIGURE 7.5
Hydrogen bonds in water (••••••••••)

elastic membrane. It's the reason why some bugs can skitter on the surface of a pond; why parlor magicians can set a steel needle afloat on water; and why water forms tight droplets and doesn't spread out on a waxy surface but does spread out on clean glass. It's also the source of the force that can make a lung collapse under certain conditions.

When a liquid's molecules are polar, they tend to jam together where the liquid meets air, as shown in Figure 7.6. Here, the polar forces of attraction that pull surface molecules downward aren't counterbalanced by forces that pull them upward. There is a net downward pull that causes the surface jam-up of molecules that is responsible for surface tension.

On a greasy or waxed surface, water forms into beads (Figure 7.7) because the inward-pulling forces of attraction in the water bead aren't matched by outward-pulling forces from the nonpolar molecules in wax or grease.

Glass, however, is a very polar material, and water spreads out on a clean, grease-free glass surface (Figure 7.8). Water molecules that are right at the glass surface find things to be more attracted to on the glass than behind them in the rest of the water sample, so the water spreads out. For the same reason, in a glass graduated cylinder or in a glass pipet, the boundary between the surface of an aqueous solution and the air curves upward at the glass walls.

Surface-Active Agents. If we replace a glass surface by a very thin, yet very polar and *flexible* membrane, such as the moist membrane of an air sac (an alveolus) in the lungs, the force with which water molecules attract the membrane ordinarily would be strong enough to make the air sac collapse. If enough air sacs do this, the lung collapses. However, a special substance is present on the surface of the air sacs that greatly reduces water's surface tension, and this protects the lungs from experiencing such a disaster. Anything that lowers the surface tension of water is called a **surface-active agent,** or a **surfactant,** for short.

All soaps and detergents are surfactants. Think of a surfactant as something that cuts the surface ''membrane'' of water to ribbons. When water contains a surfactant, its molecules can move out more readily into nonpolar environments, such as the grease and oils that act as glues to bind dirt, soil, and harmful bacteria to fabrics, skin, or cooking utensils. Water with a surfactant is an excellent cleansing agent, far better than water alone, as you already know.

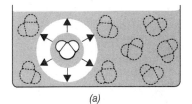
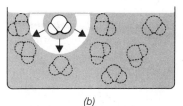

(a)

(b)

FIGURE 7.6
Suface tension. *(a)* In the interior of a sample of water, individual water molcules are attracted equally in all directions. *(b)* At the surface, nothing in the air counterbalances the downward pull that the surface molecules in water feel.

FIGURE 7.7
Water forms beads on a waxed or greasy surface. Nothing in the surface has enough polarity to attract water molecules to make the water spread out. The net inward pull created by water molecules at the surface creates the bead, because this shape minimizes the total area of the droplet.

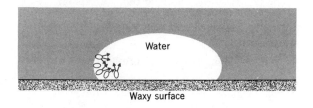

Water

Waxy surface

FIGURE 7.8
The interaction of water with the very polar surface of glass. *(a)* Water droplets spread on glass is response to polar sites in the glass. *(b)* When the glass surface is vertical, as in a graduated cylinder, pipet, or buret, water climbs the glass wall for a short distance to form a meniscus. In measuring volumes with these devices, read the bottom of the meniscus.

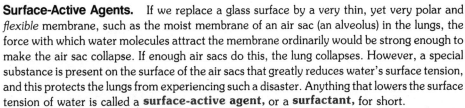
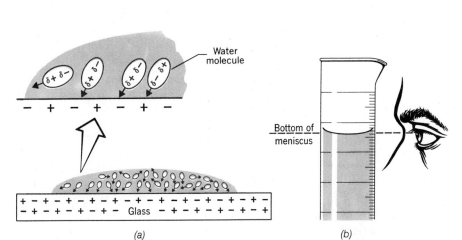

(a)

(b)

All of the chemical reactions of digestion use water as a reactant.

The lungs have a natural surfactant built into the membranes of the air sacs, and in some situations this surfactant is depleted or defective and the lungs collapse.

Bile, one of the digestive juices, contains an extremely powerful surfactant called a bile salt. Without it, it would be difficult for our systems to digest the fats and oils in our diets or to wash them from the particles of other kinds of food. The bile salts are also needed to aid in the absorption of some of the relatively nonpolar vitamins from the intestinal tract. Because the bile salts are manufactured and delivered in bile from the gall bladder, an operation that removes this organ creates special dietary problems with which health professionals have to deal.

7.2 THE TYPES OF HOMOGENEOUS MIXTURES

The sizes of the particles that are intimately mixed with a solvent determine a number of properties that are distinct from the chemical properties of these particles.

There are three chief kinds of **homogeneous mixtures,** mixtures in which any small sample has the same composition and properties as any other sample of the same size taken anywhere else in the mixture. These are *solutions, colloidal dispersions,* and *suspensions,* and all types are found in the body. They differ fundamentally in the sizes of the particles involved, and the differences in size alone can cause interesting and important changes in properties.

Solutions.   As we learned in Section 5.5, a **solution** is a homogeneous mixture in which the particles of both solvent and solutes have sizes of atoms, or ordinary ions and molecules. They have formula weights of no more than a few hundred and diameters in the range of 0.1 to 1 nm.

$1 \text{ nm} = 10^{-9} \text{ m} = 1 \text{ nanometer}$

We usually think of solutions as being liquids, but in principle the solvent can be in any state—solid, liquid, or gas—and so can the solute. Table 7.1 has a list of the several combinations that can form a solution. Solutions are generally transparent—you can see through them—but they often are colored. Solutes do not settle out of solutions under the influence of gravity, and they can't be separated from solutions by filter paper. The blood carries many substances in solution. These include the sodium ion, Na^+, and the chloride ion, Cl^-, as well as molecules of glucose, the chief sugar in blood.

Colloidal Dispersions.   A **colloidal dispersion** is a homogeneous mixture in which the dispersed particles are very large clusters of ions or molecules or are actually **macromolecules** that have formula weights in the thousands and hundreds of thousands. The dispersed particles have diameters in the range of 1 nm to 1000 nm. Table 7.2 gives several examples of colloidal dispersions, and they include many familiar substances such as whipped cream,

Macromolecule means an extremely large molecule.

TABLE 7.1
Solutions

Kinds	Common Examples
Gas in a liquid	Carbonated beverages (carbon dioxide in water)
Liquid in a liquid	Vinegar (acetic acid in water)
Solid in a liquid	Sugar in water
Gas in a gas	Air
Liquid in a gas[a]	
Solid in a gas[a]	
Gas in a solid	Alloy of palladium and hydrogen
Liquid in a solid	Toluene in rubber (e.g., rubber cement)
Solid in a solid	Carbon in iron (steel)

[a] True examples probably do not exist.

TABLE 7.2
Colloidal Systems

Type	Dispersed Phase[a]	Dispersing Medium[b]	Common Examples
Foam	Gas	Liquid	Suds, whipped cream
Solid foam	Gas	Solid	Pumice, marshmallow
Liquid aerosol	Liquid	Gas	Mist, fog, clouds, certain air pollutants
Emulsion	Liquid	Liquid	Cream, mayonnaise, milk
Solid emulsion	Liquid	Solid	Butter, cheese
Smoke	Solid	Gas	Dust in smog
Sol	Solid	Liquid	Starch in water, jellies,[c] paints
Solid sol	Solid	Solid	Black diamonds, pearls, opals, metal alloys

[a] The colloidal particles constitute the *dispersed phase*.

[b] The continuous matter into which the colloidal particles are scattered is called the *dispersing medium*.

[c] Sols that adopt a semisolid, semirigid form (e.g., gelatin desserts, fruit jellies) are called **gels.**

milk, dusty air, jellies, and pearls. The blood also carries many substances in colloidal dispersions, including a variety of proteins.

When colloidal dispersions are in a fluid state — liquid or gas — the dispersed particles, although large, are not large enough to be trapped by ordinary filter paper during filtration. They are large enough, however, to reflect and scatter light (Figure 7.9). Light scattering by a colloidal dispersion is called the **Tyndall effect,** after British scientist John Tyndall (1820–1893). This effect is responsible for the milky, partly obscuring character of smog, or the way sunlight sometimes seems to stream through a forest canopy.

The large, dispersed particles in a fluid colloidal dispersion eventually settle out under the influence of gravity, but this process can take time which, depending on the system, can range from many hours to many decades! One of the factors that keeps the particles dispersed is that they are constantly buffeted about by molecules of the solvent. Evidence for this buffeting can be seen by looking at the colloidal system under a good microscope. You can't actually see the colloidal particles, but you can see the light scintillations caused as they move erratically and unevenly about. This motion of colloidal particles is called the **Brownian movement.**

In the most stable colloidal systems, all of the particles bear like electrical charges. In living systems, it's common for colloidally dispersed proteins to be like this, for example.

Solutions do not exhibit the Tyndall effect because the solute particles are too small.

Robert Brown (1773–1858), an English botanist, first observed this phenomenon when he saw the trembling of particles inside grains of pollen that he viewed with a microscope.

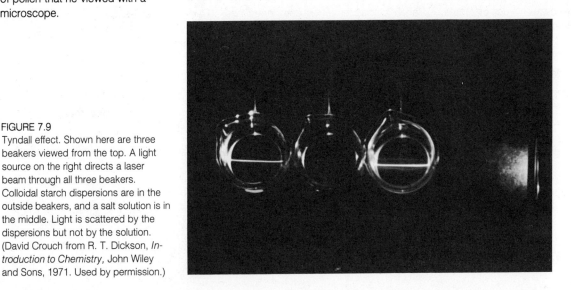

FIGURE 7.9
Tyndall effect. Shown here are three beakers viewed from the top. A light source on the right directs a laser beam through all three beakers. Colloidal starch dispersions are in the outside beakers, and a salt solution is in the middle. Light is scattered by the dispersions but not by the solution. (David Crouch from R. T. Dickson, *Introduction to Chemistry,* John Wiley and Sons, 1971. Used by permission.)

(Other dissolved species of opposite charge, such as small ions, balance the charges on the colloidal particles.) Because like-charged particles repel each other, they cannot come together and coalesce into particles that suddenly are very much larger and cannot resist separating under the influence of gravity.

Some colloidal dispersions are stabilized by protective colloids. **Emulsions** are colloidal dispersions of two liquids in each other — such as oil and vinegar — and some can be stabilized by *emulsifying agents*. For example, mayonnaise is stabilized by egg yolk whose protein molecules coat the microdroplets of olive oil or corn oil and prevent them from merging into drops large enough to rise to the surface.

Suspensions.   In **suspensions,** the dispersed or suspended particles are over 1000 nm in average diameter, and they separate under the influence of gravity. They are large enough to be trapped by filter paper. A suspension such as clay in water has to be stirred constantly to keep it from separating. Because of this fact, a suspension is always on the borderline between a homogeneous mixture and one that is heterogeneous (one not uniform throughout). The blood, while it is moving, is a suspension, besides being a solution and a colloidal dispersion. Suspended in circulating blood are its red and white cells and its platelets.

See Table 7.3 for a summary of the chief features of solutions, colloidal dispersions, and suspensions.

7.3   WATER AS A SOLVENT

Water dissolves best those substances whose ions or molecules can strongly attract water molecules and form solvent cages.

When something like solid table salt, NaCl, is put into any potential solvent, the solvent molecules bombard the crystal surfaces, and this kinetic action tends to dislodge ions (Figure 7.10). However, the individual ions cannot leave their environment in the crystal, where each

TABLE 7.3
Characteristics of Three Homogeneous Mixtures — Solutions, Colloidal Dispersions, and Suspensions

Particle Sizes Become Larger		
Solutions	Colloidal Dispersions	Suspensions
All particles are on the order of atoms, ions, or small molecules (0.1 – 1 nm)	Particles of at least one component are large clusters of atoms, ions, or small molecules, or are very large ions or molecules (1 – 1000 nm)	Particles of at least one component may be individually seen with a low-power microscope (over 1000 nm)
Most stable to gravity	Less stable to gravity	Unstable to gravity
Most homogeneous	Also homogeneous, but borderline	Homogeneous only if well-stirred
Transparent (but often colored)	Often translucent or opaque, but may be transparent	Often opaque, but may appear translucent
No Tyndall effect	Tyndall effect	Not applicable (suspensions cannot be transparent)
No Brownian movement	Brownian movement	Particles separate unless system is stirred
Cannot be separated by filtration	Cannot be separated by filtration	Can be separated by filtration
Homogeneous	to	Heterogeneous

FIGURE 7.10
The hydration of ions helps ionic substances dissolve in water.

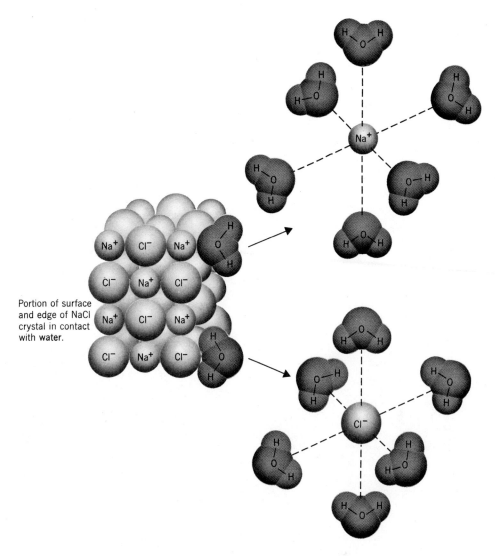

Portion of surface and edge of NaCl crystal in contact with water.

is surrounded with oppositely charged ions as nearest neighbors, unless something else substitutes for this environment. Opposite charges attract each other too strongly. This situation is ideal for water.

Hydration of Ions.   Because water molecules are very polar, they can surround sodium ions, letting their $\delta-$ sites on oxygen point toward the positively charged ion, as seen in Figure 7.10. Similarly, chloride ions can also attract water molecules (Figure 7.10), which become oriented so that their $\delta+$ sites on hydrogen point toward the negatively charged Cl^- ions.

No longer do chloride ions have Na^+ ions as nearest neighbors, but they have several water molecules performing the same service — helping to provide Cl^- with an environment of opposite charge. No longer do sodium ions have chloride ions as nearest neighbors, but water molecules take their place. This phenomenon whereby water molecules are attracted to solute particles is called **hydration,** and hydration results in a cage of water molecules about each solute particle in solution.

Of all of the common solvents, only water has molecules both polar enough and small enough to form effective solvent cages around ions. Of course, in order for water to do this, water molecules must give up some of their attractions for each other. Only ionic substances or compounds made of very polar molecules can break up the hydrogen-bonded network between water molecules. Thus the formation of a solution isn't just the separation of the

Hydr- is from the Greek *hydōr,* water.

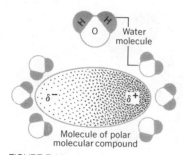

FIGURE 7.11
The hydration of a polar molecule helps polar molecular substances to dissolve in water.

solute particles from each other. It's also, to some extent, the separation of solvent molecules from each other.

Polar molecular compounds dissolve in water, too, and Figure 7.11 shows how water hydrates their molecules. In many compounds hydrated ions occur even in their crystalline forms, as we'll study next.

Hydrates.   If we let water evaporate from an aqueous solution of any one of several substances, the crystalline residue contains intact water molecules that are held within the crystals in *definite* proportions. Such water-containing solids are called **hydrates,** and they are true compounds because they obey the law of definite proportions. However, we write their formulas to emphasize that intact water molecules are present. Thus the formula of the pentahydrate of copper(II) sulfate is written as $CuSO_4 \cdot 5H_2O$, where a raised dot separates the two parts of the formula. Table 7.4 lists a number of other common hydrates.

The water indicated by the formula of a hydrate is called the **water of hydration.** It usually can be driven away from the hydrate by heat, and when this is done the residue is sometimes called the **anhydrous form** of the compound. For example:

$$CuSO_4 \cdot 5H_2O(s) \longrightarrow CuSO_4(s) \qquad + 5H_2O(g)$$

Copper(II) sulfate   Copper(II) sulfate   (as steam)
pentahydrate (deep   (anhydrous form is
blue crystals)   nearly white)

Many anhydrous forms readily take up water and re-form their hydrates. Plaster of paris, for example, although not completely anhydrous, contains relatively less water than gypsum.

TABLE 7.4
Some Common Hydrates

Formulas	Names	Decomposition Modes and Temperatures[a]	Uses
$(CaSO_4)_2 \cdot H_2O$	Calcium sulfate hemihydrate (plaster of paris)	$-H_2O$ (163)	Casts, molds
$CaSO_4 \cdot 2H_2O$	Calcium sulfate dihydrate (gypsum)	$-2H_2O$ (163)	Casts, molds, wallboard
$CuSO_4 \cdot 5H_2O$	Copper(II) sulfate pentahydrate (blue vitriol)	$-5H_2O$ (150)	Insecticide
$MgSO_4 \cdot 7H_2O$	magnesium sulfate heptahydrate (epsom salt)	$-6H_2O$ (150)	Cathartic in medicine
		$-7H_2O$ (200)	Used in tanning and dyeing
$Na_2B_4O_7 \cdot 10H_2O$	Sodium tetraborate decahydrate (borax)	$-8H_2O$ (60) $-10H_2O$ (320)	Laundry
$Na_2CO_3 \cdot 10H_2O$	Sodium carbonate decahydrate (washing soda)	$-H_2O$ (33.5)	Water softener
$Na_2SO_4 \cdot 10H_2O$	Sodium sulfate decahydrate (glauber's salt)	$-10H_2O$ (100)	Cathartic
$Na_2S_2O_3 \cdot 5H_2O$	Sodium thiosulfate pentahydrate (photographer's hypo)	$-5H_2O$ (100)	Photographic developing

[a] Loss of water is indicated by the minus sign before the symbol, and the loss occurs at the temperature in °C that is given in parentheses.

When we mix plaster of paris with water, it soon sets into a hard, crystalline mass according to the following reaction:

$$(CaSO_4)_2 \cdot H_2O + 3H_2O \longrightarrow 2CaSO_4 \cdot 2H_2O$$

Plaster of paris   Gypsum

(Notice that the first 2 in gypsum's formula is a *coefficient* for the *entire* formula.)

Some compounds in their anhydrous forms are used as drying agents or desiccants. A **desiccant** is a substance that removes moisture from air by forming a hydrate. Any substance that can do this is said to be **hygroscopic.** Anhydrous calcium chloride, $CaCl_2$, is a common desiccant, and in humid air it draws enough water to form a liquid solution. Any substance this active as a desiccant is also said to be **deliquescent.** Thus calcium chloride is often used to dehumidify damp basements.

Usually, when both ions of an ionic compound carry charges of two or three units, the compound isn't very soluble in water. The ions find more stability by remaining in the crystal than they can replace by accepting solvent cages.

Saturated Solutions Revisited. At each temperature, each potential solid solute and most liquids have a limit to the extent to which they will dissolve in water. Hundreds of ionic compounds are only slightly soluble, and hundreds seemingly do not dissolve at all. In Section 5.6 we discussed such limits, where we referred to them as the *solubilities* of substances in water, to the grams of solute necessary to prepare a saturated solution in 100 g of water at a specified temperature. We can now refine this definition of a saturated solution. A **saturated solution** is one in which there is a dynamic equilibrium between the undissolved and the dissolved solute. We can represent this as follows:

$$Solute_{undissolved} \rightleftharpoons Solute_{dissolved} \tag{7.1}$$

We must be careful here to recall what we learned about hydrates earlier and to realize that what we call the solute$_{undissolved}$ in this equilibrium expression is often the *hydrate* of the solute, and not its anhydrous form. We may begin the preparation of a saturated solution by using an anhydrous compound, but once we have the solution the equilibrium might well be between the dissolved (and hydrated) particles and the undissolved *hydrate* of what we used. In any case, in a saturated solution there is coming and going as solute particles leave the undissolved state and go into solution (the forward change) and others, at the same rate, leave the dissolved state and return to the undissolved condition (the reverse change).

With this in mind we can look a bit more closely at the way that temperature affects the solubilities of many substances. Our purpose for going into this detail is not only to become more familiar with the concept of solubility — although it's important — but to extend our study of the vastly more important concept of dynamic equilibrium.

The Effect of Temperature on Solubility. The only circumstance in which one could test the effect of temperature on solubility is an experiment in which the solvent being used is actually the solution that is already saturated in the solute in question. The question at issue here is: Does heating a saturated solution permit more undissolved material to dissolve or less? The answer is that most but by no means all solids have increased solubilities at higher temperatures. The reason is that most solids dissolve endothermically when the solution into which they are dissolving is at or very near the point of being saturated. They require heat to go into solution. For solutes that dissolve endothermically, we can refine equilibrium expression 7.1 by introducing an energy term, the *heat of solution.*

$$Solid_{undissolved} + solution + heat\ of\ solution \rightleftharpoons more\ concentrated\ solution$$

The heat of solution is the difference between the energy used to break up the solute and the energy released as the solvent cages form. Figure 7.12 shows what this means on a progress of reaction diagram.

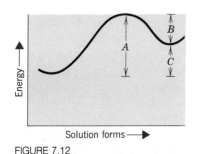

FIGURE 7.12
Energy relationships in a saturated solution at equilibrium for solid solutes that dissolve endothermically.
A = energy cost to break up the solute crystal
B = energy released as the solvent cages form
C = the net energy cost of forming the solution

When we heat a saturated solution of anything that dissolves endothermically, the stress we place on this equilibrium is absorbed, as required by Le Chatelier's principle, by a shift of the equilibrium to the right. Only a shift in this direction absorbs the additional heat. Thus more solute dissolves, and the solution becomes even more concentrated. As more and more solute particles move out into the solution at the higher temperature, the rate of their return to the undissolved state also picks up. Eventually, the two rates again become equal (although higher), and equilibrium is restored (assuming that undissolved solute is still present, of course).

A number of ionic compounds have solubilities that decrease with temperature. We already mentioned one in Chapter 5, $Ca(OH)_2$, but most of these compounds are sulfates, and most can exist as hydrates. Both the sulfate ion and the hydroxide ion strongly hydrogen-bond to water. One can only speculate, but perhaps higher temperatures interfere too much with this hydrogen-bonding to give these anions any advantage in stability by being out in solution instead of in the crystalline form.

As we learned in Section 5.5, it's sometimes possible to prepare a supersaturated solution by cooling a saturated solution that contains no undissolved solute. Without the undissolved solute already present to which dissolved solute particles can return, they sometimes simply remain in solution. The system is unstable, of course, as we mentioned on page 113. Just by adding a seed crystal, we can give the excess in solution a place to go, and the excess leaves the dissolved state.

The solubilities of gases in liquids always decrease with increasing temperature, as we saw in Table 5.1 on page 114. This is because the dissolving of gases in liquids is always an exothermic process. The following equilibrium represents this situation:

$$\text{Gas}_{\text{undissolved}} + \text{solution} \rightleftharpoons \text{more concentrated solution} + \text{heat of solution}$$

It must shift to the left in favor of undissolved gas when heat is added because, in accordance with Le Chatelier's principle, it's the only way that the stress — additional heat — can be absorbed by the system.

7.4 SOLUTIONS OF GASES IN WATER

Pressure, temperature, and sometimes reactions of gases with water affect the solubility of a gas in an aqueous solution.

Because we depend completely on the air we breathe to supply us with oxygen, we must give some time to the study of the factors that affect the solubilities of gases in water.

The Effect of Pressure on Gas Solubility.   Pressure is a factor that affects solubility only if the solute is a gas. As seen in Figure 7.13, the solubilities of two typical gases are directly proportional to the applied pressure. The equilibrium expression is

$$\text{gas}_{\text{undissolved}} + \text{solvent} \rightleftharpoons \text{solution} \qquad (7.2)$$

This equilibrium shifts to the right with increasing pressure because only such a change can absorb the volume-squeezing stress of extra pressure. (This is yet another illustration of Le Chatelier's principle.) Similarly, if we reduce the pressure above a liquid that has a dissolved gas, we have created a volume-expanding stress, and now equilibrium 7.2 shifts to the left. Dissolved gas now leaves the solution. If we apply both suction and heat to a solution of a gas, we very rapidly de-gas the solution.

William Henry (1775 – 1836) was the first to notice that gas solubility is directly proportional to gas pressure, so we now call this relationship **Henry's law** or the **pressure – solubility law.**

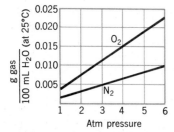

FIGURE 7.13
The solubilities of oxygen and nitrogen in water versus pressure.

Pressure–Solubility Law (Henry's law). The concentration of a gas in a liquid at any given temperature is directly proportional to the partial pressure of the gas on the solution.

Stated in the form of an equation, Henry's law says

$$C_g = k_g P_g$$

where k_g is a constant of proportionality, C_g is concentration and P_g is partial pressure.

The value of k_g *for a given gas at a given temperature* is a constant, and it does not change with the partial pressure. Because of this we can quickly change this equation to one that is easier to use in calculations, because it lets us avoid actually having to know k_g. The useful equation for Henry's law is as follows, when the gas stays the same and where the subscripts 1 and 2 refer to two different partial pressures:

$$\frac{C_1}{P_1} = \frac{C_2}{P_2} \qquad \text{(at constant temperature)} \tag{7.3}$$

(The two sides of 7.3 equal each other because they each equal the same Henry's law constant. This transformation is identical in type to what we did with Charles' law on page 136.)

EXAMPLE 7.1   USING HENRY'S LAW

Problem: At 760 mm Hg and 20 °C, the solubility of oxygen in water is 0.00430 g O_2/100 g H_2O. When air is itself saturated with water, the partial pressure of oxygen is 156 mm Hg. How many grams of oxygen dissolve in 100 g of water when the water is saturated with air and is in equilibrium with air that is saturated with water vapor?

Solution: The best idea is to collect the data to see what we have.

$$C_1 = 0.00430 \text{ g/100 g} \qquad C_2 = ?$$
$$P_1 = 760 \text{ mm Hg} \qquad P_2 = 156 \text{ mm Hg}$$

Now we can use Equation 7.3.

$$\frac{0.00430 \text{ g/100 g}}{760 \text{ mm Hg}} = \frac{C_2}{156 \text{ mm Hg}}$$

When we solve this for C_2, we get

$$C_2 = 0.00430 \text{ g/100 g} \times \frac{156 \text{ mm Hg}}{760 \text{ mm Hg}}$$

$$= 0.000883 \text{ g/100 g} \qquad \text{(answer)}$$

PRACTICE EXERCISE 1   How many grams of nitrogen dissolve in 100 g of water when the water is saturated with air and is in equilibrium with air that is saturated with water vapor? The partial pressure of nitrogen in air that is itself saturated with water vapor is 586 mm Hg. The solubility of pure nitrogen in water at 760 mm Hg is 0.00190 g/100 g H_2O.

The pressure–solubility relationship for solutions of gases in water becomes particularly relevant in dealing with people exposed to possible decompression sickness (the bends), as discussed in Special Topic 7.1.

Solubilities of Gases that React with Water.   A number of important gases are far more soluble in water than oxygen or nitrogen. They include carbon dioxide, sulfur dioxide, and ammonia. At 20 °C, only 0.0043 g of oxygen dissolves in 100 g of water, but carbon dioxide is nearly 40 times as soluble (0.169 g/100 g), sulfur dioxide is nearly 2500 times as soluble (10.6 g/100 g), and ammonia is a whopping 120,000 times as soluble (51.8 g/100 g) in water as oxygen. Behind these figures is the chemical fact that these gases do not just dissolve in water. A fraction of what dissolves reacts with water. When water contains these gases, the following chemical equilibria are involved in addition to the physical equilibria:

$$CO_2(aq) + H_2O \rightleftharpoons H_2CO_3(aq)$$
Carbonic acid

$$SO_2(aq) + H_2O \rightleftharpoons H_2SO_3(aq)$$
Sulfurous acid

$$NH_3(aq) + H_2O \rightleftharpoons NH_4^+(aq) + OH^-(aq)$$
Ammonium   Hydroxide
ion   ion

The forward reactions of these equilibria help draw the gases into solution and make them more soluble.

These gases are further assisted in dissolving in water because their molecules can form hydrogen bonds with water molecules. Figure 7.14 illustrates how hydrogen bonds exist when ammonia is the solute.

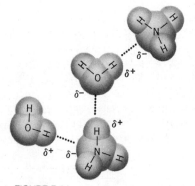

FIGURE 7.14

Hydrogen bonds (••••••) between molecules of ammonia and water help to keep ammonia in solution

Gas Tension.   In professional health care circles, the term *gas tension* is often used in referring to the solubility of a gas in water, particularly when the discussion is about the respiratory gases, oxygen and carbon dioxide, and their solubility in blood. **Gas tension** is defined as the partial pressure of a gas over a solution with which it is in equilibrium. It is an indirect measure of how much gas is in solution, because the more there is in solution the more there will be above the solution exerting a partial pressure. In other words, a high gas tension means a high availability of the gas from the solution.

In the fluids of the average human cell during rest, the oxygen tension is about 30 mm Hg. In the lungs, it is slightly over 100 mm Hg. Because gases tend to diffuse from a higher to a lower pressure, oxygen for this reason alone naturally tends to move from the lungs to the cells. (Of course, there are other mechanisms at work, too.) Carbon dioxide, on the other hand, has a gas tension in cells at rest of about 50 mm Hg, but in the lungs its gas tension is about 40 mm Hg. Thus this gas, a waste product, naturally tends to migrate from cells to the lungs where it can then be discharged in exhaled air. (Other mechanisms are also at work in the process of removing CO_2 from the body.)

7.5 PERCENTAGE CONCENTRATION

The number of grams of solute in 100 g of solution is the percentage concentration of the solution.

Many times in the lab, test tube tests have to be performed to detect some chemical. A solution is generally prepared for use in making the test. Such solutions are often referred to as **reagents,** and the test might consist of the addition of a few drops of the reagent to the substance being tested. We then look for some noticeable change that tells us that the substance we are trying to detect is present. The change might be the evolution of a gas, and we can either see the fizzing of the gas or detect its odor (if it has one). You can tell, for example, if a piece of rock is a carbonate rock such as limestone by dropping a few drops of hydrochloric acid on its surface. If fizzing occurs, and no significant change in odor accompanies it, the rock probably is a carbonate rock. The sample being tested is the rock; the reagent is a solution of hydrochloric acid; the reactant in this reagent is $HCl(aq)$; the odorless gas that causes the fizzing is CO_2; and the equation for the positive test is

$$CaCO_3 + 2HCl(aq) \longrightarrow CaCl_2(aq) + CO_2(g) + H_2O$$

Calcium   Hydrochloric   Calcium   Carbon
carbonate   acid   chloride   dioxide

There are hundreds of tests such as this where chemicals are mixed without elaborate care concerning precise stoichiometry on a mole basis. Therefore these tests do not have to be carried out with reagents for which we know molar concentrations. It wouldn't hurt to know this, but to obtain such information sometimes adds to the work without returning any special gain. Nonetheless, we do have to have some indication of the strengths or concentrations of the reagents, and here is where a number of percentage concentration expressions have been developed.

We now use percent *instead of* percentage *to conform to common usage.*

Weight/Weight Percent. The **weight/weight percent (w/w%) concentration** of a solution is the number of grams of solute in 100 g of the solution. For example, a 10.0% (w/w) glucose solution has a concentration of 10.0 g of glucose in 100 g of solution. To make 100 g of this solution, you would mix 10.0 g of glucose with 90.0 g of the solvent for a total mass of 100 g.

EXAMPLE 7.2   USING WEIGHT/WEIGHT PERCENTS

Problem: How many grams of 0.900% (w/w) NaCl contain 0.250 g of NaCl?

Solution: The concentration term, 0.900% (w/w), gives us the following two conversion factors:

$$\frac{0.900 \text{ g NaCl}}{100 \text{ g NaCl soln}} \quad \text{and} \quad \frac{100 \text{ g NaCl soln}}{0.900 \text{ g NaCl}}$$

These two ratios are just equivalent ways of understanding the concentration, and we should remind ourselves that any expression of a concentration in any units can be expressed as either of two ratios, such as we have done here. Next, we multiply the given, 0.250 g NaCl, by the second factor so that the final units will be g NaCl soln.

$$0.250 \text{ g NaCl} \times \frac{100 \text{ g NaCl soln}}{0.900 \text{ g NaCl}} = 27.8 \text{ g NaCl soln}$$

Thus 27.8 g of 0.900% (w/w) NaCl contains 0.250 g of NaCl.

EXAMPLE 7.3 PREPARING WEIGHT/WEIGHT PERCENT SOLUTIONS

Problem: A special kind of saline solution, called isotonic saline, is sometimes used in medicine. Its concentration is 0.90% NaCl (w/w). How would you prepare 750 g of such a solution?

Solution: Once again, we have to translate the label on the bottle, 0.90% (w/w) NaCl, into conversion factors.

$$\frac{0.90 \text{ g NaCl}}{100 \text{ g NaCl soln}} \quad \text{and} \quad \frac{100 \text{ g NaCl soln}}{0.90 \text{ g NaCl}}$$

What we basically have to determine is the number of grams of NaCl that we must weigh out and dissolve in water to make the final mass equal to 750 g. To find this number of grams of NaCl, we multiply the mass of the NaCl solution by the first conversion factor.

$$750 \text{ g NaCl soln} \times \frac{0.90 \text{ g NaCl}}{100 \text{ g NaCl soln}} = 6.8 \text{ g NaCl} \quad \text{(rounded from 6.75)}$$

Thus if we dissolve 6.8 g NaCl in water and add enough water to make the final mass equal to 750 g, we can write the label to read 0.90% (w/w) NaCl.

PRACTICE EXERCISE 2   Sulfuric acid can be purchased from a chemical supply house as a solution that is 96.0% (w/w) H_2SO_4. How many grams of this solution contain 9.80 g of H_2SO_4 (or 0.100 mol)?

PRACTICE EXERCISE 3   How many grams of glucose and how many grams of water are needed to prepare 500 g of 0.250% (w/w) glucose?

Sometimes weight/weight percent solutions are prepared by diluting a more concentrated solution, much as dilute solutions whose concentrations are in moles/liter are made by diluting more concentrated solutions. The equation that can be used to make the necessary calculations is similar to the one we derived in Section 5.7, and we will simply give the equation here. We leave the proof as an exercise, and some Review Exercises will require its use.

$$g_{\text{concd soln}} \times \text{percent (w/w)}_{\text{concd soln}} = g_{\text{dil soln}} \times \text{percent (w/w)}_{\text{dil soln}}$$

Volume/Volume Percent.   When a concentration is expressed as a **volume/volume percent,** it gives us the number of volumes of one substance that are present in 100 volumes of the mixture. This expression is often used for solutions of gases in each other or for solutions of liquids in each other. For example, the concentration of oxygen in air is 21% (v/v), which means that there are 21 volumes of oxygen in 100 volumes of air. The unit used for volume can be any unit, as long as we use the same unit for the constituent as for the solution. The units cancel when we deal with true percentages. Working problems involving volume/volume percents involve the same kinds of steps we used for problems of weight/weight percents.

Weight/Volume Percent.   Although this is a fairly common concentration expression, particularly among some health-care areas, a weight/volume percent isn't a true percent because the units don't cancel. When a concentration is given as a **weight/volume percent (w/v%),** it means the number of grams of solute in 100 mL of the solution. Thus a 0.90% (w/v) NaCl solution has 0.90 g of NaCl in every 100 mL of the solution.

Clinical report sheets sometimes specify units of gm/dL (or gm/dl) for the concentration of something in blood or urine. This means *grams per deciliter*, and because 1 dL = 100 mL, a concentration of, for example, 2.50 gm/dL is identical to 2.50% (w/v) as we have defined it.

Weight/volume percent problems are handled through conversion factors just as we did for weight/weight percent problems.

Milligram Percent, Parts per Million, and Parts per Billion. You'll occasionally encounter some special concentration expressions that are handy when the solutions are very dilute. **Milligram percent** means the number of milligrams of solute in 100 mL (1 dL) of the solution. For very dilute solutions, the concentration might be given in **parts per million (ppm),** which means the number of parts (in any unit) in a million parts (the same unit) of the solution. Parts per million might be interpreted as grams per million grams or pounds per million pounds. One ppm is analogous to one penny in a million pennies ($10,000) or 1 minute in a million minutes (about two years).

When water is the solvent, for which the volume of a sample in milliliters and the mass of the sample in grams are numerically the same (at least to two significant figures), it can be shown that 1 ppm is identical with 1 mg/L.

Parts per billion (ppb) similarly means parts per billion parts, such as grams per billion grams. This expression is used for extremely dilute systems. One ppb is like one penny in $10 million (a billion pennies), or like two drops of a liquid in a full, 33,000-gallon tank car.

Because of the many ways the term *percent* can be taken, there is a trend away from using it, which should be encouraged. Instead, the explicit units are given. Thus instead of referring to a concentration of, say, 10.0% (w/v) KCl, the label or the report should read 10.0 g KCl/100 mL or, better (because it takes less space), 10.0 g KCl/dL. When no units are given, just a percent, until you find out otherwise you have to interpret the percent as weight/weight if the solute is a solid when pure and as a volume/volume percent if the solute is a liquid when pure. Usually, however, the labels on fluids of clinical and medical importance are clear and, as always, *labels should be carefully read.*

7.6 OSMOSIS AND DIALYSIS

The migration of ions and molecules through membranes is an important mechanism for getting nutrients inside cells and waste products out.

Some properties of solutions and colloidal dispersions depend not so much on the chemical identities of the solutes but on nothing more than their concentrations. Such properties are called **colligative properties** (after the Greek *kolligativ,* depending on number and not on nature), and two examples of colligative properties are the depression of the freezing point and the elevation of the boiling point. For example, solutions generally have lower melting points and higher boiling points than their pure solvents. Thus aqueous solutions freeze not at 0 °C but at slightly lower temperatures. They also boil not at 100 °C but at slightly higher temperatures (both compared at 760 mm Hg).

The effects aren't very large unless the concentrations are very large, but they are related to concentrations, not chemical identities. For example, if we prepare two solutions, one that has 1.0 mole of NaCl in 1000 g of water and the other with 1.0 mole of KBr in 1000 g of water, each freezes at −3.4 °C (and not at 0 °C), and each boils at 101 °C (at 760 mm Hg). Both the freezing and the boiling points are the same, despite the difference in solutes, because the ratios of the moles of ions to the moles of water in both solutions are identical. Otherwise, the identities of the solutes are immaterial.

The temperature of a mixture made from 33 g NaCl and 100 g ice is about −22 °C (−6 °F).

When the concentrations are very high, the effects can be quite large and important. The whole basis for the use of an antifreeze mixture in an automobile radiator is the large depression of the freezing point of the radiator fluid caused by the presence of the antifreeze. A 50-50 mixture (vol/vol) of antifreeze in water gives protection to about −40 °F.

Time does not permit us to delve more deeply into these particular colligative properties. Instead, we will study a colligative property of the highest importance at the molecular level of life, osmosis, and a closely related phenomenon, dialysis. As we will see, osmosis and dialysis are quite sensitive even to low concentrations of solutes.

Osmosis.   Cells in the body are enclosed by cell membranes, and on both sides of cell membranes there are aqueous systems that involve substances in solution and colloidal dispersion. Materials and water have to be able to move through cell membranes in either direction so that nutrients can enter cells and wastes can leave.

The movements of Na⁺ and K⁺ ions into and out of cells is controlled by active transport mechanisms.

One of the means for moving ions and molecules of ordinary size through cell membranes is called **active transport.** *Active* refers to the active involvement of specialized materials in the membrane itself. These accept and pass on ions and molecules by endothermic chemical reactions. We can do no more than mention active transport here. The other means of passing things through membranes is spontaneous dialysis, and the simplest form of dialysis is the phenomenon called osmosis.

Membranes of cells are **semipermeable.** This means that they can let some but not all kinds of molecules and ions pass through. Cellophane, for example, is a synthetic, semipermeable membrane. When it is in contact with an aqueous solution, only water molecules and other small molecules and ions can migrate through it. Molecules of colloidal size are stopped. Evidently, there are ultrafine pores in cellophane that are just large enough to let small particles through but too small for larger particles.

Some membranes have pores so small that only water molecules can get through them. Because ions are hydrated, their effective sizes are apparently too large for the pores. No other molecules can pass either. A semipermeable membrane that is so selective that only the solvent molecules can get through is called an **osmotic membrane.**

Essentially no purely osmotic membranes occur naturally.

When two solutions that have different concentrations of particles are separated by an osmotic membrane, osmosis occurs. **Osmosis** is the net migration of solvent from the solution that has the lower concentration of solute into the solution that has the higher concentration. If it could continue long enough, the concentrated solution would become dilute enough and the other solution (by losing solvent) would become concentrated enough so that the two concentrations become equal. This way of thinking about the *direction* of the net flow of water in osmosis, incidentally, is perhaps the best way to remember how it goes. The net flow is such as to make the concentrated solution more dilute.

Figure 7.15 shows why there is a net flow in one direction in osmosis. It shows the special case in which pure water is on one side of the membrane. Water molecules can move in *both* directions, but when solute particles are on one side, they get in the way. Therefore water molecules are prevented from leaving as frequently from that side as they are able to come in from the opposite side where no solute particles get in the way. Therefore more water molecules enter the concentrated solution than leave it, and this solution becomes increasingly diluted. Eventually, the column of water shown in Figure 7.15 will exert a high enough back-pressure to prevent any further rise.

FIGURE 7.15
Osmosis and osmotic pressure *(a)* In the beaker, *A*, there is pure water and in the tube, *B*, there is a solution. An osmotic membrane closes the bottom of the tube. *(b)* A microscopic view at the osmotic membrane shows how solute particles interfere with the movements of water molecules from *B* to *A*, but not from *A* to *B*. *(c)* The level in *A* has fallen and that in *B* has risen because of osmosis. *(d)* To prevent osmosis, a back pressure would be needed, and the exact amount of pressure is the osmotic pressure of the solution in *B*.

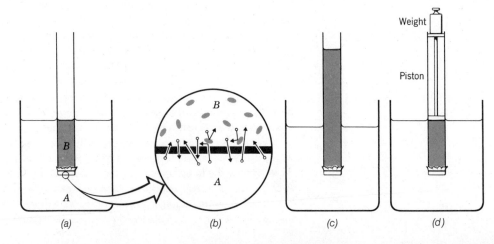

Π is the Greek capital letter pi.

Osmotic Pressure. The exact back pressure necessary to prevent osmosis is called the **osmotic pressure** of the solution, and its symbol is Π.

The value of osmotic pressure is directly proportional to the molar concentration of the particles in the solution (at least at relatively low molarities), and the equation for osmotic pressure is almost identical to the ideal gas equation, $PV = nRT$. For osmotic pressure, Π,

$$\Pi V = nRT$$

If we rearrange terms, and remember that $n/V =$ molarity (moles per liter), then

$$\Pi = \frac{n}{V} \times RT$$

Or,

$$\Pi = MRT \qquad (7.4)$$

Even in relatively dilute solutions, the osmotic pressure can be very high, as the following example illustrates.

EXAMPLE 7.4   CALCULATING OSMOTIC PRESSURE

Problem: A dilute solution, 0.100 M sugar in water, is separated from pure water by an osmotic membrane. What is its osmotic pressure at a temperature of 25 °C or 298 K?

Solution: $\Pi = \dfrac{nRT}{V} = \dfrac{0.100 \text{ mol}}{1000 \text{ mL}} \times 6.24 \times 10^4 \dfrac{\text{mm Hg mL}}{\text{mol K}} \times 298 \text{ K}$

$= 1.86 \times 10^3$ mm Hg

The solution has an osmotic pressure of 1.86×10^3 mm Hg. This means that a solution with this concentration can support a column of mercury 1.86×10^3 mm high, over 6 feet.

PRACTICE EXERCISE 4   What is the osmotic pressure of a 0.900 M glucose solution at 25 °C?

Water's density (1.00 g/mL) is much less than mercury's (13.6 g/mL), so the water column is 13.6 times higher.

If, instead of mercury, the column in Example 7.4 had been water (or the dilute solution), the column supported would have been 25.3 m (83.0 ft) high. Thus a relatively dilute solution, when separated from pure water, can be driven to a column height of several dozen feet. This phenomenon is one of the factors in the rise of sap in tall trees.

Osmolarity. The osmotic pressure of a solution has to be understood not as something that the solution is actually exerting, like some hand pushing on a surface. Instead, osmotic pressure is a potential that is directly related to the concentration of the solute particles. This potential is fully realized only in the very special circumstance in which an osmotic membrane separates the solution from pure water.

The particles can be ions, molecules, or macromolecules. When the solute is an ionic compound such as sodium chloride, the concentration of particles — ions in this situation — is twice the molar concentration of the salt given by the label on the bottle. For example, 0.10 M NaCl has a concentration of $2 \times (0.10) = 0.20$ mole of all ions per liter, because NaCl breaks up into two ions for each formula unit that goes into solution. Therefore the osmotic pressure of 0.10 M NaCl is twice as large as that of 0.10 M glucose, which does not break up into ions. For an ionic compound such as Na_2SO_4, for which three ions are released for each formula unit that dissolves — two Na^+ and one SO_4^{2-} — the concentration of particles in a 0.10 M solution is $3 \times (0.10) = 0.30$ mole of all ions per liter.

Because *molarity* doesn't reveal enough about a solution when we think about its osmotic pressure, scientists use the concept of **osmolarity** to express the concentration of all osmotically active particles in the solution. Thus a solution that is 0.10 M NaCl has a molarity of 0.10 mol/L but an osmolarity of 0.20 mol/L. The osmolarity of 0.10 M Na$_2$SO$_4$ is 0.30 mol/L. The concentration term in Equation 7.4, M, must refer to the osmolarity of the solution. The abbreviation used in labeling osmolarities is **Osm.** Thus 0.0125 Osm means that the osmolarity of this solution is 0.0125 mol/L of all of the particles that contribute to the osmotic pressure.

PRACTICE EXERCISE 5 Assuming that any *ionic* solutes in this exercise break up completely into their constituent ions when they dissolve in water, what is the osmolarity of each solution?

(a) 0.010 M NH$_4$Cl (which ionizes as NH$_4^+$ and Cl$^-$)
(b) 0.005 M Na$_2$CO$_3$ (which ionizes as 2Na$^+$ and CO$_3^{2-}$)
(c) 0.100 M fructose (a sugar and a molecular substance)
(d) A solution that contains both fructose and NaCl with concentrations of 0.050 M fructose and 0.050 M NaCl

---

Dialysis.   **Dialysis** is a phenomenon like osmosis except that in dialysis not only water molecules but also ordinary-sized ions and molecules can move through the membrane. A dialyzing membrane can be thought of as having larger pores than an osmotic membrane. Cell membranes are largely dialyzing membranes, but such membranes include substances that are able selectively to block the migration of even some small particles and to allow others to pass through.

Dialysis produces a net migration of water only if the fluid on one side of the dialyzing membrane has a higher concentration in colloidal substances than the other. Colloidal-sized particles are blocked by dialyzing membranes, so they get in the way of the movements of smaller particles through the membrane. The net flow of fluid in dialysis, as in osmosis, is from the side that has the lower concentration of colloidal substances to the side with the higher concentration. This imbalance in concentration that is related to colloidally dispersed materials causes a **colloidal osmotic pressure,** which is similar to osmotic pressure in its meaning. In the next section we present some situations involving life at the molecular level where osmotic pressure relationships are very critical and depend on the colloidal osmotic pressure of blood.

7.7   DIALYSIS AND THE BLOODSTREAM

When the osmotic pressure of blood varies too much, the result can be shock or harmful damage to red blood cells.

The body tries to maintain the concentrations of all of the substances that circulate in blood within fairly narrow limits. To do this, the body has a thirst mechanism for bringing in water and the machinery of diuresis for letting water leave. A number of hormones are involved in maintaining the integrity of blood, and in this section we will look briefly at just two situations that can arise when the osmotic pressure of blood changes too dramatically.

Shock Syndrome.   One feature of the shock syndrome is a dramatic increase in the permeability of the blood capillaries to colloidal-sized particles, particularly protein molecules. When these leave the blood, the colloidal osmotic pressure of blood falls, which is another way of saying that the concentration of colloidal materials in blood falls. The blood, in effect, becomes less concentrated and therefore less able to take up water from the spaces that surround the blood capillaries. A sufficient drop in the colloidal osmotic pressure of blood makes a net loss of water from blood possible. Such a loss of water means a loss of blood volume, and this upsets the elaborate mechanisms that bring nutrients to the brain cells and that carry wastes away. The result to the nervous system is called shock. When a person goes into shock, one of the many problems, but one that lies close to the central cause, is that blood capillaries become temporarily more permeable to the loss of macromolecules from blood.

Hemolysis.   Millions of red blood cells circulate in the bloodstream, and their membranes behave as dialyzing membranes. Within each red cell there is an aqueous fluid that contains dissolved and colloidally dispersed substances (Figure 7.16a). Even though the dispersed particles are too large to dialyze, they contribute to the colloidal osmotic pressure. Therefore they help to determine the direction in which dialysis occurs between the inside and the outside of the red cell. For example, if red cells are placed in pure water, enough fluid will migrate into the cells to make them burst open, as seen in Figure 7.16b. The rupturing of red cells is called **hemolysis,** and we say that the cells hemolyze. On the other hand, if we put red cells into a solution with an osmolarity higher than that inside the cells, dialysis occurs in the opposite direction—out of the cell and into the solution. Now the cells, losing fluid volume, shrivel and shrink, and this is called **crenation** (Figure 7.16c).

In some clinical situations, body fluids need replacement or nutrients have to be given by intravenous drip. It is very important that the osmolarity of the solution being added this way closely matches that of the fluid inside the red cells. Otherwise, hemolysis or crenation will occur.

Two solutions of equal osmolarity are called **isotonic solutions.** If one has a lower osmotic pressure than the other, the first is said to be **hypotonic** with respect to the second. A hypotonic solution has a lower osmolarity than the one to which it is compared. Red cells hemolyze if placed in a hypotonic environment, including pure water.

If one of the two solutions has a higher osmotic pressure than the other, the first is said to be **hypertonic** compared to the second. Thus 0.14 M NaCl is hypertonic with respect to 0.10 M NaCl. Red cells undergo crenation when they are in a hypertonic environment.

A 0.9% (w/w) NaCl solution—called **physiological saline solution**—is isotonic with respect to the fluid inside a red cell. Any solution that is administered into the bloodstream of a patient has to be isotonic in this way.

All of the topics we have studied in this and the preceding section are important factors in the operation of artificial kidney machines, which are discussed in Special Topic 7.2.

Red cell

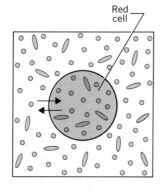 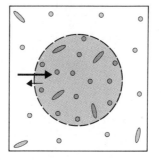

FIGURE 7.16
Dialysis. (a) The red cell is in an isotonic environment. (b) Hemolysis is about to occur because the red cell is swollen by extra fluids brought into it from its hypotonic environment. (c) The red cell experiences crenation when it is in a hypertonic environment.

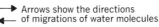

⊙ Ordinary-sized ions and molecules

⬭ Macromolecules

→ Arrows show the directions
← of migrations of water molecules

(a)   (b)   (c)

SPECIAL TOPIC 7.2 HEMODIALYSIS

The kidneys cleanse the bloodstream of nitrogen waste products such as urea and other wastes. If the kidneys stop working efficiently or are removed, these wastes build up in the blood and threaten the life of the patient. The artificial kidney is one remedy for this.

The overall procedure is called *hemodialysis* — the dialysis of blood — and the figure on the lower left shows how it works. The bloodstream is diverted from the body and pumped through a long, coiled cellophane tube that serves as the dialyzing membrane. (The blood is kept from clotting by an anticlotting agent such as heparin.) A solution called the *dialysate* circulates outside of the cellophane tube. This dialysate is very carefully prepared not only to be isotonic with blood but also to have the same concentrations of all of the essential substances that should be left in solution in the blood. When these concentrations match, the rate at which such solutes migrate out of the blood equals the rate at which they return. In this way several key equilibria are maintained, and there is no net removal of essential components. The figure on the lower right shows how this works. The dialysate, however, is kept very low in the concentrations of the wastes, so the rate at which they leave the blood is greater than the rate at which they can get back in. In this manner, hemodialysis slowly removes the wastes from the blood.

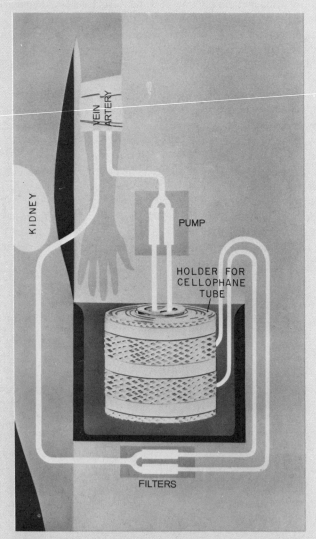

Schematic of an artificial kidney machine (Courtesy Artificial Organs Division, Travenol Laboratories, Inc.)

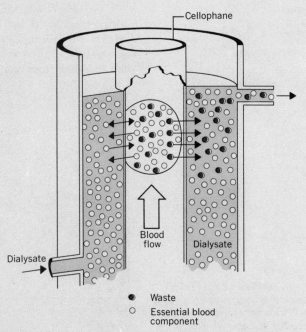

Waste molecules move out of the blood faster than they can return, but essential substances leave the blood and return at equal rates.

SUMMARY

Water The higher electronegativity of oxygen over hydrogen and the angularity of the water molecule make water a very polar compound — so polar that hydrogen bonds exist between the molecules. The geometry of water is successfully predicted by the valence-shell electron-pair repulsion (VSEPR) theory, one that also works for ammonia and methane. The high polarity of water explains many of its unusual thermal properties, such as its high heats of fusion and vaporization, its high surface tension, and its ability to dissolve ionic and polar molecular compounds.

Hydrogen bonds When hydrogen is covalently bonded to atoms of any of the three most electronegative elements (O, N, or F), its partial positive charge is large enough to be attracted rather strongly to the partial negative charge on an atom of O, N, or F on a nearby molecule. This force of attraction is the hydrogen bond. The most common errors are to think of this bond as the bond within a molecule of hydrogen, or to consider it as a covalent bond within some molecule. It's neither. The hydrogen bond is a force of attraction between the $\delta+$ on H in H—O, H—N, or H—F to the $\delta-$ of another O, N, or F.

Hydration The attraction of water molecules to ions or to polar molecules leads to a loose solvent cage that shields the ions or molecules from each other. This phenomenon is called hydration. Sometimes water of hydration is present in a crystalline material in a definite proportion to the rest of the formula unit, and such a substance is a hydrate. Heat converts most hydrates to their anhydrous forms. And some anhydrous forms serve as drying agents — desiccants.

Solutions Ions and molecules of ordinary size, if soluble in water at all, form solutions — homogeneous mixtures that neither gravity nor filtration can separate. The solubilities of most solids increase with temperature, because the process of their dissolving is usually endothermic. (More energy is needed to break up the crystal than is recovered as the solvent cages form about the ions or molecules.)

Gas solubilities Gases dissolve in water exothermically, so the addition of heat to an aqueous solution of a gas drives the gas out of solution. The solubility of a gas is directly proportional to its partial pressure in the space above the solution (Henry's law). Some gases do more than mechanically dissolve in water; part of what dissolves reacts with water to form soluble species.

Colloidal dispersions Large clusters of ions or molecules or macromolecules do not form true solutions but colloidal dispersions. These can reflect and scatter light (Tyndall effect), experience the Brownian movement, and (in time) succumb to the force of gravity (if the medium is fluid). Protective colloids sometimes stabilize these systems. If the dispersed particles grow to an average diameter of about 1000 nm, they slip over into the category of suspended matter and such systems must be stirred to maintain the suspension.

Percent concentration A variety of concentration expressions have been developed to provide ways of describing a concentration without going into molar concentrations. These include weight/weight percents, volume/volume percents, and hybrid descriptions that aren't true percentages — weight/volume percent, milligram percent, parts per million, and parts per billion.

Osmosis and dialysis When a semipermeable membrane separates two solutions or dispersions of unequal osmolarities, a net flow occurs in the direction that, if continued, would produce solutions of identical osmolarities. When the membrane is osmotic, only the solvent can migrate, and the phenomenon is osmosis. The back pressure needed to prevent osmosis is called the osmotic pressure, and it's directly proportional to the concentration of all particles of solute that are osmotically active — ions, molecules, and macromolecules.

When macromolecules are present, their particular contribution to the osmotic pressure is called the colloidal osmotic pressure of a solution. It is this factor that operates when the membrane is a dialyzing membrane. The permeability of blood capillaries changes temporarily when a person experiences shock, and macromolecules leave the blood. Their departure results in the loss of water, too, and the blood volume decreases.

Solutions of matched osmolarity are isotonic. Otherwise, one is hypertonic (more concentrated) with respect to the other, and the other is hypotonic (less concentrated). Only isotonic solutions, or those that are nearly so, should be administered intravenously.

KEY TERMS

The following terms were emphasized in this chapter, and their meanings should be mastered before continuing your study.

active transport	colloidal osmotic pressure	emulsion	homogeneous mixture
anhydrous form	crenation	gas tension	hydrate
Brownian movement	deliquescent	gel	hydration
colligative property	desiccant	hemolysis	hydrogen bond
colloidal dispersion	dialysis	Henry's law	hygroscopic

hypertonic

hypotonic

isotonic solution

macromolecule

osmolarity (Osm)

osmosis

osmotic membrane

osmotic pressure

parts per billion (ppb)

parts per million (ppm)

percent, milligram

vol/vol

wt/vol

wt/wt

physiological saline solution

pressure–solubility law

reagent

saturated solution

semipermeable

solution

surface active agent

surface tension

surfactant

suspension

tetrahedral

Tyndall effect

valence-shell electron-pair repulsion theory (VSEPR)

water of hydration

SELECTED REFERENCES

1 F. H. Stillinger. "Water Revisited." *Science*, July 25, 1980, page 45.

2 J. Walker. "Why Do Particles of Sand and Mud Stick Together When They Are Wet?" *Scientific American*, January 1982, page 174.

3 J. Sarquis. "Colloidal Systems." *Journal of Chemical Education*, August 1980, page 602.

4 T. Tanaka. "Gels." *Scientific American*, January 1981, page 124.

5 J. T. Riley. "Appetizing Colloids." *Journal of Chemical Education*, February 1980, page 153.

6 R. S. Treptow. "Le Chatelier's Principle Applied to the Temperature Dependence of Solubility." *Journal of Chemical Education*, June 1984, page 499.

7 K. J. Mysels. "Solvent Tension or Solvent Dilution." *Journal of Chemical Education*, January 1978, page 21.

8 G. B. Kolata. "Dialysis After Nearly a Decade." *Science*, May 2, 1980, page 473.

REVIEW EXERCISES

The answers to Review Questions that require a calculation and whose numbers are marked with an asterisk are given in an Appendix. The answers to the remaining Review Exercises are given in the *Study Guide* that accompanies this book.

VSEPR Theory

7.1   Describe in your own words how the VSEPR theory explains the bond angle in the water molecule.

7.2   If the bond angle in water were 180° instead of 104.5°, water would be nonpolar. Explain.

7.3   The boron trifluoride molecule, BF_3, has very polar B—F bonds. However, the molecule is nonpolar, and BF_3 is a gas at room temperature (b.p. = −100 °C). Which of the following geometries is most likely to be correct for BF_3? Explain how one can tell.

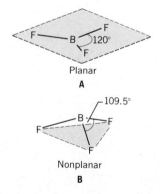

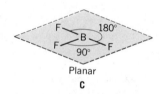

Planar

C

7.4   The electron-dot structure for BF_3 is as follows:

F:B:F
F̈

Notice that the central atom has only three pairs of electrons, not four, in its valence shell. Explain how the VSEPR theory leads to a prediction of the very geometry deduced in Review Exercise 7.3.

7.5   Both a regular planar structure and the tetrahedral structure for methane, CH_4, would have zero polarities. Explain.

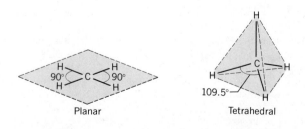

7.6   Explain why the methane molecule is tetrahedral, not planar.

7.7 The central atom, N, in NH_3 holds three atoms just like the central atom, B, in BF_3 (Review Exercise 7.3). Why, then, doesn't NH_3 have a planar geometry like BF_3?

Hydrogen Bond

7.8 The hydrogen molecule, H—H, does not become involved in hydrogen bonding.
   (a) What kind of bond occurs in a hydrogen molecule?
   (b) Why can't this molecule become involved in hydrogen bonding?

7.9 The methane molecule, CH_4, does not become involved in hydrogen bonding. Why not?

7.10 Solid sodium hydroxide, NaOH, includes two kinds of chemical bonds. Which ones are they, and how do they differ?

7.11 Draw a structure of two water molecules. Write in $\delta+$ and $\delta-$ symbols where they belong. Then draw a correctly positioned dotted line between two molecules to symbolize a hydrogen bond.

7.12 Hydrogen bonds exist between two molecules of ammonia, NH_3. Draw the structures of two ammonia molecules. (You don't have to try to duplicate their tetrahedral geometry.) Put $\delta+$ and $\delta-$ signs where they should be located. Then draw a dotted line that correctly connects two points to represent a hydrogen bond.

7.13 The hydrogen bond between two molecules of ammonia must be much weaker than the hydrogen bond between two molecules of water.
   (a) How do boiling point data suggest this?
   (b) What does this suggest about the relative sizes of the $\delta+$ and the $\delta-$ sites in molecules of water and ammonia?
   (c) Why are the $\delta+$ and the $\delta-$ sites different in their relative amounts of fractional electric charge when we compare molecules of ammonia and water?

7.14 If it takes roughly 100 kcal/mol to break the covalent bond between O and H in H_2O, about how many kilocalories per mole are needed to break the hydrogen bonds in a sample of liquid water?

7.15 Explain in your own words how hydrogen bonding helps us understand each of the following.
   (a) The high heats of fusion and vaporization of water
   (b) The high surface tension of water

7.16 Explain in your own words why water forms tight beads on a waxy surface but spreads out on a clean glass surface.

7.17 What does a surfactant do to water's surface tension?

7.18 What common household materials are surfactants?

7.19 What surfactant is involved in digestion? What juice supplies it? How does it aid digestion?

7.20 Bile comes from the gall bladder, and when a patient has the gall bladder removed, he or she is put on a diet that is relatively low in fats and oils. Why?

Homogeneous Mixtures

7.21 What does it mean when we describe a solution or a colloidal dispersion as *homogeneous*?

7.22 What is the basis for distinguishing among solutions, colloidal dispersions, and suspensions?

7.23 Which of the three kinds of homogeneous mixtures
   (a) Can be separated into its components by filtration?
   (b) Exhibits the Tyndall effect?
   (c) Shows observable Brownian movements?
   (d) Has the smallest particles of all kinds?
   (e) Is likeliest to be the least stable at rest over time?

7.24 What kinds of particles make the most stable colloidal dispersions? Explain.

7.25 The blood is simultaneously a solution, a colloidal dispersion, and a suspension. Explain.

7.26 Why won't a solution give the Tyndall effect?

7.27 What causes the Brownian movement?

7.28 What simple test could be used to tell if a clear, colorless solution contained substances in colloidal dispersion?

7.29 What is an emulsion? Give some examples.

7.30 What is a sol? Give some examples.

7.31 What is a gel? Give an example.

Aqueous Solutions

7.32 In a crystal of sodium chloride the sodium ions are surrounded by oppositely charged ions (Cl^-) as nearest neighbors. What replaces this kind of electrical environment for sodium ions when sodium chloride dissolves in water?

7.33 When we say that a chloride ion in water is *hydrated*, what does this mean? (Make a drawing as part of your answer.)

7.34 Carbon tetrachloride, CCl_4, is a liquid, and its molecules are tetrahedral, like those of methane, CH_4. This liquid does not dissolve in water. Why won't water let CCl_4 molecules in?

7.35 We have to distinguish between *how fast* something dissolves in water and *how much* can dissolve to make a saturated solution. The speed with which we can dissolve a solid in water increases if we (a) crush the solid to a powder, (b) stir the mixture, or (c) heat the mixture. Use the kinetic theory as well as the concept of forward and reverse processes to explain these facts.

7.36 Assuming that solid, undissolved sodium chloride is present, the rates of what two changes are equal in a saturated solution of sodium chloride in water? Write an equilibrium expression.

7.37 Suppose that you do not know and do not have access to a reference in which to look up the solubility of sodium nitrate, $NaNO_3$, in water at room temperature. Yet you need a solution that you know beyond doubt is saturated. How can you make such a saturated solution and know that it is saturated?

7.38 Explain why the solubility of a solid or a liquid in water generally increases with increasing temperature.

7.39 Ammonium chloride dissolves in water endothermically. Suppose that you have a saturated solution of this compound, that its temperature is 30 °C, and that undissolved solute is present. Write the equilibrium expression for this saturated solution, and use Le Chatelier's principle to predict what will happen if you cool the system to 20 °C.

Hydrates

7.40 Calcium sulfate dihydrate, $CaSO_4 \cdot 2H_2O$, loses all of its water of hydration at a temperature of 163 °C. Write the equation for this reaction.

7.41 Why are hydrates classified as compounds and not as wet mixtures?

7.42 When water is added to anhydrous copper(II) sulfate, the pentahydrate of this compound forms. Write the equation.

7.43 Anhydrous sodium sulfate is hygroscopic. What does this mean? Does this property make it useful as a desiccant?

7.44 Sodium hydroxide is sold in the form of small pellets about the size and shape of split peas. It is a very deliquescent substance. What can happen if you leave the cover off of a bottle of sodium hydroxide pellets?

7.45 When 6.29 g of the hydrate of compound M was strongly heated to drive off all of the water of hydration, the residue — the anhydrous form, M — had a mass of 4.97 g. What number should x be in the formula of the hydrate, $M \cdot xH_2O$? The formula weight of M is 136.

*7.46 When all of the water of hydration was driven off of 4.25 g of a hydrate of compound Z, the residue — the anhydrous form, Z — had a mass of 2.24 g. What is the formula of the hydrate (using the symbol Z as part of it)? The formula weight of Z is 201.

Gas Solubilities

7.47 The solubility of methane, the chief component in bunsen burner gas, in water at 20 °C and 1.0 atm is 0.025 g/L. What will be its solubility at 1.5 atm?

*7.48 At 20 °C the solubility of nitrogen in water is 0.0150 g/L when the partial pressure of the nitrogen is 580 mm Hg. What is its solubility when the partial pressure is raised to 800 mm Hg?

7.49 Explain why carbon dioxide is more soluble in water than is oxygen.

7.50 Explain why ammonia is much more soluble in water than nitrogen.

7.51 Using Le Chatelier's principle, explain why the solubility of a gas in water should increase with increasing partial pressure of the gas.

7.52 When the gas tension of CO_2 in blood is described as 30 mm Hg, what specifically does this mean?

7.53 If in one region of the body the gas tension of oxygen over blood is 80 mm Hg and in a second region it is 50 mm Hg, which region (the first or the second) has a higher concentration of oxygen in the blood itself?

Percent Concentrations

7.54 If a solution has a concentration of 1.2% (w/w) NaOH, what two conversion factors can we write based on this value?

7.55 A solution bears the label, 1.5% (w/v) NaCl. What two conversion factors can be written for this value?

7.56 The concentration of a pollutant in water is reported as 1.5 ppm. What is the concentration of this pollutant in units of mg/L?

7.57 A solution of wood alcohol in water is described as 12% (v/v). What two conversion factors are possible from this value?

7.58 How many grams of solute are needed to prepare each of the following solutions?
(a) 500 g of 0.900% (w/w) NaCl
(b) 250 g of 1.25% (w/w) $NaC_2H_3O_2$
(c) 100 g of 5.00% (w/w) NH_4Cl
(d) 500 g of 3.50% (w/w) Na_2CO_3

*7.59 Calculate the number of grams of solute needed to make each of the following solutions.
(a) 100 g of 0.500% (w/w) NaI
(b) 250 g of 0.500% (w/w) NaBr
(c) 500 g of 1.25% (w/w) $C_6H_{12}O_6$ (glucose)
(d) 750 g of 2.00% (w/w) H_2SO_4

7.60 How many grams of solute have to be weighed out to make each of the following solutions?
(a) 125 mL of 10.0% (w/v) NaCl
(b) 250 mL of 2.00% (w/v) KBr
(c) 500 mL of 1.50% (w/v) $CaCl_2$
(d) 750 mL of 0.900% (w/v) NaCl

*7.61 In order to prepare the following solutions, how many grams of solute are required?
(a) 250 mL of 5.00% (w/v) $Mg(NO_3)_2$
(b) 500 mL of 1.00% (w/v) NaBr
(c) 100 mL of 2.50% (w/v) KI
(d) 50.0 mL of 3.35% (w/v) $Ca(NO_3)_2$

7.62 How many milliliters of methyl alcohol have to be used to make 750 mL of 10.0% (v/v) aqueous methyl alcohol solution?

*7.63 A sample of 500 mL of 5.00% (v/v) aqueous ethyl alcohol contains how many milliliters of pure ethyl alcohol?

7.64 A chemical supply room has supplies of the following solutions: 5.00% (w/w) NaOH, 1.00% (w/w) Na_2CO_3, and 2.50% (w/v) glucose. If the densities of these solutions can be taken to be 1.00 g/mL, how many milliliters of the appropriate solution would you have to measure out to obtain the following quantities?
(a) 3.50 g of NaOH
(b) 0.250 g of Na_2CO_3
(c) 0.500 g of glucose
(d) 0.100 mol of NaOH
(e) 0.100 mol of glucose ($C_6H_{12}O_6$)

*7.65 The stockroom has the following solutions: 2.00% (w/w) KOH, 0.500% (w/w) HCl, and 0.900% (w/v) NaCl. Assuming that the densities of these solutions are all 1.00 g/mL, how many milliliters of the appropriate solution have to be measured out

to obtain the following quantities of solutes?
(a) 0.220 g of KOH   (b) 0.150 g of HCl
(c) 0.100 mol of NaCl   (d) 0.100 mol of KOH

7.66 A student needed 50.0 mL of 10.0% (w/w) aqueous sodium acetate, $NaC_2H_3O_2$. (The density of this solution is 1.05 g/mL.) Only the trihydrate of this compound, $NaC_2H_3O_2 \cdot 3H_3O$, was available, and the student knew that the water of hydration would just become part of the solvent once the solution was made. How many grams of the trihydrate would have to be weighed out to prepare the needed solution?

*7.67 How many grams of $Na_2SO_4 \cdot 10H_2O$ have to be weighed out to prepare 100 mL of 10.0% (w/w) Na_2SO_4 in water? (The density of this solution is 1.09 g/mL.)

7.68 A student has to prepare 500 g of 1.25% (w/w) NaOH. The stock supply of NaOH is in the form of 5.00% (w/w) NaOH. How many grams of the stock solution have to be diluted to make the desired solution?

*7.69 A 10.0% (w/w) HCl solution is available from the stockroom. How many grams of this solution have to be weighed out to prepare, by dilution, 250 g of 0.500% (w/w) HCl? If the density of the 10.0% solution is 1.05 g/mL, how many milliliters would provide the grams of the concentrated solution that are called for?

7.70 Concentrated hydrochloric acid is available as 11.6 M HCl. The density of this solution is 1.18 g/mL.
(a) Calculate the percent (w/w) of HCl in this solution.
(b) How many milliliters of this concentrated acid have to be taken to prepare 500 g of a solution that is 10.0% (w/w) HCl?

*7.71 Commercial nitric acid comes in a concentration of 16.0 mol/L. The density of this solution is 1.42 g/mL.
(a) Calculate the percent (w/w) of nitric acid, HNO_3, in this solution.
(b) How many milliliters of the concentrated acid have to be taken to prepare 250 g of a solution that is 10.0% (w/w) HNO_3?

Osmosis and Dialysis

7.72 If a solution that contains 1.00 mol of glucose in 1000 g of water freezes at -1.86 °C, what is the freezing point of a solution that contains 1.00 mol of glycerol in 1000 g of water? (Both are compounds that do not break up into ions when they dissolve.)

7.73 A solution that contains 1.00 mol of glucose in 1000 g of water has a normal boiling point of 100.5 °C. Another solution contains 1.00 mol of an unknown compound in 1000 g of water

has a normal boiling point of 101.0 °C. What is the likeliest explanation for the higher boiling point of the second solution?

7.74 Explain in your own words and drawings how osmosis gives a net flow of water from pure water into a solution on the other side of an osmotic membrane.

7.75 In general terms, how does an osmotic membrane differ from a dialyzing membrane?

7.76 Explain in your own words why the osmotic pressure of a solution should depend only on the concentration of its solute particles and not on their chemical properties.

7.77 The equation for osmotic pressure (Equation 7.4) shows that this pressure is directly proportional to the Kelvin temperature. Use the kinetic model of molecules and ions in motion and other aspects of the general kinetic theory to explain why the osmotic pressure should increase with an increase in temperature.

7.78 Why is the osmolarity of 1 M NaCl not the same as its molarity?

7.79 Which has the higher osmolarity, 0.10 M NaCl or 0.080 M Na_2SO_4? Explain.

*7.80 Which solution has the higher osmotic pressure, 10% (w/w) NaCl or 10% (w/w) NaI? Both NaCl and NaI break up in water in the same way—two ions per formula unit.

7.81 Solution A consists of 0.5 mol of NaCl, 0.1 mol of $C_6H_{12}O_6$ (glucose, a molecular substance), and 0.05 mol of starch (a colloidal, macromolecular substance), all in 1000 g of water. Solution B is made of 0.5 mol of NaBr, 0.1 mol of $C_6H_{12}O_6$ (fructose, a molecular substance related to glucose), and 0.005 mol of starch all in 1000 g of water. Which solution, if either, has the higher osmotic pressure? Explain.

7.82 What is the osmotic pressure (in mm Hg) of a 0.0100 M solution in water of a molecular substance at 25 °C?

*7.83 Calculate the osmotic pressure in mm Hg of a 0.0100 M solution in water at 25 °C of a compound that breaks up into two ions per formula unit when it dissolves.

7.84 What happens to red blood cells in hemolysis?

7.85 Physiological saline solution has a concentration of 0.9% (w/w) NaCl.
(a) Is a solution that is 1.1% (w/w) NaCl described as hypertonic or hypotonic with respect to physiological saline solution?
(b) What would happen, crenation or hemolysis, if a red blood cell were placed in (1) 0.5% (w/w) NaCl? (2) In 1.5% (w/w) NaCl?

7.86 Explain how the loss of macromolecules from the blood can lead to the increased loss of water from blood and a reduction in blood volume.

Chapter 8
Acids, Bases, and Salts

A highway snowblower is pumping sodium carbonate onto a nitric acid spill near downtown Denver following the rupture of a 20,000-gallon tank car in April, 1983. We learn how various bases can neutralize acids in this chapter.

8.1 SOURCES OF IONS

The principal ion-producers in water are acids, bases, and salts.

Almost all water contains dissolved ions, whether the water is in lakes and rivers or in the fluids of living systems. Almost all experiments done in the lab, and almost all clinical analyses involve substances that give ions in solution. Sometimes, the tiniest imbalances in the concentrations of certain ions in cells or in the blood cause the gravest medical emergencies.

We begin here a rather extensive study of ions, a study that spreads out over more than one chapter. In this chapter we will learn about the principal families of ionic compounds — acids, bases, and salts — and their chief reactions in aqueous systems.

Most of the chemical properties of ionic compounds are the properties of their individual ions. Right at the start, therefore, let's be sure we know how to distinguish among atoms, molecules, and ions. All three are extremely tiny kinds of particles, but atoms and molecules are *always* electrically neutral, and ions are *always* electrically charged. The symbols used for atoms or molecules and their corresponding ions are very similar, and in the long run this makes it easier to study them; but watch out for the differences. For example, Cu^{2+} and Cu are as different as night and day. Cu^{2+} is an ion; its hydrated form in water is bright blue. We need a trace of it in our bodies, but only a trace, because this ion is a dangerous poison. Cu, on the other hand, is the copper atom, the particle in copper metal, which is used to make pennies and electrical wires.

Another example is the difference between Na^+ and Na. Na^+, the sodium ion, is present almost everywhere in the body, and it has almost no chemical reactions. Na, the sodium atom of metallic sodium, reacts violently with water to give hydrogen gas (which usually ignites as it forms) plus a caustic solution of sodium hydroxide, household lye. The equation for this reaction is as follows:

$$2Na(s) + 2H_2O \longrightarrow 2NaOH(aq) + H_2(g)$$

Sodium   Sodium   Hydrogen
(metal)   hydroxide

Chlorine

Atom $:\overset{\cdot\cdot}{\underset{\cdot\cdot}{Cl}}\cdot$

Ion $:\overset{\cdot\cdot}{\underset{\cdot\cdot}{Cl}}:^-$

Molecule $:\overset{\cdot\cdot}{\underset{\cdot\cdot}{Cl}}:\overset{\cdot\cdot}{\underset{\cdot\cdot}{Cl}}:$

Level means concentration when we study solutions or other mixtures.

People in health-care fields often speak of the "sodium level of the blood," but what they *always* mean is the "sodium *ion* level of the blood." There are no metallic *atoms* of any element in the body, only their corresponding ions. (Dental fillings and the like don't count.)

$H \cdot$ = hydrogen atom
H^+ = hydrogen ion, a bare proton

Self-Ionization of Water. Before we study the families of compounds that supply ions in water, we have to learn about the ions that can form just from water molecules when they collide with each other. (See Figure 8.1.) When two water molecules collide powerfully enough, a transfer of a proton, H^+, occurs. Two ions form in a 1 : 1 ratio, the **hydronium** ion, H_3O^+, and the **hydroxide ion**, OH^-. Any reaction in which ions form from neutral molecules

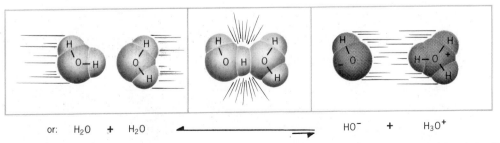

or: H_2O + H_2O HO^- + H_3O^+

FIGURE 8.1
The self-ionization of water. Extremely violent collisions are necessary, and at equilibrium the fraction of all collisions occurring at any instant that are energetic enough is very small.

FIGURE 8.2
Progress of reaction diagram for the self-ionization of water. The large quantity of energy needed to convert two moles of water into one mole each of hydronium ion and hydroxide ion means that these ions are present in very low concentrations at equilibrium.

is called **ionization,** and this self-ionization of water is actually the forward reaction in a chemical equilibrium:

$$2H_2O \rightleftharpoons H_3O^+(aq) + OH^-(aq)$$

Because we said it takes a powerful collision, you might have guessed that the energy of activation for the forward reaction is very high. The percentage of all of the collisions between water molecules that have this energy (or more), and that are correctly lined up, is very low. In other words, the forward reaction isn't favored, and at 25 °C the concentration of each product ion is only 1×10^{-7} mol/L. The reverse reaction — the transfer of a proton from H_3O^+ back to OH^- — is relatively easy. Its energy of activation is very low. Figure 8.2 gives a progress of reaction diagram for this equilibrium.

Although concentrations of 1×10^{-7} mol/L may seem too small to mention, life itself hinges on holding the concentrations of H_3O^+ and OH^- ions in body fluids at about 1×10^{-7} mol/L. We have already learned about the possibilities of shifting an equilibrium, and this one is no less susceptible to shifts. Some are life-threatening. Almost all of what we will be studying about acids, bases, and salts is essential to the study of how the body controls the equilibrium for the self-ionization of water and, by this, controls the concentrations of hydronium and hydroxide ions in body fluids.

Notice that in pure water the hydronium and hydroxide ions are always present in identical concentrations, and it is around changes in this feature, caused by acids, bases, and some salts, that we organize the families of ion-producing compounds.

The Arrhenius Acids and Bases.   A fairly large family of compounds called **acids** makes the concentration of hydronium ions in aqueous systems higher than that of hydroxide ions. Members of the family of **bases** or **alkalies** make the concentration of hydroxide ion greater than that of hydronium ion. The existence of ions such as these in water was first proposed by Svante Arrhenius (1859–1927), a Swedish scientist. He actually spoke of hydrogen ions, H^+, not hydronium ions, H_3O^+, but he had no way of knowing then that bare protons, which is what H^+ represents, are always piggybacked on something else in solution. Although protons can be *transferred* from one place to another, they have no independent existence as separate entities in solution. As explained in Special Topic 8.1, H^+ is always held by a covalent bond — a coordinate covalent bond — to a water molecule or something else.

Arrhenius's supposition about the hydrogen ion, H^+, isn't functionally too wide of the mark. It is so easily available (by a transfer process) from H_3O^+, that scientists to this day commonly use expressions such as *proton, hydrogen ion,* and *hydronium ion* interchangeably. We will use *hydrogen ion* as a convenient nickname for *hydronium ion* ourselves, and we will employ the symbol H^+ as a simpler way of writing H_3O^+. With these understandings, we can summarize the **Arrhenius definitions of acids and bases** as follows.

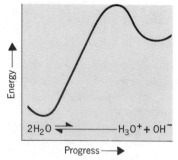

In 1903, Arrhenius won the third Nobel prize in chemistry for his theory about ions.

Arrhenius Acids and Bases

Acids are hydrogen ion producers in water.
Bases are hydroxide ion producers in water.

These definitions let us now make an important distinction among three kinds of aqueous solutions — acidic, basic, and neutral. We will use brackets, [ ], about a formula to specify that what they enclose refers to *molar concentration.*

Acidic solutions:	$[H^+] > [OH^-]$
Neutral solutions:	$[H^+] = [OH^-]$
Basic solutions:	$[H^+] < [OH^-]$

If we use electron-dot symbolism such as we employed in Section 4.5, we can visualize the formation of the water molecule as follows, where we use small x's to identify the valence-shell electrons of the oxygen atom. (We'll ignore the geometrical aspects for the moment.)

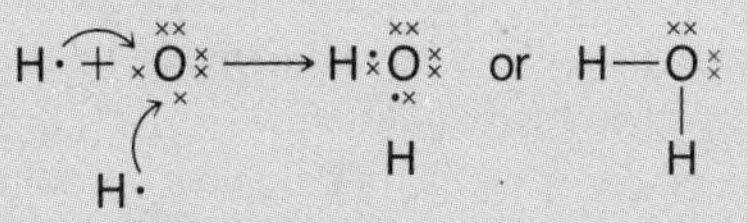

$$H \cdot + \overset{\times\times}{\underset{\times}{O}}{\times} \longrightarrow H \overset{\times\times}{\underset{\bullet\times}{O}}{\times} \quad or \quad H-\overset{\times\times}{\underset{|}{O}}{\times}$$
$$H\cdot \qquad\qquad H \qquad\qquad H$$

In the water molecule, oxygen still has two unshared pairs of electrons. If we now imagine a bare proton, H^+, coming in to one of these pairs, we can see how a coordinate covalent bond can form:

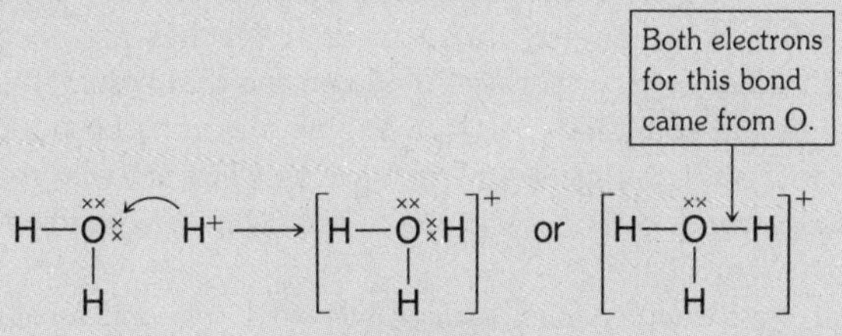

$$H-\overset{\times\times}{\underset{|}{O}}{\times} \quad H^+ \longrightarrow \left[H-\overset{\times\times}{\underset{|}{O}}{\times}H\right]^+ \quad or \quad \left[H-\overset{\times\times}{\underset{|}{O}}H\right]^+$$
$$H \qquad\qquad\qquad H \qquad\qquad\qquad H$$

Hydronium ion

> Both electrons for this bond came from O.

As the accompanying figure shows, the hydronium ion assumes an approximately tetrahedral geometry in accordance with VSEPR theory.

Once the coordinate covalent bond forms, the system completely forgets the origins of the bonds, and the O—H bonds in H_3O^+ are exactly equivalent — and all are relatively weak. Any of the three H's can break away as H^+ and transfer to some other particle that has an unshared pair of electrons and can function as a proton-acceptor. Because the energy needed for such a transfer is relatively small, we are justified in thinking of H_3O^+ as if it were the equivalent of H^+ and use their names and formulas interchangeably. As we will often see, this simplifies many equations.

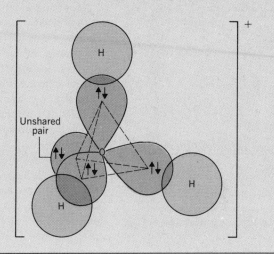

The third major family of ion-producing substances is the salts. **Salts** are ionic compounds in which the positive ion is a metal ion or any other positive ion except H^+, and the negative ion is any except OH^-. All salts are crystalline solids at room temperature, and many can be made in aqueous solution by the reaction of an acid with a base. When the acid and base are mixed in the right mole ratio, the resulting solution is neutral and $[H^+] = [OH^-]$. Therefore this and similar reactions are called **acid – base neutralizations.** For example, table salt, sodium chloride, is the product of the reaction of hydrochloric acid and sodium hydroxide. The base, sodium hydroxide, neutralizes the acid. (Or, with equal validity, we could say that the acid neutralizes the base.) The equation for this reaction is

$$HCl(aq) \quad + NaOH(aq) \longrightarrow NaCl(aq) + H_2O$$

Hydrochloric	Sodium	Sodium
acid	hydroxide	chloride

We will return to the family of the salts at the end of this chapter after we have studied the nature of acids and bases in more detail.

Electrolytes.   One property that is common to the aqueous solutions of many acids, bases, and salts is that they conduct electricity. Electrocardiograms can be obtained by attaching wires to the *outside* of the body only because the fluids in the skin and inside the body contain ions that can conduct very small electrical currents.

Electricity in metals is a flow of electrons. In a complete circuit, there is a battery or generator that forces the electrons to move, and the energy of this flow of electrons constitutes electrical energy. If the circuit is broken, the flow stops. The break might be an air gap at a switch, or it might be a gap filled with pure water, which is a nonconductor like air. Either way, there is no flow of electricity. If we add ions to the water, however, we close the gap again. The question now is, How can ions do this?

Unlike the situation in a metal, in a solution of ions subjected to an electric current, electrons do not move directly through the solution. Instead the dissolved ions move. (As we said earlier, the word *ion* is from the Greek *ienai,* which means to move.) Let's see how ions can give the effect of transporting electrons through a solution.

Figure 8.3 shows a typical setup. The plates or wires that dip into the solution are called **electrodes.** The battery forces electrons to one of these electrodes, called the **cathode** and it becomes electron-rich. Positive ions in solution naturally are attracted to the cathode, because opposite charges attract, and this is why positive ions are called **cations.** The electrons that make the cathode electron-rich are "pumped" from the other electrode, called the **anode,** which becomes electron-poor. Naturally, negative ions are attracted to the anode, which is why negative ions are called **anions.**

When a cation—a positive ion—arrives at the cathode, it picks an electron off of the cathode's electron-rich surface. When an anion—a negative ion—reaches the anode, it deposits an electron at the anode's electron-poor surface. You can see that if something is taking electrons from one electrode and something (else) is putting electrons on the other electrode, the net effect is the same as if the electrons themselves were actually moving through the solution. This is how ion-producers in water make the solution a conductor of electricity.

The passage of an electric current through a fluid is called **electrolysis,** and substances that permit electrolysis are called **electrolytes.** For electrolysis to happen, ions must be mobile. When ions are immobilized in the solid state, no electrolysis occurs. If the solid, ionic compound is heated until it melts, then the ions become mobile, and molten salts conduct electricity. Thus the term *electrolyte* can refer either to a solution of ions or to the pure, solid ionic compound. (The term does not apply to metals. Metals that conduct electricity are called *conductors.*)

Electrolytes are not equally good at enabling the flow of electricity. A "good" electrolyte is one that even in a quite dilute solution helps a strong electrical current to flow. It's one that can readily supply the carriers, the ions. Good electrolytes, in this sense, are called **strong electrolytes,** which means that in aqueous solutions essentially 100% of their formula units have broken up or separated into ions. Sodium hydroxide, hydrochloric acid, and sodium chloride are all strong electrolytes as are most of the common bases, acids, and water-soluble salts.

A **weak electrolyte** is a substance that breaks up into ions in water to a small percentage. A typical example is acetic acid, the acid that makes vinegar sour. Although 1 M NaCl is an excellent conductor, 1 M acetic acid is a poor conductor. Aqueous solutions of ammonia are also poor conductors, so aqueous ammonia is classified as a weak electrolyte.

Many substances are **nonelectrolytes,** whether they are in the liquid state or in solution. They do not conduct ordinary currents of electricity (e.g., household currents) at all. Pure water is an example of a nonelectrolyte, and alcohol and gasoline are others. We can summarize the relationships we have just studied as follows; be sure to notice the emphasis on *percentage* ionization as the feature that dominates these definitions.

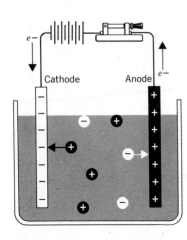

● Cation   ⊖ Anion

FIGURE 8.3
Electrolysis. Cations, positive ions, migrate to the cathode and remove electrons. Anions, negative ions, migrate to the anode and deposit electrons. The effect is a closed circuit.

Strong electrolyte	One that is strongly ionized in water — a high percentage ionization
Weak electrolyte	One that is weakly ionized in water — a low percentage ionization
Nonelectrolyte	One that does not ionize in water — essentially zero percentage ionization

8.2 THE COMMON AQUEOUS ACIDS AND BASES

Hydrochloric acid, sulfuric acid, and phosphoric acid are common acids and sodium hydroxide is a common base.

It's a good idea in the lab *never* to taste a chemical unless the instructor says that it's all right.

Because one of the ions, the hydronium ion, is present in aqueous solutions of all acids, all acids share some common properties. They have tart tastes, for example. The tartness of lemon juice is caused by citric acid. Acids also change the color of litmus, a dye, from blue to red, so when this dye is soaked into porous paper strips — litmus paper — it is easily used to test to see if a solution is acidic. We'll now survey some common substances that are acids.

The concentrated $HCl(aq)$ of commerce is 12 M, which is 37% (w/w) in HCl — a saturated solution of $HCl(g)$ in water. Handle it very carefully. It can cause several chemical burns.

Hydrochloric Acid.   You can buy hydrochloric acid, $HCl(aq)$, in hardware and plumbing supply stores (where it's likely to be called *muriatic acid*). Handle it very carefully and avoid letting either the liquid or its sharp, stinging fumes touch your skin, clothing, or eyes.

Hydrochloric acid is made by dissolving the gas, hydrogen chloride, $HCl(g)$, in water. $HCl(g)$ reacts promptly and essentially completely with the water to give a solution of hydronium ions and chloride ions. It is this solution to which the name *hydrochloric acid* belongs.

$$HCl(g) + H_2O \longrightarrow \underbrace{H_3O^+(aq) + Cl^-(aq)}$$

Hydrogen
chloride   Hydrochloric acid, HCl (aq)

When we use the formula $HCl(aq)$ in an equation or when we speak of hydrochloric acid, we always mean the separated ions, $H_3O^+(aq)$ and $Cl^-(aq)$.

Figure 8.4 shows how a collision between a molecule of hydrogen chloride and a water molecule allows the transfer of H^+ from $H—Cl(g)$ to H_2O. As explained in the figure, the curved arrows signify the movements of particles or of electron pairs. Equilibrium arrows of unequal length are used to indicate that the substances to which the longer arrow points are favored. (Later, we'll learn about a way of using numbers to tell us by how much something is favored in an equilibrium.) The covalent bond in $H—Cl(g)$ is relatively weak, weaker than the

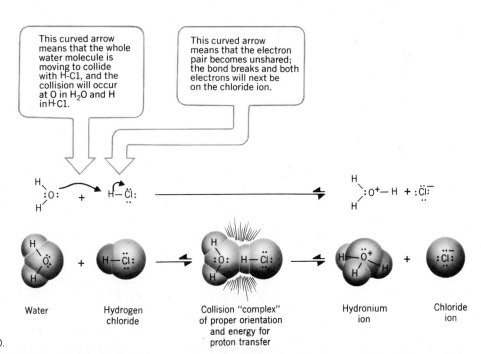

FIGURE 8.4
The reaction of $HCl(g)$ with water to make hydrochloric acid, $HCl(aq)$. The proton initially bound to Cl easily transfers to the O of a molecule of H_2O.

H—O bonds in H_3O^+, so the transfer occurs readily. Like most reactions that generate stronger bonds from weaker bonds, this reaction is exothermic.

Hydrochloric acid is called a **monoprotic acid** because the ratio of H_3O^+ ions to Cl^- ions is 1 to 1. Other monoprotic acids are those that can be made by dissolving the hydrogen halide gases in water — hydrofluoric acid, HF(aq), hydrobromic acid, HBr(aq), and hydriodic acid, HI(aq).

Hydrofluoric acid is unusual because its ions, H_3O^+ and F^-, attract each other so strongly in solution that it behaves as if it were a weak acid. Otherwise, these HX(aq) acids, where X = a halogen atom, are all strong acids. A **strong acid** ionizes essentially 100% in solution. A **weak acid** ionizes only to a few percent or less in solution. The terms *strong* and *weak* have the same connection to percentage ionization when they are applied to acids as when they are applied to electrolytes. All strong acids are strong electrolytes. All weak acids are weak electrolytes.

The general symbol for all aqueous hydrohalogen acids is HX(aq), where X may be F, Cl, Br, or I.

PRACTICE EXERCISE 1   Following the pattern that we used for the equilibrium in hydrochloric acid — the top equation in Figure 8.4 — write the expressions for the equilibria in HBr(aq) and HI(aq).

PRACTICE EXERCISE 2   We described hydrofluoric acid, HF(aq), as a weak acid. What is the equilibrium expression for the equilibrium present in this substance? Pay particular attention to the direction in which the longer of the two equilibrium arrows points.

The concentrated HNO_3(aq) of commerce is 16 M, which is 71% (w/w) in HNO_3. Handle it very carefully. It not only causes severe chemical burns, it turns the skin yellow on even a very brief contact.

Nitric Acid.   Another important monoprotic acid is nitric acid, HNO_3(aq). We can indicate the equilibrium in nitric acid as follows. The use of the curved arrows is explained in Figure 8.4.

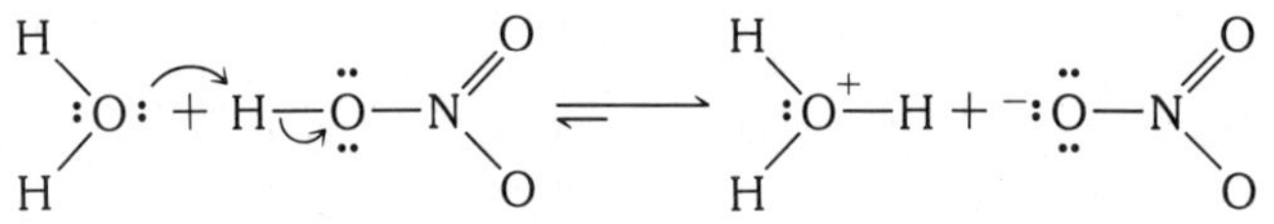

or

$$H_2O \ + \ HNO_3 \ \rightleftharpoons \ H_3O^+ \ + \ NO_3^-$$

Nitric acid   Hydronium ion   Nitrate ion

When nitric acid is selected for use in the lab, the main purpose is to use the nitrate ion, which is an oxidizing agent, rather than the hydronium ion.

Acetic Acid.   Acetic acid is a typical weak monoprotic acid, and also a typical organic acid. **Organic compounds,** which number in the millions, are the compounds of carbon other than the oxides of carbon, the cyanides, or the carbon compounds that are related to earth-like substances such as carbonate rocks and other carbonates. Compounds that are not organic compounds are called **inorganic compounds.**

The acetic acid molecule, $HC_2H_3O_2$, has four hydrogen atoms, but only one is attached to an oxygen atom. Only this one can transfer to a water molecule, and a powerful collision is needed. (See Figure 8.5.) The H—O bond in acetic acid is considerably stronger than the

Vinegar is about 5% (w/w) acetic acid in water.

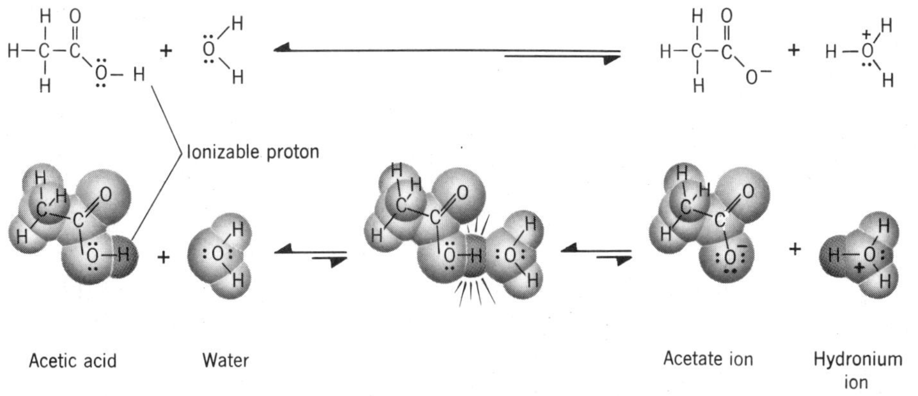

FIGURE 8.5
The ionization of acetic acid in water. Only a small fraction of the collisions are energetic enough at equilibrium to make a proton transfer from O in acetic acid to O in H$_2$O.

Acetic acid   Water   Acetate ion   Hydronium
                                                                                             ion

H—Cl bond in HCl(g), so acetic acid is a weaker acid than hydrochloric acid. We can represent the equilibrium present in aqueous acetic acid as follows:

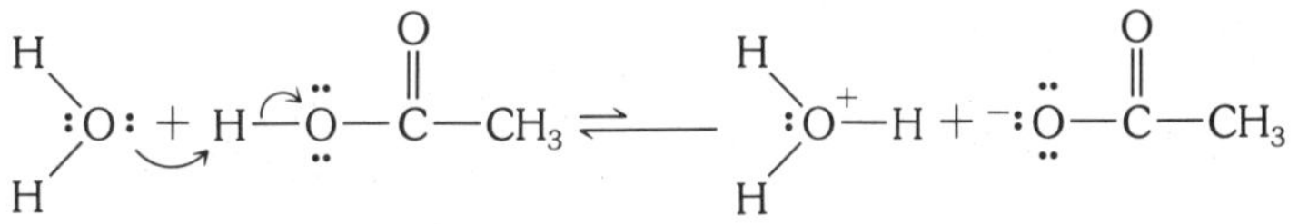

Or,

$$H_2O + H-C_2H_3O_2 \rightleftharpoons H_3O^+ + C_2H_3O_2^-$$

             Acetic acid   Acetate ion

We will continue to use HC$_2$H$_3$O$_2$ as our symbol for acetic acid in this chapter and the next. Just remember that only one hydrogen is active in acid-base reactions, and that acetic acid is monoprotic.

Sulfuric acid must be handled very carefully. See Table 8.1.

Sulfuric Acid.   Sulfuric acid, H$_2$SO$_4$(aq), is the only common aqueous **diprotic acid,** one that gives two hydronium ions per formula unit in solution. The ionization of the first proton is so easy that sulfuric acid is a strong acid.

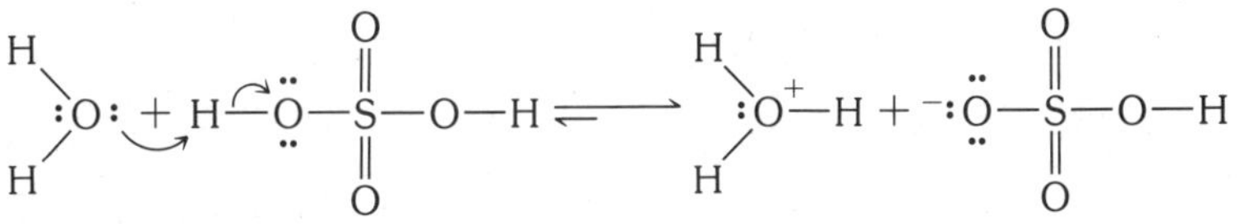

             Sulfuric acid   Hydrogen sulfate
                                                         ion

or,

$$H_2O + H_2SO_4 \rightleftharpoons H_3O^+ + HSO_4^-$$

             Sulfuric acid   Hydrogen sulfate
                                                         ion

The hydrogen sulfate ion is also an acid, but the transfer of H$^+$ away from it to a water molecule requires the movement of a positively charged particle away from one that already is oppositely charged. Although this is harder than the transfer of the first H^+ from H_2SO_4, it still happens.

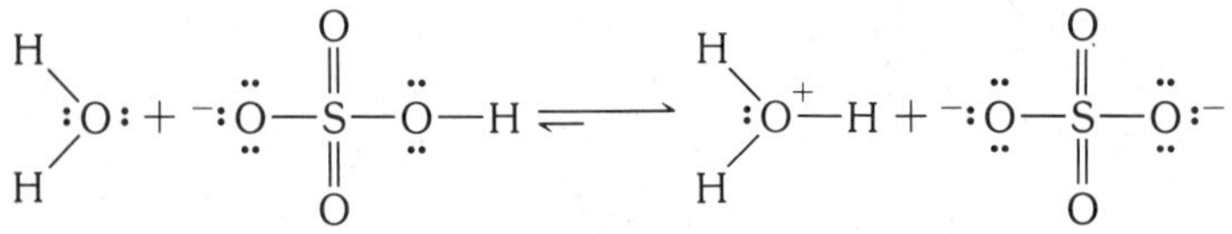

or,

$$H_2O \; + HSO_4^- \quad\rightleftharpoons\quad H_3O^+ \quad + SO_4^{2-}$$

Hydrogen sulfate   Sulfate ion
ion

Phosphoric Acid.   Phosphoric acid, H_3PO_4, is the only common, inorganic, **triprotic acid** — one that can release three H^+ ions. Like the ionization of sulfuric acid, that of phosphoric acid occurs in steps, each one more difficult than the previous.

Salts of phosphoric acid are important fertilizers. One salt, Na_3PO_4, or trisodium phosphate, is a powerful cleanser, but wear rubber gloves when using it.

$$H_2O + H_3PO_4(aq) \rightleftharpoons H_3O^+(aq) + H_2PO_4^-(aq)$$

Phosphoric   Dihydrogen
acid   phosphate ion

$$H_2O + H_2PO_4^-(aq) \rightleftharpoons H_3O^+(aq) + HPO_4^{2-}(aq)$$

Monohydrogen
phosphate ion

$$H_2O + HPO_4^{2-}(aq) \rightleftharpoons H_3O^+(aq) + PO_4^{3-}(aq)$$

Phosphate ion

Phosphoric acid is classified as a *moderate* acid. Its percentage ionization is roughly 27% in a dilute solution, which is too low to be a strong acid but high enough to make it a good conductor of electricity in water. It is important to learn the names and formulas of the three ions available from phosphoric acid, because the phosphate ion system occurs widely in the body.

Carbonic Acid.   Carbonic acid, H_2CO_3, is a weak diprotic acid that is unusual because it is unstable. However, the fact that it is unstable becomes an important property when the body has to manage one of the respiratory gases, carbon dioxide. As we learned in Section 7.4, when carbon dioxide dissolves in water, some of it reacts with water to form carbonic acid.

$$CO_2(aq) + H_2O \rightleftharpoons H_2CO_3(aq)$$

Carbonic acid

Then a small fraction of the carbonic acid molecules ionizes.

$$H_2CO_3(aq) + H_2O \rightleftharpoons H_3O^+ + HCO_3^-(aq)$$

Bicarbonate ion

The bicarbonate ion ionizes to a very slight extent in water.

$$HCO_3^-(aq) + H_2O \rightleftharpoons H_3O^+ + CO_3^{2-}(aq)$$

Carbonate ion

Special Topic 8.2 describes the molecular structures of carbonic acid and its anions.
The bicarbonate ion is the principal form in which waste carbon dioxide is transported from cells of the body where it is made to the lungs for disposal by means of exhaled air. In the lungs, reactions occur that change the bicarbonate ion back to carbonic acid, which breaks

SPECIAL TOPIC 8.2 THE CARBONIC ACID SYSTEM

In the molecule of carbonic acid, the two hydrogens are held by oxygens. However, they are held less strongly than the two hydrogens on oxygen in the water molecule because of the group in the middle of the H_2CO_3 molecule. This is the carbon-oxygen double bond, $C=O$, and it is an electro-negative group. Its oxygen atom makes it so, and this group can pull some electron density away from the oxygen atoms of the H—O groups in carbonic acid. This weakens the H—O bonds, so the hydrogens more easily transfer to water molecules during a collision such as the following:

Carbonic acid ⇌ Bicarbonate ion

Now that there is a negative charge (on the bicarbonate ion), the remaining hydrogen is much harder to release, but the following collision, if it is violent enough, can bring about the transfer of a proton as indicated.

Bicarbonate ion ⇌ Carbonate ion

Carbonic acid.

Bicarbonate ion.

Carbonate ion.

down under the influence of an enzyme into carbon dioxide and water. The carbon dioxide is then expelled.

The enzyme, called carbonic anhydrase, which breaks down carbonic acid in the lungs, only changes the rate of the decomposition of an already unstable acid from rapid to extremely rapid. You observe this decomposition at what seems to be a very rapid rate everytime you open a soft drink. When you release the internal pressure in any carbonated beverage by opening its container, dissolved carbon dioxide leaves the solution. And more continues to come as carbonic acid in the beverage breaks down. All this action involves the reversal of several equilibria, and it is triggered simply by reducing the pressure above the solution of CO_2 in water.

Club soda and beverages such as Perrier water are little more than solutions of carbon dioxide in water. Traces of salts are also present.

PRACTICE EXERCISE 3 Sulfurous acid, $H_2SO_3(aq)$, is a moderately strong, diprotic acid, but like carbonic acid it is also unstable. When sulfur dioxide dissolves in water, some of it reacts with the water to generate sulfurous acid, just like the reaction of CO_2 with water. The equilibrium is

$$SO_2(aq) + H_2O \rightleftharpoons H_2SO_3(aq)$$

Write the equilibrium equations for the successive ionizations of this weak acid, and give proper attention to the way in which the longer arrows point as well as to their relative lengths.

TABLE 8.1
Common Acids[a]

Acid	Formula	Percentage Ionization
Strong Acids		
Hydrochloric acid	HCl	>90
Hydrobromic acid	HBr	>90
Hydriodic acid	HI	>90
Nitric acid	HNO_3	>90
Sulfuric acid[b]	H_2SO_4	>60 (in 0.05 M solution)
Moderate Acids		
Phosphoric acid	H_3PO_4	27
Sulfurous acid[c]	H_2SO_3	20
Weak Acids		
Nitrous acid[c]	HNO_2	1.5
Acetic acid	$HC_2H_3O_2$	1.3
Carbonic acid[c]	H_2CO_3	0.2

[a] Data are for 0.1 M solutions of the acids in water at room temperature.

[b] *Concentrated* sulfuric acid (99%) is particularly dangerous not only because it is a strong acid but also because it is a powerful dehydrating agent. This action generates considerable heat at the reaction site, and at higher temperatures sulfuric acid becomes even more dangerous. Moreover, concentrated sulfuric acid is a thick, viscous liquid that does not wash away from skin or fabric very quickly.

[c] An unstable acid.

Table 8.1 summarizes the common aqueous acids. You should memorize the names and formulas of all of the strong and moderate acids on this list. There are very few of them, and once they are learned, you can be fairly certain that any unfamiliar acid you encounter will be a weak acid. It's easier to learn a few strong and moderate acids than several hundred weak acids.

The Common Aqueous Bases.   Dilute aqueous solutions of the common bases have a bitter, soapy taste, and they all turn red litmus blue. These are properties of the hydroxide ion. Table 8.2 lists the common bases.

TABLE 8.2
Common Bases

Base	Formula	Solubility[a]	Percentage Ionization
Strong Bases			
Sodium hydroxide	NaOH	109	>90 (0.1 M solution)
Potassium hydroxide	KOH	112	>90 (0.1 M solution)
Calcium hydroxide (The saturated, aqueous solution is called limewater)	$Ca(OH)_2$	0.165	100 (saturated solution)
Magnesium hydroxide (The saturated, aqueous solution is called milk of magnesia)	$Mg(OH)_2$	0.0009	100 (saturated solution)
Weak Base			
Aqueous ammonia	NH_3	89.9	1.3 (18 °C)[b]

[a] Solubilities are in grams of solute per 100 g of water at 20 °C except where otherwise noted.

[b] The ionization referred to here is the reaction:

$$NH_3(aq) + H_2O \rightleftharpoons NH_4^+(aq) + OH^-(aq)$$

Two important **strong bases** — those that ionize nearly 100% in water — are sodium hydroxide, NaOH, and potassium hydroxide, KOH.

$$NaOH(s) \xrightarrow{\text{water}} Na^+(aq) + OH^-(aq)$$

$$KOH(s) \xrightarrow{\text{water}} K^+(aq) + OH^-(aq)$$

Two other strong bases are of group IIA metals. These are magnesium hydroxide, $Mg(OH)_2$, and calcium hydroxide, $Ca(OH)_2$. They ionize essentially 100% in water, but as the data in Table 8.2 show, they are so insoluble in water that even saturated solutions provide only very dilute solutions of hydroxide ions.

$$Ca(OH)_2(s) \rightleftharpoons Ca^{2+}(aq) + 2OH^-(aq)$$

$$Mg(OH)_2(s) \rightleftharpoons Mg^{2+}(aq) + 2OH^-(aq)$$

Whereas both sodium and potassium hydroxides can be prepared in solutions concentrated enough to give severe chemical burns to the skin, calcium and magnesium hydroxide are very mild. Calcium hydroxide is a component of one commercial antacid tablet, and a slurry of magnesium hydroxide in water, called milk of magnesia, is used as an antacid and a laxative without posing any danger of chemical burn.

The only common weak base is a solution of ammonia in water called aqueous ammonia. When ammonia dissolves in water, a small percentage of its molecules react with water, as we learned in Section 7.4.

You'll sometimes see aqueous ammonia called "ammonium hydroxide," but it's better not to use this term. NH_4OH is unknown as a pure compound.

$$\underset{\text{Ammonia}}{NH_3(aq)} + H_2O \rightleftharpoons \underset{\text{Ammonium ion}}{NH_4^+(aq)} + OH^-(aq)$$

A dilute solution (about 5%) of ammonia in water is sold as household ammonia in supermarkets. It's a good cleaning agent, but watch out for its fumes.

The names and formulas of the bases in Table 8.2 should also be memorized.

8.3 THE CHEMICAL PROPERTIES OF AQUEOUS ACIDS AND BASES

Because they all contribute a common ion, H_3O^+, the common aqueous acids have a number of similar chemical properties.

In this section we will study the reactions of the hydronium ion, H_3O^+, and we'll use the symbol H^+ for this ion in most of the equations that we will write. The anions of the acids that furnish H_3O^+ have their own chemical reactions, of course, but we'll not be concerned about them here. Before we study any particular reactions, we will learn how to write a special kind of equation that is helpful when we work largely with ions instead of molecules or other formula units.

Molecular and Net Ionic Equations.   When we want to study the reactions of just one ion, it is useful to work with a new kind of equation called the *net ionic equation*. We'll develop such an equation for the reaction of hydrochloric acid with sodium hydroxide to illustrate it and to show its value.

We say *molecular equation* even though some of the chemicals in the equation might be ionic.

The conventional equation for a reaction is called a **molecular equation,** whenever it shows all of the substances in the formulas we would need if we wanted to plan an actual experiment — their full chemical formulas. The molecular equation for the reaction of hydrochloric acid and sodium hydroxide is

$$HCl(aq) + NaOH(aq) \longrightarrow NaCl(aq) + H_2O$$

However, as we now know, HCl(aq) is really $H^+(aq)$ and $Cl^-(aq)$; and NaOH(aq) is actually $Na^+(aq)$ and $OH^-(aq)$. We know this because we know the acid is a *strong* acid and the base is a *strong* base, so both must be essentially fully ionized in solution. We don't have nonionized molecules of HCl or NaOH in the solution. We also know that NaCl(aq) is fully ionized because it's a strong electrolyte. Besides, when we put (aq) by the formula of a *salt*, it has to mean that the salt is in *solution*. No salt goes into solution in water without separating almost entirely into its ions — at least, we'll use this as a working assumption. The fourth formula in the equation is that of water, and because water is a nonelectrolyte, its molecules aren't separated into ions (except to that infinitesimal degree involved in its self-ionization).

With these facts about the status of the reactants and products in the solution, we can expand the molecular equation into what is called the **ionic equation.** We do this by replacing anything we know is present as ions by the actual formulas of these ions. Thus the ionic equation for our example is

$$\underbrace{H^+(aq) + Cl^-(aq)}_{\text{These came from HCl }(aq)} + \underbrace{Na^+(aq) + OH^-(aq)}_{\text{These came from NaOH }(aq)} \longrightarrow \underbrace{Na^+(aq) + Cl^-(aq)}_{\text{These came from NaCl }(aq)} + \underbrace{H_2O}_{\text{Not ionized}}$$

Sometimes we don't cancel, but only reduce in number. If, for example, an equation has
$$\ldots + 4H_2O \longrightarrow \ldots + 2H_2O$$
we can simplify it to
$$\ldots + 2H_2O \longrightarrow \ldots$$

To write an ionic equation we actually do no more than a scratch-paper operation, because what we do next is cancel all of the formulas that appear identically on opposites sides of the arrow. Notice that nothing happens chemically either to $Na^+(aq)$ or to $Cl^-(aq)$, so there is little reason in letting them remain in any equation where we want to give full attention just to the actual chemical event. These ions, $Na^+(aq)$ and $Cl^-(aq)$, do serve one function; they contribute to overall electrical neutrality for their respective compounds. However, in a chemical sense in this reaction they are nothing more than *spectator particles*.

We cancel the spectator particles — whether they are ions or molecules — from the ionic equation, and then rewrite what remains as the **net ionic equation.** As you can see, this equation is beautifully simple:

$$H^+(aq) + OH^-(aq) \longrightarrow H_2O(aq)$$

This is one of the most important net ionic equations we will study. It represents the neutralization of all strong aqueous acids by metal hydroxides in water.

For a net ionic equation to be balanced, two conditions must be met. There must be a **material balance** and an **electrical balance.** We have material balance when the numbers of atoms of each element, regardless of how they are chemically present, are the same on both sides of the arrow. We have electrical balance when the algebraic sum of the charges left of the arrow equals the sum of the charges to the right.

EXAMPLE 8.1   WRITING A NET IONIC EQUATION

Problem: Sulfuric acid is the most important acid in industrial use, and sometimes it has to be neutralized by sodium hydroxide. The reaction can be carried out to produce sodium sulfate, $Na_2SO_4(aq)$, and water. Write the molecular, ionic, and net ionic equations.

Solution: We always start with a molecular equation, and to write it we put down the formulas of the reactants and products in the conventional manner, and then we balance. The molecular equation is

$$H_2SO_4(aq) + 2NaOH(aq) \longrightarrow Na_2SO_4(aq) + 2H_2O$$

Next, we have to recall the following facts:

$H_2SO_4(aq)$ means $2H^+(aq) + SO_4^{2-}(aq)$ (This is a strong, fully ionized acid.)

2NaOH(aq) means 2Na$^+$(aq) + 2OH$^-$(aq)   (This is a strong, fully ionized metal hydroxide.)

Na$_2$SO$_4$(aq) means 2Na$^+$(aq) + SO$_4$$^{2-}$($aq$)   (This is a salt, and (aq) tells us that it's in solution; hence, it's fully ionized.)

2H$_2$O means 2H$_2$O   (No breaking up occurs with this nonelectrolyte.)

These facts let us transform the molecular equation into the following ionic equation:

$$2H^+(aq) + SO_4^{2-}(aq) + 2Na^+(aq) + 2OH^-(aq) \longrightarrow 2Na^+(aq) + SO_4^{2-}(aq) + 2H_2O$$

Formulas must be of the same physical state before they can be canceled. We could not cancel, for example, HCl(g) by HCl(aq), because their states are different.

Next we identify those particles that appear identically on opposite sides of the arrow — the spectators — and cancel them:

$$2H^+(aq) + \cancel{SO_4^{2-}(aq)} + \cancel{2Na^+(aq)} + 2OH^-(aq) \longrightarrow \cancel{2Na^+(aq)} + \cancel{SO_4^{2-}(aq)} + 2H_2O$$

This leaves us with

$$2H^+(aq) + 2OH^-(aq) \longrightarrow 2H_2O$$

We have both a material and an electrical balance, but we should note that we can divide all of the coefficients by 2 and convert them to their smallest whole numbers. Thus the final ionic equation is

$$H^+(aq) + OH^-(aq) \longrightarrow H_2O$$

In other words, the only chemical event that occurs when we mix sodium hydroxide and sulfuric acid in the ratios of the molecular equation is the reaction of H$^+$(aq) with OH$^-$(aq). As we said earlier, this is the neutralization of any strong acid by a metal hydroxide.

PRACTICE EXERCISE 4   Write the molecular, the ionic, and the net ionic equation for the neutralization of nitric acid by potassium hydroxide. A water-soluble salt, KNO$_3$(aq), and water form.

The Reaction of Strong Acids with Metal Bicarbonates.   One of the important reactions used by the body to control the acid–base balance in the blood is the reaction of the bicarbonate ion with the hydrogen ion. All metal bicarbonates react the same way with strong, aqueous acids. The reactions produce carbon dioxide, water, and a salt. For example, sodium bicarbonate and hydrochloric acid react as follows:

$$HCl(aq) + NaHCO_3(aq) \longrightarrow CO_2(g) + H_2O + NaCl(aq)$$

Potassium bicarbonate and hydrobromic acid give a similar reaction:

$$HBr(aq) + KHCO_3(aq) \longrightarrow CO_2(g) + H_2O + KBr(aq)$$

What actually forms initially is not CO$_2$ and H$_2$O but H$_2$CO$_3$, carbonic acid. However, almost all of it promptly decomposes to CO$_2$ and H$_2$O.

Notice that the salt that forms and whose formula has to be written in the molecular equation is always a combination of the cation of the bicarbonate (Na$^+$ or K$^+$ in our examples) and the anion of the acid (Cl$^-$ or Br$^-$ in our examples). Let's be sure we can write the formula of the salt that forms in these reactions before we continue.

EXAMPLE 8.2   WRITING THE FORMULA OF THE SALT THAT FORMS WHEN A BICARBONATE REACTS WITH AN ACID

Problem: What salt forms when lithium bicarbonate reacts with nitric acid?

Solution: Using the name, lithium bicarbonate, we can write its formula — $LiHCO_3$ — so the cation has to be Li^+. The anion furnished by nitric acid, $HNO_3(aq)$, is the nitrate ion, NO_3^-. Therefore we *must* combine one NO_3^- ion with one Li^+ ion to figure out the formula of the salt. The 1 to 1 ratio is required by the necessity of electrical neutrality in the salt. Hence, the salt's formula is $LiNO_3$.

PRACTICE EXERCISE 5   What is the formula of the salt that forms when potassium bicarbonate reacts with sulfuric acid? Assume the salt is a sulfate and not a hydrogen sulfate.

Now that we can write the formulas of salts that form when bicarbonates and strong acids react, we'll get back to writing equations for these reactions. We'll treat all metal bicarbonates as ionized in water to the metal ion and the bicarbonate ion.

EXAMPLE 8.3   WRITING EQUATIONS FOR THE REACTIONS OF BICARBONATES WITH STRONG ACIDS

Problem: What are the molecular and the net ionic equations for the reaction of potassium bicarbonate with hydriodic acid? (Assume the salt that forms is soluble in water.)

Solution: Using what we learned in Example 8.2, the salt must be a combination of the potassium ion, K^+, and the iodide ion, I^-. The salt is KI. Now we can write the molecular equation:

$$KHCO_3(aq) + HI(aq) \longrightarrow CO_2(g) + H_2O + KI(aq)$$

To prepare the ionic equation, we analyze each of the formulas in the molecular equation.

$KHCO_3(aq)$ means $K^+(aq)$ and $HCO_3^-(aq)$	(As we were told.)
$HI(aq)$ means $H^+(aq) + I^-(aq)$	(Because this is a fully ionized acid.)
$KI(aq)$ means $K^+(aq) + I^-(aq)$	(Because we treat all water-soluble salts as fully ionized.)
$CO_2(g)$ and H_2O stay the same	(Neither is ionized.)

Using these facts, we expand the molecular equation into the ionic equation.

$$[K^+(aq) + HCO_3^-(aq)] + [H^+(aq) + I^-(aq)] \longrightarrow CO_2(g) + H_2O + [K^+(aq) + I^-(aq)]$$

The $K^+(aq)$ and the $I^-(aq)$ cancel from each side of the arrow. Draw in the cancel lines yourself. This leaves the following net ionic equation:

$$H^+(aq) + HCO_3^-(aq) \longrightarrow CO_2(g) + H_2O$$

It is balanced both materially and electrically.

SPECIAL TOPIC 8.3 CARBONATED MEDICATIONS

Besides having aspirin, Alka-Seltzer and similar tablets contain a solid acid, citric acid, and solid sodium bicarbonate. The way that these tablets respond when they're dropped into water illustrates the importance of water as a solvent in the reactions of ions. In the crystalline materials, ions are not free to move, but as soon as these tablets hit the water and the ions become mobile, the ions start to react.

Citric acid is a triprotic acid, and we can represent it as H_3Cit, where *Cit* stands for the citrate ion, an ion with a charge of 3−. The hydrogens made available by citric acid as it is in solution react with the bicarbonate ions that become free to move around when sodium bicarbonate dissolves. This reaction gives the CO_2 that fizzes out of solution as this gas forms:

$$H^+(aq) + HCO_3^-(aq) \longrightarrow CO_2(g) + H_2O$$

The equation produced by Example 8.3

$$H^+(aq) + HCO_3^-(aq) \longrightarrow CO_2(g) + H_2O$$

is the same net ionic equation for the reaction of all metal bicarbonates with all strong, aqueous acids. Because this reaction destroys the hydrogen ions of the acid, it must also be called an acid-neutralization reaction. In fact, the familiar "bicarb" used as a home remedy for acid stomach is nothing more than sodium bicarbonate. Stomach acid is roughly 0.1 *M* HCl, and the bicarbonate ion neutralizes this acid by the reaction we have just studied. An overdose of "bicarb" must be avoided because it can cause a medical emergency involving the respiratory gases. Another use of sodium bicarbonate is as an isotonic solution given intravenously to neutralize acid in the blood. For still another use, see Special Topic 8.3.

PRACTICE EXERCISE 6 Write the molecular, ionic, and net ionic equations for the reaction of sodium bicarbonate with sulfuric acid in which sodium sulfate, $Na_2SO_4(aq)$, is one of the products.

The Reaction of Strong Acids with Carbonates. Metal carbonates are attacked by strong acids—that is, by hydrogen ions—to give carbon dioxide, water, and a salt. (As before, the salt is made up of the cation of the carbonate and the anion of the acid.) These are the same kinds of products that bicarbonates give, but the stoichiometry is different. Mole for mole, the CO_3^{2-} ion can neutralizes twice as much H^+ as the HCO_3^- ion, as we'll see in the next example.

EXAMPLE 8.4 WRITING EQUATIONS FOR THE REACTIONS OF METAL CARBONATES WITH STRONG, AQUEOUS ACIDS

Problem: Sodium carbonate, Na_2CO_3, neutralizes hydrochloric acid and forms sodium chloride, carbon dioxide, and water. Write the molecular, ionic, and net ionic equations for this reaction. Assume that the reaction occurs in an aqueous solution.

Solution: We first write the formulas into a conventional, molecular equation and balance it.

$$2HCl(aq) + Na_2CO_3(aq) \longrightarrow CO_2(g) + H_2O + 2NaCl(aq)$$

Then we analyze each of the formulas in this equation to see how to use them in the ionic equation.

$2HCl(aq)$ means $2H^+(aq) + 2Cl^-(aq)$ (The acid is strong and fully ionized.)

$Na_2CO_3(aq)$ means $2Na^+(aq) + CO_3{}^{2-}(aq)$ (This is a salt and it is written with "(aq)". So, it's in solution and fully ionized.)

$2NaCl(aq)$ means $2Na^+(aq) + 2Cl^-(aq)$ (This soluble salt is treated, like all salts dissolved in water, as fully ionized.)

$CO_2(g)$ and H_2O remain unchanged

Now we can expand the molecular equation to the ionic equation.

$$[2H^+(aq) + 2Cl^-(aq)] + [2Na^+(aq) + CO_3{}^{2-}(aq)]$$
$$\longrightarrow CO_2(g) + H_2O + [2Na^+(aq) + 2Cl^-(aq)]$$

Na^+ and Cl^- are spectator ions in this reaction.

We can cancel the $2Na^+(aq)$ and the $2Cl^-(aq)$ from both sides of the equation, which leaves the following net ionic equation:

$$2H^+(aq) + CO_3{}^{2-}(aq) \longrightarrow CO_2(g) + H_2O$$

Notice in Example 8.4 that one carbonate can neutralize two hydrogen ions, twice as many as are neutralized by a bicarbonate ion. The net ionic equation that we devised in Example 8.4,

$$2H^+(aq) + CO_3{}^{2-}(aq) \longrightarrow CO_2(g) + H_2O$$

is the same for the reactions of all of the carbonates of the group IA metals with all strong, aqueous acids.

PRACTICE EXERCISE 7   Write the molecular, ionic, and net ionic equations for the reaction of aqueous potassium carbonate, $K_2CO_3(aq)$, with sulfuric acid to give potassium sulfate, $K_2SO_4(aq)$, a water-soluble salt, and the other usual products.

Adding a few drops of strong acid to a rock sample is a field test for carbonate rocks. A positive test is an odorless fizzing reaction.

Most carbonates are water-insoluble compounds. Calcium carbonate, $CaCO_3$, for example, is the chief substance in limestone and marble. Despite its insolubility in water, calcium carbonate reacts readily with strong, aqueous acids (with their hydrogen ions, of course).

$$CaCO_3(s) + 2HCl(aq) \longrightarrow CO_2(g) + H_2O + CaCl_2(aq)$$

The net ionic equation is

$$CaCO_3(s) + 2H^+(aq) \longrightarrow CO_2(g) + H_2O + Ca^{2+}$$

For water-insoluble compounds, we have to write their entire formulas in net ionic equations.

PRACTICE EXERCISE 8   Dolomite, a limestone-like rock, contains both calcium and magnesium carbonates. Magnesium carbonate is attacked by nitric acid. The salt that forms is water-soluble. Write the molecular, ionic, and net ionic equations for this reaction.

The Reaction of Strong Acids with Metal Hydroxides.   We have already discussed the reaction of strong acids with hydroxides when we learned how to write net ionic equations. Only the group IA hydroxides are very soluble in water, so their reactions in solution with strong, aqueous acids have net ionic equations that differ slightly from those of the water-insoluble metal hydroxides. If we let M stand for a group IA metal, we can write the reactions of aqueous solutions of their hydroxides with a strong acid such as hydrochloric acid by the following general equation:

$$MOH(aq) + HCl(aq) \longrightarrow MCl(aq) + H_2O$$

or, as the net ionic equation:

$$OH^-(aq) + H^+(aq) \longrightarrow H_2O$$

If we now use M for any metal in group IIA (except beryllium), the equations are:

$$M(OH)_2(s) + 2HCl(aq) \longrightarrow MCl_2(aq) + 2H_2O$$

or,

$$M(OH)_2(s) + 2H^+(aq) \longrightarrow M^{2+}(aq) + 2H_2O$$

PRACTICE EXERCISE 9   When milk of magnesia is used to neutralize hydrochloric acid (stomach acid), solid magnesium hydroxide in the suspension reacts with the acid. Write the molecular and net ionic equations for this reaction.

The Reaction of Ammonia with Strong, Aqueous Acids.   An aqueous solution of ammonia is an excellent reagent for neutralizing acids. We learned in Section 4.7 how an unshared pair of electrons on nitrogen in ammonia can form a coordinate covalent bond to H^+ furnished by an acid. This neutralizes the acid. For example,

$$NH_3(aq) + HCl(aq) \longrightarrow NH_4Cl(aq)$$

or,

$$NH_3(aq) + H^+(aq) \longrightarrow NH_4^+(aq)$$

All ammonium salts are soluble in water, so they liberate NH_4^+ ions in aqueous solutions. Many biochemicals have ammonia-like molecules that also neutralize hydrogen ions.

PRACTICE EXERCISE 10   Write the molecular and net ionic equations for the reaction of aqueous ammonia with (a) HBr(aq) and (b) $H_2SO_4(aq)$.

The Reaction of Active Metals with Strong, Aqueous Acids.   Nearly all metals are attacked more or less readily by the hydrogen ion in solution. The products are generally hydrogen gas and a salt made of the cation from the metal and the anion from the acid. Zinc, for example, reacts with hydrochloric acid as follows:

$$Zn(s) + 2HCl(aq) \longrightarrow H_2(g) + ZnCl_2(aq)$$

The net ionic equation is

$$Zn(s) + 2H^+(aq) \longrightarrow H_2(g) + Zn^{2+}(aq)$$

Aluminum is also attacked by acids. For example, its reaction with nitric acid can be written as follows:

$$2Al(s) + 6HNO_3(aq) \longrightarrow 2Al(NO_3)_3(aq) + 3H_2(g)$$

The net ionic equation is

$$2Al(s) + 6H^+(aq) \longrightarrow 2Al^{3+}(aq) + 3H_2(g)$$

PRACTICE EXERCISE 11   Write the molecular and the net ionic equations for the reaction of magnesium with hydrochloric acid.

Remember, oxidation is a loss of electrons and reduction is a gain of electrons.

Activity Series of the Metals.   Metals differ greatly in their tendencies to react with aqueous hydrogen ions. When they do, atoms of the metal are oxidized because they lose electrons and become metal ions. The electrons are transferred to protons taken from H_3O^+ ions, and these protons are reduced and made electrically neutral. Two H atoms combine and emerge as a molecule of hydrogen gas, H_2. The group IA metals such as sodium and potassium include the most reactive metals of all. They not only reduce protons taken from hydronium ions, they also reduce protons taken from water. No acid need be present. The following reaction of sodium metal with water is extremely violent, and it should never be attempted except by an experienced chemist working with safety equipment, including a fire extinguisher. (The hydrogen gas generally ignites spontaneously as it forms.)

$$2Na(s) + 2H_2O \longrightarrow 2NaOH(aq) + H_2(g)$$

If this reaction is violent in water, it's even more violent in aqueous acids.

Gold, silver, and platinum, in contrast, are stable not only toward water but also toward hydronium ions.

With such a vast difference in the reactivities of metals toward acids, it shouldn't be surprising that the metals can be arranged in an order of reactivity. The result is called the **activity series** of the metals, and it's given in Table 8.3. Atoms of any metal above hydrogen in the series can transfer electrons to H^+, either from H_2O or from H_3O^+, to form hydrogen gas, and the metal atoms change to metal ions. Tin and lead, however, react only very slowly. The metals below hydrogen in the activity series do not transfer electrons to H^+.

The rate of the reaction of an acid with a metal depends on the acid as well as the metal. When compared at the same molar concentrations, strong acids react far more rapidly than weak acids, as Figure 8.6 shows. These differences reflect the differences in percentage ionizations, because the actual reaction, as we have said, is with the hydrogen ion. When the concentration of hydrogen ion is high, as it can be when the acid is strong, the reaction is vigorous. In 1 M HCl, the concentration of $H^+(aq)$ is also 1 M, because for each HCl one $H^+(aq)$ is released. However, in 1 M $HC_2H_3O_2$, acetic acid (a weak acid), the actual concentration of $H^+(aq)$ is closer to 0.004 M, which is about 1/250 as much. No wonder the liveliness of the reaction pictured in Figure 8.6c, the reaction of zinc with 1 M acetic acid, is much less than in Figure 8.6a, the reaction with 1 M HCl. In part b of Figure 8.6, the reaction is with 1 M H_3PO_4, a moderate acid, and the vigor of the reaction is somewhere in between that of the other two.

TABLE 8.3
The Activity Series of the Common Metals

Greatest tendency to become ionic

Decreasing tendency to become ionic		Potassium Sodium	React violently with water
		Calcium	Reacts slowly with water
	React with hydrogen ions to liberate H_2	Magnesium Aluminum Zinc Chromium	React very slowly with steam
		Iron Nickel Tin Lead	
		HYDROGEN	
	Do not react with hydrogen ions	Copper Mercury Silver Platinum Gold	

Least tendency to become ionic

FIGURE 8.6
Relative hydrogen ion concentrations and the reactivity of zinc. (a) The acid is HCl(aq), a strong, fully ionized acid. (b) The acid is $H_3PO_4(aq)$, a moderate, partly ionized acid. (c) The acid is acetic acid, $HC_2H_3O_2(aq)$, a weak, poorly ionized acid. Although the molar concentrations of the acids are the same in each tube, the actual molar concentrations of their hydrogen ions are greatly different, being highest in (a), where the bubbles of hydrogen are evolving the most vigorously, next highest in (b), and lowest in (a).

(a)   (b)   (c)

8.4   BRØNSTED ACIDS AND BASES

Acid–base reactions are proton transfers between any donors and acceptors of H⁺, not just between H_3O^+ and OH^-.

As we have seen, many substances besides metal hydroxides can neutralize hydrogen ions. Both the carbonates and bicarbonates as well as ammonia do this. And we will see that many substances besides the strong acids can neutralize hydroxide ions. Acetic acid does this very well, for example. The idea of acids and bases is clearly useful, but the Arrhenius view needs broadening.

Johannes Brønsted (1879–1947), a Danish chemist, called attention to the central feature of acid–base reactions. This is the relocation or transfer of one tiny particle, the proton, H^+. The acid provides it and something else accepts it. So Brønsted suggested that we call anything that supplies protons an acid and anything that accepts them a base. Clearly, substances such as $HCl(aq)$ and $H_2SO_4(aq)$—typical Arrhenius acids—qualify as suppliers of protons. And the Arrhenius base, OH^-, is obviously something that accepts protons. But acetic acid, a weak acid that has a low percentage ionization, also supplies protons if mixed with a strong base. And ammonia, as we learned in the previous section, neutralizes acids just as well as does OH^-. These considerations gelled into the following definitions of **Brønsted acids and bases.**

Brønsted Acids and Bases

Acids   proton donors.
Bases   proton acceptors.

According to these definitions, even in the gaseous state HCl is an acid. Gaseous hydrogen chloride reacts with gaseous ammonia to form solid ammonium chloride (Figure 8.7):

$$HCl(g) + NH_3(g) \longrightarrow NH_4Cl(s)$$

FIGURE 8.7
The gases $HCl(g)$ and $NH_3(g)$ are escaping from their aqueous solutions in the bottles and reacting above to form solid NH_4Cl, which appears as a white smoke. (From J. E. Brady and J. R. Holum, *Fundamentals of Chemistry,* 2nd ed., John Wiley and Sons, Inc., New York, 1984. Used by permission.)

Here, a proton transfers from a molecule of HCl directly to a molecule of NH_3. No hydronium ions are involved.

Even the reaction of gaseous hydrogen chloride with water that produces hydrochloric acid illustrates the Brønsted concept.

$$HCl(g) + H_2O \longrightarrow H_3O^+(aq) + Cl^-(aq)$$

The acid here — the Brønsted acid — is HCl(g). The Brønsted base is water. A proton transfers from the acid to the base.

Conjugate Acid–Base Pairs.   In working with the Brønsted concept, it's useful to consider all acid–base reactions as involving chemical equilibria. Thus when acetic acid dissolves in water, a solution forms that turns blue litmus red. This tells us that hydronium ions are present in excess of hydroxide ions. By other tests, however, we can learn that in, say, 1 M $HC_2H_3O_2$, we don't have 1 mol H^+/L, which we should have if acetic acid were 100% ionized. Instead we have about 0.004 mol H^+/L. Only a small percentage — 0.4% — of all the acetic acid molecules are ionized. What is present is a chemical equilibrium in which the reactants are heavily favored.

$$HC_2H_3O_2(aq) + H_2O \rightleftharpoons H_3O^+(aq) + C_2H_3O_2^-(aq)$$

In the forward reaction, the acid — the proton donor — is the acetic acid molecule, and the base — the proton acceptor — is the water molecule. However, the reverse reaction is also a proton-transfer. In the reverse reaction, the reaction that reads from right to left, the acid is clearly the hydronium ion, because it donates a proton. And the base is the acetate ion, $C_2H_3O_2^-(aq)$, because it accepts a proton. In other words, in this equilibrium we can identify two Brønsted acids and two Brønsted bases. Each base is related to one of the two acids. Thus the acetate ion, $C_2H_3O_2^-$, is related to and comes from acetic acid, $HC_2H_3O_2$. The other base, H_2O, is related to the other acid, H_3O^+. Pairs of particles like H_2O and H_3O^+ or $C_2H_3O_2^-$ and $HC_2H_3O_2$, whose formulas differ by just one H^+, are called **conjugate acid–base pairs.** Thus H_2O is the conjugate base of H_3O^+, and H_3O^+ is the conjugate acid of H_2O. Similarly, $C_2H_3O_2^-$ is the conjugate base of $HC_2H_3O_2$, and $HC_2H_3O_2$ is the conjugate acid of $C_2H_3O_2^-$. It will be useful, given the name or formula of one member of a conjugate acid–base pair, to be able to write the formula of the other, so let's study some examples.

EXAMPLE 8.5   WRITING THE FORMULA OF A CONJUGATE ACID

Problem: Ammonia, NH_3, accepts protons, H^+, when it neutralizes acids, as we learned in the previous section. What is the conjugate acid of this Brønsted base?

Solution: All we have to do is change NH_3 by one H^+. When we do this we have to add not just the H but also the + charge. We add the charge algebraically. The conjugate acid is NH_4^+.

EXAMPLE 8.6   WRITING THE FORMULA OF A CONJUGATE ACID

Problem: The phosphate ion, PO_4^{3-}, is a Brønsted base. What is the formula of its conjugate acid?

Solution: When we add H^+ to PO_4^{3-} we get HPO_4^{2-}, the conjugate acid of PO_4^{3-}. The algebraic sum of 1+ and 3– is 2–, the charge on the conjugate acid.

EXAMPLE 8.7   WRITING THE FORMULA OF A CONJUGATE BASE

Problem: What is the conjugate base of nitrous acid, HNO_2, a weak acid?

Solution: When we take H^+ away from HNO_2, we're left with NO_2^-, the conjugate base. (When we take a charge of $1+$ from a particle with a charge of 0, the remaining charge is $1-$.)

EXAMPLE 8.8   WRITING THE FORMULA OF A CONJUGATE BASE

Problem: The anion, $H_2PO_4^-$, is a weak proton-donor — a weak Brønsted acid. What is its conjugate base?

Solution: We have to remove H^+ from $H_2PO_4^-$, both the H and a net of one $+$ charge. This leaves us with HPO_4^{2-}. (When we take $1+$ away from $1-$, the result is $2-$.)

PRACTICE EXERCISE 12   Write the formulas of the conjugate acids of the following particles.
(a) NO_3^- (b) SO_3^{2-} (c) CO_3^{2-} (d) SO_4^{2-} (e) Cl^- (f) H_2O (g) OH^-

PRACTICE EXERCISE 13   Write the formulas of the conjugate bases of the following particles.
(a) HCO_3^- (b) HPO_4^{2-} (c) H_2SO_4 (d) HSO_4^- (e) HBr (f) H_3O^+ (g) H_2O

Strong and Weak Brønsted Acids and Bases.   What we are leading up to is the study of Brønsted acid – base interactions in body fluids. We want to be able to sense, at least in a qualitative way, if a particular species poses a threat to the acid – base balance of the blood by being too strong a Brønsted acid or base. Or we might want to understand how other species are able to stand guard over the very delicate acid – base balance by neutralizing excesses of either acids or bases.

There is one important major generalization about the equilibria that involve conjugate acids and bases. *The position of the equilibrium when conjugate acids and bases interact always favors the weaker acid/base side.* For this to be applied, we obviously have to learn how to tell what is stronger and what is weaker. Two simple, logical rules are all we need:

If an acid is strong, its conjugate base is weak.

If an acid is weak, its conjugate base is strong.

Now we have to learn what is meant by *strong* and *weak* in the Brønsted theory.

Strong and Weak Acids and Bases — Brønsted Theory

A **strong acid** is a strong proton donor.
A **strong base** is a strong proton acceptor or proton binder.
A **weak acid** is a weak proton donor, a poor proton donor.
A **weak base** is a weak proton acceptor, a poor proton binder.

The starting point for using these rules is your knowledge of the list of strong acids. Let's work an example to show how this list enables us to figure out if a particular species is strong or weak.

EXAMPLE 8.9   DEDUCING IF AN ION OR MOLECULE IS A WEAK BRØNSTED ACID

Problem: Lactic acid is the acid responsible for the tart taste of sour milk. Is lactic acid a strong or a weak acid?

Solution: The list of strong acids does not include lactic acid. Hence, it is a weak acid. It's as simple as that (and we'd err very seldom).

EXAMPLE 8.10   DEDUCING IF AN ION OR MOLECULE IS A STRONG OR A WEAK BRØNSTED BASE

Problem: Is the bromide ion a strong or a weak Brønsted base?

Solution: When the question deals with a potential *base,* we have to find the answer in a roundabout fashion. We accept this because the alternative would be to memorize a rather extensive list of the stronger Brønsted bases. Here's how to go about it. Pretend that the potential base actually functions as a base, so write the formula of its conjugate acid. If Br^- were to be a base, its conjugate acid would be HBr, which, in water, is hydrobromic acid. Now comes the crucial question. Is hydrobromic acid a strong acid? We have to know the list, and HBr is on the list of *strong* acids, so we know it easily gives up a proton. Therefore we know that what remains when the proton so readily leaves HBr, Br^-, has to be a poor proton binder. So our answer is that Br^- is a weak Brønsted base.

EXAMPLE 8.11   DEDUCING IF AN ION OR A MOLECULE IS A STRONG OR A WEAK BRØNSTED BASE

Problem: Is the phosphate ion, PO_4^{3-}, a strong or a weak Brønsted base? Using the strategy described in Example 8.10, we pretend that this ion actually is a base—a proton acceptor. So we give it a proton, and write the result, the conjugate acid. The conjugate acid of PO_4^{3-} is HPO_4^{2-}. This Brønsted acid isn't on our list of strong acids, so we conclude it's a weak acid. This means that PO_4^{3-} is a good proton binder (holding the proton as HPO_4^{2-}). A good proton binder is a strong base, so our answer to the question is that PO_4^{3-} is a strong base.

PRACTICE EXERCISE 14   Classify the following particles as strong or as weak Brønsted acids.

(a) HSO_3^- (b) HCO_3^- (c) $H_2PO_4^-$

PRACTICE EXERCISE 15   Classify the following ions as strong or as weak Brønsted bases.

(a) I^- (b) NO_3^- (c) CN^- (d) NH_2^-

The strongest Brønsted base we can have in an aqueous solution (at least at any appreciable concentration) is the hydroxide ion. Any stronger base reacts with water, takes a proton, and changes to the conjugate acid. For example, the oxide ion, O^{2-}, which is the conjugate *base* of OH^-, is a much stronger base than OH^-. If we add it to water in the form of sodium oxide, the following reaction occurs very exothermically, and none of oxide ions in Na_2O are in the solution. They have all changed to hydroxide ions.

$$Na_2O(s) \quad + H_2O \longrightarrow 2NaOH(aq)$$

Sodium oxide   Sodium hydroxide

The strongest Brønsted acid we can have in an aqueous solution is the hydronium ion. Any stronger acid reacts with water, gives up a proton, and changes to its own conjugate base. Thus hydrogen chloride, H—Cl(g), is actually a stronger proton-donor than H_3O^+. As we already know, when we bubble H—Cl(g) into water, the following reaction occurs — a typical Brønsted acid-base reaction:

$$HCl(g) + H_2O \longrightarrow H_3O^+(aq) + Cl^-(aq)$$

Evidently, the hydronium ion holds the proton better than Cl is able to in H—Cl(g).

Now that we can make reasonable predictions of relative acid or base strengths, let's see how we can use this skill in predicting reactions. Keep in mind that in all proton-transfer reactions, the stronger acid and base always tend to react to give the weaker acid and base. We can condense this to a rule of thumb: *The stronger always give way to the weaker in acid–base reactions.*

EXAMPLE 8.12   PREDICTING WHICH SUBSTANCES ARE FAVORED IN AN ACID–BASE EQUILIBRIUM

Problem: If we add hydrochloric acid to an aqueous solution of sodium cyanide, NaCN, will HCN and NaCl form to any significant extent? (If they do, the evolving HCN, hydrogen cyanide, might kill anyone who mixes these substances. Hydrogen cyanide is a very dangerous poison.)

Solution: Because we are dealing with HCl(aq), the reagent actually consists of $H_3O^+(aq)$ and $Cl^-(aq)$. Because NaCN, a salt, is in solution, we are dealing with $Na^+(aq)$ and $CN^-(aq)$. Sodium ions and chloride ions would be spectator ions. The question, therefore, is, Does the following reaction occur?

$$H_3O^+(aq) + CN^-(aq) \xrightarrow{\ ?\ } H_2O + HCN(aq)$$

Remembering the rule that the "stronger give way to the weaker" in these proton-transfer reactions, we have to identify the acids and bases, and then infer what's stronger and what's weaker.

When we look for the conjugate pairs, we can see them as follows:

Conjugate acid-base pair

$$H_3O^+(aq) + CN^-(aq) \longrightarrow H_2O + HCN(aq)$$

Conjugate acid-base pair

Each pair must have an acid and each must have a base, so let's write in these labels (and, to reduce clutter, omit the lines that have served to connect conjugate pairs).

$$H_3O^+(aq) + CN^-(aq) \longrightarrow H_2O + HCN(aq)$$
   Acid   Base   Base   Acid

Now we decide which of the two acids is stronger. We know that H_3O^+ is the strongest acid species we can have in water. (We also know that because HCN isn't on the list of strong acids, it must be weak.) So we modify our labels with this new information.

$$H_3O^+(aq) + CN^-(aq) \longrightarrow H_2O + HCN(aq)$$
   Stronger   Base   Base   Weaker
   acid   acid

Because the conjugate of a stronger acid must be a weaker base, and the conjugate of a weaker acid must be a stronger base, we can modify the remaining labels as follows:

$$H_3O^+(aq) + CN^-(aq) \longrightarrow H_2O \quad + HCN(aq)$$

| Stronger | Stronger | Weaker | Weaker |
| acid | base | base | acid |

Finally, we can tell that the reaction, as written, must occur, because the stronger are always replaced by the weaker in acid–base reactions. What we actually would have is an equilibrium in which the products of the forward reaction (as we draw the equilibrium expression) are favored.

$$H_3O^+(aq) + CN^-(aq) \rightleftharpoons H_2O \quad + HCN(aq)$$

| Stronger | Stronger | Weaker | Weaker |
| acid | base | base | acid |

PRACTICE EXERCISE 16   When the meat preservative, sodium nitrite, $NaNO_2$, enters the stomach and encounters the hydrochloric acid in gastric juice, can nitrous acid, HNO_2, be produced? Write the equilibrium expression for any net ionic interactions, using the longer arrow to point toward the favored products. (Nitrous acid is suspected of being a cause of cancer, but no evidence presently exists that it actually causes cancer in humans.)

Relative Acid/Base Strengths and Weaknesses — A Qualitative Look.   The term *weak* in *weak acid* is quite a broad term, and the acids on the long list of weak acids are not equal in weakness. Even from the limited data in Table 8.1 you can see that carbonic acid is a weaker acid than acetic acid. Table 8.4 lists several Brønsted acids and bases in their orders of increasing strength.

Anything in the column of acids of Table 8.4 above the hydronium ion is so strong a proton donor that in water it is essentially 100% ionized into the hydronium ion and the

TABLE 8.4
Relative Strengths of Some Brønsted Acids and Bases

Brønsted Acid		Brønsted Base	
Name	Formula	Name	Formula
Perchloric acid	$HClO_4$	Perchlorate ion	ClO_4^-
Hydrogen iodide	HI	Iodide ion	I^-
Hydrogen bromide	HBr	Bromide ion	Br^-
Sulfuric acid	H_2SO_4	Hydrogen sulfate ion	HSO_4^-
Hydrogen chloride	HCl	Chloride ion	Cl^-
Nitric acid	HNO_3	Nitrate ion	NO_3^-
HYDRONIUM ION	H_3O^+	WATER	H_2O
Hydrogen sulfate ion	HSO_4^-	Sulfate ion	SO_4^{2-}
Phosphoric acid	H_3PO_4	Dihydrogen phosphate ion	$H_2PO_4^-$
Acetic acid	$HC_2H_3O_2$	Acetate ion	$C_3H_3O_2^-$
Carbonic acid	H_2CO_3	Bicarbonate ion	HCO_3^-
Ammonium ion	NH_4^+	Ammonia	NH_3
Bicarbonate ion	HCO_3^-	Carbonate ion	CO_3^{2-}
WATER	H_2O	HYDROXIDE ION	OH^-
Methyl alcohol	CH_3OH	Methoxide ion	CH_3O^-
Ammonia	NH_3	Amide ion	NH_2^-
Hydrogen	H_2	Hydride ion	H^-

Increasing acid strength ↑   Increasing base strength ↓

conjugate base. In the column of bases, anything below the hydroxide ion is too strong a base to exist in water. It takes a proton from a water molecule and generates the OH^- ion as well as the conjugate acid of the base.

All acids above water in the column of acids in Table 8.4 are almost quantitatively neutralized by the hydroxide ion. All bases below water in the column of bases react substantially with hydronium ion and change over to their conjugate acids plus water. (Below the acetate ion, the extent of this proton-transfer is essentially 100%.)

As the heading to this unit indicated would happen, all that we have done in this small unit is to give a qualitative idea about the strengths and weaknesses of acids and bases. In Section 8.6 we will return to this topic to describe a quantitative way of telling the relative strengths of acids and bases.

The Ammonium Ion as a Brønsted Acid.

The ammonium ion occupies a special place in our study because many biochemicals such as proteins have a molecular part that is very much like this ion. Although the ammonium ion is a weak acid (Table 8.4), it still is a Brønsted acid, and it can neutralize the hydroxide ion. When we add sodium hydroxide to a solution of ammonium chloride, the following reaction occurs:

$$NaOH(aq) + NH_4Cl(aq) \longrightarrow NH_3(aq) + H_2O + NaCl(aq)$$

The net ionic equation is

$$OH^-(aq) + NH_4^+(aq) \longrightarrow NH_3(aq) + H_2O$$

This reaction neutralizes the hydroxide ion, and it leaves a solution of the weaker base, NH_3. (If the initial solution is concentrated enough, the final solution has a strong odor of ammonia.)

In some medical emergencies, when the blood has become too alkaline or too basic, an isotonic solution of ammonium chloride is administered by intravenous drip. Its ammonium ions can neutralize some of the base in the blood and bring the acid/base balance back to normal.

8.5 SALTS

A very large number of ionic reactions can be predicted from a knowledge of the solubility rules of salts.

Salts are ionic compounds whose cations are any except H^+ and whose anions are any except OH^-. All are crystalline solids at room temperature, because forces of attraction between ions in crystals are very strong. A **simple salt** is one that is made of just *two* kinds of oppositely charged ions. Examples are NaCl, $MgBr_2$, and $CuSO_4$. *Mixed salts* are those that have three or more different ions. Alum, used in water purification, is an example: $K_2SO_4 \cdot Al_2(SO_4)_3 \cdot 24H_2O$. As the formula of alum illustrates, the salt family includes hydrates. Some salts of practical value are given in Table 8.5.

Formation of Salts.

In the laboratory, salts are obtained whenever an acid is used in any of the following ways. We summarize and review these methods here and show their similarities.

$$Acid + metal \longrightarrow a\ salt + H_2$$
$$Acid + metal\ hydroxide \longrightarrow a\ salt + H_2O$$
$$Acid + metal\ carbonate \longrightarrow a\ salt + H_2O + CO_2$$
$$Acid + metal\ bicarbonate \longrightarrow a\ salt + H_2O + CO_2$$

If the salt is soluble in water, we have to evaporate the solution to dryness to isolate it. However, sometimes the salt precipitates. To predict when to expect this, as well as to show

TABLE 8.5
Some Salts and Their Uses

Formula and Name	Uses
$BaSO_4$ Barium sulfate	Used in the "barium cocktail" given prior to X-raying the gastrointestinal tract
$(CaSO_4)_2 \cdot H_2O$ Calcium sulfate hemihydrate (plaster of paris)	Plaster casts. Wall stucco. Wall plaster
$MgSO_4 \cdot 7H_2O$ Magnesium sulfate heptahydrate (epsom salt)	Purgative
$AgNO_3$ Silver nitrate	Antiseptic and germicide. Used in eyes of infants to prevent gonorrheal conjunctivitis. Photographic film sensitizer
$NaHCO_3$ Sodium bicarbonate (baking soda)	Baking powders. Effervescent salts. Stomach antacid. Fire extinguishers
$Na_2CO_3 \cdot 10H_2O$ Sodium carbonate decahydrate (soda ash, sal soda, washing soda)	Water softener. Soap and glass manufacture
$NaCl$ Sodium chloride	Manufacture of chlorine, sodium hydroxide. Preparation of food
$NaNO_2$ Sodium nitrite	Meat preservative

still another way of making salts, and simply to add some "rules of thumb" to our general knowledge of nature, we turn to the solubility rules for salts. Quite arbitrarily, but in rough agreement with somewhat standard assumptions, we say that a salt is soluble in water if it can form a solution with a concentration of at least 3 to 5% (w/w). When a *counter ion* is referred to in the following rules, it means the unnamed ion that is part of the salt. For example, in the lithium salt, LiCl, the counter ion is the chloride ion.

Solubility Rules for Salts in Water

1. All lithium, sodium, potassium, and ammonium salts are soluble regardless of the counter ion.

2. All nitrates and acetates are soluble, regardless of the counter ion.

3. All chlorides, bromides, and iodides are soluble, except when the counter ion is lead, silver, or mercury(I).

4. Salts not in the above categories are generally insoluble or, at best, only slightly soluble.

There are exceptions to these rules, but we will seldom be wrong in applying them. One of the many applications of these rules is to predict possible reactions between salts.

Salts from Double Decomposition Reactions. In addition to the reactions we have already studied that make salts by acid-base neutralizations, we can also make salts by a "change of partners" reaction called **double decomposition.** The best way to see how this works is by a worked example.

EXAMPLE 8.13 PREDICTING DOUBLE DECOMPOSITION REACTIONS OF SALTS

Problem: What happens if we mix aqueous solutions of sodium sulfate and barium nitrate?

Solution: Because, by the solubility rules, both sodium sulfate and barium nitrate are soluble in water, their solutions contain their separated *ions*. When we pour the two solutions together, four ions experience attractions and repulsions. Hence, we have to examine each possible combination of oppositely charged ions to see which, if any, makes a water-insoluble salt. If we find one, then we can write an equation for the reaction that produces this salt. Here are the possible combinations when Ba^{2+}, NO_3^-, Na^+, and SO_4^{2-} ions intermingle in water.

$$Ba^{2+} + 2NO_3^- \xrightarrow{?} Ba(NO_3)_2(s)$$

This possibility is obviously out, because barium ions and nitrate ions do not precipitate together from water. ("All nitrates are soluble.")

$$2Na^+ + SO_4^{2-} \xrightarrow{?} Na_2SO_4(s)$$

This possibility is also out. ("All sodium salts are soluble.")

$$Na^+ + NO_3^- \xrightarrow{?} NaNO_3(s)$$

No. Again, "All sodium salts are soluble."

$$Ba^{2+} + SO_4^{2-} \xrightarrow{?} BaSO_4(s)$$

Yes. Barium sulfate, $BaSO_4$, is not in any of the categories of water-soluble salts. Therefore we predict it is insoluble.

A slurry of barium sulfate in flavored water is the barium "cocktail" a patient drinks before having an X ray taken of the gastrointestinal tract.

Because we predicted that $BaSO_4$ can form a precipitate, we can write a molecular equation, and we'll use some connector lines to show how partners exchange—how *double* decomposition occurs.

$$Ba(NO_3)_2(aq) + Na_2SO_4(aq) \longrightarrow 2NaNO_3(aq) + BaSO_4(s)$$

The net ionic equation, however, is a better way to describe what happens:

$$Ba^{2+}(aq) + SO_4^{2-}(aq) \longrightarrow BaSO_4(s)$$

The sodium and nitrate ions are only spectators. To obtain the solid barium sulfate, we would filter the mixture and collect this compound on the filter. If we also wanted the sodium nitrate, we would evaporate the clear filtrate to dryness.

PRACTICE EXERCISE 17   If solutions of sodium sulfide, Na_2S, and copper(II) nitrate, $Cu(NO_3)_2$, are mixed, what if anything will happen chemically? Write a molecular and a net ionic equation for any reaction.

One of the many uses of the solubility rules is to understand what it means for water to be called *hard water* and what it means to *soften* such water. These are discussed in Special Topic 8.4.

Reactions of Ions — A Summary. Our study of the reaction of sodium sulfate and barium nitrate in Example 8.13 illustrated the power of knowing just a few facts for the sake of predicting an enormous number of others with a high probability of success. The following facts summarize those that should now be well-learned.

1. The solubility rules of the salts (because then we can assume that all of the other salts are insoluble).
2. The five strong acids in Table 8.1 (because then we can assume that all of the other acids, including organic acids, are weak).

Ground water that contains magnesium, calcium, or iron ions at a high enough level to interact with ordinary soap to form scum is called **hard water**. In **soft water** these "hardness ions" — Ca^{2+}, Mg^{2+}, Fe^{2+}, or Fe^{3+} — are either absent or are present in extremely low concentrations. The anions that most frequently accompany the hardness ions are SO_4^{2-}, Cl^-, and HCO_3^-.

Hard water in which the principal anion is the bicarbonate ion is called **temporary hard water**. Hard water in which the chief negative ions are anything else is called **permanent hard water**. When temporary hard water is heated near its boiling point, as in hot boilers, steam pipes, and instrument sterilizers, the bicarbonate ion breaks down to the carbonate ion. This ion forms insoluble precipitates with the hardness ions. Their carbonate salts form, come out of solution, and deposit as scaly material that can even clog the equipment, as the accompanying photograph illustrates. The equations for these changes are as follows.

The breakdown of the bicarbonate ion:

$$2HCO_3^-(aq) \longrightarrow CO_3^{2-}(aq) + CO_2(g) + H_2O$$

The formation of the scaly precipitate (using the calcium ion to illustrate):

$$CO_3^{2-}(aq) + Ca^{2+}(aq) \longrightarrow CaCO_3(s)$$

Water-Softening Procedures Hard water can be softened in various ways. Most commonly, excess soap is used. Some scum does form, but then the extra soap does the cleansing work. To avoid the scum altogether, softening agents are added before the soap is used. One common water-softening chemical is sodium carbonate decahydrate, known as washing soda. Its carbonate ions take out the hardness ions as insoluble carbonates by the kind of reaction for which we wrote the previous net ionic equation.

Another home water-softening agent is household ammonia — 5% (w/w) NH_3. We've already learned about the following equilibrium in such a solution:

$$NH_3(aq) + H_2O \rightleftharpoons NH_4^+(aq) + OH^-(aq)$$

In other words, aqueous ammonia has some OH^- ions, and the hydroxides of the hardness ions are not soluble in water. Therefore when aqueous ammonia is added to hard

Calcium carbonate deposits in a 2-in. hot water pipe after two years of service in northeastern New Jersey. (Courtesy of The Permutit Company, a division of Sybron Corporation.)

water, the following kind of reaction occurs (illustrated using the magnesium ion this time):

$$Mg^{2+}(aq) + 2OH^-(aq) \longrightarrow Mg(OH)_2(s)$$

As hydroxide ions are removed by this reaction, more are made available from the ammonia-water equilibrium. (A loss of OH^- ion from this equilibrium is a stress, and the equilibrium shifts to the right in response, as we'd predict using Le Chatelier's principle.)

Still another water-softening technique is to let the hard water trickle through zeolite, a naturally-occurring porous substance that is rich in sodium ions. When the hard water is in contact with the zeolite, sodium ions go into the water and the hardness ions leave solution and attach themselves to the zeolite. Later, the hardness ions are flushed out of the zeolite by letting water that is very concentrated in sodium chloride trickle through the spent zeolite, and this zeolite is ready for reuse. Synthetic, ion-exchange materials are also used to soften water by roughly the same principle.

Perhaps the most common strategy in areas where the water is quite hard is to use synthetic detergents instead of soap. Synthetic detergents do not form scum and precipitates with the hardness ions.

3. The first two strong bases of Table 8.2 (because then we can assume that all of the rest are either weak or are too insoluble in water to matter much).

To summarize, we expect ions to react with each other if any one of the following possibilities is predicted.

1. A gas forms that (mostly) leaves the solution — it could be hydrogen (from the action of acids on metals), or carbon dioxide (from acids reacting with carbonates or bicarbonates).

2. An un-ionized, molecular compound forms that remains in solution — it could be water (from acid – base neutralizations), or a weak acid (by the action of H^+ on the acetate ion), or ammonia (by the reaction of OH^- on NH_4^+)

3. A precipitate forms — some water-insoluble salt or one of the water-insoluble hydroxides.

PRACTICE EXERCISE 18   When a solution of hydrochloric acid is mixed in the correct molar proportions with a solution of sodium acetate, $NaC_2H_3O_2$, essentially all of the hydronium ion concentration vanishes. What happens and why? Write the net ionic equation.

PRACTICE EXERCISE 19   What, if anything, happens chemically when each pair of solutions is mixed? Write net ionic equations for any reactions that occur.

(a) NaCl and $AgNO_3$
(b) $CaCO_3$ and HNO_3
(c) KBr and NaCl

The Common Ion Effect.   The solubility rules of salts work to predict the solubility of an individual salt when this salt is to be the lone solute in solution. In nature and in living systems, however, solutions this simple seldom occur. Usually, two or more electrolytes are present, and it becomes important to consider any changes in the solubility of one compound that might be caused by the presence of another. If the other solutes provide *different* ions entirely, then the solubility rules work. But if some other solute contributes one ion that is *common* to the salt whose solubility we are studying, then the solubility of the latter is reduced. This reduction in solubility of one salt by the addition of a common ion is called the **common ion effect.** Let's see how it works.

Suppose that we have a *saturated* solution of sodium chloride. The following equilibrium exists.

$$NaCl(s) \rightleftharpoons Na^+(aq) + Cl^-(aq)$$

What happens if we now pour into this solution some concentrated hydrochloric acid, a fully ionized acid? By using concentrated acid we can quickly increase the concentration of the chloride ion in the solution, and this ion is *common* to the original solute, NaCl. By increasing the concentration of Cl^-, we place a stress on the equilibrium. In accordance with Le Chatelier's principle, the equilibrium has to shift to absorb this stress, and it has to shift to the left. Only by running the reverse reaction of the equilibrium can the system reduce the concentration of dissolved Cl^- and so reduce the stress. But this has to result in the precipitation of more solid sodium chloride, and this is exactly what happens, as the photos of Figure 8.8 show. After the activity has quieted, we still have a saturated solution, but we also have more solid NaCl, and we have a lower concentration of dissolved Na^+ ions.

Special Topic 8.5 describes an interesting common ion effect that leads to some of the symptoms of urinary calculus disease (kidney stones).

8.6 ACID AND BASE IONIZATION CONSTANTS

Regardless of how we try to change the concentration of just one species in a chemical equilibrium, the equilibrium adjusts automatically so that the value of the equilibrium constant doesn't change.

Thus far we have deliberately kept our study of the strengths and weakness of acids and bases as well as the solubilities of salts at a qualitative, nonnumerical level. To develop a deeper

In medicine a *calculus* is an abnormal nonliving aggregation of mineral salts in a framework of matrix or organic materials. Urinary calculi, commonly called kidney stones or bladder stones depending on location, include calcium and magnesium salts. Why they develop in some people and not in others is not known. However, something goes wrong in the way that the system manages its calcium and magnesium ions.

Body fluids contain some of the ions of phosphoric acid. They consist mostly of $H_2PO_4^-$ and HPO_4^{2-}, but some PO_4^{3-} are also present. The salts of calcium or magnesium with the phosphate ion, PO_4^{3-}, are insoluble in water. Thus if too high a level of either calcium or magnesium ion in body fluids develops, the excess Ca^{2+} or Mg^{2+} ions begin to form insoluble matter with phosphate ions. The following equilibrium is established in certain areas of the body:

$$CaPO_4(s) \rightleftharpoons Ca^{2+}(aq) + PO_4^{3-}(aq)$$

Should the calcium or magnesium levels rise further, this equilibrium shifts more to the left, and this is an illustration of the common ion effect. The insoluble phosphates can develop into urinary calculi that become larger and larger over a period of time until they are so large that they become life-threatening.

Gallstones include different salts, but their formation follows the same general principle. Mineral deposits that form in some joints in the condition known as gout are explained in a similar fashion, except that the negative ions come from uric acid, a breakdown product of nitrogen compounds called the nucleic acids.

understanding of these equilibria and of Le Chatelier's principle, we have to look at some quantitative relationships.

The Law of Mass Action. In 1867, C. M. Guldburg and Peter Waage, two Norwegian scientists, discovered a relationship about the molar concentrations of the species in an equilibrium that we now call the **law of mass action.** The simplest statement of it is the following equation:

$$\frac{[C]^c[D]^d}{[A]^a[B]^b} = K_{eq} \tag{8.1}$$

The constant K_{eq} is called the **equilibrium constant.** The brackets denote concentrations in moles per liter. The letters correspond to reactants and products in the following generalized equilibrium:

$$aA + bB \rightleftharpoons cC + dD \tag{8.2}$$

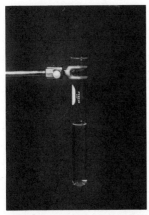
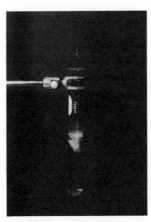
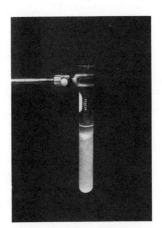
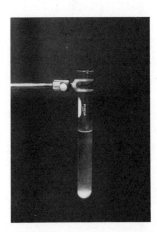

FIGURE 8.8
The common ion effect. At the start there is a saturated solution of sodium chloride—first frame. When concentrated hydrochloric acid is added—the second frame—a white precipitate of sodium chloride appears, grows in quantity (third frame), and finally settles (last frame).

For example, the equilibrium in the ionization of acetic acid is written as follows:

$$HC_2H_3O_2(aq) + H_2O \rightleftharpoons H_3O^+(aq) + C_2H_3O_2^-(aq)$$

The law of mass action expression for this equilibrium is the following equation:

$$\frac{[H_3O^+][C_2H_3O_2^-]}{[HC_2H_3O_2][H_2O]} = K_{eq} \tag{8.3}$$

To keep the expression for K_{eq} as simple as possible, we omit the designations of state such as (aq) or (l).

If we add sodium acetate, $NaC_2H_3O_2$, to a solution in which this equilibrium exists, we cause an increase in the value of $[C_2H_3O_2^-]$. Guldberg and Waage were among the first chemists to realize that a rise in $[C_2H_3O_2^-]$ causes changes in the molar concentrations of all of the other species in the equilibrium. There is nothing mysterious about these adjustments in concentrations. The common ion effect tells us, for example, that as $[C_2H_3O_2^-]$ rises, $[H_3O^+]$ must fall because the equilibrium must shift from right to left to absorb as much of the added $C_2H_3O_2^-$ as possible.

Acid and Base Ionization Constants. When we consider dilute solutions of weak acids and bases, the equation for the equilibrium constant can be simplified. This is because the molar concentration of water, $[H_2O]$, changes so little as the equilibrium becomes established that it is essentially a constant. Let's see why. In pure water, the molar concentration of water is 55.6 mol/L. For all practical purposes, and to a precision of three significant figures, this value for $[H_2O]$ is not changed when a *weak* acid (or base) is present. In this situation the fact that the solute is a *weak* acid (or base) means that only a very small percentage of the solute molecules ionize, so only the same very small percentage of the solvent molecules change. This is why we can treat $[H_2O]$ as a constant in Equation 8.3. Now if we multiply both sides of Equation 8.3 by the same constant, $[H_2O]$, we get:

The molar concentration of water is 55.6 mol/L because 1 L of water is the same as 1000 mL, and this equals 1000 g of water, so in 1 L of water we have:

$$1000 \text{ g } H_2O \times \frac{1 \text{ mol } H_2O}{18.0 \text{ g } H_2O}$$

$$= 55.6 \text{ mol } H_2O$$

$$\frac{[H_3O^+][C_2H_3O_2^-]}{[HC_2H_3O_2][H_2O]} \times [H_2O] = K_{eq} \times [H_2O] = K_a \tag{8.4}$$

The new constant, K_a, the product of multiplying one constant (K_{eq}) by another, ($[H_2O]$), is called the **acid ionization constant.** Each proton-donating substance, whether it is a molecular compound such as acetic acid or an ion such as HSO_4^- or HCO_3^-, has its own value for its acid ionization constant. We can generalize if we let HA represent any acid species — molecular or ionic. Then the equilibrium for its ionization is

$$HA \rightleftharpoons H^+ + A^-$$

The corresponding acid ionization constant is

$$K_a = \frac{[H^+][A^-]}{[HA]} \tag{8.5}$$

When working with aqueous bases, we have the following general equilibrium, where B is any proton acceptor (base):

$$B + H_2O \longrightarrow BH^+ + OH^- \tag{8.6}$$

The **base ionization constant,** K_b, is defined by the following equation.

$$K_b = \frac{[BH^+][OH^-]}{[B]} \tag{8.7}$$

One important use of the values of K_a and K_b is that they tell us not only if a particular acid or base is weak, but also how weak. For example, when K_a is a very small number, such as it is for acetic acid (1.8×10^{-5}), the acid must be a weak acid. The only way that K_a can be small is for the quantities in the *numerator* of Equation 8.5, $[H^+]$ and $[A^-]$, to be small relative to the value of the denominator, $[HA]$. Chemists have found, for example, that when $[HC_2H_3O_2] = 0.100$ mol/L, the experimental values for $[H^+]$ and $[C_2H_3O_2^-]$ are 1.3×10^{-3} mol/L for each. In terms of percentage ionization, these values mean that only 1.3% of the acetic acid molecules are ionized in $0.100\ M$ acetic acid. With still weaker acids, the values of K_a are even smaller and the percentage ionizations in water are still less. Thus we can make a simple generalization—weak acids have low K_a values.

Similarly, the smaller the value of K_b, the weaker is the base. Several values for K_a and K_b are given in Table 8.6.

Equilibrium constants and acid or base ionization constants are higher in solutions that have higher temperatures. The ionizations—the forward reactions of the equilibria—are endothermic changes, so the addition of heat shifts the equilibria to the right in favor of the products. This is why values of K_{eq}, K_a, or K_b, are reported for the particular temperatures at which they are true.

TABLE 8.6
Acid and Base Ionization Constants

Compound	Equilibrium[a]	Ionization Constant[b]
Acids		K_a
Sulfuric acid	$H_2SO_4 \rightleftharpoons H^+ + HSO_4^-$	very large
	$HSO_4^- \rightleftharpoons H^+ + SO_4^{2-}$	1.0×10^{-2}
Phosphoric acid	$H_3PO_4 \rightleftharpoons H^+ + H_2PO_4^-$	7.1×10^{-3}
	$H_2PO_4^- \rightleftharpoons H^+ + HOP_4^{2-}$	6.3×10^{-8}
	$HPO_4^{2-} \rightleftharpoons H^+ + PO_4^{3-}$	4.5×10^{-13}
Hydrofluoric acid	$HF \rightleftharpoons H^+ + F^-$	6.3×10^{-4}
Acetic acid	$HC_2H_3O_2 \rightleftharpoons H^+ + C_2H_3O_2^-$	1.8×10^{-5}
Ascorbic acid (Vitamin C)	$H_2C_6H_6O_6 \rightleftharpoons H^+ + HC_6H_6O_6$	7.9×10^{-5}
Carbonic acid	$H_2CO_3 \rightleftharpoons H^+ + HCO_3^-$	4.5×10^{-7}
	$HCO_3^- \rightleftharpoons H^+ + CO_3^{2-}$	4.7×10^{-11}
Hydrocyanic acid	$HCN \rightleftharpoons H^+ + CN^-$	6.2×10^{-10}
Bases		K_b
Phosphate ion	$PO_4^{3-} + H_2O \rightleftharpoons HOP_4^{2-} + OH^-$	2.2×10^{-2}
Carbonate ion	$CO_3^{2-} + H_2O \rightleftharpoons HCO_3^- + OH^-$	2.1×10^{-4}
Ammonia	$NH_3 + H_2O \rightleftharpoons NH_4^+ + OH^-$	1.8×10^{-5}
Monohydrogen phosphate ion	$HOP_4^{2-} + H_2O \rightleftharpoons H_2PO_4^- + OH^-$	1.6×10^{-7}
Bicarbonate ion	$HCO_3^- + H_2O \rightleftharpoons H_2CO_3 + OH^-$	2.6×10^{-8}
Dihydrogen phosphate ion	$H_2PO_4^- + H_2O \rightleftharpoons H_3PO_4 + OH^-$	1.4×10^{-12}

[a] When we have quantitative data for ionization constants, we drop all attempts to use arrows of unequal lengths to indicate relative strengths.

[b] At 25 °C. Data are from E. H. Martell and R. M. Smith, *Critical Stability Constants* (New York: Plenum Press, 1974). The value of K_a for the first ionization of ascorbic acid, a diprotic acid, is from R. C. Weast, *Handbook of Chemistry and Physics*, 64th edition (Cleveland: CRC Publishing Co., 1983)

SUMMARY

Ionization of water Trace concentrations of hydronium ions, H_3O^+, and hydroxide ions, OH^-, are always present in water. In neutral water, their molar concentrations are equal (and very low). In writing equations, we usually write H_3O^+ as H^+, calling the latter either the hydrogen ion or the proton. In explaining these reactions, however, it's usually necessary to use the correct formula, H_3O^+.

Arrhenius theory Acids are substances that liberate hydrogen ions in water and bases are substances that produce hydroxide ions in water. Salts are ionic compounds that involve any other ions. Strong acids or bases in the Arrhenius theory are those that break up into ions to the extent of a high percent, up to 90 or 100%. The five most common strong acids are hydrochloric, hydrobromic, hydriodic, sulfuric, and nitric acid. All are monoprotic except sulfuric acid, which is diprotic.

Electrolytes Salts in their molten states and the aqueous solutions of all soluble salts and of all strong acids and strong, soluble bases conduct electricity, and are called strong electrolytes. Solutions of weaker acids or of slightly soluble strong bases are weak electrolytes. All molecular substances, unless they change into ions by reacting with water, are nonelectrolytes. Pure water is a nonelectrolyte.

Brønsted theory A Brønsted acid is any that can donate a proton, and a Brønsted base is any that can accept a proton. A strong Brønsted base is able to accept and bind a proton strongly. In water, the strongest Brønsted base is the hydroxide ion. Water is a weak Brønsted base and the chloride ion in water is an even weaker base. A strong Brønsted acid has a strong ability to give up or donate a proton. The hydronium ion is an example of a strong Brønsted acid, and gaseous hydrogen chloride is even stronger. A weak Brønsted acid has a weak or poor ability to donate a proton. It holds its proton instead. Acetic acid is a weak Brønsted acid, and the water molecule is even weaker.

Aqueous acids The strong aqueous acids react with:

> metals, to give the salt of the metal and hydrogen
> metal hydroxides, to give a salt and water
> metal carbonates, to give a salt, carbon dioxide, and water
> metal bicarbonates, to give a salt, carbon dioxide, and water.

A solution of an acid is neutralized when any sufficiently strong proton-binding species is added in the correct mole proportion to make the concentration of hydrogen ion and hydroxide ion equal (and very small).

Carbonic acid and carbonates Carbonic acid, H_2CO_3, is both a weak acid and an unstable acid. When it is generated in water by the reaction of any stronger acid with a bicarbonate or a carbonate salt, virtually all of the carbonic acid decomposes to carbon dioxide and water, and most of the carbon dioxide fizzes out. The carbonate ion and the bicarbonate ion are both Brønsted bases, and the bicarbonate ion is involved in carrying waste carbon dioxide from cells, where it is made, to the lungs.

Aqueous bases Any ion that can accept a proton from the hydronium ion in water qualifies as a base. These include the hydroxide ion, the bicarbonate ion, the carbonate ion, and ammonia plus any of the anions of weak acids. The conjugate base of a weak acid is a strong base, and the conjugate acid of a weak base is a strong acid. Conjugate acid–base pairs differ in formula only by a single unit of H^+.

Ammonia and the ammonium ion Ammonia is a strong base toward H_3O^+ but a weak base toward H_2O. The ammonium ion is a strong acid toward OH^- but a weak acid toward H_2O. Ammonia can neutralize strong acids and the ammonium ion can neutralize strong bases.

Salts The chemical properties of salts in water are the properties of their individual ions. If the anion of the salt is the conjugate base of a weak acid—as HCO_3^- is the conjugate base of H_2CO_3—then the salt can neutralize strong acids. Thus bicarbonates, carbonates, acetates, and the salts of other organic acids supply Brønsted bases— their anions. If the cation of the salt is the conjugate acid of a weak base—as NH_4^+ is the conjugate acid of NH_3—then the salt supplies a Brønsted acid in water.

Salts can be produced by any of the reactions of strong acids that were studied (and summarized, above) as well as by double decomposition reactions. The solubility rules for salts are guides for the prediction of their reactions. If a combination of oppositely charged ions can lead to an insoluble salt, an un-ionized species that stays in solution, or a gas, then the ions react. If a different compound that can furnish an ion that is common to an ion of a salt already in solution is added to this solution, the solubility of the salt might be reduced enough to force it out of solution (common ion effect).

Acid and base ionization constants The values of K_a and K_b for acids and bases, respectively, disclose the relative strengths of these substances. These values increase with temperature. The equations for them are modifications of the law of mass action equations that relate to their ionization equilibria.

KEY TERMS

The following terms were emphasized in this chapter. They should be mastered before you continue the study of acids and bases in the next chapter.

acid, Arrhenius	acid-base neutralization	acid ionization constant, K_a	alkali
acid, Brønsted	acidic solution	activity series	anion

anode

Arrhenius theory of acids
   and bases

base, Arrhenius

base, Brønsted

base ionization constant, K_b

basic solution

cathode

cation

common ion effect

conjugate acid-base pairs

diprotic acid

double decomposition

electrical balance

electrode

electrolysis

electrolyte

equilibrium constant, K_{eq}

hydronium ion

hydroxide ion

inorganic compound

ionic equation

ionization

law of mass action

material balance

molecular equation

monoprotic acid

net ionic equation

neutral solution

nonelectrolyte

organic compound

salt

simple salt

strong acid

strong base

strong Brønsted acid

strong Brønsted base

strong electrolyte

triprotic acid

weak acid

weak base

weak Brønsted acid

weak Brønsted base

weak electrolyte

SELECTED REFERENCE

1 P. A. Giguere. "The Great Fallacy of the H^+ Ion and the True
Nature of H_3O^+." *Journal of Chemical Education,* September
1979, page 571.

REVIEW EXERCISES

The answers to Review Exercises that require a calculation and that
are marked with an asterisk are found in an Appendix. The answers
to the other Review Exercises are found in the *Study Guide* that
accompanies this book.

Atoms, Ions, and Molecules

8.1 What families of compounds are the principal sources of ions in
aqueous solutions?

8.2 Review the differences between atoms and ions by answering
the following questions.
  (a) Are there any atoms that have more than one nucleus? If
  so, give an example.
  (b) Are there any ions with more than one nucleus? If so, give
  an example.
  (c) Are there any ions that are electrically neutral? If so, give
  an example.
  (d) Are there any atoms that are electrically charged? If so,
  give an example.
  (e) Sometimes scientists speak of the "potassium level" in the
  blood. What specifically do they mean?

8.3 Write the equilibrium equation for the self-ionization of water,
and label the ions that are present.

Arrhenius Acids, Bases, and Salts

8.4 What features do the common aqueous acids have in common?

8.5 In the context of acid–base discussion, what are two other
names that we can use for *proton?*

8.6 How did Arrhenius define an acid? A base?

8.7 Acids have a set of common reactions, and so do bases, but not
salts. Explain.

8.8 Tell whether each of the following solutions is acidic, basic, or
neutral.
  (a) $[H^+] = 6.2 \times 10^{-6}$ mol/L and $[OH^-] = 1.6 \times 10^{-9}$
  mol/L
  (b) $[H^+] = 1.0 \times 10^{-7}$ mol/L and $[OH^-] = 1.0 \times 10^{-7}$
  mol/L
  (c) $[H^+] = 1.36 \times 10^{-8}$ mol/L and $[OH^-] = 7.35 \times 10^{-7}$
  mol/L

8.9 Salts are all crystalline solids at room temperature. Why do you
suppose this is?

Electrolytes

8.10 The word *electrolyte* can be understood in two ways. What are
they? Give examples.

8.11 To which electrode do cations migrate?

8.12 The anode has what electrical charge, positive or negative?

8.13 The electrode that is negatively charged attracts what kinds of
ions, cations or anions?

8.14 Explain in your own words how the presence of cations and
anions in water enables the system to conduct electricity.

8.15 When NaOH(s) is dissolved in water, the solution is an excellent
conductor of electricity, but when methyl alcohol is dissolved in
water, the solution won't conduct electricity at all. What does
this behavior suggest about the structural natures of NaOH and

methyl alcohol, whose structure is given below? (Notice that both appear to have OH groups in their formulas.)

$$\begin{array}{c} H \\ | \\ H-C-O-H \\ | \\ H \end{array}$$

Methyl alcohol

8.16 If a water-soluble compound breaks up entirely into ions as it dissolves in water, do we call it a weak or a strong electrolyte?

8.17 In the liquid state, tin(IV) chloride, $SnCl_4$, is a nonconductor. What does this suggest about the structural nature of this compound?

Aqueous Acids and Bases

8.18 What is the difference between hydrochloric acid and hydrogen chloride?

8.19 In which species is the covalent bond to hydrogen stronger, in $HCl(g)$ or in $H_3O^+(aq)$? How do we know?

8.20 We can represent an acid by the general symbol HA. Its equilibrium expression, then, is

$$HA \rightleftharpoons H^+ + A^-$$

On the basis of the way that the equilibrium arrows are drawn, is this acid weak or strong?

8.21 Is $HC_2H_3O_2$ a mono-, di-, tri-, or tetraprotic acid? (What is its name?)

8.22 What are the names and the formulas of the aqueous solutions of the four hydrohalogen acids?

8.23 If we represent all diprotic acids by the symbol H_2A, write the equilibrium expressions for the two separate ionization steps.

8.24 Would the ionization of the second proton from a diprotic acid occur with greater ease or with greater difficulty than the ionization of the first proton? Explain.

8.25 Compare the structures of nitrous acid, HNO_2, and nitric acid, HNO_3.

$$H-O-N=O \qquad \qquad H-O-N\!\!\underset{O}{\overset{O}{\diagup}}$$

Nitrous acid   Nitric acid

Nitrous acid is a much weaker acid than nitric acid. How does the extra oxygen in the structure of nitric acid help to explain this? (*Hint:* Study Special Topic 8.2.)

8.26 Which is the stronger acid in water, sulfurous acid or sulfuric acid?

$$\begin{array}{c} O \\ \| \\ H-O-S-O-H \end{array} \qquad \begin{array}{c} O \\ \| \\ H-O-S-O-H \\ \| \\ O \end{array}$$

Sulfurous acid   Sulfuric acid

Explain this along the lines of argument useful for Review Exercise 8.25.

8.27 Write the equilibrium expression for the solution of carbon dioxide in water that produces some carbonic acid.

8.28 Write the equilibrium expressions for the successive steps in the ionization of carbonic acid.

8.29 Magnesium hydroxide is practically insoluble in water, and yet it is classified as a strong base. Explain.

8.30 Ammonia is very soluble in water, and yet it is called a weak base. Explain.

8.31 What are the names and formulas of two bases that are both strong and are capable of forming relatively concentrated solutions in water?

8.32 When carbon dioxide is bubbled into pure water to form a solution, it takes only time and the help of a little warming to drive essentially all of it out of solution again. When this gas is bubbled into aqueous sodium hydroxide, however, it is completely trapped by a chemical reaction. If we assume that the reaction involves CO_2 and NaOH in a mole ratio of 1 to 1, what is the molecular equation for this trapping reaction?

8.33 What is meant by *aqueous ammonia*? Why don't we call it "ammonium hydroxide"?

8.34 Write the names and the formulas of the five strong acids that we have studied.

8.35 What are the four strong bases — both the names and formulas? Which are quite soluble in water?

Net Ionic Equations

8.36 Consider the following net ionic equation:

$$2H^+(aq) + Cu(s) + NO_3^-(aq) \longrightarrow$$
$$Cu^{2+}(aq) + NO_2(g) + H_2O$$

(a) Does it have material balance?
(b) Does it have electrical balance?

8.37 Complete and balance the following molecular equations, and then write the net ionic equations.
(a) $HNO_3(aq) + NaOH(aq) \rightarrow$
(b) $HCl(aq) + K_2CO_3(aq) \rightarrow$
(c) $HBr(aq) + CaCO_3(s) \rightarrow$
(d) $HNO_3(aq) + NaHCO_3(aq) \rightarrow$
(e) $HI(aq) + NH_3(aq) \rightarrow$
(f) $HNO_3(aq) + Mg(OH)_2(s) \rightarrow$
(g) $HBr(aq) + Zn(s) \rightarrow$

8.38 Complete and balance the following molecular equations, and then write the net ionic equation.
(a) $KOH(aq) + H_2SO_4(aq) \rightarrow$
(b) $Na_2CO_3(aq) + HNO_3(aq) \rightarrow$
(c) $KHCO_3(aq) + HCl(aq) \rightarrow$
(d) $MgCO_3(s) + HI(aq) \rightarrow$
(e) $NH_3(aq) + HBr(aq) \rightarrow$
(f) $Ca(OH)_2(s) + HCl(aq) \rightarrow$
(g) $Al(s) + HCl(aq) \rightarrow$

8.39 What are the net ionic equations for the following reactions of strong, aqueous acids? (Assume that all reactants and products are soluble in water.)
(a) With metal hydroxides

(b) With metal bicarbonates

(c) With metal carbonates

(d) With aqueous ammonia

8.40 Write net ionic equations for the reactions of all of the water-insoluble group IIA carbonates, where you use $MCO_3(s)$ as their general formula, with hydrochloric acid (chosen so that all of the products are soluble in water).

8.41 If we let $M(OH)_2(s)$ represent the water-insoluble group IIA metal hydroxides, what is the general net ionic equation for all of their reactions with nitric acid (chosen so that all of the products are soluble in water)?

8.42 If we let $M(s)$ represent either calcium or magnesium metal, what net ionic equation represents the reaction of either with hydrochloric acid?

8.43 Sodium and potassium in group IA are higher in the activity series than calcium and magnesium in group IIA.

(a) What does it mean to be *higher* in the activity series?

(b) If you check back to Figure 3.5b, on page 65, you will see that sodium and potassium have lower ionization energies than calcium and magnesium. In what way does this fact correlate with their higher position in the activity series of the metals?

8.44 Zinc metal reacts more rapidly with which acid, 1 M nitric acid or 1 M acetic acid?

*8.45 How many moles of sodium bicarbonate can react quantitatively with 0.250 mol of HCl?

8.46 How many moles of potassium hydroxide can react quantitatively with 0.400 mol of H_2SO_4 (assuming that both H^+ in H_2SO_4 are neutralized).

*8.47 How many grams of sodium carbonate does it take to neutralize 4.60 g of HCl?

8.48 How many grams of calcium carbonate react quantitatively with 6.88 g of HNO_3?

*8.49 How many grams of sodium bicarbonate does it take to neutralize all of the acid in 25.4 mL of 1.15 M H_2SO_4?

8.50 How many grams of potassium carbonate will neutralize all of the acid in 36.8 mL of 0.550 M HCl?

*8.51 How many milliliters of 0.246 M NaOH are needed to neutralize the acid in 32.4 mL of 0.224 M HNO_3?

8.52 How many milliliters of 0.108 M KOH are needed to neutralize the acid in 16.4 mL of 0.116 M H_2SO_4?

*8.53 For an experiment that required 12.0 L of dry CO_2 gas (as measured at 740 mm Hg and 25 °C), a student let 5.00 M HCl react with marble chips, $CaCO_3$.

(a) Write the molecular and net ionic equations for this reaction.

(b) How many grams of $CaCO_3$ and how many milliliters of the acid are needed?

8.54 How many liters of dry CO_2 gas are generated (at 750 mm Hg and 20 °C) by the reaction of $Na_2CO_3(s)$ with 250 mL of 6.00 M HCl? Write the molecular and the net ionic equations for the reaction, and calculate how many grams of Na_2CO_3 are needed.

Brønsted Acids and Bases

8.55 Acetic acid is a weak acid. Arrhenius would select which species in aqueous acetic acid as specifically being the acid species? Which species would Brønsted pick?

8.56 In aqueous ammonia, which species in equilibrium would Arrhenius say is the base? Which species would Brønsted call the most abundant base?

8.57 Write the formulas of the conjugate acids of the following:
(a) HSO_4^- (b) HCO_3^- (c) I^- (d) NO_2^-

8.58 What are the formulas of the conjugate acids of the following?
(a) HSO_3^- (b) Br^- (c) H_2O (d) $C_2H_3O_2^-$

8.59 What are the conjugate bases of the following? Write their formulas.
(a) H_2CO_3 (b) $H_2PO_4^-$ (c) NH_4^+ (d) OH^-

8.60 Write the formulas of the conjugate bases of the following:
(a) NH_3 (b) HNO_2 (c) HSO_3^- (d) H_2SO_3

8.61 Which member of each pair is the stronger Brønsted base?
(a) Br^- or HCO_3^- (b) $H_2PO_4^-$ or HSO_4^-
(c) NO_2^- or NO_3^-

8.62 Which member of each pair is the stronger Brønsted base?
(a) NH_3 or NH_2^- (b) OH^- or H_2O
(c) HS^- or S^{2-}

8.63 Study each pair and decide which is the stronger Brønsted acid.
(a) $H_2PO_4^-$ or HOP_4^{2-} (b) H_2SO_3 or HSO_3^-
(c) NH_4^+ or NH_3

8.64 Which member of each pair is the stronger Brønsted acid?
(a) H_2CO_3 or HCl   (b) H_2O or OH^-
(b) HSO_4^- or HSO_3^-

8.65 If sodium phosphate and sodium hydrogen sulfate solutions are mixed in equimolar amounts of their solutes, the following ionic equilibria is established (only the arrowheads have been omitted):

$$HPO_4^{2-}(aq) + SO_4^{2-} \rightleftharpoons PO_4^{3-}(aq) + HSO_4^-(aq)$$

Put the arrowheads at the correct ends of the proper lines to complete this equilibrium expression.

8.66 Aspirin is a weak acid. We can represent it as H(*Asp*), and it has a sodium salt that we can symbolize as Na(*Asp*). When the sodium salt of aspirin is given as a medication and it encounters gastric juice, which contains HCl(aq), the following ionic equilibrium is established (at least temporarily). The arrowheads are not shown. You are to supply them.

$$(Asp)^-(aq) + H_3O^+(aq) \rightleftharpoons H(Asp)(aq) + H_2O$$

8.67 Suppose you are handed a test tube and told that it contains a concentrated solution of either ammonium chloride or potassium chloride. An aqueous solution of one of the substances we have studied in this chapter could be added to the unknown solution as a test for deciding which of the two solutes is present. What is this test reagent, and what would you observe as a result of the test if the unknown contained ammonium chloride?

Salts

8.68 Write the names and formulas of three compounds that, by reacting with hydrochloric acid, give a solution of potassium chloride. Write the molecular equations for these reactions.

8.69 Write the names and formulas of three compounds that will give a solution of lithium bromide when they react with hydrobromic acid. Write the molecular equations for these reactions.

8.70 Which of the following compounds are insoluble in water (as we have defined solubility)?
(a) NaOH   (b) NH_4Br (c) Hg_2Cl_2
(d) $Ca_3(PO_4)_2$ (e) KBr   (f) Li_2SO_4

8.71 Which of the following compounds are insoluble in water?
(a) $(NH_4)_2SO_4$ (b) KNO_2 (c) LiCl
(d) AgCl   (e) $Mg_3(PO_4)_2$ (f) $NaNO_3$

8.72 Identify the compounds that do not dissolve in water.
(a) NH_4NO_3 (b) $BaCO_3$ (c) $PbCl_2$
(d) K_2CO_3 (e) $LiC_2H_3O_2$ (f) Na_2SO_4

8.73 Which of the following compounds do not dissolve in water?
(a) Li_2CO_3 (b) $Na_2Cr_2O_7$ (c) NH_4I
(d) AgBr   (e) K_2CrO_4 (f) $FeCO_3$

8.74 Assume you have separate solutions of each compound in the pairs below. Predict what happens chemically when the two solutions of a pair are poured together. If no reaction occurs, state so. If there is a reaction, write its net ionic equation.
(a) LiCl and $AgNO_3$ (b) $NaNO_3$ and $CaCl_2$
(c) KOH and H_2SO_4 (d) $Pb(NO_3)_2$ and KCl
(e) NH_4Br and K_2SO_4 (f) Na_2S and $CuSO_4$
(g) K_2SO_4 and $Ba(NO_3)_2$ (h) NaOH and HI
(i) K_2S and $NiCl_2$ (j) $AgNO_3$ and NaCl
(k) $LiHCO_3$ and HBr   (l) $CaCl_2$ and KOH

8.75 If you have separate solutions of each of the compounds given below and then mix the two of each pair together, what (if anything) happens chemically? If no reaction occurs, state so, but if there is a reaction write its net ionic equation.
(a) H_2S and $CdCl_2$ (g) Na_2S and $Ni(NO_3)_2$
(b) KOH and HBr   (h) NaOH and HBr
(c) Na_2SO_4 and $BaCl_2$ (i) $Hg(NO_3)_2$ and KCl
(d) $Pb(C_2H_3O_2)_2$ and Li_2SO_4 (j) $NaHCO_3$ and HI
(e) $Ba(NO_3)_2$ and KCl   (k) KBr and NaCl
(f) $NaHCO_3$ and H_2SO_4 (l) $Pb(NO_3)_2$ and Na_2CrO_4

8.76 Soap is a mixture of the sodium salts of certain organic acids. One is sodium stearate, which we can represent as Na(*Ste*).
(a) Write the equilibrium expression for a saturated solution of this salt in water.
(b) What would happen to this equilibrium if a concentrated solution of sodium chloride were added to it?
(c) The NaCl solution need not be concentrated. Sea water is about 3% (w/w) NaCl, and soap doesn't work well when sea water is used. Suggest a reason.

Acid and Base Ionization Constants

8.77 Write the mass action expression for the following equilibrium:
$$2H_2 + O_2 \rightleftharpoons 2H_2O$$

8.78 Hydrogen and iodine, both in the gaseous state, react to establish the following equilibrium:
$$H_2(g) + I_2(g) \rightleftharpoons 2HI(g)$$
(a) Write the mass action expression for this equilibrium.
(b) In which direction would this equilibrium shift if the partial pressure of hydrogen were increased? (*Note:* This is the same as increasing the molar concentration of the hydrogen.)
(c) An increase in the partial pressure of hydrogen would cause what kind of change (if any) to the value of K_{eq} for this reaction?

8.79 Nitrous acid, HNO_2, is a weak monoprotic acid.
(a) Write the equilibrium expression for its solution in water.
(b) Write the equation for K_a for this acid.
(c) At 25 °C, the K_a value for nitrous acid is 7.1×10^{-4}. Hydrofluoric acid, HF, has a K_a value of 6.8×10^{-4} at the same temperature. Which is the stronger acid, HNO_2 or HF?
(d) If some concentrated sodium nitrite solution, $NaNO_2(aq)$, were added to a solution of nitrous acid, in what way (if any) would the value of K_a change?

8.80 One of the anions present in cellular fluid is $HPO_4{}^{2-}$.
(a) Write the equilibrium expression for the situation in which this anion acts as a Brønsted base and takes *one* proton from a water molecule.
(b) Write the equation that expresses the law of mass action for this equilibrium.
(c) Write the equation that expresses K_b for this system.
(d) The value of K_b for the monohydrogen phosphate ion is 1.6×10^{-7}. The value of K_b for the bicarbonate ion is 2.6×10^{-8}. Which is the stronger base, $HPO_4{}^{2-}$ or $HCO_3{}^-$?

8.81 As we have learned, saturated solutions also involve equilibria. For example, the equilibrium present in a saturated solution of AgCl is
$$AgCl(s) \rightleftharpoons Ag^+(aq) + Cl^-(aq)$$
The mass action law for this equilibrium is
$$K_{eq} = \frac{[Ag^+(aq)][Cl^-(aq)]}{[AgCl(s)]}$$
If both sides of this equation are multiplied by [AgCl(s)], one of the new terms will be $[AgCl(s)] \times K_{eq}$. This turns out to be a constant, a new constant called the K_{sp} for AgCl.
(a) Explain how we can know that the product, $[AgCl(s)] \times K_{eq}$, must have a constant value (at a given temperature). (*Hint:* Show that the value of [AgCl(s)] is a constant. What are its units?)
(b) Each salt has its own value of K_{sp}. For example, the K_{sp} for $BaSO_4$ is 6.3×10^{-7}. Write the equation for the K_{sp} expression for $BaSO_4$.
(c) The value of K_{sp} for AgCl is 1.8×10^{-10}. Assuming that this value and the value for the K_{sp} of $BaSO_4$ (part b) relate to the same temperature, which is more soluble in water, AgCl or $BaSO_4$? How can you tell?

Chapter 9
Acidity: Detection, Control, Measurement

If this train of 33,000-gallon tank cars carried water at a pH of 7, the entire amount of water would contain only about 1 gram of H⁺ ion. The pH concept, a topic in this chapter, helps us describe such trace concentrations easily.

9.1   DESCRIBING LOW ACIDITIES

Very low levels of H⁺ are more easily described and compared in terms of pH values than as molar concentrations.

The brackets in [H⁺] signify that the concentration is specifically moles per liter.

At the molecular level of life we must have almost perfect control over the acid-base balance of body fluids. Blood, for example, must have a molar concentration of hydrogen ion, [H⁺], of 4.47×10^{-8} mol/L or extremely close to this. If the value of [H⁺] goes up to 10×10^{-8} mol/L or drops to 1×10^{-8} mol/L, death is very near. Yet we can change the acidity of pure water far more than these figures represent by adding only one drop of concentrated hydrochloric acid or one drop of concentrated sodium hydroxide to a liter.

So critical is the acid–base balance of the blood that a special vocabulary has been created to describe small shifts toward acidity or basicity. If the acidity of the blood increases, the condition is called **acidosis.** Acidosis is characteristic of untreated diabetes and emphysema, and other conditions. If the acidity of the blood decreases, the condition is called **alkalosis.** An overdose of bicarbonate, exposure to the low partial pressure of oxygen at high altitudes, or prolonged hysteria can cause alkalosis. In their more advanced stages, acidosis and alkalosis are medical emergencies because they interfere with the smooth working of **respiration** — the physical and chemical apparatus that brings oxygen in, uses it, and then removes waste carbon dioxide. We need much more background before we can understand this complex subject, but important parts of the preparation are given in this chapter.

The Ion Product Constant of Water.   We begin by returning for a quantitative look at the self-ionization of water. As we learned in Section 8.1, the equilibrium expression for this (using H⁺ for H_3O^+) is

$$H_2O \rightleftharpoons H^+(aq) + OH^-(aq)$$

According to the law of mass action (Section 8.6), we can write the following mass action equation for this equilibrium.

$$K_{eq} = \frac{[H^+][OH^-]}{[H_2O]} \tag{9.1}$$

We also learned that the values of [H⁺] and [OH⁻] in pure water are very low, 1.0×10^{-7} mol/L each at 25 °C. This is so low that the formation of these ions from water molecules has no effect on the value of $[H_2O]$, even if we round to seven significant figures. The value of $[H_2O]$ is thus a constant, and if we multiply the expression of Equation 9.1 on both sides by $[H_2O]$, we obtain a new constant:

$K_{eq} \times [H_2O]$ is one constant multiplied by another.

$$K_{eq} \times [H_2O] = \frac{[H^+][OH^-]}{[H_2O]} \times [H_2O] = \text{a new constant}$$

The new constant is called the **ion product constant of water,** and its symbol is $\boldsymbol{K_w}$.

Units are never included in quoted values for K_w or K_{eq}.

$$K_w = [H^+][OH^-] \tag{9.2}$$

In accordance with the law of mass action, no matter how we change the value of, say, [H⁺] by adding either acid or base to a solution, the value of the other concentration term in Equation 9.2 adjusts, and the *product* of the two terms remains equal to the constant, K_w. The

K_w at Various Temperatures

Temperature (°C)	K_w
0	1.5×10^{-15}
10	3.0×10^{-15}
20	6.8×10^{-15}
30	1.5×10^{-14}
40	3.0×10^{-14}

only method for changing K_w is to change the temperature, as data in the margin show.

At 25 °C (and to two significant figures),

$$[H^+] = 1.0 \times 10^{-7} \text{ mol/L}$$

and

$$[OH^-] = 1.0 \times 10^{-7} \text{ mol/L}$$

Therefore at 25 °C,

$$K_w = (1.0 \times 10^{-7})(1.0 \times 10^{-7})$$
$$= 1.0 \times 10^{-14} \text{ (at 25 °C)} \qquad (9.3)$$

In all of our work, we will assume a temperature of 25 °C.

With the value of K_w given by Equation 9.3, we can calculate the value of one of the two concentration terms, $[H^+]$ or $[OH^-]$, if we know the other.

EXAMPLE 9.1 **USING THE ION PRODUCT CONSTANT OF WATER**

Problem: The value of $[H^+]$ of blood (when measured at 25 °C, not body temperature) is 4.5×10^{-8} mol/L. What is the value of $[OH^-]$, and is the blood acidic, basic, or neutral?

Solution: We simply use the value of $[H^+]$ in the equation for K_w.

$$K_w = 1.0 \times 10^{-14} = (4.5 \times 10^{-8}) \times [OH^-]$$
$$[OH^-] = \frac{1.0 \times 10^{-14}}{4.5 \times 10^{-8}}$$
$$= 2.2 \times 10^{-7} \text{ mol/L}$$

Because the value of $[H^+]$ is less than the value of $[OH^-]$, the blood is (very slightly) basic.

PRACTICE EXERCISE 1   For each of the following values of $[H^+]$, calculate the value of $[OH^-]$ and state whether the solution is acidic, basic, or neutral.

(a) $[H^+] = 4.0 \times 10^{-9}$ mol/L
(b) $[H^+] = 1.1 \times 10^{-7}$ mol/L
(c) $[H^+] = 9.4 \times 10^{-8}$ mol/L

The pH Concept.   When we compare numbers such as we found in Example 9.1—4.5×10^{-8} versus 2.2×10^{-7}—to see which is larger, we have to look in two places for each number. We have to compare the exponents of 10 and then the numbers before the 10s. Yet numbers this small (and this close together) are often encountered in dealing with the relative acidities of body fluids such as blood. To make these kinds of comparisons easier, a Danish biochemist, S.P.L. Sorenson (1868–1939), invented the concept of pH.

There are two ways of writing the equation that defines pH. We will put the equation that involves logarithms into Special Topic 9.1. All of our needs will be met with the following closely related exponential form of the definition of pH:

$$[H^+] = 1 \times 10^{-pH} \qquad (9.4)$$

In other words, the **pH** of a solution is the negative power (the p in pH) to which the number 10 must be raised to express the molar concentration of a solution's hydrogen ions (hence, the H in pH).

The exponential definition of pH is

$$[H^+] = 1.00 \times 10^{-pH}$$

By taking the common logarithm of both sides of this equation, we get:

$$\log[H^+] = -pH$$

After multiplying both sides by -1, we obtain the usual form of this relationship, which is the logarithmic definition of pH:

$$pH = -\log[H^+]$$

In this form, we can use logarithms to calculate pH from values of $[H^+]$, or we can convert values of pH into values of $[H^+]$. For example, if $[H^+] = 6.5 \times 10^{-7}$ mol/L, then

$$
\begin{aligned}
pH &= -[\log(6.5 \times 10^{-7})] \\
&= -[\log 6.5 + \log 10^{-7}] \\
&= -[0.81 + (-7)] \\
&= -[-6.19] \\
pH &= 6.19
\end{aligned}
$$

If we are given the information that the pH of some digestive juice is 8.20, then

$$8.20 = -\log[H^+]$$

or

$$
\begin{aligned}
\log[H^+] &= -8.20 \\
&= (-9) + 0.80 \\
[H^+] &= [\text{antilog}(0.80)] \times 10^{-9} \\
&= 6.3 \times 10^{-9} \text{ mol/L}
\end{aligned}
$$

If you have a pocket calculator with the $+/-$, the log, and the 10^x keys, then working these problems becomes very easy. For example, to use the same problems, if we know that $[H^+] = 6.5 \times 10^{-7}$ mol/L, enter 6.5×10^{-7} into the calculator, and then hit the $\boxed{\log}$ button. If you don't know how your calculator handles this, consult its manual. However, most calculators have a key labeled *EE* or *EXP* that is used to enter the exponential part of a number such as the given value of $[H^+]$. Just remember that 6.5×10^{-7} reads "6.5 *times* 10 to the minus 7." The key *EE* (or *EXP*) substitutes for the phrase ". . . . times ten to the . . ." in this reading. Therefore the correct sequence of hitting the calculator keys is

$$\boxed{6}\ \boxed{.}\ \boxed{5}\ \boxed{EE}\ \boxed{+/-}\ \boxed{7}\ \boxed{\log}$$

After this sequence the calculator reads $-6.1870866\ 00$. This is the logarithm of 6.5×10^{-7}. Because the pH is the negative of this logarithm, the pH is, after correctly rounding, 6.19.

For the other problem we used above — find the value of $[H^+]$ if the pH is 8.20, simply enter 8.20 and change the sign to $-$, so -8.20 shows on the calculator screen. Now hit the 10^x button. The screen will now read either .0000000063 or 6.3×10^{-9}. If your result comes out as 0.0000000063, you can get the calculator to reexpress this exponentially by hitting the *ENG* key (meaning Engineering Notation). Try the following Practice Exercises to see if you can work these problems with your calculator.

1. What are the pH values of the solutions that have the following values of $[H^+]$?
   (a) 4.3×10^{-4} mol/L
   (b) 1.4×10^{-7} mol/L
   (c) 8.9×10^{-14} mol/L
2. What are the molar concentrations of hydrogen ion in solutions that have the following values of pH?
   (a) 2.76   (b) 6.90   (c) 7.35   (d) -1.00

Answers

1. (a) 3.37   (b) 6.85   (c) 13.05
2. (a) 1.74×10^{-3} mol/L   (b) 1.26×10^{-7} mol/L
   (c) 4.47×10^{-8} mol/L   (d) 1×10^1 mol/L

In pure water at 25 °C, $[H^+] = 1.0 \times 10^{-7}$ mol/L. Therefore the pH of pure water at 25 °C is 7.00. Thus a pH of 7.00 corresponds to a neutral solution at 25 °C.[1]

There is an analogous concept for expressing low concentrations of OH^-, the **pOH** of a solution:

$$[OH^-] = 1 \times 10^{-pOH} \tag{9.5}$$

Values of pOH are seldom used, but when they are a simple relationship between pH and pOH

---

[1] The 7 in 7.00 comes from the *exponent* in 1.0×10^{-7}, so it actually does nothing more than set off a decimal point when we rewrite the number as 0.00000010. Hence the 7 in the pH value of 7.00 can't be counted as a significant figure. A pH value of 7.00 therefore has just two significant figures, those that follow the decimal point, just as there are but two significant figures in the value of the molar concentration of H^+, 1.0×10^{-7} mol/L. To repeat, the number of significant figures in any value of pH is the number of figures that *follow* the decimal point.

exists. Notice that if we insert the pH and the pOH expressions for $[H^+]$ and $[OH^-]$ into the equation for the ion product constant of water, Equation 9.2, we get

$$(1.0 \times 10^{-pH})(1.0 \times 10^{-pOH}) = 1.0 \times 10^{-14} \qquad \text{(at 25 °C)}$$

A review of exponents is in Appendix A.

Recall that when we multiply numbers that involve exponents we *add* the exponents. Therefore this equation must mean that the following relationship is true at 25 °C:

$$pH + pOH = 14.00 \qquad \text{(at 25 °C)} \qquad (9.6)$$

Because pH occurs as a *negative* exponent in Equation 9.4, it takes a pH value that is *lower* than 7.00 for the solution to be acidic, and a value *higher* than 7.00 for it to be basic. In pH terms, then, we have the following definitions of acidic, basic, and neutral solutions when their temperatures are 25 °C.

Acidic solution	$pH < 7.00$	
Neutral solution	$pH = 7.00$	(9.7)
Basic solution	$pH > 7.00$	

When the value of $[H^+]$ is 1 mol/L or higher, the pH concept isn't used. The exponents would no longer be negative, so there would be none of the confusion that Sorenson addressed when he invented pH.

Table 9.1 gives the correlations of pH, $[H^+]$, $[OH^-]$, and pOH values for the entire useful range of pH, 0 to 14. The pH values of several common substances are shown in Figure 9.1.

One of the very deceptive features of the pH concept is that the actual hydrogen ion concentration changes greatly — by a factor of 10 — for each change of only one unit of pH.

TABLE 9.1
pH, $[H^+]$, $[OH^-]$, and pOH

pH	$[H^+]$	$[OH^-]$	pOH	
0	1	1×10^{-14}	14	
1	1×10^{-1}	1×10^{-13}	13	
2	1×10^{-2}	1×10^{-12}	12	
3	1×10^{-3}	1×10^{-11}	11	Acidic Solutions
4	1×10^{-4}	1×10^{-10}	10	
5	1×10^{-5}	1×10^{-9}	9	
6	1×10^{-6}	1×10^{-8}	8	
7	1×10^{-7}	1×10^{-7}	7	Neutral Solution
8	1×10^{-8}	1×10^{-6}	6	
9	1×10^{-9}	1×10^{-5}	5	
10	1×10^{-10}	1×10^{-4}	4	Basic Solutions
11	1×10^{-11}	1×10^{-3}	3	
12	1×10^{-12}	1×10^{-2}	2	
13	1×10^{-13}	1×10^{-1}	1	
14	1×10^{-14}	1×10^{-0}	0	

Concentrations are in mol/L at 25 °C.

FIGURE 9.1
The pH scale and the pH values of several common substances.

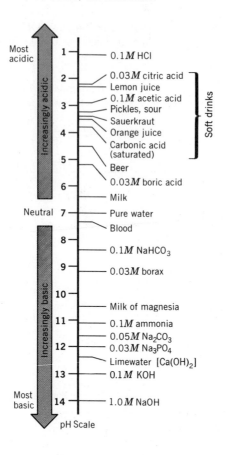

In mass, 1 mol of H⁺ has a mass of only 1.0 g.

For example, if the pH of a solution is zero (meaning that $[H^+] = 1 \times 10^0$ mol/L), only 1 L of water is needed to contain 1 mol of H^+. When the pH is 1, however, then 10 L of water (about the size of an average wastebasket) is needed to hold 1 mol of H^+. At a pH of 5, it takes a large railroad tank car full of water to include just 1 mol of H^+. If the pH of the water flowing over Niagara Falls, New York, were 10 (which, of course, it isn't), an entire 1-hour flowage would be needed for 1 mol of H^+ to pass by. And at a pH of 14, the volume that would hold 1 mol of H^+ is about a quarter of the volume of Lake Erie (one of the Great Lakes). You can see that seemingly small changes in pH numbers signify enormous changes in real concentrations of hydrogen ions.

Another point about pH to be emphasized, is that it refers to the molar concentration of *hydrogen ions*, not to the molar concentration of any particular solute that contributes these ions. Only with dilute solutions of strong, 100-percent ionized acids is there a simple correlation. For example, a solution that is 0.01 M HCl, which is the same as 1×10^{-2} mol HCl(aq)/L, must also have a molar concentration of H^+ ions of 1×10^{-2} mol/L. This is because each molecule of HCl that went into the solution breaks up into one H^+ ion and one Cl^- ion. Therefore because $[H^+] = 1 \times 10^{-2}$ mol/L, the pH is simply 2.0. Similarly, a solution that is 0.0001 M HNO_3 has a pH of 4.0, because 0.0001 is the same as 1×10^{-4}, and because HNO_3 is a strong acid and fully ionized. Each molecule of HNO_3 actually means one ion of H^+.

When the solute is a weak acid, only a small percentage of its molecules are ionized at equilibrium, so no simple correlation exists between the concentration of the weak acid and the pH of the solution. It is possible to calculate the pH from the values of the molar concentration of the acid and its K_a, but we will not need to develop this particular calculating skill.

The exponential definition of pH we have used has the number 1.0 before the times sign: 1.0×10^{-pH}. What if the pH is, say, 4.56? If we insert this into Equation 9.4, this pH value means that

$$[H^+] = 1.00 \times 10^{-4.56} \text{ mol/L}$$

A solution at pH 4.56 has 10 times the concentration of H^+ as one at a pH of 5.56.

Although this is a perfectly good number, we normally reexpress exponentials so that the exponents are whole numbers. We can tell from the number as it stands that the value of $[H^+]$ is between 1.0×10^{-4} and 1.0×10^{-5} mol/L, but we can't tell exactly where between these limits without using logarithms. Although logarithms are not hard to use, particularly when pocket calculators have both log x and 10^x functions, our needs in the area of pH are not this extensive. All we need is an ability to *interpret* pH values that are not whole numbers rather than an ability to carry out logarithmic calculations. For example, a pH of 7.35 is between 7.00 and 8.00, so the value of $[H^+]$ has to be between 1.0×10^{-7} and 1.0×10^{-8} mol/L. Of much more importance to us is the ability to recognize that a pH of 7.35 corresponds to a (slightly) basic solution.

PRACTICE EXERCISE 2

A patient suspected of having taken an overdose of barbiturates was found to have blood with a pH of 7.10.

(a) Between what limits of values for the molar concentration of H^+ does this pH correspond?
(b) Is the blood slightly basic, acidic, or is it neutral?
(c) If the pH of blood normally is 7.35, is the patient experiencing acidosis or alkalosis?
(d) What is the pOH of this patient's blood?

PRACTICE EXERCISE 3

The patient in Practice Exercise 2 had been admitted to a teaching hospital, and the attending physician asked a group of students which of two isotonic reagents, sodium bicarbonate or ammonium chloride, might be administered by intravenous drip to restore the patient's blood pH to a normal value of 7.35. Which solution could be used? Explain how it would work using a net ionic equation.

Phenolphthalein (fee-noll-THAY-lean).

Acid-Base Indicators. A number of organic dyes called **acid-base indicators** can be used to indicate if a solution is acidic, basic, or neutral. For example, a dye called phenolphthalein has a bright pink color at a pH above 10.0 and is colorless below a pH of 8.2. In the range of 8.2 to 10.0 phenolphthalein undergoes a gradual change from colorless to pale pink to deep pink. Another acid–base indicator dye is litmus, which is blue above a pH of about 8.5 and red below a pH of 4.5. Each indicator has its own pH range and set of colors, and Figure 9.2 gives just a few examples.

Some commercial pH test papers contain several indicator dyes in the same test strip, which makes possible a whole spectrum of colors according to the pH color code on the container.

When the solutions to be tested for pH are themselves highly colored, we can't use indicators. Moreover, we often need more than a rough idea of pH, which is all that indicators can actually give. In such situations there are a variety of commercial pH meters (Figure 9.3) that come equipped with specially designed electrodes which can be dipped into the solution to be tested. With a good pH meter, pH values can be read to the second decimal place.

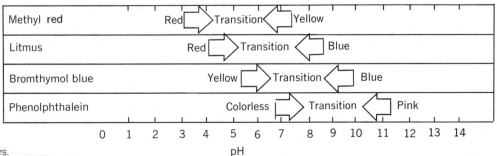

FIGURE 9.2
Some common acid–base indicators.

FIGURE 9.3
A pH meter.

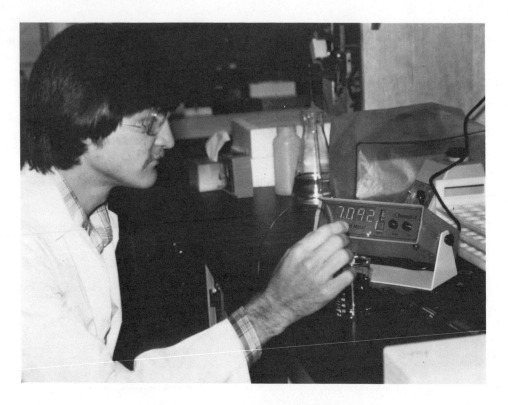

9.2   THE EFFECT OF DISSOLVED SALTS ON THE pH OF A SOLUTION

Aqueous solutions can be made basic or acidic just from the presence of salts that can hydrolyze.

Hydrolysis of Salts.   In many laboratory situations, an aqueous solution of some salt is prepared only to have the solution turn out to be either acidic or basic. If we weren't aware this could happen, we might unknowingly prepare a salt solution that could be more corrosive to metals than expected, or could harm living things, or could make food or drink or aqueous medications unfit for human use, or could contain a catalyst (e.g., hydrogen ion) that might affect a reaction rate.

"Hydrolysis" is from the Greek *hydro,* water, and *lysis,* loosening or breaking — breaking or loosening by water.

Many salts react to some extent with water to upset its 1 to 1 mole ratio of $[H^+]$ to $[OH^-]$, and this reaction is called the **hydrolysis of salts.** The effect of hydrolysis is to make the solution have a pH that is lower than 7 or higher, and usually the effect is caused by just one of the ions supplied by the salt. The reaction of a specific ion with water to affect the pH can be referred to as the hydrolysis of the ion itself.

Predicting If Salts Hydrolyze.   We want a simple strategy to predict if a salt can make a solution acidic or basic, and it would help our analysis to have a scheme for classifying salts. We will think of a salt in terms of its parents, that is, the acid and the base that could be used to make it by a neutralization reaction. For example, the salt NaCl would be formed in the reaction of its parent acid and base, NaOH with HCl:

$$NaOH + HCl \longrightarrow NaCl + H_2O$$

Similarly, the salt sodium acetate, $NaC_2H_3O_2$, would be formed in the reaction of the base NaOH and the acid $HC_2H_3O_2$ (acetic acid):

$$NaOH + HC_2H_3O_2 \longrightarrow NaC_2H_3O_2 + H_2O$$

In general, in these acid-base reactions, the base could be either strong or weak — for example, NaOH (a strong base) or NH_3 (a weak base) — and the acid could be either strong or weak; for example, HCl (a strong acid) or $HC_2H_3O_2$ (a weak acid). To determine if a salt hydrolyzes, we have to examine each ion of the salt and ask if it can react with water to generate either H^+ or OH^-. It is at this step of the analysis that we must consider the salt's acid–base parents most closely.

Salts of Strong Acids and Strong Bases.   A typical example of this kind of salt is NaCl, the salt of the strong base NaOH and the strong acid HCl. We have to decide whether either the Na^+ or Cl^- ion in the solution undergoes a reaction with water — that is, whether they have a tendency to hydrolyze.

You learned earlier that a strong acid, such as HCl, has a very weak conjugate base, Cl^-. In fact, the chloride ion is such a weak base that it has no tendency to capture protons from water molecules, so in a solution of a salt that provides Cl^- as its anion, there is no tendency for the anion to react with water, or hydrolyze. The chloride ion, therefore, does not affect the pH of the solution. Neither do the anions of other strong monoprotic acids, for example, Br^-, I^-, and NO_3^-, whose parent acids are HBr, HI, and HNO_3.

Next, we have to consider the cation in our NaCl solution, Na^+. Some metal ions do react with water and affect the acidity of a solution. However, the metal ions that do this are very small and very highly charged. Cations from group IA do not fit this description, and it is safe to assume that none of them reacts with water. The same applies to all the metal ions from Group IIA except Be^{2+}, which is a very tiny ion. (These are facts you will need to remember to help you decide whether salts hydrolyze.) In a solution of NaCl, therefore, neither the Na^+ ion nor the Cl^- ion is able to react with water. As a result, neither one affects the acidity of the solution, so the solution is neutral, and we would expect its pH to be 7.00.

As a general rule, we can state that *salts that are formed from a strong acid and a strong base do not hydrolyze in solution.*

Salts of Weak Acids and Strong Bases.   A typical example of this type of salt is sodium acetate, $NaC_2H_3O_2$. Its acid–base parents are NaOH and $HC_2H_3O_2$. A solution of $NaC_2H_3O_2$ contains the ions Na^+ and $C_2H_3O_2^-$. You've just learned that the sodium ion does not hydrolyze, so it has no effect on the pH of the solution. However, the acetate ion is the conjugate base of acetic acid, $HC_2H_3O_2$. Acetic acid isn't on our list of strong acids, so we can infer that it is a weak acid. Because $HC_2H_3O_2$ is weak, we can anticipate that the $C_2H_3O_2^-$ ion is a strong enough base to react with water to produce at least some OH^-:

$$C_2H_3O_2^-(aq) + H_2O \rightleftharpoons HC_2H_3O_2(aq) + OH^-(aq)$$

Thus the acetate ion does affect the pH of the solution, and because OH^- ion is formed in the reaction, the solution is expected to be slightly basic with a pH greater than 7.

In general, *a solution of a salt that is formed by the reaction of a weak acid and a strong base will be basic because of the hydrolysis of the anion — the conjugate base of the weak acid.*

Salts of Strong Acids and Weak Bases.   Here we have salts similar to NH_4Cl, which would form in the reaction of HCl (a strong acid) and ammonia (a weak Brønsted base). In a solution of NH_4Cl, the ions are NH_4^+ and Cl^-. We have already seen that Cl^- doesn't hydrolyze, because it is the very weak conjugate base of a strong acid. On the other hand, NH_4^+ is the conjugate acid of the weak base NH_3. Because NH_3 is weak, we can expect that its conjugate acid will be strong enough to react to some extent with water to form some hydronium ions, and therefore make the solution acidic:

$$NH_4^+(aq) + H_2O \longrightarrow NH_3(aq) + H_3O^+(aq)$$

TABLE 9.2
Hydrolysis of Salts

Category of Salt	Acidity or Basicity or Aqueous Solution	Example
Strong acid—strong base	Neutral	NaCl, KBr, NaI, CaBr$_2$
Weak acid—strong base	Basic	NaC$_2$H$_3$O$_2$, KCN
Strong acid—weak base	Acidic	NH$_4$Cl, AlCl$_3$, Cu(NO$_3$)$_2$
Weak acid—weak base	Varies with salt	NH$_4$C$_2$H$_3$O$_2$

We can now make another generalization: *A solution of a salt that is formed by the reaction of a strong acid with a weak base will be acidic because of the hydrolysis of the cation—the conjugate acid of the weak base.*

Salts of Weak Acids and Weak Bases.   A typical example of this kind of salt is ammonium acetate, NH$_4$C$_2$H$_3$O$_2$, which would be formed in the reaction of the weak acid HC$_2$H$_3$O$_2$ and the weak base NH$_3$. Salts in this category may or may not affect the pH of the solution. Both the cation and the anion have a tendency to hydrolyze, but the effect of the salt on the pH depends on whether the cation is better able to generate H$^+$ than the anion can generate OH$^-$. These salts have to be examined on a case-by-case basis.

A summary of the four kinds of salts we have discussed and their effects on the pH of a solution is given in Table 9.2.

EXAMPLE 9.2   PREDICTING IF A SALT HYDROLYZES AND ITS EFFECT ON pH

Problem: Sodium carbonate is sometimes called soda ash, and it is used to manufacture soap and glass. Does it hydrolyze, and if so, is the solution acidic or basic?

Solution: We can consider Na$_2$CO$_3$ to be the salt of NaOH and H$_2$CO$_3$ (or the anion of this acid, HCO$_3^-$). NaOH is a strong base, but neither H$_2$CO$_3$ nor HCO$_3^-$ is on our list of strong acids, so we can infer that either is a weak acid. Na$_2$CO$_3$ is therefore the salt of a strong base and a weak acid, so we can expect that only the anion, CO$_3^{2-}$, will hydrolyze:

$$CO_3^{2-}(aq) + H_2O \longrightarrow HCO_3^-(aq) + OH^-(aq)$$

This reaction generates some OH$^-$ ions, so the solution will be basic. (In fact, 0.100 *M* Na$_2$CO$_3$ has a pH of 11.63.)

Notice that in Example 9.2 we didn't need a table of Brønsted bases to make our prediction. We used our knowledge of just a few facts—which acids and bases are strong acids and bases in water—to figure out what we needed to know. Let's work another example to practice using facts about strong aqueous acids and bases to decide if a given salt hydrolyzes.

EXAMPLE 9.3   PREDICTING IF A SALT HYDROLYZES

Problem: Will a solution of ammonium bromide, NH$_4$Br, be acidic, basic, or neutral?

Solution: This salt gives NH$_4^+$ and Br$^-$ ions in water. We've already learned that Br$^-$ ions do not hydrolyze. The ammonium ion, NH$_4^+$, is the conjugate acid of a weak base, so we should expect it to be somewhat acidic. Therefore in water the following equilibrium should exist:

$$NH_4^+(aq) + H_2O \rightleftharpoons NH_3(aq) + H_3O^+(aq)$$

Because this reaction generates some hydronium ions, the solution can be expected to be acidic.

PRACTICE EXERCISE 4   Determine without the use of tables if each ion can hydrolyze. If it can, write the equilibrium expression and predict if the ion tends to make its solution acidic or basic.
(a) CO_3^{2-} (b) HPO_4^{2-}
(c) S^{2-} (d) Mg^{2+}
(e) NO_2^- (f) F^-

PRACTICE EXERCISE 5   Does potassium acetate, $KC_2H_3O_2$, hydrolyze, and if it does, is its aqueous solution acidic or basic?

PRACTICE EXERCISE 6   Does potassium bromide, KBr, hydrolyze, and if so, is its aqueous solution acidic or basic?

PRACTICE EXERCISE 7   Ammonium nitrate, NH_4NO_3, is a nitrogen fertilizer. What effect will the application of an aqueous solution of this fertilizer have on the pH of soil? Will it tend to raise the pH, lower it, or leave it unchanged?

We mentioned earlier that a small metal ion with a high enough electrical charge can react with water. One of these is the aluminum ion, Al^{3+}, which exists in water as a hexahydrate, $Al(H_2O)_6^{3+}$. The $3+$ charge on the central metal ion so strongly attracts the $\delta-$ charges on the polar molecules of the water of hydration that O—H bonds in these molecules are weakened. A properly oriented collision of high enough energy occurs often enough between this hydrated ion and water molecules to generate hydronium ions at equilibrium:

The arrows signify a force of attraction between $\delta-$ on O and the central, positively charged cation.

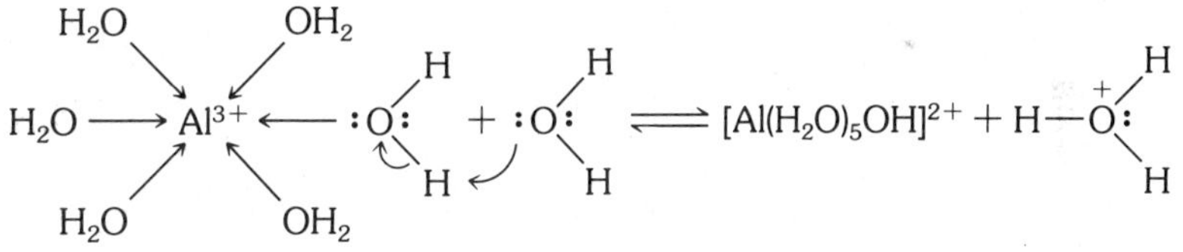

A solution that is 0.1 M $AlCl_3$ has a pH of about 3, the same as 0.1 M $HC_2H_3O_2$, acetic acid.

9.3 BUFFERS. PREVENTING LARGE CHANGES IN pH

The pH of a solution can be held relatively constant if it contains a buffer — a weak base and its conjugate acid.

$\dfrac{10^{-4}}{10^{-7}} = 1000$

It takes just a little strong acid or strong base to cause a large change in the pH of a solution. The addition of only one drop of concentrated hydrochloric acid to a liter of pure water drops the pH of the system from 7 to 4, a change of three units in pH but a 1000-fold change in acidity. If such a change occurred in the bloodstream, you would die. The pH of blood can't be allowed to change by more than 0.2 to 0.3 pH units. We'll learn how the body controls the pH of its fluids in this section.

Certain combinations of solutes, called **buffers,** can prevent large changes in pH from happening when strong acids or bases are added to an aqueous solution. One part of the buffer system can neutralize protons, H^+, and the other part can neutralize hydroxide ions, OH^-. Fluids that contain buffers are said to be *buffered* against the changes in pH that these ions otherwise cause. The blood and other body fluids include some buffers, and much of the body's work in maintaining its acid-base balance depends on these substances. There are several life-threatening situations in which the blood is forced to accept or to retain either more acid or more base than it normally has to handle. Both acidosis and alkalosis are prevented by the activities of buffers. Let's see what they are and how they work.

The Phosphate Buffer.   The principal buffer at work inside cells is called the **phosphate buffer.** It consists of the pair of ions, HPO_4^{2-} and $H_2PO_4^-$, the monohydrogen and the dihydrogen phosphate ions. Notice that $H_2PO_4^-$ is the conjugate acid of HPO_4^{2-}, so $H_2PO_4^-$ is the member of this pair that is better able to donate a proton to the hydroxide ion. Any added OH^- shifts the following equilibrium to the right, and neutralizes the OH^-. This keeps the pH from increasing.

$$H_2PO_4^-(aq) + OH^-(aq) \rightleftharpoons HPO_4^{2-}(aq) + H_2O \qquad (9.8)$$

The conjugate base to $H_2PO_4^-$, the HPO_4^{2-} ion, is a good proton acceptor, so it stands ready, by its presence in the phosphate buffer, to neutralize any added H^+. When H^+ enters, the following equilibrium shifts to the right, the H^+ is neutralized, and the pH is kept from decreasing.

$$HPO_4^{2-}(aq) + H^+(aq) \rightleftharpoons H_2PO_4^-(aq) \qquad (9.9)$$

The Carbonate Buffer.   The principal buffer in blood is called the **carbonate buffer,** and it consists of the conjugate pair, H_2CO_3 and HCO_3^-, carbonic acid and the bicarbonate ion. Carbonic acid, which is by far the stronger Brønsted acid of the two, stands ready to neutralize OH^-, and thus prevent an increase in pH should this strong base enter the system. If an increase in the OH^- level of the blood should occur, the following equilibrium shifts to the right, which neutralizes the OH^- and prevents alkalosis.

H_2CO_3 occurs in blood because some dissolved CO_2 reacts with water to form it.

$$H_2CO_3(aq) + OH^-(aq) \rightleftharpoons HCO_3^-(aq) + H_2O \qquad (9.10)$$

If the level of H^+ ion increases in the blood, then the following equilibrium, which uses HCO_3^- as a base, shifts to the right. This neutralizes the H^+ and prevents acidosis.

$$HCO_3^-(aq) + H^+(aq) \rightleftharpoons H_2CO_3(aq) \qquad (9.11)$$

When acid is neutralized by a shift of equilibrium 9.11 to the right, the level of carbonic acid, H_2CO_3, in the blood rises. We have already learned that this is an unstable acid, and not much can exist as such in solution. When the level of H_2CO_3 rises, the following equilibrium shifts to the right to produce carbon dioxide and water.

$$H_2CO_3(aq) \rightleftharpoons CO_2(aq) + H_2O \qquad (9.12)$$

By using the symbol $CO_2(aq)$ instead of $CO_2(g)$, we have shown that the carbon dioxide is still in solution. However, when the blood moves through the very thin capillaries in the alveoli of the lungs, where gaseous carbon dioxide can escape the body, the following change takes place, and $CO_2(g)$ now leaves the system. Because the gas leaves, we won't write the change as an equilibrium.

$$CO_2(aq) \longrightarrow CO_2(g) \qquad (9.13)$$

FIGURE 9.4

The irreversible neutralization of H^+ through the loss of CO_2 is one way the carbonate buffer system handles acidosis. It is the last step, the change of dissolved CO_2 into gaseous CO_2, which is exhaled, that draws all of the equilibria to the right and makes H^+ "disappear" into H_2O.

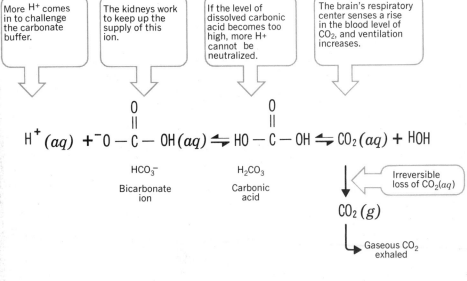

More H^+ comes in to challenge the carbonate buffer.

The kidneys work to keep up the supply of this ion.

If the level of dissolved carbonic acid becomes too high, more H^+ cannot be neutralized.

The brain's respiratory center senses a rise in the blood level of CO_2, and ventilation increases.

$$H^+(aq) + {}^-O-C-OH(aq) \rightleftharpoons HO-C-OH \rightleftharpoons CO_2(aq) + HOH$$

HCO_3^-
Bicarbonate ion

H_2CO_3
Carbonic acid

Irreversible loss of $CO_2(aq)$

$CO_2(g)$

Gaseous CO_2 exhaled

This change tends to pull or shift all of the preceding equilibria of the carbonate buffer system to the right. In other words, the blood brings to bear *two* mechanisms for handling protons. It neutralizes them by the work of the carbonate buffer, and it uses the process of ventilation to make this neutralization permanent.

Ventilation is the circulation of air into and out of the lungs. The brain has a site called the respiratory center that monitors the level of $CO_2(aq)$ in the blood. When this level rises, the brain instructs the breathing apparatus to step it up — to breathe more rapidly and deeply — a response called **hyperventilation.** This increases the rate at which a gas such as carbon dioxide can leave the lungs, and this response pulls all of the carbonate equilibria in their acid-neutralizing directions.

The successive shifts of equilibria 9.11 and 9.12 which are caused by the change described by Equation 9.13 have the important overall effect that one H^+ ion is permanently neutralized because one CO_2 molecule leaves the body. All of these steps are summarized in Figure 9.4, where you can see that the H^+ to be neutralized ends up in a molecule of water, and the ability of this water molecule to form depends on the loss of the CO_2 molecule from the body.

Hyperventilation is not rapid-fire panting with no pauses between intake and outgo. This kind of breathing is called Kussmaul breathing.

One of many lessons we can draw from this is that anything that interferes with the loss of CO_2 inhibits the neutralization of the H^+ ion and causes acidosis. **Hypoventilation**— slow and shallow breathing — occurs involuntarily in people with emphysema. They are unable to breathe deeply enough, and they struggle with acidosis as well. Similarly, people with asthma or pneumonia, or those who have taken an overdose of narcotics or barbiturates, experience unusual difficulties in managing the acid–base balance of the blood either because their lungs or their respiratory centers are not working properly.

There is much more to say about the various ways in which metabolism and respiration interact with the buffer systems in the blood, and we will return to this topic in considerable detail in a later chapter when we have learned more about metabolism and about the chemistry of the blood.

9.4 EQUIVALENTS AND MILLIEQUIVALENTS OF IONS

The electrical neutrality of an aqueous solution requires that the number of milliequivalents of cations equals the number of milliequivalents of anions.

In some situations it isn't as important to know how many moles of a compound are involved as how many moles of electrical charges. We'll study one of these situations in this section.

The term *equivalent* is analogous to the term *mole*. It's a unit for a specific quantity of a specified substance.

Equivalents.   The counterpart to the mole for this purpose is called the *equivalent*. One **equivalent,** abbreviated **eq,** of an ion is the number of grams of the ion that corresponds to Avogadro's number — one mole — of electrical charges. For example, when the charge is unity, either $1+$ or $1-$, it takes Avogadro's number of ions to have Avogadro's number of electrical charges. Thus 1 eq for ions such as Na^+, K^+, Cl^-, or Br^- is the same as the molar mass of each ion. The molar mass of Na^+ is 23.0 g Na^+/mol, and 1 eq of $Na^+ = 23.0$ g of Na^+. This much sodium ion, 23.0 g of Na^+, contributes Avogadro's number of positive charges. In analogy to the molar mass, we have a comparable term for equivalents — the *equivalent mass* — but it's usually called the **equivalent weight** of some species. Thus the equivalent weight of the sodium ion is 23.0 g Na^+/eq.

When an ion has a double charge, either $2+$ or $2-$, then the mass of one equivalent equals the molar mass divided by 2. For example, 1 mol of CO_3^{2-} ion $= 60.0$ g of CO_3^{2-}, so 1 eq of CO_3^{2-} ion $= 30.0$ g of CO_3^{2-} ion. The extension of this to ions of higher charges should now be obvious. The equivalent weight of an ion is its formula weight divided by its charge. Table 9.3 gives equivalent weights for a number of ions.

The advantage of the concept of the equivalent is the simplicity of the 1 to 1 ratio. Regardless of the amounts of charges on the individual ions in a solution, we can always be sure that for every equivalent of positive charge there has to be one equivalent of negative charge. In any solution,

$$eq \text{ of cations} = eq \text{ of anions}$$

Milliequivalents.   Because the concentrations of most ions in body fluids are very low, the numbers used to express them are awkward unless they are given in relatively small subunits. For example, the level of potassium ion, K^+, in the blood is only about 0.0040 mol/L, which is the same as 0.0040 eq/L. However, by taking advantage of the SI prefix, milli-, we can work with numbers in units of **milliequivalents, meq,** which are easier to remember and speak.

The counterpart to the milliequivalent is the millimole (mmol). 1 mol = 1000 mmol

$$1 \text{ eq} = 1000 \text{ meq}$$

Therefore the K^+ ion level in the blood, 0.0040 eq/L, can be stated less awkwardly as 4.0 meq/L, because

$$0.0040 \, \frac{\cancel{eq}}{L} \times \frac{1000 \text{ meq}}{1 \, \cancel{eq}} = 4.0 \text{ meq/L}$$

Inside the back cover of this book is a handy list of the normal ranges of values of the concentrations of several components of blood, and you will see that some are given in units of meq/L.

The Anion Gap.   The ions in blood that contribute the highest levels of concentration and charge are Na^+, Cl^- and HCO_3^-. For example, the concentration of Na^+ normally is in the range of 136–145 meq/L; of Cl^-, 95–109 meq/L; and of HCO_3^-, 21–29 meq/L. In

TABLE 9.3
Equivalents of Ions

Ion	g/mol	g/eq
Na^+	23.0	23.0
K^+	39.1	39.1
Ca^{2+}	40.1	20.1
Mg^{2+}	24.3	12.2
Al^{3+}	27.0	9.0
Cl^-	35.5	35.5
HCO_3^-	61.0	61.0
CO_3^{2-}	60.0	30.0
SO_4^{2-}	96.1	48.1

contrast, the levels of K^+, Ca^{2+}, and Mg^{2+} ions are on the order of only 2 to 5 meq/L each. The blood also carries varying concentrations of negatively charged ions of organic acids, such as the anions (the conjugate bases) of acetic acid, citric acid, and many others.

The levels of the organic anions tend to rise in several metabolic disturbances such as diabetes or kidney disease, but measuring these anions is difficult. Their levels, however, can be estimated by calculating a quantity known as the **anion gap,** which can be defined by the following equation:

$$\text{anion gap} = \frac{\text{meq of Na}^+}{L} - \left(\frac{\text{meq of Cl}^-}{L} + \frac{\text{meq of HCO}_3^-}{L} \right) \quad (9.14)$$

To determine the anion gap in a patient's blood, a sample is analyzed for the concentrations in meq/L of Na^+, Cl^-, and HCO_3^-, which as we said are the most abundant ions and are also relatively easy to analyze. Then the concentration data are fed into Equation 9.14. For example, suppose that analyses found the following data: $Na^+ = 137$ meq/L; $Cl^- = 100$ meq/L; and $HCO_3^- = 28$ meq/L. Then, the anion gap is found by:

$$\text{Anion gap} = 137 \, \frac{\text{meq}}{L} - \left(100 \, \frac{\text{meq}}{L} + 28 \, \frac{\text{meq}}{L} \right)$$
$$= 9 \text{ meq/L}$$

The normal range for the anion gap is 5 to 14 meq/L, so in our example the anion gap of 9 meq/L falls within the normal range. This 9 meq/L is accounted for by the presence of unmeasured ions of low concentration.[2] If metabolic disturbances cause the levels of organic anions to rise, the body must retain cations in the blood and excrete some of the more common anions such as Cl^- and HCO_3^- to maintain the absolute requirement that the blood be electrically neutral. In other words, negative organic ions tend to expel other negative ions but retain whatever positive ions are available. This is how the anion gap widens in metabolic disturbances that generate organic anions. The anion gap routinely rises above 14 meq/L in untreated diabetes.

You can see that by using rather easily measured data on the meq/L concentrations of Na^+, Cl^-, and HCO_3^-, and calculating the anion gap from these data, the clinical chemist can inform the health care professionals of any unusual buildups in anions which indicate possible disease. In severe exercise, the anion gap rises above normal, too, but it goes back down again in time. Thus an above normal anion gap has to be interpreted in the light of other facts.

9.5 ACID-BASE TITRATIONS

An analyst doing an acid-base titration tries to have the number of equivalents of acid and base come out exactly equal at the end point.

The pH of a solution tells us indirectly what the concentration of hydronium ions is. It does not disclose the **neutralizing capacity** of the solution — its capacity to neutralize a strong base. The 1 mol of acetic acid in 1 L of 1 M $HC_2H_3O_2$ can neutralize 1 mol of NaOH, yet this solution has an actual quantity of H_3O^+ of only about 0.004 mol. This low concentration of hydronium ion is caused by the weakness of the acidity of acetic acid. When sodium hydroxide is added, however, its hydroxide ions not only take H^+ from H_3O^+ but also from $HC_2H_3O_2$. The hydroxide ion is this strong a proton acceptor. Thus the neutralizing capacity of 1 M acetic acid is considerably greater than suggested by its concentration of hydronium ions.

---

[2] The anion gap would be even wider in our example if we added in the other measurable cations — K^+, Ca^{2+}, and Mg^{2+} — which, as we said, occur at low levels of about 2 to 5 meq/L each in the blood. However, offsetting this would be the inclusion of anions of low levels such as the sulfate ion and ions of the phosphoric acid system. In other words, the anion gap actually is the sum of the undetermined anions subtracted from the sum of the undetermined cations.

The careful measurement of the concentration of a standard solution is called standardization. We say that we standardize the solution.

Titration.   We use an indicator or a pH meter to measure, indirectly, the value of $[H_3O^+]$ in a solution. The procedure used for measuring the total neutralizing capacity of a solution is called *titration*. The apparatus for titration is shown in Figure 9.5. When **titration** is used for an acid–base analysis, a carefully measured volume of the solution of unknown acidity (or basicity) is placed in a beaker or a flask. A very small amount of an acid–base indicator such as phenolphthalein is added. Then a **standard solution**—one whose concentration is accurately known—of the neutralizing reagent is added portion by portion from a piece of glassware called a *buret* (Figure 9.5). This addition is continued until the acid–base indicator signals by a change in color that the unknown has been exactly neutralized.

Equivalence Points and End Points.   With a carefully selected indicator, the color change in an acid–base titration occurs when all of the available hydrogen ions have reacted with all of the available hydroxide ions. This point in a titration is called the **equivalence point.**

A well-chosen indicator is one whose color matches exactly the color that it would give if it were added to a solution that consists solely of the *salt* that is produced by the titration — and is at the same concentration of this salt that results from the actual titration. Remember that some salts can hydrolyze, so the equivalence point in a titration isn't necessarily at a pH of 7.00. The pH of a sodium acetate solution, for example, is above 7.00 (how much above depends on the concentration of the salt), and this is the salt produced by the titration of acetic acid by sodium hydroxide.

Whether or not the indicator has been well-chosen, the analyst has little choice but to stop the titration when the indicator's color changes. This stopping point is called the **end point** of the titration. In a well-run titration, of course, the end point and the equivalence point would coincide.

Many analysts use the concept of an equivalent in acid-base neutralizations because of the simple 1 to 1 matching of the moles of the ions in the equation for an acid-base neutralization:

$$H^+(aq) + OH^-(aq) \longrightarrow H_2O$$

FIGURE 9.5
The apparatus for titration. By manipulating the stopcock, the analyst controls the rate at which the solution in the buret is added to the flask below.

The definition of an equivalent when we work with an acid or base is related to the coefficients in this equation. One equivalent of an acid is the mass of the acid that can donate 1 mol of H^+ for neutralization by OH^-. For all monoprotic acids, the mass of 1 eq equals the mass of 1 mol. Similarly, for all group IA hydroxides such as NaOH or KOH, the mass of 1 eq of the base equals the mass of 1 mol.

For all diprotic acids, or of such bases as $Ca(OH)_2$, the molar mass has to be divided by 2 to find the mass of 1 eq. For example, 1 mol of H_2SO_4 can deliver 2 mol of H^+, so 0.5 mol of this acid can supply 1 mol of H^+. In other words, 1 mol $H_2SO_4 = 2$ eq H_2SO_4. The molar mass of H_2SO_4 is 98.1 g/mol, so

$$\frac{98.1 \text{ g } H_2SO_4}{\text{mol } H_2SO_4} \times \frac{1 \text{ mol } H_2SO_4}{2 \text{ eq } H_2SO_4} = 49.1 \text{ g } H_2SO_4/\text{eq } H_2SO_4$$

Thus 1 eq of H_2SO_4 has a mass of 49.1 g. Table 9.4 gives the mass of 1 eq of several common acids and bases, and you can see how these masses are easily related to the molar masses of the compounds.

Normality. The concept of equivalents as applied to acids and bases lets us write the following simple condition that must be met at the equivalence point in a titration, where the subscripts a and b designate the acid and the base, respectively:

$$eq_a = eq_b$$

Or, to use the more convenient subunit,

$$meq_a = meq_b$$

Titrations, of course, are carried out with *solutions* so what we now need is a vocabulary for describing the concentration of a solution in terms not of moles per liter but equivalents per liter. We used *molarity* for moles per liter. For equivalents per liter, we use the term **normality,** and the abbreviation **N**.

$$\text{Normality } (N) = eq/L = meq/mL \qquad (9.15)$$

$$\frac{eq}{L} = \frac{1000 \text{ meq}}{1000 \text{ mL}} = \frac{meq}{mL}$$

The simple calculation in the margin shows that the ratio of eq/L is numerically identical to the ratio meq/mL. If we rearrange parts of Equation 9.15, we can obtain expressions for meq for

TABLE 9.4
Equivalents of Acids and Bases

Substance	g/mol	g/eq
HCl	36.5	36.5
HBr	80.9	80.9
HI	128	128
HNO_3	63.0	63.0
H_2SO_4	98.1	49.1
H_3PO_4	98.0	32.7
NaOH	40.0	40.0
KOH	56.1	56.1
$Ca(OH)_2$	74.1	37.1
$Mg(OH)_2$	58.3	29.2
$NaHCO_3$	84.0	84.0
Na_2CO_3	106	53.0

the acid and for the base:

$$meq_a = mL_a \times N_a$$

and

$$meq_b = mL_b \times N_b$$

Because the quantities on the left of the equals signs in these two equations equal each other at the equivalence point, the quantities on the right sides must also equal each other. Thus we have the following equation that relates measured quantities to the relationship that must be true at the equivalence point of an acid–base titration.

Many analysts insist that the simplicity of this equation justifies the extra effort to learn about equivalent weights of acids and bases and about normality.

$$mL_a \times N_a = mL_b \times N_b \qquad (9.16)$$

This equation can be used for the calculations in all acid–base titrations.

The base used in a titration doesn't have to be a metal hydroxide. For example, sodium carbonate neutralizes acids by the following reaction:

$$2H^+ + CO_3^{2-} \longrightarrow CO_2 + H_2O$$

Because each CO_3^{2-} ion neutralizes $2H^+$ ions, the mass of 1 eq of any carbonate is half of its molar mass.

EXAMPLE 9.4   CALCULATING WITH NORMALITIES

Problem: How many grams of sulfuric acid are needed to prepare 250 mL of 0.150 N H_2SO_4?

Solution: The value of 0.150 N H_2SO_4 means the following ratios are true for this solution and serve as possible conversion factors.

$$\frac{0.150 \text{ eq } H_2SO_4}{1000 \text{ mL } H_2SO_4 \text{ soln}} \quad \text{or} \quad \frac{1000 \text{ mL } H_2SO_4 \text{ soln}}{0.150 \text{ eq } H_2SO_4}$$

Before we can calculate the number of grams of H_2SO_4, we have to find out how many equivalents of H_2SO_4 are in 250 mL of this solution. To find this, we multiply the given, 250 mL, by the first of the above conversion factors.

$$250 \text{ mL } H_2SO_4 \text{ soln} \times \frac{0.150 \text{ eq } H_2SO_4}{1000 \text{ mL } H_2SO_4 \text{ soln}} = 0.0375 \text{ eq } H_2SO_4$$

We have already shown that

$$1 \text{ eq } H_2SO_4 = 49.1 \text{ g } H_2SO_4$$

This relationship can be used to construct the following two conversion factors:

$$\frac{1 \text{ eq } H_2SO_4}{49.1 \text{ g } H_2SO_4} \quad \text{or} \quad \frac{49.1 \text{ g } H_2SO_4}{1 \text{ eq } H_2SO_4}$$

Now, if we multiply 0.0375 eq of H_2SO_4 by the second of these two factors, we will obtain the answer in the correct unit, g.

$$0.0375 \text{ eq } H_2SO_4 \times \frac{49.1 \text{ g } H_2SO_4}{1 \text{ eq } H_2SO_4} = 1.84 \text{ g } H_2SO_4$$

Thus if we dissolve 1.84 g of H_2SO_4 in a little water and then add more water to make the final volume equal to 250 mL, we can put 0.150 N H_2SO_4 on the label of the bottle.

PRACTICE EXERCISE 8   How many grams of NaOH are needed to prepare 250.0 mL of 0.1250 N NaOH?

EXAMPLE 9.5   DOING TITRATION CALCULATIONS

Problem: A grocery store manager received complaints that a shipment of vinegar was weak, that the vinegar seemed to be too dilute. He sent a sample to a testing laboratory where the analyst took a 10.1 mL sample of the vinegar and titrated it with 0.133 N NaOH. It took 34.9 mL of the standard base. What was the normality of the vinegar sample? (It should be between 0.7 and 0.8 N.)

Solution: It's a good idea to assemble the data in one place first.

$$mL_a = 10.1 \text{ mL} \qquad mL_b = 34.9 \text{ mL}$$
$$N_a = ? \qquad N_b = 0.133$$

Therefore

$$(10.1 \text{ mL}) \times (N_a) = (34.9 \text{ mL}) \times (0.133)$$
$$N_a = 0.460 \ N$$

Obviously the vinegar had been watered. It's only about half as concentrated as it should be.

PRACTICE EXERCISE 9   If it takes 24.3 mL of 0.110 N HCl to neutralize 25.5 mL of freshly prepared sodium hydroxide solution, what is the normality of the base?

SUMMARY

pH   The self-ionization of water produces an equilibrium in which the ion product constant, K_w, the product of the molar concentrations of hydrogen and hydroxide ions, is 1.0×10^{-14} (at 25 °C). If acids or bases are added, the value of K_w stays the same but individual values of $[H^+]$ and $[OH^-]$ adjust. A simple way to express very low values for these molar concentrations is by a value of pH, where $[H^+] = 1 \times 10^{-pH}$. When, at 25 °C, pH < 7, the solution is acidic. When pH > 7, the solution is basic. To measure the pH of a solution we use indicators — dyes whose colors change over a narrow range of pH — or we use a pH meter.

Hydrolysis of salts   A dissolved salt affects the pH of the solution if one of its ions reacts more than the other with water to generate extra H^+ or OH^- ions. Thus neither of the ions of a salt of a strong acid and a strong base hydrolyzes to change the pH of the solution. But the salt of a strong base and a weak acid (e.g., $NaC_2H_3O_2$) makes a solution slightly basic, because the anion but not the cation can react slightly with water to generate OH^- ions. The salt of a strong acid and a weak base (e.g., NH_4Cl) makes a solution slightly acidic, because the cation but not the anion reacts slightly with water to generate hydrogen ions. When the cation of a salt is an ion with $3+$ charge and a small radius (e.g., Al^{3+}), it reacts with water to generate hydrogen ions and make the solution acidic. Sometimes metal ions with just $2+$ charge hydrolyze this way, too. When a salt is made of a weak base and a weak acid (e.g., $NH_4C_2H_3O_2$), its effect on the pH of the solution has to be determined on a case-by-case basis.

Buffers   Solutions that contain something that can neutralize OH^- ion (such as a weak acid) and something else that can neutralize H^+ ion (such as the conjugate base of the same weak acid) are buffered against changes in pH when either additional base or acid is added. The phosphate buffer, which is present in the fluids inside cells of the body, consists of HPO_4^{2-} (to neutralize H^+) and $H_2PO_4^-$ (to neutralize OH^-). The carbonate buffer, which is the chief buffer in blood, consists of HCO_3^- (to neutralize H^+) and H_2CO_3 (to neutralize OH^-). The supply of H_2CO_3 comes from the reaction of dissolved CO_2 with water. Carbonic acid, however, tends to break down in the lungs where the CO_2 is expelled. When metabolism or some deficiency in respiration produces or retains H^+ at a rate faster than the blood buffer can neutralize them, the lungs try to remove CO_2 at a faster rate (hyperventilation). Overall, for each molecule of CO_2 exhaled, one proton is neutralized.

Equivalents and milliequivalents   An equivalent of an ion is the number of grams of the ion that carry Avogadro's number of positive or negative charges. It's calculated by dividing the molar mass of the ion by the size of the charge on the ion. Often the subunit, the milliequivalent is used, where 1 eq = 1000 meq. Unmeasured anion concentrations in body fluids can be estimated by subtracting the meq/L values of Cl^- and HCO_3^- from the meq/L value of Na^+. The difference, the anion gap, rises above the range of 5 to 14 meq/L in conditions that produce excessive levels of organic anions.

Acid – base titration   The concentration of an acid or a base in water can be determined by titrating the unknown solution with a standard solution of what can neutralize it. The indicator is selected to have its color change occur at whatever pH the final solution would have if it were made from the salt that forms by the neutralization. The unknown concentration can be calculated from the volumes of the acid and base used and the normality of the standard solution by the equation

$$mL_a \times N_a = mL_b \times N_b$$

where the normality, N, means the concentration in equivalents per liter, or what is numerically identical, the milliequivalents per milliliter. The equivalent weight of an acid is its formula weight divided by the number of available hydrogen ions per formula unit. For a base, the equivalent weight is its formula weight divided by the number of hydrogen ions each formula unit can neutralize.

KEY TERMS

The following terms were emphasized in this chapter, and they should be mastered before continuing. They are particularly relevant to any study of the acid – base balance or the electrolyte balance of the body.

acidic solution	end point	indicator, acid – base	pH
acidosis	equivalence point	ion product constant of water, K_w	pOH
alkalosis	equivalent, eq		phosphate buffer
anion gap	equivalent weight	milliequivalent, meq	respiration
basic solution	hydrolysis of salts	neutralizing capacity	standard solution
buffer	hyperventilation	neutral solution	titration
carbonate buffer	hypoventilation	normality	ventilation

SELECTED REFERENCES

1 Linda Janusek. "Metabolic Acidosis." *Nursing84,* July 1984, page 44.

2 Linda Janusek. "Metabolic Alkalosis." *Nursing84,* August 1984, page 60.

3 Dolores Taylor. "Respiratory Alkalosis." *Nursing84,* November 1984, page 44.

4 Linda Glass and Cheryl Jenkins. "The Ups and Downs of Serum pH." *Nursing83,* September 1983, page 34.

5 Jeffery Fisher. "Measurement of pH." *American Laboratory,* June 1984, page 54.

6 Doris Kolb. "Acids and Bases." *Journal of Chemical Education,* July 1978, page 459.

7 Doris Kolb. "The pH Concept." *Journal of Chemical Education,* January 1979, page 49.

REVIEW EXERCISES

The answers to Review Exercises that require a calculation and that are marked with an asterisk are found in an Appendix. The answers to the other Review Exercises are found in the *Study Guide* that accompanies this book.

Ion Product Constant of Water

9.1   Write the equation that defines K_w.

9.2   What is the value of K_w at 25 °C?

9.3   The higher the temperature, the higher is the value of K_w. Why should there be this trend?

*9.4   At the temperature of the human body, 37 °C, the concentration of hydrogen ion in pure water is 1.56×10^{-7} mol/L. What is the value of K_w at 37 °C?

9.5   "Heavy water" or deuterium oxide, D_2O, is used in nuclear power plants. It self-ionizes like water, and at 20 °C there is a concentration of D^+ ion of 3.0×10^{-8} mol/L. What is the value of K_w for heavy water at 20 °C?

pH

9.6   What equation defines pH in exponential terms?

9.7   The average pH of urine is about 6. Is this acidic, neutral, or basic?

9.8   The pH of gastric juice is in the range of 1.5 to 3.5. Is gastric juice acidic, basic, or neutral?

9.9   What is the pH of 0.01 M HCl(aq), assuming 100% ionization?

9.10 What is the pOH of 0.01 M NaOH(aq), assuming 100% ionization. What is the pH of this solution?

9.11 Explain why a pH of 7.00 corresponds to a neutral solution at 25 °C.

9.12 Following surgery, a patient experienced persistent vomiting and the pH of his blood became 7.56. (Normally it is 7.35.) Has the blood become more alkaline or more acidic? Is the patient experiencing acidosis or alkalosis?

9.13 A patient brought to the emergency room following an overdose of aspirin was found to have a pH of 7.20 for the blood. (Normally the pH of blood is 7.35.) Has the blood become more acidic or more basic? Is the condition acidosis or alkalosis?

9.14 A certain brand of beer has a pH of 5.0. What is the concentration of hydrogen ion in moles per liter? Is the beer slightly acidic or basic?

9.15 The pH of a soft drink was found to be 4.5. Between what limits of molar concentration does the value of $[H^+]$ occur?

9.16 A solution of a monoprotic acid was prepared with a molar concentration of 0.10 M. Its pH was found to be 1.0. Is the acid a strong or a weak acid? Explain.

9.17 The pH of a solution of a monoprotic acid was found to be 4.56, whereas its molar concentration was 0.010 M. Is this acid a strong or a weak acid? Explain.

9.18 When a soil sample was stirred with pure water, the pOH of the water changed to 6.10. Did the soil produce an acidic or a basic reaction with the water?

9.19 A patient entered the emergency room of a hospital after three weeks on a self-prescribed low-carbohydrate, high-fat diet and the regular use of the diuretic, acetazolamide. (A diuretic promotes the formation of urine and thus causes the loss of body fluid.) The pOH of the patient's blood was 6.80. What was the pH, and was the condition acidosis or alkalosis?

Hydrolysis of Salts

9.20 Explain in your own words why a solution of sodium acetate, $NaC_2H_3O_2$, is slightly basic, not neutral.

9.21 Predict whether each of the following solutions is acidic, basic, or neutral.
(a) K_2SO_4 (b) NH_4NO_3 (c) $KHCO_3$
(d) $FeCl_3$ (e) Li_2CO_3

9.22 Predict whether each of the following solutions is acidic, neutral, or basic.
(a) KNO_3 (b) Na_2HPO_4 (c) K_3PO_4
(d) $Cr(NO_3)_3$ (e) $KC_2H_3O_2$

Buffers

9.23 When we say the blood is *buffered,* what does this mean in general terms (without specifying specific substances)?

9.24 What two chemical species make up the chief buffer in blood?

9.25 Write the equations that show how the chief buffer system in the blood works to neutralize hydroxide ion and hydrogen ion.

9.26 Explain in your own words, using equations as needed, how the loss of a molecule of CO_2 at the lungs permanently neutralizes a hydrogen ion.

9.27 In high-altitude sickness, the patient *involuntarily* overbreathes and expels CO_2 from the body at a faster than normal rate. This results in an *increase* in the pH of the blood.
(a) Is this condition alkalosis or acidosis?
(b) Why should excessive loss of CO_2 result in an increase in the pH of the blood? (*Note:* Such a patient should be returned to lower elevations as soon as possible. It helps to rebreathe one's own air, as by breathing into a paper sack, because this helps the system retain CO_2.)

9.28 A patient with emphysema *involuntarily* hypoventilates. (In other words, the respiratory system is not working properly.) This leads to a decrease in the pH of the blood.
(a) Is this alkalosis or acidosis?
(b) Why should hypoventilation under these circumstances lead to a decrease in the pH of the blood?

9.29 What two ions constitute the phosphate buffer in cellular fluids?

9.30 Based on what should be predicted about the abilities of the anions of the phosphate buffer to hydrolyze, should one expect cellular fluids to be slightly alkaline or slightly acidic?

Equivalents and Milliequivalents of Ions

9.31 The concentration of potassium ion in blood serum is normally in the range of 0.0035 to 0.0050 mol K^+/L. Express this range in units of milliequivalents of K^+ per liter.

9.32 The concentration of calcium ion in blood serum is normally in the range of 0.0042 to 0.0052 eq Ca^{2+}/L. Express this range in units of milliequivalents of Ca^{2+} per liter.

*9.33 The level of chloride ion in blood serum is normally quoted as 100 to 106 meq/L. How many grams and how many milligrams constitute 106 meq of Cl^-?

9.34 The sodium ion level in the blood is normally 135 to 145 meq/L. How many grams and how many milligrams of sodium ion constitute 135 meq of Na^+?

*9.35 The potassium ion level of blood serum normally does not exceed 0.196 g of K^+ per liter. How many milliequivalents of K^+ ion are in 0.196 g of K^+?

9.36 The magnesium ion level in plasma normally does not exceed 0.0243 g of Mg^{2+}/L. How many milliequivalents of Mg^{2+} are in 0.0243 g of Mg^{2+}?

*9.37 The analysis of the blood from a young man recovering from polio found 137 meq of Na^+/L, 34 meq of HCO_3^-/L, and 93 meq of Cl^-/L. Calculate the anion gap. Does it suggest a serious disturbance in his metabolism?

9.38 A patient on a self-prescribed diet consisting essentially only of protein was found to have the following blood analyses after two weeks of the diet: Na^+, 174 meq/L; Cl^-, 135 meq/L; and HCO_3^-, 20 meq/L. Calculate the anion gap. Does it suggest a disturbance in metabolism?

Acid – Base Titrations[3]

9.39 What does it mean to have a *standard* solution of, say, HCl(aq)?

9.40 In a titration, how does one know when the end point is reached?

9.41 What steps does an analyst take to ensure that the end point and the equivalence point in a titration occur together?

9.42 Give an example of a titration in which the equivalence point has a pH that is equal to 7. (Give a specific example of an acid and a base that produce such a solution.)

9.43 Give a specific example an acid and a base that, when titrated together, produce a solution with a pH greater than 7.

9.44 At the equivalence point in a titration the pH is less than 7. Give a specific example of an acid and a base that, when titrated together, would produce this result.

9.45 Explain why 1 mol of H_2SO_4 contains 2 eq of H_2SO_4.

9.46 What choices of conversion factors are available when we have been given the following information?
(a) 36.46 g HCl = 1 eq HCl
(b) The concentration of H_2SO_4 is 0.115 N.

9.47 Given the known relationships between eq and meq as well as between the liter and the milliliter, prove that when we know the normality of a solution we not only know eq/L but we also know meq/mL—that these ratios are numerically identical.

9.48 Given that 1 mol = 1000 mmol (where mmol means milli-mole), prove that a concentration in mol/L is numerically the same as the concentration in mmol/mL.

*9.49 Calculate the mass in grams of one equivalent of each of the following compounds. Calculate to the second decimal place.
(a) $HClO_4$, perchloric acid (a very strong acid)
(b) HCO_2H, formic acid (a monoprotic acid)
(c) $Al(OH)_3$, aluminum hydroxide (the active ingredient in Am-phojet, an antacid)
(d) $H_3C_6H_5O_7$, citric acid (a triprotic acid whose anions are involved in energy metabolism in the body)

9.50 What is the mass in grams of one equivalent of each of the following compounds? (Do the calculations to the second deci-mal place.)
(a) $H_2C_2O_4 \cdot 2H_2O$, oxalic acid dihydrate (its anions are present in the cell sap of many plants; it's a diprotic acid)
(b) $MgCO_3$, magnesium carbonate (an antacid)
(c) $HC_3H_5O_3$, lactic acid (a monoprotic acid and the cause of the sour taste of sour milk)
(d) $NaAl(OH)_2CO_3$, dihydroxyaluminum sodium carbonate (the ingredient in Rolaids, a stomach antacid). Assume that each formula unit neutralizes *four* H^+ ions.

*9.51 How many equivalents are in each of the following samples? (If you did Review Exercises 9.49, you can use the formula weights calculated there for this Review Exercise. Notice that four significant figure precision is indicated.)
(a) 100.5 g $HClO_4$

(b) 25.68 g HCO_2H
(c) 28.35 g $Al(OH)_3$
(d) 500.0 mg $H_3C_6H_5O_7$ How many meq are also present?

9.52 Calculate the number of equivalents in each of the following samples. (If you did Review Exercise 9.50, you can use the formula weights calculated there for this Review Exercise. No-tice that four significant figure precision is indicated.)
(a) 3.150 g $H_2C_2O_4 \cdot 2H_2O$
(b) 28.35 g $MgCO_3$
(c) 110.4 g $HC_3H_5O_3$
(d) 56.70 g $NaAl(OH)_2CO_3$

*9.53 How many milliequivalents are in the following samples?
(a) 0.150 eq acetic acid
(b) 0.00550 eq oxalic acid
(c) 4.56 eq citric acid
(d) 0.450 eq Na^+

9.54 The following samples have how many milliequivalents?
(a) 0.00050 eq sulfuric acid   (b) 0.109 eq Cl^-
(c) 0.021 eq CO_3^{2-} (d) 0.035 eq K^+

*9.55 Individual aqueous solutions were prepared that contained the following substances. Calculate the molarity and the normality of each solution.
(a) 7.292 g of HCl in 500.0 mL of solution
(b) 16.18 g of HBr in 500.0 mL of solution

9.56 What are the molarity and the normality of each of the follow-ing solutions?
(a) 25.58 g of HI in 500.0 mL of solution
(b) 9.808 g of H_2SO_4 in 500.0 mL of solution

*9.57 If 20.00 g of a monoprotic acid in 100.0 mL of solution gives a concentration of 0.5000 N, what is the formula weight of the acid?

9.58 A solution with a concentration of 0.2500 N could be made by dissolving 4.000 g of a base in 250.0 mL of solution. What is the equivalent weight of this base?

*9.59 How many grams of each solute are needed to prepare the following solutions?
(a) 1000 mL of 0.2500 N HCl
(b) 750.0 mL of 0.1150 N HNO_3
(c) 500.0 mL of 0.01500 N H_2SO_4

9.60 To prepare each of the following solutions would require how many grams of the solute in each case?
(a) 100.0 mL of 0.1000 N HBr
(b) 250.0 mL of 0.1000 N H_2SO_4
(c) 500.0 mL of 1.250 N Na_2CO_3

*9.61 In standardizing a sodium carbonate solution, 21.45 mL of this solution was titrated to the end point with 18.78 mL of 0.1018 N HCl.
(a) Calculate the normality of the sodium carbonate solution.
(b) How many grams of sodium carbonate are in each liter?

9.62 A freshly prepared solution of sodium hydroxide was standard-ized with 0.1148 N H_2SO_4.
(a) If 18.32 mL of the base was neutralized by 20.22 mL of the acid, what was the normality of the base?
(b) How many grams of NaOH were in each liter?

---

[3] For all of the calculations in Review Exercises that follow, round atomic weights to their *second* decimal places before calculating formula weights.

Chapter 10
Radioactivity and Nuclear Chemistry

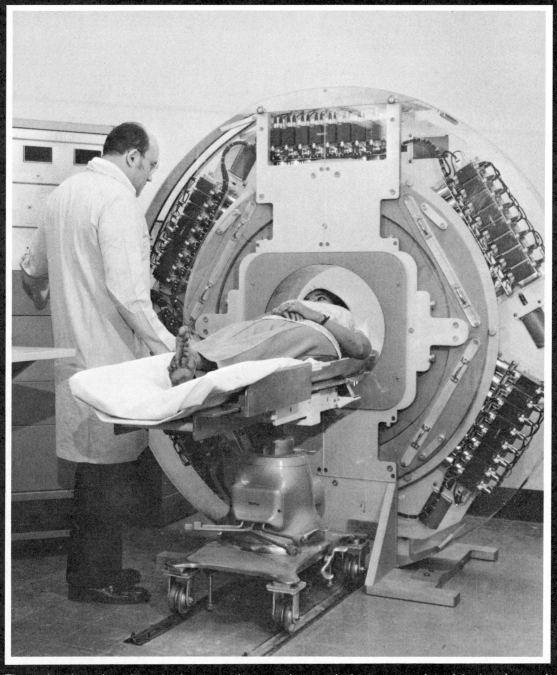

The patient is undergoing a brain scan using a positron emission tomograph to detect and record the radiation emitted by radioactive monosaccharides being metabolized in the brain. In this chapter we'll learn about radioactivity and its many uses.

10.1   ATOMIC RADIATIONS

Unstable atomic nuclei eject high-energy radiations as they change to more stable nuclei.

Some atomic nuclei are unstable and the isotopes with such nuclei are **radioactive,** meaning that they emit streams of high-energy radiations. Each radioactive isotope is called a **radionuclide.** Radiations from these isotopes as well as X rays can cause grave harm to human life, but when carefully used their potential benefits far outweigh their possible harm.

Radioactive Decay.   Radioactivity was discovered in 1896 when a French physicist, A. H. Becquerel (1852–1908), happened to store some photographic plates in a drawer that contained samples of uranium ore. The film became fogged, meaning that when developed the picture looked like a photograph of fog. Becquerel might have blamed the accident on faulty film or careless handling, but a mysterious radiation called X rays had recently been discovered by a German scientist, Wilhelm Roentgen (1845–1923). X rays were known to be easily capable of penetrating the packaging of unexposed film and fogging it. Becquerel's film proved to have been fogged by a natural source of radiation that resembled X rays. These radiations, he soon found, were emitted by any compound of uranium, but most intensely by uranium metal itself.

Several years later, two British scientists, Ernest Rutherford (1871–1937) and Frederick Soddy (1877–1956), explained radioactivity in terms of events that take place in unstable atomic nuclei. Such nuclei undergo small disintegrations called **radioactive decay,** and they hurl tiny particles into space or emit a powerful radiation like X rays but called gamma radiation. The nuclei that remain after decay almost always are those of an entirely different element, so decay is usually accompanied by the **transmutation** of one element or isotope into another. The natural sources of radiation on our planet emit one or more of three kinds — alpha radiation, beta radiation, and gamma radiation. We receive another kind called **cosmic radiation** from the sun and outer space (see Special Topic 10.1).

Alpha Radiation.   **Alpha radiation** consists of particles that move with a velocity almost one-tenth the velocity of light. Each particle, called an **alpha particle,** is a tiny cluster of two protons and two neutrons, and therefore these particles are the nuclei of helium atoms (Figure 10.1). They are the largest of the decay particles and they have the greatest charge, so when alpha particles travel in air, they soon lose their energy (and charge) by collisions with the air molecules. Alpha particles cannot penetrate even thin cardboard or the outer layer of dead cells on the skin. Exposure to an intense dose of alpha radiation, however, causes a severe burn.

The most common isotope of uranium, uranium-238 or $^{238}_{92}U$, is an alpha emitter. By

Becquerel shared the 1903 Nobel prize in physics with Pierre and Marie Curie.

Roentgen won the 1901 Nobel prize in physics, the first to be awarded.

Nobel prizes in chemistry were awarded to both Rutherford (1908) and Soddy (1921).

The special symbols for isotopes were introduced in Section 3.1. Thus in $^{238}_{92}U$, 238 is the mass number and 92 is the atomic number.

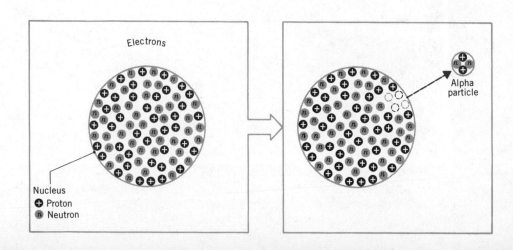

FIGURE 10.1
Emission of an alpha particle.

losing two protons, an atom of this isotope changes its atomic number from 92 to 90, and by losing four units of mass number (two protons + two neutrons), its mass number changes from 238 to 234. The result is that uranium-238 transmutes into an isotope of thorium, $^{234}_{90}$Th.

Beta Radiation.   **Beta radiation** consists of a stream of particles called **beta particles,** which are actually electrons. These electrons are produced *within* the nucleus and then emitted. Because they have less charge and are more than 7000 times smaller than alpha particles, they penetrate matter more easily. These properties make it possible for beta particles to travel much farther in air than alpha particles. However, beta particles usually do not penetrate the skin, although those from particular sources can. Each radionuclide emits its radiations at specific values of energy, and only the highest-energy beta particles can reach internal organs from outside of the body.

When a beta particle is emitted, the nucleus in effect converts a neutron into a proton (Figure 10.2). This does not change the mass number, but it makes the atomic number *increase* by 1 unit. For example, thorium-234, $^{234}_{90}$Th, is a beta emitter, and when it ejects a beta particle, it changes to an isotope of protactinium, $^{234}_{91}$Pa. Accompanying the loss of a beta particle is gamma radiation.

> Losing one electron from the nucleus is like adding one proton without changing the mass number.

Gamma Radiation.   **Gamma radiation** does not consist of particles. Instead it is a form of energy like X rays or ultraviolet rays, but with more energy. Like X rays, gamma radiation is quite penetrating and very dangerous. It easily travels through the entire body.

The composition and symbols of the three radiations studied thus far are summarized in Table 10.1.

Nuclear Equations.   Nuclear reactions are quite unlike chemical reactions because the latter lead to no changes in atomic nuclei. Therefore **nuclear equations** are different from chemical equations in important ways. Like all equations, nuclear equations are intended to convey the maximum amount of detailed information possible in a very abbreviated and balanced form. And for nuclear reactions, the critical information concerns changes in atomic numbers, mass numbers, and identities of radionuclides.

In nuclear equations the alpha particle is symbolized as $^{4}_{2}$He, and although it is positively charged, the charge is omitted from the symbol. The particle soon picks up electrons, anyway, from the matter through which it is passing, and it becomes a neutral atom of helium. The beta

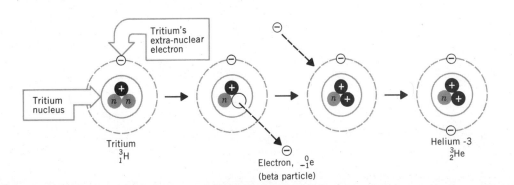

FIGURE 10.2
Emission of a beta particle.

TABLE 10.1

Radiations from Naturally Occurring Radionuclides

Radiation	Composition	Mass Number	Electrical Charge	Symbols
Alpha radiation	Helium nuclei	4	2+	$^4_2$He or α
Beta radiation	Electrons	0	1−	$^0_{-1}$e or β
Gamma radiation	X-ray type of energy	0	0	$^0_0\gamma$ or γ

It's proper to think of the electron as having an atomic number of −1.

particle has the symbol $^0_{-1}$e because its mass number is 0 and its charge is 1−. Gamma radiation is symbolized simply by γ (or, sometimes, by $^0_0\gamma$). Both its mass number and charge are 0.

A nuclear equation is balanced when the sums of the mass numbers on either side of the arrow are equal and when the sums of the atomic numbers are equal. The alpha decay of uranium-238 is represented by the following equation, for example.

$$^{238}_{92}U \longrightarrow\ ^{234}_{90}Th +\ ^4_2He$$

Notice that the sums of the atomic numbers agree: $92 = 90 + 2$. And the sums of the mass numbers agree: $238 = 234 + 4$.

The beta decay of thorium-234, which also emits gamma radiation, is represented by the following nuclear equation:

$$^{234}_{90}Th \longrightarrow\ ^{234}_{91}Pa +\ ^0_{-1}e + \gamma$$

The sums of the atomic numbers agree: $90 = 91 + (-1)$.

EXAMPLE 10.1   BALANCING NUCLEAR EQUATIONS

Problem: Cesium-137, $^{137}_{55}$Cs, is one of the radioactive wastes that forms during fission in a nuclear power plant or an atomic bomb explosion. This radionuclide decays by emitting both beta and gamma radiation. Write the nuclear equation for this decay.

Solution: First we set up as much of the nuclear equation as we can and leave blanks for any information that we have to figure out. Thus our partial nuclear equation is

$$^{137}_{55}Cs \longrightarrow\ ^0_{-1}e +\ ^0_0\gamma +\ \underline{\underline{\quad}}\ \underline{\ }$$

Mass number needed
Atomic symbol needed
Atomic number needed

The first thing to do is to figure out the atomic symbol for the isotope that forms. For this we need the atomic number, because then we can look up the symbol in a table. We will find the atomic number by using a step-by-step process, but with just a little experience, you will be able to do this by inspection. We find the atomic number by using the fact that the sum of the atomic numbers on one side of the nuclear equation must equal the sum on the other side. Therefore letting x be the atomic number,

$$55 = -1 + 0 + x$$
$$x = 56$$

Now we can use the periodic table to find that the atomic symbol that goes with element 56 is Ba (barium). Next, to find out which particular isotope of barium forms, we use the fact that the sum of the mass numbers on one side of the nuclear equation must equal the sum on the other side. Letting y equal the mass number of the barium isotope,

$$137 = 0 + 0 + y$$

$$y = 137$$

The balanced nuclear equation therefore is

$$^{137}_{55}\text{Cs} \longrightarrow {}_{-1}^{0}e + {}_{0}^{0}\gamma + {}^{137}_{56}\text{Ba}$$

EXAMPLE 10.2

BALANCING NUCLEAR EQUATIONS

The radium used in cancer therapy was held in a thin, hollow gold or platinum needle to retain the alpha particles and all the decay products.

Problem: Until the 1950s, radium-226 was widely used as a source of radiation for cancer treatment. It is an alpha and a gamma emitter. Write the equation for its decay.

Solution: We have to look up the atomic number of radium, which turns out to be 88, so the symbol we'll use for this radionuclide is $^{226}_{88}\text{Ra}$. When one of its atoms loses an alpha particle, $^{4}_{2}\text{He}$, it loses 4 units in mass number — from 226 to 222. And it loses 2 units in atomic number — from 88 to 86. Thus the new radionuclide has a mass number of 222 and an atomic number of 86. We have to look up the atomic symbol for element number 86, which turns out to be Rn, for radon. Now we can assemble the nuclear equation:

$$^{226}_{88}\text{Ra} \longrightarrow {}^{222}_{86}\text{Rn} + {}^{4}_{2}\text{He} + \gamma$$

PRACTICE EXERCISE 1

Iodine-131 has long been used in treating cancer of the thyroid. This radionuclide emits beta and gamma rays. Write the nuclear equation for this decay.

PRACTICE EXERCISE 2

Plutonium-239 is a by-product of the operation of nuclear power plants. It can be isolated from used uranium fuel and made into fuel itself or into atomic bombs. It's a powerful alpha and gamma emitter, and is one of the most dangerous of all known substances. Write the equation for its decay.

Half-Life. Some radionuclides are much more stable than others, and we use the concept of a half-life to describe the differences. The **half-life** of a radionuclide, symbolized as $t_{1/2}$, is the time it takes for half of the atoms in a sample of a single, pure isotope to decay. (The atoms that decay don't just vanish, of course. They change into different isotopes.) Table 10.2 gives several half-lives.

The half-life of uranium-238 is 4.51×10^9 years, which means that an initial 100 g of this radionuclide would have only 50.0 g of uranium-238 left after 4.51×10^9 years. Strontium-90, a by-product of nuclear power plants, is a beta emitter with a half-life of 28.1 years. Figure 10.3 shows graphically how an initial supply of 40 g is reduced successively by units of one-half for each half-life period. At the end of eight half-life periods (224.8 years, from 8×28.1), only 0.3 g of strontium-90 remain in the sample.

The shorter the half-life, the more decay events per mole per second occur in the isotope. Mole for mole, it's generally much safer to be near a sample that decays very slowly than one that decays very rapidly.

An extremely long half-life is typical of the radionuclides that head a radioactive disintegration series.

Radioactive Disintegration Series. The decay of one radionuclide often produces another radionuclide rather than a stable isotope. Sometimes one after another unstable isotope forms and decays in a long series, called a **radioactive disintegration series,** before a stable isotope forms. Uranium-238 is at the head of one of four such series that are known, and Figure 10.4 displays the entire succession that ends in a stable isotope of lead.

TABLE 10.2
Typical Half-Life Periods

Element	Isotope	Half-Life Period	Radiations
Naturally Occurring Radionuclides			
Potassium	$^{40}_{19}K$	1.3×10^9 years	Beta, gamma
Neodymium	$^{144}_{60}Nd$	5×10^{15} years	Alpha
Radon[a]	$^{222}_{86}Rn$	3.832 days	Alpha
Radium[a]	$^{226}_{88}Ra$	1590 years	Alpha, gamma
Thorium	$^{230}_{90}Th$	8×10^4 years	Alpha, gamma
Uranium	$^{238}_{92}U$	4.51×10^9 years	Alpha
Synthetic Radionuclides			
Hydrogen (tritium)	$^{3}_{1}H$	12.26 years	Beta
Oxygen	$^{15}_{8}O$	124 seconds	Positron[b]
Phosphorus	$^{32}_{15}P$	14.3 days	Beta
Technetium	$^{99m}_{43}Tc$	6.02 hours	Gamma
Iodine	$^{131}_{53}I$	8.07 days	Beta
Cesium	$^{137}_{55}Cs$	30 years	Beta
Strontium	$^{90}_{38}Sr$	28.1 years	Beta
Americium	$^{243}_{95}Am$	7.37×10^3 years	Alpha

[a] Although short-lived, radon-222 and radium-226 are found in nature because the uranium-238 disintegration series continuously produces them.

[b] The positron has a mass number of 0 and a charge of 1+. (It's sometimes called a positive electron.)

10.2   IONIZING RADIATIONS—DANGERS AND PRECAUTIONS

Atomic radiations create unstable ions and radicals in tissue, and this can lead to cancer, mutations, tumors, or birth defects.

The source of danger to living things posed by atomic radiations is their ability to generate strange, unstable, highly reactive particles as they plow through living tissues. The *chemical* properties of these alien particles make them enemies of life. Because these particles are

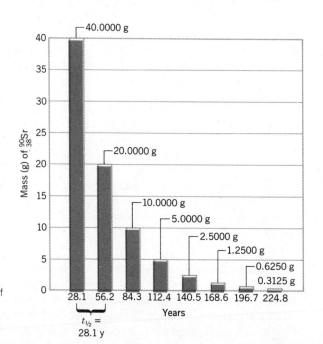

FIGURE 10.3
Each half-life period reduces the quantity of a radionuclide by a factor of two. Shown here is the pattern for strontium-90, a radioactive pollutant with a half-life of 28.1 years.

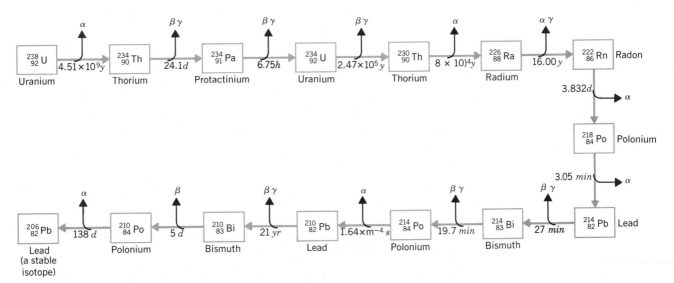

FIGURE 10.4
The uranium-238 radioactive disintegration series.

frequently ions, atomic radiations and X rays are called **ionizing radiations.** Alpha particles, for example, can generate ions from water molecules by the following reaction:

$$H—\overset{\cdot\cdot}{\underset{\cdot\cdot}{O}}—H + \text{high energy } \alpha \text{ particle} \longrightarrow [H—\overset{\cdot}{\underset{\cdot\cdot}{O}}—H]^+ + e^- + \text{lower-energy } \alpha \text{ particle}$$

The oxygen atom no longer has its outer octet, and the water molecule-ion, $[H—\overset{\cdot}{\underset{}{O}}—H]^+$, has sufficient energy from the collision that formed it to suffer the breaking of one of its covalent bonds.

$$[H—\overset{\cdot}{\underset{\cdot\cdot}{O}}—H]^+ \longrightarrow H^+ + \overset{\cdot}{\underset{\cdot\cdot}{:O}}—H$$
$$\text{Hydroxyl radical}$$

A proton forms plus a hydroxyl radical, a particle new to our study. It's a neutral particle with an unpaired electron and without an octet for its oxygen atom. Any particle with an unpaired electron is designated a **radical,** and with few exceptions, radicals are very reactive chemicals.

> Sometimes the term used is *free radical.*

The new ions and radicals cause chemical reactions among the more stable substances around them, and they alter them in ways foreign to metabolism. If this happens to the cell's genetic substances—genes and chromosomes—there could be subsequent changes that lead to cancer, or to a tumor, or to a genetic mutation. If they occur in a sperm cell, an ovum, or a fetus, the result might be a birth defect. Prolonged and repeated exposures to *low* levels of radiation are more likely to induce these problems than bursts of high-level radiation. The latter, of course, can kill outright, or at least render the target cell reproductively dead. This is why high doses of radiations are used in cancer treatment.

> Technical terms are:
>
> *Carcinogen,* a cancer causer
> *Tumorogen,* a tumor causer
> *Mutagen,* a mutation causer
> *Teratogen,* a birth defect causer

There is believed to be no **threshold exposure,** no minimum level of exposure below which radiation can cause no injury. All radiations that penetrate the skin or enter the body on food or through the lungs are considered harmful, and the damage can accumulate over a working lifetime. Medical personnel who work with radionuclides, X rays, or gamma-ray emitters are expected to wear devices that automatically record the exposure. The data accumulated from these devices are periodically logged into a permanent record book, and when the maximum permissible dose has been attained, the worker has to be transferred. Cells do have a capacity for self-repair, and some exposures carry very low risks. But no exposure is entirely risk-free. The widespread and routine use of X rays for public health screenings has long been curtailed.

> No technology in any medical field is entirely risk-free. We take risks when we believe that the benefits outweigh them.

Radiation Sickness.   Because the primary site of radiation damage is in the cell's genetic apparatus, the first symptoms of exposure to radiation occur in tissues whose cells divide most frequently; for example, the cells in bone marrow. Because these make white blood cells, an early sign of radiation damage is a fall in the white cell count of the blood. Cells in the intestinal tract also divide frequently, and even moderate exposure to X rays or gamma rays (as in cobalt-ray therapy for cancer) produces intestinal disorders.

Radiation sickness is a set of symptoms brought about by exposure to high-intensity radiation. Thus even if you initially survived a massive radiation exposure, you wouldn't be free of problems. The symptoms of radiation sickness include nausea, vomiting, a drop in the white cell count, diarrhea, dehydration, prostration, hemorrhaging, and the loss of hair. Radiations in sharp bursts are used to halt the spread of cancer in the affected tissue. Sometimes the symptoms of radiation sickness appear when radiation is used this way.

Protection from Radiation.   Dangers of direct exposure to radiation are minimized by taking advantage of some of the general properties of all radiations. Radiations move in straight lines from their origins. From an exposed sample of radioactive material, radiations stream in all directions much as light leaves a light source, spreading out. Alpha and beta rays are the easiest to stop (Table 10.3). Gamma radiations and X rays are stopped effectively only by particularly dense substances. Lead is the most common material for shielding gamma or X rays. Its effectiveness is apparent from the data in Table 10.3. A vacuum is least effective, of course, and air isn't much better. Low-density materials such as cardboard, plastic, and aluminum are poor shielders, but concrete works well if it is thick (and it's much cheaper than lead). Thus by a careful choice of a shielding material, protection can be obtained against radiations.

TABLE 10.3

Penetrating Abilities of Some Common Radiations

Type of Radiation	Common Sources	Approximate Depth of Penetration of Radiation into:		
		Dry Air	Tissue	Lead
Alpha rays	Radium-226 Radon-222 Polonium-210	4 cm	0.05 mm[b]	0
Beta rays	Tritium Strontium-90 Iodine-131 Carbon-14	6 to 300 cm[a]	0.06 to 4 mm[b]	0.005 to 0.3 mm
		Thickness to Reduce Initial Intensity by 10%		
Gamma rays	Cobalt-60 Cesium-137 Decay products of radium-226	400 m	50 cm	30 mm
X rays Diagnostic Therapeutic		120 m 240 m	15 cm 30 cm	0.3 mm 1.5 mm

Data from J. B. Little. *The New England Journal of Medicine,* Vol. 275, 1966, pages 929–938.

[a] The range of beta particles in air is about 12 ft/MeV. Thus a 2-MeV beta ray has a range of about 24 ft in air.

[b] To penetrate the protective layer of skin (0.07 mm thick), the approximate energy needed is 7.5 MeV for alpha particles and 70 keV for beta particles.

Another strategy to minimize exposure when radiations are used in medical diagnosis (such as in taking X rays) is to use fast film. With fast film, the *time* of exposure is kept as low as possible.

Still another way to get protection is to move away from the source. Because the radiations spread out, forming a cone of rays, fewer rays can strike a unit of surface of the receiving area if it is moved farther away. The area of the base of such a cone increases with the square of the distance. Hence the radiation intensity I on a unit area diminishes with the square of the distance d from the source. This is the **inverse square law** of radiation intensity.

Inverse Square Law The intensity of radiation is inversely proportional to the square of the distance from the source.

$$I \propto \frac{1}{d^2} \qquad\qquad (10.1)$$

This law holds strictly only in a vacuum, but it holds closely enough when the medium is air to make good estimates. If we move from one location, a, to another location, b, then the variation of Equation 10.1 that we can use to compare the intensities at the two different places, I_a and I_b, is given by the following equation:

$$\frac{I_a}{I_b} = \frac{d_b{}^2}{d_a{}^2} \qquad\qquad (10.2)$$

EXAMPLE 10.3   USING THE INVERSE SQUARE LAW OF RADIATION INTENSITY

Problem: At 1.0 m from a radioactive source the radiation intensity was measured as 30 units. If the operator moves away to a distance of 3.0 m, what will be the radiation intensity?

Solution: As usual, it's a good idea to assemble the data.

$$I_a = 30 \text{ units} \qquad d_a = 1.0 \text{ m}$$
$$I_b = ? \qquad d_b = 3.0 \text{ m}$$

Now we can use Equation 10.2:

$$\frac{30 \text{ units}}{I_b} = \frac{(3.0 \text{ m})^2}{(1.0 \text{ m})^2}$$

Solving for I_b, we get

$$I_b = 30 \text{ units} \times \frac{1.0 \text{ m}^2}{9.0 \text{ m}^2}$$
$$= 3.3 \text{ units}$$

Thus tripling the distance cut the intensity almost by a factor of 10.

PRACTICE EXERCISE 3   If the intensity of radiation is 25 units at a distance of 10 m, what does the intensity become if you move to a distance of 0.50 m?

PRACTICE EXERCISE 4   If you are receiving an intensity of 80 units of radiation at a distance of 6.0 m, to what distance would you have to move to reduce this intensity by half, to a value of 40 units?

Natural Background Radiation.   Shielding materials and the advantage that distance gives us in protection from radiations can never reduce our exposure completely. We are constantly exposed to **background radiation,** the radiations given off, for example, by the naturally occurring radionuclides, by radioactive pollutants, cosmic rays, and medical X rays. About 50 of the roughly 350 isotopes of all elements in nature are radioactive. On the average, the top 15 cm of soil on our planet has 1 g of radium per square mile. Radioactive materials are in the soils and rocks on which we walk and which we use to make building materials. They are in the food we eat, the water we drink, and the air we breathe. Radiations enter our bodies with every X ray taken of us. They come in cosmic ray showers.

The background radiation varies widely from place to place on our planet, and only estimates (which vary widely with the estimator) are possible. Table 10.4 gives the averages of radiations from various sources for the U.S. population. We'll discuss the unit of *dose equivalent* used in this table, the millirem (mrem), shortly. For comparison purposes, a dose of 500 rem (500,000 mrem) given to the individuals in a large population would cause the deaths of half of them in 30 days. In relation to this, the natural background is very small. At higher altitudes, the background is higher because incoming cosmic rays, which contribute to the background, have had less opportunity to be shielded by the earth's atmosphere. The background is 170 mrem, for example, for the population of Denver, Colorado, at an elevation of 1 mile.

Radioactive Pollutants.   All operating nuclear power plants in the United States today employ atomic fission. **Fission** is the disintegration of a large atomic nucleus into small fragments following neutron capture. Fission releases additional neutrons, radioactive isotopes, and enormous yields of heat. In the light-water reactors in use today, the heat generates steam, which drives electric turbines.

The uranium-235 isotope is the only naturally occurring radionuclide that spontaneously undergoes fission when it captures a slow-moving and relatively low-energy neutron. Its nucleus captures the neutron and then splits apart. It can split in a number of ways that give different products. We'll give the equation for just one mode of splitting:

$$^{235}_{92}U + ^{1}_{0}n \xrightarrow[\text{capture}]{\text{Neutron}} {}^{236}_{92}U \xrightarrow{\text{Fission}} {}^{139}_{56}Ba + {}^{94}_{36}Kr + 3{}^{1}_{0}n + \gamma + \text{heat}$$

More neutrons are released than are used up. Hence, one fission event produces particles that can initiate more than one new fission. In other words, a **nuclear chain reaction** takes place (Figure 10.5). A reactor that operates on this principle is called a *burner reactor,* because the uranium fuel is used up.

TABLE 10.4

Average Radiation Doses Received Annually by the U.S. Population

Source	Mrem Dose
Natural background (whole-body dose equivalent)	100
Medical sources	
Diagnostic X rays	20–50
Therapeutic X rays	3–5
Other sources	
Fallout (from atmospheric testing, 1954–1962)	1.5 (5 mrem maximum in early 1960s)
Radioactive pollutants introduced into the environment from	
nuclear power plants	<1
Miscellaneous	2.0[a]

Data from J. B. Little, *The New England Journal of Medicine,* Vol. 275, 1966, pages 929–938; *Technology Review,* October/November 1970; and *Nuclear Power and the Public,* H. Foreman, ed., University of Minnesota Press, 1970.

[a] The one-pack-a-day smoker's lungs get at least 40 mrem per year more dose equivalent. Naturally occurring radon, a radioactive, gaseous decay product of naturally occurring radium, decays to radioactive lead that deposits on earth and foliage (including broad-leaf tobacco plants).

FIGURE 10.5
Fission is initiated when a nucleus of uranium-235 captures a neutron. The new nucleus splits apart, and more neutrons are released. These can initiate subsequent fission events, and unless this is prevented or at least tightly controlled, the whole mass of uranium-235 will detonate—as in the explosion of an atomic bomb. One method of control is to keep the mass of uranium-235 low, below what is called the critical mass. Now, enough neutrons escape before being captured by neighboring nuclei.

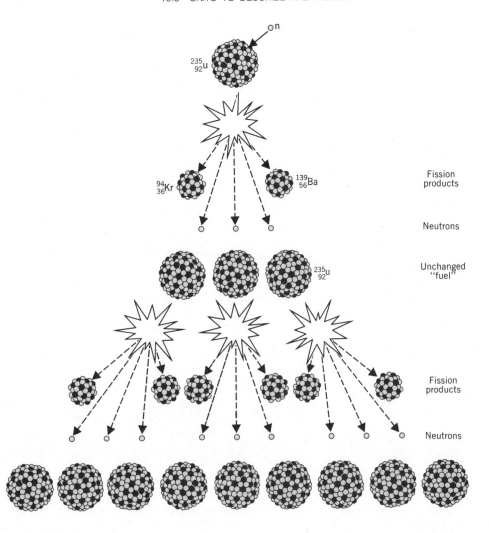

The new isotopes produced by fission are radioactive, and their decay leads to some of the radioactive pollutants of chief concern—principally strontium-90 (a bone-seeking element in the calcium family), iodine-131 (a thyroid-gland-seeker), and cesium-137 (a group IA radionuclide that goes wherever Na^+ or K^+ can go). The U.S. government has set limits to the release of each radioactive isotope into the air and into the cooling water of nuclear power plants. Those plants that operate in compliance with these standards expose people living near them to an extra dose of no more than 5% of the dose they normally receive from background radiations.

One of the most vexing problems of nuclear energy has been that of permanently storing long-lived radioactive wastes. Most are now in temporary storage at nuclear power plants. They must be sequestered from all human contact for at least 1000 years, and scientists are seeking deep geologic formations out of all contact with mining operations or underground water supplies into which these wastes can be placed.

10.3 UNITS TO DESCRIBE AND MEASURE RADIATIONS

Units have been devised to describe the activity of a radioactive sample, the energies of its radiations, and the energies they can deliver to tissue.

A number of units exist for a variety of measurements of radiations, and each one has been invented to serve in the answer to a particular question.

Marie Curie is one of two scientists to win two Nobel prizes in a field of science — a share of the physics prize in 1903 and the chemistry prize in 1911.

The Curie (Ci) and the Becquerel (Bq).   The curie and the becquerel are units of activity that were devised to help answer the question, "How *active* is a sample of a radionuclide?" The **curie, Ci,** is named after Marie Sklodowska Curie (1867 – 1934), a Polish scientist who, while working in France, discovered the element radium. One curie (1 Ci) is the number of radioactive disintegrations that occur per second in a 1.0 g sample of radium — 3.7×10^{10} disintegrations/second:

$$1 \text{ Ci} = 3.7 \times 10^{10} \text{ disintegrations/s}$$

This is an intensely active rate, so fractions of the Ci, such as the millicurie (mCi, 10^{-3} Ci), the microcurie (μCi, 10^{-6} Ci), and the picocurie (pCi, 10^{-12} Ci) are often used.

The **becquerel, Bq,** is the SI unit of activity, but it has achieved essentially no popularity among U.S. scientists (a situation which might change, of course).

$$1 \text{ Bq} = 1 \text{ disintegration/s}$$

Therefore

$$1 \text{ Ci} = 3.7 \times 10^{10} \text{ Bq}$$

The Roentgen (R).   The roentgen is a unit of exposure of X ray or gamma radiation, and it was devised to serve in answering the question, "How *intense* is the exposure to X ray or gamma ray radiation?" One **roentgen** of either of these radiations, in passing through 1 cm^3 of dry air at normal temperature and pressure, generates ions with a total charge of 2.1×10^9 units. If a large population were exposed to 650 roentgens, half would die in one to four weeks. (The rest would have radiation sickness.)

The Gray (Gy) and the Rad (D).   The gray and the rad are units that were invented to answer the question, "How much *energy* is absorbed by a unit mass of tissue or other materials?" The SI unit used for this is the **gray, Gy,** after a British radiologist, Harold Gray. It equals 1 joule (J) of energy absorbed per kilogram of tissue. (The joule is the SI unit of energy; 1 J = 4.184 cal):

$$1 \text{ Gy} = 1 \text{J/kg}$$

Rad comes from *radiation absorbed dose.*

An older unit, one perhaps more widely used, is called the **rad, D.**

$$1 \text{ D} = 10^{-2} \text{ Gy}$$

Because the joule is an extremely small amount of energy, both the gray and the rad are also very small quantities. Nevertheless, only 600 rad of gamma radiation would be lethal to most people. It's not the quantity of energy that matters, however, as much as it is the formation of unstable radicals and ions that this energy triggers. A 600 rad dose delivered to water causes the ionization of only one molecule for every 36 million present, but the ions or radicals thus produced begin a cascade of harmful reactions when they form in critical regions of the cell.

The roentgen and the rad are close enough in magnitude to say that, from a health standpoint, they are equivalent. Thus one roentgen of gamma radiation from a cobalt-60 source, a radionuclide often used in cancer treatment, equals 0.96 rad in muscle tissue and 0.92 rad in compact bone.

The Rem.   The rem is a unit that satisfies the need for a unit of absorbed dose that is additive for different radiations and different target tissues. A dose of one rad of gamma radiation is not biologically the same as a dose of one rad of beta radiation or of neutrons. Thus the rad doesn't serve as a good basis for comparison or for equivalences. The rem fits this need. The **rem** is the unit of *dose equivalent.* To convert rads to rems, the dose in rads from some radiation is multiplied by a factor that takes into account biologically significant properties of the radiation. One rem of any given radiation is the dose that has, in a human, the effect of one roentgen.

Rem comes from *roentgen equivalent for man.*

The rem, like the rad, is a small quantity in terms of energy, but significant in terms of danger. Even millirem quantities of radiation should be avoided, and when this isn't possible the workers must wear monitoring devices that allow the day-to-day exposures to be calculated.

The Electron-Volt.   Those who work with X rays or gamma radiation use an old unit of energy to describe the energies of these radiations. This unit is the **electron-volt, eV,** defined as follows:

$$1 \text{ eV} = 1.602 \times 10^{-19} \text{ J}$$

(Originally, the electron-volt was viewed as the energy an electron receives when it is accelerated by a voltage of 1 volt.)

You might guess, with an exponent of -19 in the above equation which defined the electron-volt, that this is an extremely small amount of energy. Therefore multiples of the electron-volt are common, such as the kiloelectron-volt (1 keV = 10^3 eV) and the megaelectron volt (MeV = 10^6 eV). X rays that are used for diagnosis are typically 100 keV or less. The gamma radiations from cobalt-60 that are used in cancer treatment have energies of 1.2 and 1.3 MeV. Beta radiation of 70 keV or more can penetrate the skin, but alpha particles (which are much larger than beta particles) need energies of more than 7 MeV to do this. Alpha radiation from radium-226 has an energy of 5 MeV. The cosmic radiation that enters our outer atmosphere has energies ranging from 200 MeV to 200 GeV (1 GeV = 1 gigaelectron-volt = 10^9 eV).

The higher their energy, the more penetrating are X rays and gamma rays.

Instruments for Detecting Ionizing Radiations.   People who work around radioactive sources cannot avoid receiving some exposure, and it's important that they keep a log of how much exposure they accumulate. A device for measuring exposure is called a *dosimeter,* and one common type is a film badge that contains photographic film, which becomes fogged by radiations. The degree of fogging, which is related to the exposure, can be measured.

Ionizing radiations also affect substances called phosphors — salts with traces of rare earth metal ions that scintillate when struck by radiation. These scintillations — brief, spark-like flashes of light — can be translated into doses of radiation received. Devices based on this technology are called *scintillation counters.*

Evacuated tubes that hold a gas at very low pressure and are fitted with two electrodes are used in devices such as the Geiger counter. This is a type of *ionization counter,* and Figure 10.6 shows how the tube itself is constructed. The tube, called a Geiger-Müller tube, is especially useful in measuring such beta and gamma radiation that has enough energy to penetrate the window. When a pulse of this radiation enters the rarified gaseous atmosphere in the tube, it creates ions that conduct a brief pulse of electricity which shoots from one electrode to the other. This pulse signals the apparatus to record a "count."

Inside the screen of a color TV tube there is a coating that includes various phosphors that glow with different colors when struck by the focused electron beam in the tube.

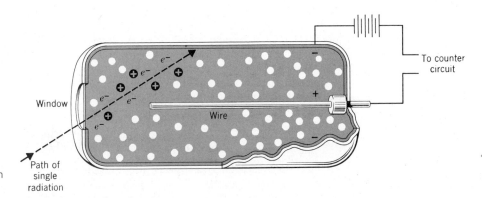

FIGURE 10.6
The basic features of a gas-ionization radiation detection tube such as used in a Geiger-Müller counter.

10.4   SYNTHETIC RADIONUCLIDES

Most radionuclides used in medicine are made by bombarding other atoms with high-energy particles.

Radioactive decay is only one way by which transmutations occur. They can also be caused by bombarding atoms with high-energy particles, and several hundred isotopes that do not occur naturally have been made this way. Several have been used successfully in medicine both in diagnosis and in treatment.

Various bombarding particles are employed, including alpha particles, neutrons, and gaseous protons. The first transmutation caused by bombardment was observed by Rutherford. It was the conversion of nitrogen-14 into oxygen-17 by alpha-particle bombardment. Rutherford let alpha particles from a naturally radioactive source travel through a tube that contained nitrogen-14. He soon detected that another radiation was being generated, one far more penetrating than the alpha radiation that was used in this experiment, and he found it to consist of high-energy protons. He also discovered that atoms of oxygen-17 now existed in the tube. To explain all of these observations, Rutherford reasoned that alpha particles had plowed right through the electron clouds of nitrogen-14 atoms and buried themselves in their nuclei. The strange new nuclei, called *compound nuclei,* evidently had too much energy to exist for long. They were excited nuclei of fluorine-18, and to rid themselves of excess energy each ejected a proton, which left behind an atom of oxygen-17. The equation is

> These are gaseous protons, true subatomic particles, and not hydronium ions.

$$\underset{\substack{\text{Alpha}\\\text{particle}}}{{}^{4}_{2}\text{He}} + \underset{\substack{\text{Nitrogen}\\\text{nucleus}}}{{}^{14}_{7}\text{N}} \longrightarrow \underset{\substack{\text{Fluorine}\\\text{nucleus}}}{{}^{18}_{9}\text{F}^{*}} \longrightarrow \underset{\substack{\text{Oxygen}\\\text{nucleus}}}{{}^{17}_{8}\text{O}} + \underset{\text{Proton}}{{}^{1}_{1}p}$$

(The asterisk by the symbol for fluorine-18 signifies that the particle is a high-energy, compound nucleus.) Oxygen-17 is a rare but nonradioactive isotope of oxygen. Usually, transmutations caused by bombardments produce radioactive isotopes of other elements.

> Remember that the energy of a moving object increases with the *square* of its velocity.
>
> $KE = (1/2)mv^2$

Electrically charged bombarding particles such as the alpha particle and the proton can be made to move with greater velocity and therefore with greater energy when they are attracted by opposite charge. Devices that accomplish this constitute some of the multimillion dollar hardware of atomic research — machines that function as particle accelerators with such names as cyclotrons, betatrons, synchrotrons, and many others. They have made possible the synthesis of dozens of new radionuclides when their ultrahigh-energy beams have been focused on selected targets.

Certain isotopes of uranium in atomic reactors eject neutrons, and although neutrons can't be accelerated (they are electrically neutral), they have sufficient energy to serve as bombarding particles. Being neutral is an advantage, because they aren't repelled either by the electrons that surround an atom or by the nucleus. One of the very important applications of neutron bombardment is the synthesis of molybdenum-99 from molybdenum-98. Using ${}^{1}_{0}n$ as the symbol for the neutron, the equation is

$$\,^{98}_{42}\text{Mo} + \,^{1}_{0}n \longrightarrow \,^{99}_{42}\text{Mo} + \gamma$$

As we will learn in the next section, the decay of molybdenum-99 leads to one of the most commonly used radionuclides in medicine.

10.5   MEDICAL APPLICATIONS OF RADIONUCLIDES

A radionuclide to be used in medical diagnoses ought to have a short half-life, emit only gamma radiation, and be easily eliminated.

Radionuclides have chemical properties identical to those of the stable isotopes of the same element. For example, strontium-90 is a beta and gamma emitter, but *chemically* it is identical to strontium-88, the most abundant stable isotope of this element. Although iodine-131 is a

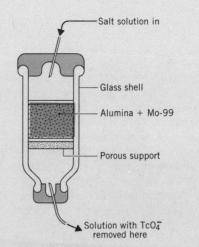
beta emitter, its chemical properties are identical to those of iodine-127, the only stable isotope of this element.

What these facts suggest is that radionuclides are not selected for any unique chemical properties but for their radiations. Their actual chemistry is also a factor, because when they are used in the body, radionuclides must be chemically compatible with the living system. Moreover, their *chemistry* is what guides them naturally to desired tissues.

Radiology: the science of radioactive substances and of X rays.
Radiologist: a specialist in radiology who also usually has a medical degree.

General Principles for the Selection of Radionuclides for Medical Uses.   Exposing anyone to any radiation entails some risks, because prolonged exposure can produce cancer. No such exposure is permitted unless the expected benefit from finding and treating a dangerous disease is thought to be greater than the risk. To minimize the risks, the radiologist uses radionuclides that, as much as possible, have the following properties.

1.   The radionuclide should have a half-life that is short. (Then it will decay *during* the diagnosis when the decay gives some benefit, and as little as possible will decay later, when the radiations are of no benefit.)

2.   The product of the decay of the radionuclide should have little if any radiation of its own. (Either the product should be a stable isotope or it should have a very long half-life.) It should also be quickly eliminated.

3.   The half-life of the radionuclide must be long enough to make it possible to prepare and administer it to the patient.

4.   If the radionuclide is to be used for diagnosis, it should decay by penetrating radiation entirely — meaning gamma radiation, if possible. (Nonpenetrating radiations, such as alpha and beta radiation, add to the risk by causing internal damage without contributing to the detection of the radiation externally. For uses in *therapy,* as in cancer therapy, nonpenetrating radiation is preferred because a radionuclide well-placed in cancerous tissue *should* cause damage to such tissue.)

5.   The diseased tissue either should concentrate the radionuclide, giving a "hot spot" where the diseased area exists, or it should do the opposite and reject the radionuclide, making the diseased area a "cold spot" insofar as external detectors are concerned.

Technetium-99*m*.   Technetium 99*m* is a radionuclide produced by the decay of molybdenum-99, whose synthesis we described at the end of Section 10.4. (See also Special Topic 10.2.) You already know that gamma radiation often accompanies the emission of other

X rays are generated by bombarding a metal surface with high-energy electrons. These can penetrate the metal atom far enough to knock out one of its low-level orbital electrons, such as a 1s electron. This creates a "hole" in the electron configuration, and orbital electrons at higher levels begin to drop down. In other words, the creation of this "hole" leads to electrons changing their energy levels. The difference between two of the lower levels corresponds to the energy of an X ray, which is emitted.

The refinement of X-ray techniques and the development of powerful computers made possible the generation of a diagnostic technology called computerized tomography, or CT for short. A. M. Cormack (United States) and G. N. Hounsfield (England) shared the 1979 Nobel prize in medicine for their work in the development of this technology. The instrument includes a large array of carefully positioned and focused X-ray generators. In the procedure called a CT scan, this array is rotated as a unit around the body or the head of the patient. Extremely brief pulses of X rays are sent in from all angles across one cross section of the patient. (See the accompanying photo.)

The changes in the X rays that are caused by internal organs or by tumors are sent to a computer that processes the data and delivers a picture of the cross section. It's like getting a picture of the inside of a cherry pit without cutting open the cherry. The CT scan is widely used for locating tumors and cancers (see the other accompanying figure).

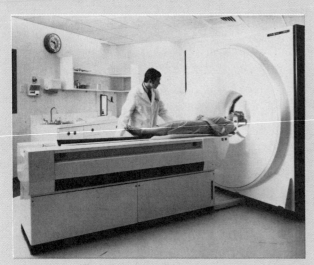

CT scanner

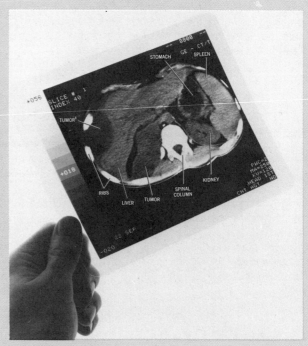

CT scan showing two tumors.

radiations, but with molybdenum-99 the gamma radiation comes after a pause. Molybdenum-99 first emits a beta particle:

$$\,^{99}_{42}\text{Mo} \longrightarrow \,^{99m}_{43}\text{Tc} + \,^{0}_{-1}e$$

The other decay product is a metastable form of technetium-99—hence the m in 99m. Metastable means poised to move toward greater stability. Technetium-99m decays by emitting gamma radiation with an energy of 143 keV:

$$\,^{99m}_{43}\text{Tc} \longrightarrow \,^{99}_{43}\text{Tc} + \gamma$$

Technetium-99m almost ideally fits the criteria for a radionuclide intended for diagnostic work. Its half-life is short—6.02 h. Its decay product, technetium-99, has a very long

A number of synthetic radionuclides emit positrons. These are particles that have the same small mass as an electron but carry one unit of *positive* charge. (They're sometimes called a positive electron.) A positron forms by the conversion of a proton into a neutron, as follows:

$$^1_1p \longrightarrow {}^1_0n + {}^0_1e$$

Proton (in atom's nucleus)   Neutron (stays in nucleus)   Positron (is emitted)

Positrons, when emitted, last for only a brief interval before they collide with an electron. The two particles annihilate each other, and in so doing their masses convert entirely into energy in the form of two tiny bursts of gamma radiation (511 keV). The gamma radiation formed in this way is called annihilation radiation.

$$^{\;0}_{-1}e + {}^0_1e \longrightarrow 2\,{}^0_0\gamma$$

Electron   Positron   Gamma radiation

The two bursts leave the collision site in almost exactly opposite directions.

To convert this science into a medically useful technology requires that a positron-emitting nuclide be made part of a molecule with a chemistry that will carry it into the particular tissue whose study is desired. Once the molecule gets in, the tissue now has a gamma radiator *on the inside*. Thus instead of sending X rays through the body as in a CT scan (Special Topic 10.3), the radiation originates right within the site being monitored. The overall procedure is called positron emission tomography, or PET for short.

Three positron emitters are often used—oxygen-15, nitrogen-13, and carbon-11. Glucose, for example, can be made in which one carbon atom is carbon-11 instead of the usual carbon-12. Glucose can cross the blood-brain barrier and get inside brain cells. If some part of the brain is experiencing abnormal glucose metabolism, this will be reflected in the way in which positron-emitting glucose is handled, and gamma-radiation detectors on the outside can pick up the differences. The use of the PET scan has led to the discovery that glucose metabolism in the brain is altered in schizophrenia and manic depression.

PET technology is being used to study a number of neuropsychiatric disorders, including Parkinson's disease. When a drug labeled with carbon-11 is used, its molecules go to those parts of the brain that have nerve endings that release dopamine. A PET scan then discloses the dopamine-releasing potential of the patient. This potential becomes impaired in Parkinson's disease.

By labeling blood platelets with a positron emitter, scientists can follow the development of atherosclerosis in even the tiniest of human blood vessels. Blood flow in the heart can be monitored without having to insert a catheter.

---

half-life—212,000 years—so it has an activity too low to be of much concern. (Technetium-99 decays to a stable isotope of ruthenium, $^{99}_{44}Ru$.) The half-life of technetium-99m, although short, is still long enough to allow time to prepare and administer it. It decays entirely by gamma radiation, which means that *all* of the radiation gets to a detector to signal where the radionuclide is in the body. Finally, a variety of chemically combined forms of technetium-99m have been developed that permit either hot spots or cold spots to form.

One form of technetium-99m is the pertechnetate ion, TcO_4^-. It behaves in the body very much like the chloride ion, so it tends to go where chloride ions go and to be eliminated by the kidneys. It is used to assess the degree of normal functioning of the kidneys and other organs such as the liver, the spleen, and the lungs. It is no longer widely used, as it once was, to spot tumors or cancerous tissue. The CT scan (Special Topic 10.3), the PET scan (Special Topic 10.4), and NMR imaging (Special Topic 10.5) have largely taken over this function.

Iodine-131 and Iodine-123.   The thyroid gland, the only user of iodine in the body, takes iodide ion and makes the hormone thyroxin. When either an underactive or an overactive thyroid is suspected, one technique is to let the patient drink a glass of flavored water that contains some radioactive iodine as I^-. By placing radiation-detection equipment near the thyroid gland, the radiologist can tell how well this gland takes up iodide ion from circulation. For diagnostic purposes, iodine-123 is popular because it has a short half-life (13.3 h) and it emits only gamma radiation (159 keV). Iodine-131 also has a short half-life (8 days), but it emits both beta particles (600 keV) and gamma radiation (mostly 360 keV). Certain types of thyroid cancer have been treated with this radionuclide. The ability of this gland to concentrate iodide ion is so good that if a small whole-body dose of iodine-131 is given, nearly 1000 times this much dose concentrates in the thyroid.

The CT scan subjects a patient to large numbers of short bursts of X rays. The PET scan exposes the patient to gamma radiation that is generated on the inside. Thus both technologies carry the usual risks that attend ionizing radiations, and they are used when the potential benefits from correct diagnoses far outweigh such risks. The NMR imaging technology operates without these dangers—at least, none has been discovered thus far. The principal developer of the hardware for NMR imaging has been Raymond Damadian.

NMR stands for nuclear magnetic resonance. Atomic nuclei that have odd numbers of protons and neutrons behave as if they were tiny magnets—hence the *nuclear magnetic* part of NMR. The nucleus of ordinary hydrogen (which has no neutrons and 1 proton) constitutes the most abundant nuclear magnet in living systems, because hydrogen atoms are parts of water molecules, and all biochemicals. Nuclear magnets spin about an axis much as the electron spins about its axis. When molecules that have spinning nuclear magnets are in a strong magnetic field and are simultaneously bathed with properly tuned radiofrequency radiation (which is of very low energy), the nuclear magnets flip their spins. (This is the *resonance* part of NMR.) As they resonate, they emit electromagnetic energy that is biologically harmless—it's also in the low-energy radiofrequency region of the spectrum—and this energy is picked up by detectors. The data are fed into computers that produce an image much like that of a CT or a PET scan—but without having subjected the patient to ultra-high energy electromagnetic radiation such as X rays or gamma rays. The NMR images are actually sophisticated plots of the distributions of the spinning nuclear magnets—of hydrogen atoms, for example.

NMR imaging has proven to be especially useful for studying soft tissue, the sort of tissue least well studied by X rays. Different soft tissues have different population densities of water molecules or fat molecules (which are loaded with hydrogen atoms). And tumors and cancerous tissue have their own water inventories. The pelvic area, packed with soft-tissue organs, is particularly amenable to NMR imaging. Natural contrasts are built in—by the pelvic fat, the water in the bladder, and the gases in the bowel—and NMR imaging is being increasingly used to aid in the diagnosis of prostate and bladder cancers. Distinguishing between actual cancer in the prostate and benign hypertrophy is one possibility.

Cobalt-60.   Cobalt-60 emits beta radiation (315 keV) and gamma radiation (2.819 MeV), and it has a half-life of 5.3 years. Gram for gram, cobalt-60 has a source intensity more than 200 times that of pure radium, which until the 1950s had been widely used in treating cancer. The sample of cobalt-60 is placed in a lead container many centimeters thick that has an opening pointed toward the cancerous site. All beta radiations are shielded by a thin piece of aluminum.

Linear accelerators are a relatively recent development in X-ray therapy. These devices can generate X rays for therapeutic uses with energies in the range of 6 to 12 MeV, which are much more powerful than the gamma rays from cobalt-60.

Other Medically Useful Radionuclides.   Indium-111 ($t_{1/2} = 2.8$ days; gamma emitter, 173 and 247 keV) has been found to be a good labeler of blood platelets.

Gallium-67 ($t_{1/2} = 78$ h; gamma emitter; 1.003 MeV) is used in the diagnosis of Hodgkin's disease, lymphomas, and bronchogenic carcinoma.

Phosphorus-32 ($t_{1/2} = 14.3$ days; beta emitter; 1.71 MeV) in the form of the phosphate ion has been used to treat a form of leukemia, a cancer of the bone that affects white cells in the blood. Because the phosphate ion is part of the hydroxyapatite mineral in bone, this ion is a bone-seeker.

SUMMARY

Atomic radiations Radionuclides in nature emit alpha radiation (helium nuclei), beta radiation (electrons), and gamma radiation (high-energy, X-ray-like radiation). This radioactive decay causes transmutation. The penetrating abilities of the radiations are a function of the sizes of the particles, their charges, and the energies with which they are emitted. Gamma radiations, which have no associated mass or charge, are the most penetrating. The decay of one radionuclide doesn't always produce a stable nuclide. Uranium-238 is at the head of a radioactive disintegration series that involves several intermediate radionuclides until a stable isotope of lead forms. Each decay can be described by a nuclear equation in which mass numbers and atomic numbers on either side of the arrow must balance. To describe how stable a radionuclide is we use its half-life, and the shorter this is, the more radioactive is the radionuclide.

Ionizing radiations — dangers and precautions When radiations travel in matter, they create unstable ions and radicals that have chemical properties dangerous to health. Intermittent exposure can lead to cancer, tumors, mutations, or birth defects. Intense exposures cause radiation sickness and death. Intense exposures focused on cancer tissue are used in cancer therapy. To guard against the hazards of ionizing radiations, the use of distance, fast film, and dense shielding material are the best strategies. According to the inverse square law, the intensity of radiation falls off with the square of the distance from the source. Complete protection, however, is not possible because of the natural background radiation that now includes traces of radioactive pollutants.

Units of radiation measurement To describe activity we use the curie, Ci, or the becquerel, Bq. To describe the intensity of exposure to X rays or gamma radiation, we use the roentgen. The gray (Gy) or the rad (D) is used to indicate how much energy has been absorbed by a unit mass of tissue (or other matter). To put the damage that different radiations can cause when they have the same values of rads (or grays) on a comparable and additive basis, we use

the rem. Finally, to describe the energy possessed by a radiation, we use the electron-volt. Diagnostic X rays are on the order of 100 keV. Radiations used in cancer treatment are in the low MeV range. To measure radiations there are devices such as film badges, scintillation counters, and ionization counters (Geiger-Müller tubes).

Radionuclides in medicine A number of synthetic radionuclides have been made by bombarding various isotopes with alpha radiation, neutrons, or accelerated protons. For diagnostic uses, the radionuclide should have a short half-life (but not so short that it decays before any benefit can be obtained). It should decay by gamma radiation only, and it should be chemically compatible with the organ or tissue so that either a hot spot or a cold spot appears. Its decay products should be as stable as possible, and capable of being eliminated from the body. Technetium-99m is almost ideal for diagnostic work, particularly for assessing the ability of an organ or tissue to function. Radionuclides of iodine (I-123 and I-131) are used in diagnosing or treating thyroid conditions. Gallium-67, indium-111, and phosphorus-32 are a few of the many other radionuclides used in diagnosis. Cobalt-60 has powerful gamma radiation, and is used in cancer treatment. Linear accelerators also provide high-energy (6 to 12 MeV) radiation for cancer therapy.

KEY TERMS

The following terms were emphasized in this chapter, and they constitute perhaps the most elementary vocabulary for the use of radiations in medicine that can be devised. Health care professionals should be well-versed in these terms and their meanings.

alpha particle	electron-volt (eV)	nuclear equation	radionuclide
alpha radiation	fission	rad (D)	rem
background radiation	gamma radiation	radiation sickness	roentgen
becquerel (Bq)	gray (Gy)	radical	threshold exposure
beta particle	half-life	radioactive	transmutation
beta radiation	inverse square law	radioactive decay	
cosmic radiation	ionizing radiation	radioactive disintegration series	
curie (Ci)	nuclear chain reaction		

SELECTED REFERENCES

1 Mary Berger and Karl Hubner. "Hospital Hazards: Diagnostic Radiation." *American Journal of Nursing,* August 1983, page 1155.

2 National Research Council—National Academy of Sciences. Committee on the Biological Effects of Ionizing Radiations. *The Effects on Populations of Exposure to Low Levels of Ionizing Radiation.* Washington, D.C., The Council, 1980.

3 C. Carl Jaffe. "Medical Imaging." *American Scientist,* November–December, 1982, page 576.

4 L. R. Prosnitz, D. S. Kapp, and J. B. Weissberg. "Radiotherapy." *New England Journal of Medicine,* Part 1, September 29, 1983, page 771. Part 2, October 6, 1983, page 834.

5 G. W. Beebe. "Ionizing Radiations and Health." *American Scientist.* January–February 1982, page 35.

6 Carol B. Jankowski. "Radiation Emergency." *American Journal of Nursing.* January 1982, page 90.

7 A. C. Upton. "The Biological Effects of Low-Level Ionizing Radiation." *Scientific American,* February 1982, page 41.

8 M. M. Ter-Pogassian, M. E. Raichle, and B. E. Sobel. "Positron-Emission Tomography." *Scientific American,* October 1980, page 171.

REVIEW EXERCISES

The answers to the Review Exercises that require a calculation or the balancing of an equation and that are marked with an asterisk are found in an Appendix. The answers to the other Review Exercises are found in the *Study Guide* that accompanies this book.

Radioactivity and Radiations

10.1 If a sample is described as *radioactive,* what specifically do we know about it?

10.2 The forerunners of chemists were called alchemists, and in ancient times one of their quests was for a way to change a base metal such as lead into gold. Today what technical word would be used to describe such a change if it were successful?

10.3 The film that Becquerel put in his desk drawer was wrapped in fairly heavy paper. In view of what we know today, his discovery of radioactivity depended on the emission of which radiation from the uranium ore sample?

10.4 When we write nuclear equations, which symbols are used for each of the following?
(a) The alpha particle
(b) The beta particle
(c) A gamma ray

10.5 The energy of an alpha particle is often higher than that of beta or gamma rays. Why is it, then, the least penetrating of the radiations?

10.6 The loss of an alpha particle changes the radionuclide's mass number by how many units? Its atomic number by how many units?

10.7 Why does the loss of a beta particle not change the radionuclide's mass number but *increases* its atomic number?

10.8 What happens to the mass number and to the atomic number of a radionuclide if it is only a gamma emitter?

10.9 Are *all* radioactive decays also transmutations?

10.10 If electrons do not exist in the nucleus, how can one originate in a nucleus in beta decay?

Nuclear Equations

*10.11 Write the symbols of the missing particles in the following nuclear equations.
(a) $^{245}_{96}Cm \rightarrow ^{4}_{2}He +$ _____
(b) $^{22}_{9}F \rightarrow ^{0}_{-1}e +$ _____

10.12 Complete the following nuclear equations by writing the symbols of the missing particles.
(a) $^{220}_{86}Rn \rightarrow ^{4}_{2}He +$ _____
(b) $^{140}_{56}Ba \rightarrow ^{0}_{-1}e +$ _____

*10.13 Write a balanced nuclear equation for each of the following changes.
(a) Alpha emission from einsteinium-252
(b) Beta emission from magnesium-28
(c) Beta emission from oxygen-20
(d) Alpha and gamma emission from californium-251

10.14 Give the nuclear equation for each of the following radioactive decays.
(a) Beta emission from bismuth-211
(b) Alpha and gamma emission from plutonium-242
(c) Beta emission from aluminum-30
(d) Alpha emission from curium-243

Half-Lives

10.15 Lead-214 is in the uranium-238 disintegration series. Its half-life is 19.7 min. Explain in your own words what being in this series means and what *half-life* means.

10.16 Which would be more dangerous to be near, a radionuclide with a short half-life that decays by alpha emission only or a radionuclide with the same half-life but that decays by beta and gamma emission? Explain.

*10.17 A 12.00 ng sample of technetium-99*m* will still have how many nanograms of this radionuclide left after 4 half-life periods? (This is about a day.)

10.18 If a patient is given 9.00 ng of iodine-123 (half-life 13.3 h), how many nanograms of this radionuclide remain after 12 half-life periods (about a week)?

Dangers of Ionizing Radiations

10.19 We have ions in every fluid of the body. Why, then, are ionizing radiations dangerous?

10.20 What is a chemical *radical,* and why is it chemically reactive?

10.21 Radiations are teratogenic agents. What does this mean?

10.22 Plutonium-239 is a carcinogen. What does this mean?

10.23 What two properties of ionizing radiations are exploited in strategies for providing radiation protection?

10.24 Atomic radiations are said to have no threshold exposure. What does this mean?

10.25 How is it that the same agent—radiations from a radionuclide—can be used both to cause cancer and to cure it?

10.26 The inverse square law tells us that if we double the distance from a radioactive source, we will reduce the radiation intensity that we receive by a factor of what number?

10.27 What general property of radiations is behind the inverse square law?

10.28 List as many factors as you can that contribute to the background radiation.

10.29 Why does a trip in a high-altitude jet plane increase one's exposure to background radiation?

10.30 Name three isotopes made in nuclear power plants that are particularly hazardous to health, and explain in what specific ways they endanger various parts of the body.

10.31 In general terms, how does fission differ from radioactive decay?

10.32 What fundamental aspect of fission makes it possible for it to proceed as a chain reaction?

10.33 The reactors of nuclear power plants are fitted with many movable rods. (Some are made of carbon, for example.) These rods are effective in capturing neutrons. To turn down a reactor or to turn it off, these rods are pushed into the regions where fission occurs. How can their presence control the rate of fissioning?

*10.34 A radiologist discovered that at a distance of 1.80 m from a radioactive source, the intensity of radiation was 140 millirad. How far should the radiologist move away to reduce the exposure to 2 millirad?

10.35 Using a Geiger-Müller counter, a radiologist found that in a 20-min period the dose from a radioactive source would measure 40 millirad at a distance of 10.0 m. How much dose would be received in the same time by moving to a distance of 1.00 m?

Units of Radiation Measurement

10.36 What SI unit is used to describe the activity of a radioactive sample?

10.37 A hospital purchased a sample of a radionuclide rated at 1.5 mCi. What does this rating mean?

10.38 Give the name of the unit that describes the intensity of an exposure to X rays.

10.39 What is the name and symbol of the SI unit used in describing how much energy a given mass of tissue receives from exposure to radiation? What is the name and symbol of the older, common unit?

10.40 Approximately how many rads would kill half of a large population within four weeks, assuming that each individual received this much?

10.41 From a health-protection standpoint, how do the roentgen and the rad compare in their potential danger?

10.42 We cannot add a 1 rad dose to some organ from gamma radiation to a 1 rad dose to the same organ from neutron radiation and say that the total biologically effective dose is 2 rads. Why not?

10.43 How is the problem implied by the previous Review Exercise resolved?

10.44 In units of mrem, what is the average natural background radiation received by the U.S. population, exclusive of medical sources, radioactive pollutants, and fallout?

10.45 What is the name of the energy unit used to describe the energy associated with an X ray or a gamma ray?

10.46 In the unit traditionally used (Review Exercise 10.45), how much energy is associated with diagnostic X rays?

10.47 Why should diagnostic radiations ideally be of much lower energy than radiations used in therapy — in cancer treatment, for example?

10.48 In general terms, how does a film badge dosimeter work?

10.49 In your own words, how does a Geiger-Müller counter work? Why doesn't it detect alpha radiation?

Synthetic Radionuclides

*10.50 When manganese-55 is bombarded by protons, the neutron is one product. What else is produced? Write a nuclear equation.

10.51 To make indium-111 for diagnostic work, silver-109 is bombarded with alpha particles. What forms if the nucleus of silver-109 captures one alpha particle? Write the nuclear equation for this capture.

*10.52 The compound nucleus that forms when silver-109 captures an alpha particle (previous Review Exercise) decays directly to indium-111, plus *two* other identical particles. What are they? Write the nuclear equation for this decay.

10.53 To make gallium-67 for diagnostic work, zinc-66 is bombarded with accelerated protons. When a nucleus of zinc-66 captures a proton, the nucleus of what isotope forms? Write the nuclear equation.

*10.54 The isotope that forms when zinc-66 captures a proton (previous Review Exercise) is unstable in a novel way (novel at least to our study). This nucleus is able to capture one of its own electrons. When it does, what new nucleus forms? Write the equation for this kind of nuclear event — called electron capture.

10.55 When fluorine-19 is bombarded by alpha particles, both a neutron and a nucleus of sodium-22 form. Write the nuclear equation, including the compound nucleus that is the intermediate.

*10.56 When boron-10 is bombarded by alpha particles, nitrogen-13 forms and a neutron is released. Write the equation for this reaction.

10.57 When nitrogen-14 is bombarded with deuterons, ^2_1H, oxygen-15 and a neutron form. Write the equation for this reaction.

*10.58 What bombarding particle could change aluminum-27 into phosphorus-32 and a proton? Write the equation.

10.59 What bombarding particle can change sulfur-32 into phosphorus-32 and a proton? Write the equation.

Medical Applications of Radiations

10.60 Why is it desirable to use a radionuclide of short half-life in diagnostic work, when we know that even small samples of such isotopes can be very active?

10.61 The possible dangers related to the high activity of a sample of a radionuclide of short half-life can be minimized by taking advantage of this activity. What can be done with such a radionuclide that could not be done as easily with one of very long half-life?

10.62 We know that gamma radiation is the most penetrating of all natural radiations. Why, then, is a diagnostic radionuclide that emits only gamma radiation preferable to one that gives, say, only alpha radiation?

10.63 Why is iodine-123 better for diagnostic work than iodine-131?

10.64 Why did cobalt-60 replace radium for cancer treatment?

10.65 In what chemical form should phosphorus-32 be used to facilitate its seeking bone tissue? Explain.

CHAPTER 11
Introduction to Organic Chemistry

The marvel of it is that the atoms making up these people ultimately came from such simple things as water, carbon dioxide, air, and a few salts. A study of how this can be has a new beginning in this chapter as we introduce organic compounds.

11.1 ORGANIC AND INORGANIC COMPOUNDS

Because most compounds of carbon — organic compounds — are molecular, not ionic, they have relatively low boiling points, melting points, and solubilities in water.

A mineral is a solid with a definite formula that can be obtained from the earth's crust.

Organic compounds are compounds of carbon, and there are more of these compounds than of all the other elements combined, except hydrogen. The name itself, implying *organism,* arose when scientists believed that organic compounds could be made only by and within living organisms. Prior to 1828 no one had succeeded in making an organic compound in the laboratory either directly from the elements or from minerals in the earth's crust. In fact, scientists who tried to overcome this difficulty eventually concluded that a law of nature made the task impossible in principle. After all, what we cannot do is just as significant in science as what we can. For example, we cannot create or destroy energy, and this natural limitation is labeled a scientific law. Similarly, the inability to synthesize an organic compound in the laboratory led to a theory known as the **vital force theory,** which declared that a catalyst-like agent — a vital force — is essential to this synthesis. And only living things, according to the theory, possessed the vital force. Organic compounds, therefore, were those that only living things could make, and **inorganic compounds** were all the rest, those not requiring the vital force for their preparation.

Vita- is from a Latin root meaning "life."

Wöhler's Experiment.   In 1828, Friedrich Wöhler (1800–1882) succeeded in making urea, a white solid that can be obtained from urine and that *everyone* regarded as an organic compound. His synthesis was the wholly unexpected result of an attempt to prepare the inorganic compound, ammonium cyanate, NH_4NCO. He prepared an aqueous solution that contained the ammonium ion, NH_4^+, and the cyanate ion, NCO^-, and evaporated it to dryness in full expectation that these oppositely charged ions would be forced to aggregate as a crystalline solid.

Wöhler obtained a white powder, but it had none of the expected properties of the salt or of any compound of ammonia. Instead of throwing the solid away, Wöhler analyzed it and found that it was urea. Evidently, the heat that was added to the system to remove the water caused the following reaction:

Urea is the chief nitrogen waste from the body. It is also manufactured from ammonia and used as a commercial fertilizer.

$$
NH_4NCO \xrightarrow{\text{heat}}
\begin{array}{ccccc}
 & H & O & H & \\
 & | & \| & | & \\
H-\!\!\!& N & -C- & N & \!\!\!-H
\end{array}
$$

Ammonium   Urea
cyanate

Other syntheses of organic compounds from minerals quickly followed Wöhler's discovery, and the vital force theory was soon dead. Roughly 6 million organic compounds are known, and all of them have been made or could in principle be made from substances of mineral origins.

Some Differences Between Organic and Inorganic Compounds.   Relatively few inorganic compounds contain carbon. Those that do are mostly the carbonates, bicarbonates, and cyanides of metal ions, but there are a few other types that need not concern us. As we have said, carbon is always present in organic compounds where its atoms, covalently bound together, form the structural "skeletons" of organic molecules. Atoms of several other nonmetal elements can be appended to these skeletons, also by covalent bonds. Such nonmetals include hydrogen, oxygen, nitrogen, sulfur, and the halogens.

Besides differences in composition, there are variations in bond types, which help to explain important differences in some properties. Covalent bonds are far more prevalent among organic compounds than among inorganic compounds, but ionic bonds are present more frequently among inorganic compounds. In other words, most organic compounds are molecular and most inorganic compounds are ionic. Because forces of attraction between polar molecules are generally much less than between oppositely charged ions, most organic compounds melt at temperatures well below those at which ionic compounds melt. A great number of organic compounds are liquids, not solids, at room temperature, but all ionic compounds are solids and melt generally well above 400 °C. Most organic compounds also have normal boiling points below 400 °C, but most inorganic compounds have very high boiling points. The exceptions are the inorganic compounds that are not ionic, such as a large number of gases (e.g., hydrogen halides, carbon dioxide, and the oxides of sulfur) and some inorganic liquids (e.g., water).

Relatively few organic compounds dissolve in water, but many inorganic compounds are water-soluble. Organic molecules often are not polar enough to be hydrated by water. However, the organic compounds that do dissolve in water are particularly important at the molecular level of life, because water is the central solvent in living things.

Typically, ionic compounds melt and boil well above 350 °C.

11.2 STRUCTURAL FEATURES OF ORGANIC MOLECULES

Organic molecules have flexible chains or rings of carbon atoms, and they almost always have a functional group.

The uniqueness of carbon among the elements is that its atoms can bond to each other successively many times and still form equally strong bonds to atoms of other nonmetals. A typical molecule in the familiar plastic, polyethylene, has hundreds of carbon atoms covalently joined in succession, and each carbon binds enough hydrogen atoms to fill out its full complement of four bonds.

Only a short segment of a typical molecule of polyethylene is shown here.

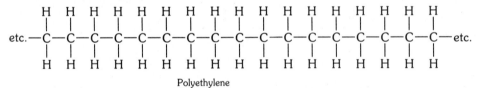

Polyethylene

Recall that each line in a structural formula stands for one shared pair of electrons. Each carbon, therefore, has four lines, meaning four shared pairs of electrons for a total of eight outside-level electrons. In all organic compounds, carbon *always* has four bonds from it, *never* five and *never* three bonds.

Straight Chains and Branched Chains.   A succession of carbon atoms that are bonded covalently together, as shown in the segment of the polyethylene molecule, is called a **straight chain.** Pentane, a constituent of gasoline, consists of molecules with straight chains that are just five carbons long. It is important to realize that "straight" has a very limited, technical meaning when used in this context. It means the absence of carbon branches. "Straight chain" means that one carbon follows another and no additional carbons are joined to this system at intermediate points.

A molecule of 2-methylpentane illustrates a **branched chain,** which means it has one (or more) carbon atoms joined to carbons that are *between* the ends of some parent chain.

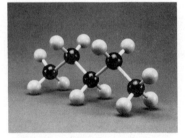

Pentane.

2-Methylpentane.

Pentane
(straight chain)

2-Methylpentane
(branched chain)

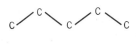

Pentane skeleton

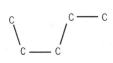

2-Methylpentane skeleton

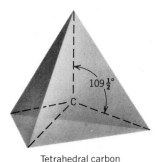

Tetrahedral carbon

When you compare the photographs of the ball-and-stick models of the molecules of pentane and 2-methylpentane with their structural formulas, be sure to notice that the written (or printed) structures disregard the correct bond angles at carbon. A carbon that has four single bonds has a tetrahedral geometry with bond angles of 109.5°. The ball-and-stick models faithfully show the correct angles at each carbon, but the printed symbols do not. The point here is that it is perfectly all right to let bond angles be "understood" unless there is some important reason to the contrary. And this isn't the only point we have to understand when we use printed structural formulas. We'll learn about another one next.

Free Rotation at Single Bonds.   Pieces of either straight chain or branched chain molecules that are connected by *single* bonds have a property called **free rotation.** This means that such pieces can be set into rotation with respect to each other around the single bonds as the result of collisions with other molecules. In an actual sample of pentane, for example, not all of the molecules are in the fully extended form shown in Figure 11.1. Pentane molecules are kinked and twisted into an almost infinite number of contorted forms, called **conformations.** Models of just a few of those of pentane are shown in Figure 11.1.

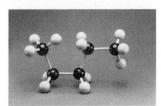

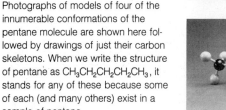

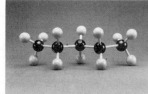

FIGURE 11.1
Free rotation at single bonds. Photographs of models of four of the innumerable conformations of the pentane molecule are shown here followed by drawings of just their carbon skeletons. When we write the structure of pentane as $CH_3CH_2CH_2CH_2CH_3$, it stands for any of these because some of each (and many others) exist in a sample of pentane.

The physical and chemical properties of pentane are the net, overall results of the effects of the whole collection of all its molecular conformations on whatever physical agent or chemical reactant has been applied to observe the property. In other words, it would seem that we could forget about the phenomena of free rotation and different conformations. However, when we study how the catalysts of the body work — its enzymes — it will be important to remember that in organic molecules there are bond angles, that about single bonds there is free rotation, and that molecules possess some flexibility and can change from one conformation to another. We will see that molecular *shape* is as important to understanding events at the molecular level of life as anything else.

Condensed Structural Formulas.   Thus far we have shown every bond in a structural formula as a straight line, but we have also seen how useful it is to leave some molecular features (e.g., bond angles) to the informed imagination. We can also leave most of the bonds in a structural formula to the imagination as well. All we *must* remember is that every carbon in a structural formula must have four bonds. With this in mind, we can group the hydrogen atoms that a particular carbon holds all together to one side or the other of its symbol.

Whenever a carbon holds three hydrogens, we can simplify the system by writing CH_3. (We could also write H_3C, but you don't see this as often.) Just remember that these three hydrogen atoms are individually joined to the carbon. The simplest example of doing this occurs with the structural formula of ethane:

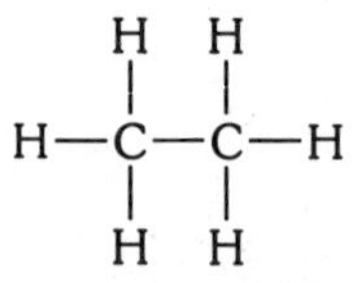

Ethane

When a carbon holds two hydrogen atoms, we can write the system either as CH_2 or (less often seen) as H_2C.

When a carbon holds just one hydrogen, we can write it as CH or (less commonly seen) as HC.

When we condense a full or expanded structure following these guides, the result is called a **condensed structure,** but because this is almost always the kind of structure used when discussing a compound, it is simply referred to as the substance's **structure.**

EXAMPLE 11.1   CONDENSING A FULL STRUCTURAL FORMULA

Problem: Condense the structural formula for 2-methylpentane.

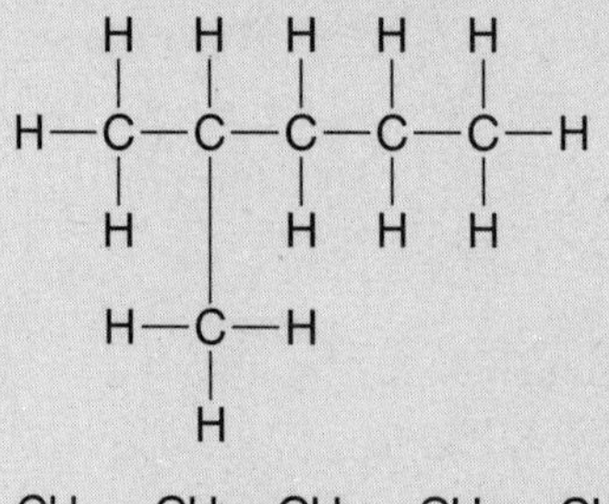

Solution: $CH_3-CH-CH_2-CH_2-CH_3$
                       |
                      CH_3

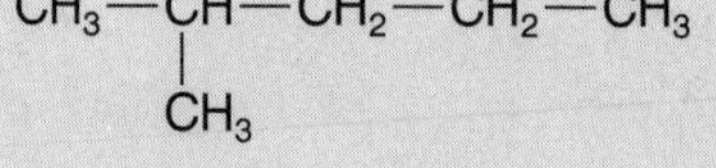

PRACTICE EXERCISE 1   Condense the following expanded structural formulas.

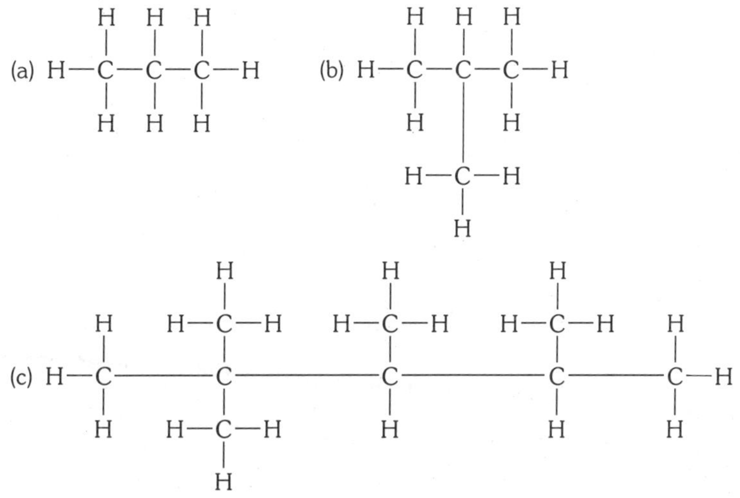

$$\text{(a)} \quad \begin{array}{ccc} & H & H & H \\ & | & | & | \\ H-&C-&C-&C-H \\ & | & | & | \\ & H & H & H \end{array} \qquad \text{(b)} \quad \begin{array}{ccc} & H & H & H \\ & | & | & | \\ H-&C-&C-&C-H \\ & | & | & | \\ & H & C & H \\ & & H-\!\!-\!\!H \\ & & | \\ & & H \end{array}$$

(c)

Because of free rotation, we have to be able to interpret zigzags. For example,

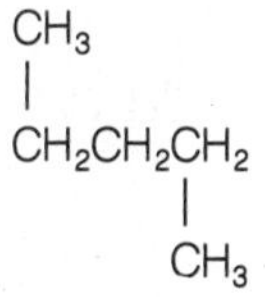

$$\begin{array}{c} CH_3 \\ | \\ CH_2CH_2CH_2 \\ | \\ CH_3 \end{array}$$

is the same molecule as $CH_3CH_2CH_2CH_2CH_3$

There is still another useful simplification in writing condensed structures. Whenever a *single* bond appears on a *horizontal* line, we need not write a straight line to represent it, we can leave such a bond understood. We cannot do this for bonds that are not on a horizontal line. For example, we can write the condensed structure of 2-methylpentane, which we considered in Example 11.1, as follows. Notice that the vertically oriented bond is shown by a line:

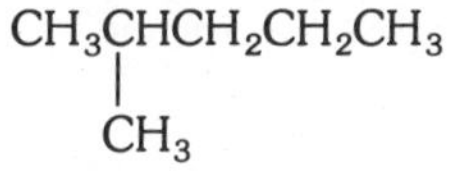

$$\underset{\underset{CH_3}{|}}{CH_3CHCH_2CH_2CH_3} \qquad \text{2-Methylpentane}$$

PRACTICE EXERCISE 2   Rewrite the condensed structures that you drew for the answers to Practice Exercise 1 and let the appropriate carbon-carbon single bonds be left to the imagination.

PRACTICE EXERCISE 3   Just to be certain you are comfortable with condensed structures, expand each of the following to make them full, expanded structures with no bonds left to the imagination.

(a) CH_3CH_3 (b) $\underset{\underset{CH_3}{|}}{\overset{\overset{CH_3}{|}}{CH_3CHCHCH_3}}$ (c) $\underset{\underset{CH_3}{|}}{\overset{\overset{CH_2CH_3}{|}}{CH_3CH_2CCH_2CH_2CH_3}}$

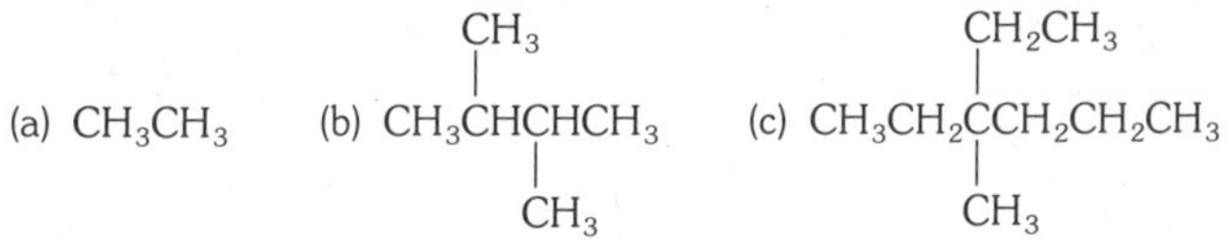

One very important skill that you have to develop in order to use condensed structures is the ability to recognize when an error has been made. The most common error is a violation of the rule that carbon must have just four bonds, no more and no fewer. Do the next Practice Exercise to develop this ability.

PRACTICE EXERCISE 4   Which of the following structures cannot represent real compounds?

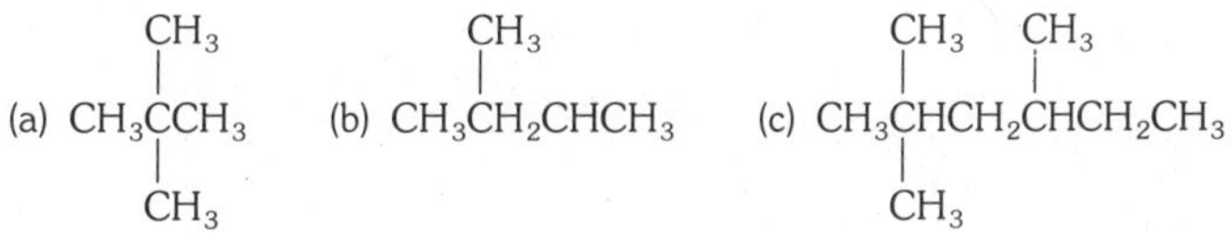

(a) $\underset{\underset{CH_3}{|}}{\overset{\overset{CH_3}{|}}{CH_3CCH_3}}$ (b) $\underset{\underset{CH_3}{|}}{CH_3CH_2CHCH_3}$ (c) $\underset{\underset{CH_3}{|}}{\overset{\overset{CH_3\ \ CH_3}{|\ \ \ \ |}}{CH_3CHCH_2CHCH_2CH_3}}$

There is still another simplification. Sometimes, two or three identical groups that are attached to the same carbon can be grouped inside a set of parentheses. For example,

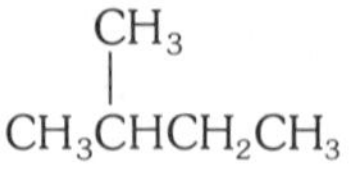

$$
\begin{array}{c}
CH_3 \\
| \\
CH_3CHCH_2CH_3
\end{array}
\qquad \text{can be written as} \qquad (CH_3)_2CHCH_2CH_3
$$

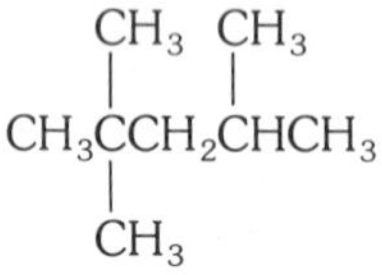

$$
\begin{array}{c}
CH_3 \quad CH_3 \\
| \qquad | \\
CH_3CCH_2CHCH_3 \\
| \\
CH_3
\end{array}
\qquad \text{can be written as} \qquad (CH_3)_3CCH_2CH(CH_3)_2
$$

We will not use this simplification very often, but you will see it in many references and you should be aware of it.

When atoms other than carbon and hydrogen are present in a molecule, there is no major new problem in writing condensed structures. Remember that *every* oxygen or sulfur atom must have two bonds, *every* nitrogen must have three, and *every* halogen atom must have just one. (These rules apply, of course, when these atoms carry no electrical charge and the structures involved are of molecules, not ions.) Another rule about condensed structures is that carbon-carbon double and triple bonds are never left to the imagination. Study the following examples that illustrate these rules.

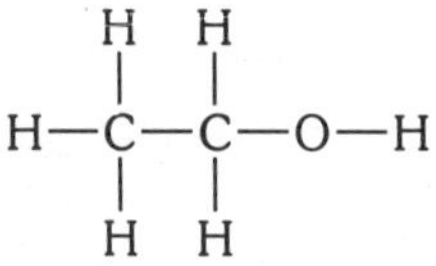

O and N can have more bonds in certain *ions,* for example, H_3O^+ and NH_4^+.

$$
\begin{array}{c}
H \quad H \\
| \quad | \\
H-C-C-O-H \\
| \quad | \\
H \quad H
\end{array}
\quad \text{condenses to} \quad CH_3-CH_2-O-H \quad \text{or to} \quad CH_3CH_2OH
$$

Ethyl alcohol (in alcoholic drinks)

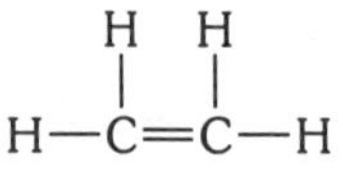

$$
\begin{array}{c}
H \quad H \\
| \quad | \\
H-C{=}C-H
\end{array}
\quad \text{condenses to} \quad CH_2{=}CH_2 \quad \text{or to} \quad H_2C{=}CH_2
$$

Ethylene (the raw material for making polyethylene)

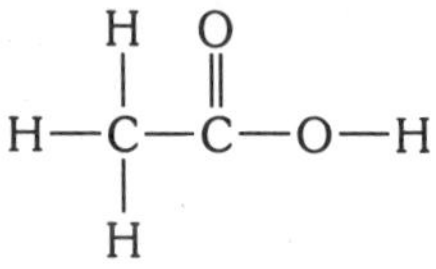

$$
\begin{array}{c}
H \quad O \\
| \quad \| \\
H-C-C-O-H \\
| \\
H
\end{array}
\quad \text{condenses to} \quad
\begin{array}{c}
O \\
\| \\
CH_3-C-O-H
\end{array}
\quad \text{or to} \quad
\begin{array}{c}
O \\
\| \\
CH_3COH
\end{array}
$$
$$\text{and often to } CH_3CO_2H \quad \text{or to} \quad CH_3COOH$$

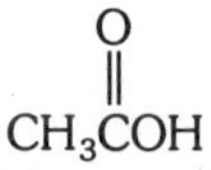

Acetic acid (in vinegar)

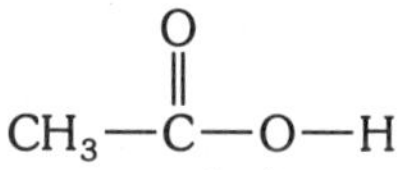

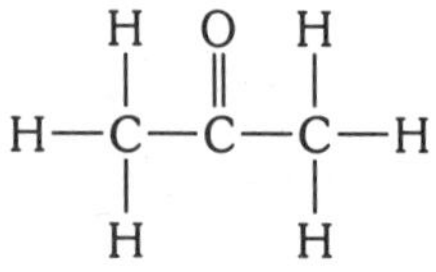

$$
\begin{array}{c}
H \quad O \quad H \\
| \quad \| \quad | \\
H-C-C-C-H \\
| \qquad | \\
H \qquad H
\end{array}
\quad \text{condenses to} \quad
\begin{array}{c}
O \\
\| \\
CH_3-C-CH_3
\end{array}
\quad \text{or to} \quad
\begin{array}{c}
O \\
\| \\
CH_3CCH_3
\end{array}
$$

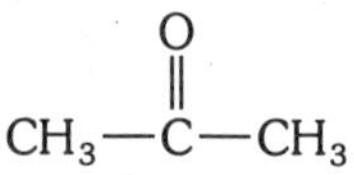

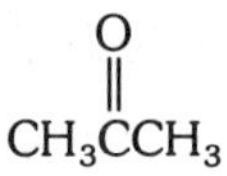

Acetone (nail polish remover)

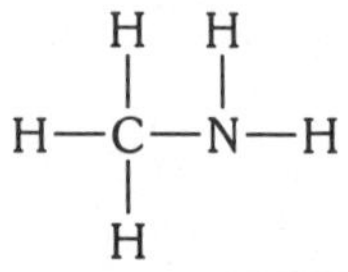

$$
\begin{array}{c}
H \quad H \\
| \quad | \\
H-C-N-H \\
| \\
H
\end{array}
\quad \text{condenses to} \quad CH_3-NH_2 \quad \text{or to} \quad CH_3NH_2
$$

Methylamine (in decaying fish)

Functional Groups and the Organization of Organic Chemistry.   Pieces of molecules that include nonmetals other than carbon and hydrogen or that have double or triple bonds are the specific sites in organic molecules that chemicals most often attack. These molecular pieces are called **functional groups,** because they are the chemically functioning locations. The other pieces of molecules that consist only of carbon and hydrogen and only single bonds are called the **nonfunctional groups.**

Although over 6 million organic compounds are known, there are only a handful of functional groups, and each one serves to define a family of organic compounds. Our study of organic chemistry will be organized around just a few of these families — those outlined in Table 11.1. One important family is called the alcohol family. We have learned, for example, that ethyl alcohol is CH_3CH_2OH. Its molecules have the —OH group attached to a chain of two carbons. But *chain length* is not what determines what family a compound is in; it only determines the name of the specific family member, as we will see later. The chain can be any length imaginable, and the substance will be in the alcohol family provided that somewhere on the chain there is an —OH group attached to a carbon from which only single bonds extend. Some simple examples of alcohols are:

$$CH_3\!-\!OH \qquad CH_3CH_2\!-\!OH \qquad CH_3CH_2CH_2\!-\!OH \qquad CH_3CHCH_3$$

Methyl   Ethyl   Propyl $|$
alcohol   alcohol   alcohol   OH

Isopropyl
alcohol

Because these all have the same functional group, they exhibit the same kinds of chemical reactions. When just one of these reactions is learned, it applies to all members of the family — literally to thousands of compounds. In fact, we will often summarize a particular reaction for an organic family by using a general family symbol. All alcohols, for example, can be symbolized by the symbol R—OH, where R— stands for a carbon chain — any chain of whatever length. All alcohols, for instance, react with sodium metal as follows:

> R is from the German word *Radikal,* which we translate here to mean *group* as in a group of atoms.

$$2R\!-\!O\!-\!H + 2Na \longrightarrow 2R\!-\!ONa + H_2$$

If we wanted to write the specific example of this reaction that involves, say, ethyl alcohol, all we have to do is replace R— by CH_3CH_2—.

$$2CH_3CH_2\!-\!OH + 2Na \longrightarrow 2CH_3CH_2\!-\!ONa + H_2$$

11.3 ISOMERISM

Compounds can have identical molecular formulas but different structures.

Ammonium cyanate and urea, the chemicals of Wöhler's important experiment, both have the molecular formula, CH_4N_2O. The atoms are just organized differently:

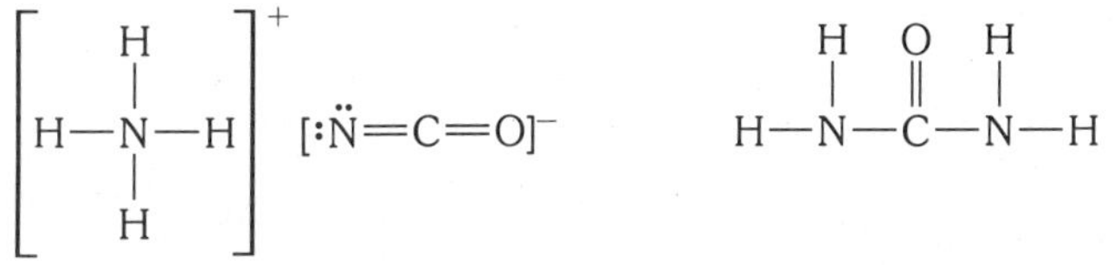

$$\left[\begin{array}{c} H \\ | \\ H\!-\!N\!-\!H \\ | \\ H \end{array}\right]^{+} \; [:\ddot{N}\!=\!C\!=\!O]^{-} \qquad\qquad \begin{array}{c} H \quad\; O \quad\; H \\ | \quad\;\; || \quad\;\; | \\ H\!-\!N\!-\!C\!-\!N\!-\!H \end{array}$$

Ammonium cyanate   Urea
CH_4N_2O CH_4N_2O

TABLE 11.1
Some Important Families of Organic Compounds

Family	Molecular Features	Example
Hydrocarbons	Only C and H present Subfamilies: 　Alkanes: only single bonds 　Alkenes: some double bonds 　Alkynes: some triple bonds 　Aromatic: benzene ring present	CH_3CH_3, ethane $CH_2{=}CH_2$, ethene $HC{\equiv}CH$, ethyne benzene
Alcohols	$-\overset{\mid}{\underset{\mid}{C}}-OH$ as in R—O—H	CH_3CH_2OH, ethyl alcohol
Ethers	$-\overset{\mid}{\underset{\mid}{C}}-O-\overset{\mid}{\underset{\mid}{C}}-$ as in R—O—R'	$CH_3CH_2OCH_2CH_3$, diethyl ether
Thioalcohols (mercaptans)	$-\overset{\mid}{\underset{\mid}{C}}-S-H$ as in R—S—H	CH_3SH, methyl mercaptan
Disulfides	$-\overset{\mid}{\underset{\mid}{C}}-S-S-\overset{\mid}{\underset{\mid}{C}}-$ as in R—S—S—R'	CH_3SSCH_3, dimethyl disulfide
Aldehydes	$-\overset{O}{\overset{\|}{C}}-H$ as in $R-\overset{O}{\overset{\|}{C}}-H$	$CH_3\overset{O}{\overset{\|}{C}}H$, acetaldehyde
Ketones	$-\overset{\mid}{\underset{\mid}{C}}-\overset{O}{\overset{\|}{C}}-\overset{\mid}{\underset{\mid}{C}}-$ as in $R-\overset{O}{\overset{\|}{C}}-R'$	$CH_3\overset{O}{\overset{\|}{C}}CH_3$, acetone
Carboxylic acids	$-\overset{O}{\overset{\|}{C}}-O-H$ as in $R-\overset{O}{\overset{\|}{C}}-O-H$	$CH_3\overset{O}{\overset{\|}{C}}OH$, acetic acid
Esters	$-\overset{O}{\overset{\|}{C}}-O-C$ as in $R-\overset{O}{\overset{\|}{C}}-O-R'$	$CH_3\overset{O}{\overset{\|}{C}}OCH_2CH_3$, ethyl acetate
Phosphate esters	$-\overset{\mid}{\underset{\mid}{C}}-O-\overset{O}{\overset{\|}{\underset{OH}{P}}}-O-H$ as in $R-O-\overset{O}{\overset{\|}{\underset{OH}{P}}}-O-H$	$CH_3O\overset{O}{\overset{\|}{\underset{OH}{P}}}OH$, methyl phosphate
Diphosphate esters	$R-O-\overset{O}{\overset{\|}{\underset{OH}{P}}}-O-\overset{O}{\overset{\|}{\underset{OH}{P}}}-O-H$	$CH_3O\overset{O}{\overset{\|}{\underset{HO}{P}}}O\overset{O}{\overset{\|}{\underset{OH}{P}}}OH$, methyl diphosphate
Triphosphate esters	$R-O-\overset{O}{\overset{\|}{\underset{OH}{P}}}-O-\overset{O}{\overset{\|}{\underset{OH}{P}}}-O-\overset{O}{\overset{\|}{\underset{OH}{P}}}-O-H$	$CH_3O\overset{O}{\overset{\|}{\underset{HO}{P}}}O\overset{O}{\overset{\|}{\underset{OH}{P}}}O\overset{O}{\overset{\|}{\underset{OH}{P}}}OH$, methyl triphosphate
Amines	$-NH_2$ as in $R-NH_2$ $-NH-$ as in $R-NH-R'$ $-\overset{\mid}{\underset{R}{N}}-$ as in $R-\overset{\mid}{\underset{R}{N}}-R'$	CH_3NH_2, methylamine $CH_3NHCH_2CH_3$, methylethylamine $CH_3\overset{}{\underset{CH_3}{N}}CH_3$, trimethylamine
Amides	$-\overset{O}{\overset{\|}{C}}-NH_2$ as in $R-\overset{O}{\overset{\|}{C}}-NH_2$	$CH_3\overset{O}{\overset{\|}{C}}NH_2$, acetamide

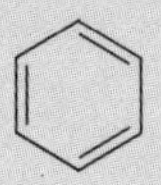

[Amides can also be of the types:

$R-\overset{O}{\overset{\|}{C}}-NH-R'$ and $R-\overset{O}{\overset{\|}{C}}-N(R)_2$]

TABLE 11.2
Properties of Two Isomers—Ethyl Alcohol and Dimethyl Ether

Property	Ethyl Alcohol	Dimethyl Ether
Structure	CH_3CH_2OH	CH_3OCH_3
Boiling point	78.5 °C	−24 °C
Melting point	−117 °C	−138.5 °C
Density (25 °C)	0.79 g/mL (a liquid)	2.0 g/L (a gas)
Solubility in water	Soluble in all proportions	Slightly soluble

"Isomer" has Greek roots—*isos,* the same, and *meros,* parts; in other words, "the same parts" (but put together differently).

The names in parentheses are the common names of these compounds. The letter *n* stands for *normal,* meaning the straight-chain isomer. *Neo* signifies *new,* as in a new isomer.

Compounds that have identical molecular formulas but different structures are called **isomers** of each other, and the existence of isomers is a phenomenon called **isomerism.** Isomerism is one reason why there are so many organic compounds.

There are three isomers of pentane, and all share the formula C_5H_{12}.

$$CH_3CH_2CH_2CH_2CH_3$$

Pentane
(*n*-pentane)

$$CH_3\overset{\overset{\displaystyle CH_3}{|}}{C}HCH_2CH_3$$

2-Methylbutane
(isopentane)

$$CH_3\overset{\overset{\displaystyle CH_3}{|}}{\underset{\underset{\displaystyle CH_3}{|}}{C}}CH_3$$

2,2-Dimethylpropane
(neopentane)

The larger the number of carbon atoms per molecule, the larger is the number of isomers. For example, someone has figured out that roughly 6.25×10^{13} isomers are possible for $C_{40}H_{82}$. (Very few have actually been prepared. It would take nearly 200 billion years to make each one at the rate of one per day.)

The isomers of pentane or of $C_{40}H_{82}$ all belong to the same organic family, and their chemical properties are quite similar. Often, however, isomers are in different families. For example, there are two ways to organize the atoms in C_2H_6O, as seen near the top of Table 11.2. One isomer is ethyl alcohol and the other is dimethyl ether. The latter is in the ether family, which has the general formula of R—O—R'. (The prime sign, ', signifies only that the two R— groups need not be identical.) Ethyl alcohol and dimethyl ether are radically different, as the data in Table 11.2 show. At room temperature, the former is a liquid and the latter is a gas. Chemically, they are also very different as we should expect when they have different functional groups. We learned in the previous section that ethyl alcohol reacts with sodium. No member of the ether family shows this reaction. It is quite common for isomers to vary this much, and this is one major reason why we use structural rather than molecular formulas in organic chemistry (and biochemistry). Only the structure lets us see at a glance just what family a compound belongs to.

EXAMPLE 11.2   RECOGNIZING ISOMERS

Problem: Which pair of structures represents a pair of isomers?

1. $CH_3\overset{\overset{\displaystyle CH_3}{|}}{C}H—\overset{\overset{\displaystyle CH_3}{|}}{C}HCH_2CH_2CH_2CH_3$ and $CH_3CH_2CH_2CH_2\overset{\overset{\displaystyle CH_3}{|}}{C}H—\overset{\overset{\displaystyle CH_3}{|}}{C}HCH_3$

2. $CH_3—O—CH_2CH_3$ and $CH_3CH_2—O—CH_3$

3. $CH_3\overset{\overset{\displaystyle CH_3}{|}}{C}HCH_2CH_2CH_3$ and $CH_3CH_2\overset{\overset{\displaystyle CH_3}{|}}{C}HCH_2CH_3$

4. $\overset{\overset{\displaystyle CH_3}{|}}{C}H_2CH_3$ and $CH_3CH_2CH_3$

Solution: Unless you spot a difference that rules out isomerism immediately, the first step is to see if the molecular formulas are the same. If they aren't, the two structures are not isomers, and if they are, the two might be identical or they might be isomers. In this problem, the members of each pair share the same molecular formula.

Pair 1: C_9H_{20} Pair 2: C_3H_8O Pair 3: C_6H_{14} Pair 4: C_3H_8

Next, to see if a particular pair represents isomers, we try to find at least one structural difference. If we can't, the two structures are identical; they are just oriented differently on the page, or their chains are twisted into different conformations. Don't be fooled by an "east-to-west" versus a "west-to-east" type of difference. The difference must be *internal* within the structure. (Whether you face east or west you're the same person!)

Pair 1 is an example of this east versus west difference in orientation. These two structures are identical. Their internal sequences are the same.

Pair 2 are similarly identical; they're just oriented differently on the page.

Pair 3 are isomers. In the first, a CH_3 group joins a five-carbon chain at the second carbon of the chain, and in the second, this group is attached at the third carbon.

Pair 4 are identical. The two structures differ only in the conformations of their chains.

PRACTICE EXERCISE 5   Examine each pair to see if the members are identical, are isomers, or are different in some other way.

(a) $H—O—CH_3$ and $CH_3—O—H$

(b) $CH_3—NH—CH_3$ and $CH_3—CH_2—NH_2$

(c)
$$\overset{\displaystyle CH_2CH_3}{\overset{|}{CH_2CH_2\underset{|}{\underset{\displaystyle CH_3}{C}HCH_3}}} \quad \text{and} \quad CH_3CH_2CH_2CH_2\overset{\displaystyle CH_3}{\overset{|}{C}}HCH_3$$

(d) $CH_2{=}CHCH_2CH_3$ and $CH_3CH{=}CHCH_3$

(e)
$$CH_3CH_2\overset{\displaystyle O}{\overset{||}{C}}OH \quad \text{and} \quad HO\overset{\displaystyle O}{\overset{||}{C}}CH_3$$

11.4 HYBRID ORBITALS AND MOLECULAR SHAPE

Because covalent bonds can form only where atomic orbitals — regular or hybrid — can overlap, molecules acquire particular bond angles and overall shapes.

We have already mentioned how important the shapes of molecules can be at the molecular level of life. What we will develop here is a theory that ties together molecular geometry and molecular orbitals, something which the VSEPR theory (Section 7.1) did not do.

Let's begin with the hydrogen sulfide molecule, H_2S, because the theory of its bonds is quite simple and the S—H bond occurs in enzymes in the body. Sulfur is in group VIA, and its electron configuration is

The atomic number of sulfur is 16, so we have to find places for 16 electrons.

$$S \qquad 1s^2 2s^2 2p^6 3s^2 3p_x^2 3p_y^1 3p_z^1$$

Thus a sulfur atom has two half-filled $3p$ orbitals, and their axes make an angle of $90°$ with each other, as seen in Figure 11.2. Therefore when two hydrogen atoms approach a sulfur atom to form the S—H bonds, the only *effective* way in which their $1s$ orbitals can overlap with sulfur's $3p$ orbitals is along the two perpendicular axes — also shown in Figure 11.2.

FIGURE 11.2
Molecular orbitals in hydrogen sulfide, H_2S.

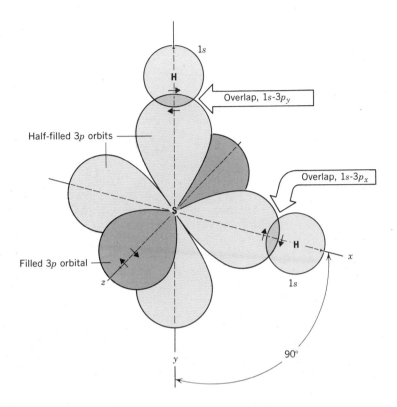

Hydrogen sulfide is one of many disagreeable scents downwind from oil refineries, plywood mills, and rotten eggs.

Only by this approach can *maximum* overlapping result, and therefore only in this way can the strongest bonds form. Hence, the two resulting S—H bonds should make an angle of 90°. The actual bond angle is 92°, which gives nice support to the theory that *the chief cause of a particular bond angle is the angle at which the axes of the atomic orbitals of the central atom cross.*

What works well for H_2S ought to work equally well for H_2O, because oxygen is in the same family as sulfur. However, the angle between the two O—H bonds is 104.5°, not 90°. A similar difficulty arises when we try to apply what worked for H_2S to NH_3. The electron configuration of nitrogen, atomic number 7, is

$$N \qquad 1s^2 2s^2 2p_x^1 2p_y^1 2p_z^1$$

The nitrogen atom has three half-filled $2p$ orbitals, and their axes cross at angles of 90°. Therefore the angle between any two N—H bonds in NH_3 should be 90°, too, but it isn't. The bond angle in NH_3 is 107.3°, not an acceptable agreement between theory and fact.

Extending what worked so well with H_2S to H_2O works poorly enough, but it becomes a total disaster when applied to CH_4, whose central atom is carbon with the electron configuration

$$C \qquad 1s^2 2s^2 2p_x^1 2p_y^1$$

Here we have only two half-filled $2p$ orbitals, an empty $2p$ orbital (the $2p_z$), and a filled $2s$ orbital. There is no way we can bring four hydrogen atoms up to these orbitals of carbon and emerge with bond angles of 109.5° between all of the pairs of bonds in CH_4. These mismatches between simple theory and fact led scientists to conclude that the atomic orbitals of the *isolated* atoms of O, N, and C are not the actual atomic orbitals used when these atoms become involved in single bonds.

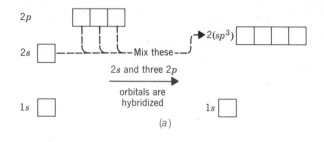

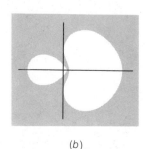

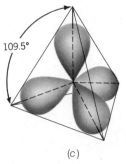

(b)   (c)

FIGURE 11.3

*sp*³ Hybridization. *(a)* The empty atomic orbitals at level 2 are mixed to give four new, level-2, *sp*³ orbitals that correspond to identical energies. How many electrons go into these hybrid orbitals depends on the

atomic number of the central atom. *(b)* The cross-sectional shape of an *sp*³ hybrid orbital. *(c)* How four equivalent *sp*³ hybrid orbitals are arranged in space about the central atom. (Only the larger lobes are shown.)

Hybrid Orbitals.   Theoreticians have found that when some atoms form bonds, some of their orbitals automatically undergo a subtle but important change. The result is as if they undergo a mixing of shapes that leads to new orbitals with new shapes *and new axes* that have different directions in space than the original orbitals had. This mixing occurs because it leads to better overlapping, stronger bonds, and therefore more stable molecules. The mixing of atomic orbitals is called **orbital hybridization.** This is a term that borrows the word *hybrid* from biology, because the new **hybrid orbitals,** like hybrids anywhere, have some resemblances to their "parents."

One way to form hybrid atomic orbitals is to mix all of the orbitals at level 2 — the 2s and the three 2p orbitals. An important rule about hybridization is that the total number of orbitals is conserved. If we mix four orbitals, we obtain four new ones.

There are just a few ways of hybridizing orbitals, and each has a name. The mixing of an s orbital with three p orbitals is called *sp*³ hybridization, and the new hybrid orbitals are named **sp³ hybrid orbitals.** Figure 11.3 shows how we can visualize the process, and in parts *(b)* and *(c)* of the figure we see the shapes of the *sp*³ orbitals. The four are identical in shape, and their axes make angles of 109.5°, which agrees precisely with VSEPR theory. Notice that each hybrid orbital has two lobes, like p orbitals, but one lobe is much larger.

Atoms of carbon, nitrogen, and oxygen can all form *sp*³ orbitals, but the number of electrons in them depends on which atom is involved. The new electron configurations for the *bonding* states of C, N, and O are as follows whenever these atoms participate in the formation of just single bonds:

C	$1s^2 2(sp^3)^1 2(sp^3)^1 2(sp^3)^1 2(sp^3)^1$	Four half-filled orbitals
N	$1s^2 2(sp^3)^2 2(sp^3)^1 2(sp^3)^1 2(sp^3)^1$	Three half-filled orbitals
O	$1s^2 2(sp^3)^2 2(sp^3)^2 2(sp^3)^1 2(sp^3)^1$	Two half-filled orbitals

These bonding states are illustrated in Figure 11.4.

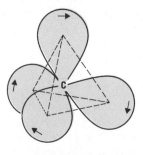

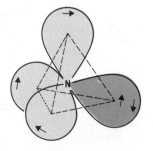

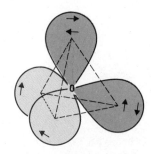

FIGURE 11.4

The bonding states for atoms of carbon, nitrogen, and oxygen when they hold other atoms or groups by *single* bonds.

Carbon
Half-filled orbitals, 4
Filled orbitals, 0

Nitrogen
Half-filled orbitals, 3
Filled orbitals, 1

Oxygen
Half-filled orbitals, 2
Filled orbitals, 2

FIGURE 11.5
The sigma bonds in methane.

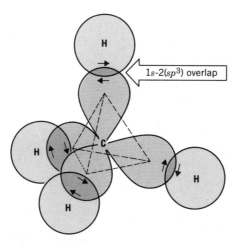

Figure 11.5 shows how we can imagine the formation of a molecule of methane from one carbon atom in its bonding state and four hydrogen atoms. A $1s$ orbital of a hydrogen atom, which holds one electron, overlaps with an sp^3 bonding-state orbital of carbon, also with one electron, to form the C—H bond.

Figure 11.6 shows how the C—H bonds and the C—C bond arise in ethane, C_2H_6. The C—C bond forms by the overlapping of two sp^3 orbitals, one supplied by each carbon atom and each holding one electron.

These single bonds, whether C—H or C—C, are called **sigma bonds,** or **σ bonds.** The extent of the overlapping of the orbitals that make up the bond — the source of the strength of the bond — is scarcely affected by any rotation of the two groups that are held by the bond. This is why free rotation around a sigma bond costs almost no energy (which is why it is easy or *free*).

The sp^3 hybrid orbitals permit the formation of stronger bonds when overlapping occurs than the original s or p orbitals of carbon, because their larger lobes extend a little farther from the nucleus. The atomic nuclei therefore do not have to get quite as close to each other. Their mutual repulsion is therefore less, and the system is more stable.

Figure 11.7 shows how the sigma bonds in NH_3 form when nitrogen's three half-filled sp^3 hybrid orbitals are used. We should expect bond angles in NH_3 of 109.5° — the tetrahedral angle — and this is in good agreement with the observed angle of 107.3°. (Agreement such as this lends support to the theory.)

FIGURE 11.6
The sigma bonds in ethane. Notice that the degree of overlapping between the sp^3 orbitals that form the carbon-carbon bond would not be affected by rotating one CH_3 group with respect to another.

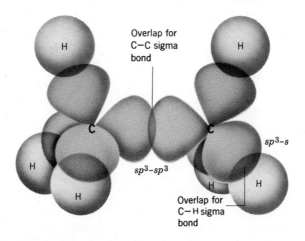

FIGURE 11.7
The molecular orbitals in ammonia, *(a)*
and in water, *(b)*. The unshared pairs of
electrons in these molecules are in
level-2 *sp*³ hybrid orbitals.

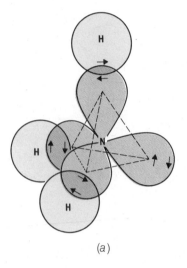

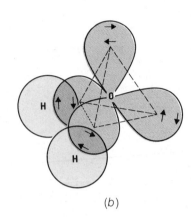

(a)   *(b)*

Figure 11.7 also carries the VSEPR theory to the water molecule and shows how the O—H bonds form when oxygen's two half-filled *sp*³ orbitals are used. Again, we would expect a bond angle of 109.5°, and this is in acceptable agreement with the observed angle of 104.5°. What is believed to squeeze the angle to the lower value are the electron clouds of the two filled *sp*³ orbitals on oxygen.

SUMMARY

Organic and inorganic compounds Most organic compounds are molecular and the majority of inorganic compounds are ionic. Molecular and ionic compounds differ in composition, in types of bonds, and in several physical properties.

Structural features of organic molecules The ability of carbon atoms to join to each other many times in succession both in straight chains and branched chains accounts in a large measure for the existence of several million organic compounds. When groups within a molecule are joined by a single bond, they can rotate relative to each other about this bond. Full structural formulas of organic compounds are usually condensed by grouping the hydrogens attached to a carbon immediately next to this carbon; by letting single bonds on a horizontal line be understood; and by leaving bond angles and conformational possibilities to the informed imagination. The families of organic compounds are organized around functional groups; these are parts of molecules at which most of the chemical reactions occur.

Nonfunctional parts of molecules can sometimes be given the general symbol R—, as in R—O—H, the general symbol for all alcohols.

Isomerism Differences in the *conformations* of carbon chains do not create new compounds, but differences in the organizations of parts do. Isomers are compounds with identical molecular formulas but different structures.

Hybrid orbitals and sigma bonds When they participate in the formation of covalent bonds, atoms of carbon, nitrogen, and oxygen use hybrid atomic orbitals instead of the orbitals associated with the isolated atoms. When an *s* orbital and three *p* orbitals of the same main level mix (hybridize), four new, equivalent *sp*³ hybrid orbitals form whose axes point to the corners of a regular tetrahedron. When such an orbital overlaps with another like it or with an ordinary atomic orbital to form a molecular orbital, the new bond is called a sigma bond. Free rotation can occur around a sigma bond, because this action does not appreciably diminish the extent of the overlap.

KEY TERMS

The following terms were emphasized in this chapter. They will be used often in succeeding chapters, so be sure to master them before continuing.

branched chain	conformation	functional group	inorganic compound
condensed structure	free rotation	hybrid orbital	isomerism

isomers

nonfunctional group

orbital hybridization

sigma bond (σ bond)

sp^3 hybrid orbital

straight chain

structure

vital force theory

SELECTED REFERENCE

1 T. O. Lipman. "Wöhler's Preparation of Urea and the Fate of Vitalism." *Journal of Chemical Education*, August 1964, p. 452.

REVIEW EXERCISES

The answers to these Review Exercises are in the *Study Guide* that accompanies this book.

Organic and Inorganic Compounds

11.1 Why are the compounds of carbon generally called *organic* compounds?

11.2 State the vital force theory in your own words.

11.3 What led scientists to adopt the vital force theory (prior to 1828)?

11.4 Describe what Wöhler did that made the vital force theory highly questionable.

11.5 What kind of bond between atoms predominates among organic compounds?

11.6 Which of the following compounds are inorganic?
(a) CH_3OH (b) CO (c) CCl_4
(d) $NaHCO_3$ (e) K_2CO_3

11.7 Are the majority of all compounds that dissolve in water ionic or molecular? Inorganic or organic?

11.8 Explain why very few organic compounds conduct electricity either in an aqueous solution or as molten materials.

11.9 Each compound described below is either ionic or molecular. State which it most likely is, and give one reason.
(a) The compound is a colorless gas at room temperature.
(b) This compound dissolves in water. When hydrochloric acid is added, the solution fizzes and an odorless, colorless gas is released, which can extinguish a burning flame.
(c) This compound melts at 300 °C, and it burns in air.
(d) This compound melts at 675 °C, and it becomes white when heated.
(e) This compound is a liquid that does not dissolve in water.

Structural Formulas

11.10 One can write the structure of butane, lighter fluid, as follows.

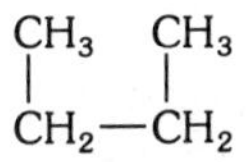

$$\begin{array}{cc} CH_3 & CH_3 \\ | & | \\ CH_2 & \!\!\!-CH_2 \end{array} \quad \text{Butane}$$

Are butane molecules properly described as straight chain or as branched chain, in the sense in which we use these terms? Explain.

11.11 Which of the following structures are possible, given the numbers of bonds that various atoms can form?
(a) $CH_2CH_2CH_3$
(b) $CH_3{=}CHCH_2CH_3$
(c) $CH_3CH{=}CH_2CH_2CH_3$

11.12 Write full (expanded) structures for each of the following *molecular* formulas. Remember how many bonds the various kinds of atoms must have. In some you will have to use double or triple bonds. (*Hint:* A trial-and-error approach will have to be used.)
(a) CH_5N (b) CH_2Br_2
(c) $CHCl_3$ (d) C_2H_6
(e) CH_2O_2 (f) CH_2O
(g) NH_3O (h) C_2H_2
(i) N_2H_4 (j) HCN
(k) C_2H_3N (l) CH_4O

11.13 Write, neat condensed structures of the following.

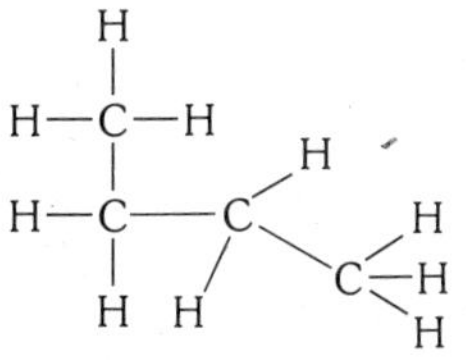

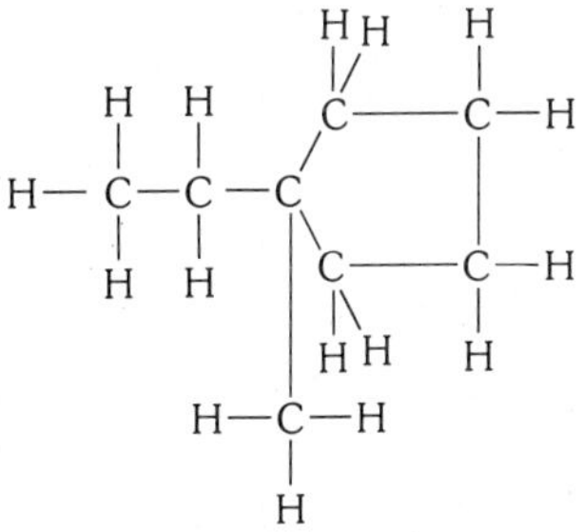

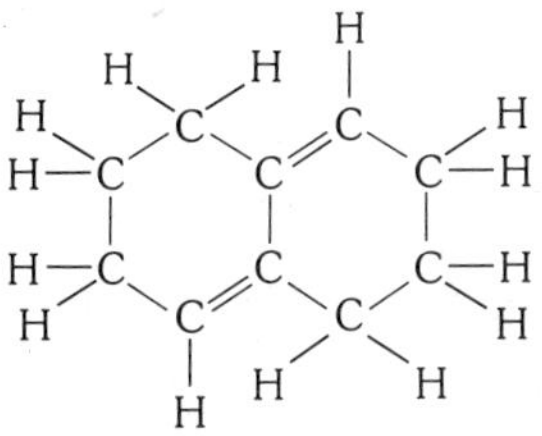

Isomers

11.14 Decide whether the members of each pair are identical, are isomers, or are unrelated.

(a) 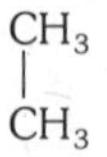 CH_3 and CH_3—CH_3
   |
   CH_3

(b) 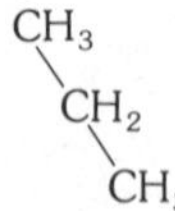 CH_3 and CH_2
   $\diagdown$
   CH_2 CH_3 CH_3
   |
   CH_3

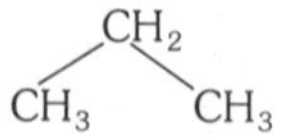

(c) CH_3CH_2—OH and $CH_3CH_2CH_2$—OH

(d) CH_3CH=CH_2 and CH_2——CH_2
   $\diagdown$ $\diagup$
   CH_2

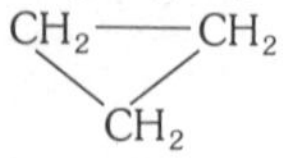

(e) H—$\overset{\overset{\displaystyle O}{\|}}{C}$—$CH_3$ and CH_3—$\overset{\overset{\displaystyle O}{\|}}{C}$—H

(f) CH_3CHCH_3 and CH_3CH
   | |
   CH_3 CH_3

(g) $CH_3CH_2CH_2$—NH_2 and CH_3CH_2—NH—CH_3

(h) $CH_3CH_2\overset{\overset{\displaystyle O}{\|}}{C}$—$O$—$H$ and H—O—$\overset{\overset{\displaystyle O}{\|}}{C}CH_2CH_3$

(i) H—$\overset{\overset{\displaystyle O}{\|}}{C}$—$O$—$CH_2CH_3$ and CH_3CH_2—$\overset{\overset{\displaystyle O}{\|}}{C}$—$O$—$H$

(j) H—$\overset{\overset{\displaystyle O}{\|}}{C}$—$O$—$CH_2CH_2OH$ and $HOCH_2CH_2$—$\overset{\overset{\displaystyle O}{\|}}{C}$—$O$—$H$

(k) $CH_3\overset{\overset{\displaystyle O}{\|}}{C}CH_2CH_3$ and $CH_3CH_2\overset{\overset{\displaystyle O}{\|}}{C}CH_3$

(l) 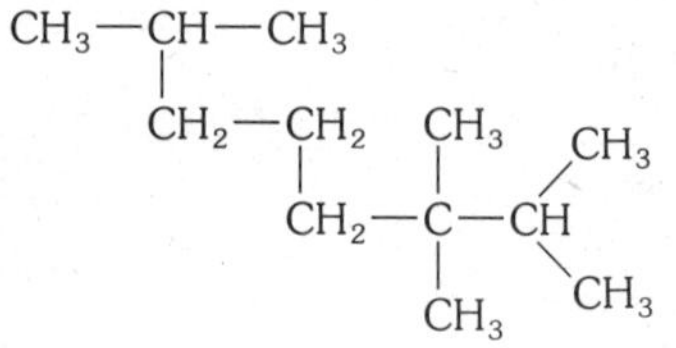 CH_3—CH—CH_3
       |
       CH_2—CH_2 CH_3 CH_3
       | | $\diagup$
       CH_2—C—CH and
            | $\diagdown$
            CH_3 CH_3

   CH_3 CH_3 CH_3
   | | |
   CH_3—CH—CH_2—CH_2—CH_2—C——CH—CH_3
                                |
                                CH_3

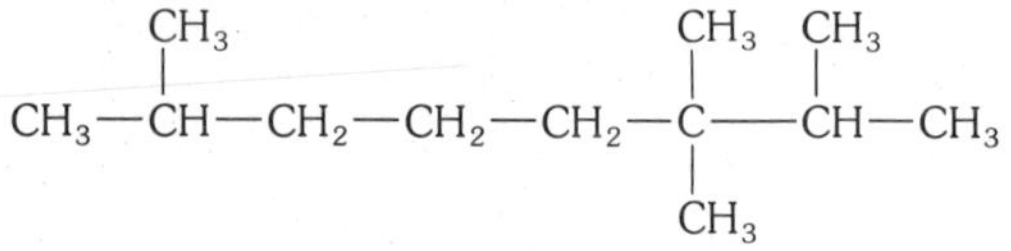

(m) CH_3—NH—$\overset{\overset{\displaystyle O}{\|}}{C}$—$CH_3$ and $CH_3CH_2\overset{\overset{\displaystyle O}{\|}}{C}NH_2$

(n) H—O—O—H and H—O—H

Families of Organic Compounds

11.15 Name the family to which each compound belongs.

(a) CH_3CH=CH_2 (b) $HOCH_2CH_2CH_3$

(c) CH_3CH_2SH (d) CH_3C≡CH

(e) $CH_3CH_2\overset{\overset{\displaystyle O}{\|}}{C}$—$O$—$CH_3$ (f) $CH_3CH_2\overset{\overset{\displaystyle O}{\|}}{C}H$

(g) $CH_3CH_2CH_2\overset{\overset{\displaystyle O}{\|}}{C}OH$ (h) $CH_3\overset{\overset{\displaystyle O}{\|}}{C}CH_2CH_2CH_3$

(i) $CH_3CH_2CH_2NH_2$ (j) CH_3—O—CH_2CH_3

11.16 Name the families to which the compounds in Practice Exercise 11.14 belong. (A few belong to more than one family.)

Hybrid Orbitals and Sigma Bonds

11.17 Why is the topic of molecular shape important at the molecular level of life?

11.18 What is an sp^3 hybrid orbital? What orbitals are used to make one, and what is its general shape? In what way is it like a p orbital as well as like an s orbital?

11.19 What kinds of orbitals overlap when a C—H bond forms in methane?

11.20 What kinds of orbitals overlap when the C—C bond forms in ethane?

11.21 What is a sigma bond?

11.22 What kind of orbital holds the *unpaired* electrons in a molecule of ammonia? In a water molecule?

11.23 Phosphorus is in the same family as nitrogen and forms the compound phosphine, PH_3, whose structure is similar to that of ammonia. The bond angles in phosphine are 93.7°. What kinds of atomic orbitals on phosphorus are most likely used in forming the P—H bonds? Explain.

Chapter 12
Saturated Hydrocarbons: Alkanes and Cycloalkanes

Candles affect moods and evoke memories. They're also made of alkanes, the family of compounds studied in this chapter.

12.1   THE HYDROCARBONS

Hydrocarbons can be saturated or unsaturated, open-chain or cyclic, and all are insoluble in water.

Petroleum and natural gas are substances that consist almost entirely of a complex mixture of molecular compounds called hydrocarbons. **Hydrocarbons** are made from the atoms of just two elements, carbon and hydrogen, and the covalent bonds between the carbon atoms can be single, double, or triple. These possibilities define the various kinds of hydrocarbons, which are outlined in Figure 12.1. **Aliphatic compounds** are any that have no benzene ring (or a similar system), and **aromatic compounds** are those with such rings. (This distinction will be explained further in the next chapter.)

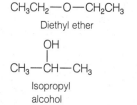

Diethyl ether

Isopropyl
alcohol

Saturated and Unsaturated Compounds.   One useful way to classify organic compounds is based on the presence or absence of multiple bonds. If the molecules of a compound involve only single bonds, it is called a **saturated compound.** If one or more double or triple bonds are present, the substance is said to be an **unsaturated compound.** Table 12.1, for example, shows the structures of several **alkanes,** hydrocarbons whose molecules have only single bonds and which, therefore, are saturated. The alkenes and alkynes (Figure 12.1), which we will study in the next chapter, are *unsaturated* hydrocarbons. Alkenes have double bonds, and alkynes have triple bonds.

Saturated compounds occur in other families, too, such as diethyl ether, once a common anesthetic, and isopropyl alcohol (rubbing alcohol).

Ring Compounds.   Besides straight-chain or branched-chain systems, which we have already introduced in the previous chapter, there is another kind of carbon skeleton called the carbon **ring.** This is an arrangement of three or more carbon atoms into a closed cycle, and ring compounds are often called cyclic compounds. For example, cyclohexane molecules have a ring of six carbon atoms, and the prefix *cyclo-* tells us that the six atoms implied by *hex* in this

FIGURE 12.1
There are several kinds of hydrocarbons. (The circles in the structures for benzene and naphthalene will be explained in the next chapter.)

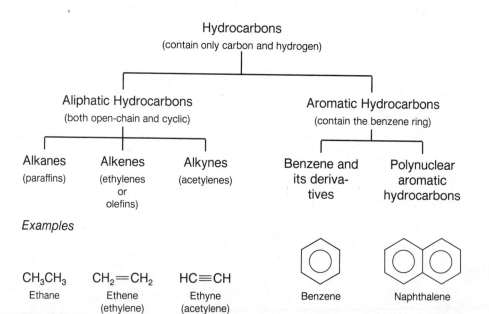

TABLE 12.1
Straight-Chain Alkanes

IUPAC Name	Number of Carbon Atoms	Molecular Formula[a]	Structure	Boiling Point (°C)	Melting Point (°C)	Density (g/mL, 20 °C)
Methane	1	CH_4	CH_4	−161.5	−182.5	
Ethane	2	C_2H_6	CH_3CH_3	−88.6	−183.3	
Propane	3	C_3H_8	$CH_3CH_2CH_3$	−42.1	−189.7	
Butane	4	C_4H_{10}	$CH_3CH_2CH_2CH_3$	−0.5	−138.4	
Pentane	5	C_5H_{12}	$CH_3CH_2CH_2CH_2CH_3$	36.1	−129.7	0.626
Hexane	6	C_6H_{14}	$CH_3CH_2CH_2CH_2CH_2CH_3$	68.7	−95.3	0.659
Heptane	7	C_7H_{16}	$CH_3CH_2CH_2CH_2CH_2CH_2CH_3$	98.4	−90.6	0.684
Octane	8	C_8H_{18}	$CH_3CH_2CH_2CH_2CH_2CH_2CH_2CH_3$	125.7	−56.8	0.703
Nonane	9	C_9H_{20}	$CH_3CH_2CH_2CH_2CH_2CH_2CH_2CH_2CH_3$	150.8	−53.5	0.718
Decane	10	$C_{10}H_{22}$	$CH_3CH_2CH_2CH_2CH_2CH_2CH_2CH_2CH_2CH_3$	174.1	−29.7	0.730

[a] The molecular formulas of the open-chain alkanes fit the general formula C_nH_{2n+2}, where n = the number of carbon atoms per molecule.

More than 2 billion pounds of cyclohexane are made annually in the United States, with over 90% being used to make nylon.

name are in a ring. The -ane ending tells us that cyclohexane is a hydrocarbon of the alkane family — a hydrocarbon with only sngle bonds. Quite often cyclic alkanes are called *cycloalkanes*. The simplest cycloalkane is cyclopropane, which was once an important anesthetic.

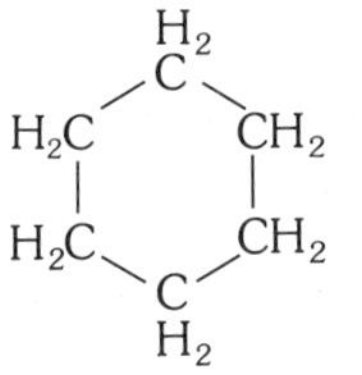

Cyclohexane

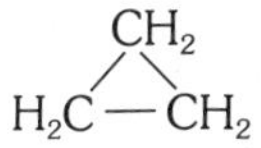

Cyclopropane

Because the bond angle at a triple bond is 180°, a ring has to be quite large to have a triple bond, and cycloalkynes are rare.

Cyclic compounds can have double bonds as in cyclohexene. (Remember that carbon-carbon double bonds, whether in open chains or in rings, are always shown by two lines. They are never "understood.") Rings, of course, can carry substituents, as in ethylcyclohexane, and not all of the ring atoms have to be carbon atoms. They can be oxygen, nitrogen, or sulfur, too, and cyclic compounds with ring atoms other than carbon are called **heterocyclic compounds.** A simple example is tetrahydropyran, which has the basic ring system widely present among molecules of carbohydrates.

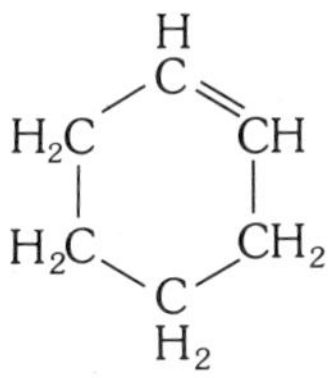

Cyclohexene

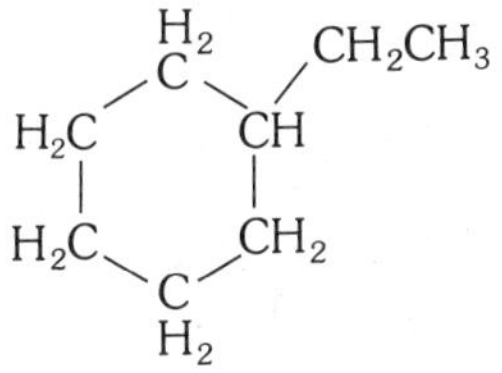

Ethylcyclohexane

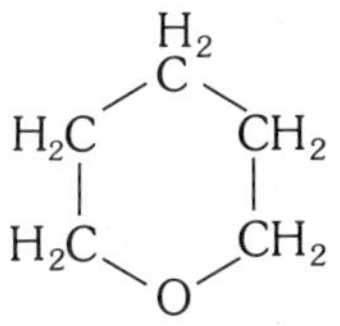

Tetrahydropyran

In structures of cyclic compounds, the ring system itself is usually represented simply by a polygon, a many-sided figure. For example, a square can represent cyclobutane. The photograph of the ball-and-stick model of cyclobutane and its progressively more condensed structures show what is left to the imagination when just a square is used. At each corner, we

TABLE 12.2
Some Cycloalkanes

IUPAC Name	Structure	Boiling Point (°C)	Melting Point (°C)	Density (20 °C)
Cyclopropane	△	−33	−127	1.809 g/L (0 °C)
Cyclobutane	 □	−13.1	−80	0.7038 g/L (0 °C)
Cyclopentane	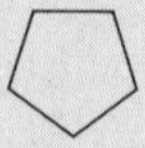 ⬠	49.3	−94.4	0.7460 g/mL
Cyclohexane	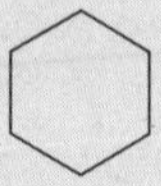 ⬡	80.7	6.47	0.7781 g/mL
Cycloheptane	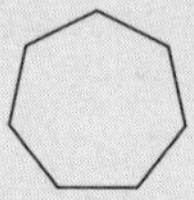 ⬭	118.5	−12	0.8098 g/mL
Cyclooctane	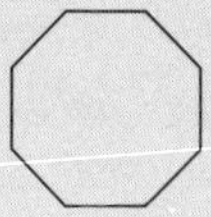 ⯃	149	14.3	0.8349 g/mL

have to understand that there is a CH_2 group. Each line in the square is a carbon-carbon single bond. (See also Table 12.2.)

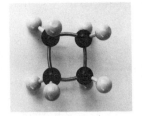

Cyclobutane.

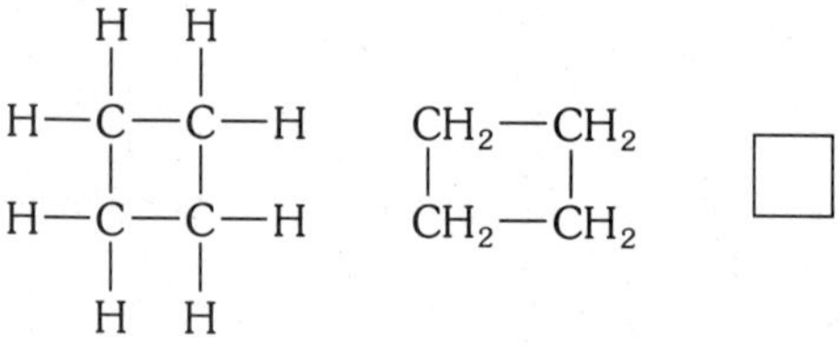

Three ways to represent the structure of cyclobutane

The photograph of the model of methylcyclopentane and its progressively more condensed structures further illustrate the use of a geometric figure to represent a ring.

Methylcyclopentane.

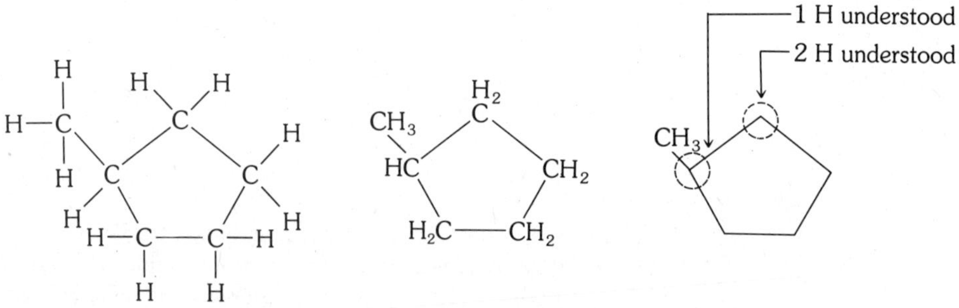

1 H understood
2 H understood

Three ways to represent the structure of methylcyclopentane

EXAMPLE 12.1   UNDERSTANDING CONDENSED STRUCTURES OF RING COMPOUNDS

Problem: To make sure that you can read a condensed structure when it includes a polygon for a ring system, expand this structure, including its side chains.

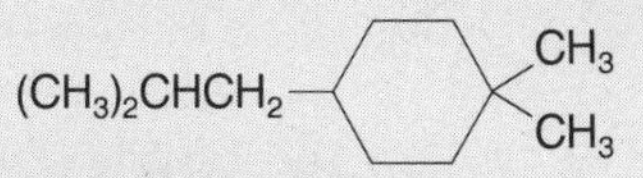

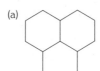

Solution:

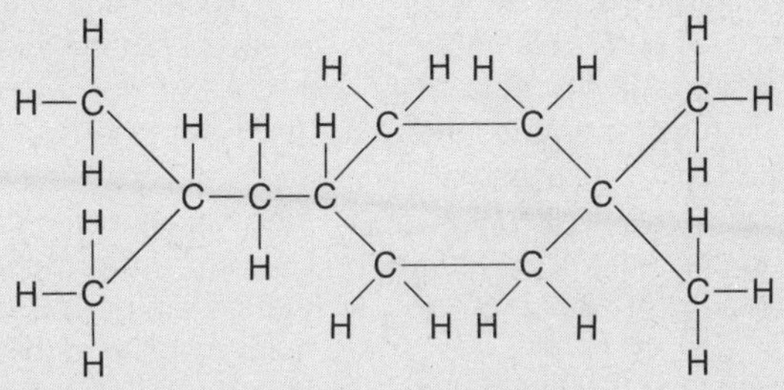

Note especially how we can tell how many hydrogens have to be attached to a ring atom. We need as many as required to fill out a set of four bonds from each carbon. This example also shows a situation (on the right side of the ring) where no bonding room is left for holding an H atom.

PRACTICE EXERCISE 1   Expand each of the following structures.

(a)   (b)

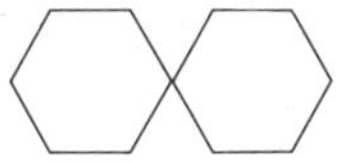

PRACTICE EXERCISE 2   Condense the following structure.

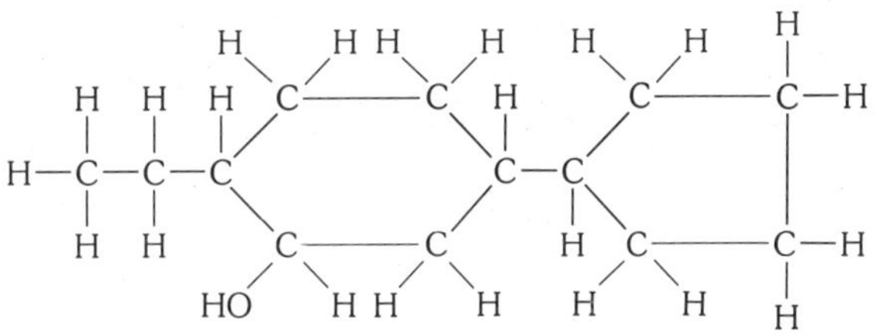

When we use polygons to represent saturated rings that have six or more ring atoms, we gloss over one feature of such molecules. The ring atoms that make up the rings of this size do not all lie in the same plane. We will discuss this situation for six-membered, saturated rings in Special Topic 12.1.

Physical Properties of Hydrocarbons.   All of the different kinds of hydrocarbons can be treated as one group in a discussion of their physical properties. Both the carbon-carbon and the carbon-hydrogen bonds are almost entirely nonpolar, so hydrocarbon molecules have very

SPECIAL TOPIC 12.1   THE BOAT AND CHAIR FORMS OF CYCLOHEXANE

The inside angle of a regular hexagon is 120°, so the six carbons of the cyclohexane ring cannot lie in the same plane and also have the tetrahedral geometry that requires bond angles of 109.5°. To resolve this, the cyclohexane ring adopts a nonplanar shape that is called the **chair form,** shown in the accompanying drawings.

There is another form, called the **boat form** of cyclohexane, in which the bond angles can also be 109.5°, and enough flexibility exists in the ring to permit it to flex and twist from the chair to the boat form. As you can see in the scale models below, however, the electron clouds that surround the hydrogen atoms are closer to each other in the boat form than in the chair form. For example, the electron clouds marked **a** nudge each other in the scale model of the boat form, but are as far from each other as possible in the chair form. Similarly, the electron clouds marked **b** are closer to each other in the boat than in the chair form. Because electron clouds repel each other, the boat form is less stable than the chair form.

If a molecule in a chair form does get twisted into its less stable boat form, it has two ways to recover greater stability. It can twist back to the original chair or twist at another place to go into an alternative but energetically

equivalent chair form. In an actual sample of cyclohexane, the two chair forms exist in equilibrium, and they constantly flip-flop back and forth. In a sense, the flat hexagonal structure that we usually draw for cyclohexane is an average of these two forms, and in most situations we can ignore the true geometry of the six-membered ring.

Saturated, heterocyclic rings of six atoms that include either oxygen or nitrogen also exist in chair forms, and we will encounter the six-membered ring that has one oxygen atom among the carbohydrates.

None of these factors matters when we deal with the chemical properties of cyclohexane, and they matter very little with monosubstituted cyclohexanes. However, when the six-membered ring holds two or more substituents, as it does among the carbohydrates, the electron clouds of these substituents dominate the system's choice of which chair form is the more stable. In other words, they dominate the system's choice of molecular shape, and we have already mentioned that the ability of an enzyme to catalyze a reaction is extraordinarily sensitive to molecular shape. For this reason, the geometry of the six-membered ring is very important at the molecular level of life.

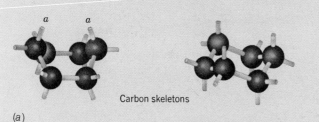

Carbon skeletons

(a)

Boat form

Chair form

(b)

Scale models

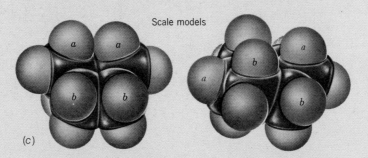

(c)

Twist this
end up ⟶

Then twist
this end down ⟶

(d)

The chair and boat forms of cyclohexane. *(a)* The carbon skeletons in ball-and-stick models. *(b)* Line drawings of these skeletons. *(c)* Space-filling models that show the effective spaces occupied by the electron clouds associated with the atoms and that cannot be easily penetrated by other electron clouds. *(d)* This shows how one chair form can be twisted into the other chair form.

Petroleum ether boils over the range of 30 °C to 60 °C and diethyl ether boils at 35 °C.

little if any overall polarity. Hence, hydrocarbons are insoluble in water, but they dissolve well in nonpolar solvents. Indeed, many are themselves nonpolar solvents. For example, *petroleum ether* is a mixture of alkanes that have four to six carbon atoms per molecule. It isn't an ether at all, but a mixture of *alkanes,* and this mixture happens to boil roughly around the boiling point of diethyl ether. Cigarette lighter fluid is similar to petroleum ether in its solvent action, and you may have used it to clean grease spots from a garment. *Ligroine* is another, higher boiling mixture of alkanes whose molecules have six to ten carbon atoms each. Gasoline is a similar mixture of alkanes. It is a good solvent for tar, but if you ever use it to clean tar from surfaces, be sure to keep all flames away.

The vapors of hydrocarbon solvents catch fire very easily, so be careful if you use these solvents for any purpose. In the right proportion in air, hydrocarbon vapors explode when ignited.

Grease and tar are relatively nonpolar materials, and what we have illustrated here is a very useful rule of thumb of predicting whether a solvent can dissolve some substance. It's called the **like-dissolves-like** rule, where "like" refers to a likeness in polarity. Polar solvents, such as water, are good for dissolving polar or ionic substances such as sugar or salt, because polar molecules or ions can attract water molecules around themselves, form solvent cages, and thus intermingle with water molecules. Nonpolar solvents such as gasoline do not dissolve sugar or salt because the nonpolar solvent molecules cannot be attracted to polar molecules or ions and form solvent cages.

Notice in Table 12.1 how the boiling points of the straight-chain alkanes increase with chain length. Because of their low polarity, the hydrocarbons that have one to four carbons per molecule are generally gases at or near room temperature. Hydrocarbons that have from five to about 16 carbon atoms per molecule are usually liquids at room temperature. When alkanes have approximately 18 or more carbon atoms per molecule, the substances are waxy solids at room temperature. Paraffin wax, for example, is a mixture of alkanes whose molecules have 20 and more carbon atoms.

Most candles are made from paraffin.

Before moving on, let's pause to reflect on what we have done here in relating physical properties to structural features. We have introduced the first "map sign" in our study of organic and biochemical compounds. This "map sign" is that substances whose molecules are entirely *or even mostly* hydrocarbon-like can be expected to be insoluble in water but soluble in nonpolar solvents. Thus when we see an unfamiliar structure, we can tell more or less at a glance if it is mostly hydrocarbon-like. If it is, we can predict with considerable confidence that the compound is not soluble in water. For example, look at the structure of cholesterol. You probably know that it can form solid deposits in blood capillaries, and that these deposits can

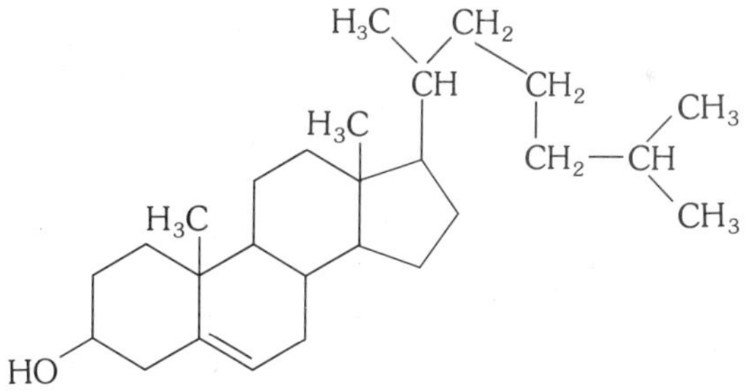

Cholesterol

even close capillaries. Consequently, the heart has to work harder, and this can cause a heart attack. Now notice that virtually the entire cholesterol molecule is hydrocarbon-like. It has only one polar group, the —OH or alcohol group, and this isn't enough to make cholesterol sufficiently polar to dissolve either in water or in blood (which is mostly water). Thus by learning one very general fact — one "map sign" — we don't have to memorize a long list of separate (but similar) facts about an equally long list of separate compounds that occur at the molecular level of life. With the molecular "map sign" in hand, we can look at the structures of hundreds of complicated compounds and confidently predict particular properties such as the likelihood of their being soluble in water.

EXAMPLE 12.2   PREDICTING PHYSICAL PROPERTIES FROM STRUCTURES

Problem: Study the following two structures and tell which is the structure of the more water-soluble compound.

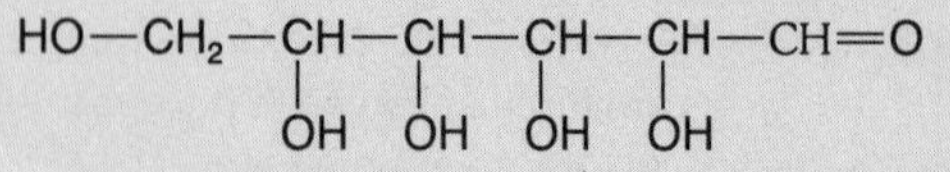

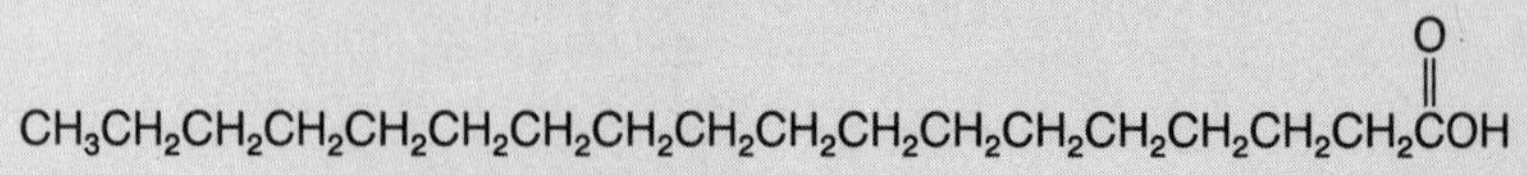

$$HO-CH_2-CH-CH-CH-CH-CH=O$$
$$\qquad\qquad OH\quad OH\quad OH\quad OH$$

$$CH_3CH_2CH_2CH_2CH_2CH_2CH_2CH_2CH_2CH_2CH_2CH_2CH_2CH_2CH_2CH_2CH_2CH_2\overset{\displaystyle O}{\overset{\|}{C}}OH$$

Solution: The first structure has several polar —OH groups, but the second is almost entirely hydrocarbon-like. Hence, the first should be (and is) much more soluble in water. The first compound is glucose (in one of its forms), the chief sugar in the bloodstream. The second is stearic acid, which forms when we digest fats in the diet, and which does not dissolve in water.

PRACTICE EXERCISE 3   Which of the following is more soluble in gasoline?

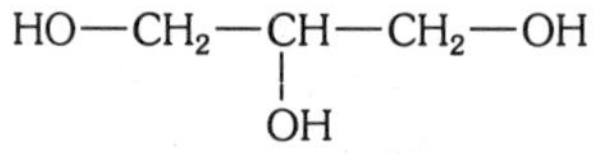

$$HO-CH_2-CH-CH_2-OH \qquad CH_3-CH_2-CH-CH_2-OH$$
$$\qquad\qquad OH \qquad\qquad\qquad\qquad CH_3$$

Glycerol 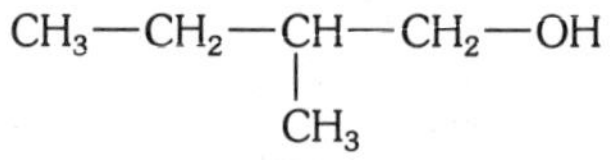   2-Methyl-1-butanol

---

12.2   THE NOMENCLATURE OF ALKANES

The IUPAC name of a compound shows the number of carbon atoms in the parent chain, the kinds and locations of substituents on this chain, and the family to which the compound belongs.

"Nomenclature" is from the Latin *nomen,* name, + *calare,* to call. Wealthy Romans had slaves called *nomenclators* who were to remind their owners of the names of important people who approached them on the street.

In chemistry, **nomenclature** refers to the rules used to name compounds. In the early years of organic chemistry, compounds were most often named after their natural sources. For example, *urea* is one of the substances in urine. *Formic acid*—from the Latin, *formica,* meaning ant—is present in the stinging juice of ants. For a time this practice worked, but chemists eventually realized that letting names refer to origins rather than molecular structures meant that they would be committing themselves to memorizing hundreds and thousands of names. It would be far easier to have a small number of rules that could be used to devise the name of any compound just by letting word units and numbers designate features of its molecular structure. In this way, a chemist could always write one (and only one) name from a structure and also write a structure from a name. These are the two goals of organic nomenclature.

The International Union of Pure and Applied Chemistry, or IUPAC, is the organization that supervises the development of rules of nomenclature, and all scientific societies in the world belong to it. Its rules are known as the **IUPAC rules.** Before we turn to them, there is something to be said for the continued use of many nonsystematic names, which are often called *common names.* When the IUPAC rules are used to name complicated molecules, the names can become very long and hard to use in the everyday conversations of chemists and engineers. The common names are short and easy to pronounce. For example, it is much easier to say *sucrose* for table sugar than α-D-glucopyranosyl β-D-fructofuranoside. You can see why we will want to learn some common names, too, but you will also see that even such names have some logic and system to them.

IUPAC Rules for Naming Alkanes and Cycloalkanes. In the IUPAC rules, the last syllable in the name of a compound is used to designate the name of the family to which the compound belongs. The names of all of the saturated hydrocarbons, for example, end in *-ane*. Hydrocarbons with double bonds have names that end in *-ene*, and those with triple bonds have names that end in *-yne*. The rules for the alkanes are as follows.

1. Use the ending *-ane* for all alkanes (and cycloalkanes).

2. Determine what is the longest continuous chain of carbons in the structure, and let this be the *parent chain* for naming purposes.

   For example, view the branched-chain alkane,

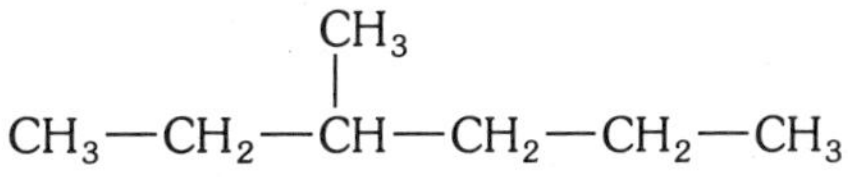

$$CH_3 - CH_2 - \overset{\overset{\displaystyle CH_3}{|}}{CH} - CH_2 - CH_2 - CH_3$$

   as coming from

$$CH_3 - CH_2 - CH_2 - CH_2 - CH_2 - CH_3$$

   by replacing a hydrogen atom on the third carbon from the left with a CH_3 group.

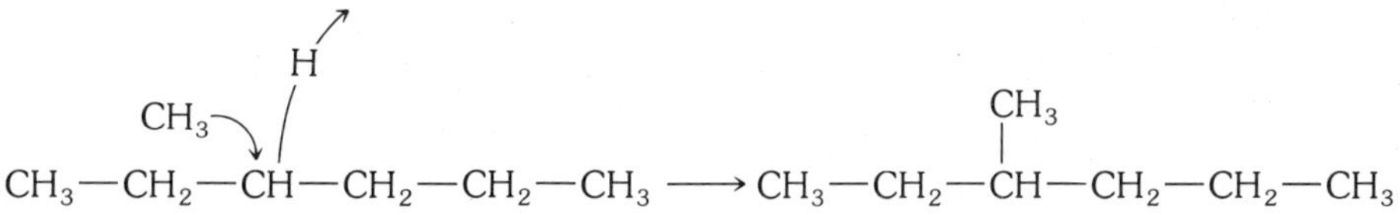

$$CH_3 - CH_2 - \overset{\overset{\displaystyle CH_3 \searrow \overset{\displaystyle H \nearrow}{|}}{}}{CH} - CH_2 - CH_2 - CH_3 \longrightarrow CH_3 - CH_2 - \overset{\overset{\displaystyle CH_3}{|}}{CH} - CH_2 - CH_2 - CH_3$$

We won't need to know the prefixes for the higher alkanes.

3. Attach a prefix to *-ane* that specifies the number of carbon atoms *in the parent chain*. The prefixes through C-10 are as follows (and these must be memorized). The names in Table 12.1 illustrate their use.

meth-	1 C	hex-	6 C
eth-	2 C	hept-	7 C
prop-	3 C	oct-	8 C
but-	4 C	non-	9 C
pent-	5 C	dec-	10 C

   For example, the parent chain of our example has six carbons, so the corresponding alkane is called hexane — *hex* for six carbons and *-ane* for being in the alkane family. The branched-chain compound whose name we are devising is regarded as a derivative of this parent, hexane.

4. Assign numbers to each carbon of the parent chain, starting from whichever of its ends gives the location of the first branch the lower of two possible numbers.

   For example, the correct direction for numbering our example is from left to right.

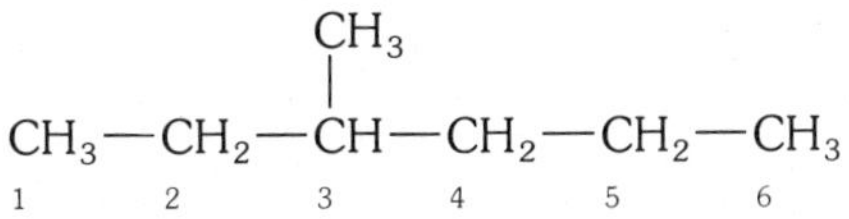

$$\underset{1}{CH_3} - \underset{2}{CH_2} - \underset{3}{\overset{\overset{\displaystyle CH_3}{|}}{CH}} - \underset{4}{CH_2} - \underset{5}{CH_2} - \underset{6}{CH_3} \quad \text{(Correct direction of numbering)}$$

Had we numbered from right to left, the carbon holding the branch would have had a higher number.

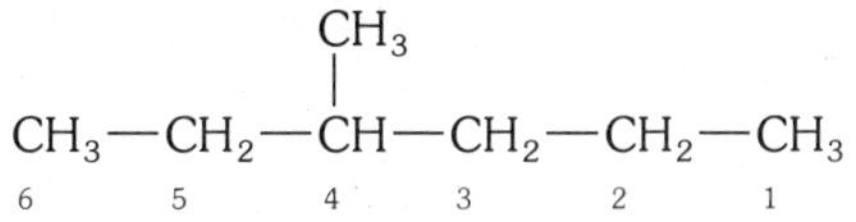

$$\underset{6}{CH_3} - \underset{5}{CH_2} - \underset{4}{\overset{\overset{\displaystyle CH_3}{|}}{CH}} - \underset{3}{CH_2} - \underset{2}{CH_2} - \underset{1}{CH_3} \quad \text{(Incorrect direction of numbering)}$$

5. Determine the correct name for each branch (or for any other atom or group). We must now pause and learn the names of a few such groups.

Any branch that consists only of carbon and hydrogen and has only single bonds is called an **alkyl group,** and the names of all alkyl groups end in *-yl*. Think of an alkyl group as an alkane minus one of its hydrogen atoms. For example,

$$ H-\overset{\overset{\displaystyle H}{|}}{\underset{\underset{\displaystyle H}{|}}{C}}-H \xrightarrow{\text{Remove one H}} H-\overset{\overset{\displaystyle H}{|}}{\underset{\underset{\displaystyle H}{|}}{C}}- \quad \text{or} \quad CH_3- $$

Methane   **Methyl**

$$ H-\overset{\overset{\displaystyle H}{|}}{\underset{\underset{\displaystyle H}{|}}{C}}-\overset{\overset{\displaystyle H}{|}}{\underset{\underset{\displaystyle H}{|}}{C}}-H \xrightarrow{\text{Remove one H}} H-\overset{\overset{\displaystyle H}{|}}{\underset{\underset{\displaystyle H}{|}}{C}}-\overset{\overset{\displaystyle H}{|}}{\underset{\underset{\displaystyle H}{|}}{C}}- \quad \text{or} \quad CH_3-CH_2- $$

Ethane   **Ethyl**

Two alkyl groups can be obtained from propane because the middle position is not equivalent to either of the end positions.

$$ H-\overset{\overset{\displaystyle H}{|}}{\underset{\underset{\displaystyle H}{|}}{C}}-\overset{\overset{\displaystyle H}{|}}{\underset{\underset{\displaystyle H}{|}}{C}}-\overset{\overset{\displaystyle H}{|}}{\underset{\underset{\displaystyle H}{|}}{C}}-H \xrightarrow[\text{from either end}]{\text{Remove one H}} H-\overset{\overset{\displaystyle H}{|}}{\underset{\underset{\displaystyle H}{|}}{C}}-\overset{\overset{\displaystyle H}{|}}{\underset{\underset{\displaystyle H}{|}}{C}}-\overset{\overset{\displaystyle H}{|}}{\underset{\underset{\displaystyle H}{|}}{C}}- \quad \text{or} \quad CH_3-CH_2-CH_2- $$

Propane   **Propyl**

$$ H-\overset{\overset{\displaystyle H}{|}}{\underset{\underset{\displaystyle H}{|}}{C}}-\overset{\overset{\displaystyle H}{|}}{\underset{\underset{\displaystyle H}{|}}{C}}-\overset{\overset{\displaystyle H}{|}}{\underset{\underset{\displaystyle H}{|}}{C}}-H \xrightarrow[\text{from the middle}]{\text{Remove one H}} H-\overset{\overset{\displaystyle H}{|}}{\underset{\underset{\displaystyle H}{|}}{C}}-\overset{\displaystyle C}{\underset{\underset{\displaystyle H}{|}}{}}-\overset{\overset{\displaystyle H}{|}}{\underset{\underset{\displaystyle H}{|}}{C}}- \quad \text{or} \quad CH_3-\overset{|}{C}H-CH_3 $$

Propane   **Isopropyl**

Two alkyl groups can similarly be obtained from butane.

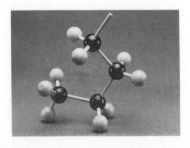

$$ H-\overset{\overset{\displaystyle H}{|}}{\underset{\underset{\displaystyle H}{|}}{C}}-\overset{\overset{\displaystyle H}{|}}{\underset{\underset{\displaystyle H}{|}}{C}}-\overset{\overset{\displaystyle H}{|}}{\underset{\underset{\displaystyle H}{|}}{C}}-\overset{\overset{\displaystyle H}{|}}{\underset{\underset{\displaystyle H}{|}}{C}}-H \xrightarrow[\text{from either end}]{\text{Remove one H}} H-\overset{\overset{\displaystyle H}{|}}{\underset{\underset{\displaystyle H}{|}}{C}}-\overset{\overset{\displaystyle H}{|}}{\underset{\underset{\displaystyle H}{|}}{C}}-\overset{\overset{\displaystyle H}{|}}{\underset{\underset{\displaystyle H}{|}}{C}}-\overset{\overset{\displaystyle H}{|}}{\underset{\underset{\displaystyle H}{|}}{C}}- $$

Butane   **Butyl**

or $ CH_3-CH_2-CH_2-CH_2- $

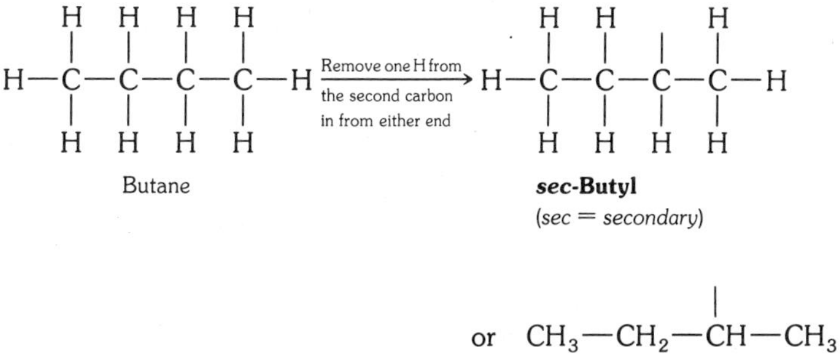

Butane

Remove one H from the second carbon in from either end →

sec-Butyl
(sec = secondary)

or $CH_3—CH_2—\overset{|}{CH}—CH_3$

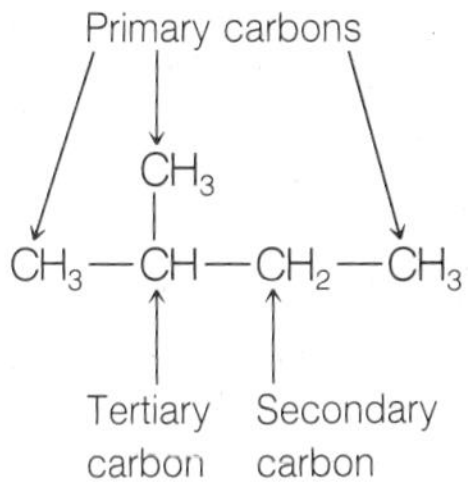

Primary carbons

CH_3

$CH_3—\overset{}{CH}—\overset{}{CH_2}—CH_3$

Tertiary   Secondary
carbon   carbon

This is called the *secondary* butyl group (abbreviated *sec*-butyl) because the open bonding site is at a **secondary carbon,** a carbon which is directly attached to just two other carbons. A **primary carbon** is one which is directly attached to just one other carbon. The open bonding site in the butyl group is at a primary carbon atom, for example. A **tertiary carbon** is one that directly holds three other carbons. We will encounter a tertiary carbon in a group we will soon study.

Butane is the first alkane to have an isomer. Its common name is isobutane, and we can derive two more alkyl groups from it.

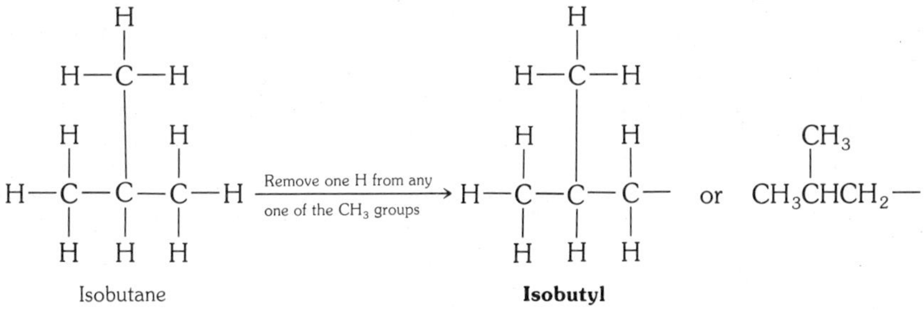

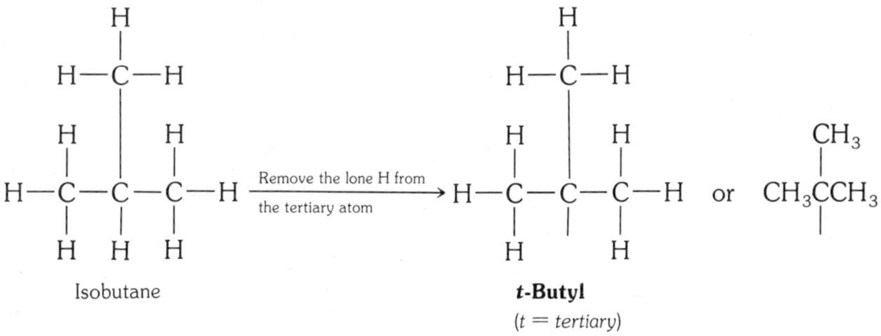

Isobutane

Remove one H from any one of the CH_3 groups →

Isobutyl   or $CH_3\overset{CH_3}{\overset{|}{CH}}CH_2—$

Isobutane

Remove the lone H from the tertiary atom →

t-Butyl   or $CH_3\overset{CH_3}{\overset{|}{C}}CH_3$
(t = tertiary)

Notice that the open bonding site in the *tertiary*-butyl group (abbreviated *t*-butyl) occurs at a tertiary carbon.

The names and structures of these alkyl groups must now be learned. If you have access to ball-and-stick models, make models of each of the parent alkanes and then remove hydrogen atoms to generate the open bonding sites and the alkyl groups. The *Study Guide* that accompanies this book has exercises that provide drills in recognizing alkyl groups when they are positioned in different ways on the page.

The prefix *iso-* in the name of an alkyl group, such as in isopropyl or isobutyl, has a special meaning. It can be used to name any alkyl group that has the following general features:

Here is another way to condense a structure. Thus

$CH_3(CH_2)_3CH_3$ represents $CH_3CH_2CH_2CH_2CH_3$.

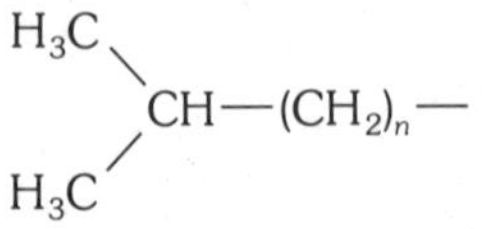

$$H_3C \diagdown \atop H_3C \diagup CH-(CH_2)_n-$$

$(n = 0, 1, 2, 3, \text{etc.})$
$n = 0$, isopropyl $n = 2$, isopentyl
$\ \ \ = 1$, isobutyl $\ \ \ = 3$, isohexyl

Notice that in each of these names there is a word fragment (for example, *-prop-*, *-but-*, etc.) associated with a number of carbon atoms. Here, it specifies the *total* number of carbons in the alkyl group. Thus the isopropyl group has three carbons and the isohexyl has six.

Before we return to the IUPAC rules for alkanes, we will introduce the names that are used for several other substituents.

—F	fluoro	—I	iodo
—Cl	chloro	—NO₂	nitro
—Br	bromo	—NH₂	amino

Now let us continue with the IUPAC rules.

6. Attach the name of the alkyl group or other substituent to the name of the parent as a prefix. Place the location number of the group in front of the resulting name and separate the number from the name by a hyphen.

    Returning to our original example, its name is 3-methylhexane.

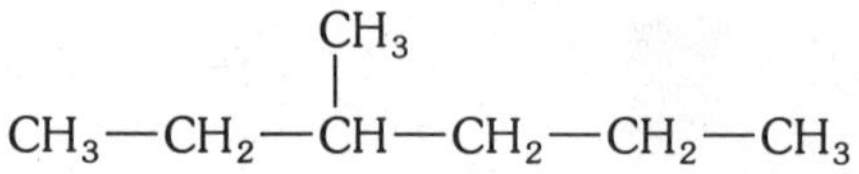

$$\overset{\displaystyle CH_3}{\underset{\displaystyle |}{CH_3-CH_2-CH-CH_2-CH_2-CH_3}}$$
3-Methylhexane

7. When two or more groups are attached to the parent, name each and locate each with a number. Always use *hyphens* to separate numbers from words in the IUPAC names of compounds.

    Here's an example of applying this rule.

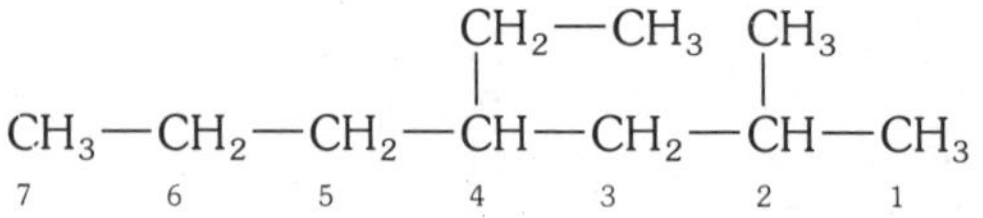

$$\overset{\displaystyle CH_2-CH_3 \quad CH_3}{\underset{\displaystyle 7 \quad\ 6 \quad\ 5 \quad\ 4 \quad\ 3 \quad\ 2 \quad\ 1}{CH_3-CH_2-CH_2-CH-CH_2-CH-CH_3}}$$

Correct name: 4-ethyl-2-methylheptane
Also correct: 2-methyl-4-ethylheptane

The IUPAC rules permit us to assemble the names of the substituents either alphabetically or in their order of increasing complexity.

8. When two or more substituents are identical, use such prefixes as di- (for 2), tri- (for 3), tetra- (for 4), and so forth; and specify the location number of *every* group. Always separate a number from another number in a name by a *comma*. For example,

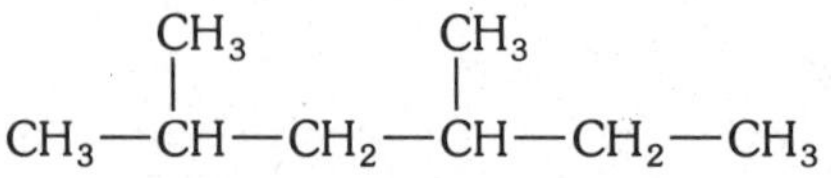

$$\overset{\displaystyle CH_3 \qquad\quad CH_3}{CH_3-CH-CH_2-CH-CH_2-CH_3}$$

Correct name: 2,4-dimethylhexane
Incorrect names: 2,4-methylhexane
3,5-dimethylhexane
2-methyl-4-methylhexane

9. When identical groups are on the *same* carbon, repeat the number of this carbon in the name. For example,

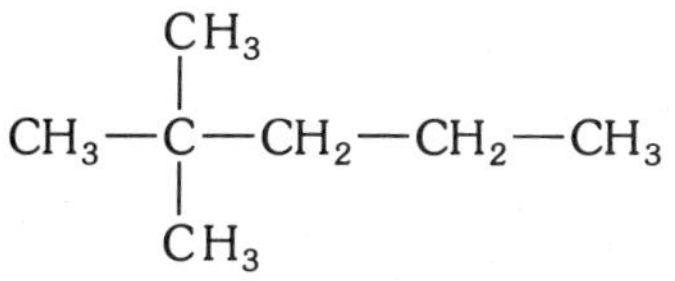

$$CH_3-\underset{\underset{\displaystyle CH_3}{|}}{\overset{\overset{\displaystyle CH_3}{|}}{C}}-CH_2-CH_2-CH_3$$

Correct name: 2,2-dimethylpentane
Incorrect names: 2-dimethylpentane
                  2,2-methylpentane
                  4,4-dimethylpentane

Another example:

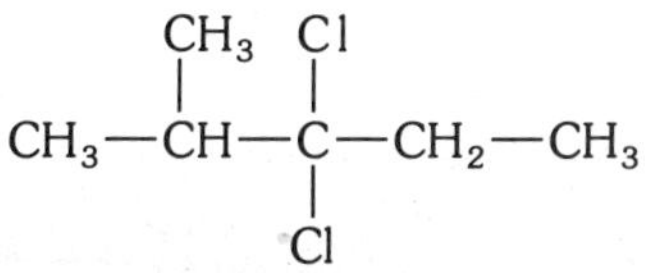

$$CH_3-CH-\underset{\underset{\displaystyle Cl}{|}}{\overset{\overset{\displaystyle CH_3 \quad Cl}{| \quad |}}{C}}-CH_2-CH_3$$

Correct name: 3,3-dichloro-2-methylpentane

Notice that the names of nonalkyl substituents are assembled first in the final name so a compound such as our previous example is viewed as a derivative of the hydrocarbon, 2-methylpentane.

10. To name a cycloalkane, place the prefix *cyclo* before the name of the straight-chain alkane that has the same number of carbon atoms as the ring.

11. When necessary, give numbers to the ring atoms by giving location 1 to a ring position that holds a substituent and numbering around the ring in whichever direction reaches the nearest substituent first. For example,

No number is needed when the ring has only one group, Thus

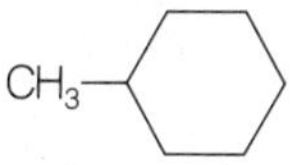

$$CH_3-\bigcirc$$

is named methylcyclohexane, not 1-methylcyclohexane.

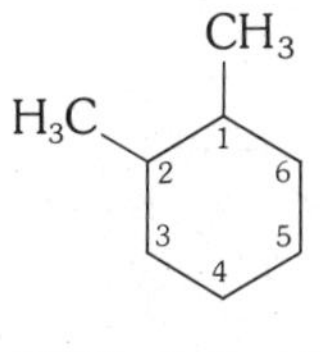

1,2-Dimethylcyclohexane 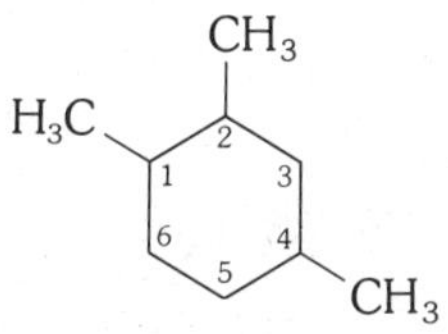   1,2,4-Trimethylcyclohexane

These are not all of the IUPAC rules for alkanes, but they will cover all of our needs. Study the following examples of correctly named compounds. Be sure to notice that in choosing the parent chain we sometimes have to go around a corner as the chain zigzags on the page.

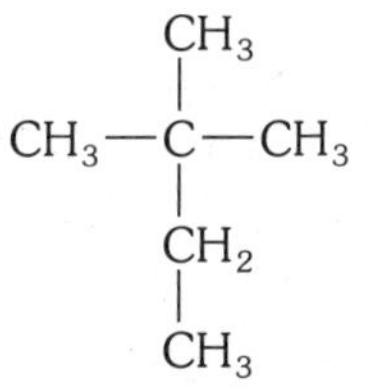

2,2-Dimethylbutane
Not 2-methyl-2-ethylpropane

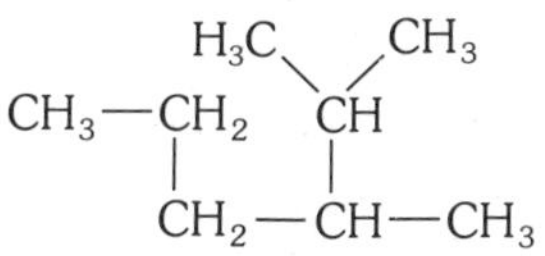

2,3-Dimethylhexane
Not 2-isopropylpentane

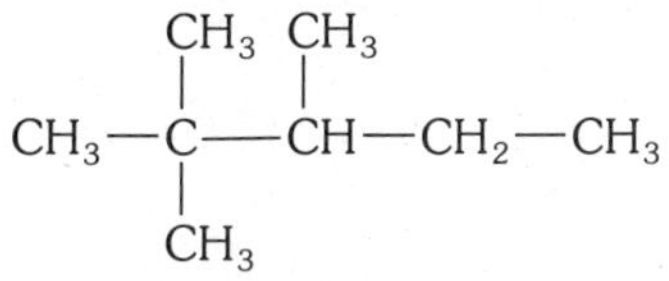

2,2,3-Trimethylpentane
Not 2,3-trimethylpentane
Not 2-*t*-butylbutane

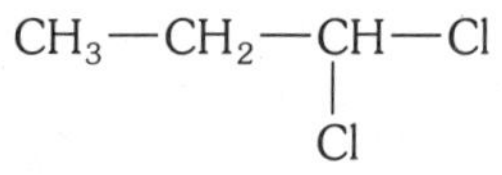

1,1-Dichloropropane
Not 3,3-dichloropropane
Not 3,3-chloropropane
Not 1-dichloropropane
Not 1,1-chloropropane

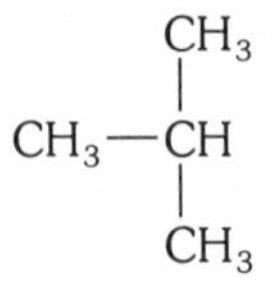

$$CH_3-\overset{\overset{\displaystyle CH_3}{|}}{\underset{\underset{\displaystyle CH_3}{|}}{CH}}$$

$$CH_3CH_2CH_2CHCH_2CHCH_3$$
$$CH_3-\overset{\overset{\displaystyle CH_3}{|}}{\underset{\underset{\displaystyle CH_3}{|}}{C}}-CH_3$$

2-Methylpropane

Not 1,1-dimethylethane

Not isobutane (which is its common name)

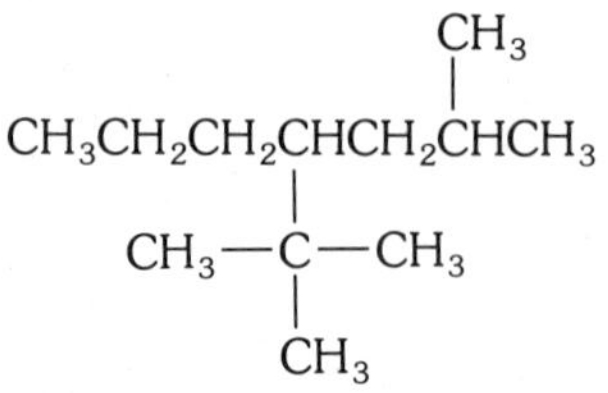

2-Methyl-4-*t*-butylheptane

Not 4-*t*-butyl-6-methylheptane

But 4-*t*-butyl-2-methylheptane is acceptable

EXAMPLE 12.3   USING THE IUPAC RULES TO NAME AN ALKANE

Problem: What is the IUPAC name for the following compound?

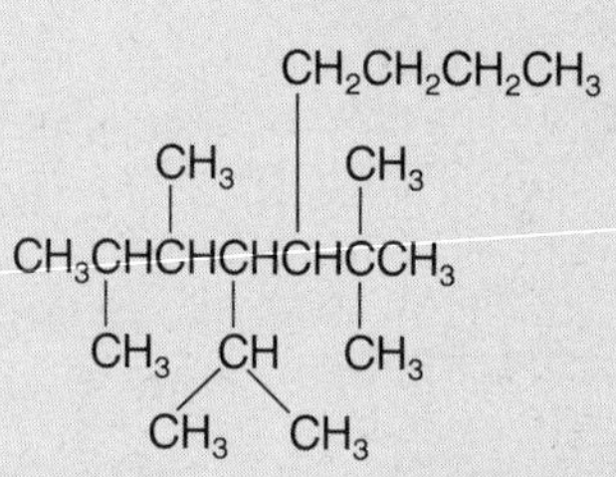

$$CH_2CH_2CH_2CH_3$$
$$CH_3CHCHCHCHCCH_3$$

Solution: The compound is an alkane because it is a hydrocarbon with only single bonds, so the ending to the name is -ane. The next step is to find the longest chain even if we have to go around corners. This chain is nine carbons long, so the name of the parent alkane is *nonane*. We have to number the chain from left to right, as follows, in order to reach the first branch with the lower number.

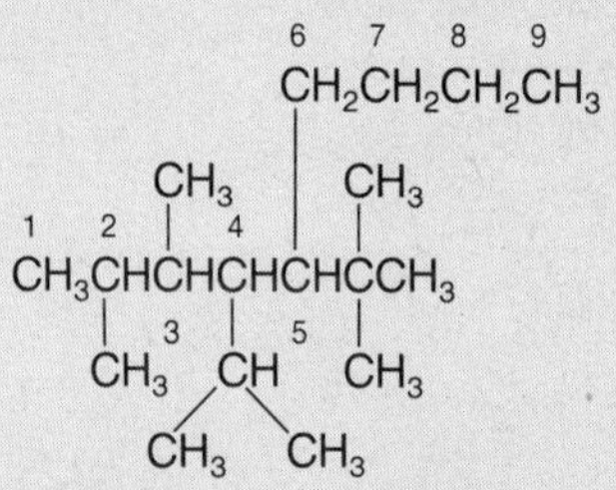

$$\overset{6\quad 7\quad 8\quad 9}{CH_2CH_2CH_2CH_3}$$
$$\underset{1\quad 2\quad\ 3\quad\ 4\ \ 5}{CH_3CHCHCHCHCCH_3}$$

At carbons 2 and 3 there are the one-carbon methyl groups. At carbon 4, there is a three-carbon isopropyl group (not the propyl group, because the bonding site is at the middle carbon of the three-carbon chain). At carbon 5, there is a four-carbon *t*-butyl group. (It has to be this particular butyl group because the bonding site is at a tertiary carbon.) We will assemble these names in their order of increasing carbon content to make the final name.

2,3-Dimethyl-4-isopropyl-5-*t*-butylnonane

Comma separates two numbers

Hyphens separate numbers from words

No hyphen, no comma, no space

PRACTICE EXERCISE 4   Write the IUPAC names of the following compounds.

(a) CH_3-CH_2
         $CH-CH_3$
     CH_2-CH_2
     CH_3

(b)
                        CH_3
         H_3C CH_3-C-CH_3
              CH —— $CH-CH_2-CH_2-CH_3$
         CH_3-CH
              CH_3

(c)
         CH_3 CH_3 CH_3
$CH_3-CH_2-CH-CH-CH-CH_2-CH-CH_3$
                   $CH_2-CH_2-CH_2-CH_3$

(d)
                   Br
$Cl-CH_2-CH-CH_2-I$

(e)
                   $CH_3-CH-CH_3$
$CH_3-CH_2-CH_2-CH-CH-CH_2-CH_2-CH_3$
         CH_3-C-CH_3
              CH_3

(f)
              CH_3
$NO_2-CH_2-C-CH_2-NO_2$
              CH_3

(g)
                   CH_2-Cl
$CH_3-CH_2-CH-CH-CH_3$
         CH_3

(h)
         CH_3 CH_2-CH_3
$CH_3-CH-CH_2-CH_2-CH-CH-CH_3$
                        CH_3

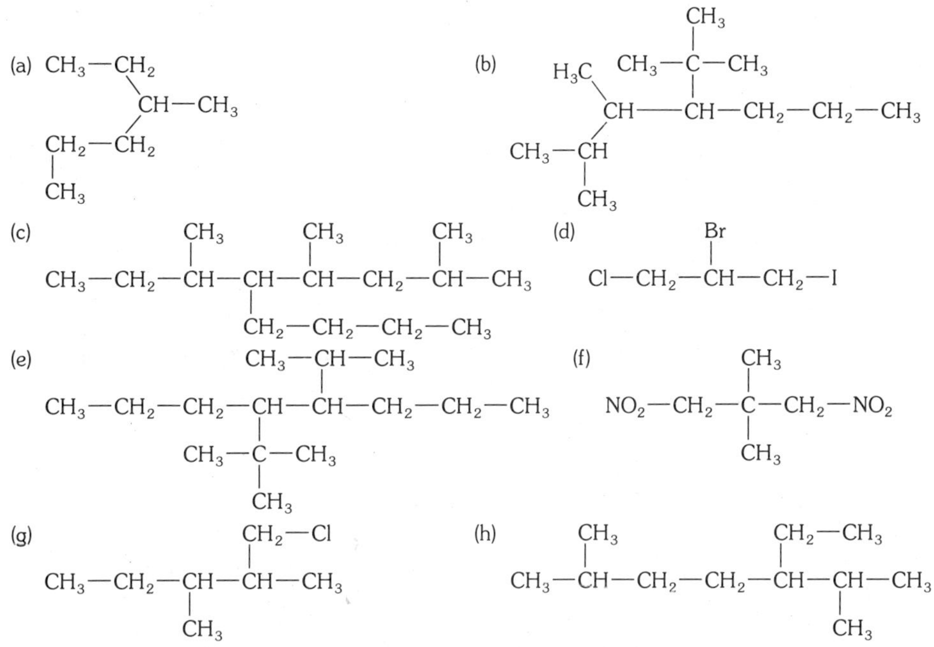

PRACTICE EXERCISE 5   Write the condensed structures of the following compounds.

(a) 1-Bromo-2-nitropentane
(b) 2,2,3,3,4,4-Hexamethyl-5-isopropyloctane
(c) 2,2-Diiodo-3-methyl-4-isopropyl-5-sec-butyl-6-t-butylnonane
(d) 1-Chloro-1-bromo-2-methylpropane
(e) 5,5-Di-sec-butyldecane

PRACTICE EXERCISE 6   Examine the structure of part (e) of Practice Exercise 5. Underline each primary carbon, draw an arrow to each secondary carbon, and circle each tertiary carbon.

Common Names.   We mentioned earlier that chemists often use *common names* for many compounds, and they actually are commonly used, even in some of the technical references and textbooks you might use in such subjects as nutrition or drugs and medicinals. For example, straight-chain alkanes sometimes have the prefix *n*- before the name. This *n*- stands for *normal,* meaning that the straight-chain isomer is the *normal* isomer. Thus the common name of what the IUPAC rules call butane is *n*-butane. (In fact, the IUPAC took over as many of the names in common use at the time as it could when it devised the rules of alkane nomenclature.)

The names of the alkyl groups are often used to make common names of alcohols, amines, and halogen derivatives of hydrocarbons. Here are some examples involving halogen compounds.

Structure	Common Name	IUPAC Name
CH_3Cl	methyl chloride	chloromethane
CH_3CH_2Br	ethyl bromide	bromoethane

Structure	Common Name	IUPAC Name
$CH_3CH_2CH_2Br$	propyl bromide	1-bromopropane
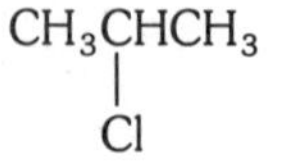 CH_3CHCH_3 $\mid$ Cl	isopropyl chloride	2-chloropropane
$CH_3CH_2CH_2CH_2Cl$	butyl chloride	1-chlorobutane
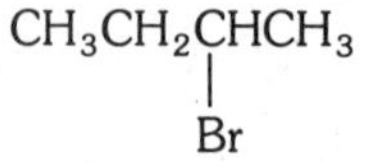 $CH_3CH_2CHCH_3$ $\mid$ Br	*sec*-butyl bromide	2-bromobutane
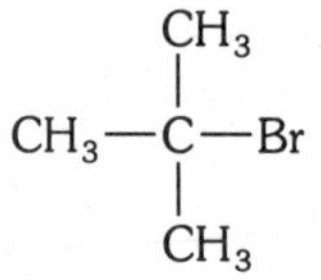 $\quad\quad CH_3$ $\quad\quad\mid$ $CH_3—C—Br$ $\quad\quad\mid$ $\quad\quad CH_3$	*t*-butyl bromide	2-bromo-2-methylpropane
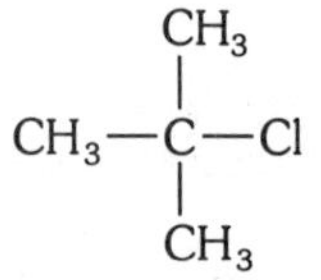 $\quad\quad CH_3$ $\quad\quad\mid$ $CH_3—C—Cl$ $\quad\quad\mid$ $\quad\quad CH_3$	*t*-butyl chloride	2-chloro-2-methylpropane

---

PRACTICE EXERCISE 7   Give the common names of the following compounds.

(a) $ClCH_2CH_3$ (b) $BrCH_2CH_2CH_2CH_3$ (c) 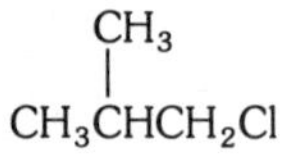 CH_3CHCH_2Cl with CH_3 branch   (d) 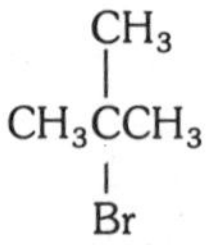 CH_3CCH_3 with CH_3 branch and Br

$$\text{(c)} \quad CH_3\underset{\mid}{\overset{CH_3}{C}}HCH_2Cl \qquad \text{(d)} \quad CH_3\underset{\underset{Br}{\mid}}{\overset{\overset{CH_3}{\mid}}{C}}CH_3$$

12.3   CHEMICAL PROPERTIES OF ALKANES

Alkanes can burn and can give substitution reactions with the halogens, but they undergo almost no other reaction.

The chemistry of the alkanes and cycloalkanes is quite simple. Very few chemicals react with them. This is why they are nicknamed the *paraffins* or the paraffinic hydrocarbons, after the Latin *parum affinis,* meaning "little affinity" or "little reactivity." The alkanes are not chemically attacked by water, by strong acids such as sulfuric or hydrochloric acid, by strong bases such as sodium hydroxide, by active metals such as sodium, by strong oxidizing agents such as the permanganate ion or the dichromate ion, or by any of the reducing agents. Even when molecules have functional groups, most of their alkane-like parts ride unchanged through reactions involving the functional groups (which, of course, is why alkyl groups are called nonfunctional groups).

> Mineral oil is a safe laxative (when used with care) because it is a mixture of high-formula-weight alkanes that undergo no chemical reactions in the intestinal tract.

Among the few reactions of alkanes are combustion and chlorination (a reaction with chlorine). Alkanes are also attacked by fluorine and bromine as well as by very hot nitric acid. We will look briefly at combustion and chlorination.

Combustion.   Virtually all organic compounds burn, and the saturated hydrocarbons are no exception. We burn them as fuel to obtain energy. Bunsen burner gas is mostly methane. Liquified propane is used as fuel in areas where gas lines have not been built. Gasoline, diesel fuel, jet fuel, and heating oil are all mixtures of hydrocarbons, mostly alkanes.

If enough oxygen is available, the sole products of the complete combustion of *any hydrocarbon,* not just alkanes, are carbon dioxide and water. To illustrate, using propane,

$$CH_3CH_2CH_3 + 5O_2 \longrightarrow 3CO_2 + 4H_2O + 531 \text{ kcal/mol propane}$$

If insufficient oxygen is present, some carbon monoxide forms. The same products — carbon

SPECIAL TOPIC 12.2 WHY EQUATIONS FOR ORGANIC REACTIONS CANNOT ALWAYS BE BALANCED

Chemical reactions involve the breaking and reforming of bonds. Because organic molecules have several bonds of similar strengths, several products can sometimes form, even when we are trying to make just one of them. The substances we don't want are called the **by-products,** and the reactions that produce by-products are called **side reactions.** The reaction that produces the largest relative quantity of product is called the **main reaction.** Naturally, the chemist hopes this is the reaction that produces the desired product. In any case, most organic reactions produce a set of products, and this mixture must then be separated. This often takes more time and effort than any other aspect of the synthesis.

As an example of a reaction that produces a mixture, we can use the chlorination of ethane. If we were to mix 1 mol of chlorine with 1 mol of ethane, naively trying to prepare 1 mole of ethyl chloride, we would obtain a mixture of mono-, di-, tri- and possibly still more highly chlorinated molecules (plus HCl).

$$CH_3CH_3 + Cl_2 \xrightarrow{\text{ultraviolet light}} CH_3CH_2Cl + CH_3CHCl_2 + ClCH_2CH_2Cl + CH_3CCl_3 + ClCH_2CHCl_2 + \text{etc.} + HCl$$

It would be foolish to write coefficients in front of any of the products or to change the coefficients in front of the reactants—which, as they stand, say that a 1 : 1 molar ratio was taken—in an effort to balance the equation. If the mixture were to be completely separated, then the molar percentages of the various products could be reported, but seldom is such a thorough separation performed when just one product is sought.

Most organic reactions pose this kind of balancing problem, so the equation written is the equation for the main reaction only. Often it is balanced, but for just one reason—to provide a basis for the selection of the relative numbers of moles of the reactants to be mixed at the start. In nearly all of the reactions that we will study, side reactions occur, which we generally will ignore.

dioxide and water—are also obtained by the complete combustion of any organic compound that consists only of carbon, hydrogen, and oxygen (e.g., the alcohols).

Chlorination. In the presence of ultraviolet radiation or at a high temperature, alkanes react with chlorine, and organochlorine compounds plus hydrogen chloride form. An atom of hydrogen in the alkane is replaced by an atom of chlorine, and this kind of replacement of one atom or group by another is called a *substitution reaction*. For example,

$$CH_4 + Cl_2 \xrightarrow[\text{or heat}]{\text{Ultraviolet light}} \underset{\text{Methyl chloride}}{CH_3Cl} + HCl$$

The hydrogen atoms in methyl chloride can also be replaced by chlorine, so as methyl chloride starts to form it competes with the methane that hasn't reacted yet. This is how some methylene chloride forms when the above reaction is carried out. In fact, when 1 mol of CH_4 and 1 mol of Cl_2 are mixed and made to react, several reactions eventually occur, and a mixture of four chlorinated methanes, hydrogen chloride, plus some unreacted methane is the result. It isn't possible to write a balanced equation, so we represent it by a flow of symbols.

The numbers in parentheses are the boiling points of the compounds. Notice how they increase with formula weight.

$$\underset{\substack{\text{Methane}\\(-162\ ^\circ\text{C})}}{CH_4} + Cl_2 \xrightarrow[\searrow HCl]{} \underset{\substack{\text{Methyl}\\\text{chloride}\\(-24\ ^\circ\text{C})}}{CH_3Cl} \xrightarrow[\searrow HCl]{Cl_2} \underset{\substack{\text{Methylene}\\\text{chloride}\\(40\ ^\circ\text{C})}}{CH_2Cl_2} \xrightarrow[\searrow HCl]{Cl_2} \underset{\substack{\text{Chloroform}\\(61\ ^\circ\text{C})}}{CHCl_3} \xrightarrow[\searrow HCl]{Cl_2} \underset{\substack{\text{Carbon}\\\text{tetrachloride}\\(77\ ^\circ\text{C})}}{CCl_4}$$

For a discussion of the use of unbalanced reaction sequences to represent organic reactions, see Special Topic 12.2.

Methylene chloride, chloroform, and carbon tetrachloride are all used as nonpolar solvents, and chloroform has been used as an anesthetic. Chlorinated solvents must be handled in well-ventilated areas, and they should not be allowed to spill on the skin. They are rapidly absorbed through the skin, and both chloroform and carbon tetrachloride can cause liver damage. They should never be poured down the sink.

Bromine reacts with methane by the same kind of substitution as chlorine. Iodine does not react. Fluorine combines explosively with most organic compounds at room temperature and complex mixtures form.

The higher alkanes can also be chlorinated. Ethyl chloride, plus more highly chlorinated products, form by the chlorination of ethane, and ethyl chloride (b.p. 12.5 °C) is used as a local anesthetic. When it is sprayed on the skin, it evaporates very rapidly, and this cools the area enough to prevent the transmission of pain signals during minor surgery at the site.

When propane is chlorinated, both propyl chloride and isopropyl chloride form in roughly equal amounts.

$$CH_3CH_2CH_3 + Cl_2 \xrightarrow{\text{ultraviolet light}} CH_3CH_2CH_2Cl + CH_3\overset{\overset{\displaystyle Cl}{|}}{C}HCH_3 + HCl$$

Propane   Propyl   Isopropyl
                                          chloride   chloride

Besides these products, some higher chlorinated compounds also form.

PRACTICE EXERCISE 8   (a) How many monochloro derivatives of butane are possible? Give both their common and IUPAC names. (Consider only the monochloro compounds with the formula C_4H_9Cl.) (b) How many monochloro derivatives of isobutane are possible? Write both their common and IUPAC names.

SUMMARY

Hydrocarbons The carbon frameworks of hydrocarbon molecules can be straight chains, branched chains, or rings. When only single bonds occur, the substance is said to be saturated; otherwise it is unsaturated. Regardless of how much we condense a structure, each carbon atom must always have four bonds. Most single bonds need not be shown, but carbon-carbon double and triple bonds are always indicated by two or three lines.

Being nonpolar compounds, the hydrocarbons are all insoluble in water, and many mixtures of alkanes are common, nonpolar solvents. The rule *like dissolves like* lets us predict solubilities.

Nomenclature of alkanes In the IUPAC system, a compound's family is always indicated in the name of a compound by the ending; the number of carbons in the parent chain, by a prefix; and the

locations of side chains or groups are specified by numbers assigned to the carbons of the parent chain. Alkane-like substituents are called alkyl groups, and the names and formulas of those having from one to four carbon atoms must be learned. Common names are still popular, particularly when the IUPAC names are long and cumbersome to use in conversation.

Chemical properties of alkanes Alkanes and cycloalkanes are generally unreactive at room temperature toward concentrated acids and bases, toward oxidizing and reducing agents, toward even the most reactive metals, and toward water. They burn, giving off carbon dioxide and water, and in the presence of ultraviolet light (or at a high temperature) they are attacked by chlorine, bromine, and nitric acid. Fluorine reacts explosively, but iodine does not react with alkanes.

KEY TERMS

The following terms were emphasized in this chapter, and they will be used often in subsequent chapters. Be sure you have mastered them before continuing.

aliphatic compound	*t*-butyl group	IUPAC rules	ring compound
alkane	ethyl group	like-dissolves-like rule	saturated compounds
alkyl group	heterocyclic compound	methyl group	secondary carbon
aromatic compound	hydrocarbon	nomenclature	tertiary carbon
butyl group	isobutyl group	primary carbon	unsaturated compound
sec-butyl group	isopropyl group	propyl group	

REVIEW EXERCISES

The answers to these Review Exercises are in the *Study Guide* that accompanies this book.

Saturated and Unsaturated Compounds

12.1 Which of the following structures represent unsaturated compounds?

(a)

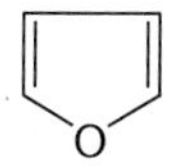

Furan

(b)

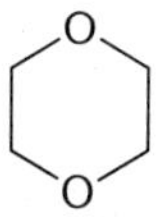

Dioxane

(c)

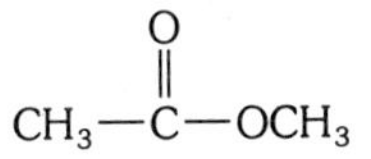

$$CH_3-\overset{\overset{\displaystyle O}{\|}}{C}-OCH_3$$

Methyl acetate

(d)

$$CH_3-\overset{\overset{\displaystyle O-CH_3}{|}}{\underset{\underset{\displaystyle O-CH_3}{|}}{C}}-O-CH_3$$

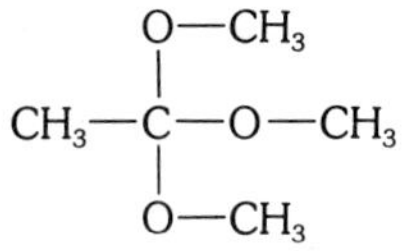

Methyl orthoacetate

12.2 Which compounds are saturated?

(a)

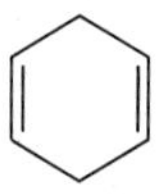

1,4-Cyclohexadiene

(b) $CH_3-C\equiv N$

Acetonitrile

(c) 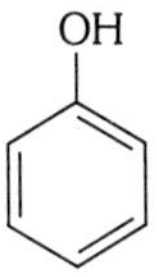 OH

Phenol

(d)

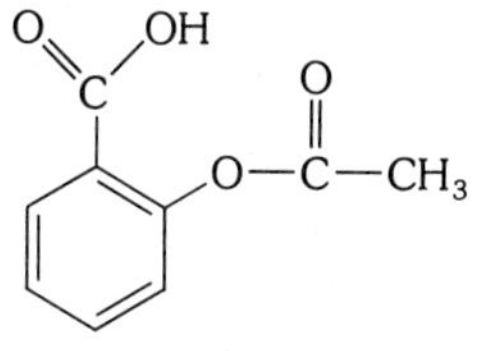

Aspirin

Structures for Cyclic Compounds

12.3 Expand the following structure of nicotinamide, one of the B-vitamins.

Nicotinamide

12.4 Expand the structure of thiamine, vitamin B_1. Notice that one nitrogen has a positive charge and therefore it has four bonds.

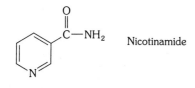

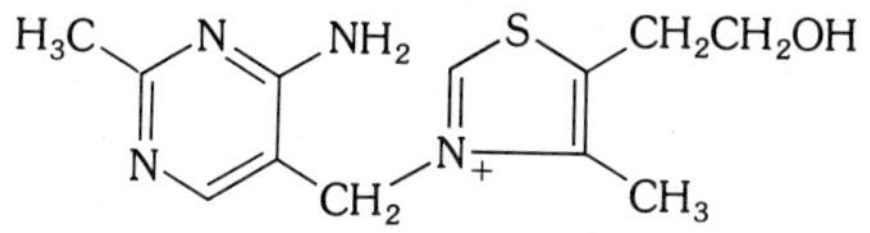

Physical Properties and Structure

12.5 Which compound must have the higher boiling point? Explain.

$$CH_3CH_2CH_3 \qquad CH_3CH_2CH_2CH_2CH_3$$
$$\text{A} \qquad\qquad\qquad \text{B}$$

12.6 Which compound must be less soluble in petroleum ether? Explain.

$$ClCH_2CH_2CH_2CH_2CH_2Cl \quad HOCH_2CH_2CH_2CH_2CH_2OH$$
$$\text{A} \qquad\qquad\qquad\qquad \text{B}$$

12.7 Suppose that you were handed two test tubes containing colorless liquids, and you were told that one contained pentane and the other held hydrochloric acid. How could you use just water to tell which tube contained which compound without carrying out any chemical reaction?

12.8 Suppose that you were given two test tubes and were told that one held methyl alcohol, CH_3OH, and the other held hexane. How could water be used to tell these substances apart without carrying out any chemical reaction?

Nomenclature

12.9 There are five isomers of C_6H_{14}. Write their condensed structures and their IUPAC names.

12.10 Which of the isomers of hexane (Review Exercise 12.9) has the common name, *n*-hexane? Write its structure.

12.11 Which of the hexane isomers (Review Exercise 12.9) has the common name, isohexane? Write its structure.

12.12 There are nine isomers of C_7H_{16}. Write their condensed structures and their IUPAC names.

12.13 Write the condensed structures of the isomers of heptane (Review Exercise 12.12) that have the following names.
(a) *n*-Heptane   (b) Isoheptane

12.14 Write the IUPAC names of the following compounds.

(a)

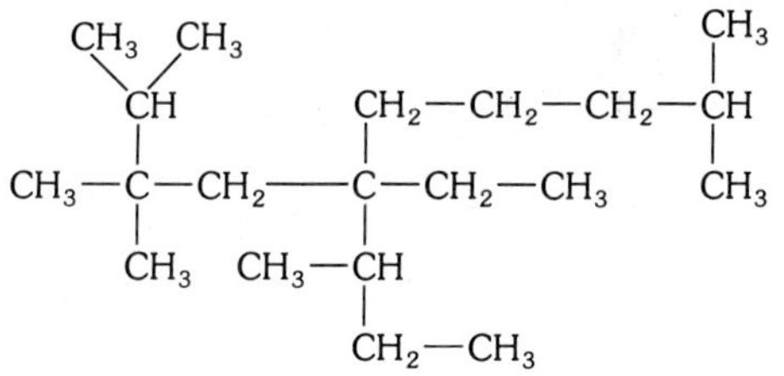

$$CH_3-\overset{\overset{\displaystyle CH_3\ \ CH_3}{\overset{\displaystyle |\ \ \ \ \ |}{CH}}}{\underset{\underset{\displaystyle CH_3\ \ CH_3-CH}{\underset{\displaystyle |}{CH_2-CH_3}}}{C}}-CH_2-\overset{}{C}-CH_2-CH_3 \qquad CH_2-CH_2-CH_2-\overset{\overset{\displaystyle CH_3}{|}}{\underset{\underset{\displaystyle CH_3}{|}}{CH}}$$

(b)

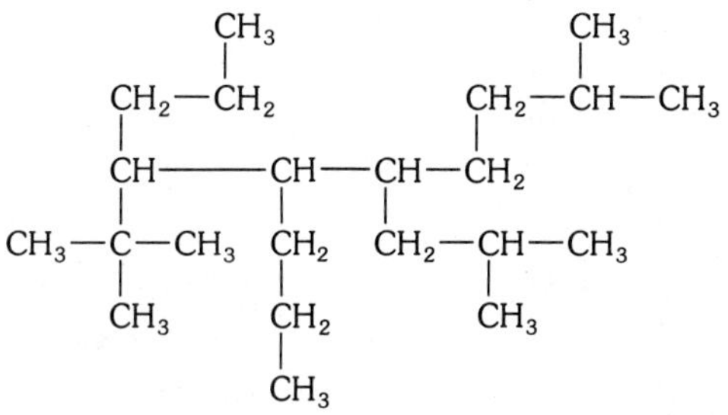

12.15 Write condensed structures for the following compounds.
   (a) Isobutyl bromide   (b) Ethyl iodide
   (c) Propyl chloride   (d) *t*-Butylcyclohexane

12.16 Write condensed structures for the following compounds.
   (a) *sec*-Butyl chloride   (b) *n*-Butyl iodide
   (c) Isopropyl bromide   (d) Isohexyl bromide

12.17 The following are incorrect efforts at naming certain compounds. What are the most likely condensed structures and correct IUPAC names?
   (a) 1,6-Dimethylcyclohexane
   (b) 2,4,5-Trimethylhexane
   (c) 1-Chloro-*n*-butane
   (d) Isopropane

12.18 The following names cannot be the correct names, but it is still possible to write structures from them. What are the correct IUPAC names and the condensed structures?
   (a) 1-Chloroisobutane
   (b) 2,4-Dichlorocyclopentane

   (c) 2-Ethylbutane
   (d) 1,3-6-Trimethylcyclohexane

Reactions of Alkanes

12.19 Write the balanced equation for the complete combustion of octane, a component of gasoline.

12.20 Gasohol is a mixture of ethyl alcohol, CH_3CH_2OH, in gasoline. Write the equation for the complete combustion of ethyl alcohol.

12.21 What are the formulas and common names of all of the compounds that can be made from methane and chlorine?

12.22 There are two isomers of $C_2H_4Cl_2$. What are their structures and IUPAC names?

12.23 When propane reacts with chlorine, besides the two isomeric monochloropropanes, there are some dichloropropanes that also form. Write the structures and the IUPAC names for all of these possible dichloropropanes.

Chapter 13
Unsaturated Hydrocarbons

Silos, such as those standing by the buildings, may eventually be replaced at a fraction of the cost by enormous plastic bags here seen holding hay silage — about 150 tons per bag — at a beef feeding operation near Moses Lake, Washington. Many plastics are made from alkenes, which we'll study in this chapter.

13.1 TYPES OF UNSATURATED HYDROCARBONS

The alkenes and cycloalkanes can exhibit geometric isomerism because there is no free rotation at the double bond or in a ring.

Unsaturation in hydrocarbons occurs as double bonds, triple bonds, or as benzene rings. The double bond is very common among biochemicals, particularly the fats and oils — the lipids. The triple bond is very uncommon in nature, and later in the chapter we will use Special Topic 13.1 to discuss it. The benzene ring and rings similar to it occur among proteins and nucleic acids, and it is widely present in pharmaceuticals. This chapter is mostly about the alkene system and the properties of the carbon-carbon double bond, sometimes called the -ene function.

Table 13.1 shows the structures and some physical properties of several alkenes. As with the alkanes, the first four are gases at room temperature, and all are much less dense than water. Like all hydrocarbons, all alkenes are insoluble in water and soluble in nonpolar solvents.

Isomerism Among the Alkenes. Alkenes can exist as isomers in any one of three ways. They can have different carbon skeletons, as illustrated by 1-butene and 2-methylpropene. They can have the same skeletons but differ in the locations of their double bonds, as in 1-butene and 2-butene. Finally, they can have the same skeletons and the same locations of the double bond, but they differ in geometry, as in the

$$CH_2{=}CHCH_2CH_3 \qquad CH_2{=}\overset{\overset{\displaystyle CH_3}{|}}{C}{-}CH_3 \qquad CH_3{-}CH{=}CH{-}CH_3$$

1-Butene   2-Methylpropene   2-Butene

TABLE 13.1
Properties of Some 1-Alkenes

Name (IUPAC)	Structure	Boiling Point (°C)	Melting Point (°C)	Density (in g/mL at 10 °C)
Ethene	$CH_2{=}CH_2$	−104	−169	—
Propene	$CH_2{=}CHCH_3$	−48	−185	—
1-Butene	$CH_2{=}CHCH_2CH_3$	−6	−185	—
1-Pentene	$CH_2{=}CHCH_2CH_2CH_3$	30	−165	0.641
1-Hexene	$CH_2{=}CHCH_2CH_2CH_2CH_3$	64	−140	0.673
1-Heptene	$CH_2{=}CHCH_2CH_2CH_2CH_3$	94	−119	0.697
1-Octene	$CH_2{=}CHCH_2CH_2CH_2CH_2CH_3$	121	−102	0.715
1-Nonene	$CH_2{=}CHCH_2CH_2CH_2CH_2CH_2CH_3$	147	−81	0.729
1-Decene	$CH_2{=}CHCH_2CH_2CH_2CH_2CH_2CH_2CH_3$	171	−66	0.741
Cyclopentene		44	−135	0.722
Cyclohexene	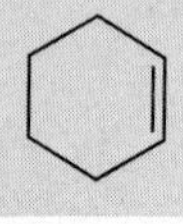	83	−104	0.811

FIGURE 13.1
The geometry at a carbon-carbon
double bond.

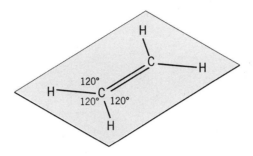

situation with 2-butene, which we will discuss next. We have to explore this matter of geometry, because the abilities of cellular catalysts (enzymes) to function are very sensitive to the shapes of molecules.

Molecular geometry is more important at the double bond than at the single bond, because there is no free rotation of the groups that the double bond holds together. As seen in Figure 13.1, all of the atoms in ethene lie in the same plane, and the lines that join them intersect at angles of 120°. If we replace a hydrogen at each end of the double bond in ethene by a methyl group, the result is 2-butene, but we can make this change in two ways. In one, the methyl groups are on the same side of the double bond, and in the other they are on opposite sides, *and we cannot twist one to make it equivalent to the other.* These two geometric forms of 2-butene represent molecules of different compounds, not just two interconvertible conformations of the same molecule.

The *side* of a double bond is not the same as the *end* of a double bond.

Same side

$$\text{C}=\text{C}$$ One end

cis-2-Butene
(b.p. 3.7 °C)

trans-2-Butene
(b.p. 0.9 °C)

Substances that have identical molecular formulas and identical orders of attachment of their atoms but differ in geometry are called **geometric isomers.** The phenomenon of the existence of such isomers is called **geometric isomerism.** When two atoms or groups are on the same side of the double bond, they are said to be *cis* to each other. When they are on opposite sides, they are said to be *trans* to each other. (Sometimes geometric isomerism is called *cis-trans isomerism.*) These cis or trans designations can be made parts of the names of the isomers, as the examples of *cis-* and *trans-2-butene* show.

Not all alkenes can exist as cis-trans isomers. When the two atoms or groups *at either end* of the double bond are identical — for example, both are H or both are CH_3 — there is nothing special toward which a group at the other end of the double bond can be uniquely cis or trans, and there can be no geometric isomerism. For example, in 1-butene, the two H atoms on one end of the double bond, are identical, so the two ways of writing 1-butene, below,

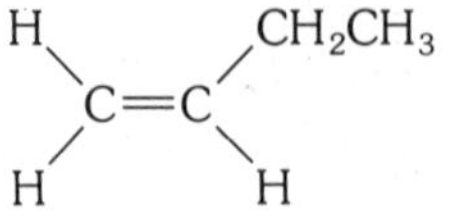

H   CH₂CH₃   H   H
 \   /   \   /
  C==C   is the same as   C==C
 /   \   /   \
H   H   H   CH₂CH₃

1-Butene   1-Butene

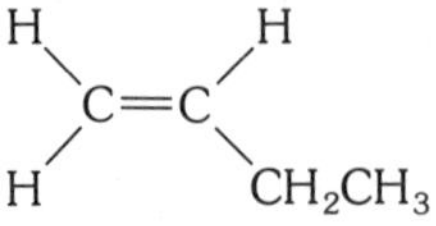

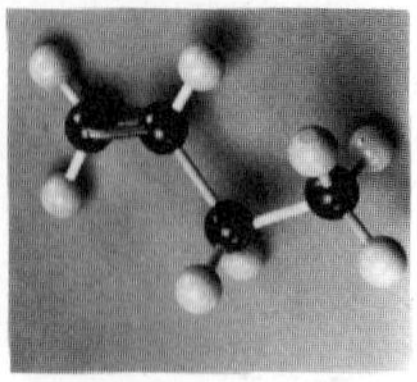

1-Butene.

represent the same substance. If we simply flop the *whole* first structure over, top to bottom, we get the second. But this does not reorganize the bonds to make anything different. There are no cis-trans isomers of 1-butene.

Cis-trans isomerism also occurs when the atoms involved at the ends of the double bond are halogen atoms or other groups, for example:

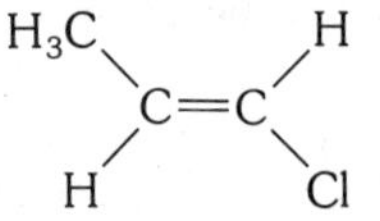

H₃C   H   H₃C   Cl
 \   /   \   /
  C======C   C======C
 /   \   /   \
H   Cl   H   H

trans-1-Chloro-1-propene 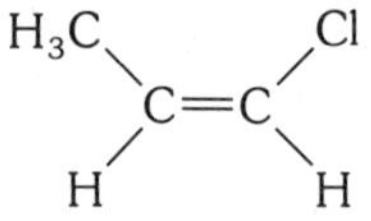   *cis*-1-Chloro-1-propene

---

EXAMPLE 13.1   WRITING THE STRUCTURES OF CIS AND TRANS ISOMERS

Problem: Write the structures of the cis and trans isomers, if any, of the following alkene.

CH₃CH==CHCH₂CH₃
2-Pentene

Solution: First, write a carbon-carbon double bond without any attached groups. Spread the single bonds at the carbon atoms at angles of roughly 120°. Draw two of these partial structures.

\   /   \   /
 C==C   C==C
/   \   /   \

Then attach the two groups that are at one of the ends of the double bond. Attach them *identically* to make identical partial structures.

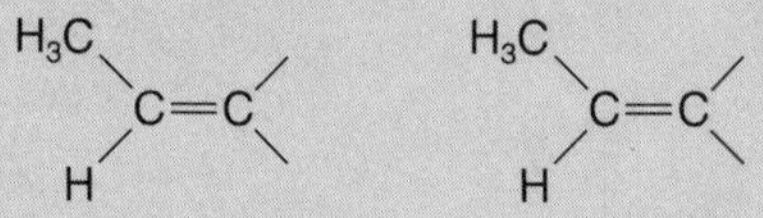

H₃C   H₃C
   \   /   \   /
    C==C   C==C
   /   \   /   \
  H   H

Finally, at the other end of the double bond, draw the other two groups, only this time be sure that they are switched in their relative positions.

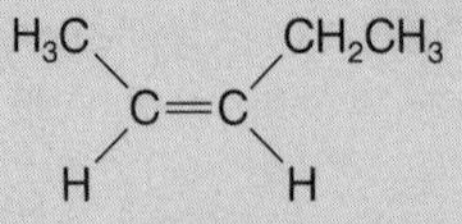

H₃C   CH₂CH₃   H₃C   H
   \   /   \   /
    C====C   C====C
   /   \   /   \
  H   H   H   CH₂CH₃

cis-2-Pentene 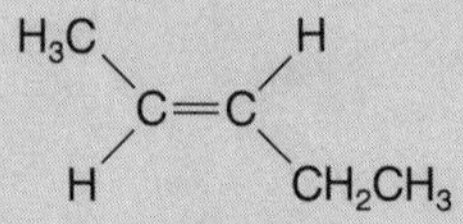   *trans*-2-Pentene

Be sure to check to see if the two structures are geometric *isomers* and not two identical structures that are merely flip-flopped on the page.

PRACTICE EXERCISE 1   Write the structures of the cis and trans isomers, if any, of the following compounds.

(a) $CH_3CH_2C{=}CHCH_3$ (b) $ClCH{=}CHCl$ (c) $CH_3C{=}CH_2$ (d) $ClC{=}CHBr$
$\quad\quad\quad\quad |$ $\quad\quad\quad\quad\quad\quad\quad\quad\quad\quad\quad\quad\quad\quad\quad\quad |$ $\quad\quad |$
$\quad\quad\quad\quad CH_3$ $\quad\quad\quad\quad\quad\quad\quad\quad\quad\quad\quad\quad\quad\quad\quad CH_3$ $\quad\quad Cl$

Geometric Isomerism Among Ring Compounds.   The double bond is not the only source of restricted rotation; the ring is another. For example, two geometric isomers of 1,2-dimethylcyclopropane are known, and neither can be twisted into the other without breaking the ring open. This costs too much energy to occur spontaneously even at quite high temperatures.

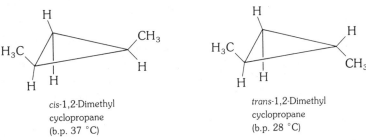

cis-1,2-Dimethyl
cyclopropane
(b.p. 37 °C)

trans-1,2-Dimethyl
cyclopropane
(b.p. 28 °C)

This kind of cis-trans difference involving ring compounds occurs often among molecules of carbohydrates.

13.2 NOMENCLATURE OF ALKENES

The IUPAC names of alkenes end in -ene, and the double bond takes precedence over side chains in numbering the parent chain.

The IUPAC rules for naming alkenes and cycloalkenes are as follows.

1.   Use the ending, -ene for all alkenes and cycloalkenes.

2.   As a prefix to this ending, count the number of carbon atoms in the longest sequence *that includes the double bond.* Then use the same prefix that would be used if the compound were saturated.

For example:

$$\overset{1}{CH_2}$$
$$\underset{2\ 3}{CH_3CH_2}\overset{\|}{C}\underset{4}{CH_2}\underset{5}{CH_2}\underset{6}{CH_2}CH_3$$

The parent chain has
six carbons, not seven.

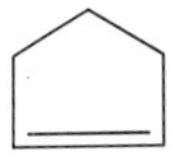

Cyclopentene
(complete name)

3.   For open-chain alkenes, number the parent chain from whichever end gives the lower number to the first carbon of the double bond to be reached.

This rule gives precedence to the numbering of the double bond over the location of the first substituent on the parent chain. For example,

$$\overset{CH_3}{\underset{|}{}}$$
$$\underset{5\ \ 4\ \ 3\ \ 2\ \ 1}{CH_3CHCH_2CH{=}CH_2}$$ The double bond is at position 1, not 4.
Not $\underset{1\ \ 2\ \ 3\ \ 4\ \ 5}{\phantom{CH_3CHCH_2CH{=}CH_2}}$

4-Methyl-1-pentene
(complete name)
Not: 2-methyl-4-pentene

4. For cycloalkenes, always give position 1 to one of the two carbons at the double bond. To decide which carbon gets this number, the ring atoms are to be numbered from carbon 1 *through the double bond* in whichever direction reaches a substituent first.

   For example, the numbers inside the ring represent the correct numbering, not the numbers outside the ring.

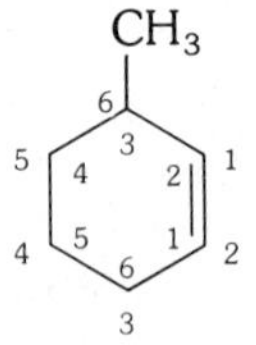

3-Methylcyclohexene (complete name)
Not: 6-Methylcyclohexene

5. To the name begun with rules 1 and 2, place the number that locates the first carbon of the double bond as a prefix, and separate this number from the name by a hyphen.

6. If substituents are on the parent chain or ring, complete the name obtained by rule 5 by placing the names and location numbers of the substituents as prefixes.

   Remember to separate numbers from numbers by commas, but use hyphens to connect a number to a word.

A few correctly named examples follow. Some have the common names that are given in parentheses. The ending -*ylene* characterizes the common names of open-chain alkenes.

The common name, ethylene, is also allowed by the IUPAC as the name for ethene.

$$CH_2{=}CH_2 \qquad CH_3CH{=}CH_2$$

Ethene   Propene[1]
(Ethylene)   (Propylene)

$$CH_3{-}\overset{\displaystyle CH_3}{\underset{}{C}}{=}CH_2$$

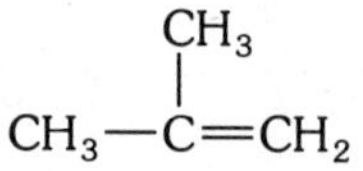

2-Methylpropene[2]
(Isobutylene)

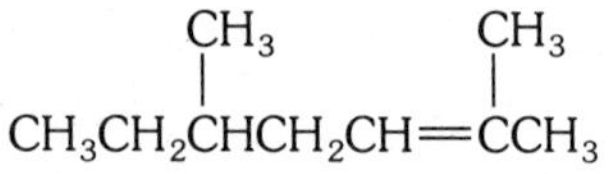

$$CH_3CH_2\overset{\displaystyle CH_3}{\underset{}{C}}HCH_2CH{=}\overset{\displaystyle CH_3}{\underset{}{C}}CH_3$$

2,5-Dimethyl-2-heptene

$$CH_3CH_2CH_2\overset{\displaystyle \overset{CH_3}{\underset{}{C}}HCH_2CH_2CH_3}{\underset{}{C}}{=}CH_2$$

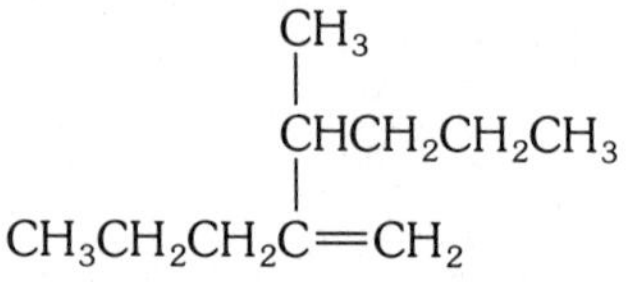

2-Propyl-3-methyl-1-hexene
or 3-Methyl-2-propyl-1-hexene

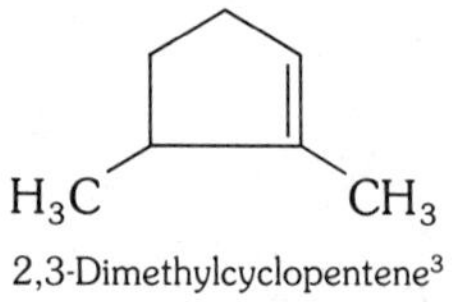

2,3-Dimethylcyclopentene[3]

$$Cl(CH_2)_6CH{=}CH_2$$

8-Chloro-1-octene

EXAMPLE 13.2   NAMING AN ALKENE

Problem: Write the name of the following alkene.

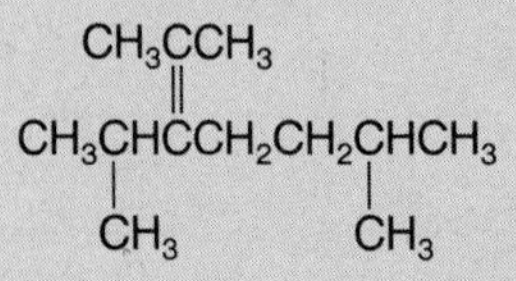

$$CH_3\overset{CH_3CCH_3}{\underset{\underset{CH_3}{|}}{CH_3CHCCH_2CH_2}}\overset{}{\underset{\underset{CH_3}{|}}{CHCH_3}}$$

[1] Propene, not 1-propene, because by rule 3 there is no 2-propene.
[2] 2-Methylpropene, not 2-methyl-1-propene, because the 1 isn't needed.
[3] We don't include a number for the double bond's location, because by rule 4 it can only be at position 1.

Solution: First, identify the longest chain *that includes the double bond,* and number it to let one of the carbons of the double bond have the lower number.

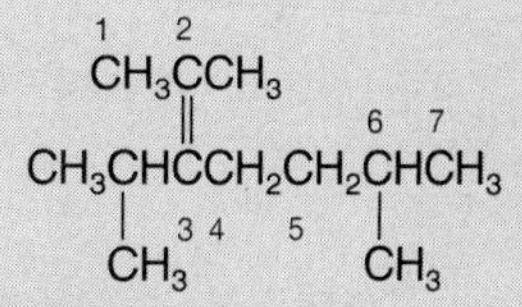

$$
\begin{array}{c}
\overset{1}{C}H_3\overset{2}{C}CH_3 \\
\| \\
CH_3\overset{}{C}HCCH_2CH_2\overset{6}{C}H\overset{7}{C}H_3 \\
\mid \quad {}^{3}{}^{4}\quad {}^{5}\quad \mid \\
CH_3 \qquad\qquad CH_3
\end{array}
$$

We can see that the parent alkene is 2-heptene. It holds two methyl groups (positions 2 and 6) and one isopropyl group (position 3). The names and location numbers are assembled as follows:

2,6-Dimethyl-3-isopropyl-2-heptene

A comma   Hyphens separate
separates   numbers from names
two numbers

PRACTICE EXERCISE 2   Write the IUPAC names for the following compounds. (Ignore cis or trans designations.)

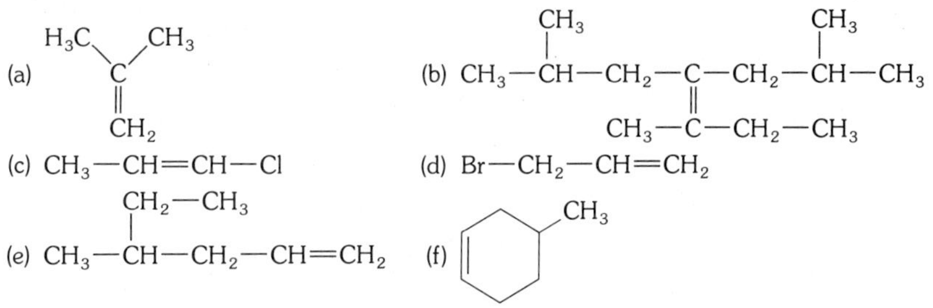

(a)
$$
\begin{array}{c}
H_3C \qquad CH_3 \\
\diagdown\ \diagup \\
C \\
\| \\
CH_2
\end{array}
$$

(b)
$$
\begin{array}{c}
CH_3 \qquad\qquad\qquad CH_3 \\
\mid \qquad\qquad\qquad\quad \mid \\
CH_3-CH-CH_2-C-CH_2-CH-CH_3 \\
\| \\
CH_3-C-CH_2-CH_3
\end{array}
$$

(c) $CH_3-CH{=}CH-Cl$

(d) $Br-CH_2-CH{=}CH_2$

(e)
$$
\begin{array}{c}
CH_2-CH_3 \\
\mid \\
CH_3-CH-CH_2-CH{=}CH_2
\end{array}
$$

(f)

PRACTICE EXERCISE 3   Write condensed structures for each of the following.

(a) 4-Methyl-2-pentene
(b) 3-Propyl-1-heptene
(c) 3,3-Dimethyl-4-chloro-1-butene
(d) 2,3-Dimethyl-2-butene

When a compound has two double bonds, it is named as a *diene* with two numbers in the name that specify the locations of the double bonds. For example,

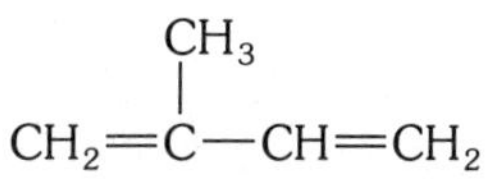

$$
\begin{array}{c}
CH_3 \\
\mid \\
CH_2{=}C-CH{=}CH_2
\end{array}
$$

2-Methyl-1,3-butadiene

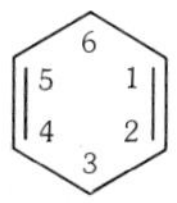

1,4-Cyclohexadiene

This pattern can be easily extended to *trienes, tetraenes,* and so forth.

13.3   ADDITION REACTIONS OF THE CARBON–CARBON DOUBLE BOND

The carbon–carbon double bond adds H_2, Cl_2, Br_2, HX, H_2SO_4, and H_2O, and it is attacked by strong oxidizing agents, including ozone.

The **addition reaction** is the most typical reaction of a carbon–carbon double bond. In this reaction, the pieces of some reactant molecule become attached at opposite ends of the double bond as the double bond changes to a single bond. Thus all additions to the double bond have the following features:

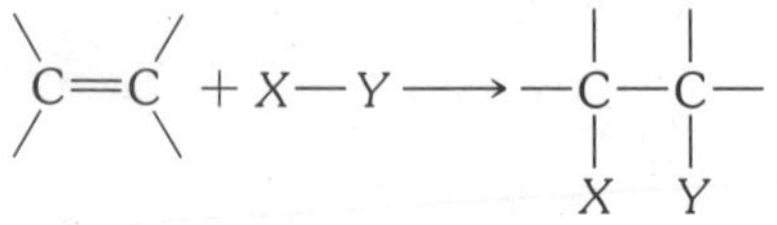

$$\overset{\diagdown}{\underset{\diagup}{}}C=C\overset{\diagup}{\underset{\diagdown}{}} + X-Y \longrightarrow -\overset{|}{\underset{X}{C}}-\overset{|}{\underset{Y}{C}}-$$

We will study several specific examples of the generalized reactant, $X-Y$, but first we should note that this kind of reaction indicates one bond of a double bond is different from the other. One of the two breaks open, but the other holds. After we have studied the reactions of the double bond, we will study its structural details and explain both its reactivity and cis-trans isomerism.

The Addition of Hydrogen — Hydrogenation.   In the presence of a powdered metal catalyst, and under both pressure and heat, hydrogen adds to double bonds as follows:

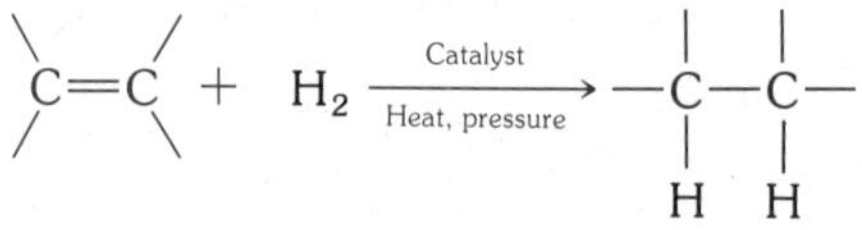

$$\overset{\diagdown}{\underset{\diagup}{}}C=C\overset{\diagup}{\underset{\diagdown}{}} + \ H_2 \ \xrightarrow[\text{Heat, pressure}]{\text{Catalyst}} \ -\overset{|}{\underset{H}{C}}-\overset{|}{\underset{H}{C}}-$$

The addition of hydrogen is often called the reduction of the double bond.

Specific examples are

$$CH_2{=}CH_2 + H_2 \xrightarrow[\text{Heat, pressure}]{\text{Catalyst}} \underset{\text{Ethene}}{CH_2{-}CH_2} \ \text{ or } \ \underset{\text{Ethane}}{CH_3CH_3}$$

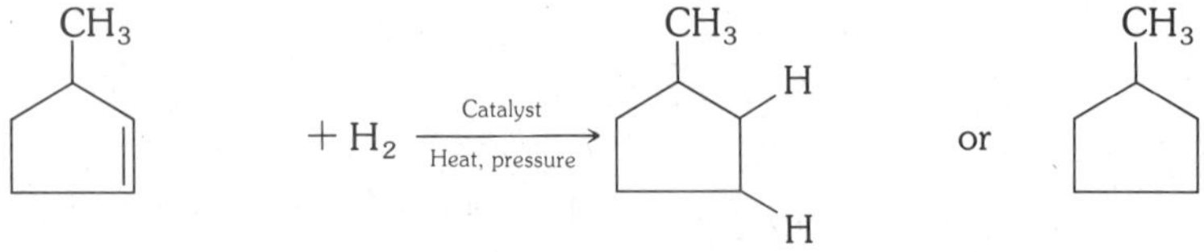

3-Methylcyclopentene   Methylcyclopentane

Although molecular hydrogen is not a reactant in body cells, there are substances in the body that can donate hydrogen to alkene groups, and this kind of reaction is very important at several stages of metabolism.

EXAMPLE 13.3   WRITING THE STRUCTURE OF THE PRODUCT OF THE ADDITION OF HYDROGEN TO A DOUBLE BOND

Problem: Write the structure of the product of the following reaction:

$$\underset{}{CH_3CH_2\overset{\overset{CH_3}{|}}{C}{=}\overset{\overset{CH_3}{|}}{C}CH_2CH_2CH_3} + H_2 \xrightarrow[\text{Heat, pressure}]{\text{Catalyst}} \ ?$$

Solution: The only change occurs at the double bond, and all of the rest of the structure goes through the reaction unchanged. *This is true of all of the addition reactions we will study.* Therefore copy the structure of the alkene just as it is, except leave only a single bond where the double bond was. Then increase by one the number of hydrogens at each carbon of the original double bond. The structure of the product can be written as

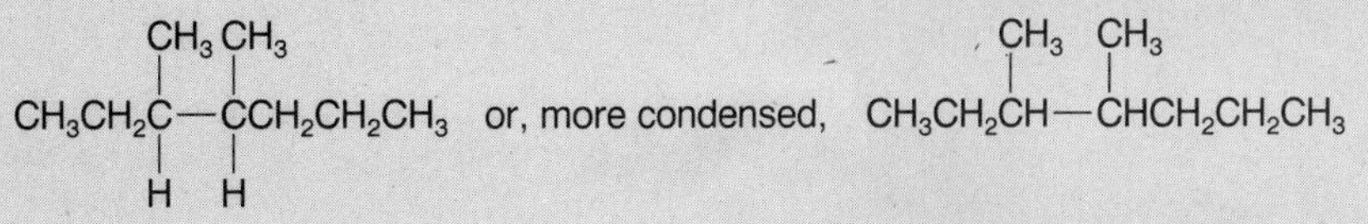

$$CH_3CH_2\overset{\overset{\displaystyle CH_3}{|}}{\underset{\underset{\displaystyle H}{|}}{C}}-\overset{\overset{\displaystyle CH_3}{|}}{\underset{\underset{\displaystyle H}{|}}{C}}CH_2CH_2CH_3 \quad \text{or, more condensed,} \quad CH_3CH_2\overset{\overset{\displaystyle CH_3}{|}}{CH}-\overset{\overset{\displaystyle CH_3}{|}}{CH}CH_2CH_2CH_3$$

Study Example 13.3 very carefully. It illustrates the strategy we will use throughout all of the chapters on organic chemistry. We need a studying strategy, because there are too many *specific* reactions involving particular compounds to memorize as individual "things-in-themselves." Yet, one of the goals of our study of organic reactions is to be able to predict the outcomes of specific reactions. The time-tested way to handle this problem is to focus the memory work on the *types* of reactions and then apply this knowledge to figure out a specific example. The *types* of reactions, such as hydrogenation, not their specific illustrations, constitute the fundamental chemical properties of functional groups—the "map signs" that each functional group stands for. We have just learned one such "map sign" for the carbon-carbon double bond. This functional group can be made to add hydrogen, and when it does, the double bond becomes a single bond, and each carbon of this functional group picks up another hydrogen atom. This statement of chemical fact is what has to be learned, but learned more by working out specific examples than straight memorizing. Remember, there will be no point in learning the chemical facts about organic substances unless they can be applied among biochemical substances later on.

PRACTICE EXERCISE 4   Write the structures of the products, if any, of the following.

(a) $CH_3CH{=}CH_2 + H_2 \xrightarrow[\text{Heat, pressure}]{\text{Catalyst}}$

(b) $CH_3CH_2CH_3 + H_2 \xrightarrow[\text{Heat, pressure}]{\text{Catalyst}}$

(c) 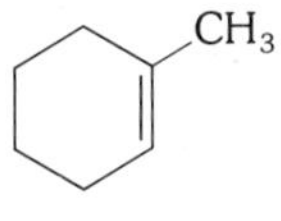 $+ H_2 \xrightarrow[\text{Heat, pressure}]{\text{Catalyst}}$

(d) $CH_3(CH_2)_7CH{=}CH(CH_2)_7CO_2H + H_2 \xrightarrow[\text{Heat, pressure}]{\text{Catalyst}}$

The Addition of Chlorine or Bromine—Chlorination or Bromination. Both chlorine, Cl_2, and bromine, Br_2, add rapidly to the carbon–carbon double bond at room temperature without the need of any catalyst. Iodine does not add, and fluorine reacts explosively with almost any organic compound to give a mixture of products.

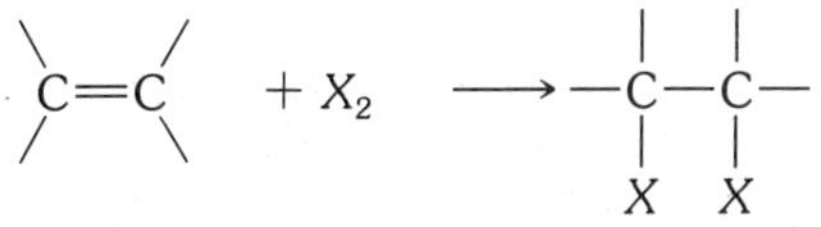

$$\overset{\diagdown}{\diagup}C{=}C\overset{\diagup}{\diagdown} + X_2 \longrightarrow -\overset{|}{\underset{|}{C}}-\overset{|}{\underset{|}{C}}-$$
$$\qquad\qquad\qquad X\;\;X$$

$$X = Cl \text{ or } Br$$

Specific examples are

$$CH_3CH{=}CH_2 + Br_2 \longrightarrow CH_3\underset{\underset{Br}{|}}{CH}-\underset{\underset{Br}{|}}{CH_2}$$

Propene   1,2-Dibromopropane

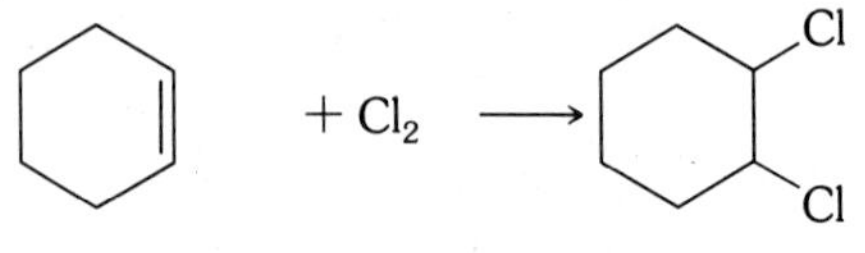

Cyclohexene $+ Cl_2 \longrightarrow$ 1,2-Dichlorocyclohexane

EXAMPLE 13.4   WRITING THE STRUCTURE OF THE PRODUCT OF THE ADDITION OF CHLORINE OR BROMINE TO A CARBON–CARBON DOUBLE BOND

Problem: What compound forms in the following situation?

$$CH_3{-}CH{=}CH{-}CH_3 + Br_2 \longrightarrow ?$$

Solution: This problem is very similar to that of the addition of hydrogen. We rewrite the alkene except that a single bond is left where the double bond was.

$$CH_3{-}CH{-}CH{-}CH_3 \quad \text{(Incomplete)}$$

Now we attach a bromine atom by a single bond to each carbon of the original double bond, and we have the answer.

$$CH_3{-}\underset{\underset{Br}{|}}{CH}-\underset{\underset{Br}{|}}{CH}{-}CH_3 \quad \text{2,3-Dibromobutane}$$

PRACTICE EXERCISE 5   Complete the following equations by writing the structures of the products. If no reaction occurs, write "no reaction."

(a) $CH_3{-}\underset{\underset{CH_3}{|}}{C}{=}CH_2 + Br_2 \longrightarrow$

(b) $CH_3CH_2CH_2CH_3 + Cl_2 \longrightarrow$

(c) $CH_2{=}CHCH_2CH_3 + Cl_2 \longrightarrow$

(d) $CH_3{-}CH{=}CH{-}CH_3 + H_2 \xrightarrow[\text{Heat, pressure}]{\text{Catalyst}}$

Use a good fume hood and protective gloves when dispensing bromine.

In the lab, the addition of bromine is easier to carry out than that of chlorine, because bromine is a liquid and chlorine is a gas. Handling a gas requires special equipment, but liquids can be poured. Both elements are dangerous, however. Pure bromine causes painful skin burns. To reduce the hazard of working with it, a dilute solution of bromine in a solvent such as dichloromethane or carbon tetrachloride is used.

A change in color occurs as bromine adds to a double bond. Bromine is dark brown, but the dibromoalkanes are virtually colorless. Therefore an unknown organic compound that rapidly decolorizes bromine at room temperature quite likely has a carbon–carbon double bond. The bromine color must disappear rapidly *without being accompanied by the evolution of hydrogen bromide gas*. If this happens, the decolorization was caused by a substitution, not an addition reaction. Remember that alkanes react (when heated or exposed to UV light) with chlorine and bromine, too, but their reaction generates HX. For example,

$$CH_3CH_2CH_2CH_2CH_2CH_3 + Br_2 \longrightarrow$$
$$CH_3CH_2CH_2CH_2\underset{\underset{Br}{|}}{C}HCH_3 + \text{other products} + HBr$$

HBr(g) also generates whitish fumes when it contacts humid air.

It's easy to test to see if HBr evolves; it will turn moist blue litmus paper red when a strip is held in the mouth of the test tube. This substitution reaction does not occur unless sunlight is absorbed by the mixture of the alkane and the bromine, so in the lab, if the reaction occurs at all, it doesn't happen instantaneously.

The Addition of Hydrogen Chloride, Hydrogen Bromide, and Sulfuric Acid. The carbon-carbon double bond readily adds gaseous H—Cl or H—Br and concentrated sulfuric acid, H—OSO$_3$H (which is the same as H$_2$SO$_4$). The pattern is the same in all these additions.

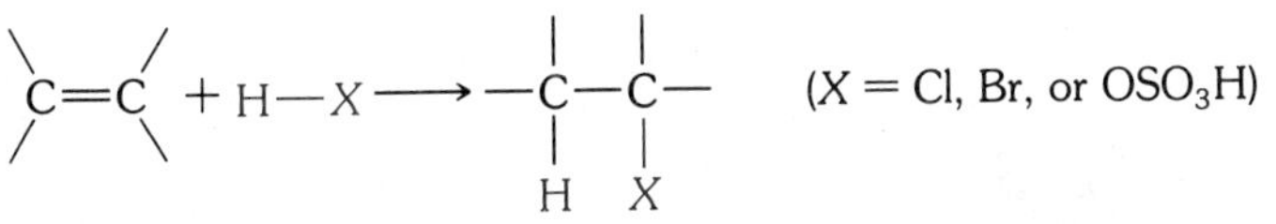

$$\overset{\backslash\quad/}{\underset{/\quad\backslash}{C=C}} + H-X \longrightarrow -\overset{|\quad|}{\underset{\underset{H\quad X}{|\quad|}}{C-C}}- \qquad (X = Cl, Br, or OSO_3H)$$

The reactants bring in a new complication, because their molecules are not symmetrical, like H—H or Br—Br or Cl—Cl. The two portions of H—Cl, H—Br, or H—OSO$_3$H that add are not identical, so when the double bond is also unsymmetrical, the structure of the product depends on which way the addition occurs. The term *unsymmetrical* has a very restricted meaning in this context. An *unsymmetrical double bond* is one whose two carbon atoms hold unequal numbers of hydrogen atoms. For example, 2-butene, CH$_3$CH=CHCH$_3$, has a symmetrical double bond, but 1-butene, CH$_2$=CHCH$_2$CH$_3$, and propene, CH$_3$CH=CH$_2$, have unsymmetrical double bonds. Thus H—Cl might possibly add in either of the two following ways to propene, or possibly both modes of addition might occur to give a mixture of products.

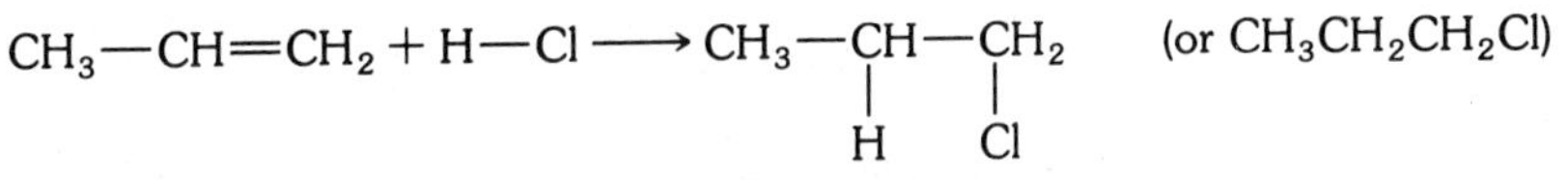

$$CH_3-CH=CH_2 + H-Cl \longrightarrow CH_3-\overset{|}{\underset{\underset{H}{|}}{C}}H-\overset{|}{\underset{\underset{Cl}{|}}{C}}H_2 \qquad (or\ CH_3CH_2CH_2Cl)$$

Propene · · · 1-Chloropropane (very little forms)

or

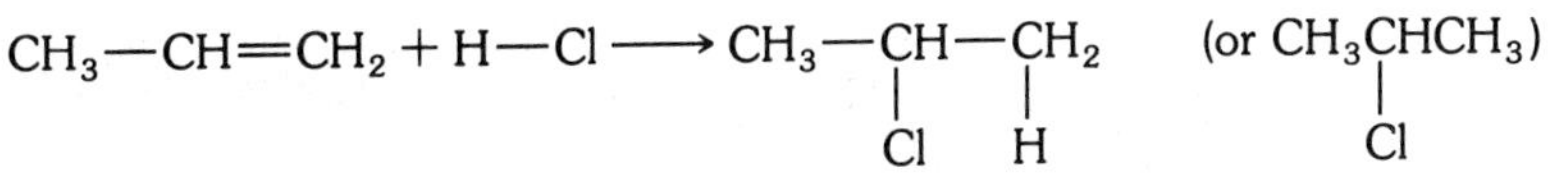

$$CH_3-CH=CH_2 + H-Cl \longrightarrow CH_3-\overset{|}{\underset{\underset{Cl}{|}}{C}}H-\overset{|}{\underset{\underset{H}{|}}{C}}H_2 \qquad (or\ CH_3CHCH_3)\underset{\ \ \ Cl}{}$$

Propene · · · 2-Chloropropane (the major product)

The actual product is largely 2-chloropropane, and very little of its isomer, 1-chloropropane, forms. In other words, the reactant, H—Cl, adds to the unsymmetrical double bond selectively.

Vladimer Markovnikov (1838–1904), a Russian chemist, was the first to notice that unsymmetrical alkenes add unsymmetrical reactants in one direction, and his work is honored by a rule named for him — **Markovnikov's rule.**[4]

"Them that has, gits" applies here, too.

> **Markovnikov's Rule** When an unsymmetrical reactant of the type H—G adds to an unsymmetrical alkene, the carbon with the greater number of hydrogens gets one more hydrogen.

The following examples illustrate Markovnikov's rule in action.

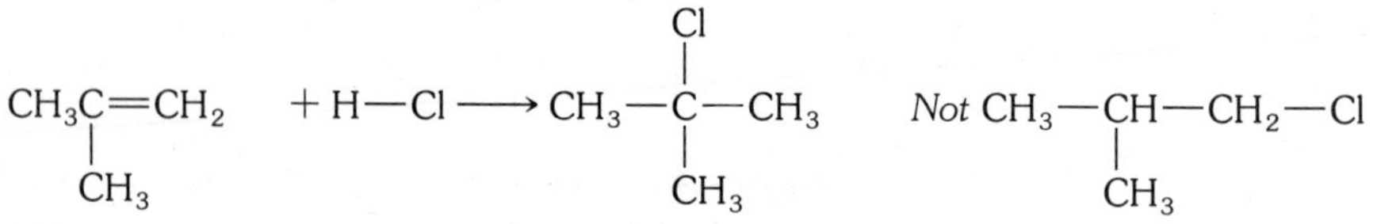

$$CH_3C{=}CH_2 \quad + H{-}Cl \longrightarrow CH_3{-}\underset{CH_3}{\overset{Cl}{\underset{|}{\overset{|}{C}}}}{-}CH_3 \qquad Not\ CH_3{-}\underset{CH_3}{\overset{|}{CH}}{-}CH_2{-}Cl$$

2-Methylpropene   *t*-Butyl chloride

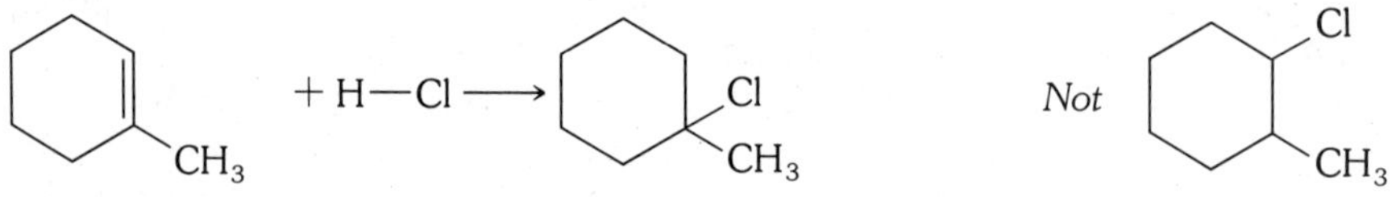

1-Methylcyclohexene   + H—Cl ⟶   1-Chloro-1-methylcyclohexane   *Not*

When an alkene is mixed with concentrated sulfuric acid, which must certainly be regarded as a very polar solvent, the hydrocarbon dissolves and heat evolves. An alkane does not behave this way at all but merely forms a separate layer that floats on the sulfuric acid. (See Figure 13.2.) The alkene dissolves because it reacts by an addition reaction to form an alkyl hydrogen sulfate. For example,

The sodium salts of long-chain alkyl hydrogen sulfates are detergents, for example, $CH_3(CH_2)_{11}OSO_3Na$.

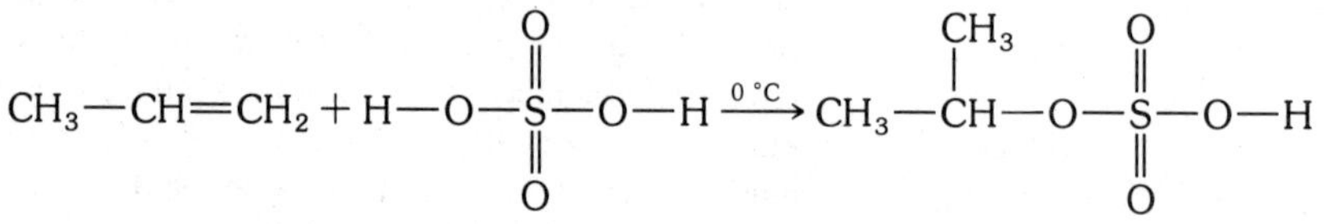

$$CH_3{-}CH{=}CH_2 + H{-}O{-}\overset{O}{\underset{O}{\overset{\|}{\underset{\|}{S}}}}{-}O{-}H \xrightarrow{0\ °C} CH_3{-}\underset{}{\overset{CH_3}{\overset{|}{CH}}}{-}O{-}\overset{O}{\underset{O}{\overset{\|}{\underset{\|}{S}}}}{-}O{-}H$$

Propene   Sulfuric acid   Isopropyl hydrogen sulfate

Alkyl hydrogen sulfates are also very polar compounds, so as their molecules form, they move smoothly into the polar sulfuric acid layer. The heat generated by the reaction and the strongly acidic nature of the mixture causes side reactions to occur that generate black by-products. All of these properties make concentrated sulfuric acid another test reagent to distinguish between an alkene and an alkane. As the photos in Figure 13.2 show, the alkene turns black but the alkane is unaffected by the presence of concentrated sulfuric acid.

When the carbon atoms at the double bond have identical numbers of hydrogen atoms, Markovnikov's rule cannot be applied. Unsymmetrical reactants add in both of the two possible directions, and a mixture of products forms. For example,

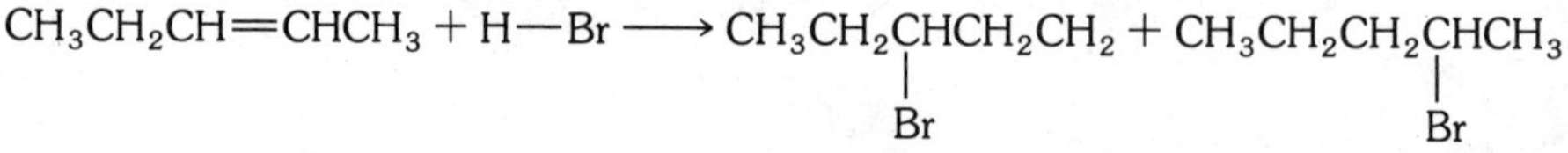

$$CH_3CH_2CH{=}CHCH_3 + H{-}Br \longrightarrow CH_3CH_2\underset{Br}{\overset{|}{CH}}CH_2CH_2 + CH_3CH_2CH_2\underset{Br}{\overset{|}{CH}}CH_3$$

2-Pentene   3-Bromopentane   2-Bromopentane

---

[4] When hydrogen bromide is used, it is important that no peroxides, compounds of the type R—O—O—H or R—O—O—R, be present. Traces of peroxides commonly form in organic liquids that are stored in contact with air for long periods. When peroxides are present, the addition of H—Br occurs in the direction opposite to that predicted by Markovnikov's rule. Peroxides catalyze this anti-Markovnikov addition only of H—Br, not of H—Cl or H—OSO₃H. Having mentioned this complication, we will do nothing more with it.

FIGURE 13.2
The effect of concentrated sulfuric acid on an alkane and an alkene. In each photo, the tube with the alkene is on the left and the tube with the alkane is on the right. Both are clear liquids *(left),* but soon after concentrated sulfuric acid was added to both, the alkene turned black *(right).* The alkane floats unaffected on the surface of the acid.

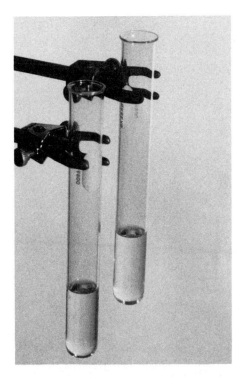
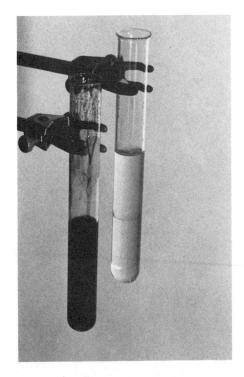

(For a reminder of why we don't try to balance an equation such as this, check back to Special Topic 12.2.)

The Addition of Water.   Water adds to the carbon – carbon double bond but only if an acid catalyst (or the appropriate enzyme) is present. The product is an alcohol. Water alone, or aqueous bases have no effect on alkenes whatsoever.

$$\underset{\text{Alkene}}{\diagdown\!\!C\!=\!\!C\diagup} + H\!-\!OH \xrightarrow[\text{Heat}]{H^+} \underset{\text{Alcohol}}{-\overset{\displaystyle|}{\underset{\displaystyle H}{C}}\!-\!\overset{\displaystyle|}{\underset{\displaystyle OH}{C}}\!-}$$

Specific examples are:

$$CH_2\!=\!CH_2 + H\!-\!OH \xrightarrow[\substack{240\,°C \\ \text{(closed vessel)}}]{10\%\ H_2SO_4} CH_3\!-\!CH_2\!-\!OH$$

Ethene   Ethyl alcohol

$$CH_3\!-\!\overset{\displaystyle \overset{CH_3}{|}}{C}\!=\!CH_2 + H\!-\!OH \xrightarrow[25\,°C]{10\%\ H_2SO_4} CH_3\!-\!\overset{\displaystyle \overset{CH_3}{|}}{\underset{\displaystyle \underset{OH}{|}}{C}}\!-\!CH_3 \quad \textit{Not } CH_3\!-\!\overset{\displaystyle \overset{CH_3}{|}}{CH}\!-\!CH_2\!-\!OH$$

2-Methyl   *t*-Butyl alcohol   (very little forms)
propene   (major product)

As you can see, Markovnikov's rule applies to this reaction, too. One H in H—OH goes to the carbon with the greater number of hydrogens, and the —OH goes to the other carbon of the double bond. Notice in the last example, and in all previous examples of addition reactions, that the carbon skeleton does not change. Although this is not always true, it will be in all of the examples we will use as well as in all of the Practice and Review Exercises.

EXAMPLE 13.5   USING MARKOVNIKOV'S RULE

Problem: What product forms in the following situation?

$$CH_3-CH=CH_2 + H_2O \xrightarrow[\text{Heat}]{H^+} ?$$

Solution: As in all the addition reactions we are studying, the carbon skeleton can be copied over intact, except that a single bond is shown where the double bond was.

$$CH_3-CH-CH_2 \quad \text{(Incomplete)}$$

To decide which carbon of the original double bond gets the H atom from H—OH, we use Markovnikov's rule—the H atom has to go to the CH_2 end. The —OH unit from H—OH goes to the other carbon. The structure of the product is

$$CH_3-CH-CH_2 \quad \text{or, more condensed,} \quad CH_3CHCH_3$$
$$\;\;\;\;\;\;\; | \;\;\; | \; |$$
$$\;\;\;\;\;\;\; OH \;\; H \; OH$$
Isopropyl alcohol

Be sure at this stage to see if each carbon in the structure of the product has four bonds. If not, you can be certain that some mistake has been made. This is always a useful way to avoid at least some of the common mistakes made in solving a problem such as this.

PRACTICE EXERCISE 6   Write structures for the product(s), if any, that would form under the conditions shown. If no reaction occurs, write "no reaction."

(a) $CH_2=CHCH_2CH_3 + HCl \longrightarrow$

(b) $CH_2=C-CH_3 + HBr \longrightarrow$
$\;\;\;\;\;\;\;\;\; |$
$\;\;\;\;\;\;\;\;\; CH_3$

(c) $CH_3-CH=C$ $+ H_2O \xrightarrow[\text{Heat}]{H^+}$

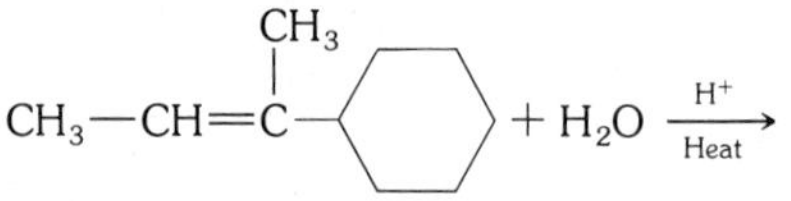

(d) 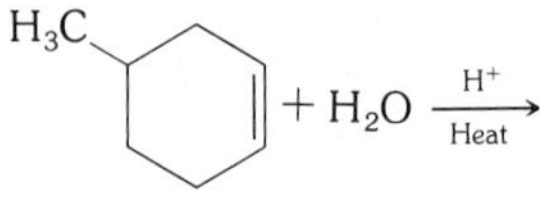 $+ H_2O \xrightarrow[\text{Heat}]{H^+}$ (What *mixture* forms?)

(e) 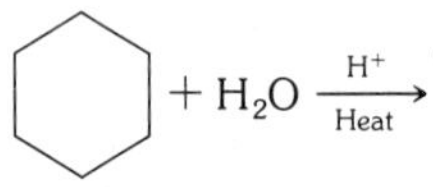 $+ H_2O \xrightarrow[\text{Heat}]{H^+}$

The Oxidation of the Carbon–Carbon Double Bond.   Hot solutions of potassium permanganate ($KMnO_4$) and potassium dichromate ($K_2Cr_2O_7$) vigorously oxidize molecules at carbon–carbon double bonds. The reaction begins as an addition reaction, but continues beyond to the point where the molecule is split apart. We will not study any of the details or learn how to predict products. We just have to take note of the fact that at the carbon–carbon double bond a molecule is susceptible to strong oxidizing agents. The products can be ketones, carboxylic acids, carbon dioxide, or mixtures of these.

The permanganate ion is intensely purple in water, and as it oxidizes double bonds it changes to manganese dioxide, MnO_2, a brownish, sludge-like, insoluble solid. In other words, an easily observed change occurs that makes of a potassium permanganate solution still another reagent for distinguishing between an alkane and an alkene. An alkane gives no reaction with permanganate ion, but when an alkene is stirred with aqueous permanganate, the purple color gives way to the brownish precipitate.

The dichromate ion is bright orange in water, and when it acts as an oxidizing agent, it changes to the bright green chromium(III) ion, Cr^{3+}. Hence, aqueous potassium (or sodium) dichromate also can be used as a test reagent to find out if a substance has an easily oxidized group.

So far we have studied four tests for distinguishing an alkane and an alkene — the bromine test, the concentrated sulfuric acid test, and the use of aqueous permanganate or aqueous dichromate.

Ozone, O_3, a pollutant in smog, is dangerous because it is a very powerful oxidizing agent that attacks biochemicals wherever they have carbon–carbon double bonds. Because such bonds occur in the molecules of all cell membranes, you can see that exposure to ozone must be kept very low. Even a concentration in air of only 1 part per million parts (1 ppm) warrants the declaration of a smog emergency condition.

Ozone destroys any vegetation that has the green pigment, chlorophyll, because chlorophyll contains double bonds.

13.4 HOW ADDITION REACTIONS OCCUR

The carbon–carbon double bond is a site that can accept a proton from an acid and change into a carbocation.

Several of the reactants that add to a double bond are proton donors — HCl, HBr, $HOSO_3H$, for example. Even in the addition of water, the initial attack is by the acid catalyst, not the water molecule. The double bond is a proton acceptor. One of its pairs of electrons can pivot away from one carbon and form an electron-pair bond to a proton. Let's use the symbol H—G to represent any proton donor. (In H—Cl, G is Cl. In H_2SO_4, G is OSO_3H. And in dilute acid, the proton-donating species is H_3O^+, so G here is H_2O.) Using this general symbol, we can write the first step in an addition reaction as follows:

What once were two electrons in a double bond are now the electrons of this bond.

$$CH_3-CH{=}CH_2 + H-\overset{..}{G}: \longrightarrow CH_3-\overset{+}{C}H-CH_2-H + :\overset{..}{\underset{..}{G}}:^-$$

Propene   Isopropyl carbocation

The curved arrows show how one of the two electron pairs of the double bond in propene swings away from the central carbon to strike the H end of the proton-donor and remove the proton. The electrons, of course, do not move through space. This switching occurs during a collision between the two reactants. The organic cation that forms is an example of a **carbocation,** a species in which a carbon atom has only a sextet of electrons, not an octet. The isopropyl carbocation, not the propyl carbocation, $CH_3CH_2CH_2^+$ is the one that forms — for a reason we will soon study.

Carbocations are particularly unstable cations. All of their reactions are geared to the recovery of an outer octet of electrons for carbon. The instant that a carbocation forms, it becomes very attractive to an electron-rich species, and just such particles emerge when H—Cl, H—Br, or H—OSO_3H attack a double bond. When these acids give up protons, $:\overset{..}{\underset{..}{Cl}}:^-$, $:\overset{..}{\underset{..}{Br}}:^-$, or $^-:\overset{..}{O}SO_3H$ remain. As we said, we can use $:\overset{..}{\underset{..}{G}}:^-$ to represent all of them.

The newly formed carbocation reacts with $:\ddot{G}:^-$ in the next step of the addition.

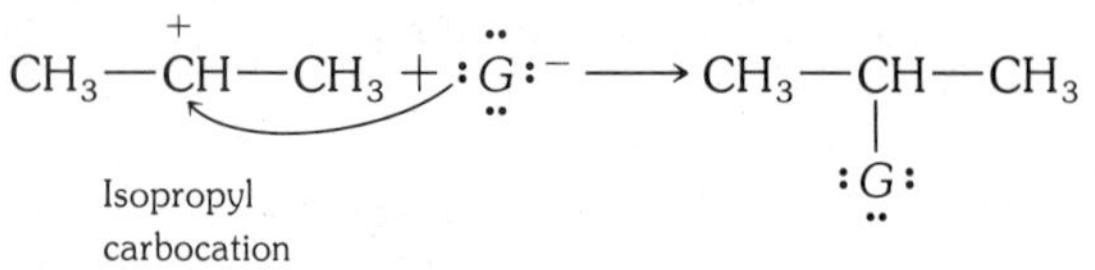

$$CH_3-\overset{+}{C}H-CH_3 + :\ddot{G}:^- \longrightarrow CH_3-\underset{\underset{\ddot{\,}}{\overset{|}{:\ddot{G}:}}}{C}H-CH_3$$

Isopropyl
carbocation

This restores the octet to carbon.

The question now is: Why does the isopropyl and not the propyl carbocation form? The answer would explain Markovnikov's rule. Unstable though any carbocation is, it helps if its positively charged center is surrounded as much as possible by neighboring electron clouds. When the positive charge is on the middle carbon, $CH_3-\overset{+}{C}H-CH_3$, the electron clouds of *two* alkyl groups crowd around it. When the charge is on the end carbon, $CH_3-CH_2-CH_2^+$, the electron cloud of only one alkyl group is nearby. Hence, this carbocation is less stable, and *it is therefore not the one that forms*. In general, when carbon has to carry a positive charge, however briefly, the order of stability of the carbocations is as follows. And when there is an option, the most stable carbocation is produced.

The R groups are alkyl groups, and they need not be identical.

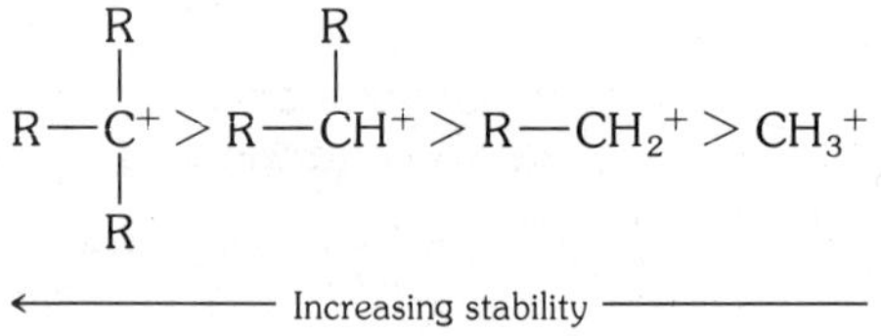

$$\underset{\underset{R}{\overset{|}{R}}}{\overset{\overset{R}{\overset{|}{}}}{R-C^+}} > \underset{\overset{R}{\overset{|}{}}}{R-CH^+} > R-CH_2^+ > CH_3^+$$

$\longleftarrow$ ———— Increasing stability ————— $\longrightarrow$

This order of stability determines the direction of the addition of an unsymmetrical reactant to an unsymmetrical double bond, because it determines which of the two possible carbocations preferably forms in the first step.

When water adds to an alkene under acid catalysis, the first step is a proton transfer from the acid catalyst, not from a water molecule. The water molecule holds its protons much too strongly. This is why the acid catalyst is necessary.

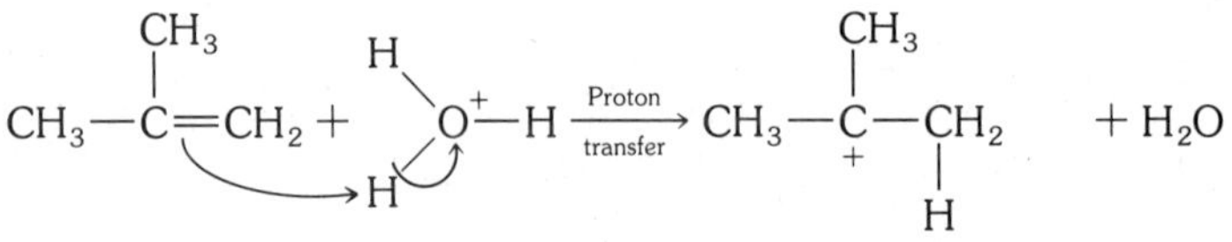

$$CH_3-\underset{\overset{|}{CH_3}}{\overset{\overset{CH_3}{\overset{|}{}}}{C}}=CH_2 + \overset{H}{\underset{H}{\overset{}{O^+}}}-H \xrightarrow[\text{transfer}]{\text{Proton}} CH_3-\underset{\underset{H}{\overset{|}{}}}{\overset{\overset{CH_3}{\overset{|}{}}}{C}}-\overset{+}{C}H_2 \quad + H_2O$$

t-Butyl carbocation
(three alkyl groups on C^+)

The other possible carbocation, the less stable ion that does *not* form, is

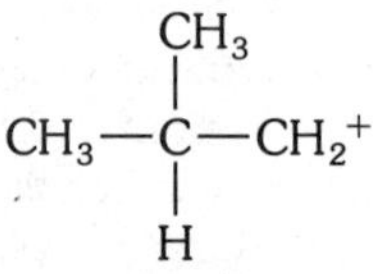

$$CH_3-\underset{\underset{H}{\overset{|}{}}}{\overset{\overset{CH_3}{\overset{|}{}}}{C}}-CH_2^+$$

Isobutyl carbocation
(one alkyl group on C^+)

The *t*-butyl carbocation quickly attracts a water molecule, which has two unshared pairs of electrons on oxygen. One pair now becomes shared as a bond forms to carbon. The product is a hydronium ion in which one H has been replaced by an alkyl group.

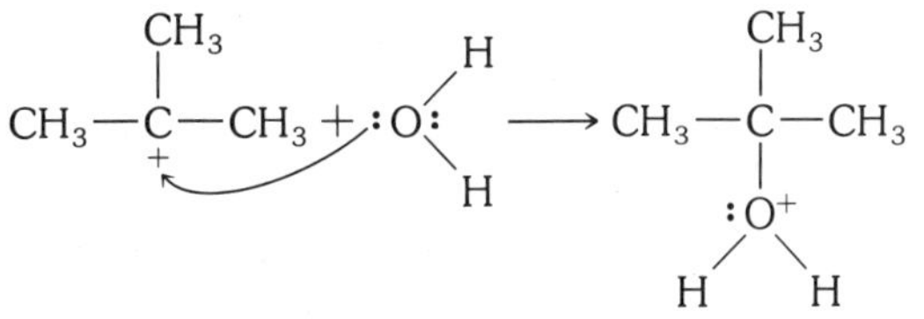

$$\underset{\underset{+}{\overset{\overset{CH_3}{|}}{CH_3-C-CH_3}}}{} + :\overset{\overset{H}{\diagup}}{\underset{\underset{H}{\diagdown}}{O}}: \longrightarrow \underset{\underset{\underset{H\quad H}{:O^+}}{\overset{CH_3}{|}}}{CH_3-C-CH_3}$$

Now the positive charge is on the oxygen atom, but this is acceptable because oxygen also has an octet. In the last step, the catalyst, H_3O^+ is recovered by another proton transfer, this time from the alkyl-substituted hydronium ion.

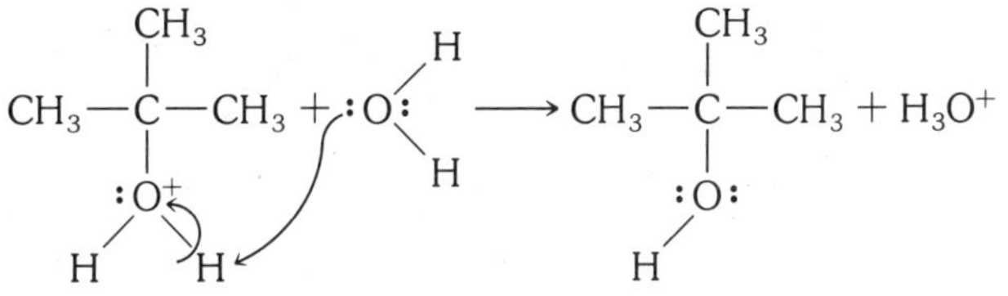

$$\underset{\underset{H\quad H}{:O^+}}{\overset{\overset{CH_3}{|}}{CH_3-C-CH_3}} + :\overset{\overset{H}{\diagup}}{\underset{\underset{H}{\diagdown}}{O}}: \longrightarrow \underset{\underset{H}{:O:}}{\overset{\overset{CH_3}{|}}{CH_3-C-CH_3}} + H_3O^+$$

An alkyl-substituted   *t*-Butyl   Recovered
hydronium ion   alcohol   catalyst

PRACTICE EXERCISE 7   Write the condensed structures for the two carbocations that could conceivably form if a proton became attached to each of the following alkenes. Circle the carbocation that is preferred. If both are reasonable, state that they are. Then write the structures of the alkyl chlorides that would form by the addition of hydrogen chloride to each alkene.

(a) $CH_3-CH_2-CH{=}CH_2$ (b) $CH_3-\overset{\overset{CH_3}{|}}{C}{=}CH_2$ (c)

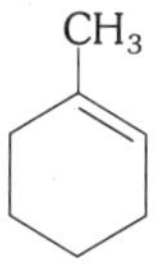

(d) $CH_3-CH{=}CH-CH_3$ (e) $CH_3-CH{=}CH-CH_2-CH_3$

(f) The addition of water to 2-pentene (part e) gives a mixture of alcohols. What are their structures? Why is the formation of a mixture to be expected here but not when water adds to propene?

13.5 THE POLYMERIZATION OF ALKENES

Hundreds to thousands of alkene molecules can join together to make one large molecule of a polymer.

Under a variety of conditions, many hundreds of ethene molecules can reorganize their bonds, join together, and change into one large molecule.

$$nCH_2{=}CH_2 \xrightarrow[\text{Trace of } O_2]{\text{Heat, pressure}} {+}CH_2-CH_2{+}_n \qquad (n = \text{a large number})$$

Ethylene   Polyethylene
(Ethene)   (repeating unit)

''Polymer'' has Greek roots: *poly,* many, and *meros,* parts.

The product is an example of a **polymer,** a substance of very high formula weight whose molecules have repeating structural units. The repeating unit in polyethylene is CH_2-CH_2, and one of these after another is joined together into an extremely long chain. Chain-branching reactions also occur during the formation of the polymer, so the final product includes both straight- and branched-chain molcules. The starting material for making a polymer is called a **monomer,** and the reaction is called **polymerization.**

As we said, a variety of conditions can be used to cause a monomer to polymerize. The catalyst, for example, can be one that generates positively charged intermediates — carbocations. A proton-donor can convert ethylene into the ethyl carbocation:

$$G-H + CH_2=CH_2 \longrightarrow H-CH_2-CH_2^+ + G^-$$

Acid   Ethylene   Ethyl
catalyst   carbocation

This cation then attacks an unchanged molecule of the alkene:

$$CH_3-CH_2^+ + CH_2=CH_2 \longrightarrow CH_3-CH_2-CH_2-CH_2^+$$

But this new and longer carbocation then attacks another molecule of alkene:

$$CH_3-CH_2-CH_2-CH_2^+ + CH_2=CH_2 \longrightarrow CH_3-CH_2-CH_2-CH_2-CH_2-CH_2^+$$

Polymerization is an example of a *chemical chain reaction* because the product of one step initiates the next step.

You can begin to see how this works. Still another carbocation has been produced, and it can attack still another molecule of the alkene. Thus the chain grows with each step by one repeating unit, $-CH_2-CH_2-$, until it picks up some stray anion — an anion from the catalyst, for example — and the chain stops growing. Actually, the catalyst should not be called by this term, because it is consumed. The term *promoter* is better. Some chains grow longer than others, and chain-branching reactions also occur, so the final polymer is a mixture of molecules. All are very large, however, and they all have the same general feature. This is why we can write the structure of a polymer simply by writing its repeating unit, as we did for polyethylene.

When propene (common name, propylene) polymerizes, methyl groups appear on alternate carbons of the main chain.

$$-CH_2-\overset{\overset{\displaystyle CH_3}{|}}{CH}-CH_2-\overset{\overset{\displaystyle CH_3}{|}}{CH}-CH_2-\overset{\overset{\displaystyle CH_3}{|}}{CH}-CH_2-\overset{\overset{\displaystyle CH_3}{|}}{CH}-CH_2-\overset{\overset{\displaystyle CH_3}{|}}{CH}-etc \quad or, \quad \left(CH_2-\overset{\overset{\displaystyle CH_3}{|}}{CH}\right)_n$$

Polypropylene

Polymerizations generally take place in orderly fashions such as this.

Because polypropylene and polyethylene are fundamentally alkanes, they have all of the chemical inertness of this family. These polymers are, therefore, popular raw materials for making containers that must be inert to food juices and to fluids used in medicine — for making refrigerator boxes and bottles, containers for chemicals, sutures, catheters, various drains, and wrappings for aneurysms.

The idea of a polymer is important in our study because many biochemicals are polymers; for example, proteins, some carbohydrates, and the nucleic acids (the chemicals of heredity). However, we must note that polymers are extremely important commercial substances and we encounter them in our daily lives all the time. Table 13.2 contains a list of just a few examples of polymers for which the monomers are substituted alkenes. Dienes also polymerize, and natural rubber is a polymer of a diene called isoprene (Table 13.2).

TABLE 13.2
Some Polymers of Substituted Alkenes

Polymer	Monomer	Uses
Polyvinyl chloride (PVC)	CH_2=CHCl	Bottles and other containers; insulation; plastic pipe
Saran	CH_2=CCl_2 and CH_2=CHCl	Packaging film; fibers; tubing
Teflon	F_2C=CF_2	Nonsticking surfaces for pots and pans
Orlon	CH_2=CH—C≡N	Fabrics
Polystyrene	C_6H_5—CH=CH_2	Foam plastics and molded items
Lucite	CH_2=$\overset{\overset{\textstyle CH_3}{\textstyle \mid}}{C}$—$CO_2CH_3$	Coatings; windows; molded items
Natural Polymer		
Natural rubber	CH_2=$\overset{\overset{\textstyle CH_3}{\textstyle \mid}}{C}$—CH=$CH_2$ (Isoprene)	Tires; hoses; boots

13.6 THE PI BOND

The carbon–carbon double bond consists of one sigma bond and one pi bond.

One of the two bonds at a carbon–carbon double bond is not like the other, and by studying this difference we can learn why cis-trans isomers exist, why alkenes are so readily attacked by oxidizing agents, and why alkenes easily undergo addition reactions that are begun by the transfer of a proton.

***sp²* Hybridization.**   The carbon atoms at the double bond cannot use sp^3 orbitals because they hold three other groups, not four. Different hybrid orbitals are used, and these emerge from the mixing or hybridizing of the $2s$ and just *two* of the $2p$ atomic orbitals, as shown in Figure 13.3. The third $2p$ orbital is left unchanged, but only for a moment. Because one s and two p orbitals are mixed, the new hybrid orbitals are called ***sp²* orbitals.** Each carbon of the double bond develops three of them, and it also has one unhybridized $2p$ orbital. Thus there are four orbitals at each carbon with one electron apiece.

The shapes and the distributions of the sp^2 orbitals at carbon are shown in Figure 13.4. The axes of these orbitals lie in the same plane, and between them are angles of $120°$. The axis of the unhybridized $2p$ orbital is perpendicular to the plane and intersects it at the point where the three sp^2 orbitals intersect.

FIGURE 13.3
sp^2 Hybridization at carbon. *(a)* The empty atomic orbitals of carbon at main energy levels 1 and 2. *(b)* The new, but still empty orbitals after sp^2 hybridization of the atomic orbitals has occurred. *(c)* The valence state of sp^2 hybridized carbon.

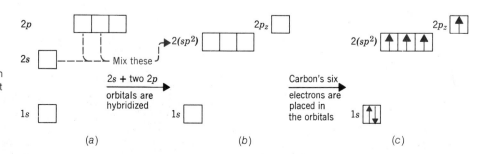

FIGURE 13.4
The shapes and arrangements of the orbitals at carbon when it is sp^2 hybridized. (a) Cross section of one sp^2 hybrid orbital. (b) The axes of the three sp^2 hybrid orbitals lie in the same plane and intersect at an angle of 120°. (c) A perspective view of the three sp^2 hybrid orbitals plus the unhybridized $2p$ orbital at carbon. (The smaller lobes of the sp^2 orbitals are omitted here.)

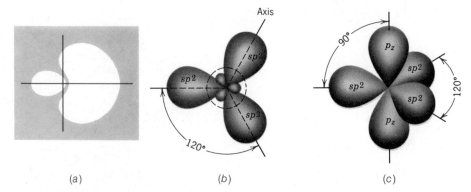

(a)   (b)   (c)

Figure 13.5 illustrates how a molecular orbital forms from the overlapping of two sp^2 orbitals, one from each carbon at the double bond. Two electrons are in this molecular orbital and are shared by the two carbons. The result is a sigma bond and, as with any ordinary single, covalent bond, we would expect free rotation about it. However, at one stopping point during such a rotation, the two $2p$ orbitals become perfectly aligned, side by side. They are so close to each other that they now overlap *side by side* to create a new kind of molecular orbital, one with two banana-shaped halves, as shown in Figure 13.6. This orbital now holds another shared pair of electrons, and it locks free rotation out of the system. To continue with free rotation, the bond generated by this molecular orbital would have to break, and this costs too much energy. Thus the formation of this second bond, called a **pi bond,** or **π-bond,** explains why there is no free rotation at a double bond and why cis-trans isomers of alkenes are possible.

The electrons in the pi bond, sometimes called the **pi electrons,** are not concentrated as much *between* the carbon nuclei as those in the sigma bond. The pi electrons are a little farther away; they are perched above and below the plane in which the carbon nuclei are found; therefore they are more open to attack by anything with an electron-seeking nature — protons, for example, or oxidizing agents. The pi electrons are what give alkenes such a rich chemistry compared with the alkanes.

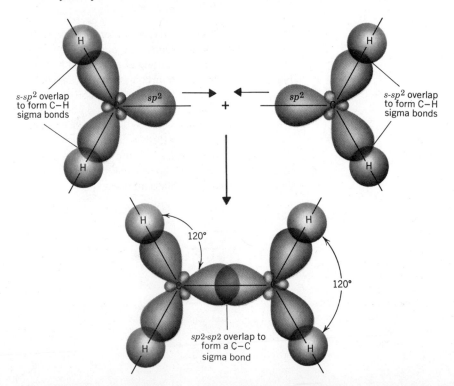

FIGURE 13.5
The sigma bond network is ethene. (The 2p orbital at each carbon is omitted.)

FIGURE 13.6

Formation of the pi bond in ethene. *(a)* This is the situation after the sigma bonds have formed but before the 2*p* orbitals have overlapped. *(b)* This shows the side-to-side overlap of the 2*p* orbitals that creates the pi bond. *(c)* One sigma bond (heavy line) and one pi bond (with two banana-shaped halves) are the two bonds of the double bond. *(d)* This is an end-on view of the pi bond.

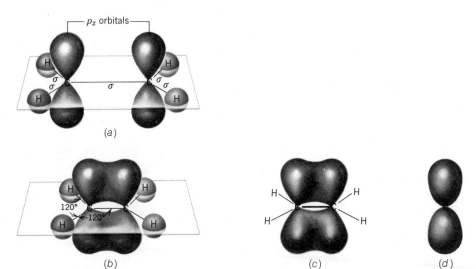

(a)

(b)   (c)   (d)

The triple bond of an alkyne has two pi bonds and a sigma bond. Special Topic 13.1 discusses this further.

13.7 AROMATIC COMPOUNDS

The benzene ring undergoes substitution reactions instead of addition reactions despite a high degree of unsaturation.

Until well into this century, the structure of benzene was a problem. Its formula, C_6H_6, indicates that it is quite unsaturated, because the ratio of hydrogens to carbons is much lower than in hexane, C_6H_{14} or in cyclohexane, C_6H_{12}. And, in fact, benzene does have a reaction that unsaturated compounds give. It adds hydrogen, and the product is cyclohexane. However, extremely rigorous conditions of pressure and temperature are necessary.

$$C_6H_6 + 3H_2 \xrightarrow[\substack{\text{High pressure and} \\ \text{temperature}}]{\text{Catalyst}} C_6H_{12}$$

Benzene   Cyclohexane

Because cyclohexane forms, the six carbons of a benzene molecule must also form a six-membered ring, but now the locations of the hydrogens become a problem. It is well-known that all six of benzene's hydrogens are equivalent. For example, it is possible to replace one by Cl to make chlorobenzene, C_6H_5Cl, and no isomeric monochlorobenzenes form. In other words, it doesn't matter which of the six hydrogens in benzene is replaced; the same chlorobenzene forms. Therefore it seems most reasonable to put one H on each of the ring carbon atoms.

The older structure, **2**, is still widely used to represent benzene, although it is usually abbreviated further:

1
(Incomplete)

2
(Older structure for benzene)

The trouble with the incomplete structure identified as **1** is that each carbon has only three bonds from it, not four.

The triple bond forms from a series of steps very similar to those of the double bond. The $2s$ orbital of carbon and just one of its $2p$ orbitals mix or hybridize, which leaves two of the $2p$ orbitals of carbon unchanged. See the accompanying figure below.

The new hybrid orbitals are called sp hybrid orbitals. In shape they are similar to sp^2 and sp^3 hybrid orbitals, but their axes point in exactly opposite directions, not at angles of 109.5° or 120°. Their axes are also at right angles to the axes of the $2p$ orbitals, as seen in the second series of drawings below. In this series, we see how the sigma bonds

form in the simplest member of the alkyne family, ethyne (common name, acetylene).

In the last drawing below, the two pi orbitals are indicated. They result from the side-to-side overlapping of the $2p$ orbitals. Because the triple bond has pi bonds like the double bond, alkynes give the same kinds of addition reactions as alkenes. Some examples are the following, and notice that the triple bond can add *two* molecules of a reactant. Usually, it is possible to control the reaction so that only one molecule adds.

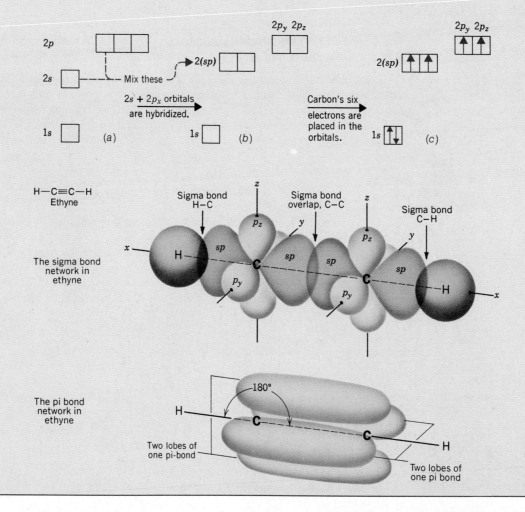

$$CH_3C\equiv CH + H_2 \xrightarrow[\text{Heat, pressure}]{\text{Special catalyst}} CH_3CH=CH_2 \xrightarrow{\text{more } H_2} CH_3CH_2CH_3$$

Propyne   Propene   Propane

2-Chloropropene   2,2-Dichloropropane

H—C≡C—H
Ethyne

The sigma bond network in ethyne

The pi bond network in ethyne

For decades, scientists made carbon have four bonds in the structure of benzene simply by writing in three double bonds, as seen in structure **2**. The difficulty with this, however, is it says that benzene is a triene, a substance with three alkene groups per molecule. Trienes are known compounds, and they have the same kinds of reactions as mono-enes (e.g., ethene) and dienes. Trienes react readily at room temperature with oxidizing agents, and they easily add bromine, chlorine, and sulfuric acid. In sharp and dramatic contrast, benzene gives no reaction whatsoever at room temperature with concentrated sulfuric acid, potassium permanganate, bromine, or chlorine. It isn't that benzene has no chemical properties, as we will soon see. It's that benzene is in a class by itself.

Because the three double bonds indicated in structure **2** are misleading in a chemical sense, scientists often represent benzene simply by a hexagon with a circle inside, structure **3.**

3
Benzene

The ring, which is often referred to as the *benzene ring,* is planar. All of the atoms of benzene lie in the same plane, and all of the bond angles are 120°, as seen in the scale model of Figure 13.7.

The Characteristic Reactions of Benzene. Benzene undergoes **substitution reactions** far more readily than addition reactions. This property is what sets benzene and all compounds that have benzene rings apart from other unsaturated substances. Any substances whose molecules have benzene rings and that give substitution reactions at the ring instead of addition reactions are called **aromatic compounds.** This term is a holdover from the days when most of the known compounds of benzene actually had aromatic fragrances, but now it is not necessarily related to odor. Although oil of wintergreen and vanillin do have pleasant fragrances, aspirin does not. Yet all three have the benzene ring and all are classified as aromatic compounds.

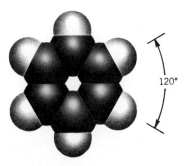

FIGURE 13.7
Scale model of a molecule of benzene. It shows the relative volumes of space occupied by the electron clouds of the atoms.

Oil of wintergreen

Vanillin

Aspirin

In a substitution reaction of a benzene ring, a hydrogen atom on the ring is replaced by another atom or group. For example, benzene gives a substitution reaction with chlorine or bromine, provided that iron or an iron halide catalyst is present. (Without the catalyst, there is no reaction.)

$$C_6H_6 + Cl_2 \xrightarrow[\text{FeCl}_3]{\text{Fe or}} C_6H_5-Cl + H-Cl$$
Benzene   Chlorobenzene

$$C_6H_6 + Br_2 \xrightarrow[\text{FeBr}_3]{\text{Fe or}} C_6H_5-Br + H-Br$$
Bromobenzene

Benzene also reacts, by substitution, with sulfur trioxide when it is dissolved in concentrated sulfuric acid. (Recall that alkenes react exothermically with concentrated sulfuric acid by *addition*.)

$$C_6H_6 + SO_3 \xrightarrow[\text{Room temperature}]{\text{H}_2\text{SO}_4 \text{ (concd)}} C_6H_5-\overset{\overset{\displaystyle O}{\parallel}}{\underset{\underset{\displaystyle O}{\parallel}}{S}}-O-H$$

Benzenesulfonic acid

Benzenesulfonic acid is about as strong an acid as hydrochloric acid, and it is a raw material for the synthesis of aspirin.

Benzene reacts with warm, concentrated nitric acid when it is dissolved in concentrated sulfuric acid. We will represent nitric acid as $HO-NO_2$, instead of HNO_3, because it loses the HO group during the reaction.

$$C_6H_6 + HO-NO_2 \xrightarrow[\text{50-55 °C}]{\text{H}_2\text{SO}_4 \text{ (concd)}} C_6H_5-NO_2 + H_2O$$

Nitrobenzene

A reaction very similar to this occurs when nitric acid is spilled on the skin. Proteins usually include substituted benzene rings, only the rings are more reactive than benzene itself. They undergo a reaction with nitric acid at body temperature to produce a yellow substance that stains the skin (and that eventually wears off).

1-Phenylpropane is an example of an aromatic compound that has an aliphatic side chain. (In the IUPAC, the group C_6H_5-, derived from benzene by removing one H atom, is called the **phenyl** group.) So stable is the benzene ring toward oxidizing agents, that alkylbenzenes such as 1-phenylpropane are attacked by hot permanganate at the side chain and not at the ring. Benzoic acid can be made as follows:

$$C_6H_5-CH_2CH_2CH_3 \xrightarrow{\text{Hot KMnO}_4} C_6H_5-\overset{\overset{\displaystyle O}{\parallel}}{C}-OH \quad (+CO_2 + H_2O + MnO_2)$$

1-Phenylpropane Benzoic acid

In many compounds, the benzene ring can be attacked by oxidizing agents. This is particularly true of rings that hold the $-OH$ or the $-NH_2$ groups. However, we cannot take the time to go more than this into the chemistry of aromatic compounds. Our chief goals have been to call attention to the existence of benzene rings and how they are represented, and then to point out by a few examples of their chemical properties how unlike alkenes they are.

Naming Derivatives of Benzene. The names of several monosubstituted benzenes are straightforward. The substituent is indicated by a prefix to the word *benzene*. For example,

All of these compounds are oily liquids.

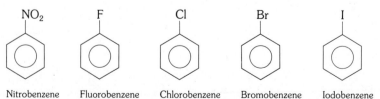

Nitrobenzene   Fluorobenzene   Chlorobenzene   Bromobenzene   Iodobenzene

Other derivatives of benzene have common names that are always used.

Phenol was the first antiseptic used by British surgeon Joseph Lister (1827–1912), the discoverer of antiseptic surgery.

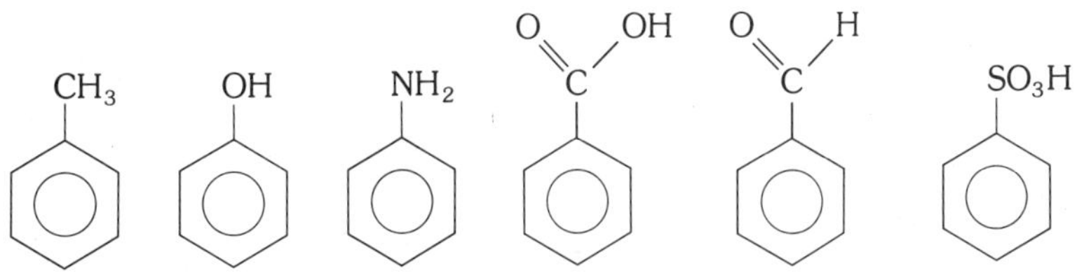

Toluene   Phenol   Aniline   Benzoic acid   Benzaldehyde   Benzene-sulfonic acid

When two or more groups are attached to the benzene ring, both what they are and where they are must be specified. One common way to indicate where two groups are in relationship to each other is by the prefixes *ortho-*, *meta-*, and *para-*, which usually are abbreviated *o-*, *m-* and *p-*, respectively. When two groups are *ortho* to each other, they are located 1,2- relative to each other, as in 1,2-dichlorobenzene, which commonly is called *o*-dichlorobenzene. In *m*-dichlorobenzene, the two chlorine atoms are in a 1,3- relationship. And in *p*-dichlorobenzene, the substituents are 1,4- relative to each other.

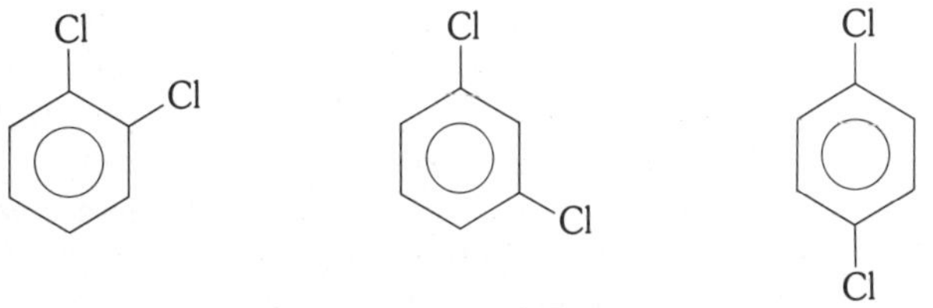

Ortho or 1,2
o-Dichlorobenzene   Meta or 1,3
m-Dichlorobenzene   Para or 1,4
p-Dichlorobenzene

A disubstituted benzene is often named as a derivative of a monosubstituted benzene when the latter has a common name such as toluene or aniline, and *o-*, *m-*, or *p-* is used to specify the relative positions. For example,

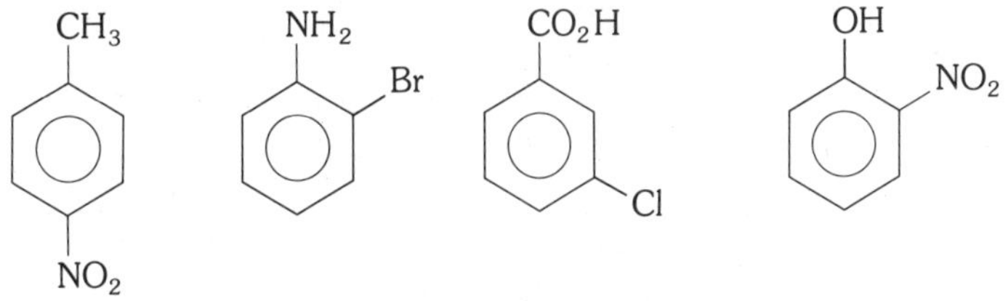

p-Nitrotoluene   *o*-Bromoaniline   *m*-Chlorobenzoic acid   *o*-Nitrophenol

When three or more groups are on a benzene ring, we can no longer use the *ortho, meta,* or *para* designations; we have to use numbers assigned to ring positons. For example,

TNT is an important explosive. TNB is an even better explosive than TNT, but it is more expensive to make.

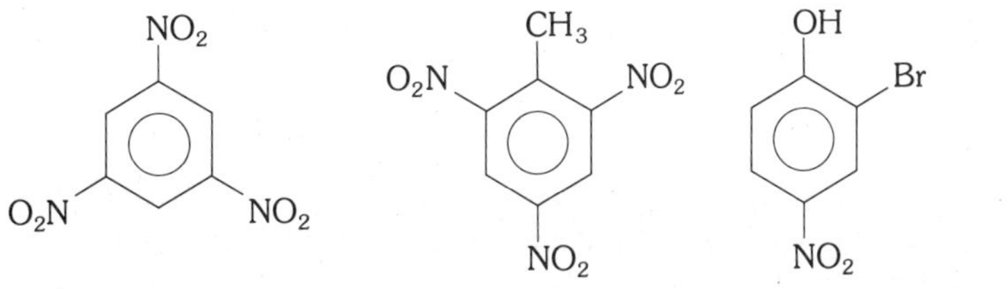

1,3,5-Trinitrobenzene (TNB)   2,4,6-Trinitrotoluene (TNT)   2-Bromo-4-nitrophenol

FIGURE 13.8
The molecular orbital model of benzene. (a) The sigma bond framework. (b) The six 2p orbitals of the ring carbon atoms. (c) The double-doughnut shaped region formed by the side-by-side overlapping of the six 2p orbitals.

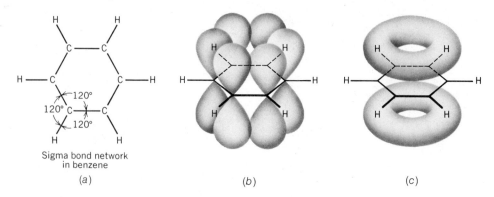

Sigma bond network
in benzene
(a)

(b)

(c)

13.8   THE PI ELECTRONS IN BENZENE

The cyclic, pi-electron network of a benzene ring strongly resists reactions that would disrupt it.

Part (a) of Figure 13.8 shows the sigma bond network in a molecule of benzene. Because each carbon atom holds three other atoms, not four, it is sp^2 hybridized, and it has an unhybridized 2p orbital. The axes of the 2p orbitals are parallel to each other and perpendicular to the plane of the ring, as seen in part (b) of Figure 13.8. The novel and important feature of the structure of benzene is that these six 2p orbitals overlap side to side *all around the ring*. They don't just pair off as in ethene and form three isolated double bonds. The result is a large, circular, double-doughnut shaped space above and below the plane, as seen in part (c) of the figure. Six electrons are in this space, and we can refer to them as the pi electrons of the benzene ring.

The six pi electrons are in three energy sublevels within the double-doughnut shaped space, and they enjoy considerably more room and freedom of movement than if they were in isolated double bonds. The electrons are said to be *delocalized*. When electrons have more room, the system is more stable. The delocalization of the pi electrons, in fact, explains much of the unusual stability of benzene and its resistance to addition reactions. When a hydrogen atom held by a ring carbon is replaced by another group, the closed-circuit pi electron network is not broken up. But if an addition were to occur, the ring system would no longer be that of benzene but, instead, of a cyclic diene. For example,

Benzene   5,6-Dichloro-
                 1,3-cyclohexadiene

It costs the system far more energy to react this way than to react by a substitution reaction, so the benzene ring strongly resists addition reactions.

SUMMARY

Alkenes   The lack of free rotation at a double bond makes geometric (cis-trans) isomers possible, but they exist only when the two groups are not identical at *either* end of the double bond. Cyclic compounds also exhibit cis-trans isomerism.

Alkenes and cycloalkenes are given IUPAC names by a set of rules very similar to those used to name their corresponding saturated forms. However, the double bond takes precedence both in selecting and in numbering the main chain (or ring). The first unsaturated carbon encountered in moving down the chain or around the ring through the double bond must have the lower number.

Addition reactions Several compounds add to the carbon–carbon double bond—H_2, Cl_2, Br_2, HCl, HBr, $HOSO_3H$, and H_2O (when an acid catalyst is present). The kinds of products that can be made are outlined in the accompanying chart of the reactions of alkenes. When both the alkene and the reactant are unsymmetrical, the addition proceeds according to Markovnikov's rule—the end of the double bond that already has the greater number of hydrogens gets one more. The double bond is vigorously attacked by strong oxidizing agents such as the permanganate ion, the dichromate ion, and ozone.

How additions occur The carbon–carbon double bond serves as a proton-acceptor toward strong proton-donors. The proton becomes attached to one carbon at the double bond, using the pair of electrons in the pi bond to make the sigma bond to this hydrogen, and the other carbon becomes positively charged. This carbocation then accepts an electron-rich particle to complete the formation of the product.

Polymerization of alkenes The polymerization of an alkene is like an addition reaction. The alkene serves as the monomer, and one alkene molecule adds to another, and so on until a long chain with a repeating unit forms—the polymer molecule.

The pi bond Following sp^2 hybridization at carbon, the sp^2 orbitals participate in the formation of the sigma bond framework. Then the side-to-side overlap of the unhybridized $2p$ orbitals creates the pi bond, a double-banana shaped space above and below the plane of the double bond that holds a pair of electrons. The cost in energy to undo this overlap in order that free rotation might occur is too high, and there is no free rotation at a double bond. However, the pi electrons are not held as strongly as the sigma electrons, and they are responsible for most of the chemistry of a double bond.

Aromatic properties When aromatic compounds undergo reactions at the benzene ring, substitutions rather than additions occur. In this way, the close-circuit pi electron network of the ring remains unbroken. This network forms when six unhybridized $2p$ orbitals of the ring carbon atoms overlap side by side to form a double-doughnut shaped space found above and below the plane of the ring. There are three sublevels in this space, and the pi electrons enjoy considerable room and freedom of motion. Except when the ring holds —OH or —NH_2 groups, it is very resistant to oxidation.

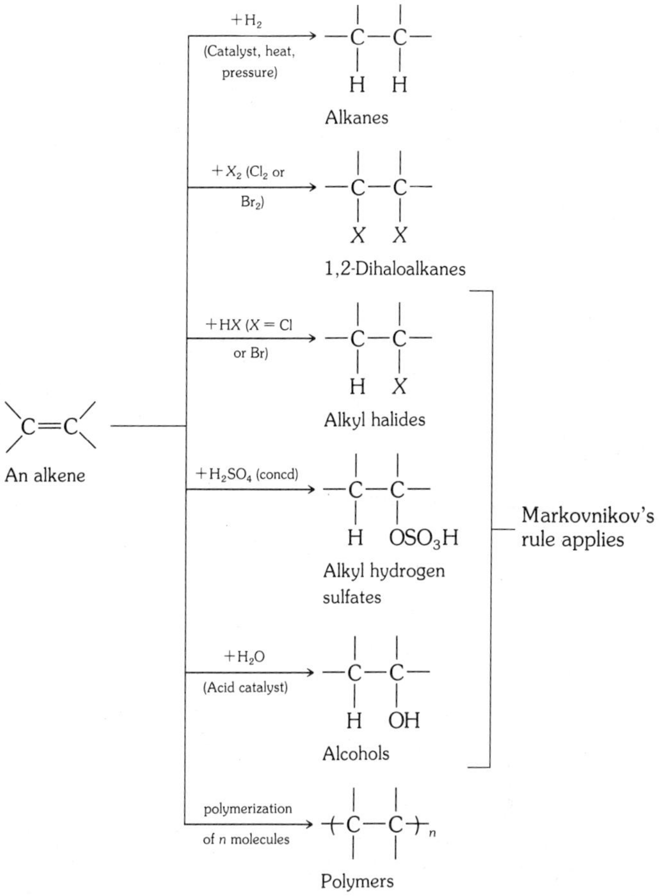

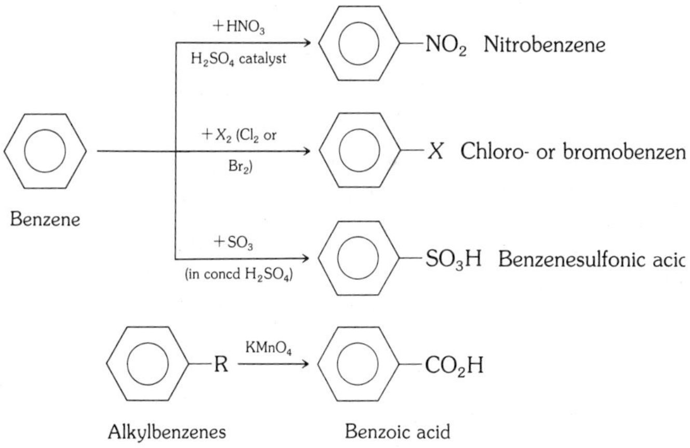

KEY TERMS

These terms will be used again and they should now be mastered.

addition reaction	geometric isomers	pi bond	sp^2 orbitals
aromatic compound	Markovnikov's rule	pi electrons	substitution reaction
carbocation	monomer	polymer	
geometric isomerism	phenyl group	polymerization	

SELECTED REFERENCES

1 F. W. Harris. "Introduction to Polymer Chemistry." *Journal of Chemical Education*, November 1981, page 837

2 "A Close-Up Look at Polymers." *Chemistry*, June 1978. An issue with the following articles:

   C. E. Carraher, Jr. "What Are Polymers?" page 6.

A. E. Tonelli. "Sizes and Shapes of Giant Molecules," page 11.

S. W. Shalaby and E. M. Pearce. "The Role of Polymers in Medicine and Surgery," page 17.

H. F. Mark. "Polymers for the Future," page 28.

REVIEW EXERCISES

The answers to these Review Exercises are in the *Study Guide* that accompanies this book.

Cis-Trans Isomerism

13.1   Which of the following pairs of structures represent identical compounds or isomers?

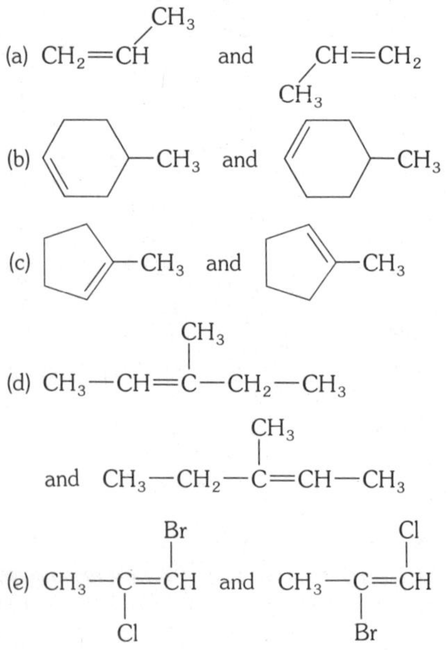

(a) $CH_2=CH$ CH_3 and CH_3 $CH=CH_2$

(b) ⬡—CH_3 and ⬡—CH_3

(c) ⬠—CH_3 and ⬠—CH_3

(d) $CH_3-CH=\overset{\overset{\displaystyle CH_3}{|}}{C}-CH_2-CH_3$

and $CH_3-CH_2-\overset{\overset{\displaystyle CH_3}{|}}{C}=CH-CH_3$

(e) $CH_3-\overset{\overset{\displaystyle Br}{|}}{\underset{\underset{\displaystyle Cl}{|}}{C}}=CH$ and $CH_3-\overset{\overset{\displaystyle Cl}{|}}{\underset{\underset{\displaystyle Br}{|}}{C}}=CH$

13.2   Write the structures of the cis and trans isomers, if any, of each compound.

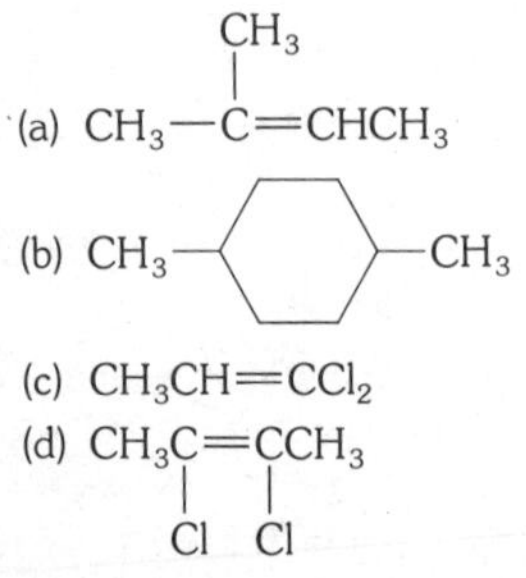

(a) $CH_3-\overset{\overset{\displaystyle CH_3}{|}}{C}=CHCH_3$

(b) CH_3-⬡$-CH_3$

(c) $CH_3CH=CCl_2$

(d) $CH_3\overset{}{C}=\overset{}{C}CH_3$ with Cl Cl

13.3   Identify which of the following compounds can exist as cis and trans isomers and write the structures of these isomers.

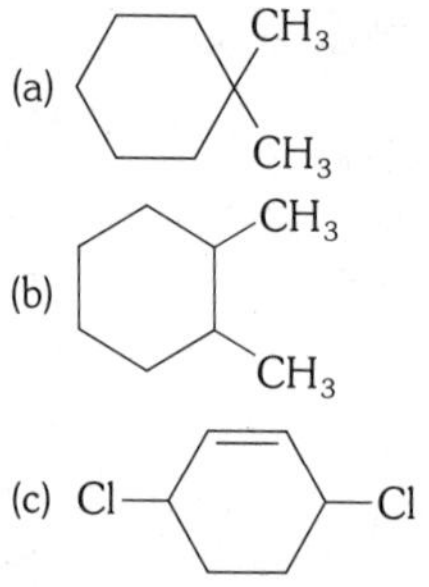

(a) ⬡ with two CH_3

(b) ⬡ with two CH_3

(c) $Cl-$⬡$-Cl$

Nomenclature

13.4   Write the condensed structures of the following compounds.
(a) Propylene
(b) Isobutylene
(c) *cis*-2-Hexene
(d) 3-Chloro-2-pentene
(e) 1,2-Diethylcyclohexene
(f) 2,4-Dimethylcyclohexene

13.5   Write the IUPAC names of the following compounds.
(a) $CH_3(CH_2)_7CH=CH_2$
(b) $Cl-CH=CHCH\overset{\overset{\displaystyle CH_3}{|}}{C}CH_3$
(c) $CH_3CH_2CH_2\overset{\overset{\displaystyle CH_2}{||}}{C}CH_2\overset{\overset{\displaystyle CH_3}{|}}{C}HCH_2CH_3$
(d) $CH_3\overset{\overset{\displaystyle CH_3}{|}}{\underset{\underset{\displaystyle CH_3}{|}}{C}}CH=\overset{\underset{\underset{\displaystyle CH_3}{|}}{}}{C}H$

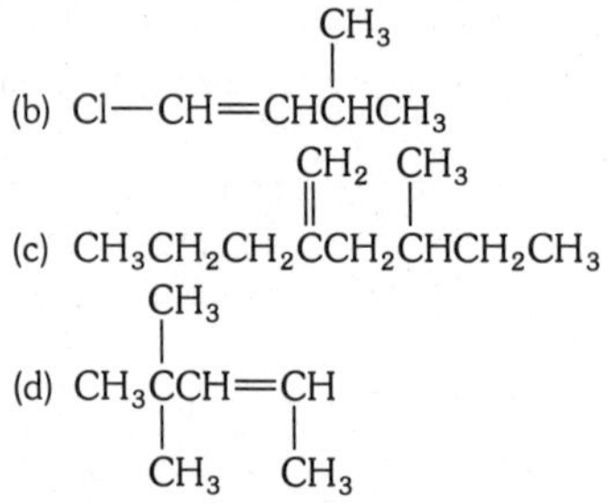

13.6   Write the condensed structures and the IUPAC names for all of the isomeric pentenes, C_5H_{10}. Include cis and trans isomers.

13.7   Write the condensed structures and the IUPAC names for all of the isomeric methylcyclopentenes.

13.8   Write the condensed structures and IUPAC names for all of the isomeric pentynes. The IUPAC rules for naming alkynes are identical with the rules for naming alkenes, except that the name ending is *-yne*, not *-ene*.

13.9 Write the condensed structures and the IUPAC names for all of the isomeric dimethylcyclohexenes. Include the cis and trans isomers.

13.10 Write the condensed structures and the names for all of the open chain dienes with the molecular formula C_5H_8. (Note, there can be two double bonds from the same carbon.)

Reactions of the Carbon–Carbon Double Bond

13.11 Write equations for the reactions of 2-methylpropene with the following reactants.
(a) Cold, concentrated H_2SO_4
(b) H_2 (Ni, heat, pressure)
(c) H_2O, (H^+ catalyst)
(d) H—Cl
(e) H—Br
(f) Br_2

13.12 Write equations for the reactions of 1-methylcyclopentene with the reactants listed in Review Exercise 13.11.

13.13 Write equations for the reactions of 2-methyl-2-butene with the compounds given in Review Exercise 13.11.

13.14 Write equations for the reactions of 1-methylcyclohexene with the reactants listed in Review Exercise 13.11. (Do not attempt to predict if cis or trans isomers form.)

13.15 Ethane is insoluble in concentrated sulfuric acid, but ethene dissolves readily. Write an equation to show how ethene is changed into a substance polar enough to dissolve in concentrated sulfuric acid, which, of course, is highly polar.

13.16 1-Hexene, C_6H_{12}, has several isomers, and most but not all of them decolorize bromine, dissolve in concentrated sulfuric acid, and react with potassium permanganate. Show the structures of at least two isomers of C_6H_{12} that give *none* of these reactions.

13.17 One of the raw materials for the synthesis of nylon, adipic acid, can be made from cyclohexene by oxidation using potassium permanganate. The balanced equation for the first step in which the potassium salt of adipic acid forms is as follows:

$$3C_6H_{10} + 8KMnO_4 \longrightarrow$$
$$3K_2C_6H_8O_4 + 8MnO_2 + 2KOH + 2H_2O$$

How many grams of potassium permanganate are needed for the oxidation of 10.0 g of cyclohexene, assuming that the reaction occurs exactly and entirely as written?

13.18 Referring to Review Exercise 13.17, how many grams of the potassium salt of adipic acid can be made if 24.0 g of $KMnO_4$ are used in accordance with the equation given?

How Addition Reactions Occur

13.19 When 1-butene reacts with hydrogen chloride, the product is 2-chlorobutane, not 1-chlorobutane. Explain.

13.20 When 4-methylcyclohexene reacts with hydrogen chloride, the product consists of a mixture of roughly equal quantities of 3-chloro-1-methylcyclohexane and 4-chloro-1-methylcyclohexane. Explain why substantial proportions of *both* isomers form.

Polymerization

13.21 Rubber cement can be made by mixing some polymerized 2-methylpropene with a solvent such as toluene. When the solvent evaporates, a very tacky and sticky residue of the polymer remains, which soon hardens and becomes the glue. Write the structure of the polymer of 2-methylpropene (called polyisobutylene) in two ways. (The structure is quite regular, like polypropylene.)
(a) One that shows four repeating units, one after the other.
(b) The condensed structure.

13.22 Safety glass is made by sealing a thin film of polyvinyl acetate between two pieces of glass. If the glass shatters, its pieces remain glued to the film and cannot fly about. Using four vinyl acetate units, write part of the structure of a molecule of polyvinyl acetate. Also write its condensed structure. The structure of vinyl acetate is as follows:

$$\begin{array}{c} O \\ \parallel \\ OCCH_3 \\ | \\ CH_2{=}CH \end{array}$$

Vinyl acetate

When it polymerizes, its units line up in a regular way as in the polymerization of propene.

13.23 Gasoline is mostly a mixture of alkanes. However, when a sample of gasoline was shaken with aqueous potassium permanganate, a brown precipitate of MnO_2 appeared and the purple color of the permanganate ion disappeared. What kind of hydrocarbon was evidently also in this sample?

13.24 If gasoline rests for months in the fuel line of some engine, the line slowly accumulates some sticky material. This can clog the line and also make the carburetor work poorly or not at all. In view of your answer to Review Exercise 13.23, what is likely to be happening, chemically?

The Pi Bond

13.25 In your own words, using your own drawings, describe what is meant by sp^2 hybridization at carbon.

13.26 Using your own drawing, show how the axes of the four orbitals at an sp^2 hybridized carbon atom (three sp^2 hybrid orbitals plus one $2p$ orbital) are arranged. Indicate the angles.

13.27 If the valence-shell electron-pair repulsion theory were applied to the electron-occupied orbitals of an sp^2 hybridized carbon atom, what angles would the axes of these orbitals make with each other? How does your answer compare with the angles used in the answer to Review Exercise 13.26?

13.28 Why couldn't just the sigma bond between the two carbon atoms of a double bond prevent free rotation of the groups held by this bond?

13.29 Using drawings of your own making, explain how the pi bond prevents free rotation of the groups held by a double bond.

13.30 Which bonding electrons are more directly *between* the two carbon atoms in ethene, the sigma or the pi electrons?

Aromatic Properties

13.31 Dipentene has a very pleasant, lemon-like fragrance, but it is not classified as an aromatic compound. Why?

Dipentene

13.32 Sulfanilamide, the structurally simplest of the sulfa drugs, has no odor at all, but it is still classified as an aromatic compound. Explain.

Sulfanilamide

13.33 Write equations for the reactions, if any, of benzene with the following compounds.
(a) Sulfur trioxide (in concentrated sulfuric acid)
(b) Concentrated nitric acid (in concentrated sulfuric acid)
(c) Hot sodium hydroxide solution
(d) Hydrochloric acid
(e) Chlorine (alone)
(f) Hot potassium permanganate
(g) Bromine in the presence of $FeBr_3$

13.34 Write the structure of any compound that would react with hot potassium permanganate to give benzoic acid.

13.35 Terephthalic acid is one of the raw materials for making Dacron, a polymer that is popular for making synthetic fabrics.

Terephthalic acid

What hydrocarbon with the formula C_8H_{10} could be changed to terephthalic acid by the action of hot potassium permanganate? (Write its structure.)

Names of Aromatic Compounds

13.36 Write the condensed structure of each compound.
(a) Toluene   (b) Aniline
(c) Phenol   (d) Benzoic acid
(e) Benzaldehyde   (f) Nitrobenzene

13.37 Give the condensed structures of the following compounds.
(a) o-Nitrobenzenesulfonic acid
(b) p-Chlorotoluene
(c) m-Bromonitrobenzene
(d) 2,4-Dinitrophenol

The Bonds in Benzene

13.38 Describe in your own words, making your own drawings, the ways in which the bonds in benzene form.

13.39 What is it about the delocalization of benzene's pi electrons that makes the benzene ring relatively stable?

13.40 Explain why benzene strongly resists addition reactions and gives substitution reactions instead.

Predicting Reactions

13.41 Write the structures of the products to be expected in the following situations. If no reaction is to be expected, write "no reaction." To work this kind of exercise, you have to be able to do three things.
(a) *Classify* a specific organic reactant into its proper family.
(b) *Recall* the short list of chemical facts about the family. (If there is no match-up between this list and the reactants and conditions specified by a given problem, assume that there is no reaction.)
(c) *Apply* the recalled chemical fact, which might be some "map sign" associated with a functional group, to the specific situation.

Study the next two examples before continuing.

EXAMPLE 13.6 PREDICTING REACTIONS

Problem: What is the product, if any, of the following?

$$CH_3CH_2CH_2CH_3 + H_2SO_4 \longrightarrow$$

Solution: We note first that the organic reactant is an alkane, so we next turn to the list of chemical properties about all alkanes that we learned. With this family, of course, the list is very short. Except for combustion and halogenation, we have learned no reactions for alkanes, and we assume therefore that there aren't any others, not even with sulfuric acid. Hence, the answer is "no reaction."

EXAMPLE 13.7 PREDICTING REACTIONS

Problem: What is the product, if any, in the following situation?

$$CH_3CH=CHCH_3 + H_2O \xrightarrow{\text{acid catalyst}} ?$$

Solution: We first note that the organic reactant is an alkene, so we review our mental "file" of reactions of the carbon–carbon double bond.
1. Alkenes add hydrogen (in the presence of a metal catalyst and under heat and pressure) to form alkanes.
2. They add chlorine and bromine to give 1,2-dihaloalkanes.
3. They add hydrogen chloride and hydrogen bromide to give alkyl halides.
4. They add sulfuric acid to give alkyl hydrogen sulfates.
5. They add water in the presence of an acid catalyst to give alcohols.
6. They are attacked by strong oxidizing agents.
7. They polymerize.

These are the chief chemical facts about the carbon–carbon double bond that we have studied, and we see that the list includes a reaction with water in the presence of an acid catalyst. We remember that in all addition reactions the double bond changes to a single bond and the pieces of the molecule that adds end up on the carbons at this bond. We also have to remember Markovnikov's rule to tell us which pieces of the water molecule go to which carbon. However, in this specific example, Markovnikov's rule does not apply (the alkene is symmetrical).

Answer:
$$CH_3CH_2\underset{\underset{OH}{|}}{CH}CH_3$$

Now work the following parts. (Remember that C_6H_6 stands for benzene and that C_6H_5— is the phenyl group.)

(a) $CH_3CH_2CH{=}CH_2 + H_2 \xrightarrow{\text{Ni, heat, pressure}}$

(b) $CH_3CH{=}CHCH_3 + H_2O \xrightarrow{\text{H}^+}$

(c) $C_6H_6 + Br_2 \xrightarrow{\text{FeBr}_3}$

(d) $CH_3CH_2CH_3 + \text{concd } H_2SO_4 \longrightarrow$

(e) 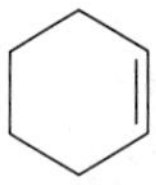 ⬡ $+ H_2O \xrightarrow{\text{H}^+}$

(f) $CH_3CH{=}CHCH_2CH_3 + H_2 \xrightarrow{\text{Ni, heat, pressure}}$

(g) 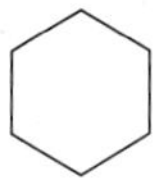 ⬡ $+ \text{concd } H_2SO_4 \longrightarrow$

(h) $C_6H_5{-}CH{=}CH{-}C_6H_5 + H_2O \xrightarrow{\text{H}^+}$

(i) 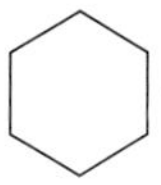 ⬡ $+ H_2 \xrightarrow{\text{Ni, heat pressure}}$

(j) $CH_3CH_2CH_2CH_3 + O_2 \xrightarrow{\text{Complete combustion}}$

(Balance the equation.)

(k) $C_6H_6 + H_2O \xrightarrow{\text{H}^+}$

13.42 Write the structures of the products in the following situations. If no reaction is to be expected, write "no reaction."

(a) $CH_3CH{=}CH_2 + H_2O \xrightarrow{\text{H}^+}$

(b) $CH_2{=}CH{-}CH_2{-}CH{=}CH_2 + 2H_2 \xrightarrow{\text{Ni, heat pressure}}$

(c) $+CH_2{-}CH_2{+}_n + H_2SO_4 \text{ (concd)} \longrightarrow$

(d) $CH_3{-}\underset{\underset{CH_3}{|}}{C}{=}\underset{\underset{CH_3}{|}}{C}{-}CH_3 + HCl \longrightarrow$

(e) 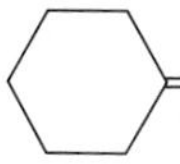 ⬡${=}CH_2 + HBr \longrightarrow$

(f) 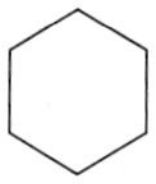 ⬡ $+ Br_2 \longrightarrow$

(g) $C_6H_6 + Cl_2 \xrightarrow{\text{FeCl}_3}$

(h) $C_6H_6 + NaOH(aq) \longrightarrow$

(i) $C_6H_5{-}CH_3 + O_2 \xrightarrow{\text{Complete combustion}}$ (Balance the equation.)

(j) 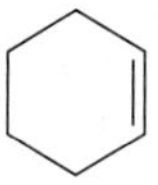 ⬡ $+ NaOH(aq) \longrightarrow$

(k) $CH_2{=}CH{-}CH{=}CH_2 + 2Cl_2 \longrightarrow$

(l) $C_6H_5{-}CH{=}CH_2 + Br_2 \longrightarrow$

Chapter 14
Alcohols, Thioalcohols, Phenols, and Ethers

Alcohols and thioalcohols differ in the replacement of O by S, but what a difference to the skunk, whose chief defense is a liquid mixture of thioalcohols, among nature's most vile smelling compounds.

14.1 OCCURRENCE, TYPES, AND NAMES OF ALCOHOLS

In the alcohol family, molecules have the —OH group attached to a saturated carbon.

The alcohol system is one of the most widely occurring in nature. It is present, for example, in all carbohydrates, all proteins, all nucleic acids, and in one of the building blocks of fats and oils. Wood alcohol, rubbing alcohol, beverage alcohol, and antifreezes are all different members of the alcohol family.

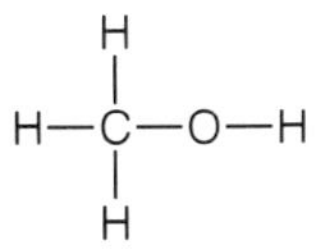

In the **alcohols,** the —OH group is covalently held by a saturated carbon atom, one from which only single bonds extend. When this conditon is met, the —OH group is called the **alcohol group.** Table 14.1 includes a number of simple alcohols. As you can see by the following structures, the —OH group also occurs in the family of the phenols, where it is attached to a benzene ring, and in the family of the carboxylic acids, where it is attached to a carbon that has a double bond to oxygen.

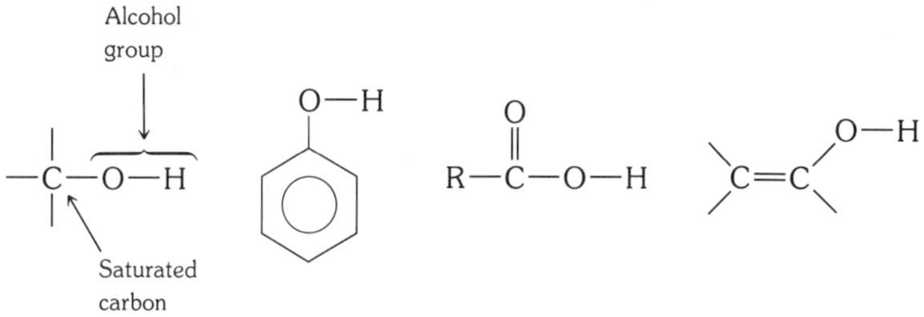

Methanol, the simplest alcohol

When the —OH is attached to an alkene group, the system — called the *enol* system (ene + ol) — is unstable. Most of our attention will be given to alcohols in this chapter, with some study of phenols. We will study the carboxylic acids in a later chapter.

PRACTICE EXERCISE 1   Classify the following as alcohols, phenols, or carboxylic acids.

(a) 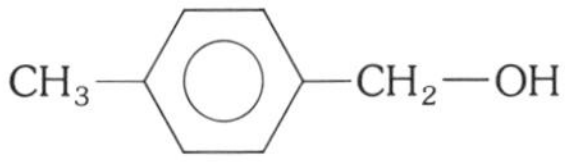 CH_3—⟨benzene ring⟩—CH_2—OH   (b)  CH_3—⟨benzene ring⟩—OH

(c) 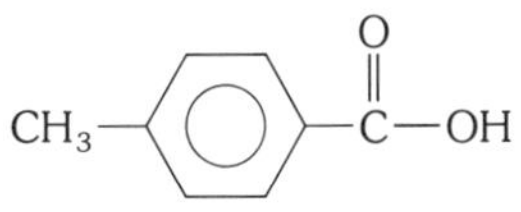 CH_3—⟨benzene ring⟩—$\overset{O}{\overset{\|}{C}}$—OH   (d) CH_2=CH—CH_2—OH

(e) $CH_3CH_2CH_2CH_2OH$ (f) 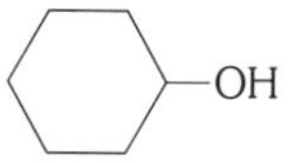 ⟨cyclohexane ring⟩—OH

Subclasses of Alcohols.   The alcohol group is classified as being primary (1°), secondary (2°), or tertiary (3°) according to the kind of carbon to which it is attached. When the —OH group is held by a 1° carbon, one that has only one carbon atom directly joined to it, the alcohol is a **primary alcohol.** In a **secondary alcohol,** the —OH group is held by a 2° carbon. When the —OH group is joined to a 3° carbon, the alcohol is a **tertiary alcohol.** We

Pronounce 1° as *primary,* 2° as *secondary,* and 3° as *tertiary.*

TABLE 14.1
Some Common Alcohols

Name[a]	Structure	Boiling Point (°C)
Methanol	CH_3OH	65
Ethanol	CH_3CH_2OH	78.5
1-Propanol	$CH_3CH_2CH_2OH$	97
2-Propanol	CH_3CHCH_3 　　$\vert$ 　　OH	82
1-Butanol	$CH_3CH_2CH_2CH_2OH$	117
2-Butanol (sec-butyl alcohol)	$CH_3CH_2CHCH_3$ 　　　$\vert$ 　　　OH	100
2-Methyl-1-propanol (isobutyl alcohol)	CH_3 $\vert$ CH_3CHCH_2OH	108
2-Methyl-2-propanol (t-butyl alcohol)	CH_3 $\vert$ CH_3COH $\vert$ CH_3	83
1,2-Ethanediol (ethylene glycol)	$CH_2\!-\!CH_2$ 　$\vert$　　$\vert$ 　OH　OH	197
1,2-Propanediol (propylene glycol)	$CH_3CH\!-\!CH_2$ 　　$\vert$　　$\vert$ 　　OH　OH	189
1,2,3-Propanetriol (glycerol)	$CH_2\!-\!CH\!-\!CH_2$ 　$\vert$　　$\vert$　　$\vert$ 　OH　OH　OH	290

[a] The IUPAC names with the common names in parentheses.

will see the need for these subclasses when we learn that not all alcohols respond in the same way to oxidizing agents.

The R— groups in 2° and 3° alcohols don't have to be the same.

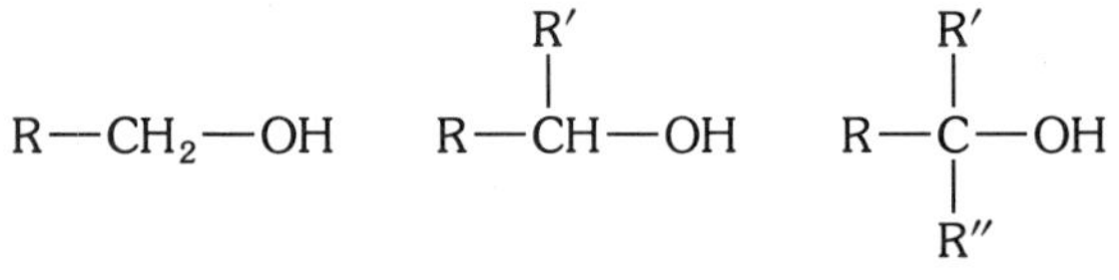

$$R\!-\!CH_2\!-\!OH \qquad R\!-\!\overset{R'}{\underset{}{CH}}\!-\!OH \qquad R\!-\!\overset{R'}{\underset{R''}{\overset{\vert}{\underset{\vert}{C}}}}\!-\!OH$$

Primary alcohol　　　Secondary alcohol　　　Tertiary alcohol

Many substances classified as alcohols have more than one —OH group per molecule. When only one is present, the alcohol is a **monohydric alcohol.** Often the monohydric alcohols are referred to as the *simple alcohols.* When two —OH groups occur in the same molecule, as in 1,2-ethanediol (ethylene glycol), the substance is a **dihydric alcohol** or a **glycol.** A **trihydric alcohol** is one whose molecules have three alcohol groups, such as 1,2,3-propanetriol (glycerol). Many substances, particularly among the various sugars, have several —OH groups per molecule. One important structural restriction is that almost no stable system is known in which one carbon holds more than one —OH group. If such should form during a reaction, it breaks up (more will be said about this in the next chapter). Special Topic 14.1 describes several important, individual alcohols.

The following system, the 1,1-diol, is unstable; only rare examples are known.

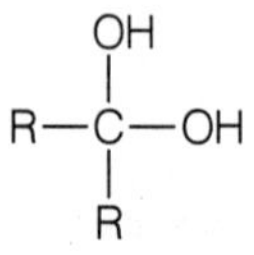

$$R\!-\!\overset{OH}{\underset{R}{\overset{\vert}{\underset{\vert}{C}}}}\!-\!OH$$

1,1-Diols

Methanol (methyl alcohol, wood alcohol). When taken internally in enough quantity, methanol causes either blindness or death. In industry, it is used as the raw material for making formaldehyde (which is used to make polymers), as a solvent, and as a denaturant (poison) for ethanol. It is also used as the fuel in Sterno, as well as in burners for fondue pots.

Most methanol is made by the reaction of carbon monoxide with hydrogen under high pressure and temperature:

$$2H_2 + CO \xrightarrow[\substack{\text{Temp.} = 350-400\ °C \\ \text{Catalyst} = ZnO-Cr_2O_3}]{3000\ \text{lb/in.}^2} CH_3-OH$$

Ethanol (ethyl alcohol, grain alcohol). Some ethanol is made by the fermentation of sugars, but most is synthesized by the addition of water to ethene in the presence of a catalyst. A 70% (v/v) solution of ethanol in water is used as a disinfectant.

In industry, ethanol is used as a solvent and to prepare pharmaceuticals, perfumes, lotions, and rubbing compounds. For these purposes, the ethanol is adulterated by poisons that are very hard to remove in order that the alcohol cannot be sold or used as a beverage. (Nearly all countries derive revenue by taxing potable, or drinkable, alcohol.)

2-Propanol (isopropyl alcohol). 2-Propanol is a common substitute for ethanol for giving back rubs. It is twice as toxic as ethanol, and in solutions with concentrations from 50 to 99% (v/v) it is used as a disinfectant.

1,2-Ethanediol (ethylene glycol), and **1,2-Propanediol** (propylene glycol). Ethylene and propylene glycol are the chief components in permanent-type antifreezes. Their great solubility in water and their very high boiling points make them ideal for this purpose. An aqueous solution that is roughly 50% (v/v) in either glycol does not freeze until about −40 °C (−40 °F).

1,2,3-Propanetriol (glycerol, glycerin). Glycerol, a colorless, syrupy liquid with a sweet taste, is freely soluble in water and insoluble in nonpolar solvents. It is one product of the digestion of the fats and oils in our diets. Because it has three —OH groups per molecule, each capable of hydrogen-bonding to water molecules, glycerol can draw moisture from humid air. It is sometimes used as a food additive to help keep foods moist.

Sugars. All carbohydrates consist of polyhydroxy compounds, and we will study them in a later chapter.

$$\begin{array}{ccc} CH_2-CH_2 & CH_2-CH-CH_2 & CH_2-CH-CH-CH-CH-\overset{\overset{\displaystyle O}{\|}}{C}-H \\ |\quad\ \ | & |\quad\ |\quad\ \ | & |\quad\ \ |\quad\ \ |\quad\ \ |\quad\ \ | \\ OH\ \ OH & OH\ \ OH\ \ OH & OH\ \ OH\ \ OH\ \ OH\ \ OH \end{array}$$

1,2-Ethanediol   1,2,3-Propanetriol   Glucose — a sugar
(ethylene glycol)   (glycerol)   (open-form of molecule)

PRACTICE EXERCISE 2   Classify each of the following as monohydric or dihydric. For each found to be monohydric, classify it further as 1°, 2°, or 3°. If the structure is too unstable to exist, state so.

(a) $CH_3-\underset{\underset{\displaystyle OH}{|}}{CH}-CH_3$

(b) (cyclohexene)—OH

(c) $CH_3-\underset{\underset{\displaystyle OH}{|}}{\overset{\overset{\displaystyle OH}{|}}{C}}-CH_3$

(d) (cyclohexane with two OH groups)

(e) $HO-CH_2-\underset{\underset{\displaystyle CH_3}{|}}{\overset{\overset{\displaystyle CH_3}{|}}{C}}-CH_3$

(f) (benzene)—CH_2-OH

(g) $CH_3-\underset{\underset{\displaystyle CH_3}{|}}{\overset{\overset{\displaystyle CH_3}{|}}{C}}-OH$

(h) $CH_3-CH_2-\underset{\overset{\displaystyle CH_3}{|}}{CH}-OH$

(i) $CH_3-\underset{\underset{\displaystyle OH}{|}}{\overset{\overset{\displaystyle OH}{|}}{C}}-OH$

The Nomenclature of Alcohols.   Among the simple alcohols, the common names are most often used. To write a common name, simply write the word *alcohol* after the name of the alkyl group involved. For example:

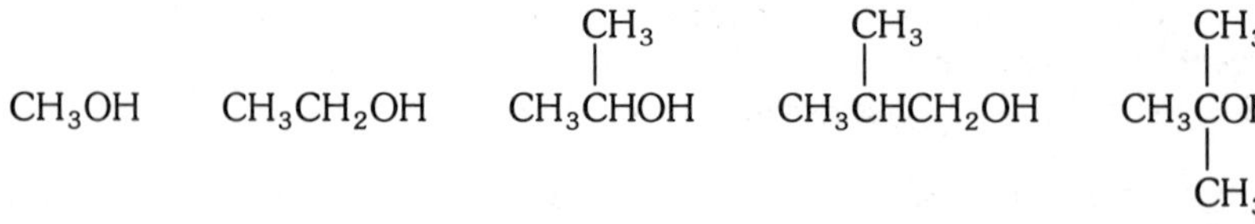

$$CH_3OH \qquad CH_3CH_2OH \qquad CH_3\overset{\overset{\displaystyle CH_3}{|}}{C}HOH \qquad CH_3\overset{\overset{\displaystyle CH_3}{|}}{C}HCH_2OH \qquad CH_3\overset{\overset{\displaystyle CH_3}{|}}{\underset{\underset{\displaystyle CH_3}{|}}{C}}OH$$

| Methyl alcohol | Ethyl alcohol | Isopropyl alcohol | Isobutyl alcohol | *t*-Butyl alcohol |

When the alkyl group has no common name, then the IUPAC system has to be used, and the rules are similar to those for naming alkanes. The full name of the alcohol is based on the idea of a *parent alcohol* that has substituents on its carbon chain. The rules are as follows.

1.   Determine the parent alcohol by selecting the longest chain of carbons *that includes the carbon atom to which the* — *OH group is attached.* Name the parent alcohol by changing the name ending of the alkane that corresponds to this chain from *-e* to *-ol*. Examples are:

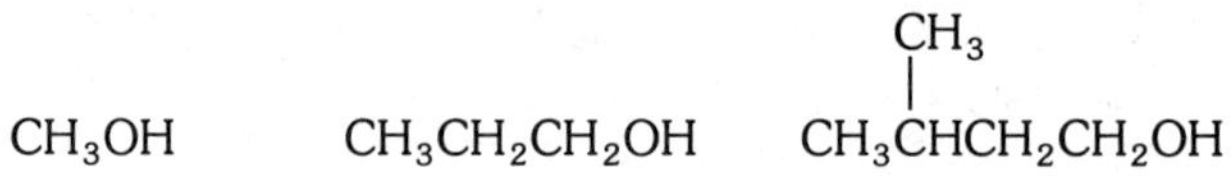

$$CH_3OH \qquad CH_3CH_2CH_2OH \qquad CH_3\overset{\overset{\displaystyle CH_3}{|}}{C}HCH_2CH_2OH$$

| Methanol (parent alkane is methane) | Propanol (parent alkane is propane) | A substituted butanol (parent alkane is butane) |

2.   Number the parent chain from whichever end gives the carbon that holds the — OH group the lower number. Be sure to notice this departure from the IUPAC rules for numbering chains for alkanes, where the location of the first branch determined the direction of numbering. With alcohols, *the location of the* — *OH group takes precedence over alkyl groups or halogen atoms in deciding the direction of numbering.* Examples are:

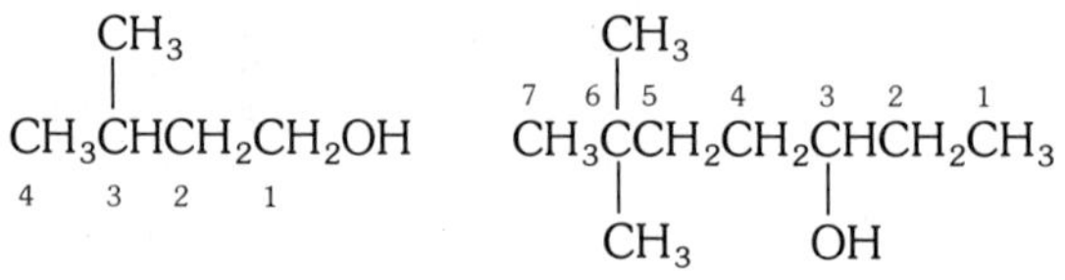

$$\overset{\overset{\displaystyle CH_3}{|}}{CH_3CHCH_2CH_2OH}$$
$${}_4_3_2_1$$

3-Methyl-1-butanol
(*Not* 2-methyl-4-butanol)

$$\overset{7\ \ 6\ |\ 5\ \ \ \ 4\ \ \ 3\ \ \ 2\ \ \ 1}{CH_3\underset{\underset{\displaystyle CH_3}{|}}{C}CH_2CH_2\underset{\underset{\displaystyle OH}{|}}{C}HCH_2CH_3}$$

6,6-Dimethyl-3-heptanol
(*Not* 2,2-dimethyl-5-heptanol)

No number is needed to specify the location of the —OH group in the names methanol and ethanol.

3.   Write the number that locates the — OH group in front of the name of the parent, and separate this number from the name of the parent by a hyphen. Then, as prefixes to what you have just written, assemble the names of the substituents and their location numbers. Use commas and hyphens in the usual way. See the examples above that followed rule 2 for illustrations. If a halogen atom is present, place its part in the final name first. For example,

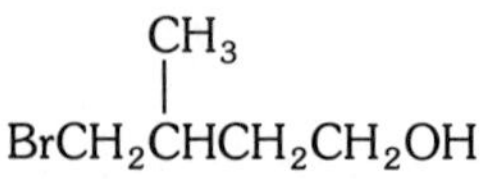

$$BrCH_2\overset{\overset{\displaystyle CH_3}{|}}{C}HCH_2CH_2OH$$

4-Bromo-3-methyl-1-butanol
(*Not* 3-methyl-4-bromo-1-butanol)

The order in which the names of the alkyl groups appear in the final name is not important. Either an alphabetical ordering or an ordering according to increasing carbon content is acceptable, as we will show in Example 14.1.

4.   When two or more —OH groups are present, use name endings such as *-diol* (for two —OH groups), *-triol,* and so forth. Then place the two, three, or more location numbers immediately in front of the name of the parent alcohol. For example,

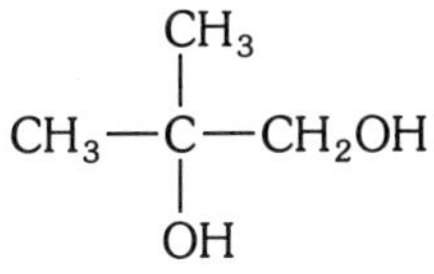

$$CH_3-\underset{\underset{\displaystyle OH}{\displaystyle |}}{\overset{\overset{\displaystyle CH_3}{\displaystyle |}}{C}}-CH_2OH \qquad\qquad CH_3-\underset{\underset{\displaystyle OH}{\displaystyle |}}{CH}-\underset{\underset{\displaystyle OH}{\displaystyle |}}{CH}-CH_2OH$$

2-Methyl-1,2-propanediol 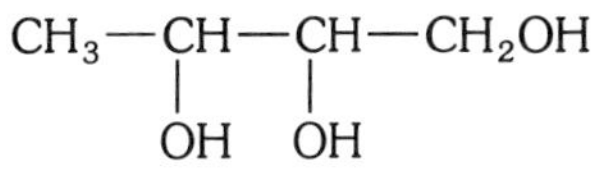   1,2,3-Butanetriol

5.   If no parent alcohol name is possible or convenient, then the —OH group can be treated as another substituent, and it is named *hydroxy.* For example,

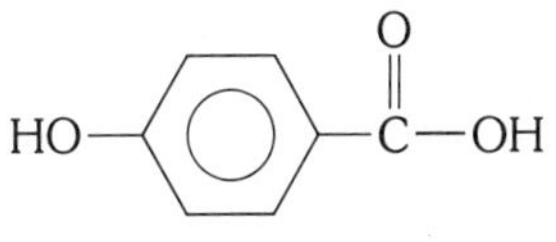

$$HO-\underset{}{\bigcirc}-\overset{\overset{\displaystyle O}{\displaystyle \|}}{C}-OH$$

4-Hydroxybenzoic acid

EXAMPLE 14.1   USING THE IUPAC RULES TO NAME AN ALCOHOL

Problem: What is the name of the following compound?

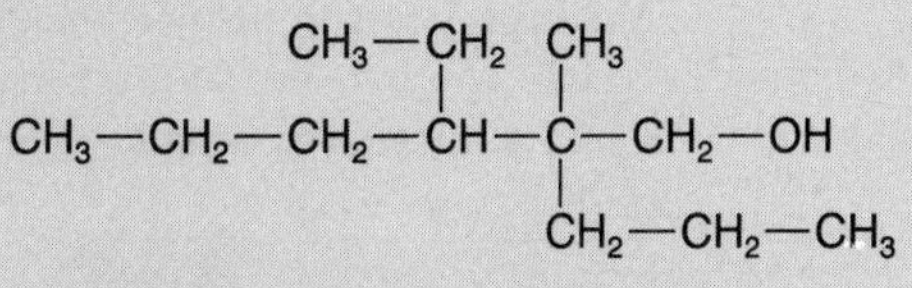

$$CH_3-CH_2-CH_2-\underset{\underset{\displaystyle CH_2-CH_2-CH_3}{\displaystyle |}}{CH}-\underset{\underset{}{}}{\overset{\overset{\displaystyle CH_3-CH_2\quad CH_3}{\displaystyle |\qquad\quad|}}{C}}-CH_2-OH$$

Solution: The longest chain *that includes the carbon atom to which the —OH group is attached* has six carbons in it, so the name of the parent alcohol is hexanol. (We can't use the longest chain of all, a chain of eight, because it doesn't include the —OH group.) We have to number the chain from right to left because this gives the carbon that holds the OH group the lower number.

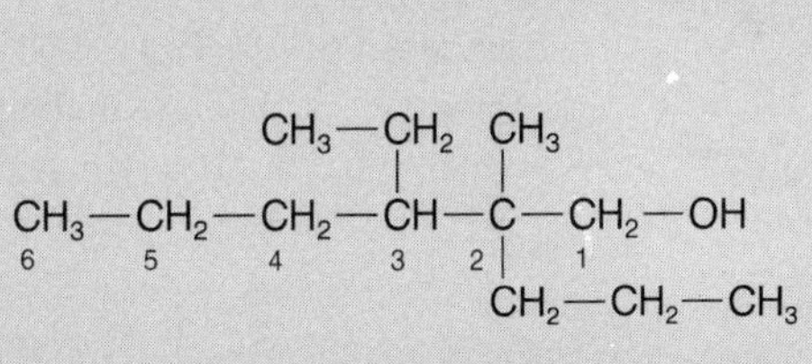

$$\underset{6}{CH_3}-\underset{5}{CH_2}-\underset{4}{CH_2}-\underset{3}{CH}-\underset{2}{C}-\underset{1}{CH_2}-OH$$

Thus the final name must end in -1-hexanol. Carbon 2 holds both a methyl and a propyl group, and carbon 3 has an ethyl group, so the final name, arranging the alkyl groups alphabetically, is

3-Ethyl-2-methyl-2-propyl-1-hexanol

Also acceptable, with the alkyl groups arranged in their order of increasing complexity, is

2-Methyl-3-ethyl-2-propyl-1-hexanol

PRACTICE EXERCISE 3   Name the following compounds by the IUPAC system.

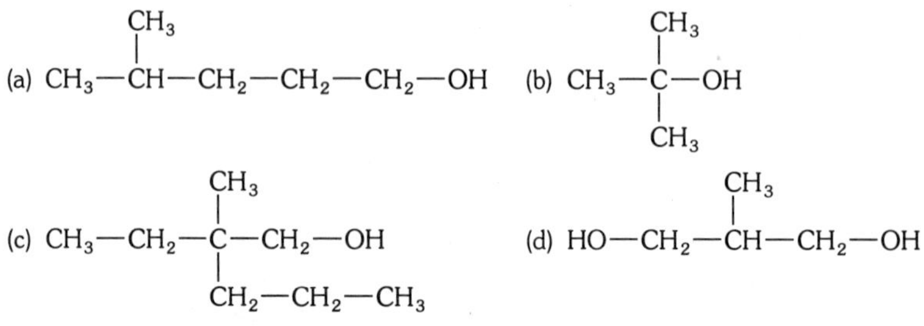

$$\text{(a) } CH_3-\underset{\underset{CH_3}{|}}{CH}-CH_2-CH_2-CH_2-OH \qquad \text{(b) } CH_3-\underset{\underset{CH_3}{|}}{\overset{\overset{CH_3}{|}}{C}}-OH$$

$$\text{(c) } CH_3-CH_2-\underset{\underset{CH_2-CH_2-CH_3}{|}}{\overset{\overset{CH_3}{|}}{C}}-CH_2-OH \qquad \text{(d) } HO-CH_2-\underset{\underset{}{}}{\overset{\overset{CH_3}{|}}{CH}}-CH_2-OH$$

14.2   PHYSICAL PROPERTIES OF ALCOHOLS

Hydrogen bonding dominates the physical properties of alcohols.

Alcohol molecules are quite polar, and they can both donate and accept hydrogen bonds. The effect of this can be seen in Table 14.2, which gives the boiling points of some alkanes (which are nonpolar substances) and the boiling points of alcohols of comparable formula weights.

The difference in boiling point of 154 °C between ethane and methyl alcohol shows how great is the force of attraction between molecules of the alcohol. And, yet, water with a lower formula weight has an even higher boiling point than methyl alcohol. Figure 14.1 shows how three hydrogen bonds can extend from just one water molecule to neighbors, but as you can see in this figure, only two such bonds can extend from molecules of monohydric alcohols. Molecules of dihydric alcohols, such as ethylene glycol (Table 14.2), can have four hydrogen bonds that extend between neighboring molecules, and ethylene glycol boils nearly 200 °C higher than butane. Thus boiling point data show the importance of the forces of attraction between molecules when —OH groups are present.

Only three elements, F, O, and N, have atoms that are electronegative enough to participate significantly in hydrogen bonds.

PRACTICE EXERCISE 4   1,2-Propanediol (propylene glycol) and 1-butanol have similar formula weights. In which of these two compounds is hydrogen bonding between molecules more extensive and stronger? How do the data in Table 14.1 support the answer?

Solubilities of Alcohols.   The —OH group greatly enhances the ability of a substance to dissolve in water. Thus methane is insoluble in water but methyl alcohol dissolves in all

TABLE 14.2
The Influence of the Alcohol Group on Boiling Points

Name	Structure	Formula Weight	Boiling Point (°C)	Difference in Boiling Point
Ethane	CH_3CH_3	30	−89	154
Methyl alcohol	CH_3OH	32	65	
Propane	$CH_3CH_2CH_3$	44	−42	120
Ethyl alcohol	CH_3CH_2OH	46	78	
Butane	$CH_3CH_2CH_2CH_3$	58	0	197
Ethylene glycol	$HOCH_2CH_2OH$	62	197	

FIGURE 14.1
Hydrogen bonding in water *(a)*
and in alcohols *(b)*.

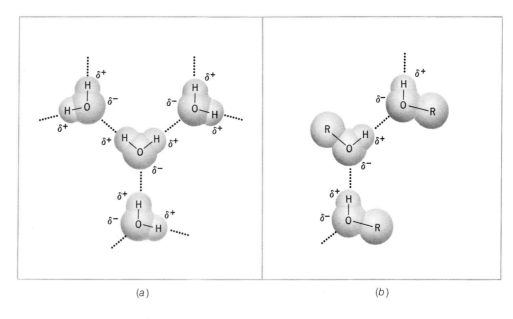

(a)   (b)

proportions. Methyl alcohol dissolves because its molecules can donate and accept hydrogen bonds and so it can slip into the hydrogen-bonding network in water, as seen in Figure 14.2.

As the size of a monohydric alcohol molecule increases, its molecules become more and more alkane-like. In 1-decanol, for example,

$$CH_3CH_2CH_2CH_2CH_2CH_2CH_2CH_2CH_2CH_2OH$$
1-Decanol

the small, water-like OH group is overwhelmed by the long hydrocarbon chain. Water molecules have no attraction for this part of the molecule, and the flexings and twistings of this chain interfere too much with the hydrogen-bonding networks in water for water to let these molecules into solution. This alcohol, and most with five or more carbons are insoluble in water. However, they do dissolve in such nonpolar solvents as diethyl ether, gasoline, and petroleum ether.

We need to be aware of these facts, because some substances in cells must be in solution and others must stay out of solution. The presence or absence of OH groups or of long hydrocarbon groups are major factors in determining relative solubilities. Moreover, the body

FIGURE 14.2
How a short-chain alcohol dissolves in water. *(a)* The alcohol molecule can take the place of a water molecule in the hydrogen bonding network of water. *(b)* An alkane molecule cannot break into the hydrogen bonding network in water, so the alkane can't dissolve.

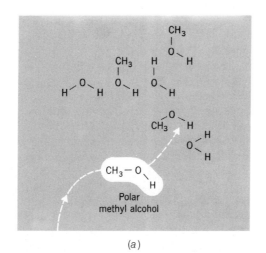

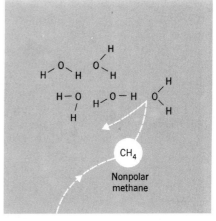

(a)   (b)

handles many toxic substances by having special enzymes in the liver attach —OH groups to their molecules so that the toxic chemicals can more easily be carried by the blood to the kidneys, where they are passed out via the urine.

14.3 CHEMICAL PROPERTIES OF ALCOHOLS

The loss of water (dehydration) and the loss of hydrogen (oxidation) are two very important reactions of alcohols.

Alcohols react with both inorganic and organic compounds, but we will study only the inorganic reactants in this chapter. Before continuing, however, we have to point out some properties that the alcohols do not have, but which might be assumed because of their —OH groups.

When an alcohol dissolves in water, it doesn't raise or lower the pH.

The alcohols are *not* donors of hydroxide ions. Furthermore, they are not donors of hydrogen ions, either. The alcohols are very similar to water in these respects. Water, as you know, does not ionize — at least not to more than an extent of roughly 0.0000001%. In other words, alcohols are neither acids nor bases in the Arrhenius sense, but are neutral compounds.

When some alcohols undergo dehydration, their carbon skeletons rearrange, but we won't study any examples.

The Dehydration of Alcohols to Alkenes.   The action of heat and a strong acid catalyst on an alcohol molecule causes it to lose a water molecule, and a carbon–carbon double bond emerges. This reaction is called the dehydration of an alcohol, and it results in an alkene. The pieces needed to make up a water molecule — one H and one OH — come from *adjacent* carbons.

In general:

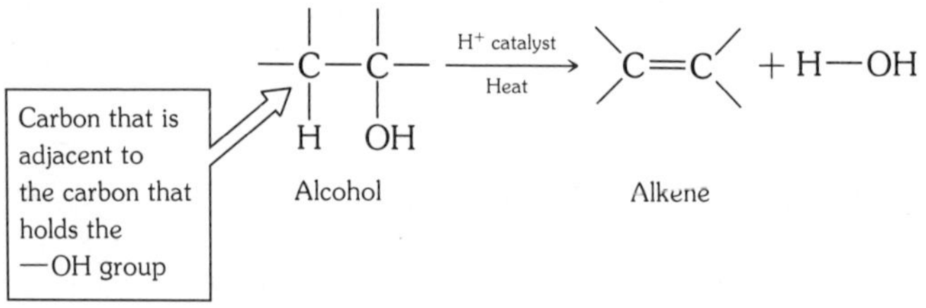

Specific examples are as follows.

$$CH_3CH_2OH \xrightarrow[170-180\ °C]{Concd\ H_2SO_4} CH_2{=}CH_2 + H_2O$$
$$\text{Ethanol} \qquad\qquad\qquad \text{Ethene}$$

$$\underset{\text{2-Butanol}}{CH_3CH_2\overset{\overset{\textstyle OH}{|}}{C}HCH_3} \xrightarrow[100\ °C]{60\%\ H_2SO_4} \underset{\substack{\text{2-Butene}\\ \text{(chief product)}}}{CH_3CH{=}CHCH_3} + \underset{\substack{\text{1-Butene}\\ \text{(by-product)}}}{CH_3CH_2CH{=}CH_2} + H_2O$$

The second example illustrates another fact about the dehydration of an alcohol. Water can split out in two directions from 2-butanol, because its molecules have hydrogen atoms on *two* carbons that are adjacent to the carbon that holds the —OH group. Therefore two alkenes are possible. Some molecules of 2-butanol split out water to give 2-butene, and other molecules react to give 1-butene. However, whenever two or more alkenes seem to be possible in a dehydration, generally just one alkene is the major product — the one whose molecules have the greatest number of alkyl groups attached to the double bond. Chemists have learned that the more substituted an alkene is, the more stable it is. There are two alkyl groups at the double bond in 2-butene (two methyl groups) but only one alkyl group (an ethyl group) in 1-butene. Therefore 2-butene is the more stable alkene and it is the major product, and 1-butene, the less stable alkene, is a by-product. This complication is not a problem in

When a strong acid is added to an alcohol, the first chemical event is the ionization of the acid. The reaction is exactly analogous to the ionization of a strong acid when it is added to water—a proton transfers from the acid to the oxygen atom of a solvent molecule. Because sulfuric acid is often used as the catalyst in the dehydration of an alcohol, we will use it to illustrate these reactions. Its ionization in water establishes the following equilibrium (with the products very strongly favored).

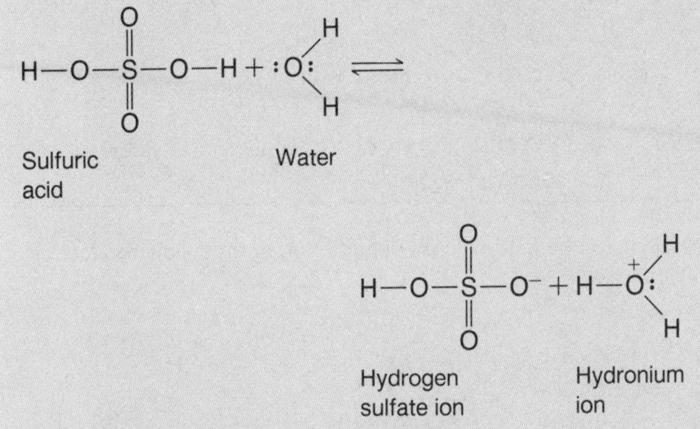

H—O—S—O—H + :O: ⇌

Sulfuric acid   Water

H—O—S—O⁻ + H—O:⁺

Hydrogen sulfate ion   Hydronium ion

Similarly, in an alcohol:

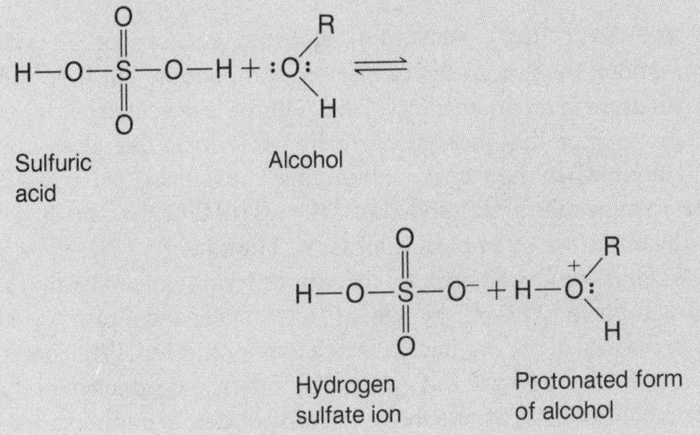

H—O—S—O—H + :O: ⇌

Sulfuric acid   Alcohol

H—O—S—O⁻ + H—O:⁺

Hydrogen sulfate ion   Protonated form of alcohol

All of the covalent bonds to the oxygen atoms in either the hydronium ion or in the protonated form of the alcohol are weak including, in the latter, *the covalent bond to carbon.* The weakness of this bond is caused by the work of the catalyst. As ions and molecules bump into each other, some of the ions of the protonated form of the alcohol break up, as we can illustrate using the protonated form of ethyl alcohol.

CH_3CH_2—O:⁺ ⇌ $CH_3CH_2^+$ + :O:

Protonated form of ethyl alcohol   Ethyl carbocation

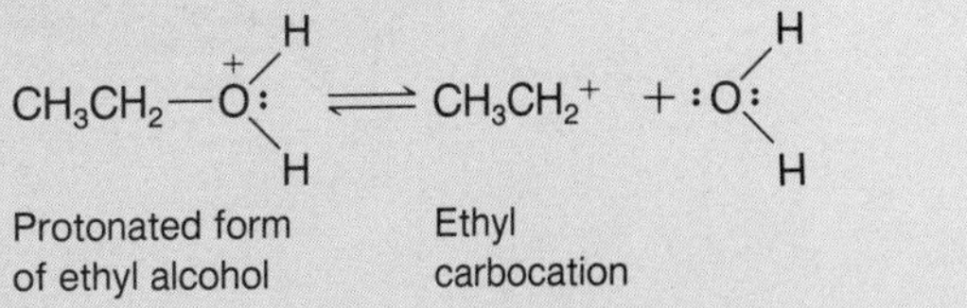

All carbocations are unstable, because a carbon atom does not handle anything less than an octet very well. The octet for carbon in the ethyl carbocation is restored when a proton from the carbon adjacent to the site of the positive charge transfers to a proton acceptor. As this proton transfers, the electron pair of its covalent bond to carbon pivots in to form the second bond of the emerging double bond. It is a smooth, synchronous operation. One acceptor for the proton is the hydrogen sulfate ion, and its acceptance of a proton restores the catalyst, as follows.

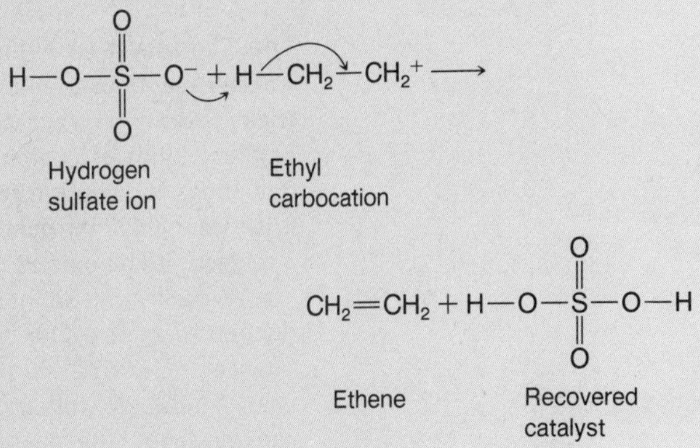

H—O—S—O⁻ + H—CH_2—CH_2^+ ⟶

Hydrogen sulfate ion   Ethyl carbocation

CH_2=CH_2 + H—O—S—O—H

Ethene   Recovered catalyst

Enzymes are exceedingly selective in what they do and how they control reactions.

alcohol dehydrations among body chemicals, because enzymes direct the reactions in very specific ways. Special Topic 14.2 explains how the acid catalyzes the dehydration of an alcohol.

EXAMPLE 14.2   WRITING THE STRUCTURE OF THE ALKENE THAT FORMS WHEN AN ALCOHOL UNDERGOES DEHYDRATION

Problem: What is the product of the dehydration of isobutyl alcohol?

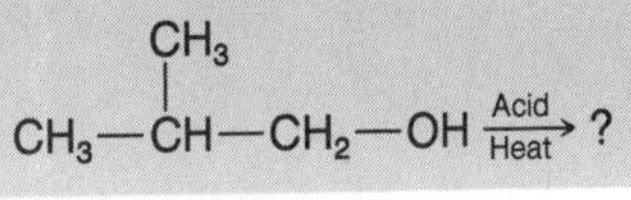

$$CH_3\text{—}CH\text{—}CH_2\text{—}OH \xrightarrow[\text{Heat}]{\text{Acid}} ?$$

with CH_3 branch on the CH

Solution: All we have to do is rewrite the given structure of the alcohol, except that we leave out the —OH group from its carbon, and we omit one H from a carbon that is adjacent to the carbon that holds the —OH group. Doing this to isobutyl alcohol leaves:

$$CH_3-\overset{\overset{\displaystyle CH_3}{|}}{C}-CH_2 \quad \text{(Incomplete answer)}$$

Between the two adjacent carbons that lost the H and the OH groups we now have to write in another bond to make a double bond:

$$CH_3-\overset{\overset{\displaystyle CH_3}{|}}{C}=CH_2 \quad \text{2-Methylpropene, the answer}$$

Each carbon now has four bonds, unlike the situation in the incomplete answer.

PRACTICE EXERCISE 5   Write the structures of the alkenes that can be made by the dehydration of the following alcohols.

(a) $CH_3CH_2CH_2OH$ (b) $CH_3\underset{\underset{\displaystyle OH}{|}}{CH}CH_3$ (c) $CH_3-\overset{\overset{\displaystyle CH_3}{|}}{\underset{\underset{\displaystyle CH_3}{|}}{C}}-OH$ (d) 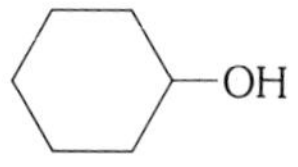 $\langle hexagon \rangle$—OH

The Oxidation of Primary and Secondary Alcohols.   Redox reactions or electron-transfer reactions are common among many families of organic compounds, but it is often tricky to determine exactly which atom in an organic molecule gains or loses electrons in one of these kinds of chemical changes. For this reason, chemists sometimes use alternative definitions of oxidation and reduction when they work with organic substances. An *oxidation* is the loss of H or the gain of O by a molecule, and a *reduction* is the loss of O or the gain of H by a molecule. The oxidations of alcohols are examples of losses of H atoms.

We will not take the time to learn how to balance the equations for organic oxidation-reduction reactions. Our needs will be met simply by the use of unbalanced "reaction sequences." Our chief interest is in the fate of the organic molecule that is oxidized. What does it change into? We will simplify even further and use the symbol (O) for any oxidizing agent that can bring about the reaction. Typical oxidizing agents that work are potassium permanganate, $KMnO_4$, and either sodium or potassium dichromate, $Na_2Cr_2O_7$ or $K_2Cr_2O_7$. In the body, there are other substances that remove hydrogen from alcohols, and we will learn about them later.

When an alcohol reacts with an oxidizing agent and undergoes the loss of hydrogen, molecular hydrogen, H_2, does not itself form. Instead, the hydrogen atoms end up in the form of a water molecule where the oxygen atom comes from the oxidizing agent. One H comes from the —OH group of the alcohol, and the other H comes from the carbon that has been holding the —OH group.

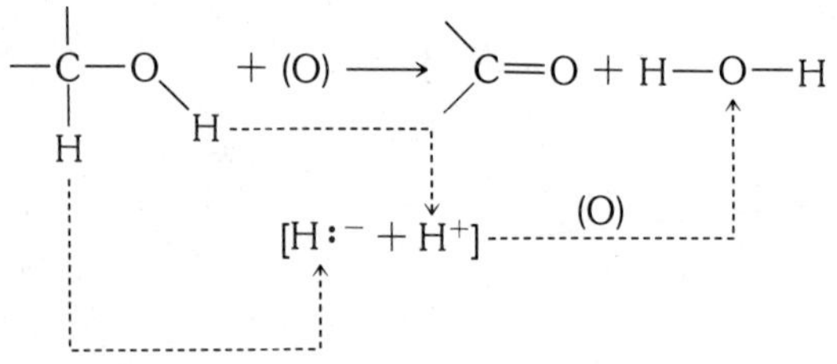

$$-\overset{|}{\underset{\underset{\displaystyle H}{|}}{C}}-O{\diagdown}_H + (O) \longrightarrow {\diagdown}C=O + H-O-H$$

$$[H\!:^- + \overset{+}{H}] ------ (O)$$

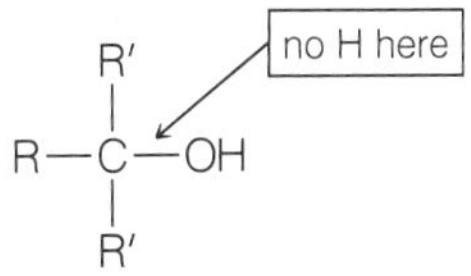

R—C—OH

no H here

$$R-\overset{\overset{R'}{|}}{\underset{\underset{R'}{|}}{C}}-OH$$

3° alcohol system

$$R-\overset{\overset{\displaystyle O}{\parallel}}{C}-H$$

Aldehyde

$$R-\overset{\overset{\displaystyle O}{\parallel}}{C}-OH$$

Carboxylic acid

$$R-\overset{\overset{\displaystyle O}{\parallel}}{C}-R'$$

Ketone

Only 1° and 2° alcohols can be oxidized by this kind of loss of hydrogen; 3° alcohol molecules do not have an H atom on the carbon that holds the —OH group, so 3° alcohols cannot be oxidized in this way.

The organic product of the oxidation of a 1° or a 2° alcohol has a carbon-oxygen double bond, but the particular family that forms depends on the subclass of the alcohol used. The product can be an aldehyde, a carboxylic acid, or a ketone, all of which have carbon-oxygen double bonds.

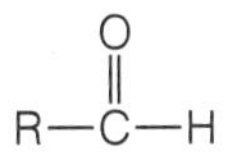

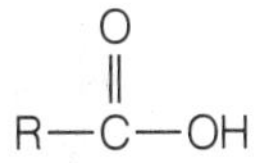

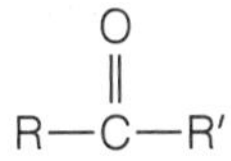

In laboratory experiments that use conventional oxidizing agents such as potassium permanganate, 1° alcohols are oxidized first to aldehydes and then to carboxylic acids. It is difficult to stop the oxidation of a 1° alcohol at the aldehyde stage, because aldehydes oxidize more easily than alcohols. In body cells, of course, this is not a problem, because different enzymes are needed for each step. However, the oxidation of a 1° alcohol is usually carried out with enough oxidizing agent to take the oxidation all the way to the carboxylic acid.

In general,

$$RCH_2OH \xrightarrow[\text{(Enzyme-catalyzed)}]{(O)} R-\overset{\overset{\displaystyle O}{\parallel}}{C}-H + H_2O$$

1° Alcohol   Aldehyde

$$RCH_2OH \xrightarrow[\substack{\text{(by KMnO}_4\text{,}\\ \text{for example)}}]{(O)} (R-\overset{\overset{\displaystyle O}{\parallel}}{C}-H) \xrightarrow{\text{More (O)}} R-\overset{\overset{\displaystyle O}{\parallel}}{C}-O^-K^+ + MnO_2 + KOH$$

Aldehyde   Salt of a   Manganese
                  carboxylic   dioxide
                  acid

$$\downarrow \text{Add } H^+$$

$$R-\overset{\overset{\displaystyle O}{\parallel}}{C}-OH + K^+$$

Carboxylic acid

As the oxidation proceeds, the deep purple color of the permanganate ion gives way to the appearance of a brown precipitate of manganese dioxide. When (O) is the dichromate ion, its bright orange color changes to the bright green color of the Cr^{3+} ion as the oxidation takes place.

$$RCH_2OH + Na_2Cr_2O_7 \xrightarrow{H^+} (R-\overset{\overset{\displaystyle O}{\parallel}}{C}-H) \xrightarrow{\text{more } Cr_2O_7^{2-}} R-\overset{\overset{\displaystyle O}{\parallel}}{C}-OH + Cr^{3+}$$

The following are some specific examples of making aldehydes from 1° alcohols.

$$CH_3CH_2CH_2OH \xrightarrow{Cr_2O_7^{2-},\ H^+} CH_3CH_2\overset{\overset{\displaystyle O}{\parallel}}{C}H$$

1-Propanol   Propanal
(b.p. 97 °C)   (b.p. 55 °C)

$$CH_3CH_2CH_2CH_2OH \xrightarrow{Cr_2O_7^{2-},\ H^+} CH_3CH_2CH_2\overset{\overset{\displaystyle O}{\parallel}}{C}H$$

1-Butanol   Butanal
(b.p. 118 °C)   (b.p. 82 °C)

These oxidations can be stopped at the aldehyde stage because the aldehydes boil so much lower than their parent alcohols that they can be boiled out of the reaction mixture as soon as they form. This removes them from the oxidizing agent, so they are not further oxidized. However, it is easy to make the oxidation of a 1° alcohol go to the carboxylic acid stage, as in the next example. All it takes is enough oxidizing agent.

$$CH_3CH_2CH_2OH \xrightarrow{Cr_2O_7^{2-}, H^+} CH_3CH_2\overset{\displaystyle O}{\overset{\displaystyle \|}{C}}OH$$

1-Propanol   Propanoic acid

EXAMPLE 14.3 **WRITING THE STRUCTURE OF THE PRODUCT OF THE OXIDATION OF A PRIMARY ALCOHOL**

Problem: What aldehyde and what carboxylic acid could be made by the oxidation of ethyl alcohol?

Solution: First, write the structure of the given alcohol.

$$CH_3—CH_2—OH$$

Then either cross out or erase the H on the OH group, and reduce by one the number of H atoms on the carbon that holds the OH group. When we do this to ethyl alcohol, we have

$$CH_3—CH—O \quad \text{(Incomplete structure)}$$

Finally, we make the carbon-oxygen bond a double bond. Thus the aldehyde that forms is

$$CH_3—CH{=}O \quad \text{or} \quad CH_3—\overset{\displaystyle O}{\overset{\displaystyle \|}{C}}—H \quad \text{(Ethanal)}$$

The second way of writing the structure of the product might make it easier to solve the second part of the problem: write the structure of the carboxylic acid that can be made from ethyl alcohol. This is most easily written by inserting an oxygen atom between the carbon atom of the C=O group in ethanal and the H atom attached to it:

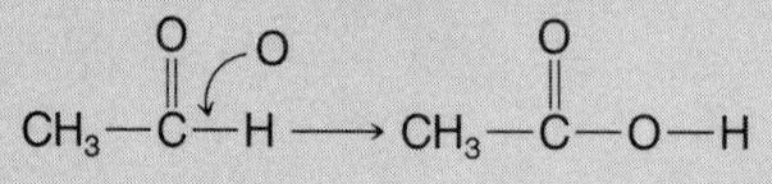

$$CH_3—\overset{\displaystyle O}{\overset{\displaystyle \|}{C}}\overset{O}{\frown}H \longrightarrow CH_3—\overset{\displaystyle O}{\overset{\displaystyle \|}{C}}—O—H \quad \text{(Acetic acid)}$$

PRACTICE EXERCISE 6   Write the structures of the aldehydes and carboxylic acids that can be made by the oxidation of the following alcohols.

(a) $CH_3—\overset{\displaystyle CH_3}{\overset{\displaystyle |}{CH}}—CH_2—OH$ (b) 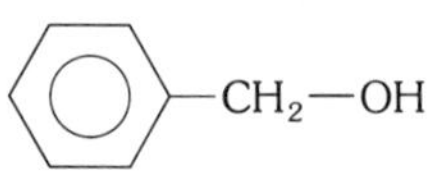 ⬡—$CH_2—OH$ (c) $CH_3—OH$

Secondary alcohols are oxidized to ketones. Because ketones strongly resist further oxidation, they are easily made by this reaction.

In general, for 2° alcohols:

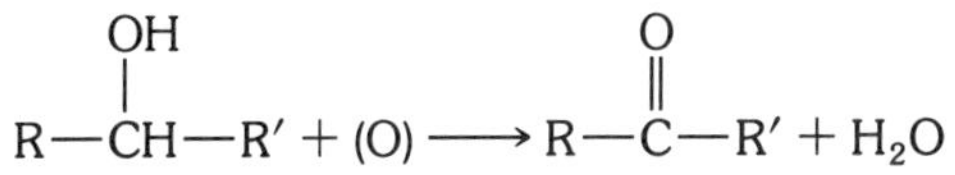

$$\underset{\text{2° alcohol}}{R-\overset{\overset{\displaystyle OH}{|}}{C}H-R'} + (O) \longrightarrow \underset{\text{Ketone}}{R-\overset{\overset{\displaystyle O}{\|}}{C}-R'} + H_2O$$

Specific examples are:

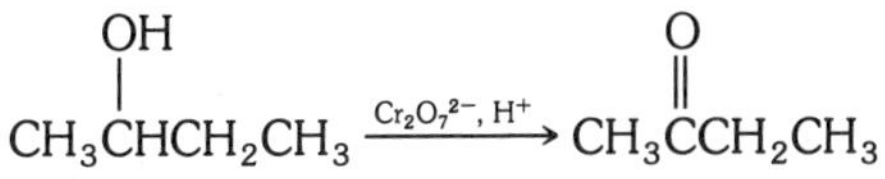

$$\underset{\text{2-Butanol}}{CH_3\overset{\overset{\displaystyle OH}{|}}{C}HCH_2CH_3} \xrightarrow{Cr_2O_7{}^{2-},\ H^+} \underset{\text{Butanone}}{CH_3\overset{\overset{\displaystyle O}{\|}}{C}CH_2CH_3}$$

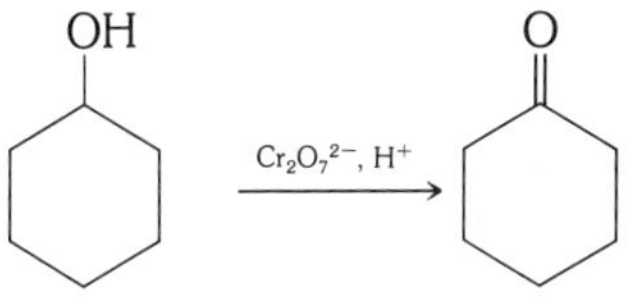

Cyclohexanol $\xrightarrow{Cr_2O_7{}^{2-},\ H^+}$ Cyclohexanone

---

EXAMPLE 14.4 WRITING THE STRUCTURE OF THE PRODUCT OF THE OXIDATION OF A SECONDARY ALCOHOL

Problem: What ketone forms when 2-propanol is oxidized?

Solution: We work this essentially as we worked Example 14.3. We start with the structure of the given alcohol, and then we cross out or erase the H on the —OH group and the H on the carbon that is holding this —OH group.

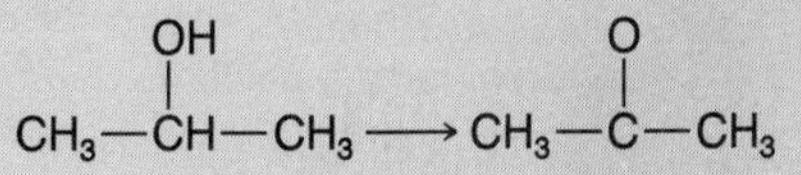

$$CH_3-\overset{\overset{\displaystyle OH}{|}}{C}H-CH_3 \longrightarrow CH_3-\overset{\overset{\displaystyle O}{|}}{C}-CH_3 \quad \text{(Incomplete answer)}$$

Then we make the bond to oxygen a double bond:

The name chemists always use for propanone is *acetone*.

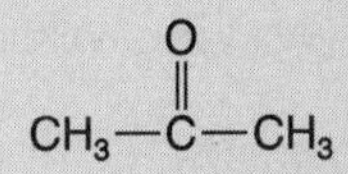

$$CH_3-\overset{\overset{\displaystyle O}{\|}}{C}-CH_3 \quad \text{(Acetone, or propanone)}$$

PRACTICE EXERCISE 7 Write the structures of the ketones that can be made by the oxidation of the following alcohols.

$$\text{(a) } CH_3-\overset{\overset{\displaystyle OH}{|}}{C}H-CH_2CH_3 \quad \text{(b) } \overset{\overset{\displaystyle OH}{|}}{\underset{}{C}}H-CH_3 \quad \text{(c) } \bigcirc\!\!-OH$$

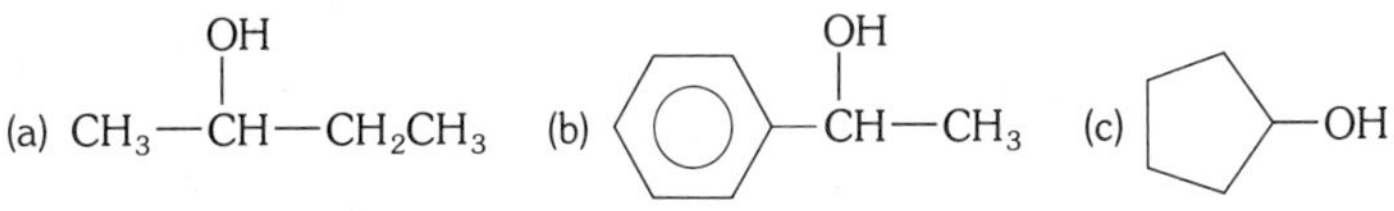

PRACTICE EXERCISE 8 What are the products of the oxidation of the following alcohols? If the alcohol is a 1° alcohol, show the structures of both the aldehyde and the carboxylic acid that could be made, depending on the conditions. If the alcohol cannot be oxidized, write "no reaction."

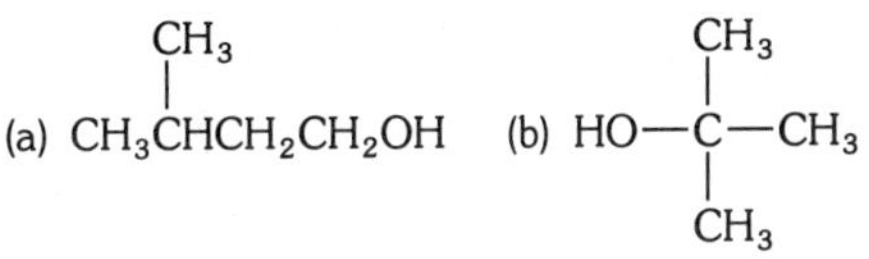

$$\text{(a) } CH_3\overset{\overset{\displaystyle CH_3}{|}}{C}HCH_2CH_2OH \quad \text{(b) } HO-\overset{\overset{\displaystyle CH_3}{|}}{\underset{\underset{\displaystyle CH_3}{|}}{C}}-CH_3$$

$$
\begin{array}{ccc}
& \text{CH}_3 & \text{CH}_3 \quad \text{OH} \\
& | & | \quad\quad | \\
\text{(c)} \quad \text{CH}_3\text{CCH}_2\text{OH} & & \text{(d)} \quad \text{CH}_3\text{CH}-\text{CH}-\text{CH}_3 \\
& | & \\
& \text{CH}_3 &
\end{array}
$$

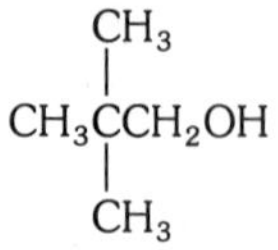

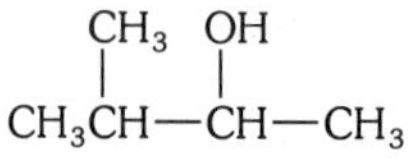

14.4   THIOALCOHOLS AND DISULFIDES

Both the —SH group, an easily oxidized group, and the —S—S— system, an easily reduced group, are present in molecules of proteins.

Mercaptan is a contraction of mercury capturer. Compounds with —SH groups form precipitates with mercury ions.

Alcohols, R—O—H, can be viewed as alkyl derivatives of water, H—O—H. Similar derivatives of hydrogen sulfide, H—S—H, are also known, and they are in the family called the **thioalcohols** or, more commonly, the **mercaptans.**

$$
\begin{array}{ccc}
\text{R—S—H} & \text{R—S—R}' & \text{R—S—S—R}' \\
\text{Thioalcohols} & \text{Thioethers} & \text{Disulfides} \\
\text{(mercaptans)} & &
\end{array}
$$

The —SH group is variously called the thiol group, the mercaptan group, or the sulfhydryl group.

Lower-formula-weight thioalcohols are present in and are responsible for the respect usually accorded skunks.

Table 14.3 gives the IUPAC names and structures of a few thioalcohols. (We will not study the IUPAC nomenclature as a separate topic because it is straightforward, and because our interest in thioalcohols is limited just to one reaction.) Some very important properties of proteins depend on the presence of the —SH group, which is contributed to protein structure by one of the building blocks of proteins, the amino acid called cysteine (Table 14.3).

Dialkyl derivatives of water, the ethers (R—O—R'), have their sulfur counterparts, too, the thioethers. (You can see that the prefix *thio-* indicates the replacement of an oxygen atom by a sulfur atom.) Although proteins also have the thioether system, our studies will not require that we learn anything of the chemistry of this group, and we will say no more about it.

The Oxidation of the —SH Group.   The one reaction of thioalcohols that will be important in our study of proteins is oxidation. Thioalcohols can be oxidized to **disulfides,** compounds whose molecules have two sulfur atoms joined by a covalent bond, R—S—S—R'.

TABLE 14.3
Some Common Thioalcohols

Name	Structure	Boiling Point (°C)
Methanethiol	CH_3SH	6
Ethanethiol	CH_3CH_2SH	36
1-Propanethiol	$CH_3CH_2CH_2SH$	68
1-Butanethiol	$CH_3CH_2CH_2CH_2SH$	98
Cysteine (a monomer for proteins)	$^+NH_3-\overset{\displaystyle }{CH}-\overset{\overset{\text{O}}{\|\|}}{C}-O^-$ $\underset{}{CH_2-SH}$	(Solid)

In general:

$$R—S—H + H—S—R + (O) \longrightarrow R—S—S—R + H_2O$$

Two molecules of   One molecule
a thioalcohol   of a disulfide

Specific example:

$$2CH_3SH + (O) \longrightarrow CH_3—S—S—CH_3 + H_2O$$

Methanethiol   Dimethyl disulfide

EXAMPLE 14.5 WRITING THE PRODUCT OF THE OXIDATION OF A THIOALCOHOL

Problem: What is the product of the oxidation of ethanethiol?

Solution: Because the oxidation of the SH group generates the —S—S— group, begin simply by writing this group down.

$$—S—S—$$

Then attach the alkyl group from the thioalcohol, one on each sulfur atom. Ethanethiol furnishes ethyl groups, so the answer is:

$$CH_3CH_2—S—S—CH_2CH_3 \quad \text{(Diethyl disulfide)}$$

If the problem had called for an equation, then we would have to use the coefficient of 2 for the ethanethiol.

$$2CH_3CH_2SH + (O) \longrightarrow CH_3CH_2—S—S—CH_2CH_3 + H_2O$$

However, this coefficient isn't necessary when all we are asked to do is to write the structure of the disulfide that can be made by the oxidation of ethanethiol. Always remember that balancing an equation comes *after* you have written the correct formulas for the reactants and products.

The Reduction of Disulfides. The sulfur-sulfur bond in disulfides is very easily reduced, and the products are molecules of the thioalcohols from which the disulfide could be made. We will use the symbol (H) to represent any reducing agent that can do the task, just as we used (O) for an oxidizing agent.

In general:

$$R—S—S—R + 2(H) \longrightarrow R—S—H + H—S—R$$

One molecule   Two molecules of
of disulfide   a thioalcohol

Specific example:

$$CH_3CH_2—S—S—CH_2CH_2CH_3 + 2(H) \longrightarrow$$
$$CH_3CH_2—S—H + H—S—CH_2CH_2CH_3$$

EXAMPLE 14.6   WRITING THE PRODUCT OF THE REDUCTION OF A DISULFIDE

Problem: What forms when the following disulfide reacts with a reducing agent?

$$CH_3—S—S—CH_2CH_3$$

Solution: The easiest approach is simply to split the disulfide molecule between the two sulfur atoms:

$$CH_3—S\!+\!S—CH_2CH_2 \dashrightarrow CH_3—S + S—CH_2CH_3 \quad \text{(Incomplete)}$$

Then attach one H to each sulfur to make the —SH groups.

$$CH_3—S—H + H—S—CH_2CH_3 \quad \text{(The answers)}$$
Methanethiol   Ethanethiol

PRACTICE EXERCISE 9   Complete the following equations by writing the structures of the products that form. If no reaction occurs (insofar as we have studied organic chemistry), write "no reaction."

(a) $CH_3—S—S—CH_3 + (H) \longrightarrow ?$ (b) $CH_3—\underset{\underset{SH}{|}}{CH}—CH_3 + (O) \longrightarrow ?$

(c) 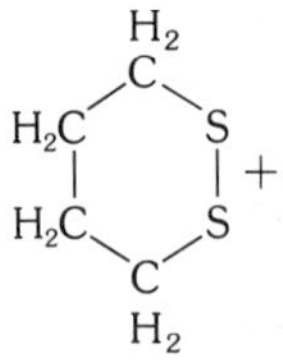 $+ (H) \longrightarrow ?$ (d) 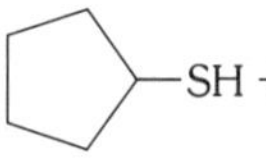 $—SH + (O) \longrightarrow ?$

14.5 PHENOLS

Phenols are weak acids that can neutralize sodium hydroxide, and their benzene rings are easily oxidized.

In **phenols,** the —OH group is attached to a benzene ring, and phenols are widespread in nature and in commerce. The simplest member of this family is called phenol, and it is a raw material for making aspirin. Vanillin, a flavoring agent, is both an aldehyde and a phenol. Tyrosine, one of the amino acids the body uses to make proteins, is also a phenol. Special Topic 14.3 describes some other phenols.

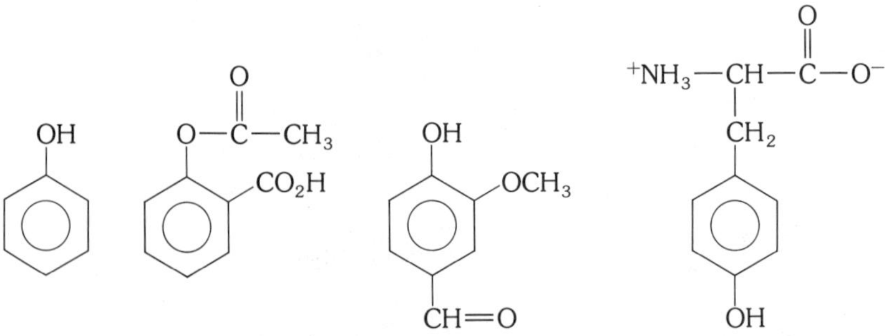

Phenol   Aspirin   Vanillin   Tyrosine

SPECIAL TOPIC 14.3 SOME IMPORTANT PHENOLS

Phenol (carbolic acid). Phenol, as we have already noted, was the first antiseptic to be used by Joseph Lister (1827–1912), an English surgeon. It kills bacteria by denaturing their proteins—that is, by destroying the abilities of their proteins to function normally. (Interestingly, phenol does not denature nucleic acids, the chemicals of heredity.) However, phenol is dangerous to healthy tissue—it is a general protoplasmic poison—and other antiseptics have been developed since Lister's time.

BHA and **BHT.** BHA and BHT are widely used as antioxidants in gasoline, lubricating oils, rubber, edible fats and oils, and in materials used for packaging foods that might turn rancid. Being phenols and therefore easily oxidized themselves, they act by interfering with oxidations of the materials they are designed to protect.

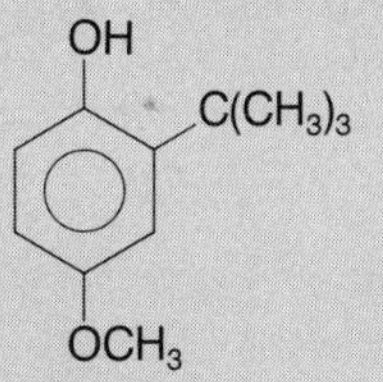

BHA (butylated hydroxy anisole)

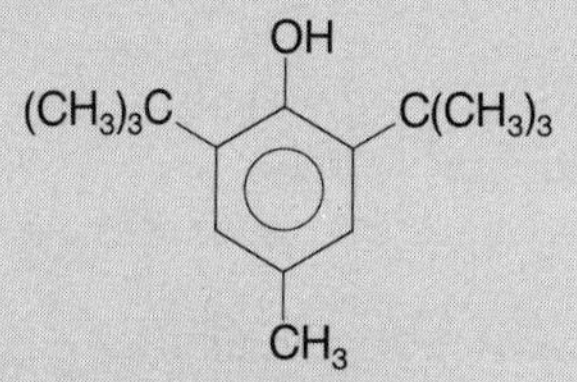

BHT (butylated hydroxy toluene)

Unlike alcohols, phenols cannot be dehydrated. Another difference between alcohols and phenols is that phenols are weak acids. They can neutralize sodium hydroxide. For example:

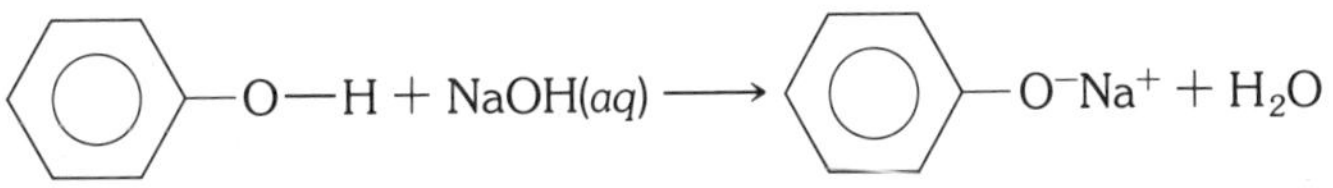

Phenol Sodium phenoxide

The benzene ring in a phenol is easily oxidized, but the oxidations produce complex mixtures that include some highly colored compounds. Even phenol crystals that are left exposed to air slowly turn dark because of the oxidation caused by the oxygen in the air. This is in sharp contrast to benzene itself, which strongly resists oxidation. In the body, special hydroxylating enzymes steer the oxidations of the benzene ring in phenols to specific products, and we will encounter such reactions later.

14.6 ETHERS

The ethers are almost as chemically unreactive as the alkanes.

Ethers are compounds in whose molecules two organic groups are joined to the same oxygen atom, and their family structure is R—O—R′. The two groups can be almost anything except that the carbon joined to the central oxygen atom cannot be the one in C=O. The first three compounds given below are ethers, but methyl acetate is in the ester family, not the ether family. (We will study esters in a later chapter.)

$CH_3CH_2—O—CH_2CH_3$

Diethyl ether

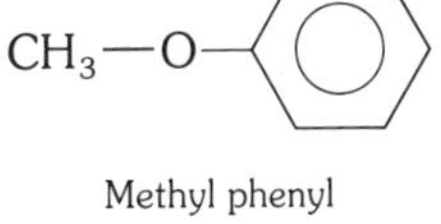

Methyl phenyl ether

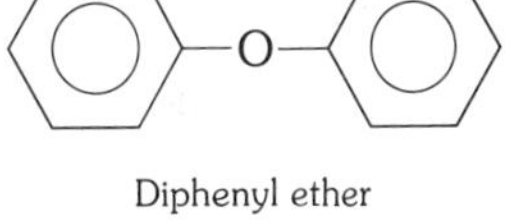

Diphenyl ether

$$CH_3—\overset{\overset{\displaystyle O}{\|}}{C}—O—CH_3$$

Methyl acetate (an ester, not an ether)

Table 14.4 gives some examples of ethers, and three are described in Special Topic 14.4, including methyl *t*-butyl ether, which, in preliminary tests on humans conducted in 1985, rapidly dissolves gallstones.

The common names of ethers are made by naming the groups attached to the oxygen and adding the word *ether,* as illustrated in Table 14.4.

Diethyl Ether ("ether"). Diethyl ether is a colorless, volatile liquid with a pungent, somewhat irritating odor, and it was once widely used as an anesthetic. It acts as a depressant for the central nervous system and somewhat of a stimulant for the sympathetic system. It exerts an effect on nearly all tissues of the body.

Because mixtures of ether and air in the right proportions can explode, anesthesiologists avoid using it as an anesthetic whenever possible.

Divinyl Ether (vinethene). Divinyl ether is another anesthetic, but it also forms an explosive mixture in air. Its anesthetic action is more rapid than that of diethyl ether.

Methyl _t_-Butyl Ether. The standard treatment for stones in the gallbladder or the gallbladder duct has been surgical removal of the gallbladder. Medical scientists have long sought a safe way to dissolve the stones where they are, and they have known that a chief constituent of these stones is cholesterol. Cholesterol, as we have already learned, is mostly hydrocarbon-like, so the search has been for a relatively nonpolar solvent for cholesterol that could therefore dissolve the stones without causing serious side effects, and that would be at least as safe as the surgical treatment. In 1985, medical scientists at the Mayo Clinic found that methyl _t_-butyl ether worked rapidly and well when used with four patients, and more studies are in progress. This ether is less toxic than diethyl ether, it can be inserted into the gallbladder by a catheter, it remains a liquid at body temperature, and it works in just a few hours with no evidence of disagreeable side effects.

Compound	Boiling Point (°C)
Pentane	36
Diethyl ether	35
1-Butanol	118

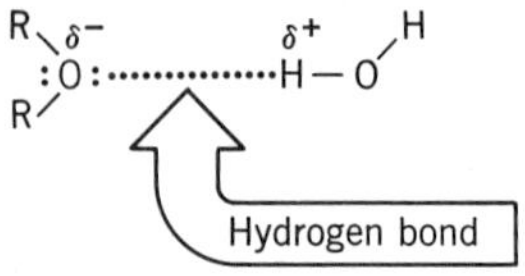

FIGURE 14.3
An ether molecule can accept a hydrogen bond from a water molecule. (The δ^- end of a hydrogen bond is the acceptor end.)

Properties of Ethers. Because the ether group cannot donate hydrogen bonds, the boiling points of simple ethers are more like those of the alkanes of comparable formula weights than those of the alcohols. However, the oxygen of the ether group can accept hydrogen bonds (Figure 14.3), so ethers are more soluble in water than alkanes. For example, both 1-butanol and its isomer, diethyl ether, dissolve in water to the extent of about 8 g/100 mL, but pentane is completely insoluble in water.

Chemically, ethers are similar to alkanes. At room temperature, they do not react with strong acids, bases, oxidizing, or reducing agents. Like all organic compounds, they burn. We will learn no reactions of ethers, but we must be able to recognize this group and then remember that it is not very reactive, particularly in the environment within the body.

The Synthesis of Ethers from Alcohols. Ethers can be synthesized in many ways, and we will look briefly at one method because it relates to some of the chemistry we will study in the next chapter.

TABLE 14.4
Some Ethers

Common Name	Structure	Boiling Point (°C)
Dimethyl ether	CH_3OCH_3	−23
Methyl ethyl ether	$CH_3OCH_2CH_3$	11
Methyl _t_-butyl ether	$CH_3OC(CH_3)_3$	55.2
Diethyl ether	$CH_3CH_2OCH_2CH_3$	34.5
Dipropyl ether	$CH_3CH_2CH_2OCH_2CH_2CH_3$	91
Methyl phenyl ether	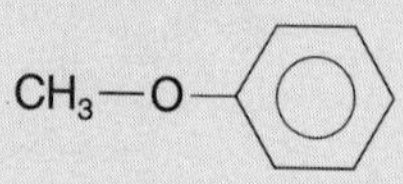 $CH_3-O-\bigcirc$	155
Diphenyl ether	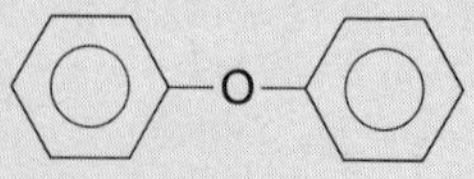 $\bigcirc-O-\bigcirc$	259
Divinyl ether	$CH_2{=}CHOCH{=}CH_2$	29

We learned earlier in this chapter that alcohols can be dehydrated by the action of heat and an acid catalyst to give alkenes. The precise temperature that works best for this reaction has to be determined for each alcohol, because if the temperature is not set correctly, a different kind of dehydration occurs. Water splits out *between* two alcohol molecules rather than from one molecule, and the product is an ether.

In general,

$$R-O-H + H-O-R \xrightarrow{\text{Acid catalyst}} R-O-R + H_2O$$

Two alcohol molecules   Ether

A specific example:

Earlier we learned that concd H_2SO_4 acts on ethanol to give ethene when the temperature is 170 °C.

$$2CH_3CH_2OH \xrightarrow[140\ °C]{H_2SO_4} CH_3CH_2-O-CH_2CH_3 + H_2O$$

Ethyl   Diethyl ether
alcohol

The dehydration that produces an ether usually requires a lower temperature than the dehydration that gives an alkene, but no further details are necessary. Our interest is simply in the *possibility* of making an ether from an alcohol as well as in what ether can be made from a given alcohol.

EXAMPLE 14.7   WRITING THE STRUCTURE OF AN ETHER THAT CAN FORM FROM AN ALCOHOL

Problem: If the conditions are right, 1-butanol can be converted to an ether. What is the structure of this ether?

Solution: What this questions asks is to complete the following reaction:

$$CH_3CH_2CH_2CH_2OH \xrightarrow[\text{Heat}]{\text{Acid catalyst}} ?$$

As usual, the structure of the starting material gives us most of the answer. We know that the ether must get its organic groups from the alcohol, so to write the structure of the ether we write one O atom with two bonds from it.

$$-O-$$

Then we attach the alkyl group of the alcohol, one to each of the bonds.

$$CH_3CH_2CH_2CH_2-O-CH_2CH_2CH_2CH_3$$

Of course, we need *two* molecules of the alcohol to make one molecule of this ether, but always remember, *we balance an equation after we have written the correct formulas for reactants and products.* Remembering that water is the other product, the balanced equation is

$$2CH_3CH_2CH_2CH_2OH \xrightarrow[\text{Heat}]{\text{Acid catalyst}} CH_3CH_2CH_2CH_2-O-CH_2CH_2CH_2CH_3 + H_2O$$

1-Butanol   Dibutyl ether

PRACTICE EXERCISE 10   Write the structures of the ethers to which the following alcohols can be converted.

(a) CH_3OH (b) $CH_3CH_2CH_2OH$ (c) 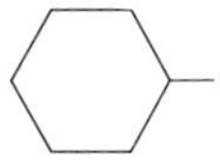 ⬡—OH

SUMMARY

Alcohols   The alcohol system has an —OH group attached to a saturated carbon. The IUPAC names of simple alcohols end in *-ol*, and their chains are numbered to give precedence to the location of the —OH group. The common names have the word *alcohol* following the name of the alkyl group.

Alcohol molecules hydrogen bond to each other and to water molecules. By the action of heat and an acid catalyst they can be dehydrated internally to give carbon–carbon double bonds or externally to give ethers. Primary alcohols can be oxidized to aldehydes and to carboxylic acids. Secondary alcohols can be oxidized to ketones. Tertiary alcohols cannot be oxidized (without breaking up the carbon chain). The —OH group in alcohols does not function either as an acid or a base in the Arrhenius sense.

Thioalcohols   The thioalcohols or mercaptans, R—S—H, are

easily oxidized to disulfides, R—S—S—R, and these are easily reduced back to the original thioalcohols.

Phenols   When the —OH group is attached to a benzene ring, the system is the phenol system, and it is now acidic enough to neutralize strong bases.

Ethers   The ether system, R—O—R′, does not react at room temperature or body temperature with strong acids, bases, oxidizing agents or reducing agents. It can accept hydrogen bonds but cannot donate them.

Reactions studied   Without attempting to present balanced equations, or even all of the inorganic products, we can summarize the reactions studied in this chapter as follows.

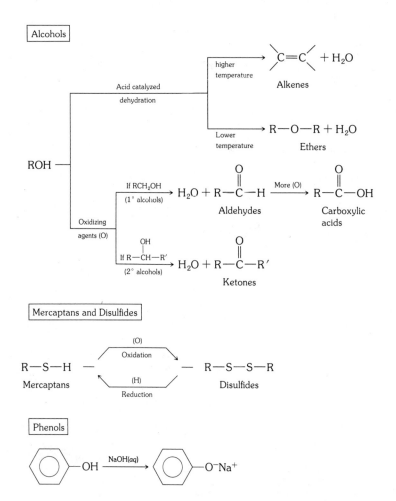

KEY TERMS

The following terms were emphasized in this chapter. They will be used often in the future, so be sure that you master them before continuing.

alcohol	ether	phenols	thioalcohol
alcohol group	glycol	primary alcohol	trihydric alcohol
dihydric alcohol	mercaptan	secondary alcohol	
disulfide	monohydric alcohol	tertiary alcohol	

SELECTED REFERENCE

1 G. E. Vaillant. *The Natural History of Alcoholism.* Harvard University Press, Cambridge, MA, 1983.

REVIEW EXERCISES

The answers to all Review Exercises are in the *Study Guide* that accompanies this book.

Functional Groups

14.1 Name the functional groups to which the arrows point in the structure of eugenol, an oily liquid sometimes used as a substitute for oil of cloves in formulating fragrances and perfumes.

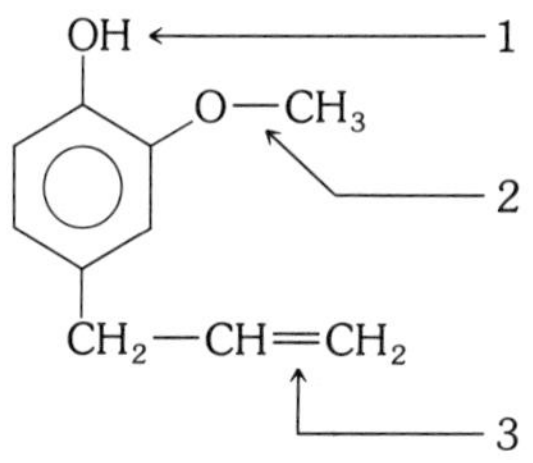

Eugenol

14.2 What are the functional groups to which the arrows point in the structure of geranial, a constituent of oil of lemon grass?

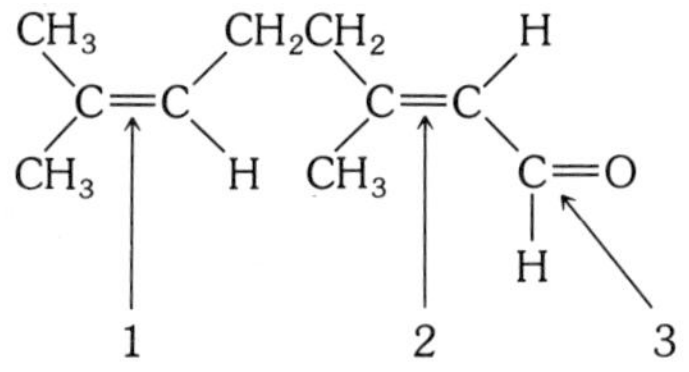

Geranial

14.3 Name the functional groups to which the arrows point in the structure of cortisone, a drug used in treating certain forms of arthritis. If a group is an alcohol, state if it is a $1°$, $2°$, or $3°$ alcohol.

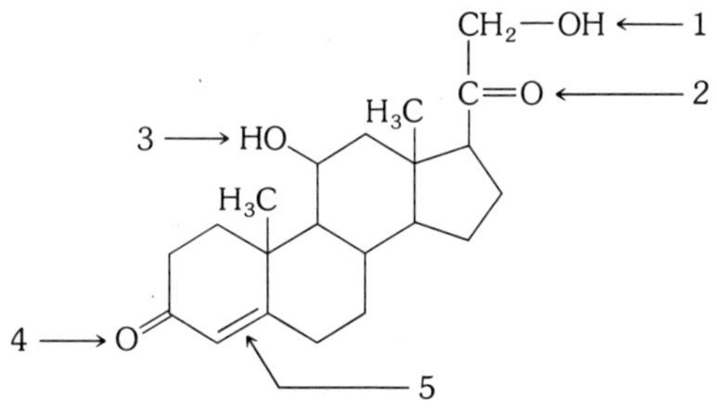

Cortisone

14.4 Give the names of the functional groups to which the arrows point in prostaglandin E_1, one of a family of compounds that are smooth muscle stimulants.

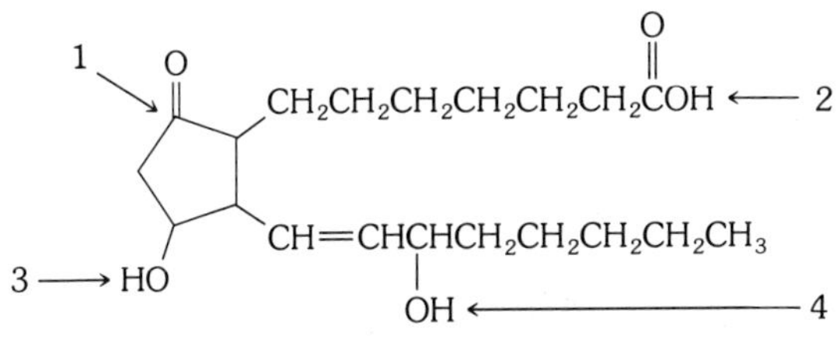

Prostaglandin E_1

Structures and Names

14.5 Write the structure of each compound.
(a) Isobutyl alcohol (b) Isopropyl alcohol
(c) Propyl alcohol   (d) Glycerol

14.6 Write the structures of the following compounds.
(a) Methyl alcohol   (b) t-Butyl alcohol
(c) Ethyl alcohol   (d) Butyl alcohol

14.7 Give the common names of the following compounds.

(a) CH_3CHCH_3 (b) $HOCH_2CH_2OH$
    |
    OH

        CH_3
        |
(c) CH_3-C-OH (d) $HOCH_2CHCH_2OH$
        | |
        CH_3 OH

14.8 What are the common names of the following compounds?
(a) CH_3CH_2OH (b) $CH_3CH_2CH_2OH$
(c) $HOCH_2CHCH_3$ (d) $HOCH_2CH_2CH_2CH_3$
        |
        CH_3

14.9 What is the structure and the IUPAC name of the simplest, stable dihydric alcohol?

14.10 Give the structure and the IUPAC name of the simplest, stable trihydric alcohol.

14.11 Give the IUPAC names for the compounds listed in Review Exercise 14.5.

14.12 What are the IUPAC names for the compounds given in Review Exercise 14.6?

14.13 Provide the IUPAC names for the compounds in Review Exercise 14.7.

14.14 What are the IUPAC names for the structures given in Review Exercise 14.8?

14.15 Write the IUPAC name of the following compound.

        CH_3 CH_3
        | |
$CH_3CH_2CHCH-CHCH_3$
        |
        CH_2OH

14.16 Write the IUPAC name of the following compound.

        OH
        |
  CH_3 CH_2 $CH_3CH_2CH_3$
     | | |
$CH_3CH-CH-C-CHCH_2$
        | |
    CH_3CCH_3 CH_2CH_3
        |
       CH_3

Physical Properties

14.17 Draw a figure that illustrates a hydrogen bond between two molecules of ethyl alcohol. Use a dotted line to represent this bond, and write in the $\delta-$ and $\delta+$ symbols where they belong.

14.18 When ethyl alcohol dissolves in water, its molecules slip into the hydrogen bond network in water. Draw a figure that illustrates this. Use dotted lines for hydrogen bonds, and place $\delta-$ and $\delta+$ symbols where they belong.

14.19 Arrange the following compounds in their order of increasing boiling points. Place the letter symbol of the compound that has the lowest boiling point on the left end of the series, and arrange the remaining letters in the correct order.

    OH
    |
$CH_3CHCH_2CH_3$ $CH_3CH_2CH_2CH_2CH_2CH_3$
     A   **B**

   HO   OH
    | |
CH_3CH-CH_2 CH_3-O-CH_3
    C   **D**

   ___ $<$ ___ $<$ ___ $<$ ___

Lowest   Highest
b.p.   b.p.

14.20 Arrange the following compounds in their order of increasing solubility in water. Place the letter symbol of the compound with the least solubility on the left end of the series, and arrange the remaining letters in the correct order.

HO   OH   OH
 | | |
$CH_3CHCHCHCH_3$ $CH_3CH_2CHCH_2CH_3$ $CH_3CH_2CH_2CH_2CH_3$
 |
OH
 A   **B**   **C**

  ___ $<$ ___ $<$ ___

Least   Most
soluble   soluble

Chemical Properties of Alcohols

14.21 Write the structures of the alkenes that form when the following alcohols undergo acid-catalyzed dehydration.

     OH
     |
(a) $CH_3CHCH_2CH_3$

     CH_3
     |
(b) $CH_3CCH_2CH_2CH_3$
     |
     OH

(c) [cyclohexane ring with CH_3 and OH on same carbon]

(d) [benzene ring]$-CH_2CH_2OH$

(e) [benzene ring]$-CHCH_2-$[benzene ring]
           |
           OH

(f) CH_3-[cyclohexane ring]$-OH$

14.22 Write the structures of the products of the oxidation of the alcohols listed in Review Exercise 14.21. If the alcohol is a 1° alcohol, give the structures of both the aldehyde and the carboxylic acid that could be made by varying the quantities of the oxidizing agent.

14.23 Write the structure of any alcohol that could be dehydrated to give each alkene. In some cases, more than one alcohol would work.

(a) $CH_2{=}CH_2$ (b) $CH_3CH{=}CH_2$

(c) 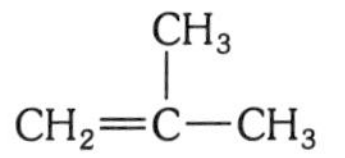 $CH_2{=}\overset{\overset{\displaystyle CH_3}{|}}{C}{-}CH_3$ (d) $CH_3CH{=}CHCH_3$

14.24 Write the structures of any alcohols that could be used to prepare the following compounds by an oxidation.

(a) 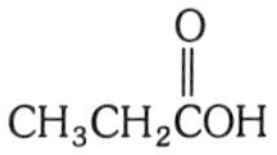 $CH_3CH_2\overset{\overset{\displaystyle O}{||}}{C}OH$ (b) $CH_3\overset{\overset{\displaystyle O}{||}}{C}CH_2CH_2CH_3$

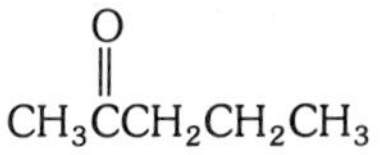

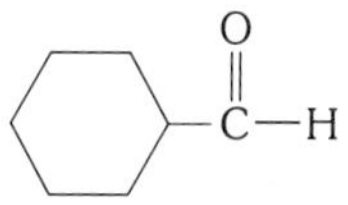

(c) $\langle\ \rangle{-}\overset{\overset{\displaystyle O}{||}}{C}{-}H$ (d) 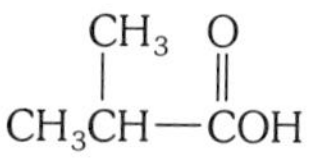 $CH_3\overset{\overset{\displaystyle CH_3}{|}}{C}H{-}\overset{\overset{\displaystyle O}{||}}{C}OH$

14.25 Suppose you were given an unknown liquid and told that it is either *t*-butyl alcohol or butyl alcohol. You're also told that when a few drops of the unknown are shaken with dilute potassium permanganate, the purple color of the MnO_4^- ion disappears and a brown precipitate forms. Which alcohol is it? Explain, and include a reaction sequence for the reaction that occurs.

14.26 An unknown is either pentane or 2-butanol. When it is shaken with $Na_2Cr_2O_7$ and some acid, the orange color of the dichromate ion disappears and the solution becomes green. What is the unknown? Explain, including an equation.

Thioalcohols and Disulfides

14.27 Although we did not systematically develop the rules for naming thioalcohols, the patterns in Table 14.3 make these rules fairly obvious. Write the structures of the following compounds.
(a) Isopropyl mercaptan (b) 1-Butanethiol
(c) Dimethyl disulfide (d) 1,2-Ethanedithiol

14.28 Complete the following reaction sequences by writing the structures of the organic products that form.
(a) $CH_3SH + (O) \longrightarrow$

(b) 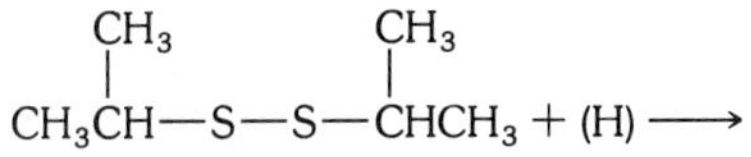 $CH_3\overset{\overset{\displaystyle CH_3}{|}}{C}H{-}S{-}S{-}\overset{\overset{\displaystyle CH_3}{|}}{C}HCH_3 + (H) \longrightarrow$

(c) $\langle\ \rangle{-}SH + (O) \longrightarrow$

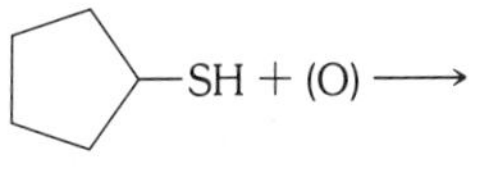

(d) $CH_3{-}S{-}S{-}CH_2CH_2CH_3 + (H) \longrightarrow$

14.29 Propane, methanethiol, and ethanol have similar formula weights, but propane boils at $-42\ °C$, methanethiol at $6\ °C$, and ethanol at $78\ °C$. What do these data suggest about hydrogen bonding in the thioalcohol family? Does it occur at all? Are the hydrogen bonds as strong as those in the alcohol family?

Phenols

14.30 What is one chemical difference between phenol and cyclohexanol? Write an equation.

14.31 The cresols are three members of the phenol family, and they are isomers. In each structure there is a $-OH$ group attached to the benzene ring in toluene. Write the structures of these isomers, and give each a common name (using the designations *o*, *m*, and *p*). (Disinfectants sometimes use a mixture of these cresols as germ killers.)

14.32 An unknown was either **A** or **B**.

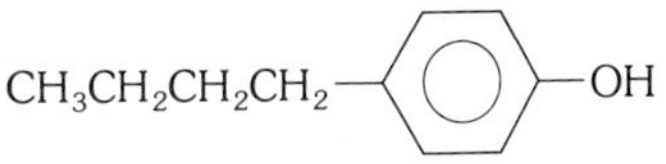

$CH_3CH_2CH_2CH_2{-}\langle O \rangle{-}OH$

A

$CH_3CH_2CH_2{-}\langle O \rangle{-}CH_2OH$

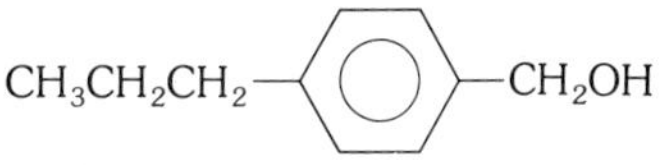

B

The unknown dissolved in aqueous sodium hydroxide, but not in water. Which compound was it? How can you tell? (Write an equation.)

14.33 An unknown was either **I** or **II**.

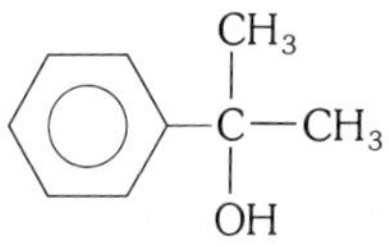

$\langle O \rangle{-}\overset{\overset{\displaystyle CH_3}{|}}{\underset{\underset{\displaystyle OH}{|}}{C}}{-}CH_3 \qquad HO{-}\langle O \rangle{-}\overset{\overset{\displaystyle CH_3}{|}}{C}H{-}CH_3$

I 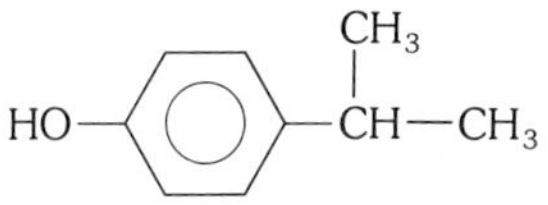 **II**

When it was shaken with aqueous KOH, no reaction occurred. Which was the unknown? How can you tell?

Ethers

14.34 Which of the following compounds have the ether function and which do not?

(a) $CH_3{-}O{-}CH_2CH_2{-}O{-}CH_3$ (b) 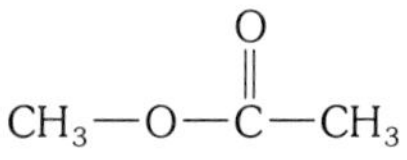 $CH_3{-}O{-}\overset{\overset{\displaystyle O}{||}}{C}{-}CH_3$

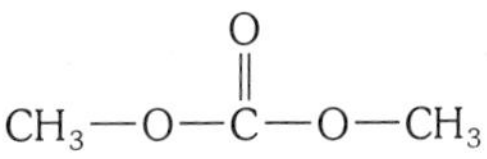

(c) $CH_3{-}O{-}\overset{\overset{\displaystyle O}{||}}{C}{-}O{-}CH_3$ (d) 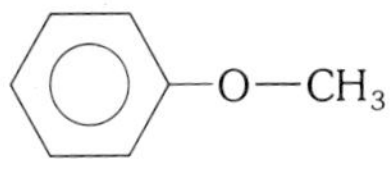 $\langle O \rangle{-}O{-}CH_3$

(e)  [structure: tetrahydropyran ring with O]

(f) $CH_3-O-O-CH_3$

(g) $CH_2=CH-O-CH_2CH_3$

(h) 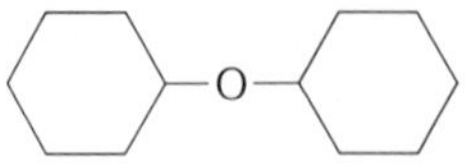 [structure: two cyclohexyl rings joined by O]

14.35 What are the structures of the ethers into which the following alcohols can be changed?

(a) $CH_3\overset{\underset{\displaystyle |}{OH}}{C}HCH_3$

(b) $CH_3\overset{\underset{\displaystyle |}{CH_3}}{C}HCH_2OH$

(c) 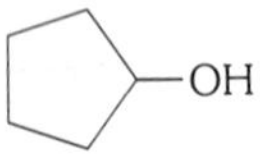 [cyclopentyl]—OH

(d) 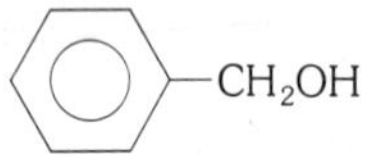 [benzene ring]—CH_2OH

14.36 What alcohols would be needed to prepare the following ethers? Write their structures.
(a) CH_3-O-CH_3
(b) $CH_3CH_2CH_2CH_2-O-CH_2CH_2CH_2CH_3$

14.37 If an equimolar mixture of methanol and ethanol were heated with concentrated sulfuric acid under conditions suitable for making the ether system, what organic products would be isolated from this mixture after the reaction is over?

14.38 What happens, chemically, when the following compound is heated with aqueous sodium hydroxide?

$$CH_3CH_2-O-CH_2CH_3$$

Organic Reactions

14.39 Write the structure of the principal organic product that would be expected to form in each of the following situations. If no reaction occurs, state so.

   If the reaction is the oxidation of a primary alcohol, give the structure of the *aldehyde* only. If the conditions are sulfuric acid and heat and the reactant is an alcohol, write the structure of the *alkene* when the coefficient of the alcohol is given as "1." If it is given as "2," write the structure of the *ether*. (This violates our rule that balancing an equation is the *last* step in writing a reaction, but we need a signal to tell what kind of a reaction is meant.)

(a) $(CH_3)_2CHOH \xrightarrow[H^+]{Cr_2O_7{}^{2-}}$

(b) $CH_3CH_2CH_2CH_3 + H_2SO_4 \longrightarrow$

(c) $CH_3CH_2CH_2\overset{\underset{\displaystyle |}{}}{C}(CH_3)_2 \xrightarrow{MnO_4{}^-}$ with OH

(d) $CH_3CH_2CH_2OH + NaOH(aq) \longrightarrow$

(e) $2CH_3OH \xrightarrow[Heat]{H_2SO_4}$

(f) $CH_3CH_2\overset{\underset{\displaystyle |}{OH}}{C}HCH_2CH_3 + (O) \longrightarrow$

(g) $CH_3-O-CH_3 + (O) \longrightarrow$

(h) $CH_3CH=CHCH_3 + H_2O \xrightarrow[Heat]{H^+}$

(i) $CH_3CH_2CH_2OH \xrightarrow[Heat]{H_2SO_4}$

(j) $CH_3\overset{\underset{\displaystyle |}{OH}}{C}HCH_2CH_3 + (O) \longrightarrow$

14.40 Following the same directions as given in Review Exercise 14.39, write the structure of the principal organic product that would be expected to form in each situation. If no reaction occurs, write "no reaction."

(a) $CH_3CH=CH_2 + H_2 \xrightarrow[Pressure]{Ni, heat}$

(b) $CH_3\overset{\underset{\displaystyle |}{OH}}{C}CH_2CH_3 + (O) \longrightarrow$ with CH_3 below

(c) 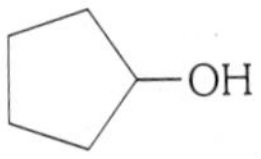 [cyclopentyl]—OH $\xrightarrow[Heat]{H_2SO_4}$

(d) $CH_3CH_2OCH_3 + HCl(aq) \longrightarrow$

(e) 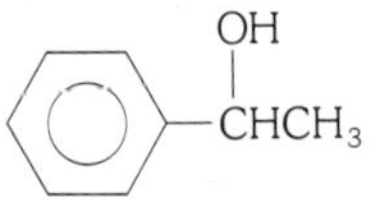 [benzene ring]—$\overset{\underset{\displaystyle |}{OH}}{C}HCH_3 \xrightarrow[Heat]{dil\ H_2SO_4}$

(f)  [cyclohexane ring with OH and CH$_3$] $+ (O) \longrightarrow$

(g) $CH_3-O-CH_2CH=CH_2 + H_2 \xrightarrow[Pressure]{Ni, heat}$

(h)  [cyclohexyl]—$CH_2CH_3 + H_2O \longrightarrow$

(i) $CH_3\overset{\underset{\displaystyle |}{CH_3}}{\underset{\underset{\displaystyle OH}{|}}{C}}CH_3 \xrightarrow[Heat]{H_2SO_4}$

(j) $CH_3CH_2CH_2OH + (O) \longrightarrow$

(k) $CH_3-O-CH_2\overset{\underset{\displaystyle |}{OH}}{C}HCH_3 + (O) \longrightarrow$

Chapter 15
Aldehydes and Ketones

Mirrors have always fascinated people. Some mirrors are of polished steel, as here. Others are thin coatings of silver on glass which are deposited by a chemical reaction studied in this chapter.

15.1 STRUCTURAL FEATURES AND NAMES

Molecules of both aldehydes and ketones contain the carbonyl group, C=O, and when it holds at least one hydrogen atom, it becomes the aldehyde group.

A knowledge of some of the properties of aldehydes and ketones is essential to an understanding of the carbohydrates. All simple sugars are either polyhydroxy aldehydes or polyhydroxy ketones, and many intermediates in metabolism are aldehydes or ketones. Special Topic 15.1 describes just a few of them.

The Carbonyl Group.   Both aldehydes and ketones contain the carbon-oxygen double bond, and this particular system has its own special name, the **carbonyl group** (pronounced carbon-EEL).

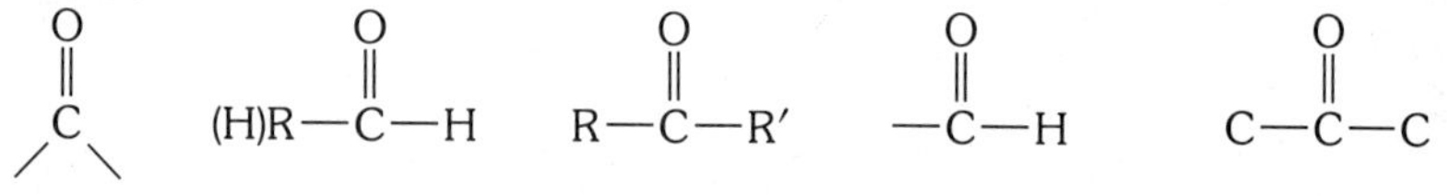

| Carbonyl group | Aldehydes | Ketones | Aldehyde group | Ketone system |

Special Topic 15.2 describes the nature of the double bond in a carbonyl group.

Notice that the carbonyl group in aldehydes must have at least one H atom attached. Thus **aldehydes** are compounds whose molecules all have the **aldehyde group,** which is often condensed and written as —CH=O or as —CHO. Attached to this system must be either a carbon atom (and its associated group), or a hydrogen atom. The simplest aldehyde, H_2C=O, methanal (formaldehyde), has two hydrogens on the carbonyl carbon atom. Table 15.1 shows some other relatively simple aldehydes.

In **ketones,** the carbonyl carbon must be flanked on both sides by carbon atoms. When this condition is met, the carbonyl group is sometimes called the **keto group.** Several ketones are listed in Table 15.2. Sometimes the keto group is condensed in structures as —CO—, and you have to learn how to read this as consisting of a carbon-oxygen double bond, and that the other bonds both go away from the *carbon* atom of this group.

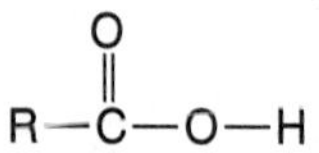

The carbonyl group also occurs in carboxylic acids, and their salts, esters, anhydrides, and amides—substances we will study in the next two chapters.

Carboxylic
acids

TABLE 15.1
Aldehydes

Name	Structure	Boiling Point (°C)	Solubility in Water
Methanal	CH_2=O	−21	Very soluble
Ethanal	CH_3CH=O	21	Very soluble
Propanal	CH_3CH_2CH=O	49	16 g/dL (25 °C)
Butanal	$CH_3CH_2CH_2CH$=O	76	4 g/dL
Benzaldehyde	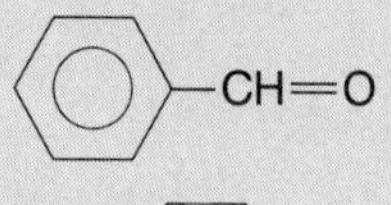 ⬡—CH=O	178	0.3 g/dL
Vanillin	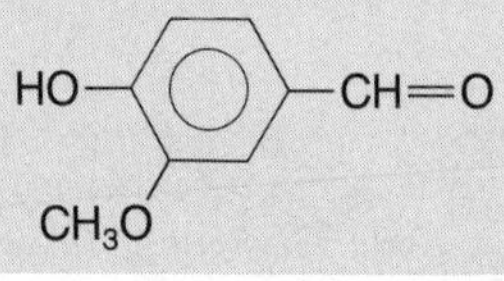 HO—⬡—CH=O CH_3O	285	1 g/dL

Formaldehyde Pure formaldehyde is a gas at room temperature, and it has a very irritating and distinctive odor. It is quite soluble in water, so it is commonly marketed as a solution called formalin (37% w/w) to which some methanol is added. In this and more dilute forms, formaldehyde was once commonly used as a disinfectant and as a preservative for biological specimens. (Concern over formaldehyde's potential hazard to health has caused these uses to decline.) Most formaldehyde today is used to make various plastics such as Bakelite.

Acetone Acetone is valued as a solvent. It not only dissolves a wide variety of organic compounds, it also is miscible with water in all proportions. Nail polish remover is generally acetone. Should you ever use "superglue," it would be a good idea to have some acetone (nail polish remover) handy because superglue can cause your fingers to stick together so tightly that it takes a solvent such as acetone to get them unstuck.

Acetone is a minor by-product of metabolism, but in some situations (e.g., untreated diabetes) enough is produced to give breath the odor of acetone.

Some Aldehydes and Ketones in Metabolism The aldehyde group and the keto group occur in many compounds at the molecular level of life. The following are just a few examples of substances with the aldehyde group.

$$HOCH_2-CH-CH-CH-CH-\overset{\displaystyle O}{\overset{\|}{C}}-H$$
$$\quad\quad\ \ \ |\ \ \ \ \ |\ \ \ \ \ |\ \ \ \ \ |$$
$$\quad\quad\ \ \ OH\ \ OH\ \ OH\ \ OH$$

Glucose, a product of
the digestion of
sugars and starch

$$H-\overset{\displaystyle O}{\overset{\|}{C}}-CH-CH_2-OPO_3{}^{2-}$$
$$\quad\quad\ \ \ |$$
$$\quad\quad\ \ OH$$

Glyceraldehyde 3-phosphate,
an intermediate in glucose
metabolism

Pyridoxal, one of
the vitamins (B_6)

Just a few of the many substances at the molecular level of life that contain the keto group are the following.

$$CH_3-\overset{\displaystyle O}{\overset{\|}{C}}-CO_2{}^-$$

Pyruvate ion, a
product of the
metabolism of
glucose and fructose

$$CH_3-\overset{\displaystyle O}{\overset{\|}{C}}-CH_2-CO_2{}^-$$

Acetoacetate ion, a product
of the metabolism of long-
chain carboxylic acids, and
present in blood at elevated
levels in diabetes

$$HO-CH_2-\overset{\displaystyle O}{\overset{\|}{C}}-CH_2-O-PO_3{}^{2-}$$

Dihydroxyacetone phosphate,
an intermediate in the
metabolism of
glucose and fructose

Estrone, a female sex hormone

Polarization
of the carbonyl
group

Physical Properties of Aldehydes and Ketones. Because oxygen is much more electronegative than carbon, there is a permanent $\delta-$ on the carbonyl group's oxygen atom and a permanent $\delta+$ charge on its carbon atom. In other words, the carbonyl group is permanently polarized, and aldehydes and ketones are moderately polar compounds. Their molecules are attracted to each other, but not as strongly as they would be if they had —OH groups instead. When comparing substances of nearly the same formula weights but from different families, as seen in Table 15.3, aldehydes and ketones boil higher than alkanes, but lower than alcohols.

In Section 13.6, we learned that the carbon-carbon double bond consists of one sigma bond and one pi bond. The carbon-oxygen double bond is exactly like this, except that an oxygen atom has replaced a carbon atom. Both the carbon atom and the oxygen atom of the carbonyl group are sp^2 hybridized, and the accompanying figure shows how, in methanal, the carbon-oxygen sigma bond forms by the overlap of two such hybrid orbitals. The figure also shows how the pi bond results from the side-to-side overlap of two unhybridized p orbitals.

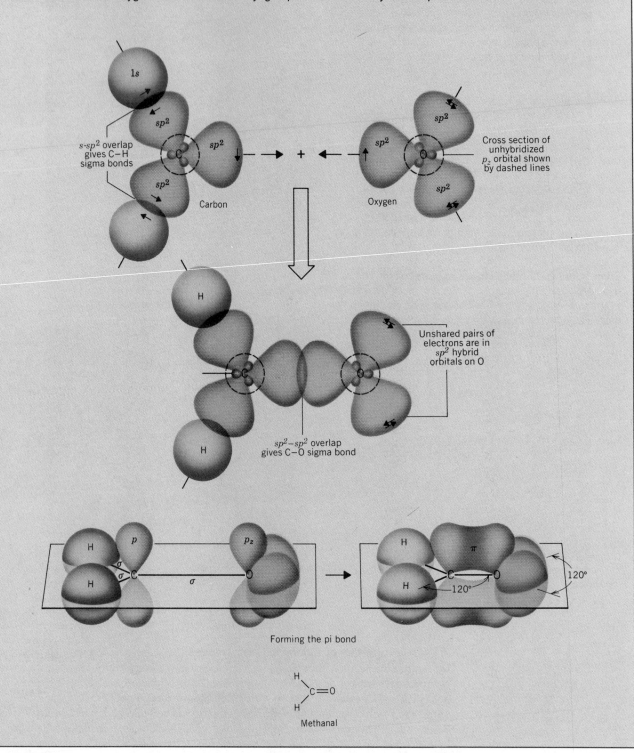

TABLE 15.2
Ketones

Name	Structure	Boiling Point (°C)	Solubility in Water
Propanone (acetone)	CH_3CCH_3 (with =O above C)	56	Very soluble
Butanone	$CH_3CCH_2CH_3$ (with =O above C)	80	33 g/dL (25 °C)
2-Pentanone	$CH_3CCH_2CH_2CH_3$ (with =O above C)	102	6 g/dL
3-Pentanone	$CH_3CH_2CCH_2CH_3$ (with =O above C)	102	5 g/dL
Cyclopentanone	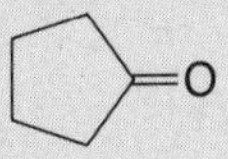 (cyclopentane ring)=O	129	Slightly soluble
Cyclohexanone	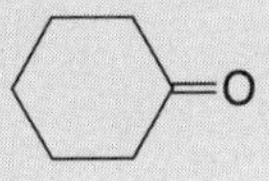 (cyclohexane ring)=O	156	Slightly soluble

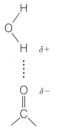

Hydrogen bond (· · · ·)
between a water molecule
and a carbonyl group

We see, again, that the ability of molecules to form hydrogen bonds, which alcohols have, bears strongly on physical properties.

Although the carbonyl group cannot donate hydrogen bonds, it can accept them. Therefore the aldehyde group and the keto group provide some help in bringing organic molecules into solution in water. As seen in the data of Tables 15.1 and 15.2, the low-formula-weight aldehydes and ketones are relatively soluble in water.

Nomenclature of Aldehydes.   The common names of the simple aliphatic and aromatic aldehydes are quite important, because they are commonly used—much more so than the IUPAC names. What is easy about the common names is that they all (well, nearly all) end in *-aldehyde*. The prefixes to this are taken from the prefixes of the carboxylic acids to which the aldehydes are easily oxidized, so this is a good place to study the common names of both acids and aldehydes.

TABLE 15.3
Boiling Point versus Structure

Compound	Boiling Point (°C)
$CH_3CH_2CH_2CH_3$	0
$CH_3CH_2CH=O$	49
CH_3CCH_3 (with =O above C)	57
$CH_3CH_2CH_2OH$	98
$HOCH_2CH_2OH$	198

The common names of the simple carboxylic acids have been known for centuries, and they were based on some natural sources of the acids. Thus the one-carbon acid is called formic acid because it is present in the stinging fluid of ants — and the Latin root for ants is *formica*. The prefix in formic acid is *form-*, so the one-carbon aldehyde is called *formaldehyde*. Here are the four simplest carboxylic acids and their common names together with the structures and names of their corresponding aldehydes.

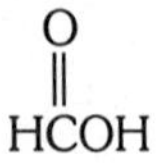

$$\underset{\text{Formic acid}}{\overset{\overset{\displaystyle O}{\parallel}}{HCOH}} \qquad \underset{\text{Formaldehyde}}{\overset{\overset{\displaystyle O}{\parallel}}{HCH}}$$

$$\underset{\text{Acetic acid}}{\overset{\overset{\displaystyle O}{\parallel}}{CH_3COH}} \qquad \underset{\text{Acetaldehyde}}{\overset{\overset{\displaystyle O}{\parallel}}{CH_3CH}}$$

L. *acetum*, vinegar

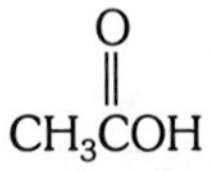

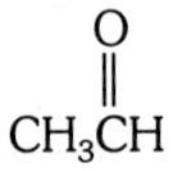

$$\underset{\text{Propionic acid}}{\overset{\overset{\displaystyle O}{\parallel}}{CH_3CH_2COH}} \qquad \underset{\text{Propionaldehyde}}{\overset{\overset{\displaystyle O}{\parallel}}{CH_3CH_2CH}}$$

Gr. *proto*, first, and *pion*, fat

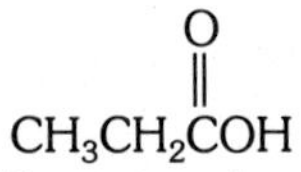

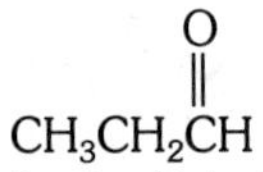

$$\underset{\text{Butyric acid}}{\overset{\overset{\displaystyle O}{\parallel}}{CH_3CH_2CH_2COH}} \qquad \underset{\text{Butyraldehyde}}{\overset{\overset{\displaystyle O}{\parallel}}{CH_3CH_2CH_2CH}}$$

L. *butyrum*, butter

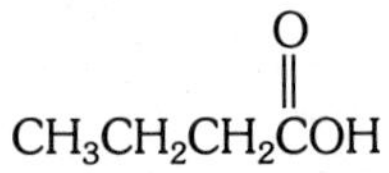

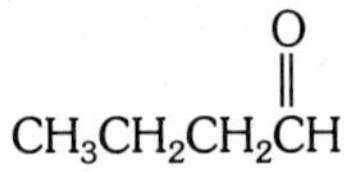

In the aromatic series we have the following examples:

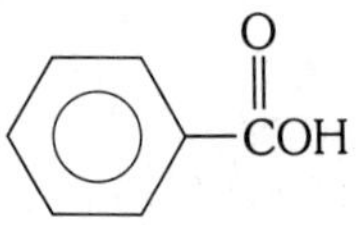

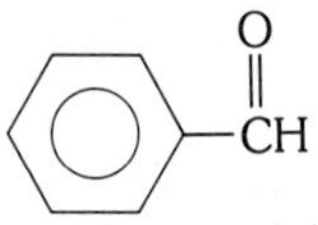

Benzoic acid   Benzaldehyde

The IUPAC rules for naming aldehydes are very similar to those for the alcohols, and they are based on the idea of a *parent aldehyde*.

1. Select as the parent aldehyde the longest chain *that includes the carbon atom of the aldehyde group.*

    The parent aldehyde in the following structure is a five-carbon aldehyde.

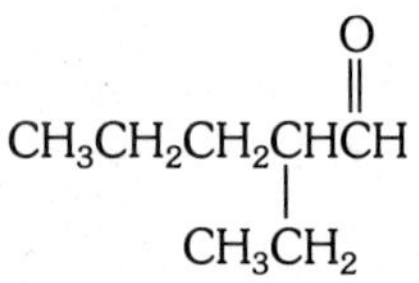

$$\overset{\overset{\displaystyle O}{\parallel}}{CH_3CH_2CH_2CHCH} \\ \underset{\displaystyle CH_3CH_2}{|}$$

There is a longer chain, one of six carbons, but it doesn't include the carbon of the aldehyde group, so this longer chain can't be used in the parent name.

2. Name the parent by changing the *-e* ending of the corresponding alkane to *-al*.

    In the example shown with rule 1, the alkane that corresponds to the correct chain is pentane, so the name of the parent aldehyde in this structure is pentanal.

3. Number the chain to give the carbon atom of the carbonyl group number 1.

    This is another instance of the precedence accorded the aldehyde group. Regardless of where other substituents on the parent chain occur, they have to take whatever numbers they receive following the assignment of 1 to the carbonyl carbon atom.

4. Assemble the rest of the name in the same way that was used in naming alcohols, except do not include "1" to specify the location of the aldehyde group.

The carbonyl carbon can't have any other number but 1, so we don't include this number. Thus using the example begun under rule 1,

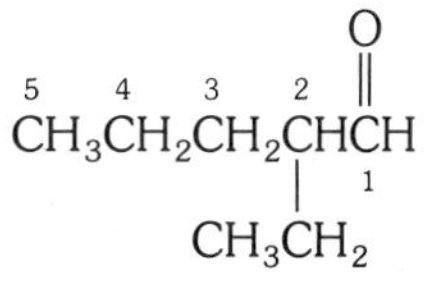

$$
\begin{array}{c}
\hspace{4.2em} \overset{\displaystyle O}{\underset{\displaystyle \parallel}{}} \\
\overset{5}{CH_3}\overset{4}{CH_2}\overset{3}{CH_2}\overset{2}{CH}\overset{}{CH} \\
\hspace{4.2em} | \hspace{1em} {}^{1} \\
\hspace{3.5em} CH_3CH_2
\end{array}
$$

2-Ethylpentanal
Not: 2-ethyl-1-pentanal

EXAMPLE 15.1   WRITING THE IUPAC NAME OF AN ALDEHYDE

Problem: What is the IUPAC name of the following compound?

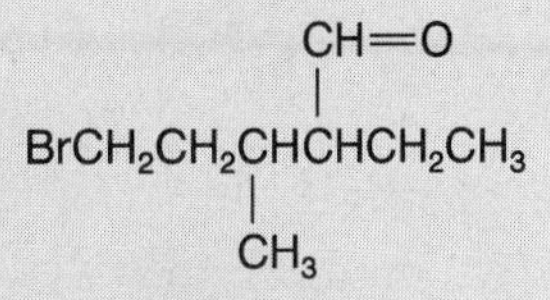

$$
\begin{array}{c}
CH{=}O \\
| \\
BrCH_2CH_2CHCHCH_2CH_3 \\
| \\
CH_3
\end{array}
$$

Solution: First, we identify the parent aldehyde. The longest chain that includes the carbon atom of the aldehyde group has five carbons, so the name of the parent aldehyde is *pentanal*. Next, we number the chain beginning with the carbon atom of the aldehyde group.

$$
\begin{array}{c}
\overset{1}{CH}{=}O \\
| \\
\overset{5}{Br}\overset{}{CH_2}\overset{4}{CH_2}\overset{3}{CH}CHCH_2CH_3 \\
\hspace{3em}| \hspace{0.3em}{}_{2} \\
\hspace{2em}CH_3
\end{array}
$$

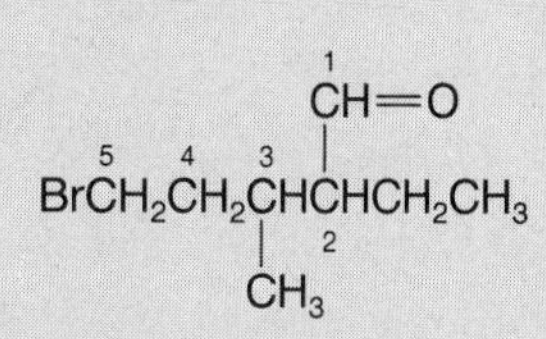

At position 2 there is an ethyl group; at 3, a methyl group; and at 5, a bromo group. Remember that we put the bromo group first in the name. We can organize the names of the two alkyl groups in their order of increasing carbon content. Thus the name of this aldehyde is

5-bromo-3-methyl-2-ethylpentanal

(Also acceptable is 5-bromo-2-ethyl-3-methylpentanal.)

PRACTICE EXERCISE 1   Write the IUPAC names of the following aldehydes.

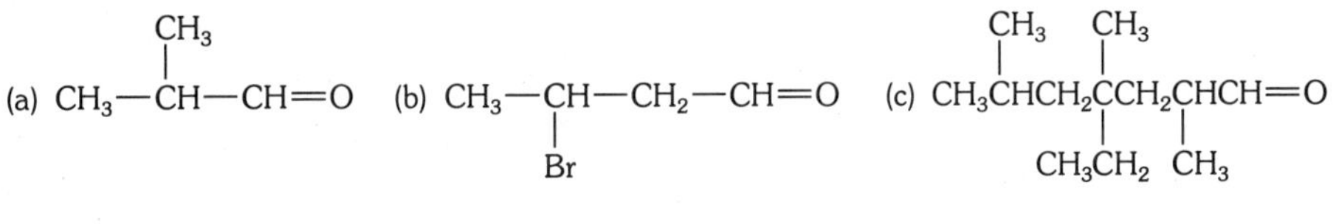

$$
\text{(a)} \quad
\begin{array}{c}
CH_3 \\
| \\
CH_3{-}CH{-}CH{=}O
\end{array}
\qquad
\text{(b)} \quad
\begin{array}{c}
CH_3{-}CH{-}CH_2{-}CH{=}O \\
| \\
Br
\end{array}
\qquad
\text{(c)} \quad
\begin{array}{c}
CH_3 \quad CH_3 \\
| \quad\;\; | \\
CH_3CHCH_2CCH_2CHCH{=}O \\
| \quad\;\; | \\
CH_3CH_2 \; CH_3
\end{array}
$$

PRACTICE EXERICSE 2   What is wrong with the name 2-isopropylpropanal?

Nomenclature of Ketones.   The IUPAC rules for naming ketones are identical to those of the aldehydes, except for one obvious change. The name of the parent ketone must end in -*one* (not -*al*). Otherwise, the location of the carbonyl group takes precedence in numbering the chain, as with aldehydes.

Pronouce -one as own.

EXAMPLE 15.2   WRITING THE IUPAC NAME OF A KETONE

Problem: What is the IUPAC name of the following ketone?

$$CH_3-CH-\underset{\underset{CH_3-CH_2-CH_2}{|}}{\overset{\overset{CH_3}{|}}{C}}-\overset{\overset{CH_3}{|}\ \ \overset{O}{\|}}{C}-CH_3$$

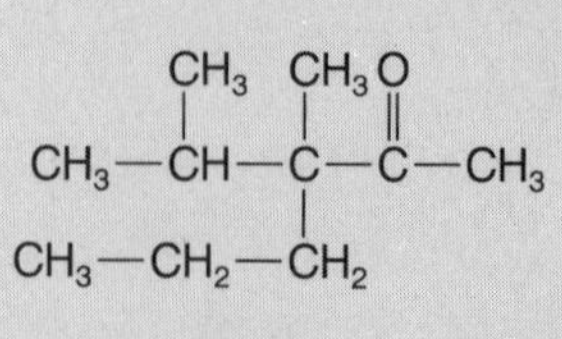

Solution: There are two chains that have six carbon atoms, but we have to use the one that has the carbon atom of the carbonyl group. We number this chain to give the location of the carbonyl group the lower number.

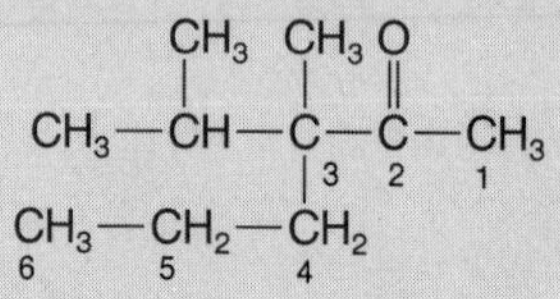

$$CH_3-CH-\underset{\underset{\underset{6\ \ \ \ 5\ \ \ \ 4}{CH_3-CH_2-CH_2}}{|}}{\overset{\overset{CH_3}{|}}{\underset{3}{C}}}-\overset{\overset{CH_3}{|}\ \ \overset{O}{\|}}{\underset{2}{C}}-\underset{1}{CH_3}$$

The parent ketone is therefore 2-hexanone. At carbon 3 its chain has a methyl group plus an isopropyl group, so the full name of this compound is

3-Methyl-3-isopropyl-2-hexanone

(Also acceptable is 3-isopropyl-3-methyl-2-hexanone.)

PRACTICE EXERCISE 3   Write the IUPAC names for the following ketones.

(a) 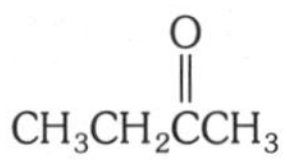 $CH_3CH_2\overset{\overset{O}{\|}}{C}CH_3$ (b) 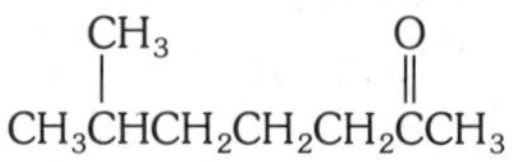 $CH_3\overset{\overset{CH_3}{|}}{C}HCH_2CH_2CH_2\overset{\overset{O}{\|}}{C}CH_3$ (c)

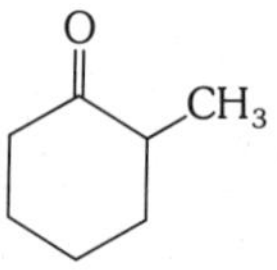

Quite often the simpler ketones are given common names that are made by naming the two alkyl groups attached to the carbon atom of the carbonyl group and then following these names by the word ketone. For example

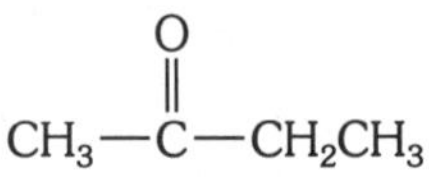

$$CH_3-\overset{\overset{O}{\|}}{C}-CH_2CH_3 \qquad CH_3CH_2-\overset{\overset{O}{\|}}{C}-CH_2CH_3 \qquad CH_3-\overset{\overset{O}{\|}}{C}-CH_3$$

Methyl ethyl ketone 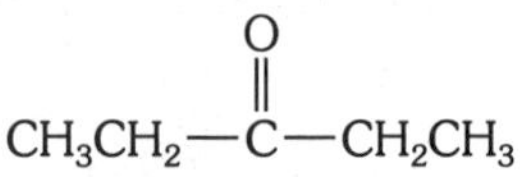   Diethyl ketone 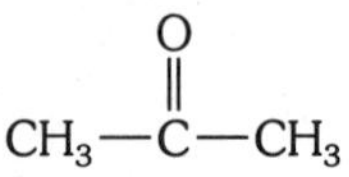   (Dimethyl ketone)
                                                       Acetone

The name *acetone* stems from the fact that this ket*one* can be made by heating the calcium salt of *acet*ic acid.

As we have already noted, almost all chemists and biochemists use the name *acetone* for dimethyl ketone or propanone.

PRACTICE EXERCISE 4   Write the structures of the following ketones.

(a) Ethyl isopropyl ketone   (b) Methyl phenyl ketone
(c) Dipropyl ketone   (d) Di-*t*-butyl ketone

15.2 THE OXIDATION OF ALDEHYDES AND KETONES

The aldehyde group is easily oxidized to the carboxylic acid group, but the ketone system is difficult to oxidize.

We learned in the previous chapter that the oxidation of a 1° alcohol to an aldehyde requires care, because aldehydes are themselves easily oxidized, but that much less care is needed to oxidize a 2° alcohol to a ketone.

$$RCH_2OH + \quad (O) \quad \longrightarrow \quad R-\overset{\overset{\displaystyle O}{\|}}{C}-H + H_2O$$

1° Alcohol   Oxidizing   Aldehyde
agent

$$\overset{\overset{\displaystyle OH}{|}}{RCHR'} \quad + \quad (O) \quad \longrightarrow \quad R-\overset{\overset{\displaystyle O}{\|}}{C}-R' + H_2O$$

2° Alcohol   Ketone

The ease with which the aldehyde group is oxidized by even mild reactants has led to some simple test tube tests for aldehydes that we will study next.

Tollens' Test or the Silver Mirror Test.   One very mild oxidizing agent consists of an alkaline solution of the diammonia complex of the silver ion, $Ag(NH_3)_2{}^+$, and this reagent is called the **Tollens' reagent.** It reacts with an aldehyde to oxidize it to a carboxylic acid (in the form of its anion).

$$RCH{=\!=}O + 2Ag(NH_3)_2{}^+ + 3OH^- \longrightarrow RCO_2{}^- + 2Ag + 2H_2O + 4NH_3$$

When this reaction occurs in a test tube that has been thoroughly cleaned and made free of any greasy deposit, the metallic silver that forms as one product emerges as a beautiful silver mirror. The appearance of this mirror provides dramatic evidence that the reaction occurred, so this reagent provides chemists with a simple test tube test that can be used to tell if an unknown compound is an aldehyde or a ketone. The test is called the **Tollens' test** (or, sometimes, the silver mirror test). A positive Tollens' test is the formation of metallic silver in any form — as a mirror in a clean test tube or as a grayish precipitate otherwise. Glucose, a carbohydrate that we will learn more about later, gives a positive Tollens' test.

The Tollens' reagent is prepared by adding sodium hydroxide to dilute silver nitrate. This causes silver oxide, Ag_2O, to precipitate, but then dilute ammonia is added. Molecules of ammonia are able to pull silver ions out of the solid silver oxide and form the soluble diammonia complex ion, $Ag(NH_3)_2{}^+$. Thus the silver oxide dissolves, and now the compound suspected of being an aldehyde or a ketone is added and the mixture is warmed in a hot water bath for a few minutes. If the unknown is an aldehyde (and if the test tube was clean), the silver mirror slowly takes form. This is the technology used to make mirrors in general. The reagent has to be freshly made, because it deteriorates on standing.

Benedict's Test.   Benedict's reagent, another mild oxidizing agent, consists of the copper(II) ion that has been made into a complex ion so it can remain dissolved in a basic solution. The complexing agent is the citrate ion from citric acid. (We won't need to know more details here, but Cu^{2+} forms a precipitate of $Cu(OH)_2$ in a basic solution, and the citrate ion protects the copper(II) ion from this happening.)

When an easily oxidized compound is added to a test tube that contains some Benedict's reagent, and the solution is warmed, Cu^{2+} ions are reduced to Cu^+ ions, and these cannot be protected from OH^- ions by the citrate system. The newly formed Cu^+ ions are pulled out of the surrounding citrate ions, and are changed by the base into a precipitate of copper(I) oxide, Cu_2O. The Benedict's reagent has a brilliant blue color caused by the copper(II) ion, but Cu_2O has a brick-red color. Therefore the visible evidence of a positive **Benedict's test** is the disappearance of a blue color and the appearance of a reddish precipitate. Three systems, all of which occur among various carbohydrates, give positive Benedict's tests:

A carbon atom immediately adjacent to a carbonyl group is often called an alpha (α) carbon:

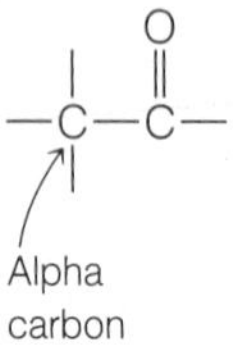

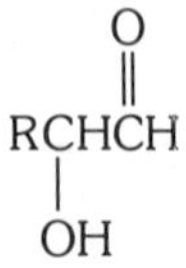

$$\overset{\overset{\displaystyle O}{\|}}{-\underset{\underset{\displaystyle Alpha\ carbon}{\uparrow}}{C}-C-}$$

$$\overset{\overset{\displaystyle O}{\|}}{RCH}\underset{\underset{\displaystyle OH}{|}}{C}H$$
α-Hydroxy aldehyde

$$\overset{\overset{\displaystyle O}{\|}}{RC}-\overset{\overset{\displaystyle O}{\|}}{C}H$$
α-Keto aldehyde

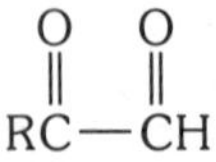

$$\overset{\overset{\displaystyle O}{\|}}{RCH}\underset{\underset{\displaystyle OH}{|}}{C}R'$$
α-Hydroxy ketone

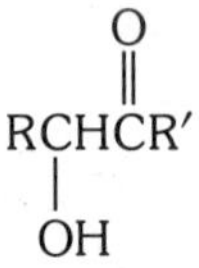

Glucose, for example, is an α-hydroxy aldehyde, and Benedict's test is a common method for detecting glucose in urine. In certain conditions, such as diabetes, the body cannot prevent some of the excess glucose in the blood from going into the urine, so testing the urine for its glucose concentration has long been used in medical diagnosis.

Clinitest tablets are a convenient solid form of Benedict's reagent. They contain all of the needed reactants in their solid forms, and to test for glucose a few drops of urine are mixed with a tablet. As the tablet dissolves, the heat needed for the test is generated. The color that develops is compared with a color code on a chart provided with the tablets. (It should be said that specialists in the control of diabetes prefer to monitor the carbohydrate status of a person with diabetes by testing the concentration of glucose in the patient's blood instead of its concentration in the urine. However, not all patients can or are willing to manage blood tests, particularly when they are needed frequently.) Other tests for glucose are based on enzyme-catalyzed reactions, and we will learn about them later.

15.3   THE REDUCTION OF ALDEHYDES AND KETONES

Aldehydes and ketones are reduced to alcohols when hydrogen adds to their carbonyl groups.

Aldehydes are reduced to 1° alcohols and ketones are reduced to 2° alcohols by a variety of conditions, and we will study two methods — the direct addition of hydrogen and reduction by hydride ion transfer.

Catalytic Hydrogenation.   Under heat and pressure and in the presence of a finely divided metal catalyst, the carbonyl groups of aldehydes and ketones add hydrogen.

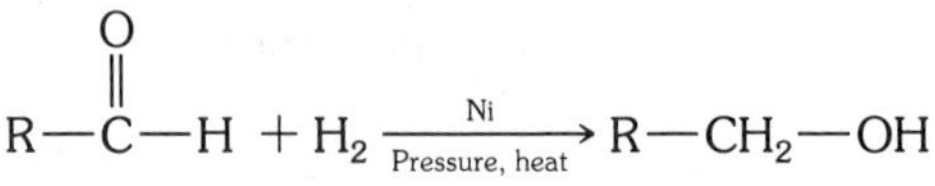

$$\underset{\text{Aldehyde}}{R-\overset{\overset{\displaystyle O}{\|}}{C}-H} + H_2 \xrightarrow[\text{Pressure, heat}]{\text{Ni}} \underset{\text{1° alcohol}}{R-CH_2-OH}$$

$$\underset{\text{Ketone}}{R-\overset{\overset{\displaystyle O}{\|}}{C}-R'} + H_2 \xrightarrow[\text{Pressure, heat}]{\text{Ni}} \underset{\text{2° alcohol}}{R-\overset{\overset{\displaystyle OH}{|}}{C}H-R'}$$

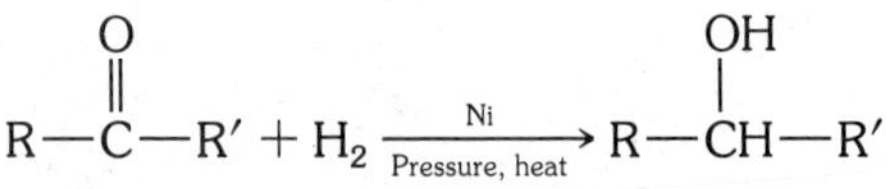

Specfic examples include:

$$CH_3CH_2CH_2CH{=}O + H_2 \xrightarrow[\text{Pressure, heat}]{\text{Ni}} CH_3CH_2CH_2CH_2OH$$

Butanal   1-Butanol

$$\underset{\text{Acetone}}{CH_3\overset{\displaystyle O}{\overset{\|}{C}}CH_3} + H_2 \xrightarrow[\text{Pressure, heat}]{\text{Ni}} \underset{\text{2-Propanol}}{CH_3\overset{\displaystyle OH}{\overset{|}{C}}HCH_3}$$

Reduction by Hydride Ion Donation.   The hydride ion, $H:^-$, is a powerful reducing agent, and before we see how it works with aldehydes and ketones we should learn more about how it can be supplied and some of its other properties.

Certain metal hydrides such as sodium borohydride, $NaBH_4$, and lithium aluminum hydride, $LiAlH_4$, are common sources of the hydride ion in experimental work in chemistry, but neither is found in the body. Body cells obtain hydride ions, instead, from molecules of organic donors of $H:^-$. We say a *donor* of hydride ions, because free hydride ions cannot exist in water. The hydride ion is a very strong proton-acceptor, and it reacts with water as follows to give hydrogen gas and the hydroxide ion:

$$H:^- + H{-}\overset{\cdot\cdot}{\underset{\cdot\cdot}{O}}{-}H \longrightarrow H{-}H + :\overset{\cdot\cdot}{\underset{\cdot\cdot}{O}}{-}H^-$$

Hydride ion

However, an organic molecule that can donate a hydride ion passes it *directly* to the acceptor molecule, and the ion itself never comes into contact with the water present in body cells.

The carbonyl group is an excellent acceptor of the hydride ion. We'll represent an organic hydride-ion donor by the symbol $M:H$, where M refers to a *metabolite* — a chemical intermediate in metabolism. When an aldehyde or ketone group accepts a hydride ion, the following reaction occurs.

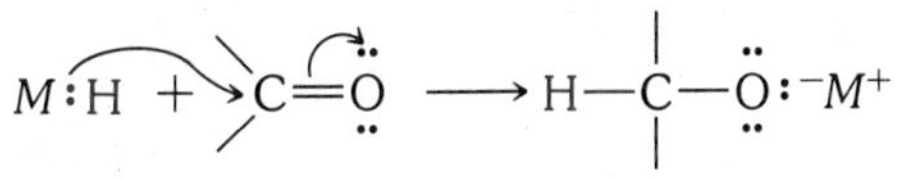

$$M{:}H + \overset{\diagdown}{\underset{\diagup}{C}}{=}\overset{\cdot\cdot}{\underset{\cdot\cdot}{O}} \longrightarrow H{-}\overset{|}{\underset{|}{C}}{-}\overset{\cdot\cdot}{\underset{\cdot\cdot}{O}}{:}^-M^+$$

Hydride   Aldehyde   Anion of
donor   or ketone   an alcohol

The anion of an alcohol is a stronger proton-acceptor than a hydroxide ion, so in the instant when the newly formed anion emerges, it takes a proton from a water molecule. Thus the final organic product is an alcohol.

$${-}\overset{|}{\underset{|}{C}}{-}\overset{\cdot\cdot}{\underset{\cdot\cdot}{O}}{:}^- + H{-}O{-}H \longrightarrow H{-}\overset{|}{\underset{|}{C}}{-}O{-}H + :\overset{\cdot\cdot}{\underset{\cdot\cdot}{O}}{-}H^-$$

Anion of   Alcohol
alcohol

The overall effect is the same as if direct hydrogenation had been used on the aldehyde or ketone. As we have already noted, another name for *hydrogenation* is *reduction,* and when a carbonyl carbon atom accepts the pair of electrons carried by the hydride ion, it *gains* this pair and so is reduced. (Remember, reduction is a gain of electrons.)

One of the many examples in cells of reduction by hydride-ion donation is the reduction of the keto group in the pyruvate ion to form the lactate ion, a step in the metabolism of glucose.

$M:H$ in the body often is a B-vitamin unit in an enzyme.

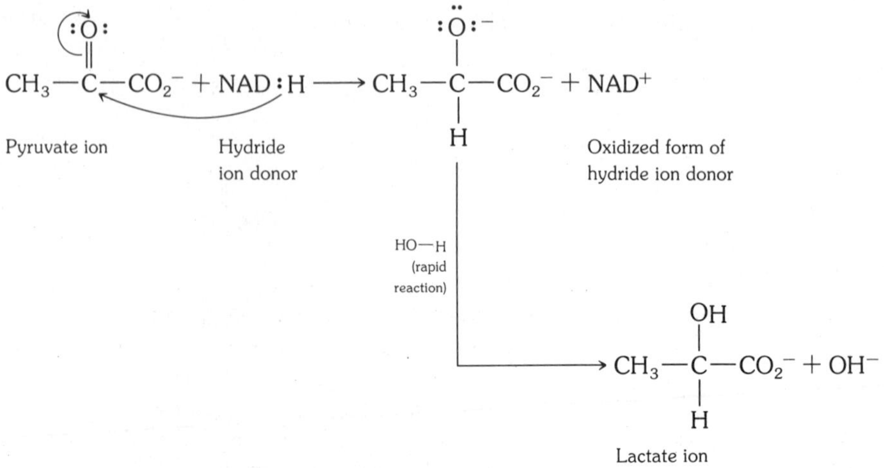

$$CH_3-\overset{\overset{\displaystyle :O:}{\|}}{C}-CO_2^- + NAD\!:\!H \longrightarrow CH_3-\overset{\overset{\displaystyle :\ddot{O}:^-}{|}}{\underset{\underset{\displaystyle H}{|}}{C}}-CO_2^- + NAD^+$$

Pyruvate ion   Hydride
                     ion donor

Oxidized form of
hydride ion donor

HO—H
(rapid
reaction)

$$\longrightarrow CH_3-\overset{\overset{\displaystyle OH}{|}}{\underset{\underset{\displaystyle H}{|}}{C}}-CO_2^- + OH^-$$

Lactate ion

NAD is a molecular unit present in several enzymes and made from a B-vitamin.

(In this sequence, NAD stands for *nicotinamide adenine dinucleotide,* a substance which we will discuss further in a later chapter. Right now, all we need to know about NAD is that in what is called its reduced form, NAD:H, it is a good hydride-ion donor.)

EXAMPLE 15.3   WRITING THE STRUCTURE OF THE PRODUCT OF THE REDUCTION OF AN ALDEHYDE OR KETONE

Problem: What is the product of the reduction of propanal?

Solution: All of the action is at the carbonyl group; it changes to an alcohol group. Therefore all we have to do is copy over the structure of the given compound, change the double bond to a single bond, and supply the two hydrogen atoms—one to the oxygen atom of the original carbonyl group and one to the carbon atom.

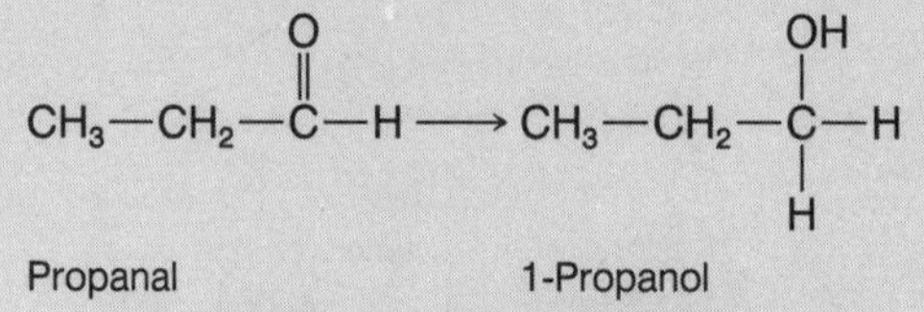

$$CH_3-CH_2-\overset{\overset{\displaystyle O}{\|}}{C}-H \longrightarrow CH_3-CH_2-\overset{\overset{\displaystyle OH}{|}}{\underset{\underset{\displaystyle H}{|}}{C}}-H \qquad (or,\ CH_3CH_2CH_2OH)$$

Propanal   1-Propanol

PRACTICE EXERCISE 5   Write the structures of the products that form when the following aldehydes and ketones are reduced.

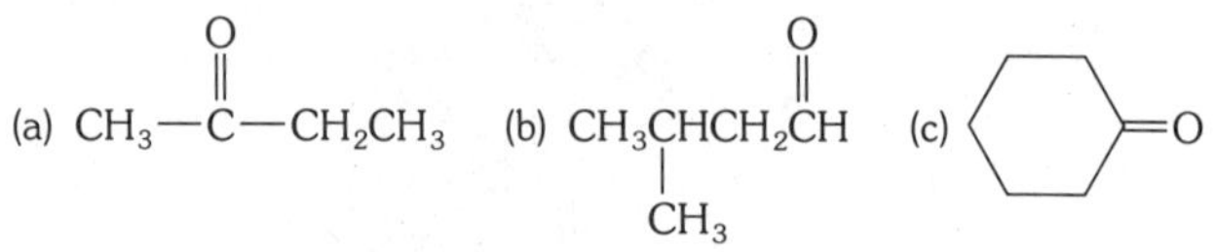

(a) $CH_3-\overset{\overset{\displaystyle O}{\|}}{C}-CH_2CH_3$ (b) $CH_3\overset{\overset{\displaystyle O}{\|}}{\underset{\underset{\displaystyle CH_3}{|}}{CH}}CH_2\overset{\overset{\displaystyle O}{\|}}{C}H$ (c) ⬡=O

15.4 THE REACTIONS OF ALDEHYDES AND KETONES WITH ALCOHOLS

1,1-Diethers—acetals or ketals—form when aldehydes or ketones react with alcohols in the presence of an acid or enzyme catalyst.

This section is background to our later study of carbohydrates whose molecules have the functional groups introduced here. We will study the simplest possible examples of these groups now so that carbohydrate structures will be easier to understand.

Hemiacetals from Aldehydes and Alcohols.   When a solution of an aldehyde in an alcohol is prepared, molecules of the alcohol add to molecules of the aldehyde and the following equilibrium becomes established:

The hemiacetal system:

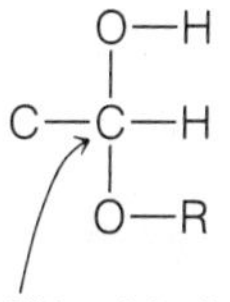

$$
\begin{array}{c}
O-H \\
| \\
C-C-H \\
/ \quad | \\
O-R
\end{array}
$$

This originally was the carbon atom of an aldehyde group.

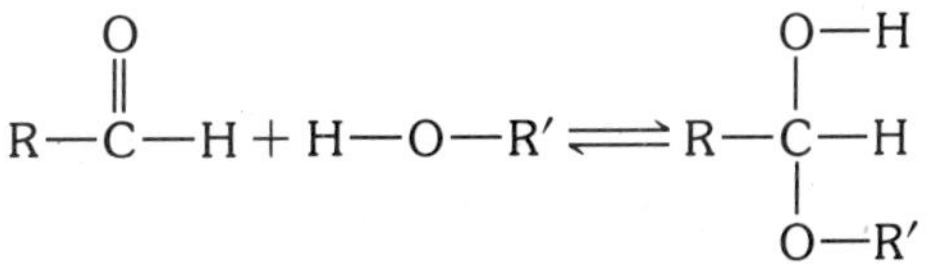

$$
\underset{\text{Aldehyde}}{R-\overset{\displaystyle O}{\overset{\|}{C}}-H} + \underset{\text{Alcohol}}{H-O-R'} \rightleftharpoons \underset{\text{Hemiacetal}}{R-\overset{\displaystyle O-H}{\underset{\displaystyle O-R'}{\overset{|}{\underset{|}{C}}}}-H}
$$

The product is called a **hemiacetal,** and in hemiacetal molecules there is always a carbon atom that holds both an —OH group and an —OR group. When these two groups are this close to each other, they so modify each others' properties that it's useful to place the whole system into its own separate family.

In the formation of a hemiacetal, the —O—R part of the alcohol molecule *always* ends up attached to the carbon atom of the original carbonyl group, and the H— atom of the alcohol always goes to the carbonyl oxygen.

Except among carbohydrates, a hemiacetal is generally an unstable compound. If we try to isolate and purify one, it breaks back down. The addition reaction that made the hemiacetal reverses itself and nothing but the original aldehyde and alcohol are obtained. Thus hemiacetals generally exist only in a solution in which the equilibrium shown just above has formed. We're interested in them because they are intermediate structures on the way to 1,1-diethers — acetals — which are stable enough to be isolated, and because among carbohydrates, hemiacetals do occur.

The relative ease with which hemiacetals break back down means that the hemiacetal system is a site of structural weakness in an organic molecule — even among carbohydrates. For this reason, we have to learn how to recognize the system when it occurs in a structure, so let's work an example.

EXAMPLE 15.4   IDENTIFYING THE HEMIACETAL SYSTEM

Problem: Which of the following structures has the hemiacetal system? Draw an arrow pointing to any carbon atoms that were initially the carbon atoms of aldehyde groups.

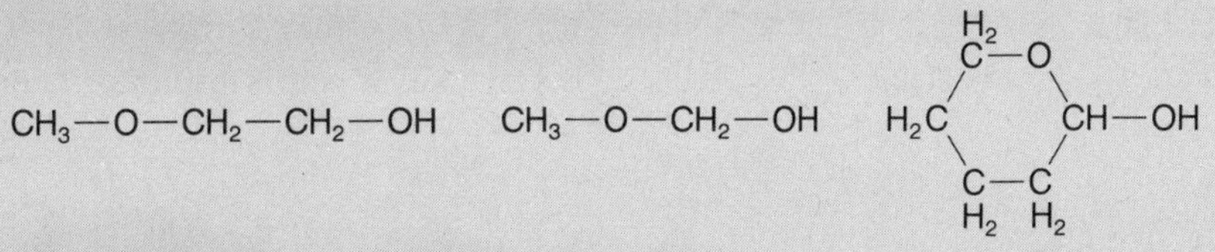

$$CH_3-O-CH_2-CH_2-OH \qquad CH_3-O-CH_2-OH \qquad \begin{array}{c} \overset{H_2}{C}-O \\ H_2C \qquad CH-OH \\ C-C \\ H_2 \quad H_2 \end{array}$$

1   2   3

Solution: To have a hemiacetal system, the molecule must have a carbon to which are attached one —OH group, and one —O—C unit. Notice that in structure **1**, there is an —OH group and an —O—C unit, but they are not joined to the *same* carbon. Therefore **1** is not a hemiacetal. It just has an ordinary ether group and an alcohol group.

In structure **2**, the —OH and the —O—C are joined to the same carbon, so **2** is a hemiacetal. Similarly, in structure **3**, the carbon on the far right corner of the ring holds both an —OH group and a —O—C unit, and **3** is also a hemiacetal—a cyclic hemiacetal.

Structure **3** shows the way in which the hemiacetal system occurs in many carbohydrates, as a cyclic hemiacetal.

The ring system of **3** also occurs in glucose.

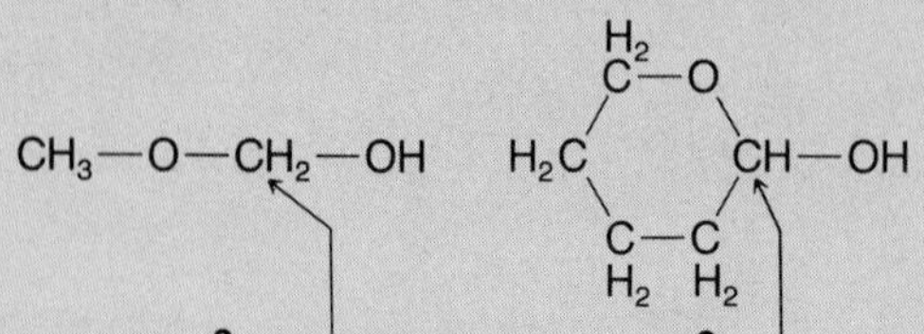

$$CH_3-O-CH_2-OH \qquad$$

2   3

These carbons were initially part of aldehyde groups.

PRACTICE EXERCISE 6   Identify the hemiacetals among the following structures, and draw arrows that point to the carbon atoms that initially were part of the carbonyl groups of parent aldehydes.

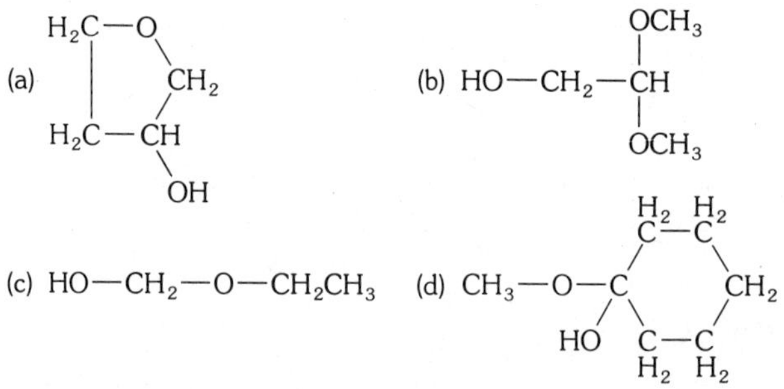

(a)

(b) $HO-CH_2-CH$ with OCH_3 above and OCH_3 below

(c) $HO-CH_2-O-CH_2CH_3$ (d) CH_3-O-C (ring)

Another skill that will be useful in our study of carbohydrates is the ability to write the structure of a hemiacetal that could be made from a given aldehyde and alcohol.

EXAMPLE 15.5   WRITING THE STRUCTURE OF A HEMIACETAL GIVEN ITS PARENT ALDEHYDE AND ALCOHOL

Problem: Write the structure of the hemiacetal that is present at equilibrium in a solution of propanal in ethanol.

Solution: The best way to start is to rewrite the structure of the aldehyde, but show only one bond from carbon to oxygen:

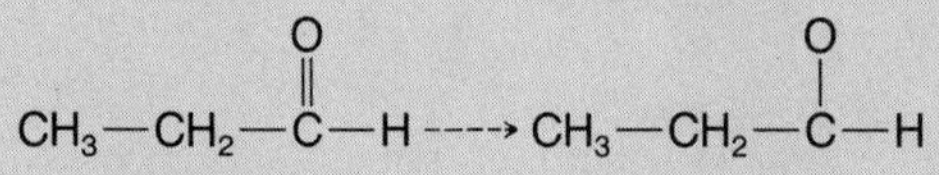

$$CH_3-CH_2-\overset{\overset{\textstyle O}{\|}}{C}-H \dashrightarrow CH_3-CH_2-\overset{\overset{\textstyle O}{|}}{C}-H$$

Propanal   (Incomplete)
(given)

Then look at the structure of the given alcohol, CH_3-CH_2-O-H. The H on its oxygen atom ends up on the oxygen atom of the developing structure:

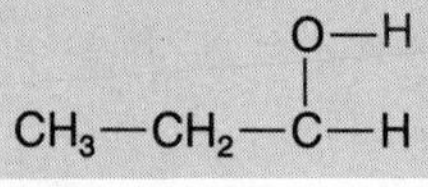

$$CH_3-CH_2-\overset{\overset{\textstyle O-H}{|}}{C}-H \qquad (Incomplete)$$

Finally, the entire remainder of the alcohol molecule is attached by its oxygen atom to the carbon atom that presently has the —O—H group in our developing structure:

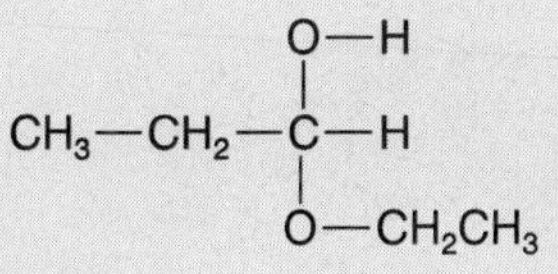

$$
\begin{array}{c}
\text{O—H} \\
| \\
\text{CH}_3\text{—CH}_2\text{—C—H} \\
| \\
\text{O—CH}_2\text{CH}_3
\end{array}
\qquad \text{(Final answer)}
$$

We can leave the answer in this form, or we can rewrite it to condense it a little more:

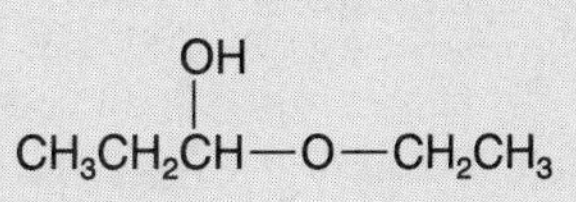

$$
\begin{array}{c}
\text{OH} \\
| \\
\text{CH}_3\text{CH}_2\text{CH—O—CH}_2\text{CH}_3
\end{array}
$$

PRACTICE EXERCISE 7   Write the structures of the hemiacetals that are present in the equilibria that involve the following pairs of compounds.

(a) Ethanal and methanol   (b) Butanal and ethanol
(c) Benzaldehyde and 1-propanol   (d) Methanal and methanol

Still another skill that will be useful in studying carbohydrates is the ability to write the structures of the aldehyde and alcohol that are liberated by the breakdown of a hemiacetal.

EXAMPLE 15.6   WRITING THE BREAKDOWN PRODUCTS OF A HEMIACETAL

Problem: What aldehyde and alcohol form when the following hemiacetal breaks down?

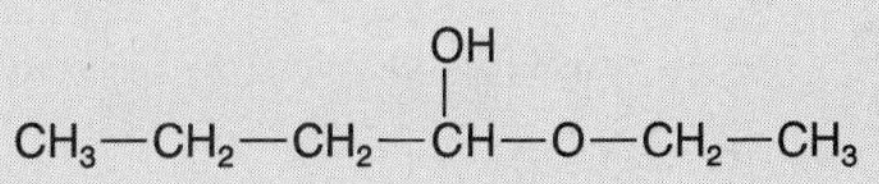

$$
\begin{array}{c}
\text{OH} \\
| \\
\text{CH}_3\text{—CH}_2\text{—CH}_2\text{—CH—O—CH}_2\text{—CH}_3
\end{array}
$$

Solution: This is the kind of problem where the ability to pick out the carbon atom of the original carbonyl group is especially helpful. Remember that to find this carbon we look for one that holds both a —OH group and a —O—R system. When we find this carbon, we break its bond to the —O—R unit and separate the pieces. (This bond was the bond that formed when the alcohol added to the aldehyde to form the hemiacetal system.) *Do not break any other bond in the given hemiacetal.*

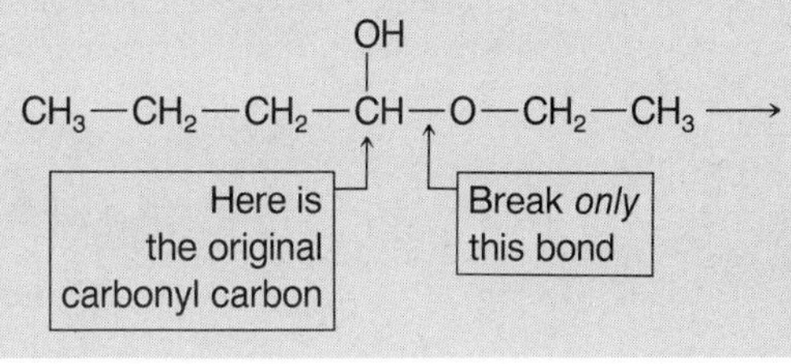

$$
\begin{array}{c}
\text{OH} \\
| \\
\text{CH}_3\text{—CH}_2\text{—CH}_2\text{—CH}\overset{\uparrow}{—}\text{O—CH}_2\text{—CH}_3 \longrightarrow
\end{array}
$$

Here is
the original
carbonyl carbon

Break *only*
this bond

(completed on next page)

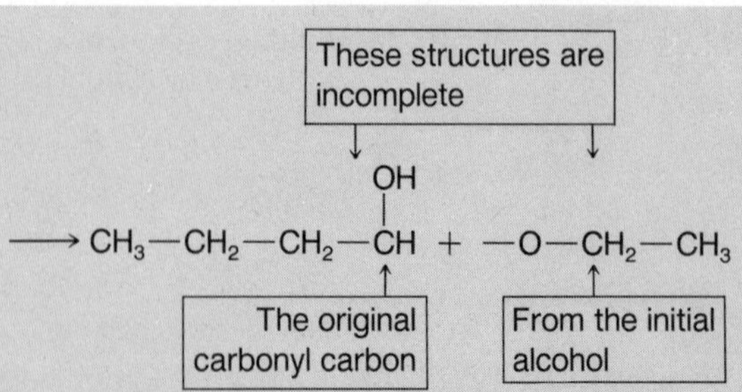

These structures are incomplete

$$OH$$
$$\longrightarrow CH_3-CH_2-CH_2-CH \ + \ -O-CH_2-CH_3$$

The original carbonyl carbon

From the initial alcohol

Now all we have to do is fix the structure of the fragment that has the original carbonyl carbon into a structure with an actual carbonyl group. We do this by moving the H atom on the oxygen over to the O atom of the other fragment. (This gives us the original alcohol molecule.) Then we write a carbon-oxygen double bond to make the carbonyl group, and we have the original aldehyde. The answer, then, is

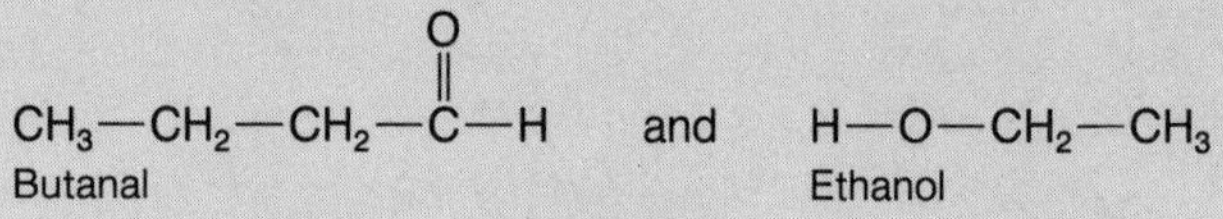

$$O$$
$$CH_3-CH_2-CH_2-C-H \quad \text{and} \quad H-O-CH_2-CH_3$$
Butanal   Ethanol

PRACTICE EXERICISE 8   Write the structures of the breakdown products of the following hemiacetals.

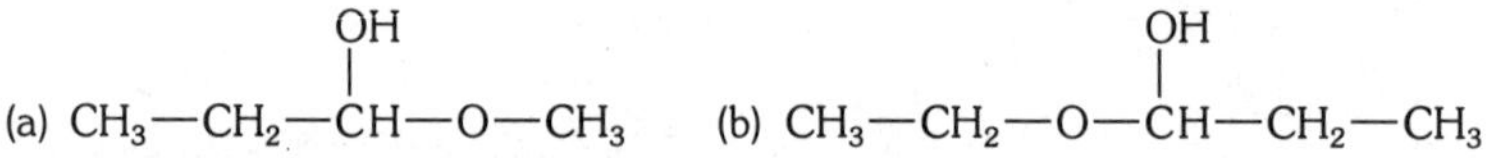

$$OH$$
(a) $CH_3-CH_2-CH-O-CH_3$

$$OH$$
(b) $CH_3-CH_2-O-CH-CH_2-CH_3$

Hemiketals from Ketones and Alcohols.   Ketones behave very much like aldehydes toward alcohols. They add alcohols to form equilibria that contain a structure almost identical in type to that of a hemiacetal, only now it's called a **hemiketal.** Like the hemiacetal, the hemiketal has a carbon atom — originally the carbonyl carbon atom of the parent ketone — which holds both —OH and —O—R.

The hemiketal system:

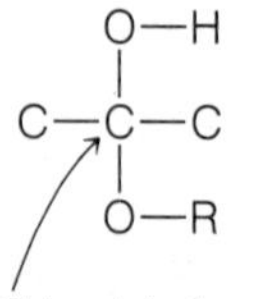

$$O-H$$
$$C-C-C$$
$$O-R$$

This originally was the carbon atom of a keto group.

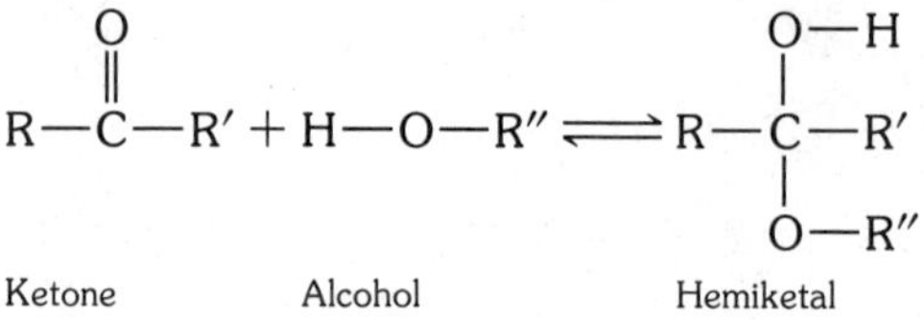

$$O \qquad\qquad\qquad O-H$$
$$R-C-R' + H-O-R'' \rightleftharpoons R-C-R'$$
$$O-R''$$

Ketone   Alcohol   Hemiketal

Alcohols do not add as readily to ketones as they do to aldehydes, so the position of the equilibrium favors hemiketal formation less than corresponding equilibria favor hemiacetals. Among carbohydrates, however, fructose (levulose) has a relatively stable hemiketal system.

The kinds of problems that we studied under the hemiacetals are exactly analogous to those of the hemiketals. Try the following exercises to prove this.

PRACTICE EXERCISE 9   Examine the following structures and decide if any represent hemiketals. Draw an arrow pointing to any carbon atoms that initially came from the carbonyl groups of ketones.

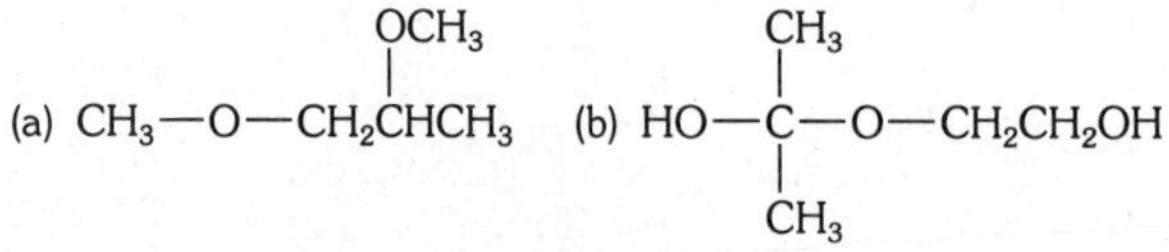

$$OCH_3$$
(a) $CH_3-O-CH_2CHCH_3$

$$CH_3$$
(b) $HO-C-O-CH_2CH_2OH$
$$CH_3$$

PRACTICE EXERCISE 10   Write the structure of the hemiketal that would be present at equilibrium in a solution of acetone in methanol.

PRACTICE EXERCISE 11   What ketone and alcohol would be needed to prepare (in an equilibrium) the following hemiketal?

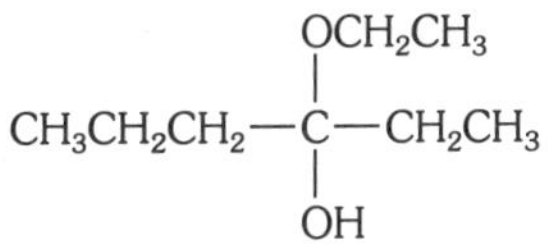

$$CH_3CH_2CH_2-\underset{\underset{OH}{|}}{\overset{\overset{OCH_2CH_3}{|}}{C}}-CH_2CH_3$$

Acetals and Ketals.   Hemiacetals and hemiketals are special kinds of alcohols, and they resemble alcohols in one important property. They can be converted into ethers — special kinds of 1,1-diethers called **acetals** and **ketals.** The overall change that leads to an acetal is as follows:

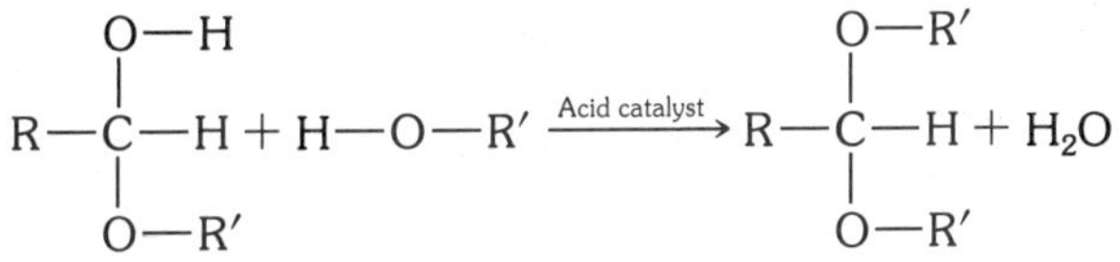

$$R-\underset{\underset{O-R'}{|}}{\overset{\overset{O-H}{|}}{C}}-H + H-O-R' \xrightarrow{\text{Acid catalyst}} R-\underset{\underset{O-R'}{|}}{\overset{\overset{O-R'}{|}}{C}}-H + H_2O$$

Hemiacetal   Acetal

Hemiketals give the identical kind of reaction, but the products are called ketals. Unlike hemiacetals and hemiketals, both acetals and ketals are stable compounds that can be isolated and stored.

The difference between the formation of an acetal and an ordinary ether is that acetals form more readily. As a rule, when two functional groups are very close to each other in a molecule, each modifies the properties of the other in some way. Here, the O—R′ group makes the nearby —OH group much more reactive toward forming an ether.

An acetal is a 1,1-diether. This doesn't mean that we have to number the chain this way. The 1,1-designation here means only that the two ether groups come to the same carbon. Ketals are also 1,1-diethers. Such common carbohydrates as sucrose (table sugar), lactose (milk sugar), and starch have the 1,1-diether system.

Acetals and ketals are stable, as we said, but only if they are kept out of contact with aqueous acids. In water, acids (or enzymes) catalyze the hydrolysis of these compounds. They break back down to give their parent alcohols and carbonyl compounds — aldehydes or ketones. This hydrolysis of acetals and ketals is their only chemical reaction we need to study, and it is the chemistry of the digestion of carbohydrates. Before we go into it, let's be sure we can recognize the acetal or ketal system when it occurs in a structure.

EXAMPLE 15.7   RECOGNIZING THE ACETAL OR KETAL SYSTEMS AND WHICH OF THEIR CARBON ATOMS CAME FROM PARENT CARBONYL CARBONS

Problem: Examine each structure to see if it is an acetal or a ketal. If it is, identify the carbon atom that was the carbonyl carbon atom of the parent aldehyde or ketone.

$$CH_3-O-CH_2-O-CH_3 \qquad CH_3-O-\underset{\underset{}{}}{\overset{\overset{O-CH_3}{|}}{C}H}-CH_3$$

4 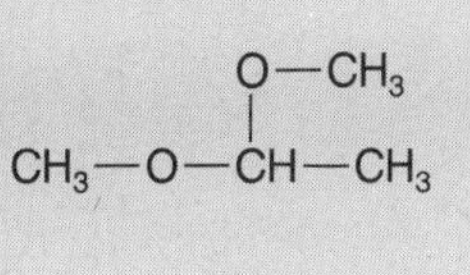   5 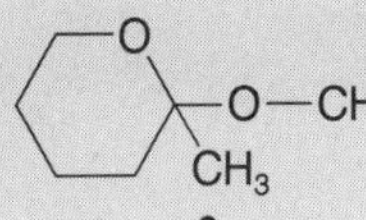   6

Solution: We have to find one carbon that holds two —O—R types of groups. Structure **4** has such a carbon in its central CH_2 unit. This carbon was initially the carbonyl carbon atom of an aldehyde (methanal), because it also holds at least one H atom. Structure **5** similarly has such a carbon, in the —CH— unit, and it also came from an aldehyde group (because it has at least one H atom). In structure **6,** we can also find a carbon that holds two —O—C networks. This carbon lacks an H-atom, however, so it must have come from the carbonyl group of a ketone system.

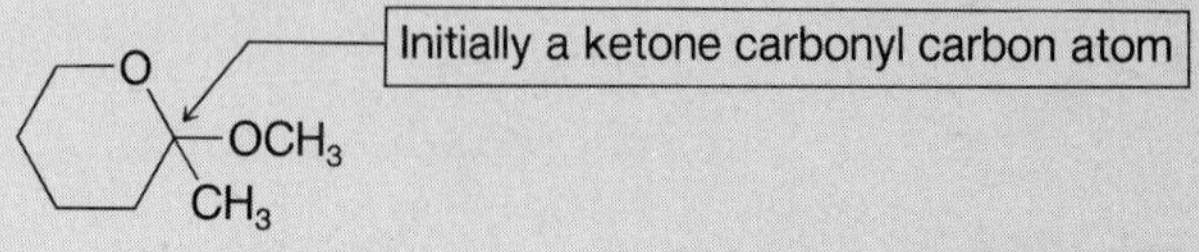

Initially a ketone carbonyl carbon atom

PRACTICE EXERCISE 12   Which of the following, if any, is an acetal or a ketal? For any that is either, identify the carbon atom which came originally from the carbonyl group of a parent aldehyde or ketone.

(a) CH_3—O—CH_2—CH_2—O—CH_3 (b) CH_3CH_2—O—$\underset{\underset{CH_3}{|}}{\overset{\overset{CH_3}{|}}{C}}$—O—$CH_2CH_3$

Because acetals and ketals can be hydrolyzed, we have to be able to examine the structures of such compounds and write the structures of their parent alcohols and aldehydes (or ketones) — the hydrolysis products. We'll do a worked example to show how this can be done.

EXAMPLE 15.8   WRITING THE STRUCTURES OF THE PRODUCTS OF THE HYDROLYSIS OF ACETALS OR KETALS

Problem: What are the products of the following reaction?

$$\underset{CH_3—\overset{\overset{O—CH_3}{|}}{CH}—O—CH_3 + H_2O}{} \xrightarrow{\text{Acid catalyst}} ?$$

Solution: The best way to proceed is to find the carbon atom in the structure that holds *two* oxygen atoms. This carbon is the carbonyl carbon atom of the parent aldehyde (or ketone). Break both of its bonds to these oxygen atoms. *Do not break any other bonds.* Separate the fragments for further processing.

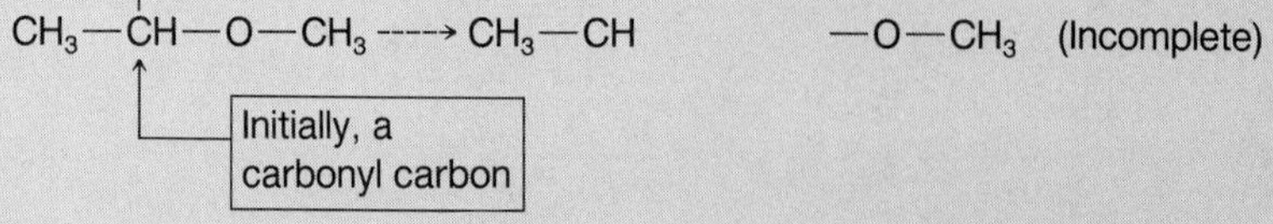

$$CH_3—\overset{\overset{O—CH_3}{|}}{CH}—O—CH_3 \dashrightarrow CH_3—CH \qquad \begin{array}{c} —O—CH_3 \\ —O—CH_3 \end{array} \quad \text{(Incomplete)}$$

Initially, a carbonyl carbon

Next, we finish writing the carbonyl group where it belongs — at the identified carbon atom. Then we place H atoms on the oxygen atoms of the other fragments to finish writing the structures of the alcohol molecules that also form. The final products of the hydrolysis of the given acetal are:

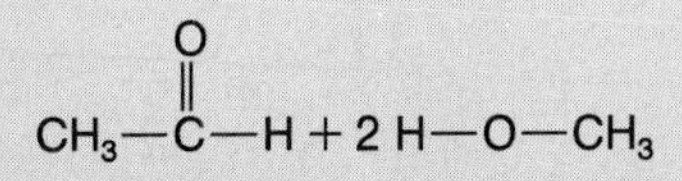

$$CH_3-\overset{\overset{\displaystyle O}{\|}}{C}-H + 2\ H-O-CH_3$$

Notice that the oxygen atom for the new carbonyl group comes from the molecule of water that acted to hydrolyze the acetal. The two hydrogen atoms needed to complete the structures of the two alcohol molecules also come from the water molecule.

PRACTICE EXERCISE 13   Write the structures of the aldehydes (or ketones) and the alcohols that are obtained by hydrolyzing the following compounds. If they do not hydrolyze like acetals or ketals, write "no reaction."

(a) $CH_3-O-CH_2-O-CH_3$

(b) $CH_3-O-CH_2-CH_2-O-CH_2-CH_3$

(c) $CH_3-\overset{\overset{\displaystyle CH_3}{|}}{CH}-\overset{\overset{\displaystyle O-CH_3}{|}}{\underset{\underset{\displaystyle CH_3}{|}}{C}}-O-CH_3$

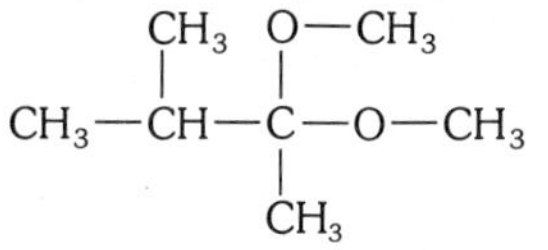

SUMMARY

Naming aldehydes and ketones The IUPAC names of aldehydes and ketones are based on a parent compound, one with the longest chain that includes the carbonyl group. The names of aldehydes end in -al and of ketones in -one, and the chains are numbered so as to give the carbonyl carbons the lower of two possible numbers.

Physical properties of aldehydes and ketones The carbonyl group confers moderate polarity, which gives aldehydes and ketones higher boiling points and solubilities in water than hydrocarbons but lower boiling points and solubilities in water than alcohols (that have comparable formula weights).

Chemical properties of aldehydes and ketones Aldehydes are easily oxidized to carboxylic acids, but ketones resist oxidation. Aldehydes give a positive Tollens' test and ketones do not. α-Hydroxy aldehydes and ketones give the Benedict's test.

When an aldehyde or a ketone is dissolved in an alcohol, some of the alcohol adds to the carbonyl group of the aldehyde or ketone. An equilibrium is thus formed that includes molecules of a hemiacetal (or hemiketal). The chart at the end of this Summary outlines the chemical properties of aldehydes and ketones we have studied.

Hemiacetals and hemiketals Hemiacetals and hemiketals are usually unstable compounds that exist only in an equilibrium involving the parent carbonyl compound and the parent alcohol (which generally is the solvent). Hemiacetals and hemiketals readily break back down to their parent carbonyl compounds and alcohols. When an acid catalyst is added to the equilibrium, a hemiacetal or hemiketal reacts with more alcohol to form an acetal or ketal.

Acetals and ketals Acetals and ketals are 1,1-diethers that are stable in aqueous base or in water, but not in aqueous acid. Acids catalyze the hydrolysis of acetals and ketals, and the final products are the parent aldehydes (or ketones) and alcohols.

Aldehydes

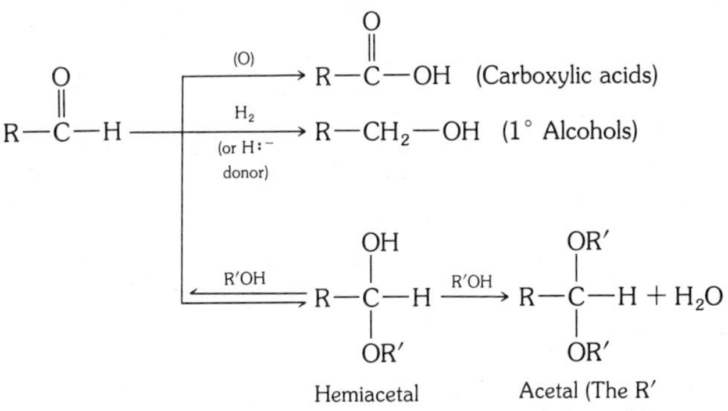

Hemiacetal   Acetal (The R′ groups need not be the same.)

Ketones

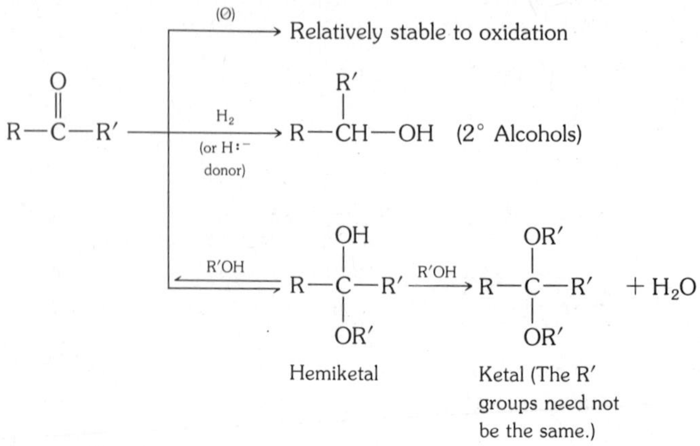

$$R-\overset{\overset{\displaystyle O}{\|}}{C}-R'$$

$$\xrightarrow{(O)} \text{Relatively stable to oxidation}$$

$$\xrightarrow[\substack{\text{(or H:}^-\\ \text{donor)}}]{H_2} R-\overset{\overset{\displaystyle R'}{|}}{CH}-OH \quad (2° \text{ Alcohols})$$

$$\xrightarrow[R'OH]{R'OH} R-\overset{\overset{\displaystyle OH}{|}}{\underset{\underset{\displaystyle OR'}{|}}{C}}-R' \xrightarrow{R'OH} R-\overset{\overset{\displaystyle OR'}{|}}{\underset{\underset{\displaystyle OR'}{|}}{C}}-R' \quad + H_2O$$

Hemiketal   Ketal (The R' groups need not be the same.)

Acetals and Ketals

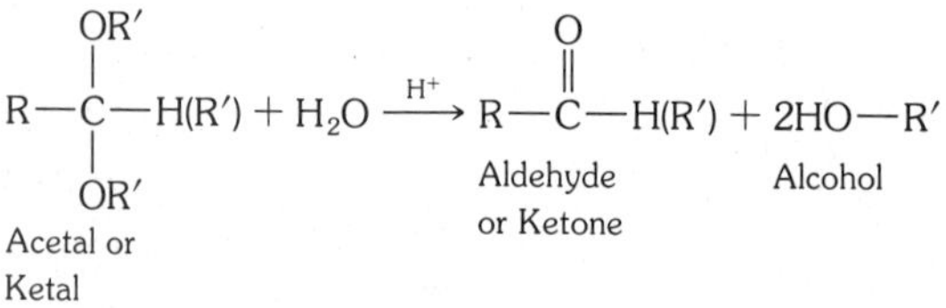

$$R-\overset{\overset{\displaystyle OR'}{|}}{\underset{\underset{\displaystyle OR'}{|}}{C}}-H(R') + H_2O \xrightarrow{H^+} R-\overset{\overset{\displaystyle O}{\|}}{C}-H(R') + 2HO-R'$$

Acetal or Ketal   Aldehyde or Ketone   Alcohol

KEY TERMS

Most of the Key Terms involve functional groups that we'll often see again. Be sure you know these terms before continuing.

acetal	Benedict's test	ketal	Tollens' test
aldehyde	carbonyl group	keto group	
aldehyde group	hemiacetal	ketone	
Benedict's reagent	hemiketal	Tollens' reagent	

REVIEW EXERCISES

The answers to these Review Exercises are in the *Study Guide* that accompanies this book.

NAMES AND STRUCTURES

15.1   What is the *structural* difference between an aldehyde and a ketone?

15.2   Write the structure of each of the following compounds.
   (a) 3-Methylpentanal
   (b) 2,3-Dibromocyclopentanone
   (c) 1-Phenyl-1-butanone
   (d) Dibutyl ketone
   (e) Butane-2,3-dione

15.3   What are the structures of the following compounds?
   (a) 2-Cyclohexylcyclohexanone
   (b) 2-Ethylbutanal
   (c) Di-*sec*-butyl ketone
   (d) 1,3-Diphenyl-2-propanone
   (e) 1,3,5-Cyclohexanetrione

15.4   Although we can write structures that correspond to the following names, when we do, we find that the names aren't proper. How should these compounds be named in the IUPAC system?
   (a) 6-Methylcyclohexanone
   (b) 1-Hydroxy-1-propanone (Give common name.)
   (c) 1-Methylbutanal
   (d) 2-Methylethanal
   (e) 2-Propylpropanal

15.5   The following names can be used to write structures, but the names turn out to be improper. What should be their IUPAC names?
   (a) 2-*sec*-Butylbutanal
   (b) 1-Phenylethanal

(c) 4,5-Dimethylcyclopentanone

(d) 1-Hydroxyethanal (Give common name.)

(e) 1-Butanone

15.6 If the IUPAC name of $CH_3-\overset{O}{\overset{\|}{C}}-CH_2-\overset{O}{\overset{\|}{C}}-H$ is 3-ketobutanal, what is the IUPAC name of the following compound?

$$H-\overset{O}{\overset{\|}{C}}-CH-\overset{CH_3}{\underset{|}{CH}}-\overset{O}{\overset{\|}{C}}-CH_3$$
$$\underset{|}{\overset{}{CH_3}}$$

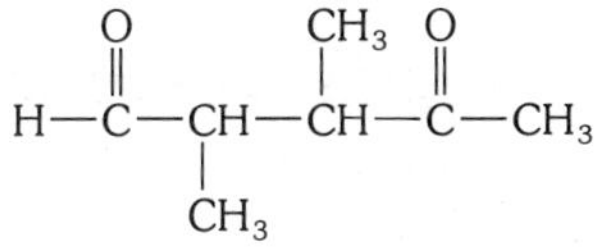

15.7 We can name compound **A** as 2-methylformylcyclohexane. Taking a clue from this, how might compound **B** be named?

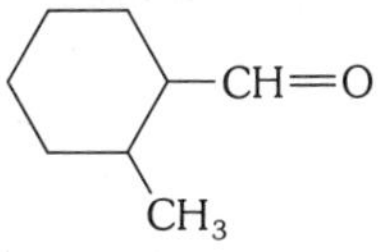

A 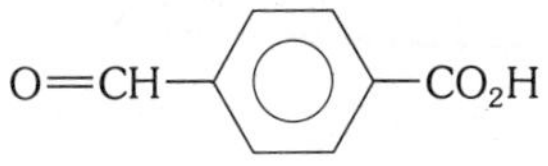   **B**

15.8 Write the IUPAC names of the following compounds.

(a) 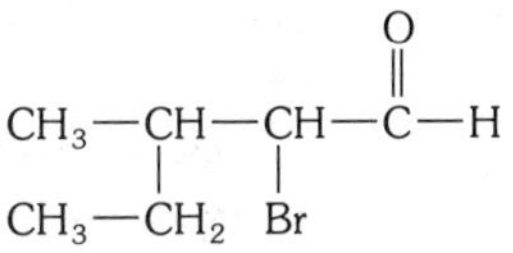 $CH_3-\overset{}{\underset{\underset{CH_3-CH_2}{|}}{CH}}-\overset{}{\underset{\underset{Br}{|}}{CH}}-\overset{O}{\overset{\|}{C}}-H$

(b) 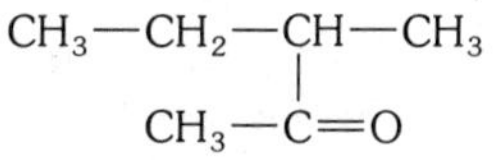 $CH_3-CH_2-\overset{}{\underset{\underset{CH_3-C=O}{|}}{CH}}-CH_3$

(c)

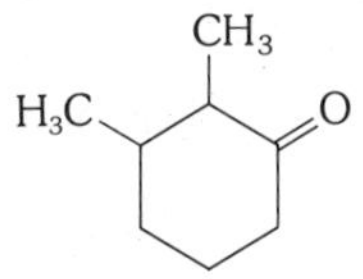

(d)

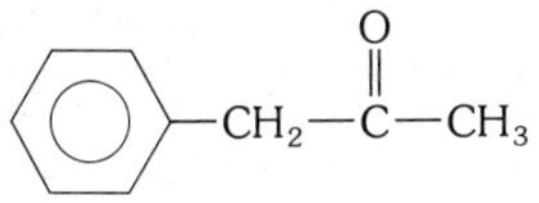

(e)

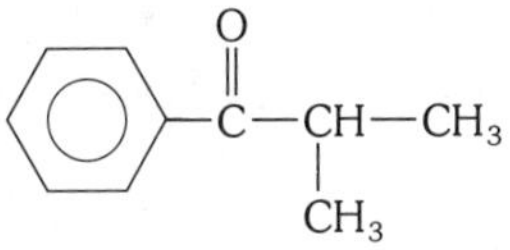

15.9 What are the IUPAC names of the following compounds?

(a) 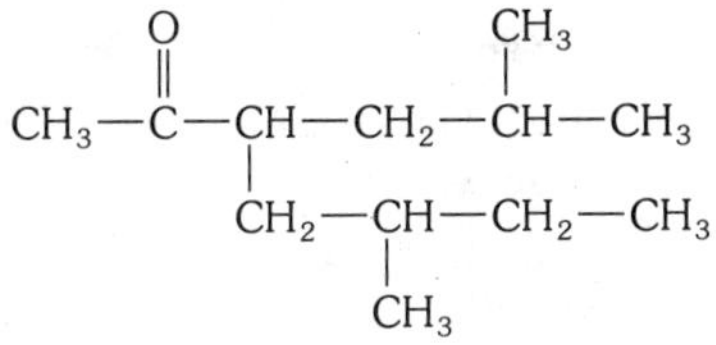 $CH_3-\overset{O}{\overset{\|}{C}}-\overset{}{\underset{\underset{CH_2-CH-CH_2-CH_3}{|}}{CH}}-CH_2-\overset{CH_3}{\underset{|}{CH}}-CH_3$
$$\underset{CH_3}{}$$

(b) 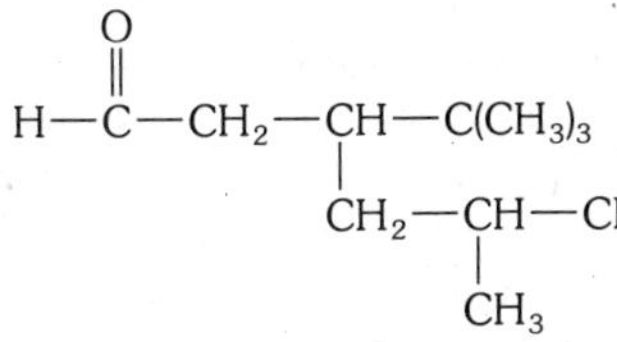 $H-\overset{O}{\overset{\|}{C}}-CH_2-\overset{}{\underset{\underset{CH_2-CH-CH_3}{|}}{CH}}-C(CH_3)_3$
$$\underset{CH_3}{}$$

(c)

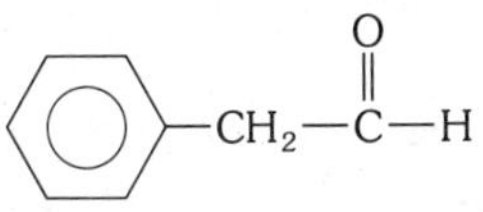

(d)

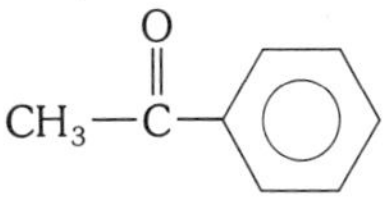

(e)

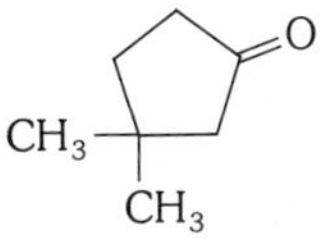

15.10 If the common name of $CH_3CH_2CH_2CH_2\overset{O}{\overset{\|}{C}}OH$ is valeric acid, what is the most likely common name of the following compound?

$$CH_3CH_2CH_2CH_2CH=O$$

15.11 If the common name of this compound

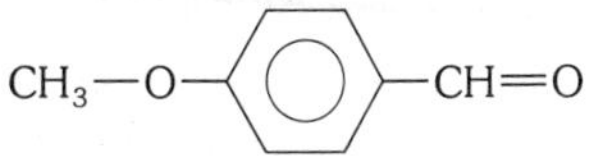

is anisaldehyde, then what is the most likely common name of the following compound?

Physical Properties of Aldehydes and Ketones

15.12 Arrange the following compounds in their order of increasing boiling points. Do this by placing the letters that identify them in the correct order, starting with the lowest-boiling compound and moving in order to the highest-boiling compound.

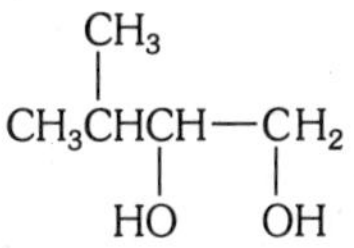

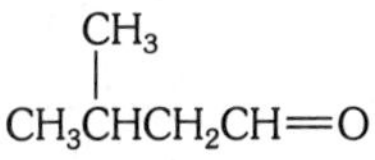

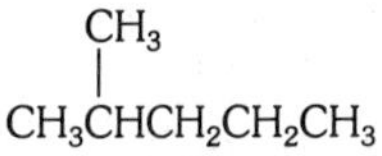

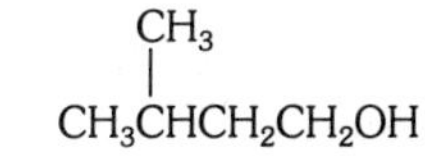

15.13 Arrange the following compounds in their order of increasing boiling points. Do this by placing the letters that identify them in the correct order, starting with the lowest-boiling compound on the left in the series and moving to the highest-boiling compound.

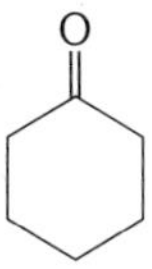

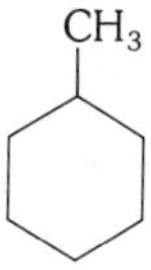

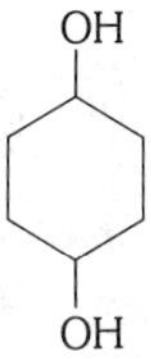

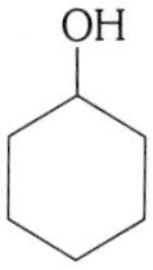

15.14 Reexamine the compounds of Review Exercise 15.12, and arrange them in their order of increasing solubility in water.

15.15 Arrange the compounds of Review Exercise 15.13 in their order of increasing solubility in water.

15.16 Draw the structure of a water molecule and an acetone molecule and align them on the page to show how the acetone molecule can accept a hydrogen bond from the water molecule. Use a dotted line to represent this hydrogen bond and place $\delta+$ and $\delta-$ symbols where they are appropriate.

15.17 Draw the structures of molecules of methanol and ethanal, and align them on the page to show how a hydrogen bond (which you are to indicate by a dotted line) can exist between the two. Place $\delta+$ and $\delta-$ symbols where they are appropriate.

Oxidation of Alcohols and Aldehydes

15.18 What are the structures and the IUPAC names of the aldehydes and ketones to which the following compounds can be oxidized?

(a) $HO-\overset{\overset{\displaystyle CH_3}{|}}{C}HCH_2CH_3$ (b) $HO-\hspace{-2pt}\bigcirc$

(c) $\bigcirc\hspace{-4pt}-CH_2OH$ (d) $CH_3\overset{\overset{\displaystyle CH_3}{|}}{\underset{\underset{\displaystyle CH_3}{|}}{C}}CH_2OH$

15.19 Examine each of the following compounds to see if it can be oxidized to an aldehyde or to a ketone. If it can, write the structure and the IUPAC name of the aldehyde or ketone.

(a) CH_3CH_2OH (b) $CH_3\overset{\overset{\displaystyle OH}{|}}{C}H-OH$

(c) cyclohexane with CH_3 and OH (d) $CH_3\overset{\overset{\displaystyle OH}{|}}{C}H\overset{\overset{\displaystyle O}{||}}{C}H_2COH$

15.20 An unknown compound, C_3H_6O, reacted with permanganate ion to give $C_3H_6O_2$, and the same unknown also gave a positive Tollens' test. Write the structures of C_3H_6O and $C_3H_6O_2$.

15.21 An unknown compound, $C_3H_6O_2$, could be oxidized easily by permanganate ion to $C_3H_4O_3$, and it also gave a positive Benedict's test. Write structures for $C_3H_6O_2$ and $C_3H_4O_3$.

15.22 Which of the following compounds can be expected to give a positive Benedict's test? All are intermediates in metabolism.

(a) $HOCH_2\overset{\overset{\displaystyle O}{||}}{C}\underset{\underset{\displaystyle HO}{|}}{C}H$ (b) $HOCH_2\overset{\overset{\displaystyle O}{||}}{C}CH_2OH$

(c) $CH_3\overset{\overset{\displaystyle O}{||}}{C}\underset{\underset{\displaystyle OH}{|}}{C}H_2\overset{\overset{\displaystyle O}{||}}{C}OH$ (d) $CH_3\overset{\overset{\displaystyle O}{||}}{C}CH_2\overset{\overset{\displaystyle O}{||}}{C}OH$

15.23 Which of the following compounds give a positive Benedict's test? (Most are intermediates in metabolism.)

(a) $HOCH_2\overset{\overset{\displaystyle O}{||}}{C}H-\underset{\underset{\displaystyle HO}{|}}{C}\underset{\underset{\displaystyle OH}{|}}{C}H$ (b) $HOCH_2CH_2\overset{\overset{\displaystyle O}{||}}{C}CH_3$

(c) $CH_3\overset{\overset{\displaystyle O}{||}}{C}-\overset{\overset{\displaystyle O}{||}}{C}H$ (d) $HO\overset{\overset{\displaystyle O}{||}}{C}-\overset{\overset{\displaystyle O}{||}}{C}CH_3$

15.24 Why is the presence of ammonia required in the Tollens' reagent?

15.25 What is the function of the citrate ion in the Benedict's reagent?

15.26 What is the formula of the precipitate that forms in a positive Benedict's test?

15.27 Clinitest tablets are used for what?

15.28 What is one practical commercial application of the Tollens' test?

15.29 Neither the lactate ion nor the pyruvate ion gives a positive Tollens' test. When the body metabolizes the lactate ion, it oxidizes it to the pyruvate ion, $C_3H_3O_3^-$. Using these facts, write the structure of the lactate ion.

15.30 One of the steps in the metabolism of fats and oils in the diet is the oxidation of the following compound:

$$CH_3\overset{\overset{\displaystyle OH}{|}}{C}HCH_2CO_2^-$$

Write the structure of the product of this oxidation.

Reduction of Aldehydes and Ketones

15.31 The hydride ion, as we learned, reacts as follows with water:

$$H:^- + H-OH \longrightarrow H_2 + OH^-$$

The hydride ion reacts in a similar way with $H-O-CH_3$. Write the net ionic equation for this reaction.

15.32 Based on our strategies for figuring out if a particular negative ion is a strong Brønsted base, how can we tell if CH_3-O^- is a strong or a weak Brønsted base?

15.33 Consider the reaction that occurs when a hydride ion transfers from some hydride ion donor (which we can write as $M:H$) to ethanal. (We will see several examples of this kind of reaction in our later study of metabolism.)
(a) Write the structure of the organic anion that forms when this transfer occurs to ethanal.
(b) What is the net ionic equation of the reaction of this anion with water?
(c) What is the IUPAC name of the organic product of this reaction with water?

15.34 If a hydride ion donor ($M : H$) transfers its hydride ion to a molecule of acetone:
(a) What is the structure of the organic ion that forms?
(b) What happens to this anion in the presence of water? (Write a net ionic equation.)
(c) What is the IUPAC name of the organic product of this reaction with water?

15.35 The metabolism of aspartic acid, an amino acid, occurs by a succession of steps, one of which is indicated as follows.

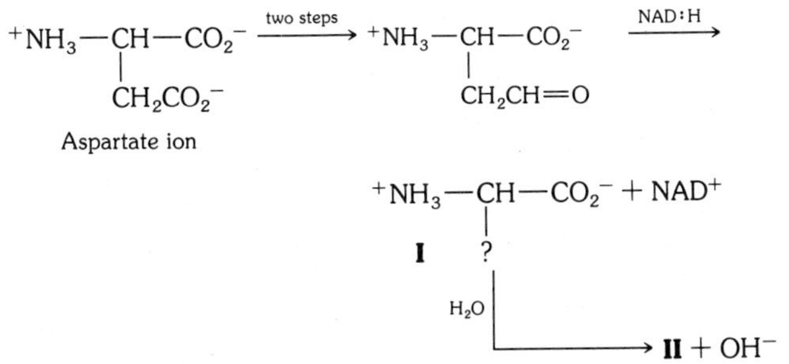

$$^{+}NH_3{-}CH{-}CO_2^{-} \xrightarrow{\text{two steps}} {}^{+}NH_3{-}CH{-}CO_2^{-} \xrightarrow{NAD:H}$$
$$\qquad\quad |\qquad\qquad\qquad\qquad\qquad\quad |$$
$$\qquad CH_2CO_2^{-}\qquad\qquad\qquad\quad CH_2CH{=}O$$

Aspartate ion

$$^{+}NH_3{-}CH{-}CO_2^{-} + NAD^{+}$$
$$\qquad\qquad |$$
$$\mathbf{I}\quad ?$$
$$\qquad\qquad | \;\; H_2O$$
$$\qquad\qquad \longrightarrow \mathbf{II} + OH^{-}$$

Complete the structure of **I**, and write the structure of **II**.

15.36 One of the steps the body uses to make long-chain carboxylic acids involves a reaction similar to the following reaction.

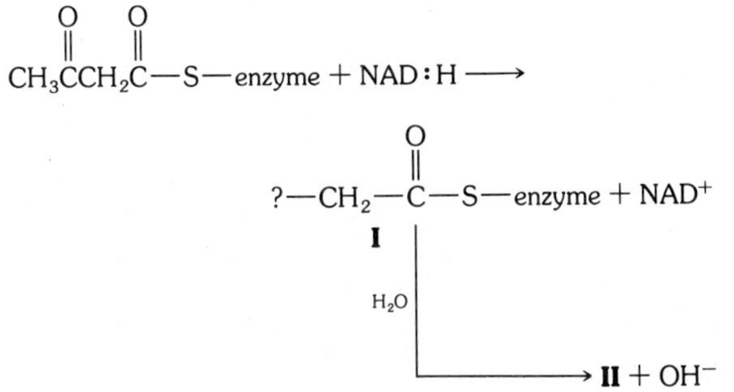

$$\underset{O}{\overset{O}{CH_3C}}CH_2\underset{O}{\overset{O}{C}}{-}S{-}enzyme + NAD:H \longrightarrow$$

$$? {-}CH_2{-}\underset{O}{\overset{O}{C}}{-}S{-}enzyme + NAD^{+}$$
$$\mathbf{I}$$
$$\qquad | \;\; H_2O$$
$$\qquad \longrightarrow \mathbf{II} + OH^{-}$$

Complete the structure of **I** and write the structure of **II**.

15.37 Write the structures of the aldehydes or ketones that could be used to make the following compounds by reduction (hydrogenation).

(a)  [cyclopentane ring with OH]

(b) 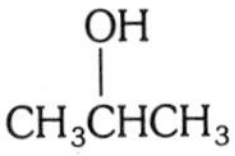 CH_3CHCH_3 with OH

(c) 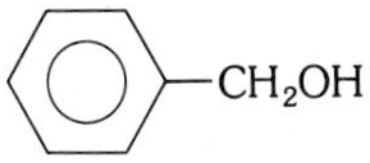 [benzene ring]$-CH_2OH$

(d) 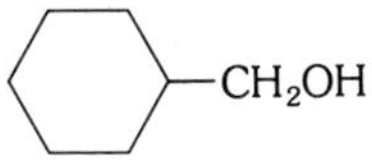 [cyclohexane ring]$-CH_2OH$

Hemiacetals and Acetals. Hemiketals and Ketals

15.38 Examine each structure and decide if it represents a hemiacetal, hemiketal, acetal, ketal, or something else.

(a) 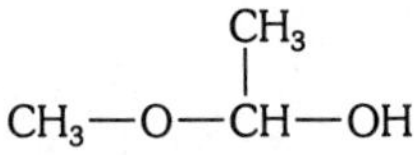 $CH_3{-}O{-}\underset{CH_3}{\overset{\;\;CH_3}{CH}}{-}OH$

(b) 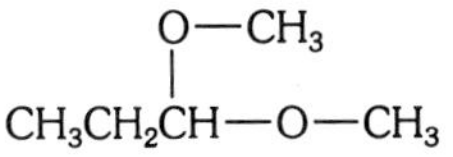 $CH_3CH_2\underset{\overset{|}{O{-}CH_3}}{CH}{-}O{-}CH_3$

(c) 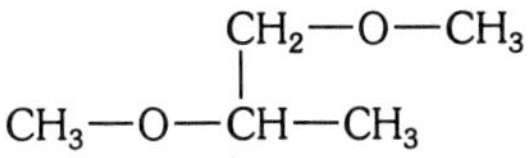 $CH_3{-}O{-}\underset{\overset{|}{CH_2{-}O{-}CH_3}}{CH}{-}CH_3$

(d) 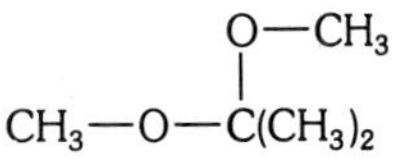 $CH_3{-}O{-}\underset{\overset{|}{O{-}CH_3}}{C}(CH_3)_2$

15.39 Examine each structure and decide if it represents a hemiacetal, hemiketal, acetal, ketal, or something else.

(a) 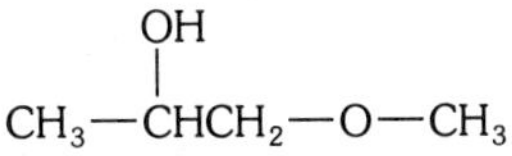 $CH_3{-}\underset{\overset{|}{OH}}{CH}CH_2{-}O{-}CH_3$

(b) $HOCH_2OCH_3$

(c) 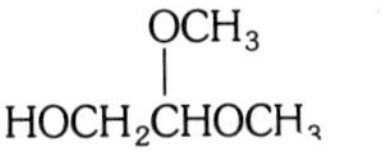 $HOCH_2\underset{\overset{|}{OCH_3}}{CH}OCH_3$

(d) 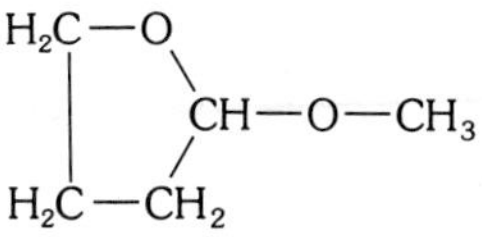 [five-membered ring with $H_2C{-}O$, $CH{-}O{-}CH_3$, $H_2C{-}CH_2$]

15.40 Either an aldehyde or a ketone could be used to make each of the following compounds by hydrogenation. Write the structure of the aldehyde or ketone suitable in each part.

(a) 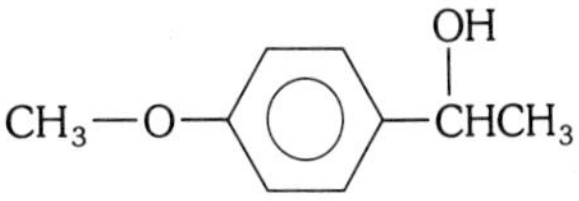 $CH_3{-}O{-}$[benzene ring]$-\underset{\overset{|}{OH}}{CH}CH_3$

(b) 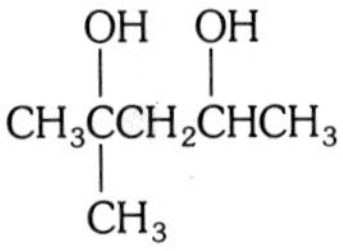 $CH_3\underset{\overset{|}{CH_3}}{C}CH_2\underset{\overset{|}{OH}}{CH}CH_3$ (with OH OH)

(c) $CH_3CH_2OCH_2CH_2OH$

(d) 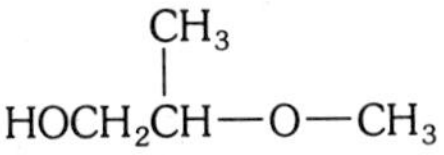 $HOCH_2\underset{\overset{|}{CH_3}}{CH}{-}O{-}CH_3$

15.41 Write the structures of the hemiacetals and the acetals that can form between ethanal and the following alcohols.
(a) Methanol   (b) Ethanol

15.42 What are the structures of the hemiketals and the ketals that can form between acetone and the following alcohols?
(a) Methanol   (b) Ethanol

15.43 Write the structure of the hydroxyaldehyde (a compound having both the alcohol group and the aldehyde group in the same molecule) from which the following hemiacetal forms in a ring-closing reaction. (You may leave the chain of the open-chain compound coiled somewhat.)

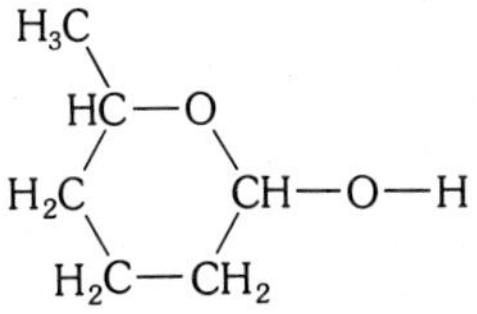

[six-membered ring: $H_3C{-}HC{-}O$, H_2C, $CH{-}O{-}H$, $H_2C{-}CH_2$]

15.44 One form in which a glucose molecule exists is given by the following structure. (*Note:* The atoms and groups that are attached to the carbon atoms of the six-membered ring must be seen as projecting *above* or *below* the ring.)

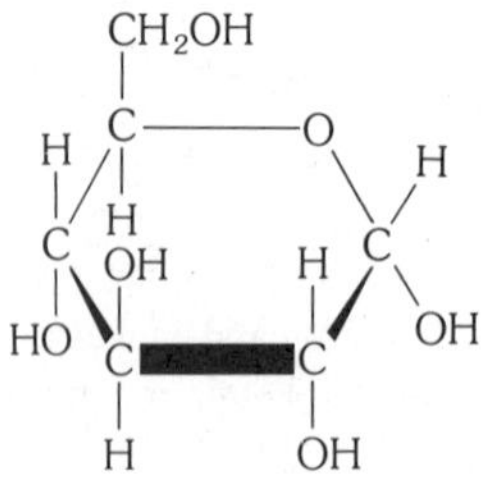

(a) Draw an arrow that points to the hemiacetal carbon.
(b) Write the structure of the open-chain form that has a free aldehyde group. (You may leave the chain coiled.)

15.45 Write the structure of a hydroxy ketone (a molecule that has both the —OH group and the keto group) from which the following hemiketal forms in a ring-closing reaction. (You may leave the chain of the open-chain compound somewhat coiled.)

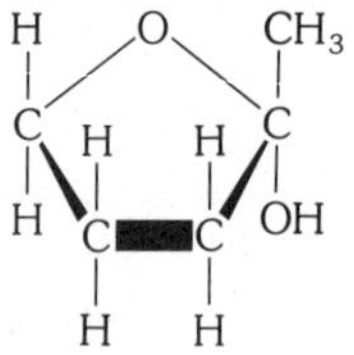

15.46 Fructose occurs together with glucose in honey, and it is sweeter to the taste than table sugar. One form in which a fructose molecule can exist is given by the following structure.

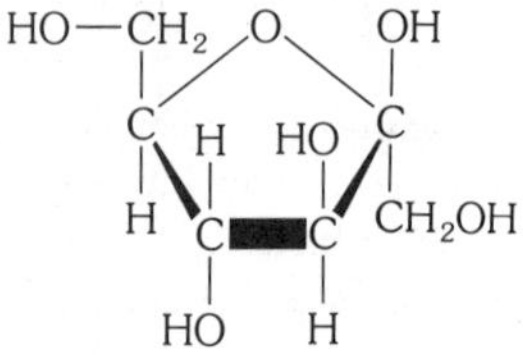

(a) Draw an arrow to the carbon of the hemiketal system that came initially from the carbon atom of a keto group.
(b) In water, fructose exists in an equilibrium with an open-chain form of the given structure. This form has a keto group in the same molecule as five —OH groups. Draw the structure of this open-chain form (leaving the chain coiled somewhat as it is in the structure that was given).

15.47 The digestion of some carbohydrates is simply their hydrolysis catalyzed by enzymes. Acids catalyze the same kind of hydrolysis of acetals and ketals. Write the structures of the

products, if any, that form by the action of water and an acid catalyst on the following compounds.

(a) 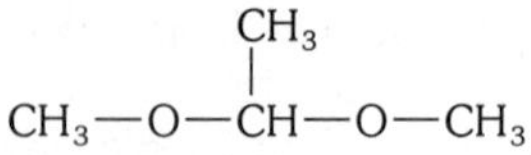 $CH_3—O—\underset{\underset{CH_3}{|}}{CH}—O—CH_3$

(b) 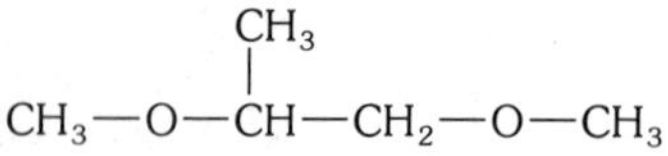 $CH_3—O—\underset{\underset{CH_3}{|}}{CH}—CH_2—O—CH_3$

(c) 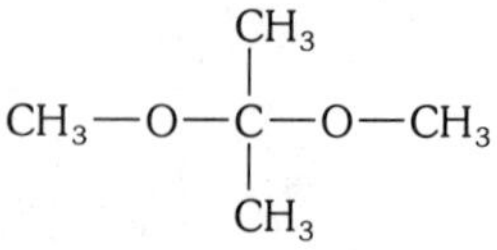 $CH_3—O—\underset{\underset{CH_3}{|}}{\overset{\overset{CH_3}{|}}{C}}—O—CH_3$

(d)

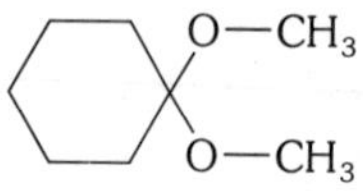

15.48 What are the structures of the products, if any, of the action of water that contains a trace of acid catalyst on the following compounds?

(a)

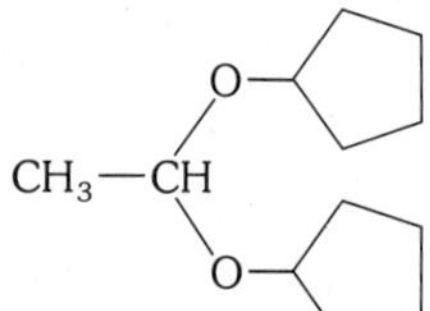

(b)

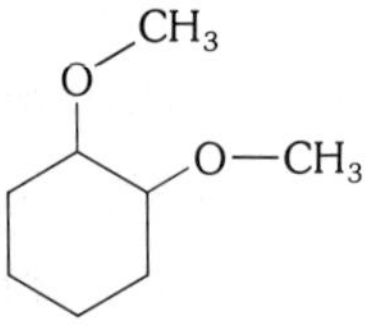

(c)

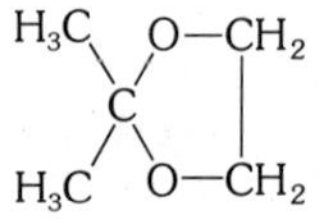

(d) 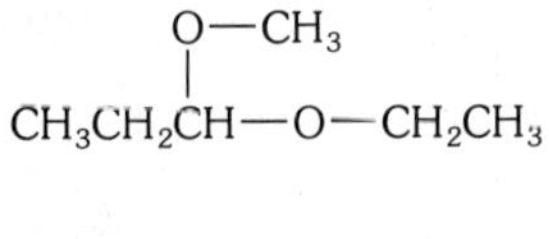 $CH_3CH_2\overset{\overset{O—CH_3}{|}}{CH}—O—CH_2CH_3$

15.49 Complete the following reaction sequences by writing the structures of the organic products that form. If no reaction occurs, write "no reaction." (Reviewed here are reactions of earlier chapters, too.)

(a) $CH_3CH{=}CHCH_3 + H_2 \xrightarrow[\text{Heat, pressure}]{\text{Ni}}$

(b) $CH_3\overset{\overset{OH}{|}}{CH}CH_3 + (O) \longrightarrow$

(c) $CH_3—O—\overset{\overset{OCH_3}{|}}{CH}CH_3 + H_2O \xrightarrow{H^+}$

(d) $CH_3\overset{\overset{O}{||}}{CH} + H_2 \xrightarrow[\text{Heat, pressure}]{\text{Ni}}$

(e) $CH_3CH_2\overset{\overset{O}{||}}{CH} + (O) \longrightarrow$

$$\text{(f)} \quad CH_3-\overset{\overset{\displaystyle CH_3}{|}}{\underset{\underset{\displaystyle CH_3}{|}}{C}}-OH + (O) \longrightarrow$$

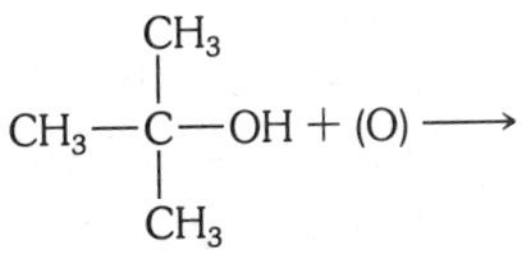

$$\text{(g)} \quad CH_3OH + CH_3CH_2\overset{\overset{\displaystyle O}{\|}}{C}H \rightleftharpoons$$

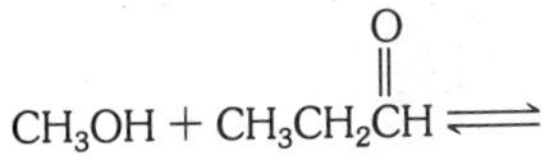

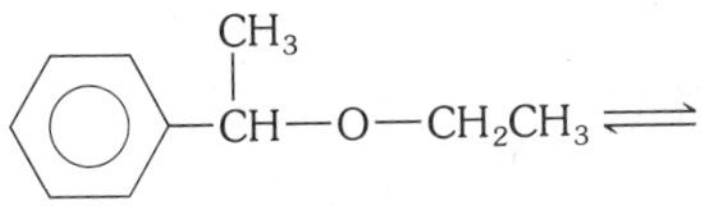

(h) $CH_3CH_2-O-CH_2CH_3 + (O) \longrightarrow$

$$\text{(i)} \quad \langle\bigcirc\rangle-\overset{\overset{\displaystyle CH_3}{|}}{C}H-O-CH_2CH_3 \rightleftharpoons$$

$$\text{(j)} \quad CH_3\overset{\overset{\displaystyle O}{\|}}{C}H + 2CH_3OH \xrightarrow{\text{Acid catalyst}}$$

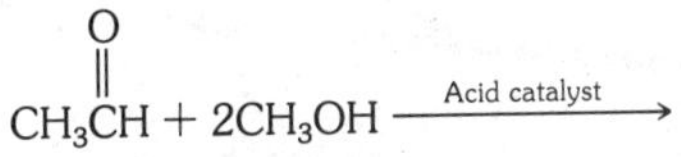

15.50 Write the structures of the organic products that form in each of the following situations. If no reaction occurs, write "no reaction." (Some of the situations constitute a review of earlier chapters.)

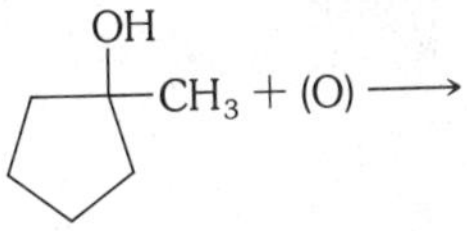

$$\text{(a)} \quad \langle\bigcirc\rangle-\overset{\overset{\displaystyle O}{\|}}{C}H + (O) \longrightarrow$$

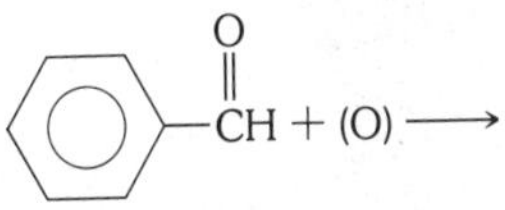

(b) [cyclopentane with OH and CH₃] $+ (O) \longrightarrow$

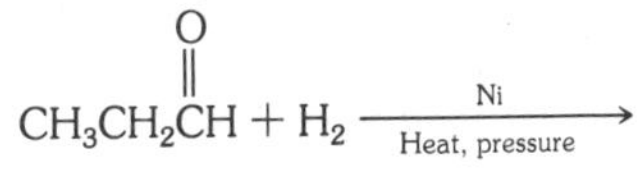

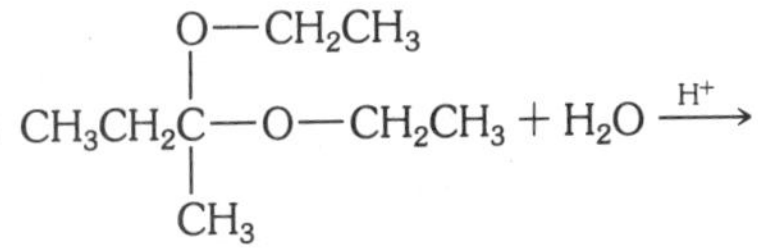

$$\text{(c)} \quad CH_3\overset{\overset{\displaystyle O}{\|}}{C}H + CH_3CH_2CH_2OH \rightleftharpoons$$

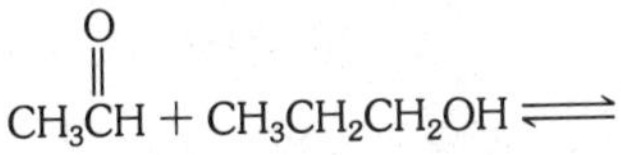

(d) $CH_3-O-CH_2CH_2-O-CH_3 + H_2O \xrightarrow{H^+}$

$$\text{(e)} \quad CH_3CH_2-O-\overset{\overset{\displaystyle OH}{|}}{\underset{\underset{\displaystyle CH_3}{|}}{C}}-CH_3 \rightleftharpoons$$

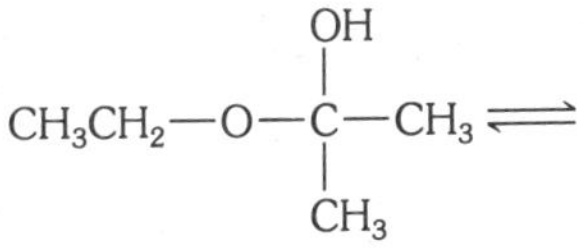

$$\text{(f)} \quad CH_3CH_2\overset{\overset{\displaystyle O}{\|}}{C}H + H_2 \xrightarrow[\text{Heat, pressure}]{Ni}$$

$$\text{(g)} \quad CH_3CH_2\overset{\overset{\displaystyle O-CH_2CH_3}{|}}{\underset{\underset{\displaystyle CH_3}{|}}{C}}-O-CH_2CH_3 + H_2O \xrightarrow{H^+}$$

$$\text{(h)} \quad \langle\bigcirc\rangle-\overset{\overset{\displaystyle OH}{|}}{C}HCH_3 + (O) \longrightarrow$$

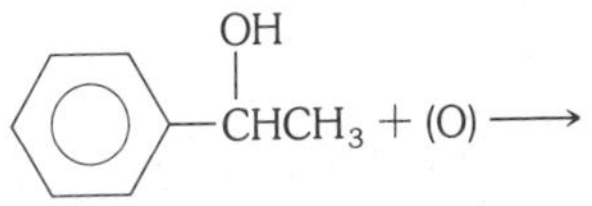

(i) $CH_3-O-CH_2CH=CH_2 + H_2 \xrightarrow[\text{Heat, pressure}]{Ni}$

$$\text{(j)} \quad CH_3CH_2\overset{\overset{\displaystyle O}{\|}}{C}H + 2CH_3OH \xrightarrow{H^+}$$

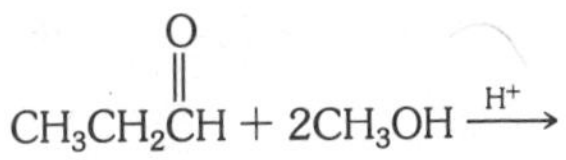

15.51 Catalytic hydrogenation of compound A (C_3H_6O) gave B (C_3H_8O). When B was heated strongly in the presence of sulfuric acid, it changed to compound C (C_3H_6). The acid-catalyzed addition of water to C gave compound D (C_3H_8O); and when D was oxidized, it changed to E (C_3H_6O). Compounds A and E are isomers, and compounds B and D are isomers. Write the structures of A–E.

15.52 When compound F ($C_4H_{10}O$) was gently oxidized, it changed to compound G (C_4H_8O), but vigorous oxidation changed F (or G) to compound H ($C_4H_8O_2$). Action of hot sulfuric acid on F changed it to compound I (C_4H_8). The addition of water to I (in the presence of an acid catalyst) gave compound J ($C_4H_{10}O$), a compound that could not be oxidized. Compounds F and J are isomers. Write the structures of F–J.

Chapter 16
Carboxylic Acids and Esters

Dacron, a polyester described in this chapter, is often used to make thin but strong fabrics for sails.

16.1 OCCURRENCE, NAMES, AND PHYSICAL PROPERTIES OF ACIDS

The carboxylic acids are polar compounds whose molecules form strong hydrogen bonds to each other.

"Carboxyl" comes from carbonyl + hydroxyl.

The two principal types of organic acids are the carboxylic acids and the sulfonic acids. The latter are much less common than the former, and we will not study them.

$$\begin{matrix} & O \\ & \parallel \\ - & C - O - H \end{matrix} \quad or \quad -CO_2H \quad or \quad -COOH \qquad \begin{matrix} & O \\ & \parallel \\ - & S - O - H \\ & \parallel \\ & O \end{matrix} \quad or \quad -SO_3H$$

Carboxyl group   Sulfonic acid group

In **carboxylic acids,** the carbonyl carbon holds a hydroxyl group and either another carbon atom or a hydrogen atom. A number of specific examples are given in Table 16.1. The acids with straight, alkane-like chains are often called the **fatty acids,** because they are products of the digestion of fats (and oils) in the diet.

The simplest acid, the first one listed in Table 16.1, is formic acid, and it has a sharp, irritating odor. This acid is responsible for the sting of certain ants and of the nettle plant. The next acid, acetic acid, gives tartness to vinegar, where its concentration is 4 to 5%. Butyric acid causes the odor of rancid butter. Valeric acid gets its name from the Latin *valerum,* meaning to be strong. What is strong about valeric acid is its odor. Other acids with vile odors are caproic, caprylic, and capric acids, which get their names from the Latin *caper,* meaning "goat"—a reference to odor.

Some acids have two carboxyl groups. The simplest such dicarboxylic acid is oxalic acid, which gives the tart taste to rhubarb. The tartness of citrus fruits is caused by citric acid. Lactic acid gives the tart taste to sour milk. These facts plus information in Table 16.1 on the origins of other acids show how widely occurring carboxylic acids are in nature.

Citric acid is a tricarboxylic acid.

$$\begin{matrix} & O & O \\ & \parallel & \parallel \\ HO - & C - C - OH \end{matrix} \qquad \begin{matrix} CH_2CO_2H \\ | \\ HO - C - CO_2H \\ | \\ CH_2CO_2H \end{matrix} \qquad \begin{matrix} CH_3CHCO_2H \\ | \\ OH \end{matrix}$$

Oxalic acid   Citric acid   Lactic acid

All carboxylic acids are weak Brønsted acids, and they exist in the form of their anions both in basic solutions and in their salts. It is largely as their anions that they occur in living cells and body fluids.

$$\begin{matrix} & O \\ & \parallel \\ R - & C - O^- \end{matrix} \quad or \quad RCO_2^- \quad or \quad RCOO^-$$

Symbols of anions of carboxylic acids

Names of Carboxylic Acids and Their Anions.   In all of the examples of naturally occurring acids we described, the common names were used. Several additional examples are in Table 16.1, where the IUPAC names for most are shown in parentheses. The IUPAC rules for naming carboxylic acids are similar to those for the aldehydes, including the requirements that the parent acid have the longest chain that includes the carbonyl group and that the carbon atom of this group be given number 1. Once the parent acid is identified, change the ending of the name of the alkane that has the same number of carbons (the parent alkane) from *-e* to *-oic acid.*

Use the name of the acid, not the name of the alkane, to devise the name of the anion of the acid.

TABLE 16.1
Carboxylic Acids

n	Structure	Name[a]	Origin of Name	M.Pt. (°C)	B.Pt. (°C)	Solubility (in g/100 g water at 20 °C)	K_a (25 °C)
Straight-chain saturated acids, $C_nH_{2n}O_2$							
1	HCO_2H	Formic acid (methanoic acid)	L. *formica*, ant	8	101	∞	1.8×10^{-4} (20 °C)
2	CH_3CO_2H	Acetic acid (Ethanoic acid)	L. *acetum*, vinegar	17	118	∞	1.8×10^{-5}
3	$CH_3CH_2CO_2H$	Propionic acid (Propanoic acid)	L. *proto, pion* first fat	−21	141	∞	1.3×10^{-5}
4	$CH_3(CH_2)_2CO_2H$	Butyric acid (Butanoic acid)	L. *butyrum*, butter	−6	164	∞	1.5×10^{-5}
5	$CH_3(CH_2)_3CO_2H$	Valeric acid (Pentanoic acid)	L. *valere*, to be strong (valerian root)	−35	186	4.97	1.5×10^{-5}
6	$CH_3(CH_2)_4CO_2H$	Caproic acid (Hexanoic acid)	L. *caper*, goat	−3	205	1.08	1.3×10^{-5}
7	$CH_3(CH_2)_5CO_2H$	Enanthic acid (Heptanoic acid)	Gr. *oenanthe*, vine blossom	−9	223	0.26	1.3×10^{-5}
8	$CH_3(CH_2)_6CO_2H$	Caprylic acid (Octanoic acid)	L. *caper*, goat	16	238	0.07	1.3×10^{-5}
9	$CH_3(CH_2)_7CO_2H$	Pelargonic acid (Nonanoic acid)	Pelargonium geranium	15	254	0.03	1.1×10^{-5}
10	$CH_3(CH_2)_8CO_2H$	Capric acid (Decanoic acid)	L. *caper*, goat	32	270	0.015	1.4×10^{-5}
12	$CH_3(CH_2)_{10}CO_2H$	Lauric acid (Dodecanoic acid)	Laurel	44	—	0.006	—
14	$CH_3(CH_2)_{12}CO_2H$	Myristic acid (Tetradecanoic acid)	Myristica (nutmeg)	54	—	0.002	—
16	$CH_3(CH_2)_{14}CO_2H$	Palmitic acid (Hexadecanoic acid)	Palm oil	63	—	0.0007	—
18	$CH_3(CH_2)_{16}CO_2H$	Stearic acid (Octadecanoic acid)	Gr. *stear*, solid	70	—	0.0003	—
Miscellaneous carboxylic acids							
	$C_6H_5CO_2H$	Benzoic acid	Gum benzoin	122	249	0.34 (25 °C)	6.5×10^{-5}
	$C_6H_5CH{=}CHCO_2H$	Cinnamic acid (*trans* isomer)	Cinnamon	132	—	0.04	3.7×10^{-5}
	$CH_2{=}CHCO_2H$	Acrylic acid	L. *acer*, sharp	13	141	soluble	5.6×10^{-5}
	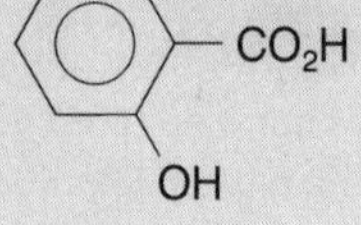 (benzene ring)—CO_2H / OH	Salicylic acid	L. *salix*, willow	159	211	0.22 (25 °C)	1.1×10^{-3} (19 °C)

[a] In parentheses below each common name is the IUPAC name.

If the acid is in the form of its anion, change the ending of the name of the parent *acid* from *-ic* to *-ate* (and omit the word *acid*). This rule for naming carboxylate anions applies both to the IUPAC names and to the common names. For example,

Pronounce "oate" as "oh-ate."

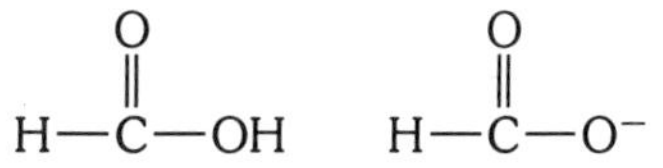

$$\underset{\substack{\text{Methanoic acid} \\ \text{(Formic acid)}}}{\text{H}-\overset{\displaystyle\overset{\text{O}}{\|}}{\text{C}}-\text{OH}} \qquad \underset{\substack{\text{Methanoate ion} \\ \text{(Formate ion)}}}{\text{H}-\overset{\displaystyle\overset{\text{O}}{\|}}{\text{C}}-\text{O}^-}$$

EXAMPLE 16.1   NAMING A CARBOXYLIC ACID AND ITS ANION

Problem: The following carboxylic acid has the common name of isovaleric acid. What is the IUPAC name of this acid and its sodium salt?

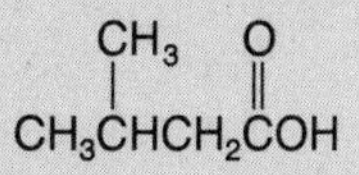

$$\overset{\text{CH}_3}{\underset{}{\text{CH}_3\text{CHCH}_2\overset{\displaystyle\overset{\text{O}}{\|}}{\text{C}}\text{OH}}}$$

Solution: The longest chain that includes the carboxyl group has four carbon atoms, so the parent acid is named by changing the name *butane* — the parent alkane — to *butanoic acid*. Then we have to number the chain, starting with the carboxyl group's carbon.

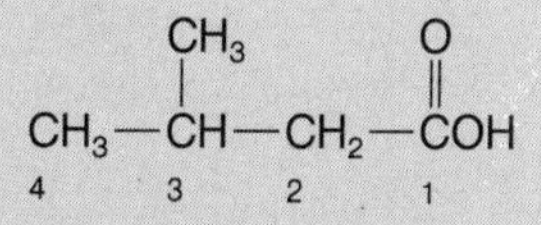

$$\underset{4}{\text{CH}_3}-\underset{3}{\overset{\overset{\text{CH}_3}{|}}{\text{CH}}}-\underset{2}{\text{CH}_2}-\underset{1}{\overset{\displaystyle\overset{\text{O}}{\|}}{\text{C}}\text{OH}}$$

The methyl group is at position 3, so the IUPAC name of this acid is 3-methylbutanoic acid.
  To name its anion, we drop *-ic acid* from the name and add *-ate*. Therefore the name of the anion is 3-methylbutanoate, and the name of the sodium salt of this acid is sodium 3-methylbutanoate. (The common name is sodium isovalerate.)

PRACTICE EXERCISE 1   What are the IUPAC names of the following compounds?

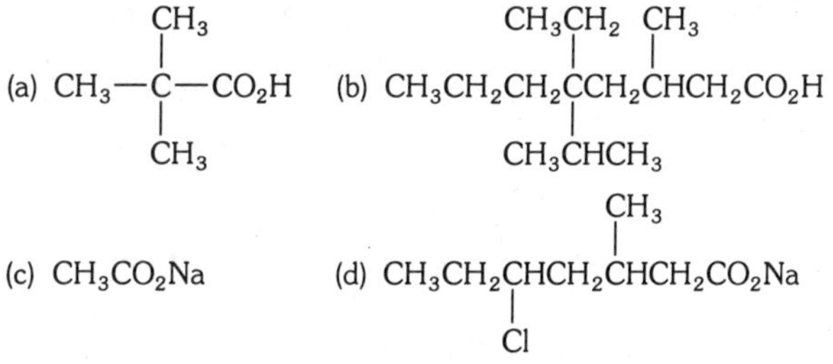

(a) $\text{CH}_3-\overset{\overset{\text{CH}_3}{|}}{\underset{\underset{\text{CH}_3}{|}}{\text{C}}}-\text{CO}_2\text{H}$

(b) $\text{CH}_3\text{CH}_2\text{CH}_2\overset{\overset{\text{CH}_3\text{CH}_2}{|}}{\text{C}}\text{CH}_2\overset{\overset{\text{CH}_3}{|}}{\text{CH}}\text{CH}_2\text{CO}_2\text{H}$
   $\overset{|}{\text{CH}_3\text{CHCH}_3}$

(c) $\text{CH}_3\text{CO}_2\text{Na}$

(d) $\text{CH}_3\text{CH}_2\text{CHCH}_2\overset{\overset{\text{CH}_3}{|}}{\text{CH}}\text{CH}_2\text{CO}_2\text{Na}$
   $\underset{\text{Cl}}{|}$

PRACTICE EXERCISE 2   If the IUPAC name of HOCCH_2COH (with two C=O groups) is propanedioic acid (and not 1,3-propanedioic acid), what must be the IUPAC name for $\text{HOCCH}_2\text{CH}_2\text{CH}_2\text{COH}$ (with two C=O groups)?

PRACTICE EXERCISE 3

Oleic acid is actually the *cis* isomer. The name of the *trans* isomer is elaidic acid.

If the IUPAC name of $CH_3CH{=}CHCH_2CH_2CO_2H$ is 4-hexenoic acid, what must be the IUPAC name of the following acid, which is commonly called *oleic acid*—one of the products of the digestion of almost any edible vegetable oil or animal fat. (The name of the straight-chain alkane with 18 carbon atoms is octadecane.)

$$CH_3CH_2CH_2CH_2CH_2CH_2CH_2CH_2CH{=}CHCH_2CH_2CH_2CH_2CH_2CH_2CH_2CO_2H$$

Oleic acid (common name)

Hydrogen Bonding in Carboxylic Acids.   Carboxylic acids have higher boiling points than alcohols of comparable formula weights. The reason is that molecules of carboxylic acids form hydrogen-bonded pairs:

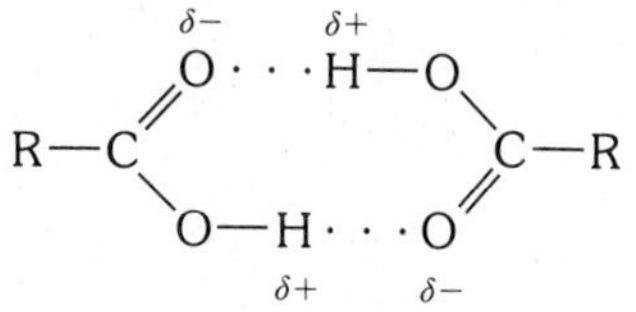

This makes the *effective* formula weight of a carboxylic acid much higher than its calculated formula weight, and therefore the boiling point is higher.

The lower-formula-weight carboxylic acids (C_1—C_4) are soluble in water largely because the carboxyl group has *two* oxygen atoms that can accept hydrogen bonds from water molecules. In addition, the carboxyl group has the —OH group that can donate hydrogen bonds.

Remember, we're interested in how structure affects solubility in water because water is the fluid medium in the body.

16.2   THE ACIDITY OF CARBOXYLIC ACIDS

The carboxylic acids are weak Brønsted acids toward water but strong Brønsted acids toward the hydroxide ion.

Aqueous solutions of carboxylic acids contain the following species in equilibrium:

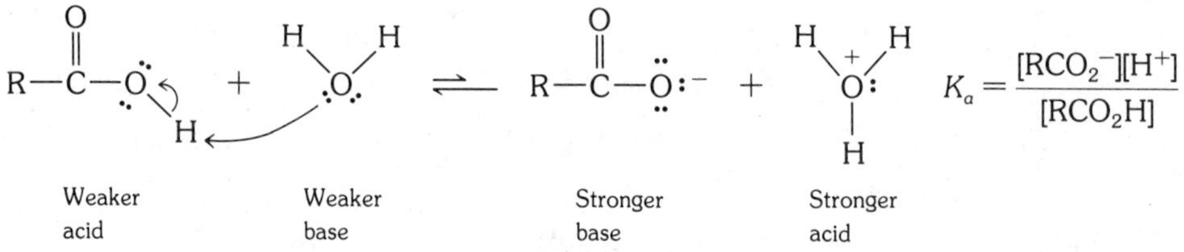

$$K_a = \frac{[RCO_2^-][H^+]}{[RCO_2H]}$$

The K_a values of several carboxylic acids are given in Table 16.1, and you can see that most are on the order of 10^{-5}. Thus the carboxylic acids are weak acids toward water, and their percentage ionizations are low. For example, in a $1M$ solution at room temperature, acetic acid is ionized only to about 0.5%. Of course, this is vastly greater than the ionization of alcohols in water. Alcohols have K_a values on the order to 10^{-16}, so they are several billion times weaker acids than carboxylic acids. In fact, we don't think of alcohols as being acids at all.

Neutralization of Carboxylic Acids by Strong Bases.   The hydroxide ion, the carbonate ion, and the bicarbonate ion are much stronger bases than a water molecule. Therefore when any of these ions is supplied to a carboxylic acid (in a 1 : 1 mole ratio), the acid is neutralized. This reaction is extremely important at the molecular level of life, because the carboxylic acids we normally produce by metabolism must be neutralized. Otherwise, the pH of body fluids, such as the blood, would fall too low (and we would die).

The organic product of the neutralization of a carboxylic acid is the anion of the acid, a carboxylate ion, RCO_2^-. The inorganic products vary with the base used. Acid neutralization occurs rapidly at room temperature. With hydroxide ion, the reaction is as follows:

$$RCO_2H + OH^- \longrightarrow RCO_2^- + H-OH$$

| Stronger | Stronger | Weaker | Weaker |
| acid | base | base | acid |

With bicarbonate ion, the chief base in the buffer systems of the blood, the following reaction occurs:

$$RCO_2H + HCO_3^- \longrightarrow RCO_2^- + H_2O + CO_2$$

Some specific examples are as follows:

$$CH_3-CO_2H + OH^- \longrightarrow CH_3-CO_2^- + H_2O$$

Acetic acid   Acetate ion

The stearate ion is one of several organic ions in soap.

$$CH_3(CH_2)_{16}CO_2H + OH^- \longrightarrow CH_3(CH_2)_{16}CO_2^- + H_2O$$

Stearic acid   Stearate ion
(insoluble in water)   (soluble in water)

$$C_6H_5CO_2H + HCO_3^- \longrightarrow C_6H_5CO_2^- + H_2O + CO_2$$

Benzoic acid   Benzoate ion
(insoluble   (soluble in
in water)   water)

PRACTICE EXERCISE 4   Write the structures of the carboxylate anions that form when the following carboxylic acids are neutralized.

(a) $CH_3CH_2CO_2H$ (b) 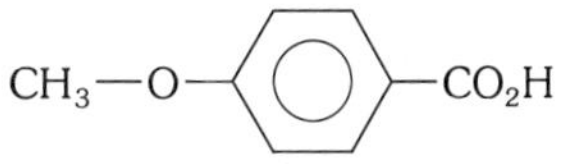 $CH_3-O-\langle\!\!\bigcirc\!\!\rangle-CO_2H$ (c) $CH_3CH{=}CHCO_2H$

The purified salts that involve carboxylate anions and metal ions are genuine salts— assemblies of oppositely charged ions—and therefore all of these salts are solids at room temperature. Table 16.2 gives a few examples of the sodium salts of some carboxylic acids. Like all sodium salts, they are soluble in water but completely insoluble in such nonpolar solvents as ether or gasoline. Several are used as decay inhibitors in foods and bottled beverages, as described in Special Topic 16.1.

Carboxylate Ions as Bases.   Because carboxylate ions are the anions of *weak* acids, they themselves must be relatively good bases, especially toward a strong proton donor such as the hydronium ion. At room temperature, the following neutralization of a strong acid by a carboxylate ion occurs virtually instantaneously. It is the most important reaction of the carboxylate ion that we will study, because it makes the carboxylate ion group a neutralizer of excess acid at the molecular level of life.

$$R-CO_2^- + H-\overset{+}{O}\!\!\diagdown\!\!\diagup^H_H \longrightarrow R-CO_2H + H-OH$$

| Stronger | Stronger | Weaker | Weaker |
| base | acid | acid | base |

TABLE 16.2
Some Sodium Salts of Carboxylic Acids

Common Name[a]	Structure	Melting Point (°C)	Solubility	
			Water	Ether
Sodium formate (Sodium methanoate)	HCO_2Na	253	Soluble	Insoluble
Sodium acetate (Sodium ethanoate)	CH_3CO_2Na	323	Soluble	Insoluble
Sodium propionate (Sodium propanoate)	$CH_3CH_2CO_2Na$	—	Soluble	Insoluble
Sodium benzoate	$C_6H_5CO_2Na$	—	66 g/100 mL	Insoluble
Sodium salicylate	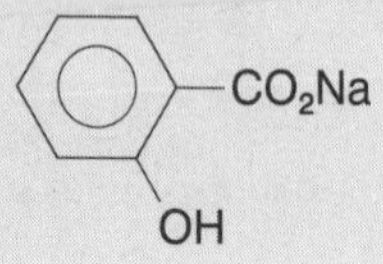 (structure: benzene ring with $-CO_2Na$ and OH)	—	111 g/100 mL	Insoluble

[a] The IUPAC names are in parentheses.

For example,

These rapid proton transfers don't require any heating, either.

$$C_6H_5CO_2^- + H_3O^+ \longrightarrow C_6H_5CO_2H + H_2O$$

Benzoate ion (soluble in water)   Benzoic acid (insoluble in water)

$$CH_3(CH_2)_{16}CO_2^- + H_3O^+ \longrightarrow CH_3(CH_2)_{16}CO_2H + H_2O$$

Stearate ion (soluble in water)   Stearic acid (insoluble in water)

The Carboxylic Acid Group as a Solubility "Switch".   In these reactions, we have noted the solubilities in water of several species to draw attention to a very important property that the carboxylic acid group gives to a molecule. It can be used to switch on or off the solubility in water of any substance that contains this group. By simply trying to increase the pH of a solution — by adding a strong base — a water-insoluble carboxylic acid almost instantly dissolves, because it changes to its carboxylate ion. Similarly, by trying to decrease the pH of a solution — by adding a strong acid — a water-soluble carboxylate anion instantly changes to its much less soluble, free carboxylic acid form. In other words, by suitably adjusting the pH of an aqueous solution we can make a substance with a carboxyl group more soluble or less soluble in water. The body often takes advantage of this kind of solubility "switch."

PRACTICE EXERCISE 5   Write the structures of the organic products of the reactions of the following compounds with dilute hydrochloric acid at room temperature.

(a)  CH_3-O-(benzene ring)$-CO_2^-K^+$ (b) $CH_3CH_2CO_2^-Li^+$ (c) $(CH_3CH=CHCO_2^-)_2Ca^{2+}$

16.3 THE CONVERSION OF CARBOXYLIC ACIDS TO ESTERS

Carboxylic acids can be used directly or indirectly to make esters from alcohols.

The carboxylic acids are the parent compounds for several families that collectively are called the **acid derivatives.** These include the **acid chlorides,** the **anhydrides** (both common

SPECIAL TOPIC 16.1 SOME IMPORTANT CARBOXYLIC ACIDS AND SALTS

Acetic Acid Most people experience acetic acid directly in the form of its dilute solution in water called vinegar. Because blood is slightly alkaline, acetic acid circulates as the acetate ion, and we will meet this species many times when we study metabolic pathways. The acetate ion, in fact, is one of the major intermediates in the metabolism of carbohydrates, lipids, and proteins.

Acetic acid is also an important industrial chemical, and more than 3 billion pounds (23 billion moles) are manufactured each year in the United States. Acetate rayon is just one consumer product made using acetic acid.

Propanoic Acid and Its Salts Propanoic acid occurs naturally in Swiss cheese with a concentration that can be as high as 1%. Its sodium and calcium salts are food additives used in baked goods and processed cheese to retard the formation of molds or the growth of bacteria. (On ingredient labels, these salts are listed under their common names, sodium or calcium propionate.)

Sorbic Acid and the Sorbates Sorbic acid, or 2,4-hexadienoic acid, $CH_3CH{=}CHCH{=}CHCO_2H$, and its sodium or potassium salts are added in trace concentrations to a variety of foods to inhibit the growth of molds and yeasts. The sorbates often appear on ingredient lists for fruit juices, fresh fruits, wines, soft drinks, sauerkraut and other pickled products, and some meat and fish products. For food products that usually are wrapped, such as cheese and dried fruits, solutions of sorbate salts are sometimes sprayed onto the wrappers.

Sodium Benzoate Traces of sodium benzoate inhibit molds and yeasts in products that normally have pH values below 4.5 or 4.0. (The sorbates work better at slightly higher pH values—up to 6.5.) You'll see sodium benzoate on ingredient lists for beverages, syrups, jams and jellies, pickles, salted margarine, fruit salads, and pie fillings. Its concentration is low—0.05 to 0.10%—and neither benzoic acid nor the benzoate ion accumulates in the body.

and mixed anhydrides), the **esters,** and the **amides.** They are called acid *derivatives* because they can be made from the acids and they can be hydrolyzed back to the acids.

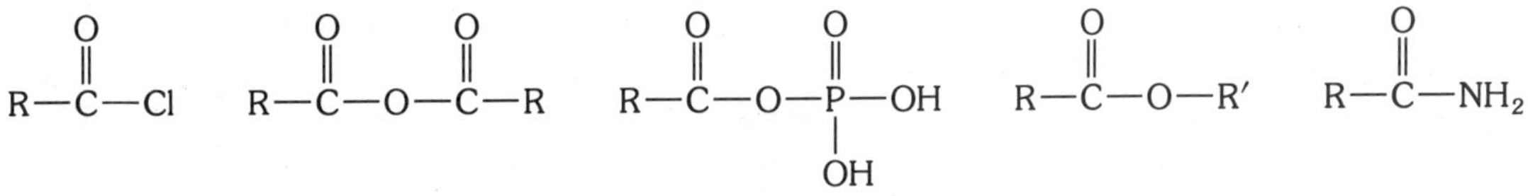

Acid chlorides — Acid anhydrides — Mixed anhydrides with phosphoric acid — Esters — Amides

Neither the acid chlorides nor the simple acid anhydrides occur in the body, because these compounds react too readily with water and with alcohol groups. We include them in our study because they are often used in the lab to make esters (e.g., aspirin), and because their reactions that lead to esters are simple examples of **acyl group transfer reactions.** An **acyl group** is a carboxylic acid minus its —OH group.

Acyl group — Acetyl group — Benzoyl group

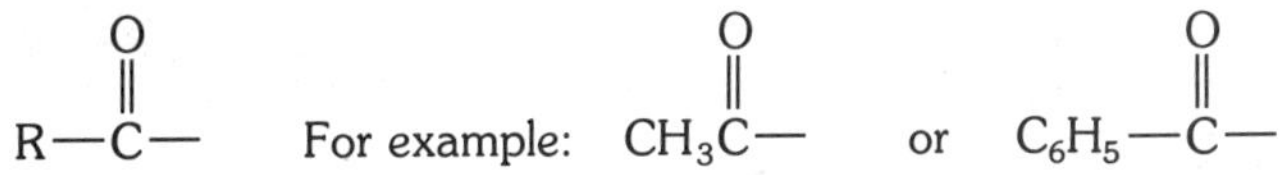

Thus the acyl group is in the molecules of all carboxylic acids, all esters, all acid chlorides, anhydrides (and amides).

Acid Chlorides. Acid chlorides react with water to give the parent acids and hydrogen chloride (which, because of the excess water, forms as hydrochloric acid). In this reaction an acyl group transfers from —Cl to —OH.

Acid chloride — Carboxylic acid

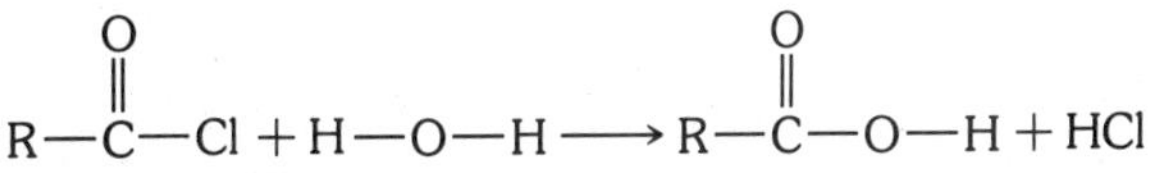

With alcohols, the reaction is the transfer of an acyl group to —OR and an ester is the organic product.

$$R-\underset{\underset{\text{Acid}}{\text{chloride}}}{\overset{\overset{O}{\|}}{C}}-Cl + \underset{\text{Alcohol}}{H-O-R'} \longrightarrow R-\underset{\text{Ester}}{\overset{\overset{O}{\|}}{C}}-O-R' + HCl$$

For example,

We'll learn how to name esters soon, but we will not develop the rules for naming acid chlorides or acid anhydrides.

$$CH_3-\underset{\underset{\text{Acetyl}}{\text{chloride}}}{\overset{\overset{O}{\|}}{C}}-Cl + \underset{\underset{\text{Ethyl}}{\text{alcohol}}}{H-O-CH_2CH_3} \longrightarrow CH_3-\underset{\underset{\text{Ethyl acetate}}{\text{(ethyl ethanoate)}}}{\overset{\overset{O}{\|}}{C}}-O-CH_2CH_3 + HCl$$

This is an example of a reaction called an **esterification**—the synthesis of an ester. We say that the alcohol is *esterified*. We could also say that the acid has been esterified by this method.

Acid Anhydrides (Common).   When a carboxylic acid anhydride reacts with an alcohol, an acyl group transfers from a carboxylate group to —OR. This reaction occurs with roughly the same ease as the reaction of an acid chloride with an alcohol. Acid anhydrides react with alcohols as follows:

$$R-\underset{\underset{\text{Acid}}{\text{anhydride}}}{\overset{\overset{O}{\|}}{C}}-O-\overset{\overset{O}{\|}}{C}-R + \underset{\text{Alcohol}}{H-O-R'} \longrightarrow R-\underset{\text{Ester}}{\overset{\overset{O}{\|}}{C}}-O-R' + \underset{\underset{\text{Carboxylic}}{\text{acid}}}{H-O-\overset{\overset{O}{\|}}{C}-R}$$

Phenols, like alcohols, can also be esterifed by acid anhydrides. When the acid anhydride is acetic anhydride and phenol is used, one product is aspirin.

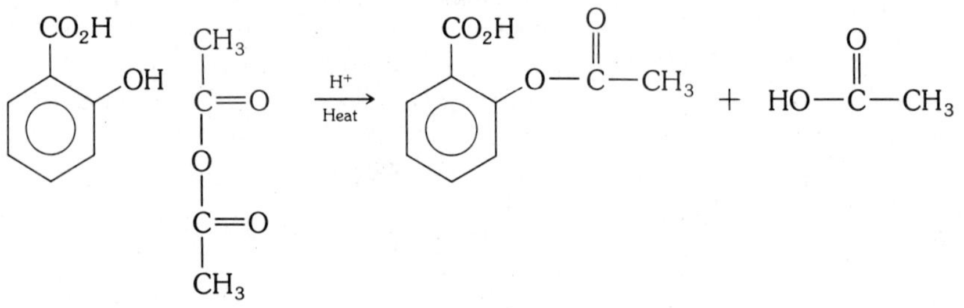

Both salicylic acid and acetic anhydride are common, readily available organic chemicals.

Salicylic acid   Acetic anhydride   Acetyl salicylic acid (aspirin)

Direct Esterification of Acids and Alcohols.   When a solution of a carboxylic acid in an alcohol is heated, and a strong acid catalyst is present, the following species become involved in an equilibrium.

$$R-\underset{\underset{\text{Carboxylic}}{\text{acid}}}{\overset{\overset{O}{\|}}{C}}-O-H + \underset{\text{Alcohol}}{H-O-R'} \underset{}{\overset{H^+}{\rightleftharpoons}} R-\underset{\text{Ester}}{\overset{\overset{O}{\|}}{C}}-O-R' + H-OH$$

When the alcohol is in *excess*, Le Chatelier's principle operates and the equilibrium shifts so much to the right that this a good method for making an ester. This synthesis of an ester is called *direct esterification*. Some specific examples are as follows:

$$CH_3-\overset{\overset{O}{\|}}{C}-O-H + H-O-CH_2CH_3 \xrightarrow{H^+} CH_3-\overset{\overset{O}{\|}}{C}-O-CH_2CH_3 + H_2O$$

Acetic acid   Ethyl alcohol   Ethyl acetate
                   (large excess)

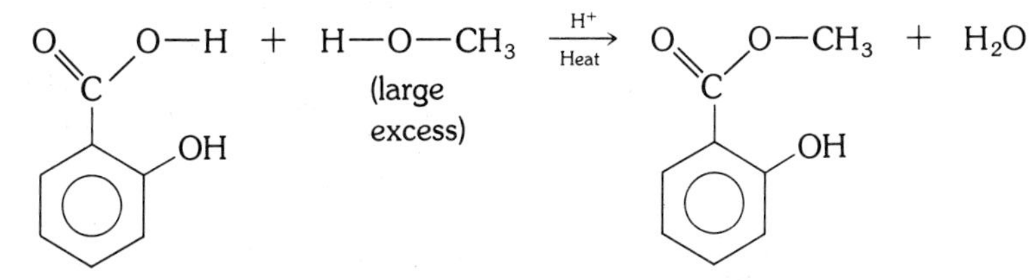

Salicylic acid   Methyl   Methyl salicylate
                      alcohol   (oil of wintergreen)

EXAMPLE 16.2   **WRITING THE STRUCTURE OF A PRODUCT OF DIRECT ESTERIFICATION**

Problem: What is the ester that can be made from benzoic acid and methyl alcohol?

Solution: Write the structures of the two reactants. Sometimes it helps to let the —OH groups "face" each other:

$$C_6H_5-\overset{\overset{O}{\|}}{C}-O-H + H-O-CH_3$$
Benzoic acid   Methyl alcohol

To obtain the pieces of the water molecule, the other product of direct esterification, we have to remove the —OH group from the acid and the proton on oxygen from the alcohol. This leaves us with the following fragments:

$$C_6H_5-\overset{\overset{O}{\|}}{C} \quad and \quad -O-CH_3$$

Now all that is left to do is join these fragments.

$$C_6H_5-\overset{\overset{O}{\|}}{C}-O-CH_3 \quad \text{(The product of the esterification)}$$
Methyl benzoate

Although it is not important in predicting correct structures of products, always erase the —OH group from the carboxylic acid, not the alcohol. This will make it easier to learn a reaction coming up in the next chapter.

PRACTICE EXERCISE 6   Write the structures of the esters that form by the direct esterification of acetic acid by the following alcohols.

(a) Methyl alcohol   (b) Propyl alcohol   (c) Isopropyl alcohol

PRACTICE EXERCISE 7   Write the structures of the esters that can be made by the direct esterification of ethyl alcohol by the following acids.

(a) Formic acid   (b) Propionic acid   (c) Benzoic acid

The Acid Catalyst in Direction Esterification.   Without the acid catalyst, direct esterification would proceed very slowly. The catalyst works by converting what is a very poor leaving group in the acid into one that, because it is so stable, is a good leaving group. The leaving group of the acid itself would have to be the hydroxide ion, OH^-, a strong base. But the acid catalyst donates a proton to the —OH group of the acid, and this sets up the molecule for the departure not of OH^- but of the far more stable H_2O, as follows:

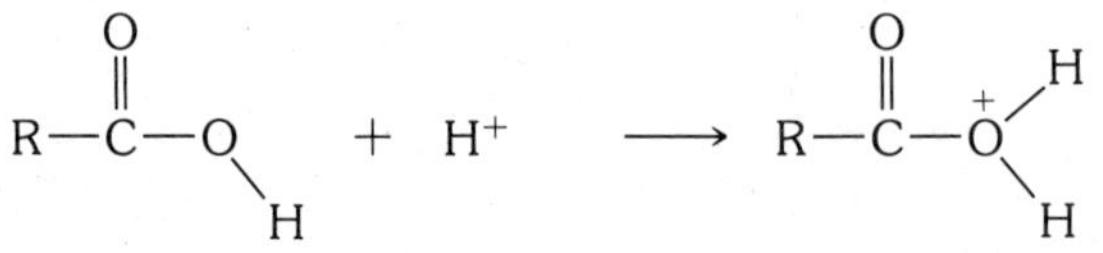

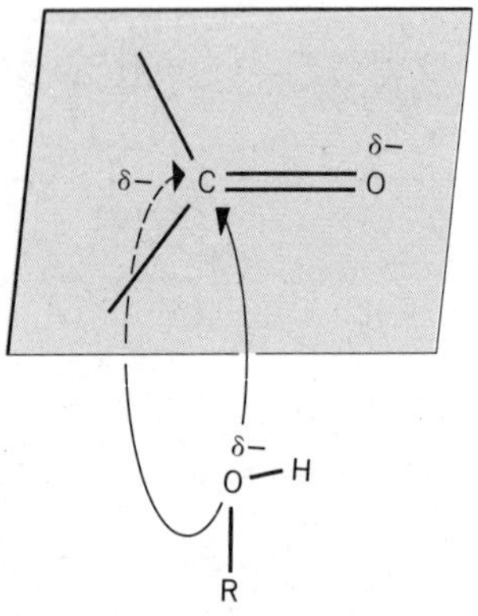

Catalyst   Protonated form of the carboxylic acid

In its protonated form, as shown, the bond from the carbonyl carbon atom to the potential leaving group (water) is weak in comparison to this bond in the unprotonated form of the carboxylic acid. The extra positive charge in the protonated acid draws the electrons of this bond away from the carbonyl carbon. Moreover, this carbon now has a larger partial positive charge, $\delta+$, so besides being wide open to attack (Figure 16.1), it attracts the alcohol molecule more strongly. For these reasons, the alcohol more readily attacks the carboxylic acid when the strong acid catalyst is used.

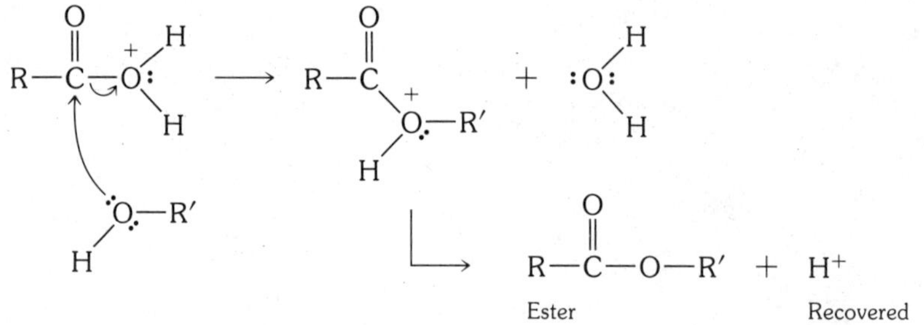

FIGURE 16.1
An alcohol molecule can attack a carbonyl carbon atom from either side of the planar carbonyl group. The other groups already attached to the carbonyl carbon atom interfere with this attack very little. Therefore, when one of these other groups is a particularly stable leaving group, such as Cl or a carboxylate group, the carbonyl compound is especially reactive toward alcohols (and water, too).

Ester   Recovered catalyst

16.4 OCCURRENCE, NAMES, AND PHYSICAL PROPERTIES OF ESTERS

Esters are moderately polar compounds.

The functional group of an ester is the central structural feature of all of the edible fats and oils as well as a number of constituents of body cells. Be sure to be able to recognize this group and to pick out what we will call the *ester linkage* — the single bond between the carbonyl carbon atom and the oxygen atom that holds the ester's alkyl group. This linkage is where an ester breaks apart when it reacts with water.

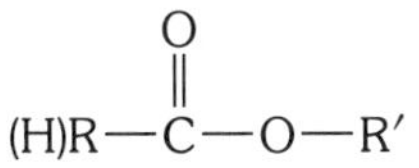

$$\underset{\substack{\text{Two general formulas for}\\\text{esters}}}{(H)R-\overset{\displaystyle O}{\overset{\|}{C}}-O-R' \quad \text{or} \quad (H)RCO_2R'}$$

Ester linkage

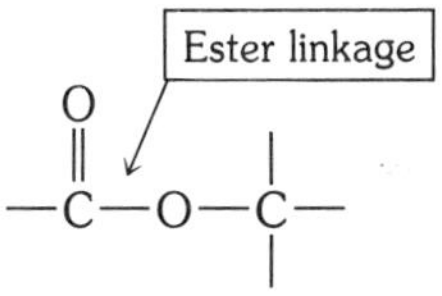

$$\underset{\substack{\text{Ester group}\\\text{(carbonyl-oxygen-carbon)}}}{-\overset{\displaystyle O}{\overset{\|}{C}}-O-\overset{|}{\underset{|}{C}}-}$$

One interesting feature about acids and their esters is that the low-formula-weight acids have vile odors, but their esters have some of the most pleasant fragrances in all of nature, as the information in Table 16.3 shows. Special Topic 16.2 describes some important esters. Table 16.4 gives physical properties of several common esters.

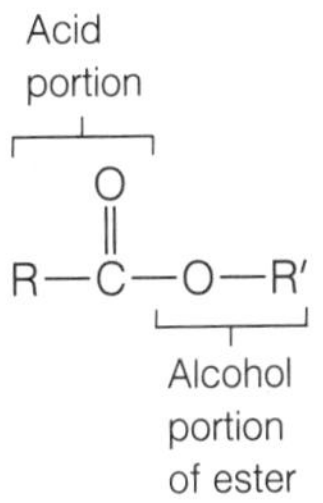

Naming Esters. Common names and IUPAC names are devised in very similar ways. For the moment, simply ignore the R′ group of an ester — the part supplied by the alcohol — and focus your attention on the acid portion. Pretend you are naming the *anion* of the acid. Remember that in both the common and the IUPAC names, this anion is named by changing the *-ic* ending of the name of the parent acid to *-ate*. Thus salts of acetic acid (ethanoic acid) are called acetate salts (common name) or ethanoate salts (IUPAC). Similarly, esters of this acid are called acetate esters (common) or ethanoate esters (IUPAC). Thus the acid portions of esters and the anions of these acids are named in the same way.

Once you have the name of the acid portion of the ester, simply write the name of the alkyl group in the ester's alcohol portion in front of this name (as a separate word). Here are some examples that show the pattern. (The IUPAC names are in parentheses.)

Acid portion

$$R-\overset{\displaystyle O}{\overset{\|}{C}}-O-R'$$

Alcohol portion of ester

Ester	Name of Parent Acid	Name of Acid Portion of Ester	Alkyl Group in Ester	Name of Ester
$CH_3-\overset{O}{\overset{\|}{C}}-O-CH_3$	Acetic acid (ethanoic acid)	Acetate (ethanoate)	Methyl	Methyl acetate (methyl ethanoate)
$CH_3CH_2-O-\overset{O}{\overset{\|}{C}}-H$	Formic acid (methanoic acid)	Formate (methanoate)	Ethyl	Ethyl formate (ethyl methanoate)

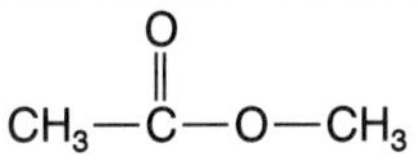

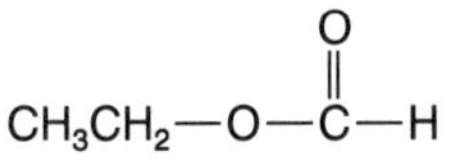

TABLE 16.3
Fragrances or Flavors of Some Esters

Name	Structure	Source or Flavor
Ethyl formate	$HCO_2CH_2CH_3$	Rum
Isobutyl formate	$HCO_2CH_2CH(CH_3)_2$	Raspberries
Pentyl acetate	$CH_3CO_2CH_2CH_2CH_2CH_2CH_3$	Bananas
Isopentyl acetate	$CH_3CO_2CH_2CH_2CH(CH_3)_2$	Pears
Octyl acetate	$CH_3CO_2(CH_2)_7CH_3$	Oranges
Ethyl butyrate	$CH_3CH_2CH_2CO_2CH_2CH_3$	Pineapples
Pentyl butyrate	$CH_3CH_2CH_2CO_2(CH_2)_4CH_3$	Apricots
Methyl salicylate	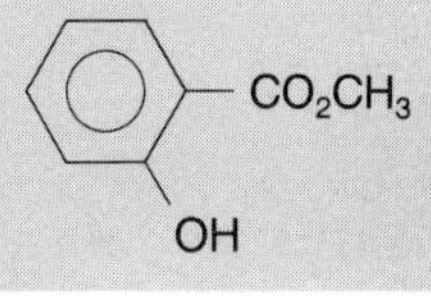 ⬡—CO_2CH_3 OH	Oil of wintergreen

EXAMPLE 16.3 WRITING IUPAC NAMES FOR ESTERS

Problem: What is the IUPAC name for the following ester?

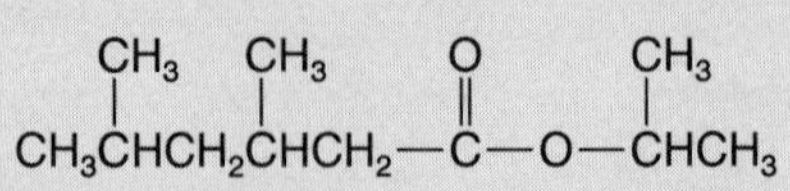

$$
\begin{array}{ccccc}
CH_3 & CH_3 & O & CH_3 \\
| & | & \| & | \\
CH_3CHCH_2CHCH_2\!-\!C\!-\!O\!-\!CHCH_3
\end{array}
$$

Solution: First, figure out what part of this structure came initially from a parent alcohol. This is the group attached to an oxygen atom by a *single* bond, but is not the carbonyl carbon atom. In this ester, the alcohol part is the isopropyl group on the right in the structure. Now we can figure out the name of the parent acid, the part of the ester system that includes the carbonyl group. Its chain is six carbons long, and it has methyl groups on carbons 3 and 5 (numbering from the carbonyl carbon). Therefore the name of the parent acid is 3,5-dimethylhexanoic acid. We have to change the *-ic acid* ending in this name to *-ate* and we have to add the name *isopropyl* to the result (as a separate word). This ester is named

Isopropyl 3,5-dimethylhexanoate   (The answer)

PRACTICE EXERCISE 8 Write the IUPAC names of the following esters.

$$
CH_3
$$
(a) $CH_3CH_2CO_2CH_3$ (b) $CH_3CH_2CHCH_2CO_2CH_2CH_2CH_3$

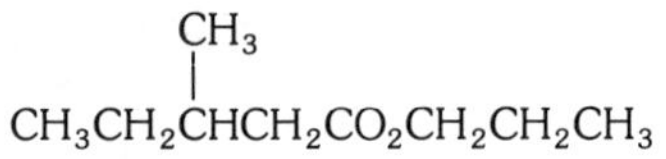

PRACTICE EXERCISE 9 Using the patterns developed, write the common names of the following esters.

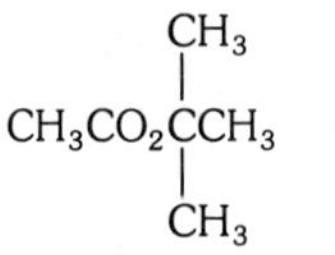

$$
CH_3
$$
(a) $CH_3CO_2CCH_3$ (b) $CH_3CH_2CH_2CO_2CH_2CH_3$
$$
CH_3
$$

The Effect of the Ester Group on Physical Properties. The ester group cannot donate hydrogen bonds, because it has no H—O (or H—N) group. The effect of this is to give esters of the lower-formula-weight alcohols, such as methyl and ethyl alcohol, lower boiling points than the parent acids of the esters. For example, even though methyl acetate has a higher formula weight than acetic acid, this ester boils at 57 °C but the parent acid, acetic acid, boils at 118 °C. The ester group, because it has oxygen atoms, can *accept* hydrogen bonds, however. This allows the lower-formula-weight esters to be relatively soluble in water.

16.5 SOME REACTIONS OF ESTERS

Ester molecules are broken apart by water in the presence of either acids or bases.

When we digest fats and oils, ester groups are hydrolyzed.

Hydrolysis of Esters. An ester reacts with water to give the carboxylic acid and the alcohol from which the ester could be made. This reaction is called the hydrolysis of an ester, and a

TABLE 16.4
Esters of Carboxylic Acids

Common Name[a]	Structure	Melting Point (°C)	Boiling Point (°C)	Solubility (in g/100 g water, 20 °C)
Ethyl esters of straight-chain carboxylic acids, $RCO_2C_2H_5$				
Ethyl formate (Ethyl methanoate)	$HCO_2C_2H_5$	−79	54	Soluble
Ethyl acetate (Ethyl ethanoate)	$CH_3CO_2C_2H_5$	−82	77	7.35 (25 °C)
Ethyl propionate (Ethyl propanoate)	$CH_3CH_2CO_2C_2H_5$	−73	99	1.75
Ethyl butyrate (Ethyl butanoate)	$CH_3(CH_2)_2CO_2C_2H_5$	−93	120	0.51
Ethyl valerate (Ethyl pentanoate)	$CH_3(CH_2)_3CO_2C_2H_5$	−91	145	0.22
Ethyl caproate (Ethyl hexanoate)	$CH_3(CH_2)_4CO_2C_2H_5$	−68	168	0.063
Ethyl enanthate (Ethyl heptanoate)	$CH_3(CH_2)_5CO_2C_2H_5$	−66	189	0.030
Ethyl caprylate (Ethyl octanoate)	$CH_3(CH_2)_6CO_2C_2H_5$	−43	208	0.007
Ethyl pelargonate (Ethyl nonanoate)	$CH_3(CH_2)_7CO_2C_2H_5$	−45	222	0.003
Ethyl caproate (Ethyl decanoate)	$CH_3(CH_2)_8CO_2C_2H_5$	−20	245	0.0015
Esters of acetic acid, CH_3CO_2R				
Methyl acetate	$CH_3CO_2CH_3$	−99	57	24.4
Ethyl acetate	$CH_3CO_2CH_2CH_3$	−82	77	7.39 (25 °C)
Propyl acetate	$CH_3CO_2CH_2CH_2CH_3$	−93	102	1.89
Butyl acetate	$CH_3CO_2CH_2CH_2CH_2CH_3$	−78	125	1.0 (22 °C)
Miscellaneous esters				
Methyl acrylate (Methyl propenoate)	$CH_2{=}CHCO_2CH_3$		80	5.2
Methyl benzoate	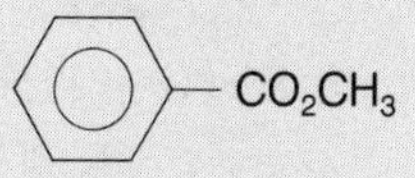 (phenyl)—CO_2CH_3	−12	199	Insoluble
Methyl salicylate	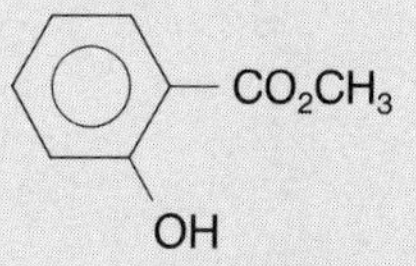 (phenyl)—CO_2CH_3 with OH	−9	223	Insoluble
Acetylsalicylic acid	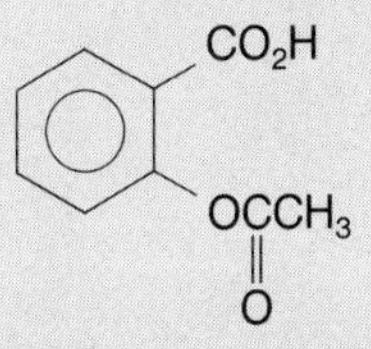 (phenyl) with CO_2H and $OCCH_3$ ($\overset{\parallel}{O}$)	135		
Natural waxes	$CH_3(CH_2)_nCO_2(CH_2)_nCH_3$	n = 23–23: carnauba wax = 25–27: beeswax = 14–15: spermaceti		

[a] IUPAC names are in parentheses.

Esters of *p*-Hydroxybenzoic Acid—the Parabens Several alkyl esters of *p*-hydroxybenzoic acid—referred to as *parabens* on ingredient labels—are used to inhibit molds and yeasts in cosmetics, pharmaceuticals, and food.

Salicylates Certain esters and salts of salicylic acid are analgesics—pain suppressants—and antipyretics—fever reducers. The parent acid, salicylic acid, is itself too irritating to the stomach for these uses, but sodium salicylate and acetylsalicylic acid (aspirin) are commonly used. Methyl salicylate, a pleasant-smelling oil, is used in liniments, and it readily migrates through the skin.

Sodium
salicylate

Acetyl
salicylic acid
(aspirin)

Methyl
salicylate
(oil of
wintergreen)

FIGURE 16.2
The knitted tubing for this graft in an operative site
is made of Dacron fibers.

Dacron Dacron, a polyester of exceptional strength, is widely used in making fabrics and film backing for recording tapes. (Actually, the name *Dacron* applies just to the fiber form of this polyester. When it is cast as a thin film, it is called *Mylar*.) Dacron fabrics have been used in surgery to repair or replace segments of blood vessels, as seen in Figure 16.2.

The formation of Dacron and many other polyfunctional polymers starts with two difunctional monomers, *aAa* and *bBb*, which are able to react with each other in such a way that a small molecule, *ab*, splits out, and the monomer fragments join end to end to make a very long polymer molecule. In principle, the polymerization can be represented as follows:

$aAa + bBb + aAa + bBb$
$\quad\quad + aAa + bBb + \ldots etc. \longrightarrow$
$\quad\quad -A-B-A-B-A-B- \ldots etc. + n(ab)$
$\quad\quad\quad\quad\quad$ A copolymer

Because two monomers are used, the reaction is called *copolymerization.*

To make Dacron, one monomer is ethylene glycol, which has two alcohol groups. The other monomer is dimethyl terephthalate, which has two ester groups. One reaction of esters that we will not study in detail is that they can be made to react with alcohols to replace the original alkyl group in the ester by another. An alcohol molecule splits out. When ethylene glycol and dimethyl terephthalate polymerize, the *ab* molecule that splits out is methyl alcohol, and thus the terephthalate ester has become an ester of ethylene glycol instead of methyl alcohol. The copolymerization proceeds as follows.

etc.) C ($-OCH_3 + H$) $-OCH_2CH_2O-$ ($H + CH_3O-$) C C ($-OCH_3 + H$) $-OCH_2CH_2O-$ ($H +$ etc.

Ethylene
glycol

Dimethyl
terephthalate

etc.) C (OCH_2CH_2O- C C)$_n$ $O-CH_2CH_2O-$ etc. $+ nCH_3OH$

(Repeating unit)
Dacron/Mylar

strong acid catalyst is generally used. (In the body, an enzyme acts as the catalyst.) In general,

$$
\underset{\substack{}}{R-\overset{\overset{\textstyle O}{\|}}{C}-O-R'} + H-OH \xrightarrow[\text{Heat}]{H^+} R-\overset{\overset{\textstyle O}{\|}}{C}-O-H + H-O-R'
$$

Specific examples are:

$$
\underset{\text{Ethyl acetate}}{CH_3-\overset{\overset{\textstyle O}{\|}}{C}-O-CH_2CH_3} + H_2O \xrightarrow[\text{Heat}]{H^+} \underset{\text{Acetic acid}}{CH_3-\overset{\overset{\textstyle O}{\|}}{C}-OH} + \underset{\text{Ethyl alcohol}}{HO-CH_2CH_3}
$$

$$
\underset{\text{Methyl benzoate}}{CH_3-O-\overset{\overset{\textstyle O}{\|}}{C}-\bigcirc} + H_2O \xrightarrow[\text{Heat}]{H^+} \underset{\substack{\text{Methyl}\\\text{alcohol}}}{CH_3-OH} + \underset{\text{Benzoic acid}}{HO-\overset{\overset{\textstyle O}{\|}}{C}-\bigcirc}
$$

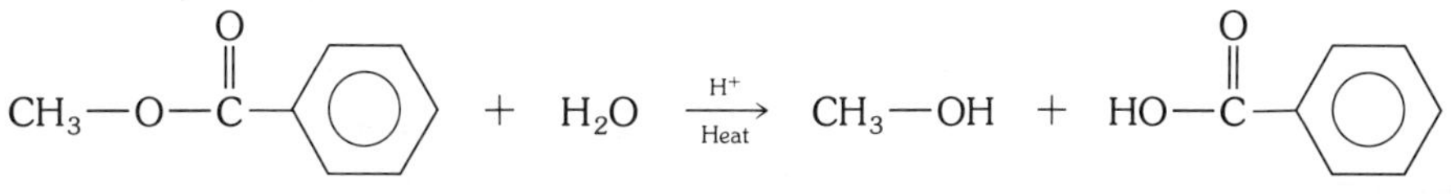

If you want to avoid a mistake that students often make, notice that the *only* bond that breaks in the hydrolysis of an ester is the single covalent bond that joins the carbon atom of the carbonyl group to the oxygen atom. We have called this the "ester bond." Notice also that the products are always the "parents" of the ester and that the names of these parents are strongly implied in the name of the ester itself. Thus methyl benzoate hydrolyzes to *methyl* alcohol and *benzo*ic acid. Now let's work an example.

EXAMPLE 16.4   PREDICTING THE PRODUCTS OF AN ESTER HYDROLYSIS

Problem: What are the products of the hydrolysis of the following ester?

$$
CH_3-\overset{\overset{\textstyle O}{\|}}{C}-O-CH_2CH_2CH_3
$$

Solution: The crucial step is locating the ester bond, the carbonyl-to-oxygen bond. It doesn't matter in which direction this bond happens to point on the page:

$$
CH_3CH_2CH_2-O-\overset{\overset{\textstyle O}{\|}}{C}-CH_3 \quad \text{or} \quad CH_3-\overset{\overset{\textstyle O}{\|}}{C}-O-CH_2CH_2CH_3
$$

Ester bond
(carbonyl-to-oxygen bond)

As we said, this bond is the only bond that breaks in ester hydrolysis, so break it—erase it and separate the fragments. If the ester were written as follows:

$$
CH_3CH_2CH_2-O-\overset{\overset{\textstyle O}{\|}}{C}-CH_3 \dashrightarrow CH_3CH_2CH_2-O + \overset{\overset{\textstyle O}{\|}}{C}-CH_3
$$

On the other hand, if the ester were written in the other direction:

$$
CH_3-\overset{\overset{\textstyle O}{\|}}{C}-O-CH_2CH_2CH_3 \dashrightarrow CH_3-\overset{\overset{\textstyle O}{\|}}{C} + O-CH_2CH_2CH_3
$$

Both the —OH and the H— are supplied by H—OH, the other reactant.

Either way gives the same results. Next we attach the pieces of the water molecule so as to make the "parents" of the ester. We attach —OH to the carbonyl carbon and we put H— on the oxygen atom of the other fragment. The products therefore are the following—propyl alcohol and acetic acid.

$$HOCH_2CH_2CH_3 + HO\overset{\displaystyle O}{\overset{\displaystyle \|}{-C}}-CH_3$$

PRACTICE EXERCISE 10

Write the structures of the products of the hydrolysis of the following esters.

(a) $CH_3\overset{\displaystyle O}{\overset{\displaystyle \|}{-O-C}}-CH_3$ (b) $CH_3CH_2\overset{\displaystyle O}{\overset{\displaystyle \|}{-C}}-O-\overset{\displaystyle CH_3}{\overset{\displaystyle |}{CH}}CH_3$ (c) $CH_3\overset{\displaystyle CH_3}{\overset{\displaystyle |}{CH}}-\overset{\displaystyle O}{\overset{\displaystyle \|}{C}}-O-CH_2CH_2CH_3$

The hydrolysis of an ester is the reverse of the direct formation of the ester from the carboxylic acid and the alcohol. It involves the formation of a chemical equilibrium. Therefore if we were to take the ester and water in a 1:1 mole ratio, we could not convert all of the ester molecules into their parent acid and alcohol molecules. Some of the ester molecules and some of the water molecules would still be present. To make sure that all of the ester is hydrolyzed, we have to take an excess of water. In accordance with Le Chatelier's principle, this excess of one reactant shifts the equilibrium in favor of making the products.

Saponification. If a strong base instead of a strong acid is used to promote the breakup of an ester, the products are the salt of the parent acid and the parent alcohol. This reaction of an ester with strong, aqueous alkali is called **saponification,** and it requires a full mole (not just a catalytic trace) of base for each mole of ester bonds. The base *promotes* the reaction but, unlike a true catalyst, it is used up as the reaction proceeds. No equilibrium forms, because one product is the *anion* of the parent acid; and such anions cannot be converted directly into esters by heating with alcohols.

L. *sapo*, soap, and *onis*, to make. Ordinary soap is made by the saponification of the ester groups in fats and oils.

In general:

$$R\overset{\displaystyle O}{\overset{\displaystyle \|}{-C}}-O-R' + OH^- \xrightarrow{Heat} R\overset{\displaystyle O}{\overset{\displaystyle \|}{-C}}-O^- + H-O-R'$$

Specific examples are as follows. (Assume that OH^- comes from NaOH or KOH.)

$$CH_3\overset{\displaystyle O}{\overset{\displaystyle \|}{-C}}-O-CH_2CH_3 + OH^- \underset{Heat}{\rightleftharpoons} CH_3\overset{\displaystyle O}{\overset{\displaystyle \|}{-C}}-O^- + HO-CH_2CH_3$$

Ethyl acetate Acetate ion Ethyl alcohol

$$CH_3-O\overset{\displaystyle O}{\overset{\displaystyle \|}{-C}}-\bigcirc + OH^- \xrightarrow{Heat} CH_3-OH + {}^-O\overset{\displaystyle O}{\overset{\displaystyle \|}{-C}}-\bigcirc$$

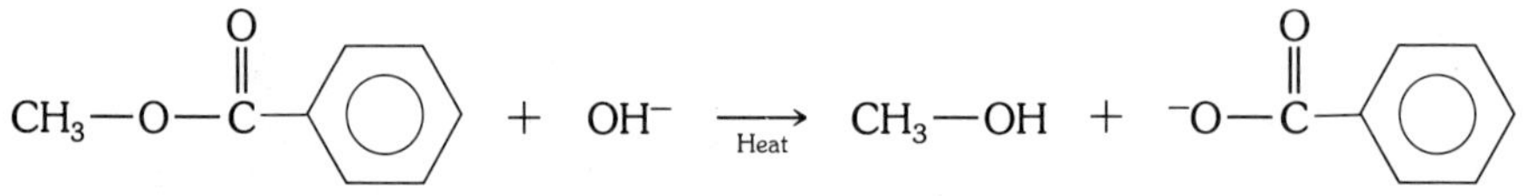

Methyl benzoate Methyl alcohol Benzoate ion

EXAMPLE 16.5   WRITING THE STRUCTURES OF THE PRODUCTS OF SAPONIFICATION

Problem: What are the products of the saponification of the following ester?

$$CH_3CH_2\overset{\displaystyle O}{\overset{\displaystyle \|}{C}}-O-CH_3$$

Solution: Remember that saponification is very similar to ester hydrolysis. Therefore first break (erase) the ester bond — *break only this bond* — and separate the fragments:

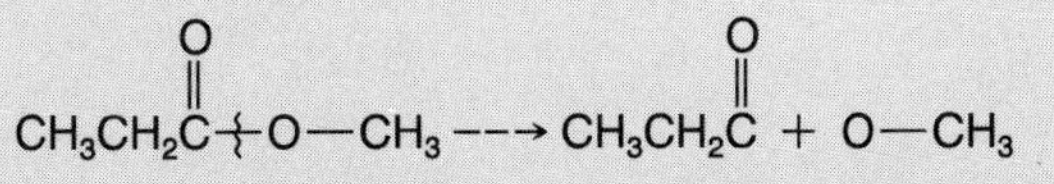

$$CH_3CH_2\overset{\overset{\displaystyle O}{\|}}{C} \!+\! O\!-\!CH_3 \dashrightarrow CH_3CH_2\overset{\overset{\displaystyle O}{\|}}{C} + O\!-\!CH_3 \quad \text{(Incomplete)}$$

Now change the fragment that has the carbonyl group into the *anion* of a carboxylic acid. Do this by attaching $-O^-$ to the carbonyl carbon atom. Then attach an H atom to the oxygen atom of the other fragment to make the alcohol molecule:

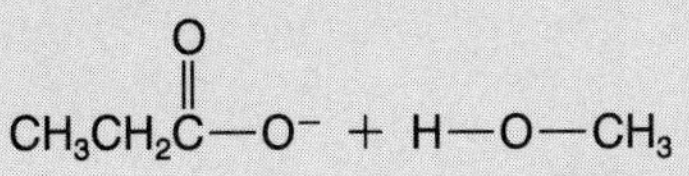

$$CH_3CH_2\overset{\overset{\displaystyle O}{\|}}{C}\!-\!O^- + H\!-\!O\!-\!CH_3 \quad \text{(The products)}$$

PRACTICE EXERCISE 11   Write the structures of the products of the saponification of the following esters.

(a) 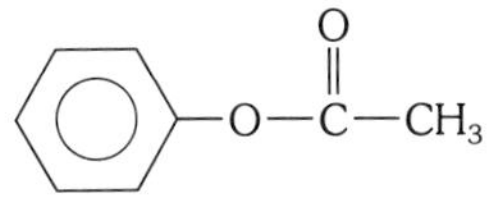 (b)

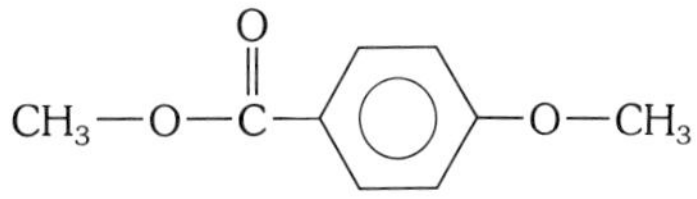

16.6 ORGANOPHOSPHATE ESTERS AND ANHYDRIDES

The most widely distributed kinds of esters and anhydrides in living organisms are the anions of esters of phosphoric acid, diphosphoric acid, and triphosphoric acid.

Phosphoric acid appears in several forms and anions in the body, but the three fundamental parents of all these forms are phosphoric acid, diphosphoric acid, and triphosphoric acid.

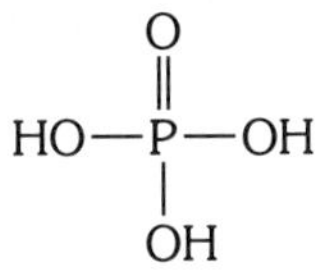

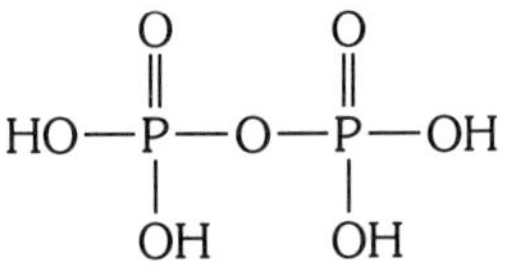

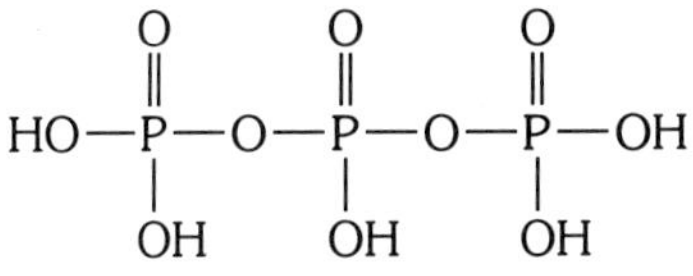

These are all polyprotic acids, but at the slightly alkaline pH values of the fluids of the body, they cannot exist as free acids. Instead they occur as a mixture of negative ions.

Monophosphate Esters.   If you look closely at the structure of phosphoric acid, you can see that part of it resembles a carboxyl group.

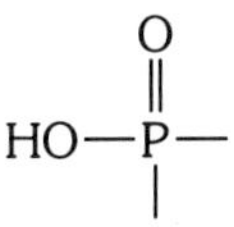

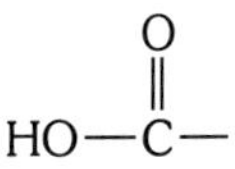

It isn't surprising therefore that esters of phosphoric acid exist and that they are structurally similar to esters of carboxylic acids.

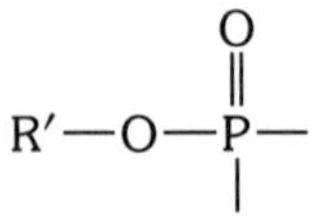

$$R'-O-\overset{\displaystyle O}{\underset{|}{\overset{\|}{P}}}- \qquad \text{compares to} \qquad R'-O-\overset{\displaystyle O}{\overset{\|}{C}}-$$

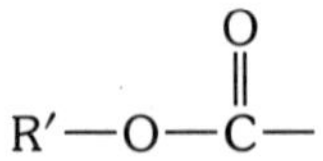

Part of a
phosphate
ester

Part of a
carboxylate
ester

One large difference between a phosphate ester and a carboxylate ester is that a phosphate ester is still a diprotic acid. Its molecules still carry two proton-donating —O—H groups. Therefore, depending on the pH of the medium, a phosphate ester can exist in any one of three forms, and usually there is an equilibrium mixture of all three.

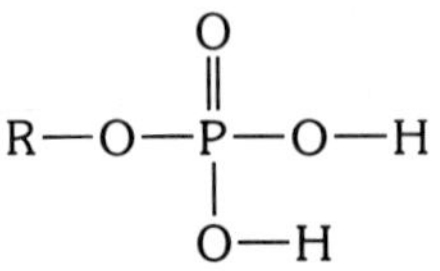

$$R-O-\overset{\displaystyle O}{\underset{\underset{\displaystyle O-H}{|}}{\overset{\|}{P}}}-O-H \qquad R-O-\overset{\displaystyle O}{\underset{\underset{\displaystyle O-H}{|}}{\overset{\|}{P}}}-O^- \qquad R-O-\overset{\displaystyle O}{\underset{\underset{\displaystyle O^-}{|}}{\overset{\|}{P}}}-O^-$$

Phosphate ester
(as a diprotic acid)

—favored at low pH

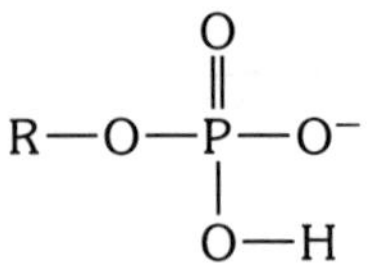

Phosphate ester
(as a singly
ionized species)

—favored at pH
values just
below 7

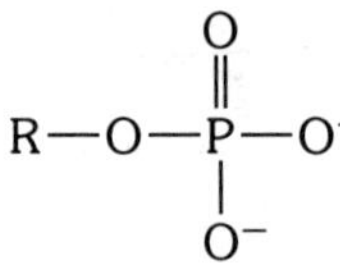

Phosphate ester
(as a doubly
ionized species)

—favored at pH
values above 7

At the pH of most body fluids (just slightly more than 7), phosphate esters exist mostly as the doubly ionized species — as the di-negative ion. All forms, however, are generally soluble in water, and one reason that the body converts so many substances into their phosphate esters might be to improve their solubilities in water.

Diphosphate Esters.   A diphosphate ester actually has three functional groups — the phosphate ester group, the proton-donating —OH group, and the phosphoric anhydride system.

phosphoric anhydride system

$$R-O-\overset{\displaystyle O}{\underset{\underset{\displaystyle OH}{|}}{\overset{\|}{P}}}-O-\overset{\displaystyle O}{\underset{\underset{\displaystyle OH}{|}}{\overset{\|}{P}}}-OH$$

ester group

proton-donating groups

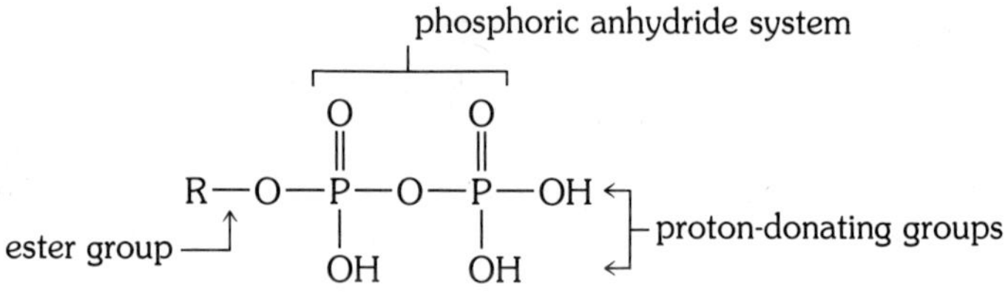

Diphosphate ester

Notice the similarity of part of the structure of a diphosphate ester to that of an anhydride of a carboxylic acid:

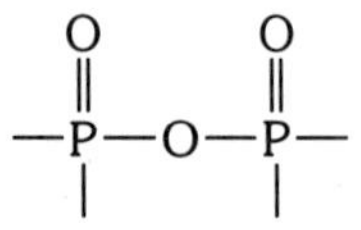

$$-\overset{\displaystyle O}{\underset{|}{\overset{\|}{P}}}-O-\overset{\displaystyle O}{\underset{|}{\overset{\|}{P}}}- \qquad -\overset{\displaystyle O}{\overset{\|}{C}}-O-\overset{\displaystyle O}{\overset{\|}{C}}-$$

Part of the
diphosphate
system

The phosphoric
anhydride group

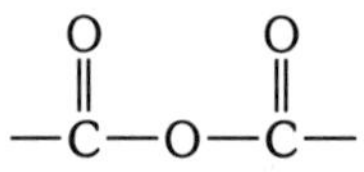

Part of a
carboxylic acid
anhydride

The carboxylic
anhydride group

One of the many diphosphate esters in the body is called adenosine diphosphate, or ADP. We will show its structure as its triply charged anion, because it exists largely in this fully ionized form at the pH of most body fluids.

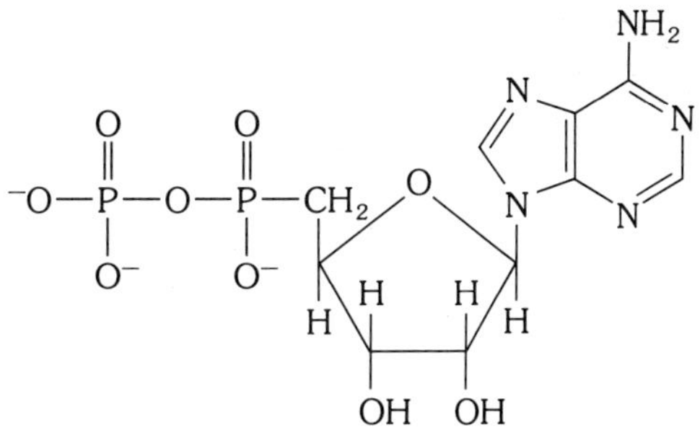

Adenosine diphosphate, ADP
(fully ionized form)

Notice the phosphoric anhydride system in ADP. This structural unit is one of the body's chief means of storing chemical energy, and it is worthwhile to ask why. The source of the internal energy in the triply charged anion of ADP is the tension up and down the anhydride chain. This central chain bears oxygen atoms with full negative charges, and these charges repel each other. This internal repulsion primes the phosphoric anhydride system for breaking apart exothermically when it is attacked by a suitable reactant. Molecules with alcohol groups are examples of such reactants in the body, but these reactions require enzymes for catalysis. Without a catalyst, the triply charged ADP anion repels electron-rich species that might otherwise be attracted to the phosphorus atoms in the phosphoric anhydride chain. A successful attack by an alcohol group — an attack that breaks apart the anhydride — requires that the oxygen atom of the alcohol group strike a phosphorus atom.

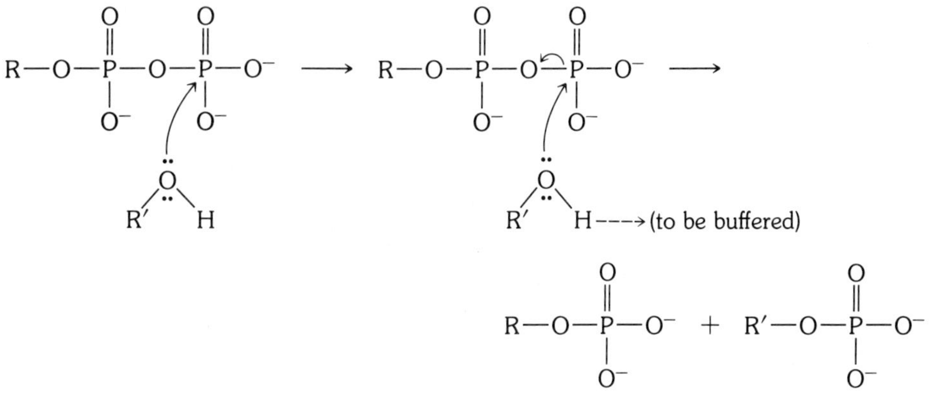

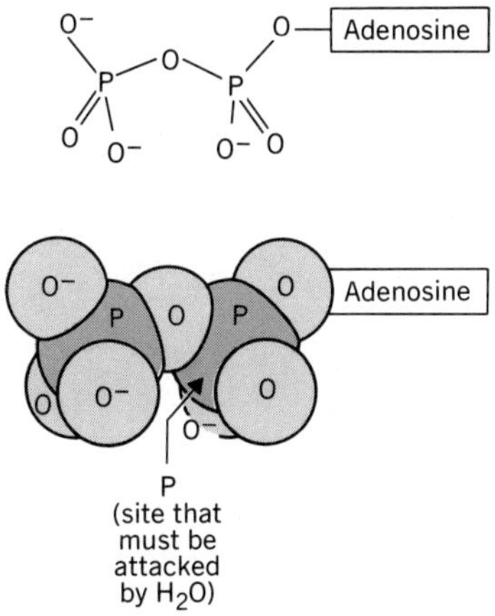

FIGURE 16.3
The oxygen atoms in the phosphoric anhydride system of ADP screen the phosphorus atoms. The negative charges on these oxygen atoms deflect incoming, electron-rich particles such as molecules of an alcohol or of water. Therefore, this kind of anhydride system reacts very slowly with these reactants, unless a special catalyst such as an enzyme is also present.

As seen in Figure 16.3, however, the phosphorus atoms in the chain are buried within a clutch of negatively charged oxygen atoms that repel the alcohol molecule. Thus the internal tension cannot be relieved by this kind of reaction *unless an enzyme for the reaction is present.* You have probably already guessed that *the body exerts control over energy-releasing reactions of diphosphates by its control of the enzymes for these reactions.*

Water could make the same kind of exothermic attack on a diphosphate ester as an alcohol, but *the body has no enzymes inside cells that catalyze this reaction.* Hence, energy-rich diphosphates can exist in cells despite the abundance of water.

Triphosphate Esters. Adenosine triphosphate or ATP is the most common and widely-occurring member of a small family of energy-rich triphosphate esters. Because the triphosphates have two phosphoric anhydride systems in each molecule, on a mole-for-mole basis the triphosphates are among the most energy-rich substances in the body.

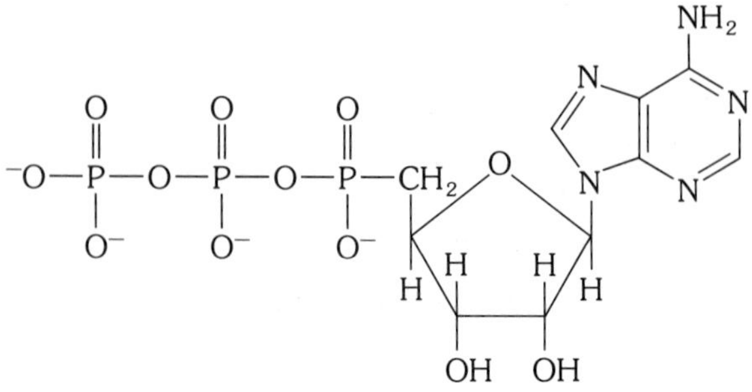

Adenosine triphosphate, ATP
(in its fully neutralized form)

Triphosphates are much more widely used in cells as sources of energy than are the diphosphates. For example, we can write the overall reaction for the contraction of a muscle as follows, where we use the symbol P_i to stand for the set of inorganic phosphate ions — mostly $H_2PO_4^-$ and HPO_4^{2-} — that are produced by the breaking up of ATP and that are present at equilibrium at body pH.

$$\text{Relaxed muscle} + \text{ATP} \xrightarrow{\text{enzyme}} \text{Contracted muscle} + \text{ADP} + P_i$$

Thus muscular work requires ATP, and if the body's supply of ATP were used up *with no way to remake it,* we'd soon lose all capacity for such work. The resynthesis of ATP from ADP and inorganic phosphate ion is one of the major uses of the chemical energy in the food we eat. What we have studied here is *how* these tri- and diphosphates can be energy-rich and, at the same time, not be destroyed by uncatalyzed reactions in the body. Later, we will study more details of how a cell makes these phosphate systems and how it uses them.

SUMMARY

Acids and their salts The carboxyl group, CO_2H, is a polar group that confers moderate water solubility to a molecule without preventing its solubility in nonpolar solvents. This group is very resistant to oxidation and reduction. Carboxylic acids are strong proton donors toward hydroxide ions, whereas alcohols are not. Toward water, carboxylic acids are weak acids. Therefore their conjugate bases, the carboxylate anions, are good proton acceptors toward the hydronium ions of strong acids.

Salts of carboxylic acids are ionic compounds, and the potassium or sodium salts are very soluble in water. Hence, the carboxyl group is one of nature's important "solubility switches." An insoluble acid becomes soluble in base, but it is thrown out of solution again by the addition of acid.

The derivatives of acids — acid chlorides, anhydrides, and esters — can be made from the acids and are converted back to the acids by reacting with water. We can organize the reactions we have studied for the carboxylic acids as follows:

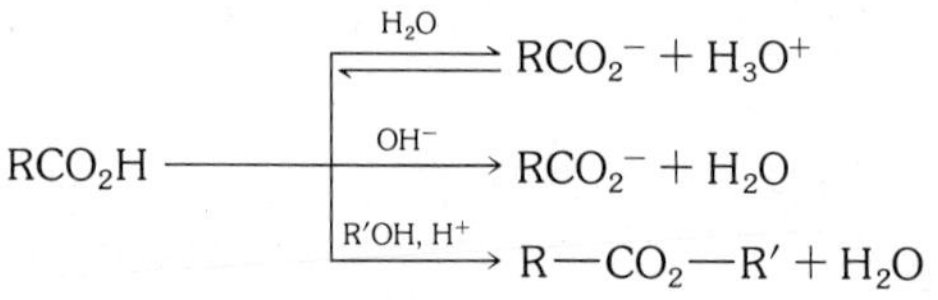

Esters We can organize the reactions used to make esters and the reactions of esters as follows:

Esterifications

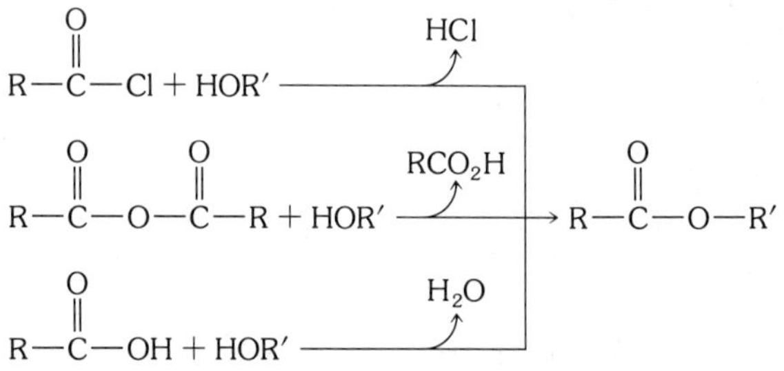

Reactions of Esters:

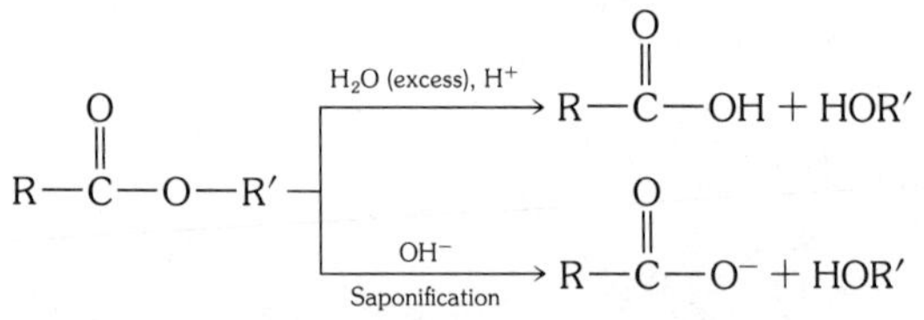

Esters and anhydrides of the phosphoric acid system Esters of phosphoric acid, diphosphoric acid, and triphosphoric acid occur in living systems largely as anions, because these esters are also polyprotic acids. In addition, those of di- and triphosphoric acid are phosphoric anhydrides. These anhydrides are energy-rich compounds. Their reactions with water or alcohols are very exothermic, but the reactions are also very slow unless a catalyst (an enzyme) is present.

KEY TERMS

The following terms will occur in subsequent chapters, and they should be learned well before you continue.

acid anhydride	acyl group	carboxylic acid	fatty acid
acid chloride	acyl group transfer reaction	ester	saponification
acid derivative	amide	esterification	

REVIEW EXERCISES

The answers to these Review Exercises are in the *Study Guide* that accompanies this book.

Structures and Names of Carboxylic Acids and Their Salts

16.1   What is the structure of the carboxyl group, and in what way does it differ from the functional group in an alcohol? In a ketone?

16.2   Fatty acids are carboxylic acids obtained from what substances?

16.3   What is the common name of the acid in vinegar? In sour milk?

16.4   Write the structures of the following substances.
   (a) Butyric acid   (b) Acetic acid
   (c) Formic acid   (d) Benzoic acid

16.5   What are the structures of the following?
   (a) 2,2-Dimethylpropanoate ion
   (b) 4-Bromo-3-methylhexanoic acid
   (c) Pentanedioic acid
   (d) 2-Butenoate ion

16.6   Write the IUPAC names of the following compounds.

(a) $CH_3CHCHCH_3$ with CO_2H and CH_3CH_2 substituents

(b) $HO_2CCH-C-CH_2CH_2$ with CH_3, CH_3, CH_3, CH_3CHCH_3 substituents

(c) CH_3CHCO_2K with CH_3 substituent

(d) NaO_2CC-CH_2 with CH_3, CH_3, CH_3 substituents

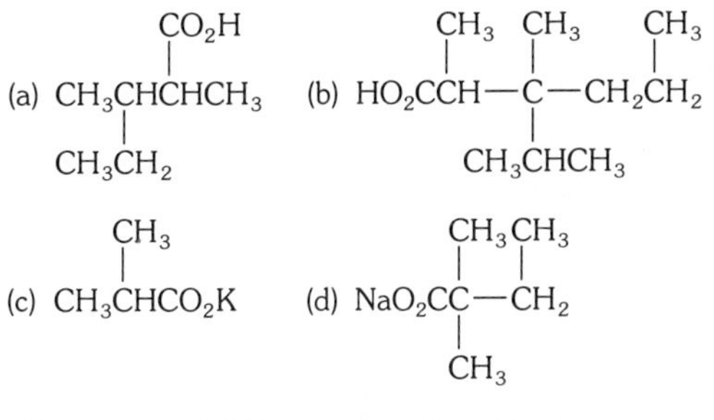

16.7   What are the IUPAC names of the following compounds?

(a) $CH_3CH-CHCO_2H$ with Br and CH_3 substituents

(b) CH_3CHCO_2Na with OH substituent

(c) $CH_3CHCH_2CO_2H$ with OH substituent

(d) $CH_3CHCHCH_2C-CHCH_3$ with CH_3, $CH_3CHCH_2CH_3$, CO_2H, $CH_3CH_2CHCH_3$, CH_3 substituents

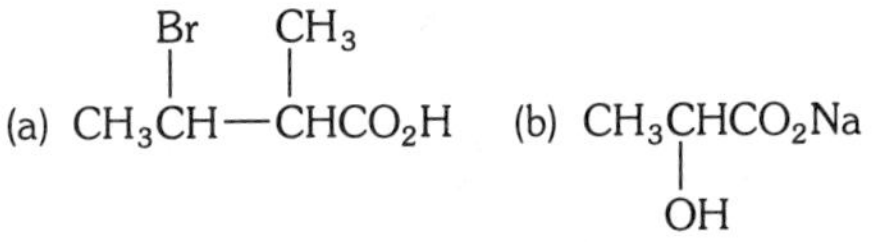

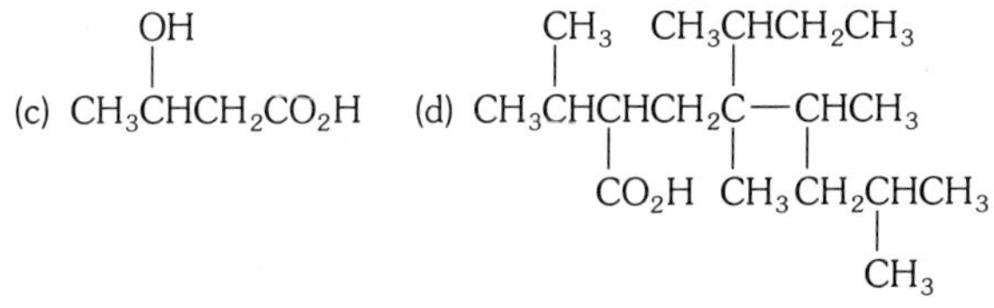

16.8   One of the *ketone bodies* whose concentration in blood rises in unchecked diabetes has the following structure.

$$CH_3CCH_2CO_2H$$
with carbonyl O

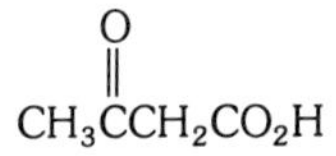

If its IUPAC name is 3-ketobutanoic acid and its common name is acetoacetic acid, what are the IUPAC and common names for its sodium salt?

16.9   The tricarboxylic acid cycle is one of the major metabolic sequences of reactions in the body. One of the acids in this series of reactions is commonly called α-ketoglutaric acid, which has the following structure.

$$HO_2CCCH_2CH_2CO_2H$$
with carbonyl O

What is the IUPAC name of this acid?

Physical Properties of Carboxylic Acids

16.10   Draw a figure that shows how two acetic acid molecules can pair in a hydrogen-bonded form.

16.11   The hydrogen bond system in formic acid includes an array of molecules, one after the other, each carbonyl oxygen of one molecule attracted to the HO group of the next molecule in line. Represent this linear array of hydrogen-bonded molecules of formic acid by a drawing.

16.12 Arrange the following compounds in their order of increasing solubility in water. Do this by arranging their identifying letters in a row in the correct order, placing the letter of the least soluble on the left.

$$CH_3CO_2H \qquad CH_3(CH_2)_5CO_2H \qquad CH_3(CH_2)_6CH_3$$
$$\text{A} \qquad\qquad \text{B} \qquad\qquad\qquad \text{C}$$

16.13 Arrange the following compounds in their order of increasing boiling points by arranging their identifying letters in a row in the correct order. Place the letter of the lowest-boiling compound on the left.

$$HO_2CCO_2H \qquad CH_3CO_2H \qquad CH_3OH \qquad CH_3CH_2CH_3$$
$$\text{A} \qquad\qquad \text{B} \qquad\quad \text{C} \qquad\quad \text{D}$$

Carboxylic Acids as Weak Acids

16.14 Write the equation for the equilibrium that is present in an aqueous solution of acetic acid.

16.15 What is the equation for the equilibrium that is present in a solution of formic acid in water?

16.16 Arrange the following compounds in their order of increasing acidity by writing their identifying letters in a row in the correct sequence. (Place the letter of the least acidic compound on the left.)

$$CH_3CO_2H \qquad HNO_3 \qquad CH_3CH_2OH \qquad \langle\!\bigcirc\!\rangle\!-\!OH$$
$$\text{A} \qquad\qquad \text{B} \qquad\qquad \text{C} \qquad\qquad\qquad \text{D}$$

16.17 Give the order of increasing acidity of the following compounds. Arrange their identifying letters in the order that corresponds to their acidity, with the letter of the least acidic compound on the left.

$$\langle\!\bigcirc\!\rangle\!-\!CH_2\!-\!OH \qquad CH_3\!-\!\langle\!\bigcirc\!\rangle\!-\!OH$$
$$\text{A} \qquad\qquad\qquad\qquad \text{B}$$

$$\langle\!\bigcirc\!\rangle\!-\!CO_2H \qquad H_2SO_4$$
$$\text{C} \qquad\qquad\qquad \text{D}$$

16.18 Write the net ionic equation for the complete reaction, if any, of sodium hydroxide with the following compounds at room temperature.

(a) $CH_3CH_2CO_2H$ (b) $HO_2C\!-\!\langle\!\bigcirc\!\rangle\!-\!CH_3$

(c) CH_3CH_2OH

16.19 What are the net ionic equations for the reactions of the following compounds with aqueous potassium hydroxide at room temperature?

(a) $\langle\!\bigcirc\!\rangle\substack{-CO_2H \\ -CO_2H}$ (b) $HOCH_2\overset{\displaystyle O}{\overset{\|}{C}}CH_2CH_2CO_2H$

(c) $\langle\!\bigcirc\!\rangle\!-\!OH$

Salts of Carboxylic Acids

16.20 Which compound, A or B, is more soluble in water? Explain.

$$CH_3(CH_2)_8CO_2H \qquad CH_3(CH_2)_8CO_2Na$$
$$\text{A} \qquad\qquad\qquad \text{B}$$

16.21 Which compound, A or B, is more soluble in ether? Explain.

$$CH_3\!-\!\langle\!\bigcirc\!\rangle\!-\!CO_2H \qquad CH_3\!-\!\langle\!\bigcirc\!\rangle\!-\!CO_2K$$
$$\text{A} \qquad\qquad\qquad\qquad \text{B}$$

16.22 Suppose you added 0.1 mol of hydrochloric acid to an aqueous solution that contained 0.1 mol of the compound given in each of the following parts. If any reaction occurs rapidly at room temperature, write its net ionic equation.

(a) $CH_3\!-\!\langle\!\bigcirc\!\rangle\!-\!CO_2^-$ (b) CH_3CO_2H

(c) $^-O_2CCH_2CH_2CO_2^-$

16.23 Suppose that you have each of the following compounds in a solution in water. What reaction, if any, would occur rapidly at room temperature if an equimolar quantity of hydrochloric acid were added? Write net ionic equations.
(a) $CH_3CH_2CO_2H$
(b) $CH_3CH_2CH_2CO_2^-$
(c) $HOCH_2CH_2CH_2CO_2^-$

Esterification and Reactivity

16.24 What are the structures of the reactants needed to make methyl propanoate from methanol and each of the following kinds of starting materials?
(a) An acid chloride
(b) A carboxylic acid anhydride
(c) By direct esterification

16.25 In order to prepare ethyl benzoate, what are the structures of the reactants needed for each kind of approach?
(a) By direct esterification
(b) From an acid chloride
(c) From an acid anhydride

16.26 The reaction of methyl alcohol with acetyl chloride, $CH_3\overset{\overset{\textstyle O}{\|}}{C}Cl$, is rapid.
(a) What is the structure of the organic product?
(b) How is the speed of this reaction explained?

16.27 Ethyl alcohol reacts rapidly with acetic anhydride,

$CH_3\overset{\overset{\textstyle O}{\|}}{C}-O-\overset{\overset{\textstyle O}{\|}}{C}CH_3$, to give ethyl acetate and acetic acid. How can the very rapid rate of the reaction be explained?

16.28 What are the structures of the products of the esterification by methyl alcohol of each compound?
(a) Formic acid
(b) 2-Methylbutanoic acid
(c) *p*-Chlorobenzoic acid
(d) Oxalic acid, $HO\overset{\overset{\textstyle O}{\|}}{C}-\overset{\overset{\textstyle O}{\|}}{C}OH$. (Show the esterification of both of the carboxyl groups.)

16.29 When propanoic acid is esterified by each of the following compounds, what are the structures of the esters that form?
(a) Ethanol
(b) 2-Methyl-1-propanol
(c) Phenol
(d) $HOCH_2CH_2OH$ (1,2-ethanediol). Show the esterification of both alcohol groups.

16.30 Explain by means of equations how H^+ works as a catalyst in the direct esterification of acetic acid by ethyl alcohol.

16.31 The first of the following two reactions is very difficult to cause, but the second occurs fairly readily. What is a logical explanation for this difference?

(1) $H_2O + CH_3CH_2-\overset{\overset{\textstyle H}{|}}{\overset{+}{O}}-CH_2CH_3 \longrightarrow$

$CH_3CH_2-\overset{+}{O}\overset{\diagup H}{\diagdown H} + \overset{H}{\diagdown}O-CH_2CH_3$

(2) $H_2O + CH_3\overset{\overset{\textstyle O}{\|}}{C}-\overset{\overset{\textstyle H}{|}}{\overset{+}{O}}-CH_2CH_3 \longrightarrow$

$CH_3\overset{\overset{\textstyle O}{\|}}{C}-\overset{+}{O}\overset{\diagup H}{\diagdown H} + \overset{H}{\diagdown}O-CH_2CH_3$

16.32 Suppose that a way could be found to remove H_2O as rapidly as it is produced in direct esterification. What would this do to the equilibrium in this reaction, shift it to the right (favoring the ester) or to the left (favoring the carboxylic acid and the alcohol)? Explain.

16.33 How do we explain the fact that esters react much more slowly than acid chlorides with water?

Structures and Physical Properties of Esters

16.34 Write the structures of the following compounds.
(a) Methyl formate   (b) Ethyl benzoate

16.35 What are the structures of the following compounds?
(a) Isopropyl propanoate   (b) Isobutyl 2-methylbutanoate

16.36 Arrange the following compounds in their order of increasing boiling points. Do this by placing their identifying letters in a row, starting with the lowest-boiling compound on the left.

$CH_3CH_2CH_2CH_2CO_2H$ $CH_3CH_2OCH_3$
A   **B**

$CH_3CH_2CO_2CH_3$ $CH_3CH_2CO_2CH_2CH_3$
C   **D**

16.37 Arrange the following compounds in their order of increasing solubilities in water by placing their identifying letters in the correct sequence, beginning with the least soluble on the left.

$CH_3(CH_2)_4CO_2Na$ $CH_3(CH_2)_2CO_2H$
A   **B**

$CH_3(CH_2)_4CO_2CH_3$ $CH_3(CH_2)_6OCH_3$
C   **D**

Reactions of Esters

16.38 Write the equation for the acid-catalyzed hydrolysis of each compound. If no reaction occurs, write "no reaction."
(a) $CH_3\overset{\overset{\textstyle CH_3}{|}}{CH}-O-\overset{\overset{\textstyle O}{\|}}{C}CH_3$ (b) $CH_3\overset{\overset{\textstyle }{}}{\underset{\overset{\textstyle |}{CH_3}}{CH}}\overset{\overset{\textstyle O}{\|}}{C}-O-CH_3$
(c) $CH_3-O-CH_2CO_2H$ (d) $HOCH_2\overset{\overset{\textstyle O}{\|}}{C}-O-CH_3$

16.39 What are the equations for the acid-catalyzed hydrolyses of the following compounds? If no reaction occurs, write "no reaction."

(a) $CH_3CH_2-O-\overset{\overset{\textstyle O}{\|}}{C}-\bigcirc$

(b) $CH_3CH_2\overset{\overset{\textstyle O}{\|}}{C}-O-\bigcirc$

(c) $CH_3CH_2-O-\bigcirc-\overset{\overset{\textstyle O}{\|}}{C}CH_3$

(d) $CH_3-O-\overset{\overset{\textstyle O}{\|}}{C}CH_2CH_2\overset{\overset{\textstyle O}{\|}}{C}-O-CH_3$

16.40 The digestion of fats and oils involves the complete hydrolysis of molecules such as the following. What are the structures of its hydrolysis products?

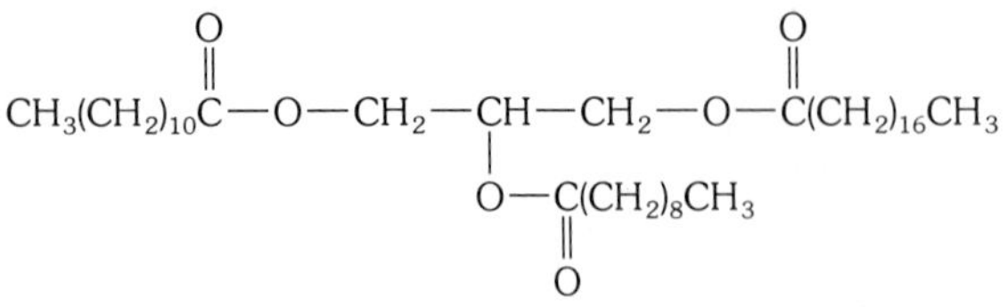

$$CH_3(CH_2)_{10}\overset{\overset{\displaystyle O}{\|}}{C}-O-CH_2-\underset{\underset{\displaystyle \overset{\|}{O}}{\underset{\displaystyle C(CH_2)_8CH_3}{|}}}{CH}-CH_2-O-\overset{\overset{\displaystyle O}{\|}}{C}(CH_2)_{16}CH_3$$

16.41 Cyclic esters are known compounds. What is the structure of the product of the hydrolysis of the following compound?

$$\begin{array}{c} H_2C-O \\ H_2C \qquad C=O \\ H_2C-CH_2 \end{array}$$

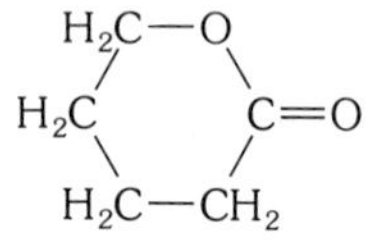

16.42 What are the structures of the products of the saponification of the compounds in Review Exercise 16.38?

16.43 What forms, if anything, when the compounds of Review Exercise 16.39 are subjected to saponification by aqueous KOH? Write their structures.

16.44 What are the products of the saponification of the compound given in Review Exercise 16.40? (Assume that aqueous NaOH is used.)

16.45 Write the structure of the organic ion that forms when the compound of Review Exercise 16.41 is saponified.

16.46 A pharmaceutical chemist needed to prepare the ethyl ester of an extremely expensive and rare carboxylic acid in order to test this form of the drug for its side effects. Direct esterification had to be used. How could the conversion of all of the acid to its ethyl ester be maximized? Use RCO_2H as a symbol for the acid in any equations you write.

16.47 Write the steps in the mechanism of the acid-catalyzed hydrolysis of methyl acetate. (Remember, this is the exact reverse of the acid-catalyzed, direct esterification of acetic acid by methyl alcohol.)

Phosphate Esters and Anhydrides

16.48 Write the structures of the following compounds.
(a) Monoethyl phosphate
(b) Monomethyl diphosphate
(c) Monopropyl triphosphate

16.49 State one apparent advantage to the body of its converting many compounds into phosphate esters.

16.50 What part of the structure of ATP is particularly responsible for its being described as an *energy-rich* compound? Explain.

16.51 Why is ATP harder to hydrolyze than acetyl chloride?

Review of Organic Reactions

16.52 Complete the following reaction sequences by writing the structures of the organic products. If no reaction occurs, state so. (These constitute a review of this and earlier chapters on organic chemistry.)

(a) $CH_3\overset{\overset{\displaystyle O}{\|}}{C}H \xrightarrow{K_2Cr_2O_7}$

(b) $CH_3CH_2CH_3 + H_2SO_4 \longrightarrow$

(c) $CH_3OH + CH_3CO_2H \xrightarrow[\text{Heat}]{H^+}$

(d) $CH_3CH_2\underset{\underset{\displaystyle OH}{|}}{CH}CH_3 \xrightarrow[\text{Heat}]{KMnO_4}$

(e) $CH_3\underset{\underset{\displaystyle OH}{|}}{CH}CH_3 \xrightarrow[\text{Heat}]{H_2SO_4}$

(f) $CH_3O\overset{\overset{\displaystyle O}{\|}}{C}\underset{\underset{\displaystyle CH_3}{|}}{CH}CH_3 + H_2O \xrightarrow{H^+}$

(g) $CH_3CH_2CH{=}CH_2 + HCl(g) \longrightarrow$

(h) $CH_3\underset{\underset{\displaystyle O-CH_3}{|}}{CH}-O-CH_3 + H_2O \xrightarrow{H^+}$

(i) $CH_3\underset{\underset{\displaystyle CH_3}{|}}{CH}\overset{\overset{\displaystyle O}{\|}}{C}-O-CH_3 + NaOH(aq) \longrightarrow$

(j) $CH_3-O-CH_2CH_2\overset{\overset{\displaystyle O}{\|}}{C}CH_3 + H_2O \longrightarrow$

(k) $CH_3CH_2\overset{\overset{\displaystyle O}{\|}}{C}-Cl + H_2O \longrightarrow$

(l) $CH_3CH_2CH_2CO_2H + NaOH \longrightarrow$

16.53 Write the structures of the organic products, if any, that form in the following situations. If no reaction occurs, state so. Some of these constitute a review of the reactions of earlier chapters.

(a) $CH_3CH_2CO_2^- + HCl(aq) \longrightarrow$

(b) 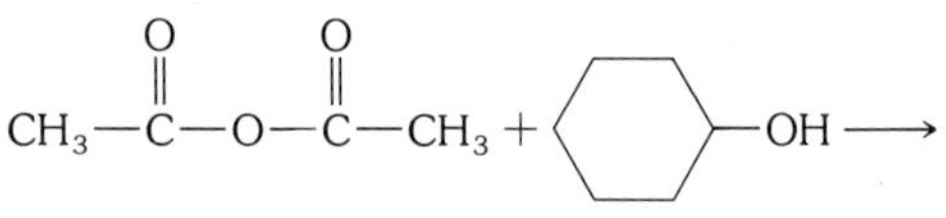 $CH_3-\overset{\overset{\displaystyle O}{\|}}{C}-O-\overset{\overset{\displaystyle O}{\|}}{C}-CH_3 + \langle\!\!\!\bigcirc\!\!\!\rangle-OH \longrightarrow$

(c) $CH_3CH_2-O-CH_2CH_2\overset{\overset{\displaystyle O}{\|}}{C}CH_3 + NaOH \longrightarrow$

(d) $CH_3CH_2CH_2-O-\overset{\overset{\displaystyle O}{\|}}{C}-CH_2CH_2-O-\overset{\overset{\displaystyle O}{\|}}{C}-CH_3$
$+ H_2O \xrightarrow{H^+}$
(Excess)

(e) 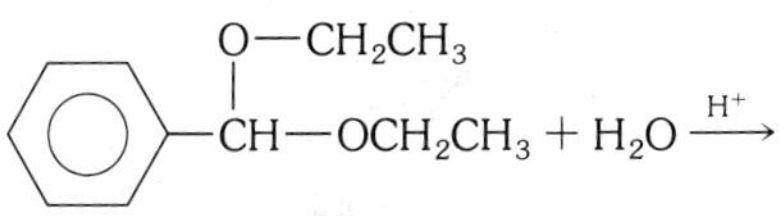 $\langle\bigcirc\rangle\overset{\displaystyle O-CH_2CH_3}{\underset{\displaystyle |}{CH}}-OCH_2CH_3 + H_2O \xrightarrow{H^+}$

(f) $CH_3\overset{\displaystyle OH}{\underset{\displaystyle |}{C}}CH_2CH_3 \xrightarrow{KMnO_4}$
$\underset{\displaystyle |}{}$
$\underset{\displaystyle CH_3}{}$

(g) $CH_3CH_2CO_2H + CH_3\overset{\displaystyle CH_3}{\underset{\displaystyle |}{CHOH}} \xrightarrow[\text{Heat}]{H^+}$

(h) $CH_3(CH_2)_9CH_3 + NaOH \longrightarrow$

(i) $CH_3-O-\overset{\displaystyle O}{\overset{\displaystyle \|}{C}}CH_2CH_2-O-\overset{\displaystyle O}{\overset{\displaystyle \|}{C}}CH_2CH_3$
$+ NaOH(aq) \longrightarrow$
(Excess)

(j) $CH_3(CH_2)_6CO_2H + NaOH(aq) \longrightarrow$

(k) $^-O_2CCH_2CH_2CH_2CO_2^- + HCl(aq) \longrightarrow$
(Excess)

(l) $CH_3OH + HO_2CCH_2CH_2CH_2CO_2H \xrightarrow[\text{Heat}]{H^+}$
(Excess)

Chapter 17
Amines and Amides

Nylon is strong, light-weight, and it won't mildew. It's a fabric to which many sky divers trust their lives. Nylon is a polyamide, and we'll study the amide function in this chapter.

17.1 OCCURRENCE, NAMES, AND PHYSICAL PROPERTIES OF AMINES

The amino group, $-NH_2$, has some of the properties of ammonia, including the ability to be involved in hydrogen bonding.

Both the amino group and its protonated form occur in living things in molecules of proteins, enzymes, and genes. When a carbonyl group is attached to nitrogen, we have the amide system, and this also occurs widely in living things.

$$-NH_2 \qquad -NH_3^+ \qquad \overset{\displaystyle O}{\underset{\displaystyle |}{\overset{\displaystyle \|}{-C}}} \!\! -\!\! \overset{|}{N}-$$

| Amino group | Protonated amino group | Amide system |

Structural Features of Amines.   The **amines** are organic relatives of ammonia in which one, two, or all three of the hydrogen atoms on an ammonia molecule have been replaced by a hydrocarbon group. Some examples are:

$$CH_3NH_2 \qquad CH_3NHCH_3 \qquad CH_3\overset{\displaystyle CH_3}{\overset{\displaystyle |}{N}}CH_3 \qquad CH_3NHCH_2CH_3$$

Methylamine   Dimethylamine   Trimethylamine   Methylethylamine

These are the common, not the IUPAC names.

Several amines are listed in Table 17.1. All of these are classified as *amines*, and all are basic, like ammonia. It's quite important to realize that for a compound to be an amine, its molecules must not only involve a nitrogen with three bonds, but also that none of these bonds can be to a carbonyl group. If such a system is present — a carbonyl-nitrogen bond — the substance is an **amide**. Thus the structure given by **1** is an amide, not an amine. However, the structure given by **2** is not an amide, because there is no carbonyl-nitrogen bond. Instead, **2** has two functional groups — a keto group and an amino group. The chemical difference is that amines are basic

TABLE 17.1
Amines

Common Name	Structure	Boiling Point (°C)	Solubility in Water	K_b (at 25 °C)
Methylamine	CH_3NH_2	−8	Very soluble	4.4×10^{-4}
Dimethylamine	CH_3NHCH_3	8	Very soluble	5.3×10^{-4}
Trimethylamine	CH_3NCH_3 $\quad\ \ |$ $\quad\ \ CH_3$	3	Very soluble	0.5×10^{-4}
Ethylamine	$CH_3CH_2NH_2$	17	Very soluble	5.6×10^{-4}
Diethylamine	$CH_3CH_2NHCH_2CH_3$	55	Very soluble	9.6×10^{-4}
Triethylamine	$CH_3CH_2NCH_2CH_3$ $\qquad\quad |$ $\qquad\quad CH_2CH_3$	89	14 g/dL	5.7×10^{-4}
Propylamine	$CH_3CH_2CH_2NH_2$	49	Very soluble	4.7×10^{-4}
Aniline	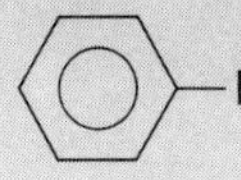 ⬡$-NH_2$	184	4 g/dL	3.8×10^{-10}

and amides are not. Another difference is that the carbon-nitrogen bond in amines can't be broken by water, but the carbonyl-nitrogen bond in amides can.

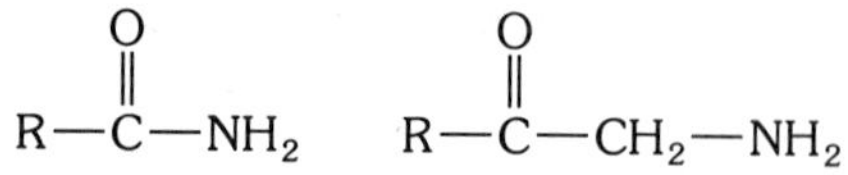

$$R—\overset{\overset{\displaystyle O}{\|}}{C}—NH_2 \qquad R—\overset{\overset{\displaystyle O}{\|}}{C}—CH_2—NH_2$$

$$\mathbf{1} \qquad\qquad\qquad \mathbf{2}$$

If one or more of the groups attached directly to nitrogen in an amine is a benzene ring, then the amine is an *aromatic* amine. Otherwise, it is classified as an *aliphatic* amine. Thus benzylamine is an aliphatic amine and aniline, N-methylaniline, and N,N-dimethylaniline are all aromatic amines.

| Aniline | N-Methylaniline | N,N-Dimethylaniline | Benzylamine |

Naming the Amines.   The common names of the simple, aliphatic amines are made by writing the names of the alkyl groups attached to nitrogen in front of the word *amine* (and leaving no space). We have already seen how this works. Here are three more examples.

$$\underset{\text{Isobutylamine}}{CH_3\overset{\overset{\displaystyle CH_3}{|}}{C}HCH_2NH_2} \qquad \underset{\text{Ethylisopropylamine}}{CH_3\overset{\overset{\displaystyle CH_3}{|}}{C}HNHCH_2CH_3} \qquad \underset{\text{Methylethylpropylamine}}{CH_3CH_2\overset{\overset{\displaystyle CH_3}{|}}{N}CH_2CH_2CH_3}$$

In the IUPAC system, the —NH_2 group is called the *amino group*. Thus isobutylamine has the IUPAC name of 1-amino-2-methylpropane. We will not develop IUPAC names for amines having two or more groups attached to the nitrogen atom.

PRACTICE EXERCISE 1   Give common names for the following compounds.

(a) $(CH_3)_2NCH(CH_3)_2$ (b) 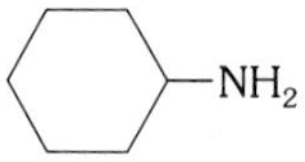 —NH_2 (c) $(CH_3)_2CHCH_2NHC(CH_3)_3$

PRACTICE EXERCISE 2   Write the structures of the following compounds.

(a) *t*-Butyl-*sec*-butylamine

(b) *p*-Nitroaniline

(c) *p*-Aminobenzoic acid (the PABA of sun-screening lotions)

Heterocyclic Amines.   Both proteins and genes are rich in nitrogen-containing heterocyclic rings. For example, the amino acid proline has a saturated ring that includes a nitrogen atom. Unsaturated rings are present in other amino acids such as tryptophan.

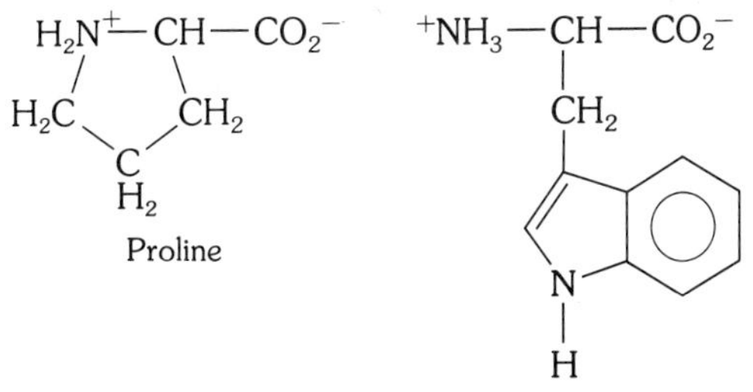

$$H_2\overset{+}{N}-CH-CO_2^-$$

Proline

$$^+NH_3-CH-CO_2^-$$

Tryptophan

In molecules of genes, heterocyclic amine systems generally involve either the pyrimidine ring or the purine ring system.

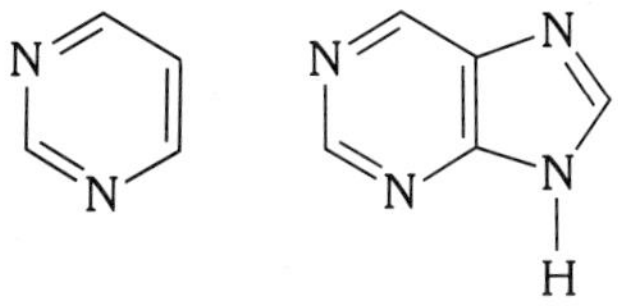

Pyrimidine   Purine

Compound	Boiling Point (°C)
CH_3CH_3	−89
CH_3NH_2	−6
CH_3OH	65

Physical Properties of Amines. You may recall that we sometimes use boiling point data to tell us something about forces between molecules. If you study the data in the table in the margin, you can see that when compounds of similar formula weights are compared, the boiling points of amines are higher than those of alkanes but lower than those of alcohols. This suggests that the forces of attraction between molecules are stronger in amines than in alkanes, but they are weaker in amines than in alcohols.

We can understand these trends in terms of hydrogen bonds. When a hydrogen atom is bound to oxygen or nitrogen but not to carbon, the system can donate and accept hydrogen bonds. However, nitrogen has a lower electronegativity than oxygen, so the polarity of the N—H bond in amines is weaker than the polarity of the O—H bond in alcohols. Consequently, the N—H system develops weaker hydrogen bonds than the O—H system. As a result, amine molecules can't attract each other as strongly as alcohol molecules, so amines boil lower than alcohols (of comparable formula weights). But amine molecules, nevertheless, do develop some hydrogen bonds (Figure 17.1), which alkane molecules cannot do, so amines boil higher than alkanes (of comparable formula weights).

Even though hydrogen bonding that involves amines is weaker than that which involves alcohols, it has a very important function at the molecular level of life among proteins and nucleic acids, where it stabilizes the special molecular shapes of these substances.

Hydrogen bonding also helps amines to be much more soluble in water than alkanes, as Figure 17.1 also depicts.

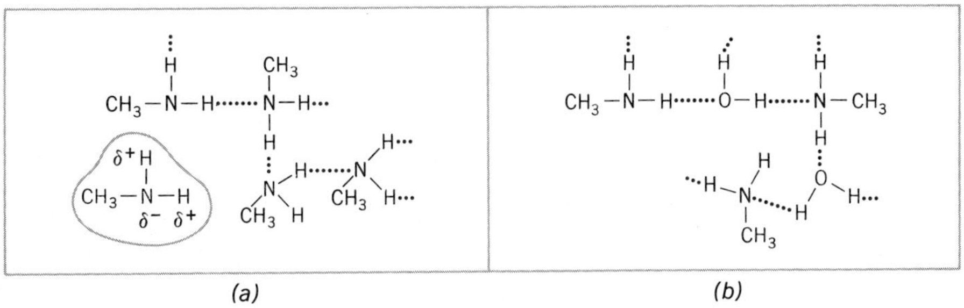

FIGURE 17.1
Hydrogen bonds in (a) amines and in (b) aqueous solutions of amines.

(a)   (b)

The processes in the body that lead to the sensations of odor or taste begin with *chemical* reactions.

Odor isn't actually a *physical* property, but we should note that the amines with lower formula weights smell very much like ammonia. Just above these amines in the homologous series, the odors of amines become very "fishy."

17.2 CHEMICAL PROPERTIES OF AMINES

The amino group is a proton acceptor, and the protonated amino group is a proton donor.

We will examine two chemical properties of amines that will be particularly important to our study of biochemicals — the basicity of amines, in this section, and their conversion to amides, in the next.

Basicity of Amines.   Because amines are related to ammonia, it shouldn't be too surprising that amines are basic. When either ammonia or any water-soluble amine is dissolved in water, the following equilibrium becomes established:

$$R-\overset{..}{N}H_2 + H_2O \rightleftharpoons R-NH_3^+ + OH^-$$

Amine (or
ammonia,
when R = H)

Protonated
amine (or
ammonium ion
when R = H)

How much the products are favored in this equilibrium is usually expressed by the **base ionization constant, K_b,** which is defined as follows:

Remember, the brackets [   ] denote the moles-per-liter concentration of the compound or ion that the brackets enclose.

$$K_b = \frac{[RNH_3^+][OH^-]}{[RNH_2]}$$

Table 17.1 gives the K_b values for several amines. The K_b of ammonia is 1.8×10^{-5}, so you can see that most amines have K_b values slightly larger than that of ammonia. Remember, the larger the K_b value, the *stronger* is the base, because a larger value means that the terms in the numerator, including [OH⁻], have to be larger. Therefore, the aliphatic amines are generally slightly stronger bases than ammonia, and their aqueous solutions are basic solutions. In other words — and this is the important conclusion for us — compounds with aliphatic amino groups tend to raise the pH of an aqueous solution. Any factor that can affect the pH of an aqueous system is vitally important at the molecular level of life, as we have often emphasized.

Compounds with amino groups can also neutralize hydronium ions. The following acid-base neutralization occurs rapidly and essentially completely at room temperature.

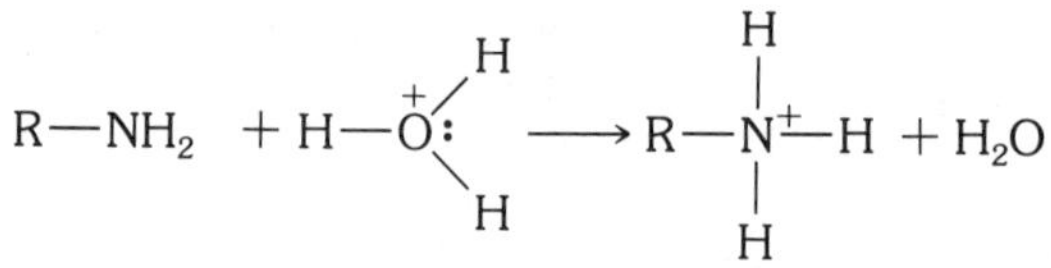

$$R-NH_2 + H-\overset{+}{\underset{H}{O}}{:}\overset{H}{} \longrightarrow R-\overset{\overset{H}{|}}{\underset{\underset{H}{|}}{N}}{}^+H + H_2O$$

Amine (or
ammonia,
when R = H)

Hydronium
ion

Protonated
amine (or
ammonium ion
when R = H)

Tetraalkylammonium ions:

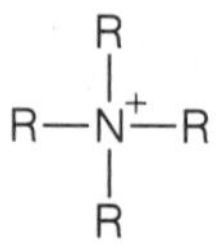

$$R-\overset{\overset{R}{|}}{\underset{\underset{R}{|}}{N}}{}^+R$$

are also known species, but they can't be basic because they have no unshared pair of electrons on nitrogen.

For example,

$$CH_3NH_2 + HCl(aq) \longrightarrow CH_3NH_3^+Cl^- + H_2O$$

Methylamine   Hydrochloric
acid

Methylammonium
chloride

It doesn't matter if the nitrogen atom in an amine bears one, two, or three hydrocarbon groups. The amine can still neutralize strong acids, because the reaction involves just the unshared pair of electrons on the nitrogen, not any of the bonds to the other groups.

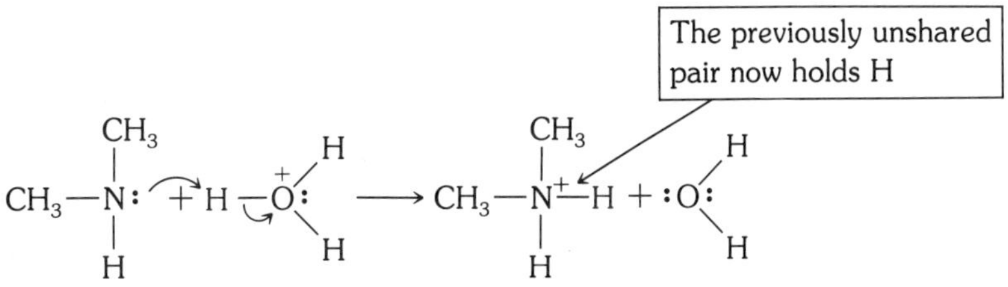

> The previously unshared pair now holds H

Dimethyl-amine   Hydronium ion   Dimethyl-ammonium ion

EXAMPLE 17.1   WRITING THE STRUCTURE OF THE PRODUCT WHEN AN AMINE IS NEUTRALIZED BY A STRONG ACID

Problem: What organic cation forms when hydrochloric acid (or any strong acid) neutralizes each of the following amines?

(a) $CH_3CH_2NH_2$ (b) $CH_3CH_2NCH_3$ (with CH_3 below)   (c) ring $N-CH_3$

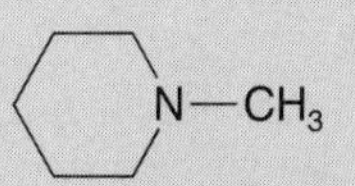

Solution: All we have to do is increase the number of H atoms attached to the nitrogen atom by one, and then we have to write a positive sign to show the charge. Thus the answers are as follows:

(a) $CH_3CH_2NH_3^+$ (b) $CH_3CH_2\overset{+}{N}HCH_3$ (with CH_3 below)   (c) ring $\overset{+}{N}H-CH_3$

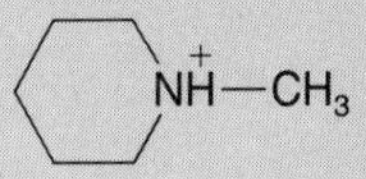

PRACTICE EXERCISE 3   What are the structures of the cations that form when the following amines react completely with hydrochloric acid?

(a) aniline   (b) trimethylamine   (c) $NH_2CH_2CH_2NH_2$

Some amine salts are internal salts, like all of the amino acids, the building blocks of proteins.

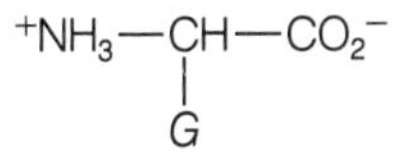

$$^+NH_3-CH-CO_2^-$$
with G below

General formula of all amino acids. G = an organic group

Properties of Amine Salts.   A protonated amine and an associated anion make up an organic salt called an **amine salt.** Table 17.2 gives some examples and, like all salts, they are crystalline solids at room temperature. In addition, like the salts of the ammonium ion, nearly all amine salts of strong acids are soluble in water *even when the parent amine is not.* Amine salts are much more soluble in water than amines because the *full* charges carried by the ions of an amine salt can be much better hydrated by water molecules than the amine itself, where only the small, partial charges of polar bonds occur.

Protonated amines neutralize the hydroxide ion and revert to amines in the following manner (where we show only skeletal structures):

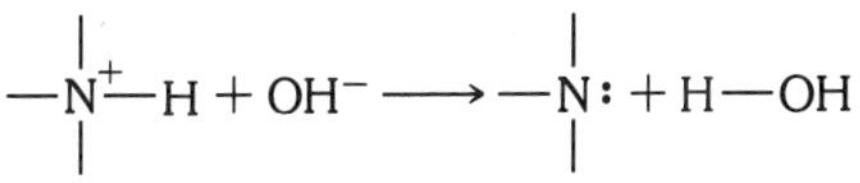

$$-\overset{+}{N}-H + OH^- \longrightarrow -N: + H-OH$$

Protonated amine   Amine

TABLE 17.2
Amine Salts

Name	Structure	Melting Point (°C)
Methylammonium chloride	$CH_3NH_3^+Cl^-$	232
Dimethylammonium chloride	$(CH_3)_2NH_2^+Cl^-$	171
Dimethylammonium bromide	$(CH_3)_2NH_2^+Br^-$	134
Dimethylammonium iodide	$(CH_3)_2NH_2^+I^-$	155
Tetramethylammonium hydroxide (a base as strong as KOH)	$(CH_3)_4N^+OH^-$	130–135 (decomposes)

For example,

$$CH_3NH_3^+ + OH^- \longrightarrow CH_3\overset{..}{N}H_2 + H_2O$$

$$CH_3CH_2\overset{+}{N}H_2CH_3 + OH^- \longrightarrow CH_3CH_2\overset{..}{N}HCH_3 + H_2O$$

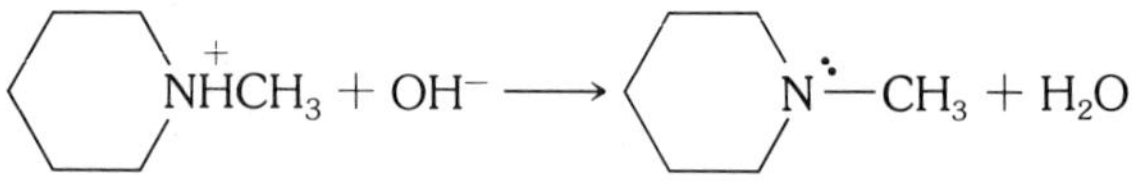

$$\bigcirc\overset{+}{N}HCH_3 + OH^- \longrightarrow \bigcirc\overset{\cdot\cdot}{N}\!\!-CH_3 + H_2O$$

The Amino Group as a Solubility Switch.   We have just learned that the amino group is a good proton acceptor, and that its protonated form is not only a good proton donor but is also more soluble in water than the amine. These properties make the amino group an excellent "solubility switch." This means that when a compound is an amine, Its solubility in water can be turned on simply by adding enough strong acid to protonate the amine. Triethylamine, for example, is insoluble in water, but we can switch on its solubility merely by adding a strong acid such as hydrochloric acid. The amine dissolves as its protonated form is produced.

The other important solubility switch that we have studied involved the carboxylic acid group:

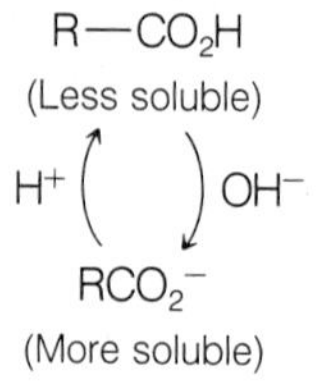

R—CO₂H
(Less soluble)

$$H^+ \Big(\updownarrow\Big) OH^-$$

RCO₂⁻
(More soluble)

$$(CH_3CH_2)_3N\!:\; + HCl(aq) \longrightarrow (CH_3CH_2)_3\overset{+}{N}H\;Cl^-$$

Triethylamine
(water-insoluble)   Triethylammonium chloride
                                           (water-soluble)

We can just as quickly and easily bring the amine back out of solution again simply by adding a strong base, such as the hydroxide ion. Protonated amines, as we just learned, are good proton donors.

$$(CH_3CH_2)_3\overset{+}{N}H\; + OH^- \longrightarrow (CH_3CH_2)_3N + H_2O$$

Triethylammonium
ion
(water-soluble)   Triethylamine
                                   (water-insoluble)

The significance of this "switching" relationship is that the solubilities of complex compounds that have the amine function can be changed almost instantly, simply by adjusting the pH of the medium.

One application of the property we have just studied occurs in medicine when certain drugs have to be administered. A number of amines obtained from the bark, roots, leaves, flowers, or fruit of various plants are useful drugs. These naturally-occurring, acid-neutralizing, physiologically active amines are called **alkaloids,** and morphine, codeine, and quinine are just three typical examples.

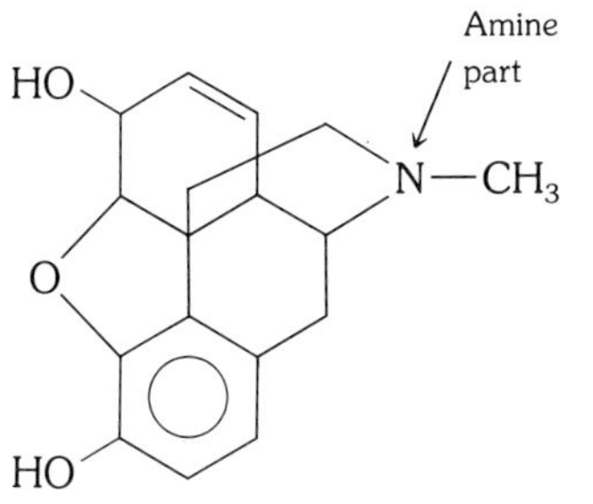

Morphine 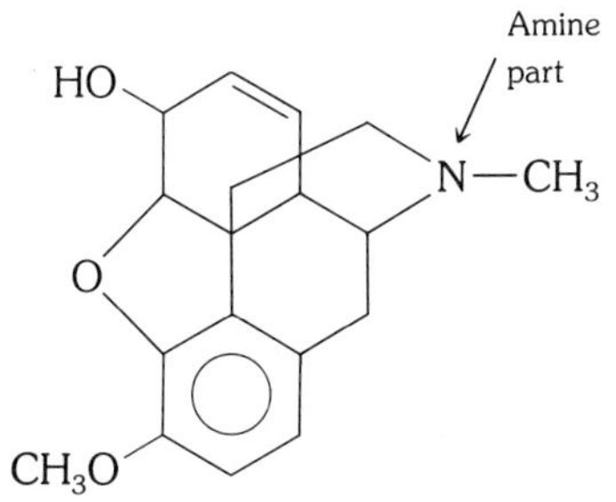   Codeine 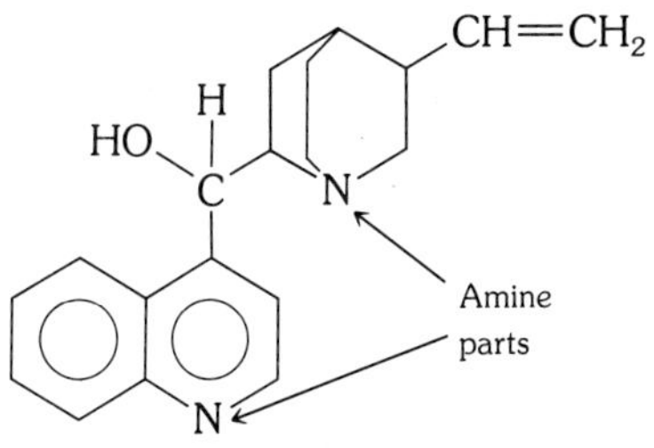   Quinine

To make it easier to administer alkaloidal drugs in the dissolved state, they often are prepared as their water-soluble amine salts. For example, morphine, a potent sedative and painkiller, is often given as morphine sulfate, the salt of morphine and sulfuric acid. Quinine, an antimalarial drug, is available as quinine sulfate. Codeine, sometimes used in cough medicines, is often present as codeine phosphate. Special Topic 17.1 tells about a few other physiologically active amines, most of which are also prepared as their amine salts.

EXAMPLE 17.2

Amphetamine sulfate (also known as benzedrine sulfate) is the form in which this drug is administered. It can be prescribed as an anorexigenic agent—one that reduces the appetite.

WRITING THE STRUCTURE OF THE PRODUCT OF DEPROTONATING THE CATION OF AN AMINE SALT

Problem: The protonated form of amphetamine is shown below. What is the structure of the product of its reaction with hydroxide ion?

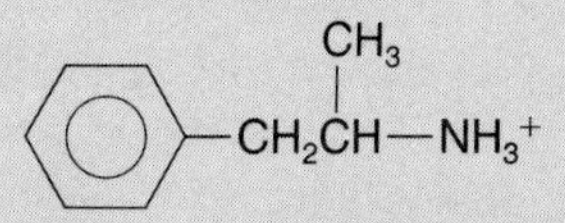

Solution: Because OH^- removes just one H^+ from the protonated amine's cation, all we have to do is reduce the number of H's on the nitrogen by one and cancel the positive charge. The answer, therefore is as follows—the structure of amphetamine itself.

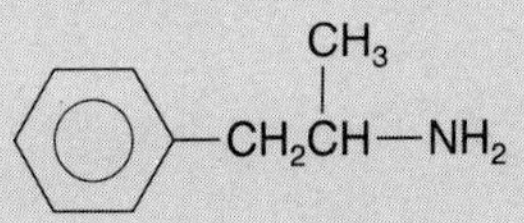

(The structure of amphetamine, given here, appears to be identical with that of Dexedrine, shown in Special Topic 17.1. However, there is an important difference that we will explore in the next chapter.)

PRACTICE EXERCISE 4

Write the structures of the products after the following protonated amines have reacted with OH^- in a 1:1 mole ratio.

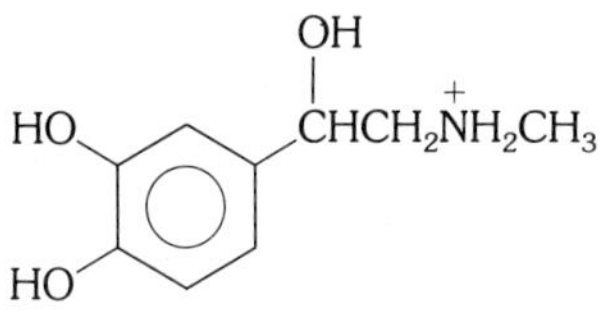

Epinephrine (adrenaline), a hormone given here in its protonated form. As the chloride salt in a 0.1% solution, it is injected in some cardiac failure emergencies. (See also Special Topic 17.1.)

Epinephrine and norepinephrine are two of the many hormones in our bodies. We will study the nature of hormones in a later chapter, but we can use a definition here. **Hormones** are compounds the body makes in special glands to serve as chemical messengers. In response to a stimulus somewhat unique for each hormone, such as fright, food odor, sugar ingestion, and others, the gland secretes its hormone into circulation. The hormone then moves to some organ or tissue where it activates a particular metabolic series of reactions that constitute the biochemical response to the initial stimulus. Maybe you have heard the expression, "I need to get my adrenalin flowing." Adrenalin — or epinephrine, its technical name — is made by the adrenal gland. If you ever experience a sudden fright, a trace of epinephrine immediately flows and the results include a strengthened heartbeat, a rise in blood pressure, and a release of glucose into circulation from storage — all of which get the body ready to respond to the threat.

Norepinephrine has similar effects, and because these two hormones are secreted by the adrenal gland, they are called **adrenergic agents.**

HO

HO — (ring) — CHCH$_2$NHCH$_3$ with OH

Epinephrine

HO

HO — (ring) — CHCH$_2$NH$_2$ with OH

Norepinephrine

Several useful drugs mimic epinephrine and norepinephrine, and all are classified as *adrenergic drugs.* Most of them, like epinephrine and norepinephrine, are related structurally to β-phenylethanolamine (which is not a formal name, obviously). In nearly all of their uses, these drugs are prepared as dilute solutions of their amine acid salts. Several of the β-phenylethanolamine drugs are even more structurally like epinephrine and norepinephrine, because they have the structural features of 1,2-dihydroxybenzene. This compound is commonly called *catechol,* so the catechol-like adrenergic drugs are called the **catecholamines.** Synthetic epinephrine (an agent in Primatene Mist), Ethylnorepinephrine, and Isoproterenol are examples.

OH

(ring) — CHCH$_2$NH$_2$

β-Phenylethanolamine

HO

HO — (ring) — CHCHCHNH$_2$ with OH and CH$_2$CH$_3$

Ethylnorepinephrine
(used for treating asthma
in children)

HO

HO — (ring) — CHCH$_2$NHCH(CH$_3$)$_2$ with OH

Isoproterenol
(used in treating emphysema
and asthma)

Another family of physiologically active amines is the β-phenylethylamines. For example, dopamine (which is also a catecholamine) is the compound the body uses to make norepinephrine. Its synthetic form is used to treat shock associated with severe congestive heart failure.

The amphetamines are a family of β-phenylethylamines that include Dexedrine ("speed") and Methedrin ("crystal," "meth"). The amphetamines can be legally prescribed as stimulants and antidepressants, and sometimes they are prescribed for weight-control programs. However, millions of these "pep pills" or "uppers" are sold illegally, and this use of amphetamines constitutes a major drug abuse problem. The dangers of overuse include suicide, belligerence and hostility, paranoia, and hallucinations.

HO

HO — (ring) — CH$_2$CH$_2$NH$_2$

Dopamine

(ring) — CH$_2$CHNH$_2$ with CH$_3$

Dexedrine

(ring) — CH$_2$CHNHCH$_3$ with CH$_3$

Methedrin

In a later chapter we will discuss the mechanisms by which these drugs and the naturally occurring hormones work.

Hallucinogens are drugs that cause illusions of time and place, make unreal experiences or things seem real, and distort the qualities of things.

(b)

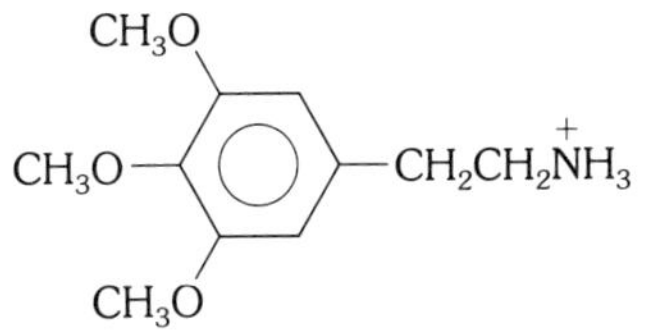

$$CH_3O$$

$$CH_3O \text{—} \underset{}{\bigcirc} \text{—} CH_2CH_2\overset{+}{N}H_3$$

$$CH_3O$$

Mescaline, a mind-altering hallucinogen shown here in its protonated form. It is isolated from the mescal button, a growth on top of the peyote cactus. Indians in the southwestern United States have used it in religious ceremonies.

17.3 AMIDES OF CARBOXYLIC ACIDS

Amides are neutral nitrogen compounds that can be hydrolyzed to carboxylic acids and ammonia (or amines).

The amide bond is called the *peptide bond* in the chemistry of proteins.

The carbonyl-nitrogen bond is sometimes called the **amide bond,** because it is the bond that forms when amides are made, and it is the bond that breaks when amides are hydrolyzed. As the following general structures show, an amide can be derived either from ammonia or from amines. Those derived from ammonia itself are often referred to as *simple* amides.

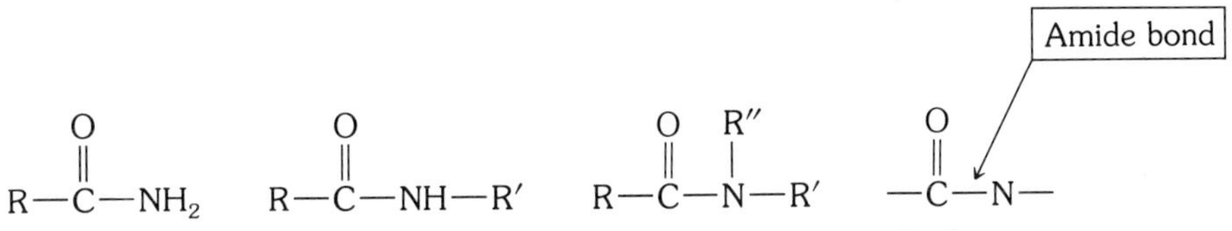

$$\underset{\substack{\text{Amides of}\\\text{ammonia}\\\text{(Simple amides)}}}{R\text{—}\overset{\overset{\textstyle O}{\|}}{C}\text{—}NH_2} \qquad \underset{\text{Amides of amines}}{R\text{—}\overset{\overset{\textstyle O}{\|}}{C}\text{—}NH\text{—}R' \qquad R\text{—}\overset{\overset{\textstyle O}{\|}}{C}\text{—}\overset{\overset{\textstyle R''}{|}}{N}\text{—}R'} \qquad \underset{\text{Amide group}}{\text{—}\overset{\overset{\textstyle O}{\|}}{C}\text{—}N\text{—}}$$

Amide bond

We study the amide system because all proteins are essentially polyamides whose molecules have regularly spaced amide bonds.

Table 17.3 lists several low-formula-weight amides. Their molecules are quite polar, and when they have an H atom bonded to N, they can both donate and accept hydrogen bonds. These forces are so strong in amides that all simple amides except methanamide are solids at room temperature. When we study proteins, we'll see how the hydrogen bond is involved in

TABLE 17.3
Amides of Carboxylic Acids

IUPAC Name	Structure	Melting Point (°C)
Methanamide	$HCONH_2$	3
N-Methylmethanamide	$HCONHCH_3$	−5
N,N-Dimethylmethanamide	$HCON(CH_3)_2$	−61
Ethanamide	CH_3CONH_2	82
N-Methylethanamide	$CH_3CONHCH_3$	28
N,N-Dimethylethanamide	$CH_3CON(CH_3)_2$	−20
Propanamide	$CH_3CH_2CONH_2$	79
Butanamide	$CH_3CH_2CH_2CONH_2$	115
Pentanamide	$CH_3CH_2CH_2CH_2CONH_2$	106
Hexanamide	$CH_3CH_2CH_2CH_2CH_2CONH_2$	100
Benzamide	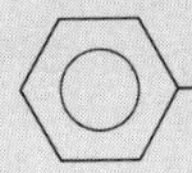 $\bigcirc\text{—}CONH_2$	133

stabilizing the shapes of protein molecules, shapes that are as important to the functions of proteins as anything else about their structures.

Naming Amides.   Our needs for systematic names do not go beyond those of the simple amides, and they are based on the names of the parent acids. The common names of simple amides are made by replacing *-ic acid* by *-amide,* and their IUPAC names are devised by replacing *-oic acid* by *-amide.* In the following examples, notice how we can condense the structure of the amide group.

$$CH_3CNH_2 \quad or \quad CH_3CONH_2$$

Acetamide (common name)
Ethanamide (IUPAC name)

$$CH_3CH_2CH_2CNH_2 \quad or \quad CH_3CH_2CH_2CONH_2$$

Butyramide (common name)
Butanamide (IUPAC name)

$$C_6H_5{}^- = \langle\bigcirc\rangle -$$

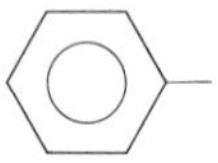

The simplest aromatic amide is called benzamide, $C_6H_5CONH_2$, where C_6H_5 signifies the phenyl group.

PRACTICE EXERCISE 5   Write the IUPAC names of the following amides.

(a) $CH_3CH_2CHCH_2CH_2CONH_2$
       |
       CH_3

(b) $CH_3CH_2CHCONH_2$
       |
       CH_3CH_2

Amides as Neutral Compounds.   One reason for creating a separate family for the amides apart from the amines is that, unlike amines, amides are not proton-acceptors. They're not proton donors, either. Amides are neutral in an acid-base sense. In other words — and this is the important point — the amide group does not affect the pH of an aqueous system.

The electronegative carbonyl group on the nitrogen atom causes the acid-base neutrality of amides. Both an amide and an amine have an unshared pair of electrons on nitrogen, but in the amide this pair is drawn back so tightly by the electron-withdrawing ability — the electronegativity — of the carbonyl group that the nitrogen atom cannot accept and hold a proton.

The oxygen atom of the carbonyl group is what makes the whole group electronegative.

Making Amides from Amines by Acyl Group Transfer Reactions.   Amides can be made from amines just as esters can be made from alcohols. Either acid chlorides or acid anhydrides react smoothly with ammonia or amines to give amides. (The amine, of course, must have at least one hydrogen atom on nitrogen, because one hydrogen has to be replaced as the amide forms.) We can illustrate these reactions using ammonia.

$$R-\overset{O}{\overset{||}{C}}-Cl + 2NH_3 \longrightarrow R-\overset{O}{\overset{||}{C}}-NH_2 + [NH_4{}^+ + Cl^-]$$

Acid chloride   Amide   Ammonium chloride

$$R-\overset{O}{\overset{||}{C}}-O-\overset{O}{\overset{||}{C}}-R + 2NH_3 \longrightarrow R-\overset{O}{\overset{||}{C}}-NH_2 + [NH_4{}^+ + {}^-O-\overset{O}{\overset{||}{C}}-R]$$

Acid anhydride   Amide   Ammonia salt of the
                                                                                                         carboxylic acid

These reactions are further examples of the *acyl group transfer reaction* we learned about in the previous chapter. In the formation of an amide from an acid chloride, the acyl group in

SPECIAL TOPIC 17.2 NYLON, A POLYAMIDE

The term *nylon* is a coined name that applies to any synthetic, long-chain, fiber-forming polymer with repeating amide linkages. One of the most common members of the nylon family, nylon-66, is made from 1,6-hexanediamine and hexanedioic acid.

$$NH_2CH_2CH_2CH_2CH_2CH_2CH_2NH_2$$

1,6-Hexanediamine
(Hexamethylenediamine)

Hexanedioic acid
(Adipic acid)

Repeating unit in nylon-66

(The "66" means that each monomer has six carbon atoms.) To be useful as a fiber-forming polymer, each nylon-66 molecule should contain from 50 to 90 of each of the monomer units. Shorter molecules form weak or brittle fibers.

When molten nylon resin is being drawn into fibers, newly emerging strands are caught up on drums and stretched as they cool. Under this tension, the long polymer molecules within the fiber line up side by side, overlapping each other, to give a finished fiber of unusual strength and beauty. Part of nylon's strength comes from the innumerable hydrogen bonds that extend between the polymer molecules and that involve their many regularly spaced amide groups.

Nylon is more resistant to combustion than wool, rayon, cotton, or silk, and it is as immune to insect attack as Fiberglas. Molds and fungi to not attack nylon molecules. In medicine, nylon is used in specialized tubing, and as velour for blood contact surfaces. Nylon sutures were the first synthetic sutures and are still used.

the acid chloride transfers from the Cl atom to the N atom of the amine (or ammonia). An acyl group can transfer from an acid anhydride to N, also.

In the body, instead of ordinary acid chlorides and anhydrides as sources of the acyl group, there are other kinds of acyl carrier molecules. In making proteins from amino acids, for example, the acyl portions of amino acids—aminoacyl units—are held by carrier molecules.

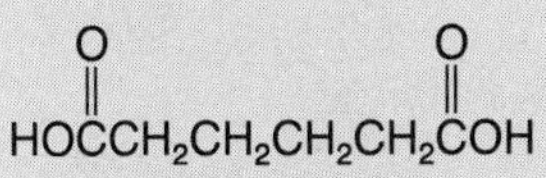

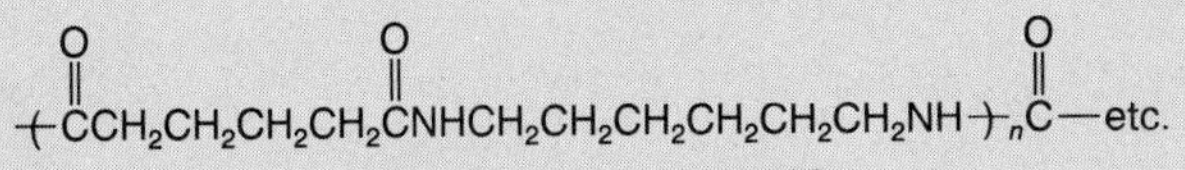

G is some organic group, but not necessarily an alkyl group. Hence, the symbol R is not used here.

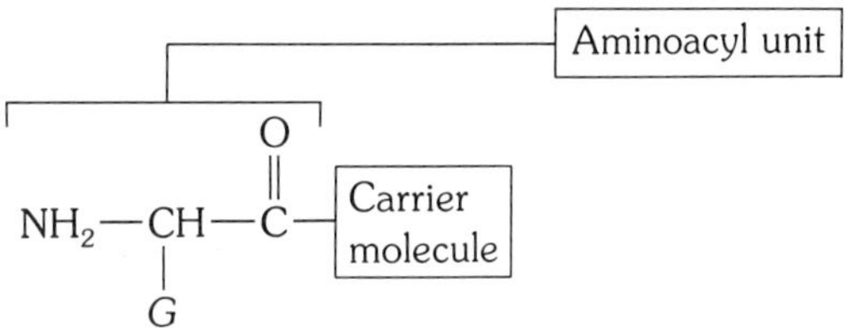

When a cell makes an amide bond, it transfers an aminoacyl group from its carrier molecule to the nitrogen atom of the amino group on another molecule. The carrier molecule is released to be reused.

The molecule given here as NH_2—R can be another aminoacyl group that is bound to another carrier molecule.

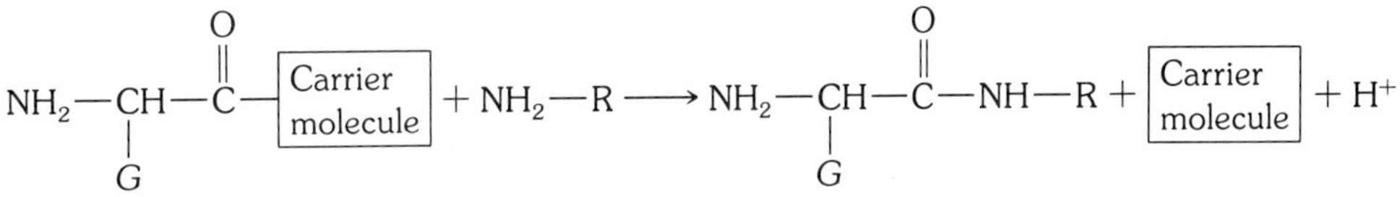

This is the aspect of making amides—aminoacyl transfers—that is of greatest interest as we prepare for our upcoming study of biochemistry. The skill we should carry forward is the ability to figure out the structure of the amide that can be made from ammonia (or some amine) and a carboxylic acid regardless of the exact nature of the acyl transfer agent. The kind of question you should be able to answer is illustrated in the next worked example.

EXAMPLE 17.3   WRITING THE STRUCTURE OF AN AMIDE THAT CAN BE MADE FROM THE ACYL GROUP OF AN ACID AND AN AMINE

Problem: What amide can be made from the following two substances, assuming that a suitable acyl group transfer process is available?

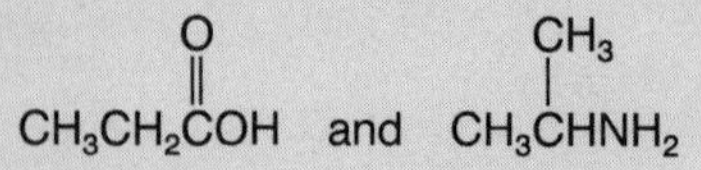

$$\underset{CH_3CH_2COH}{\overset{O}{\overset{\|}{}}} \quad and \quad \underset{CH_3CHNH_2}{\overset{CH_3}{\overset{|}{}}}$$

Solution: The best way to proceed is to write the skeleton of the amide system and then build on it. It doesn't matter how we orient this skeleton:

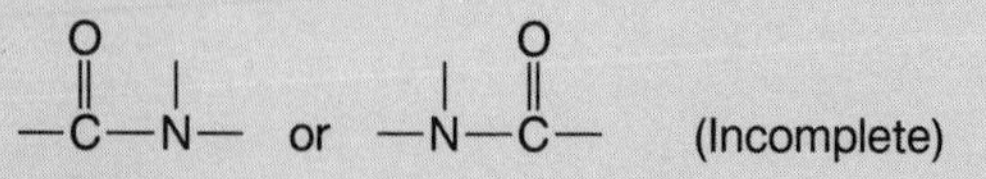

$$\overset{O}{\overset{\|}{-C-N-}} \quad or \quad \overset{O}{\overset{\|}{-N-C-}} \quad \text{(Incomplete)}$$

Then we look at the acid and see what has to be attached to the carbon atom of this skeleton, and we write it in:

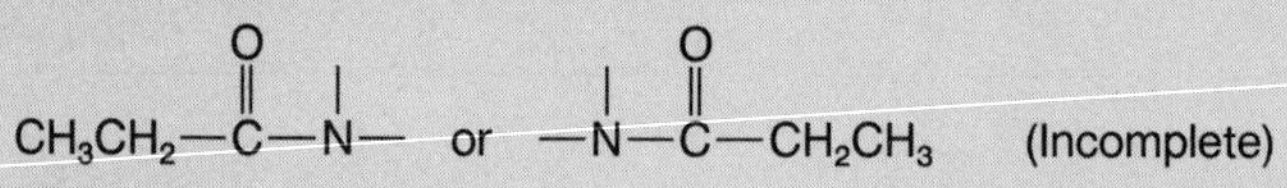

$$CH_3CH_2-\overset{O}{\overset{\|}{C}}-N- \quad or \quad -N-\overset{O}{\overset{\|}{C}}-CH_2CH_3 \quad \text{(Incomplete)}$$

Then we examine the given amine to see what organic group it carries, and we attach it to the N atom. (If there are *two* organic groups on N in the amine, we have to attach both, of course.)

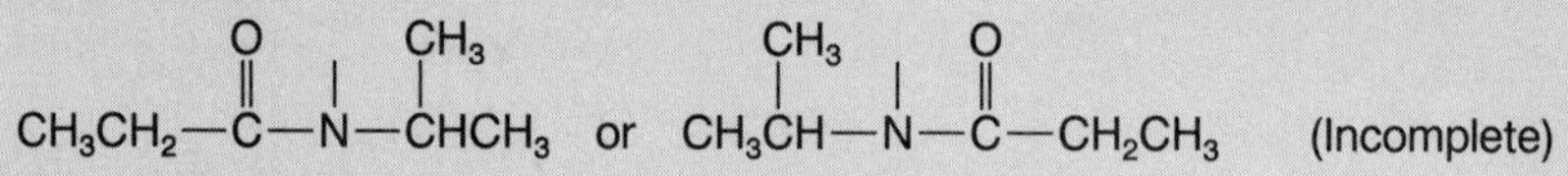

$$CH_3CH_2-\overset{O}{\overset{\|}{C}}-\overset{CH_3}{\overset{|}{N}}-CHCH_3 \quad or \quad CH_3\overset{CH_3}{\overset{|}{CH}}-\overset{}{N}-\overset{O}{\overset{\|}{C}}-CH_2CH_3 \quad \text{(Incomplete)}$$

Finally, of the two H atoms on N in the amine, one survives, and our last step is to write it in. The final answer is:

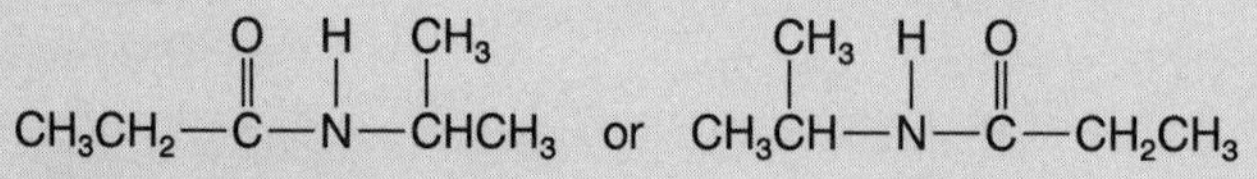

$$CH_3CH_2-\overset{O}{\overset{\|}{C}}-\overset{H}{\overset{|}{N}}-\overset{CH_3}{\overset{|}{CH}}CH_3 \quad or \quad CH_3\overset{CH_3}{\overset{|}{CH}}-\overset{H}{\overset{|}{N}}-\overset{O}{\overset{\|}{C}}-CH_2CH_3$$

These structures, of course, are identical.

PRACTICE EXERCISE 6   What amides, if any, could be made by suitable acyl group transfer reactions from the following pairs of compounds?

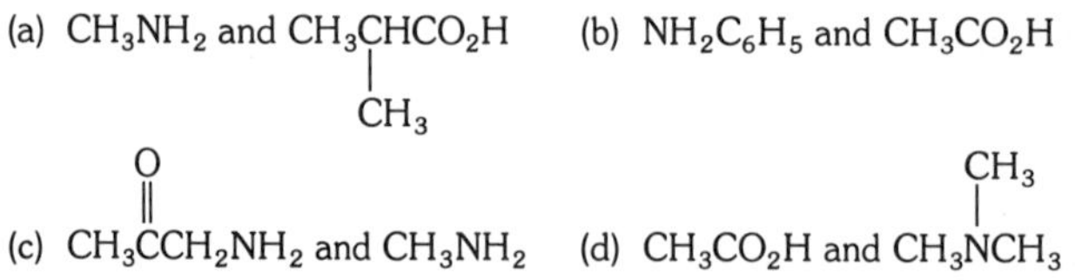

(a) CH_3NH_2 and $\underset{\underset{CH_3}{|}}{CH_3CHCO_2H}$ (b) $NH_2C_6H_5$ and CH_3CO_2H

(c) $\overset{O}{\overset{\|}{CH_3CCH_2NH_2}}$ and CH_3NH_2 (d) CH_3CO_2H and $\underset{\underset{CH_3}{|}}{CH_3NCH_3}$

The Hydrolysis of Amides.   The only reaction of amides we will study is their hydrolysis, a reaction in which the amide bond breaks and we obtain the amide's parent acid and amine (or ammonia). Either acids or bases promote this reaction, and enzymes catalyze it in the body.

The hydrolysis of the amide bond is all that is involved in the overall chemistry of the digestion of proteins.

When an acid promotes the hydrolysis of an amide, one of the products, the amine, neutralizes the acid. (This is why we don't say that the acid *catalyzes* the hydrolysis. Catalysts, by definition, are reaction promotors that are not used up.) Thus instead of obtaining the amine itself, we get the salt of the amine. For example,

$$R-\overset{\underset{\parallel}{O}}{C}-NH-CH_3 + H-OH + HCl(aq) \longrightarrow R-\overset{\underset{\parallel}{O}}{C}-OH + CH_3-\overset{\overset{\textstyle H}{|}}{\underset{\underset{\textstyle H}{|}}{N^+}}-H\ Cl^-$$

On the other hand, if we use a base to promote this hydrolysis, then the carboxylic acid that forms neutralizes the base, and we get the salt of the carboxylic acid. For example,

$$R-\overset{\underset{\parallel}{O}}{C}-NH-CH_3 + NaOH(aq) \longrightarrow R-\overset{\underset{\parallel}{O}}{C}-O^-Na^+ + CH_3NH_2$$

The hydrolysis of an amide can occur without any promoter; it is just much slower this way. Moreover, when enzymes catalyze this hydrolysis, they are not used up by the reaction. Therefore we'll write amide hydrolysis as a simple reaction with water to give the free carboxylic acid and the free amine. Here are some examples:

$$R-\overset{\underset{\parallel}{O}}{C}-NH_2 + H_2O \longrightarrow R-\overset{\underset{\parallel}{O}}{C}-OH + NH_3$$

$$R-\overset{\underset{\parallel}{O}}{C}-NH-R' + H_2O \longrightarrow R-\overset{\underset{\parallel}{O}}{C}-OH + NH_2-R'$$

$$R-\overset{\underset{\parallel}{O}}{C}-\overset{\overset{\textstyle R''}{|}}{N}-R' + H_2O \longrightarrow R-\overset{\underset{\parallel}{O}}{C}-OH + H-\overset{\overset{\textstyle R''}{|}}{N}-R'$$

EXAMPLE 17.4   WRITING THE PRODUCTS OF THE HYDROLYSIS OF AN AMIDE

Problem: Acetophenetidin (phenacetin) has long been used in some brands of headache remedies. (APC tablets, for example, consist of aspirin, phenacetin, and caffeine.)

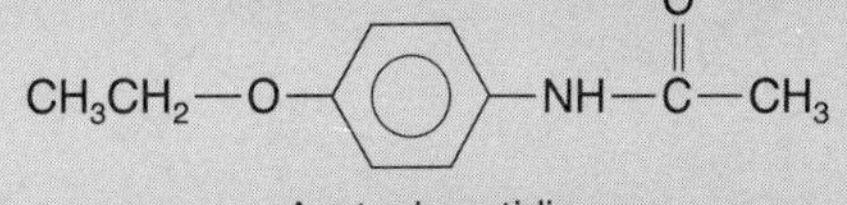

$$CH_3CH_2-O-\!\!\left\langle\!\!\bigcirc\!\!\right\rangle\!\!-NH-\overset{\underset{\parallel}{O}}{C}-CH_3$$

Acetophenetidin

If this compound is an amide, what are the products of its hydrolysis?

Solution: Acetophenetidin does have the amide bond — carbonyl to nitrogen — so it can be hydrolyzed. (The functional group on the left side of this structure is that of an ether, and ethers do not react with water.)

Because the amide bond breaks when an amide is hydrolyzed, simply erase this bond from the structure and separate the parts. *Do not break any other bond.*

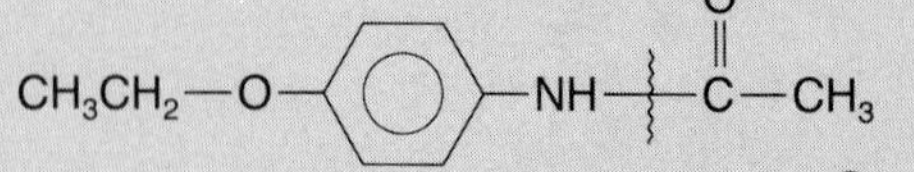

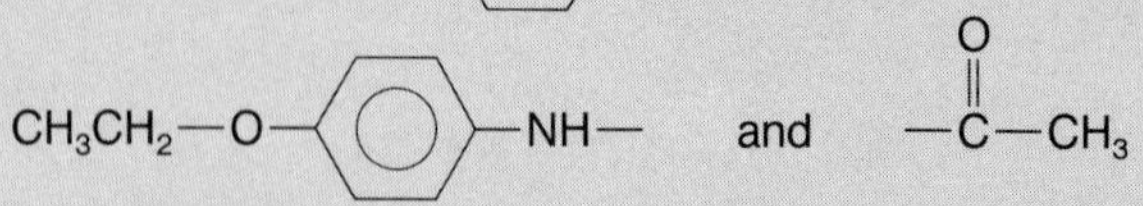

$$CH_3CH_2-O-\text{⬡}-NH \bigg| \overset{\displaystyle O}{\overset{\|}{-C}}-CH_3$$

$$CH_3CH_2-O-\text{⬡}-NH- \quad \text{and} \quad \overset{\displaystyle O}{\overset{\|}{-C}}-CH_3 \quad \text{(Incomplete)}$$

We know that the hydrolysis uses HO—H to give a carboxylic acid and an amine, so we put a HO— group on the carbonyl group in the appropriate fragment and we put H— on the nitrogen of the other fragment. The products, therefore, are

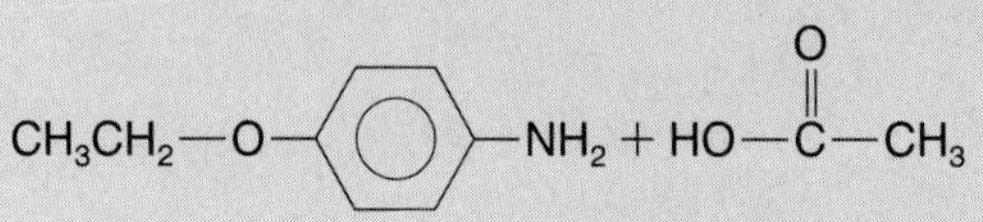

$$CH_3CH_2-O-\text{⬡}-NH_2 + HO-\overset{\displaystyle O}{\overset{\|}{C}}-CH_3$$

PRACTICE EXERCISE 7   For any compounds in the following list that are amides, write the products of their hydrolysis.

(a) 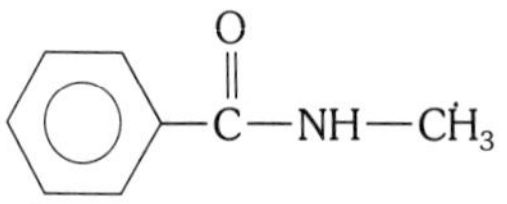 $\text{⬡}-\overset{\displaystyle O}{\overset{\|}{C}}-NH-CH_3$

(b) $\text{⬡}-\overset{\displaystyle O}{\overset{\|}{C}}-CH_2-NH_2$

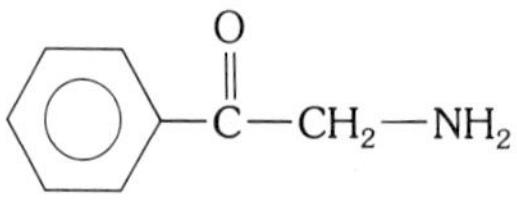

(c) $\text{⬡}-NH-\overset{\displaystyle O}{\overset{\|}{C}}-CH_3$

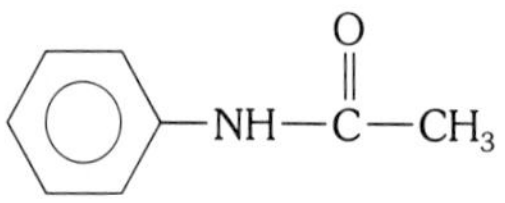

(d)  $CH_3-\overset{\displaystyle O}{\overset{\|}{C}}-NH-CH_2CH_2-NH-\overset{\displaystyle O}{\overset{\|}{C}}-CH_3$
        (Use an excess of water)

PRACTICE EXERCISE 8   The following structure illustrates some of the features of protein molecules. What are the products of the complete, enzyme-catalyzed hydrolysis (the digestion) of this substance? (A typical protein would hydrolyze to several hundred and, in some, several thousand of the kinds of molecules produced here.)

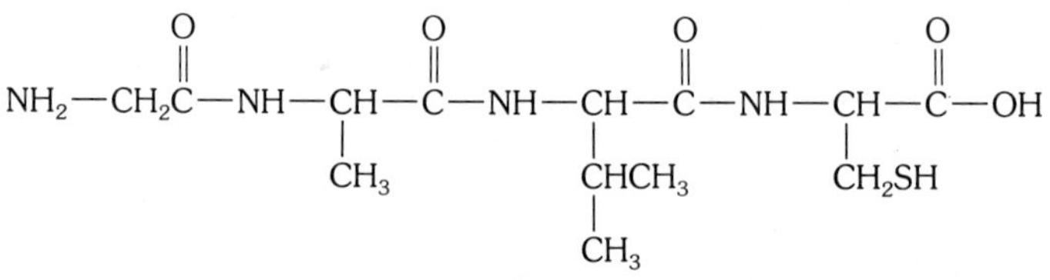

$$NH_2-CH_2\overset{\displaystyle O}{\overset{\|}{C}}-NH-\underset{\underset{\displaystyle CH_3}{|}}{CH}-\overset{\displaystyle O}{\overset{\|}{C}}-NH-\underset{\underset{\underset{\displaystyle CH_3}{|}}{\underset{\displaystyle CHCH_3}{|}}}{CH}-\overset{\displaystyle O}{\overset{\|}{C}}-NH-\underset{\underset{\displaystyle CH_2SH}{|}}{CH}-\overset{\displaystyle O}{\overset{\|}{C}}-OH$$

SUMMARY

Amines and protonated amines   When one, two, or three of the hydrogen atoms in ammonia are replaced by an organic group (other than a carbonyl group), the result is an amine. The nitrogen atom can be part of a ring, as in heterocyclic amines. Like ammonia, the amines are weak bases, but all can form salts with strong acids. The cations in these salts are protonated amines. Amine salts are far more soluble in water than their parent amines. Protonated amines are easily deprotonated by any strong base to give back the original, and usually far less soluble amine. Thus any compound with the amine function has a "solubility switch," because its solubility in an aqueous system can be turned on by adding acid (to form the amine salt) and turned off again by adding base (to recover the amine).

Amides The carbonyl-nitrogen bond, the amide bond, can be formed by letting an amine or ammonia react with anything that can transfer an acyl group (e.g., an acid chloride or an acid anhydride). Amides are neither basic nor acidic, but are neutral compounds. Amides can be made to react with water to give back their parent acids and amines. The accompanying chart summarizes the reactions studied in this chapter.

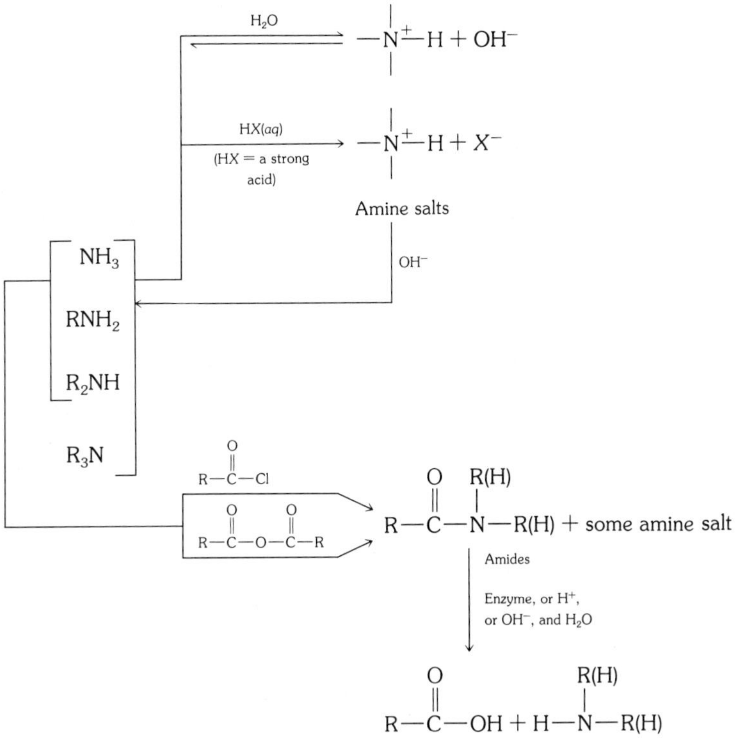

KEY TERMS

The following terms should be mastered before you continue.

alkaloid	amide bond	amine salt	base ionization constant, K_b
amide	amine		

REVIEW EXERCISES

The answers to these Review Exercises are in the *Study Guide* that accompanies this book.

Structures of Amines and Amides — Review of Functional Groups

17.1 Classify the following as aliphatic, aromatic, or heterocyclic amines or amides, and name any other functional groups, too.

(a) $CH_3-O-CH_2-\overset{\overset{O}{\|}}{C}-NH_2$

(b) $CH_3-O-\overset{\overset{O}{\|}}{C}-CH_2-NH_2$

(c)

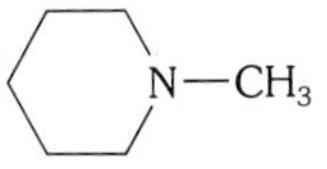

(d)

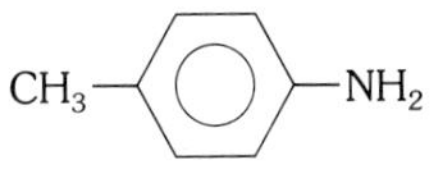

17.2 Classify each of the following as aliphatic, aromatic, or heterocyclic amines or amides. Name any other functional groups that are present.

(a)

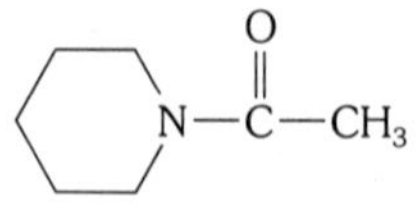

(b)

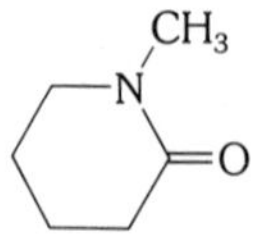

(c)

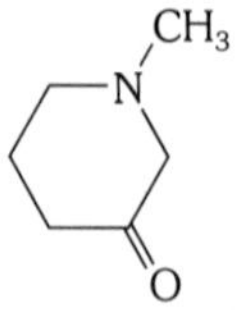

(d)

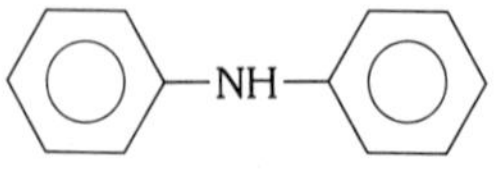

17.3 The following compounds are all very active physiological agents. Name the numbered functional groups that are present in each.

(a)

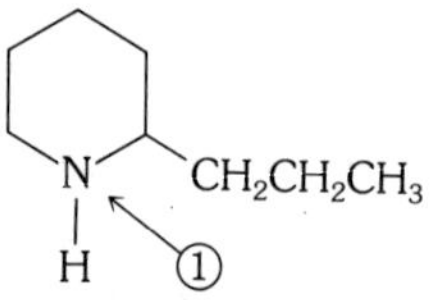

Coniine, the poison in the extract of hemlock that was involved in the death of the Greek philosopher, Socrates.

(b)

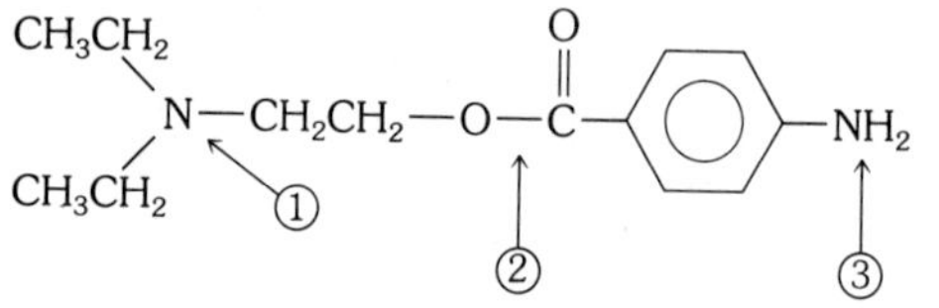

Novocaine, a local anesthetic

(c)

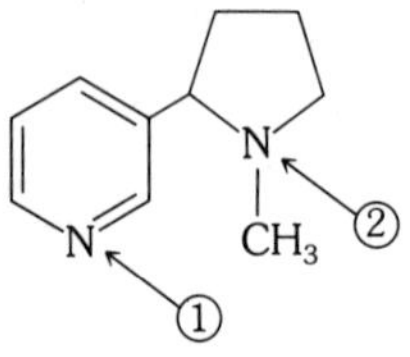

Nicotine, a poison in tobacco leaves

(d)

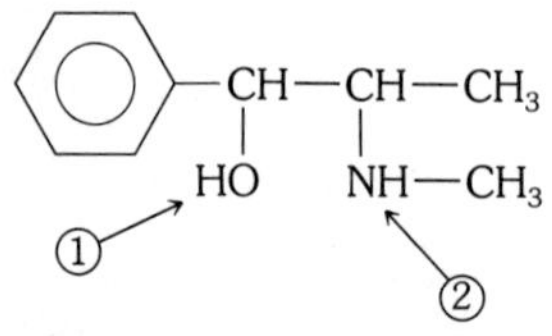

Ephedrine, a bronchodilator

17.4 Some extremely potent, physiologically active compounds are in the following list. Name the functional groups that they have.

(a)

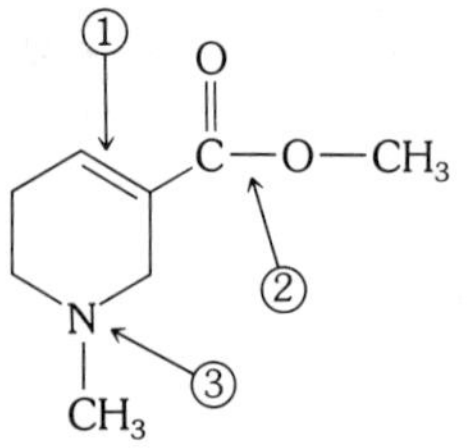

Arecoline, the most active component in the nut of the betel palm. This nut is chewed daily by millions of inhabitants of parts of Asia and the Pacific islands as a narcotic.

(b)

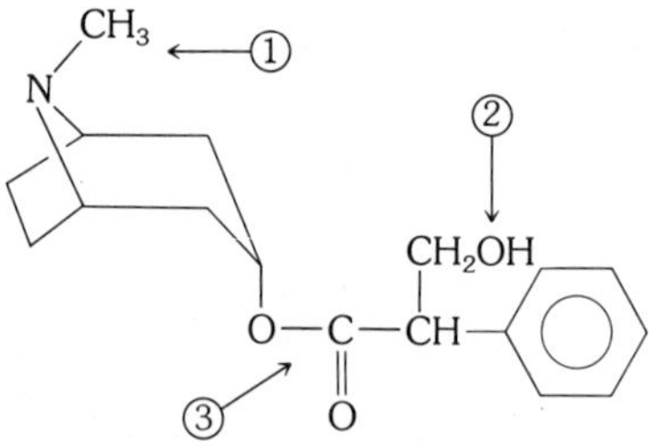

Hyoscyamine, a constituent of the seeds and leaves of henbane, and a smooth muscle relaxant. (A similar form is called atropine, a drug used to counteract nerve poisons.)

(c)

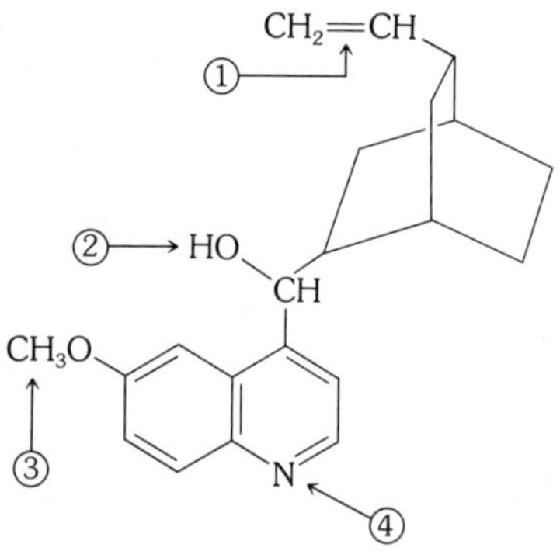

Quinine, a constituent of the bark of the chinchona tree in South America and used worldwide to treat malaria.

(d)

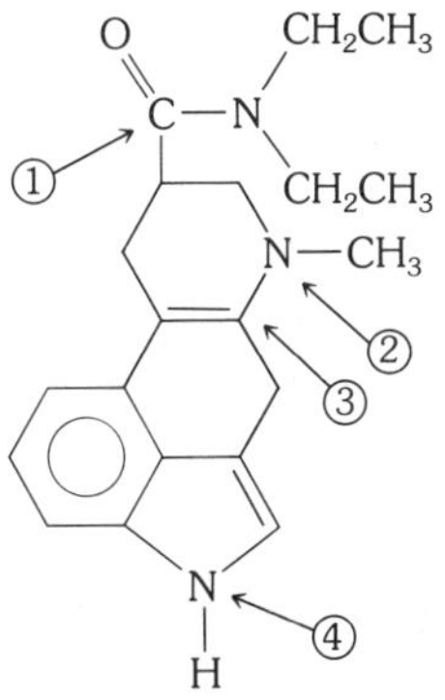

LSD (lysergic acid diethylamide), a constituent of diseased rye and a notorious hallucinogen.

Nomenclature of Amines and Amine Salts

17.5 Give the common names of the following compounds or ions.

(a)

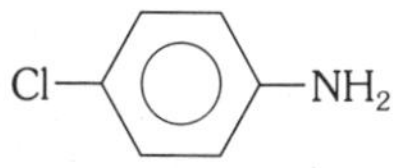

(b) 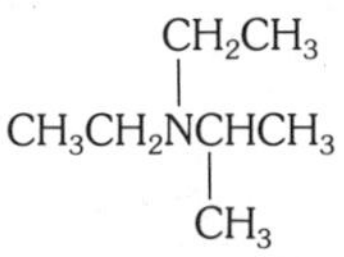 $CH_3CH_2NCHCH_3$

(c)

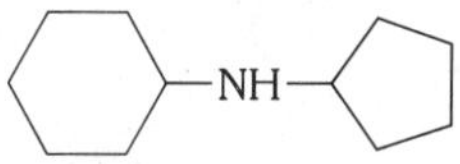

(d) 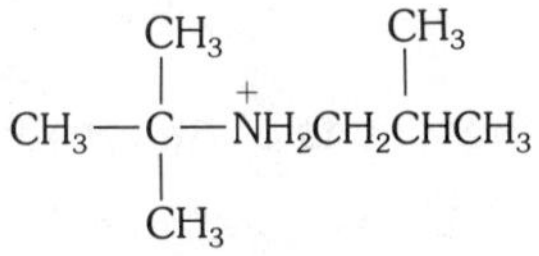 $CH_3-C-NH_2CH_2CHCH_3$

17.6 What are the common names of the following compounds?
(a) $[(CH_3)_3C]_3N$

(b)

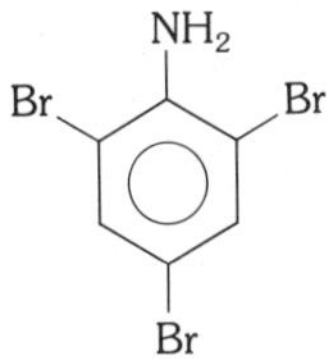

(c) $(CH_3)_2CHNHCH_2CH_2CH_2CH_3$
(d) $(CH_3CH_2)_3\overset{+}{N}H\ Cl^-$

Chemical Properties of Amines and Amine Salts

17.7 Complete the following reaction sequences by writing the structures of the organic products. If no reaction occurs, write "no reaction."
(a) $CH_3NH_2 + NaOH(aq) \rightarrow$

(b)

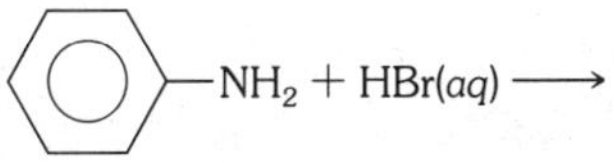

(c) $NH_4Cl(aq) + NaOH(aq) \rightarrow$

(d)

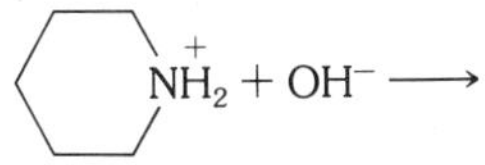

(e) $^+NH_3CH_3 + OH^- \rightarrow$

(f)

17.8 Write the structures of the organic products that form in each situation. Assume that all reactions occur at room temperature. (Some of the named compounds are described in Review Exercises 17.3 and 17.4.) If no reaction occurs, write "no reaction."

(a)

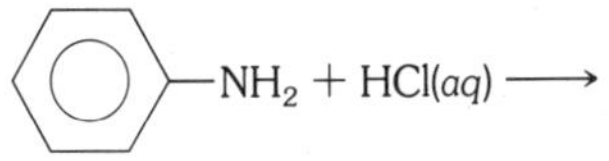

(b)

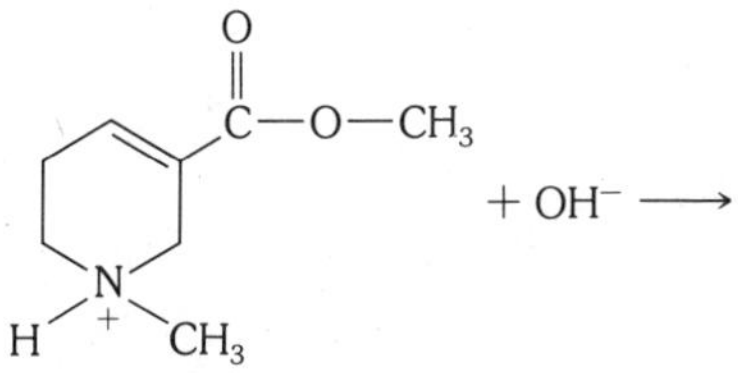

Protonated form of arecoline

(c)

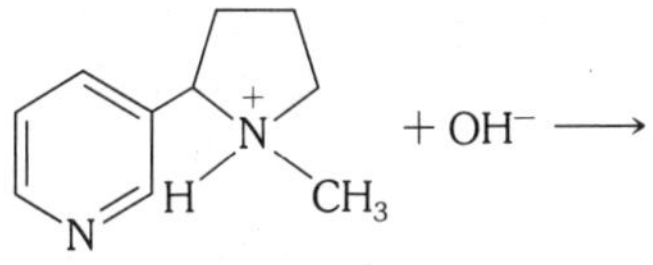

Protonated form of nicotine

(d)

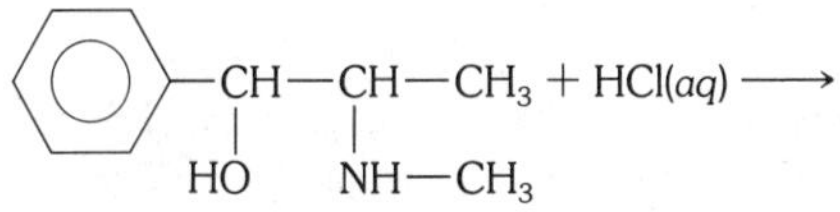

Ephedrine

(e)

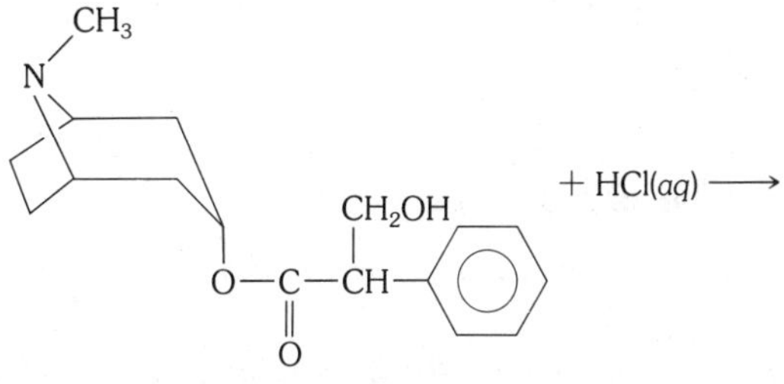

Hyoscamine

17.9 Which is the stronger base, A or B? Explain.

$$CH_3CH_2\overset{\overset{\displaystyle O}{\|}}{C}NHCH_3 \qquad CH_3\overset{\overset{\displaystyle O}{\|}}{C}H\overset{\overset{\displaystyle O}{\|}}{C}CH_3$$
$$\underset{\displaystyle NH_2}{|}$$

A   B

17.10 Which is the stronger proton acceptor, A or B? Explain.

$$CH_3-\overset{\overset{\displaystyle CH_3}{|}}{N}-CH_3 \qquad CH_3-\overset{\overset{\displaystyle CH_3}{|}}{\overset{+}{N}}-CH_3$$
$$\underset{\displaystyle CH_3}{|}$$

A   B

Names and Structures of Amides

17.11 What are the IUPAC names of the following compounds?

(a) $CH_3-\overset{\overset{\displaystyle CH_3}{|}}{\underset{\underset{\displaystyle CH_3}{|}}{C}}-\overset{\overset{\displaystyle O}{\|}}{C}-NH_2$

(b) $CH_3\overset{\overset{\displaystyle CH_3}{|}}{C}H\overset{}{C}H-\overset{\overset{\displaystyle O}{\|}}{C}-NH_2$
$$\underset{\displaystyle Br}{|}$$

17.12 If the common name of hexanoic acid is caproic acid, what is the common name of its simple amide?

17.13 Write the structure of *N,N*-dimethylbenzamide.

17.14 What is the structure of butanediamide?

17.15 What is the structure of lysergic acid? The structure of its *N,N*-diethylamide was given in Review Exercise 17.4, part (d).

17.16 What is the structure of the amide between ephedrine, given in part (d) of Review Exercise 17.3, and acetic acid?

Synthesis of Amides

17.17 Write the equations for two ways to make acetamide using ammonia as one reactant.

17.18 What are two different ways to make *N*-methylacetamide if methylamine is one reactant? Write the equations.

17.19 Examine the following acyl group transfer reaction.

$$NH_2CH_2\overset{\overset{\displaystyle O}{\|}}{C}NH\overset{}{C}H\overset{\overset{\displaystyle O}{\|}}{C}-\boxed{\substack{\text{Carrier}\\\text{molecule}}}_1 + NH_2\overset{}{C}H\overset{\overset{\displaystyle O}{\|}}{C}-\boxed{\substack{\text{Carrier}\\\text{molecule}}}_2 \longrightarrow$$
$$\underset{\displaystyle CH_3}{|} \qquad\qquad \underset{\underset{\displaystyle CH_3}{|}}{CHCH_3}$$

$$NH_2CH_2\overset{\overset{\displaystyle O}{\|}}{C}NH\overset{}{C}H\overset{\overset{\displaystyle O}{\|}}{C}NH\overset{}{C}H\overset{\overset{\displaystyle O}{\|}}{C}-\boxed{\substack{\text{Carrier}\\\text{molecule}}}_2 + \boxed{\substack{\text{Carrier}\\\text{molecule}}}_1$$
$$\underset{\displaystyle CH_3}{|} \qquad \underset{\underset{\displaystyle CH_3}{|}}{CHCH_3}$$

(a) Which specific acyl group transferred? (Write its structure.)

(b) How many amide bonds are in the product?

17.20 If the following anhydride were mixed with ammonia, what organic products that are not salts can form? Write their structures.

$$CH_3-\overset{\overset{\displaystyle O}{\|}}{C}-O-\overset{\overset{\displaystyle O}{\|}}{C}-CH_2CH_3$$

Reactions of Amides

17.21 What are the products of the hydrolysis of the following compounds? (If no hydrolysis occurs, state so.)

(a) $CH_3NH\overset{\overset{\displaystyle O}{\|}}{C}CH_2CH_3$

(b) $CH_3NHCH_2\overset{\overset{\displaystyle O}{\|}}{C}CH_3$

(c) $CH_3\overset{\overset{\displaystyle O}{\|}}{C}NCH_3$
$$\underset{\displaystyle CH_3}{|}$$

(d) $CH_3\overset{}{C}H\overset{\overset{\displaystyle O}{\|}}{C}NHCH_3$
$$\underset{\displaystyle CH_3}{|}$$

17.22 Write the structures of the products of the hydrolysis of the following compounds. If no reaction occurs, state so. If more than one bond is subject to hydrolysis, be sure to hydrolyze all of them.

(a) $NH_2CH_2\overset{\overset{\displaystyle O}{\|}}{C}NHCH_2\overset{\overset{\displaystyle O}{\|}}{C}OH$

(b) $CH_3\overset{\overset{\displaystyle CH_3}{|}}{N}CH_2\overset{\overset{\displaystyle O}{\|}}{C}NHCH_2CH_2\overset{\overset{\displaystyle O}{\|}}{C}NHCH_3$

(c)

$$H_2C-NH$$
$$H_2C \qquad C=O$$
$$H_2C-CH_2$$

(d)

$$\underset{\overset{|}{CH_3}}{CH_3N}CH_2CH_2NH\overset{\overset{O}{\|}}{C}CH_2CH_2\overset{\overset{O}{\|}}{C}NHCH_3$$

Review of Organic Reactions

17.23 What are all of the functional groups we have studied that can be changed by each of the following reactants? Write the equations for the reactions, using general symbols such as ROH or RCO₂H and so forth to illustrate these reactions, and name the organic families to which the reactants and products belong.
(a) Water, either with an acid or an enzyme catalyst.
(b) Hydrogen (or a hydride ion donor) and any needed catalysts and special conditions.
(c) An oxidizing agent represented by (O), such as $Cr_2O_7{}^{2-}$ or $MnO_4{}^-$, but not ozone or oxygen used in combustion.

17.24 We have described three functional groups that typify those involved in the chemistry of the digestion of carbohydrates, fats and oils, and proteins. What are the names of these groups and to which type of food does each belong?

17.25 Write the structures of the organic products that would form in the following situations. If no reaction occurs, state so. These constitute a review of nearly all of the organic reactions we have studied, beginning with Chapter 12.

(a) $CH_3CH_2CO_2H + NaOH(aq) \rightarrow$

(b) $CH_3-\hspace{-4pt}\bigcirc\hspace{-4pt}- + MnO_4{}^-(aq) \longrightarrow$

(c) $CH_3\overset{\overset{O}{\|}}{C}CH_3 + H_2 \xrightarrow[\text{Heat, pressure}]{Ni}$

(d) $CH_3O\overset{\overset{O}{\|}}{C}CH_3 + H_2O \xrightarrow[\text{Heat}]{H^+}$

(e) $CH_3-\overset{\overset{O}{\|}}{C}-Cl + 2CH_3\underset{\overset{|}{NH_2}}{C}HCH_3 \rightarrow$

(f) $CH_3(CH_2)_8CH_3 + NaOH(aq) \rightarrow$

(g) $CH_3CH_2\underset{\overset{|}{OCH_3}}{C}HOCH_3 + H_2O \xrightarrow[\text{Heat}]{H^+}$

(h) $\bigcirc\hspace{-4pt}-\overset{\overset{O}{\|}}{C}-H + (O) \longrightarrow$

(i) $CH_3CH_2\overset{\overset{O}{\|}}{C}-O-CH_2CH_3 + NaOH(aq) \xrightarrow[\text{Heat}]{}$

(j) $CH_3-\overset{\overset{O}{\|}}{C}-H + 2CH_3OH \xrightarrow[\text{Heat}]{H^+}$

(k) $CH_3\overset{\overset{O}{\|}}{C}CH_3 + (O) \rightarrow$

(l) $CH_3-O-\underset{\overset{|}{OCH_3}}{C}HCH_2CH_2CH_3 + NaOH(aq) \rightarrow$

(m) $CH_3CH_2CO_2H + CH_3OH \xrightarrow[\text{Heat}]{H^+}$

(n) $CH_3CH_2CH_2NH_2 + HCl(aq) \xrightarrow[25\ °C]{}$

(o) $H\overset{\overset{O}{\|}}{C}NH_2 + H_2O \xrightarrow[\text{Heat}]{}$

(p) $CH_3SH + (O) \rightarrow$

17.26 What are the structures of the organic products that form in the following situations? (If there is no reaction, state so.) These reactions review most of the chemical properties of functional groups we have studied, beginning with Chapter 12.

(a) $\bigcirc\hspace{-4pt}-\overset{\overset{O}{\|}}{C}-O-CH_2CH_3 + NaOH(aq) \xrightarrow[\text{Heat}]{}$

(b) $\bigcirc\hspace{-4pt}-\overset{\overset{O}{\|}}{C}-H + 2CH_3OH \xrightarrow[\text{Heat}]{H^+}$

(c) $CH_3(CH_2)_3CH_3 + Cr_2O_7{}^{2-} \rightarrow$

(d) $CH_3\overset{\overset{O}{\|}}{C}CH_2CH_3 + (O) \rightarrow$

(e) $HO\overset{\overset{O}{\|}}{C}CH_2CH_2CH_3 + NaOH(aq) \xrightarrow[25\ °C]{}$

(f) $CH_3-\underset{\overset{|}{CH_3}}{\overset{\overset{|}{OCH_2CH_3}}{C}}-O-CH_2CH_3 + H_2O \xrightarrow{H^+}$

(g) $\pentagon + NaOH(aq) \longrightarrow$

(h) $CH_3-O-\overset{\overset{O}{\|}}{C}CH_2CH_2\overset{\overset{O}{\|}}{C}-O-CH_3 + H_2O \xrightarrow{\text{Enzyme}}$
(Excess)

(i) $NH_2CH_2CH_2CH_2NH_2 + HCl(aq) \xrightarrow[25\,°C]{}$

(Excess)

(j) $CH_3-O-CH_2CH_2\overset{\displaystyle O}{\overset{\|}{C}}-H + (O) \rightarrow$

(k) $CH_3CH_2OH + CH_3(CH_2)_4CO_2H \xrightarrow[Heat]{H^+}$

(l) $CH_3CH_2-O-\overset{\displaystyle CH_3}{\overset{|}{CH}}-O-CH_2CH_3 + H_2O \xrightarrow[Heat]{H^+}$

(m) $CH_3CH_2\overset{\displaystyle O}{\overset{\|}{C}}CH_2CH_3 + H_2 \xrightarrow[Heat,\ pressure]{Ni}$

(n) $2CH_3NH_2 + CH_3\overset{\displaystyle O}{\overset{\|}{C}}-O-\overset{\displaystyle O}{\overset{\|}{C}}CH_3 \rightarrow$

(o) $CH_3NH\overset{\displaystyle O}{\overset{\|}{C}}CH_2CH_2\overset{\displaystyle O}{\overset{\|}{C}}NHCH_3 + H_2O \xrightarrow{Enzyme}$

(Excess)

(p) $CH_3S-SCH_3 + 2(H) \rightarrow$

Chapter 18
Optical Isomerism

Part of the light reflecting from a glass surface is polarized, and it obscures what's behind the glass, as seen in the top photo. When Polaroid film filters polarized light, the view is much clearer, as seen in the bottom photo. An unusual application of polarized light is described in this chapter.

18.1   TYPES OF ISOMERISM

Structural isomers and stereoisomers are the two broad classes of isomers.

Compounds that have identical molecular formulas can be different in two general ways, as structural isomers or as stereoisomers. **Structural isomers** generally have major differences in their molecular frameworks. For example, butane and isobutane, both C_4H_{10}, have different chains. Ethanol and dimethyl ether, both C_2H_6O, have different functional groups. Each of these pairs illustrates structural isomerism.

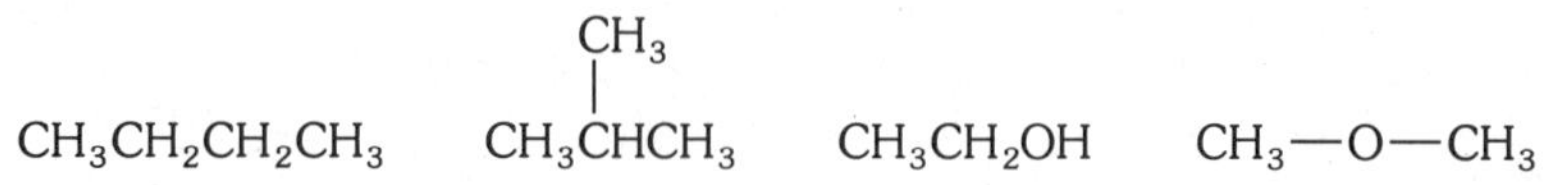

$$CH_3CH_2CH_2CH_3 \qquad \underset{\overset{|}{CH_3}}{CH_3CHCH_3} \qquad CH_3CH_2OH \qquad CH_3\!-\!O\!-\!CH_3$$

Butane   Isobutane   Ethanol   Dimethyl ether

Stereoisomers differ only in some three-dimensional or geometric way. They have the same basic atom-to-atom connections, the same skeletons, and identical functional groups, but yet they still are different compounds. There are two kinds of stereoisomers, and we have already studied one — the geometric or cis-trans isomers, illustrated by *cis-* and *trans-2*-butene.

cis-2-Butene   *trans*-2-Butene

In this chapter we'll study the second kind of stereoisomer, called optical isomers. These are isomers with the most subtle difference in structure but with some very dramatic differences in chemical behavior at the molecular level of life. One such difference is illustrated by the bitter-sweet story of asparagine.

Asparagine is a building block for making proteins in the body.

Asparagine is a white solid that was first isolated in 1806 from the juice of asparagus. When it's from this source, asparagine has a bitter taste. Its molecular formula is $C_4H_8N_2O_3$, and its structure — at least its pattern of atom-to-atom connections — is given by structure **1.**

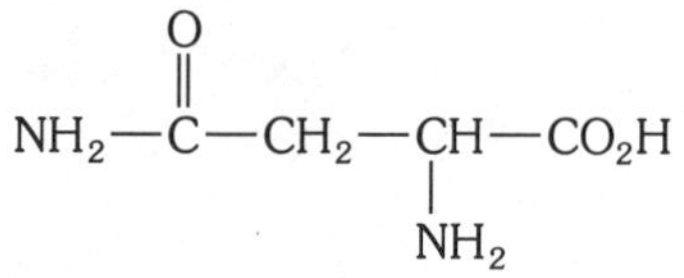

$$NH_2\!-\!\overset{\overset{\displaystyle O}{\|}}{C}\!-\!CH_2\!-\!\underset{\underset{\displaystyle NH_2}{|}}{CH}\!-\!CO_2H$$

1 Asparagine

Vetch is a member of a genus of herbs, some of which are useful as fodder for cattle.

In 1886, a chemist isolated from sprouting vetch a white substance with the same molecular formula *and structure*, but it has a sweet taste. To have names for these, the one isolated from asparagus is now called L-asparagine, and the one from vetch sprouts is called D-asparagine.

Chemists have long worked with the principle of *one-substance – one-structure*. If two samples of matter have identical physical and chemical properties, they have to be identical at the level of their individual formula units. If two samples differ in even one way in their fundamental properties, there has to be at least one difference in the way that their molecules are put together. Taste is a chemical sense, so the two samples of asparagine do have one difference in chemical property. Under the one-substance – one-structure law, the molecules of D- and L-asparagine must be structurally different in at least one way. This difference is very subtle, and it arises from a peculiar lack of symmetry in their molecules that makes possible two different relative configurations of their molecular parts.

The letters D and L will acquire more specific meaning in the next chapter. Consider them to be only labels now.

18.2 MOLECULAR CHIRALITY

The molecules of many substances have a handedness like that of the left and the right hands.

Two partial ball-and-stick molecular models of asparagine are shown in part (a) of Figure 18.1. Examine each to make sure that it faithfully represents structure **1,** the structure of asparagine. To aid in the following discussion, these models have been simplified as shown in part (b) of the figure. Notice again how alike the two molecular structures are. Notice particularly that the same four groups are attached to a central carbon atom, and there is no cis-trans kind of difference between them. Yet the fact remains that one represents a bitter-tasting compound, and the other is of a sweet-tasting compound. In some way therefore these two similiar structures have to be different, and the difference isn't just something that can be removed by rotating groups around single bonds. We'll learn about the difference in this section.

Superimposability.   To understand how the two asparagine structures are different, we first have to learn how to tell if two structures are identical. For two structures to be identical, it must be possible to superimpose them. This is a manipulation you can only do to completion in

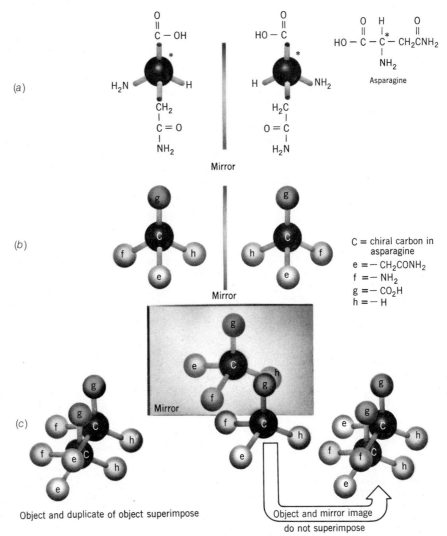

FIGURE 18.1
The two stereoisomers of asparagine.
(a) Two ways of joining the four groups to the carbon marked by the asterisk.
(b) Simplified representations of the models in part (a). (c) What is in front of the mirror is identical with the model on the left in part (b). What is seen in the mirror as the image is identical with the model on the right in part (b). You'll mentally have to spin the mirror image 120° counterclockwise about the bond from C to g to see that they are the same. The object and its mirror image do not superimpose, so they can't be identical. Instead, they are enantiomers.

C = chiral carbon in asparagine
e = $- CH_2CONH_2$
f = $- NH_2$
g = $- CO_2H$
h = $- H$

Object and duplicate of object superimpose

Object and mirror image do not superimpose

In some references you'll see the word *superposition* used for *superimposition*.

your mind, but working with molecular models is a great help. **Superimposition** is mentally blending one molecular model with another so that the two would coincide in every atom and bond simultaneously if the operation could, in fact, be completed. The left-hand side of part *(c)* of Figure 18.1 shows how it works when two identical models of one of the asparagines is used (and it also shows why the full blending can only be done in the mind). When working with actual models, it's fair to twist parts about single bonds as much as the models allow in order to try to get superimposition to work, but it's not legal to break any bonds and remake them in other arrangements. To repeat, the fundamental criterion of identity of two structures is that they must pass the test of superposition. The models of the two different asparagines, redrawn from part *(b)* to the right side of part *(c)* of Figure 18.1, fail this test. The next question is, "Just exactly how do they differ?"

There is one significant way in which the two asparagine models are related — one is the mirror image of the other. By this we mean that if you held up either model to a mirror and made an exact duplicate *of what you see in the mirror,* this second model would, in fact, correspond to the *other* asparagine. Each asparagine model is the mirror image of the other, and yet the model and its mirror image do not superimpose. Substances, like the two asparagines, whose molecules are related as object to mirror image but that cannot be superimposed are called **enantiomers** of each other.

Enantiomers are just special kinds of isomers. Like all isomers, they have identical molecular formulas, but unlike structural isomers they also have identical atom-to-atom sequences. The fundamental difference between enantiomers lies only in their configurations. As we said, isomers that differ only in a configurational way are called stereoisomers, so enantiomers are examples of stereoisomers. To distinguish them from cis-trans isomers, which are also members of the broad family of stereoisomers, remember what makes cis-trans isomers different is some lock against free rotation — either a double bond or a ring. Stereoisomers that are not geometric isomers are called **optical isomers.** Thus the two main families of stereoisomers are geometric isomers and optical isomers. (See also Figure 18.2.) Enantiomers constitute one way in which optical isomers can occur — the only way that we'll study apart from a Special Topic later in the chapter.

Lack of free rotation is not the problem with the asparagine enantiomers. Instead, *molecules of enantiomers have opposite configurations,* and our next task is to see what this means.

Configurational Differences Between Enantiomers.   In Figure 18.3, we have redrawn the abbreviated versions of the two asparagine enantiomers that were given in part *(b)* of Figure 18.1. Notice again that these two models are related as an object to its mirror image. (Imagine that the mirror stands between the two and is perpendicular to the page.) You're now going to let your eyes make a special scanning trip around each molecule. Imagine that the bond from C to H in each model is the steering column of the steering wheel of a car and that the remaining three groups — **e, f,** and **g** — are distributed around the perimeter of the steering wheel. Taking each model in turn, the scanning trip we have in mind is to move your eyes from **g** to **e** and then to **f.** When you do this with the asparagine model on the left in

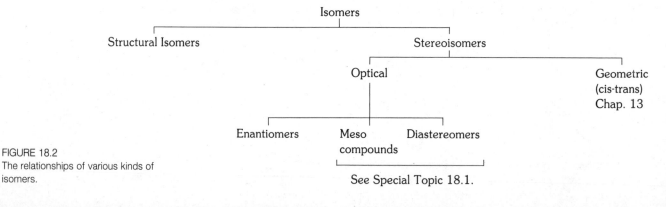

FIGURE 18.2
The relationships of various kinds of isomers.

FIGURE 18.3
When the two stereoisomers of asparagine are viewed down the same axis, the C-H bond axis, the remaining three groups at the central carbon atom have opposite configurations.

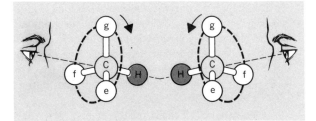

Figure 18.3, your eyes move clockwise. But to make the identical trip — **g** to **e** to **f** — in the model on the right, your eyes move counterclockwise. This is what we mean when we say that the molecules of the two asparagines have opposite configurations. The four groups in one asparagine that are attached to the central carbon atom are the same four groups as in the other asparagine, but they are configured oppositely in space.

The Chirality of Enantiomers.   We have to remind ourselves here that any object has a mirror image. Spheres, cubes, broom handles, water glasses, and so on, can all be reflected in a mirror. It's only when the model of the object and the model of the mirror image can't be superimposed that we call the two enantiomers. Your two hands are like a pair of enantiomers, if you disregard small differences such as wrinkles, scars, rings, and fingerprints. Now place your left hand in front of a mirror near its edge. Place your other hand just off the edge of the mirror and notice that the reflection of your left hand in the mirror is the same as your right hand (disregarding, as we said, the small differences). Your right hand is the mirror image reflection of your left hand. Next, try to superimpose the two hands. Because the mirror image of your left hand is your real right hand, use your two hands to see if they superimpose. If you put them palm to palm, it seems as if all the fingers do superimpose. But remember, you have to carry this blending through to completion (in your mind), and when you do, the palm of one hand comes out on the back side of the other. The palms won't superimpose when the fingers seem to. And if you try to get the palms to come out right, then the fingers come out all wrong. Left and right hands, although related as object to mirror image, don't superimpose. They are related as enantiomers.

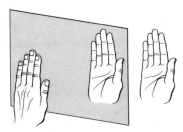

The image of the left hand is just like the right hand in the relative orientations of the fingers, thumb, and palm.

The two hands are related as an object to its mirror image, but they can't be superimposed.

Notice, now, how the two hands have opposite configurations. Look at them from the same angle (as we did with the asparagine models in Figure 18.3), say, both palms facing you. To make the trip from thumb to little finger, touching every other finger on the way, you have to go one direction for one hand and in exactly the opposite direction with the other. Thus the thumb and the four fingers have opposite configurations in the two hands.

The experiment with the hands has been used for decades to teach about configurational differences of enantiomers, and this is why we say that molecules of different enantiomers have different *handedness*. The technical term for this configurational property is **chirality** (from the Greek word for hand, which is *cheir*). We say that the two enantiomers of a pair have opposite chirality, meaning opposite handedness. We also say that the molecules of any given enantiomer are **chiral** — they possess handedness or chirality.

The opposite of chiral is **achiral.** An achiral molecule is one that is symmetrical enough so that its model and the model of its mirror image do superimpose. The methane molecule is achiral, for example. Some examples of larger achiral objects include a cube, a sphere, a broom handle, and a water glass.

Some Differences that Chirality Makes.   One asparagine enantiomer tastes sweet and the other tastes bitter — surely a large difference. We have already mentioned that taste is a chemical sense. The details are far from fully known, but the initiation of the sensation of tasting one flavor or another probably begins with a chemical reaction that is catalyzed by an enzyme. The enzymes involved, *like all enzymes,* themselves consist of very large, chiral molecules with molecular surfaces that are different for each enzyme.

The substance whose reaction an enzyme catalyzes is called the **substrate** for that enzyme. An enzyme works by letting a molecule of the substrate come and temporarily fit into the contours on the enzyme's surface. This idea of fitting can be illustrated by a return to our hands, only now we'll add gloves. Gloves, like hands, are chiral, and a glove fits well only to its matching hand. The left hand fits well into the left glove, not the right glove. Now suppose that an enzyme responsible ultimately for the sensation of a sweet taste is like a glove for the left hand. This means that only those substrate molecules that have the matching handedness can interact with this enzyme. Substrate molecules of the opposite handedness can't fit to this enzyme.

We can now shift back from this analogy of hands and gloves to chiral molecules with the aid of Figure 18.4. To make it easier, simple geometric forms have been used to create two enantiomers, and indentations that match these forms are part of the enzyme surface. One enantiomer can fit to the enzyme surface, but no matter how you turn the model of the other enantiomer you can't get it to fit to the same enzyme surface. There is, evidently, a different enzyme whose surface chirality matches the other asparagine enantiomer that lets us know that this other enantiomer has a different taste. The phenomenon of chirality is absolutely central to this difference. And the different chemical properties that relate to the taste of asparagine are illustrated in countless ways by events at the molecular level of life. We'll see time after time that differences in geometry and configuration are fully as important as functional groups to the chemical reactions of life.

If the enzyme molecule weren't chiral, it couldn't discriminate between enantiomers. In fact, *enantiomers have identical chemical properties toward all reactants whose molecules are achiral* — reactants such as H_2O, NaOH, HCl, Cl_2, NH_3, and H_2. Using our analogy, a broom handle, which isn't chiral, fits just as easily to the left hand as to the right. It can't discriminate between the hands. In like manner, the molecules of an achiral reactant can't tell the difference between molecules of enantiomers.

To summarize, enantiomers react differently toward reactants whose molecules are chiral but identically toward achiral reactants. What about physical properties? With one exception (to be discussed shortly), enantiomers have identical physical properties — identical melting points, boiling points, densities, and solubilities in common (i.e., achiral) solvents such as water, ether, gasoline, or ethanol. This *must* be so, because the molecules of two enantiomers *must* have identical molecular polarities — they have, after all, identical intranuclear distances and bond angles. Returning once again to our analogy with the hands, you can see that the distance, say, from the tip of the thumb to the tip of the little finger is the same in both hands. You can pick any other such intra-hand distance that you please, and it is the same in both hands. Similarly, the angle between any two fingers is the same in both hands. In like manner, if we pick any distance between atoms or any bond angle in one asparagine enantiomer, we will find it to be the same in the other. When all these distances and angles are identical in the two enantiomers, their molecules as a whole must have identical polarities. This is responsible for the identical physical properties that we mentioned.

FIGURE 18.4
Since an enzyme is chiral, it can accept substrate molecules of only one of a pair of enantiomers. To illustrate this difference, we have used simple geometric forms. On the left, the enzyme can acept as a substrate the molecule of one enantiomer. On the right, the same enzyme can't accept a molecule of the other enantiomer, because the shapes don't match.

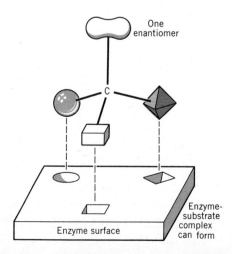
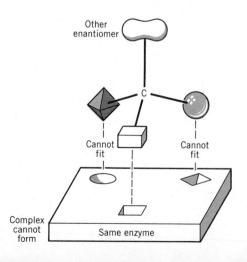

Deciding if Molecules Are Chiral.   Because enantiomers have such different chemical properties at the molecular level of life, where chiral enzymes are the catalysts everywhere, it's important that we be able to recognize when a potential substrate is chiral. In all of the examples of enantiomers that we'll encounter, their molecules always have at least one carbon to which *four different groups* are attached. A carbon that holds four different atoms or groups is called a **chiral carbon,** because its presence usually (although not always) guarantees that a molecule with such a carbon is itself chiral.

The asparagine molecule has one (and just one) chiral carbon — and when a molecule has only one chiral carbon, we can be absolutely certain that the molecule is itself chiral. Here, then, is one way to predict if a substance consists of chiral molecules — we look for a chiral carbon. If we find one, we can be certain that the molecules are chiral. If we find more than one, we have to be careful, but this brings us to a complication that we'll use Special Topic 18.1 to examine. Probably 99.99% of all examples of substances that have two or more chiral carbons also have chiral molecules, but a few exceptions do exist where the molecule is achiral despite having chiral carbon atoms. (These are the *meso compounds* discussed in Special Topic 18.1.)

There are some substances — both organic and inorganic — whose molecules have no chiral carbon atoms and yet are chiral molecules, but we won't encounter any of these in our study.

EXAMPLE 18.1   IDENTIFYING CHIRAL CARBONS

Problem: Amphetamine, which we introduced in Example 17.2, exists as a pair of enantiomers. One of them has its own name — Dexedrine. Find the chiral carbon in amphetamine, and list the four groups attached to it.

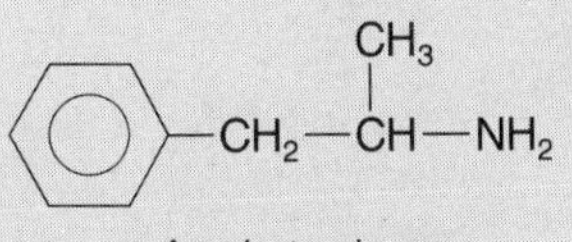

Amphetamine

Solution: There is one chiral carbon in amphetamine, the one labeled with the asterisk:

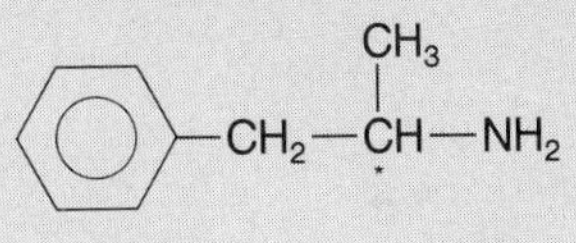

The four groups are:

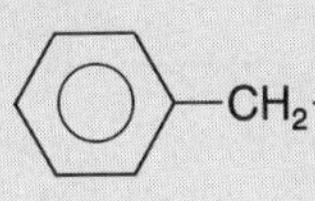

PRACTICE EXERCISE 1   Place an asterisk next to each chiral carbon in the following structures.

(a) 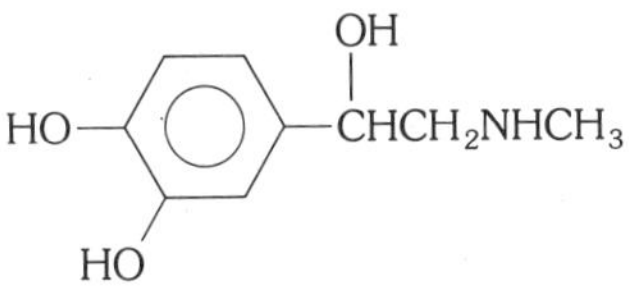
Epinephrine, a hormone (see special Topic 17.1)

(b) CH_3CHCO_2H
      |
     OH
Lactic acid, the sour constituent in sour milk

(c) $CH_3CHCHCO_2^-$
      | |
    HO NH_3^+
Threonine, one of the amino acid building blocks of proteins

(d) $HOCH_2CH—CH—CH—CH{=}O$
              | | |
             OH OH OH
Ribose, a sugar unit at the molecular level of heredity

When a molecule has two or more chiral carbons, as in parts (c) and (d) of Practice Exercise 1, then it becomes useful to judge if the chiral carbons are *different*. When used in this context, *different* means that the sets of four atoms or groups at the various chiral carbons have at least one difference. Two chiral carbons are said to be *different* if the set of four groups at one is not duplicated by the set at the other. Whenever the chiral carbons in a molecule are different in this sense — as they were in parts (c) and (d) of Practice Exercise 1 — then the substance can exist in the forms of 2^n optical isomers, where n is the number of different chiral carbons. These 2^n optical isomers occur as half as many *pairs* of enantiomers. We'll see this illustrated in the next worked example.

EXAMPLE 18.2   **JUDGING IF CHIRAL CARBONS ARE DIFFERENT AND CALCULATING THE NUMBER OF OPTICAL ISOMERS.**

Problem: The threonine molecule, part (c) of Practice Exercise 1, has two chiral carbons, labeled here by asterisks.

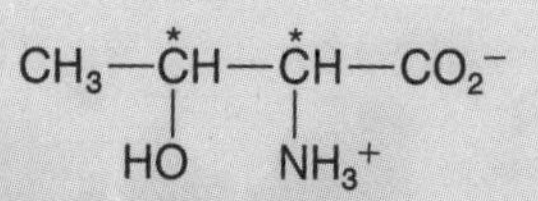

$$CH_3 - \overset{*}{C}H - \overset{*}{C}H - CO_2^-$$
$$\qquad\quad | \qquad\quad |$$
$$\qquad\quad HO \qquad NH_3^+$$

Threonine

Are these chiral carbons different? If so, how many optical isomers of threonine are there?

Solution: Make a list of the sets of four different groups at each chiral carbon and compare the sets. If they're not identical in every respect, then the two chiral carbons are different.

At one chiral carbon:
CH_3, H, OH, $-CH-CO_2^-$
$\qquad\qquad\qquad\quad |$
$\qquad\qquad\qquad\quad NH_3^+$

At the other chiral carbon:
CH_3-CH-, H, CO_2^-, NH_3^+
$\qquad\quad |$
$\qquad\quad HO$

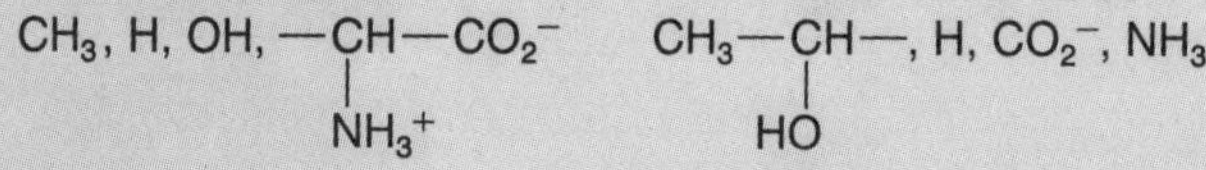

The sets are obviously different, so $n = 2$ is the number of different chiral carbons in a threonine molecule. Therefore, $2^n = 2^2 = 4$, the number of optical isomers of threonine. These occur as half of 4 or 2 pairs of enantiomers. The complete set has the following structures. One pair of enantiomers is on the left. Just imagine that the mirror is between them and is perpendicular to the page. The other pair of enantiomers is on the right.

Only L-threonine works as a building block for making proteins in the body. There is no enzyme that can accept any of the other optical isomers as substrates.

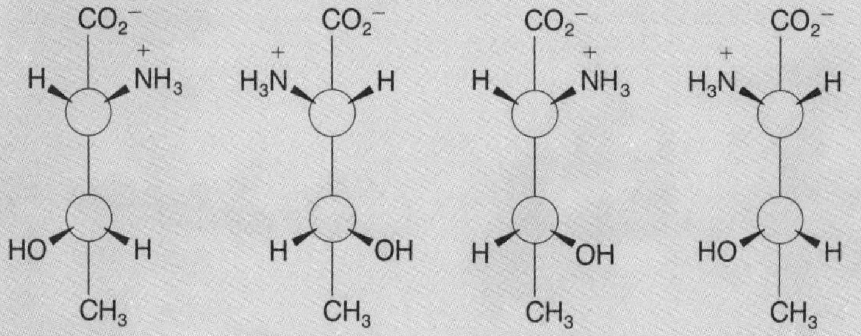

The meaning of the experimental values given for the symbol $[\alpha]_D^{26}$ is explained in the next section.

D-Threonine
$[\alpha]_D^{26} +28.3°$

L-Threonine
$[\alpha]_D^{26} -28.3°$

D-Allothreonine
$[\alpha]_D^{26} -9.6°$

L-Allothreonine
$[\alpha]_D^{26} +9.6°$

PRACTICE EXERCISE 2   Examine the structure of ribose that was given in part (d) of Practice Exercise 1. (a) How many different chiral carbons does it have? (b) How many optical isomers are there of this structure? (Only one is actually the ribose that can be used by the body.) (c) How many pairs of enantiomers correspond to this structure?

In Example 18.2, the structures of the four optical isomers of threonine were given. There were four of them, and they occurred as two pairs of enantiomers. D-threonine is the enantiomer of L-threonine, and D-allothreonine is the enantiomer of L-allothreonine. Thus all four belong to the same set of stereoisomers. The members of such a set that are *not* related as enantiomers are called **diastereomers.** Thus either D- or L-threonine is a diastereomer of either D- or L-allothreonine.

Unlike enantiomers, diastereomers do have many differences in physical properties. This is because the set of intramolecular distances and bond angles in one diastereomer isn't exactly duplicated in any other diastereomer. Hence, molecules of one diastereomer should be expected to have at least slightly different polarities than those of any other diastereomer. You can see an illustration of one such difference in physical property in the values of the specific rotation of the diastereomers of the threonine/allothreonine system.

Another kind of optical isomer occurs among the stereoisomers of tartaric acid. Tartaric acid is 2,3-dihydroxybutanedioic acid, and it has two chiral carbons. However, these two chiral carbons are identical.

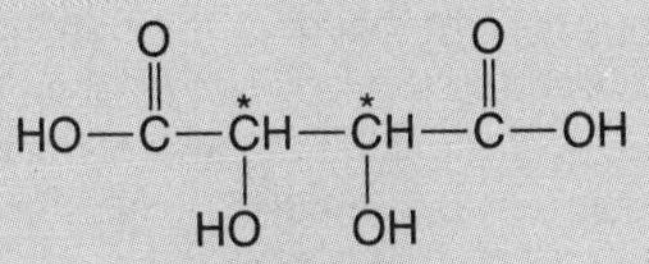

Tartaric acid

Each has the same set of attached atoms or groups as the other: CO_2H, H, HO, and $HOCHCO_2H$. There is no simple equation for calculating how many optical isomers exist of a compound that has two or more identical chiral carbons. Tartaric acid happens to have three optical isomers:

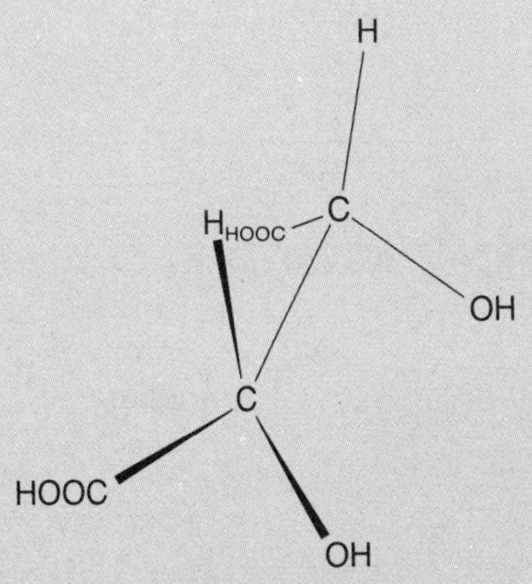

meso-Tartaric acid
$[\alpha]_D^{20} = 0$
m.p. 140 °C

You can see that two are related as enantiomers, because their models are related as an object to its mirror image and they can't be superimposed. There is no enantiomer for the third stereoisomer, the one labeled *meso*-tartaric acid. If you made a mirror image of this model (can you draw it?), and rotated it 180° about the correct axis (can you find it?), the model and its mirror image would superimpose. So here is an example of an optical isomer with chiral carbons that is optically inactive.

Remember that the molecule as a *whole* has to be chiral for the compound to be able to affect plane-polarized light. We still call *meso*-tartaric acid an optical isomer even though it is itself optically inactive, because it belongs to a set of stereoisomers that does include optically active forms. Optical isomers that are achiral are classified as **meso compounds,** and this is why the achiral form of tartaric acid is called *meso*-tartaric acid.

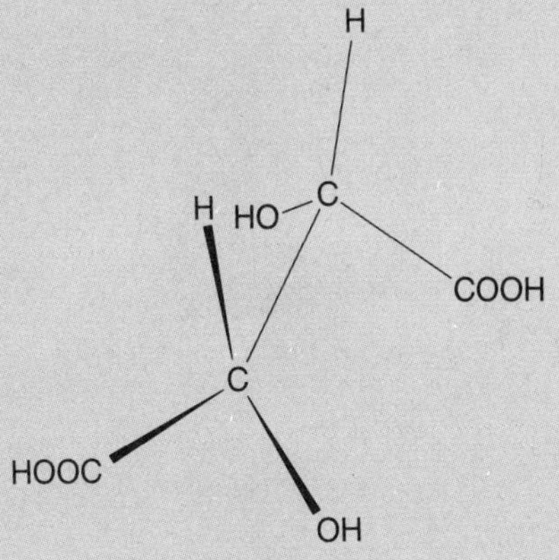

D-Tartaric acid
$[\alpha]_D^{20} = -11.98°$
m.p. 170 °C

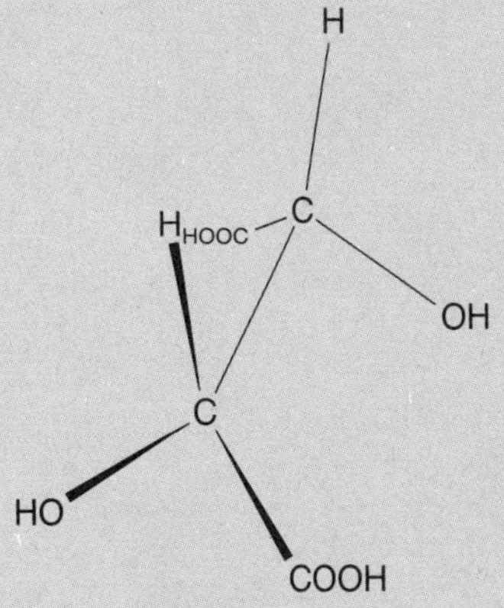

L-Tartaric acid
$[\alpha]_D^{20} = +11.98°$
m.p. 170 °C

PRACTICE EXERCISE 3   Write the structure of 2,3-butanediol and place an asterisk by each carbon that is chiral. Are they *different* chiral carbons?

18.3   OPTICAL ACTIVITY

Enantiomers affect polarized light in equal and opposite ways when they are compared under identical conditions.

We mentioned earlier that there is one physical property in which enantiomers differ, and to describe it we first have to learn something about polarized light.

Polarized Light.   Light is electromagnetic radiation in which the intensities of the electric and magnetic fields set up by the light source oscillate in a regular way. In ordinary light, these oscillations occur equally in all directions about the line that defines the path of the light ray.

Certain materials, such as the polarizing film used to make the lenses of Polaroid sunglasses, affect ordinary light in a special way. Polarizing film interacts with the oscillating electrical field of any light passing through it to make this field oscillate in just one plane. The light that emerges is now **plane-polarized light.** (See Figures 18.5 and Figure 18.6a.) If we look at some object through polarizing film and then place a second film in front of the first, we can rotate one film until the object no longer can be seen, as shown in Figure 18.6b and Figure 18.7. If we now rotate one film by 90°, we'll see the object at maximum brightness again. The first film seems to act as a lattice fence, forcing any light that goes through it to vibrate only in the direction allowed by the long spaces between the slats. This light then moves on to the molecular slats of the second film. If its slats are perpendicular to those of the first, the light has no freedom to oscillate, and it cannot get through. At intermediate angles, fractional amounts of light can go through the second polarizing film. Only when the slats of the two films are *parallel* to each other can the light leaving the first film slip easily through the second.

> You can try this out using two Polaroid sunglass lenses. The lenses of these glasses reduce glare by cutting out the plane-polarized light produced when sunlight reflects from a plane surface such as a road, a snowfield, or a lake.

Optical Rotation.   When a solution of D-asparagine in water is placed in the path of plane-polarized light, the plane of polarization of the light is rotated. Any substance that can rotate the plane of plane polarized light is said to be **optically active.** We have not actually explained how this phenomenon happens — we are unable to do so. We have only reported that it does happen. It's because of this physical property that the term *optical isomerism* was first applied to compounds that were found to be optically active.

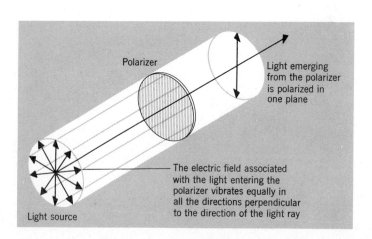

FIGURE 18.5
When light passes through polarizing film, it becomes polarized light.

FIGURE 18.6
The principal working parts of a polarimeter and how optical rotation is measured.

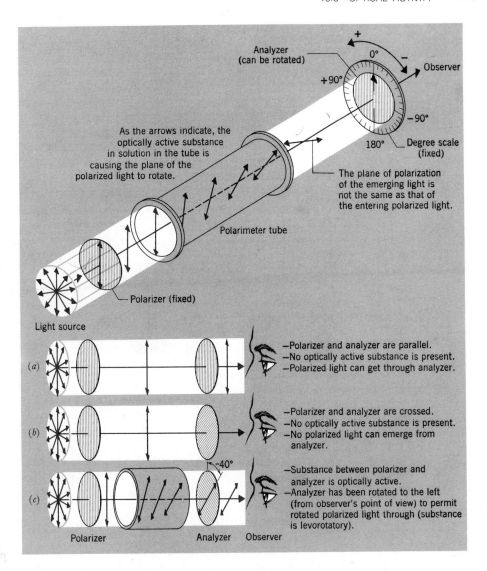

Analyzer (can be rotated)

As the arrows indicate, the optically active substance in solution in the tube is causing the plane of the polarized light to rotate.

The plane of polarization of the emerging light is not the same as that of the entering polarized light.

Polarimeter tube

Polarizer (fixed)

Light source

+90°   0°   +   −   Observer
180°   −90°   Degree scale (fixed)

(a)
—Polarizer and analyzer are parallel.
—No optically active substance is present.
—Polarized light can get through analyzer.

(b)
—Polarizer and analyzer are crossed.
—No optically active substance is present.
—No polarized light can emerge from analyzer.

(c) −40°
—Substance between polarizer and analyzer is optically active.
—Analyzer has been rotated to the left (from observer's point of view) to permit rotated polarized light through (substance is levorotatory).

Polarizer   Analyzer Observer

The Polarimeter.   The instrument used to detect and measure optical activity is called a **polarimeter.** (See the top part of Figure 18.6.) Its principal working parts consist of a polarizer for the light, a tube for holding solutions in the path of the polarized light, an analyzer (actually, just another polarizing device), and a circular scale for measuring the number of degrees of rotation. When the "slats" of the polarizer and the analyzer are parallel and the tube contains no optically active material, the polarized light goes through the solution to emerge with maximum intensity. Let's assume that we start with this parallel orientation of analyzer and polarizer.

When a solution of one pure enantiomer is placed in the light path, the plane-polarized light encounters molecules of just one chirality. They cause the plane of oscillation of the polarized light to be rotated. The plane of oscillation of the polarized light that leaves the solution is now no longer parallel with the analyzer, as shown in Figure 18.6c. Consequently, not as much light gets through the solution, so now the intensity of the observed light is less than the original maximum. To restore the original maximum intensity, the analyzer has to be rotated to the right or to the left a definite number of degrees until the analyzer is parallel with the light *that emerges from the tube.* The operator, looking *toward* the light source, might find that rotating the analyzer to the right restores the light intensity with fewer degrees of rotation than rotating the analyzer to the left. When such a rightward rotation works, the degrees are

FIGURE 18.7
When two polarizing films are
"crossed," no light can get through to
the observer.

Latin, *dextro*, right, and *levo*, left.

recorded as positive, and the optically active substance is said to be **dextrorotatory.** If the fewer degrees of rotation are found by a leftward or counterclockwise rotation, then the degrees are recorded as negative. In this case, the substance is said to be **levorotatory.** In Figure 18.6c, the reading is $\alpha = -40°$, where α stands for the **optical rotation,** the *observed* number of degrees of rotation for the solution.

The observed number of degrees of rotation varies both with the temperature and the frequency of the light that is used. There is no simple, direct relationship, however, so both of these data simply have to be recorded. For example, $\alpha_D^{20} = -40°$ means that the solution had a temperature of 20 °C and that the so-called D-light of a sodium vapor lamp was the light source. (This is the intense yellow light given by a sodium vapor lamp such as used in some street lights.)

Specific Rotation.   The observed rotation that a compound can cause at a given temperature of its solution and under a given light frequency is also a function of the population of its molecules that encounter the light. At a given tube length, we can vary this population by changing the concentration, and α is proportional to the concentration. On the other hand, at a given concentration, the population of molecules that can affect the light varies with the length of the tube, and α is also proportional to the tube length. Because both factors involve a simple, direct relationship, it's possible to construct a ratio-like number that takes these two factors into account. This is done by a new quantity called the **specific rotation** of a compound — the number of degrees of rotation per unit concentration per unit of path length. The symbol for specific rotation is $[\alpha]$ and the defining equation is:

If $\alpha \propto c$
and $\alpha \propto \ell$
then $\alpha \propto c \times \ell$
or $\alpha = (\text{const}) \times c \times \ell$
Therefore

$(\text{const}) = \dfrac{\alpha}{c\ell} = [\alpha]$

$$[\alpha]_\lambda^t = \frac{100\alpha}{c\ell} \tag{18.1}$$

where

c = concentration in grams per 100 mL of solution
ℓ = length of the light path in the solution in decimeters (1 dm = 10 cm)
α = observed rotation in degrees (plus or minus)
λ = wavelength of light
t = temperature of the solution in degrees Celsius

L-asparagine is the more common form.

The value of $[\alpha]_D^{20}$ for D-asparagine is $+5.42°$; for L-asparagine, $[\alpha]_D^{20} = -5.42°$. Here, then, is an illustration of the only physical difference between a pair of enantiomers. All of their physical properties are identical, including the number of degrees of specific rotation, except the *direction* of rotation. One enantiomer is always dextrorotatory and the other is always levorotatory by the same number of degrees.

It is important to realize that optical activity and optical isomerism are not the same. Optical activity refers to a phenomenon we can observe with the right kind of instrument. We can *see* the number of degrees of rotation. Optical isomerism is a factor in our explanation of optical activity. We *infer* optical isomerism (and other things) from the observation of optical activity. It's possible to have a sample of a substance that consists entirely of chiral molecules and is still optically inactive. All we have to do to make such a sample is mix two enantiomers in an equimolar ratio. The activity of the molecules of one enantiomer to rotate the plane of polarized light, say, clockwise, is canceled by the molecules of the other enantiomer. Such 50:50 mixtures of enantiomers, called **racemic mixtures,** are common. It's both interesting and significant that if a racemic mixture of the asparagine enantiomers were given in the diet, the body would use one and excrete the other entirely unused. Naturally-occurring glucose is dextrorotatory, and it's often named (+)-glucose to distinguish it from its synthetic enantiomer (−)-glucose, which the body can't use. Naturally-occurring fructose, another sugar, is levorotatory and is often named (−)-fructose.

This is why glucose is often called *dextrose* and fructose is sometimes called *levulose.*

The specific rotation of an optically active compound is an important physical constant, comparable to its melting point or its boiling point. If we want to know the concentration of, say, (+)-glucose in water, all we have to do is measure the observed rotation in a tube of known path length. Then we can calculate the concentration of the solution by Equation 18.1, because we know $[\alpha]$, α, and ℓ.

SUMMARY

Optical activity   Optical activity is a natural phenomenon detected by means of a polarimeter if polarized light that passes through a substance or its solution undergoes a rotation in its plane of polarization. Substances that do this are called optically active.

Optical isomerism   At the molecular level, optically active compounds consist of chiral molecules. Models of such molecules and models of their mirror images do not superimpose. Substances whose molecules are related as object to mirror image that do not superimpose are enantiomers. Almost always, an enantiomer molecule has a chiral carbon, one that holds four different atoms or groups. If a molecule has n different chiral carbons, then the number of optical isomers is 2^n, and these occur as half as many pairs of enantiomers. Each enantiomer is an example of one kind of optical isomer, which places it in the broad family of stereoisomers. A 50:50 mixture of enantiomers, a racemic mixture, is optically inactive.

Properties of enantiomers   Enantiomers are identical in every physical property except the sign of their specific rotation. They are also identical in every chemical respect provided that the molecules or ions of the reactant are achiral. When the reactant particles are chiral, then one enantiomer reacts differently with it than the other enantiomer, a phenomenon always observed when an enzyme is involved.

Specific rotation   The specific rotation of an optically active substance is what its observed rotation is at one unit of concentration (1g/100 mL) in one unit of path length (1 dm). It varies, but not in a simple way, with the temperature and the wavelength of the light used. Values of specific rotation can be used to determine concentrations of optically active substances.

KEY TERMS

The following terms were emphasized in this chapter, and they should be well-learned before continuing into succeeding chapters where they will be encountered again.

achiral	chiral carbon	dextrorotatory	levorotatory
chiral	chirality	enantiomer	optical isomers

optical activity

optical rotation

plane polarized light

polarimeter

racemic mixture

specific rotation

stereoisomers

structural isomers

substrate

superimposition

SELECTED REFERENCES

1 E. L. Eliel. "Stereochemistry Since LeBel and van't Hoff." *Chemistry*, January/February 1976, page 6.

2 G. B. Kauffman and R. D. Myers. "The Resolution of Racemic Acid." *Journal of Chemical Education*, December 1975, page 777.

REVIEW EXERCISES

The answers to the Review Exercises marked with an asterisk are in an Appendix. The answers to the other Review Exercises are in the *Study Guide* that accompanies this book.

Structural Isomers and Stereoisomers

18.1   What are the structures of the simplest chloroalkanes that can exhibit structural isomerism?

18.2   Draw the structures of the simplest chloroalkenes that can exist as a set of stereoisomers. (Be sure to draw these carefully to show the geometric relationships.)

18.3   Classify the following pairs of structures as structural isomers or as stereoisomers.

(a)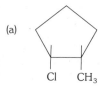

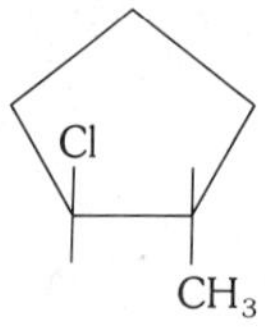

(b)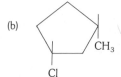

18.4   What is the "one-substance–one-structure" principle?

Optical Isomers

18.5   The general structure of several optical isomers that include glucose is that of 2,3,4,5,6-pentahydroxyhexanal:

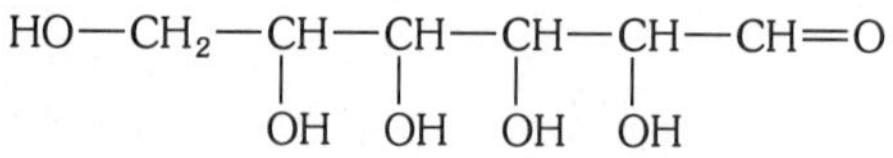

$$HO-CH_2-CH-CH-CH-CH-CH=O$$
$$\qquad\quad |\quad\ |\quad\ |\quad\ |$$
$$\qquad\quad OH\ \ OH\ \ OH\ \ OH$$

(a) Place an asterisk by each chiral carbon.

(b) How many of these chiral carbons qualify as *different* chiral carbons?

(c) How many optical isomers of this compound are possible?

(d) These optical isomers occurs as how many *pairs* of enantiomers?

18.6   One of the building blocks of nucleic acids used in transmitting genetic information from a cell nucleus to the place in a cell where enzymes are made is adenosine monophosphate, AMP.

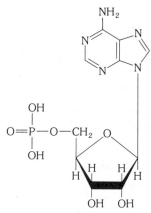

Adenosine monophosphate, AMP

(a) Place an asterisk by each chiral carbon in AMP.

(b) How many optical isomers can be calculated for AMP? (Only one is usable by the cell.)

18.7   One of the approximately 20 building blocks of protein molecules is glycine: $^+NH_3-CH_2-CO_2^-$. Does glycine have optical isomers? How can you tell?

18.8   One of the very important intermediate substances in the body's energy-producing metabolism is an anion of citric acid.

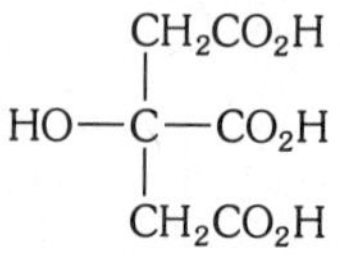

$$CH_2CO_2H$$
$$\qquad\ |$$
$$HO-C-CO_2H$$
$$\qquad\ |$$
$$CH_2CO_2H$$

Does citric acid have optical isomers? How can you tell?

18.9 Cholesterol is involved in a beneficial way in making some of the sex hormones and in a harmful way by plugging small capillaries.

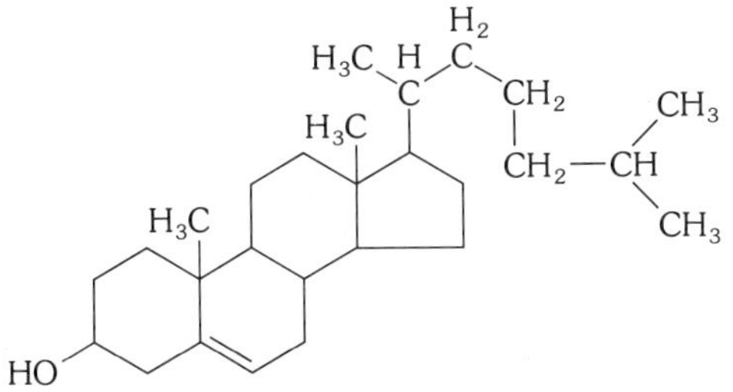

Cholesterol

(a) Place an asterisk by each carbon in the cholesterol molecule that is chiral.
(b) How many optical isomers can be calculated for cholesterol? (Cholesterol is just one of these.)

18.10 Vitamin D_2 is one of a few members of the vitamin D family of substances.

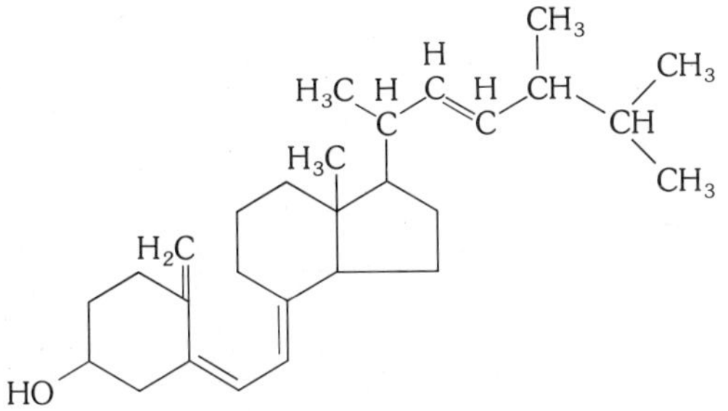

Vitamin D_2

(a) Put an asterisk by each of the chiral carbons in the structure of vitamin D_2.
(b) How many optical isomers are there of this compound?

Properties of Enantiomers

18.11 The melting point of (−)-cholesterol is 148.5 °C. What is the melting point of (+)-cholesterol? How can we know this without actually making the measurement?

18.12 Explain why enantiomers should have identical physical properties (except for the sign of specific rotation).

18.13 Explain why enantiomers cannot have different chemical properties toward reactants whose molecules or ions are achiral. (Use an analogy if you wish.)

18.14 Explain why enantiomers have different chemical properties toward reactants whose molecules or ions are chiral. (Use an analogy if you wish.)

Specific Rotation

18.15 Pantothenic acid was once called vitamin B_3.

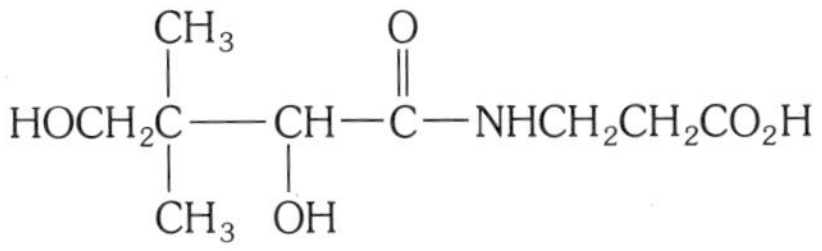

Pantothenic acid

Only its dextrorotatory form can be used by the body, and its specific rotation is $[\alpha]_D^{25} = +37.5°$. In a tube 1.00 dm long and at a concentration of 1.00 g/dL, what is the *observed* optical rotation of the levorotatory enantiomer of (+)-pantothenic acid?

18.16 Ascorbic acid is also known as vitamin C. Only its dextrorotatory enantiomer can be used by the body.

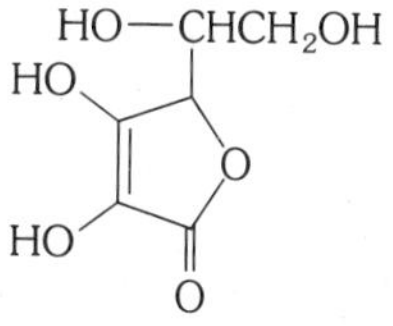

Ascorbic acid

Its specific rotation is $[\alpha]_D^{25} = +21°$. What is the specific rotation of (−)-ascorbic acid?

18.17 Explain why the observed rotation is proportional to the concentration of the optically active compound and to the length of the tube through which the polarized light travels.

18.18 What becomes of the optical activity of a substance if it consists of a mixture of equal numbers of moles of its two enantiomers? Explain.

*18.19 A solution of sucrose (table sugar) in water at 25 °C in a tube that is 10.0 cm long gives an observed rotation of +2.00°. The specific rotation of sucrose in water at this temperature and the same wavelength of light is +66.4°. What is the concentration of the sucrose solution in g/dL?

18.20 Quinine sulfate, an antimalarial drug, has a specific rotation in water at 17 °C of −214°. If a solution of quinine sulfate in this solvent in a 1.00 dm tube and under the same conditions of temperature and wavelength has an observed rotation of −10.4°, what is its concentration in g/dL?

*18.21 Strychnine and brucine are structurally similar compounds that have extremely bitter tastes, and both are very poisonous. The specific rotation in chloroform at 20 °C of brucine is −127° and that of strychnine under identical conditions is −139°. If a solution of one of these in chloroform at a concentration of 1.68 g/dL in a tube 1.00 dm long gave an observed rotation of $-2.34°$, which of the two compounds was it? Do the calculation needed for the answer.

18.22 Corticosterone and cortisone are two substances used to treat arthritis. Under identical conditions of solvent and tempera- ture, the specific rotation of corticosterone is $+223°$ and that of cortisone is $+209°$. If a solution of one of these at a concentration of 1.48 g/dL and in a tube 1 dm long has an observed rotation of $+3.10$, which compound was in the solution? Do the calculation.

Chapter 19
Carbohydrates

The summer sunshine, caught by photosynthesis in this corn being harvested, is now chemical energy in corn starch and other materials. People make a kind of starch, too, as we'll learn in this chapter.

19.1   BIOCHEMISTRY — AN OVERVIEW

Building materials, energy, and information are basic essentials for life.

Biochemistry is the systematic study of the chemicals of living systems, their organization, and the principles of their participation in the processes of life.

A Molecular Basis of Life.   The molecules of living systems are lifeless, yet life has a molecular basis. When they are isolated from their cells or when they are studied in living environments, the chemicals at the foundation of life obey all of the known laws of chemistry and physics. In isolation, however, not one single compound of a cell has life. The cell's intricate organization is as important to life as the chemicals themselves. The cell is the smallest unit of matter that lives and that can make a new cell exactly like itself. A change in its environment — a change in temperature, pH, pressure, or the appearance of hostile substances — can quickly render a cell nothing more than a lifeless bag of chemicals.

An ancient writer said that we once were dust and to dust we will return. Compared with dust and ashes scattered to the four winds, an organism is a finely ordered, high-energy system. The marvelous interlude between dust and dust that we call life endures only as long as the organism can maintain this dynamic organization of parts and defend it against the many natural tendencies to disintegrate and decay.

Cornstarch, potato starch, table sugar, and cotton are all carbohydrates.

Vital Needs: Materials, Energy, and Information.   Whether it be a plant or an animal, all life has three absolutely vital needs: materials, energy, and information. Our purpose in the remainder of this book is to study the molecular basis of meeting these three needs and how one organism, the human body, uses the substances that supply them. As we look in a general way at life at the subcellular and molecular levels, we will also learn of several chemical dangers constantly and almost always met by a living system.

We will begin in the next three chapters to study the organic materials of life, starting with the three great classes of foodstuffs: carbohydrates, lipids, and proteins. We use their molecules to build and run ourselves and to try to stay in some state of repair. Proteins are particularly important in both the structures and functions of cells, whether of plants or animals. Plants rely heavily on carbohydrates for cell walls, and animals obtain considerable energy from carbohydrates made by plants. Lipids (fats and oils) serve many purposes and are rich in energy.

Meat is rich in protein.

Butter, lard, margarine, and corn oil are all lipids.

What Controls Enzymes Controls Life.   Because of their central catalytic role in regulating chemical events in cells, we will study a special family of proteins, the enzymes.

Enzymes are actually a part of the molecular basis of information, and some kind and form of information are essential to life. Without plans or blueprints, materials and energy could combine to produce only rubble and rubbish. Monkeys swinging hammers would only reduce a stack of lumber to splinters. Carpenters, using no more raw energy, can make the same lumber into a building, because they have both a plan and experience. Enzymes, however, do not originate the plans of a living system. They only help to carry them out. The blueprint for an individual member of a species is encoded in the molecular structures of a family of substances called nucleic acids. Nucleic acids direct the synthesis of a cell's enzymes, and each individual has a unique set. Therefore before we study how enzymes work, we will study enzyme makers and see how different species can take essentially the same raw materials and energy and make unique enzymes.

Individual genes are sections of molecules of a polymer called DNA, which is one kind of nucleic acid.

Supplying and Processing Materials.   To supply materials for any use — parts, energy, or information — each organism has basic nutritional needs. These include not just organic materials, but also minerals, water, and oxygen. We'll survey these aspects of the biochemistry of life, too.

All of the materials taken in the diet have to be processed by the digestive system and distributed by the bloodstream and the lymph if cells are to obtain what they need in forms suitable for that species. Therefore we will study the fluid networks that process materials, distribute them, and carry wastes away, with a special emphasis on the molecular basis of using oxygen and releasing carbon dioxide.

Energy for Living. How can gulps of air, a peanut butter sandwich, and a glass of milk be used by the body to make or repair, say, muscle cells? How do these materials translate into such a whirlwind of energy as an active little child? Or an athlete? The molecular basis of energy for life forms the final broad topic of our study—the last four chapters. And as we study biochemical energetics and its enzymes and metabolic pathways, we'll have numerous occasions to peer deeply into the molecular basis of some disorders and diseases.

We have an exciting trip ahead of us. In the preceding chapters we have slowly and carefully built a solid foundation of chemical principles. It's been like a mountain-climbing trip where the route for a large part of the trek is through country with few grand vistas but still with a beauty of its own. Now we're moving to elevations where the vistas begin to open. It's like a new beginning, and we start it with a study of the first of the three chief classes of food materials, the carbohydrates.

19.2 MONOSACCHARIDES

The structure of glucose is the key to the structures of most carbohydrates.

The oxidized and reduced forms of polyhydroxy aldehydes and ketones as well as certain amino derivatives are also in the family of carbohydrates.

Carbohydrates are polyhydroxy aldehydes, polyhydroxy ketones, or substances that yield these by simple hydrolysis. Carbohydrates that cannot be hydrolyzed to compounds with simpler molecules are called **monosaccharides** or sometimes **simple sugars.** Monosaccharides have the general formula $(CH_2O)_n$, and they can be classified as follows, where you will notice the characteristic name ending of virtually all carbohydrates, -ose.

1. The number of carbons in one molecule, or the value of n in $(CH_2O)_n$.

If the number of carbons is	The monosaccharide is a
3	triose
4	tetrose
5	pentose
6	hexose, etc.

2. The nature of the carbonyl group. If an aldehyde group is present, the monosaccharide is an **aldose.** If a keto group is present, the monosaccharide is a **ketose.**

A hexose that has an aldehyde group is called an **aldohexose.** Glucose is an example. A **ketohexose** possesses a keto group and a total of six carbons. Fructose is a ketohexose. You can see that we can combine word parts to make descriptive names.

Other Families of Carbohydrates. Carbohydrates that can be hydrolyzed to two monosaccharides are called **disaccharides.** Sucrose, maltose, and lactose are common examples. Starch and cellulose are called **polysaccharides,** because when one of their molecules reacts with water, it gives hundreds of monosaccharide molecules.

Oligosaccharide molecules yield 3 to 10 monosaccharide molecules when they are hydrolyzed.

Another way of classifying carbohydrates is by their ability to give positive tests with Tollens' and Benedict's reagents. When the tests are positive, something in the reagent (e.g., Ag^+ or Cu^{2+}) is reduced, so carbohydrates that give such positive tests are called **reducing carbohydrates.** All monosaccharides and nearly all disaccharides are reducing carbohydrates. Sucrose (table sugar) is not, and neither are the polysaccharides.

Aldohexoses.   If we measure importance by relative abundance, then the hexoses in general and the aldohexoses in particular are the most important saccharides. Before we study the major aldohexoses, we should note that two trioses, glyceraldehyde and dihydroxyacetone, occur in metabolism, and two aldopentoses, ribose and 2-deoxyribose, are essential to the nucleic acids, the chemicals of heredity.

$$
\underset{\text{Glyceraldehyde}}{\overset{\displaystyle \text{O}}{\underset{\text{OH}}{\text{HOCH}_2\overset{|}{\text{CH}}\overset{\|}{\text{CH}}}}}
\qquad
\underset{\text{Dihydroxyacetone}}{\overset{\displaystyle \text{O}}{\text{HOCH}_2\overset{\|}{\text{C}}\text{CH}_2\text{OH}}}
\qquad
\underset{\text{Ribose}}{\overset{\displaystyle \text{O}}{\text{HOCH}_2\underset{\text{OH}}{\overset{|}{\text{CH}}}—\underset{\text{OH}}{\overset{|}{\text{CH}}}—\underset{\text{OH}}{\overset{|}{\text{CH}}}—\overset{\|}{\text{CH}}}}
\qquad
\underset{\text{2-Deoxyribose}}{\overset{\displaystyle \text{O}}{\text{HOCH}_2\underset{\text{OH}}{\overset{|}{\text{CH}}}—\underset{\text{OH}}{\overset{|}{\text{CH}}}\text{CH}_2\overset{\|}{\text{CH}}}}
$$

Fructose has a much sweeter taste than glucose or sucrose.

Among the important hexoses are two aldohexoses, **glucose** and **galactose,** and a ketohexose, **fructose.** All have sweet tastes. Like all of the monosaccharides, they are white, crystalline substances with very polar molecules. They dissolve freely in water but not in any of the nonpolar solvents. The several —OH groups in each molecule form quite strong hydrogen-bonding networks between molecules that simply won't allow molecules of nonpolar solvents to get between them.

Glucose.   If we count all of its combined forms, (+)-glucose is perhaps the most abundant organic species on earth. It's the building block for molecules of **cellulose,** a polysaccharide that makes up about 10% of all of the tree leaves of the world (on a dry weight basis), about 50% of the woody parts of plants, and nearly 100% of cotton. Glucose is also a building block for **starch,** another polysaccharide in many of our foods. Glucose and fructose are the major components of honey. Glucose is also commonly found in plant juices. It is by far the most common carbohydrate in blood. In fact, glucose is often called **blood sugar,** although this more general term actually applies to the mixture of all of the carbohydrates in blood. In human blood there is roughly 100 mg of glucose per 100 mL of volume.

Glucose is also called corn sugar, because it can be made by the hydrolysis of cornstarch.

Glucose normally does not appear in the urine, at least not for sustained periods of time. If it does, there is almost certainly some disorder (e.g., diabetes mellitus), so the detection and measurement of glucose in the urine is a common clinical analysis.

Structure of Glucose.   The molecular formula of glucose is $C_6H_{12}O_6$. It forms a pentaacetate ester, so five of its six oxygens are in —OH (alcohol) groups, which can form esters. It is easily oxidized to a C_6 monocarboxylic acid by agents such as Tollens' reagent that do not oxidize alcohol groups, so the sixth oxygen is in an aldehyde group. Under strong, forcing conditions, glucose can be reduced to a straight-chain derivative of hexane, so the six carbons in glucose must be in a straight chain. The five —OH groups must be strung out, one on each of five carbons, because 1,1-diols are not stable. These data tell us that the structure of glucose is that of 2,3,4,5,6-pentahydroxyhexanal.

A 1,1-diol consists of the following system.

$$
—\overset{|}{\underset{|}{\text{C}}}—\text{OH}
$$
$$
\text{OH}
$$

$$
\underset{\text{OH}}{\overset{|}{\text{CH}_2}}—\underset{\text{OH}}{\overset{|}{\text{CH}}}—\underset{\text{OH}}{\overset{|}{\text{CH}}}—\underset{\text{OH}}{\overset{|}{\text{CH}}}—\underset{\text{OH}}{\overset{|}{\text{CH}}}—\overset{\displaystyle \text{O}}{\overset{\|}{\text{C}}}—\text{H}
$$

Basic structure of all
aldohexoses, including glucose

Mutatrotation.   Carbons 2, 3, 4, and 5 in the glucose chain are all chiral and all different. We learned in the previous chapter that the number of optical isomers of a compound whose molecules have n different chiral carbons is 2^n. Therefore 2,3,4,5,6-pentahydroxyhexanal must have $2^4 = 16$ optical isomers, or eight pairs of enantiomers. (+)-Glucose is one of these 16; galactose is another.

Although glucose is optically active, when we try to measure its optical activity, it behaves in a very strange way. A *freshly prepared* solution of (+)-glucose has a specific rotation of $[\alpha]_D^{20} = +113°$. As this solution ages, however, its specific rotation changes until it stabilizes at a value of $+52°$. We'll call a glucose solution with this specific rotation an "aged glucose solution."

By a special method of recovery (which we'll not discuss), we can recover crystalline (+)-glucose from the aged solution. It's the same glucose in every respect with which we began these experiments. And a freshly prepared solution of this recovered glucose again shows a specific rotation of $+113°$. This new solution ages in the identical way until its specific rotation stabilizes at $+52°$. This cycle can be repeated as often as we please.

It's possible to change the method of recovering the solute in an aged glucose solution and obtain a slightly different crystalline compound. Its freshly prepared solution has a specific rotation of $+19°$. The specific rotation of this solution also changes with time, and it finally stops at the same value that was observed for the other aged solution, $+52°$! Using this second method of recovering the solute, we can repeat this cycle as often as we please, too.

To summarize, from the original aged solution we can use one method to recover the solute and get back glucose with a specific rotation of $+113°$. With the second recovery technique, we get a glucose with a specific rotation of $+19°$. We can interconvert these forms through the aged solution as often as we please. This change in the optical rotation with time shown by a solution of an optically active substance, a substance that can be recovered from an aged solution without any other apparent change, is called **mutarotation.** All of the hexoses and most of the disaccharides mutarotate. Let us now see what is behind it.

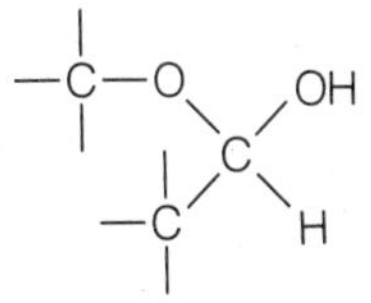

The Cyclic Forms of Glucose. Three forms of the glucose molecule exist, given by structures **1, 2,** and **3.** Structure **2** is the open-chain form, 2,3,4,5,6-pentahydroxyhexanal, except that it is shown in a coiled configuration. The other two forms, α-glucose (**1**) and β-glucose (**3**), are cyclic compounds — cyclic hemiacetals. α-Glucose is the form that glucose molecules are in when a freshly prepared aqueous solution has a specific rotation of $+113°$.

Hemiacetal system

The H at C-2 and C-5 and the OH at C-3 do not stick *inside* the ring. They stick *above or below the plane* of the ring.

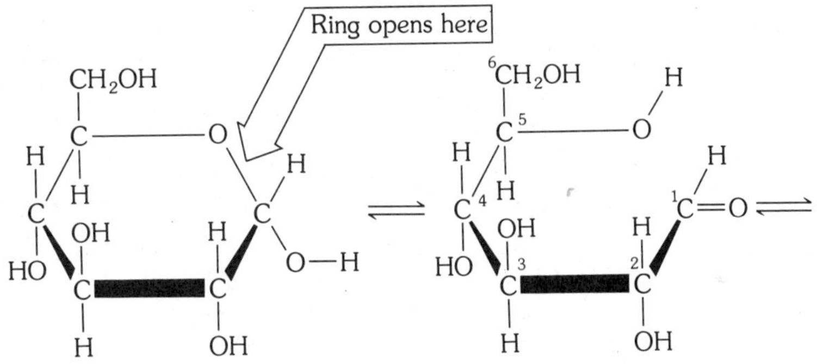

α-Glucose

1

Open form
(polyhydroxyaldehyde)

2

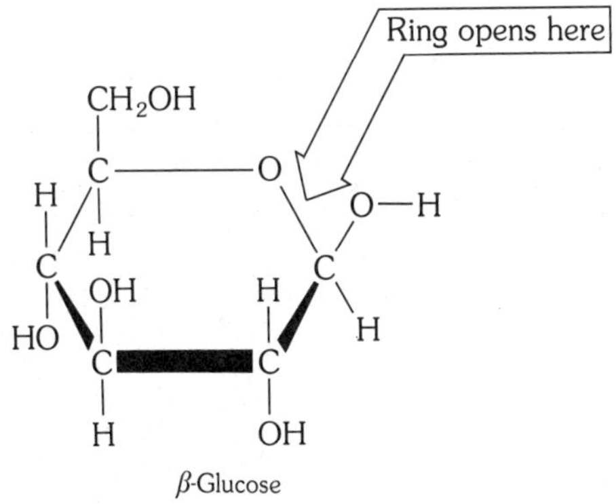

β-Glucose

3

When α-glucose is first dissolved in water, all of its molecules are in the cyclic form. Remember, however, that hemiacetals are unstable, and glucose in its cyclic forms is a hemiacetal. First one molecule of **1** and then another opens up into the open form that has the aldehyde group and all of the alcohol groups. The —OH at C-5 is the alcohol group that participates with the aldehyde group in making a cyclic hemiacetal, and *both are in the same molecule,* structure **2.** Molecules of **2,** as soon as they emerge, can and do reclose. The —OH group at C-5 merely has to add across the double bond of the aldehyde group. This is illustrated in Figure 19.1. An equilibrium is beginning to be established.

While a glucose molecule is in its open form, **2,** rotation about any of the single bonds becomes possible. The aldehyde group, for example, can pivot from one orientation to another, as seen near the left side of Figure 19.1. If an open form does this *and then immediately ring-closes,* the new cyclic hemiacetal system of β-glucose emerges.

To keep a complicated situation as simple as possible, use the CH_2OH group as a reference, and notice that in both of the cyclic forms of glucose it sticks upward from the plane of the ring at the same time that the oxygen atom in the ring is in the upper right-hand corner. Now notice that the hemiacetal —OH group at C-1 projects upward in β-glucose and downward in α-glucose. (The projections of all of the other —OH groups are the same in the two forms.) Thus the two cyclic forms of glucose differ only in the orientation of the —OH group at C-1. With this in mind, let's review what happens during mutarotation.

As we said, when α-glucose is freshly dissolved in water, it has a specific rotation of $+113°$. But its molecules open and close, because the hemiacetal system easily breaks apart and reforms. Some of the newly formed open-chain molecules reclose as α-glucose, and some as β-glucose. These events take place during mutarotation, whether we start with α- or β-glucose, until one grand, dynamic equilibrium involving all three forms of glucose is established. And it has to be the same equilibrium regardless of which form of glucose is used to

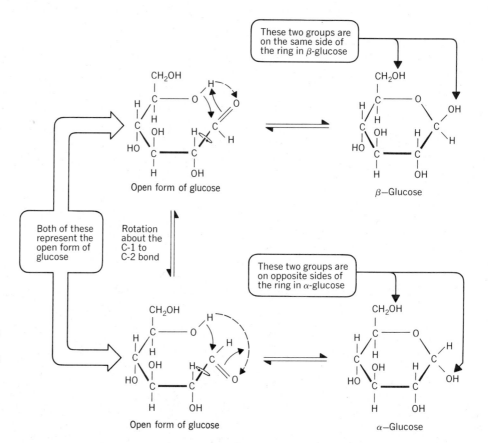

FIGURE 19.1

The α- and β-forms of glucose arise from the same intermediate, the open-chain form. Depending on how the aldehyde group. CH=O, is pointing when the ring closes, one ring form or the other results. These ring-openings and closing can be seen in computer-driven animation with *Prelab Studies for General, Organic, and Biological Chemistry* by Sandra L. Olmsted and Richard D. Olmsted. See page xi.

These two groups are on the same side of the ring in β-glucose

These two groups are on opposite sides of the ring in α-glucose

Both of these represent the open form of glucose

Rotation about the C-1 to C-2 bond

Open form of glucose

β—Glucose

Open form of glucose

α—Glucose

start it, because we obtain an aged solution that has the identical specific rotation, $+52°$. We can express the equilibrium in words as follows:

$$\alpha\text{-Glucose} \rightleftharpoons \text{open-chain form of glucose} \rightleftharpoons \beta\text{-glucose}$$

At equilibrium, the solute molecules are 36% α-glucose, 64% β-glucose, and scarcely a trace ($<0.05\%$) of the open-chain form.

One of the methods of recovering the solute from an aged solution succeeds in getting just α-glucose molecules to start the formation of a crystal. The molecules of β-glucose can't fit to this crystal and help it grow, so they remain in solution. But the loss of molecules of α-glucose from the equilibrium puts a stress on it, and the equilibrium shifts to replace the lost molecules. In this way, all of the β-glucose molecules eventually get changed to α-glucose molecules and nestle into the growing crystals. The other method of crystallization succeeds in getting crystals started that include just the β-glucose molecules, so eventually all of the α-glucose molecules get switched over to the β-form and join the growing crystals.

Molecules of the open-chain form never crystallize. They occur only in the solution. Of course, they are the ones attacked by Tollens' or Benedict's reagents, and as they are removed they are replaced by a steady shifting of the equilibrium from closed forms to the open form. This is why glucose, despite the fact that in the solid state it is in either one cyclic form or the other, gives the chemical properties of a pentahydroxy aldehyde (and why it's still all right to define a monosaccharide as a polyhydroxy aldehyde — or ketone — rather than as a cyclic hemiacetal).

Before continuing, you should now pause to learn how to write the cyclic forms of glucose. Figure 19.2 outlines the steps in mastering this that have worked well for many students. Even though six-membered rings (as was described in Special Topic 12.1, page 290) are not *flat* hexagons, the relative orientations of the groups around the ring, whether they project to the same side of the ring as the CH_2OH group or to the opposite side, are faithfully shown by these kinds of drawings.

This is another example of Le Chatelier's principle in action.

1. First write a six-membered ring with an oxygen in the upper right-hand corner.

2. Next "anchor" the —CH_2OH on the carbon to the left of the oxygen. (Let all the —Hs attached to ring carbons be "understood")

3. Continue in a *counterclockwise* way around the ring, placing the —OHs first down, then up, then down.

4. Finally, at the last site on the trip, how the last —OH is positioned depends on whether the alpha or the beta form is to be written. The alpha is "down," the beta "up."

    If this detail is immaterial, or if the equilibrium mixture is intended, the structure may be written as.

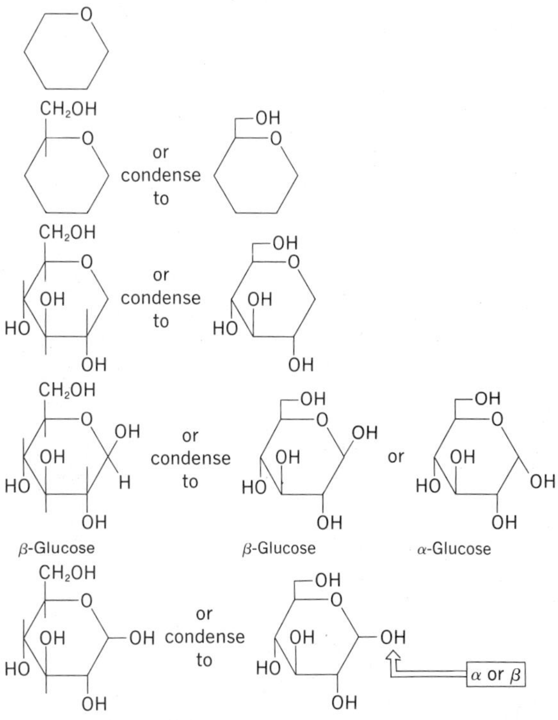

FIGURE 19.2
How to draw the cyclic forms of glucose in a highly condensed way.

Lactose is milk sugar.

Galactose.   Galactose is an aldohexose and an optical isomer of glucose. It occurs in nature mostly as structural units in larger molecules such as the disaccharide lactose. Galactose differs from glucose only in the orientation of the C-4 —OH group. Like glucose, it is a reducing sugar, it mutatrotates, and it exists in solution in three forms—alpha, beta, and open.

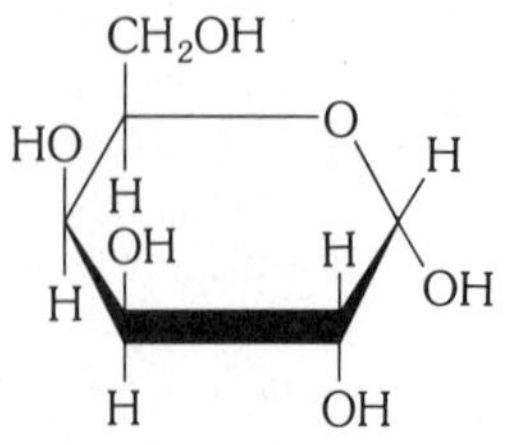

α-Galactose 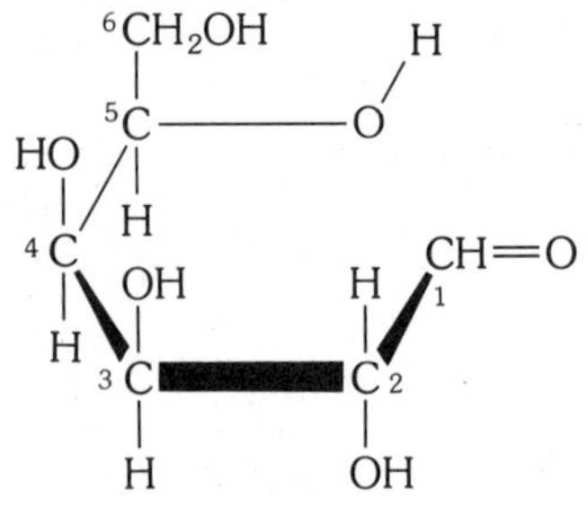   Galactose   β-Galactose
                              (open form)

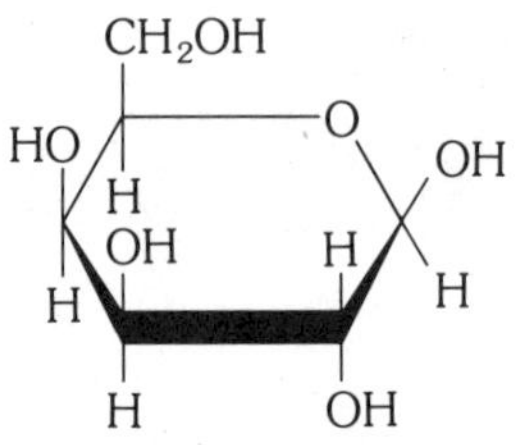

Fructose.   Fructose, the most important ketohexose, is found together with glucose and sucrose in honey and in fruit juices. It, too, can exist in more than one form, including cyclic hemiketals. We'll show only one.

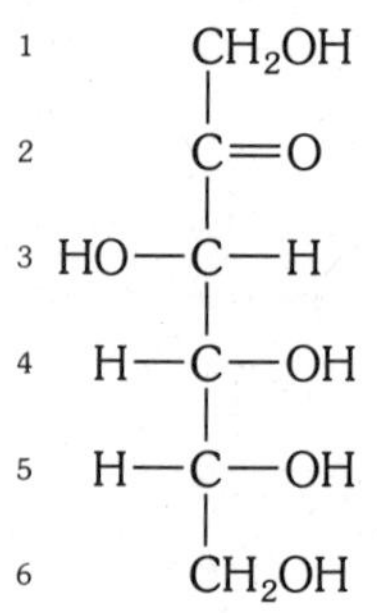

Fructose   β-Fructose
(open-form)   (cyclic hemiketal form)

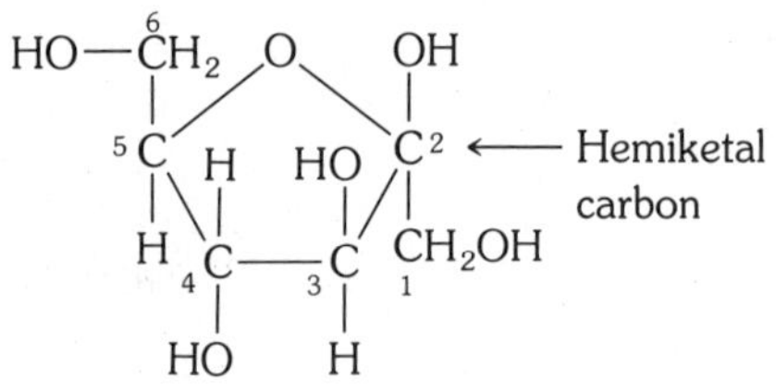

An old name for fructose is *levulose*, after its strong levorotatory power.

Fructose is strongly levorotatory with a specific rotation of $[\alpha]_D^{20} = -92.4°$. Because its molecules have the α-hydroxyketone system, fructose is a reducing sugar. Mono- and diphosphoric acid esters of fructose are important compounds in the metabolism of carbohydrates.

Deoxycarbohydrates.   A deoxycarbohydrate is one with a molecule that lacks a —OH group where normally such a group is expected. Thus 2-deoxyribose is the same as ribose except that there is no —OH group at C-2, just two Hs instead. Both ribose and 2-deoxyribose, as we noted earlier, are building blocks of nucleic acids, and each of these aldopentoses can exist in three forms, two cyclic hemiacetals and an open form.

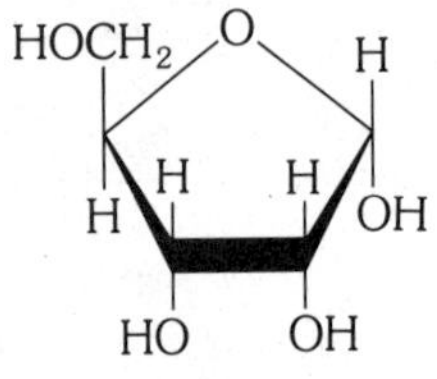

α-Ribose 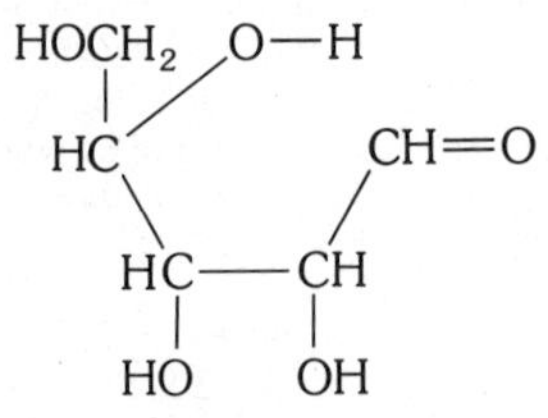   Ribose   β-Ribose
                          (open form)

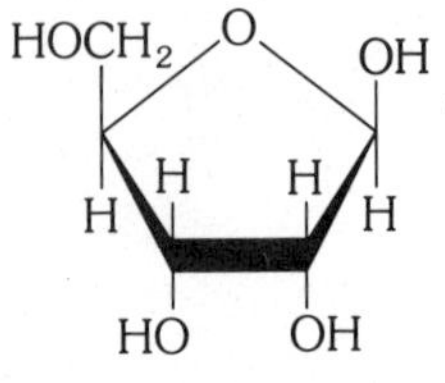

α-2-Deoxyribose

2-Deoxyribose
(open form)

β-2-Deoxyribose

19.3   D- AND L-FAMILIES OF CARBOHYDRATES

When the chiral carbon farthest from the carbonyl group of a monosaccharide has the same configuration as (+)-glyceraldehyde, the monosaccharide is in the D-family.

In this section we will study the question of the actual orientations or configurations of the chiral carbons in the monosaccharides.

Absolute Configuration.   To simplify this study, we'll retreat from the complexities of (+)-glucose and go back to the simplest aldose, glyceraldehyde. The structures of the two enantiomers of glyceraldehyde are shown in Figure 19.3. Both of these compounds are known, and it is also known that (+)-glyceraldehyde actually has the absolute configuration given in this figure. **Absolute configuration** refers to the actual arrangement in space about each chiral center in a molecule. When we know the absolute configuration of (+)-glyceraldehyde, we also know that of (−)-glyceraldehyde. The two are known to be enanatiomers, so the structure of (−)-glyceraldehyde has to be the mirror image of (+)-glyceraldehyde. (See also Figure 19.3.)

Chemists have used the absolute configurations of the enantiomers of glyceraldehyde to devise configurational or optical families for the rest of the carbohydrates. Any compound that has a configuration like that of (+)-glyceraldehyde and can be related to it by known reactions is said to be in the **D-family.** For example, (−)-glyceric acid is in the D-family because it can be made from D-(+)-glyceraldehyde by an oxidation that doesn't disturb any of the four bonds to

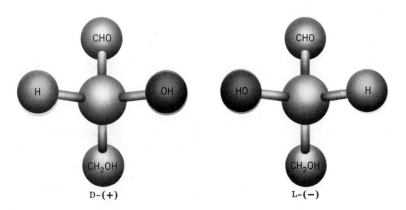

FIGURE 19.3
The absolute configurations of the
enantiomers of glyceraldehyde.

the chiral carbon, as the following equation shows:

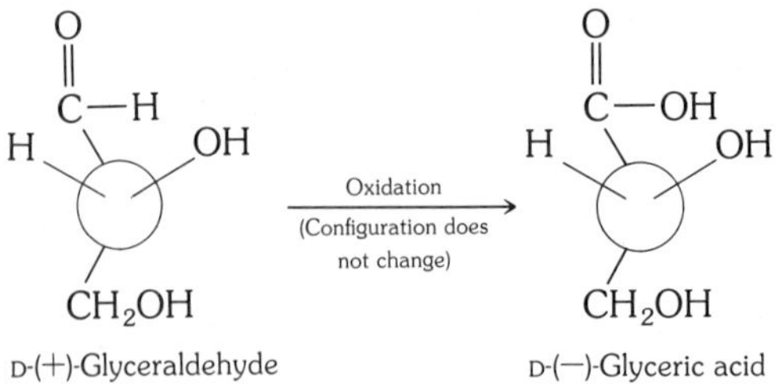

D-(+)-Glyceraldehyde   D-(−)-Glyceric acid

When the molecules of a compound are the mirror images of an enantiomer in the D-family, the compound is in the **L-family.**

The letters D and L are only family names. They have nothing to do with actual signs of rotation. There exists, in fact, no way to tell from the sign of rotation if a compound is in the D- or L-family. These letters signify something about absolute configurations only.

Emil Fischer (1852–1919), the "father" of carbohydrate chemistry, won the second Nobel prize in chemistry in 1902.

Fischer Projection Formulas.   When a molecule has several chiral centers, it becomes quite difficult for most people to make a perspective three-dimensional drawing of the structure. Therefore to be able to represent absolute configurations on paper without asking that anyone be an artist, we follow a set of rules that let us depict what we want. The rules let us project onto a plane surface the three-dimensional configuration of each chiral carbon in a molecule, and the resulting structure is called a Fischer projection formula.

Rules for Writing Fischer Projection Formulas

1. Visualize the molecule with its main carbon chain vertical and with the bonds that hold the chain together projecting to the rear at each chiral carbon. Carbon-1 is at the top.

2. Mentally flatten the structure, chiral carbon by chiral carbon, onto a plane surface. (See Figures 19.4 and 19.5.)

3. In the projected structure, represent each chiral carbon as either the intersection of two lines or conventionally as C.

4. The horizontal lines at a chiral center actually represent bonds that project *forward,* out of the plane of the paper.

5. The vertical lines at a chiral center actually represent bonds that project rearward, behind the plane.

A Fischer projection formula can have more than one intersection of lines, each representing a chiral carbon, as seen in Figure 19.5. Always remember that at each chiral carbon, a horizontal line is a bond coming toward you and a vertical line is a bond going away from you.

Once we have one plane projection structure, it's easy to draw the mirror image, as we saw in Figures 19.4 and 19.5. We can easily test for superimposition, too, provided we strictly heed one important additional rule. We may never (mentally) lift a Fischer projection formula out of the plane of the paper. We may only slide it and rotate it within the plane. This rule is necessary because if we turn a Fischer projection formula out of the plane and over, we actually make groups that project in one direction project oppositely, but the operation will not show this reversal.

FIGURE 19.4
The relationships of the perspective (three-dimensional) drawings of D-(+)-glyceraldehyde and L-(−)-glyceraldehyde to their corresponding Fischer projection formulas.

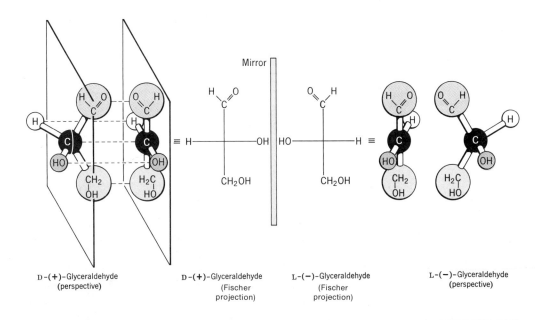

D-(+)-Glyceraldehyde
(perspective)

D-(+)-Glyceraldehyde
(Fischer projection)

L-(−)-Glyceraldehyde
(Fischer projection)

L-(−)-Glyceraldehyde
(perspective)

EXAMPLE 19.1   WRITING FISCHER PROJECTION FORMULAS

Problem: Write the Fischer projection formulas for the optical isomers of glyceric acid.

$$HOCH_2CHCO_2H$$
$$|$$
$$OH$$

Glyceric acid

Solution: There are three carbons in the chain, but only the center carbon is chiral. Therefore glyceric acid has two optical isomers, two enantiomers. We represent its chiral carbon by the intersection of two perpendicular lines, and we make two of these, one for each enantiomer:

$+$ $+$ (Incomplete)

Then we attach the other two nonchiral carbons. According to the rules, we have to put C-1, the carbon with the carbonyl group in CO_2H, at the top:

CO_2H CO_2H
$+$ $+$ (Incomplete)
CH_2OH CH_2OH

We know that the —OH (alcohol) group at C-2 can be either on the right or the left, so we finish the Fischer projection formulas:

CO_2H CO_2H
H—OH   HO—H   (Complete)
CH_2OH CH_2OH

These are the two enantiomers of glyceric acid. The one on the left is D-glyceric acid, and the other is L-glyceric acid.

FIGURE 19.5
The four aldotetroses in their perspective and Fischer projection formulas. There are two different chiral carbons, so there are $2^2 = 4$ optical isomers that occur as two pairs of enantiomers, those of erythrose and those of threose.

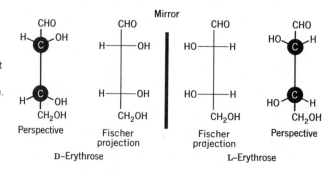

PRACTICE EXERCISE 1　Fischer projection formulas that correspond to the various optical isomers of tartaric acid are given below. (a) Which are identical? (b) Which are related as enantiomers? (c) One is a *meso* compound (Special Topic 18.1). Which is it?

CO₂H	CO₂H	CO₂H	HO₂C	CO₂H
H—OH	H—OH	H—OH	HO—H	HO—H
HO—H	HO—H	H—OH	H—OH	HO—H
CO₂H	CO₂H	CO₂H	HO₂C	CO₂H
(a)	(b)	(c)	(d)	(e)

PRACTICE EXERCISE 2　Write Fischer projection formulas for the optical isomers of the following compound:

$$HOCH_2CH-CH-CO_2H$$
$$\qquad\quad |\quad\ \ |$$
$$\qquad\quad OH\ \ OH$$

D- and L-Families of Monosaccharides.　Monosaccharides are assigned to the D-family or the L-family according to the projection of the —OH group at the chiral carbon farthest from the carbonyl group. The compound is in the D-family when this —OH group projects to the right *in a Fischer projection formula* oriented so that the carbonyl group is at or near the top. When this —OH group projects to the left, the substance is in the L-family. We have

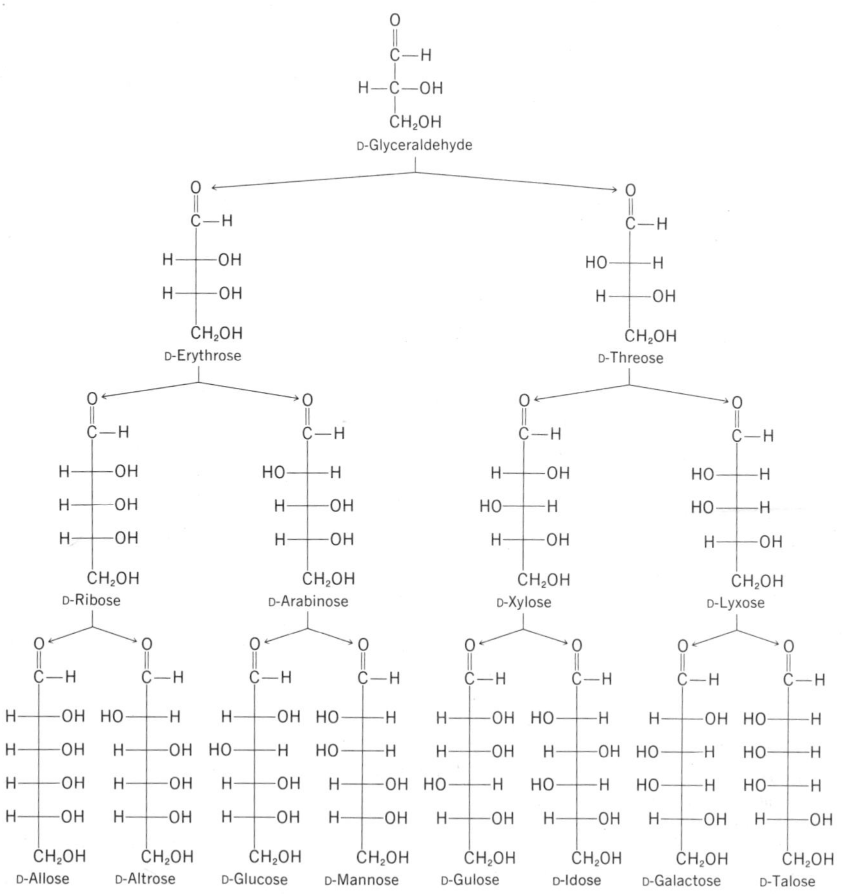

FIGURE 19.6
The D-family of the aldoses through the aldohexoses. Notice that in all of them, the —OH group on the chiral center that is farthest from the carbonyl group projects to the right. Each pair of arrows points to a pair of aldoses whose configurations are identical except at C-2.

already illustrated these rules by D- and L-glyceraldehyde (Figure 19.4), and by the enantiomers of threose and erythrose in Figure 19.5. It doesn't matter how the other —OH groups at the other chiral carbons project. Membership in the D- or L-family is determined solely by the projection of the —OH on the chiral carbon farthest from the carbonyl carbon.

The nutritionally important carbohydrates are all in the D-family. Therefore throughout the rest of this book, if the family membership of a carbohydrate isn't given, assume that the D-family is meant.

Notice that D-glucose and D-galactose differ only in the projection of the —OH group at C-4.

Figure 19.6 gives the Fischer projection formulas of all of the aldoses in the D-family from the aldotriose through the aldohexoses. There are eight D-aldohexoses. The enantiomers of these constitute eight L-aldohexoses (not shown). In all, therefore, there are 16 optical isomers of the aldohexoses, as we calculated earlier.

19.4 DISACCHARIDES

The disaccharides are glycosides that can be hydrolyzed to monosaccharides.

The aldoses, as we have just seen, are hemiacetals. Like all hemiacetals, they react with alcohols in the presence of a catalyst to give acetals and water. For example, methanol can be made to react with glucose to give either an α- or a β-acetal, depending on how the new $—OCH_3$ group becomes oriented at the ring. If it's on the same side of the ring as our reference CH_2OH group, then we have the β-form. If it is on the opposite side, then we have the α-form.

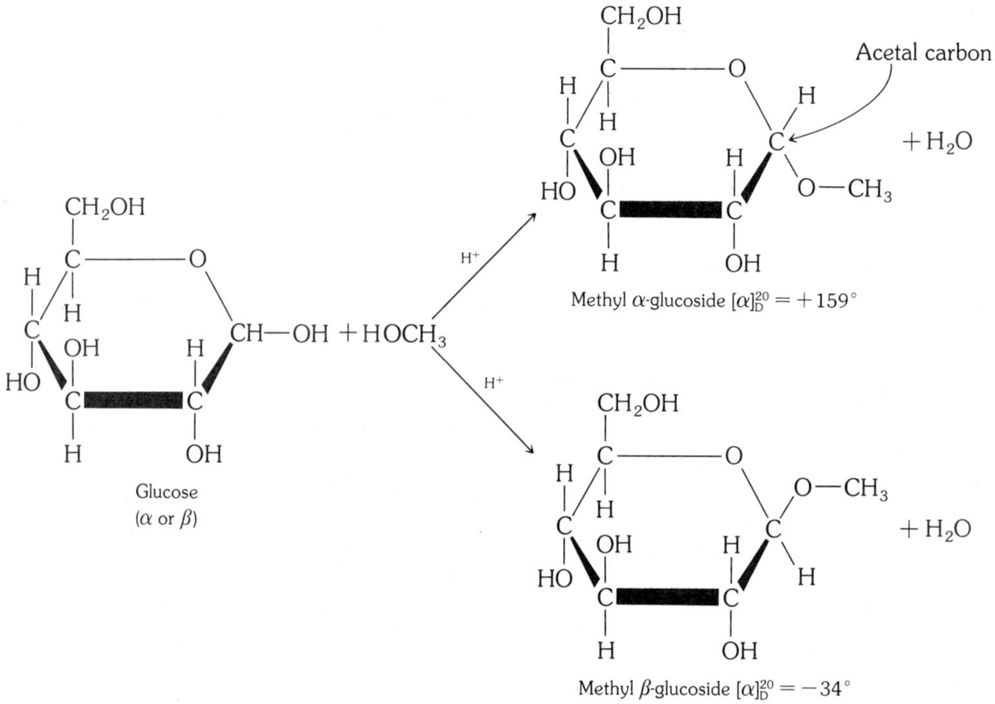

Methyl α-glucoside $[\alpha]_D^{20} = +159°$

Methyl β-glucoside $[\alpha]_D^{20} = -34°$

Glucose
(α or β)

Glucoside comes from a combination of *gluc*ose + gly*coside*. There are also galactosides, fructosides, and so forth.

Sugar acetals have a special name—**glycosides,** in general, and when they are made from glucose, then **glucosides.** These cyclic acetals, like all acetals, are stable enough to isolate, but they readily react with water when an acid catalyst (or the appropriate enzyme) is present. Thus the glycosides have rings that do not spontaneously open and close in water, so they do not mutarotate and do not give positive Tollens' or Benedict's tests. Our reason for even mentioning them is that all of the di- and polysaccharides are glycosides of a special type. The only difference is that they aren't methyl glycosides, made from a very simple alcohol. They are made from more complex alcohols—from one of the alcohol groups of another sugar molecule. Thus the disaccharides, like all acetals, react readily with water in the presence of an acid or enzyme catalyst.

The three nutritionally important disaccharides are **maltose, lactose,** and **sucrose.** All are in the D-family. We'll first show their relationships to monosaccharides by word-equations, and then we'll look more closely, but briefly, at their structures.

You can see why glucose is of such central interest to carbohydrate chemists.

$$\text{Maltose} + H_2O \longrightarrow \text{glucose} + \text{glucose}$$
$$\text{Lactose} + H_2O \longrightarrow \text{glucose} + \text{galactose}$$
$$\text{Sucrose} + H_2O \longrightarrow \text{glucose} + \text{fructose}$$

Maltose.   Maltose or malt sugar does not occur widely as such in nature, although it is present in germinating grain. It occurs in corn syrup, which is made from cornstarch, and it forms from the partial hydrolysis of starch. As the first equation above indicated, maltose is made from two glucose units. They are joined by an acetal oxygen bridge that in carbohydrate chemistry is called a **glycosidic link.**

An $\alpha(1 \rightarrow 4)$ glycosidic link

Hemiacetal system

If the —OH group on the far right projected downward instead of upward, the structure would be that of α-D-maltose.

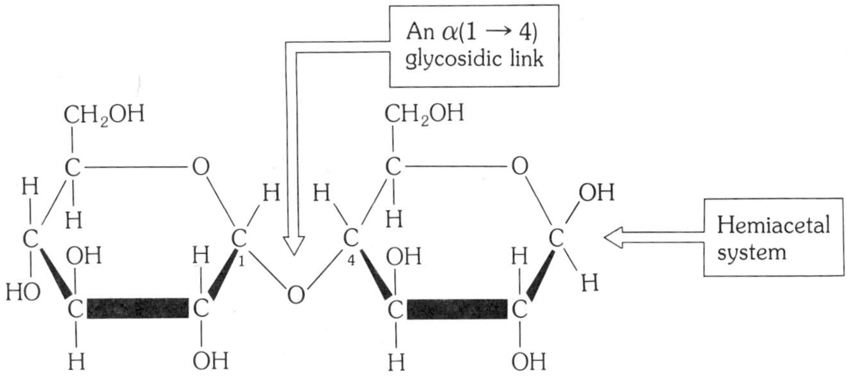

β-D-maltose

In the structure of maltose, the glycosidic link proceeds from C-1 of one glucose unit to C-4 of the other unit, and this relationship is symbolized by saying that the glycosidic link is $(1 \rightarrow 4)$. Moreover, the bond to O from C-1 points in the *alpha* direction, so the full description is that the glycosidic link is $\alpha(1 \rightarrow 4)$. Had the glycosidic link pointed in the beta direction and therefore been $\beta(1 \rightarrow 4)$, the resulting molecule would not have been maltose but a different disaccharide instead, cellobiose. This may seem to be a trifle, but the difference is that we can digest maltose but not cellobiose. Just this difference in *geometry* bars humans from an enormous potential food source, for nature could supply much of it from the polysaccharide, cellulose. We have an enzyme, maltase, that catalyzes the digestion (the hydrolysis) of maltose. We have no similar enzyme for cellobiose (or, for that matter, cellulose), although some organisms do.

One feature of maltose you should be sure to notice is that the glucose unit on the 4-side of a $(1 \rightarrow 4)$ glycosidic link still has a hemiacetal group. This part of the maltose molecule therefore can open and close, so maltose can exist in three forms — α-, β-, and the open form. In all three, the $\alpha(1 \rightarrow 4)$ glycosidic link to the other glucose unit is present. The ring-opening action occurs only at the hemiacetal part. Maltose therefore mutarotates. It also is a reducing sugar, giving positive tests with Tollens' and Benedict's reagents.

Lactose.   Lactose or milk sugar occurs in the milk of mammals — 4 to 6% in cow's milk and 5 to 8% in human milk. It is obtained commercially as a by-product in the manufacture of cheese. Like maltose, lactose is a glycoside — actually, a galactoside. From C-1 of its galactose unit there is a $\beta(1 \rightarrow 4)$ glycosidic link to C-4 of a glucose unit. The glucose unit therefore still has a free hemiacetal system, so lactose mutarotates and is a reducing sugar.

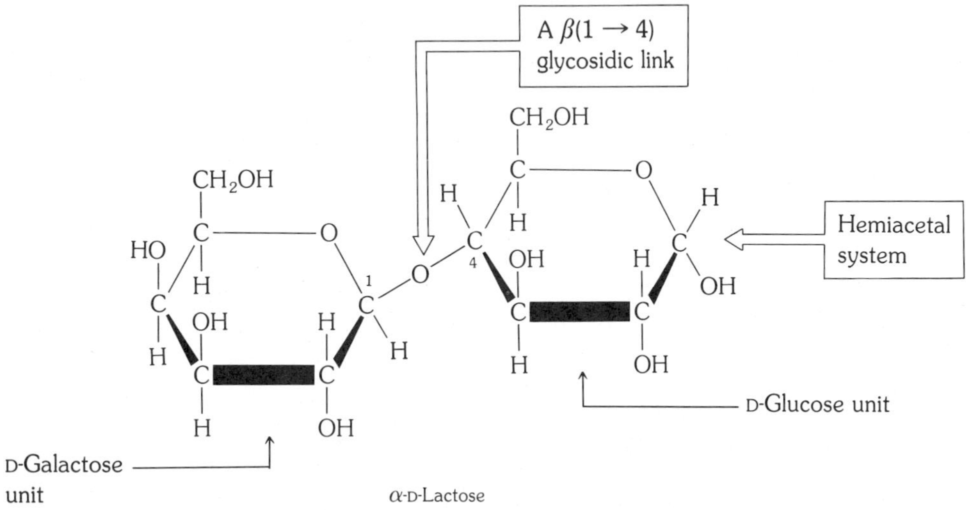

A $\beta(1 \rightarrow 4)$ glycosidic link

Hemiacetal system

D-Glucose unit

D-Galactose unit

α-D-Lactose

Beet sugar and cane sugar are identical compounds — sucrose.

Sucrose.   Sucrose, our familiar table sugar, is obtained from sugarcane or from sugar beets. Structurally, it is made from a glucose and a fructose unit between which is an acetal oxygen bridge. It has no hemiacetal or hemiketal group, so neither ring in sucrose can open and close spontaneously in water. Hence, sucrose neither mutarotates nor gives positive tests with Tollens' or Benedict's reagents. It's our only common nonreducing disaccharide.

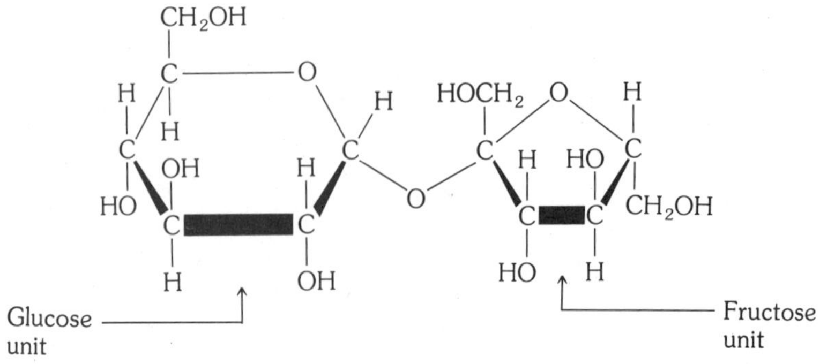

D-Sucrose

The 50 : 50 mixture of glucose and fructose that forms when sucrose is hydrolyzed is called *invert sugar,* and it makes up the bulk of the carbohydrate in honey. (The sign of specific rotation inverts from $+$ to $-$ when sucrose, $[\alpha]_D^{20} + 66.5°$, changes to invert sugar, $[\alpha]_D^{20} - 19.9°$ and this inversion of the sign is the origin of the term *invert* sugar.)

19.5   POLYSACCHARIDES

Starch, glycogen, and cellulose are all polyglucosides.

Much of the glucose a plant makes by photosynthesis goes to make cellulose and other substances that it needs to build its cell walls and its rigid fibers. Some glucose also is stored for the food needs of the plant. Free glucose, however, is too soluble in water for storage, so most is converted to a much less soluble form, starch. This polymer of glucose is particularly abundant in plant seeds.

We use plant starch for food. During the digestion of starch, we hydrolyze it to glucose, and we store what the body does not need right away. Normally we don't excrete any excess glucose. Instead, our bodies convert it to a starch-like polymer, glycogen, or to fat. In this section we will study the structures and some of the properties of the three types of glucose polymers — starch, glycogen, and cellulose.

Starch.   The basic structure of starch is shown in Figure 19.7. **Starch** is actually a mixture of two kinds of polymers of α-glucose. One is linear, **amylose,** and it has $\alpha(1 \rightarrow 4)$ glucosidic links exclusively. Its long molecules coil in a helix as shown in Figure 19.8. The other polyglucoside polymer in starch, **amylopectin,** is branched and it has both $\alpha(1 \rightarrow 4)$ and $\alpha(1 \rightarrow 6)$ glucosidic links. The latter form bridges that link the C-1 ends of long amylose molecules to C-6 positions of glucose units in other long amylose chains, as seen in Figure 19.7. There are hundreds of such links per molecule, so amylopectin is heavily branched, and the branches prevent any coiling of the polymer. This leaves many more —OH groups exposed to water than in amylose, so amylopectin tends to be somewhat more soluble in water than amylose. However, neither dissolves well. The "solution" is actually a colloidal dispersion, because it gives the Tyndall effect.

So-called *soluble* starch is partially hydrolyzed starch, and its smaller molecules more easily dissolve in water.

FIGURE 19.7
The glucose polymers in starch—amylose and amylopectin. Depending on the origin of the starch, the formula weight varies from 50,000 to several million. (A formula weight of 1 million corresponds to about 6000 glucose units per polymer molecule.)

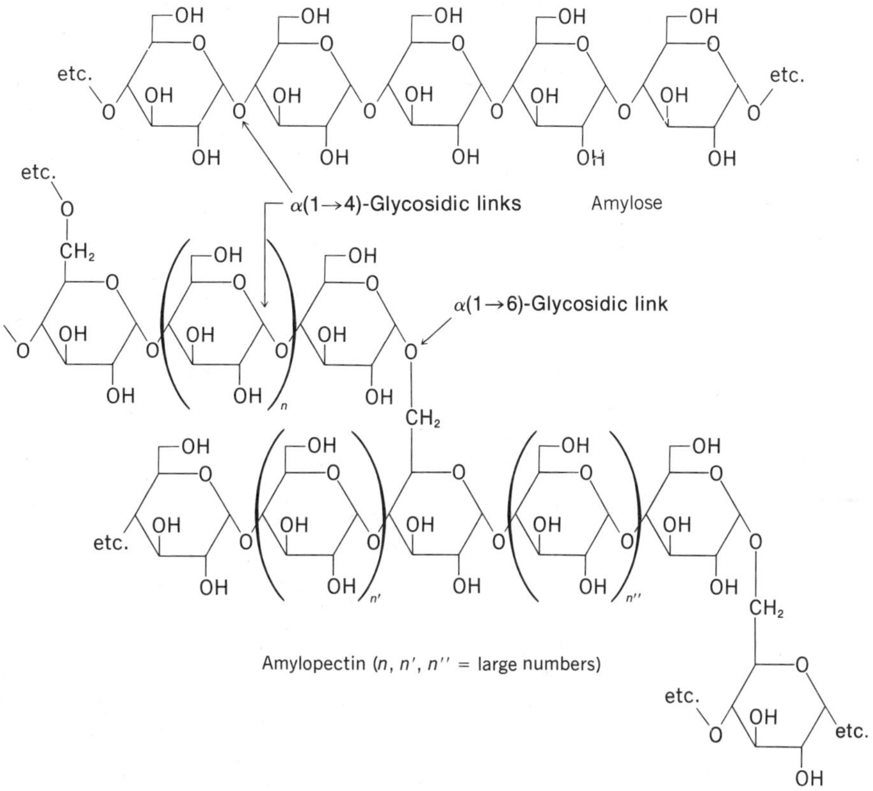

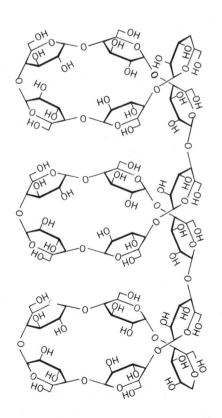

α(1→4)-Glycosidic links   Amylose

α(1→6)-Glycosidic link

Amylopectin (n, n', n'' = large numbers)

FIGURE 19.8
Amylose molecules coil somewhat as indicated here.

FIGURE 19.9

The hydrolysis of amylopectin. *(a)* The circles, either light or dark, represent individual glucose units in a molecule of amylopectin. *(b)* After partial hydrolysis, many maltose molecules have been freed (the pairs of light circles). The remaining, highly branched residue (all dark circles) is a dextrin. [After P. Bernfeld. *Advances in Enzymology,* vol. 12, 1951, p. 390.]

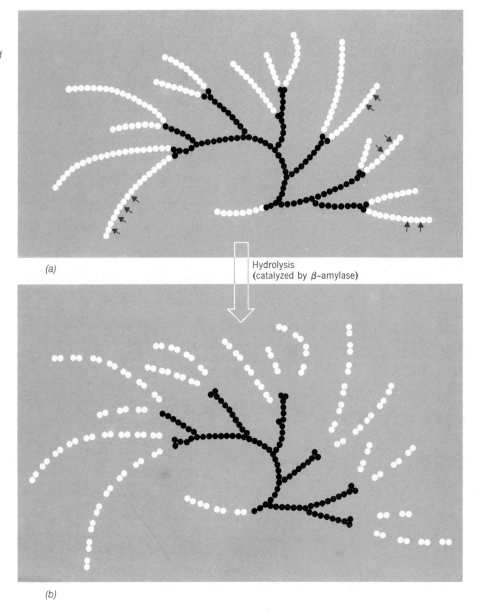

(a)

Hydrolysis
(catalyzed by β–amylase)

(b)

Natural starches are about 10 to 20% amylose and 80 to 90% amylopectin. Neither is a reducing sugar and neither gives a positive Tollens' or Benedict's test. One unique test that starch does give is called the **iodine test** for starch, and it can detect extremely minute traces of starch.[1] When a drop of iodine reagent is added to starch, an intensely purple color develops as the iodine molecules become trapped within the vast network of starch molecules. In a starch sample undergoing hydrolysis, this network gradually breaks up so the system slowly loses its ability to give the iodine test.

The glycosidic links in starch are easily hydrolyzed in the presence of acids or the appropriate enzyme, which humans have. Thus the complete digestion of starch gives us only glucose. The partial hydrolysis of starch produces smaller polymer molecules that make up a

[1] The starch-iodine reagent is made by dissolving iodine, I_2, in aqueous sodium iodide, NaI. Iodine by itself is very insoluble in water, but iodine molecules combine with iodide ions to form the triiodide ion, I_3^-. Molecular iodine is readily available from this ion if some reactant is able to react with it.

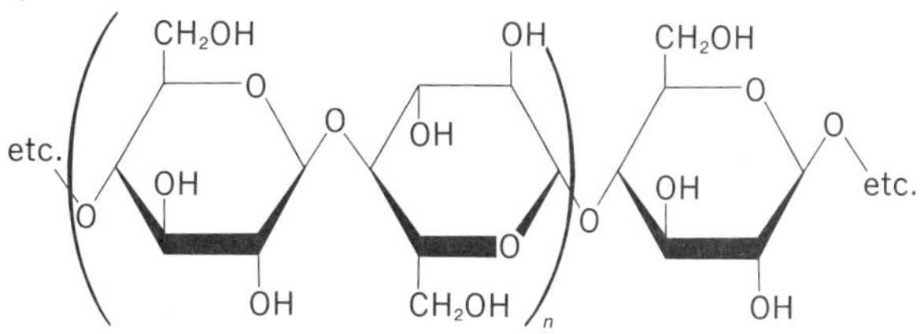

FIGURE 19.10
Cellulose, a linear polymer of β-D-glucose. The value of n varies from 1000 to 13,000 (or 2000 to 26,000 glucose units) in different varieties of cotton. The strength of a cotton fiber comes in part from the thousands of hydrogen bonds that can exist between parallel and overlapping cellulose molecules.

substance known as **dextrin,** which has been used to manufacture mucilage and paste. Figure 19.9 illustrates the progressive stages in the hydrolysis of amylopectin.

Glycogen.   The molecules of **glycogen** are essentially like those of amylopectin, perhaps even more branched. The formula weights of various samples of glycogen have been reported in the range of 300,000 to 100,000,000, which correspond roughly to 1700 to 600,000 glucose units per molecule.

Glycogen is sometimes called animal starch.

We store glucose as glycogen principally in the liver and in muscle tissue. A network of enzymes controls the formation of glycogen when glucose is in a plentiful supply after a meal containing carbohydrates. A different set of enzymes handles the debranching and the hydrolysis of glycogen when glucose molecules are needed between meals or during other periods of fasting. Some of these enzymes are controlled by hormones. We'll learn more details when we study the metabolism of carbohydrates in a later chapter.

Cellulose.   Cellulose, unlike starch or glycogen, is a polymer of the beta form of glucose. Its molecules (Figure 19.10) are unbranched, so they bear a resemblance to amylose, but all of the bridges are $\beta(1 \rightarrow 4)$ instead of $\alpha(1 \rightarrow 4)$. As we have noted, humans have no enzyme that can catalyze the hydrolysis of a beta-glucosidic link, so none of the huge supply of cellulose in the world, or the cellobiose that could be made from it, is nutritionally useful to us. Many bacteria have this enzyme, however, and some strains dwell in the stomachs of cattle and other animals. Bacterial action converts cellulose in hay and other animal feed into molecules that the larger animals can then use.

Cellobiose is to cellulose what maltose is to starch.

SUMMARY

Carbohydrates   Carbohydrates are polyhydroxy aldehydes or ketones or glycosides of these. Those that can't be hydrolyzed are the monosaccharides, which in pure forms exist as cyclic hemiacetals or cyclic hemiketals that can mutarotate and that are reducing sugars.

Monosaccharides   The three nutritionally important monosaccharides are glucose, galactose, and fructose—all in the D-family. Glucose is the chief carbohydrate in blood. Galactose, which differs from glucose only in the orientation of the —OH at C-4, is obtained (together with glucose) from the hydrolysis of lactose. Fructose, a reducing ketohexose, differs from glucose only in the location of the carbonyl group. It's at C-2 in fructose and at C-1 in glucose.

Disaccharides   The disaccharides are glycosides whose molecules break up into two monosaccharide molecules when they react with water. Maltose is made of two glucose units joined by an $\alpha(1 \rightarrow 4)$ glucosidic link. In a molecule of lactose (milk sugar)—a galactoside — a galactose unit joins a glucose unit by a $\beta(1 \rightarrow 4)$ oxygen bridge. In sucrose (cane or beet sugar), there is an oxygen bridge from C-1 of a glucose unit to C-2 of a fructose unit. Both maltose and lactose retain hemiacetal systems, so both mutarotate and are reducing

sugars. They also exist in α- and β-forms. Sucrose is a nonreducing disaccharide. The digestion of these disaccharides is by the following reactions:

$$\text{Maltose} + H_2O \longrightarrow \text{glucose} + \text{glucose}$$

$$\text{Lactose} + H_2O \longrightarrow \text{glucose} + \text{galactose}$$

$$\text{Sucrose} + H_2O \longrightarrow \text{glucose} + \text{fructose}$$

Polysaccharides   Three important polysaccharides of glucose are starch (a plant product), glycogen (an animal product), and cellulose (a plant fiber). In molecules of each, $(1 \rightarrow 4)$ glycosidic links occur. They're alpha bridges in starch and glycogen and beta bridges in cellulose. In the molecules of the amylopectin portion of starch as well as in glycogen, numerous $\alpha(1 \rightarrow 6)$ bridges also occur. No polysaccharide gives a positive test with Tollens' or Benedict's reagents. Starch gives the iodine test. As starch is hydrolyzed its molecules successively break down to dextrins, maltose, and finally, glucose. Humans have enzymes that catalyze the hydrolysis of $\alpha(1 \rightarrow 4)$ glucosidic links, but not the $\beta(1 \rightarrow 4)$ glucosidic links of cellulose.

Absolute configuration Fischer projection structures of open-chain forms are made according to a set of rules. The carbon chain is positioned vertically with any carbonyl group as close to the top as possible. At each chiral carbon, this chain projects rearward. Any groups on bonds that appear horizontal project forward. If the —OH group of a carbohydrate that is farthest from the carbonyl group projects to the right in a Fischer projection structure, the carbohydrate is in the D-family. If this —OH group projects to the left, the substance is in the L-family.

KEY TERMS

The following terms were stressed in this chapter, and they should be mastered before continuing.

absolute configuration	cellulose	glycogen	monosaccharide
aldohexose	dextrin	glycoside	mutarotation
aldose	D-family, L-family	glycosidic link	polysaccharide
amylopectin	disaccharide	iodine test	reducing carbohydrate
amylose	fructose	ketohexose	simple sugar
biochemistry	galactose	ketose	starch
blood sugar	glucose	lactose	sucrose
carbohydrate	glucoside	maltose	

SELECTED REFERENCE

1 N. Sharon. "Carbohydrates." *Scientific American,* November 1980, page 90.

REVIEW EXERCISES

The answers to these Review Exercises are in the *Study Guide* that accompanies this book.

Biochemistry

19.1 Substances in the diet must provide raw materials for what three essentials for life?

19.2 What are the three broad classes of foods?

19.3 What two kinds of compounds are most involved in the molecular basis of information? Which one carries the genetic "blueprints"?

19.4 What is as important to the life of a cell as the chemicals that make it up or that it receives?

Carbohydrate Terminology

19.5 Examine the following structures and identify by letter(s) which structure(s) fit(s) each of the labels. If a particular label is not illustrated by any structure, state so.

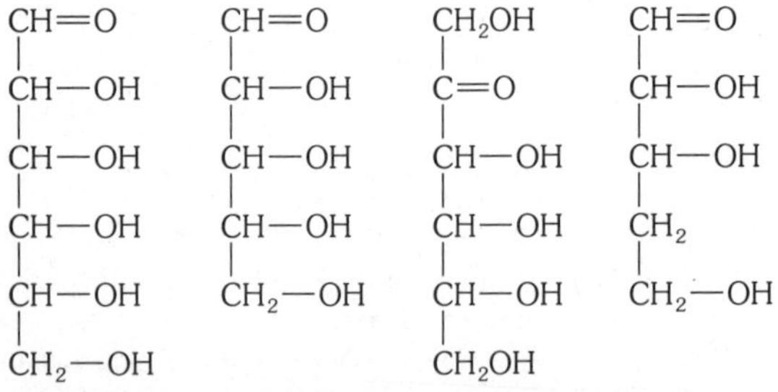

```
   CH=O        CH=O        CH2OH       CH=O
    |           |           |           |
   CH—OH       CH—OH        C=O         CH—OH
    |           |           |           |
   CH—OH       CH—OH       CH—OH        CH—OH
    |           |           |           |
   CH—OH       CH—OH       CH—OH        CH2
    |           |           |           |
   CH—OH       CH2—OH      CH—OH        CH2—OH
    |                       |
   CH2—OH                  CH2OH

     A           B           C           D
```

(a) Ketose(s)   (b) Deoxy sugar(s)
(c) Aldohexose(s) (d) Aldopentose(s)

19.6 Write the structure (open-chain form) that illustrates
(a) any aldopentose
(b) any ketotetrose

19.7 What is the structure and the common name of the simplest aldose?

19.8 What is the structure and the common name of the simplest ketose?

19.9 A sample of 0.0001 mol of a carbohydrate reacted with water in the presence of a catalyst and 1 mol of glucose was produced. Classify this carbohydrate as a mono-, di-, or polysaccharide.

19.10 An unknown carbohydrate failed to give a positive Benedict's test. Classify it as a reducing or a nonreducing carbohydrate.

Monosaccharides

19.11 What is the name of the most abundant carbohydrate in (a) blood and in (b) corn syrup?

19.12 What is the name of the ketose present in honey?

19.13 Write the structure (open-chain form) of any carbohydrate that has all of the following properties. It forms a tetraacetate ester. It is oxidized without loss of carbon or hydrogen by Tollens' reagent to give a carboxylic acid. It can be reduced by a series of steps to pentane.

19.14 An unknown carbohydrate could be reduced by a series of steps to butane. It gave a positive Benedict's test, and its molecules had just one chiral carbon. What is a structure consistent with these facts?

19.15 What is the structure of the open form of the following cyclic hemiacetal? (Write the open form with its chain coiled in the same way it is coiled in the closed form.)

19.16 Examine the following structure. If you judge that it is either a cyclic hemiketal or a cyclic hemiacetal, write the structure of the open-chain form (coiled in like manner as the chain of the ring).

19.17 Mannose mutarotates like glucose. Mannose is identical with glucose except that in the cyclic structures the —OH at C-2 in mannose projects on the same side of the ring as the CH$_2$OH group. Write the structures of the three forms of mannose that are in equilibrium after mutarotation gives a steady value of specific rotation. Identify which corresponds to α-mannose and which to β-mannose.

19.18 Allose is identical with glucose except that in its cyclic forms the —OH group at C-3 projects on the opposite side of the ring from the CH$_2$OH group. Allose mutarotates like glucose. Write the structures of the three forms of allose that are in equilibrium after mutarotation gives a final value of specific rotation. Which structures are α- and β-allose?

19.19 If less than 0.05% of all galactose molecules are in their open-chain form at equilibrium in water, how can galactose give a strong, positive Tollens' test, a test for the aldehyde group?

19.20 At equilibrium, after mutarotation, a glucose solution consists of 36% α-glucose and 64% β-glucose (and just a trace of the open form). Suppose that in some enzyme-catalyzed process the beta-form is removed from this equilibrium. What becomes of the other forms of glucose?

19.21 Study the cyclic form of fructose on page 456 again. If its designation as β-fructose signifies a particular relationship between the CH$_2$OH group at C-5 and the —OH group at C-2, what must be the cyclic formula of α-fructose?

19.22 Is the following structure that of α-fructose, β-fructose, or something else? Explain.

19.23 Write the cyclic structure of α-3-deoxyribose and draw an arrow that points to its hemiacetal carbon.

19.24 Could 4-deoxyribose exist as a cyclic hemiacetal with a five-membered ring (one of whose atoms is O)? Explain.

Absolute Configurations

19.25 Write the Fischer projection formula of L-glucose. (Refer to Figure 19.6.)

19.26 What is the Fischer projection formula of D-2-deoxyribose? (Refer to Figure 19.6.)

19.27 Suppose that the aldehyde group of D-glyceraldehyde is oxidized to a carboxylic acid group, and that this is then converted to a methyl ester under conditions that do not touch any of the four bonds to C-2. What is the Fischer projection formula of this methyl ester? To what family, D or L, does it belong? Explain.

19.28 Sorbose has the structure given below. It is made by the fermentation of sorbitol, and hundreds of tons of sorbose are used each year to make vitamin C. In what configurational family, D or L, is sorbose?

19.29 Sorbitol, C$_6$H$_{14}$O$_6$, is found in the juices of many fruits and berries (e.g., pears, apples, cherries, and plums). It can be made by the addition of hydrogen to D-glucose:

$$C_6H_{12}O_6 + H_2 \xrightarrow[\text{Heat, pressure}]{\text{Ni}} C_6H_{14}O_6$$

D-Glucose   D-Sorbitol

Write the Fischer projection formula of D-sorbitol.

19.30 The magnesium salt of D-gluconic acid (Glucomag) is used as an antispasmodic and to treat dysmenorrhea. D-gluconic acid forms by the mild oxidation of D-glucose:

$$C_6H_{12}O_6 + (O) \longrightarrow C_6H_{12}O_7$$

D-Glucose   D-Gluconic acid

What is the Fischer projection formula of D-gluconic acid?

Glycosides

19.31 Using cyclic structures, write the structures of ethyl α-gluco-side and ethyl β-glucoside. Are these two enantiomers or are they some other kind of stereoisomers?

19.32 What are the structures of methyl α-galactoside and methyl β-galactoside? Could these be described as cis-trans isomers? Explain.

Disaccharides

19.33 What are the names of the three nutritionally important disac-charides?

19.34 What is invert sugar?

19.35 Why isn't sucrose a reducing sugar?

19.36 Examine the following structure and answer the questions about it.

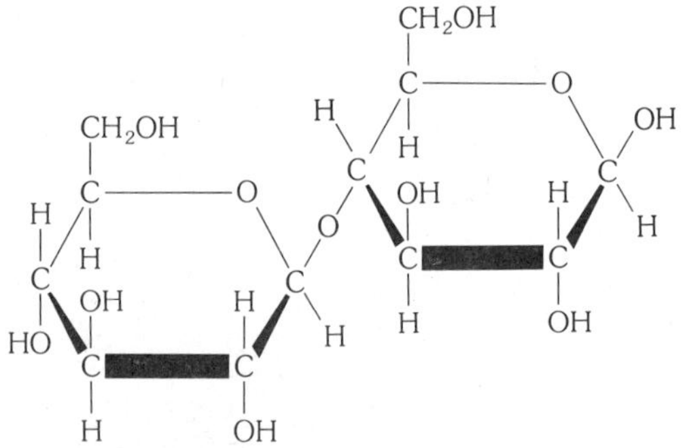

(a) Does it have a hemiacetal system? Where? (Draw an arrow to it or circle it.)
(b) Does it have an acetal system? Where? (Circle it.)
(c) Does this substance give a positive Benedict's test? Explain.
(d) In what specific structural way does it differ from mal-tose?
(e) Write the cyclic structures and names of the products of the acid-catalyzed hydrolysis of this compound. (This compound, incidentally, is cellobiose.)

19.37 Trehalose is a disaccharide found in young mushrooms and yeast, and it is the chief carbohydrate in the hemolymph of certain insects. On the basis of its structural features, answer the questions below.

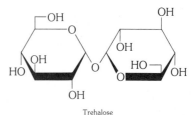

Trehalose

(a) Is trehalose a reducing sugar? Explain.
(b) Can trehalose mutarotate? Explain.
(c) Identify, by name only, the products of the hydrolysis of trehalose.

19.38 Maltose has a hemiacetal system. Write the structure of mal-tose in which this group has changed to the open form.

19.39 When lactose undergoes mutarotation, one of its rings opens up. Write this open form of lactose.

Polysaccharides

19.40 Name the polysaccharides that give only D-glucose when they are completely hydrolyzed.

19.41 What is the main structural difference between amylose and cellulose?

19.42 How are amylose and amylopectin alike structurally?

19.43 How are amylose and amylopectin different structurally?

19.44 Why can't humans digest cellulose?

19.45 What is the iodine test? Describe the reagent, state what it is used to test for, and what is seen in a positive test.

19.46 How do amylopectin and glycogen compare structurally?

19.47 How does the body use glycogen?

Chapter 20
Lipids

The kitten won't wait for butter, and doesn't prefer it to cream either. Cream is rich in butterfat, which pioneers separated from the water and proteins of cream using a simple churn such as on the woman's right. Then the butterfat was salted and blended as shown here. We'll learn about fats and oils in this chapter.

20.1 WHAT LIPIDS ARE

The edible fats and oils consist of hydrocarbon-like molecules that are esters of long-chain fatty acids and glycerol.

Lipids are defined not by their structures so much as by an experimental operation—extraction using a nonpolar solvent. When undecomposed plant or animal material is crushed and ground with a nonpolar solvent such as ether, whatever dissolves is classified as a **lipid.** Among the many substances that won't dissolve are carbohydrates, proteins, other very polar organic substances, inorganic salts, and water.

Extraction means to shake or stir a mixture with a solvent that dissolves just part of the mixture.

Types of Lipids. One of the major classes of lipids — the **saponifiable lipids** — consists of compounds whose molecules have one or more groups that can be hydrolyzed or saponified. A number of families are in this class, including the waxes, the neutral fats, the phospholipids, and the glycolipids. The other principal class of lipids — the **nonsaponifiable lipids** — lack groups that can be hydrolyzed or saponified. The steroids, including many sex hormones, are in this class. The chart in Figure 20.1 outlines the many kinds of lipids.

Butterfat and salad oils are neutral fats.

Cholesterol and several sex hormones are steroids.

Waxes. The waxes, the simplest of the saponifiable lipids, occur as protective coatings on fruit and leaves as well as on fur, feathers, and skin. Nearly all **waxes** are esters of long-chain monohydric alcohols and long-chain monocarboxylic acids in each of which there is an even number of carbons. As many as 26 to 34 carbon atoms can be incorporated in each of the alcohol and the acid units, which makes the waxes almost totally hydrocarbon-like.

An acyl group has the general structure:

$$R-\overset{\overset{\displaystyle O}{\|}}{C}-$$

$$R-O-\overset{\overset{\displaystyle O}{\|}}{C}-R'$$

| Alcohol unit | Fatty acyl unit |

Components of waxes

Any particular wax, such as beeswax, consists of a mixture of similar compounds that share the feature shown above. In molecules of lanolin (wool fat), however, the alcohol portion is contributed by steroid alcohols, which have large ring systems that we'll study in Section 20.4. Waxes exist in sebum, a secretion of human skin that helps to keep the skin supple.

Lanolin is used to make cosmetic skin lotions.

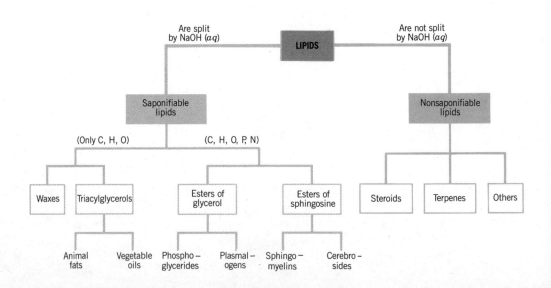

FIGURE 20.1
Lipid families.

PRACTICE EXERCISE 1   One particular ester in beeswax can be hydrolyzed to give a straight-chain primary alcohol with 26 carbons and a straight-chain carboxylic acid with 28 carbons. Write the structure of this ester.

Triacylglycerols.   The most abundant group of lipids is the neutral fats. These include lard (hog fat), tallow (beef fat), butterfat — all animal fats — as well as such plant oils as olive oil, cottonseed oil, corn oil, peanut oil, soybean oil, coconut oil, and linseed oil. These neutral fats — so named because their molecules carry no ionic (or otherwise highly polar) sites — are all **triacylglycerols.** An older, still widely used name for them is **triglycerides.** Their molecules consist of esters between glycerol and long-chain monocarboxylic acids called **fatty acids.**

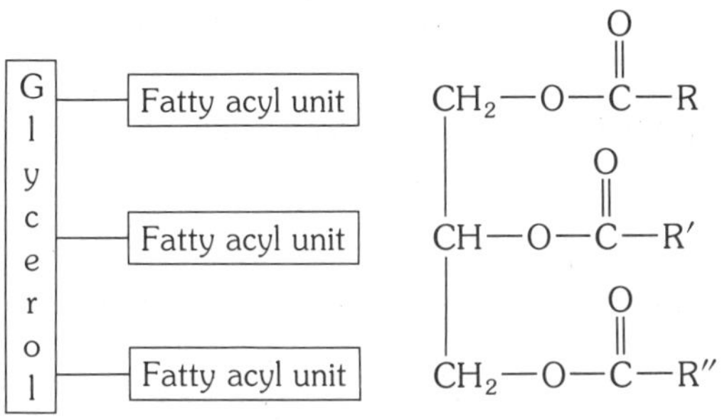

Components of neutral fats and oils (triacylglycerols)

The three acyl units in a typical triacylglycerol are not identical. They usually are from three different fatty acids. Thus the fats and oils of any given source are not pure in the sense that all of their molecules are identical. Instead, they are mixtures of different molecules that share common structural features. The differences among the molecules are in the specific fatty acids that are incorporated. Although we can't write, for example, the structure of cottonseed oil, we can describe a typical molecule such as that given by structure **1.**

$$CH_2-O-\overset{\displaystyle O}{\overset{\|}{C}}(CH_2)_7CH=CH(CH_2)_7CH_3$$
$$CH-O-\overset{\displaystyle O}{\overset{\|}{C}}(CH_2)_{14}CH_3$$
$$CH_2-O-\overset{\displaystyle O}{\overset{\|}{C}}(CH_2)_7CH=CHCH_2CH=CH(CH_2)_4CH_3$$

1

Notice that the middle carbon of the glycerol unit in **1** is a chiral carbon.

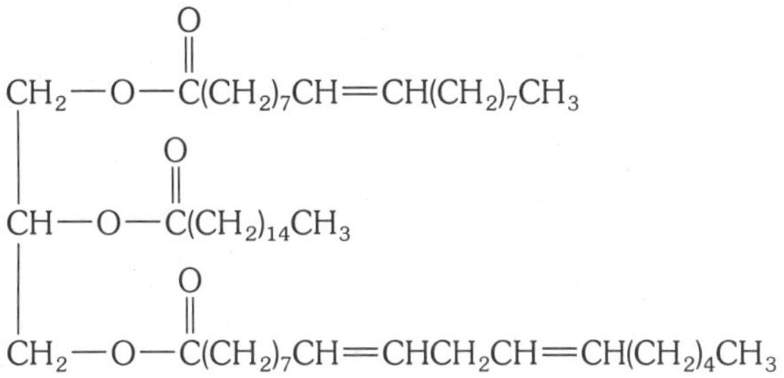

In a particular kind of fat or oil, certain fatty acids tend to predominate, certain others are either absent or are present in trace amounts, and virtually all of the molecules are triacylglycerols. Data for several fats and oils are listed in Table 20.1.

Fatty Acids.   The fatty acids obtained from the lipids of most plants and animals share the following features.

1.   They are usually monocarboxylic acids, RCO_2H.

2.   The R— group is usually an unbranched chain.

3.   The number of carbon atoms is almost always even.

4.   The R— group can be saturated, or it can have one or more double bonds, which are cis.

TABLE 20.1
Fatty Acids Obtained from Neutral Fats And Oils

Type	Fat or Oil	Average Composition of Fatty Acids (%)					
		Myristic Acid	Palmitic Acid	Stearic Acid	Oleic Acid	Linoleic Acid	Others
Animal Fats	Butter	8–15	25–29	9–12	18–33	2–4	a
	Lard	1–2	25–30	12–18	48–60	6–12	b
	Beef tallow	2–5	24–34	15–30	35–45	1–3	b
Vegetable Oils	Olive	0–1	5–15	1–4	67–84	8–12	
	Peanut	—	7–12	2–6	30–60	20–38	
	Corn	1–2	7–11	3–4	25–35	50–60	
	Cottonseed	1–2	18–25	1–2	17–38	45–55	
	Soybean	1–2	6–10	2–4	20–30	50–58	c
	Linseed	—	4–7	2–4	14–30	14–25	d
Marine Oils	Whale	5–10	10–20	2–5	33–40	—	e
	Fish	6–8	10–25	1–3	—	—	e

a Also, 3–4% butyric acid, 1–2% caprylic acid, 2–3% capric acid, 2–5% lauric acid.

b Also, linolenic acid, 1%.

c Also, linolenic acid, 5–10%.

d Also, linolenic acid, 45–60%.

e Large percentages of other highly unsaturated fatty acids.

The most abundant saturated fatty acids are palmitic acid, $CH_3(CH_2)_{14}CO_2H$, and stearic acid, $CH_3(CH_2)_{16}CO_2H$, which have 16 and 18 carbons, respectively. Refer back to Table 16.1 for the other saturated fatty acids used to make triacylglycerols. (In this table, they are the acids that follow acetic acid and that have an even number of carbons, but their acyl units are present in only relatively small amounts in triacylglycerols.)

The most prevalent of the unsaturated fatty acids, listed in Table 20.2, are oleic, linoleic, and linolenic acids, all with 18-carbon skeletons. Oleic acid is the most abundant and most widely distributed fatty acid in nature. The double bonds in these acids have a cis geometry.

Notice from the data in Table 20.1 how the animal fats and the vegetable oils differ. The latter incorporate more double bonds per molecule than the former. The vegetable oils are said to be *polyunsaturated,* because the acyl units from oleic and linoleic acids occur more often in these oils than they do in the animal fats. The saturated fatty acyl units of palmitic and stearic acids are far more common in animal fats than in vegetable oils. This is why the animal fats are likely to be solids at room temperature and the vegetable oils are liquids. As you can see from the data in Table 20.2, adding alkene double bonds to fatty acyl units lowers the melting points of the acids (and of the triacylglycerols that are made from such unsaturated acids).

PRACTICE EXERCISE 2   Write the structure of oleic acid in a way that correctly shows the geometry of the alkene group — a cis relationship for the carbons in the units that are attached to the carbons of the double bond.

The properties of the fatty acids are those to be expected of compounds that have a carboxyl group, double bonds (in some), and long hydrocarbon chains. For example, they are insoluble in water and soluble in nonpolar solvents. They can form salts, and they can be esterified. Those that have alkene groups react with bromine and they take up hydrogen in the presence of a catalyst.

One of the unusual families of fatty acids are the *prostaglandins,* which are C-20 carboxylic acids with five-membered rings. They have a wide variety of effects in the body and are under study as potential pharmaceuticals. Special Topic 20.1 discusses them further.

TABLE 20.2
Common Unsaturated Fatty Acids

Name	Number of Double Bonds	Total Number of Carbons	Structure	Melting Point (°C)
Palmitoleic acid	1	16	$CH_3(CH_2)_5CH{=}CH(CH_2)_7CO_2H$	32
Oleic acid	1	18	$CH_3(CH_2)_7CH{=}CH(CH_2)_7CO_2H$	4
Linoleic acid	2	18	$CH_3(CH_2)_4CH{=}CHCH_2CH{=}CH(CH_2)_7CO_2H$	−5
Linolenic acid	3	18	$CH_3CH_2CH{=}CHCH_2CH{=}CHCH_2CH{=}CH(CH_2)_7CO_2H$	−11
Arachidonic acid	4	20	$CH_3(CH_2)_4CH{=}CHCH_2CH{=}CHCH_2CH{=}CHCH_2CH{=}CH(CH_2)_3CO_2H$	−50

20.2 CHEMICAL PROPERTIES OF TRIACYLGLYCEROLS

Triacylglycerols can be hydrolyzed (digested), saponified, and hydrogenated.

Hydrolysis of Triacylglycerols. Enymes in the digestive tracts of humans and animals catalyze the hydrolysis of the ester links in triacylglycerols. In general:

This hydrolysis of triacylglycerols is what happens when we digest fats and oils.

Triacylglycerol — Glycerol — Fatty acids

A specific example is

1 — Glycerol — Oleic acid — Palmitic acid — Linoleic acid

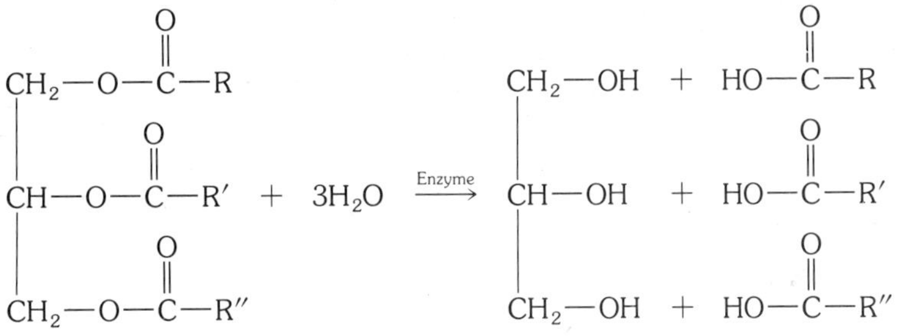

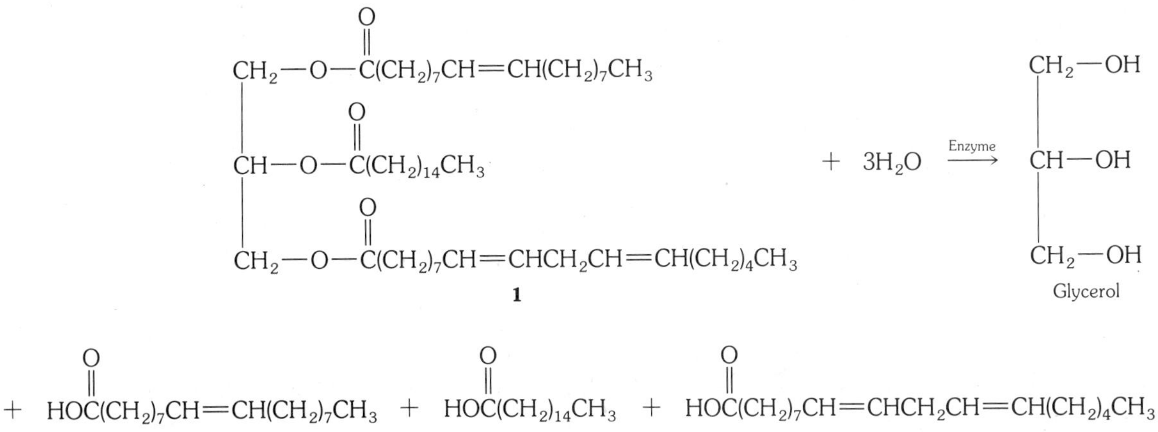

SPECIAL TOPIC 20.1   THE PROSTAGLANDINS

The prostaglandins were discovered in the mid-1930s by a Swedish scientist, Ulf von Euler (Nobel prize, 1970), but they didn't arouse much interest in medical circles until the late 1960s, when it became apparent that these compounds, which occur widely in the body, affect a large number of processes. Their general name comes from the organ, the prostate gland, from which they were first obtained. About 20 are known, and they occur in four major subclasses designated as PGA, PGB, PGE, and PGF. (A subscript is generally placed after the third letter to designate the number of alkene double bonds that occur outside of the five-membered ring.) The structures of some typical examples are shown here.

Prostaglandins are made in membranes from C-20 fatty acids such as arachidonic acid. By coiling a molecule of this acid, as shown, you can see how its structure needs only a ring closure (suggested by the dashed arrow) and three more oxygen atoms to become PGF$_2$. The oxygen atoms are all provided by molecular oxygen itself.

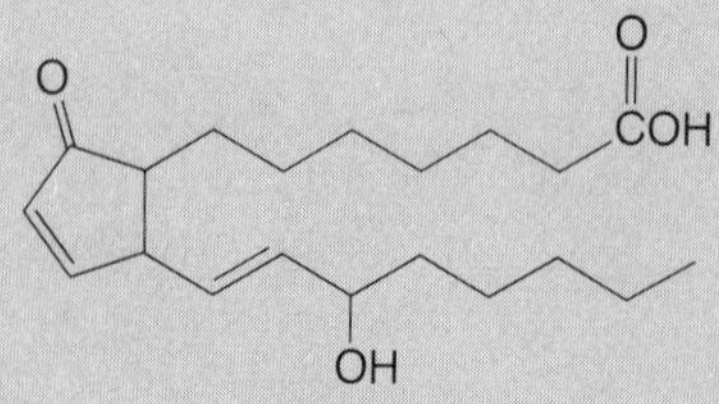

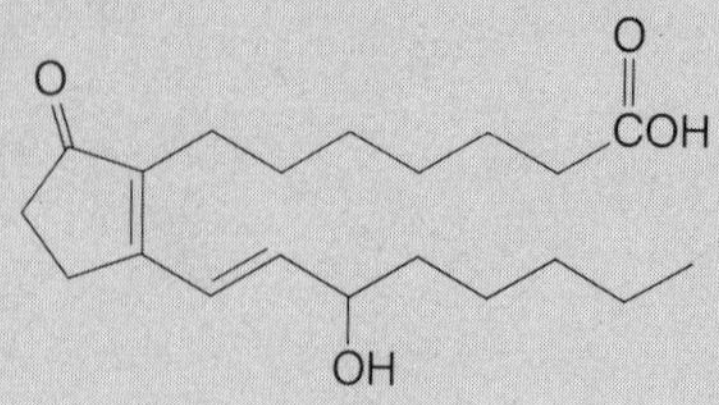

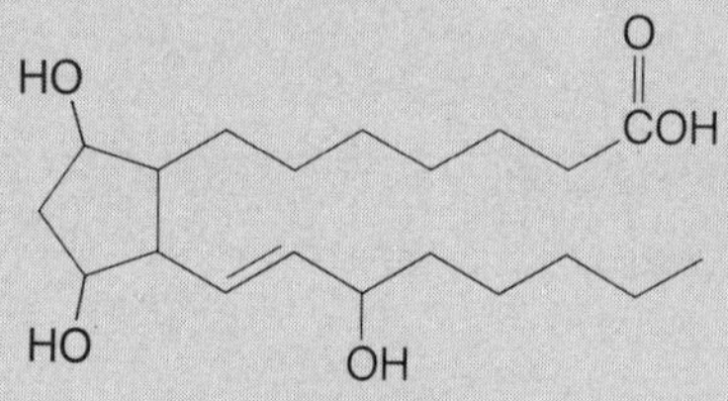

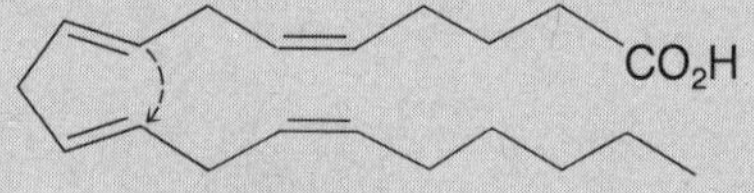

Prostaglandins as Chemical Messengers The prostaglandins work inside cells, but just how they work is not yet fully understood. One feature seems clear, however. They are not hormones, but they work together with hormones to modify the chemical messages that hormones bring to cells. Some prostaglandins stimulate inflammation in a tissue, and it is interesting that aspirin—an inflammation reducer—does exactly the opposite. Some of aspirin's effect is caused by its ability to inhibit the synthesis of prostaglandins.

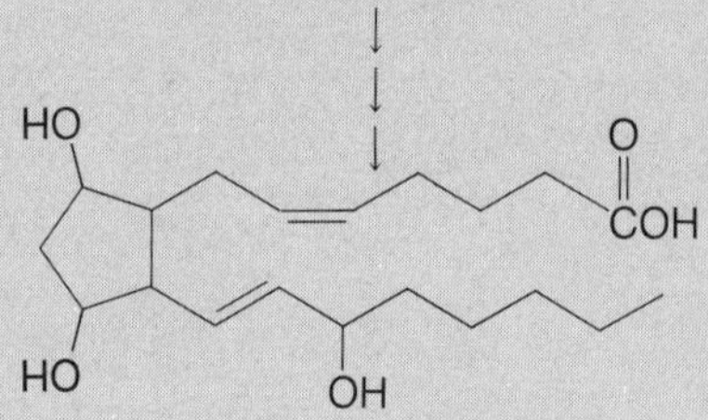

Prostaglandins as Pharmaceuticals In experiments that use prostaglandins as pharmaceuticals, they have been found to have an astonishing variety of effects. One prostaglandin induces labor at the end of a pregnancy. Another stops the flow of gastric juice while the body heals an ulcer. Other possible uses are in treating high blood pressure, rheumatoid arthritis, asthma, nasal congestion, and certain viral diseases.

Saponification of Triacylglycerols. The saponification of the ester links in triacylglycerols by the action of a strong base (e.g., NaOH or KOH) gives glycerol and a mixture of the salts of fatty acids. These salts are soaps, and how they exert their detergent action is described in Special Topic 20.2. In general:

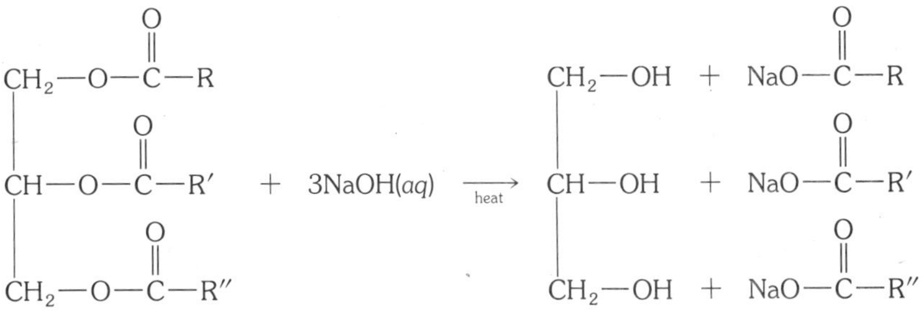

$$CH_2-O-\overset{\overset{\displaystyle O}{\|}}{C}-R$$

$$CH-O-\overset{\overset{\displaystyle O}{\|}}{C}-R' \quad + \quad 3NaOH(aq) \xrightarrow[\text{heat}]{} \quad CH-OH \quad + \quad NaO-\overset{\overset{\displaystyle O}{\|}}{C}-R'$$

$$CH_2-O-\overset{\overset{\displaystyle O}{\|}}{C}-R''$$

$$CH_2-OH \quad + \quad NaO-\overset{\overset{\displaystyle O}{\|}}{C}-R$$

$$CH_2-OH \quad + \quad NaO-\overset{\overset{\displaystyle O}{\|}}{C}-R''$$

Glycerol   Mixture of salts

PRACTICE EXERCISE 3   Write a balanced equation for the saponification of **1** with sodium hydroxide.

Hydrogenation of Vegetable Oils.   When hydrogen is made to add to some of the double bonds in vegetable oils, the oils become like animal fats, both physically and structurally. One very practical consequence is that they change from being liquids to solids at room temperature. Many people prefer solid, lard-like shortening for cooking, instead of liquid oils. Therefore the manufacturers of such products as Crisco and Spry use inexpensive, readily available vegetable oils — corn oil and cottonseed oil are commonly used — and catalytically add hydrogen to some (not all) of the alkene groups in their molecules.

Hydrogenated vegetable oils are chemically identical to animal fats.

PRACTICE EXERCISE 4   Write the balanced equation for the complete hydrogenation of the alkene links in structure **1.**

The chief lipid material in oleomargarine is produced from vegetable oils in the same way. The hydrogenation is done with special care so that the final product can melt on the tongue, a property that makes butterfat such a desirable commodity. (If all of the alkene groups in a vegetable oil were hydrogenated, instead of just some of them, the product would be just like tallow or mutton fat — relatively hard materials that would not melt on the tongue.)

The popular brands of peanut butter whose peanut oils do not separate are made by the partial hydrogenation of the oil in real peanut butter. The lipid present becomes a solid at room temperature (and therefore it can't separate).

20.3   PHOSPHOLIPIDS

Phospholipid molecules have very polar or ionic sites in addition to long hydrocarbon chains.

Phospholipids are esters either of glycerol or sphingosine. Sphingosine is a long-chain, dihydric amino alcohol with one double bond.

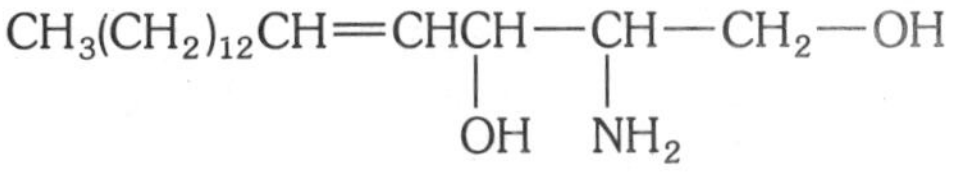

$$CH_3(CH_2)_{12}CH=CHCH-CH-CH_2-OH$$
$$\underset{OH}{|} \quad \underset{NH_2}{|}$$

Sphingosine

SPECIAL TOPIC 20.2   SOAP AND DETERGENTS

Soap Water is a very poor cleansing agent because it can't penetrate greasy substances, the "glues" that bind soil to skin and fabrics. When just a little soap is present, however, water cleans very well, especially warm water. Soap is a simple chemical, a mixture of the sodium or potassium salts of the long-chain fatty acids obtained by the saponification of fats or oils.

Detergents Soap is just one kind of detergent. All detergents are surface-active agents that lower the surface tension of water. All consist of ions or molecules that have long hydrocarbon portions plus ionic or very polar sections at one end. The accompanying structures illustrate these features and show the varieties of detergents that are available.

Although soap is manufactured, it's not called a synthetic detergent. This term is limited to detergents that are not soap, that is, not the salts of naturally-occurring fatty acids obtained by the saponification of lipids. Most synthetic detergents are salts of sulfonic acids, but others have different kinds of ionic or polar sites. The great advantage of synthetic detergents is that they work in hard water and are not precipitated by the hardness ions — Mg^{2+}, Ca^{2+}, and the two ions of iron. The anions of the fatty acids present in soap form messy precipitates with these ions. The anions of synthetic detergents do not have this property.

The accompanying figure shows how detergents work. In (a) we see the hydrocarbon tails of the detergent work their way into the hydrocarbon environment of the grease layer. ("Like dissolves like" is the principle at work here.) The ionic heads stay in the water phase, and the grease layer becomes pincushioned with electrically charged sites. In (b) we see the grease layer breaking up, aided with some agitation or scrubbing. Part (c) shows a magnified view of grease globules studded with ionic groups and, being like-charged, these globules repel each other. They also tend to dissolve in water, so they are ready to be washed down the drain.

$$CH_3(CH_2)_{14}CO_2^-Na^+$$

Soap — an anionic detergent

$$CH_3(CH_2)_8OSO_3^-Na^+$$

A sodium alkyl sulfate — an anionic detergent

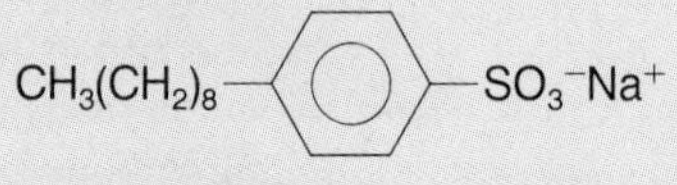

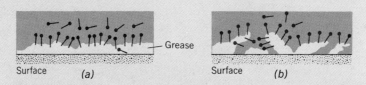

A sodium alkylbenzenesulfonate — an anionic detergent

$$CH_3(CH_2)_{11}\overset{+}{N}(CH_3)_3Cl^-$$

A trimethylalkylammonium ion — a cationic detergent

$$CH_3(CH_2)_8\text{---}(OCH_2CH_2O)_n H$$

A nonionic detergent

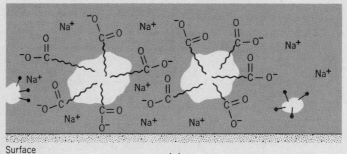

How detergents work.

Phosphoglycerides.   Molecules of **phosphoglycerides** have two ester bonds from glycerol to fatty acids plus one ester bond to phosphoric acid. The phosphoric acid unit, in turn, is joined by a phosphate ester link to a small alcohol molecule. When this link is absent, the material is called phosphatidic acid.

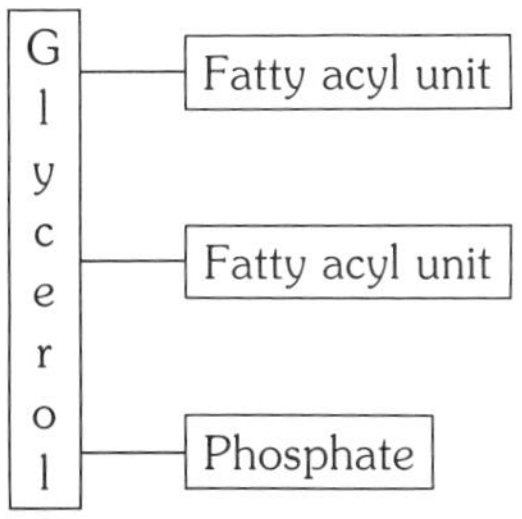

Components of
phosphatidic acid

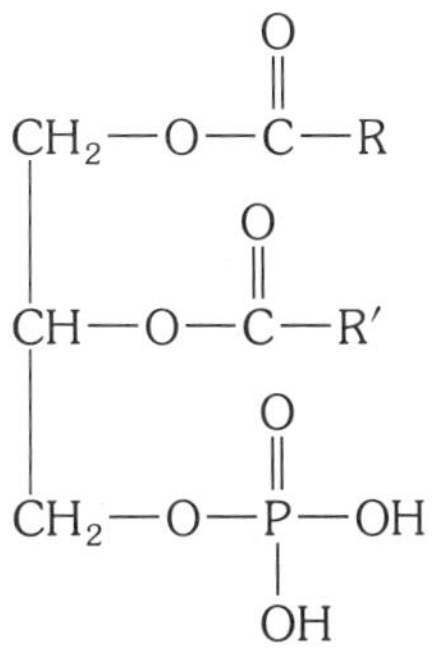

Phosphatidic
acid

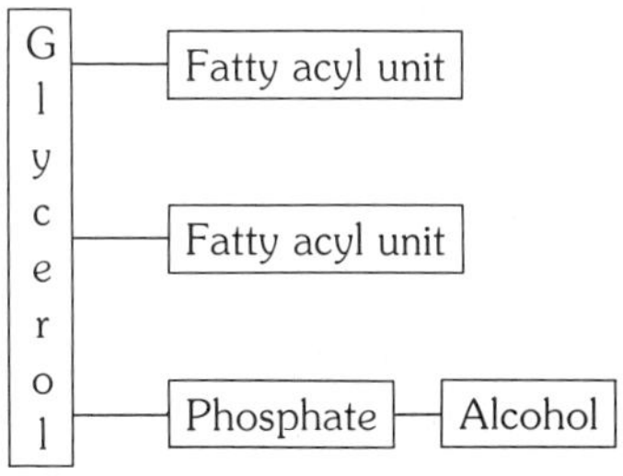

Components of
phosphoglycerides

Lecithin is from the Greek lekitos, egg yolk — a rich source of this phospholipid.

The three principal phosphoglycerides are esters that are formed between phosphatidic acid and either choline, ethanolamine, or serine to give, respectively, phosphatidylcholine (lecithin), **2,** phosphatidylethanolamine (cephalin), **3,** and phosphatidylserine, **4.**

$$\overset{+}{HOCH_2CH_2N(CH_3)_3}$$ $$HOCH_2CH_2NH_2$$ $$HOCH_2\underset{\underset{NH_3^+}{|}}{C}HCO_2^-$$

Choline
(a cation)

Ethanolamine

Serine
(an amino alcohol)

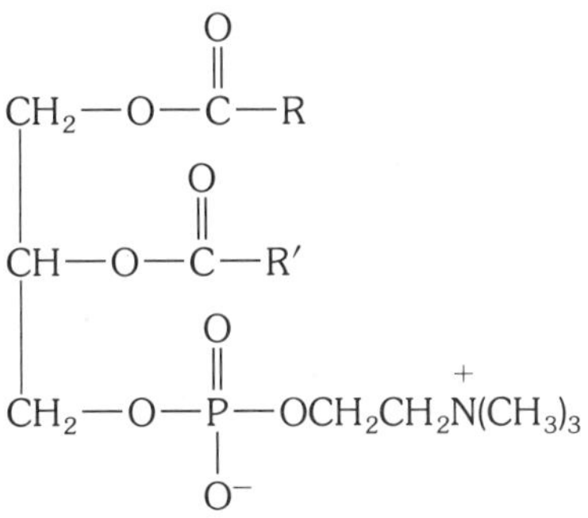

2
Phosphatidylcholine
(lecithin)

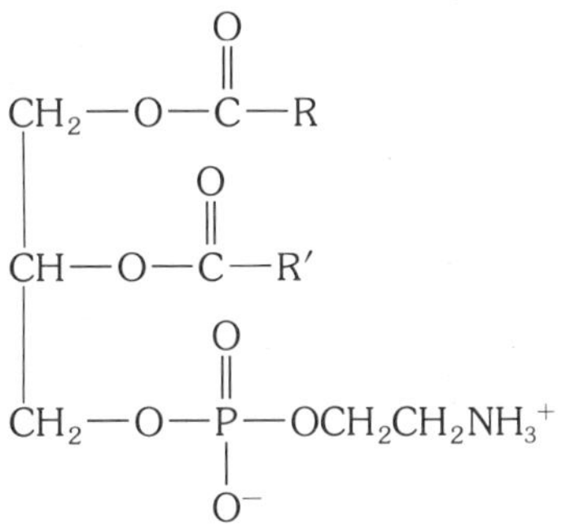

3
Phosphatidylethanolamine
(cephalin)

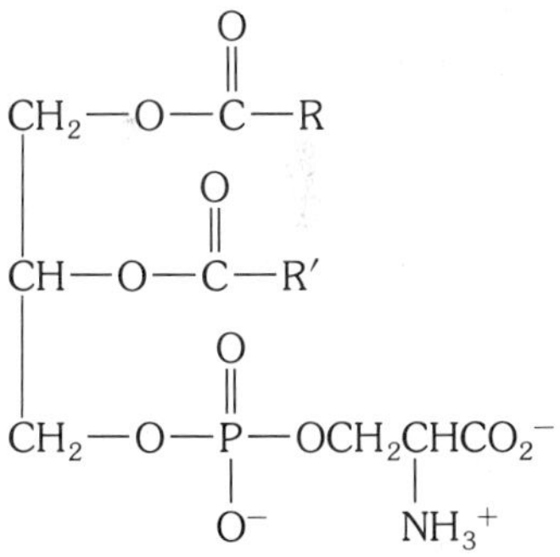

4
Phosphatidylserine

Cephalin is from the Greek kephale, head. Cephalin is found in brain tissue.

As the structures of **2, 3,** and **4** show, one part of each phosphoglyceride molecule is very polar because it carries full electrical charges. The remainder is nonpolar and hydrocarbon-like. These characteristics have important implications in understanding how phosphoglycerides are used to make cell membranes (Section 20.5).

When pure, lecithin is a clear, waxy solid that is very hygroscopic. In air, it is quickly attacked by oxygen, which makes it turn brown in a few minutes. Lecithin is a powerful emulsifying agent for triacylglycerols, and this is why egg yolks, which contain it, are used to make the emulsions found in mayonnaise, ice cream, custards, and cake dough.

Plasmalogens.   The **plasmalogens** make up another family of glycerol-based phospholipids, and they occur widely in the membranes of nerve cells and muscle cells. They differ from the other phosphoglycerides by the presence of an unsaturated ether group instead of an acyl group at one end of the glycerol unit.

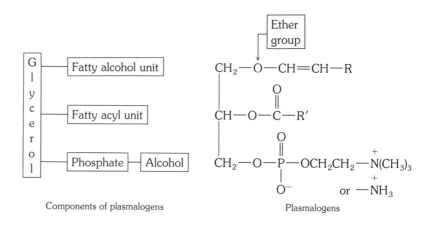

Components of plasmalogens   Plasmalogens

Sphingolipids.   The two types of sphingosine-based lipids or **sphingolipids** are the sphingomyelins and the cerebrosides, and they are also important constituents of cell membranes. The sphingomyelins are phosphate diesters involving sphingosine. Their acyl units occur as acylamido parts, and they come from unusual fatty acids that are not found in neutral fats.

The cerebrosides are not actually phospholipids. Instead they are **glycolipids,** lipids with a sugar (i.e., glycose) unit and not a phosphate ester system. The sugar unit is usually that of D-galactose, or D-glucose, or amino derivatives of these.

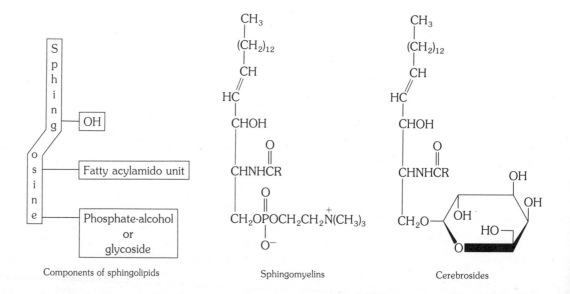

Components of sphingolipids   Sphingomyelins   Cerebrosides

20.4 STEROIDS

Cholesterol and other steroids are nonsaponifiable lipids.

Steroids are high-formula-weight aliphatic compounds whose molecules include the characteristic four-ring feature called the steroid nucleus. It consists of three six-membered rings and one five-membered ring, as seen in structure **5.** Several steroids are very active, physiologically.

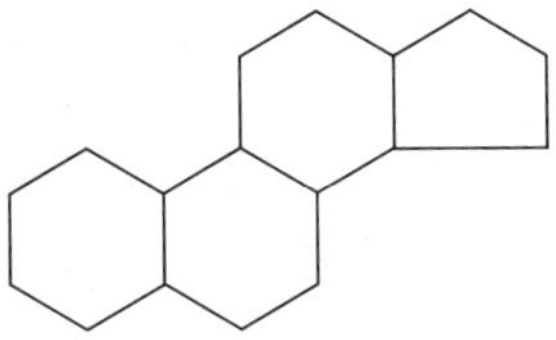

Steroid nucleus
5

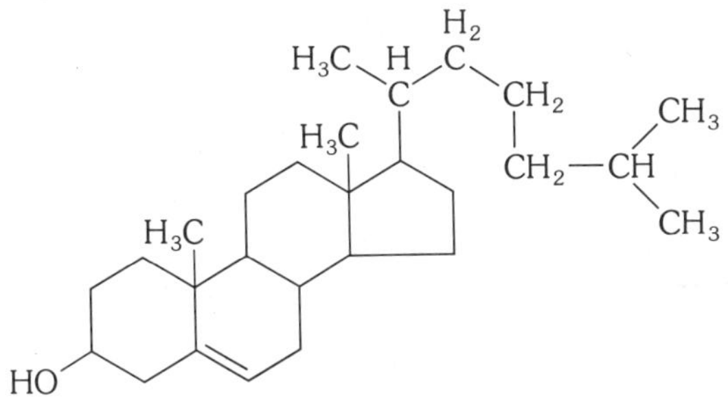

Cholesterol
(Greek: *chole,* bile; *stereos,* solid; -ol, alcohol)

Table 20.3 lists several steroids and their functions, and you can see the large range of properties they have in the body.

Steroid alcohols are called *sterols.*

Cholesterol.   Cholesterol is an unsaturated steroid alcohol that makes up a significant part of the membranes of cells and is the chief constituent in gallstones. It's the body's raw material for making the bile salts and the steroid hormones, including the sex hormones listed in Table 20.3.

Cholesterol enters the body via the diet, but about 800 mg per day is synthesized normally in the body, chiefly in the liver, from acetate units. The relationship between cholesterol and the risk of heart disease will be discussed when we study the metabolism of lipids. We need to know more about proteins, genes, and enzymes before we can discuss this relationship.

20.5 CELL MEMBRANES

Cell membranes consist of a lipid bilayer that includes molecules of proteins and cholesterol.

Cell membranes are made of both lipids and proteins. The principal lipids are the phospholipids, the glycolipids, and cholesterol. The proteins often serve as enzymes, or they act as receptor molecules for various nutrients, or for hormones or the small molecules (neurotransmitters) that move across gaps between cells in the nervous system when a nerve signal is on the move. A **receptor molecule** is one that has a unique shape that enables the molecule of a compound that it is supposed to receive, and only this molecule, to fit to it. In this way a receptor molecule "recognizes" among the hundreds of kinds of molecules that bump against it just the molecules of one compound. Once these molecules fit to their receptor, further biochemical changes occur. An enzyme in the cell might be activated, for example. Some proteins act as "pumps" for moving solutes across a cell membrane. Other proteins are formed into small channels through which certain ions can migrate.

TABLE 20.3
Important Steroids

Vitamin D₃ precursor

Irradiation of this derivative of cholesterol by ultraviolet light opens one of the rings to produce vitamin D₃. Meat products are sources of this compound.

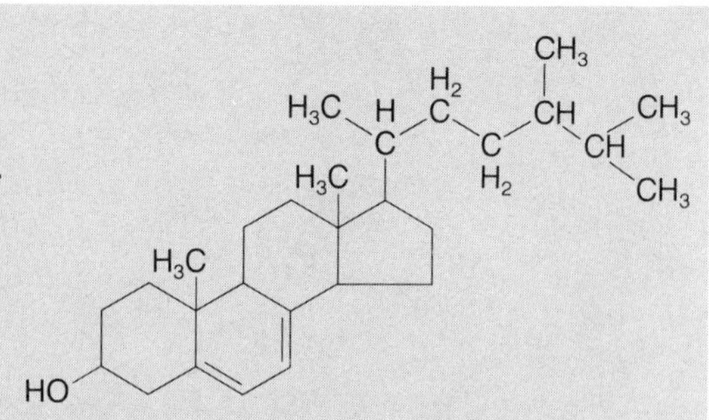

7-Dehydrocholesterol

Ultraviolet light

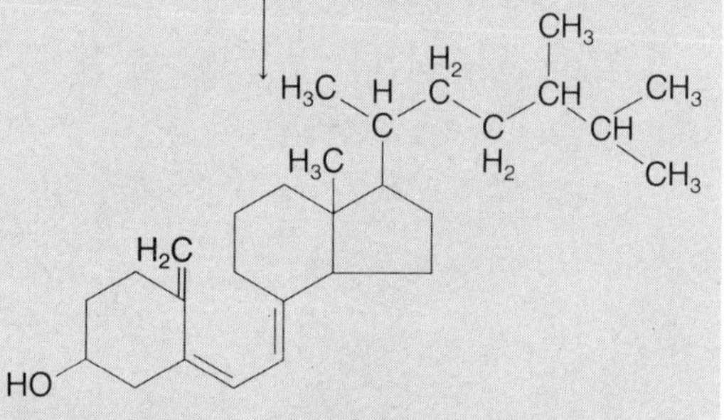

Vitamin D₃, is an antirachitic factor. Its absence leads to rickets, an infant and childhood disease characterized by faulty deposition of calcium phosphate and poor bone growth.

Vitamin D₃

Bile acid

Cholic acid is found in bile in the form of its sodium salt. This and closely related salts are the bile salts that are powerful surface active agents needed to aid in the digestion of lipids and in the absorption of vitamins A, D, E, and K from the intestinal tract.

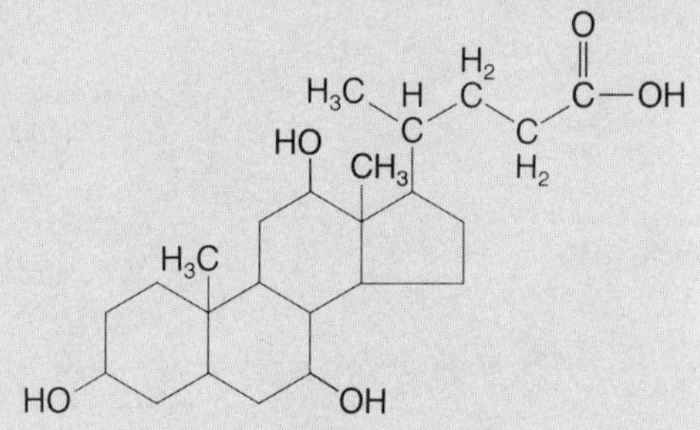

Cholic acid

Adrenocortical hormone

Cortisol is one of the 28 hormones secreted by the cortex of the adrenal gland. Cortisone, very similar to cortisol, is another such hormone. When cortisone is used to treat arthritis, the body changes much of it to cortisol by reducing a keto group to the 2° alcohol group that you see in the structure of cortisol.

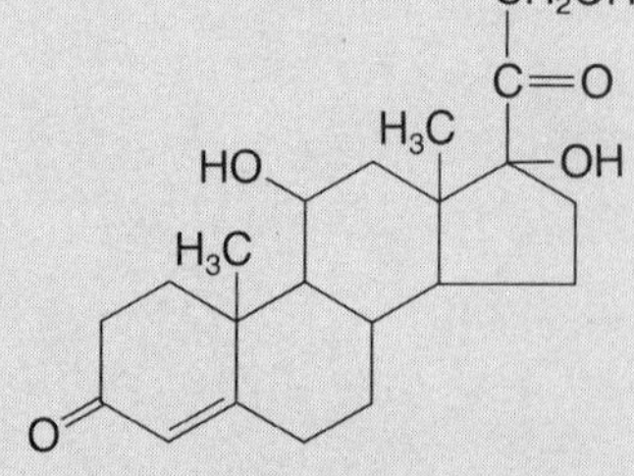

Cortisol

TABLE 20.3 (continued)

Cardiac aglycone

Digitoxigenin is found in many poisonous plants, notably digitalis, as a complex glycoside. In small doses it stimulates the vagus mechanism and increases heart tone. In larger doses it acts as a potent poison.

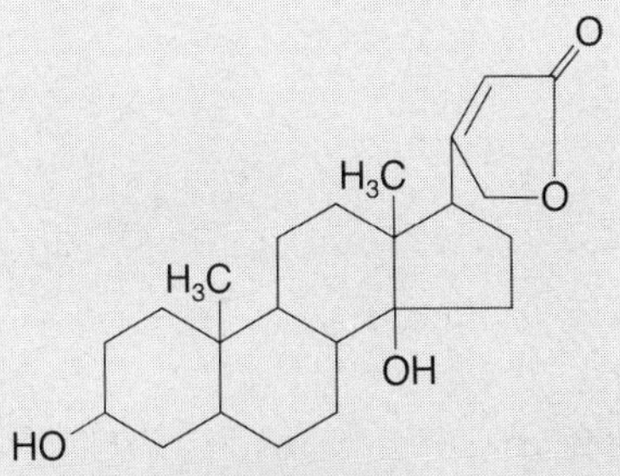

Digitoxigenin

Sex hormones

Estradiol is a human estrogenic hormone.

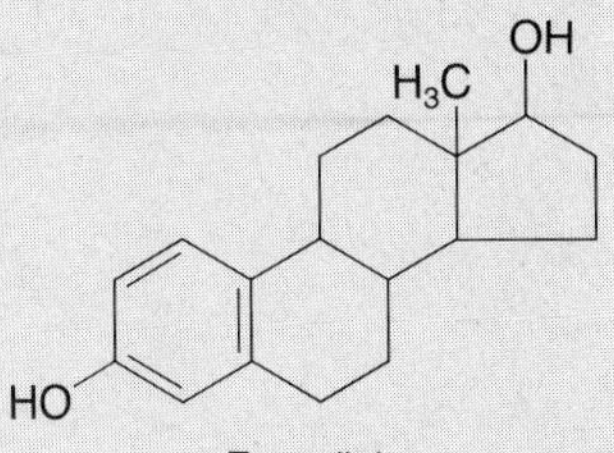

Estradiol

Progesterone, a human pregnancy hormone, is secreted by the corpus luteum.

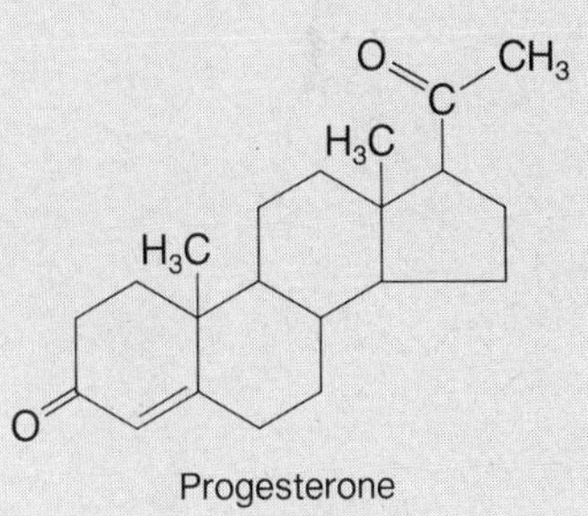

Progesterone

Testosterone, a male sex hormone, regulates the development of reproductive organs and secondary sex characteristics.

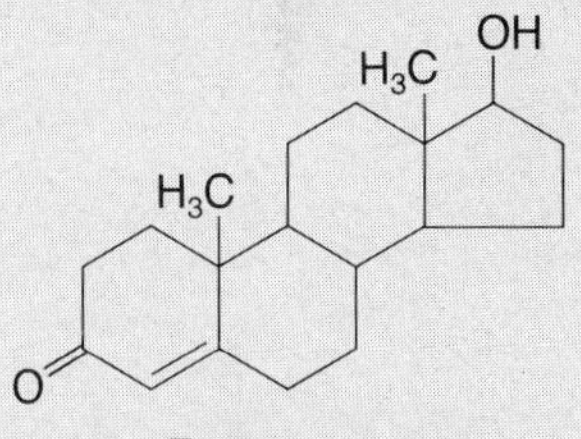

Testosterone

Androsterone is another male sex hormone.

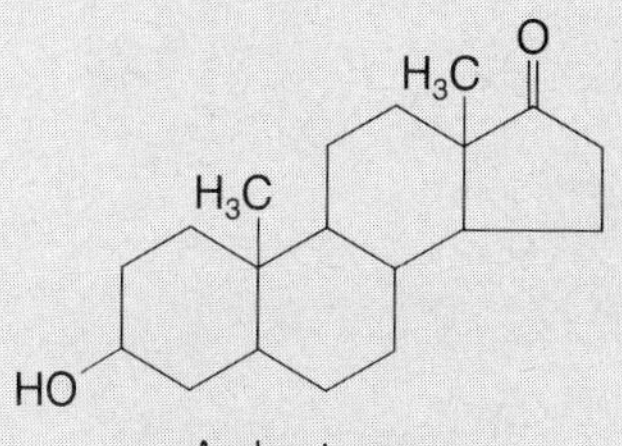

Androsterone

TABLE 20.3   (continued)

Synthetic hormones in fertility control

Most oral contraceptive pills contain one or two synthetic, hormone-like compounds. (Synthetics must be used because the real hormones are broken down in the body.)

The most widely used pill is a combination of an estrogen (not more than 30 micrograms) and a progestin (up to 4 milligrams).[a]

Synthetic estrogens

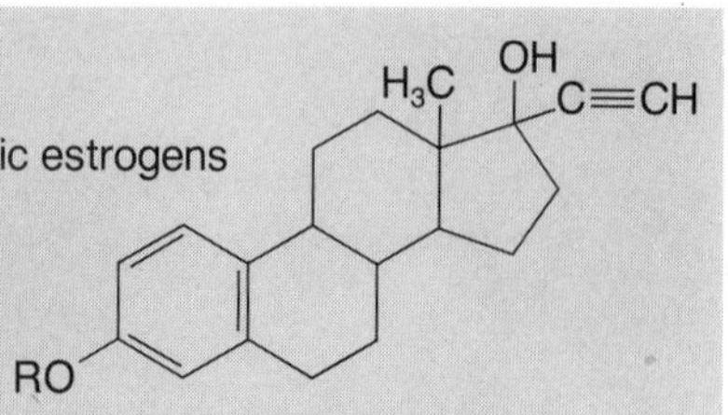

If R = H, ethynylestradiol
R = CH₃, mestranol

Synthetic progestins

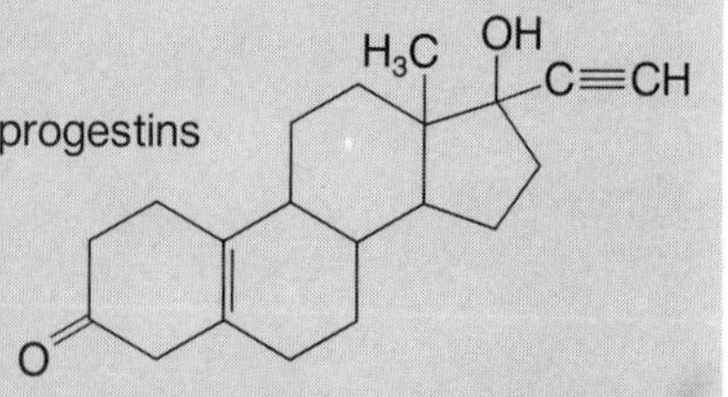

Norethynodrel

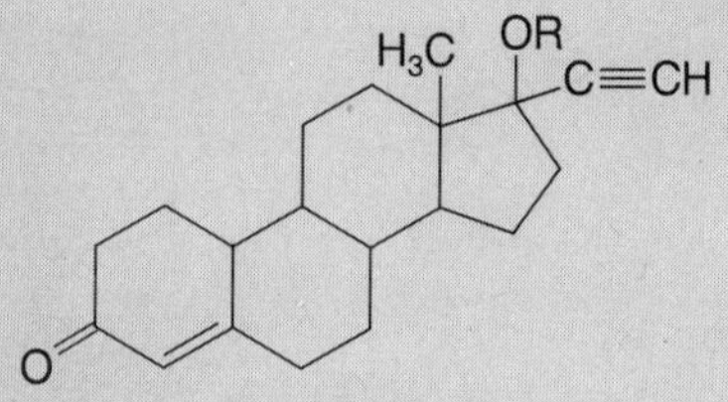

If R = H, norethindrone

R = C—CH₃, norethindrone acetate

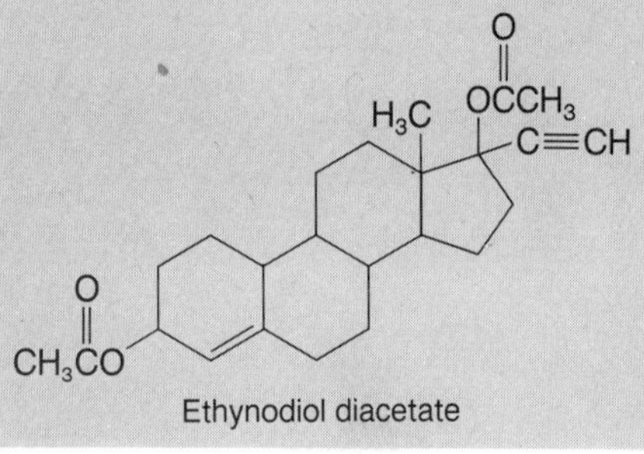

Ethynodiol diacetate

[a] The U.S. Food and Drug Administration warns that progestins must not be taken during pregnancy because they can cause severe birth defects.

Hydrophilic and Hydrophobic Groups.   The molecules in cell membranes have parts that are either very polar or have full electrical charges and other parts that are relatively nonpolar. The polar or ionic sites are called **hydrophilic groups,** because they are able to attract water molecules and they tend to associate as much as possible with the aqueous systems of cell fluids or fluids outside the cell. The hydrophilic parts of the phospholipids are the phosphate diester units with their ionic sites. In a glycolipid, the sugar unit with its many —OH groups is hydrophilic.

The nonpolar, hydrocarbon sections of membrane lipids are called **hydrophobic groups** because they do not attract water molecules nor are they attracted by them. Sub-

Hydrophilic — from the Greek *hydor,* water, and *philos,* loving. *Hydrophobic* — from the Greek *phobikos,* hating.

stances with both hydrophilic and hydrophobic groups are called **amphipathic compounds.** In an aqueous medium, hydrophobic groups tend as much as possible to avoid water molecules. They do this by "dissolving" in each other almost as if they formed a hydrocarbon solution, as we will next see.

The Lipid Bilayer of Cell Membranes. When phospholipids or glycolipids are mixed with water, their molecules spontaneously form a **lipid bilayer,** a sheet-like array that consists of two layers of lipid molecules aligned side by side, as illustrated in Figure 20.2. The hydrophobic "tails" of the lipid molecules intermingle in the center of the bilayer so that they are away from water molecules. The hydrophilic "heads" stick out into the aqueous phase, where they are hydrated.

The lipid bilayer is quite flexible. If a pin were stuck through it and then pulled out, the layer would close back spontaneously. No covalent bonds hold neighboring lipid molecules to each other. Only the net forces of attraction that we imply when we use the terms *hydrophobic* and *hydrophilic* are at work. Yet the bilayer is strong enough to be a membrane, and it is flexible enough to let things in and out. The bilayer is very selectively permeable to the ordinary migrations of ions and polar molecules, with the important exception of water molecules, which move back and forth easily.

Concentration Gradients and Active Transport. As we just indicated, the cell membrane is not an ordinary dialyzing membrane. Among its proteins are some that participate in maintaining the needed concentration gradients. A **gradient** is the existence of an unevenness in the value of some physical property throughout a system. A concentration gradient exists in a solution, for example, when one region of the solution has a higher concentration of solute than another region.

As the data in the table in the margin show, both sodium ions and potassium ions have quite different concentrations in the fluids on the inside of a cell as compared with their concentrations in the fluids on the outside. Thus between the inside and the outside of a cell there is a considerable concentration gradient for both of these ions, and *this gradient must be maintained at all costs.* However, it takes energy to maintain any kind of gradient, because the natural, spontaneous tendency in nature is for things that are in a gradient to diffuse enough to destroy the difference that makes up the gradient. Thus sodium ions naturally tend to migrate spontaneously from blood plasma and into the fluids outside the bloodstream, including the fluids inside cells. Similarly, potassium ions tend to migrate in the opposite direction. If these tendencies were allowed to take their course, we would die. The smooth operation of our bodies demands the maintenance of these (and many other) gradients.

Here is where proteins that are integral parts of cell membranes carry out a vital function. One kind of membrane protein can move sodium ions against their gradient. When too many sodium ions leak to the inside of a cell, they are "pumped" back out by a special molecular

	mmol/L	
Ion	Plasma	Cells
Na^+	136–145	10
K^+	3.5–5.0	125

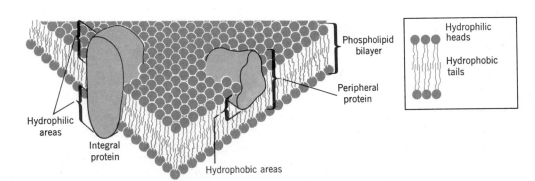

FIGURE 20.2
Cell membrane.

machinery called the sodium-potassium pump. The same pump can move potassium ions back inside a cell. This movement of any solute against its concentration gradient is an example of **active transport,** and other reactions in cells supply the chemical energy that lets it work.

SUMMARY

Lipids Lipids are ether-extractable substances in animals and plants, and they include saponifiable esters and nonsaponifiable compounds. The esters are generally of glycerol or sphingosine with their acyl portions contributed by long-chain carboxylic acids (fatty acids). Because molecules of all lipids are mostly hydrocarbon-like, lipids are soluble in nonpolar solvents but not in water. The fatty acids obtained from lipids by hydrolysis generally have long chains of even numbers of carbons, seldom are branched, and often have one or more alkene groups. The latter are generally cis.

Molecules of the waxy coatings on leaves and fruit, or in beeswax or sebum, are simple esters between long-chain monohydric alcohols and fatty acids.

Triacylglycerols Molecules of neutral fats — those without electrically charged sites or sites that are similarly polar — are esters of glycerol and a variety of fatty acids, both saturated and unsaturated. Vegetable oils have more double bonds per molecule than animal fats. The triacylglycerols can be hydrogenated, hydrolyzed (digested), and saponified.

Phosphoglycerides Molecules of the phosphoglycerides are esters both of glycerol and of phosphoric acid. A second ester bond from the phosphate unit goes to a small alcohol molecule that can also have a positively charged group. Thus this part of a phosphoglyceride is strongly hydrophilic.

Sphingomyelins Sphingomyelins are esters of sphingosine, a dihydric amino alcohol. Their molecules also have a strongly hydrophilic phosphate system.

Glycolipids Also sphingosine-based, the glycolipids use a monosaccharide instead of the phosphate-to-small alcohol unit to provide the hydrophilic section. Otherwise, they resemble the sphingomyelins.

Steroids Steroids are nonsaponifiable lipids with the steroid nucleus of four fused rings (three being C-6 rings and one a C-5 ring). Several steroids are sex hormones, and oral fertility control drugs mimic their structure and functions. Cholesterol, the raw material used by the body to make other steroids, is also manufactured by the body.

Membranes A double layer of lipid molecules — phospholipids or glycolipids plus cholesterol — make up the lipid bilayer part of a cell membrane. The hydrophobic tails of these amphipathic lipids intermingle within the bilayer, away from the aqueous phase. The hydrophilic heads are in contact with the aqueous medium.

On or in the bilayer are proteins that serve as enzymes, or as receptors, or are parts of "pumps," such as the sodium-potassium pump, that work to maintain important concentration gradients, or as channels for small ions or molecules.

KEY TERMS

The following terms were emphasized in this chapter, and they should be mastered before you continue.

active transport	hydrophilic group	phosphoglyceride	sphingolipid
amphipathic compound	hydrophobic group	phospholipid	steroid
fatty acid	lipid	plasmalogen	triacylglycerol
glycolipid	lipid bilayer	receptor molecule	triglyceride
gradient	nonsaponifiable lipid	saponifiable lipid	wax

SELECTED REFERENCES

1 Mark S. Bretscher. "The Molecules of the Cell Membrane." *Scientific American,* October 1985, page 100.

2 N. Unwin and R. Henderson. "The Structure of Proteins in Biological Membranes." *Scientific American,* February 1984, page 78.

3 J. H. Fendler. "Membrane Mimetic Chemistry." *Chemical and Engineering News,* January 2, 1984, page 25.

4 Alice Dautry-Varsat and H. F. Lodish. "How Receptors Bring Proteins and Particles into Cells." *Scientific American,* May 1984, page 52.

5 J. L. Goldstein, T. Kita, and M. S. Brown. "Defective Lipoprotein Receptors and Atherosclerosis." *New England Journal of Medicine,* August 4, 1983, page 288.

6 R. D. Keynes. "Ion Channels in the Nerve-Cell Membrane." *Scientific American,* March 1979, page 126.

7 Doris Kolb. "A Pill for Birth Control." *Journal of Chemical Education,* September 1978, page 591.

REVIEW EXERCISES

The answers to these Review Exercises are in the *Study Guide* that accompanies this book.

Lipids in General

20.1 Crude oil is soluble in ether, yet it isn't classified as a lipid. Explain.

20.2 Cholesterol has no ester group, yet we classify it as a lipid. Why?

20.3 Ethyl acetate has an ester group, but it isn't classified as a lipid. Explain.

20.4 What are the criteria for deciding if a substance is a lipid?

Waxes

20.5 One component of beeswax has the formula $C_{34}H_{68}O_2$. When it is hydrolyzed, it gives $C_{16}H_{32}O_2$ and $C_{18}H_{38}O$. Write the most likely structure of this compound.

20.6 When all of the waxes from the leaves of a certain shrub are separated, one has the formula of $C_{60}H_{120}O_2$. Its structure is either **A**, **B**, or **C**. Which is it most likely to be? Explain why the others can be ruled out.

$$CH_3(CH_2)_{56}CO_2CH_2CH_3 \qquad CH_3(CH_2)_{29}CO_2(CH_2)_{28}CH_3$$
$$\textbf{A} \qquad\qquad\qquad \textbf{B}$$

$$CH_3(CH_2)_{28}CO_2(CH_2)_{29}CH_3$$
$$\textbf{C}$$

Fatty Acids

20.7 What are the structures and the names of the two most abundant saturated fatty acids?

20.8 Write the structures and names of the unsaturated fatty acids that have 18 carbons each and that have no more than three double bonds. Show the correct geometry at each double bond.

20.9 Write the equations for the reactions of palmitic acid with (a) NaOH(*aq*) and (b) CH_3OH (when heated in the presence of an acid catalyst).

20.10 What are the equations for the reactions of oleic acid with (a) Br_2, (b) KOH(*aq*), (c) H_2 (in the presence of a catalyst and under pressure), and (d) CH_3CH_2OH (heated in the presence of an acid catalyst)?

20.11 Which of the following acids, **A** or **B**, is more likely to be obtained by the hydrolysis of a lipid? Explain.

$$\overset{\displaystyle CH_3}{\underset{\textbf{A}}{CH_3CH(CH_2)_{11}CO_2H}} \qquad \underset{\textbf{B}}{CH_3(CH_2)_{12}CO_2H}$$

20.12 Without writing structures, state what kinds of chemicals the prostaglandins are.

Triacylglycerols

20.13 Write the structure of a triacylglycerol that involves linolenic acid, oleic acid, and myristic acid, besides glycerol.

20.14 What is the structure of a triacylglycerol made from glycerol, stearic acid, oleic acid, and palmitic acid?

20.15 Write the structures of all of the products that would form from the complete digestion of the following lipid.

$$\begin{array}{l} CH_2-O-\overset{\displaystyle O}{\overset{\displaystyle \|}{C}}(CH_2)_7CH{=}CH(CH_2)_7CH_3 \\[2mm] CH-O-\overset{\displaystyle O}{\overset{\displaystyle \|}{C}}(CH_2)_{12}CH_3 \\[2mm] CH_2-O-\overset{\displaystyle O}{\overset{\displaystyle \|}{C}}(CH_2)_7CH{=}CH(CH_2)_7CH_3 \end{array}$$

20.16 Write the structures of the products that are produced by the saponification of the triacylglycerol whose structure was given in Review Exercise 20.15.

20.17 The hydrolysis of a lipid produced glycerol, lauric acid, linoleic acid, and oleic acid in equimolar amounts. Write a structure that is consistent with these results. Is there more than one structure that can be written? Explain.

20.18 The hydrolysis of 1 mol of a lipid gave 1 mol each of glycerol and oleic acid and 2 mol of lauric acid. This lipid was optically active. Write its structure. Is more than one structure possible? Explain.

20.19 What is the structural difference between the triacylglycerols of the animal fats and the vegetable oils?

20.20 Products such as corn oil are advertised as being "polyunsaturated." What does this mean in terms of the structures of the molecules that are present? And corn oil is more "polyunsaturated" than what?

20.21 What chemical reaction is used in the manufacture of oleomargarine?

20.22 Lard and butter are chemically almost the same substances, so what is it about butter that makes it so much more desirable a spread for bread than, say, lard or tallow?

Phospholipids

20.23 What are the names of the two chief kinds of phospholipids?

20.24 In structural terms, how do the phosphoglycerides and plasmalogens differ?

20.25 How are the sphingomyelins and cerebrosides different structurally?

20.26 What structural unit provides the most polar group in a molecule of a glycolipid? (Name it.)

20.27 Phospholipids are not classified as neutral fats. Explain.

20.28 Phospholipids are particularly common in what part of a cell?

20.29 What are the names of the two types of sphingosine-based lipids?

20.30 Are the sugar units that are incorporated into the cerebrosides bound by glycosidic links or by ordinary ether links? How can one tell? Which kind of link is more easily hydrolyzed (assuming an acid catalyst)?

20.31 The complete hydrolysis of 1 mol of a phospholipid gave 1 mol each of the following compounds: glycerol, linolenic acid, oleic acid, phosphoric acid, and the cation, $HOCH_2CH_2\overset{+}{N}(CH_3)_3$.
  (a) Write a structure of this phospholipid that is consistent with the information given.
  (b) Is the substance a phosphoglyceride or a sphingolipid? Explain.
  (c) Are its molecules chiral or not? How can you tell?
  (d) Is it an example of a lecithin or a cephalin? Explain.

20.32 When 1 mol of a certain phospholipid was hydrolyzed, there was obtained 1 mol each of lauric acid, oleic acid, phosphoric acid, glycerol, and $HOCH_2CH_2NH_2$.
  (a) What is a possible structure for this phospholipid?
  (b) Is it a sphingolipid or a phosphoglyceride? Explain.
  (c) Can its molecules exist as enantiomers or not? Explain.
  (d) Is it a cephalin or a lecithin? Explain.

Steroids

20.33 What is the name of the steroid that occurs as a detergent in our bodies?

20.34 What is the name of a vitamin that is made in our bodies from a dietary steroid by the action of sunlight on the skin?

20.35 Give the names of three steroidal sex hormones.

20.36 What is the name of a steroid that is part of the cell membranes in many tissues?

Cell Membranes

20.37 Describe in your own words what is meant by the *lipid bilayer* structure of cell membranes.

20.38 How do the hydrophobic parts of phospholipid molecules avoid water in a lipid bilayer?

20.39 Besides lipids, what kinds of substances are present in a cell membrane?

20.40 What kinds of forces are at work in holding a cell membrane together?

20.41 Immediately after you add a teaspoon of sugar to a hot cup of coffee, can a concentration gradient for sugar be present? Explain what this is. What happens to it, given enough time, even without stirring the mixture? What causes this change to occur? What are the chances of the original gradient ever restoring itself spontaneously?

20.42 Which has the higher level of sodium ion, plasma or cell fluid?

20.43 Does cell fluid or plasma have the higher concentration of potassium ion?

20.44 In which fluid, plasma or cell fluid, would the level of sodium ion increase if the sodium ion gradient could not be maintained?

20.45 What is meant by *active transport* in a cell membrane?

20.46 What does the sodium-potassium pump in a cell membrane do?

20.47 Name the functions that the proteins of a cell membrane can serve.

Chapter 21
Proteins

The sinuous tube in this molecular model of myoglobin shows where the main backbone is. Myoglobin is an oxygen-storage protein in muscles, and we'll learn about all of the major structural features of proteins in this chapter.

21.1   AMINO ACIDS: THE BUILDING BLOCKS OF PROTEINS

Living things select from among the molecules of about 20 α-amino acids to make the polypeptides in proteins.

Proteins, found in all cells and in virtually all parts of cells, constitute about half of the body's dry weight. They give strength and elasticity to skin. As muscles and tendons, they function as the cables that enable us to move the levers of our bones. They reinforce our teeth and bones much as thick steel rods reinforce concrete. The molecules of antibodies, of hemoglobin, and of the various kinds of albumins in our blood serve as protectors and as the long-distance haulers of substances, such as oxygen or lipids, that otherwise do not dissolve well in blood. Other proteins form parts of the communications network of our nervous system. Some proteins are enzymes, hormones, and gene regulators that direct and control all forms of repair, construction, and energy conversion in the body. No other class of compounds is involved in such a variety of functions, all essential to life. They deserve the name *protein*, taken from the Greek *proteios*, "of the first rank."

Proteins are substances of large formula weights that are made up wholly or mostly of polymers called **polypeptides.** Figure 21.1 outlines the relationship of polypeptides and nonpolypeptide units to whole proteins. The monomer units for polypeptides are **α-amino acids,** which have the general structure given by **1,** and about 20 of these amino acids are used to make polypeptides. Hundreds of **amino acid residues, 2,** each derived from one or another of the various α-amino acids, are joined together in a single polypeptide molecule, but before we can study how polypeptides are put together we must learn more about amino acids.

Hereafter, when we say "amino acid," we'll mean α-amino acid.

$$\boxed{\alpha\text{-position}}$$

$$^+NH_3-\underset{\underset{G}{|}}{CH}-\overset{\overset{O}{\|}}{C}-O^- \qquad -NH-\underset{\underset{G}{|}}{CH}-\overset{\overset{O}{\|}}{C}-$$

α-Amino acids   Amino acid residue
(general formula)

1   **2**

The same set of 20 amino acids is used by all species of plants and animals. Their structures are given in Table 21.1. In rare instances, a few others are present in certain polypeptides, and some of the 20 shown in Table 21.1 can occur in slightly modified forms.

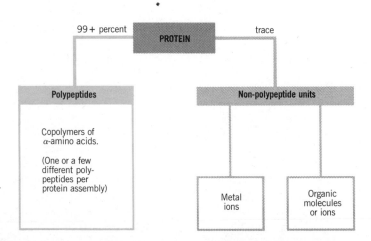

FIGURE 21.1
Components of proteins. Some proteins consist exclusively of polypeptide molecules, but most also have nonpolypeptide units such as small organic molecules or metal ions, or both.

TABLE 21.1
Amino Acids $^+NH_3$—CH—CO_2^-
                        |
                        G

Type	Side Chain, G	Name	Symbol	pI
Side chain is nonpolar	—H	Glycine	Gly	5.97
	—CH_3	Alanine	Ala	6.00
	—$CH(CH_3)_2$	Valine	Val	5.96
	—$CH_2CH(CH_3)_2$	Leucine	Leu	5.98
	—$CHCH_2CH_3$ $\;$ CH_3	Isoleucine	Ile	6.02
	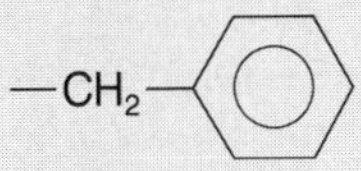 —CH_2— ⬡	Phenylalanine	Phe	5.48
	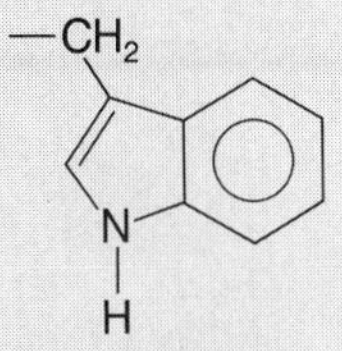 —CH_2 (indole ring)	Tryptophan	Trp	5.89
	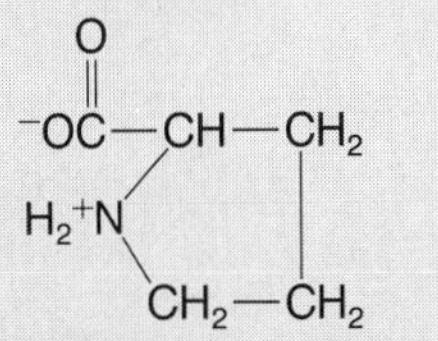 ^-OC—CH—CH_2 H_2^+N $\;$ CH_2—CH_2 (complete structure)	Proline	Pro	6.30
Side chain has a hydroxyl group	—CH_2OH	Serine	Ser	5.68
	—$CHOH$ $\;$ CH_3	Threonine	Thr	5.64
	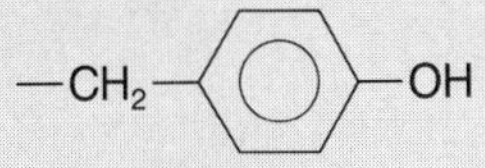 —CH_2—⬡—OH	Tyrosine	Tyr	5.66
Side chain has a carboxyl group (or an amide)	—CH_2CO_2H	Aspartic acid	Asp	2.77
	—$CH_2CH_2CO_2H$	Glutamic acid	Glu	3.22
	—CH_2CONH_2	Asparagine	Asn	5.41
	—$CH_2CH_2CONH_2$	Glutamine	Gln	5.65
Side chain has a basic amino group	—$CH_2CH_2CH_2CH_2NH_2$	Lysine	Lys	9.74
	—$CH_2CH_2CH_2NH$—C—NH_2 $\;$ NH	Arginine	Arg	10.76
	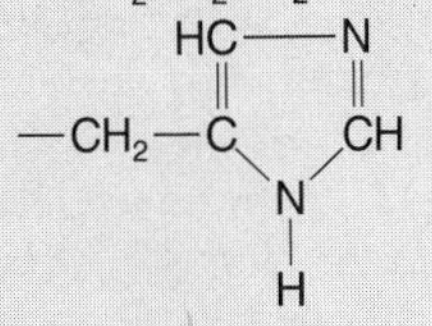 —CH_2—C (imidazole ring)	Histidine	His	7.59
Side chain contains sulfur	—CH_2SH	Cysteine	Cys	5.07
	—$CH_2CH_2SCH_3$	Methionine	Met	5.74

Structures and Properties of the α-Amino Acids.   As you can see in Table 21.1, the amino acids differ in their **side chains,** the G-groups that can be attached to the α-position in **1.**

In the solid state, amino acids exist entirely in the form shown by **1,** which is called a **dipolar ion.** Such ions actually are neutral particles that have a positive and a negative charge on different sites. Because they are neutral, dipolar ions are basically molecules, and we'll usually use this term. Because their molecules are exceedingly polar, amino acids, like salts, have melting points that are considerably higher than those of most molecular compounds, and they are insoluble in nonpolar solvents, but soluble in water.

When they are dissolved in water, amino acid particles take on structures that depend on the pH. Nearly all amino acids exist in water mostly as dipolar ions, **1,** when the pH is roughly 6 to 7. Structure **1** is actually an internally neutralized molecule. We can imagine that it started out with a regular amino group, $—NH_2$, and an ordinary carboxyl group, $—CO_2H$. But then the amino group, a proton-acceptor, took a proton from the carboxyl group, a proton donor, to give structure **1.** Of course, **1,** has its own (weaker) proton-donating group, $—NH_3^+$, and its own (also weaker) proton-accepting group, $—CO_2^-$, so these dipolar ions can neutralize acids or basis of sufficient strength, such as H_3O^+ and OH^-. (In fact, amino acids can serve as buffers.)

If we add a strong acid such as HCl (*aq*) to a solution of an amino acid initially at a pH of about 6 to 7, and we add enough acid to lower the pH to about 1, the $—CO_2^-$ groups in the molecules of structure **1** take on protons. They change to the structure given by **2,** and now they are cations that can migrate to a cathode.

$$^+NH_3—CH—CO_2^-$$

G

1

H^+ OH^- OH^- H^+

$$^+NH_3—CH—CO_2H \qquad NH_2—CH—CO_2^-$$

G   G

2 **3**

These shifts of H⁺ ions are illustrations of Le Chatelier's principle at work.

If we add enough strong base to a solution of an amino acid to raise the pH to about 11, then most of the amino acid molecules transfer protons from their $—NH_3^+$ groups to OH^- ions, and the amino acid molecules change to **3.** This is an anion, so it can migrate to an anode.

Isoelectric Points. When electrodes are immersed in an aqueous solution of an amino acid, molecules that are in the form of **2** start to migrate to the cathode. Those of form **3** begin their migration to the anode. The neutral molecules, **1,** can go nowhere. They have equal numbers of opposite charge, and so are neutral. A dipolar molecule in this condition is said to be **isoelectric.**

Of course, the solution involves a complex dynamic equilibrium, as indicated by the double arrows, above. Therefore a cation on its way to the cathode could flip a proton to some acceptor, thereby become **1** and neutral, and stop dead. In another instant, it could shed another proton, become **3,** and so turn around and head for the anode. Similarly, an anion on its way to the anode might pick up a proton, become neutral, and also stop dead. Then it might take another proton, become **2,** and turn itself around. In the meantime, an isoelectric molecule, **1,** might either donate or accept a proton, become electrically charged, and start its own migration. (It reminds one of amusement park bumper cars that move in every direction. The question is, what is the net change?)

If either **2** or **3** is in any molar excess in the solution, then there will be some *net* migration toward one electrode or the other. If the molar concentration of **2,** for example, is greater than that of **3,** some statistical net movement to the cathode will occur. Remember, however, that these equilibria can be shifted by adding acid or base. Therefore by carefully adjusting the pH, the concentrations at equilibrium can be so finely tuned that no net migration can occur. When the pH is correctly adjusted, the rates of proton exchange are such that each unit that is not **1** spends an equal amount of time as **2** and as **3.** (And the concentrations of **2** and **3** are very low.) In this way any net migration toward one electrode is blocked. The pH at which no net migration of an amino acid can occur in an electric field is called the **isoelectric point** of the amino acid, and its symbol is **pI.** Table 21.1 includes a column of the pI values of the amino acids. But what has all this to do with proteins?

As we will soon see, all proteins have groups that are, or by a change in pH, can easily become, $—NH_3^+$ or $—CO_2^-$ groups. Therefore whole protein molecules can be isoelectric, or they can be cations or anions, according to their structures and the pH of the medium. The point is that the entire electrical character of a protein can be dramatically changed simply by changing the pH of the medium. Such changes in electrical charge have serious implications for the reactions at the molecular level of life. We'll return to this concept in this and later chapters, but the discussion focuses our attention again on how important it is that an organism control the pH values of its fluids.

We will next survey the types of side chains in amino acids and how they affect the properties of polypeptides and proteins. You should memorize the structures of a minimum of five amino acids that illustrate the types we are about to study — glycine, alanine, cysteine, lysine, and glutamic acid. How to use Table 21.1 to write their structures is described in the following worked example.

EXAMPLE 21.1   **WRITING THE STRUCTURE OF AN AMINO ACID**

Problem: What is the structure of cysteine?

Solution: Doing this kind of problem depends on two things — on remembering what is common to all amino acids:

$$^+NH_3—CH—CO_2^-$$
$$|$$

and then either looking up or remembering the side chain for the particular amino acid. For cysteine, this is CH_2SH, so simply attach this group to the α-carbon. The structure of cysteine is:

$$^+NH_3—CH—CO_2^-$$
$$|$$
$$CH_2SH$$

The memorization of amino acid structures involves learning the structures of the side chains and fixing them in the mind to the name of the amino acid.

PRACTICE EXERCISE 1   Write the structures of the dipolar ionic forms of glycine, alanine, lysine, and glutamic acid.

Amino Acids with Nonpolar Side Chains.   The first amino acids in Table 21.1 have essentially nonpolar, hydrophobic side chains. When a long polypeptide molecule folds into its distinctive shape, these hydrophobic groups tend to be folded next to each other rather than next to highly polar groups or to water molecules in the solution.

Amino Acids with Hydroxyl-Containing Side Chains.   The second set of amino acids in Table 21.1 consists of those whose side chains carry alcohol or phenol groups. These are neither acidic nor basic in the cell, but they are polar and hydrophilic. They can donate and accept hydrogen bonds. As a long polypeptide chain folds into its final shape, these side chains tend to stick out into the surrounding aqueous phase to which they are attracted.

Amino Acids with Acidic Side Chains.   Molecules of aspartic and glutamic acid have extra carboxyl groups on their side chains. At the slightly basic pH of body fluids, these groups have lost their protons and they exist as $—CO_2^-$ groups. To prevent the side chain from being in this form, one has to make the solution quite acidic to force protons onto the side chain $—CO_2^-$ groups. This is why the pI values of aspartic and glutamic acid are low.

Aspartic acid and glutamic acid often occur as asparagine and glutamine in which the side chain $—CO_2H$ groups have become amide groups, $—CONH_2$, instead. These are also polar, hydrophilic groups, but they are not electrically charged. They are neither proton donors nor acceptors, so the pI values of asparagine and glutamine are higher than those of aspartic or glutamic acids.

PRACTICE EXERCISE 2   Write the structure of aspartic acid (in the manner of **1**): (a) with the side-chain carboxyl in its carboxylate form, and (b) in its amide form.

Amino Acids with Basic Side Chains.   A lysine molecule has an extra amino group that makes its side chain basic, hydrophilic, and a hydrogen-bond donor or acceptor. This group tends to exist in its protonated form, $—NH_3^+$, so a solution of lysine in water has to have some extra hydroxide ion present to prevent this, keep the side chain amino group in the unprotonated form as $—NH_2$, and thus permit the molecule as a whole to be isoelectric. This is why the pI value, 9.47, of lysine is relatively high. Arginine and histidine have similarly basic side chains.

PRACTICE EXERCISE 3   Write the structure of arginine in the manner of **1**, but with its side-chain amino group in its protonated form. (Put the extra proton on the $=NH$ unit, not the $—NH_2$ unit of the side chain.)

PRACTICE EXERCISE 4   Classify the side chain of the following amino acid as hydrophilic or hydrophobic. Does this side chain have an acidic, basic, or neutral group?

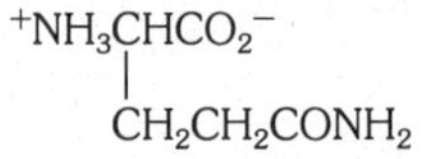

$$^+NH_3CHCO_2^-$$
$$|$$
$$CH_2CH_2CONH_2$$

Amino Acids with Sulfur-Containing Side Chains.   The side chain in cysteine has a $—SH$ group. As we studied in Section 14.4, molecules with this group are easily oxidized to disulfide systems, and disulfides are easily reduced to $—SH$ groups:

$$2RSH \underset{(H)}{\overset{(O)}{\rightleftharpoons}} R—S—S—R + H_2O$$

Cysteine and its oxidized form, cystine, are interconvertible by oxidation and reduction, a property of far-reaching importance in some proteins.

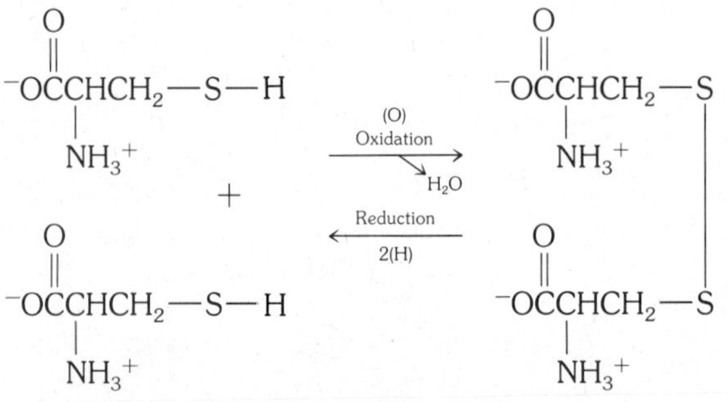

The three letter symbol for cystine is

Cys
|
Cys

Cysteine   Cystine

The **disulfide link** contributed by cystine is especially prevalent in the proteins that have a protective function, such as those in hair, fingernails, and the shells of certain crustaceans.

Optical Isomerism of Amino Acids.   The α-carbon is chiral in all of the naturally-occurring amino acids, except glycine. Therefore the chiral amino acids can exist as enantiomers, but (with very rare exceptions) all of the naturally-occurring amino acids belong to the same optical family, the L-family. What this means is illustrated in Figure 21.2. It also means that all of the proteins in our bodies, including all enzymes, are made from L-amino acids and are all chiral.

The mirror-image molecules of the L-amino acids are the D-amino acids, which can be synthesized in the laboratory.

21.2   PRIMARY STRUCTURES OF PROTEINS

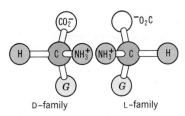

D-family   L-family

FIGURE 21.2
The two possible enantiomers of α-amino acids whose molecules have just one chiral center. The absolute configuration on the right, which is in the L-family, represents virtually all of the naturally occurring α-amino acids.

The backbones of all polypeptides of all plants and animals have a repeating series of

$$-\overset{|}{\underset{|}{N}}-\overset{|}{\underset{|}{C}}-\overset{O}{\overset{\|}{C}}- \text{ units.}$$

Overview of the Levels of Protein Structure.   There are four levels of complexity in the structures of most proteins. Disarray at any level almost always renders the protein biologically useless. The first and most fundamental level, the **primary structure** of a protein, concerns only the sequence of amino acid residues joined by covalent bonds, called *peptide bonds,* in the polypeptide(s) of the protein.

The next level, the **secondary structure,** also concerns just individual polypeptides. It entails noncovalent forces, particularly the hydrogen bond, and it consists of the particular way in which a long polypeptide strand has coiled or in which strands have either coiled together or lined up side to side.

The **tertiary structure** of a polypeptide concerns the further bending, kinking, or twisting of secondary structures. If you've ever played with a coiled door spring, you know that the coil (secondary structure) can be bent and twisted (tertiary structure). Noncovalent forces, generally hydrogen bonds, stabilize these shapes.

Finally, we deal with the final, intact protein. Its **quaternary structure** is the way in which individual polypeptides (each with all previous levels of structure), and any other molecules or ions, come together in one grand whole.

In this section, we'll study the primary structural features of polypeptides, whose central feature is the peptide bond.

The Peptide Bond.   The chief covalent bond that forms when amino acids are linked together is called the **peptide bond.** It's nothing more than an amide system, carbonyl-to-nitrogen. To illustrate a peptide bond, as well as to show how polypeptides acquire their primary structure, we will begin by putting just two amino acids together.

Suppose that glycine acts at its carboxyl end and alanine acts at its amino end such that, by a series of steps (not given in detail but indicated by a dashed arrow, $---\!\!\rightarrow$), a molecule of water splits out and a carbonyl-to-nitrogen bond is created.

Simple amides are:

$$R-\overset{O}{\overset{\|}{C}}-NH_2$$

Amide bond

How a cell causes a peptide bond to form is a major topic under the chemistry of heredity.

Peptide bond

$$^+NH_3CH_2C\!-\!O^- \;+\; H\!-\!\overset{+}{N}HCHCO^- \;\;-\!-\!-\!\rightarrow\;\; ^+NH_3CH_2C\!-\!NHCHCO^- \;+\; H_2O$$

Glycine, Gly   Alanine, Ala   A dipeptide, Gly · Ala

4

Of course, there is no reason why we could not picture the roles reversed so that alanine acts at its carboxyl end and glycine at its amino end. This results in a different dipeptide (but an isomer of the first).

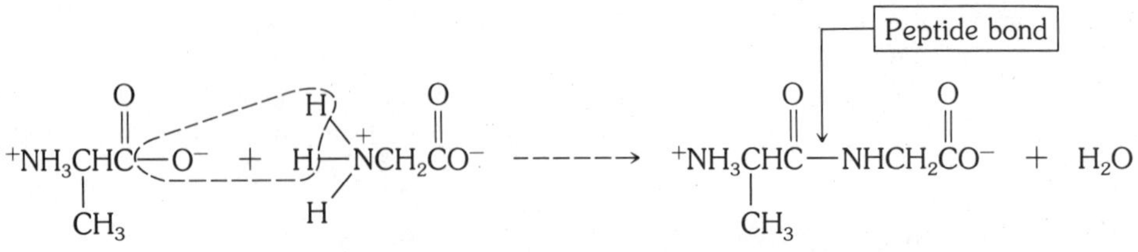

Alanine, Ala   Glycine, Gly   Another dipeptide, Ala·Gly

The product of the union of any two amino acids by a peptide bond is called a **dipeptide,** and all dipeptides have the following features:

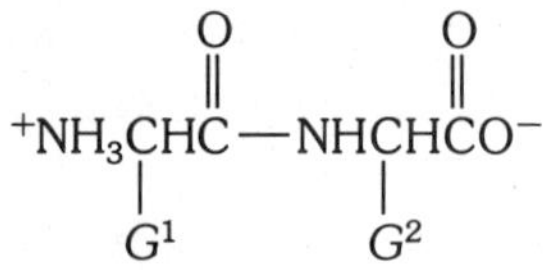

Dipeptide

EXAMPLE 21.2   WRITING THE STRUCTURE OF A DIPEPTIDE

Problem: What are the two possible dipeptides that can be put together from alanine and cysteine?

Solution: Both must have the same backbone, so we write two of these first (and we follow the convention that such backbones are always written in the N to C—left to right—direction):

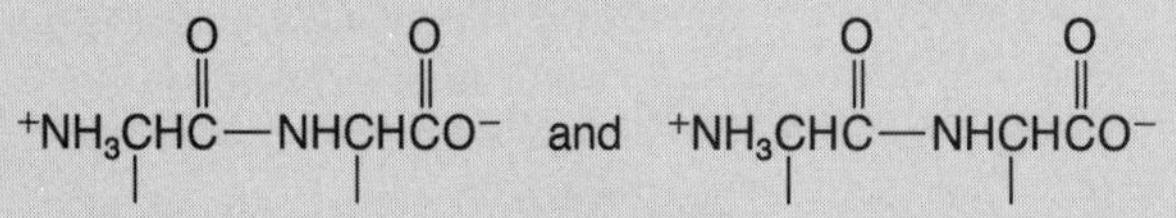

Then, either from memory or by the use of Table 21.1, we recall the two side chains, —CH$_3$ for alanine and —CH$_2$SH for cysteine. We simply attach these in their two possible orders to make the finished structures:

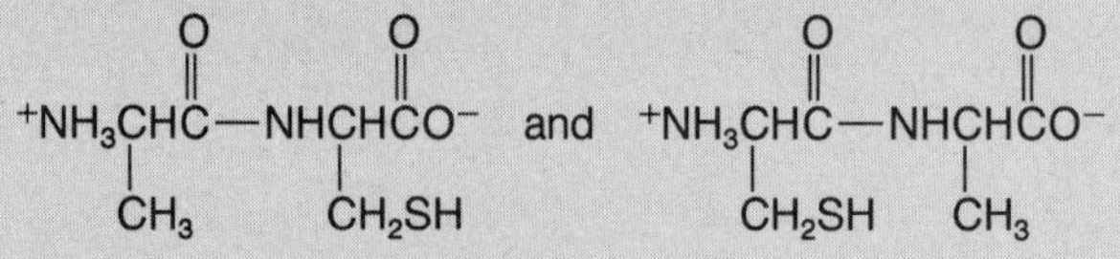

It would be worthwhile at this time simply to memorize the easy repeating sequence in a dipeptide, because it carries forward to higher peptides.

nitrogen-carbon-carbonyl-nitrogen-carbon-carbonyl

PRACTICE EXERCISE 5   Write the structures of the two dipeptides that can be made from alanine and glutamic acid.

One-letter symbols have been proposed and may become official.

To make it easier to write the structures of polypeptides, chemists use the three-letter symbols of amino acids given in Table 21.1. The convention is that a series of these three-letter symbols, each separated by a raised dot, represents a polypeptide structure, provided that the first symbol (reading left to right) is the free amino end, $^+NH_3$—, and the last symbol has the free carboxylate end, —CO_2^-. For example, the structure of the dipeptide **4** can be rewritten as Gly·Ala, and its isomer **5** as Ala·Gly. In both, the backbones are identical.

Structures **4** and **5** differ only in the sequence in which the side chains, —H and —CH_3, occur on α-carbons. This is fundamentally how polypeptides also differ, in their sequences of side chains. Of course, they also differ in their backbone lengths, so let's continue.

Dipeptides still have $^+NH_3$— and —CO_2^- groups, so a third amino acid can react at either end. In general,

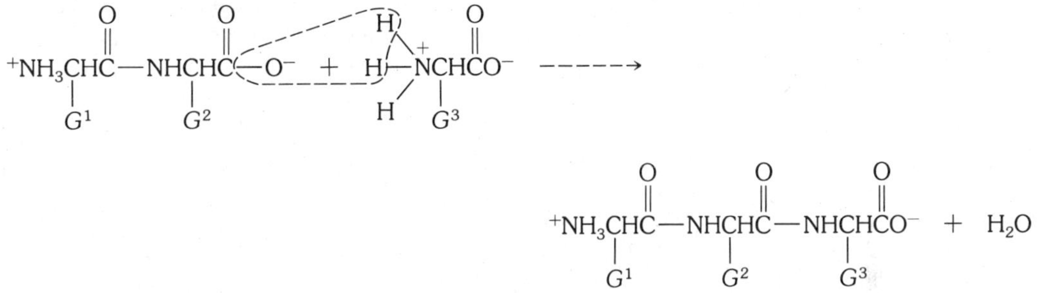

6

A tripeptide

A specific example is:

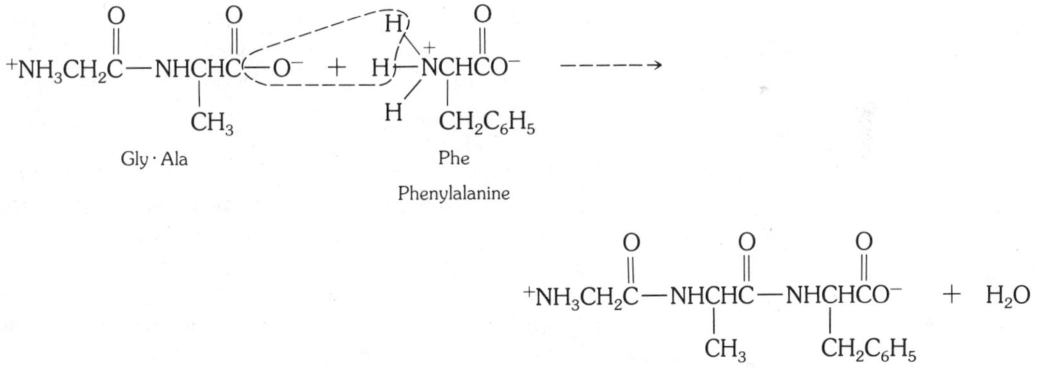

This tripeptide, Gly·Ala·Phe, is only one of six possible tripeptides that involves these three different amino acids. The set of all possible sequences for a tripeptide made from glycine, alanine, and phenylalanine is as follows:

Gly·Ala·Phe   Ala·Gly·Phe   Phe·Gly·Ala
Gly·Phe·Ala   Ala·Phe·Gly   Phe·Ala·Gly

FIGURE 21.3
The disulfide link in polypeptides. *(a)*
Two neighboring strands are joined. *(b)*
Loops within the same strand can form.

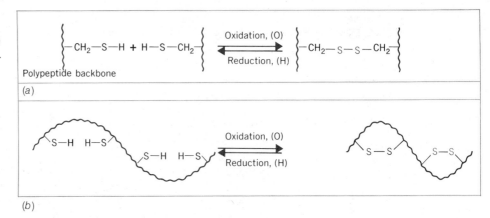

(a)

(b)

Each of these tripeptides still has groups at each end, $^+NH_3$— and —CO_2^-, that can interact with still another amino acid to make a tetrapeptide. And this product would still have the end groups from which the chain could be extended still further. You can see how a repetition of this many hundreds of times can produce a long polymer, a polypeptide, in which the backbone would have the following structure:

$$^+NH_3-\underset{|}{CH}-\overset{\overset{O}{\|}}{C}(NH-\underset{|}{CH}-\overset{\overset{O}{\|}}{C})_n NH-\underset{|}{CH}-CO_2^-$$ (*n* can be several thousand)

N-terminal   C-terminal
unit   unit

This repeating sequence, together with the sequence of side chains, constitutes the **primary structure** of all polypeptides. (Notice, for later reference, the designations N-terminal unit and C-terminal unit for the residues with the free α-NH_3^+ and the free α-CO_2^- groups, respectively.) The peptide bond is the chief covalent linkage that holds amino acid residues together in polypeptides. The only other covalent bond at the primary level is the disulfide bond.

As the number of amino acid residues increases, some used several times, the number of possible polypeptides rises rapidly. For example, if 20 different amino acids are incorporated, each used only once, there are 2.4×10^{18} possible isomeric polypeptides! (And a polypeptide with only 20 amino acid residues in its molecule is a very small polypeptide.)

Disulfide Bonds in Polypeptides.   If the —SH group on the side chain of cysteine appears on two neighboring polypeptide molecules, then mild oxidation is all it takes to link the two molecules by a disulfide bond, as shown in Figure 21.3. This cross-linking can also occur between parts of the same polypeptide molecule, in which case a closed loop results.

Some polypeptides feature both kinds of S—S cross-linking, and one example is the hormone insulin (Figure 21.4), which is considered to be a relatively simple polypeptide. You can find three disulfide bonds, one which creates a loop, and two that hold the two insulin subunits together.

21.3   SECONDARY STRUCTURES OF PROTEINS

The α-helix, the β-pleated sheet, and the triple helix are three kinds of secondary protein structures.

The α-Helix.   Once a cell puts together a polypeptide molecule, noncovalent forces of attraction between parts of the structure make the molecule twist into a particular shape. The hydrogen bond is the chief noncovalent force, and we'll see in this section how it can stabilize shapes of polypeptides.

FIGURE 21.4
Human insulin.

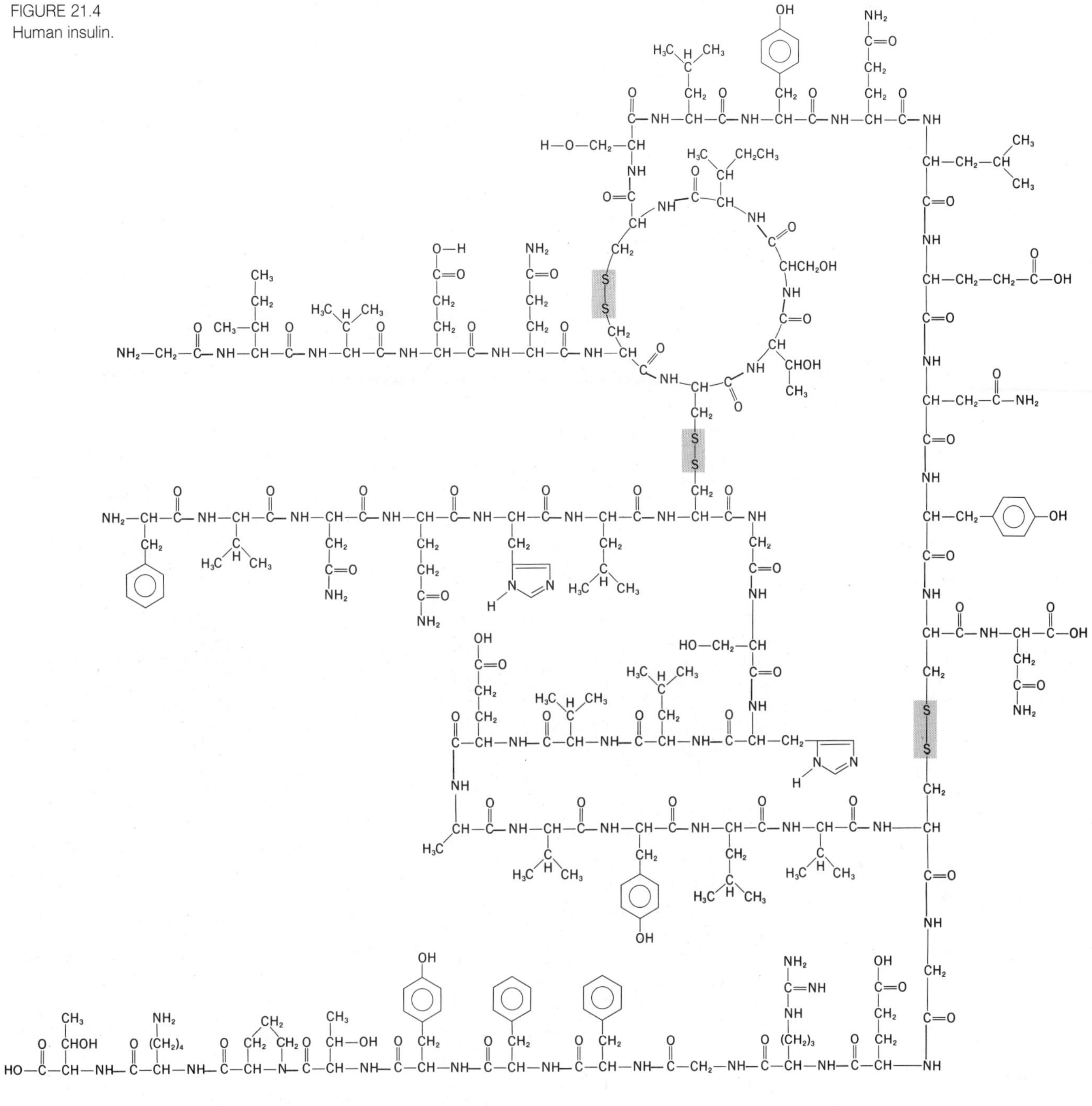

Linus Pauling won the 1954 Nobel
prize in chemistry for this work.

Using data obtained with X rays, Linus Pauling and R. B. Corey showed that a coiled configuration is one secondary structural feature, and they called it the α-helix, illustrated in Figure 21.5. In the α-helix, the polypeptide backbone coils as a right-handed screw with all of its side chains sticking to the outside. Hydrogen bonds extend from the oxygen atoms of carbonyl groups to hydrogen atoms of NH groups farther along the backbone.

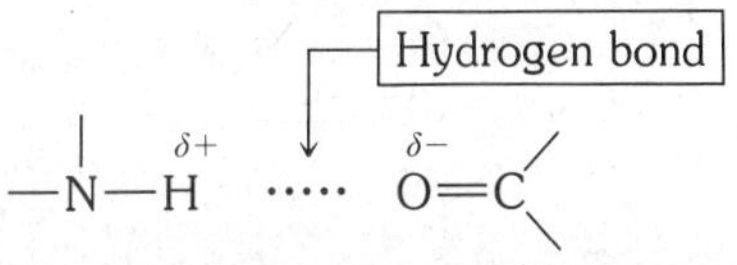

FIGURE 21.5
The α-helix. (From G. H. Haggis, D. Michie, A. R. Muir, K. B. Roberts, and P. M. B. Walker, *Introduction to Molecular Biology,* John Wiley & Sons, New York, 1964. Used by permission.)

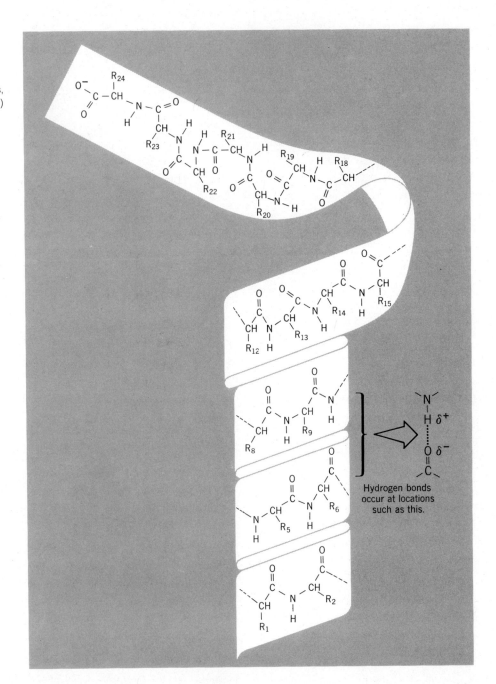

Hydrogen bonds occur at locations such as this.

The designation α was picked only because this secondary structure was the first to be identified.

Individually, a single hydrogen bond is a weak force of attraction, but when there are hundreds of them up and down a coiled polypeptide, these forces add up much as the individual "forces" accumulate that hold a zipper strongly shut. Generally, only segments of polypeptides, not entire lengths, are in an α-helix configuration.

The β-Pleated Sheet. Pauling and Corey also discovered that molecules in some proteins line up side by side, to form a sheet-like array. Hydrogen bonds hold the polypeptide molecules together. Then the entire sheet changes over into a partial accordian fold to give a pleat-like array, and this secondary structure is called the **β-pleated sheet.** (See Figure 21.6.) This structure is the dominant feature in fibroin, the protein in silk. In other proteins, these sheets seldom contribute much to the overall structure.

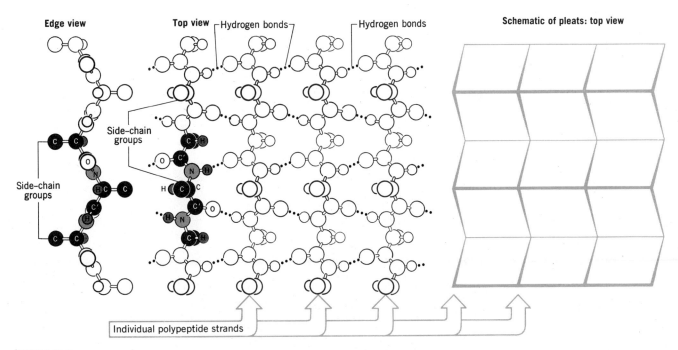

Edge view

Side-chain groups

Top view ⌐Hydrogen bonds⌐   ⌐Hydrogen bonds⌐

Side-chain groups

Side-chain groups

Schematic of pleats: top view

Individual polypeptide strands

FIGURE 21.6
The β-pleated sheet.

A collagen fibril only 1 mm in diameter can hold a mass as large as 10 kg (22 lb).

The Triple Helix. The third important secondary structure is present in a small family of proteins called the collagens. These are the proteins that give strength to bone, teeth, cartilage, tendons, and skin. The polypeptide units in collagen are called tropocollagen, and each tropocollagen molecule consists of three polypeptide chains. Each chain has about 1000 amino acid residues, which are twisted together to form a **triple helix.** Some of the residues are hydroxylated derivatives of lysine and proline that are made with the help of vitamin C *after* the initial polypeptide is made. You can see why vitamin C is essential to the formation of strong bones. In some of the types of collagen, sugar molecules are incorporated as glycosides of side-chain —OH groups.

The individual strands in tropocollagen helices are in a very open helix that is not stabilized by hydrogen bonds, but these open helices wrap around each other into a right-handed helical cable. Within the cable, hydrogen bonds help to stabilize the system. In addition, some cross-links, such as disulfide bonds, are made.

The collagen *fibril* forms when individual tropocollagen cables overlap lengthwise, as seen in Figure 21.7. The mineral deposits in bones and teeth become tied into the protein at the gaps between the heads of tropocollagen molecules and the tails of others.

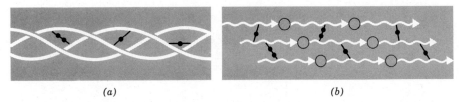

(a)   *(b)*

FIGURE 21.7
Collagen. *(a)* The tropocollagen cable is made of three polypeptide molecules, each in an α-helix (not pictured), wrapped together. *(b)* Individual tropocollagen cables, shown here by wavy lines, line up side by side with overlapping to give collagen fibrils. The circles represent gaps into which minerals can deposit in bones and teeth. The colored lines with solid color dots represent cross-links.

FIGURE 21.8
Myoglobin. The tube-like forms outline the segments that are in an α-helix. The black dots identify α-carbons. The side-chain groups at FG2, H16, H24, CD2, and CD3 are hydrophilic and are somewhat exposed to the aqueous medium. Hydrophobic groups that are somewhat tucked inside are at A7, A8, A9, A11, and A12, to note a few. (Reproduced by permission from R. E. Dickerson, "X-Ray Analysis of Proteins," in H. Neurath (ed.), *The Proteins,* Academic Press, New York; © 1964, all rights reserved.)

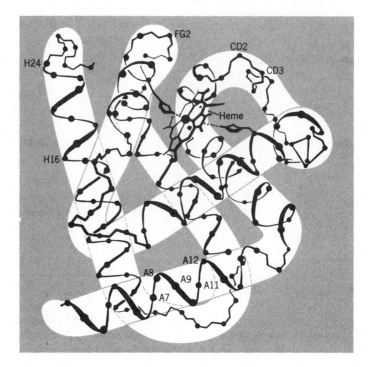

21.4 TERTIARY AND QUATERNARY STRUCTURES OF PROTEINS

Tertiary structures are the results of folding, bending, and twisting of secondary structures.

When α-helices take shape, their side chains tend to project outward where, in an aqueous medium, they would be in contact with water molecules. Even in water-soluble proteins, however, as many as 40% of the side chains are hydrophobic. Because such groups can't break up the hydrogen-bonding networks among the water molecules, the entire α-helix undergoes further twisting and folding until the hydrophobic groups, as much as possible, are tucked to the inside, away from the water, and the hydrophilic groups stay exposed to the water. Thus the final shape of the polypeptide, its **tertiary structure,** emerges in response to simple molecular forces set up by the water-avoiding and the water-attracting properties of the side chains.

The tertiary structure of myoglobin, the oxygen-storing protein in muscle tissue, is given in Figure 21.8. It consists of just one polypeptide plus a nonprotein, heme (Figure 21.9). About 75% of the myoglobin molecule is in an α-helix that is further folded as the figure shows. Virtually all of its hydrophobic groups are folded inside and its hydrophilic groups are on the outside.

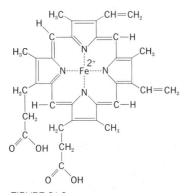

FIGURE 21.9
The heme molecule with its Fe^{2+} ion.

"Prosthetic" is from the Greek *prosthesis,* an addition.

Prosthetic Groups in Proteins.   A nonprotein, organic compound that associates with a polypeptide, such as heme in myoglobin, is called a **prosthetic group.** It is often the focus of the protein's biological purpose. Heme, for example, is the actual oxygen-carrier in myoglobin. It serves the same function in **hemoglobin,** the oxygen-carrier in blood.

Salt Bridges in Tertiary Structures.   Another force that can stabilize a tertiary structure is the attraction between a full positive and a full negative charge, each occurring on a particular side chain. For example, at the pH of body fluids, the side chains of both aspartic acid and glutamic acid carry CO_2^- groups. The side chains of lysine and arginine carry

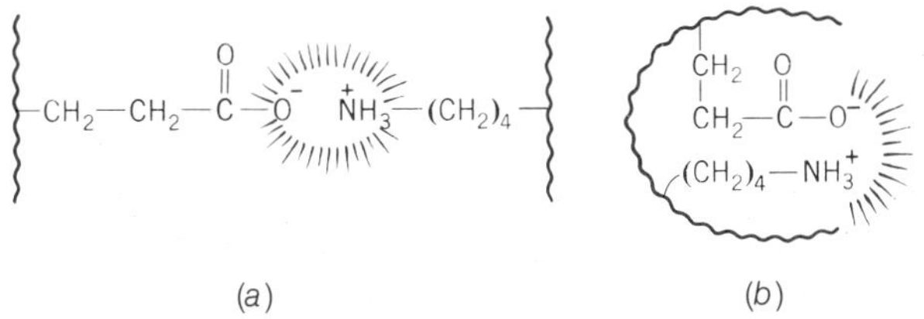

FIGURE 21.10
The salt bridge. This attraction between full and unlike charges can (a) hold one polypeptide to another or (b) stabilize a loop or coil within the same molecule.

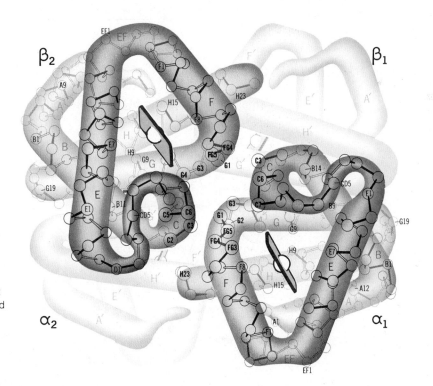

(a)   (b)

$—NH_3^+$ groups. These oppositely charged groups naturally attract each other much as do the oppositely charged ions in an ionic crystal. The attraction is called a **salt bridge.** (See Figure 21.10.)

Quaternary Structures. There are some proteins whose structures do not go beyond the tertiary level. Myoglobin in one. They are made up of single polypeptide molecules, sometimes with prosthetic groups. Many proteins, however, are aggregations of two or more polypeptides, and these aggregations constitute **quaternary structures.** For example, one molecule of the enzyme phosphorylase consists of two tightly aggregated and identical molecules of polypeptide. If the two polypeptide molecules become separated, the enzyme will not function. Individual molecules of polypeptides that make up an intact protein molecule are called the protein's *subunits.*

Hemoglobin has four subunits, two of one kind (designated α-subunits) and two of another (called the β-subunits). Each of the four subunits carries a heme molecule. (See Figure 21.11.) Salt bridges and hydrogen bonds hold the subunits together. These forces do not work unless each subunit has the appropriate primary, secondary, and tertiary structural features. If even one amino acid residue is wrong, the results can be very serious. The case of sickle-cell anemia is one example, and it is described in Special Topic 21.1.

FIGURE 21.11
Hemoglobin. Four polypeptide chains, each with one heme molecule that is represented here by the flat plates that contain spheres (Fe^{2+} ions), are nestled together. (From R. E. Dickerson and I. Geis, *The Structure and Action of Proteins,* W. A. Benjamin, Inc., Menlo Park, CA., © 1969. All rights reserved. Used by permission.)

The decisive importance of the primary structure to all other structural features of a polypeptide or its associated protein is illustrated by the grim story of sickle-cell anemia. This inherited disease is widespread among those whose roots are in central and western Africa.

In its mild form, where only one parent carries the genetic trait, the symptoms of sickle-cell anemia are seldom noticed except when the environment has a low partial pressure of oxygen, as at high altitudes. In the severe form, when both parents carry the trait, the infant usually dies by the age of 2. The problem is an impairment in blood circulation traceable to the altered shape of hemoglobin in sickle-cell anemia, particularly after the hemoglobin has delivered oxygen and is on its way back to the heart and lungs for more.

The fault at the molecular level lies in a β-subunit of hemoglobin. One of the amino acid residues should be glutamic acid but is valine, instead. Thus instead of a side-chain $-CO_2^-$ group, which is electrically charged and hydrophilic, there is an isopropyl side chain, which is neutral and hydrophobic. Normal hemoglobin, symbolized as HHb, and sickle-cell hemoglobin, HbS, therefore have different patterns of electrical charges. Both have about the same solubility in well-oxygenated blood, but HbS without its oxygen load tends to precipitate inside red cells. This distorts the cells into a telltale sickle shape. The distorted cells are harder to pump, and they sometimes clump together and plug capillaries. Sometimes they split open. Any of these events places a greater strain on the heart. The error in one side chain seems miniscule, but it is far from small in human terms.

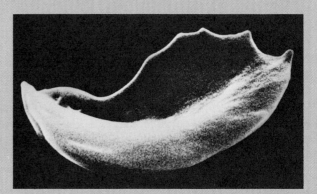

Electron micrographs of a normal red blood cell, left, and a sickle cell, right.

21.5   COMMON PROPERTIES OF PROTEINS

Even small changes in the pH of a solution can affect a protein's solubility and its physiological properties.

Although proteins come in many diverse biological types, they generally have similar chemical properties because they have similar functional groups.

Hydrolysis and Digestion.   The digestion of a protein is nothing more than the hydrolysis of its peptide bonds (amide linkages), as illustrated in Figure 21.12.

Denaturation.   It isn't necessary to hydrolyze peptide bonds to denature a protein, that is, to destroy its ability to perform its biological function. All that has to happen is some disruption of secondary or higher structural features. **Denaturation** is the disorganization of the overall molecular shape of a protein. It can occur as an unfolding or uncoiling of helices, or as the separation of subunits. (See Figure 21.13.)

Usually, denaturation is accompanied by a major loss in solubility. For example, when egg white is whipped or is heated, as when you cook an egg, the albumin molecules unfold and become entangled among themselves. The system no longer blends with water—it's insoluble—and it no longer allows light to pass through.

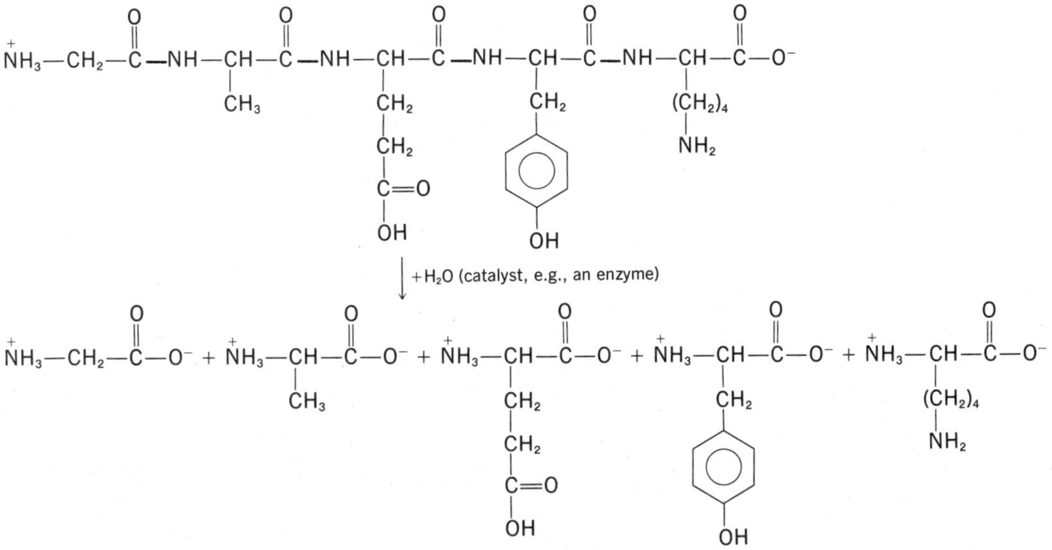

$$\overset{+}{N}H_3-CH_2-\overset{\overset{\displaystyle O}{\|}}{C}-NH-CH-\overset{\overset{\displaystyle O}{\|}}{C}-NH-CH-\overset{\overset{\displaystyle O}{\|}}{C}-NH-CH-\overset{\overset{\displaystyle O}{\|}}{C}-NH-CH-\overset{\overset{\displaystyle O}{\|}}{C}-O^-$$

with side chains: CH_3; CH_2–CH_2–$C=O$–OH; CH_2–(ring)–OH; $(CH_2)_4$–NH_2

$$\downarrow\ +H_2O \text{ (catalyst, e.g., an enzyme)}$$

$$\overset{+}{N}H_3-CH_2-\overset{\overset{\displaystyle O}{\|}}{C}-O^- + \overset{+}{N}H_3-CH-\overset{\overset{\displaystyle O}{\|}}{C}-O^- + \overset{+}{N}H_3-CH-\overset{\overset{\displaystyle O}{\|}}{C}-O^- + \overset{+}{N}H_3-CH-\overset{\overset{\displaystyle O}{\|}}{C}-O^- + \overset{+}{N}H_3-CH-\overset{\overset{\displaystyle O}{\|}}{C}-O^-$$

with side chains: CH_3; CH_2–CH_2–$C=O$–OH; CH_2–(ring)–OH; $(CH_2)_4$–NH_2

Glycine   Alanine   Glutamic acid   Tyrosine   Lysine

FIGURE 21.12
The digestion of a polypeptide, illustrated here by the hydrolysis of a pentapeptide. Only the peptide bonds (in color) break.

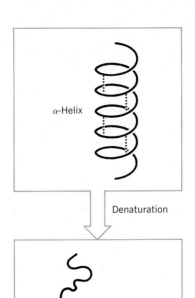

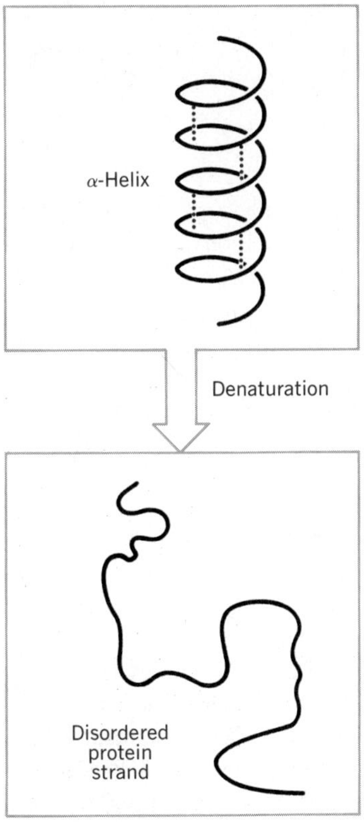

FIGURE 21.13
Denaturation of a protein happens when the protein loses secondary, tertiary, or quaternary structure.

α-Helix

Denaturation

Disordered protein strand

Table 21.2 has a list of several reagents or physical forces that cause denaturation, together with explanations of how they work. How effective a given denaturing agent is depends on the kind of protein. The proteins of hair and skin and of fur or feathers, quite strongly resist denaturation because they are rich in disulfide links.

Effect of pH on Protein Solubility.   Because some side chains as well as the end groups of polypeptides bear electrical charges, the entire molecule bears a net charge. Because these groups are either proton donors or acceptors, the net charge is easily changed by changing the pH. For example, $-CO_2^-$ groups become electrically neutral $-CO_2H$ groups when they pick up protons as a strong acid is added.

Suppose the net charge on a polypeptide is $1-$, and that one extra $-CO_2^-$ is responsible for it. When acid is added, we might imagine the following change, where the elongated shape is the polypeptide system.

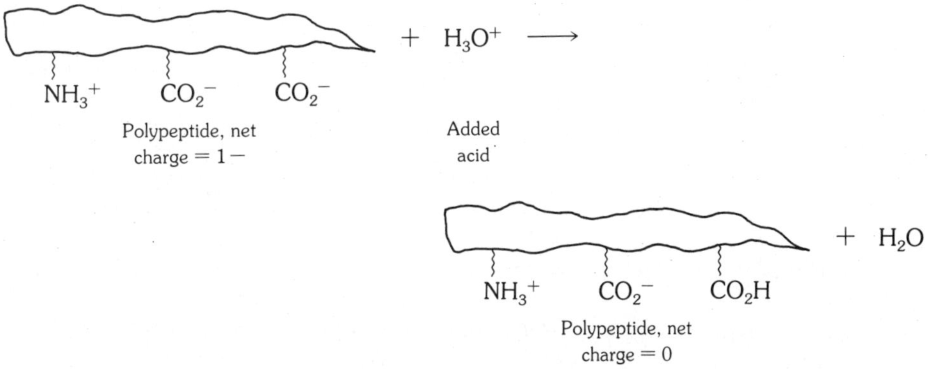

$$\text{(polypeptide)} \quad + \quad H_3O^+ \quad \longrightarrow$$

$NH_3^+ \qquad CO_2^- \qquad CO_2^-$

Polypeptide, net charge = $1-$

Added acid

$$\text{(polypeptide)} \quad + \quad H_2O$$

$NH_3^+ \qquad CO_2^- \qquad CO_2H$

Polypeptide, net charge = 0

TABLE 21.2
Sequence Rule Priorities — The *R/S* System

Denaturing Agent	How the Agent May Operate
Heat	Disrupts hydrogen bonds by making molecules vibrate too violently. Produces coagulation, as in the frying of an egg.
Microwave radiation	Causes violent vibrations of molecules that disrupt hydrogen bonds.
Ultraviolet radiation	Probably operates much as the action of heat (e.g., sunburning)
Violent whipping or shaking	Causes molecules in globular shapes to extend to longer lengths, which then entangle (e.g., beating egg white into meringue).
Detergents	Probably affect hydrogen bonds and salt bridges.
Organic solvents (e.g., ethanol, 2-propanol, acetone)	May interfere with hydrogen bonds because these solvents can form hydrogen bonds, also. Quickly denature proteins in bacteria, killing them (e.g., the disinfectant action of 70% ethanol).
Strong acids and bases	Disrupt hydrogen bonds and salt bridges. Prolonged action leads to actual hydrolysis of peptide bonds.
Salts of heavy metals (e.g., salts of Hg^{2+}, Ag^+, Pb^{2+})	Cations combine with —SH groups and form precipitates. (These salts are all poisons.)
Solutions of urea	Disrupt hydrogen bonds. (Urea, being amide-like, can form hydrogen bonds of its own.)

Now the polypeptide is isoelectric and neutral. On the other hand, a polypeptide might have a net charge of 1 +, caused by an excess of one —NH_3^+ group. The addition of OH^- can cause the following change, which makes the polypeptide isoelectric and neutral.

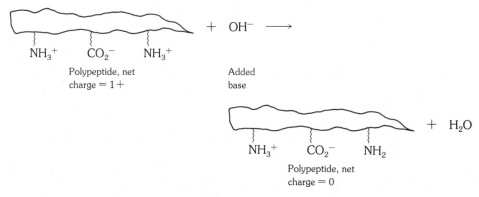

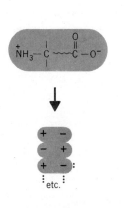

FIGURE 21.14
Several isoelectric protein molecules (top) can aggregate into very large clusters that no longer dissolve in water.

Like each amino acid, each protein has a characteristic pH, its isoelectric point, at which its net charge is zero and at which it cannot migrate in an electric field. One major significance of this is that polypeptide molecules that are neutral can aggregate and clump together to become particles of enormous size that simply drop entirely out of solution. (See Figure 21.14.) A protein is least soluble in water when the pH equals the protein's isoelectric point. Therefore whenever a protein must be in solution to work, as is true for many enzymes, the pH of the medium must be kept away from the protein's isoelectric point. Buffers in body fluids have the task of ensuring this.

An example of the effect of pH on solubility is given by casein, milk protein. Its pI value is 4.7, and when the pH of milk drops from its normal value of 6.3 – 6.6 to 4.7 as milk turns sour, the casein separates as curds. As long as the pH of milk is something *other than* the pI for casein, the protein remains colloidally dispersed.

21.6 CLASSES OF PROTEINS

Three criteria for classifying proteins are their solubility in aqueous systems, their compositions, and their biological functions.

We began this chapter with hints about the wide diversities of the kinds and uses of proteins. Now that we know about their structures, we can better understand how so many types of proteins with so many functions are possible. The following three major classifications of proteins and their several examples give substance to the chapter's introduction.

Classifying Proteins According to Solubility.   When proteins are classified according to their solubilities, two major families are the **fibrous proteins** and the **globular proteins.** The fibrous proteins are insoluble in water, and they including the following special types.

When meat is cooked, some of its collagen changes to gelatin, which makes the meat easier to digest.

1. **Collagens** — in bone, teeth, tendons, skin, and soft connective tissue. When such tissue is boiled with water, the portion of its collagen that dissolves is called gelatin.

Elastin is not changed to gelatin by hot water.

2. **Elastins** — in many places where collagen is found, but particularly in ligaments, the walls of blood vessels, and the necks of grazing animals. Elastin is rich in hydrophobic side chains. Cross-links between elastin strands are important to its recovery after stretching.

3. **Keratins** — in hair, wool, animal hooves, nails, and porcupine quills. The keratins are rich in disulfide links.

4. **Myosins** — the proteins in contractile muscle.

5. **Fibrin** — the protein of a blood clot. During clotting, it forms from its precursor, fibrinogen, by an exceedingly complex series of reactions.

Globular proteins, which are soluble in water or in water that contains certain salts, include the following.

1. **Albumins** — in egg white and in blood. In the blood the albumins serve many functions. Some are buffers. Some carry water-insoluble molecules of lipids or fatty acids. Some carry metal ions that could not be dissolved in water that is even slightly alkaline, such as Cu^{2+} ions.

2. **Globulins** — including the gamma-globulins that are part of the body's defense mechanism against diseases. The globulins need the presence of dissolved salts to be soluble in water.

Classifying Proteins According to Composition.   The nature of the prosthetic group in these proteins provides another way of classifying proteins.

1. Glycoproteins — proteins with sugar units. Gamma globulin is an example.

2. Hemoproteins — proteins with heme units such as hemoglobin, myoglobin, and certain cytochromes (enzymes that help cells use oxygen).

3. Lipoproteins — proteins that carry lipid molecules, including cholesterol.

4. Metalloproteins — proteins that incorporate a metal ion, such as many enzymes do.

5. Nucleoproteins — proteins bound to nucleic acids, such as ribosomes and some viruses.

6. Phosphoproteins — proteins with a phosphate ester to a side chain —OH group, such as in serine. Milk casein is an example.

Classifying Proteins According to Biological Function. Perhaps no other system of classification more clearly dramatizes the importance of proteins.

1. Enzymes — the body catalysts.
2. Contractile muscle — with stationary filaments, myosin, and moving filaments, actin.
3. Hormones — such as growth hormone, insulin, and others.
4. Neurotransmitters — such as the enkephalins and endorphins.
5. Storage proteins — those that store nutrients that the organism will need, such as seed proteins in grains, casein in milk, ovalbumin in egg white, and ferritin, the iron-storing protein in human spleen.
6. Transport proteins — those that carry things from one place to another. Hemoglobin and the serum albumins are examples already mentioned. Ceruloplasmin is a copper-carrying protein.
7. Structural proteins — proteins that hold a body structure together, such as collagen, elastin, keratin, and proteins in cell membranes.
8. Protective proteins — those that help the body to defend itself. Examples are the antibodies and fibrinogen.
9. Toxins — poisonous proteins. Examples are snake venom, diphtheria toxin, and clostridium botulinus toxin (a toxic substance that causes some types of food poisoning).

SUMMARY

Amino acids About 20 α-amino acids supply the amino acid residues that make up a polypeptide. All but one (glycine) are optically active and in the L-family. In the solid state or in water at a pH of roughly 6–7, most amino acids exist as dipolar ions. Their isoelectric points, the pH values of solutions in which they are isoelectric, are in this pH range of 6–7. Those with —CO_2H groups on side chains have lower pI values. Those with basic side chains have higher pI values. Several amino acids have hydrophobic side chains, but the side chains in others are strongly hydrophilic. The —SH group of cysteine opens the possibility of disulfide cross-links between polypeptide units.

Polypeptides Amino acid residues are held together by peptide (amide) bonds, so the repeating unit in polypeptides is

$$-NH-\underset{|}{CH}-CO-$$

Each residue carries its own side chain. This repeating system with a unique sequence of side chains constitutes the primary structure of a polypeptide. Once this is fashioned, the polypeptide coils and folds into higher features — secondary and tertiary — that are stabilized largely by hydrogen bonds and the water-avoiding or water-attracting properties of the side chains. The most prominent secondary structures are the α-helix, the β-pleated sheet, and the collagen triple helix.

Proteins Many proteins consist just of one kind of polypeptide. Many others have nonprotein, organic groups — prosthetic groups — or metal ions. And still other proteins — those with quaternary structure — involve two or more polypeptides whose molecules aggregate in definite ways, stabilized sometimes by salt bridges. Thus the terms *protein* and *polypeptide* are not synonyms, although for some specific proteins they turn out to be.

Because of their higher levels of structure, proteins can be denatured by agents that do nothing to peptide bonds. The acidic and basic side chains of polypeptides affect protein solubility, and when a protein is in a medium whose pH equals the protein's isoelectric point, the substance is least soluble. The amide bonds (peptide bonds) of proteins are hydrolyzed during digestion.

KEY TERMS

The following terms were stressed in this chapter, and they will be used again. Be sure that you have mastered them before you continue.

albumins	collagen	dipolar ion	fibrin
amino acid	denaturation	disulfide link	fibrous protein
amino acid residue	dipeptide	elastin	globular protein

globulins	keratin	polypeptide	salt bridge
α-helix	myosin	primary structure	secondary structure
hemoglobin	peptide bond	prosthetic group	side chain
isoelectric	pI	protein	tertiary structure
isoelectric point (pI)	β-pleated sheet	quaternary structure	triple helix

SELECTED REFERENCES

1 David Freifelder. "The Physical Structure of Protein Molecules," Chapter 5 in *Molecular Biology*. Science Books International, Boston, MA, 1983.

2 A. L. Lehninger. *Principles of Biochemistry* (Chapters 5–8), Worth Publishers, Inc., New York, NY, 1982.

3 Lubert Stryer. *Biochemistry,* 2nd ed. (Chapter 2). W. H. Freeman and Company, San Francisco, CA, 1981.

4 R. F. Doolittle. "Proteins." *Scientific American,* October 1985, page 88.

5 R. E. Doolittle. "Fibrinogen and Fibrin." *Scientific American,* December 1981, page 126.

6 A. I. Caplan. "Cartilage." *Scientific American,* October 1984, page 84.

REVIEW EXERCISES

The answers to the following Review Exercises are in the *Study Guide* that accompanies this book.

Amino Acids

21.1 What structure will nearly all of the molecules of glycine have at a pH of about 1?

21.2 In what structure will most of the molecules of alanine be at a pH of about 12?

21.3 Pure alanine does not melt, but at 290 °C it begins to char. However, the ethyl ester of alanine, which has a free —NH$_2$ group, has a low melting point, 87 °C. Write the structure of this ethyl ester, and explain this large difference in melting point.

21.4 The ethyl ester of alanine (Review Exercise 21.3) is a much stronger base — more like ammonia — than alanine. Explain this.

21.5 Which of the following amino acids has the more hydrophilic side chain? Explain.

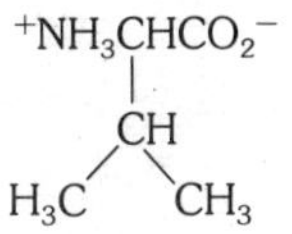

$$^+NH_3CHCO_2^-$$
$$|$$
$$CH$$
$$H_3C \quad CH_3$$
A

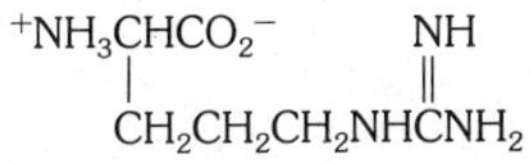

$$^+NH_3CHCO_2^-$$
$$|$$
$$CH_2CH_2CH_2NHCNH_2$$
$$\quad\quad\quad\quad\quad || $$
$$\quad\quad\quad\quad\quad NH$$
B

21.6 Which of the following amino acids has the more hydrophobic side chain? Explain.

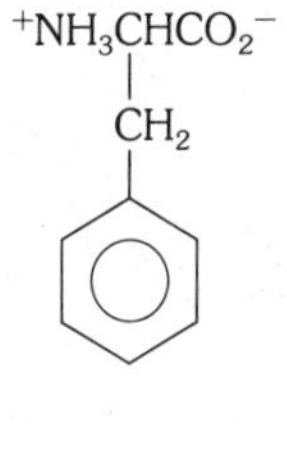

A

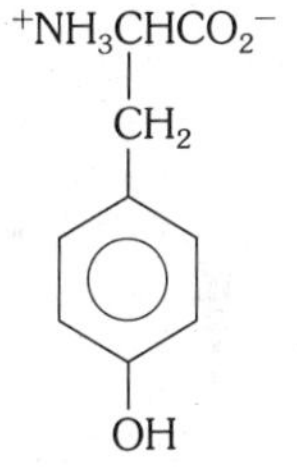

B

21.7 One of the possible forms for lysine is

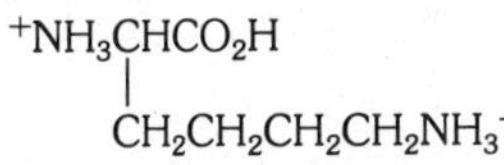

$$^+NH_3CHCO_2H$$
$$|$$
$$CH_2CH_2CH_2CH_2NH_3^+$$

(a) Is lysine most likely to be in this form at pH 1 or at pH 11? Explain.
(b) Would lysine in this form migrate to the anode, to the cathode, or not migrate at all in an electric field?

21.8 Aspartic acid can exist in the following form.

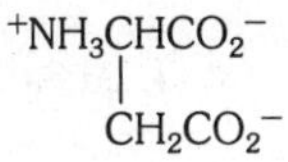

$$^+NH_3CHCO_2^-$$
$$|$$
$$CH_2CO_2^-$$

(a) Would this form predominate at a pH of 0 or a pH of 10? Explain.

(b) To which electrode, the anode or the cathode — or to neither — would aspartic acid in this form migrate in an electric field?

21.9 Complete the following Fischer projection formula to show correctly the absolute configuration of L-serine.

$$CO_2^-$$
$$+$$
$$CH_2OH$$

21.10 The molecules of all but one of the amino acids have at least one chiral carbon. Two of the amino acids have molecules with two chiral carbons. Write the structures and names of these two, and place asterisks by their chiral carbons.

Polypeptides

21.11 Write the conventional, condensed structure of the dipeptides that can be made from lysine and glycine.

21.12 What are the condensed structures of the dipeptides that can be made from cysteine and glutamic acid? (Do not use the three-letter symbols.)

21.13 Using three-letter symbols, write the structures of all of the tripeptides that can be made from lysine, glutamic acid, and alanine.

21.14 Write the structures in three-letter symbols of all of the tripeptides that can be made from cysteine, glycine, and alanine.

21.15 What is the conventional structure of Val·Ile·Phe?

21.16 The artificial sweetener in aspartame (the chief ingredient in NutraSweet), is the methyl ester of Asp·Phe. When the peptide bond in aspartame is hydrolyzed (and only this bond), one of the products cannot exist as an anion in water, even when the pH is 12. Write a conventional, condensed structure for aspartame.

21.17 Write the conventional structure for Ala·Val·Phe·Gly·Leu.

21.18 What is the conventional structure for:
Asp·Thr·Lys·Glu·Tyr?

21.19 Compare the side chains in the pentapeptide of Review Exercise 21.17 (call it A) with those in:
Lys·Glu·Asp·Thr·Ser
(which we can call B).
(a) Which of the two, A or B, is the more hydrocarbon-like?
(b) Which is probably more soluble in water? Explain.

21.20 Compare the side chains in the pentapeptide of Review Exercise 21.18, which we'll label C, with those in Phe·Leu·Gly·Ala·Val, which we can label D. Which of the two would tend to be less soluble in water? Explain.

21.21 If the tripeptide Gly·Cys·Ala were subjected to mild oxidizing conditions, what would form? Write the structure of the product using three-letter symbols.

21.22 Write the structure(s) of the product(s) of the mild reduction of the following polypeptide.

Higher Levels of Protein Structure

21.23 What kind of force makes it possible for a polypeptide to be stabilized in the shape of an α-helix?

21.24 The side-by-side alignment of polypeptides in a β-pleated sheet is maintained through the agency of what kind of force?

21.25 Give a brief description of the way in which polypeptide strands are organized in collagen.

21.26 What function does vitamin C perform in the formation of strong bones?

21.27 What factors affect the bending and folding of α-helices in the presence of an aqueous medium?

21.28 What is meant by a salt bridge? Draw structures to illustrate your answer.

21.29 In what way does hemoglobin represent a protein with quaternary structure (in general terms only)?

21.30 How do myoglobin and hemoglobin compare (in general terms only)?
(a) Structurally — at the quaternary level.
(b) Where they are found in the body.
(c) In terms of their prosthetic group(s).
(d) In terms of their functions in the body.

Properties of Proteins

21.31 What products form when the following polypeptide is completely digested?

21.32 The *partial* digestion of a polypeptide gave several products that *were not* amino acids. Some were dipeptides, for example. Which of the following compounds are possible products, and which could not possibly have formed? Explain.

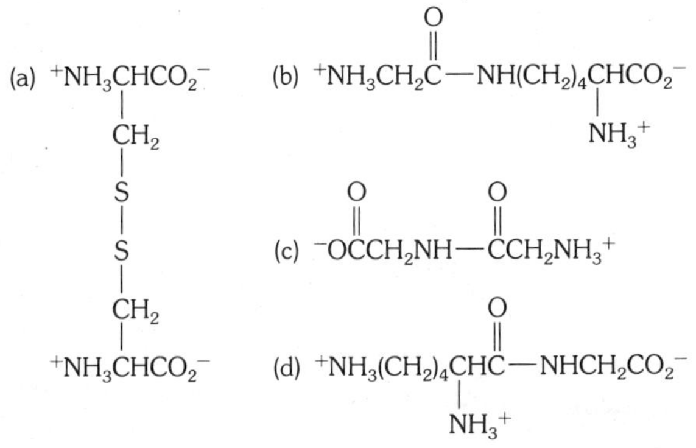

(a) $^+NH_3CHCO_2^-$
$|$
CH_2
$|$
S
$|$
S
$|$
CH_2
$|$
$^+NH_3CHCO_2^-$

(b) $^+NH_3CH_2\overset{\overset{\displaystyle O}{\|}}{C}-NH(CH_2)_4CHCO_2^-$
$|$
NH_3^+

(c) $^-O\overset{\overset{\displaystyle O}{\|}}{C}CH_2NH-\overset{\overset{\displaystyle O}{\|}}{C}CH_2NH_3^+$

(d) $^+NH_3(CH_2)_4\overset{\overset{\displaystyle O}{\|}}{C}HC-NHCH_2CO_2^-$
$|$
NH_3^+

21.33 Explain why a protein is least soluble in an aqueous medium that has a pH equal to the protein's pI value.

21.34 What is the difference between the digestion and the denaturation of a protein?

Types of Proteins

21.35 What experimental criterion distinguishes between fibrous and globular proteins?

21.36 What is the relationship between collagen and gelatin?

21.37 How are collagen and elastin alike? How are they different?

21.38 What experimental criterion distinguishes between the albumins and the globulins?

21.39 What is fibrin and how is it related to fibrinogen?

21.40 What general name can be given to a protein that carries a lipid molecule?

Chapter 22
Nucleic Acids

Whatever she's wondering, it's probably not how kids and kittens can grow on the same food. But we're ready to explore this rather astonishing fact as we study in this chapter how the same molecules can be reorganized into widely different species.

22.1 HEREDITY AND THE CELL

The chemicals associated with living things are, by and large, organized in structural units of life called cells.

We have learned that nearly *every* reaction in a living organism requires its special catalyst, and that all of these catalysts are in a family of proteins called enzymes. The set of enzymes in one organism is not exactly the same as the set of another, although many enzymes from different species are quite similar both in structure and function. In reproduction, each organism transmits the capacity to possess a unique set of enzymes to new offspring. However, the organism does not duplicate the enzymes themselves and then pass them on directly. Rather, the organism sends on the *instructions* for making the enzymes from amino acids. It does this by duplicating the compounds in a different family, the nucleic acids. These then direct the synthesis of enzymes in the offspring. Our purpose in this chapter is to study how nucleic acids do this, but we have to learn something about the biological context first.

The English scientist Robert Hooke was the first to use the word *cell* when, in 1665, he described the hollow compartments he saw with the aid of a microscope in thin slices of cork.

The Cell.   An artist's rendering of a typical cell is shown in Figure 22.1. Cells of different tissues differ widely in shape and size, so all that this figure does is display what most cells generally have.

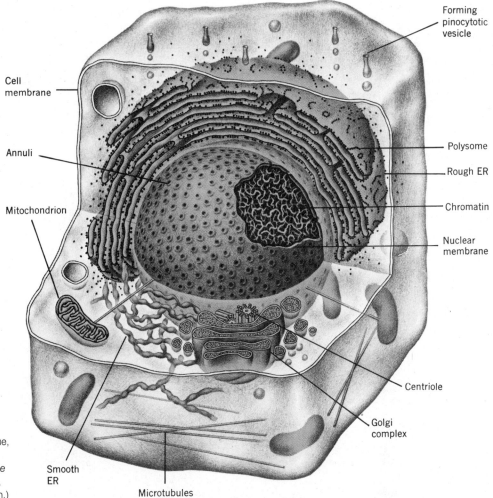

Cell membrane

Annuli

Mitochondrion

Forming pinocytotic vesicle

Polysome

Rough ER

Chromatin

Nuclear membrane

Centriole

Golgi complex

Smooth ER

Microtubules

FIGURE 22.1
A generalized animal cell. Although cells differ greatly from tissue to tissue, most have the features shown here. (From A. Nason and R. L. Dehan, *The Biological World*. John Wiley & Sons, New York, 1973. Used by permission.)

The boundary of a cell is the cell membrane, which we studied in Section 20.5. Everything enclosed by this membrane is called *protoplasm,* and it contains several discrete particles or cellular bodies called *organelles* ("little organs"). Prominent among them are the *mitochondria,* which make adenosine triphosphate (ATP) for the cell's needs for chemical energy. The endoplasmic reticulum (ER) consists of membrane-lined tubules with many convoluted branches in which fluids and chemicals move and interact. The ER generally extends from the cell membrane to the wall of the cell nucleus. All of the contents of the cell outside of the nucleus is called the *cytoplasm.* The pinocytotic vesicles in Figure 22.1 are a part of the way a cell moves materials to the outside. The Golgi complex makes lipids, collagen, and packages of enzymes. The centriole is involved in cell division. Polysomes (polyribosomes) consist mostly of nucleic acid. They help make polypeptides under the direction of the nucleic acid that is inside the cell's nucleus.

German biologists M. J. Schleiden and T. Schwann in 1839 proposed a formal theory of the cell as an organism and of plants and animals as composed of these organisms.

Chromosomes, Genes, and Heredity.
The nucleus has its own membrane, and inside is a web-like network of protein called the nuclear matrix. Twisted and intertwined filaments inside the nucleus consist of **chromatin,** and each chromatin strand is made up of nucleic acid, mostly of one kind called DNA, but some of a second kind called RNA. Associated with the nucleic acid is a mixture of polypeptides called histones. Molecules of DNA have sections that constitute individual **genes,** the fundamental units of heredity.

When cell division begins (Figure. 22.2), the chromatin strands thicken and become rod-like bodies that accept staining agents and so can be seen under a microscope. These discrete bodies are called **chromosomes.** The thickening of chromatin into chromosomes is caused by the synthesis of new nucleic acid (and associated proteins). The new chromatin is an exact copy of the old, if all goes well as it usually does.

The cycle of cell division illustrated in Figure 22.2 has now reached the prophase stage. Each gene has undergone **replication**—meaning that it has been reproduced in duplicate.

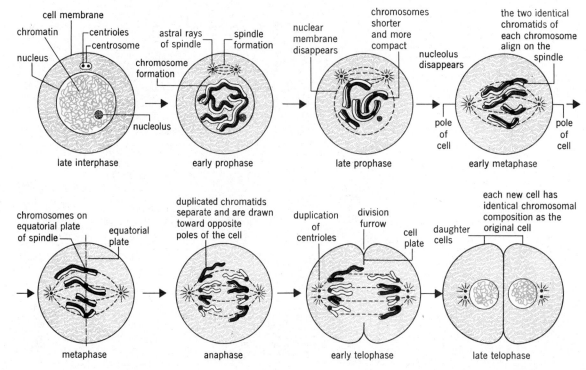

FIGURE 22.2
The major events in mitosis, showing a cell that has four chromosomes. (From J. R. McClintic, *Basic Anatomy and Physiology of the Human Body.* John Wiley & Sons, New York, 1975. Used by permission.)

Cell division now continues until the new daughter cells are complete. Thus through the replication of DNA, the genetic message of the first cell is passed to each of the two new cells.

22.2 THE STRUCTURE OF THE HEREDITARY UNITS

Genetic information is carried by the side chains of a twisted, double-stranded polymer called the DNA double helix.

Nucleic Acids. The chemicals that are the cell's primary means of storing and transmitting genetic information are polymers called **nucleic acids,** and these occur as two types nicknamed DNA and RNA. **DNA** is **deoxyribonucleic acid,** and **RNA** is **ribonucleic acid.** The monomer molecules for the nucleic acids are called **nucleotides.**

Unlike the monomers of proteins, the nucleotides can be further hydrolyzed. As outlined in Figure 22.3, the hydrolysis of a representative mixture of nucleotides produces three kinds of products: inorganic phosphate, a pentose sugar, and a group of heterocyclic amines called the **bases.** Their structures are given in Figure 22.3, but usually they are referred to by their names or their single-letter symbols, as follows.

Bases from DNA		Bases from RNA	
Adenine	A	Adenine	A
Thymine	T	Uracil	U
Guanine	G	Guanine	G
Cytosine	C	Cytosine	C

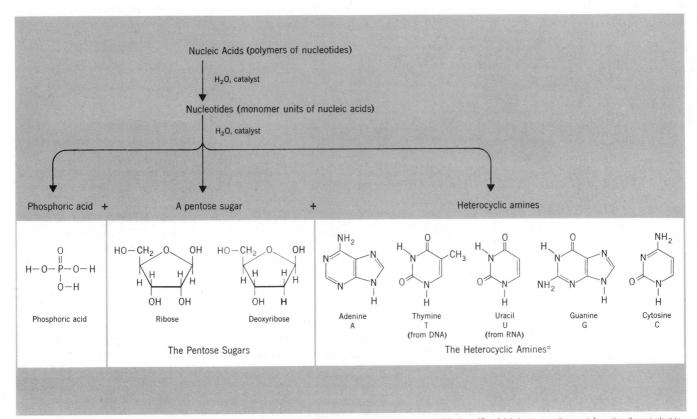

FIGURE 22.3
The hydrolysis products of nucleic acids.

* These are the five principal heterocyclic amines obtainable from nucleic acids. Others, not shown, are known to be present. While they differ slightly in structure they are informationally equivalent to one or another of the five shown here.

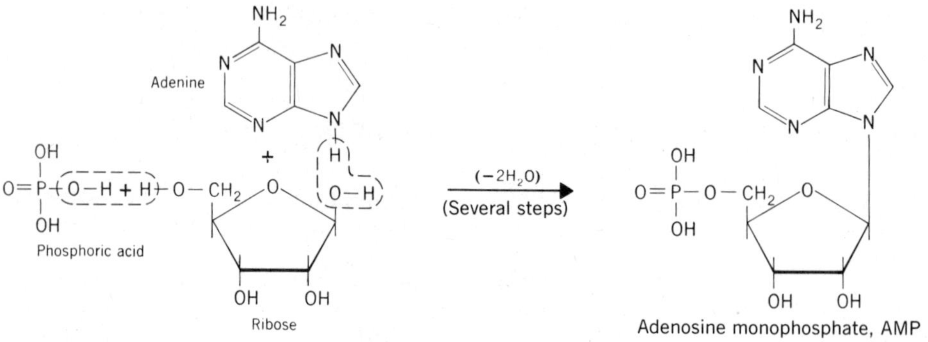

FIGURE 22.4
A typical nucleotide, AMP, and the
smaller units from which it is assembled.

You can see that three bases are common to both DNA and RNA, and that one base is different.

Another difference between DNA and RNA is the sugar unit. The pentose from the hydrolysis of RNA is ribose — hence the R in RNA — and that from DNA is deoxyribose — which lends the D to DNA.

With the products of the complete hydrolysis of nucleic acids now in mind, we can work backwards and see how a nucleic acid is built. First, we envision the formation of a typical monomer unit — a nucleotide. As seen in Figure 22.4, the pentose uses its C-5 alcohol group to make a phosphate ester. It also uses its C-1 alcohol group — actually, this alcohol unit is in a hemiacetal group — to fashion a bond to one of the bases, which is adenine (A) in the example used in Figure 22.4. Thus by the splitting out of two molecules of water (by a series of steps that we won't discuss), a nucleotide takes shape. In our example, it is adenosine monophosphate or AMP, a nucleotide for RNA, because it's made from ribose. When the —OH at C-2 is replaced by —H, and the pentose is deoxyribose, the nucleotide is one for DNA. The pattern for any nucleotide is as follows:

base
|
—phosphate—pentose—

Next, we move to a level higher and imagine how a number of nucleotides go together to make a nucleic acid. Figure 22.5 conveys the general idea. In principle, the formation of a nucleic acid polymer involves nothing more than the splitting out of water between a phosphate unit of one nucleotide and the C-3 alcohol group on the pentose of the next nucleotide in the sequence. The result is a phosphodiester system. Of course, many steps are required, each calling for its own enzyme, but when these are repeated thousands of times, a nucleic acid is the result. Thus the pattern for the backbone of a nucleic acid is alternating phosphate and pentose units, and projecting from the backbone at each pentose is one of the bases. With this in mind, we can move to simplify our structures. Figure 22.6 shows how a nucleic acid structure can be condensed. The distinctiveness of any one nucleic acid is in the sequence of the side-chain bases and in the length of the chain.

Base Pairing.   The side-chain bases have functional groups so arranged geometrically that they can fit to each other in pairs by means of hydrogen bonds, as seen in Figure 22.7. This phenomenon is called **base pairing,** and the relative locations of the functional groups and the geometries of these bases are such that in DNA G and C always form a pair, and A and T form another pair. In RNA, G and C always pair, and U and T always pair. Neither G nor C ever pairs with A, T, or U.

Crick-Watson Theory.   In 1953, Francis Crick of England and James Watson of the United States proposed a structure for DNA that made possible an understanding of how heredity works at the molecular level. Using X-ray data obtained by Rosalind Franklin, they deduced

Ring carbon number 1 is the carbon of the aldehyde group when the pentose ring is open.

Crick and Watson shared the 1962 Nobel prize in medicine and physiology with Maurice Wilkins.

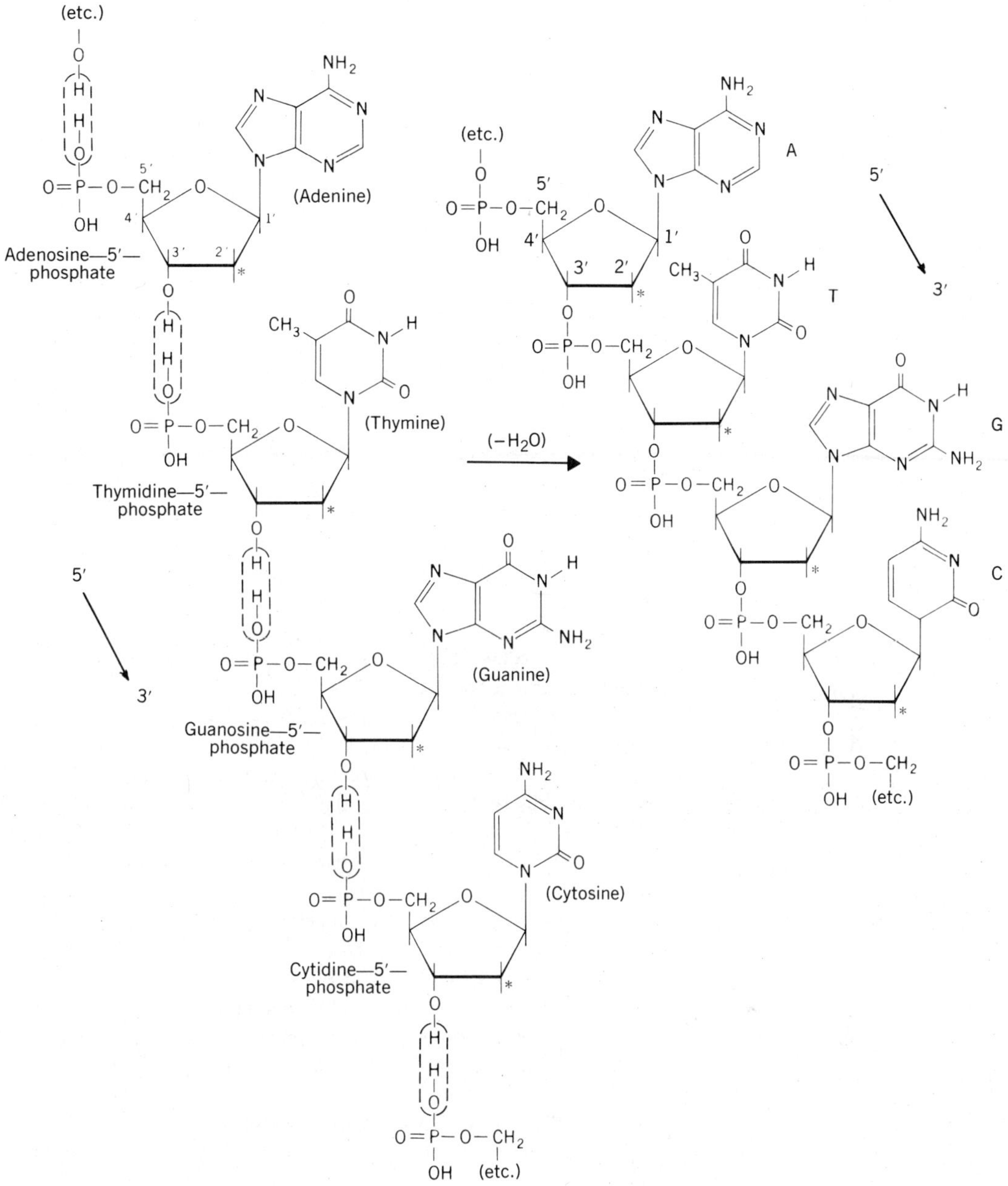

FIGURE 22.5
The relationship of a nucleic acid chain to its nucleotide monomers. On the right is a short section of a DNA strand. On the left are the nucleotide monomers from which it is made (after many steps). The colored asterisks by the pentose units identify the 2′ positions of these rings where there would be another —OH group if the nucleic acid were RNA (assuming that uracil also replaced thymine). The designation 5′ → 3′ means that the complete strand would have an unesterified —OH group on C-5′ of the first pentose unit and an unesterified C-3′ on the other end, and that the sequence of bases is written from the 5′ end to the 3′ end. Thus the sequence here is written as ATGC, not as CGTA.

that the long strands of DNA molecules occur in cell nuclei as pairs that are twisted into a right-hand helix. A key feature of this **double-helix model** of DNA is that the two intertwined strands are complementary in terms of the bases. Crick and Watson knew from the work of Erwin Chargaff that the DNA *in all species* contains A and T in a mole ratio of 1 : 1 and

FIGURE 22.6
Condensed structures of nucleic acids. Shown here is a representation of the same short segment of DNA that was given in Figure 22.5. The same representations could be used for RNA if U replaced T, and if the pentose units were understood to be ribose instead of deoxyribose.

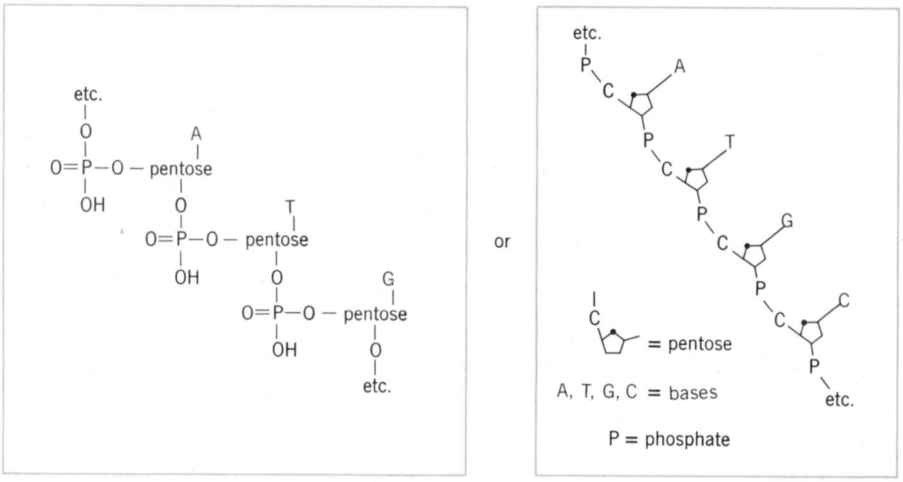

that the mole ratio of G and C also is 1 : 1. Crick and Watson made sense of these simple ratios when they found that A and T pair *between the two strands,* and that G and C also pair between the strands. Thus interstrand pairing *must* be reflected in 1 : 1 ratios of A to T and G to C. Whenever adenine (A) is one strand, then thymine (T) is immediately opposite it on the other strand. And whenever guanine (G) projects from one strand, then cytosine (C) is opposite it on the other strand. The hydrogen bonds between A and T and between G and C hold the strands together.

A molecular model of the DNA double helix is shown in Figure 22.8. The system resembles a spiral staircase in which the steps, which are perpendicular to the long axis of the spiral, consist of the base pairs. Hydrogen bonds between the pairs are centered around the long axis. Figure 22.9 is a schematic representation that points out an additional feature. The two DNA strands of the double helix run in opposite directions, as indicated in this figure by the $5' \rightarrow 3'$ specifications involving the pentose units. (By convention, the primed numbers such as $5'$ are for ring positions of the pentose ring when the pentose is part of a nucleotide. Side-chain ring positions have unprimed numbers.)

Figures 22.8 and 22.9 show only short segments of DNA double helices. They do not show that these helices are further twisted and coiled into superhelices. This looping and coiling is necessary if the cell's DNA is to fit into its nucleus. A typical human cell nucleus, for example, is only about 10^{-7} m across, but if all of its DNA double helices were stretched out, they would measure a little over 1 m, end to end.

There are 3 to 5 billion base pairs in one nucleus of a human cell.

The Replication of DNA. When DNA replicates, the cell makes an exact complementary strand for each of the two original strands, and two identical double helices emerge. A number

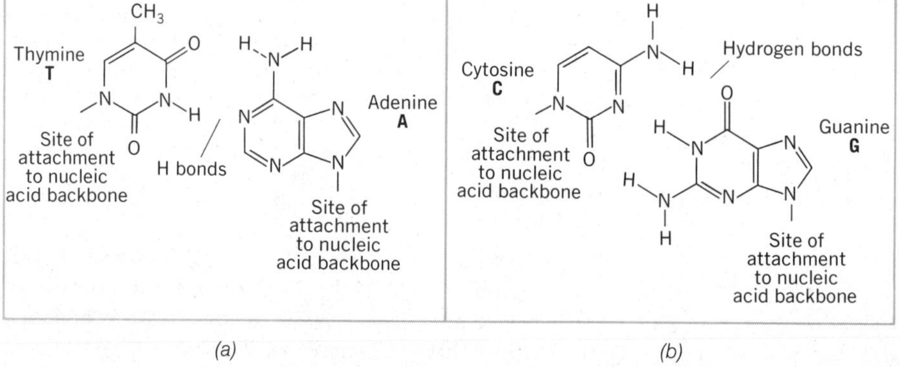

FIGURE 22.7
Hydrogen bonding between base pairs. (a) Thymine (T) and adenine (A) form one base pair between which are two hydrogen bonds. (b) Cytosine (C) and guanine (G) form another base pair between which there are three hydrogen bonds. Adenine can also base-pair to uracil (U).

(a)   (b)

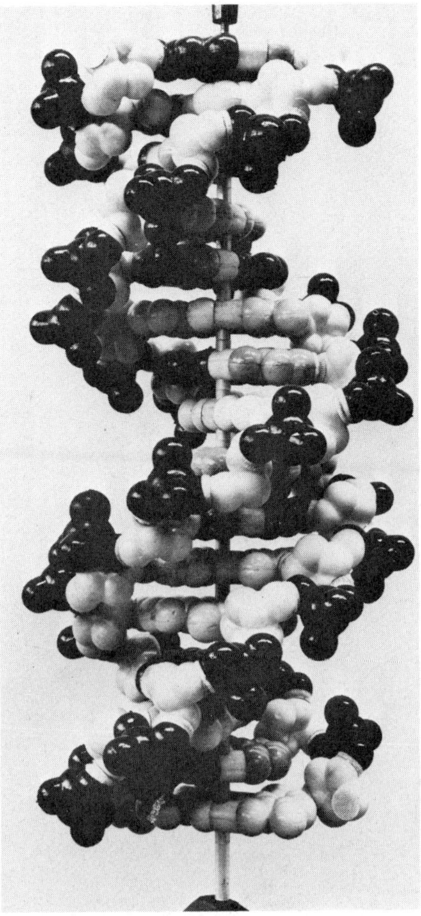

FIGURE 22.8 (left)
A molecular model of a short section of a DNA double helix. The darkest parts are phosphate groups. The lightest are deoxyribose units. What appear to be horizontal steps in a spiral staircase are the interchain base pairs.

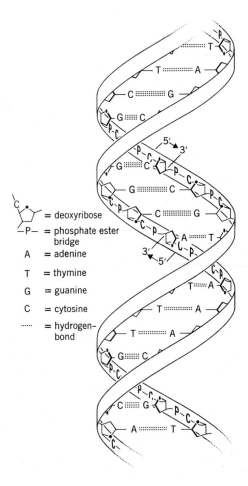

FIGURE 22.9 (right)
A schematic representation of the DNA double helix. The two spiraling strands are held side by side by the hydrogen bonds (rows of colored dots) between base pairs on opposite strands. (See the legend to Figure 22.5 for the meaning of the 5′ → 3′ designations.)

of enzymes are involved. The principal guarantee that each new strand is a complement to one of the old strands is the requirement that the only pairs allowed are A to T and C to G. A very general picture that shows how this works is in Figure 22.10. This figure explains only one of the many aspects of replication—how base pairing ensures complementarity between strands.

How replication can occur without everything becoming hopelessly tangled in the nucleus is still not fully known. In mammals, replication occurs at **replisomes,** tiny packages made mostly of protein that occur at fixed points in the nuclear matrix. Each DNA double helix forms a continuous loop that enters and leaves a replisome at many points, as seen in Figure 22.11. During replication, each of the double-helix loops slips through its own part of the replisome and is copied as it moves. It's as if each replisome were the head of a tape recorder, except that the continuous looping in and out and the duplication of the tape occur at several places at the same time. As the DNA loops move, molecules of free nucleotides, which are present in the fluids of the nucleus, hook together to make the second, complementary strand.

DNA and the Gene. In higher organisms, an individual gene is a series of sections of a single strand of DNA. A gene was once thought to be one *continuous* segment of a DNA molecule, but in the late 1970s scientists learned otherwise. Virtually all single genes are divided or split genes. Such a gene is made up of *sections* of a DNA chain that are called **exons.** Interrupting the exons and separating them are other parts of the DNA chain called **introns,** which have no direct bearing on the genetic message carried by the gene.

Exon refers to the part that is expressed and *intron* to the segments that *int*errupt the exons.

FIGURE 22.10

The accuracy of the replication of DNA is related to the exclusive pairing of A with T and G with C. The two new strands at the bottom are replicas of the original parent strand at the top.

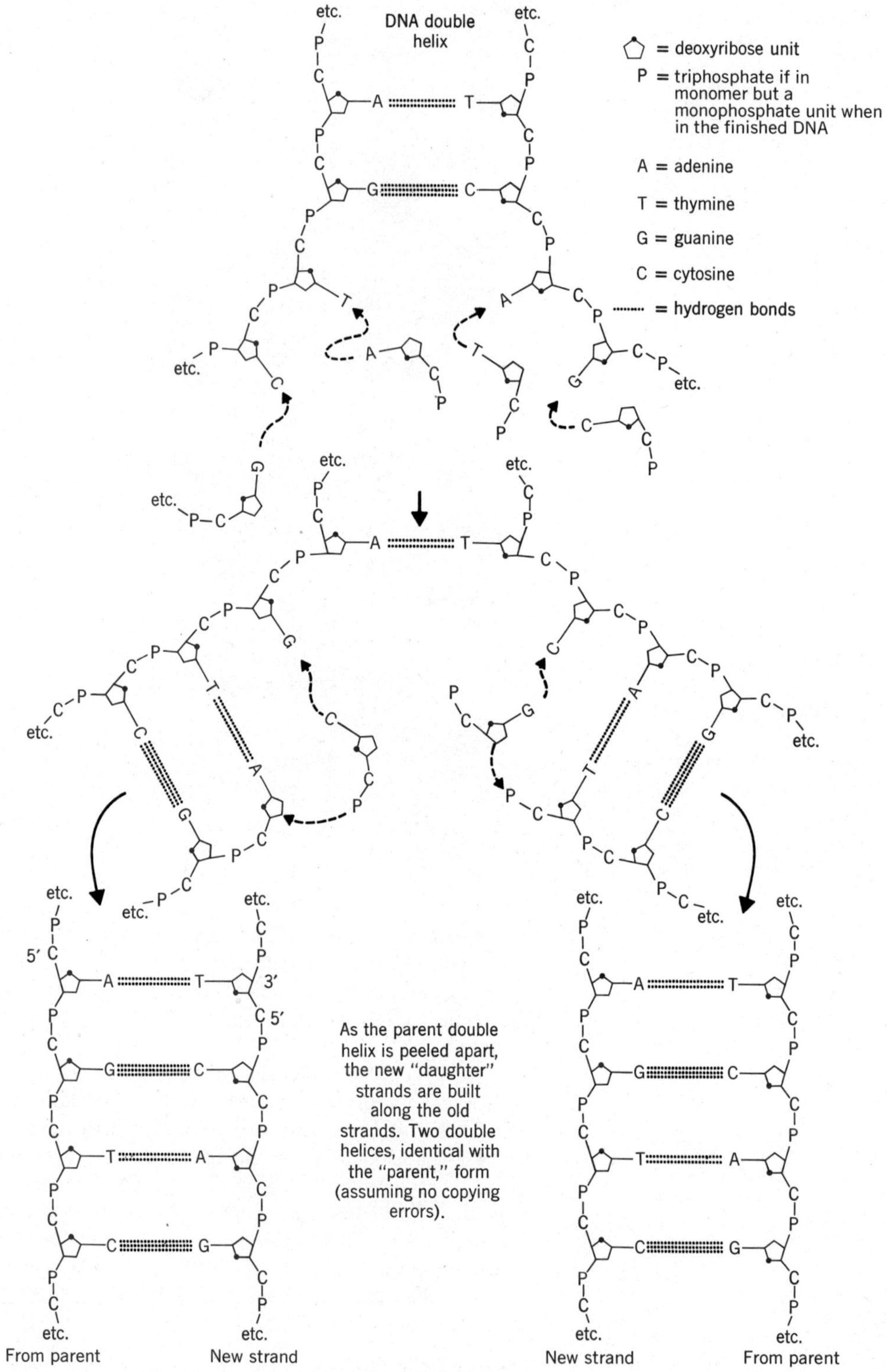

= deoxyribose unit

P = triphosphate if in monomer but a monophosphate unit when in the finished DNA

A = adenine

T = thymine

G = guanine

C = cytosine

·········· = hydrogen bonds

As the parent double helix is peeled apart, the new "daughter" strands are built along the old strands. Two double helices, identical with the "parent," form (assuming no copying errors).

From parent   New strand   New strand   From parent

We'll see later in this chapter how the exons get their message together and thus give expression to a single gene. For the present, we can consider that a gene is a particular section of a DNA strand minus all of the introns in this section. A gene, in other words, is a specific

FIGURE 22.11
At fixed sites called replisomes, which are attached to the nuclear matrix, new DNA double helices form in loops. (Adapted from D. M. Pardoll, B. Vogelstein, and D. S. Coffey, *Cell*, February 1980.)

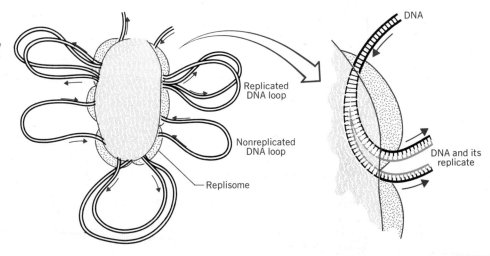

series of bases strung in a definite sequence along a DNA backbone. In humans, a single gene has between 1000 and 3000 bases.

There are no introns in human genes that code for histones or for α-interferon.

Not all DNA molecules in higher organisms involve introns. Some do have completely continuous sequences, but relatively few of such intronless genes have been identified.

22.3 RIBONUCLEIC ACIDS

The genetic code consists of a set of base triplets each of which corresponds to a specific amino acid.

Something like 10,000 to 100,000 genes are needed to account for all of the kinds of proteins in a human body.

The general scheme that relates DNA to polypeptides is illustrated in Figure 22.12. To go into more detail, we need to learn more about RNA and its various types, particularly the four that participate in expressing genes in higher organisms.

Ribosomal RNA (*r*RNA). Studding portions of the endoplasmic reticulum (Figure 22.1) are small clusters of polypeptides and nucleic acids called **ribosomes.** Their nucleic acid is called **ribosomal RNA,** abbreviated **rRNA.** Each ribosome forms from two subunits, as indicated in Figure 22.12, which come together to form a complex with messenger RNA, another type that we'll study soon.

Ribosomes are the sites of polypeptide synthesis, but the RNA in these particles does not itself direct this work. Messenger RNA does this, and it's made from the second kind of RNA of our study.

hnRNA is also called primary transcript RNA.

Heterogeneous Nuclear RNA (*hn*RNA). Just as the genetic machinery uses the side chains of DNA to ensure that an exact complement to a DNA strand is made during replication, so it uses a single DNA strand to guide the assemblage of a complementary molecule of RNA. Figure 22.13 shows how this takes place. (Remember that when RNA is made, uracil is used in the place of thymine. Thus when a DNA strand has an adenine side chain, uracil, not thymine, takes the position opposite it on newly-made RNA.) The RNA complement to a DNA section — exons plus introns plus some end-piece units that become tacked on but that need not concern us — is called **heterogeneous nuclear RNA,** abbreviated ***hn*RNA.**

At the 3' end of hnRNA molecules there is a long poly-A tail (about 200 adenosine units long), and at the other end there is a nucleotide triphosphate "cap."

The synthesis of *hn*RNA is the first general step in the expression of a gene. There are several steps within this general step, but we're interested in a relatively broad overview only. The next general step consists in processing the *hn*RNA strand to make the next kind of RNA in our study, messenger RNA.

FIGURE 22.12
The relationships of nuclear DNA to the
various RNAs and to the synthesis of
polypeptides.

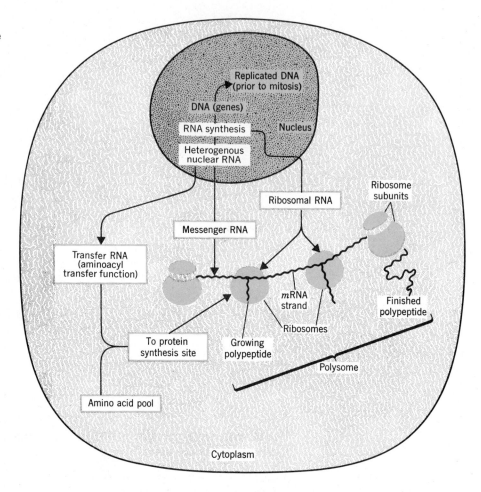

Messenger RNA (*m*RNA). The *hn*RNA molecules have large sections that were produced by the introns of the DNA. Special enzymes catalyze reactions that snip these pieces out and splice together just the units corresponding to the exons of the divided gene. (See Figure 22.14.) The result is a much shorter RNA molecule called **messenger RNA,** abbreviated **mRNA.** Now we have a sequence of bases that carries the undivided genetic message of the particular gene. This overall process that starts with the divided gene on a DNA molecule and ends with a molecule of *m*RNA is called **transcription.** It means that the gene's genetic message has been transcribed to an *m*RNA molecule.

Each group of three adjacent bases on a molecule of *m*RNA constitutes a unit of genetic information called a **codon** (taken from the word *code.*) Thus it is a series of codons strung along the *m*RNA backbone that now carries the genetic message.

Once they are made, the *m*RNAs move from the nucleus to the cytoplasm where they hook ribosomes to themselves at specific *initiation sites* on the *m*RNA chains. Such an assembly of many ribosomes along an *m*RNA chain is called a *polysome* (short for *polyribosome,* the term that is often used instead).

The ribosomes are somewhat like beads strung at regular intervals along an *m*RNA necklace.

Ribosomes are traveling packages of enzymes intimately associated with *r*RNA. Each ribosome moves along its *m*RNA chain while the *m*RNA uses its codons to guide the synthesis of a polypeptide. To complete this system, we need a mechanism for bringing individual amino acids to the polypeptide assembly site. For this task, the cell uses still another type of RNA.

FIGURE 22.13
DNA-directed synthesis of *hn*-RNA in the nucleus of a cell in a higher organism. The shaded oval on the left represents a complex of enzymes that catalyze this step. (Notice the direction of the *hn*RNA strand is opposite that of the DNA strand.)

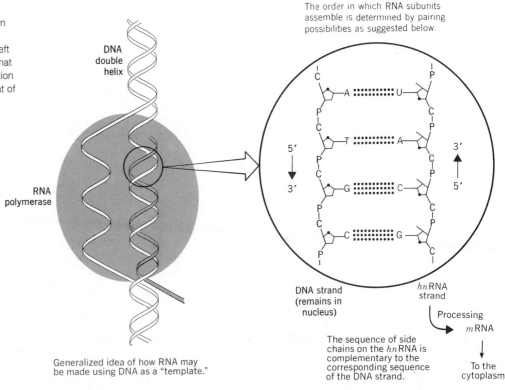

The order in which RNA subunits assemble is determined by pairing possibilities as suggested below.

DNA
double
helix

RNA
polymerase

DNA strand
(remains in
nucleus)

*hn*RNA
strand

Processing
*m*RNA

Generalized idea of how RNA may be made using DNA as a "template."

The sequence of side chains on the *hn*RNA is complementary to the corresponding sequence of the DNA strand.

To the
cytoplasm

FIGURE 22.14
The RNA made directly at a DNA strand is *hn*RNA. Only the segments made at sites *a*, *b*, *c*, and *d* — the exons — are needed to carry the genetic message to the cytoplasm. Therefore the *hn*RNA is processed and its segments that matched the introns of the gene are snipped out. Then the segments that matched the exons are rejoined to make the *m*RNA strand.

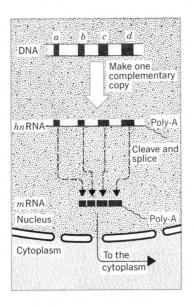

DNA

Make one
complementary
copy

*hn*RNA   Poly-A

Cleave and
splice

*m*RNA
Nucleus   Poly-A

Cytoplasm

To the
cytoplasm

Transfer RNA (*t*RNA).   The bearers of amino acyl units are molecules of the smallest of the RNAs, which typically have only 80 nucleotide units per molecule. These transfer amino acyl units from their dissolved state in the cytoplasm to the growing polypeptide molecule, so this kind of RNA is called **transfer RNA,** abbreviated ***t*RNA.** As shown in Figure 22.15, a molecule of *t*RNA has two particularly important sites. One can bind an amino acyl unit and, for a given *t*RNA, only one of the 20 amino acyl units. The other site consists of a unique set of

FIGURE 22.15
Transfer RNA (*tRNA*). These are highly schematic representations of a very convoluted molecule whose loops are stabilized by the hydrogen bonds of base pairs. *(a)* The *tRNA* molecule without its amino acyl unit. *(b)* The amino acyl group is attached at one end. *(c)* The symbol of the amino acyl—*tRNA* unit that will be used in succeeding figures.

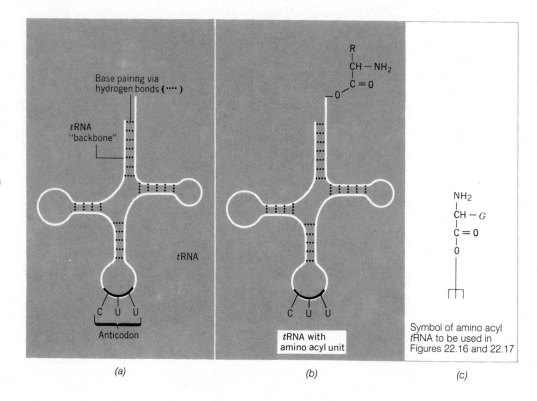

three adjacent side-chain bases that is complementary to a particularly codon. This triplet of bases on *tRNA* is called an **anticodon,** because its three bases can line up opposite the codon complementary to them on *mRNA*, using the fitting pattern allowed via hydrogen bonding.

When a molecule of *tRNA* performs its task, a unique enzyme first catalyzes a reaction that attaches an amino acyl unit. Then this molecule can, at a different place, line up against the corresponding codon of *mRNA* at a polysome. If the codon happens to be over the site on a ribosome that has the necessary enzymes, then the amino acyl unit is transferred to the growing polypeptide chain. You can see that a unique series of codons *must* let the polypeptide chain grow with an equally unique sequence of amino acid residues. The pairing of the triplets of bases between the codons and anticodons can permit only one sequence.

DNA also occurs in mitochondria, and there are some variations in the code for this DNA.

The Genetic Code. The correlation between codons and amino acids is the **genetic code.** The alphabet of this code has four letters, the four bases, A, U, G, and C. The alphabet of the polypeptide language has 20 letters, the 20 α-amino acids. Obviously, the body couldn't translate from a four-letter alphabet to a 20-letter alphabet on a one-letter to one-letter basis. This is why living systems use the four genetic letters in groups of three, as we have said.

Associated with each amino acid is at least one base triplet, one three-letter word or codon. Each codon specifies a particular amino acid. The known codon assignments are given in Table 22.1. Phenylalanine, for example, is coded either by UUU or by UUC. Alanine is coded by any one of the four triplets: GCU, GCC, GCA, or GCG. Until 1984, scientists believed that the codon assignments in Table 22.1 applied to all animals and plants. A few single-celled species have been found, however, that exhibit some variations, and we no longer can say that the genetic code is universal.

It's important to remember that a codon cannot appear on a strand of *mRNA* unless a complementary triplet of bases was on an exon unit of the original DNA strand. For example,

TABLE 22.1

Codon Assignments

First	Second				Third
	U	C	A	G	
U	Phenylalanine	Serine	Tyrosine	Cysteine	U
	Phenylalanine	Serine	Tyrosine	Cysteine	C
	Leucine	Serine	CT[a]	CT	A
	Leucine	Serine	CT	Tryptophan	G
C	Leucine	Proline	Histidine	Arginine	U
	Leucine	Proline	Histidine	Arginine	C
	Leucine	Proline	Glutamine	Arginine	A
	Leucine	Proline	Glutamine	Arginine	G
A	Isoleucine	Threonine	Asparagine	Serine	U
	Isoleucine	Threonine	Asparagine	Serine	C
	Isoleucine	Threonine	Lysine	Arginine	A
	Methionine[b]	Threonine	Lysine	Arginine	G
G	Valine	Alanine	Aspartic acid	Glycine	U
	Valine	Alanine	Aspartic acid	Glycine	C
	Valine	Alanine	Glutamic acid	Glycine	A
	Valine	Alanine	Glutamic acid	Glycine	G

[a] The codon CT is a signal codon for chain-termination.

[b] The codon for methionine, AUG, serves also as the codon form N-formyl-methionine, the chain-initiating unit in polypeptide synthesis in bacteria and mitochondria.

there could not be the UUC codon on *m*RNA unless the DNA strand had the triplet AAG, because G pairs with C and A of DNA pairs with U of *m*RNA.

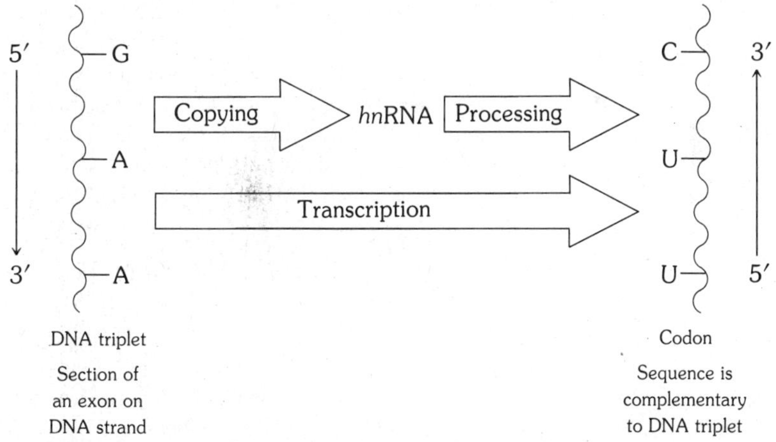

As shown above, the DNA strand and the RNA strand made directly from it run in opposite directions. To avoid confusion in writing codons on a horizontal line, scientists use the following conventions. The 5′ end of a codon is written on the left end of the three-letter symbol, and the direction, left to right, is 5′ to 3′. (See also Figure 22.5.) This is why the codon given above is written as UUC, not as CUU. Opposite this codon is the triplet GAA on the DNA strand, which is also written from the 5′ to 3′ end. To give another example, the complement to the *m*RNA codon, AAG, is the DNA triplet, CTT.

PRACTICE EXERCISE 1   Using Table 22.1, what amino acids are specified by each of the following codons on an *m*RNA molecule?

(a) CCU   (b) AGA   (c) GAA   (d) AAG

PRACTICE EXERCISE 2   What amino acids are specified by the following base triads on DNA?

(a) GGA   (b) TCA   (c) TTC   (d) GAT

22.4   *m*RNA-DIRECTED POLYPEPTIDE SYNTHESIS

_t_RNA molecules carry amino acyl groups to places at an *m*RNA strand where anticodons match codons.

In the previous section we saw how a particular genetic message can be transcribed from the exons of a divided gene to a series of codons on *m*RNA. In this section we will learn in broad terms how the next general step happens — the *m*RNA-directed assemblage of a polypeptide.

$$NH_2CHC\overset{\displaystyle O}{\overset{\displaystyle \|}{-}}$$
$$\underset{G}{|}$$

An amino acyl unit

$$^+NH_3CHCO_2^-$$
$$\underset{CH_2CH_2SCH_3}{|}$$

Methionine, Met

Translation.   The steps by which the sequence of codons on *m*RNA directs the formation of a polypeptide with a matching sequence of amino acyl groups is called **translation.** As we have already mentioned, there is at least one kind of *t*RNA molecule for each of the 20 amino acids and each amino acyl unit is carried by its own *t*RNA molecule to the polypeptide assembly site at a ribosome. We will now use the very abbreviated forklike symbol for an amino acyl-*t*RNA combination shown in part *c* of Figure 22.15. For the remainder of our study of translation, we will assume that all of the needed amino acyl-*t*RNA combinations have been assembled and are waiting like so many spare parts to be used at the assembly line.

The cell begins a polypeptide with an N-terminal methionine residue. After the end of the synthesis, the amino acyl group of methionine will be left in place only if the polypeptide is supposed to have this as its N-terminal unit. Otherwise, it will be removed, and the second amino acyl group will be the final N-terminal unit.

The principal steps in making the polypeptide are as follows.

The "P" in P site refers to the peptidyl transfer site. The A site is the amino acyl binding site.

1.   Formation of the elongation complex.   As shown in Figure 22.16, the elongation complex is made of several pieces: two subunits of the ribosome, the first amino acyl-*t*RNA unit (which is Met-*t*RNA$_1$), and the *m*RNA molecule, beginning at its first codon end. The Met-*t*RNA$_1$ comes to rest with its anticodon matched to the first codon and with the bulk of its system in contact with a portion of the ribosome's surface called the P site. This is a site where there are enzymes that can catalyze the transfer of a growing polypeptide chain to a newly arrived amino acyl unit. Now the second *t*RNA unit, *t*RNA$_2$, which holds the second amino acyl group, aa$_2$, has to find the *m*RNA codon that matches its own anticodon, and it does this at another site on the ribosome called the A site. The work of completing the elongation complex is now complete, and actual chain lengthening can start.

2.   Elongation of the polypeptide chain.   A series of repeating steps now occurs, illustrated in Figure 22.17. The methionine residue moves from its *t*RNA to the newly arrived, second amino acyl unit. This makes the first peptide bond, and it takes place by

FIGURE 22.16
Formation of the elongation complex at the beginning of the synthesis of a polypeptide. (A computer-driven animation of the overall process presented here and in Figure 22.17 can be viewed with the *Prelab Studies for General, Organic, and Biological Chemistry*, by Sandra L. Olmsted and Richard D. Olmsted. See page xi.

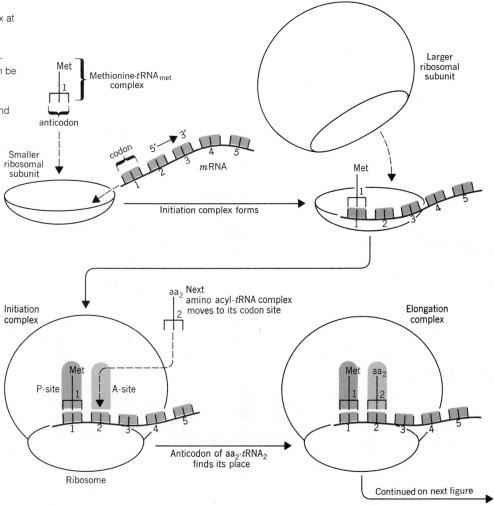

Polypeptide synthesis can occur at several ribosomes moving along the *m*RNA strand at the same time.

acyl transfer much as we discussed when we studied the synthesis of amides in an earlier chapter.

FIGURE 22.17
The elongation steps in the synthesis of a polypeptide. The dipeptidyl unit, Met-aa₂, formed by the process of Figure 22.16, and it now has a third amino acid residue, aa₃, added to it.

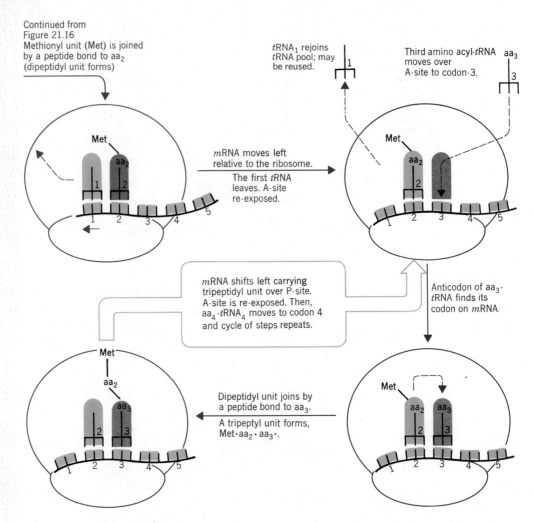

In mammals, it takes only about 1 second to move each amino acid residue into place in a growing polypeptide.

The *mRNA* unit now shifts one codon over — leftward as we have drawn it. (It's actually a relative motion of the *mRNA* and the ribosome.) This movement positions what is now a *di*peptidyl-*tRNA* unit over the P site. Now the third amino acyl-*tRNA* finds its anticodon to codon matching at the third codon of the *mRNA* strand, and it's over the recently vacated A site. Elongation now occurs; the dipeptide unit transfers to the amino group of the third amino acid. Another peptide bond forms. And a tripeptidyl system has been made. The cycle of steps can now occur again, starting with a movement of the *mRNA* chain relative to the ribosome that shifts this tripeptidyl-*tRNA* and positions it over the P site. The fourth amino acid residue is carried to the *mRNA*; the tripeptidyl unit transfers to it to make a tetrapeptidyl unit, and so forth. This cycle of steps continues until a special chain-terminating codon is reached.

3. **Termination of polypeptide synthesis.**   Once a ribosome has moved down to one of the chain-terminating codons of the *mRNA* strand — UAA, UAG, or UGA — the polypeptide synthesis is complete, and the polypeptide is released. The ribosome can be reused, and the polypeptide spontaneously acquires its higher levels of structure.

How Polypeptide Synthesis Might Be Controlled.   Because so many steps occur between the divided gene and the finished polypeptide, there are a large number of points where the cell can control the overall process. We'll briefly discuss one, and we'll have to rely heavily on Figure 22.18.

FIGURE 22.18

Repression and induction of the enzyme *β*-galactosidase in *E. coli.* The inducer is lactose whose hydrolysis requires this enzyme. The first lactose molecules to arrive bind to and remove the repressor from the operator gene. Now the structural gene is free to direct the synthesis of *m*RNA coded to make the enzyme. The repressor is made at the direction of another gene (top), the regulator gene.

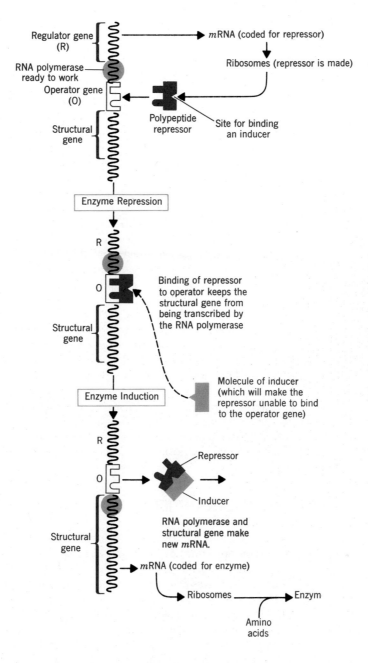

It is believed that *every* cell in the body carries exact copies of *all* of the genes of the individual. Some mechanism(s) must exist in the developed individual for keeping most of these genes switched off all of the time. It wouldn't do, for example, to have a gene that directs the synthesis of a protein-digesting enzyme busily at work in cells that would be destroyed by such an enzyme.

From studies with bacterial genes, the evidence indicates that some genes are repressed until particular events activate just those genes that the cell needs to have activated. A **repressor** molecule, probably a polypeptide made by the direction of its own gene, called a *regulator gene,* binds to a small segment of the whole DNA strand, a segment called the *operator gene.* This is shown in Figure 22.18. The binding of this repressor prevents the gene that is coded for the structure of some enzyme — the *structural gene* — from working; it prevents transcription.

Now suppose that some molecules start to appear whose processing by the cell requires the enzyme that is made under the direction of the structural gene. The first molecules to arrive function as **inducer** molecules. They combine with the repressor by fitting to it in some way, and this forces a change in the shape of the repressor. It can't fit any longer to the operator gene, so it drops off. The structural gene is free to be used to help to make the needed mRNA. For example, as illustrated in Figure 22.18, lactose is believed to be the inducer that kicks the repressor off of the gene that helps to make the enzyme needed to hydrolyze lactose. If no lactose is in the cell, the enzyme isn't needed, and the genetic machinery for making this enzyme stays switched off. Only when the enzyme is needed — signaled by the arrival of its substrate — is it manufactured.

These inhibiting activities render the *enzymes* for the various steps inactive.

Inhibition of Polypeptide Synthesis in Bacteria by Antibiotics.   Several important antibiotics kill bacteria by inhibiting the synthesis of the polypeptides that these organisms are trying to make in order to multiply. Streptomycin, for example, inhibits the initiation of polypeptide synthesis. Chloramphenicol inhibits the ability to transfer newly arrived amino acyl units to the elongating strand. The tetracyclines inhibit the binding of tRNA — aa units when they arrive at the ribosome. Actinomycin binds tightly to DNA. Erythromycin, puromycin, and cycloheximide interfere with elongation.

Radiations and Gene Damage.   Atomic radiations, particularly X rays and gamma rays, go right through soft tissue where they create unstable ions and radicals (particles with unpaired electrons). New covalent bonds might form. Even side-chain bases might be annealed together. If such events were to happen to DNA molecules, the polypeptide eventually made at their direction might be faulty. And the DNA made by replication might be seriously altered. If the initial damage to the DNA is severe enough, replication won't be possible and the cell involved is reproductively dead. This, in fact, is the *intent* when massive radiation doses are used in cancer therapy — to kill cancer cells.

Chemicals that are used in cancer therapy often mimic radiations by interfering with the genetic apparatus of cancer cells. Such chemicals are called **radiomimetic substances.**

22.5   SOME MEDICAL APPLICATIONS OF MOLECULAR BIOLOGY

Chemicals that can be used to treat viral infections, cancer, and genetic diseases are some of the fruits of molecular biology.

Simple viruses have only three genes. The most complex have about 250 genes.

Viruses.   **Viruses** are agents of infection that are made of nucleic acid usually surrounded by an overcoat of protein. Unlike a cell, a virus has either RNA or DNA, but not both. Viruses can neither synthesize polypeptides nor generate their own energy for metabolism. They are at the borderline of living systems, and are generally thought of only as unique packages of dead chemicals, except when they get inside their host cells. They then seem to be living things, because they reproduce.

The protein overcoat of some viruses includes an enzyme that catalyzes the breakdown of the cell membrane of the host cell. Thus when such a virus particle sticks to the surface of a host cell, its overcoat catalyzes the opening of a hole into the cell. Then the nucleic acid in the virus squirts into the cell, or the whole virus might move in. Each virus that works this way evidently has its own unique membrane-dissolving enzyme, so any one virus can affect only those cells in the host whose own membranes are susceptible to this enzyme. At least, this theory helps us understand how viruses are so unusually selective. For example, a virus that attacks, say, nerve cells in the spinal cord has no effect on muscle cells in the heart. And a large number of viruses exist that do not affect any kind of human cell. Many viruses attack only plants.

Once a virus particle is inside its host cell, it uses the metabolic and genetic machinery of the host to make copies of itself. When enough new virus particles are made, the host cell

SPECIAL TOPIC 22.1 INTERFERON

The term *interferon* refers to a family of similar polypeptides that are characterized chiefly by an ability to inhibit viruses. There are at least three types in humans, designated as α-, β-, and γ-interferon. The interferons are glycoproteins that have about 150 amino acid residues per molecule.

Viruses are potent stimuli for the induction of interferon synthesis in humans. When an invading virus particle first encounters a white blood cell, particularly the type made in lymph tissue, the mechanism in the cell for making the *m*RNA coded for interferon is switched on. In a short time, the cell is manufacturing and releasing interferon, which then acts as a signal to other cells to make interferon, too. The interferon activates the circulating cells, called killer cells, that attack and destroy virus particles. In this way, interferon works to inhibit a viral infection.

The discovery of interferon in 1957 came about when two scientists, Alick Isaacs and Jean Lindenmann, wondered why victims of one viral disease never seemed to come down with a second viral disease at the same time. It seemed to them that something in the first viral attack triggered a mechanism that provided protection against an attack by a different virus. The search for this "something" led to interferon. In initial clinical trials, interferon provided relief and sometimes a cure to a few patients who had rapidly-acting cancer, such as osteogenic sarcoma (bone cancer), multiple myeloma, melanoma (a form of skin cancer), breast cancer, and some forms of leukemia and lymphoma. However, it took nearly 25,000 pints of blood to make 100 mg of interferon — at a cost of a few billion dollars — so large-scale clinical tests depend on manufactured interferon.

The initial excitement that interferon would be the cure-all, free of side effects, for any kind of viral disease or any kind of cancer has faded. Yet, tests with experimental animals continue to be promising, and several medical groups are continuing their testing efforts. A number of high-tech companies are working at ways to make interferon, to separate its types from each other, and to test the individual types.

bursts and releases its load of virus particles to invade surrounding cells. In this way the virus infection spreads. Several viruses have been implicated as cancer-causing agents.

One of the many features of the body's defense against viral infections is a small family of polypeptides called the interferons. Currently, intensive research is in progress to make individual interferons and to test them in the treatment of viral diseases and cancer. Special Topic 22.1 discusses the interferons further.

Recombinant DNA Technology. Human insulin, human growth hormone, and human interferons are now being manufactured by a technology that involves the production of *recombinant DNA*. With the aid of Figure 22.19, we'll learn how this technology works. It represents one of the important advances in scientific technology of this century.

E. coli, or *Escherichia coli*, is a one-celled organism found in the human intestinal tract.

Bacteria such as *E. coli* make polypeptides using the same genetic code as humans. There are some differences in the machinery, however. For example, bacteria have DNA not only as parts of their chromosomes but also in the form of large, circular, supercoiled DNA molecules called **plasmids.** Each plasmid carries just a few genes, but several copies of a plasmid can exist in one bacterial cell. Each plasmid can replicate independently of the chromosome.

Plasmids can be isolated from bacteria, and they can be given new DNA material — a new gene — that has the base triplets for directing the synthesis of a particular polypeptide. It might even be a polypeptide that is completely alien to the bacteria, such as the subunits of human insulin, or human growth hormone, or a human interferon. The plasmids are snipped open by special enzymes, the new DNA combines with the open ends, and then the plasmid loops are reclosed. The DNA in these altered plasmids is called **recombinant DNA.**

The term *cloning* is used for the operation that places new genetic material into a cell where it becomes a part of the cell's gene pool. The new cells that follow this operation are called *clones*.

The remarkable feature of bacteria with altered plasmids made of recombinant DNA is that when the bacteria multiply, the new bacteria have within them the *altered* plasmids. And when these multiply, still more altered plasmids are made. Between their cell divisions, the bacteria manufacture the proteins for which they are genetically programmed, including the proteins specified by the recombinant DNA. In this way, bacteria can be tricked into making the *human* proteins we have mentioned. The technology isn't limited to bacteria; yeast cells work, too.

FIGURE 22.19
Recombinant DNA is made by inserting
a DNA strand coded for some protein
not made by the bacteria into the
circular DNA of the bacterial plasmid.

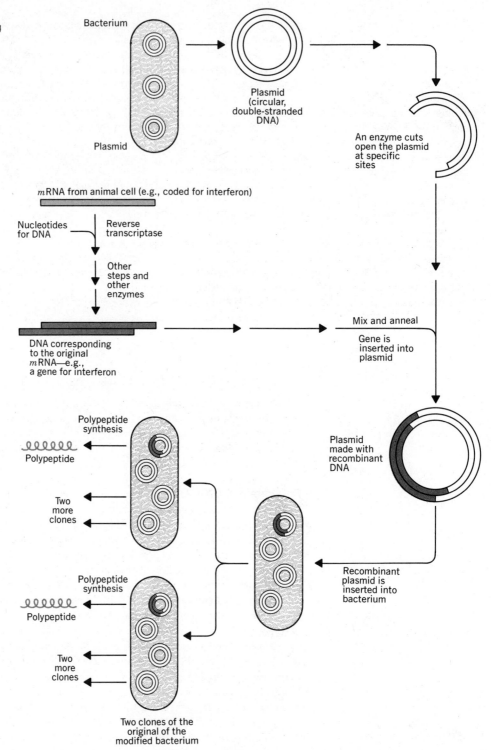

Bacterium

Plasmid

Plasmid
(circular,
double-stranded
DNA)

An enzyme cuts
open the plasmid
at specific
sites

*m*RNA from animal cell (e.g., coded for interferon)

Nucleotides
for DNA

Reverse
transcriptase

Other
steps and
other
enzymes

DNA corresponding
to the original
*m*RNA—e.g.,
a gene for interferon

Mix and anneal

Gene is
inserted into
plasmid

Plasmid
made with
recombinant
DNA

Recombinant
plasmid is
inserted into
bacterium

Polypeptide
synthesis

Polypeptide

Two
more
clones

Polypeptide
synthesis

Polypeptide

Two
more
clones

Two clones of the
original of the
modified bacterium

People who rely on the insulin of animals—cows, pigs, sheep, for example—sometimes experience allergic responses. They are also vulnerable to the availability of the pancreases of these animals. Therefore the ability to make *human* insulin has been a welcome development.

If an interferon should prove effective in treating various types of viruses, hepatitis, and possibly even certain kinds of cancer, the ability to make this unusually rare and costly substance in large quantities at low cost by recombinant DNA technology will truly be a major technological advance. In fact, just the clinical tests alone would be too costly without the availability of manufactured interferon.

Recombinant DNA technology has interacted with archeology in an interesting way. In the early 1980s scientists were able to remove segments of DNA molecules from an Egyptian mummy and from an extinct horse-like animal (a quagga) and reproduce these segments using bacteria. No segment was long enough to include a complete gene, but the technique no doubt will be developed as another tool for the study of evolutionary history.

Hereditary Diseases.   About 2000 diseases in humans are caused directly or indirectly by a defect in the genetic machinery. Cystic fibrosis is one of the most common examples. Its victims overproduce a thick mucus in the lungs and the digestive system, which clogs them and which often leads to death in children. About one person in 20 carries the defective gene associated with this disease, and it strikes about one in every 1000 newborn. The defective gene directs the synthesis of an abnormal enzyme involved in biological oxidations, an enzyme that leads to excessive oxygen consumption and the overproduction of mucus.

Another disease traceable to a defective gene is sickle-cell anemia, a condition that was discussed in a Special Topic in the previous chapter.

Albinism, the absence of pigments in the skin and the irises of the eyes, is caused by a defect in a gene that directs the synthesis of an enzyme needed to make these pigments. The pigments absorb the ultraviolet rays in sunlight, radiation that can induce cancer, so victims of albinism are more susceptible to skin cancer.

Phenylketonuria, or PKU disease, is a brain-damaging genetic disease in which abnormally high levels of the phenylketoacid called phenylpyruvic acid occur in the blood. This condition causes permanent brain damage in the newborn. Because of a defective gene, an enzyme needed to handle phenylalanine is not made, and this amino acid is increasingly converted to phenylpyruvic acid.

$$C_6H_5CH_2\overset{\overset{O}{\|}}{C}CO_2H \qquad \underset{\underset{NH_3^+}{|}}{C_6H_5CH_2CHCO_2^-}$$

Phenylpyruvic acid   Phenylalanine

There is enough phenylalanine in just one slice of bread to be potentially dangerous to a PKU infant.

PKU can be detected by a simple blood test within four or five days after birth. If the infant's diet is kept very low in phenylalanine, it can survive the critical danger period and experience no brain damage. The infant's diet should not include aspartame, which is used in low-calorie sweeteners such as Nutra Sweet, because it is hydrolyzed by the digestive processes to give phenylalanine.

$$\underset{\underset{\underset{CO_2^-}{|}}{\underset{CH_2}{|}}}{^+NH_3CH\overset{\overset{O}{\|}}{C}}-\underset{\underset{\underset{C_6H_5}{|}}{\underset{CH_2}{|}}}{NHCH\overset{\overset{O}{\|}}{C}OCH_3} + 2H_2O \longrightarrow \underset{\underset{\underset{CO_2^-}{|}}{\underset{CH_2}{|}}}{^+NH_3CHCO_2^-} + \underset{\underset{\underset{C_6H_5}{|}}{\underset{CH_2}{|}}}{^+NH_3CHCO_2^-} + CH_3OH$$

Aspartame   Aspartic acid   Phenylalanine   Methanol

SUMMARY

Hereditary information   The genetic apparatus of a cell is mostly in its nucleus and consists of chromatin, a complex of DNA, proteins, and some RNA. Strands of DNA, a polymer, carry segments that are individual genes. Chromatin replicates prior to cell division, and the duplicates segregate as the cell divides. Each new cell thereby inherits exact copies of the chromatin of the parent cell. If copying errors are made, the daughter cells (if they form at all) are mutants. They may be reproductively dead — incapable of themselves dividing — or they may transmit the mutant character to succeeding cells. The expression of this might be as a cancer, a tumor, or a birth defect. Atomic radiations, particularly X rays and gamma rays, are potent mutagens, but many chemicals, those that are radiomimetic, mimic these rays.

DNA   Complete hydrolysis of DNA gives phosphoric acid, deoxyribose, and a set of four heterocyclic amines, the bases adenine (A), thymine (T), guanine (G), and cytosine (C). The molecular backbone of the DNA polymer is a series of deoxyribose units joined by phosphodiester groups. Attached to each deoxyribose is one of the four bases. The order in which triplets of bases occur is the cell's way of storing genetic information.

A gene in higher organisms consists of successive groups of triplets — the exons — separated by introns. Thus the gene is a divided system, not a continuous series of nucleotide units. DNA strands exist in cell nuclei as double helices, and the helices are held near each other by hydrogen bonds that extend from bases on one strand to bases on the other. A always pairs with T and C always pairs with G. Using this structure and the faithfulness of base pairing, Crick and Watson explained the accuracy of replication, a process now known to occur at replisomes on the nuclear matrix in higher animals. After replication, each new double helix has one of the parent DNA strands and one new, complementary strand.

RNA   RNA is similar to DNA except that in RNA ribose replaces deoxyribose and uracil (U) replaces thymine (T). Four main types of RNA are involved in polypeptide synthesis. One is *r*RNA, which is in ribosomes. A ribosome contains both *r*RNA and proteins that have enzyme activity needed during polypeptide synthesis.

*m*RNA is the carrier of the genetic message from the nucleus to the site where a polypeptide is assembled. *m*RNA results from a chemical processing of the longer RNA strand, *hn*RNA, which is made directly under the supervision of DNA.

*t*RNA molecules are the smallest RNAs, and their function is to convey amino acyl units to the polypeptide assembly site. They recognize where they are to go by base pairing between an anticodon on *t*RNA and its complementary codon on *m*RNA. Both codon and anticodon consist of a triplet of bases.

Polypeptide synthesis   Genetic information is first transcribed when DNA directs the synthesis of *m*RNA. Each base triplet on the exons of DNA specifies a codon on *m*RNA. The *m*RNA moves to the cytoplasm to form an elongation complex with subunits of a ribosome and the first and second *t*RNA – aa unit to become part of the developing polypeptide.

The ribosome then rolls down the *m*RNA as *t*RNA – aa units come to the *m*RNA codons during the moment when the latter are aligned over the proper enzyme site of a ribosome. Elongation of the polypeptide then proceeds to the end of the *m*RNA strand or to a chain-terminating codon. After chain-termination, the polypeptide strand leaves, and it may be further modified to give it its final N-terminal amino acid residue.

The whole operation can be controlled by a feedback mechanism in which an inducer molecule removes a repressor of the gene, thus letting the gene work.

Several antibiotics inhibit bacterial polypeptide synthesis, which causes the bacteria to die.

Applications   Viruses are packages of DNA or RNA that usually are encapsulated by protein. Once they get inside their host cell, virus particles take over the cell's genetic machinery, make enough new virus particles to burst the cell, and then repeat this in neighboring cells. Some viruses are implicated in human cancer.

Recombinant DNA is DNA made from bacterial plasmids and DNA obtained from another source — DNA encoded to direct the synthesis of some desired polypeptide. The altered plasmids are reintroduced into the bacteria, where they become machinery for synthesizing this polypeptide (e.g., human insulin, or growth hormone, or interferon). Yeast cells can be used instead of bacteria for this technology.

Hereditary diseases stem from defects in DNA molecules that either prevent the synthesis of necessary enzymes or that make the enzymes in forms that won't work.

KEY TERMS

The following terms form the heart of the chemistry of heredity, and they should be made a part of the vocabulary of anyone who intends to grow professionally in the knowledge of the applications of this chemistry to medical problems.

anticodon	chromatin	deoxyribonucleic acid, DNA	gene
base (heterocyclic)	chromosome	double-helix DNA model	genetic code
base pairing	codon	exon	heterogeneous nuclear RNA, *hn*RNA

inducer	plasmid	repressor	transcription
intron	radiomimetic substance	ribonucleic acid, RNA	transfer RNA, *t*RNA
messenger RNA, *m*RNA	recombinant DNA	ribosomal RNA, *r*RNA	translation
nucleic acid	replication	ribosome	virus
nucleotide	replisome		

SELECTED REFERENCES

1 D. Freifelder. *Molecular Biology.* Science Books International, Boston, 1983.

2 A. L. Lehninger. *Principles of Biochemistry.* Worth Publishers, Inc., New York, 1982.

3 G. Felsenfeld. "DNA." *Scientific American,* October 1985, page 58.

4 J. E. Darnell, Jr. "RNA." *Scientific American,* October 1985, page 68.

5 N. O. Thorpe. *Cell Biology.* John Wiley and Sons, New York, 1984.

6 J. E. Darnell, Jr. "The Processing of RNA." *Scientific American,* October 1983, page 90.

7 R. E. Dickerson. "The DNA Helix and How It Is Read." *Scientific American,* December 1983, page 94.

8 M. Nomura. "The Control of Ribosome Synthesis." *Scientific American,* January 1984, page 102.

9 R. D. Kornberg and A. Klug. "The Nucleosome." *Scientific American,* February 1981, page 52.

10 P. Chambon. "Split Genes." *Scientific American,* May 1981, page 60.

11 W. Gilbert and L. Villa-Komaroff. "Useful Proteins from Recombinant Bacteria." *Scientific American,* April 1980, page 74.

12 C. Baglioni and T. W. Nilsen. "The Action of Interferon at the Molecular Level." *American Scientist,* July–August 1981, page 392.

13 W. F. Anderson. "Prospects for Human Gene Therapy." *Science,* October 26, 1984, page 401.

14 Rebecca L. Rawls. "Progress in Gene Therapy Brings Human Trials Near." *Chemical and Engineering News,* August 13, 1984, page 39.

15 Yvonne Baskin. "Doctoring the Genes." *Science 84,* December 1984, page 52.

REVIEW EXERCISES

The answers to these Review Exercises are in the *Study Guide* that accompanies this book.

The Cell

22.1   What is protoplasm?

22.2   What are the mitochondria, and what do they do?

22.3   The cytoplasm makes up what part of a cell?

22.4   What is the nuclear matrix?

22.5   Chromatin is made up of what kinds of substances?

22.6   What is the name of the chemical that makes up genes?

22.7   What is the relationship between a chromosome and chromatin?

22.8   What is the name of the overall process by which an exact duplicate of a gene is synthesized?

22.9   The duplication of a gene occurs in what part of the cell?

Structural Features of Nucleic Acids

22.10   What is the general name for the chemicals that are most intimately involved in the storage and the transmission of genetic information?

22.11   The monomer units for the nucleic acids have what *general* name?

22.12   What are the names of the two sugars produced by the complete hydrolysis of all of the nucleic acids in a cell?

22.13   What are the names and symbols of the four bases that are liberated by the complete hydrolysis of (a) DNA, and (b) RNA?

22.14   How are all DNA molecules structurally alike?

22.15   How do different DNAs differ structurally?

22.16   How are all RNA molecules structurally alike?

22.17   What are the principal structural differences between DNA and RNA?

22.18   When DNA is hydrolyzed, the ratios of A to T and of G to C are each very close to 1 : 1, *regardless of the species investigated.* Explain.

22.19   What is the principal noncovalent force in a DNA double helix?

22.20   What does base pairing mean, in general terms?

22.21   If the AGTCGGA sequence appeared on a DNA strand, what would be the sequence on the DNA strand opposite it in a double helix?

22.22   What does replication (of DNA) mean, in general terms?

22.23   The accuracy of replication is assured by the operation of what factors?

22.24 What is a replisome, and what is its function?

22.25 What is the relationship between a single molecule of single-stranded DNA and a single gene?

22.26 Suppose that a certain DNA strand has the following groups of nucleotides, where each lowercase letter represents a group several side chains long.

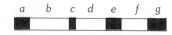

Which sections are likelier to be the introns? Why?

22.27 In general terms only, what particular contribution does a gene make to the structure of a polypeptide?

Ribonucleic Acids

22.28 What is the general composition of a ribosome, and what function does this particle have?

22.29 What is *hn*RNA, and what role does it have?

22.30 What is a codon, and what kind of nucleic acid is a continuous, uninterrupted series of codons?

22.31 Suppose that sections *x*, *y*, and *z* of the following hypothetical DNA strand are the exons of one gene.

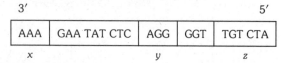

What is the structure of each of the following substances made under its direction?
(a) The *hn*RNA
(b) The *m*RNA

22.32 What is an anticodon, and on what kind of RNA is it found?

22.33 Which triplet, ATA or CGC, cannot be a codon? Explain.

22.34 Which amino acids are specified by the following codons?
(a) UUU   (b) UCC   (c) ACA   (d) GAU

22.35 What are the anticodons for the codons of Review Exercise 22.34?

Polypeptide Synthesis

22.36 What is meant by *translation,* as used in this chapter. And what is meant by *transcription?*

22.37 To make the pentapeptide, Met·Ala·Try·Ser·Tyr,
(a) What has to be the sequences of bases on the *m*RNA strand?
(b) What is the anticodon on the first *t*RNA to move into place?

22.38 In general terms, what is a repressor, and what does it do?

22.39 What does an inducer do (in general terms)?

22.40 How do some of the antibiotics work at the molecular level?

22.41 The genetic code is the key to translating between what two "languages"?

22.42 What is meant by the statement that the genetic code is universal? And is the code strictly universal?

22.43 In general terms, how do X rays cause cancer?

22.44 How do X rays and gamma rays work in cancer therapy?

22.45 What is a radiomimetic substance?

Applications of Molecular Biology

22.46 What is a virus made of?

22.47 How do at least some viruses get into their host cells?

22.48 What is a plasmid, and what is it made of?

22.49 Recombinant DNA is made from the DNA of two different kinds of sources. What are they?

22.50 Recombinant DNA technology is carried out to accomplish the synthesis of what kind of substance (in general terms)?

22.51 At the molecular level of life, what kind of defect is the fundamental cause of a heredity disease?

22.52 What is the molecular defect in PKU, and how does it cause the problems of the victims? How is it treated?

Chapter 23
Nutrition

These two chicks, of the same age, had identical diets except that the chick on the right received less tryptophan than needed for normal growth. Tryptophan-deficient diets among human babies also retards growth. Now that we've studied the raw materials for life, we will learn about the human requirements for them.

23.1 GENERAL NUTRITIONAL REQUIREMENTS

The intent of the recommended dietary allowances (RDAs) is to help meal planners provide for the nutritional needs of practically all healthy people.

Two of the great scientific triumphs of the nineteenth century were the germ theory of disease and the birth of the science of nutrition.

Nutrition can be defined in both technical and personal terms. Technically, nutrition is a field of science that investigates the identities, the quantities, and the sources of substances, called *nutrients,* that are needed for health. In personal terms, we speak of our own nutrition, the sum of the foods taken in the proper proportions at the best moments as we try to maintain a state of well-being and avoid diet-related diseases and infirmities. The science of **dietetics** is the application of the findings of the science of nutrition to feeding individual humans, whether they are ill or well.

This chapter is mostly about some of the major findings of the science of nutrition — a field that is huge and complex. There are whole departments in universities devoted to it. Universal agreement simply does not exist concerning a number of matters. Therefore we'll do little more than introduce some of the major aspects of nutrition and develop a small vocabulary in the field.

TABLE 23.1
Recommended Daily Dietary Allowances[a] of the Food and Nutrition Board,
National Academy of Sciences National Research Council, Revised 1980

	Age	Weight		Height		Protein	Fat-Soluble Vitamins		
	(years)	(kg)	(lb)	(cm)	(in.)	(g)	A (μg)[b]	D (μg)[c]	E (mg)[d]
Infants	0.0–0.5	6	13	60	24	kg × 2.2	420	10	3
	0.5–1.0	9	20	71	28	kg × 2.0	400	10	4
Children	1–3	13	29	90	34	23	400	10	5
	4–6	20	44	112	44	30	500	10	6
	7–10	28	62	132	54	36	700	10	7
Males	11–14	45	99	157	63	45	1000	10	8
	15–18	66	145	176	69	56	1000	10	10
	19–22	70	154	177	69	56	1000	7.5	10
	23–50	70	154	178	69	56	1000	5	10
	51+	70	154	178	69	56	1000	5	10
Females	11–14	46	101	157	62	46	800	10	8
	15–18	55	120	163	65	46	800	10	8
	19–22	55	120	163	65	44	800	7.5	8
	23–50	55	120	163	65	46	800	5	8
	51+	55	120	163	65	44	800	5	8
Pregnant						+30	+200	+5	+2
Lactating						+20	+400	+5	+3

[a] The allowances are intended to provide for individual variations among most normal persons as they live in the United States under usual environmental stresses. Diets should be based on a variety of common foods in order to provide other nutrients for which human requirements have been less well defined.

[b] Retinol equivalents. 1 retinol equivalent = 1 μg retinol or 6 μg β-carotene.

[c] As cholecalciferol. 10 μg cholecalciferol = 400 IU of vitamin D.

[d] α-Tocopherol equivalents. 1 mg d-α-tocopherol = 1 α-tocopherol equivalent.

When the British navy required its sailors to eat limes regularly to combat scurvy, the sailors came to be called "limeys."

When James Lind discovered in 1852 that oranges, lemons, and limes could cure scurvy, and K. Takadi in the 1880s found that a proper diet could cure beriberi, the science of nutrition was on its way. Good personal nutrition prevents a number of specific diseases and it cures several. It provides the right nutrients for whatever happens to be the metabolic or health status of the body. The **nutrients** are any chemical substances that take part in any nourishing, health-supporting or health-promoting metabolic activity. Carbohydrates, lipids, proteins, vitamins, minerals, and trace elements are nutrients. Oxygen and water are usually not on the list, although they formally qualify under the definition. (We'll be saying more about these two later, anyway.) Materials that supply one or more nutrients are called **foods.**

Recommended Dietary Allowances.   For nearly 40 years the Food and Nutrition Board of the National Research Council of the National Academy of Sciences has published what are known as the **recommended dietary allowances,** or the **RDAs.** In the judgment of this Board, these allowances are the intake levels of essential nutrients that are "adequate to meet the known nutritional needs of practically all healthy persons." The values of the RDAs, as revised in 1980, are given in Table 23.1.

TABLE 23.1   (continued)

Water-Soluble Vitamins							Minerals					
Ascorbic Acid (mg)	Folacin[e] (μg)	Niacin[f] (mg)	Riboflavin (mg)	Thiamine (mg)	Vitamin B$_6$ (mg)	Vitamin B$_{12}$ (μg)	Calcium (mg)	Phosphorus (mg)	Iodine (μg)	Iron (mg)	Magnesium (mg)	Zinc (mg)
35	30	6	0.4	0.3	0.3	0.5[g]	360	240	40	10	50	3
35	45	8	0.6	0.5	0.6	1.5	540	360	50	15	70	5
45	100	9	0.8	0.7	0.9	2.0	800	800	70	15	150	10
45	200	11	1.0	0.9	1.3	2.5	800	800	90	10	200	10
45	300	16	1.4	1.2	1.6	3.0	800	800	120	10	250	10
50	400	18	1.6	1.4	1.8	3.0	1200	1200	150	18	350	15
60	400	18	1.7	1.4	2.0	3.0	1200	1200	150	18	400	15
60	400	19	1.7	1.5	2.2	3.0	800	800	150	10	350	15
60	400	18	1.6	1.4	2.2	3.0	800	800	150	10	350	15
60	400	16	1.5	1.2	2.2	3.0	800	800	150	10	350	15
60	400	15	1.3	1.1	1.8	3.0	1200	1200	150	18	300	15
60	400	14	1.3	1.1	2.0	3.0	1200	1200	150	18	300	15
60	400	14	1.3	1.1	2.0	3.0	800	800	150	18	300	15
60	400	13	1.2	1.0	2.0	3.0	800	800	150	18	300	15
60	400	13	1.2	1.0	2.0	3.0	800	800	150	10	300	15
+20	+400	+2	+0.3	+0.4	+0.6	+1.0	+400	+400	+25	[h]	+150	+5
+40	+100	+5	+0.5	+0.5	+0.5	+1.0	+400	+400	+50	[h]	+150	+10

[e] The folacin allowances refer to dietary sources as determined by *Lactobacillus casei* assay under standard conditions for the assay.

[f] 1 niacin equivalent (NE) equals 1 mg of niacin or 60 mg of dietary tryptophan tocopherols.

[g] This is based on the average concentration of vitamin B$_{12}$ in human milk.

[h] This increased requirement cannot be met by ordinary American diets nor do many women have enough iron in existing stores. Therefore the use of supplemental iron is recommended.

A number of points and qualifications about the RDAs must be emphasized.

1. *The RDAs are not the same as the U.S. Recommended Daily Allowances (USRDA).*   The latter are set by the U.S. Food and Drug Administration, based on the RDAs, as standards for nutritional information on food labels.

2. *The RDAs are not the same as the Minimum Daily Requirements (the MDRs) for any one individual.*   The MDRs are set very close to the levels at which actual signs of deficiencies occur. The RDAs are 2 to 6 times the MDRs. Just as individuals differ greatly in height, weight, and appearance, they also differ greatly in specific biochemical needs. Therefore an effort has been made to set the RDAs far enough above average requirements so that "practically all healthy people" will thrive. Most will receive more than they need; a few will not receive enough. One of the controversies over the RDAs concerns the validity of the statistical research and analysis by which "average requirements" were determined. Moreover, some nutritionists insist that among healthy people the range of daily need for a particular nutrient can vary far more widely than the Food and Nutrition Board has determined.

It's the job of hospital dieticians to devise the large variety of diets needed by the patient population.

3. *The RDAs do not define therapeutic nutritional needs.*   People with chronic diseases such as prolonged infections or metabolic disorders; people who take certain medications on a continuing basis; and prematurely born infants all require special diets. The RDAs do cover people according to age, sex, and size, and they indicate special needs for pregnant and lactating women, but they do not include any other special needs. Therapeutic needs for water and salt rise during strenuous physical activity and prolonged exposure to high temperatures. In some areas of the United States and in many other parts of the world, intestinal parasites are common. These organisms rob the affected people of some of their food intake each day, and they also need special diets.

4. *The RDAs can (and ought to) be provided in the diet from a number of combinations and patterns of food.*   No single food contains all nutrients. People take dangerous risks with their health when they go on fad diets limited to one particular food — brown rice, gelatin, yogurt, or liquid protein are some examples. The ancient wisdom of a varied diet that includes meat, fruit, vegetables, grains, nuts, pulses (e.g., peas, beans), and dairy products may seem to be supported solely by cultural and esthetic factors. However, a varied diet also assures us of getting any trace and needed nutrients that might not yet have been discovered.

Except for water, the nutrients in Table 23.2 are arranged in their order of requirements expressed in *moles*.

Table 23.2 is a compilation of the nutrients known to be required by every adult, according to American scientists and those with the United Nations. Some of the listed nutrients are not included in official U.S. tabulations either because they are so common that a deficiency is almost impossible or because it hasn't been possible to determine the minimum requirements.

Table 23.2 includes several elements such as iron and zinc. These are known as *trace elements,* and we'll return to them in Section 23.4.

Water Needs.   Our water intake comes from the fluids we drink, the water in the foods we eat, and the water made by the oxidations of nutrients in cells. Most comes in response to the thirst mechanism. Water leaves the body via the urine, the feces, exhaled air, and perspiration. Figure 23.1 gives the relative quantities by each route.

No amount of physical training or willpower can condition anyone to go without water.

Not shown in Figure 23.1 is the water that can be lost by sensible perspiration — sweat. An activity in the hot sun of an arid desert can cause the loss of as much as 10 L of water per day along with body salts. If these losses aren't made up at a reasonable rate, the results can be heat exhaustion, heat stroke, heat cramps, and even death.

Energy Requirements.   The oxygen needs per day are given in Table 23.2, because they define overall energy demands. These demands vary, of course, with the activity, and Table 23.3 shows the ranges as they vary from the relative inactivity of sleep to heavy activity.

FIGURE 23.1
Water intake and outgo by the principal routes (excluding sensible perspiration — sweating). The dashed lines at M are the minimal volumes of urine at the maximal concentrations of solutes. "Ox" refers to water formed by the oxidation of foods. (*Source: Recommended Dietary Allowances*, 9th ed., 1980. National Academy of Sciences, Washington, D.C.).

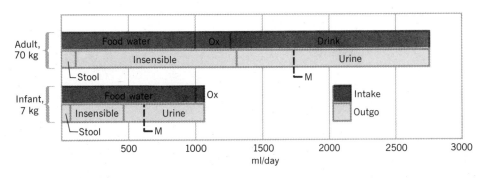

At rest, the largest user of oxygen in the body is the brain. Its mass is only about 2% of the body mass, but it consumes 20% of the oxygen used by the resting body. Much of the energy used in the brain goes to keep up concentration gradients of ions across the membranes of the billions of brain cells.

The energy obtained from food should not come exclusively either from carbohydrates alone or from fats alone. If you went on a zero-carbohydrate diet, your body would make glucose to meet its needs. (The brain derives nearly all of its energy from the breakdown of glucose.) If the body has to make glucose over a long period of time, say weeks, a slow buildup of toxic wastes would cause harm. On the other hand, if you were to go on a zero-fat diet, then some fatty acids that are essential to the body would be unavailable, and your system would not absorb the fat-soluble vitamins very well. A daily diet that includes at least 15 – 25 g of food fat (the equivalent of 2 – 4 pats of margarine or butter) and 50 – 100 g of digestible carbohydrate prevents these problems.

High protein diets, if prolonged, can make the kidneys overwork and become damaged.

The Essential Fatty Acids.   Linoleic and arachidonic acid (Table 20.2, page 475) are called the **essential fatty acids,** and they are needed for the health and growth of infants. If they are not in the diet, the victims experience skin problems and problems in the transportation of lipids in the bloodstream. Linoleic acid apparently is also essential to adults, but its widespread presence in all edible vegetable oils makes it virtually impossible for an adult not to obtain enough. Linoleic acid in adults can be used to make what arachidonic acid is needed. However, arachidonic acid is present in trace amounts in animal fats. This acid is used to make the prostaglandins that were described in Special Topic 20.1, page 476.

23.2   PROTEIN REQUIREMENTS

A protein's biological value is determined largely by its digestibility and its ability to supply the essential amino acids.

This is only about 2 ounces per day.

The total protein requirement (Table 23.2) is 56 g/day for adult men and 46 g/day for adult women. Small as these numbers are, they are still more than the bare minimum. Their intent is to allow about 30% extra to cover a wide range of variations in the protein needs of individuals. Moreover, they assume that we digest and absorb into the bloodstream only 75% of the protein in the diet. (These allowances are considered too narrow by some nutritionists.)

Actually, the idea of a *protein* requirement is somewhat misleading. What we require are certain daily intakes of specific amino acids. For several of them, their daily inclusion in the diet is so critical that they are called the *essential amino acids*. Let's see what they are.

Essential Amino Acids

Isoleucine

Leucine

Lysine

Methionine

Phenylalanine

Threonine

Tryptophan

Valine

Histidine is believed to be essential to infants.

Essential Amino Acids.   One reason that we can get by on relatively little protein is that not all of the 20 amino acids have to be present in the diet. Adults can synthesize 12 of them from parts and pieces of other molecules, including other amino acids. This leaves eight amino acids that the body can't prepare, at least not at rates rapid enough to make much difference. These eight amino acids that must be in the diet regularly, which are listed in Table 23.2 as well as in the margin, are called the **essential amino acids.**

TABLE 23.2
Adult Human Nutritional Requirements

Nutrient	Mass		Moles	
	Man[a]	Woman[b]	Man[a]	Woman[b]
Oxygen	800 g	593 g	25.0	18.5
	560 L[c]	414 L[c]		
As air, 21% oxygen	2930 L[d]	2173 L[d]		
Water[e]	2800 g	2000 g	155	111
Energy	2700 kcal	2000 kcal		
If all from glucose	675 g	500 g	3.75	2.78
If all from fat	300 g	222 g	0.338	0.25
Protein	56 g	46 g	0.47[f]	0.38[f]
Essential amino acids				
Tryptophan	0.5 g		0.0025	
Phenylalanine	2.2 g		0.0133	
Lysine	1.6 g		0.0110	
Threonine	1.0 g		0.0085	
Methionine[h]	2.2 g		0.0150	
Leucine	2.2 g		0.0168	
Isoleucine	1.4 g		0.0106	
Valine	1.6 g		0.0137	
Sodium chloride[i]	3.0 g		0.051	
Potassium	1.0 g		0.026	
Phosphorus	0.8 g		0.026	
Calcium	0.8 g		0.020	
Magnesium	0.35 g		0.015	
Essential fatty acids[j]	3.0 g		0.010	
Iron[k]	0.018 g		3.2×10^{-4}	
Vitamin C	0.045 g		2.5×10^{-4}	
Zinc	0.015 g		2.2×10^{-4}	
Niacin	0.018 g		1.5×10^{-4}	
Fluorine	0.002 g		1.0×10^{-4}	
Vitamin E[l]	0.015 g		3.5×10^{-5}	
Copper	0.0015 g		2.4×10^{-5}	
Pantothenic acid[m]	0.005 g		2.3×10^{-5}	
Manganese[m]	0.001 g		1.8×10^{-5}	
Vitamin K[l]	0.005 g		1.1×10^{-5}	
Vitamin B_6	0.002 g		1.0×10^{-5}	

     Our total daily needs for the essential amino acids add up only to about 13 g, less than half an ounce. When our tissues are actually manufacturing proteins, however, all of the amino acids that will be needed must be present and available *at the same time,* even if only a trace of any one is required. The absence of lysine, for example, when a protein that includes it has to be made, would prevent the synthesis of the protein altogether.

Coefficients of Digestibility.   Not all proteins are easily and completely digested. The variations are expressed by a **coefficient of digestibility,** defined by Equation 23.1.

TABLE 23.2 (continued)

Nutrient	Mass		Moles	
	Man[a]	Woman[b]	Man[a]	Woman[b]
Vitamin A[l]	0.0015 g		5.2×10^{-6}	
Riboflavin	0.0016 g		4.2×10^{-6}	
Thiamin	0.0014 g		4.2×10^{-6}	
Iodine	0.00012 g		9.4×10^{-7}	
Folic acid	0.0004 g		9.0×10^{-7}	
Biotin[m]	0.00015 g		6.0×10^{-7}	
Vitamin D[n]	0.00001 g		2.6×10^{-8}	
Chromium	0.000001 g		2.0×10^{-8}	
Vitamin B$_{12}$	0.0000005 g		3.7×10^{-9}	
Cobalt	0.00000002 g		3.7×10^{-9}	
Molybdenum[o]				
Vanadium[o]				
Selenium[o]				
Tin[o]				

Source: F. E. Deatherage, *Food for Life,* Plenum Publishing Corporation, New York, 1975. Used by permission.

[a] Man: Weight, 70 kg (154 lb). Height, 175 cm (5 ft 8 in.). Age, 22–50 years.

[b] Woman: Weight, 58 kg (128 lb). Height, 163 cm (5 ft 5 in.). Age, 22–50 years.

[c] At 0 °C and 760 mm Hg pressure.

[d] At 20 °C and 740 mm Hg—room conditions.

[e] Includes water consumed directly and as a component in food plus the water made by the oxidation of nutrients.

[f] Average formula weight of amino acids is 120.

[g] 0.22 g if tyrosine is available.

[h] 0.35 g if cystine is available.

[i] 2–3 g is minimum, but the salt requirement as well as the water requirement may increase in warm climates and during heavy work.

[j] As linoleic, linolenic, and/or arachidonic acid.

[k] Recommended for menstruating women; for men about 0.01 g is recommended.

[l] A number of closely related chemical compounds may serve this vitamin function. 1 IU (International Unit) of A = 0.3 μg of retinol or 0.6 μg of β-carotene. 1 IU of E = 1.0 mg of DL-α-tocopherol acetate or 0.81 mg D-α-tocopherol.

[m] Although definitely required, it is difficult to establish the amount because this nutrient's daily requirement has not been established. Sufficient amounts are in many foods.

[n] Vitamin D may not be required by the adult but is required by growing children; it is made from normal skin constituents by ultraviolet light or it may be fed in the diet. 1 IU of D = 0.025 μg of D$_3$, activated 7-dehydrocholesterol.

[o] It is known that this trace element is an integral part of vital enzyme systems, but the daily requirement has not been established. Sufficient amounts are in many foods.

$$\text{Coefficient of digestibility} = \frac{(\text{N in food eaten}) - (\text{N in feces})}{(\text{N in food eaten})} \qquad (23.1)$$

The difference of terms in the numerator:

$$(\text{N in food eaten}) - (\text{N in feces})$$

is the nitrogen that is actually absorbed by the bloodstream.

Animal proteins have higher coefficients of digestibility than those of plants or fruits. For the average animal protein this coefficient is 0.97, which means 97% digestible. The coeffi-

TABLE 23.3
Daily Energy Expenditures of Adult Men and Women in Light Occupations

Activity Category	Time (hr)	Man (70 kg)		Woman (58 kg)	
		Rate		Rate	
		kcal/min	Total (kcal)	kcal/min	Total (kcal)
Sleeping, reclining	8	1.0–1.2	540	0.9–1.1	440
Very light[a]	12	up to 2.5	1300	up to 2.0	900
Light[b]	3	2.5–4.9	600	2.0–3.9	450
Moderate[c]	1	5.0–7.4	300	4.0–5.9	240
Heavy[d]	0	7.5–12.0		6.0–10.0	
Total	24		2740		2030

Source: J. V. G. A. Durin and R. Passmore, 1967. In *Recommended Dietary Allowances,* 9th ed., 1980. National Academy of Sciences, Washington, D.C.

[a] Seated and standing activities, driving cars and trucks, secretarial work, laboratory work, sewing, ironing, playing musical instruments.

[b] Walking (on the level, 2.5–3 mph), tailoring and pressing, carpentry, electrical trades, restaurant work, washing clothes, light recreation such as golf, table tennis, volleyball, sailing.

[c] Walking at 3.5–4 mph, garden work, scrubbing floors, shopping with heavy load, moderate sports such as skiing, tennis, dancing, bicycling.

[d] Uphill walking with a load, pick-and-shovel work, lumbering, heavy sports such as swimming, climbing, football, or basketball.

cients for the proteins in fruits and fruit juices average about 0.85; for whole wheat flour, 0.79; and for whole rye flour, 0.67. Milling improves the digestibility coefficients — to 0.83 for all-purpose bread flour and 0.89 for cake flour.

Milling does the same to rice. The digestibility coefficient for brown rice is 0.75 but 0.84 for polished white rice. Milling also reduces the quantities of vitamins, minerals, and fiber in the grain product. However, these are usually added back (as in enriched flours.)

The average digestibility coefficient of the proteins in legumes and nuts is 0.78, and in vegetables the range is 0.65 to 0.74.

The Concept of Biological Value for a Protein.   The digestibility of a protein is just one factor in the quality of a protein source. If a protein is lacking or low in one or more of the essential amino acids, it is a low-quality protein. When the body gets the amino acids from such a protein but doesn't obtain enough of an essential amino acid, it has a load of amino acids it can't use. It has to get rid of them, because the body doesn't store unneeded amino acids nor does it excrete them as such. It has to break them down to be able to excrete the nitrogen (as urea). Any of the nitrogen-free amino acid breakdown products can be used by the body to generate energy, as we'll study later. But if the body isn't demanding this energy by its level of activity, then it converts the breakdown products into fat. You can get fat on a high-protein, low-activity diet, and you strain your kidneys besides.

The ratio of the nitrogen that the body retains to the total nitrogen it takes in, expressed as a percent, when the body operates under the stress of receiving not quite enough protein overall, is called the **biological value** of the protein. It is a measure of the efficiency with which the body uses the nitrogen *of the actually absorbed amino acids* furnished by the protein.

The largest single factor in a protein's biological value is its amino acid composition. What most limits this value is the extent to which the protein supplies the essential amino acids in sufficient quantities for human use. Human milk protein is the best of all proteins in terms of digestibility and biological value. Whole-egg protein is very close, and it is often taken as the reference for experimental work. When the diet is well-balanced in absorbable amino acids,

$$NH_2 - \overset{\overset{\displaystyle O}{\displaystyle \|}}{C} - NH_2$$
Urea

Infants and school children must have a positive nitrogen balance, meaning that they must ingest more nitrogen than they excrete, in order to grow.

the amount of nitrogen ingested equals the amount excreted. This condition is called **nitrogen balance.** Most 70-kg men could, in principle, be in nitrogen balance by ingesting 35 g/day of the proteins in human milk. This value is used as a reference standard in rating other proteins.

Table 23.4 summarizes some information about most of the proteins that are prominent in various diets of the world's peoples. The essential amino acid that is most poorly supplied by a given protein is called its **limiting amino acid.** These are named in the second column of Table 23.4.

The third column gives the number of grams of each food that a 70-kg man would have to digest *and absorb* per day to get the same amount of its limiting amino acid as is available from 35 g of human milk protein. Such an intake, of course, would also supply all of the other

TABLE 23.4
Comparisons of Food Proteins with Human Milk Protein

Food	Limiting Amino Acid	Food's Protein Equivalent to 35 g Human Milk Protein (g)	Digestibility Coefficient of Food's Protein	Amount of Food's Protein[a] (g)	Percent of Protein in the Food	Amount of Food Needed[b]	Kilocalories of Food Received
Wheat	Lysine	80.6	79.0	102	13.3%	767 g (1.7 lb)	2560
Corn	Tryptophan and Lysine	72.4	60.0	120	7.8%	1540 g (3.4 lb)	5660
Rice	Lysine	51.7	75.0	68.9	7.5%	919 g (2.0 lb)	3310
Beans	Valine	50.5	78.0	64.8	24.0%	270 g (0.59 lb)	913
Soybeans	Methionine and Cysteine	43.8	78.0	56.2	34.0%	165 g (0.36 lb)	665
Potatoes	Leucine	71.6	74.0	96.7	2.1%	4600 g (10.1 lb)	3500
Cassava	Methionine and Cysteine	82.4	60.0	137	1.1%	12,500 g (27.5 lb)	16,400
Eggs	Leucine	36.6	97.0	37.8	12.8%	295 g (0.65 lb)	477
Meat	Tryptophan	43.1	97.0	44.4	21.5%	206 g (0.45 lb)	295
Cow's milk	Methionine and Cysteine	43.8	97.0	45.2	3.2%	1410 g (3.1 lb)	903

Source: Data from F. E. Deatherage, *Food for Life*, Plenum Press, New York, 1975. Used by permission.

[a] The grams of protein that have to be obtained from each food source in order that the protein be nutritionally equivalent (with respect to essential amino acids) to 35 g of human milk protein, allowing for the poorer digestibility of that food's protein (i.e., its digestibility coefficient).

[b] The grams of each food that are equivalent in nutritional value (with respect to essential amino acids) to 35 g human milk protein allowing for the digestibility coefficient and the percent protein in the food.

$102 \times 0.790 = 80.6$

$767 \times 0.133 = 102$

needed but nonlimiting amino acids. For example, 80.6 g of wheat protein—not wheat, but wheat protein—would have to be digested and absorbed to obtain the lysine in 35 g of human milk protein. However, we have to adjust this for the digestibility coefficient of wheat protein, 0.790, so the 80.6 g of wheat protein means that 102 g of wheat protein (column 5) actually has to be eaten. But wheat, on the average, is only 13.3% protein (column 6), so the 70-kg man would have to eat 767 g (1.69 lb) of wheat (column 7) if his daily needs for all amino acids are to be met by wheat alone. If he does this, however, he unavoidably receives 2560 kcal of food energy, which is nearly as much total energy per day as he should have—and his diet is boring besides. Just imagine his problems if he had to get all his amino acids from potatoes—10.1 lb/day!

The data in Table 23.4 are of tremendous importance to those concerned about supplying both the protein and total calorie needs of the world's burgeoning population. Neither children nor adults can eat enough corn, rice, potatoes, or cassava per day to meet both their protein and energy needs. Proteins of eggs and meat, however, are particularly good. They are said to be **adequate proteins,** because they include all of the essential amino acids in suitable proportions to make it possible to satisfy amino acid and total nitrogen needs without excess intake of calories.

Soybeans are the best of the nonanimal sources of proteins. Cassava, a root, is an especially inadequate protein, and corn (or maize) is also poor. Unhappily, huge numbers of the world's peoples, especially those in Africa, Central and South America, India, and the countries of the Middle and Far East, rely heavily on these two foods, and there are widespread dietary deficiency diseases in these regions.

The data in Table 23.4 also provide scientific support for the long-standing practices in all major cultures of including a wide variety of foods in the daily diet. Meat and eggs can ensure adequate protein while they leave room for an attractive variety that is important to good eating habits. Milk, soybeans, and other beans also leave room for variety. By including both rice (low in lysine) and beans (low in valine) in equal proportions, about 43 g of the combination is equal in protein value to 35 g of human milk protein. Less of the combined rice-bean diet is needed than rice alone or beans alone, which means there is room for other foods that our taste buds crave. The rice and beans should, of course, be eaten fairly closely together to ensure that all essential amino acids are simultaneously available during protein synthesis. Eating just the rice early in the morning and the beans in the evening works against efficient protein synthesis.

23.3   VITAMINS

The vitamins essential to health must be in the diet because the body cannot make them.

Vitamin = "vital amine," from an early belief that vitamins might all be amines.

The term **vitamin** applies to any compound or a closely related group of compounds that satisfies the following criteria:

1.   It is organic rather than inorganic or an element.
2.   It cannot be synthesized at all (or at least in sufficient amounts) by the body and it must be in the diet.
3.   Its absence causes a specific **vitamin deficiency disease.**
4.   Its presence is essential to normal growth and health.
5.   It is present in foods in *small* concentrations, and it is not a carbohydrate, a saponifiable lipid, an amino acid, or a protein.

As we just mentioned, some vitamins occur as a set of compounds any one of which can be converted by the body into the form or forms that it needs. Such a set of related compounds, all serving as the preventer of a specific vitamin deficiency disease, are called *vitamers*.

In addition to vitamin deficiency diseases, there are at least 25 disorders classified as *vitamin-responsive inborn errors of metabolism*. These conditions arise from a genetic fault that can be partly and sometimes entirely overcome by the daily ingestion of 10 to 1000 times as much of a particular vitamin as normally should be present in the diet. Genetic faults can compromise the body's use of a vitamin at several places in the path the vitamin takes once it has been ingested. There might be impaired absorption of the vitamin, or impaired transport in the blood. As we'll learn in the next chapter, many vitamins are needed to finish the synthesis of an enzyme, so the inborn genetic error could be a defect in this use of the vitamin.

The two major classes of vitamins are the fat-soluble and the water-soluble vitamins. The former are largely hydrocarbon-like, and the latter are polar or ionic. If we ingest surpluses of any of the fat-soluble vitamins, fat tissue tends to accumulate the extra amount; such accumulations can be dangerous. Surpluses of any water-soluble vitamins from the diet leave the body in the urine.

In this section we'll study what the vitamins are, their related deficiency diseases, and their sources. In the next chapter, we will see how the body uses them at the molecular level.

"Like dissolves like" applies here, too.

Fat-Soluble Vitamins.   The fat-soluble vitamins are vitamins A, D, E, and K, and they occur in the fat fractions of living systems. Table 23.2 gives their requirements for adults. We'll see as we go along that the molecules of these vitamins either have alkene double bonds or phenol rings. Both molecular features are readily attacked by oxidizing agents. Therefore these vitamins are susceptible to destruction by prolonged exposures to the oxygen in air or to oxidizing agents, such as those that occur in fats and oils that are turning rancid because of oxidation.

Vitamin A occurs in a few closely related forms. One is retinol, a polyunsaturated alcohol.

Retinol

Some and maybe all forms of vitamin A, when taken in excess, are known to be teratogenic (they cause birth defects), so vitamin supplements with extra vitamin A should not be taken by pregnant women. In excess, vitamin A is toxic to adults. The livers of polar bears and seals are particularly rich in vitamin A, and Eskimos are careful about eating these otherwise desirable foods. Siberian huskies also have very high levels of vitamin A in their livers, and early explorers who were driven to eating their sled dogs to survive risked serious harm and death by eating too much husky liver. In early 1913, the Antarctic explorer X. Mertz almost certainly died in just this way and his companion, Douglas Mawson, barely survived.

Those suffering from excess vitamin A recover quite quickly when they stop taking vitamin A.

Many people believe that vitamin A prevents cancer, but thus far all efforts to obtain the kind of scientific proof that convinces all nutritionists have failed.

The deficiency disease for vitamin A is nyctalopia, or night blindness — impaired vision in dim light. We also need vitamin A for healthy mucous membranes.

Because the liver can make vitamin A out of β-carotene, the orange-yellow pigment in carrots and other naturally yellow foods, such foods are good sources of vitamin A.

Vitamin D exists in a number of forms. Cholecalciferol (D₃) and ergocalciferol (D₂) are two that occur naturally.

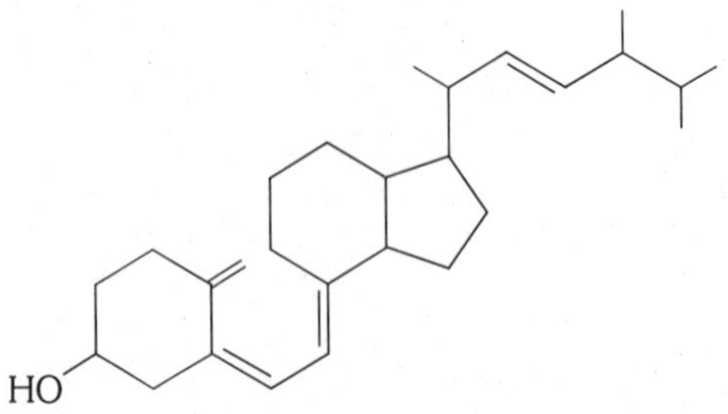

Ergocalciferol (D₂)

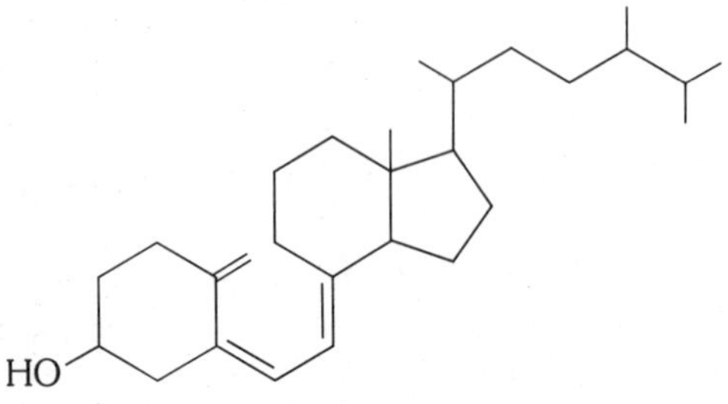

Cholecalciferol (D₃)

Poor management of Ca²⁺ metabolism contributes to a disfiguring bone condition of old age called osteoporosis.

These two are equally useful in humans, and either can be changed in the body to the slightly oxidized forms that the body uses as hormones to stimulate the absorption and uses of calcium ions and phosphate ions. Vitamin D is especially important during the years of early growth when bones and teeth, which need calcium and phosphate ions, are developing. Lack of vitamin D causes the deficiency disease known as rickets, a bone disorder.

Eggs, butter, liver, fatty fish, and fish oils such as cod-liver oil are good natural sources of vitamin D precursors. The most common source today is vitamin-D fortified milk. We are able to make some vitamin D ourselves from certain steroids that we can manufacture, provided we get enough direct sunlight on the skin. Energy absorbed from sunlight converts these steroids to the vitamin. Youngsters who worked from dawn to dusk in dingy factories or mines during the early years of the industrial revolution were particularly prone to rickets simply because they saw little if any sun.

No advantage is gained by taking large doses of vitamin D, and in sufficient excess it is dangerous. Excesses promote a rise in the calcium ion level of the blood, and this damages the kidneys and causes soft tissue to calcify and harden.

Our need for **vitamin E** is satisfied by any member of the tocopherol family. The most active member is α-tocopherol.

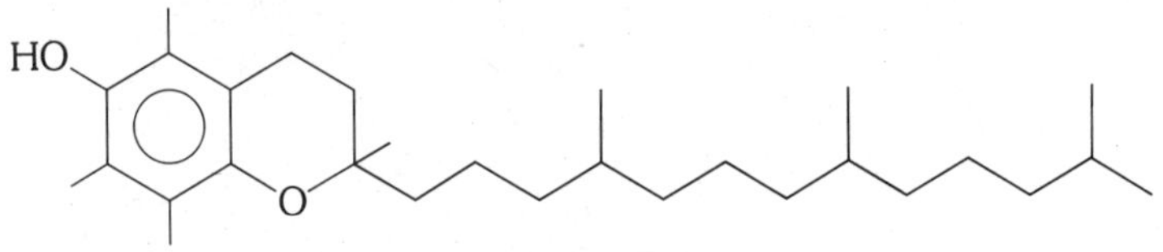

α-Tocopherol

Vitamin E detoxifies organic peroxides, those with structures such as $R-O-O-H$. These can form when oxygen attacks CH_2 groups that are adjacent to alkene double bonds, as in the many $-CH_2-CH=CH-$ units in molecules of the unsaturated fatty acids. Because the acyl units of these acids are present in the phospholipids of which cell membranes are made, vitamin E is needed to protect all membranes. In the absence of vitamin E, the activities of certain enzymes are reduced, and red blood cells hemolyze more readily. Anemia and edema are reported in infants whose feeding formulas are low in vitamin E. According to some studies, a lack of vitamin E can contribute to an increased susceptibility to sudden heart attacks, especially in males under stress.

Vitamin E occurs so widely in vegetable oils that it is almost impossible not to obtain enough of it. It is generally considered to be relatively nontoxic to humans, but because it's a fat-soluble vitamin it can accumulate in fatty tissue. Caution must be used in taking it as a supplement, but intakes in the range of 200 to 600 mg/day appear to be acceptable in humans.

Much of the vitamin K that we need is made for us by our own intestinal bacteria.

Vitamin K is the antihemorrhagic vitamin. It is present in green, leafy vegetables, and deficiencies are rare. It works as a cofactor in the blood-clotting mechanism. Sometimes women about to give birth and their newborn infants are given vitamin K to provide an extra measure of protection against possible hemorrhaging.

Water-Soluble Vitamins. The water-soluble vitamins that are recognized by the Food and Nutrition Board are vitamin C, choline, thiamin, riboflavin, niacin (nicotinamide), folacin, vitamin B_6, vitamin B_{12}, pantothenic acid, and biotin.

Collagen is the protein in bone that holds the minerals together, much as steel rods work in reinforced concrete.

Vitamin C, or ascorbic acid, prevents scurvy, a sometimes fatal disease in which collagen is not well made. Whether it prevents other diseases is the subject of enormous controversy, speculation, and research. It is an antioxidant, much as is vitamin E. A variety of studies have found that vitamin C is involved in the metabolism of amino acids, in the synthesis of some adrenal hormones, and in the healing of wounds. There are probably millions of people who believe that vitamin C in sufficient dosages — up to several grams per day — acts to prevent or to reduce the severity of the common cold. The vitamin appears to be nontoxic at these high levels.

Vitamin C is present in citrus fruits, potatoes, leafy vegetables, and tomatoes. It is destroyed by extended cooking, heating over steam tables, or prolonged exposure to air or to ions of iron or copper. Even when vitamin C is kept in a refrigerator in well-capped bottles, it slowly deteriorates.

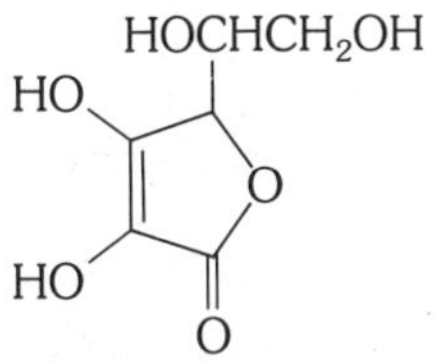

Vitamin C (ascorbic acid)

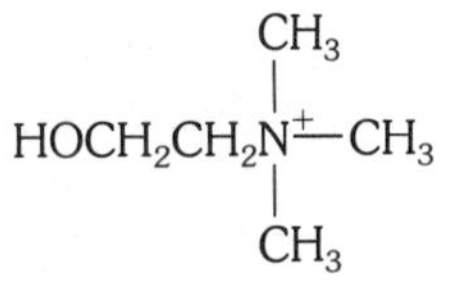

Choline

Acetylcholine is one of several neurotransmitters.

Choline is needed to make complex lipids, as we discussed in Section 20.3. Acetylcholine, which is made from choline, is one of several substances that carry nerve signals from one nerve cell to another. The body can make choline, but the Food and Nutrition Board calls it a vitamin because there are 10 species of higher animals that have dietary requirements for it. We make it too slowly to meet all of our daily needs, so having it in the diet provides protection. It occurs widely in meats, egg yolk, cereals, and legumes. No choline deficiency disease has been demonstrated in humans, but in animals the lack of dietary choline leads to fatty livers and to hemorrhagic kidney disease.

In rice, most of the thiamine is in the husk, which is lost when raw rice is milled.

Thiamine is needed for the breakdown of carbohydrates. Its deficiency disease is beriberi, a disorder of the nervous system. Good sources are lean meats, legumes, and whole (or enriched) grains. It is stable when dry, but destroyed by alkaline conditions or prolonged cooking. Thiamine is not stored, and excesses are excreted in the urine. One's daily thiamine requirement is proportional to the number of calories represented by the diet. The Food and Nutrition Board recommends 0.5 mg/1000 kcal for children and adults.

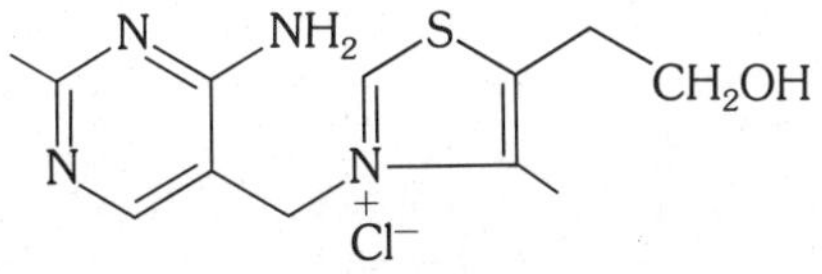

Thiamine

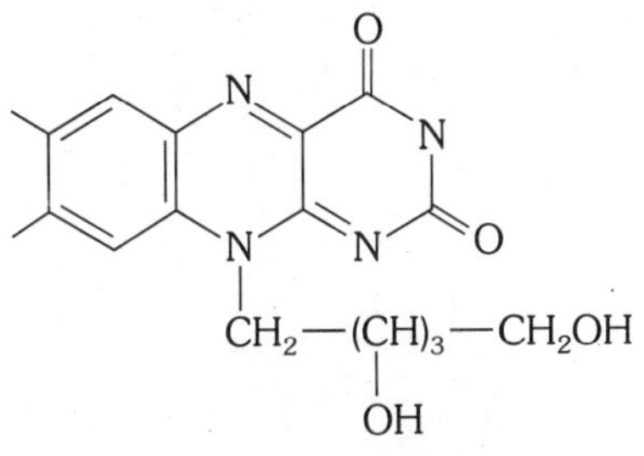

Riboflavin

Riboflavin is required by a number of oxidative processes in metabolism. Deficiencies lead to the inflammation and breakdown of tissue around the mouth, nose, and the tongue, a scaliness of the skin, and burning, itching eyes. Wound healing is impaired. The best source is milk, but certain meats (e.g., liver, kidney, and heart) also supply it. Cereals are poor sources unless they have been enriched. Little if any riboflavin is stored, and excesses in the diet are excreted. Alkaline substances, prolonged cooking, and irradiation by light destroy this vitamin.

Severe niacin deficiency can cause delerium and dementia.

We can make about 1 mg of niacin for every 60 mg of dietary tryptophan — nearly all we need.

Niacin — meaning both nicotinic acid and nicotinamide — is essential for nearly all biological oxidations. It's needed by every cell of the body every day. Its deficiency disease is pellagra, a deterioration of the nervous system and the skin. Pellagra is particularly a problem where corn (or maize) is the major item of the diet. Corn (maize) is low in niacin and the essential amino acid tryptophan, from which we are able to make some of our own niacin. Where the diet is low in tryptophan, niacin must be provided in other foods, such as enriched grains. Prolonged cooking destroys niacin.

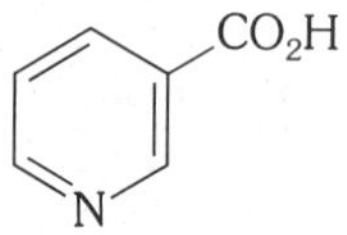

Nicotinic acid
(niacin)

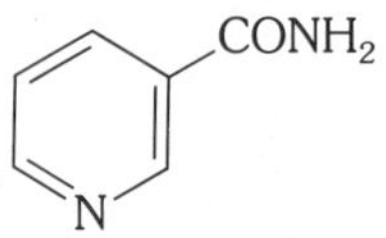

Nicotinamide
(niacinamide)

Chronic alcoholism causes folacin deficiency.

Folacin is the name used by the Food and Nutrition Board for folic acid and related compounds. Its deficiency disease is megaloblastic anemia. Several drugs, including alcohol, promote folic acid deficiency. It is needed for the synthesis of nucleic acids and heme. Good sources are fresh, leafy green vegetables, asparagus, liver, and kidney. Folacin is relatively unstable to heat, air, and ultraviolet light, and its activity is often lost in both cooking and food storage.

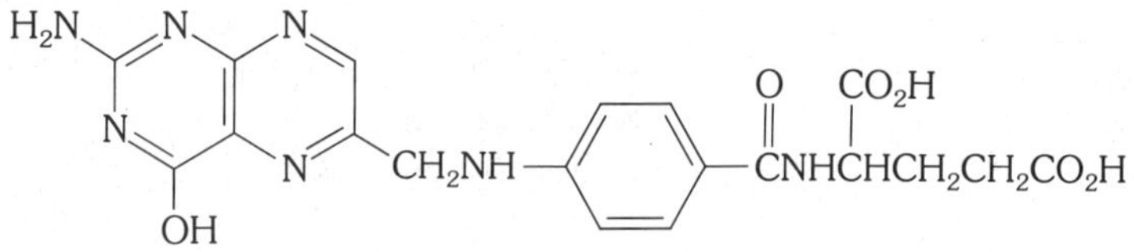

Folic acid

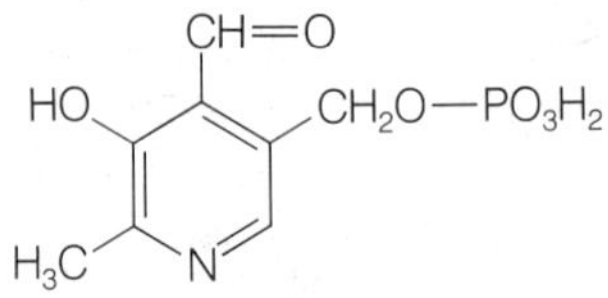

Pyridoxal phosphate

Vitamin B$_6$ activity is supplied by pyridoxine, pyridoxal, or pyridoxamine. All can be changed in the body to the active form, pyridoxal phosphate. The activities of at least 60 enzymes involved in the metabolism of various amino acids depend on pyridoxal phosphate. One deficiency disease is hypochromic microcytic anemia, and disturbances in the central nervous system also occur. The vitamin is present in meat, wheat, yeast, and corn. It is relatively stable to heat, light, and alkali.

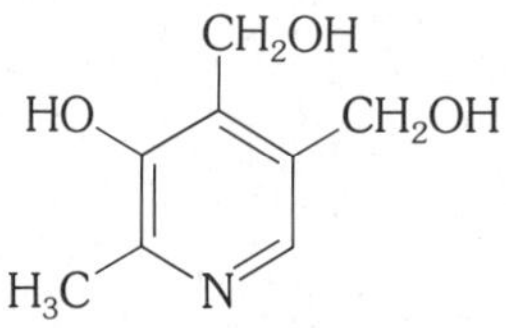

Pyridoxine

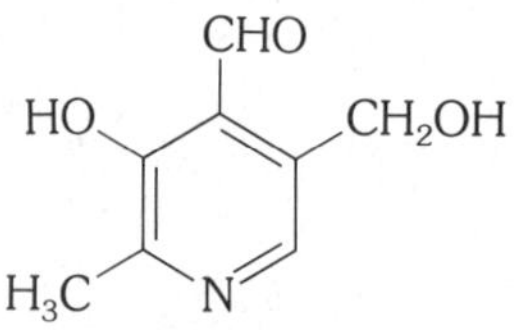

Pyridoxal

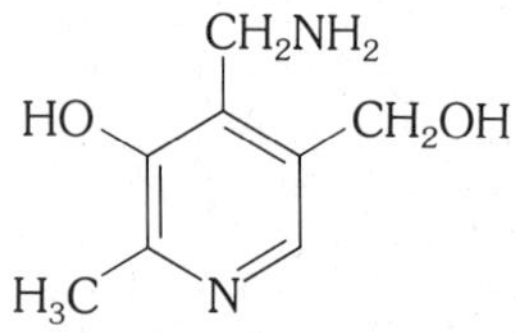

Pyridoxamine

Vitamin B$_{12}$, or cobalamin, is a specific controlling factor for pernicious anemia. This deficiency disease, however, is very rare, because it is very difficult to design a diet that lacks this vitamin. Animal products such as liver, kidney, and lean meats as well as milk products and eggs are good sources — and virtually the only sources. Thus most people who develop

B_{12} deficiency are true vegetarians or are infants born of true vegetarian mothers. The onset of any symptoms of B_{12} deficiency occurs slowly because the body stores it fairly well and because such minute traces are needed.

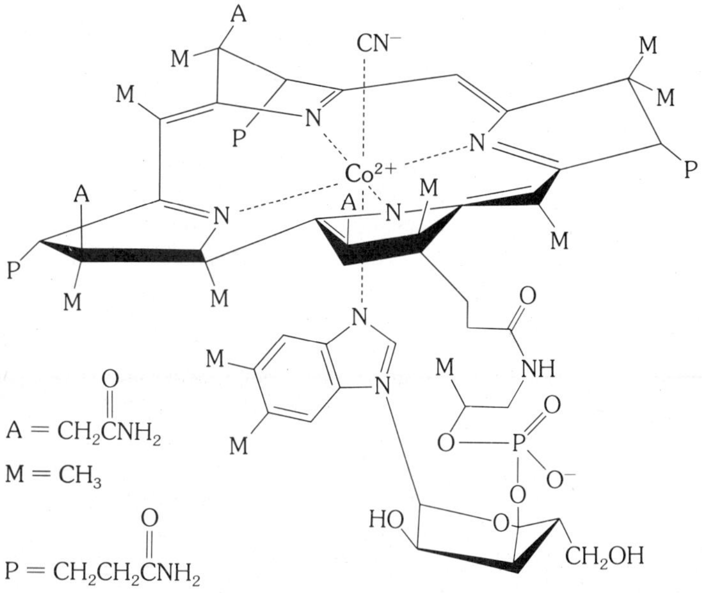

$$A = CH_2\overset{\displaystyle O}{\overset{\|}{C}}NH_2$$

$$M = CH_3$$

$$P = CH_2CH_2\overset{\displaystyle O}{\overset{\|}{C}}NH_2$$

Cyanocobalamin

Pantothenic acid is used to make a coenzyme, coenzyme A (symbol: CoASH), which the body needs to metabolize fatty acids. Signs of a deficiency disease for this vitamin have not been observed clinically in humans, but the deliberate administration of compounds that work to lower the availability of pantothenic acid in the body causes symptoms of cellular damage in vital organs. This vitamin is supplied by many foods, especially by liver, kidney, egg yolk, and skim milk.

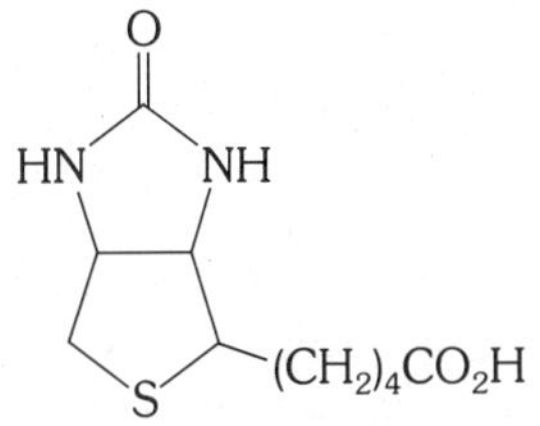

Biotin

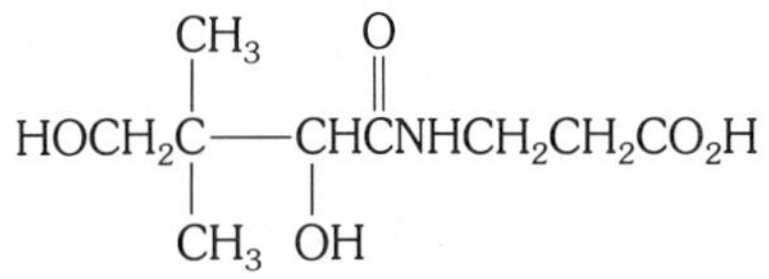

Pantothenic acid

Biotin is required for all pathways in which carbon dioxide is temporarily used as a reactant, as in the synthesis of fatty acids that we'll study in Chapter 28. Signs of biotin deficiency are hard to find, but when such a deficiency is deliberately induced, the individual experiences nausea, pallor, dermatitis, anorexia, and depression. When biotin is given again, the symptoms disappear. Our own intestinal microorganisms probably make biotin for us. Egg yolk, liver, tomatoes, and yeast are good sources.

23.4 MINERALS AND TRACE ELEMENTS IN NUTRITION

Many metal ions are necessary for the activities of enzymes.

Most of the minerals and trace elements that must be provided in the diet are metal ions, but anions are also on these lists. Whether one is classified as *mineral* or as a *trace element* is a matter of quantity. The **minerals** are needed in the diet at levels of 100 mg/day or more, usually much more. The **trace elements** are needed at levels of 20 mg/day or less, some of them much, much less.

The dietary minerals are calcium (Ca^{2+}), phosphorus (phosphate ion, P_i), magnesium (Mg^{2+}), sodium (Na^+), potassium (K^+), and chlorine (Cl^-). These are the chief electrolytes in the body. The uses of these minerals will be described elsewhere, principally in Chapter 25.

The Food and Nutrition Board recognizes 17 trace elements, all of which have been found to have various biological functions in animals, and which therefore are quite likely used in the human body. Ten of them are known to be needed by humans — fluorine (as F^-) and iodine (as I^-), and the ions of chromium, manganese, iron, cobalt, copper, zinc, selenium, and molybdenum. All are toxic in excessive amounts. They occur widely in food and drink, but some are removed by food refining and processing.

Fluorine (as fluoride ion) is essential to the growth and development of sound teeth, and the Food and Nutrition Board recommends that public water supplies be fluoridated wherever natural fluoride levels are too low.

Iodine (as iodide ion) is essential in the synthesis of certain hormones made by the thyroid gland. Iodide-ion deficient diets lead to an enlargement of the thyroid gland known as a goiter. It takes only about 1 μg (1×10^{-6} g) of I^- per kilogram of body weight each day to prevent a goiter. Seafood is an excellent source of iodide ion, but iodized salt with $75-80$ μg of iodide-ion equivalent per gram of salt is the surest way to obtain the iodine that is needed. Prior to the days of iodized salt, goiters were quite common, especially in regions where little if any fish or other seafood was in the diet.

High-fiber diets can work against the absorption of some of the trace elements.

Chromium (Cr^{3+}) is required for normal glucose metabolism, and it may be needed for the action of insulin. Chromium that occurs naturally in foods is absorbed significantly more easily than chromium given as the simple salts added to vitamin-mineral supplements. Most animal proteins and whole grains supply this trace element.

Manganese (Mn^{2+}) is required for normal nerve function, for the development of sound bones, and for reproduction. It is essential to the activities of certain enzymes. Nuts, whole grains, fruits, and vegetables supply this element, but a recommended daily allowance has yet to be set by the Food and Nutrition Board.

Iron (Fe^{2+}) is needed in heme and several enzymes. The intestines regulate how much dietary iron is absorbed, so a proper level of iron in circulation is maintained this way. A high serum iron level apparently renders an individual more susceptible to infections.

Cobalt (Co^{2+}) is part of the vitamin B_{12} molecule (cyanocobalamin, page 551). Apparently there is no other use for it in humans.

Copper (Cu^{2+}) occurs in a number of proteins and enzymes. If deficient in copper, the individual synthesizes lower-strength collagen and elastin and will tend to suffer anemia, skeletal defects, and degeneration of the myelin sheaths of nerve cells. Ruptures and aneurysms of the aorta become more likely. The structure of hair is affected, and reproduction tends to fail. Fortunately, copper occurs widely in foods, particularly in nuts, raisins, liver, kidney, certain shellfish, and legumes. An intake of just 2 mg/day assures a copper balance for nearly all people. Unfortunately, however, the copper contents of the foods in a typical American diet have declined over the last four decades. Some scientists believe that the increase in the incidence of heart disease in the United States during this period has been caused partly by copper-low diets. In one study that involved women, the lower the copper level in the blood the higher were their blood cholesterol levels. (Whether higher cholesterol levels alone are a direct cause of heart disease, however, is by no means certain. We'll discuss cholesterol and heart disease in Chapter 28.)

Zinc (Zn^{2+}) is required for the activities of several enzymes. Without sufficient zinc in the diet, an individual will experience loss of appetite and poor wound healing. Insufficient zinc in an infant's diet, which is a chronic problem in the Middle East, results in dwarfism and poor development of the gonads. If too much zinc is in the diet, and too little copper is present, the extra zinc acts to inhibit the absorption of copper. This correlation between zinc and copper is still a matter of considerable research, but it appears that the ratio of zinc to copper doesn't matter as long as sufficient copper is available. When enough zinc is present, it acts to inhibit the absorption of a rather toxic pollutant, cadmium (as Cd^{2+}), which is just below zinc in the periodic table.

Selenium is known to be essential in many animals, including humans. How much we need is not known, but too much is very toxic. A strong statistical correlation exists between high levels of selenium in livestock crops and low incidences of human deaths by heart disease. In the United States, those who live where selenium levels are high — the Great Plains between the Mississippi River and the eastern Rocky Mountains — have one-third the chance of dying from heart attack and strokes than those where levels are very low — the northeastern quarter of the United States, Florida, and the Pacific Northwest. Rats, lambs, and piglets on selenium-poor diets develop damage to heart tissue and abnormal electrocardiograms. Selenium reduces the occurrence of certain cancers in animals, and it may possibly provide a similar benefit in humans.

Molybdenum is in an enzyme required for the metabolism of nucleic acids. Deficiencies in humans are unknown, meaning that almost any reasonable diet furnishes enough.

Nickel, silicon, tin, and vanadium are possibly trace elements for humans, because deficiency diseases for these elements have been induced in experimental animals.

SUMMARY

Nutrition Good nutrition entails the ingestion of all of the substances needed for health — water, oxygen, food energy, essential amino acids and fatty acids, vitamins, minerals, and trace elements. Foods that best supply various nutrients have been identified. The Food and Nutrition Board of the National Academy of Sciences regularly publishes *recommended daily allowances* that are intended to be amounts that will meet the nutritional needs of practically all healthy people. Some people need more, most need less. But neither more nor less of anything is necessarily better. The RDAs should be obtained by a varied diet because it promotes good eating habits and because such a diet might supply a nutrient no one yet knows is essential.

Water needs The thirst mechanism leads to our chief source of water. Our principal routes of exporting water are the urine, perspiration (both sensible and insensible), and exhaled air. When excessive water losses occur during vigorous exercise, the body also loses electrolytes.

Energy Both carbohydrates and lipids should be in the diet as sources of energy. Without carbohydrates for several days, certain poisons build up in the blood as the body works to make glucose internally from amino acids. A zero-lipid diet means zero ingestion of essential fatty acids. Our oxygen requirements adjust to the caloric demands of our activities.

Protein in the diet What we need most from proteins are certain essential amino acids, and we need some extra nitrogen if we have to make the nonessential amino acids. When a diet excretes as much nitrogen as it takes in, the individual is in nitrogen balance. The most superior, balanced proteins — those that are highly digestible and that supply the essential amino acids in the right proportions — are proteins associated with animals, such as the proteins in milk and whole eggs. (Human milk protein is the standard of excellence, and whole egg protein is very close to it.)

The most important factor in the biological value of a protein is its limiting amino acid — the essential amino acid that it supplies in the lowest quantity. Another factor is the digestibility of the protein. For several reasons — poor source of an essential amino acid; low digestibility coefficients; low concentration of the protein — several foods cannot be used as exclusive or even major components of a healthy diet. Foods that have inadequate proteins include corn (maize), rice, potatoes, cassava, and manioc.

Vitamins Organic compounds, called vitamins, or sets of closely related compounds that satisfy the same need, must be in the diet in at least trace amounts or an individual will suffer from a vitamin deficiency disease. Vitamins can't be made by the body. (Essential amino acids and essential fatty acids are generally not classified as vitamins, nor are carbohydrates, proteins, or triacylglycerols and other saponifiable lipids.) Excessive amounts of the fat-soluble vitamins (A, D, E, and K) — especially A and D — must be avoided. These vitamins accumulate in fatty tissue, and in excess they can cause serious trouble. Any excesses of the water-soluble vitamins are eliminated.

Minerals and trace elements The minerals are inorganic cations and anions that are needed in the diet in amounts in excess of 100 mg/day. The trace elements are needed at levels of 20 mg/day or less. The whole body quantities of the minerals are large relative to the trace elements. (More about minerals is described in later chapters.) The trace elements that are metal ions are essential to several enzyme systems. Two anions are trace elements, F^- and I^-. Fluoride ion is needed to make strong teeth, and iodide ion is needed to make thyroid hormones and to prevent goiter.

KEY TERMS

The following terms were emphasized in this chapter, and they constitute an important beginning vocabulary for the study of nutrition.

adequate protein

biological value

biotin

choline

coefficient of digestibility

dietetics

essential amino acids

essential fatty acids

folacin

food

limiting amino acid

mineral

niacin

nitrogen balance

nutrient

nutrition

pantothenic acid

recommended dietary allowance (RDA)

riboflavin

thiamine

trace element

vitamin

vitamin A

vitamin B_6

vitamin B_{12}

vitamin C

vitamin D

vitamin deficiency disease

vitamin E

vitamin K

SELECTED REFERENCES

1 *Recommended Dietary Allowances,* 9th edition. National Academy of Sciences, Washington, D.C., 1980.

2 F. E. Deatherage. *Food for Life.* Plenum Press, New York, 1975.

3 J. Bland (ed.) *Medical Applications of Clinical Nutrition.* Keats Publishing, Inc., New Canaan, CT, 1983. Particularly, Chapter 3, by D. R. Davis, "Nutritional Needs and Biochemical Diversity."

4 H. S. Mitchell, H. J. Rynbergen, L. Anderson, and M. V. Dibble. *Nutrition in Health and Disease,* 16th edition. J. B. Lippincott Company, Philadelphia, 1976.

5 L. J. Teply, Study Chairman. *Chemistry and the Food System.* American Chemical Society, Washington, D.C., 1980.

6 H. J. Sanders. "Nutrition and Health." *Chemical and Engineering News,* March 26, 1979, page 27.

7 Patricia K. Johnston. "Getting Enough to Grow On." *American Journal of Nursing,* March 1984, page 336.

8 D. Rudman and P. J. Williams. "Megadose Vitamins." *The New England Journal of Medicine,* August 25, 1983, page 488.

9 *Chemistry of the Food Cycle — State of the Art.* A symposium consisting of 16 articles in the *Journal of Chemical Education,* April 1984.

REVIEW EXERCISES

The answers to these Review Exercises are in the *Study Guide* that accompanies this book.

Nutrition

23.1 What does the science of nutrition study?

23.2 What is meant by the term *nutrient?*

23.3 What is the relationship of nutrients to foods?

23.4 What is the relationship of the science of dietetics to nutrition?

23.5 Why are the recommended dietary allowances higher than minimum daily requirements?

23.6 What are seven situations that require special therapeutic diets?

23.7 Why should our diets be drawn from a variety of foods?

23.8 What is potentially dangerous about the following diets?
(a) A carbohydrate-free diet
(b) A lipid-free diet

Protein and Amino Acid Requirements

23.9 Because the full complement of 20 amino acids is required to make all of the body's proteins, why are less than half of this number labeled essential amino acids?

23.10 If we are able to make all of the nonessential amino acids, why should our daily protein intake include more than what is represented solely by the essential amino acids?

23.11 What does the body do with the amino acids that it absorbs from the bloodstream but doesn't use?

23.12 What is the equation that defines the coefficient of digestibility of a protein? What does the numerator in this equation stand for?

23.13 Which kind of protein is generally more fully digested by the body, the protein from an animal or a plant source?

23.14 What can be done to whole grains to improve the digestibility of their proteins? What also happens in this process that reduces the food value of the grains?

23.15 What is the most important factor in determining the biological value of a given protein?

23.16 Which specific protein has the highest coefficient of digestibility and the highest biological value for humans? Which protein comes so close to this on both counts that it is possible to use it as a substitute for research purposes?

23.17 What is meant by the *limiting* amino acid of a protein?

23.18 Why does the protein in corn have a lower biological value than the protein in whole eggs?

23.19 In one variety of hybrid corn, the limiting amino acids are lysine and tryptophan. It takes 75 g of this corn to provide the protein equivalent to 35 g of human milk protein. The coefficient of digestibility of the protein in this corn is 0.59.
  (a) In order to match the nutritional value of human milk protein, how many grams of the protein in this corn must be ingested?
  (b) To get this much protein from this variety of corn, how many grams of the corn must be eaten? The corn is only 7.6% protein.
  (c) How many kilocalories are also consumed with the quantity of corn calculated in part (b) if the corn has 360 kcal/100 g?
  (d) Could a child eat enough corn per day to satisfy all of its protein needs and still have room for any other food?

23.20 The limiting amino acid in peanuts is lysine, and 62 g of protein obtained from peanuts is equivalent to 35 g of human milk protein. The coefficient of digestibility of peanut protein is 0.78. In order to match human milk protein in nutritional value:
  (a) How many grams of peanut protein must be eaten?
  (b) How many grams of peanuts must be eaten if peanuts are 26.2% protein?
  (c) How many kilocalories are also ingested with this many grams of peanuts if peanuts have 282 kcal/100 g?
  (d) Could a child eat enough peanuts per day to satisfy its protein needs and still have room for other foods?

Vitamins

23.21 Make a table that lists each vitamin, at least one good source of each, and a serious consequence of a deficiency of each.

(Use just the information available in this book.) Set up the table with the following column heads.

Vitamin	Source(s)	Problem(s) If Deficient

23.22 Why aren't the essential amino acids listed as vitamins?

23.23 On a strict vegetarian diet — no meat, eggs, and dairy products of any sort — which one vitamin is hardest to obtain?

23.24 Why should strict vegetarians use two or more different sources of proteins?

23.25 Which are the fat-soluble vitamins?

23.26 Which vitamin is activated when the skin is exposed to sunlight?

23.27 Which vitamin acts as a hormone in its active forms?

23.28 Which vitamin has been shown to be teratogenic when used in excess?

23.29 The yellow-orange pigment in carrots, carotene, can serve as a source of the activity of which vitamin?

23.30 The vitamin needed to participate in the blood-clotting mechanism is which one?

23.31 The Food and Nutrition Board of the National Academy of Sciences recognizes which substances as the water-soluble vitamins?

23.32 Scurvy is prevented by which vitamin?

23.33 Name two vitamins that tend to be destroyed by prolonged cooking.

23.34 Which vitamin prevents beriberi?

23.35 When corn (maize) is the chief food in the diet, which vitamin is likely to be in short supply because a raw material for making it is in short supply?

Minerals and Trace Elements

23.36 What criterion makes the distinction between *minerals* and *trace elements*?

23.37 Name and give the chemical forms of the six minerals.

23.38 Make a table of the 10 trace elements and at least one particular function of each.

23.39 What can be the body's response to an iodine-deficient diet?

Chapter 24
Enzymes, Hormones, and Neurotransmitters

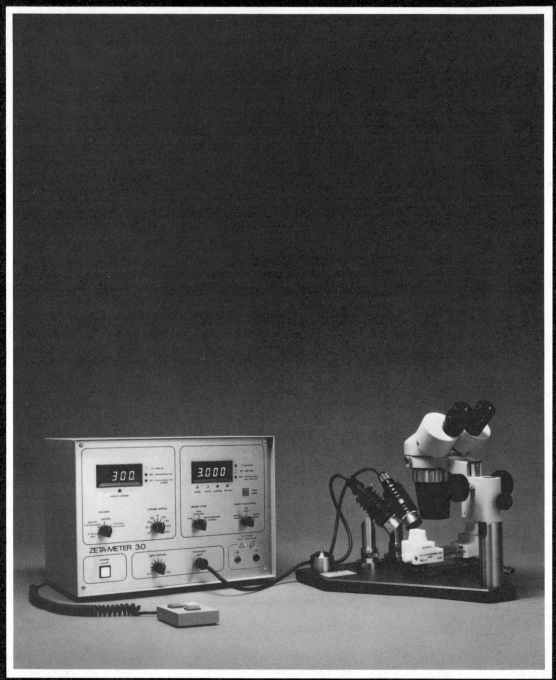

Mixtures of very similar proteins can be separated by electricity using an apparatus such as shown here mounted on the microscope stage. The procedure is electrophoresis, and it's a powerful tool in chemical analysis, as we'll learn in this chapter.

24.1 ENZYMES

The catalytic abilities of enzymes often depend on cofactors that are made from B-vitamins.

Virtually all body catalysts — enzymes — are proteins. It was once almost a dogma that *all* body catalysts are proteins, but in the early 1980s more than one team of scientists confirmed that the catalytic activity of one enzyme, ribonuclease P, is caused solely by its RNA component, not its protein. The feeling among scientists now is that more examples of nonprotein catalysts will be found. Despite these developments, it is still the case that all but a few of the thousands of catalysts in the body are proteins, and our study in this chapter will deal only with these. Let's look first at some general properties of enzymes.

All enzyme molecules and most substrate molecules are chiral: they have handedness.

Enzyme Specificity.   When we studied optical isomerism, we learned that enzyme activity occurs when molecules of the enzyme and the substrate fit together. Because each substrate consists of molecules that have their own unique shapes, each reaction needs its own specific enzyme.

Few enzymes are so specific that they catalyze just one particular reaction of one particular compound. Most enzymes possess *relative specificity*. For example, enzymes that catalyze the hydrolysis of esters usually handle a variety of esters, not just one, but they do not work well, if at all, for the hydrolysis of any other kind of functional group. Some enzymes in the digestive tract specialize in catalyzing the hydrolysis of peptide bonds that are in the interiors of polypeptide chains, but they hit only those peptide bonds that are adjacent to specific amino acid side chains. Their work splits long polypeptides into much shorter molecules. Then, another digestive enzyme goes to work exclusively on the C-terminal amino acyl units of these shorter chains, and still another enzyme works on the N-terminal units.

Rate Enhancements by Enzymes.   Enzymes, like all catalysts, affect the *rates* of reactions by providing a reaction pathway that has a lower energy of activation than the path the uncatalyzed reaction would have to use. Even small reductions of energy barriers can cause spectacular increases in rates. For example, the enzyme carbonic anhydrase (CA) catalyzes the interconversion of carbonic acid with carbon dioxide and water:

$$CO_2 + H_2O \underset{\text{Carbonic anhydrase}}{\rightleftharpoons} H_2CO_3$$

The equilibrium *must* shift to the right to make H_2CO_3 when the supply of CO_2 is high — a consequence of Le Chatelier's principle.

In actively metabolizing cells, where the supply of CO_2 is relatively high, this equilibrium shifts to the right, and each molecule of carbonic anhydrase aids in the conversion of 600,000 molecules of CO_2 *each second!* This is the fastest known rate for any enzyme-catalyzed reaction, and it is 10 million times faster than the uncatalyzed reaction. In blood that circulates into the lungs, where exhaling keeps the supply of CO_2 low, this equilibrium must shift to the left, and the same enzyme participates in this change.

It is important to realize that a catalyst, such as carbonic anhydrase, that is involved in an equilibrium speeds up the preservation of the *equilibrium*. In other words, it speeds up *both* the forward and the reverse reactions. Whether the equilibrium shifts to the right or to the left doesn't depend on the catalyst at all. It depends strictly on the relative concentrations of reactants and products; on whether other reactions are going on that feed substances into the equilibrium or that continuously remove them; and on the temperature. All the catalyst does is to make whatever shift is mandated by these conditions occur very rapidly.

Coenzymes.   The molecules of most enzymes include a nonpolypeptide component called a **cofactor,** and where the cofactor of such an enzyme is missing there is no enzymic activity. The polypeptide part of an enzyme that has a cofactor is called the **apoenzyme.**

The cofactor of some enzymes is simply a metal ion such as the Fe^{2+} ion in a family of enzymes called the cytochromes, which are involved in biological oxidations. In other enzymes, the cofactor is an organic molecule or ion, in which case the cofactor is called a **coenzyme.** Some enzymes have both a coenzyme and a metal ion cofactor.

Some of the vitamins, particularly the B vitamins, are used to make coenzymes. For example, the coenzyme thiamine diphosphate, structure **1**, is a diphosphate ester of thiamine, a B vitamin.

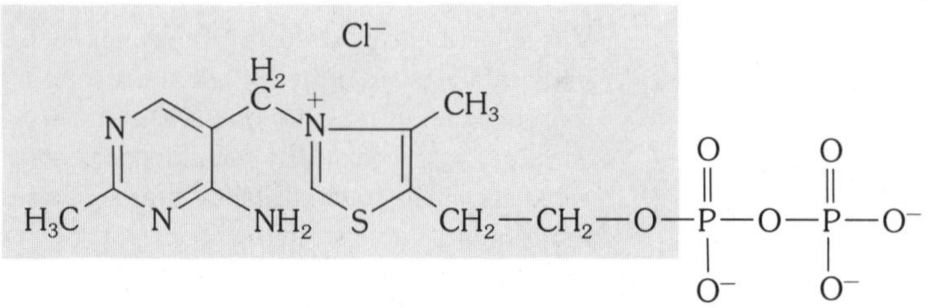

Thiamine is vitamin B_1.

1
Thiamine diphosphate

When pure, this coenzyme is a triprotic acid, but at the pH of body fluids it is ionized as shown.

Another coenzyme that incorporates a B vitamin, nicotinamide, into its structure is nicotinamide adenine dinucleotide, **2a,** which is usually referred to simply as NAD^+ (or, sometimes, just NAD). Notice that the bottom half of the NAD^+ molecule is from adenosine monophosphate, AMP, and that the upper half is almost like it except that a molecule of nicotinamide has replaced adenine.

Dinucleotide means that two nucleotides are joined together.

Another important coenzyme, nicotinamide adenine dinucleotide phosphate, **2b,** is a phosphate ester of NAD^+. It's usually referred to as $NADP^+$ (or, sometimes, just NADP). Both

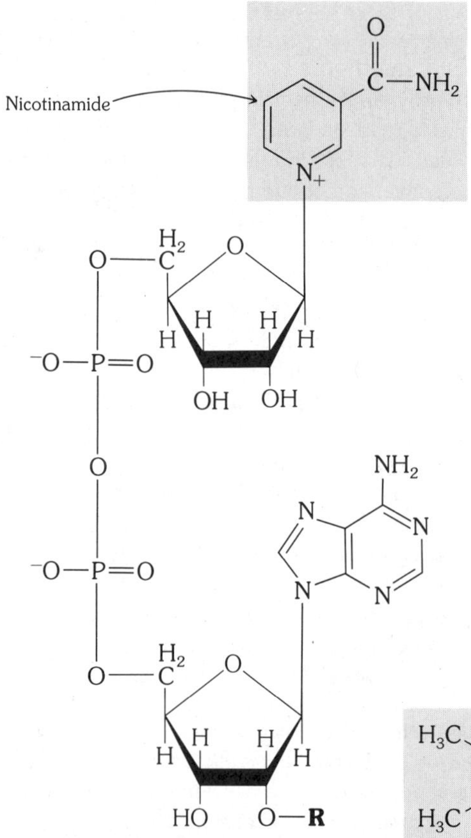

The P in $NADP^+$ refers to the extra phosphate ester unit.

2

a NAD^+ **R** $= H$

b $NADP^+$ **R** $= -OPO_3^{2-}$

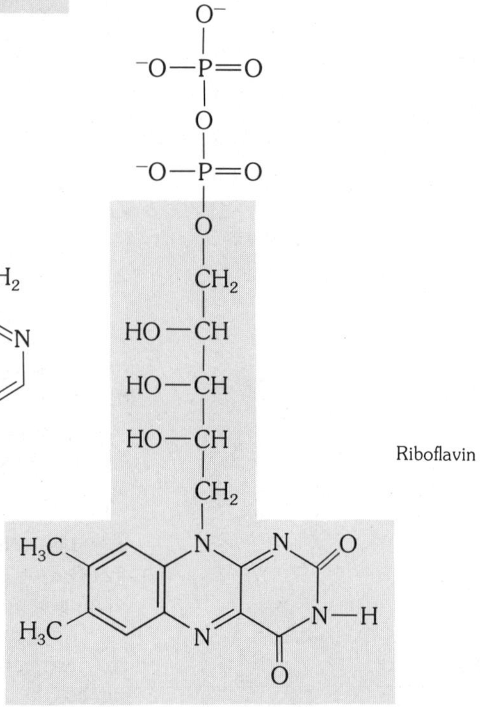

3

FMN

SPECIAL TOPIC 24.1 HOW NAD^+ AND FMN (OR FAD) PARTICIPATE IN ELECTRON TRANSFER

In the nicotinamide unit of both NAD^+ and $NADP^+$ there is a positive charge on the nitrogen atom, an atom that is also part of an aromatic ring. Despite how rich this ring is in electrons, the positive charge on N makes it an electron acceptor. Without the charge, the ring would otherwise be very similar to the ring in benzene, and it would strongly resist any reaction that disrupted its closed-circuit electron system. The positive charge on N, however, places the ring of the nicotinamide unit on an energy teeter-totter that the cell can tip either way without too much energy cost.

As seen in the equations below, when an electron-pair donor such as an oxidizable alcohol arrives at the enzyme, the donor can transfer the electron pair (as $H:^-$) to the ring. Even though the ring will temporarily lose its aromatic character, the positive charge of the ring nitrogen atom encourages this transfer of negative charge. In the next step of this metabolic pathway, the pair of electrons, still as $H:^-$, transfers to a riboflavin unit, also shown in the equations below. As this occurs, the ring of the nicotinamide unit recovers its stable, benzene-like nature.

Eventually, the pair of electrons winds up on an oxygen atom when, in the last step of this long metabolic pathway, oxygen that has been supplied by the air we breathe is reduced to water. We'll return to these steps in Chapter 26.

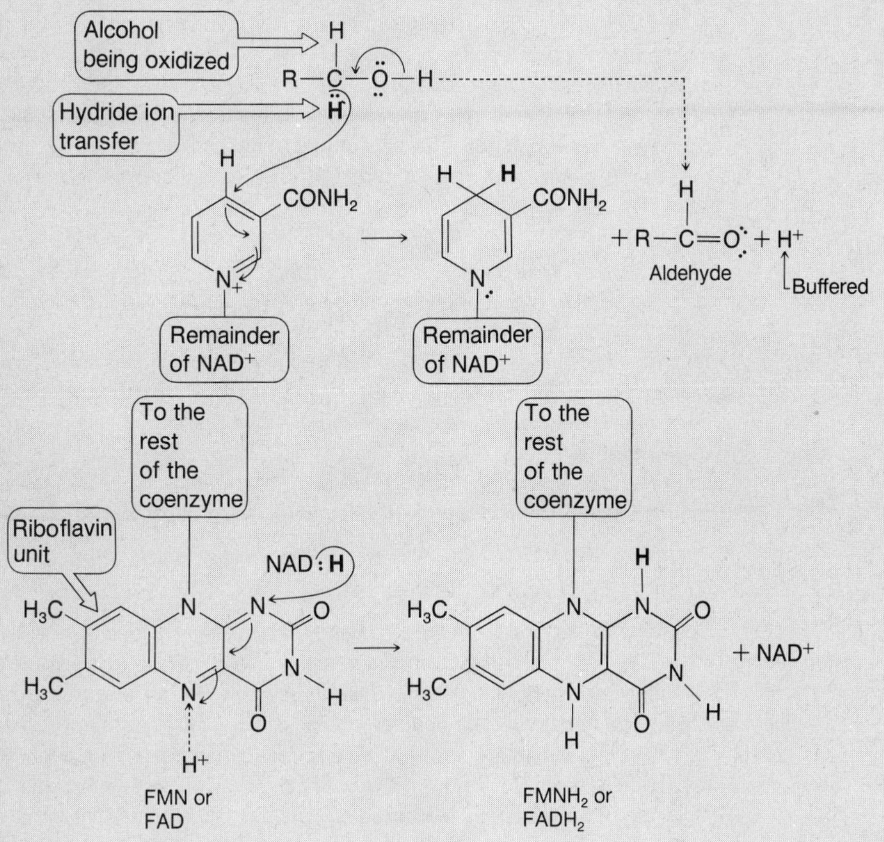

NAD^+ and $NADP^+$ are the coenzymes in enzymes needed for biological oxidation-reduction reactions.

Quite often equations that involve enzymes with recognized coenzymes are written with the symbol of the coenzyme as a reactant or as a product. For example, the body metabolizes ethyl alcohol by first oxidizing it to the aldehyde. The cofactor for the enzyme is NAD^+, which serves as the actual acceptor of the hydride ion, $H:^-$, that ethyl alcohol gives up. Although the details are reserved to a Special Topic, 24.1, the equation is written simply as follows:

$$\underset{\text{Ethyl alcohol}}{CH_3CH_2OH} + NAD^+ \longrightarrow \underset{\text{Ethanal}}{CH_3CH{=}O} + \underset{\text{Reduced form of } NAD^+}{NAD{:}H} + \underset{\text{Hydrogen ion (buffered)}}{H^+}$$

NAD^+ accepts $H\!:\!^-$ from the alcohol, and we can write this part of the above reaction by the following equation:

$$NAD^+ + H\!:\!^- \longrightarrow NAD\!:\!H$$

By accepting the *pair of electrons* in $H\!:\!^-$, NAD^+ is reduced, and $NAD\!:\!H$ (usually written simply as NADH) is referred to as the *reduced form* of NAD^+. The same kind of transfer of hydride ion can involve $NADP^+$, and its reduced form is written as NADPH. When we use the symbol of a coenzyme as if it were a chemical formula, remember that this symbol stands for the entire enzyme of which the coenzyme is a part.

We have learned earlier that a catalyst cannot be called a catalyst unless it undergoes no *permanent* change, but these examples seem to call for such a change. In the body, however, most enzymes are members of a whole series of enzymes that catalyze a chain of reactions. In such a chain, a reaction that alters an enzyme is followed at once by a reaction that changes the enzyme back again. For example, the NAD^+ that is reduced to NADH as ethyl alcohol is oxidized to ethanal, is recovered in the next step in which still another B vitamin, riboflavin, functions as a coenzyme. Flavin mononucleotide or FMN, **3,** is the coenzyme that uses riboflavin, and it can accept a pair of electrons in $NAD\!:\!H$. As also shown in Special Topic 24.1, FMN reacts with NADH and H^+ (from the buffer system present) as follows:

Riboflavin is vitamin B_2.

$$NADH + FMN + H^+ \longrightarrow NAD^+ + FMNH_2$$

Thus the NAD^+-containing enzyme is restored, but now the FMN-containing enzyme has its coenzyme in its reduced form, $FMNH_2$. We won't continue this discussion now except to say that $FMNH_2$ passes on its load of hydrogen and electrons in yet another step and so is reoxidized to FMN. The main points of our discussion here are that B vitamins are key parts of coenzymes, and that the catalytic activities of the associated enzymes involve the molecules of these vitamins directly.

Flavin mononucleotide or FAD is a near relative of FMN that contains riboflavin. Its reduced form is $FADH_2$, and it is also involved in biological oxidations.

How Enzymes Are Named and Classified.   Nearly all enzymes have names that end in *-ase*. The prefix is either from the name of the substrate or from the kind of reaction. For example, an **esterase** aids the hydrolysis of esters. A **lipase** works on the hydrolysis of lipids. A **peptidase** catalyzes the hydrolysis of peptide bonds. Similarly, an **oxidase** is an enzyme that catalyzes an oxidation, and a **reductase** handles a reduction. An **oxidoreductase** handles a redox equilibrium. A **transferase** catalyzes the transfer of a group from one molecule to another, and a **kinase** s a special transferase that handles phosphate groups.

These names are chemically informative, but not all enzymes are so named. For example, the peptidases *trypsin* and *pepsin* have old names that neither end in *-ase* nor disclose their substrates. To minimize the confusion that would result if all enzymes had such names, an International Enzyme Commission has developed a system of classifying and naming enzymes. The names of the principal reactants, separated by a colon, are written first and then the name of the kind of reaction is written as a prefix to *-ase*. For example, in moving from left to right in the following equilibrium, an amino group transfers from the glutamate ion to the pyruvate ion:

Whenever we see *-ase* as a suffix in the name of any substance or type of reaction, the word is the name of an enzyme.

This reaction, incidentally, is an example of how the body can make an amino acid—here, alanine—from other substances.

$$^-O_2CCH_2CH_2\overset{\underset{\displaystyle |}{NH_3^+}}{C}HCO_2^- + CH_3\overset{\overset{\displaystyle O}{\|}}{C}CO_2^- \rightleftharpoons {}^-O_2CCH_2CH_2\overset{\overset{\displaystyle O}{\|}}{C}CO_2^- + CH_3\overset{\underset{\displaystyle |}{NH_3^+}}{C}HCO_2^-$$

Glutamate ion   Pyruvate ion α-Ketoglutarate ion   Alanine

The systematic name for the enzyme is *glutamate : pyruvate aminotransferase.* In all but formal publications, such a cumbersome (but unambiguous) name is seldom used. This enzyme is most often referred to simply as GPT.

CK is the newer symbol for an older one, CPK (for the older name, creatine phosphokinase).

Isoenzymes. Identical reactions are often catalyzed by enzymes with identical cofactors but slightly different apoenzymes. For example, creatine kinase or CK catalyzes the transfer of a phosphate group in the following equilibrium:

ATP is adenosine triphosphate, which we described together with ADP in Section 16.6.

$$NH_2CNCH_2CO_2^- + ATP \underset{\substack{\text{Creatine} \\ \text{kinase}}}{\rightleftharpoons} {}^-OPONHCNCH_2CO_2^- + ADP$$

Creatine   Creatine phosphate

Creatine kinase consists of two polypeptide chains, but the version in skeletal muscle is slightly different from the versions made in the brain and in heart muscle. In skeletal muscle CK, the two polypeptides are identical, and we can use the letter *M* (for muscle, skeletal) to stand for them. Thus skeletal muscle CK has the symbol CK(*MM*). In brain CK, the two strands are also identical, but not the same as those in skeletal muscles. The symbol CK(*BB*) is used for this version (where *B* is for brain, of course). Heart muscle contains a third CK isoenzyme, also made of two strands. One is identical with *M* and the other with *B*, so this version is symbolized by CK(*MB*). Enzymes that catalyze the identical reaction but that have slightly different structures are called **isoenzymes,** where the prefix *iso-* carries the same idea—a certain sameness—as it does in the word *isomer.*

Some references call isoenzymes by the name *isozymes.*

The differences among isoenzymes occur in their apoenzymes, not their cofactors. Isoenzymes have an extraordinarily important function in clinical analysis and diagnosis, as we'll see later in the chapter.

24.2 THE ENZYME-SUBSTRATE COMPLEX

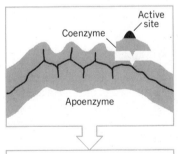

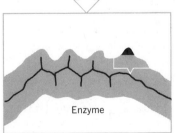

FIGURE 24.1
A coenzyme, which contributes the active site, joins to an apoenzyme in the formation of many kinds of complete enzymes.

Enzymes have one or more active sites as well as substrate-binding sites that function when enzyme-substrate complexes forms.

Substrate molecules are usually much smaller than those of the enzyme, so only a small part of the enzyme molecule is responsible for its catalytic properties. This part is called the **active site,** and the functional groups that do the catalytic work are called the catalytic groups. As we indicated in the previous section, the catalytic groups in many enzymes are provided by the coenzyme. (See also Figure 24.1.) Elsewhere on the enzyme molecule there are functional groups whose shapes and distributions of partial charges let them serve as the **binding sites.** These not only hold the substrate molecule over the active site, they help to guide it into place.

The combination that forms between the substrate molecule and the enzyme molecule during the reaction is called the **enzyme-substrate complex.** It forms and breaks down as part of the following series of chemical equilibria.

$$E + S \rightleftharpoons E{-}S \rightleftharpoons E{-}S^* \rightleftharpoons E{-}P \rightleftharpoons E + P$$

| Enzyme | Substrate | Enzyme-substrate complex | Substrate-activated $E{-}S$ complex | Enzyme-product complex | Enzyme (recovered) | Product |

The fact that an enzyme can be specific for its substrate stems from the need for the molecules of the two to be able to fit snugly together. This specificity resembles that of a key for just one tumbler lock, and so this theory of how enzymes work is sometimes called the **lock-and-key theory.**

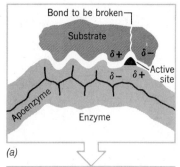

(a)

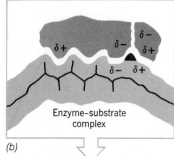

(b)

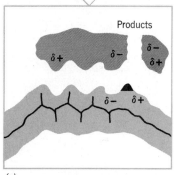

(c)

FIGURE 24.2
The lock-and-key model for enzyme action. *(a)* The enzyme and its substrate fit together to form an enzyme-substrate complex. *(b)* A reaction, such as the breaking of a chemical bond, occurs. *(c)* The product molecules separate from the enzyme. (How enzymes work can also be studied by means of a computer-driven animation in a program available in *Prelab Studies for General, Organic, and Biological Chemistry*, by Sandra L. Olmsted and Richard D. Olmsted. See page xi.

Figure 24.2, part *(a)*, shows how partial charges and complementary shapes might bring the substrate molecule directly over the active site. The figure supposes that some bond scheduled to be broken in the substrate is positioned closest to the active site. Part *(b)* of the figure also supposes, quite reasonably, that the bond-breaking reaction will radically alter the distribution of partial charges so that sites that are attracted to each other before the reaction repel each other afterward. (This might occur as new functional groups emerge in the product fragments.) Thus the product fragments quickly leave the active site, as seen in part *(c)* of Figure 24.2, and another substrate molecule can be processed.

Sometimes the approach of the substrate molecule induces the enzyme molecule to alter its shape as the enzyme-substrate complex forms. (See Figure 24.3.) According to this **induced-fit theory,** one reason for such behavior might be to strip away from the key functional groups any molecules of water that have formed solvent cages. The more exposed groups may then more readily undergo their reactions at lower energy costs.

Effect of Concentration on Enzyme-Catalyzed Reaction Rates.   What we are going to develop in this short subsection is some necessary background for the next section, the regulation of enzymes. We'll give some evidence for what was postulated by the enzyme-substrate theory — that when an enzyme-catalyzed reaction occurs, the enzyme functions temporarily as a reactant.

When two reactants are mixed, their reaction starts off with some *initial rate.* Then as the reactants are used up, the initial rate gives way to increasingly slower rates because the concentrations of the reactants decrease. We're interested here only in the *initial* rate, because this is most readily related to the easily known *initial* concentrations. What we want to learn about is how the initial rate varies with the initial concentration of the substrate.

We have to imagine a series of experiments in all of which the molar concentration of the enzyme, $[E]$, is the same. All that we'll do is vary the initial concentration of the substrate, $[S]$, from experiment to experiment. In most ordinary reactions between two reactants carried out in this way, the initial rate would double each time we doubled the initial concentration of one reactant. And in our series of experiments, we do observe something like this, but only in those trials that have *low* initial values of $[S]$. A plot of initial rates versus initial values of $[S]$ is shown in Figure 24.4, and in the part of the plot marked A a small increase in $[S]$ does cause a proportionate increase in initial rate. In succeeding experiments, however, at higher and higher initial values of $[S]$, the initial rates respond less and less until in part B of the plot the initial rates are constant *regardless of the value of* $[S]$.

The best explanation for what is happening in part B of the curve is that at a sufficiently high initial value of $[S]$, there are enough substrate molecules to saturate all of the active sites of all of the enzyme molecules. Any additional substrate molecules have to wait their turns, so to speak, and if this is true, then we are treating the enzyme itself as one of the reactants. The overall rates of such reactions are functions not only of the substrate(s) but also of the catalyst.

One of the very significant features of several enzymes is that the initial sharp rise of the curve in Figure 24.4 is delayed. For these enzymes, the plots of initial values of $[S]$ versus the initial rates of the reactions look more like the curve in Figure 24.5. It's as if there were some initial resistance to the catalysis itself, a resistance that the substrate breaks down once its initial concentration is made high enough. Then the rate takes off in the usual way with increasing initial values of $[S]$, as seen by the steeply rising part B of the curve in Figure 24.5. Eventually, the enzyme becomes saturated, and the curve levels off, just as it did in Figure 24.4.

The "lazy S" shape of the curve in Figure 24.5 is referred to as a *sigmoid shape* (after the Greek letter that corresponds to our letter S, sigma). Scientists have learned that whenever they study a new enzyme, and it gives a sigmoid plot of initial rates versus initial concentration, the enzyme is likely one that requires a preliminary activation by the substrate itself. We'll learn more about this in the next section, as well as in our study of the behavior of hemoglobin in the next chapter.

FIGURE 24.3
Induced fit theory. *(a)* A molecule of an enzyme, hexokinase, has a gap into which a molecule of its substrate, glucose, can fit. *(b)* The entry of the glucose molecule induces a change in the shape of the enzyme molecule, which now surrounds the substrate entirely. (Courtesy of T. A. Steitz, Department of Molecular Biophysics and Biochemistry, Yale University, New Haven, CT.)

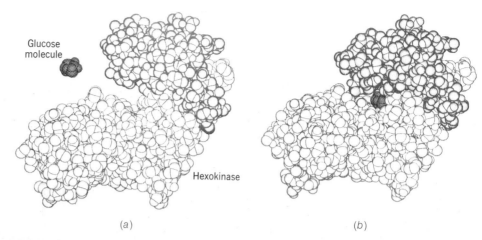

(a)   *(b)*

24.3   THE REGULATION OF ENZYMES

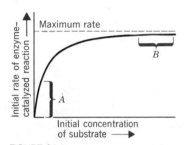

FIGURE 24.4
Initial rates of an enzyme-catalyzed reaction plotted versus the initial concentrations of substrates when the concentration of the enzyme is fixed in each experiment. The sections labeled *A* and *B* are discussed in the text.

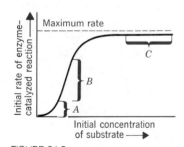

FIGURE 24.5
Initial rates of enzyme-catalyzed reactions plotted versus initial values of substrate concentrations, at fixed enzyme concentration, when an allosteric effect is observed. Sections labeled *A* and *B* of the curve — a sigmoid curve — are discussed in the text.

Allo-, other; *-steric,* space — the other space or the other site.

Enzymes are switched on and off by initiators, effectors, inhibitors, genes, poisons, hormones, and neurotransmitters.

A cell can't be doing everything at once. Some of its possible reactions have to be shut down while others occur. One way to keep a reaction switched off is to prevent its enzyme from forming, as we learned when we studied how genes can be controlled. Another (but harmful) way is to deny the cell the amino acids, vitamins, and minerals that it needs to make enzymes. Hormones and neurotransmitters are natural regulators of enzymes, and we'll learn about them in the next section. In this section we'll study several other means to control enzymes.

Allosteric Effects.   Enzymes that display sigmoid rate curves (Figure 24.5) turn out to have more than one active site, and such enzymes normally exist in inactive states. Thus their regulation requires mechanisms to activate them when needed and that then allow them to return to their inactive state when the need has passed. One such mechanism uses the substrate itself as the activator, but only after the substrate concentration has risen above some particular level. For simplicity, we'll suppose we are studying an enzyme that is made of just two polypeptide chains and that each chain bears one active site. We know that each site has its own shape, so we can pick any distinctive shape, as in Figure 24.6, to represent it.

At a sufficiently low substrate concentration, an average of only one of the two active sites will be scouted by substrate molecules. As Figure 24.6 shows, however, the shapes of the active site and the substrate molecules aren't complementary, so they can't fit together and no reaction occurs. We're in region *A* of the sigmoid rate curve (Figure 24.5), the region of the slow rate. Of course, because so little substrate is present there's little need for the reaction. However, given more substrate and more time, a sufficiently violent collision between a substrate molecule and one of the two active sites becomes very probable, and it occurs. This collisions forces the active site to "shape up," *and it also pops the other active site into shape.* This is how a sufficiently high concentration of substrate forces the complete activation of the enzyme. Once both sites acquire the right shapes, the reaction speeds up, the concentration of the substrate decreases, and as the last substrate molecules are used the active sites pop back into their inactive shapes. This phenomenon in which one active site is activated by an event that occurs at another place on the enzyme is called **allosteric activation.** It is a way to regulate how active an enzyme is according to how much its services are needed.

FIGURE 24.6
Allosteric activations. *(a)* Allosteric activation by substrate. *(b)* Allosteric activation by effector.

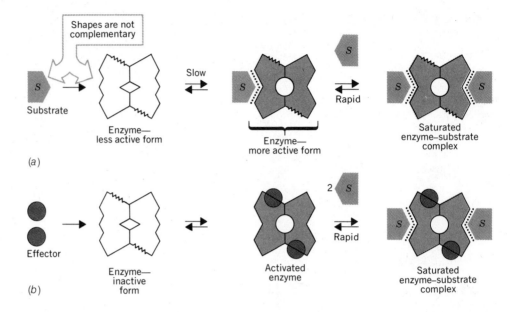

In the next chapter, we'll see how a similar allosteric effect is caused by oxygen when it interacts with hemoglobin, a nonenzyme protein, and how this enables hemoglobin to operate at 100% efficiency, or very nearly so.

Effectors.   Sometimes allosteric activation is caused by a nonsubstrate molecule called an **effector.** Part *(b)* of Figure 24.6 illustrates how *effector* molecules, represented by circles, might work by finding places on the enzyme away from the active sites. By binding at such places, they pop the active sites into their proper shapes, and now the enzyme can display the rate curve of Figure 24.4, not a sigmoid curve. The effector might, for example, be a molecule whose own metabolism cannot proceed without the *products* of the work of the enzyme that it activates.

Zymogen-Enzyme Conversion.   Some enzymes, particularly several involved in digestion, are made in inactive forms called **zymogens** or **proenzymes.** Their polypeptide strands have several more amino acid residues than the enzyme, and these extra units cover over the active site. Then, when the active enzyme is needed, a complex process is launched that clips off the extra units and, by exposing the active site, activates the enzyme. One of the functions of enterokinase, a compound released in the upper intestine when food moves in from the stomach, is to convert the zymogen, trypsinogen, into the enzyme, trypsin, which helps to digest proteins. The activation of trypsin is shown in Figure 24.7. When no food is present, there is no need for trypsin, but when food enters, then trypsin is activated.

Reversible Inhibition of Enzymes.   Some enzymes are kept in an inactive form by the work of an **inhibitor,** a substance whose molecules bind to the enzyme and prevent its active site from functioning. Two kinds of reversible enzyme inhibition are known. In one, called **allosteric inhibition,** molecules of the inhibitor bind to the enzyme somewhere other than the active site. This affects the shape of the enzyme at its active site or at its binding sites, and the enzyme-substrate complex cannot form. (See Figure 24.8*a*.)

The second kind of reversible enzyme inhibition is called **competitive inhibition.** The inhibitor is a nonsubstrate whose molecules have shapes similar enough to those of the true substrate that they compete with it for the active site. When the inhibitor molecules lock to the enzyme's active sites, but do not undergo a reaction and leave, then the enzyme has become useless to the true substrate. (See Figure 24.8*b*.)

The competitive inhibitor sometimes is the *product* of the reaction that the enzyme catalyzes, and sometimes it's the product of a reaction farther down a series of reactions. Such

In allosteric activation by a substrate, the system *starts up* when enough substrate arrives.

FIGURE 24.7
Conversion of a zymogen (proenzyme) into its active enzyme. Shown here is the activation of trypsin's zymogen. Removal of the hexapeptide segment lets the histidine (His) and serine (Ser) units come close together, and these two act together to achieve the catalysis.

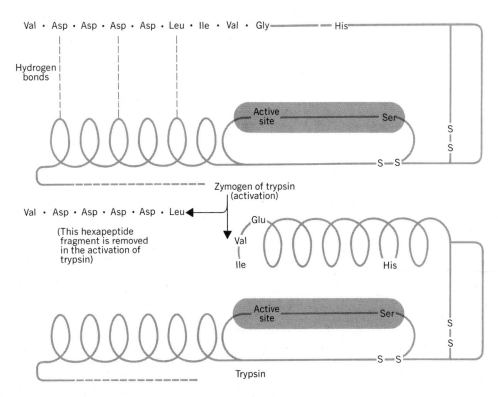

competitive inhibition is called **feedback inhibition,** because as the level of product increases, its molecules "feed back" with increasing success as inhibitors to an enzyme that helped to make them. For example, the amino acid isoleucine is made from another amino acid, threonine, by a series of steps, each with its own enzyme. We don't need any of the chemical details, except that molecules of isoleucine inhibit the enzyme for the first step, E_1:

$$\overset{+}{N}H_3CHCO_2^- \xrightarrow{E_1} \xrightarrow{E_2} \xrightarrow{E_3} \xrightarrow{E_4} \xrightarrow{E_5} \overset{+}{N}H_3CHCO_2^-$$

$$\underset{CH_3}{\overset{|}{CHOH}} \qquad\qquad \text{Inhibition of } E_1 \text{ by} \atop \text{molecules of isoleucine} \qquad \underset{CH_2CH_3}{\overset{|}{CHCH_3}}$$

Threonine   Feedback inhibition   Isoleucine

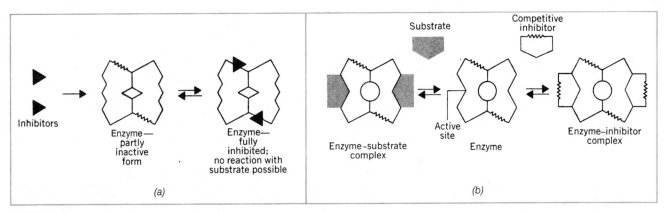

FIGURE 24.8
Enzyme inhibition. *(a)* Allosteric inhibition. *(b)* Competitive inhibition.

The beautiful feature of feedback inhibition is that the system for making a product shuts down automatically when enough is made. Then, as the cell consumes this product, it eventually uses even those product molecules that have been serving as inhibitors. The result is that when the product concentration has dropped very low, the enzyme needed to make more is released from its bondage. This phenomenon, which is very common in nature, is called **homeostasis** — the behavior of an organism to a stimulus that starts a series of events that restore the system to the original state. Some states of the body — body temperature is an example — are maintained so steadily by homeostatic mechanisms that diseases can be diagnosed by even small changes in the states. A familiar homeostatic mechanism in a home is the work of a furnace controlled by a thermostat. When the room becomes hot enough (the desired "product"), the thermostat trips and the furnace shuts off. In time, the temperature drops, the thermostat trips back, and the furnace restarts.

A competitive inhibitor doesn't have to be a product of the enzyme's own work. It can be something else the cell makes, or it could be a medication. What its molecules must do is resemble those of the normal substrate enough to bind, without reacting, to the active site of the enzyme.

An antimetabolite is called an antibiotic when it is the product of the growth of a fungus or a natural strain of bacteria.

Inhibition by Metabolites.   Antibiotics are members of a broad family of compounds called **antimetabolites,** substances that inhibit or prevent the normal metabolism of a disease-causing bacterial system. Some antimetabolites work by inhibiting an enzyme that the bacterium needs for its own growth. Both the sulfa drugs and penicillin work this way.

Inhibition by Poisons.   The most dangerous **poisons** are effective even at very low concentrations because they are powerful inhibitors of enzymes. The cyanide ion, for example, forms a strong complex with one of the metal ions that is a cofactor in an enzyme needed by cells to use oxygen. Enzymes that have —SH groups are denatured and deactivated by such heavy metal ions as Hg^{2+}, Pb^{2+}, Cu^{2+}, and Ag^+ — all poisons. Nerve gases and their weaker cousins, the organophosphate insecticides, inactivate enzymes of the nervous system.

24.4   ENZYMES IN MEDICINE

The specificity of the enzyme for its substrate and the slight differences in properties of isoenzymes provide several unusually sensitive methods of medical diagnosis.

Enzymes that normally work only inside cells are not found in the blood, except at extremely low concentrations. When cells are diseased or injured, however, their enzymes sometimes spill into the bloodstream. By detecting such enzymes in blood and measuring their levels, much can be learned about the disease or injury. Despite the enormous complexity of blood as a mixture and the low concentrations of the enzymes in it, the analyses are relatively easy to carry out. The substrate for the enzyme is used as a chemical "tweezers" to tweak just its own enzyme and nothing else. If none of the enzyme is present, nothing happens to the substrate reagent. Otherwise, the extent of the reaction with the substrate is a measure of the concentration of the enzyme. In this section, we'll learn of some examples of this medical technology.

This is the transaminase introduced on page 56.

Viral Hepatitis.   Heart, muscle, kidney, and liver tissue all contain the enzyme glutamate : pyruvate aminotransferase or GPT, which we introduced earlier. The liver, however, has about three times as much GPT as any other tissue, so the appearance of GPT in the blood generally indicates liver damage or a virus infection of the liver, such as viral hepatitis.

The level of another enzyme, glutamate : oxaloacetate aminotransferase or GOT, also increases in viral hepatitis, but the GPT level goes much higher than the GOT level. The ratio of GPT to GOT in the serum of someone with viral hepatitis is typically 1.6, compared with a level of 0.7 to 0.8 in healthy individuals. (Notice that we speak here of *ratios,* not absolute amounts, which are very low in healthy individuals.)

FIGURE 24.9
The concentrations of three enzymes in blood serum increase after a myocardial infarction. CK is creatine kinase; GOT is glutamate : oxaloacetate aminotransferase; and LD is lactate dehydrogenase.

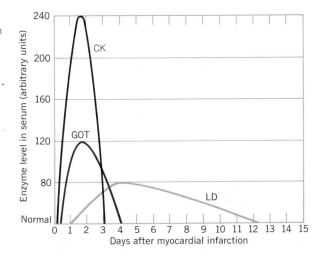

The popular term for this set of events is *heart attack*.

Myocardial Infarction. A myocardial infarction is the withering of a portion of the heart muscle following some blockage of the blood vessels that supply this muscle with oxygen and nutrients. The blockage can be caused by deposits, by hardening, or by a blood clot. If the patient survives, the withered muscle becomes scar tissue, and the outlook for a reasonably active life is generally good.

When a patient appears with the symptoms traditionally associated with a heart attack, much is at stake in making a correct diagnosis and ruling a myocardial infarction either in or out. There are decisions that have to be based on the likelihood of living or dying and on the need to be in an intensive coronary care unit. The diagnosis can now be made with an exceptionally high reliability by the analysis of the serum for several enzymes and isoenzymes.

When a myocardial infarction occurs, the levels of three enzymes that are normally confined to heart muscle cells begin to rise in blood serum. (See Figure 24.9.) These enzymes are CK (page 561), GOT (described just above), and LD, which stands for lactic dehydrogenase. LD catalyzes the oxidation of lactate to pyruvate. A clinical report from an actual patient who had a myocardial infarction is shown in Table 24.1. You can see how sharply the levels of these three enzymes rose between day I and day II. The line of data labeled "CK(*MB*)" and the data in the columns headed by "LD ISOENZYME" provided the clinching evidence for the infarction. As we learned on page 561, the CK enzyme occurs as three isoenzymes, one from skeletal muscle, CK(*MM*), one from the brain, CK(*BB*), and one from the heart, CK(*MB*). The technique that uses a chemical substrate to find the serum CK level can't tell these isoenzymes apart, because all three catalyze the reaction with the substrate. Therefore to be sure that the rise in serum CK level is caused by injury to the *heart* tissue, the clinical chemist has to analyze specifically for the CK(*MB*) isoenzyme. This is done with a technique called *electrophoresis*, which is described in Special Topic 24.2. Electrophoresis separates the individual isoenzymes, and then each is analyzed separately. As you can see in the chart, the patient's serum CK(*MB*) level did rise.

The older (and still often used) symbol for LD is LDH.

FIGURE 24.10
The lactate dehydrogenase isoenzymes. *(a)* The normal pattern of the relative concentrations of the five isoenzymes. *(b)* The pattern after a myocardial infarction. Notice the reversal in relative concentration between LD$_1$ and LD$_2$. This is the LD-flip.

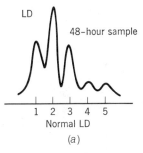

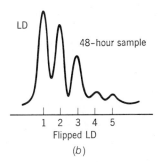

TABLE 24.1
Clinical Report: Myocardial Infarction

DAY I		DAY II		DAY III	
DATE	*5-25*	DATE	*5-26*	DATE	*5-27*
CK	*51*	CK	*552, Moderate*	CK	*399 Weak*
CK (MB)	*Negative*	CK (MB)	*Positive*	CK (MB)	*Positive*
GOT	*19*	GOT	*91*	GOT	*117*
LD	*88*	LD	*151*	LD	*247*

LD ISOENZYME		LD ISOENZYME		LD ISOENZYME	
% of Total LD activity		% of Total LD activity		% of Total LD activity	
LD_1	*28.3* %	LD_1	*33.8* %	LD_1	*37.9* %
LD_2	*32.9* %	LD_2	*32.3* %	LD_2	*32.8* %
LD_3	*19.7* %	LD_3	*16.3* %	LD_3	*15.7* %
LD_4	*11.6* %	LD_4	*9.1* %	LD_4	*7.6* %
LD_5	*7.5* %	LD_5	*8.5* %	LD_5	*6.0* %
	CB		*CB*		*B.L.*

Normal Range			LD Isoenzymes Normals	
CK	Male	5–75 mU/ml	LD_1	14–29%
	Female	5–55 mU/ml	LD_2	29–40%
			LD_3	18–28%
GOT		5–20 mU/ml	LD_4	7–17%
			LD_5	3–16%
LD		30–110 mU/ml		

INTERPRETATION: The presence of a CK(MB) and/or an LD_1 greater than LD_2 strongly suggest myocardial infarction.

One International Unit (U) of activity is the reaction under standard conditions of one micromole per minute of a particular substrate used in the test. (Data courtesy of Dr. Gary Hemphill, Clinical Laboratories, Metropolitan Medical Center, Minneapolis, MN.)

The numbered subscripts of the LD isoenzymes are simply the order in which their molecules separate during electrophoresis.

As additional confirmation of an infarction, the serum LD fraction is further separated by electrophoresis into the five LD isoenzymes, and each is individually analyzed. The *relative* concentrations of the five serum LD isoenzymes differ distinctively in a healthy person (Figure 24.10*a*) and one who has suffered an infarction (Figure 24.10*b*). Of particular importance are the relative concentrations of the LD_1 and the LD_2 isoenzymes. Normally the LD_1 level is less than that of the LD_2, but following a myocardial infarction what is called an "LD_1–LD_2 flip" occurs. The relative concentrations of LD_1 and LD_2 become reversed and the level of LD_1 rises higher than that of LD_2. When both the CK(*MB*) band and the LD_1–LD_2 flip occur, the diagnosis of a mycardial infarction is essentially 100% certain.

Molecules of proteins carry electrical charges, and both the numbers of charges and their signs depend on the pH of the medium. A procedure called *electrophoresis* that is used to separate a mixture of proteins is based on such charges and their pH-dependence.

First, a surface is prepared from a porous but mechanically sturdy solid, such as filter paper, or a gel made from a polymer such as polyacrylamide. This surface is wetted with a solution that contains ions and whose pH has been adjusted to some predetermined value. Electrical poles or plates are attached at opposite ends of this wetted surface, as seen in the accompanying figure, and electricity can flow because ions are present to carry it.

Next, a solution of a mixture of proteins is deposited as a very narrow band across the middle of the strip. The molecules of proteins — they could be isoenzymes — begin to travel because they are also electrically charged. However, when they have different *kinds* of charge (positive or negative), they are bound to travel in opposite directions. And when they have different *quantities* of charge and possibly even different masses, they travel at different rates. Thus slowly, the different proteins move out from the band where they were deposited, and they become increasingly separated, each into its own narrow zone or band. These bands can't be seen directly, but there are many techniques, including color tests, for locating them and measuring the relative amounts of the individual proteins or isoenzymes in them.

Despite being *catalytically* the same, isoenzymes can be separated by electrophoresis. For example, the *MM*, *BB*, and *MB* isoenzymes of creatine kinase can be separated, and an increase in the CK(*MB*) band in blood serum helps to diagnose a myocardial infarction. The five isoenzymes of lactate dehydrogenase, LD, can also be separated and their relative amounts determined using electrophoresis.

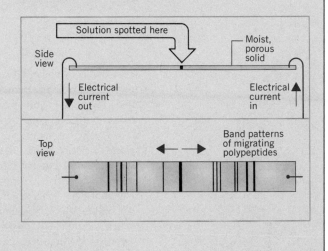

The Determination of Glucose. The detection of glucose in urine is important to people with diabetes, because in poorly managed diabetes this sugar is present in urine. One commercially available test, the Clinistix test, uses porous paper test strips impregnated with two enzymes and an aromatic compound that can be oxidized to a dye by hydrogen peroxide. If glucose is in the drop of urine used to moisten a Clinistix test strip, one of the enzymes, glucose oxidase, catalyzes a reaction that uses glucose to make hydrogen peroxide. The other enzyme, a peroxidase, then catalyzes the oxidation of the aromatic compound by hydrogen peroxide to give the dye. If the color appears in a few seconds, the urine contains glucose.

Diastix is another enzyme-based test strip for urinary glucose.

Measuring the glucose in urine is actually a rather crude way of assessing the glucose level in the blood. In managing diabetes, it is much better to measure the glucose level in blood serum directly. Figure 24.11 illustrates one method for doing this. A transparent film supports a thin layer that holds the reactants, which are the same in most respects as those in Clinistix paper. Above the reagent layer is a thicker layer made white with titanium dioxide. When a drop of serum is added to this layer, it soaks in until it reaches the reagent layer. Then the reactions that produce the dye occur. The intensity of the dye color, when viewed from the back side against the white background, is a measure of the serum glucose level.

TiO_2, titanium dioxide, is the whitening and hiding agent in paints.

Similar test slides are available to measure the serum levels of urea, triacylglycerols, bilirubin (a breakdown product of hemoglobin), and other compounds.

Other Applications of Enzymes. A large number of other applications of enzymes in medicine and technology have been developed. In some, the enzymes are physically immobilized onto the surfaces of ultra-tiny, inert plastic beads, which makes it easier to separate the products from the enzymes. Immobilized enzymes last longer, are less sensitive to temperature, and are less vulnerable to oxygen. Enzymes, for example, are used in filtering systems to remove bacteria and viruses from air and water.

FIGURE 24.11
The concentration of glucose in blood serum can be measured by this Ektachem clinical chemistry slide, seen here in cross section. (Courtesy of Eastman Kodak Company.)

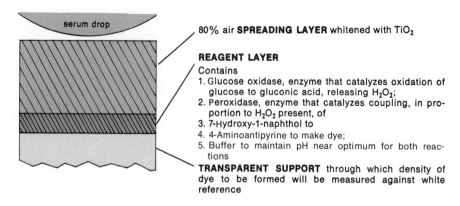

serum drop

80% air **SPREADING LAYER** whitened with TiO_2

REAGENT LAYER
Contains
1. Glucose oxidase, enzyme that catalyzes oxidation of glucose to gluconic acid, releasing H_2O_2;
2. Peroxidase, enzyme that catalyzes coupling, in proportion to H_2O_2 present, of
3. 7-Hydroxy-1-naphthol to
4. 4-Aminoantipyrine to make dye;
5. Buffer to maintain pH near optimum for both reactions

TRANSPARENT SUPPORT through which density of dye to be formed will be measured against white reference

The enzyme heparinase is immobilized onto plastic beads which are used to catalyze the breakdown of the heparin that is added to blood sent through a hemodialysis machine. The added heparin inhibits the clotting of the blood outside the body, but it has to be removed before the blood goes back into the body. The heparinase catalyzes this removal.

Analytical chemists routinely use a variety of procedures lumped under the general term, *electroanalysis*. There are special electrical conducting devices called electrodes that can be used to send a current through a solution. You already know about the electrodes of a pH meter, which are sensitive just to the concentration of hydrogen ions. There are special electrodes for other chemical species, too. Clinical chemists, for example, have at their disposal a variety of enzyme electrodes. These take advantage of the specificity of an enzyme, which is immobilized on the electrode's tip, and the sensitivity of electrical measurements. For example, the urea level of blood can be measured as a function of the concentration of the ammonium ions that are produced when urea is hydrolyzed. The tip of the urea electrode is lightly coated with a polymer that immobilizes urease, the enzyme that catalyzes the hydrolysis of urea. When this electrode is dipped into blood serum that contains urea, urea and water migrate into the polymer, the urea is promptly hydrolyzed (thanks to the urease), and the electrode helps to register the appearance of the newly formed ammonium ion. It is relatively easy to correlate the response of the electrode to the concentration of the urea.

The level of urea in the blood is called the BUN level (for blood urea nitrogen). A high BUN level indicates a kidney disorder.

Enzyme electrodes for the determination of glucose, uric acid, tyrosine, lactic acid, acetylcholine, cholesterol, and other substances have been developed.

Some enzymes are used to treat illness. For example, the enzymes urokinase and streptokinase can be used to assist in the removal of a blood clot in the lungs. These enzymes catalyze the breakdown of the fibrin in the clot, and the clot breaks up harmlessly.

About 150,000 people a year die from clots in their lungs.

Streptokinase and urokinase are also being used to reduce possible damage to heart muscle when a myocardial infarction either has just occurred or is about to take place. An occluded coronary artery can be opened up in the majority of cases if streptokinase infusion is used within 6 hours of the appearance of chest pains. When the treatment is begun within 45 to 120 minutes of the onset of symptoms, the developing infarction is halted in most of the cases. In hearts damaged enough to prevent a thorough perfusion of the enzyme into the affected areas, the treatment is relatively unsuccessful.

24.5 HORMONES AND NEUROTRANSMITTERS

Hormones carry chemical signals from endocrine glands to target cells, and neurotransmitters bring chemical signals from one nerve cell to the next.

Like any complex organism, the body is made up of many highly specialized parts. Therefore information of some sort must flow among the parts so that everything is well-coordinated. This flow is handled by chemical messengers and electrical signals.

The receptor molecules are proteins.

In the broadest terms, the messengers are molecules or ions, and they deliver messages by the act of binding to receptor molecules at cells. The cell at which a messenger acts is called the messenger's **target cell.** The messenger molecule or ion *induces* some kind of change at the cell when it binds to its receptor. The change itself is the actual "message." It might be the speeding up or slowing down of some reaction, or it might be a change in the permeability of a cell membrane to the flow of an ion or a molecule. The change induced by the messenger might even be the activation of a gene. In nerve cells, called neurons, the change could be the excitation of a neighboring neuron to cause it to send on a signal, or the change might be the opposite—the inhibition of a signal.

Greek *hormon,* arousing.

Broadly speaking, there are two kinds of chemical messengers, *primary* and *secondary.* The former bring messages and the latter pass them on. The chief primary chemical messengers are hormones and neurotransmitters. They differ not so much in *how* they work as in *at what distance* and *where* they work. (See Figure 24.12.) **Hormones** are chemicals made in specialized organs, called the **endocrine glands** (Figure 24.13), and they travel in the blood to target cells that might be as much as 15 to 20 cm away. A hormone is released into circulation when its endocrine gland receives a unique signal. The signal might be something conveyed by one of our senses, such as an odor, or it might be a stress, or a variation in the concentration of a particular substance in the blood or in another fluid. Insulin, for example, is released when the concentration of glucose in blood increases. **Neurotransmitters** are chemicals that are made in neurons and sent over very short distances to neighboring neurons.

In many examples, the mechanisms of action of hormones and neurotransmitters are very similar. To illustrate, we will study one such mechanism, one in which a cyclic nucleotide conveys the message across a cell membrane.

Cyclic refers to the *extra* ring of the phosphate diester system.

Cyclic Nucleotides.   Cyclic nucleotides, particularly 3′,5′-cyclic AMP, are important secondary chemical messengers. How cyclic AMP works is sketched in Figure 24.14. At the top of the figure we see a hormone—it could just as well be a neurotransmitter—that can combine with a receptor molecule at the surface of a cell. A "lock-and-key" mechanism is at work so that the hormone bypasses all cells that do not have a matching receptor. The complex that forms between hormone and receptor activates an enzyme, adenylate cyclase, that is an integral part of the cell membrane. This activation constitutes the specific chemical change that the hormone causes. But the process thus launched continues. The enzyme promptly catalyzes the conversion of ATP into cyclic AMP and diphosphate ion, PP_i.

FIGURE 24.12
Chemical communication in the human body. *(a)* A hormone travels from an endocrine gland, where it is made, through the bloodstream to its target cell. *(b)* A neurotransmitter travels from one nerve cell to the next nerve cell across the synapse.

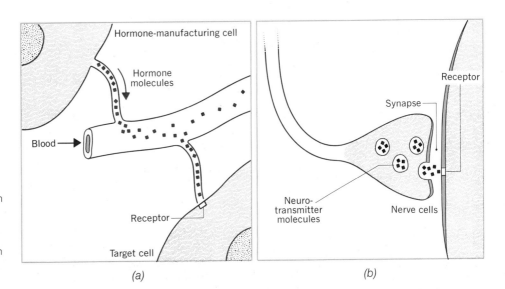

Hormone-manufacturing cell

Hormone molecules

Blood

Receptor

Target cell

(a)

Receptor

Synapse

Neuro-transmitter molecules

Nerve cells

(b)

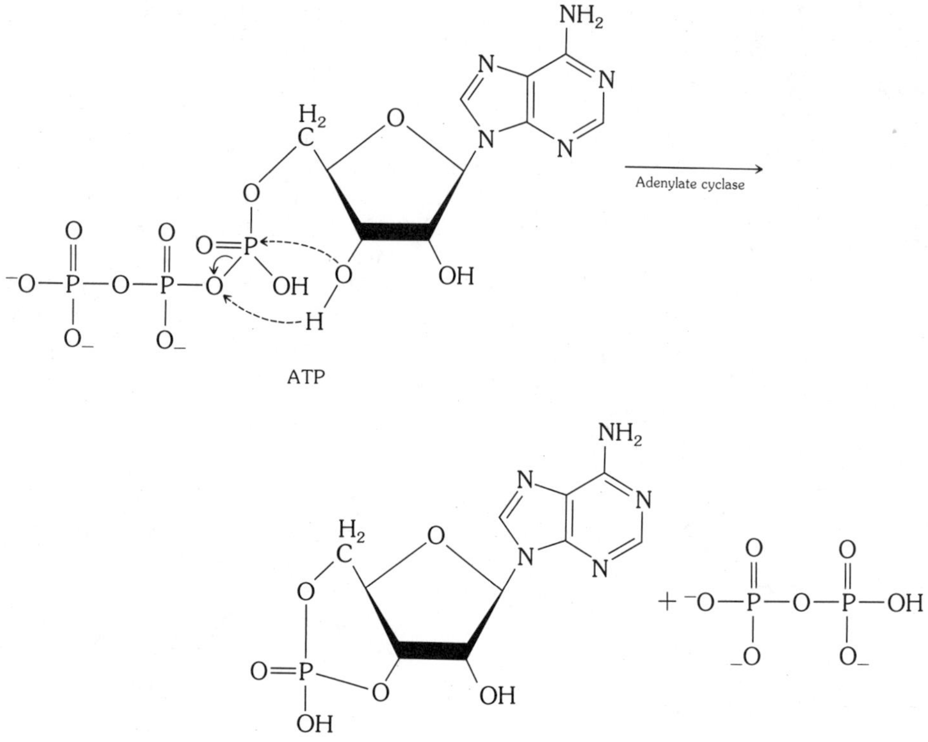

ATP

Cyclic AMP   Diphosphate ion, PP$_i$

The newly formed cyclic AMP now activates an enzyme, which, in turn, catalyzes a reaction. This last event is what the original message was all about. Finally, an enzyme called phosphodiesterase catalyzes the hydrolysis of cyclic AMP to AMP, and this shuts off the cycle. Energy-producing reactions in the cell will now remake ATP from the AMP.

Let's summarize the steps.

1. A signal releases the hormone or neurotransmitter.

2. It travels to its target cell — next door for a neurotransmitter but some farther distance away for a hormone.

3. The primary messenger molecule finds its target cell by a lock-and-key mechanism, and binds to a receptor. The resulting complex activates adenylate cyclase.

4. Adenylate cyclase catalyzes the conversion of ATP to cyclic AMP.

5. Cyclic AMP activates an enzyme inside the cell.

6. The enzyme catalyzes a reaction, one that corresponds to the primary message.

7. Cyclic AMP is hydrolyzed to AMP, which is reconverted to ATP, and the system returns to the pre-excited state.

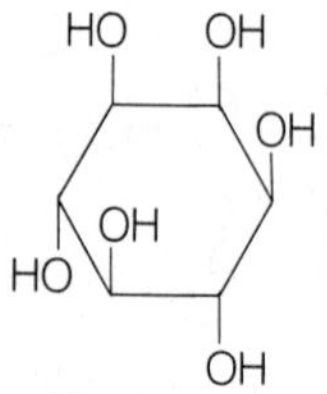

In the early 1980s, teams of scientists confirmed that a cyclic nucleotide such as cyclic AMP is not the only kind of secondary messenger. Another system involves the release of a triphosphate ester of inositol from a molecule of a glycolipid in the membrane's lipid bilayer. The pattern is similar. The hormone binds to a receptor. The complex activates an enzyme in the lipid bilayer — this time, one that catalyzes the release of inositol triphosphate from the glycolipid. This triphosphate and what it leaves behind — a diacylglycerol — work out the spreading of the chemical message inside the cell. Because our objective is to illustrate general principles rather than study all types of examples, we will say no more about the inositol triphosphate system. Because this system may be the site of action of lithium ion, which, as

Inositol

FIGURE 24.13
The endocrine glands.

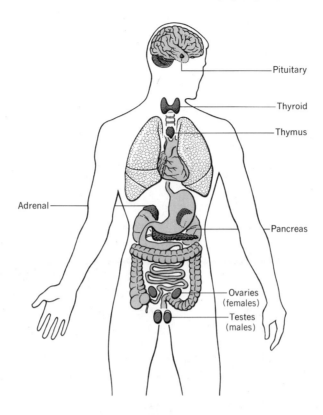

Pituitary

Thyroid

Thymus

Adrenal

Pancreas

Ovaries
(females)

Testes
(males)

lithium carbonate, is used in the treatment of manias; and because the system may also be involved as a target for the action of errant genes that cause cancer, we will certainly hear much more about it in the future.

Hormones.   The principal hormones of the human body are listed in Table 24.2. They vary widely in structure, from the relatively simple epinephrine (page 420), to the polycyclic

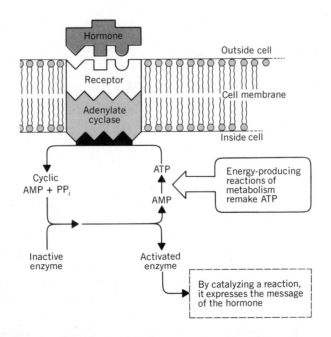

Hormone

Outside cell

Receptor

Cell membrane

Adenylate
cyclase

Inside cell

Cyclic
AMP + PP$_i$

ATP

Energy-producing
reactions of
metabolism
remake ATP

AMP

Inactive
enzyme

Activated
enzyme

By catalyzing a reaction,
it expresses the message
of the hormone

FIGURE 24.14
The activation of the enzyme, adenylate cyclase, by a hormone (or a neurotransmitter). The hormone-receptor complex activates this enzyme, which then catalyzes the formation of cyclic AMP. This, in turn, activates an enzyme inside the cell.

TABLE 24.2
The Principal Human Endocrine Glands and Their Hormones

Gland or Tissue	Hormone	Major Function of Hormone
Thyroid	Thyroxine	Stimulates rate of oxidative metabolism and regulates general growth and development
	Thyrocalcitonin	Lowers level of Ca^{2+} in blood
Parathyroid	Parathormone	Regulates the levels of calcium and phosphate ions in blood
Pancreas, β-cells	Insulin	Decreases blood glucose level
α-cells	Glucagon	Elevates blood glucose level
Adrenal medulla	Epinephrine	Elevates blood glucose level and heartbeat
Adrenal cortex	Cortisone and related hormones	Control carbohydrate, protein, mineral, salt, and water metabolism
Anterior pituitary	Thyrotropic hormone	Stimulates thyroid gland functions
	Adenocorticotropic hormone	Stimulates development and secretion of adrenal cortex
	Growth hormone	Stimulates body weight and rate of growth of skeleton
	Gonadotropic hormones	Stimulate gonads
	Prolactin	Stimulates lactation
Posterior pituitary	Oxytocin	Causes contraction of some smooth muscles
	Vasopressin	Inhibits secretion of water from the body by way of the urine
Ovary (follicle)	Estrogens	Influence development of sex organs and female characteristics
Ovary (corpus luteum)	Progesterone	Influences menstrual cycle; prepares uterus for pregnancy; maintains pregnancy
Uterus (placenta)	Estrogens and progesterone	Function in maintenance of pregnancy
Testes	Testosterone	Responsible for development and maintenance of sex organs and secondary male characteristics
Digestive system	Several gastrointestinal hormones	Integration of digestive processes

steroid hormones (page 481), to polypeptides such as insulin (page 499). We leave to other books and to other courses the study of the many ways in which individual hormones affect human well-being. However, we will examine a few hormones in greater detail in later chapters, because they are so influential in particular molecular affairs we have to study. For the rest, we deal only with some broad principles.

We've already mentioned the three principal ways by which hormones act — by activating a gene, by activating an enzyme, or by changing the permeability of a cell membrane (or the membrane of a smaller, subcellular unit) to some compound or ion. The sex hormones activate genes. Epinephrine, some thyroid hormones, and a few others activate enzymes. Insulin and human growth hormone affect permeabilities of cell membranes.

Neurotransmitters.   A partial list of neurotransmitters is given in Table 24.3. Some are nothing more than simple amino acids. Others are β-phenylethylamines or catecholamines (Special Topic 17.1, page 420), and many are polypeptides.

Each nerve cell has a fiber-like part called an *axon* that reaches to the face of the next neuron or to one of its filament-like extensions called *dendrites* (Figure 24.15). A nerve impulse consists of a traveling wave of electrical charge that sweeps down the axon as small ions migrate at different rates between the inside and the outside of the neuron. The problem is how to get this impulse launched into the next neuron so that it can continue along the length of the nerve fiber. This is solved by a *chemical* communication from one neuron to the next.

Between the terminal of an axon and the end of the next neuron, there is a very narrow, fluid-filled gap called the *synapse*. Neurotransmitters move across the synapse when the electrical wave causes them to be released from their tiny storage sacs or *vesicles* that are near the ends of axons. When neurotransmitter molecules lock to their receptors on the other side

The traveling wave of electrical charge moves rapidly, but still not as rapidly as electricity moves in electrical wires.

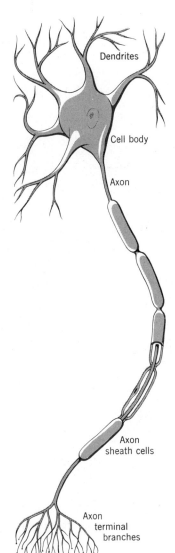

Dendrites

Cell body

Axon

Axon sheath cells

Axon terminal branches

FIGURE 24.15
One kind of neuron or nerve cell.

TABLE 24.3
Neurotransmitters

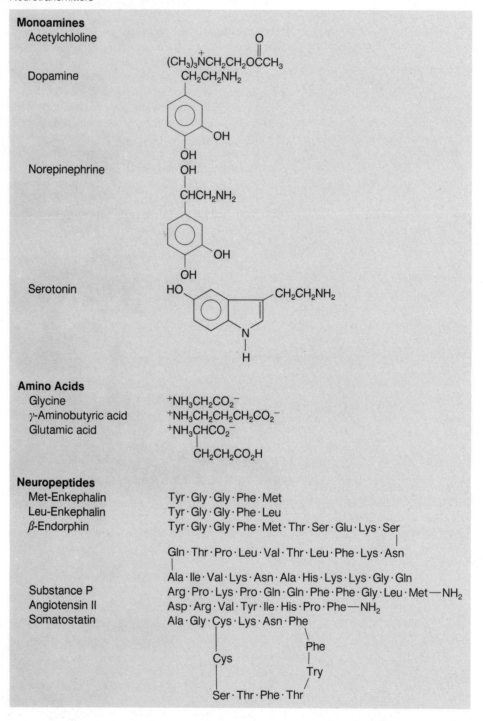

Monoamines
Acetylchloline
Dopamine
Norepinephrine
Serotonin

Amino Acids
Glycine $^+NH_3CH_2CO_2^-$
γ-Aminobutyric acid $^+NH_3CH_2CH_2CH_2CO_2^-$
Glutamic acid $^+NH_3CHCO_2^-$
           $CH_2CH_2CO_2H$

Neuropeptides
Met-Enkephalin   Tyr·Gly·Gly·Phe·Met
Leu-Enkephalin   Tyr·Gly·Gly·Phe·Leu
β-Endorphin   Tyr·Gly·Gly·Phe·Met·Thr·Ser·Glu·Lys·Ser
           Gln·Thr·Pro·Leu·Val·Thr·Leu·Phe·Lys·Asn
           Ala·Ile·Val·Lys·Asn·Ala·His·Lys·Lys·Gly·Gln
Substance P   Arg·Pro·Lys·Pro·Gln·Gln·Phe·Phe·Gly·Leu·Met—NH₂
Angiotensin II   Asp·Arg·Val·Tyr·Ile·His·Pro·Phe—NH₂
Somatostatin   Ala·Gly·Cys·Lys·Asn·Phe
           Cys   Phe
                   Try
         Ser·Thr·Phe·Thr

of the synapse, a nucleotide cyclase enzyme is activated, such as adenylate cyclase. (See Figure 24.16.) Now the formation of cyclic AMP is catalyzed, and newly formed cyclic AMP initiates whatever change is programmed by the chemicals in the target neuron. An enzyme then deactivates adenylate cyclase by catalyzing the release of the neurotransmitter molecule. If it were a hormone, it would be swept away in the bloodstream, but it's not. It's still a

FIGURE 24.16
Neurotransmission. Neurotransmitter molecules, released from vesicles of the pesynaptic neuron, travel across the synapse. At the postsynaptic neuron, they find their receptors, and the cyclic AMP system similar to that shown in Figure 24.14 becomes activated.

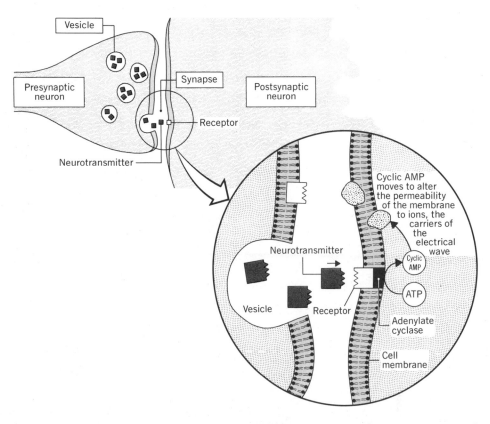

neurotransmitter in the synapse, so unless the system wants it to act again, it must be removed or deactivated.

One method used to remove the neurotransmitter is to break it up by a chemical reaction. For example, acetylcholine, a neurotransmitter in the autonomic nervous system or ANS, is catalytically hydrolyzed to choline and acetic acid. The enzyme is choline acetyltransferase.

The ANS nerves handle the signals that run the organs that have to work autonomously (without conscious effort), such as the heart and the lungs.

$$(CH_3)_3\overset{+}{N}CH_2CH_2O\overset{\displaystyle O}{\overset{\|}{C}}CH_3 + H_2O \underset{\substack{\text{choline}\\ \text{acetyltransferase}}}{\rightleftharpoons} (CH_3)_3\overset{+}{N}CH_2CH_2OH + HO\overset{\displaystyle O}{\overset{\|}{C}}CH_3$$

Acetylcholine   Choline   Acetic acid

The ANS nerves that use acetylcholine are called the *cholinergic nerves.*

Within two milliseconds (2×10^{-3} s) of the appearance of acetylcholine molecules in the synapse, they are all broken down—but not before they have served their purpose. The synapse is now cleared for a fresh release of acetylcholine from the presynaptic neuron if the signal for its release continues. If the signal does not come, then the action is shut down.

Before showing other means that are used to handle neurotransmitters that are no longer needed, we can illustrate how several poisons work by interfering with a neurotransmitter. The botulinus toxin, the extremely powerful toxic agent made by the food-poisoning botulinus bacillus, works by preventing the *synthesis* of acetylcholine. Without it, the cholinergic nerves of the ANS can't work. The nerve gases act by deactivating choline acetyltransferase, so now the signal transmitted by acetylcholine can't be turned off. It continues unabated until the heart fails, usually in a minute or two. An antidote for nerve gas poisoning, atropine, works by blocking the receptor protein for acetylcholine, so despite the continuous presence of this neurotransmitter, it isn't able to complete the signal-sending work. This tones the system down, and other processes slowly restore the system to normal. (Given the extreme speed with which nerve gases work, atropine must be used promptly.) Some organophosphate insecticides are mild nerve gases, and work in the same way. Other blockers of the receptor protein for acetylcholine are some local anesthetics such as nupercaine, procaine, and tetracaine. Drugs that block the action of a neurotransmitter are called **antagonists** to the neurotransmitter. The neurotransmitter itself is sometimes referred to as an **agonist.**

The nerves that use norepinephrine are called the *adrenergic nerves* (after an earlier name for norepinephrine, noradrenaline).

Norepinephrine, another neurotransmitter, is deactivated by being reabsorbed by the neuron that released it, where it is then degraded. (Some is also deactivated right within the synapse.) One place where norepinephrine works is in the brainstem where mood regulation is centered. Its degradation is catalyzed by enzymes called the **monoamine oxidases** or MAO. If these enzymes are themselves made inactive, then an excess of norepinephrine builds up, which can spill back into the synapse and send signals on. Such a deactivation of the MAOs is sometimes desired, particularly if the level of norepinephrine is low for any reason, and signal-sending activity dies down too much. Thus some of the antidepressant drugs, such as iproniazid, work by inhibiting the monoamine oxidases.

$$CHCH_2CH_2N(CH_3)_2$$

Amitriptyline (Elavil)

$$CH_2CH_2CH_2N(CH_3)_2$$

Imipramine (Tofranil)

$$CH_2CH_2CH_2N(CH_3)_2$$

Chlorpromazine

Haloperidol

$$^+NH_3CHCO_2^-$$
$$CH_2$$

L-DOPA

Other antidepressants, such as amitriptyline (e.g., Elavil) and imipramine (e.g., Tofranil), inhibit the reabsorption of norepinephrine by the presynaptic neuron. Without this reabsorption into the degradative hands of the monoamine oxidases, the level of norepinephrine and its signal-sending work stays high.

Norepinephrine is both a hormone and a neurotransmitter. The adrenal medulla secretes it into the bloodstream in emergencies when it must be made available to all of the nerve tissues that use it.

Dopamine, like norepinephrine, is also a monoamine neurotransmitter. It occurs in neurons of the midbrain that are involved with feelings of pleasure and arousal as well as with the control of certain movements. In schizophrenia, the neurons that use dopamine are overstimulated, because either the releasing mechanism or the receptor mechanism is overactive. Drugs commonly used to treat schizophrenia such as chlorpromazine (e.g. Thorazine) and haloperidol (Haldol) bind to dopamine receptors and thus inhibit its signal-sending work. On the other hand, certain stimulants such as the amphetamines (cf. Special Topic 17.1, page 420) work by triggering the release of dopamine into the arousal and pleasure centers of the brain. When amphetamines are abused, they cause the same kind of overstimulation associated with schizophrenia, with such resulting symptoms as delusions of persecution, hallucinations, and other disturbances of the thought processes.

In Parkinson's disease, the dopamine-using neurons in the brain have degenerated, so now an extra supply of dopamine itself is needed to help to compensate. This is why a compound called L-DOPA [*levorotatory dihydroxyphenylalanine*] is used. The neurons that still work can use it to make extra dopamine.

The normal function of some neurotransmitters is to *inhibit* signals instead of to initiate them. Gamma-aminobutyric acid (GABA) is an example, and as many as a third of the synapses in the brain have GABA available. The inhibiting work of GABA can be made even greater by mild tranquilizers such as diazepam (e.g., Valium) and chlordiazepoxide hydrochloride (Librium), as well as by ethanol. The augmented inhibition of signals reduces anxiety, affects judgment, and induces sleep. Of course, you've probably heard of the widespread abuse of Valium and Librium, to say nothing of alcohol.

Diazepam (Valium)

Chlordiazepoxide
hydrochloride (Librium)

Greek, *chorea,* dance.

In a hereditary neurological disorder called Huntington's chorea, GABA is deficient. The victims, lacking the moderating influences of GABA, suffer from speech disturbances, irregular movements, and a steady mental deterioration. Unhappily, GABA can't be administered in this disease, because it can't move out of circulation and into the regions of the brain where it works.

*En-*or *end-,* within; *kephale,* brain; *-orph-,* from morphine.

As we said, some neurotransmitters are relatively small polypeptides. One type includes the *enkephalins;* another consists of the *endorphins.* As indicated in Table 24.3, they are powerful pain inhibitors. One of them, dynorphin, is the most potent painkiller yet discovered, being 200 times stronger than morphine, an opium alkaloid that is widely used to relieve severe pain. Sites in the brain that strongly bind molecules of morphine also bind those of the enkephalins, so these natural painkillers are now often referred to as the body's natural opiates.

The enkaphalins appear to work by inhibiting the release of a pain-signaling, polypeptide neurotransmitter called *substance P,* as shown in Figure 24.17. According to one theory, when a pain-transmitting neuron is activated, it releases substance P into the synapse. However, butting against such neurons are other neurons that can release enkephalin. And these, when released, inhibit the work of substance P. In this way, the intensity of the pain signal can be toned down. This might explain the delay of pain that sometimes occurs during an emergency when the brain and the body must continue to function to escape the emergency. Substance P, according to recent findings, might be involved in the link between the nervous system and the body's immune system. It is known that some forms of arthritis

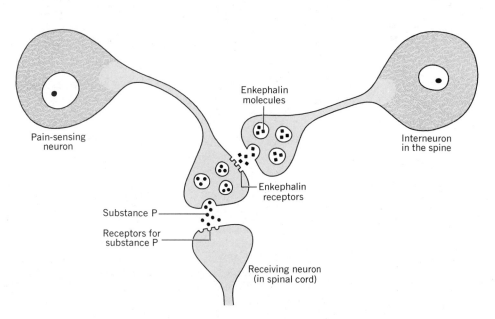

FIGURE 24.17

The inhibition of the release of Substance *P* by enkephalin to reduce the intensity of the pain signal to the brain.

flare-up under stress, and stress deeply involves the nervous system. But arthritis is generally regarded as chiefly a disease of the immune system. In tests on rats with arthritis, flare-ups of the arthritis could be induced by injections of substance P. What interests scientists about this is the possibility of controlling the severity of arthritis by somehow diminishing substance P levels in the affected joints.

One of the interesting developments in connection with the endorphins is that acupuncture — a procedure developed in China to alleviate pain — might work by stimulating the production and release of endorphins.

Many neurotransmitters can be received by more than one kind of receptor. For example, at least three types of receptors for the opiates have been identified thus far. Such receptor multiplicity may explain how some neurotransmitters have multiple effects. Thus opiates not only reduce pain, they affect emotions, and they induce sleep; and each of the opiate receptors handles a different one of these functions.

What we have done in this section is look at some *molecular* connections between conditions of the nervous system and particular chemical substances. This whole field is one of the most rapidly moving areas of scientific investigation today, and during the next several years we may expect to see a number of dramatic advances both in our understanding of what is happening and in the strategies of treating mental diseases.

SUMMARY

Enzymes Enzymes are the catalysts in cells. Some consist wholly of one or more polypeptides and others include, besides polypeptides, a cofactor — an organic coenzyme, a metal ion, or both. Some coenzymes are phosphate esters of B vitamins, and in these examples the vitamin unit usually furnishes the enzyme's active site. Because they are mostly polypeptide in nature, enzymes are vulnerable to all of the conditions that denature proteins. The name of an enzyme, which almost always ends in *-ase*, usually discloses either the identity of its substrate or the kind of reaction it catalyzes.

Some enzymes occur as small families called isoenzymes in which the polypeptide components vary slightly from tissue to tissue in the body.

An enzyme is very specific both in the kind of reaction it catalyzes and in its substrate. Enzymes make possible reaction rates that are substantially higher than the rates of uncatalyzed reactions. An enzyme has to be treated as a reactant in the initial phase of the reaction, as evidenced by the pattern for the changes in rate caused by increases in the relative initial concentrations of the substrate. At a sufficiently high initial substrate concentration, the active sites on all of the enzyme molecules become saturated by substrate molecules, so at still higher initial substrate concentrations, there is little further increase in the initial rate. Some enzymes seem to respond sluggishly to small increases in substrate concentration when the latter is very low. These display a sigmoid rate curve, and they require activation by substrate molecules before they produce their dramatic rate enhancements.

Theories of enzyme action When an enzyme-substrate complex forms, the active site is brought up to the part of the substrate that is to react. Binding sites on the enzyme guide the substrate molecule in, and sometimes a change in the conformation of the enzyme is induced by the substrate. The recognition of the enzyme by the substrate occurs as a lock-and-key model that involves complementary shapes and electrical charges.

Regulation of enzymes Enzymes that display a sigmoid rate curve can be activated allosterically either by their own substrates or by effectors. Some enzymes are activated by genes. Some enzymes that are parts of the membranes of cells (or small bodies within cells) are activated by the interaction between a hormone or a neurotransmitter and its receptor protein. Other enzymes, such as certain digestive enzymes, exist as zymogens (proenzymes) and are activated when some agent acts to remove a small part that blocks the active site.

Enzymes can be inhibited by competitive feedback that involves a product of the enzyme's action, or by a similar action by something that isn't a product. Some inhibitors act allosterically. Some of the most dangerous poisons bind to active sites and irreversibly block the work of an enzyme, or they carry enzymes out of solution by a denaturant action. Many antibiotics and other antimetabolites work by inhibiting enzymes in pathogenic bacteria.

Medical uses of enzymes The serum levels of many enzymes rise when the tissues or organs that hold these enzymes are injured or diseased. By monitoring these serum levels, and by looking for certain isoenzymes, many diseases can be diagnosed — for example, viral hepatitis and myocardial infarctions. Enzymes are also used in analytical systems that measure concentrations of substrates, such as in tests for glucose. In some analytical systems, enzymes are immobilized on electrodes where they catalyze a reaction that produces a product that the electrode then senses and measures.

Hormones Endocrine glands secrete hormones, and these primary chemical messengers travel in the blood to their target cells where they activate a gene, or an enzyme, or affect the permeability of a cell membrane. They recognize their own target cells by a lock-and-key type of recognition that involves their receptor proteins.

Neurotransmitters In response to an electrical signal, vesicles in an axon release a neurotransmitter that moves across the synapse. Its molecules bind to a receptor protein on the next neuron, and then the pattern is much like that of hormones. One common response to the formation of a neurotransmitter-receptor protein complex is the activation of adenylate cyclase, which triggers the formation of cyclic AMP. In turn, cyclic AMP sets off other events, such as the activation of an enzyme that catalyzes a reaction that is ultimately what the "signal" of the neurotransmitter was all about.

Neurotransmitters include amino acids, monoamines, and polypeptides. Some neurotransmitters *activate* some response in the next neuron, whereas others *deactivate* some activity. A number of medications work by interfering with neurotransmitters.

KEY TERMS

The following terms were emphasized in this chapter and they constitute a minimum vocabulary for anyone who is to continue into additional studies of enzymes, hormones, neurotransmitters, and drugs.

active site	cofactor	induced-fit theory	oxidoreductase
agonist	competitive inhibition	inhibitor	peptidase
allosteric activation	effector	isoenzymes	poison
allosteric inhibition	endocrine gland	kinase	proenzyme
antagonist	enzyme-substrate complex	lipase	reductase
antimetabolite	esterase	lock-and-key theory	target cell
apoenzyme	feedback inhibition	monoamine oxidase	transferase
binding site	homeostasis	neurotransmitter	zymogen
coenzyme	hormone	oxidase	

SELECTED REFERENCES

1 A. L. Lehninger. *Principles of Biochemistry.* Chapter 9, "Enzymes," and Chapter 25, "Hormones." Worth Publishers, Inc., New York, 1982.

2 L. Stryer. *Biochemistry,* 2nd ed. Chapter 6, "Introduction to Enzymes," and Chapter 35, "Hormone Action." W. H. Freeman and Company, San Francisco, 1981.

3 N. A. Daugherty. "Isoenzymes." *Journal of Chemical Education,* July 1979, page 442.

4 P. W. Carr and L. D. Bowers. *Immobilized Enzymes in Analytical and Clinical Chemistry: Fundamentals and Applications.* Wiley-Interscience, New York, 1980.

5 G. J. Davis, S. Chierchia, and A. Maseri. "Prevention of Myocardial Infarction by Very Early Treatment with Intracoronary Streptokinase." *The New England Journal of Medicine,* December 6, 1984, page 1488. *See also* J. W. Kennedy, et al. "The Western Washington Randomized Trial of Intracoronary Streptokinase in Acute Myocardial Infarction." *The New England Journal of Medicine,* April 25, 1985, page 1073 (and references given in this paper).

6 S. H. Snyder. "The Molecular Basis of Communication Between Cells." *Scientific American,* October 1985, page 132.

7 M. J. Berridge. "The Molecular Basis of Communication Within the Cell." *Scientific American,* October 1985, page 142.

8 S. H. Snyder. "Drug and Neurotransmitter Receptors in the Brain." *Science,* April 6, 1984, page 22.

9 Alice Dautry-Varsat and Harvey F. Lodish. "How Receptors Bring Proteins and Particles into Cells." *Scientific American,* May 1984, page 52.

10 R. J. Lefkowitz, M. G. Caron, and G. L. Stiles. "Mechanisms of Membrane-Receptor Regulation." *The New England Journal of Medicine,* June 14, 1984, page 1570.

11 Jean L. Marx. "A New View of Receptor Action." *Science,* April 20, 1984, page 271.

12 J. Axelrod and T. D. Reisine. "Stress Hormones: Their Interaction and Regulation." *Science,* May 4, 1984, page 452.

13 Pamela S. Zurer. "Drugs in Sports." *Chemical and Engineering News,* April 30, 1984, page 69.

14 S. H. Snyder, "Medicated Minds." *Science 84,* November 1984, page 141.

15 Y. Dunant and M. Israël. "The Release of Acetylcholine." *Scientific American,* April 1985, page 58.

16 J. S. Tompkins and M. D. Brown. "Chemonucleolysis." *Nursing 85,* July 1985, page 47.

REVIEW EXERCISES

The answers to these Review Exercises are in the *Study Guide* that accompanies this book.

Nature of Enzymes

24.1 What is meant by the term *enzyme* with respect to its (a) function and (b) composition (in general terms only)?

24.2 To what does the term *specificity* refer in enzyme chemistry?

24.3 Define and distinguish among the following terms.
(a) apoenzyme   (b) cofactor   (c) coenzyme

Coenzymes

24.4 What B vitamin is involved in the NAD^+/NADH system?

24.5 The active part of either FAD or FMN is furnished by which vitamin?

24.6 Complete and balance the following equation.

$$\underset{\overset{|}{\text{CH}_3\text{CHCH}_3}}{\overset{\text{OH}}{}} + \text{NAD}^+ \longrightarrow \underset{\text{CH}_3\text{CCH}_3}{\overset{\text{O}}{\overset{\|}{}}} + \underline{\quad} + \underline{\quad}$$

24.7 Complete and balance the following equation.

$$\underline{\quad} + \text{NADH} + \text{FMN} \longrightarrow \text{NAD}^+ + \underline{\quad}$$

Kinds of Enzymes

24.8 What is most likely the substrate for each of the following enzymes?
(a) Sucrase   (b) Glucosidase
(c) Protease   (d) Esterase

24.9 What *kind* of reaction does each of the following enzymes catalyze?
(a) An oxidase   (b) Transmethylase
(c) Hydrolase   (d) Oxidoreductase

24.10 What is the difference between lactose and lactase?

24.11 What is the difference between a hydrolase and hydrolysis?

24.12 What are isoenzymes (in general terms)?

24.13 What are the three isoenzymes of creatine kinase? Give their symbols and where they are principally found.

Theory of How Enzymes Work

24.14 What name is given to that part of an enzyme where the catalytic work is carried out?

24.15 How is enzyme specificity explained?

24.16 What is the induced-fit theory?

24.17 How does the plot of initial rate versus initial [S] at constant [E] look when (a) an allosteric effect is occurring, and (b) no allosteric effect is observed? (Draw pictures.)

24.18 What happens to the initial rate of an enzyme-catalyzed reaction as the initial concentration of the substrate is increased to the point where all enzyme molecules are saturated with substrate molecules?

24.19 If the plot of initial reaction rate versus initial substrate concentration (at constant [E]) has a sigmoid shape, what does this signify about the active site(s) on the enzyme?

24.20 How does a substrate molecule activate an enzyme whose rate curve is sigmoid?

Enzyme Activation and Inhibition

24.21 How does an effector differ from a substrate in causing allosteric activation?

24.22 What is the relationship of a zymogen to its corresponding enzyme? Give an example of an enzyme that has a zymogen.

24.23 How does competitive inhibition of an enzyme work?

24.24 Feedback inhibition of an enzyme works in what way?

24.25 Why is feedback inhibition an example of a homeostatic mechanism?

24.26 How do competitive inhibition and allosteric inhibition differ?

24.27 How do the following poisons work?
(a) CN^- (b) Hg^{2+}
(c) nerve gases or organophosphate insecticides

24.28 What are antimetabolites, and how are they related to antibiotics?

24.29 The following overall change is accomplished by a series of steps, each with its own enzyme.

$$\underset{\substack{\text{1,3-Diphosphoglycerate}\\ \text{(1,3-DPG)}}}{{}^{2-}\text{O}_3\text{POCH}_2\underset{\overset{|}{\text{OH}}}{\overset{\text{O}}{\overset{\|}{\text{CHCOPO}_3{}^{2-}}}}} \longrightarrow \longrightarrow \longrightarrow \underset{\substack{\text{2,3-Diphosphoglycerate}\\ \text{(2,3-DPG)}}}{{}^{2-}\text{O}_3\text{POCH}_2\underset{\overset{|}{\text{OPO}_3{}^{2-}}}{\text{CHCO}_2{}^-}}$$

One of the enzymes in this series is inhibited by 2,3-DPG. What kind of control is exerted by 2,3-DPG on this series? (Name it.)

24.30 If you drink enough methanol, you will become blind or die. One strategy to counteract methanol poisoning is to give the victim a nearly intoxicating drink of dilute ethanol. As the ethanol floods the same enzyme that attacks the methanol, the methanol gets a lessened opportunity to react and it is slowly and relatively harmlessly excreted. Otherwise, it is oxidized to formaldehyde, the actual poison from an overdose of methanol:

$$\underset{\text{Methanol}}{\text{CH}_3\text{OH}} \xrightarrow{\text{Dehydrogenase}} \underset{\text{Formaldehyde}}{\text{CH}_2\text{O}}$$

What kind of enzyme inhibition might ethanol be achieving here? (Name it.)

Enzymes in Medicine

24.31 If an enzyme such as CK or LD is normally absent from blood, how can a *serum* analysis for either tell anything? (Answer in general terms.)

24.32 What is the significance of the CK(*MB*) band in trying to find out if a person has had a heart attack and not just some painful injury in the chest region?

24.33 What CK band obtained by electrophoresis would increase if the injury in Review Exercise 24.32 were to skeletal muscle?

24.34 What is the LD flip, and how is it used in diagnosis?

24.35 What happens, chemically, in a positive Clinistix test for glucose? (Give the answer in terms of the steps in the series of changes that can be described by words, not equations.)

Hormones and Neurotransmitters

24.36 In what general ways do hormones and neurotransmitters resemble each other?

24.37 What function does adenylate cyclase have in the work of at least some hormones?

24.38 How is cyclic AMP involved in the work of some hormones and neurotransmitters?

24.39 After cyclic AMP has caused the activation of an enzyme inside a cell, what happens to the cyclic AMP that stops its action until more is made?

24.40 What are the names of the sites of the synthesis of (a) hormones and (b) neurotransmitters?

24.41 What does the lock-and-key concept have to do with the work of hormones and neurotransmitters?

24.42 In general terms, name three ways by which hormones work, and give an example of a hormone for each.

24.43 What happens to acetylcholine after it has worked as a neurotransmitter? What is the name of the enzyme that catalyzes this change? In chemical terms, what specifically does a nerve gas poison do?

24.44 How does atropine counter nerve gas poisoning?

Neurotransmitters and Medicine

24.45 How does a local anesthetic such as procaine affect the functioning of acetylcholine as a neurotransmitter?

24.46 What, in general terms, are the monoamine oxidases, and in what way are they important?

24.47 What does iproniazid do chemically in the neuron-signaling that is carried out by norepinephrine?

24.48 In general terms, how do antidepressants such as amitriptyline or imipramine work?

24.49 Which neurotransmitter is also a hormone, and what is the significance of this dual character to the body?

24.50 The overactivity of which neurotransmitter is thought to be one biochemical problem in schizophrenia?

24.51 How do the schizophrenia-control drugs chlorpromazine and haloperidol work?

24.52 How can the amphetamines, when abused, give schizophrenia-like symptoms?

24.53 How does L-DOPA work in treating Parkinson's disease?

24.54 Which common neurotransmitter in the brain is a signal inhibitor? How do such tranquilizers as Valium and Librium affect it?

24.55 Why is enkaphalin called one of the body's own opiates? How does it appear to work?

Chapter 25
Extracellular Fluids of the Body

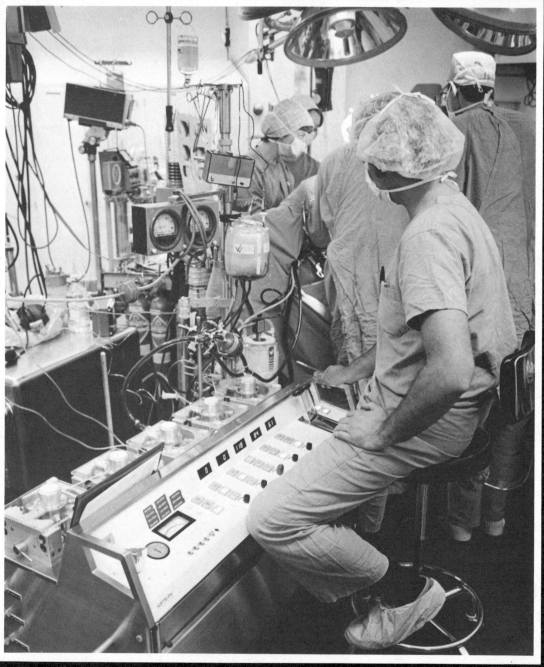

All these machines, solutions, tubes, and specialists are needed to duplicate outside the body, during open-heart surgery, what the heart-lung network of one body routinely does to maintain circulation and respiration. The chemistry of respiration is the major topic in this chapter.

25.1 DIGESTIVE JUICES

The end products of the complete digestion of the nutritionally important carbohydrates, lipids, and proteins are monosaccharides, fatty acids, glycerol, and amino acids.

Life engages two environments, the outside environment we commonly think of, and the internal environment, which we usually take for granted. When healthy, our bodies have nearly perfect control over the internal environment, and so we are able to handle large changes outside more or less well — large temperature fluctuations, chilling winds, stifling humidity, and a fluctuating atmospheric tide of dust and pollutants.

The fluids of the internal environment make up about 20% of the mass of the body.

The **internal environment** consists of all of the **extracellular fluids** — those that aren't actually inside cells. The **interstitial fluid,** the fluid in the spaces or interstices *between* cells, makes up three-quarters of the total volume of the extracellular fluids, and the blood constitutes nearly all of the rest. Other extracellular fluids are the lymph, the cerebrospinal fluid, the digestive juices, and synovial fluids.

Synovial fluids are the viscous lubricants of joints.

We will be using several chapters and many fundamental principles — those of acids, bases, buffers, and other electrolytes, as well as osmosis and dialysis — to study what goes on in the cellular fluid. In this chapter we'll concentrate on two of the extracellular fluids, the digestive juices and the blood. Because the operation of the kidneys and the formation of urine are inseparable from the chemistry of blood, we will investigate what the kidneys do, also.

The Digestive Tract.   The names and locations of the principal parts of the digestive tract are given in Figure 25.1. At various places, different solutions of enzymes — the **digestive juices** — hydrolyze the various molecules in food. The chief components of a typical meal — di- and polysaccharides, saponifiable lipids, polypeptides, and nucleic acids — cannot be absorbed directly into the rest of the body. They must first be hydrolyzed to monosaccharides, fatty acids, glycerol, amino acids, and other substances. Our strategy is to study the digestive juices in turn in the order in which they act on food.

Saliva.   The flow of **saliva** is stimulated by the sight, smell, taste, and even the thought of food. Besides water (99.5%), saliva includes a food lubricant called **mucin** (a glycoprotein) and an enzyme, α**-amylase.** This enzyme catalyzes the partial hydrolysis of starch to dextrins and maltose, and it works best at the pH of saliva, 5.8 – 7.1. Proteins and lipids pass through the mouth essentially unchanged.

Gastric Juice.   When food arrives in the stomach, the cells of the gastric glands that line the inner surface of the stomach are stimulated by hormones to release **gastric juice.** One kind of gastric gland secretes mucin, which coats the stomach to protect it against its own digestive enzymes and its acid. Mucin is continuously produced and only slowly digested. If for any reason its protection of the stomach is hindered, part of the stomach itself could be digested, and this would lead to an ulcer.

The pH of gastric juice is normally in the range of 0.9 – 2.0.

Another gastric gland secretes hydrochloric acid at a concentration of roughly 0.1 mol/L — about a million times more concentrated in hydrogen ion than the blood. A third gastric gland secretes the zymogen, **pepsinogen.** Pepsinogen is changed into **pepsin,** a proteolytic enzyme, by the action of hydrochloric acid and traces of pepsin. The optimum pH of pepsin is in the range of 1 to 1.5, which is found in the stomach fluid. Pepsin catalyzes the only important digestive work in the adult stomach, the hydrolysis of some of the peptide bonds of proteins to make relatively short polypeptides.

A *proteolytic enzyme* is one that catalyzes the digestion of proteins.

In infants, whose stomach fluid has a higher, less-acidic pH, two other components of gastric juice — rennin and gastric lipase — serve necessary functions. Rennin helps to coagulate milk protein, and when coagulated, the protein stays longer in the stomach where pepsin can work on it. (In the adult stomach, the hydrochloric acid coagulates proteins.) The lipase, which can work at the higher pH of the stomach fluid of infants, helps the infant get an early start on the digestion of lipids. In adults, this lipase awaits its action after it moves, together

FIGURE 25.1
Organs of the digestive tract.

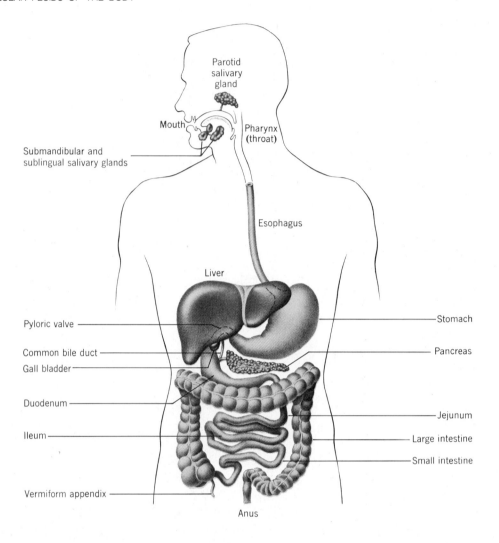

with the rest of the stomach contents, into the more alkaline environment of the upper intestinal tract.

The churning and digesting activities in the stomach produce a liquid mixture called **chyme.** Chyme is released in portions through the pyloric valve into the duodenum, the first 12 inches of the upper intestinal tract.

Pancreatic Juice. The pancreas makes a whole catalog of compounds that become involved in the digestion of starch, lipids, proteins, and small polypeptides. As soon as chyme appears in the duodenum, hormones are released that circulate to the pancreas and induce this organ to release two juices. One is almost entirely a large volume of dilute sodium bicarbonate, which neutralizes the acid in chyme. The other fluid, the one usually called **pancreatic juice,** carries enzymes or zymogens that become involved in the digestion of practically everything in food. It contributes an **α-amylase** similar to that present in saliva, a **lipase, nucleases,** and protein-digesting enzymes.

The nucleases include ribonuclease (RNAase) and deoxyribonuclease (DNAase).

The proteolytic enzymes in pancreatic juice—**trypsin, chymotrypsin,** and **carboxypeptidase**—are prepared in and leave the pancreas as zymogens. (Otherwise, if they were fully active enzymes in the pancreas itself, they would digest the proteins of this organ!) The conversion of these zymogens to active enzymes begins with the work of a "master switch" enzyme called **enterokinase,** which is released from cells that line the

In severe pancreatitis, self-digestion of the pancreas does occur.

duodenum when chyme arrives. Enterokinase catalyzes
zymogen, trypsinogen.

$$\text{Trypsinogen} \xrightarrow{\text{Enterokinase}} \text{Tr}$$

Then trypsin catalyzes the change of the other zymoge

$$\text{Procarboxypeptidase} \xrightarrow{\text{Trypsin}} \text{Car}$$

$$\text{Chymotrypsinogen} \xrightarrow{\text{Trypsin}} \text{Ch}$$

Both trypsin and chymotrypsin catalyze the hydrolysis of large polypept
Carboxypeptidase, working in from C-terminal ends of small polypeptides, carries the ac
further to amino acids and di- or tripeptides.

Bile.   In order to digest most lipids, the lipase in pancreatic juice needs the help of the
powerful detergents in bile, called the **bile salts.** These surface-active agents help to break up
the water-insoluble fatty material in chyme into extremely tiny globules. This greatly increases
the exposure of the lipids to water and lipase, and the lipids are smoothly digested. Triacylglycerols are hydrolyzed to fatty acids, glycerol, and to some monoacylglycerols.

The structure of a typical bile salt was given in Table 20.3, page 482.

Bile is a secretion that enters the duodenum from the gall bladder when the contraction
of this organ is stimulated by a hormone. (This hormone is released into circulation whenever
fatty material is in chyme.) Bile is also an avenue of excretion for the body, because it carries
away excess cholesterol as well as certain breakdown products of hemoglobin — the bile
pigments that give color to feces.

The bile salts not only aid in the digestion of lipids, they also assist in the absorption of the
fat-soluble vitamins (A, D, E, and K) from the digestive tract into the blood. This work also
carries some of the freshly delivered bile pigments back into the bloodstream, and they
eventually leave the body via the urine. Thus the bile pigments are responsible for the color of
both feces and urine.

Intestinal Juice.   The term **intestinal juice** embraces not only a secretion but also the
enzyme-rich fluids found inside certain kinds of cells that line the duodenum. As a secretion,
intestinal juice delivers enterokinase, which we just described, and an amylase. The other
enzymes of this fluid work within their cells as digestible compounds are already in migration
from the duodenum toward the bloodstream or the lymph system. An **aminopeptidase,**
working inward from N-terminal ends of small polypeptides, digests them to amino acids. The
enzymes **sucrase, lactase,** and **maltase,** handle the digestion of the dietary disaccharides
— sucrose to fructose and glucose; lactose to galactose and glucose; and maltose to glucose.
An intestinal lipase is present as well as enzymes for the hydrolysis of nucleic acids.

These intestinal cells last only about 2 days before they self-digest. They are constantly being replaced.

As fatty acids, glycerol, and monoacylglycerols migrate through the cells of the duodenal
lining, much is reconstituted into triacylglycerols, which are taken up by the lymph system
rather than the blood.

Events in the Large Intestine.   No digestive functions are performed in the large intestine.
However, large numbers of microorganisms are in residence there, and as by-products of their
own metabolism, they produce vitamins K and B, plus some essential amino acids. These are
absorbed by the body, but their contributions to overall nutrition in humans is not large.
Undigested matter, including nearly all cellulosic fiber from vegetables and whole-grain cereals
in the diet, plus the digestive secretions and water, make up the feces.

D AND THE **EXCHANGE OF NUTRIENTS**

ervous system with its
rotransmitters is the other line
f communication.

About 8% of the body's mass is
blood. In the adult, the blood
volume is 5–6 L.

The balance between the blood's pumping pressure and its colloidal osmotic pressure tips at capillary loops.

The circulatory system, Figure 25.2, is one of our two main lines of chemical communication between the external and internal environments. All of the veins and arteries together are called the **vascular compartment.** The **cardiovascular compartment** includes this plus the heart.

Functions of Blood. Blood in the pulmonary branches moves through the lungs where waste carbon dioxide in the blood is exchanged for oxygen from freshly inhaled air. The oxygenated blood then moves to the rest of the system via the systemic branches. At the intestinal tract, the blood picks up the products of digestion. Most of these are immediately monitored at the liver, which among a number of duties detects drugs and works to process them into forms that can be used or excreted. In the kidneys, the blood is purified of nitrogen wastes, particularly urea, and the blood also replenishes some of its buffer supplies. The kidneys are critical to the maintenance of the pH of the blood and the concentrations of other electrolytes.

At various endocrine glands, the blood picks up and circulates whatever hormones these glands release (and such releases are often in response to something present in the blood).

Always present in blood are white cells, which attack invading bacteria; red cells or **erythrocytes,** which carry hemoglobin; and platelets, which are needed for blood clotting

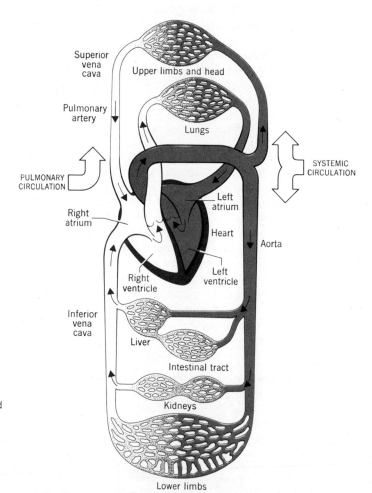

FIGURE 25.2
Human circulatory system. When oxygen-depleted venous blood (light areas) returns to the heart, it is pumped into the capillary beds of the alveoli in the lungs to reload oxygen and get rid of carbon dioxide. Then the freshly oxygenated blood (dark areas) is distributed by the arteries throughout the body, including the heart muscle.

FIGURE 25.3
Major components of blood.

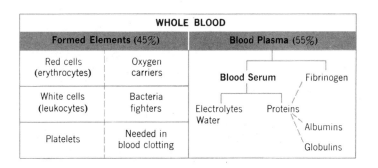

WHOLE BLOOD			
Formed Elements (45%)		**Blood Plasma (55%)**	
Red cells (erythrocytes)	Oxygen carriers	**Blood Serum** Fibrinogen	
White cells (leukocytes)	Bacteria fighters	Electrolytes Proteins Water	
Platelets	Needed in blood clotting	Albumins Globulins	

and other purposes. The blood also carries several zymogens that participate in the blood-clotting mechanism.

Composition of Blood.   The principal types of substances in whole blood are summarized in Figure 25.3. One kilogram of blood plasma contains 80 g of proteins—albumins (54–58%), globulins (40–44%), and fibrinogen (3–5%). **Albumins** help carry hydrophobic molecules such as fatty acids, other lipids, and steroid hormones, and they contribute 75–80% of the osmotic effect of the blood. Some globulins carry ions (e.g., Fe^{2+} and Cu^{2+}) that otherwise could not be soluble in a fluid with a pH slightly greater than 7. The γ-**globulins** help to protect the body against infectious disease. **Fibrinogen** is converted to an insoluble form, **fibrin,** when a blood clot forms.

Figure 25.4 shows the quantities of various components of the major body fluids. One striking difference is in the higher concentration of various proteins in blood compared with that of interstitial fluid. This is the principal reason why blood has a higher osmotic pressure than interstitial fluid.[1] The *total* osmotic pressure of blood is caused by all of the dissolved and colloidally dispersed solutes—electrolytes, organic compounds and ions, and proteins. However, the small ions and molecules can dialyze back and forth between the blood and the interstitial compartment. The large protein molecules can't do this, so it is their presence that gives to blood the higher effective osmotic pressure. This contribution to the blood's osmotic pressure made by colloidally dispersed substances is called the colloidal osmotic pressure of blood. As a consequence of the higher osmotic pressure of blood, water tends to flow into the blood from the interstitial compartment. Of course, this can't be allowed to happen everywhere and continually, or the interstitial spaces and then the cells would eventually become too dehydrated to maintain life. Before we continue, we need to survey some of the **electrolytes** in the blood and what they do.

The Electrolytes in the Blood.   The relative quantities of the electrolytes in the major body fluids are given also in Figure 25.4. The sodium ion is the chief cation in both the blood and the interstitial fluid and the potassium ion is the major cation inside cells. A sodium/potassium pump—a special protein complex that uses cellular energy—maintains these gradients. Both ions are needed to maintain osmotic pressure relationships, and both are a part of the regulatory system for acid-base balance. Both ions are also needed for the smooth working of the muscles and the nervous system.

Changes in the concentrations of sodium and potassium ion in blood can lead to serious medical emergencies, so a special vocabulary has been developed to describe various situations. We will see here how a technical vocabulary can be built on a few word parts, and we will use some of these word parts in later chapters, too. For example, we use -emia to signify "in the blood." *Hypo-* indicates a condition of a low concentration of something, and *hyper-* is the opposite, a condition of a high concentration of something. We can specify this something by a word part, too. Thus -nat- signifies sodium (from the Latin *natrium* for sodium), and -kal-

About a quarter of the plasma proteins are replaced each day.

The osmolarity of plasma is about 290 mOsm/L.

Another word part is -uria, of the urine. Thus *glucosuria* means glucose in the urine.

---

[1] As a reminder and a useful memory aid, high solute concentration means high osmotic pressure; and solvent flows in osmosis or dialysis from a region where the solute is dilute to the region where it is concentrated. The "goal" of this flow is to even out the concentrations everywhere.

FIGURE 25.4
Electrolyte composition of body fluids. (Adapted by permission from J. L. Gamble, *Chemical Anatomy, Physiology and Pathology of Extracellular Fluids,* 6th ed. Harvard University Press, Cambridge, MA, 1954.)

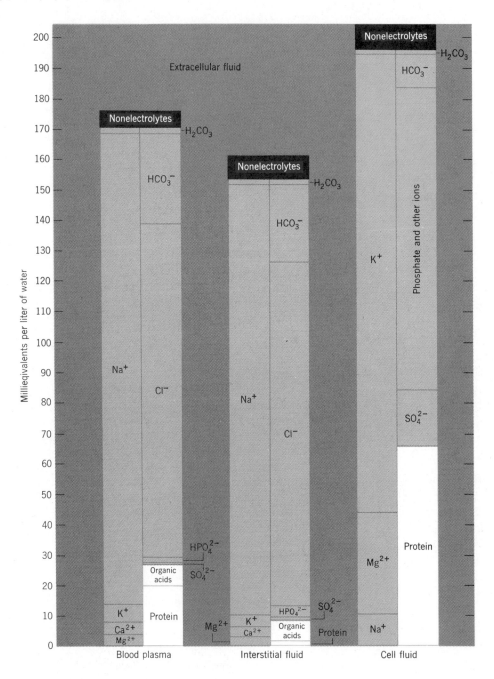

You'll see it called the ''sodium level,'' not sodium *ion* level, in nearly all references, but sodium *ion* level is always intended.

designates potassium (from the Latin *kalium* for potassium). Putting these together gives us the following terms.

Hyponatremia:	low level of sodium ion in blood
Hypokalemia:	low level of potassium ion in blood
Hypernatremia:	high level of sodium ion in blood
Hyperkalemia:	high level of potassium ion in blood

Figure 25.5 shows the normal range for the sodium ion level in blood and where the terms we have just learned would fit when the level is outside this range. Figure 25.6 does the same

thing for the potassium ion level. Table 25.1 gives the normal ranges for these cations in various units.

Because the level of potassium ion in blood is so low, small changes are particularly dangerous. Severe hyperkalemia leads to death by heart failure, and this danger exists when a person has experienced crushing injuries, or severe burns, or heart attacks. In these circumstances, cells break open and spill their contents, including their potassium ions, into general circulation. At the other extreme, severe hypokalemia, death by heart failure can also occur, and this can result from any unusual losses of body fluids, including prolonged, excessive sweating. Both body fluids *and electrolytes* have to be replaced during severe exercise.

The serum levels of sodium ions and potassium ions are regulated in tandem by the kidneys. If the intake of Na^+ is high, there will be a loss of K^+ from the body. If the intake of K^+ is high, there will be a loss of Na^+ from the body. One reason we can't tolerate sea water (3% NaCl) and will die if we drink it is that it upsets the sodium/potassium balance in the body.

If our kidneys cannot make urine or if we drink water faster than the kidneys can handle it, the sodium level of the blood decreases and we will display the signs of hyponatremia — flushed skin, fever, and a dry tongue (and a noticeable decrease in urine output). Hypernatremia occurs from excessive losses of water under circumstances in which sodium ions are not lost — as in diarrhea, diabetes, and even in some high-protein diets.

The magnesium ion is second only to the potassium ion as a cation inside cells, and it has a low but vitally important level in the blood, too. (See Table 25.2.) Inside cells, the magnesium ion participates in the activations of several enzymes. Changes in its level in the blood accompany certain conditions. For example, hypomagnesemia is seen when the kidneys aren't working properly, or in alcoholism, or in untreated diabetes. Some of its signs are muscle weakness, insomnia, and cramps in the legs or the feet. Injections of isotonic magnesium sulfate solution are sometimes used to restore the serum magnesium level. On the opposite side, hypermagnesemia can lead to cardiac arrest, and it can be brought on by the overuse of magnesium-based antacids such as milk of magnesia, $Mg(OH)_2$.

Nearly all of the body's calcium ion is in bones and teeth, but what is in solution is absolutely vital. The calcium ion is intimately involved in the mechanism whereby muscles contract — including heart muscle. Under the impetus of a nerve signal, a calcium ion enters a muscle cell and, with the help of calcium ions already there, initiates a rapid series of events leading to contraction of the muscle. Then a relaxing factor takes the excess calcium ion

Experienced backpackers use salt tablets that contain not only NaCl but also some KCl to supply K^+.

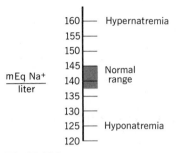

FIGURE 25.5
The sodium ion level in the blood.

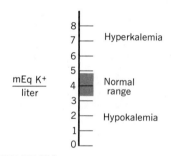

FIGURE 25.6
The potassium ion level in the blood.

TABLE 25.1
Group IA Cations in the Human Body

	Na^+	K^+
Total body	2700–3000 meq	3200 meq
Plasma level	135–145 meq/L	3.5–5.0 meq/L
Intracellular level	10 meq/L	125 meq/L
Mass of 1 meq	23.0 mg	39.1 mg

TABLE 25.2
Group IIA Cations in the Human Body

	Ca^{2+}	Mg^{2+}
Total body	6×10^4 meq (1.2 kg)	1000 meq (24 g)
Plasma level	4.2–5.2 meq/L	1.5–2.0 meq/L
Intracellular level	—	35 meq/L
Mas of 1 meq	20.0 mg	12.2 mg

outside the cell again, and the muscle relaxes. Many medications for heart conditions are directed toward the channels through which the calcium ions migrate. Acting as calcium channel blockers, these medications reduce the migration of calcium ions into the cell's interior. Although a contraction isn't prevented, it occurs with less force, and the heart works less hard.

Hypocalcemia can be brought on by vitamin D deficiency, the overuse of laxatives, an impaired activity of the thyroid gland, and even by hypomagnesemia. Hypercalcemia, on the other hand, is caused by the opposite conditions — an overdose of vitamin D, the overuse of calcium ion-based antacids, or an overactive thyroid. In severe hypercalcemia, the heart functions poorly.

As seen in Figure 25.4, the principal anion in blood and interstitial fluid is the chloride ion. Virtually none is present in cell fluid. The range of concentration of Cl^- in blood is 100 – 106 meq/L (and 1 meq of Cl^- is 35.5 mg). The chloride ion helps to maintain osmotic pressure relationships, the acid-base balance, and the distribution of water in the body. Should there be an excessive loss of Cl^- from the blood, the body must retain another kind of negative ion to keep an electrical charge balance, and the chief ion retained is HCO_3^-. Because this ion tends to raise the pH of a fluid, a condition of hypochloremia can cause alkalosis. By the same token, a rise in the level of Cl^- — hyperchloremia — could mean that HCO_3^- has to be dumped, which would mean a loss of base. Thus hyperchloremia could lead to acidosis.

You can see that the levels of none of the ions in body fluids can change without causing changes in the levels of others. Everything is tied to everything else.

The Exchange of Nutrients, Wastes, and Fluids at Tissue Cells.   The blood vessels undergo extensive branching until the narrowest tubes called the capillaries are reached. Blood enters a capillary loop (Figure 25.7) as arterial blood, but it leaves on the other side of the loop as venous blood. During the switch, fluids and nutrients leave the blood and move into the interstitial fluids and into the tissue cells themselves. *In the same volume* the fluids must return to the blood, but now they must carry the wastes of metabolism. The rate of this diffusion of fluids throughout the body is sizable, about 25 – 30 liters per second. Some fluids return to circulation by way of the lymph ducts, which are thin-walled, closed-end capillaries that bed in soft tissue.

On the arterial side of a capillary loop, the blood pressure is sufficiently high to overcome the natural tendency of fluids to move *into* the blood. Instead, water and dissolved solutes are forced out of circulation and into the surrounding tissue where exchanges of chemicals occur. On the venous side of the loop, the blood pressure is too low to prevent the natural diffusion of fluids back into the bloodstream, but by this time the fluids are carrying waste products. These relationships are illustrated in Figure 25.7, which shows how the colloidal osmotic pressure contributed by the dispersed macromolecules in blood — particularly the albumins — make the difference in determining the direction of diffusion.

When the capillaries become more permeable to blood proteins, as they do in such trauma as sudden severe injuries, major surgery, and extensive burns, the proteins migrate out of the blood. Unfortunately, this protein loss also means the loss of the colloidal osmotic pressure that helps fluids to return from the tissue areas to the bloodstream. As a result, the total volume of circulating blood drops quickly, and this drastically reduces the blood's ability to carry oxygen and to remove carbon dioxide. The drop in blood volume and the resulting loss of oxygen supply to the brain sends the victim into traumatic **shock.**

Sometimes the proteins in the blood are lost at malfunctioning kidneys. The effect, although gradual, is a slow but unremitting drop in the blood's colloidal osmotic pressure. Fluids accumulate in the interstitial regions. Because this takes place more slowly and water continues to be ingested, there is no sudden drop in blood volume as in shock. The victim appears puffy and waterlogged, a condition called **edema.**

Edema can also appear at one stage of starvation, when the body has metabolized its circulating proteins to make up for the absence of dietary proteins.

It's common among the elderly to suffer the loss of Ca^{2+} from bones, a condition called *osteoporosis.*

The lymph system makes antibodies and it has white cells that help defend the body against infectious diseases.

The prompt restoration of blood volume is mandatory in the treatment of shock.

Greek, *oidema,* swelling.

FIGURE 25.7
The exchange of nutrients and wastes at capillaries. As indicated at the top, on the arterial side of a capillary loop the blood pressure counteracts the pressure from dialysis and osmosis, and fluids are forced to leave the bloodstream. On the venous side of the loop, the blood pressure has decreased below that of dialysis and osmosis, so fluids flow back into the bloodstream. On the top right and the bottom is shown how a normal red cell distorts as it squeezes through a capillary loop. Red cells in sickle-cell anemia do not pass through as smoothly. The bottom drawing also shows how some fluids enter the lymph system.

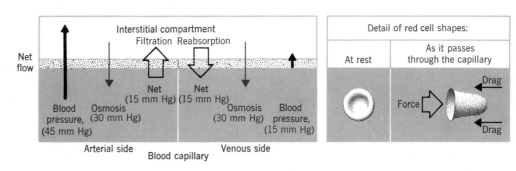

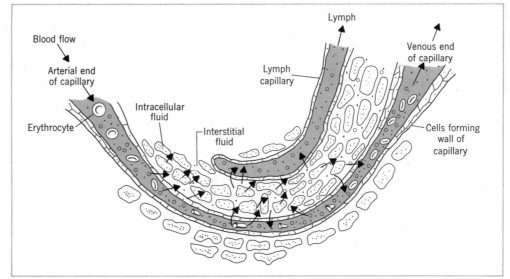

Any obstruction in the veins can also cause edema, as in varicose veins and certain forms of cancer. Now it is the venous blood pressure that rises, creating a back pressure that reduces the rate at which fluids can return to circulation from the tissue areas. The localized swelling that results from a blow is a temporary form of edema caused by injuries to the capillaries.

25.3 BLOOD AND THE EXCHANGE OF RESPIRATORY GASES

The binding of oxygen to hemoglobin is allosteric, and it is affected by the pH, pCO_2, and the pO_2 of the blood.

Each red cell carries about 2.8×10^8 molecules of hemoglobin.

The carrier of oxygen in blood is **hemoglobin,** a complex protein found inside erythrocytes. It consists of four subunits, each with one molecule of heme, an organic group that holds an iron(II) ion and that is the actual oxygen-binding unit. Two of the subunits are identical and have the symbol α. The other two, although not being α, are also identical and have the symbol β. When hemoglobin is oxygen-free, and is sometimes referred to as *deoxyhemoglobin,* the molecule has a small cavity in which a small organic anion nestles. This is the **2,3-diphosphoglycerate** ion, called simply **DPG,** and it has an important function in oxygen transport.

$$^{2-}O_3POCH_2CHCO_2^-$$
$$|$$
$$OPO_3^{2-}$$

DPG
(2,3-diphosphoglycerate)

The Oxygenation of Hemoglobin in the Lungs.   We define the **oxygen affinity** of the blood as the percent to which the blood has all of its hemoglobin molecules saturated with oxygen. A fully laden hemoglobin molecule carries four oxygen molecules and is called **oxyhemoglobin.** For the maximum efficiency in moving oxygen from the lungs to tissues that need it, all hemoglobin molecules should leave the lungs in the form of fully loaded oxyhemoglobin. Let's see what factors ensure this.

First, the partial pressure of oxygen is higher in the lungs than anywhere else in the body; $pO_2 = 100$ mm Hg in freshly inhaled air in alveoli and only about 40 mm Hg in oxygen-depleted tissues. Because of this partial pressure gradient, oxygen naturally migrates from the lungs into the bloodstream. It's as if the higher partial pressure *pushes* oxygen into the blood. Second, newly arrived oxygen actually reacts with hemoglobin to form oxyhemoglobin, so this tends to *pull* oxygen into the blood.

Another factor that helps to load hemoglobin with oxygen is an allosteric effect. Figure 25.8 shows a plot of oxygen affinity versus the oxygen partial pressure. It has the sigmoid shape that in Chapter 24 we learned to associate with the allosteric effect among enzymes. At low values of pO_2, in region A of the plot, the ability of the blood to take up oxygen rises only slowly with increases in pO_2. But eventually the oxygen affinity takes off, and rises very steeply with still more increases in pO_2—in region B of the plot. Eventually, the oxygen affinity starts to level off. It almost seems that a small "molecular dam" thwarts the efforts of oxygen molecules to be joined to hemoglobin at low partial pressure of oxygen.

What is thought to be happening is as follows. We'll represent deoxyhemoglobin by structure **1**, below. Each circle in **1** is a polypeptide subunit with its heme unit but without oxygen. The diamond figure centered within the structure represents one DPG anion. When the first O_2 molecule binds (**1** → **2**)—and it does so only weakly—it induces a change in the shape of the affected subunit, which we have represented as a change from a circle to a square:

We first learned about the allosteric effect on page 563.

This natural resistance is needed in working cells where we want no restrictions on the deoxygenation of blood.

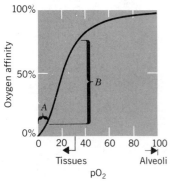

FIGURE 25.8
Hemoglobin–oxygen dissociation curve. Regions A and B are discussed in the text.

Carbon monoxide binds 150–200 times more strongly to hemoglobin than oxygen and thus prevents the oxygenation of hemoglobin and causes internal suffocation.

About 20% of a smoker's hemoglobin is more or less permanently tied up by carbon monoxide.

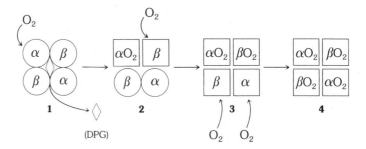

This change tends to squeeze out the DPG unit, and this "breaks the dam" we alluded to above. The next oxygen molecule enters more readily (**2** → **3**), and as its affected subunit changes its shape, the remaining two subunits also change their shapes and become very receptive to the third and fourth molecules of oxygen. These two flood in to this hemoglobin molecule (**3** → **4**) far more readily than either would bind to a completely deoxygenated hemoglobin molecule. Thus if a hemoglobin molecule accepts just one oxygen molecule, it's almost certain that it will soon accept three more to become fully oxygenated. Few if any partially oxygenated hemoglobin molecules leave the lungs.

For the sake of the remaining discussion, we'll simplify what we have just described by letting the symbol HHb represent an entire hemoglobin molecule. The first H in HHb stands for a potential hydrogen ion, and we overlook the fact that more than one is actually present in hemoglobin. (We're now also overlooking the fact that *four* molecules of O_2 bind to one of hemoglobin.) With this in mind, we can represent the oxygenation of hemoglobin as the *forward* reaction in the following equilibrium where oxyhemoglobin is represented as the anion, HbO_2^-:

$$\text{HHb} + O_2 \rightleftharpoons HbO_2^- + H^+ \qquad (25.1)$$

Hemoglobin   Oxyhemoglobin

Two facts indicated by this equilibrium are that HHb is a weak acid and that it becomes a stronger acid as it becomes oxygenated. Being stronger, it is produced in its ionized state as HbO_2^- and H^+. *The presence of H^+ in equilibrium 25.1 means that the equilibrium can be shifted one way or another simply by changing the pH,* a fact of enormous importance at the molecular level of life, as we'll see.

To understand the oxygenation of hemoglobin we have to see how various stresses can shift equilibrium 25.1 to the right. One stress, as we've already noted, is the relatively high value of pO_2 — 100 mm Hg — in the alveoli. This stress on the left side of 25.1 helps to shift the equilibrium to the right.

$$HHb + O_2 \longrightarrow HbO_2^- + H^+$$

| In the | From | In the | In the |
| red cell | air | red cell | red cell |

Another stress that isn't evident from 25.1 is the removal of H^+ *as it forms.* The newly forming H^+ ions are neutralized by the buffer system, chiefly by HCO_3^-.

$$H^+ + HCO_3^- \longrightarrow H_2CO_3$$

| In the | In the | In the |
| red cell | red cell | red cell |

This switch from the appearance of H^+ as a product to its disappearance as a reactant is called the **isohydric shift.**

Now comes one of the beautiful examples of coordinated activity in the body — the coupling *in the lungs* of the uptake of oxygen to the release of carbon dioxide, which is promptly exhaled. The neutralization of the H^+ ions produced from the uptake of O_2 produces an unstable acid, H_2CO_3. And the red cell carries the enzyme carbonic anhydrase, which promptly establishes the following equilibrium that involves this freshly made H_2CO_3:

$$H_2CO_3 \underset{}{\overset{\text{Carbonic anhydrase}}{\rightleftharpoons}} H_2O + CO_2 \qquad (25.2)$$

In the	Diffuses from
red cell	the red cell and
	is exhaled

The appearance of H_2CO_3 shifts equilibrium 25.2 to the right. Moreover, the loss of CO_2 by exhaling also shifts — pulls — this equilibrium to the right. Thus *the uptake of O_2 as hemoglobin oxygenates simultaneously aids in the release of waste CO_2.* Let's now see how this waste is picked up at cells that have produced it and is carried to the lungs; and let's also see how this works cooperatively with the *release* of oxygen at cells that need it.

The Deoxygenation of Oxyhemoglobin in Metabolizing Tissues. Consider, now, a tissue that has done some chemical work, used up some oxygen, and made some waste carbon dioxide. When a fully oxygenated red cell arrives in such a tissue, some of the events we just described reverse themselves. We can think of this reversal as beginning with the diffusion of waste CO_2 from the tissue into the blood. An impetus for this diffusion is the higher pCO_2 (50 mm Hg) in active tissue versus its value in blood (40 mm Hg). Once the CO_2 arrives in the blood, it moves inside a red cell where it encounters carbonic anhydrase. Equilibrium 25.2 is therefore promptly established, and it shifts to the left as more and more CO_2 arrives. In other words, the equilibrium now shifts to *make* H_2CO_3:

$$H_2O + CO_2 \overset{\text{Carbonic anhydrase}}{\longrightarrow} H_2CO_3 \qquad (25.2\text{-reversed})$$

From the	In the
working	red cell
tissue	

As we learned in the last chapter, carbonic anhydrase is the body's fastest-working enzyme, and it has to be fast because the red cells are always on the move.

CO_2 molecules diffuse in body fluids 30 times more easily than O_2 molecules, so the partial pressure gradient for CO_2 need not be as steep as that for O_2.

No enzyme is needed for this ionization.

Of course, this carbonic acid undergoes a small amount of ionization:

$$H_2CO_3 \longrightarrow H^+ + HCO_3^-$$

| In the red cell | In the red cell | In the red cell |

This generates hydrogen ions, and if you'll look back to equilibrium 25.1, you will see that an increase in the level of H^+ (caused by waste CO_2) can only shift equilibrium 25.1 backward:

$$H^+ + HbO_2^- \longrightarrow HHb + O_2 \qquad \text{(25.1-reverse)}$$

| Just made in the red cell at tissue | In the red cell, just arrived | In the red cell | Will diffuse into the tissue |

This reaction, another isohydric shift, not only neutralizes the acid generated by the arrival of waste CO_2, it also makes oxyhemoglobin give up its oxygen. Notice the cooperation. The tissue that needs oxygen has made CO_2 and, hence, it has indirectly made the H^+ that is required to release this needed oxygen from newly arrived HbO_2^-.

The high negative charge on DPG keeps it from diffusing through the red cell's membrane.

The deoxygenation of HbO_2^- is also aided by the DPG anions that were pushed out when HbO_2^- formed. These anions are still inside the red cell, and as soon as an O_2 molecule leaves oxyhemoglobin, a DPG anion starts to move back in. The changes in shapes of the hemoglobin subunits now operate in reverse, and all oxygen molecules smoothly leave. It's all or nothing again, and the efficiency of the unloading of oxygen is so high that if one O_2 molecule leaves, the other three follow essentially at once. DPG helps this to happen. Partially deoxygenated hemoglobin units do not slip through and go back to the lungs.

It's interesting that those who live and work at high altitudes, such as the populations in Nepal in the Himalayan Mountains, or the people in the Andes Mountains in Bolivia, have 20% higher levels of DPG in their blood and more red blood cells than those who live at sea level. The extra red cells help them carry more oxygen per mL of blood, and the extra DPG increases the efficiencies of both loading and unloading oxygen. When lowlanders take trips to high altitudes, their bodies start to build more red cells and to make more DPG so that they can function better where the partial pressure of the atmospheric oxygen is lower. Those who patiently wait during the few days that it takes for these events to occur before they set off on strenuous backpacking expeditions are far less likely to suffer high-altitude sickness, a condition that can cause death.

No conditioning at a low altitude can get the cardiovascular system ready for a low pO_2 at a high altitude.

To summarize the chemical reactions we have just studied, we can write the following:

Oxygenation:

$$HHb + O_2 \longrightarrow HbO_2^- + \cancel{H^+}$$

| In red cell | From air | Goes in red cell to tissues | In red cell |

$$\cancel{H^+} + HCO_3^- \longrightarrow \cancel{H_2CO_3}$$

| Just made | In red cell (but from tissues) | In red cell |

CA is carbonic anhydrase.

$$\cancel{H_2CO_3} \xrightarrow{CA} H_2O + CO_2$$

| | Just made in red cell | | Leaves the lungs |

Net effect of oxygenating hemoglobin:

$$\mathbf{HHb + O_2 + HCO_3^- \longrightarrow HbO_2^- + CO_2 + H_2O} \qquad \text{(25.3)}$$

| In red cell | From air | In red cell (but from tissues) | Goes in red cell to tissues | Leaves the lungs in exhaled air |

Deoxygenation:

$$CO_2 \quad + H_2O \xrightarrow{\;CA\;} \cancel{H_2CO_3}$$

Waste from   In red cell but
tissues   still by tissues

$$\cancel{H_2CO_3} \longrightarrow \cancel{H^+} + HCO_3^-$$

In red   In red   In red cell;
cell   cell   will go to lungs

$$\cancel{H^+} + HbO_2^- \longrightarrow HHb \quad + O_2$$

Just   In red   In red cell;   Goes into
made   cell; by   will return   tissue
        tissue   to lungs   needing it

Net effect of deoxygenating oxyhemoglobin:

$$\mathbf{CO_2 + H_2O + HbO_2^- \longrightarrow HHb + HCO_3^- + O_2} \qquad (25.4)$$

Waste   In red   In red   Goes in   Goes
from   cell; by   cell; will   blood to   into
tissue   tissues   go to lungs   lungs   tissue

These summarizing equations omit two features. They do not show the importance of DPG and of the concentration of H^+ in both the loading and the unloading of oxygen. Concerning the concentration of H^+, Figure 25.9 shows the plots of oxygen affinity versus pO_2 at two different values of pH, one relatively low (pH 7.2) compared with the normal value of 7.35, and the other relatively high (pH 7.6). You can recall that the pO_2 in the vicinity of oxygen-starved cells is around 30–40 mm Hg. Notice in Figure 25.9 that in this range of the partial pressure of oxygen, the blood's ability to hold oxygen is much less at the lower pH of 7.2 than it is at a pH of 7.6. In actively metabolizing cells, there is a localized drop in pH caused chiefly by the presence of the CO_2 that these cells have made. This drop in pH caused by CO_2 cannot help but to assist in the deoxygenation of HbO_2^-. Thus precisely where O_2 should be unloaded there is a chemical signal (a lower pH) that makes it happen. It's an altogether beautiful example of how a set of interrelated chemical equilibria shift in just those directions that are required for health and life. Figure 25.9 will also help us understand in what ways both acidosis and alkalosis are serious threats.

There is another chemical reaction involving waste carbon dioxide that we must now mention. Not all of the CO_2 made by metabolizing cells winds up as HCO_3^-. Some CO_2 reacts with the hemoglobin that has just been freed by deoxygenation:

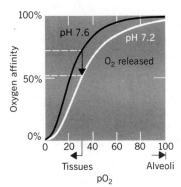

100% — pH 7.6   pH 7.2
O_2 released
50%
0%
0   20   40   60   80   100
Tissues   Alveoli
pO_2

Oxygen affinity

FIGURE 25.9
Hemoglogin–oxygen dissociation curves at two different values of the pH of blood.

$$CO_2 + HHb \longrightarrow Hb—CO_2^- + H^+ \qquad (25.5)$$

Carbaminohemoglobin

Waste   Just released   In red   Will react
from   by deoxygenation   cells   with HbO_2^-
tissues

This is actually the forward reaction of an equilibrium. The product, $Hb—CO_2^-$, is called **carbaminohemoglobin**, and it is one form in which some of the waste CO_2 travels in the blood back to the lungs. Notice in Equation 25.5, however, that this reaction also produces H^+, just as did the ionization of the H_2CO_3 that formed from other molecules of waste CO_2. Thus whether waste CO_2 is changed to H_2CO_3 (and thence, by ionization, to $HCO_3^- + H^+$) or to $Hb—CO_2^- + H^+$, either fate helps to generate hydrogen ions that are needed to react with HbO_2^- and make it unload O_2. When the red cell reaches the lungs where H^+ will now be *generated*:

$$HHb + O_2 \longrightarrow HbO_2^- + H^+ \qquad \text{(25.1, again)}$$

then the reaction of Equation 25.5 is forced to shift into reverse, because the added H^+ provides this kind of stress. Of course, this releases the CO_2 where it can be exhaled, and it gets hemoglobin ready to take on more oxygen.

Still another factor we have not mentioned thus far and that helps to unload oxygen from oxyhemoglobin is the chloride ion. Hemoglobin, HHb, binds chloride ion, and it binds it better than does oxyhemoglobin. The following equilibrium exists along with all of the others:

$$HHb(Cl^-) + O_2 \rightleftharpoons HbO_2^- + Cl^- + H^+ \qquad \text{(25.6)}$$

Therefore to *unload* oxygen (do the reverse of 25.6), chloride ion must be available *inside* the red cell. The red cell obtains it by a mechanism called the *chloride shift*. Let's see how it occurs.

As chloride ions are drawn into the red cell to react with some of the HHb that is released in active tissue, negative ions have to move out so that there is electrical charge balance. The negative ions that leave are the newly forming bicarbonate ions. From 60% to 90% of all waste CO_2 returns to the lungs as the bicarbonate ion, but most of this makes the trip *outside* red cells.

For every chloride ion that enters the red cell, one HCO_3^- ion leaves, and this switch is called the **chloride shift.** When the red cell gets back to the lungs, the various new chemical stresses make all of the equilibria, including the chloride shift, run in reverse.

If you're bewildered by all of these equilibria and how they are made to shift in the correct directions, you're almost certainly not alone among your classmates. This isn't easy material, but it is so much at the heart of so many serious medical situations that the effort to master it is very worthwhile. As you make this effort, get the key equilibria down and memorized. For each one ask, What are the stresses that can make it shift, and where in the body is each stress important — in the lungs or at actively metabolizing cells? The key stresses are the following: the relative partial pressures of O_2; the relative partial pressures of CO_2; the changes in the concentration of H^+ caused by the influx of CO_2 or by the loss of CO_2 by exhaling.

After you have studied the various equilibria from the stress point of view, and you can write all of the equilibria and discuss the influence of various stresses, then you might find Figure 25.10 a useful way to review. Follow the direction of the large U-shaped arrow that curves around the legend, and use the boxed numbers in the legend to follow the events in the figure. Notice particularly that the reactions that occur in the red cell when it is in metabolizing tissue are just the reverse of those that happen when the cell is in the lungs.

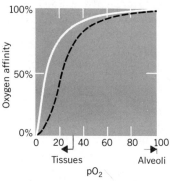

FIGURE 25.11
Myoglobin–oxygen dissociation curve — solid line. The dashed line is the curve for hemoglobin. Over the entire range of pH in tissues that might need oxygen, the oxygen affinity of myoglobin is greater than that of hemoglobin.

HMy = myoglobin

The Oxygenation of Myoglobin.   Myoglobin is a heme-containing protein in red muscle tissue such as heart muscle. Its function is to bind and store oxygen for the needs of such tissue. Unlike hemoglobin, myoglobin has only one polypeptide unit and only one heme unit per molecule. Moreover, there is no allosteric effect when it binds oxygen, as the shape of the myoglobin — oxygen dissociation curve (Figure 25.11) indicates.

Myoglobin's oxygen affinity is greater than that of hemoglobin, especially in the range of pO_2 associated with actively metabolizing tissues. Consequently, *myoglobin is able to take oxygen from oxyhemoglobin:*

$$HbO_2^- + HMy \longrightarrow HHb + MyO_2^-$$

This ability is vital to heart muscle which, as much as the brain, must have an assuredly continuous supply of oxygen. When oxymyoglobin, MyO_2^- gives up its oxygen for the cell's needs, it can at once get a fresh supply from the circulating blood. Not only does this cell now have CO_2 and H^+ available to deoxygenate HbO_2^- but it also has the superior oxygen affinity of its own myoglobin to draw more O_2 into the cell.

AT THE LUNGS

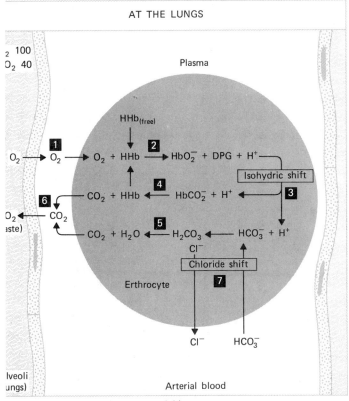

AT ACTIVE TISSUE

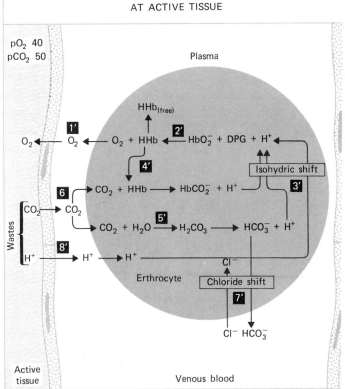

HHb = hemoglobin

HbO_2^- = oxyhemoglobin [actually, $Hb(O_2)_4$]

$HbCO_2^-$ = carbaminohemoglobin [actually, $Hb(CO_2)_4^{n-}$ where $n-$ = 1− to 4−]

DPG = diphosphoglycerate

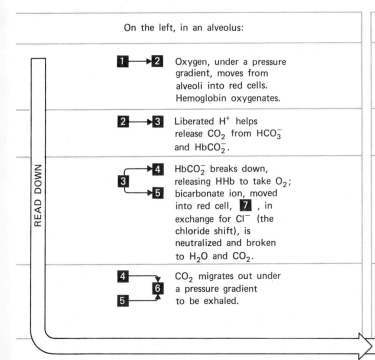

On the left, in an alveolus:

1 → 2 Oxygen, under a pressure gradient, moves from alveoli into red cells. Hemoglobin oxygenates.

2 → 3 Liberated H^+ helps release CO_2 from HCO_3^- and $HbCO_2^-$.

3 → 4, 5 $HbCO_2^-$ breaks down, releasing HHb to take O_2; bicarbonate ion, moved into red cell, **7**, in exchange for Cl^- (the chloride shift), is neutralized and broken to H_2O and CO_2.

4, 5 → 6 CO_2 migrates out under a pressure gradient to be exhaled.

READ DOWN

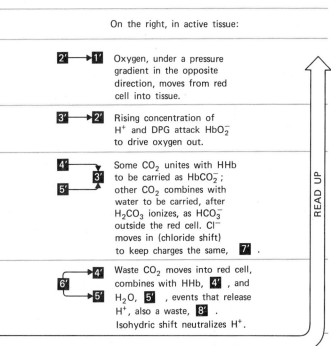

On the right, in active tissue:

2′ → 1′ Oxygen, under a pressure gradient in the opposite direction, moves from red cell into tissue.

3′ → 2′ Rising concentration of H^+ and DPG attack HbO_2^- to drive oxygen out.

4′, 5′ → 3′ Some CO_2 unites with HHb to be carried as $HbCO_2^-$; other CO_2 combines with water to be carried, after H_2CO_3 ionizes, as HCO_3^- outside the red cell. Cl^- moves in (chloride shift) to keep charges the same, **7′**.

6′ → 4′, 5′ Waste CO_2 moves into red cell, combines with HHb, **4′**, and H_2O, **5′**, events that release H^+, also a waste, **8′**. Isohydric shift neutralizes H^+.

READ UP

FIGURE 25.10

Oxygen and carbon dioxide exchange in the blood.

The Oxygenation of Fetal Hemoglobin.   The hemoglobin in a fetus is slightly different from that of an adult, and it has a higher oxygen affinity than adult hemoglobin. This helps to ensure to the fetus that it can successfully pull oxygen from the mother's oxyhemoglobin and satisfy its own needs.

25.4   ACID-BASE BALANCE OF THE BLOOD

The proper treatment of acidosis or of alkalosis depends on knowing if the underlying cause is a metabolic or a respiratory disorder.

Blood specimens are usually taken of venous blood—from veins. Arterial blood is blood in the arteries.

Acid-base balance exists when the pH of venous blood is in the range of 7.33 to 7.43. A decrease in pH—acidosis—or an increase—alkalosis—is serious and requires prompt attention, because all of the equilibria that involve H^+ in the oxygenation or the deoxygenation of blood are sensitive to pH.

In general, acidosis results either from the retention of acid or the loss of base by the body, and these can be induced by disturbances either in metabolism or in respiration. Similarly, alkalosis results either from the loss of acid or the retention of base, and some disorder in either metabolism or respiration can be the underlying cause.

A malfunction in respiration can be caused by any kind of injury to the *respiratory centers* in the brain. These are units in the brain that sense changes in the pH and pCO_2 of the blood and instruct the lungs either to breathe more rapidly or more slowly. Another cause of a malfunction of respiration is any kind of injury or disease of the lungs.

Normal values (venous blood):

$$pCO_2 = 38-50 \text{ mm Hg}$$
$$[HCO_3^-] = 22-29 \text{ meq/L}$$
$$(1.0 \text{ meq } HCO_3^- = 61 \text{ mg } HCO_3^-)$$

What we'll do in this section is study four situations, metabolic and respiratory acidosis as well as metabolic and respiratory alkalosis. We will learn how the values of pH, pCO_2, and serum $[HCO_3^-]$ change in each situation. (Normal values are given in the margin.) If your intended career is nursing—particularly surgical, critical care, or emergency room nursing—or respiration therapy, you should be aware that experienced nurses who return to study the chemistry they have missed (and now need) report that this unit is the most important unit they study in all of biological chemistry.

Metabolic Acidosis.   In **metabolic acidosis,** the lungs and the respiratory centers are working, and the problem lies somewhere in metabolism. For some reason, acids are produced faster than they are neutralized or otherwise exported. Excessive loss of base, such as from severe diarrhea, can also result in metabolic acidosis. (In diarrhea, the alkaline fluids of the duodenum leave the body, and as base migrates to replace them, there will be a depletion of base somewhere else, such as in the blood, at least for a period of time.)

As the pH of the blood falls and the molar concentration of H^+ rises, there are parallel *but momentary* increases in the values of pCO_2 and $[HCO_3^-]$. The value of pCO_2 starts to increase because the carbonate buffer, working hard to neutralize the extra H^+, manufactures CO_2:

$$\underset{\substack{\text{From} \\ \text{acidosis}}}{H^+} + HCO_3^- \longrightarrow H_2CO_3 \longrightarrow H_2O + CO_2$$

The kidneys work harder during this situation to try to keep up the supply of HCO_3^-. However, now the respiratory centers, which are sensitive to changes in pCO_2, instruct the lungs to blow the CO_2 out of the body. The lungs, in other words, hyperventilate.

As the equation given just above shows, the loss of each molecule of CO_2 means a net neutralization of one H^+ ion. Hyperventilation, however, is overdone. So much CO_2 is blown out that pCO_2 actually decreases. Thus as the blood pH decreases, so too do the values of pCO_2 (from hyperventilation) and $[HCO_3^-]$ (from the reaction with H^+). We can summarize the range of a number of clinical situations that involve metabolic acidosis as follows.

Clinical Situations of Metabolic Acidosis

(↓) means a decrease from a normal and (↑) means an increase. Some typical values are in parentheses. Note that some changes do not necessarily bring values outside the normal ranges.

Lab results:	pH↓ (7.20); pCO$_2$↓ (30 mm Hg); [HCO$_3^-$]↓ (14 meq/L)
Typical patient:	An adult comes to the clinic with a severe infection. Unknowingly, he has diabetes.
Range of causes:	Diabetes mellitus; severe diarrhea (with loss of HCO$_3^-$); kidney failure (to export H$^+$ or to make HCO$_3^-$); prolonged starvation; severe infection; aspirin overdose.
Symptoms:	Hyperventilation (because the respiratory centers have told the lungs to remove excess CO$_2$ from the blood); increased urine output (to remove H$^+$ from the blood); thirst (to replace water lost as urine); drowsiness; headache; restlessness; disorientation.
Treatment:	If the kidneys function, use isotonic HCO$_3^-$ intravenously to restore HCO$_3^-$ level, thereby neutralizing H$^+$ and raising pCO$_2$. Also, restore water. In diabetes, use insulin therapy. If the kidneys do not function, hemodialysis must be tried.

Respiratory Acidosis. In **respiratory acidosis,** either the respiratory centers or the lungs have failed, and the lungs are hypoventilating *because they can't help it.* The blood now can't help but retain CO$_2$. But the retention of CO$_2$ means the retention of its acid, H$_2$CO$_3$. The partial ionization of this acid is the source of the hydrogen ions that lower the pH and give rise to respiratory acidosis.

Clinical Situations of Respiratory Acidosis

Lab results:	pH↓ (7.10); pCO$_2$↑ (68 mm Hg); [HCO$_3^-$]↑ (40 meq/L)
Typical patient:	Chain smoker with emphysema or anyone with chronic obstructive pulmonary disease.
Range of causes:	Emphysema, pneumonia, asthma, anterior poliomyelitis, or any cause of shallow breathing such as an overdose of narcotics, barbiturates, or general anesthesia; severe head injury.
Symptoms:	Shallow breathing (which is involuntary)
Treatment:	Underlying problem must be treated; possibly hemodialysis.

An overdose of "bicarb" (NaHCO$_3$) can result from a too aggressive use of this home remedy for "heartburn."

Metabolic Alkalosis. In **metabolic alkalosis,** the system has lost acid; or it has retained base (HCO$_3^-$), or it has been given an overdose of base (e.g., antacids). Metabolic alkalosis can also be caused by a kidney-associated decrease in the serum levels of K$^+$ or Cl$^-$. The loss of these ions means the retention of Na$^+$ and HCO$_3^-$ ions, because these work in tandem and oppositely. The loss of acid could be from prolonged vomiting, which removes the gastric acid. This is followed by an effort to borrow serum H$^+$ to replace it, and the pH of the blood increases. Improperly operated nasogastric suction can also remove too much gastric acid.

Whatever the cause, the respiratory centers sense an increase in the level of base in the blood (as the level of acid drops), and they instruct the lungs to retain the most readily available neutralizer of base it has — namely CO$_2$, the source of carbonic acid:

$$CO_2 + H_2O \longrightarrow H_2CO_3$$

Retained   Carbonic acid

Then:

$$H_2CO_3 + OH^- \longrightarrow H_2O + HCO_3^-$$

From
alkalosis

Compensation by hypoventilation is obviously limited by the fundamental need of the body for some oxygen.

To help retain CO_2, the lungs hypoventilate. Thus metabolic alkalosis leads to hypoventilation. Notice carefully, hypoventilation alone cannot be used to tell if the patient has metabolic *alkalosis* or respiratory *acidosis*. Either condition means hypoventilation. But one condition—respiratory acidosis—could be treated by intravenous sodium bicarbonate, a base. Such a treatment would aggravate metabolic alkalosis. You can see that the lab data on pH, pCO_2, and $[HCO_3^-]$ must be obtained to determine which kind of condition is actually present. Otherwise, the treatment used could be just the opposite of what should be done. People working in emergency care situations get the requisite lab data rapidly, and they must be able to interpret the data on the spot.

Clinical Situations of Metabolic Alkalosis

Lab results:	pH↑ (>7.45); pCO_2↑ (>45 mm Hg); $[HCO_3^-]$↑ (>29 meq/L)
Typical patient:	Postsurgery patient with persistent vomiting.
Range of causes:	Prolonged loss of stomach contents (vomiting or nasogastric suction); overdose of bicarbonate or of medications for stomach ulcers; severe exercise, or stress, or kidney disease (with loss of K^+ and Cl^-).
Symptoms:	Hypoventilation (to retain CO_2); numbness, headache, tingling; possibly convulsions.
Treatment:	Isotonic ammonium chloride (a mild acid), intraveneously with great care. Replace K^+ loss.

Ammonium ion acts as a neutralizer as follows:

$$NH_4^+ + OH^- \longrightarrow NH_3 + H_2O$$

Respiratory Alkalosis. In **respiratory alkalosis,** the body has lost acid usually by some involuntary hyperventilation—hysterics, prolonged crying, or overbreathing at high altitudes—or by the mismanagement of a respirator. The respiratory centers have lost control, and the body expels CO_2 too rapidly. The loss of CO_2 means the loss of H_2CO_3, the substance than can neutralize base in the blood. Hence, the level of base rises; the pH rises.

Tissue that gets too little O_2 is in a state of **hypoxia.** If it gets none at all, it is in a state of **anoxia.**

An example of extreme respiratory alkalosis occurs among successful climbers of Mount Everest (8848 m, 29,030 ft), where the summit barometric pressure is 253 mm Hg and the pO_2 of the air that the climbers inhale is only 43 mm Hg (as compared to 149 mm Hg at sea level). Hyperventilation brings their arterial pCO_2 down to only 7.5 mm Hg (compared to a normal of 40 mm Hg) and the blood pH is above 7.7!

Clinical Situations of Respiratory Alkalosis

Lab results:	pH↑ (7.54); pCO_2↓ (32 mm Hg); $[HCO_3^-]$↓ (20 meq/L)
Typical patient:	Someone nearing surgery and experiencing anxiety.
Range of causes:	Prolonged crying; rapid breathing at high altitudes; hysterics; fever; disease of the central nervous system; improper management of a respirator.
Symptoms:	Hyperventilation (that can't be helped). Convulsions may occur.
Treatment:	Rebreathe one's own exhaled air (by breathing into a sack); administer carbon dioxide; treat underlying causes.

Take careful notice that hyperventilation alone cannot be used to tell what the condition is. Either metabolic *acidosis* or respiratory *alkalosis* is accompanied by hyperventilation, but the treatments are opposite in nature.

25.5 BLOOD AND THE FUNCTIONS OF THE KIDNEYS

Both filtration and chemical reactions in the kidneys help to regulate the electrolyte balance of the blood.

Diuresis is the formation of urine in the kidneys, and it is an integral part of the body's control of the levels in the blood of its electrolytes and its buffers. Figure 25.12 shows the parts of the kidneys that participate.

The net urine production is 0.6 to 2.5 L/day.

Huge quantities of fluids leave the blood by diffusion each day at the hundreds of thousands of filtering units, the glomeruli. Substances in solution but not those in colloidal dispersions (e.g., proteins) leave in these fluids. Then active transport processes in kidney cells pull all of any escaped glucose, any amino acids, and most of the fluids and electrolytes back into the blood. Most of the wastes are left in the urine that is being made.

Ur-, of the urine; *-emia*, of the blood. *Uremia* means substances of the urine present in the blood.

Urea is the chief nitrogen waste (30 g/day), but creatinine (1 – 2 g/day), uric acid (0.7 g/day), and ammonia (0.5 g/day) are also excreted with the urine. If the kidneys are injured or diseased and cannot function, wastes build up in the blood, which leads to a condition known as *uremic poisoning*.

Hormones and Diuresis. Vasopressin, a nonapeptide, is a hormone that helps to regulate the overall concentrations of substances dissolved in blood. The hypophysis, which makes vasopressin, releases this hormone when the osmotic pressure of blood rises by as little as 2%. At the kidneys, vasopressin promotes the reabsorption of water, and therefore it is often called the antidiuretic hormone (ADH). A higher-than-normal osmotic pressure (hypertonicity) means a higher concentration of solutes and colloids in blood. In such a condition, vasopressin helps the blood to retain water. By doing this, the concentrations of solutes are kept from rising still higher. In the meantime, the thirst mechanism is stimulated to bring in water to dilute the blood.

In *diabetes insipidus,* vasopressin secretion is blocked and unchecked diuresis can make from 5 to 12 liters of urine a day.

Conversely, if the osmotic pressure of blood decreases (becomes hypotonic) by as little as 2%, the hypophysis retains vasopressin. None reaches the kidneys, so the water that has left the bloodstream at the glomeruli does not return as much. Remember that a low osmotic pressure means a low concentration of solutes, so the absence of vasopressin at the kidneys when the blood is hypotonic lets urine form. This reduces the amount of water in the blood and thereby raises the concentrations of its dissolved matter. You can see that with the help of

FIGURE 25.12
The kidneys. *(a)* Principal parts. *(b)* Location of the nephrons. *(c)* Details of a nephron and its associated capillary bed. Each kidney has about a million nephrons. In the renal glomerulus, fluids and their solutes leave the blood, enter Bowman's capsule, and start to move down the tubules. The kidney now works to put most of the electrolytes and organic molecules back into the blood and to leave certain wastes in the urine. Final adjustments of concentrations are made in the distal tubule.

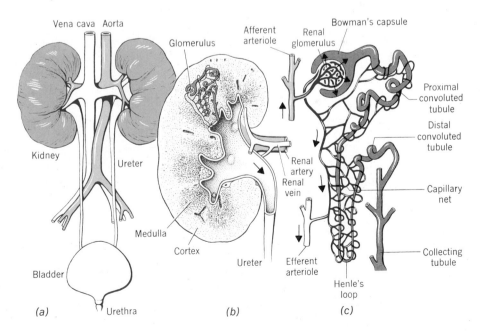

vasopressin a normal individual can vary the intake of water widely and yet preserve a stable, overall concentration of substances in blood.

Aldosterone is another hormone that affects diuresis, because it works to stabilize the sodium ion level of the blood. This steroid hormone, made in the adrenal cortex, is secreted if the sodium ion level drops. When aldosterone arrives at the kidneys, it initiates reactions that make sodium ions return from the urine being made to the blood. Of course, to keep things isotonic in the blood, the return of sodium ions also requires the return of water.

Conversely, if the sodium ion level of the blood increases, then aldosterone is not secreted, and sodium ions that have migrated out of the blood at glomeruli are permitted to stay out. They remain in the urine being made, together with some extra water.

Diuresis and the Acid–Base Balance.   The changes in breathing that affect the acid–base balance of the blood are rapid-acting. For the more routine, day-to-day maintenance of this balance, the body uses the services of the kidneys. Their cells have the ability to adjust the level of HCO_3^- in the blood and thus to assist in the management of the pH of the blood. In fact, a useful generalization is that the kidneys deal with the blood's acid-base balance most directly by controlling the *base* in the blood (HCO_3^-), and the lungs do this most directly by controlling the *acid* (CO_2, meaning H_2CO_3).

The kidneys also adjust the blood's levels of HPO_4^{2-} and $H_2PO_4^-$, the anions of the phosphate buffer. Moreover, when acidosis develops, the kidneys can put H^+ ions into the urine. Some neutralization of these ions by HPO_4^{2-} and by NH_3 takes place, but the urine becomes more acidic as acidosis continues, as we've mentioned before.

Figure 25.13 shows the various reactions that take place in the kidneys, particularly during acidosis. (The numbers in the following boxes refer to this figure.) The breakdown of metabolites, $\boxed{1}$, makes carbon dioxide, which enters the equilibrium whose formation is catalyzed by carbonic anhydrase, $\boxed{2}$. The ionization of carbonic acid, $\boxed{3}$, makes both bicarbonate ion and hydrogen ion. The bicarbonate ion goes into the bloodstream, $\boxed{4}$, but the hydrogen ion is put into the tubule, $\boxed{5}$, where urine is accumulating. This urine already contains sodium ions and monohydrogen phosphate ions, but to make step $\boxed{5}$ possible, *some* positive ion has to go with the HCO_3^-. Otherwise, there would be no net electrical balance. The kidneys have the ability to select Na^+ to go with HCO_3^- at $\boxed{4}$. The kidneys can make Na^+ travel one way and H^+ the other. Newly arrived H^+ can be buffered by HPO_4^{2-} in the developing urine, $\boxed{6}$. Moreover, the kidneys have an ability not generally found in other tissues to synthesize ammonia and use it to neutralize H^+, $\boxed{7}$. Thus the ammonium ion also appears in the urine.

If the acidosis has a metabolic origin, the level of the organic anions in the blood rises. These are the anions of organic acids that are made at accelerated rates in metabolic disorders and that are the chief cause of the pH change in metabolic acidosis. The kidneys let these anions stay in the urine, but there is a limit to how concentrated the urine can become in all of its solutes. Hence, the more that the body lets solutes stay in the urine, the more it must also let water stay. As a result, the patient with metabolic acidosis can experience a general dehydration as the system borrows water from other fluids to make urine. Usually, the thirst mechanism brings in replacement water, and the patient drinks copious amounts of water.

In alkalosis, the kidneys can put bicarbonate ion into the urine, and it no longer uses HPO_4^{2-} to neutralize H^+. Both actions raise the pH of the urine, and in *severe* alkalosis it can go over a pH of 8.

Blood Pressure and the Kidneys.   If the blood pressure drops, as in hemorrhaging, the kidneys secrete a trace of renin into the blood. Renin is an enzyme that acts on one of the globular proteins in blood, a zymogen called angiotensinogen, to convert it to another enzyme, angiotensin I. This, in turn, helps to convert still another protein in blood to angiotensin II, a neurotransmitter. It is the most potent vasoconstrictor known, and by making blood capillaries constrict, the heart has to work harder, which makes the blood pressure increase. This increase helps to ensure that some semblance of proper filtration continues at the kidneys.

Urine taken after several hours of fasting normally has a pH of 5.5 to 6.5.

In severe acidosis, the pH of urine can go as low as 4.

FIGURE 25.13
Acidification of the urine. The numbers refer to the text discussion.

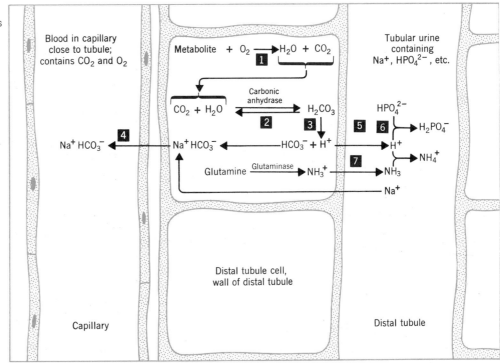

Angiotensin II also triggers the release of aldosterone, which we've already learned helps the blood to retain water. This is important because the maintenance of the overall blood volume is needed to sustain a proper blood pressure.

SUMMARY

Digestion α-Amylase in saliva begins the digestion of starch. Pepsin in gastric juice starts the digestion of proteins. In the duodenum, trypsinogen (from the pancreas) is activated by enterokinase (from the intestinal juice) and becomes trypsin, which helps to digest proteins. It also activates chymotrypsin (from chymotrypsinogen) and carboxypeptidase (from procarboxypeptidase). These also help to digest proteins. The pancreas supplies an important lipase, which, with the help of the bile salts, catalyzes the digestion of saponifiable lipids. The bile salts also aid in the absorption of the fat-soluble vitamins, A, D, E, and K.

Intestinal juice supplies enzymes for the digestion of disaccharides, nucleic acids, small polypeptides, and lipids.

The end products of the digestion of proteins are amino acids; of carbohydrates, glucose, fructose, and galactose; and of the triacylglycerols, fatty acids and glycerol. Complex lipids are also hydrolyzed, and nucleic acids yield phosphate, pentoses, and side-chain bases.

Blood Proteins in blood give it a colloidal osmotic pressure that assists in the exchange of nutrients at capillary loops. Albumins are carriers for hydrophobic molecules and serum-soluble metallic ions.

γ-Globulins help defend the body against bacterial infections. Fibrinogen is the precursor of fibrin, the protein of a blood clot.

Among the electrolytes, anions of carbonic and phosphoric acid are involved in buffers, and all ions are involved in regulating the osmotic pressure of the blood. The chief cation in blood is Na^+, and the chief cation inside cells is K^+. The balances between Na^+ and K^+ as well as between Ca^{2+} and Mg^{2+} are tightly regulated. Ca^{2+} is vital to the operation of muscles, including heart muscle. Mg^{2+} is involved in a number of enzyme systems.

The blood transports oxygen and products of digestion to all tissues. It carries nitrogen wastes to the kidneys. It unloads cholesterol and heme breakdown products at the gall bladder. And it transports hormones to their target cells. Lymph, another fluid, helps to return some substances to the blood from tissues.

Sudden failure to retain the protein in blood leads to an equally sudden loss in blood volume and a condition of shock. Slower losses of protein, as in kidney disease or starvation, lead to edema.

Respiration The relatively high pO_2 in the lungs helps to force O_2 into HHb. This creates HbO_2^- and H^+. In an isohydric shift, the H^+ is neutralized by HCO_3^-, which is returning from working tissues that make CO_2. The resulting H_2CO_3 decomposes to H_2O and CO_2, and the latter leaves during exhaling. Some of the H^+ also converts $HbCO_2^-$ to HHb and CO_2.

In deoxygenating HbO_2^- at cells that need oxygen, the influx of CO_2 makes H_2CO_3 that ionizes to HCO_3^- and H^+. The H^+ then moves (isohydric shift) to HbO_2^- and breaks it down to HHb and O_2. Both Cl^- and DPG help to ease the last of the four O_2 molecules out of oxyhemoglobin. The low oxygen affinity of blood in tissues where pCO_2 is high helps in the release of oxygen also.

In red muscle tissue, myoglobin's superior oxygen affinity ensures that such tissue can obtain oxygen from the deoxygenation of oxyhemoglobin. Fetal hemoglobin also has a superior oxygen affinity to adult hemoglobin.

Acid–base balance   The body uses the bicarbonate ion of the carbonate buffer to inhibit acidosis by irreversibly removing H^+ when the lungs release CO_2. HCO_3^- is replaced by the kidneys, which can also put excess H^+ into the urine. H_2CO_3 in the carbonate buffer works to control alkalosis by neutralizing OH^-. Metabolic acidosis, with hyperventilation, and metabolic alkalosis, with hypoventilation, arise from dysfunctions in metabolism. Respiratory acidosis, with hypoventilation, and metabolic alkalosis, with hyperventilation, occur when the respiratory centers or the lungs are not working.

Diuresis   The kidneys, with the help of hormones and changes in blood pressure, blood osmotic pressure, and concentrations of ions, monitor and control the concentrations of solutes in blood. Vasopressin tells the kidneys to keep water in the bloodstream. Aldosterone tells the kidneys to keep sodium ion (and therefore water also) in the bloodstream. A drop in blood pressure tells the kidneys to release renin, which activates a vasoconstrictor, and aldosterone, which helps to raise blood pressure and retain water. In acidosis, the kidneys transfer H^+ to the urine and replace some of the HCO_3^- lost from the blood. In alkalosis the kidneys put some HCO_3^- into urine.

KEY TERMS

The following terms were emphasized in this chapter, and they constitute a minimum vocabulary for the concepts that were studied.

acidosis, metabolic	chloride shift	globulins	mucin
acidosis, respiratory	chyme	hemoglobin	nuclease
albumin	chymotrypsin	hyperkalemia	oxygen affinity
aldosterone	digestive juice	hypernatremia	oxyhemoglobin
alkalosis, metabolic	2,3-diphosphoglycerate (DPG)	hypokalemia	pancreatic juice
alkalosis, respiratory	edema	hyponatremia	pepsin
aminopeptidase	electrolytes, blood	internal environment	pepsinogen
α-amylase	enterokinase	interstitial fluids	saliva
bile	erythrocytes	intestinal juice	shock, traumatic
bile salts	extracellular fluids	isohydric shift	sucrase
carbaminohemoglobin	fibrin	lactase	trypsin
carboxypeptidase	fibrinogen	lipase	vascular compartment
cardiovascular compartment	gastric juice	maltase	vasopressin

SELECTED REFERENCES

1 Helen Hamilton (ed.) *Monitoring Fluid and Electrolytes Precisely.* Intermed Communications, Inc., Horsham, PA, 1979.

2 R. E. Dickerson and I. Geis. *Hemoglobin.* Benjamin/Cummings Publishing Company, Menlo Park, CA, 1983.

3 Linda B. Glass and Cheryl A. Jenkins. "The Ups and Downs of Serum pH." *Nursing 83,* September 1983, page 34.

4 Chris Perkins and Hirotaka Bralley. "Metabolic Alkalosis." *Nursing 83,* January 1983, page 57.

5 Dolores L. Taylor. "Respiratory Acidosis." *Nursing 84,* October 1984, page 44.

6 Dolores L. Taylor. "Resiratory Alkalosis." *Nursing 84,* November 1984, page 44.

7 N. M. Senozan and R. L. Hunt. "Hemoglobin: Its Occurrence, Structure, and Adaptation." *Journal of Chemical Education,* March 1982, page 173.

8 M. F. Perutz. "Hemoglobin Structure and Respiratory Transport." *Scientific American,* December 1978, page 92.

9 J. B. West. "Human Physiology at Extreme Altitudes on Mount Everest." *Science,* February 24, 1984, page 223.

REVIEW EXERCISES

The answers to these Review Exercises are in the *Study Guide* that accompanies this book.

Digestion

25.1 What are the names of the two chief extracellular fluids?

25.2 Name the fluids that have digestive enzymes or digestive zymogens.

25.3 What enzymes or zymogens are there, if any, in each of the following?
  (a) Saliva   (b) Gastric juice
  (c) Pancreatic juice   (d) Bile
  (e) Intestinal juice

25.4 Name the enzymes and the digestive juices that supply them (or their zymogens) that catalyze the digestion of each of the following.
  (a) Large polypeptides   (b) Triacylglycerols
  (c) Amylose   (d) Sucrose
  (e) Di- and tripeptides   (f) Nucleic acids

25.5 What are the end products of the complete digestion of each of the following?
  (a) Proteins   (b) Carbohydrates
  (c) Triacylglycerols

25.6 What functional groups are hydrolyzed when each of the substances in Review Exercise 25.5 is digested? (Refer back to earlier chapters if necessary.)

25.7 In what way does enterokinase function as a "master switch" in digestion?

25.8 What would happen if the pancreatic zymogens were activated within the pancreas?

25.9 What services do the bile salts render in digestion?

25.10 What does mucin do (a) for food in the mouth, and (b) for the stomach?

25.11 What is the catalyst for each of the following reactions?
  (a) Pepsinogen $\rightarrow$ pepsin
  (b) Trypsinogen $\rightarrow$ trypsin
  (c) Chymotrypsinogen $\rightarrow$ chymotrypsin
  (d) Procarboxypeptidase $\rightarrow$ carboxypeptidase

25.12 Rennin does what for an infant?

25.13 Why is gastic lipase unimportant to digestive processes in the adult stomach but useful in the infant stomach?

25.14 In terms of where they work, what is different about intestinal juice compared to pancreatic juice?

25.15 What secretion neutralizes chyme, and why is this work important?

25.16 What happens to the molecules of glycerol and fatty acids that form from digestion?

25.17 In a patient with a severe obstruction of the bile duct, the feces appear clay-colored. Explain why the color is light.

25.18 When the gall bladder is surgically removed, lipids of low formula weight are the only kinds that can be easily digested. Explain.

Substances in Blood

25.19 In terms of their general composition, what is the greatest difference between blood plasma and interstitial fluid?

25.20 What is the largest contributor to the net osmotic pressure of the blood as compared to the interstitial fluid?

25.21 What is fibrinogen? Fibrin?

25.22 What services are performed by albumins in blood?

25.23 What does γ-globulin do?

25.24 In what two different regions are Na^+ and K^+ ions mostly found?

25.25 What are the major functions of Na^+ and K^+?

25.26 In hypernatremia, the sodium ion level of blood is above what value?

25.27 The sodium ion level of blood is below what value in hyponatremia?

25.28 What causes the hyperkalemia in crushing injuries?

25.29 Above what level is the blood described as hyperkalemic?

25.30 Excessive drinking of water tends to cause what condition, hyponatremia or hypernatremia?

25.31 The overuse of milk of magnesia can lead to what condition that involves Mg^{2+}?

25.32 Inside cells, what is a function that Mg^{2+} serves?

25.33 What does the calcium ion have to do with the contraction of heart muscle?

25.34 In what way do medications classified as calcium channel blockers affect the movements of calcium ions, and in what clinical circumstance is this kind of medication indicated?

25.35 What condition is brought on by an overdose of vitamin D, hypercalcemia or hypocalcemia?

25.36 Injections of magnesium sulfate would be used to correct which condition, hypomagnesemia or hypermagnesemia?

25.37 What is the principal anion in both the blood and the interstitial fluid?

25.38 Explain how hypochloremia leads to alkalosis.

25.39 What is the normal range of concentration of Cl^- in blood?

Exchange of Nutrients at Capillary Loops

25.40 What two opposing forces are at work on the arterial side of a capillary loop? What is the net result of these forces, and what does the net force do?

25.41 On the venous side of a capillary loop there are two opposing forces. What are they, what is the net result, and what does this cause?

25.42 Explain how a sudden change in the permeability of the capillaries can lead to shock.

25.43 Explain how each of the following conditions leads to edema.
(a) Kidney disease   (b) Starvation
(c) A mechanical blow

Exchange of Respiratory Gases

25.44 What are the respiratory gases?

25.45 What compound is the chief carrier of oxygen to actively metabolizing tissues?

25.46 The binding of oxygen to hemoglobin is said to be allosteric. What does this mean, and why is it important?

25.47 Write the equilibrium expression for the oxygenation of hemoglobin. In what direction does this equilibrium shift when:
(a) The pH decreases?
(b) The pO_2 decreases?
(c) The red cell is in the lungs?
(d) The red cell is in a capillary loop of an actively metabolizing tissue?
(e) CO_2 comes into the red cell?
(f) HCO_3^- ions flood into the red cell?

25.48 Using chemical equations, describe the isohydric shift when a red cell is (a) in actively metabolizing tissues and (b) in the lungs.

25.49 In what two ways does the oxygenation of hemoglobin in red cells in alveoli help to release CO_2?

25.50 In what way does waste CO_2 at active tissues help to release oxygen from the red cell?

25.51 In what way does extra H^+ at active tissue help release oxygen from the red cell?

25.52 Where is carbonic anhydrase found in the blood, and what function does it have in the management of the respiratory gases in (a) an alveolus and (b) actively metabolizing tissues?

25.53 How does DPG help in the process of oxygenating hemoglobin?

25.54 In what way is DPG involved in helping to deoxygenate HbO_2^-?

25.55 What are some changes involving the blood that occur when the body remains at a high altitude for a period of time, and how do these changes help the individual?

25.56 It has been reported that some long-distance Olympic runners have trained at high altitudes and then had some of their blood withdrawn and frozen. Days or weeks after returning to lower altitudes and just prior to a long race, they have used some of this blood to replace an equal volume of what they are carrying. This is supposed to help them in the race. How would it work?

25.57 How would the net equations for the oxygenation and the deoxygenation of blood be changed to include the function of DPG?

25.58 What are the two main forms in which waste CO_2 moves to the lungs?

25.59 How is oxygen affinity affected by pCO_2, and how is this beneficial?

25.60 What is the chloride shift and how does it aid in the exchange of respiratory gases?

25.61 In what way is the superior oxygen affinity of myoglobin over hemoglobin important?

25.62 Aquatic diving animals are known to have much larger concentrations of myoglobin in their red muscle tissue than humans. How is this important to their lives?

25.63 Fetal hemoglobin has a higher oxygen affinity than adult hemoglobin. Why is this important to the fetus?

Acid-Base Balance of the Blood

25.64 Construct a table using arrows ($\uparrow$) or ($\downarrow$) and typical lab data that summarize the changes observed in respiratory and metabolic acidosis and alkalosis. The column headings should be as follows:

Condition	pH	pCO_2	$[HCO_3^-]$

25.65 With respect to the *directions* of the changes in the values of pH, pCO_2, and $[HCO_3^-]$ in both respiratory acidosis and metabolic acidosis, in what way are the two types of acidosis the same? In what way are they different?

25.66 Hyperventilation is observed in what two conditions that relate to the acid–base balance of the blood? In one, giving carbon dioxide is sometimes used, and in the other, giving isotonic HCO_3^- can be a form of treatment. Which treatment goes with which condition and why?

25.67 In what two conditions that relate to the acid–base balance of the blood is hypoventilation observed? Isotonic ammonium chloride or isotonic sodium bicarbonate are possible treatments. Which treatment goes with which condition, and how do they work?

25.68 In which condition that relates to acid–base balance does hyperventilation have a beneficial effect? Explain.

25.69 Hyperventilation is part of the *cause* of the problem in which condition relating to the acid–base balance of the blood?

25.70 Hypoventilation is the body's way of helping itself in which condition relating to the acid–base balance of the blood?

25.71 In which condition that concerns the acid–base balance of the blood is hypoventilation part of the *problem* rather than the cure?

25.72 How can a general dehydration develop in metabolic acidosis?

25.73 Which condition, metabolic or respiratory acidosis or alkalosis, results from each of the following situations?
(a) Hysterics
(b) Overdose of bicarbonate
(c) Emphysema
(d) Narcotic overdose

(e) Diabetes

(f) Overbreathing at a high altitude

(g) Severe diarrhea

(h) Prolonged vomiting

(i) Cardiopulmonary disease

(j) Barbiturate overdose

25.74 Referring to Review Exercise 25.73, which is happening in each situation, hyperventilation or hypoventilation?

25.75 Why does hyperventilation in hysterics cause alkalosis?

25.76 Explain how emphysema leads to acidosis.

25.77 Prolonged vomiting leads to alkalosis. Explain.

25.78 Uncontrolled diarrhea can cause acidosis. Explain.

Blood Chemistry and the Kidneys

25.79 If the osmotic pressure of the blood has increased, what, in general terms, has changed to cause this?

25.80 How does the body respond to an increase in the osmotic pressure of the blood?

25.81 If the sodium ion level of the blood falls, how does the body respond?

25.82 What is the response of the kidneys to a decrease in blood pressure?

25.83 Alcohol in the blood suppresses the secretion of vasopressin. How does this affect diuresis?

25.84 In what ways do the kidneys help to reduce acidosis?

Chapter 26
Biochemical Energetics

What a surge of energy the mind calls forth from the body when there is a prize 100 meters away! We study the molecular basis of energy for living in this chapter.

26.1 ENERGY FOR LIVING

High-energy phosphates, such as ATP, are the body's means of trapping the energy of the oxidation of the products of digestion.

Catabolism is from the Greek, *cata-*, down; *ballein*, to throw or cast.

We cannot use solar energy directly, like plants. We cannot use steam energy, like a locomotive. We need *chemical* energy for living. We obtain it from food, and we use it to make high-energy molecules. Then these drive the chemical engines behind muscular work, signal-sending, and chemical manufacture in tissue. Our principal source of chemical energy is the **catabolism** (i.e., the breaking down) of carbohydrates and fatty acids, although we can also use proteins for energy.

The energy released as heat by the *combustion* of 1 mol of glucose is exactly the same as the energy that is available by any other method of converting this compound to carbon dioxide and water — 673 kcal/mol. The difference is that our cells don't *burn* glucose and so receive its energy solely as heat. We metabolize it by a number of small steps, some of which make other high-energy compounds. Thus some of the energy available from changing glucose to CO_2 and H_2O is used to run chemical reactions, and the rest of the energy is released as heat.

If we were to burn 1 mol of palmitic acid, $CH_3(CH_2)_{14}CO_2H$, a typical fatty acid, we could obtain 2400 kcal. In the body, we can also change this compound to CO_2 and H_2O, and we obtain exactly the same energy per mole. However, as with glucose, not all of this energy is released as heat. Some is used to make high-energy compounds, compounds such as the high-energy phosphates.

High-Energy Phosphates.   In Section 16.6 we first learned about the existence of energy-rich triphosphate esters. There are many of these and similar high-energy compounds involved in the mobilization of chemical energy in the body. Table 26.1 gives the names and structures of the principal phosphates we'll encounter. They are arranged in the order of their **phosphate group transfer potentials** — their relative abilities to transfer a phosphate group to an acceptor, as in the following equation:[1]

The bond in color denotes the P—O bond that breaks in an energy-releasing reaction of a high-energy phosphate.

$$R-O-PO_3^{2-} + R'-O-H \longrightarrow R-O-H + R'-O-PO_3^{2-} \qquad (26.1)$$

Higher potential   Lower potential

The numbers in the last column of Table 26.1 are measures of the actual potentials the compounds have to transfer their phosphate groups under standard conditions. (We don't need to know anything about these conditions, just that they supply a common reference.) Only those compounds from ATP and above are classified as **high-energy phosphates.** Thus the first compound in the table has the highest potential of all for transferring a phosphate group to a receptor.

We can use the positions of compounds in Table 26.1 to predict whether a particular transfer is possible. A compound with a higher phosphate group transfer potential can, in principal, always be used to make one with a lower potential (assuming that the right enzyme and any other needed reactants are available).

PRACTICE EXERCISE 1   Using Table 26.1, tell if each reaction can occur.

(a) ATP + glycerol 3-phosphate → ADP + glycerol 1,3-diphosphate
(b) ATP + glucose → ADP + glucose 1-phosphate
(c) Glucose 1-phosphate + creatine → Glucose + creatine phosphate

[1] We will usually write the phosphate group as $-PO_3^{2-}$, but its state of ionization varies with the pH of the solution. At physiological pH, the group is mostly in the singly and doubly ionized forms, $-PO_3H^-$ and $-PO_3^{2-}$.

TABLE 26.1
Some Organophosphates in Metabolism

Phosphate	Structure	Phosphate Group Transfer Potential
Phosphoenolpyruvate	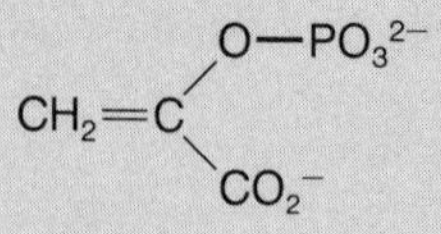 CH_2=C with O—PO_3^{2-} and CO_2^-	14.8
1,3-Diphosphoglycerate	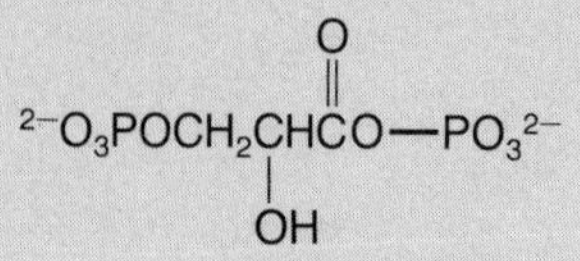 $^{2-}O_3POCH_2CHCO$—PO_3^{2-}, with =O above CO and OH below	11.8
Creatine phosphate	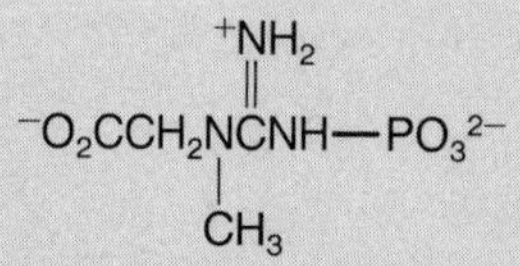 $^-O_2CCH_2NCNH$—PO_3^{2-}, with $^+NH_2$ double bond above and CH_3 below	10.3
Acetyl phosphate	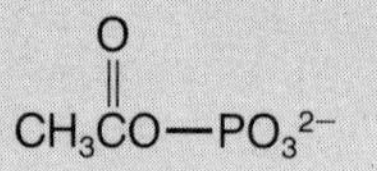 CH_3CO—PO_3^{2-}, with =O above	10.1
Adenosine triphosphate ATP	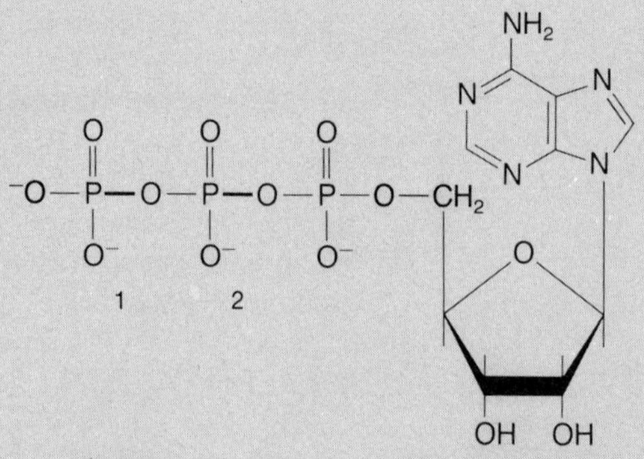	7.3 (at 1) 7.3 (at 2)
Glucose 1-phosphate	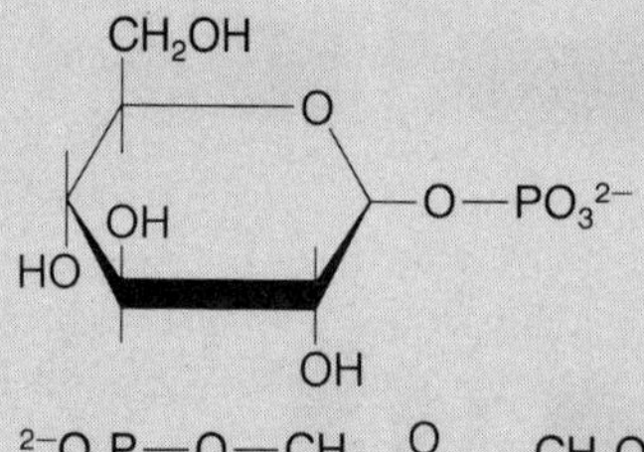	5.0
Fructose 6-phosphate	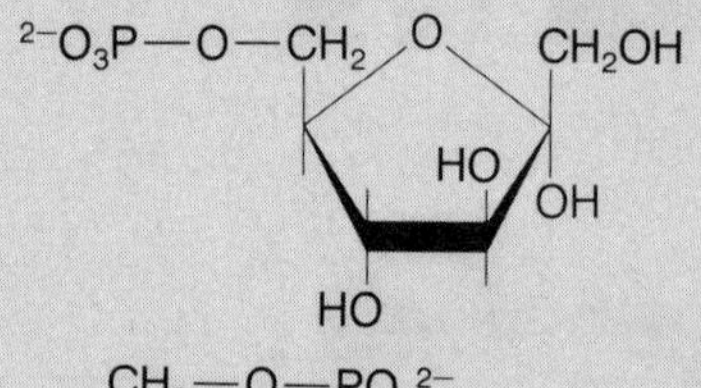	3.8
Glucose 6-phosphate	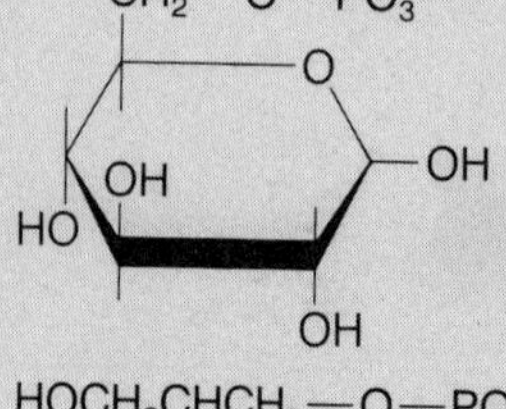	3.3
Glycerol 3-phosphate	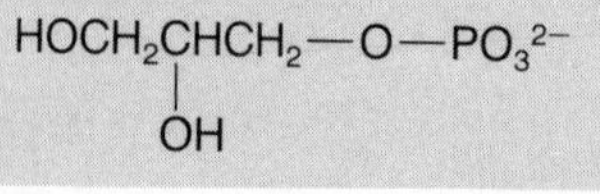 $HOCH_2CHCH_2$—O—PO_3^{2-}, with OH below	2.2

Adenosine Triphosphate (ATP).   One compound in Table 26.1 — **adenosine triphosphate** or **ATP** — is so important that we must review what we learned about it in Section 16.6. From the lowest to the highest forms of life, ATP is universally used as the principal carrier of energy for bodily functions. It is the chief means used by the body to trap energy available by oxidations. Virtually all of biochemical energetics comes down to the synthesis and uses of this compound.

Almost any energy-demanding activity of the body consumes ATP. We saw in Section 16.6, for example, that muscle contraction can be very simply written as a chemical reaction of ATP with the proteins in relaxed muscle (or with something within the proteins):

$$\text{``Relaxed'' muscle} + \text{ATP} \longrightarrow \text{``contracted'' muscle} + \text{ADP} + \text{P}_i$$

This reaction causes changes in tertiary structures of muscle proteins that cause the fibers made from these proteins to contract. Simultaneously, ATP changes to **ADP,** which is **adenosine diphosphate,** plus inorganic phosphate, P_i.

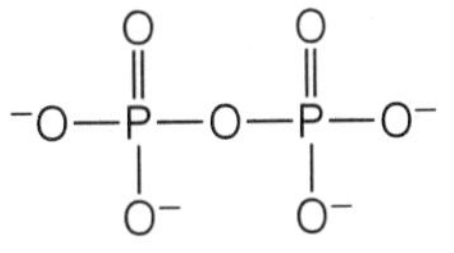

ATP has two phosphate bonds either of whose rupture can release much useful energy. Occasionally the second bond (see Table 26.1) is broken in some transfer of chemical energy, and the products then are **AMP, adenosine monophosphate,** and the inorganic diphosphate ion, PP_i.

Occasionally triphosphates other than ATP are the carriers of energy. Guanosine triphosphate, GTP, is an example that we'll encounter later in this chapter.

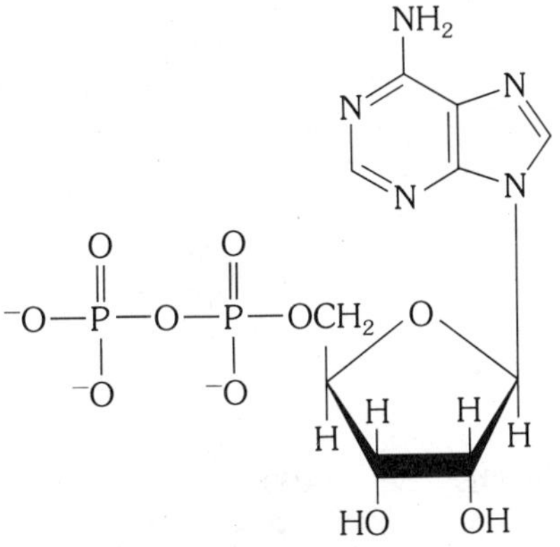

PP_i

ADP
Adenosine
diphosphate

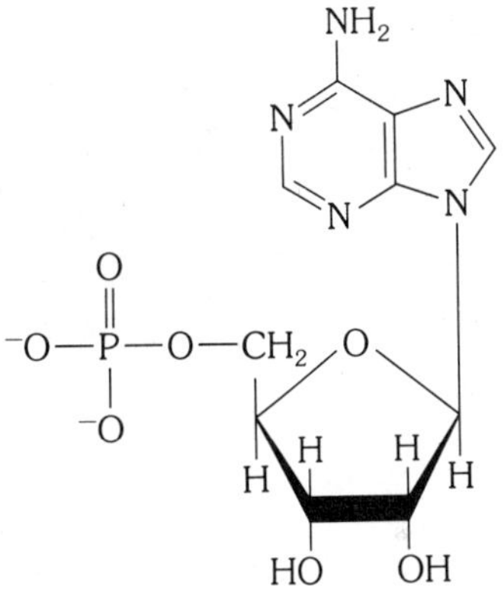

AMP
Adenosine
monophosphate

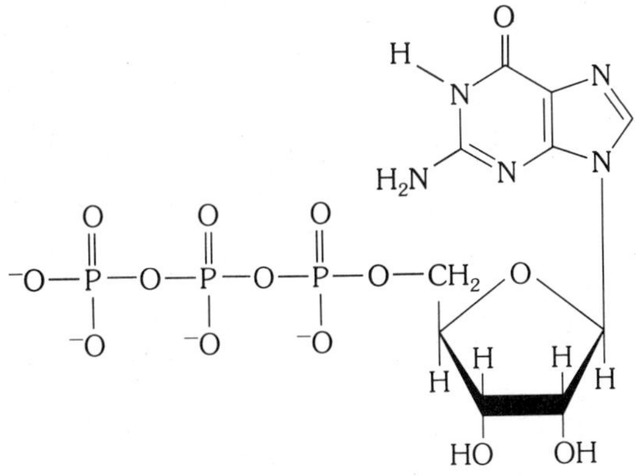

GTP
Guanosine
triphosphate

Once ATP is used, it must be remade or no more internal work can be done. One of the major purposes of catabolism is to transfer chemical energy in such a way that ATP is resynthesized from ADP and P_i. Sometimes, one of the compounds in Table 26.1 *above* ATP is first made by catabolism, and then its phosphate group is made to transfer to ADP to give ATP. When this happens, the overall ATP-synthesis is called **substrate phosphorylation** (where the substrate is ADP). The transfer is direct from an organic phosphate donor to ADP. In substrate phosphorylation, the phosphate group doesn't come directly from a phosphate ion in solution. Not that inorganic phosphate can't be pinned directly to ADP. This happens in another kind of phosphorylation called **oxidative phosphorylation** that is accomplished by an important series of reactions called the **respiratory chain.** We'll study this in the next section.

As you saw in Table 26.1, not all high-energy phosphates are triphosphates. Creatine phosphate, for example, is as important to muscle contraction as ATP. The ATP that is actually present in rested muscle can sustain muscle activity for only a fraction of a second. To provide for the immediate regeneration of ATP, muscle tissue makes and stores creatine phosphate (also known as phosphocreatine) during periods of rest. Then, as soon as some ATP is used, and ADP + P_i is made, creatine phosphate regenerates ATP. The enzyme is creatine kinase, the CK enzyme we studied in Chapter 24. What actually happens is that a rise in the supply of ADP shifts the following equilibrium to the right, to raise the concentration of ATP.

When the level of ATP drops and the levels of ADP + P_i rise, the rate of breathing is also accelerated.

$$\text{Creatine phosphate} + \text{ADP} \underset{\text{CK}}{\rightleftharpoons} \text{creatine} + \text{ATP}$$

Then, during periods of rest while ATP is made by other means, this equilibrium shifts back to the left to recharge the creatine phosphate reserves. Although muscle tissue has from 3 to 4 times as much creatine phosphate as ATP, even this reserve won't supply the high-energy phosphate needs for long—no more than what a sprinter needs for a 100 m to a 200 m sprint. For a long period of work, the body uses other methods to make ATP. We'll now take an overview of all of the pathways for ATP synthesis, and then we will look closely at the respiratory chain.

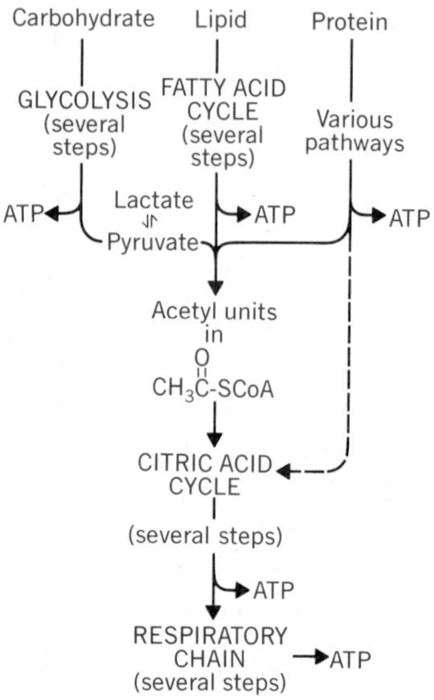

FIGURE 26.1
The major pathways for making ATP.

Major Bioenergetic Pathways: An Overview. Our first concern is what *initiates* ATP synthesis. This process is under feedback control, and if the supply of ATP is high, no more needs to be made. Only as ATP is used up is first one mechanism and then another thrown into action. Generally, it's the appearance of ADP that triggers these actions.

Figure 26.1 is a broad outline of the metabolic pathways that can generate ATP. Think of the last one mentioned, the respiratory chain, as being at the bottom of a tub, nearest the drain through which ATP will leave for some use. As in water that leaves a tub, the first to leave is at the plug. This activates the motions of water at higher levels.

By analogy, the next pathway to be thrown into action once the respiratory chain is launched is the next one from the bottom in Figure 26.1, the citric acid cycle. The chief purpose of the **citric acid cycle** is to supply the chemical needs of the respiratory chain.

The citric acid cycle also requires a "fuel," and this need is filled by an acetyl derivative of a coenzyme called coenzyme A. **Acetyl coenzyme A,** often written as acetyl CoA, is the "fuel" for the citric acid cycle. The catabolism of molecules from all three major foods — carbohydrates, lipids, and proteins — can produce acetyl coenzyme A.

From starch or glucose a pathway called **glycolysis** breaks glucose units down to the pyruvate ion. Then a short pathway converts this to acetyl coenzyme A. Glycolysis also produces some ATP. It does so independently of the respiratory chain by the use of substrate phosphorylation. Even when a cell is temporarily starved for oxygen, glycolysis can make some ATP for a while. When glycolysis has to run without oxygen, then its end product is the

The pyruvate ⇌ lactate equilibrium is discussed in Chapter 27.

lactate ion, not the pyruvate ion. Once the cell gets oxygen, however, it converts lactate to pyruvate.

$$C_6H_{12}O_6 \xrightarrow[\text{not balanced}]{\text{(Several steps;}} 2CH_3\overset{\displaystyle O}{\overset{\displaystyle \|}{C}}CO_2^- \underset{\text{(With O}_2\text{)}}{\overset{\text{(If no O}_2\text{)}}{\rightleftharpoons}} 2CH_3\overset{\displaystyle OH}{\overset{\displaystyle |}{C}}HCO_2^-$$

Glucose   Pyruvate   Lactate

$$2CH_3\overset{\displaystyle O}{\overset{\displaystyle \|}{C}}-S-CoA$$

Acetyl coenzyme A

Thus glycolysis can end either with lactate or with pyruvate, depending on the oxygen supply. We will study details of glycolysis in the next chapter.

Aerobic signifies the use of air. Anaerobic, stemming from "not air," means in the absence of the use of oxygen.

The full sequence of oxygen-consuming reactions from glucose units to pyruvate ions to acetyl CoA and on through the respiratory chain is sometimes called the **aerobic sequence** of glucose catabolism. The sequence that runs without oxygen from glucose units to lactate ions is called the **anaerobic sequence** of glucose catabolism.

Fatty acids are another source of acetyl CoA. A series of reactions called the **fatty acid cycle** breaks them down. We will study this pathway in Chapter 28.

Most amino acids can also be catabolized to acetyl units or to intermediates of the citric acid cycle itself, and we'll survey these reactions in Chapter 29.

Our interest in this chapter is in the pathways that can accept breakdown products from any food, namely, the citric acid cycle and the respiratory chain.

26.2 OXIDATIVE PHOSPHORYLATION AND THE CHEMIOSMOTIC THEORY

The flow of electrons to oxygen in the respiratory chain creates a proton gradient in mitochondria that drives the synthesis of ATP.

The term respiration refers to more than just breathing. It includes the chemical reactions that use oxygen in cells. Oxygen is reduced to water, and the following equation is the most basic statement we can write for what happens:

$$(\colon) \quad + 2H^+ + \cdot\ddot{O}\cdot \longrightarrow H-\overset{\cdot\cdot}{\underset{\cdot\cdot}{O}}-H + energy$$

Pair of   Pair of   Atom of   Molecule
electrons   protons   oxygen   of water

The gain of e^- or of e^- carriers such as $H\colon^-$ is reduction; the loss of e^- or of $H\colon^-$ is oxidation.

The electrons and protons come mostly from intermediates in the catabolism of sugars and fats. Generally, the electrons of C—H bonds are used, and sometimes the hydride ion $H\colon^-$ is the vehicle for carrying the electrons. Hydride ions can be passed directly from organic donors to organic receptors. Sometimes just electrons are passed. Remember, now, that when any molecule loses electrons it is oxidized, and the acceptor is reduced.

The chief agents of electron transfer are the **respiratory enzymes** that catalyze the reactions of the respiratory chain to which we referred when we laid out an overview of biochemical energetics. The respiratory chain is one long series of oxidation-reduction reactions. The flow of electrons from the initial donor is down an energy hill all the way to oxygen, and it is irreversible.

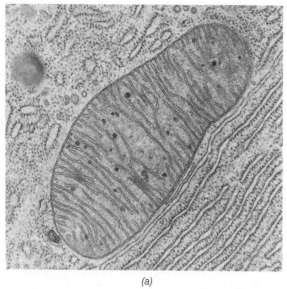

(a)

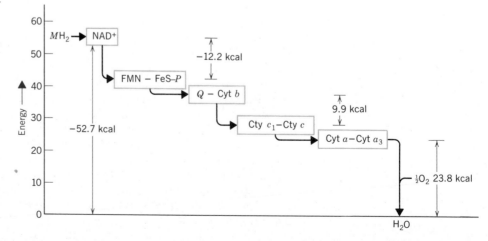

Respiratory enzymes
(in and on
inner membrane)

ATP synthesis
sites

Outside of
inner membrane

Inside of
inner membrane

Inner membrane

Outer membrane

(b)

FIGURE 26.2

A mitochondrion. *(a)* Electron micrograph ($\times$53,000) of a mitochondrion in a pancreas cell of a bat. *(b)* Perspective showing the interior. The respiratory enzymes are incorporated into the inner membrane. On the inside of this membrane are enzymes that catalyze the synthesis of ATP. (Micrograph courtesy of Dr. Keith R. Porter.)

A cell in the flight muscle of a wasp has about a million mitochondria.

Think of MH_2 as $M\!:\!H$, which becomes M when $H\!:\!^-$ and H^+ leave.

FIGURE 26.3

The enzymes of the respiratory chain and the energy drop from NAD^+ to reduced oxygen (water). The figures for energy refer to that portion of the energy change that is not unavoidably dissipated as heat. The symbols for the enzymes are as follows. (See also Figure 26.4.)

MH_2 = a metabolite; a donor of
$\quad$ ($H\!:\!^- + H^+$)
NAD^+ = enzyme with nicotinamide
$\quad$ cofactor
FMN = enzyme with riboflavin
$\quad$ phosphate cofactor
FeS—P = iron-sulfur protein
Q = enzyme with coenzyme Q
$\quad$ cofactor
Cyt = enzyme of the cytochrome
$\quad$ family

The Respiratory Enzymes. The cell's principal site of ATP synthesis is the *inner* membrane of a mitochondrion, Figure 26.2. Some tissues have thousands of these tiny organelles in the cytoplasm of a single cell. The inner membrane of a mitochondrion has built into it innumerable clusters of the respiratory enzymes. One very important property of this membrane is that it is impermeable, except at certain places, toward certain chemical species, such as H^+.

Figure 26.3 gives an outline of which enzyme interacts with which together with an indication of the favorable drops in energy that impel the overall sequence. The first enzyme is one that carries NAD^+ as the coenzyme, and it accepts $H\!:\!^-$ from a donor molecule. The donor is frequently an intermediate in the citric acid cycle. We studied how $H\!:\!^-$ transfers to NAD^+ in Section 24.1 and Special Topic 24.1. If we let MH_2 represent any metabolite that can donate $H\!:\!^-$ to NAD^+, we can write the following equation for the first step in the respiratory chain:

$$MH_2 + NAD^+ \longrightarrow M + NADH + H^+$$

The electron flow has now started toward oxygen. The electron pair has moved from MH_2 to NADH.

The next enzyme in the respiratory chain is FMN, which carries the vitamin riboflavin as part of its coenzyme. (We studied this enzyme in Section 24.1 and Special Topic 24.1.) The hydride in NADH now transfers to FMN, so we can write the following equation:

$$NADH + H^+ + FMN \longrightarrow FMNH_2 + NAD^+$$

This restores the NAD^+ enzyme, and moves the electron pair one more step down the respiratory chain.

What happens next is the transfer of just the pair of electrons while the hydrogen nucleus of $H:^-$ leaves the carrier as H^+ and *is placed into the fluid on the outside of the inner membrane.* This step begins the most critical work of the respiratory chain, the buildup of an H^+ gradient across the inner mitochondrial membrane. The electron acceptor is called the *iron-sulfur protein,* which we'll symbolize by FeS—P. The iron in FeS—P occurs as Fe^{3+}. By accepting one electron, it is reduced to the Fe^{2+} state, so we need two units of FeS—P to handle the pair of electrons now about to leave $FMNH_2$. This step can be written as:

$$FMNH_2 + 2FeS—P \longrightarrow FMN + 2FeS—P\cdot + 2H^+$$

where we use FeS—$P\cdot$ to represent the reduced form of the iron-sulfur protein in which Fe^{2+} occurs.

Before we go on, we will introduce a new way of writing these equations that helps to emphasize both the oxidation and the reduction aspects, as well as the restoration of each enzyme to its original condition. For example, the last three reactions we have studied can be represented as follows:

To the *outside* of the
inner membrane of the
mitochondrion

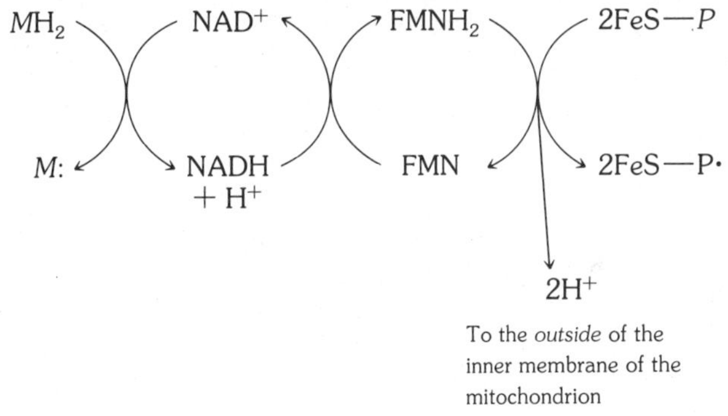

Reading from left to right, at the point where the first pair of curved arrows touch we see that as MH_2 changes to $M:$, it passes $H:^-$ to NAD^+ to change it to NADH. H^+ is also released. At the second pair of curved arrows, we read that as $NADH + H^+$ are changed back to NAD^+, the system passes one unit of $H:^- + H^+$ across to FMN to change it to $FMNH_2$. At the third pair of curved arrows, the display shows that as $FMNH_2$ changes back to FMN, it passes two electrons to two FeS—P molecules and expels two protons (H^+) into the surroundings. Let's now pause to see where we are and where we are headed.

At this stage, two electrons have moved from MH_2 to two molecules of the iron-sulfur protein, and two hydrogen ions have been added to the fluid on the *outside* of the inner mitochondrial membrane. This region now has more hydrogen ions than the inside of the inner membrane. In other words, the flow of electrons in the respiratory chain has started to establish a proton gradient that will eventually drive the synthesis of ATP. (And remember, the inner membrane is not permeable to protons except at certain specially structured places. Protons can't diffuse back just anywhere along this inner membrane.)

Think of NADH as NAD:H, in which the dots in boldface are those that move along the respiratory chain from MH_2.

Think of $FMNH_2$ as FMN:H, which becomes FMN when it loses $H:^-$ and H^+.

Visualize these enzymes and their reactions as occurring in the structure of the inner membrane itself.

What we are developing here is now called the **chemiosmotic theory,** which was first proposed in 1961 by Peter Mitchell of England (Nobel prize, 1981). As a way of seeing where we are heading, let's list the basic principles of this theory and then go back to pick up the trail of the electrons flowing in the respiratory chain.

Principles of the Chemiosmotic Theory

1. The synthesis of ATP occurs at an enzyme located on the *inside* of the inner mitochondrial membrane.

2. ATP synthesis is driven by a flow of protons that occurs from the outside to the inside of the inner membrane.

3. The flow of protons is through special ports in the membrane and down a concentration gradient of protons that exists across the inner membrane.

4. The energy that creates the proton gradient and the (+) charge gradient is provided by the flow of electrons within the respiratory chain whose enzymes make up integral packages in the inner membrane.

5. The inner membrane is a closed envelope except for special ports for the flow of protons and special transport systems that convey needed solutes into or out of the innermost mitochondrial compartment.

Getting back to the respiratory chain, its entire function is to transport electrons from a metabolite, MH_2, to oxygen while setting up the gradients of H^+ and (+) charge. As you have seen, the respiratory chain is quite complicated, and from this point on it becomes even more so. What follows next are further transfers of electrons through a series of enzymes, including several called the cytochromes. These are designated, in the order in which they participate, as cytochromes b, c_1, c, a, and a_3. Cytochromes a and a_3 are collectively known as the enzyme *cytochrome oxidase*, and this is the enzyme that catalyzes the reduction of oxygen we described earlier. Cytochrome a_3 carries a copper ion that alternates between the Cu^{2+} and the Cu^+ states.

Cyto-, cell; *-chrome,* pigment. The cytochromes are colored substances.

We will have to pass over the details of the rest of the respiratory chain. Figure 26.4 does show them, including a branch in the chain that involves an enzyme whose coenzyme is FAD.

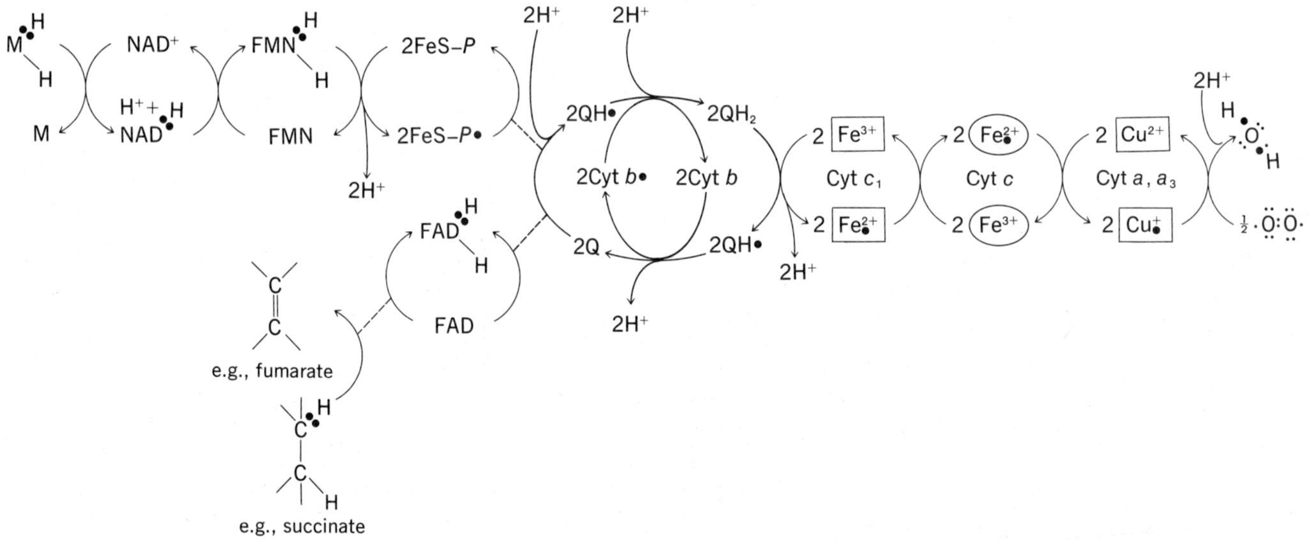

FIGURE 26.4
The respiratory chain. The boldface dots highlight the electrons that are passed along in the process from a donor to oxygen. Notice the three pairs of H^+ ions that disappear from the top side, which is inside the inner membrane, and the three pairs that appear on the bottom side, which is outside the inner membrane. (See also Figure 26.5.)

This enzyme can also help to feed electrons into the main chain, and we'll see an example of its operation in the next section.

Oxidative Phosphorylation. Figure 26.5 places the events of the respiratory chain into the context of the inner membrane, and it shows one of the portals through which protons can move back into the innermost part of the mitochondrion. As the figure as well as the previous net equation indicate, the net result of the operation of the respiratory chain, starting with MH_2 and NAD^+, can be expressed by the following equation:

$$MH_2 + 6H^+ \quad + \tfrac{1}{2}O_2 \xrightarrow{\text{Respiratory chain}} M + H_2O + 6H^+$$

From inside the inner membrane   Now on the outside of the inner membrane

Thus the oxidation of one molecule of MH_2 forces an excess of $6H^+$ ions to go outside the inner membrane. The overall change from the branch that involves FAD is to put an excess of $4H^+$ rather than $6H^+$. (This branch isn't shown in Figure 26.5.) The operation of the respiratory chain pumps these protons outside *without also putting the equivalent in negatively charged ions there, too.* Thus there are actually two gradients that are set up, a gradient of H^+ ions — more outside than inside — and a gradient of positive charge. The latter gradient could be erased either by the migration of negative ions to the outside or the migration of any kind of positive ion to the inside. Calcium ions, for example, might move, and precisely such movements of Ca^{2+} ions are intimately involved with nerve signals and muscular work. Thus this theory helps us understand events other than ATP synthesis.

Exactly *how* ATP synthesis takes place in mitochondria is still not a question settled in every detail, so we'll take just a brief look at one theory (one of two advanced by Mitchell).

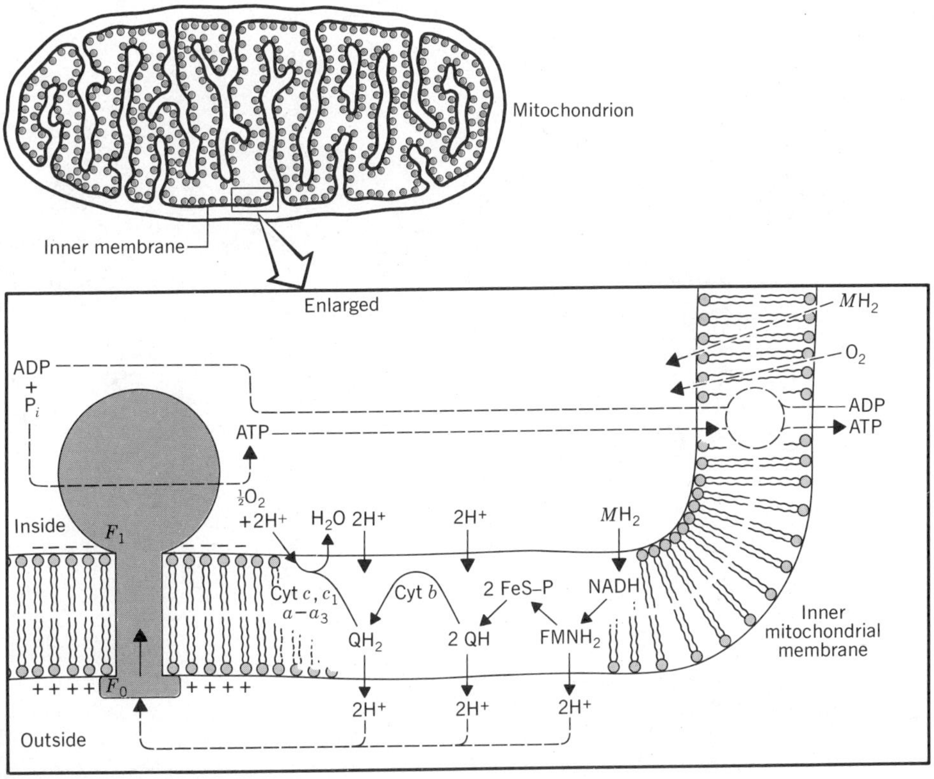

Because chemical reactions create the gradients, we have the chemi-*part of the term* chemiosmotic.

FIGURE 26.5
Cutaway of a portion of the inner mitochondrial membrane that shows one respiratory chain and one proton conduit—labeled as F_0–F_1. As indicated on the right, ADP can move inside only if ATP moves outside; the movements of these two are coupled.

This migration through a semipermeable membrane explains the -*osmotic* part of the term *chemiosmotic*.

The structure of PO_3^- is something like

and the P atom is much more exposed than it is in PO_4^{3-}.

Embedded in the inner mitochondrial membrane are complexes of proteins that form a tube through it. At least one tube exists for every respiratory chain unit. The tube is a conduit for protons, and it terminates on the inside of the inner membrane with an enzyme that catalyzes the formation of ATP from ADP and P_i. As each pair of protons flows through one of these conduits and encounters a phosphate ion, P_i, the protons convert this ion into an extremely reactive form. This might occur as the abstraction of O^{2-} from PO_4^{3-} to leave PO_3^-, an extremely unstable particle and one that would be wide open to attack by ADP. (The O^{2-} combines with $2\,H^+$ to give a molecule of water.) Figure 26.6 shows how all of this might work. Each pair of protons that moves in the gradient can thus lead to the formation of one ATP molecule. Each molecule of MH_2 that enters the chain at NAD^+ generates three pairs of protons in the gradient, so each MH_2 that donates hydrogen to NAD^+ can lead to a maximum of three ATPs. Each metabolite that donates hydrogen to FAD can lead to just two ATPs.

Because a series of oxidations establishes the proton gradient that makes this ATP synthesis possible, the overall synthesis is called **oxidative phosphorylation.** You'll also see this called *respiratory chain phosphorylation.*

Inhibitors of Oxidative Phosphorylation. A number of compounds, including some poisons and several antibiotics, interfere with the respiratory chain, with phosphorylation, or both. For example, one of the barbiturates, amytal sodium, blocks the respiratory chain between NAD^+ and cytochrome b. The powerful insecticide, rotenone, does the same thing. The cyanide ion blocks the chain at its very end, at cytochrome a_3. The antibiotic antimycin A stops the chain between cytochromes b and c.

Rotenone is a naturally occurring insecticide.

Now that we've taken a careful look at the "plug" end of the energy mobilization "tub," the respiratory chain and oxidative phosphorylation, we will move back one major step to the supplier of electrons for the respiratory chain, the citric acid cycle.

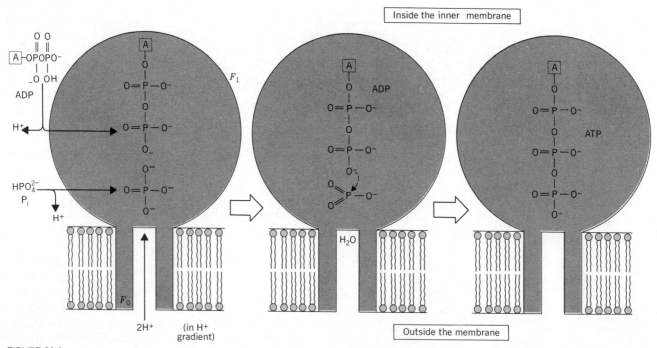

FIGURE 26.6
A theory that relates ATP synthesis to the movement of a pair of protons through the conduit in the inner mitochondrial membrane. The large shaded sphere is an enzyme that catalyzes the phosphorylation.

The idea that $2H^+$ interact with PO_4^{3-} to produce something like PO_3^- is conjecture.

26.3 THE CITRIC ACID CYCLE

Acetyl CoA is used to make the citrate ion that is then broken down bit by bit to CO_2 and units of $(H:^- + H^+)$, which funnel into the respiratory chain.

Hans Krebs won a share of the 1953 Nobel prize in medicine and physiology for his work on the citric acid cycle.

Figure 26.7 gives the reactions of the citric acid cycle, a series of reactions that break down acetyl groups.[2] The two carbon atoms of this group end up in molecules of CO_2, and the hydrogen atoms are fed into the respiratory chain. These reactions occur in the innermost part of a mitochondrion, inside the inner membrane, in what is called the *mitochondrial matrix.*

Acetyl Coenzyme A. For an acetyl group to enter the citric acid cycle it must be joined to coenzyme A, which we represent as CoA—SH.

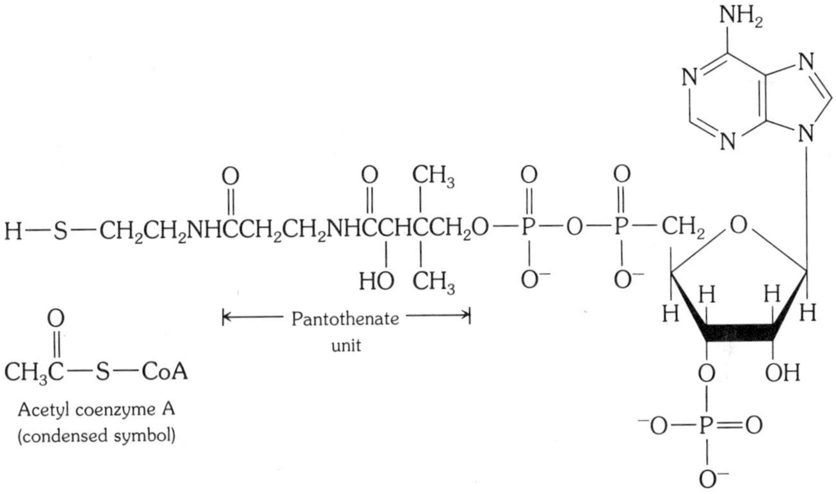

Coenzyme A (CoA—SH)

This coenzyme requires pantothenic acid, one of the vitamins. The conversion of a pyruvate ion to acetyl coenzyme A involves both a decarboxylation and an oxidation. The overall equation for this complicated step is as follows, but we'll not study any details.

$$\underset{\text{Pyruvate}}{CH_3\overset{O}{\overset{\|}{C}}—CO_2^-} + \underset{\text{Coenzyme A}}{CoA—S—H} + NAD^+ \longrightarrow \underset{\text{Acetyl coenzyme A}}{CH_3\overset{O}{\overset{\|}{C}}—S—CoA} + CO_2 + NADH$$

Acetyl coenzyme A is a thio ester, which means an ester in which an oxygen atom has been replaced by a sulfur atom. A carbon-sulfur bond is far more reactive than a carbon-oxygen bond, particularly when the carbon is part of a carbonyl group. Therefore acetyl coenzyme A is a particularly active transfer agent for the acetyl group. And just such a transfer of an acetyl group launches one turn of the citric acid cycle.

Steps in the Citric Acid Cycle. In the first step of the citric acid cycle, the acetyl group of acetyl coenzyme A transfers to oxaloacetate ion, an ion that has two carboxylate groups and a keto group. The acetyl unit adds across the keto group in a type of reaction that we have not

[2] You should be aware that the citric acid cycle goes by two other names as well—the *tricarboxylic acid cycle* and the *Krebs' cycle.* You might encounter any of these names in other references.

At physiological pH, the acids in the cycle exist largely as their anions.

studied before. The product is the citrate ion. Now begins a series of reactions by which the citrate ion is degraded bit by bit until another oxaloacetate ion is recovered. The numbers of the following steps match those in Figure 26.7.

1. Citrate is dehydrated to give the double bond of *cis*-aconitate. (This is the dehydration of an alcohol.)

2. Water adds to the double bond of *cis*-aconitate to give an isomer of citrate called isocitrate. Thus the net effect of steps 1 and 2 is to switch the alcohol group in citrate to a different carbon atom. However, this changes the alcohol from tertiary to secondary, from one that can't be oxidized to one that can.

3. The secondary alcohol group in isocitrate is dehydrogenated (oxidized) to give oxalosuccinate. NAD^+ accepts the hydrogen, so three ATPs can now be made as NAD : H passes its electron pair down the respiratory chain.

4. Oxalosuccinate loses a carboxyl group — it decarboxylates — to give α-ketoglutarate.

5. α-Ketoglutarate now undergoes a very complicated series of reactions, all catalyzed by one team of enzymes that includes coenzyme A. The results are the loss of a carboxyl group and another dehydrogenation. The hydrogen is accepted by NAD^+, so three more ATPs can now be made by the respiratory chain. The product, not shown in Figure 26.7, is the coenzyme A derivative of succinic acid. It is converted to the succinate ion, which is in Figure 26.7, in a reaction that also generates guanosine triphosphate, GTP. This is another high-energy triphosphate, which we mentioned on page 613 as being similar to ATP. GTP is able to phosphorylate ADP to make one ATP. Thus the citric acid cycle includes one substrate phosphorylation.

6. Succinate donates hydrogen to FAD (not to NAD^+), and the fumarate ion forms. Two ATPs (not three) can be made via FAD's involvement in the respiratory chain.

7. Fumarate adds water to its double bond, and malate forms.

8. The secondary alcohol in malate gives up hydrogen to NAD^+, and three more ATPs can be made. The cycle is now closed. We have remade one molecule of the carrier, oxaloacetate. It can accept another acetyl group from acetyl coenzyme A. If the demand for ATP is still high enough, another turn of the cycle can now occur.

ATP Yield in Oxidative Phosphorylation. The citric acid cycle itself makes a maximum of 12 molecules of ATP, as the data in Table 26.2 show. If we add to this the three ATPs made

TABLE 26.2
ATP Production by Oxidative Phosphorylation

Steps	Receiver of $(H:^- + H^+)$ in the Respiratory Chain	Molecules of ATP Formed
Isocitrate $\longrightarrow$ α-ketoglutarate	NAD^+	3
α-Ketoglutarate $\longrightarrow$ succinyl CoA	NAD^+	3
Succinyl CoA $\longrightarrow$ succinate (via GTP)	—	1
Succinate $\longrightarrow$ fumarate	FAD	2
Malate $\longrightarrow$ oxaloacetate	NAD^+	3
Total ATP via citric acid cycle		12 ATP
Pyruvate $\longrightarrow$ acetyl CoA	NAD^+	3
Total ATP from pyruvate via citric acid cycle		15 ATP

FIGURE 26.7
The citric acid cycle. The boxed
numbers refer to the text discussion.
The names of the enzymes for each
step are given by the arrows.

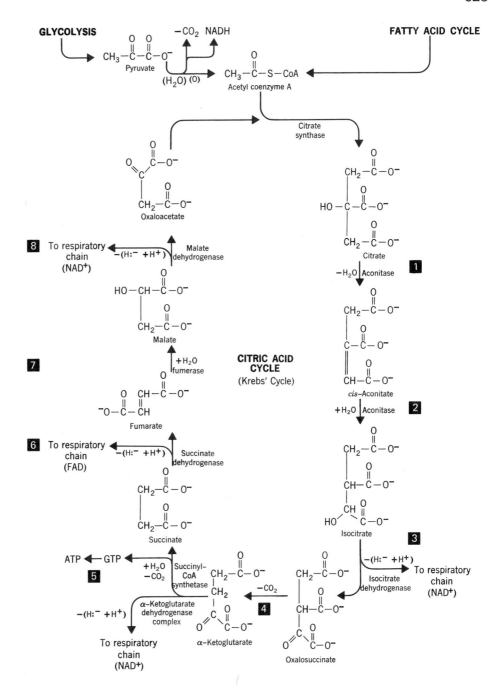

possible by the conversion of pyruvate to acetyl CoA, then 15 ATPs can be made by the
degradation of one pyruvate ion.

Keep in mind that pyruvate isn't the only raw material either for acetyl coenzyme A or for
other metabolites (MH_2) that can fuel the respiratory chain. Also remember that the energy of
the respiratory chain can be used to run other operations besides making ATP — for example,
the operation of nerves. Thus Table 26.2 has to be viewed as giving the upper limits to ATP
production from pyruvate by the respiratory chain.

SUMMARY

High-energy compounds   Organophosphates whose phosphate group transfer potentials equal or are higher than that of ATP are classified as high-energy phosphates. A lower energy phosphate can be made by phosphate transfer from a high energy phosphate — a process called substrate phosphorylation.

Respiratory chain   A series of electron-transfer enzymes called the respiratory enzymes occur together as groups called respiratory assemblies in the inner membranes of mitochondria. These enzymes process metabolites (MH_2), which often are obtained by the operation of the citric acid cycle. Either NAD^+ or FAD can accept ($H:^- + H^+$) from a metabolite. Their reduced forms, NADH or $FADH_2$, then pass on the electrons until cytochrome oxidase uses them (together with H^+) to reduce oxygen to water.

Oxidative phosphorylation   Each pair of electrons that flows from NADH in the respiratory chain is responsible for forcing three pairs of H^+ ions into the proton gradient created across the inner mitochondrial membrane. Each pair of H^+ ions that flows back at the allowed conduits of this membrane leads to the formation of one ATP from ADP and P_i. In some systems, other kinds of positive ions migrate, as in the operation of nerves. Various drugs and antibiotics can block the respiratory chain.

Citric acid cycle   Acetyl groups from acetyl coenzyme A are joined to a four-carbon carrier, oxaloacetate, to make citrate. This six-carbon salt of a tricarboxylic acid then is degraded bit by bit as ($H:^- + H^+$) is fed to the respiratory chain. Each acetyl unit leads to the synthesis of a maximum of 12 ATP molecules.

Fatty acids and glucose are important suppliers of acetyl groups. In glycolysis, glucose units are broken to pyruvate units, and the oxidative decarboxylation of pyruvate does two things. It leads to the synthesis of one NADH (which leads to the formation of 3 ATPs), and it supplies acetyl units to the citric acid cycle from which 12 more ATPs are made.

KEY TERMS

The following terms were emphasized in this chapter and they will be used in succeeding chapters. Be sure that you know them before you continue.

acetyl coenzyme A	aerobic sequence	fatty acid cycle	phosphate group transfer potential
adenosine diphosphate (ADP)	anaerobic sequence	glycolysis	
adenosine monophosphate (AMP)	catabolism	high-energy phosphate	respiratory chain
	chemiosmotic theory	oxidative phosphorylation	respiratory enzymes
adenosine triphosphate (ATP)	citric acid cycle		substrate phosphorylation

SELECTED REFERENCES

1   N. O. Thorpe. *Cell Biology.* John Wiley and Sons, Inc., 1984.

2   A. L. Lehninger. *Principles of Biochemistry.* Worth Publishers, Inc., New York, 1982.

3   L. Stryer. *Biochemistry,* 2nd ed. W. H. Freeman and Company, San Francisco, 1981.

4   P. Mitchell. "Keilin's Respiratory Chain Concept and Its Chemiosmotic Consequences." *Science,* December 7, 1979, page 1148.

5   R. E. Dickerson. "Cytochrome *c* and the Evolution of Energy Metabolism." *Scientific American,* March 1980, page 137.

6   H. E. Bent. "Reactions for Every Occasion — Energy and Exercise." *Journal of Chemical Education,* December 1978, page 796.

REVIEW EXERCISES

The answers to these Review Exercises are in the *Study Guide* that accompanies this book.

Energy Sources

26.1   What products of the digestion of carbohydrates, triacylglycerols, and the polypeptides can be used as souces of biochemical energy to make ATP?

26.2   The complete catabolism of glucose gives what products?

26.3   The identical products form when glucose is burned in open air as when it is fully catabolized in the body. How, then, do these two processes differ in their overall accomplishments?

26.4   What are the end products of the complete catabolism of fatty acids?

High-Energy Phosphates

26.5 Complete the following structure of ATP.

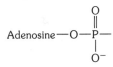

$$\text{Adenosine}-\text{O}-\overset{\overset{\displaystyle O}{\|}}{\underset{\underset{\displaystyle O^-}{|}}{P}}-$$

26.6 Write the structures of ADP and of AMP in the manner started by Review Exercise 26.5.

26.7 At physiological pH, what does the term *inorganic phosphate* stand for? (Give formulas and names.)

26.8 Why are phosphoenolpyruvate and ATP called high-energy phosphates but glycerol 3-phosphate is not?

26.9 If the organophosphate of M has a *lower* phosphate group transfer potential than *that of N*, in which direction does the following reaction tend to go spontaneously, to the left or to the right?

$$M-OPO_3{}^{2-} + N \overset{?}{\rightleftharpoons} N-OPO_3{}^{2-} + M$$

26.10 Using data in Table 26.1, tell (yes or no) if ATP can transfer a phosphate group to each of the following possible compounds. (Assume, of course, that the right enzyme is available.)
(a) Glycerol   (b) Fructose
(c) Creatine   (d) Acetic acid

26.11 All of the possible phosphate transfers in Review Exercise 26.10 are classified as *substrate* phosphorylations. What does this mean?

26.12 What is the function of creatine phosphate in muscle tissue?

26.13 Whether or not creatine phosphate is used in muscle tissue is under *feedback control*. Explain.

Overview of Metabolic Pathways

26.14 In the general area of biochemical energetics, what is the purpose of each of the following pathways?
(a) Respiratory chain   (b) Anaerobic glycolysis
(c) Citric acid cycle   (d) Fatty acid cycle

26.15 What prompts the respiratory chain to go into operation?

26.16 In general terms, the intermediates that send electrons down the respiratory chain come from what metabolic pathway that consumes acetyl groups?

26.17 Arrange the following sets of terms in sequence in the order in which they occur or take place. Place the identifying letter of the first sequence of a set to occur on the left of the row of numbers.

(a) Citric acid cycle   pyruvate   respiratory chain

   1   **2**   **3**

   glycolysis   acetyl CoA

   4   **5**

(b) Respiratory chain   fatty acid cycle

   1   **2**

   citric acid cycle   acetyl CoA

   3   **4**

26.18 The *aerobic sequence* begins with what metabolic pathway and ends with which pathway?

Respiratory Chain

26.19 What is missing in the following basic expression for what must happen in the respiratory chain?

$$\tfrac{1}{2}O_2 + 2H^+ \longrightarrow H_2O$$

26.20 What general name is given to the set of enzymes involved in electron transport?

26.21 Write the following display in the normal form of a chemical equation.

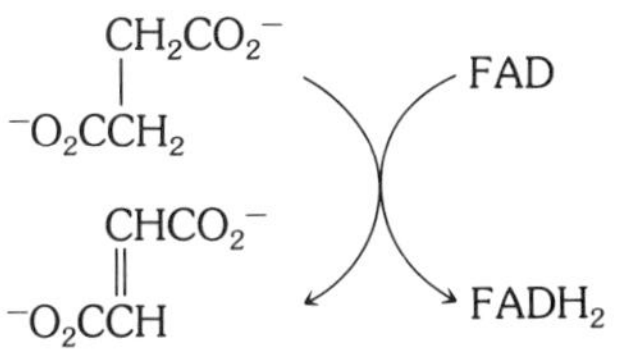

$$\underset{\underset{\displaystyle ^-O_2CCH}{|}}{\overset{\overset{\displaystyle CH_2CO_2{}^-}{|}}{}} \qquad \begin{matrix} \rightarrow FAD \\ \\ \rightarrow FADH_2 \end{matrix}$$

$$\overset{\overset{\displaystyle CHCO_2{}^-}{\|}}{}$$

(a) Which specific species is oxidized? (Write its structure.)
(b) Which species is reduced?

26.22 Write the following equation in the form of a display like that shown in Review Question 26.21.

$$\overset{\overset{\displaystyle OH}{|}}{CH_3CHCO_2{}^-} + NAD^+ \longrightarrow \overset{\overset{\displaystyle O}{\|}}{CH_3CCO_2{}^-} + NADH + H^+$$

26.23 Arrange the following in the order in which they receive and pass on electrons.

   FMN   Cyt a   NAD$^+$ FeSP

26.24 What does cytochrome oxidase do?

26.25 What is FAD, and where is it involved in the respiratory chain?

Chemiosmotic Theory

26.26 Across which cellular membrane does the respiratory chain establish a gradient of H$^+$ ions? On which side of this membrane is the value of the pH lower?

26.27 According to the chemiosmotic theory, the flow of what particles most directly leads to the synthesis of ATP?

26.28 If the inner mitochondrial membrane is broken, the respiratory chain can still operate, but the phosphorylation of ADP that normally results stops. Explain this in general terms.

26.29 Complete and balance the following equation. (Use $\tfrac{1}{2}$ as the coefficient of oxygen as shown.)

$$MH_2 + H^+ + \tfrac{1}{2}O_2 \longrightarrow$$

From inside
the inner
membrane

26.30 What is the difference between substrate and oxidative phosphorylation?

26.31 Besides a gradient of H$^+$ ions, what other gradient exists in mitochondria as a result of the operation of the respiratory chain? In terms of helping to explain how cations other than H^+ move across a membrane, what significance is this gradient?

26.32 Briefly describe the theory presented in this chapter that explains how a flow of a pair of protons across the inner mitochondrial membrane initiates the synthesis of ATP from ADP and P_i.

Citric Acid Cycle

26.33 What makes the citric acid cycle start up?

26.34 What chemical unit is degraded by the citric acid cycle? Give its name and structure.

26.35 How many times is a secondary alcohol group oxidized in the citric acid cycle?

26.36 Water adds to a carbon-carbon double bond how many times in one turn of the citric acid cycle?

26.37 The conversion of pyruvate to an acetyl unit is both an oxidation and a decarboxylation.
   (a) If *only* decarboxylation occurred, what would form from pyruvate? Write the structure of the other product in:

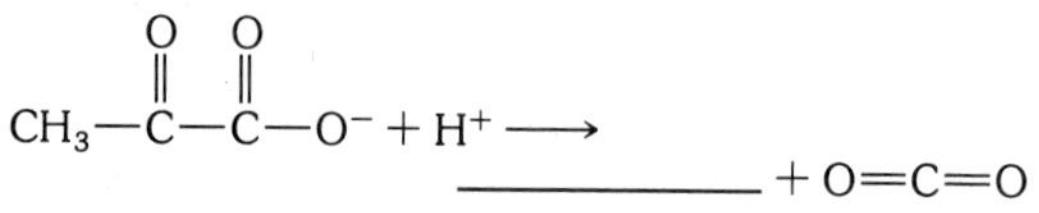

$$CH_3-\overset{\overset{\displaystyle O}{\|}}{C}-\overset{\overset{\displaystyle O}{\|}}{C}-O^- + H^+ \longrightarrow \underline{\hspace{2cm}} + O{=}C{=}O$$

   (b) If this product is oxidized, what is the name and the structure of the product of such oxidation?
   (c) Referring to Figure 26.7, which specific compound undergoes an oxidative decarboxylation similar to that of pyruvate? (Give its name.)

26.38 One cofactor in the enzyme assembly that catalyzes the oxidative decarboxylation of pyruvate requires thiamine, one of the B vitamins. Therefore in beriberi, the deficiency disease for this vitamin, the level of what substance can be expected to rise in blood serum (and for which an analysis can be made as part of the diagnosis of beriberi)?

26.39 The enzyme for the conversion of isocitrate to oxalosuccinate (Figure 26.7) is stimulated by one of these two substances, ATP or ADP. Which one is the likelier activator? Explain.

26.40 What is the maximum number of ATP molecules that can be made from the use of respiratory chain phosphorylation to break down
   (a) Pyruvate?
   (b) An acetyl group in acetyl CoA?

26.41 Glutamic acid, one of the amino acids, can be converted to α-ketoglutarate (Figure 26.7). How many ATP molecules can be made from the entry of α-ketoglutarate into the citric acid cycle?

Chapter 27
Metabolism of Carbohydrates

We can't give the molecular basis of mouth watering, but we can study how sugars are metabolized — in this chapter.

27.1 GLYCOGEN METABOLISM

Much of the body's control over the blood sugar level is handled by its regulation of the synthesis and breakdown of glycogen.

The digestion of the starch and disaccharides in the diet gives glucose, fructose, and galactose, but their catabolic pathways all converge very quickly to that of glucose itself. Galactose, for example, is changed by a few steps in the liver to glucose 1-phosphate. Fructose is changed to a compound that occurs early in glycolysis. For these reasons, this chapter concentrates on the metabolism of glucose, and we'll begin with the glucose in circulation.

Mg/dL = milligrams per deciliter, where 1 dL = 100 mL.

Blood Sugar Level. The concentration of monosaccharides in whole blood, expressed in mg/dL, is called the **blood sugar level.** Because glucose is overwhelmingly the major monosaccharide in blood, the "sugar" in blood sugar level means glucose for all practical purposes. After several hours of fasting, the normal range for the blood sugar level in whole blood, called the **normal fasting level,** is 65 – 95 mg/dL (70 – 110 mg/dL in plasma). Wide variations from the normal fasting level generally signify something wrong, so a special vocabulary exists to define such variations.

-glyc-, sugar
-emia, in blood
hypo-, under, below
hyper-, above, over
renal, of the kidneys
-uria, in urine

    Hypoglycemia is the condition of a blood sugar level *below* the normal, and **hyperglycemia** is the condition when this level is *above* the normal fasting level. If the blood sugar level goes high enough, the kidneys are unable to return to circulation all of the glucose that leaves the blood in a glomerulus. Glucose now appears in the urine and the term for this is **glucosuria.** The blood sugar level above which this happens is called the **renal threshold** for glucose, and it is in the range of about 140 – 160 mg/dL, sometimes higher.

    Whenever hypoglycemia develops rapidly, the individual experiences dizziness and may faint. People with diabetes who unknowingly take too much insulin experience this in a severe form known as *insulin shock.* Behind such responses by the central nervous system to hypoglycemia is the brain's almost utter reliance on the glucose in circulation for its routine source of minute-by-minute energy. The brain consumes about 120 g/day of glucose, and a quick onset of hypoglycemia starves the brain cells. Although they have the ability to switch over to other nutrients, brain cells can't do this very rapidly.

    Whenever hyperglycemia develops and tends to persist, something is wrong with the mechanisms for withdrawing glucose from circulation. Diabetes is a common cause of hyperglycemia.

Glycogenesis and Glycogenolysis. When the blood sugar level is high enough to meet immediate energy needs, any excess glucose is removed from circulation and put into the body's energy storage reserves. There are two ways to do this. One is to convert glucose to fat, and we'll study how this is done in the next chapter. The other is to synthesize glycogen, which we'll discuss here. (We studied glycogen on page 467.)

    Liver and muscle cells can convert glucose to glycogen by a series of steps called **glycogenesis** ("glycogen creation"). The liver holds 70 – 110 g of glycogen, and the muscles, taken as a whole, contain 170 – 250 g. When muscles need glucose, they take it back out of glycogen. When the blood needs glucose because the blood sugar level has dropped too much, the liver hydrolyzes as much of its glycogen reserves as needed and then puts the glucose into circulation. The overall series of reactions in either tissue that hydrolyzes glycogen is called **glycogenolysis.**

Hormones and the Regulation of the Blood Sugar Level. The tapping of the glycogen reserves can be initiated by several conditions, including simply a decrease in the blood sugar

FIGURE 27.1
The epinephrine "cascade."

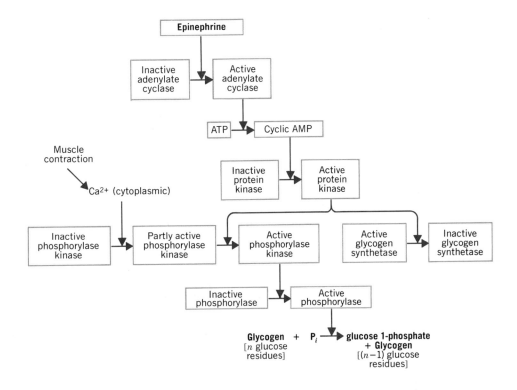

level itself. Various hormones respond to signals such as this and help to adjust the blood sugar level.

In a moment of sudden fright or danger, the adrenal medulla secretes the hormone **epinephrine.** In muscle tissue, and to some extent in the liver, epinephrine activates glycogenolysis by the steps outlined in Figure 27.1. It acts directly to stimulate the enzyme adenylate cyclase, which catalyzes the conversion of some ATP to cyclic AMP in the manner we studied in Section 24.5. Cyclic AMP then activates still another enzyme, and this, still another enzyme, and so on in a cascade of events shown in Figure 27.1. Each molecule of epinephrine, by activating adenylate cyclase, triggers the formation of dozens of molecules of cyclic AMP. *Each* of these activates a succeeding enzyme, and so on until the final enzyme appears not only suddenly but in a relatively large quantity. Thus one epinephrine molecule triggers the rapid mobilization of thousands of glucose units. And these are now ready to supply energy that the body might need to handle the emergency.

While the epinephrine "cascade" proceeds to trigger the release of glucose units, the team of enzymes called glycogen synthetase that does the *reverse* — helps to convert glucose to glycogen — has to be shut down. This is done at one of the steps shown in Figure 27.1. Notice in this figure where the enzyme called active protein kinase catalyzes the phosphorylation of *two* systems — a partly active phosphorylase kinase and active glycogen synthetase. The first action leads to further developments of the epinephrine cascade and the appearance of glucose units. The second action, the phosphorylation of glycogen synthetase, *shuts this enzyme down.* This cancels the potential competition. It's a remarkable aspect of this system.

Also indicated in Figure 27.1 is the involvement of calcium ion. When a muscle contracts, the level of Ca^{2+} in the cytoplasm increases by an exceedingly small amount, but this acts as a warning: "Get ready, we're going to need more glucose soon." Ca^{2+} changes an inactive enzyme, phosphorylase kinase, into a *partly active* form, which is then ready for final activation whenever epinephrine initiates its cascade effect.

The end product of glycogenolysis isn't actually glucose but glucose 1-phosphate. Cells that are capable of glycogenolysis also have an enzyme called phosphoglucomutase which

An estimated 30,000 molecules of glucose are released from glycogen for each molecule of epinephrine that initiates glycogenolysis.

catalyzes the conversion of glucose 1-phosphate to its isomer, glucose 6-phosphate:

Glucose 1-phosphate   Glucose 6-phosphate

Glucose is now in the form in which it must be to enter those catabolic pathways that produce ATP. *It is also in a form that can't migrate out of the cell.* Liver cells, however, have an enzyme, glucose 6-phosphatase, that can catalyze the hydrolysis of glucose 6-phosphate to glucose and inorganic phosphate.

$$\text{Glucose 6-P} + H_2O \xrightarrow{\text{glucose 6-phosphatase}} \text{Glucose} + P_i$$

The letter P is often used to represent the whole phosphate group in the structures of phosphate ester intermediates in metabolism.

Glucose itself is able to leave the liver and so help raise the blood sugar level. Thus the overall process — glycogenolysis in the liver to glucose 1-phosphate to glucose 6-phosphate to glucose — is a major supplier of glucose for the blood during periods of fasting when the blood sugar level decreases. Thus the brain depends on the liver during fasting to maintain its favorite source of chemical energy, glucose. (We are beginning to see in chemical terms how vital the liver is to the performance of other organs.) When circulating glucose is taken up by a brain cell, it is promptly trapped in the cell by being converted to glucose 6-phosphate. The same kind of trapping mechanism works at muscle tissue when its cells take up circulating glucose.

Several hereditary diseases involve the glucose–glycogen interconversion, and some are discussed in Special Topic 27.1.

Another hormone that helps to keep the blood sugar at a normal level is **glucagon.** It's a polypeptide made in the α-cells of the pancreas, and when the blood sugar level drops, these cells release glucagon. Its target tissue is the liver, where it is an excellent activator of glycogenolysis, doing its work by a cascade process which is very similar to that initiated by epinephrine. Like epinephrine, glucagon activates adenylate cyclase. Unlike epinephrine, glucagon also inhibits glycolysis, so this action helps to keep the supply of glucose up. Glucagon, also unlike epinephrine, does not cause an increase in blood pressure or pulse rate, and it is longer-acting than epinephrine.

The glucagon molecule has 29 amino acid residues.

One of the many effects of **human growth hormone** is to stimulate the release of glucagon. Growth requires energy, so the action of glucagon that helps to supply a source of energy — glucose — aids in the work of the human growth hormone. In some situations, such as a disfiguring condition known as acromegaly, there is an excessive secretion of human growth hormone that promotes too high a level of glucose in the blood. This is undesirable because a prolonged state of hyperglycemia from any cause can lead to diabetes and to some of the blood-capillary related complications of diabetes.

Acromegaly is sometimes called *giantism* because the bone structures of victims are enlarged.

Insulin, a polypeptide hormone released from the β-cells of the pancreas, has a powerful lowering effect on the blood sugar level. The release of insulin is stimulated by an increase in the blood sugar level, such as normally occurs after a meal that is rich in carbohydrates. As the insulin moves into action, it finds its receptors at the cell membranes of muscle and adipose tissue. The insulin-receptor complexes somehow make it possible for glucose molecules to move easily into the cells. Of course, this lowers the blood sugar level. Not all cells depend on insulin to take up glucose. Brain cells, for example, and cells in the kidneys, the intestinal tract, red blood cells, and in the lenses of the eyes take up glucose directly. If too much insulin gets into circulation, as from an error in insulin therapy, the individual's blood sugar level falls too low, which leads to insulin shock as we've already mentioned.

Adipose tissue is the fatty tissue that surrounds internal organs.

Life-saving first aid for someone in insulin shock is sugared fruit juice or candy to counter the hypoglycemia.

A number of inherited diseases involve the storage of glycogen. For example, in **Von Gierke's disease,** the liver lacks the enzyme glucose 6-phosphatase, which catalyzes the hydrolysis of glucose 6-phosphate. Unless this hydrolysis occurs, glucose units cannot leave the liver. They remain as glycogen in such quantities that the liver becomes very large. At the same time, the blood sugar level falls, the catabolism of glucose accelerates, and the liver releases more and more pyruvate and lactate.

In **Cori's disease,** the liver lacks an enzyme needed to catalyze the hydrolysis of 1,6-glycosidic bonds, the bonds that give rise to the many branches of a glycogen molecule.

Without this enzyme, only a partial utilization of the glucose in glycogen is possible. The clinical symptoms resemble those of Von Gierke's disease, but they are less severe.

In **McArdle's disease,** the muscles lack phosphorylase, the enzyme needed to obtain glucose 1-phosphate from glycogen (Figure 27.5). Although the individual is not capable of much physical activity, physical development is otherwise relatively normal.

In **Andersen's disease,** both the liver and the spleen lack the enzyme for putting together the branches in glycogen. Liver failure from cirrhosis usually causes death by age 2.

In an individual with a sustained hyperglycemia — as in a diabetic — some glucose combines with hemoglobin to give glycohemoglobin (or glycosylated hemoglobin). The measurement of the level of this substance is a far superior way to monitor the average blood sugar level of a diabetic than any direct measurements of blood glucose, because its level doesn't fluctuate as widely as the blood sugar level and it doesn't have to be determined as frequently.

When the β-cells of the pancreas secrete insulin, which helps to *lower* the blood sugar level, the α-cells should not at the same time release glucagon, which helps to *raise* this level. Another polypeptide hormone, **somatostatin,** which is also made in the pancreas (as well as in the hypothalamus), apparently functions to handle this problem. It acts at the pancreas to inhibit the release of glucagon as well as to slow down the release of insulin. Thus glucagon also acts as a moderator for insulin.

> Some studies are being made to use insulin and somatostatin together in the treatment of diabetes.

27.2   GLUCOSE TOLERANCE

The ability of the body to tolerate swings in the blood sugar level is essential to health.

The body's ability to handle dietary glucose without letting the blood sugar level rise too high or plummet too low is called its **glucose tolerance.** To measure this ability a **glucose tolerance test** is performed. The individual is given a drink that contains glucose — generally 75 g for an adult and 1.75 g per kilogram of body weight for children — and then the blood sugar level is checked at regular intervals. Figure 27.2 gives typical plots of this level versus time. The lower curve is that of a person with normal glucose tolerance, and the upper

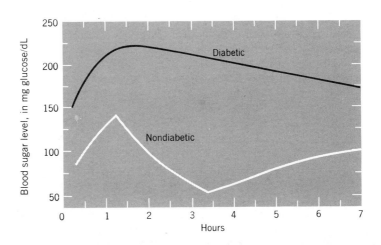

FIGURE 27.2
Glucose tolerance curves.

curve is of one whose glucose tolerance is typical of a person with diabetes. In both, the blood sugar level rises sharply at first. However, the healthy person soon manages the high level and brings it back down with the help of a normal flow of insulin and somatostatin. In the diabetic, the level comes down only very slowly and remains essentially in the hyperglycemic range throughout.

Notice that in the normal individual, the blood sugar level can sometimes drop to a mildly hypoglycemic level. Such hypoglycemia is possible also in someone who has eaten a carbohydrate-rich breakfast. With glucose pouring into the bloodstream, the release of a bit more insulin than needed can occur. This leads to the overwithdrawal of glucose from circulation, and mid-morning brings dizziness and sometimes even fainting (even falling asleep in class). The prevention isn't more sugared doughnuts but a balanced breakfast.

The subject of glucose tolerance is nowhere more discussed than in connection with diabetes. The real causes of diabetes, discussed in Special Topic 27.2, are not yet well-known, but the medical community has a working definition that leaves open the question of cause. **Diabetes** is defined clinically as a disease in which the blood sugar level persists in being much higher than warranted by the dietary and nutritional status of the individual. Invariably, a person with untreated diabetes has glucosuria, and the discovery of this condition often triggers the clinical investigations that are necessary to rule diabetes in or out.

As discussed in Special Topic 27.2, there are two broad kinds of diabetes, type I and type II. Type I diabetics are unable to manufacture insulin (at least not enough), and they need daily insulin therapy to manage their blood sugar levels. Maintaining a relatively even and normal blood sugar level is the best single strategy that such diabetics have for the prevention of some of the vascular and neural problems that can complicate their health later.

The Cori Cycle.   The strategies used by the body to maintain its blood sugar level within the normal range are, as we have seen, many and varied. They form a cycle of events called the **Cori cycle,** outlined in Figure 27.3.

Unhappily, most people with midmorning sag go into another round of sugared coffee and sugared rolls. The glucose gives a short lift, but then an oversupply of insulin restores the mild hypoglycemia of the sag.

Carl and Gerti Cori shared the 1947 Nobel prize in physiology and medicine.

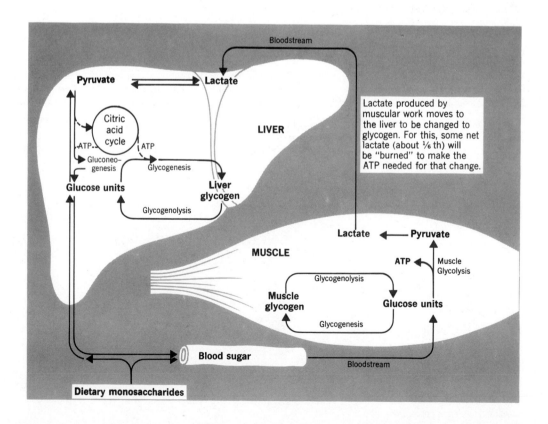

FIGURE 27.3
The Cori cycle.

FIGURE 27.4
Factors that affect the blood sugar level.

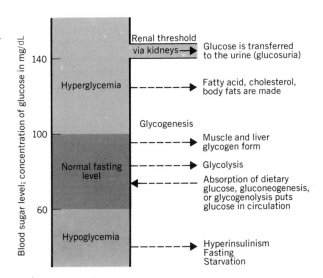

At the bottom of the figure we see glucose as it enters the bloodstream from the intestinal tract. Its molecules either stay in circulation or are soon removed by various tissues. Two are shown in the figure, those of muscle and liver. Muscle cells can trap glucose molecules and either use them in glycolysis to make ATP or to replenish the muscle's glycogen reserves. Liver cells can similarly trap glucose, and the liver is able to release glucose back into the bloodstream when the blood sugar level must be raised.

When glucose is used in glycolysis, the end product is either pyruvate or lactate, depending on the oxygen supply, as we learned in the last chapter. Either can be used to make more ATP by means of the citric acid cycle and the respiratory chain. However, when extensive anaerobic glycolysis is carried out in a tissue, then the lactate level rises considerably. Because lactate still has C—H bonds, it continues to have useful chemical energy, so instead of simply excreting excess lactate at the kidneys, the body recycles it. It converts a fraction of it—about five-sixth—to glucose, which, of course, can either be stored as glycogen or sent out into circulation again. The other one-sixth of the lactate is directly catabolized via pyruvate, acetyl CoA, the citric cycle, and the respiratory chain to make the ATP needed for the conversion of the other five-sixth to glucose. Remember that glycolysis is overall exothermic. An energy slope is descended in going from glucose to pyruvate or lactate. To go in the opposite direction, uphill, requires energy–namely, ATP or the equivalent high-energy phosphate. The synthesis of glucose from smaller molecules is called **gluconeogenesis,** and this process completes the Cori cycle. (We go into more details about gluconeogenesis later in this chapter.)

We can see from all of these processes that many factors affect the blood sugar level. Some tend to raise it and some do the opposite. Figure 27.4 summarizes them in a different kind of display. We'll now turn to a detailed study of one of these factors, the catabolism of glucose.

Neo, new; *-neogenesis,* new creation; *gluconeogenesis,* the synthesis of new glucose.

27.3 THE CATABOLISM OF GLUCOSE

Glycolysis and the pentose phosphate sequence are the chief catabolic pathways open to glucose.

Greek, *glykos,* sugar or sweet; *-lysis,* dissolution.

Glycolysis, as we noted in the previous chapter, is a series of reactions that change glucose to pyruvate or to lactate while a small but important amount of ATP is made. When a cell receives oxygen at a rate that is slower than it needs, glycolysis ends with the lactate ion, and

The name for this disorder is from the Greek *diabetes,* to pass through a siphon, and *mellitus,* honey-sweet— meaning to pass urine that contains sugar. We'll call it diabetes for short. In severe, untreated diabetes, the victim's body wastes away despite efforts to satisfy a powerful thirst and hunger. To the ancients, it seemed as if the body were dissolving from within.

In the United States there are nearly 6.5 million known cases of diabetes, and almost as many others are believed to have this disease. It ranks third behind heart disease and cancer as the cause of death.

Less than 10% of all cases of diabetes are of the severe, insulin-dependent variety in which the β-cells of the pancreas are unable to make and secrete insulin. This is **type I diabetes,** and insulin therapy is essential. Most victims contract it before they are 20, so it's sometimes referred to as *juvenile-onset diabetes.*

The rest of all those with diabetes have a form called **type II diabetes,** and most are able to manage their blood sugar levels by diet and exercise alone, without insulin injections. Their problem is actually not a lack of insulin but with a breakdown in the machinery for taking advantage of it. Often it's a reduction in the number of insulin-receptor proteins per target cell, and this ratio tends to correct itself with weight loss. (At the time of diagnosis, most type II diabetics are quite overweight.) Most get type II diabetes when they are over 40, so it's often called *adult-onset diabetes.*

Diabetes and Viruses No single cause of diabetes of either type has been firmly established, despite intensive research for decades. All that seems clear is that several factors are involved.

Starting in the 1970s, the possible interplay between viruses and the human immune system has been intensively studied. There is some evidence, for example, that the mumps virus, or even the vaccine against mumps, can trigger in some the onset of type I diabetes. Other viruses have been suspected from studies with laboratory rats and mice. The EMC or encephalomyocarditis virus destroys the β-cells of the pancreas in experimental animals. The equine encephalitis virus, the reovirus, and the *B4* strain of the Coxsackie virus all cause diabetes in animals. Thus variants of a number of viruses that infect humans also attack the β-cells in animals and cause the symptoms of diabetes.

Diabetes and the Immune System Most people who are exposed to viruses don't contract diabetes, which raises the question "Why are most people immune to whatever in the environment might cause type I diabetes?" Do some people have defective genes in their immune systems that lead to a lower-than-normal resistance to certain viruses? Do the viruses that seem to be able to cause diabetes in animals actually themselves destroy β-cells? Or do they

initiate enough changes, say, in the surface proteins of the β-cells, to induce the body itself to make antibodies against these altered proteins? If this were to happen, the antibodies made by the body could actually destroy its own β-cells. In other words, is type I diabetes caused by an autoimmune reaction?

Some toxic chemicals, including Vacor, a rat poison introduced in 1975, can destroy β-cells. Are there other chemicals that are toxic in this way among the hundreds of toxic substance in the environment? Do they directly cause the destruction of β-cells, or do they indirectly induce the body to make antibodies against its β-cells? Is it possible that there are as many causes of type I diabetes as there seem to be for the common cold?

If the problem is somehow in the immune system, then at least two conclusions might be drawn. The supposed genetic factor in diabetes that leads people to think that diabetes runs in families might be located in the genes dealing with the immune system rather than with the β-cells directly. And therefore one type of treatment might be to use drugs that suppress the immune system. Experiments with such treatments are being made. In a recent Canadian pilot study, the immunosuppressive drug cyclosporine was used. (This drug is used to suppress the immune-centered rejection of transplanted organs such as kidneys.) When the therapy was begun within a few weeks of the diagnosis of type I diabetes, and continued for a year, 16 of 30 patients became independent of insulin therapy. Controlled trials of this promising approach are now underway.

Causes of Type II Diabetes The onset of type II diabetes is much slower than that for type I, and obesity is a factor in most cases. The number of insulin-receptor proteins on the surfaces of target cells declines in obesity. Obesity usually involves a diet that is far richer in sugars and other carbohydrates than normal, and some scientists believe these substances evoke such a continuous presence of insulin that the receptor proteins literally wear out faster than they can be replaced. When the weight is reduced, particularly in connection with physical exercise, the relative numbers of receptor proteins rebound.

The nature of the diet—both quality and quantity— does appear to be a decisive factor in the onset of type II diabetes. In countries with very low levels of refined sugars in the national diets, the incidence of type II diabetes is rare.

Complications of Diabetes The immediate complications are an elevated blood sugar level, metabolic acidosis, and eventual death from coma and uremic poisoning. Insulin therapy corrects these immediate problems, but it deals less well with the longer-term complications.

One such complication, which might be at the bottom of all of the others, is a thickening of the *basement membrane* of blood capillaries, a condition called *microangio-*

pathy. This is accompanied by changes in the composition and metabolism of this membrane. (The basement membrane is the protein support structure that encases the single layer of cells of a capillary.) Microangiopathy is believed to lead to the other complications most of which involve the vascular system or the neural networks — kidney problems, gangrene of the lower limbs, and blindness. Diabetes is the leading cause of new cases of blindness in the United States, and it is the second most common cause of blindness, overall. (Ophthalmologists can detect the development of microangiopathy in the retina of the eye during an eye examination before other symptoms of diabetes are recognized.)

The change in the basement membrane might cause the complications that result from the overuse of glucose in some kinds of cells. Whereas the fat and muscle cells in a diabetic starve for glucose — in fact, they have to make their own — other cells might get too much glucose with the result that certain chemical equilibria shift unfavorably. For example, glucose is reduced by the enzyme aldose reductase to sorbitol. It's a minor reaction in cells of the lens of the eye, but in an abundance of glucose too much sorbital is made. Sorbitol, unlike glucose, tends to be trapped in lens cells, and as the sorbitol concentration rises so does the osmotic pressure in the fluid. This draws water into the lens cells, which generates pressure and leads to cataracts. The same kinds of swelling might occur in peripheral nerve cells, too, and cause them to deteriorate and no longer be able to assist in motor nerve functions. The osmotic swelling might also be the cause of poorer circulation into peripheral capillaries, which eventually opens the way for gangrene or for a breakdown of the filtration mechanisms in the kidneys.

Treatment The best single treatment of diabetes is any effort that keeps an absolutely strict control on the blood sugar level. What must at all costs be avoided are the episodes of upward surges followed by precipitous declines as either the diet or the insulin treatments are not well managed.

Among the technologies being used to deal with the problem of proper insulin injections are surgically implanted insulin pumps that are analogous to heart pacemakers. Whole pancreas transplants have been tried during the last two decades but with disappointing results. In the United States only about 18% of the slightly more than 300 such transplants have worked. Efforts to transplant pancreas tissue that contains the islet cells have similarly been unsuccessful.

In the early 1980s, however, scientists discovered that islet cells stripped of neighboring cells can be transplanted, that they need not even be inserted into the receiver's pancreas, and that they start to make insulin in a few weeks. It seems that the cells *adjacent to* islet cells are responsible for the signals that induce the body's immune-centered rejection process. Human islet cells work best, of course, but islet cells from pigs and cows also appear to be usable provided they are encapsulated in very small spheres. These spheres have microscopic holes large enough to let insulin molecules escape but not large enough to let antibodies inside. These techniques have cured type I diabetes in experimental animals. Several companies are vigorously pursuing these technologies in the conviction that type I diabetes can be *cured* by the early 1990s.

this kind of glycolysis is referred to as *anaerobic glycolysis* or the **anaerobic sequence.** The equation for this, overall, is:

$$C_6H_{12}O_6 + 2ADP + 2P_i \xrightarrow[\text{glycolysis}]{\text{Anaerobic}} 2CH_3\overset{\overset{\displaystyle OH}{|}}{C}HCO_2^- + 2H^+ + 2ATP$$

Glucose .... Lactate

Except during extensive exercise, glycolysis is operated with sufficient oxygen, and it is aerobic.

The Steps in Glycolysis. Figure 27.5 outlines the steps to pyruvate that can begin either with glucose or glycogen. Those steps that lead to fructose 1,6-diphosphate are actually up an energy hill, because they consume ATP. But this is like pushing a sled or bike up the short backside of a long hill, because the investment in energy is more than repaid by the long, downhill slide to lactate and more ATP. When glycolysis starts with glycogen instead of glucose, this initial investment in ATP is slightly smaller.

The numbers in this discussion refer to Figure 27.5.

1. Glucose is phosphorylated by ATP under catalysis by hexokinase to give glucose 6-phosphate.

FIGURE 27.5
Glycolysis.

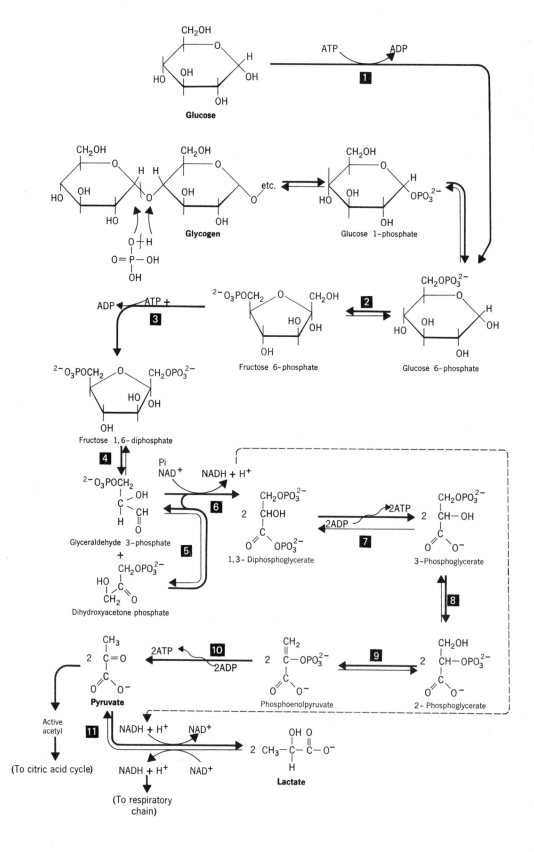

2. Glucose 6-phosphate changes to its isomer, fructose 6-phosphate. The enzyme is phosphoglucose isomerase. This may seem to be a major structural change, but it involves little more than some shifts of bonds and hydrogens, as the arrows in the following sequence show:

These equilibria shift constantly to the right as long as later reactions continuously remove products as they form.

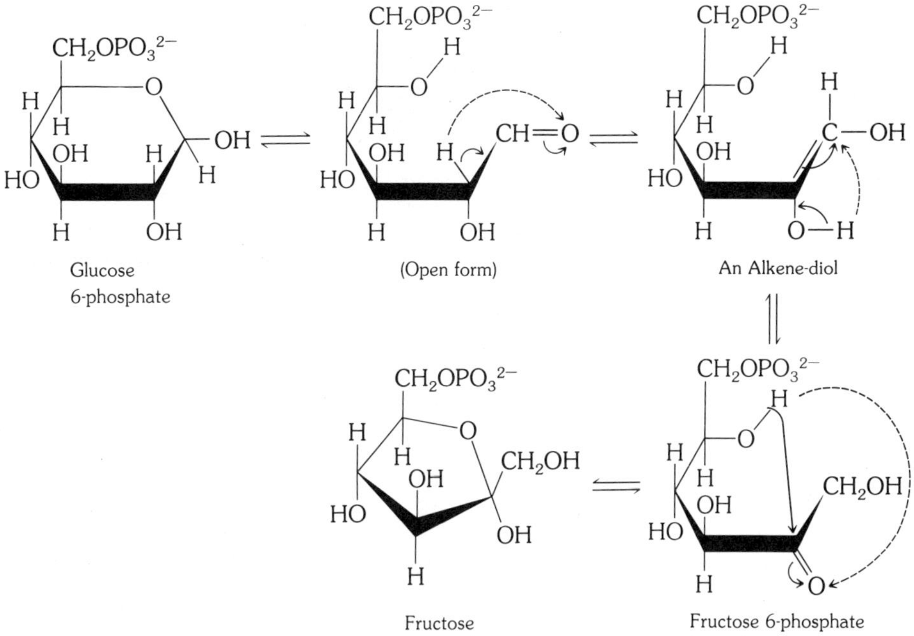

Glucose
6-phosphate

(Open form)

An Alkene-diol

Fructose
6-phosphate
(closed form)

Fructose 6-phosphate
(open form)

3. ATP phosphorylates fructose 6-phosphate to make fructose 1,6-diphosphate. The enzyme is phosphofructokinase. This step is essentially irreversible, and it ends the energy-using phase of glycolysis.

4. Fructose 1,6-diphosphate breaks apart into two triose monophosphates. The reaction is catalyzed by aldolase. We can visualize this reaction in terms of a few simple and reasonable shifts of electrons and protons:

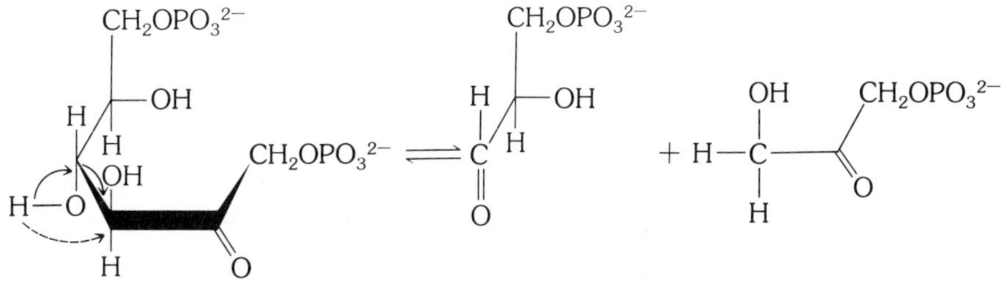

Fructose 1,6-
diphosphate
(open form)

Glyceraldehyde
3-phosphate

Dihydroxyacetone
phosphate

5. Dihydroxyacetone phosphate, in the presence of triose phosphate isomerase, changes to its isomer, glyceraldehyde 3-phosphate.

The continuous removal of glyceraldehyde 3-phosphate by its subsequent reaction shifts the dihydroxyacetone phosphate equilibrium to the right.

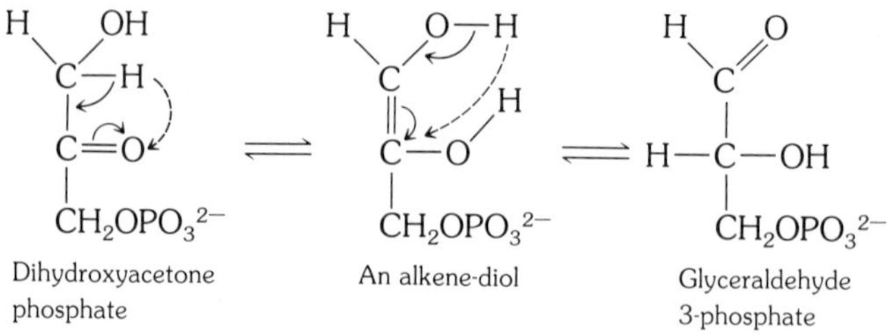

Dihydroxyacetone phosphate   An alkene-diol   Glyceraldehyde 3-phosphate

This change ensures that all of the chemical energy in glucose will be obtained, because the main path of glycolysis continues with glyceraldehyde 3-phosphate. All dihydroxyacetone shuttles through glyceraldehyde 3-phosphate.

6. Glyceraldehyde 3-phosphate is simultaneously oxidized and phosphorylated. The enzyme has NAD^+ as a cofactor, and an —SH group on the enzyme participates. Inorganic phosphate is the source of the new phosphate group. We can visualize how it happens as follows.

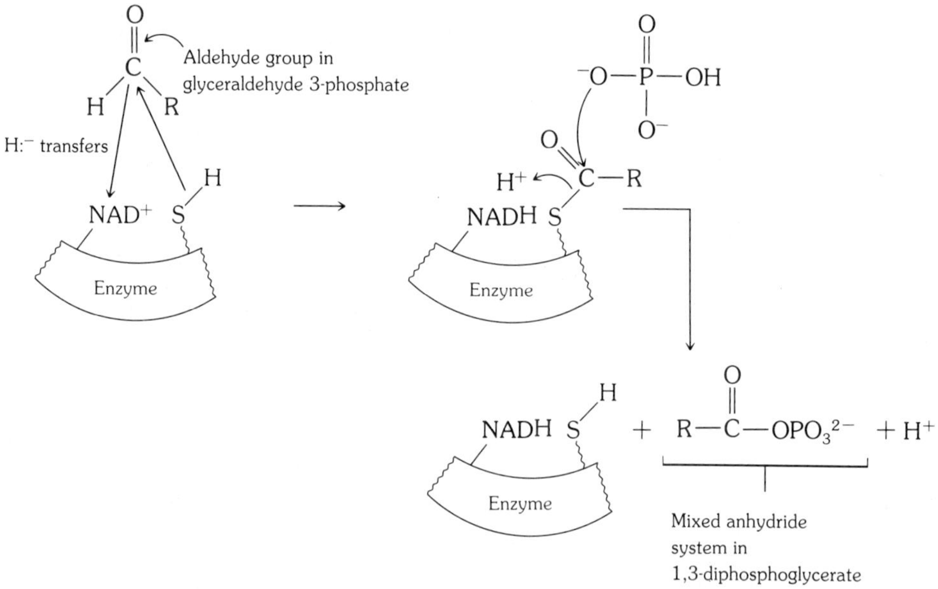

Mixed anhydride system in 1,3-diphosphoglycerate

7. 1,3-Diphosphoglycerate has a higher phosphate group transfer potential than ATP. With the help of phosphoglycerate kinase, it transfers a phosphate to ADP, so this gives us back the original investment of ATP. (Remember that each glucose molecule with six carbons is processed through *two* three-carbon molecules, so *two* ATPs are made here for each original single glucose molecule.)

8. The phosphate group in 3-phosphoglycerate shifts to the 2-position, catalyzed by phosphoglyceromutase.

9. The dehydration of 2-phosphoglycerate to phosphoenolpyruvate is catalyzed by enolase. This step has been compared to cocking a huge bioenergetic gun. A simple, low-energy reaction, dehydration, converts a low-energy phosphate into the highest

energy phosphate in all of metabolism, the phosphate ester of the enol form of pyruvate, phosphoenolpyruvate.[1]

10. More ATP is made as the phosphate group in phosphoenolpyruvate transfers to ADP. The enzyme is pyruvate kinase. The enol form of pyruvate that is stranded promptly and irreversibly rearranges to the keto form of the pyruvate ion. The instability of enols (see footnote 1) is the driving force for this entire step.

11. If the mitochondrion is running aerobically when pyruvate is made, the pyruvate ion changes to acetyl coenzyme A, from which an acetyl group can enter the citric acid cycle (or be used in another way).

No reaction in the body is truly complete until its enzyme is fully restored.

Importance of Anaerobic Glycolysis.   To restore the enzyme at step 6 that now has its coenzyme, NAD^+, in its reduced form as NADH, the hydride unit in NADH is donated directly into the respiratory chain *provided that oxygen is available to run the chain.* Glycolysis would shut down quickly if this NADH unit couldn't unload its $H:^-$ into something. As long as the $H:^-$ unit is on NADH, this enzyme is plugged and glycolysis is blocked at step 6.

If oxygen isn't available, the $H:^-$ in NADH is unloaded into the keto group in pyruvate. This group can accept $H:^-$ (along with H^+ from the buffer system), and change to the alcohol group in lactate. This is why lactate is the end product of anaerobic glycolysis. Lactate serves to store $H:^-$ made at step 6 until the cell once again becomes aerobic. The importance of anaerobic glycolysis is that the cell can continue to make some ATP even when insufficient oxygen is available to run the respiratory chain. Of course, there are limits. The longer the cell operates anaerobically and the more that lactate accumulates, the more the cell runs an increasing **oxygen debt.** Eventually the system has to slow down to let respiration bring back oxygen to metabolize lactate.

By describing the production of lactate rather than lactic acid we have obscured the generation of acid during glycolysis. However, this acid is promptly neutralized by the buffer so that the lactate ion is the product. Of course, extensive physical exercise that forces considerable tissue to operate anaerobically can overtax the buffer. A form of metabolic acidosis called *lactic acid acidosis* is the result. Hyperventilation is initiated to blow out carbon dioxide and thus help to remove acid.

Other Carbohydrates in Glycolysis.   Figure 27.6 shows how central glucose 6-phosphate is to the catabolism of all carbohydrates. It forms from any of the dietary monosaccharides.

Pentose Phosphate Pathway of Glucose Catabolism.   The biosyntheses of some substances in the body require a reducing agent. For example, fatty acids are almost entirely alkane-like, and alkanes are the most reduced types of organic compounds. The reducing agent used in the biosynthesis of fatty acids is NADPH, the reduced form of $NADP^+$.

$NADP^+$ is a phosphate derivative of NAD^+.

The body's principal route to NADPH is the **pentose phosphate pathway** of glucose catabolism. This complicated series of reactions (which we'll not study in detail) is very active in adipose tissue, where fatty acid synthesis occurs. Skeletal muscles have very little activity in this pathway.

---

[1] When a carbon atom of an alkene group holds an —OH group, the compound is called an *enol* ("ene" + "-ol"). Enols are unstable alcohols that spontaneously rearrange into carbonyl compounds:

Enol form   Carbonyl form

FIGURE 27.6
Convergence of the pathways in the metabolism of dietary carbohydrates.

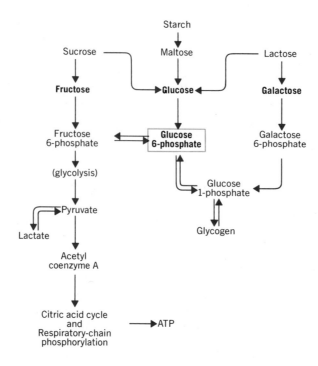

The pentose phosphate pathway also goes by the names *hexose monophosphate shunt* and *phosphogluconate pathway*.

The oxidative reactions in the pentose phosphate pathway convert a hexose phosphate into a pentose phosphate—hence, the name of the series. The overall equation for the oxidative sequence is

$$\text{Glucose 6-phosphate} + 2\text{NADP}^+ + \text{H}_2\text{O} \longrightarrow$$
$$\text{ribose 5-phosphate} + 2\text{NADPH} + 2\text{H}^+ + \text{CO}_2$$

The ribose 5-phosphate can now be used to make the pentose systems in the nucleic acids, if they are needed. If not, ribose 5-phosphate undergoes a series of isomerizations and group transfers that make up the nonoxidative phase of the pentose phosphate pathway. These reactions have the net effect of converting three pentose units into two hexose units (glucose) and one triose unit (glyceraldehyde). The hexoses can be catabolized by glycolysis, or they can be recycled to the oxidative series of the pentose phosphate pathway. Glyceraldehyde, as we'll soon see, can be converted to glucose and recycled as well. Thus glycolysis, the pentose phosphate reactions, and the resynthesis of glucose all interconnect, and specific bodily needs of the moment determine which pathway is operated.

Overall, with pentose recycle, the balanced equation for the complete oxidation of one glucose molecule via the pentose phosphate pathway is as follows:

$$6 \text{ Glucose 6-phosphate} + 12\text{NADP}^+ \xrightarrow{\text{Pentose phosphate pathway}}$$
$$5 \text{ glucose 6-phosphate} + 6\text{CO}_2 + 12\text{NADPH} + 12\text{H}^+ + \text{P}_i$$

27.4 GLUCONEOGENESIS

Some of the steps in gluconeogenesis are the reverse of steps in glycolysis.

The overall scheme of gluconeogenesis, by which glucose is made from smaller molecules, is given in Figure 27.7. Excess lactate can be used as a starting material, as we have already mentioned. Just as important, several amino acids can be degraded to molecules that can be used to make glucose, too. You'll recall that the brain normally uses circulating glucose for energy. Therefore the ability of the body to manufacture glucose from noncarbohydrate sources such as amino acids—even those from the body itself—is an important backup

FIGURE 27.7
Gluconeogenesis. The straight arrows signify steps that are the reverse of corresponding steps in glycolysis. The heavy, curved arrows denote steps that are unique to gluconeogenesis.

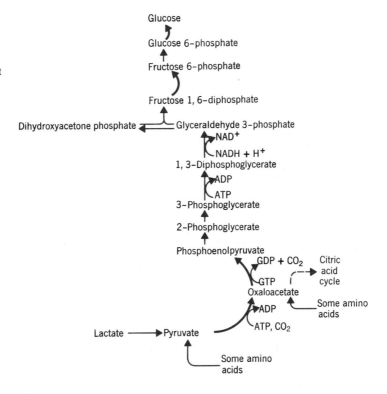

during times when glucose either isn't in the diet (starvation) or it cannot be effectively used (untreated diabetes).

Gluconeogenesis is not the exact reverse of glycolysis. There are three steps in glycolysis that cannot be directly reversed — steps 1, 3, and 10 of Figure 27.5. However, the liver and the kidneys have special enzymes that create bypasses. In the bypass that gets back and around step 10 of glycolysis — the synthesis of phosphoenolpyruvate from pyruvate — carbon dioxide is used as a reactant, whereas ATP is used to drive the steeply uphill reaction, and pyruvate carboxylase is the enzyme. Pyruvate is changed to oxaloacetate:

The enzyme for this step requires the vitamin biotin.

$$CH_3-\overset{\overset{O}{\|}}{C}-CO_2^- + CO_2 + ATP + H_2O \longrightarrow$$

Pyruvate

$$^-O_2C-CH_2-\overset{\overset{O}{\|}}{C}-CO_2^- + ADP + P_i + 2H^+$$

Oxaloacetate

Oxaloacetate then reacts with another triphosphate, guanosine triphosphate (GTP), as carbon dioxide splits out and bonds rearrange:

$$\overset{..}{:}\overset{..}{O}-\overset{\overset{O}{\|}}{C}-CH_2-\overset{\overset{O}{\|}}{C}-CO_2^- + \ ^-O-\overset{\overset{O}{\|}}{\underset{\underset{O^-}{|}}{P}}-O-\overset{\overset{O}{\|}}{\underset{\underset{O^-}{|}}{P}}-O-\overset{\overset{O}{\|}}{\underset{\underset{O^-}{|}}{P}}-O-guanosine \longrightarrow$$

Oxaloacetate

GTP

$$CO_2 + CH_2=\overset{\overset{O-PO_3^{2-}}{|}}{C}-CO_2^- + GDP$$

Phosphoenol-
pyruvate

Because each glucose to be made requires two pyruvates, and because each pyruvate uses two high-energy phosphates in gluconeogenesis, this bypass costs the equivalent of four ATPs (2ATP + 2GTP) per glucose molecule to be made.

At the reversal of step 7 (Figure 27.5), two more ATPs are used per molecule of glucose made. Thus a total of six ATPs is needed to make one glucose molecule by gluconeogenesis that starts from pyruvate, which may be compared with two ATPs that are produced by anaerobic glycolysis (that begins with glucose).

The bypasses to the reverses of steps 3 and 1 of glycolysis require only specific enzymes that catalyze the hydrolysis of phosphate ester groups, not high-energy boosters. Such enzymes are integral parts of the enzyme team for gluconeogenesis. The other steps in gluconeogenesis are run as reverse shifts in equilibria that occur in the opposite direction in glycolysis.

SUMMARY

Glycogen metabolism The regulation of glycogenesis and glycogenolysis is a part of the machinery for glucose tolerance in the body. Hyperglycemia stimulates the secretion of insulin and somatostatin, and insulin helps cells of adipose tissue to take glucose from the blood. Somatostatin helps to suppress the release of glucagon (which otherwise stimulates glycogenolysis and leads to an increase in the blood sugar level).

When glucose is abundant, the body either replenishes its glycogen reserves, or it makes fat. In danger or other stresses, epinephrine stimulates a cascade of enzyme activations that begins with the activation of adenylate cyclase and ends with the release of many glucose molecules from glycogen.

When glucose is in short supply, the body makes its own by gluconeogenesis from noncarbohydrate molecules, including several amino acids. In diabetes, some cells that are starved for glucose make their own, also. Such cells are unable to obtain glucose from circulation, so the blood sugar level is hyperglycemic to a glucosuric level. The glucose tolerance test is used to see how well the body handles an overload of glucose. In the management of diabetes, the maintenance of a reasonably steady blood sugar level in the normal range is vital. The measurement of the glycohemoglobin in circulation serves to monitor how well the normal range is kept over a long period of time.

Following strenuous exercise, when lactate is plentiful, the liver makes glucose from lactate. The many pathways that involve glycogen and glucose form a cycle of events called the Cori cycle.

Glycolysis Under anaerobic conditions, glucose can be catabolized to lactate ion. (Galactose and fructose enter this pathway, too.) By using lactate to store $H:^-$, one of the enzymes in glycolysis can be regenerated in the absence of oxygen, and glycolysis can be run to make some ATP without the involvement of the respiratory chain. Then, when the cell is aerobic again, the $H:^-$ that this enzyme must shed to work again is given directly into the respiratory chain, and pyruvate instead of lactate becomes the end product of glycolysis.

Pentose phosphate pathway The body's need for NADPH to make fatty acids is met by catabolizing glucose through the pentose phosphate pathway.

Gluconeogenesis Most of the steps in gluconeogenesis are simply the reverse of steps in glycolysis, but there are a few that require rather elaborate bypasses. Special teams of enzymes and supplies of high-energy phosphate energy are used for these. Many amino acids can be used to make glucose by gluconeogenesis.

KEY TERMS

The following terms were emphasized in this chapter and they should be mastered before continuing.

anaerobic sequence	gluconeogenesis	glycolysis	oxygen debt
blood sugar level	glucose tolerance	human growth hormone	pentose phosphate pathway
Cori cycle	glucose tolerance test	hyperglycemia	renal threshold
diabetes	glucosuria	hypoglycemia	somatostatin
epinephrine	glycogenesis	insulin	
glucagon	glycogenolysis	normal fasting level	

SELECTED REFERENCES

1 A. L. Lehninger. *Principles of Biochemistry.* Worth Publishers, Inc., New York, 1982.

2 M. Davidson. *Diabetes Mellitus: Diagnosis and Treatment.* John Wiley and Sons, New York, 1981.

3 A. L. Notkins. "The Causes of Diabetes." *Scientific American,* November 1979, page 62.

4 Alice C. Maurer. "The Therapy of Diabetes." *American Scientist,* July – August 1979, page 422.

5 G. B. Kolata. "Blood Sugar and the Complications of Diabetes." *Science,* March 16, 1979, page 1098.

6 Dolores L. Taylor. "Hyperglycemia." *Nursing83,* February 1983, page 52.

7 Dolores L. Taylor. "Hypoglycemia." *Nursing83,* March 1983, page 44.

8 J. E. Gerich. "Role of Growth Hormone in Diabetes Mellitus." *The New England Journal of Medicine,* March 29, 1984, page 848.

9 Nancy E. White and Barbara K. Miller. "Glycohemoglobin. A New Test to Help the Diabetic Stay in Control." *Nursing83,* August 1983, page 55.

10 Claire W. Surr. "New Blood-Glucose Monitoring Products." *Nursing83.* Part 1, January 1983, page 42; Part 2, February 1983, page 58.

11 C. R. Stiller, J. Dupre, M. Gent, M. R. Jenner, P. A. Keown, A. Laupacis, R. Martell, N. W. Rodger, B. V. Graffenried, and B.M.J. Wolfe. "Effects of Cyclosporine Immunosuppression in Insulin-Dependent Diabetes Mellitus of Recent Onset." *Science,* March 30, 1984, page 1362.

12 G. Bylinsky. "Closing in on a Cure for Diabetes." *Fortune,* August 6, 1984, page 70.

REVIEW EXERCISES

The answers to these Review Exercises are in the *Study Guide* that accompanies this book.

Blood Sugar

27.1 What are the end products of the complete digestion of the carbohydrates in the diet?

27.2 Why can we treat the catabolism of carbohydrates as almost entirely that of glucose?

27.3 What is meant by *blood sugar level?* By *normal fasting level?*

27.4 What is the range of concentrations in mg/dL for the normal fasting level of whole blood?

27.5 What characterizes the following conditions?
    (a) Glucosuria   (b) Hypoglycemia
    (c) Hyperglycemia   (d) Glycogenolysis
    (e) Gluconeogenesis   (f) Glycogenesis

27.6 Explain how severe hypoglycemia can lead to disorders of the central nervous system.

Hormones and the Blood Sugar Level

27.7 When epinephrine is secreted, what soon happens to the blood sugar level?

27.8 At which one tissue is epinephrine the most effective?

27.9 What does epinephrine directly activate at its target cell?

27.10 Arrange these in the correct order in which they work in epinephrine-initiated glycogenolysis.

    Phosphorylase kinase, phosphorylase, cyclic AMP,
                1   2   3

    adenylate cyclase, protein kinase
                4   5

27.11 What function does Ca^{2+} have in glycogenolysis in muscles?

27.12 What might be the result if phosphorylase and glycogen synthetase were both activated at the same time?

27.13 What switches glycogen synthetase off when glycogenolysis is activated?

27.14 What is the end product of glycogenolysis, and what does phosphoglucomutase do to it?

27.15 Why can liver glycogen but not muscle glycogen be used to resupply blood sugar?

27.16 What is glucagon, what does it do, and what is its chief target tissue?

27.17 Which is probably better at increasing the blood sugar level, glucagon or epinephrine? Explain.

27.18 How does human growth hormone manage to promote the supply of the energy needed for growth?

27.19 What is insulin, where is it released, and what is its chief target tissue?

27.20 What triggers the release of insulin into circulation?

27.21 If brain cells are not insulin-dependent cells, how can too much insulin cause insulin shock?

27.22 What is somatostatin, where is it released, and what kind of effect does it have on the pancreas?

Glucose Tolerance and the Cori Cycle

27.23 What is meant by *glucose tolerance?*

27.24 For what chief purpose is a glucose tolerance test performed, and how is it carried out?

27.25 Describe what happens when each of the following types has a glucose tolerance test.
    (a) A nondiabetic individual
    (b) A diabetic individual

27.26 Describe a circumstance in which hyperglycemia might arise in a nondiabetic individual.

27.27 In general terms, what happens to excess lactate produced in muscles during exercise?

27.28 Suppose that a sample of glucose is made using some atoms of carbon-13 in place of the common isotope, carbon-12. Suppose further that this is fed to a healthy, adult volunteer and that all of it is taken up by the muscles.
(a) Will some of the original molecules be able to go back out into circulation? Explain.
(b) Can we expect any carbon-13 compounds to end up in the liver? Explain.
(c) Can we ever expect to see carbon-13 labeled glucose molecules in circulation again? Explain.

Catabolism of Glucose

27.29 Fill in the missing substances and balance the following incomplete equation:

$$C_6H_{12}O_6 + 2ADP + \underline{\hspace{2cm}} \longrightarrow$$
Glucose

$$2C_3H_5O_3^- + 2H^+ + \underline{\hspace{2cm}}$$
Lactate

27.30 What particular significance does glycolysis have when a tissue is running an oxygen debt?

27.31 Why is the rearrangement of dihydroxyacetone phosphate into glyceraldehyde 3-phosphate important?

27.32 What happens to pyruvate under (a) aerobic conditions and (b) anaerobic conditions?

27.33 What happens to lactate when an oxygen debt is repaid?

27.34 What is the maximum number of ATPs that can be made by the complete catabolism of (a) one molecule of glucose and (b) one glucose residue in glycogen?

27.35 The pentose phosphate pathway uses $NADP^+$, not NAD^+. What forms from $NADP^+$, and how does the body use it (in general terms)?

Gluconeogenesis

27.36 In a period of prolonged fasting or starvation, what does the system do to try to maintain its blood sugar level?

27.37 Amino acids are not excreted, and they are not stored in the same way that glucose residues are stored in a polysaccharide. What probably happens to the excess amino acids in a high-protein diet of an individual who does not exercise much?

27.38 The amino groups of amino acids can be replaced by keto groups. Which amino acids could give the following keto acids that participate in carbohydrate metabolism?
(a) Pyruvic acid   (b) Oxaloacetic acid

27.39 Referring to data in the previous chapter: (a) How much ATP can be made from the chemical energy in one pyruvate? (b) The conversion of one lactate to one pyruvate transfers one $H\!:^-$ to NAD^+. How many ATP molecules can be made just from the operation of this step? (c) If all of the possible ATP that can be obtained from the complete catabolism of lactate were made available for gluconeogenesis, how many molecules of glucose could be made? (Assume that ATP can substitute for GTP.)

Chapter 28
Metabolism of Lipids

A marathon race quickly draws down the glycogen reserves and then taps fatty acids for the long haul. We'll study how fatty acids are used for energy in this chapter.

28.1   ABSORPTION AND DISTRIBUTION OF LIPIDS

Several lipoprotein complexes in the blood transport triacylglycerols, fatty acids, cholesterol, and other lipids from tissue to tissue.

The complete digestion of triacylglycerols, as we learned in Chapter 25, produces glycerol and a mixture of long-chain fatty acids. These, together with some monoacylglycerols (from incomplete digestion), leave the digestive tract, and as they migrate across the intestinal barrier they are extensively reconstituted into triacylglycerols. What is delivered to circulation consists mostly of these triacylglycerols.

Lipoprotein Complexes in Circulation.   Lipids are insoluble in water, and they are carried in blood by proteins with which they form **lipoprotein complexes.** Because defects in this system can cause heart disease, we'll look more closely at these species.

Lipoproteins are classified according to their densities, and each class has its own functions. They range in densities from 0.93 g/cm³ to 1.21 g/cm³, and the lipoproteins with the lowest density are called **chylomicrons.** They are only 2% protein or less and are put together in the liver. Let's use the scheme in Figure 28.1 to follow the migrations of lipid materials. The numbers that follow refer to the numbers in this figure. In some of the carrier units in this figure there are letters such as E, C, B-48, and B-100. They represent polypeptides, and they not only participate as lipid carriers but as the species that receptor proteins of other tissues, such as the liver, can recognize. Except for pointing out that the identities of these polypeptides change, and that B-100 is the polypeptide recognized by one key receptor, we will not call any further attention to them. Now let's get to the steps in the figure.

Remember that cholesterol is a nonsaponifiable lipid so it can't be hydrolyzed by digestion.

Where lipids from the processes of digestion, $\boxed{1}$, enter circulation, $\boxed{2}$, chylomicrons pick up any that are being delivered—triacylglycerols, cholesterol, and free fatty acids. They transport these lipids, $\boxed{3}$, and when they are in capillaries of adipose tissue, they unload some of their triacylglycerols, $\boxed{4}$. (A lipoprotein lipase catalyzes the hydrolysis of triacyglycerols. The resulting fatty acids and glycerol are absorbed by adipose tissue and reconstituted as triacylglycerols that are then stored until needed.) This process leaves chylomicron remnants, $\boxed{5}$, that are now richer in cholesterol, and the liver has special receptor proteins that recognize and help the liver absorb these remnants, $\boxed{6}$.

The liver can do any one of a number of things with cholesterol, whether the cholesterol comes from the digestive tract or is made by the liver. The liver can excrete cholesterol by transferring it to bile, which puts it into the digestive tract, and such cholesterol eventually leaves the body via the feces. The liver can also make bile salts from cholesterol, and these are needed in the digestive tract to help digest lipids. The liver can send cholesterol out into circulation either as free cholesterol or as esters of cholesterol to be picked up by cells that need the steroid nucleus for any purpose. Remember that cells everywhere use cholesterol as one raw material for building cell membranes, and endocrine glands make steroid hormones from cholesterol. So now the liver manufactures another lipoprotein complex designated the very-low-density-lipoprotein, or VLDL, for short, $\boxed{7}$. This has a slightly higher density than the chylomicrons.

Cholesterol (d = 1.05 g/cm³) is more dense than triacylglycerols (density of about 0.9 g/cm³).

The liver can make both cholesterol and triacylglycerols itself, as we'll study later in this chapter. Thus the VLDL complexes that are now put together by the liver carry cholesterol and cholesterol esters, regardless of their orgins, as well as triacylglycerols to adipose tissue and muscles, $\boxed{8}$. Triacylglycerols are mostly removed (again via hydrolysis and reconstitution). However, the lipoprotein complex that emerges, $\boxed{9}$, is now of slightly higher density, is mostly cholesterol in one form or another, and the complex is now designated as an intermediate-density-lipoprotein-complex, or IDL for short. Much of the IDL is returned to the liver, $\boxed{10}$, being helped in by special receptor proteins that recognize the B-100 polypeptide. Some of the IDL, however, escapes reabsorption by the liver and it experiences further removal of triacylglycerol. This causes a further increase in density. Now the particles are classified as low-density-lipoproteins, or LDL, $\boxed{11}$.

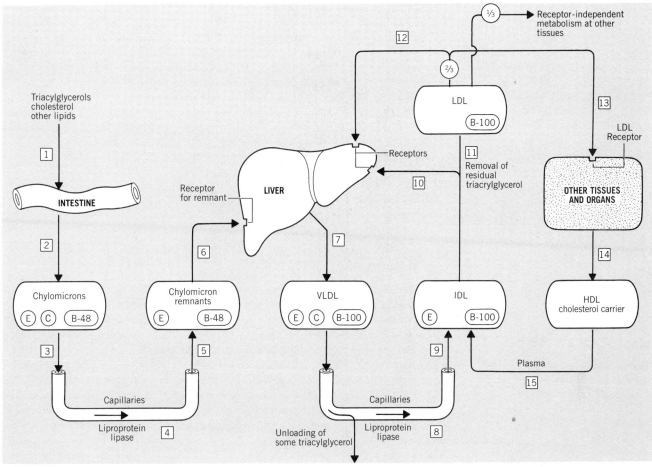

FIGURE 28.1
The transport of cholesterol and triacylglycerols by lipoprotein complexes. (Adapted by permission from J. L. Goldstein, T. Kita, and M. S. Brown, *New England Journal of Medicine,* August 4, 1983, page 289.)

J. L. Goldstein and M. S. Brown won the 1985 Nobel prize in medicine for discoveries described in this section.

A good portion of the LDL is captured by the liver by receptor proteins, and transporting cholesterol back to the liver, $\boxed{12}$, is one function of LDL. However, about a third of the LDL reaches peripheral tissues, $\boxed{13}$, including the adrenal glands. Thus another purpose of LDL is to carry cholesterol wherever cholesterol is needed to make cell membranes or to make steroid hormones. Some of these tissues carry special receptor proteins, but others can obtain cholesterol from LDL by other means.

Any leftover cholesterol must now be removed from the extrahepatic tissues (tissues other than the liver), and this job is handled by the high-density-lipoprotein complexes, or HDL, $\boxed{14}$, that the liver makes for this purpose. The chief function of the HDL is to carry cholesterol back to the liver, $\boxed{15}$.

Cholesterol and Heart Disease. The receptor proteins for the IDL and LDL have a crucial function. If they are reduced in number or are absent, there is little ability by the liver to remove excess cholesterol and export it via the bile. In some people, there is a genetic defect that bears specifically on these receptors. Two genes are involved. Those who carry two mutant genes have *familial hypercholesterolemia*—genetically caused high level of cholesterol in the blood. On even a zero-cholesterol diet, the victims have very high cholesterol levels. Their cholesterol slowly comes out of the blood at valves and other sites, reduces the dimensions of the blood capillaries, and this restricts blood flowage. Atherosclerosis has set in.

The genetic condition is described as *homozygous* when mutant genes for it come from both parents.

When just one of the two parents contributes a mutant gene for the trait, the condition is said to be heterozygous. About 1 person in 500 is heterozygous for familial hypercholesterolemia.

The heart must now work harder, and eventually arteries and capillaries in the heart itself no longer are able to bring oxygen to heart tissue. The victims generally have their first heart attacks as children, and are dead by their early 20s. People with one defective gene and one normal gene for the LDL receptor proteins generally have blood cholesterol levels that are two or three times higher than normal. Although they number only about 0.5% of all adults, they account for 5% of all heart attacks among those younger than 60.

High blood cholesterol levels also occur in many other people, even those with normal genes for the receptor proteins, and the causes have not been fully unraveled. High cholesterol foods appear to contribute, and there is some evidence that as the liver receives more and more cholesterol from the diet it loses more and more of the receptor proteins. This forces more and more cholesterol to linger in circulation. Smoking, obesity, and lack of exercise contribute to the cholesterol problem also.

28.2   STORAGE AND MOBILIZATION OF LIPIDS

Triacylglycerols are the most weight-efficient means of storing chemical energy in the body.

It's easier for the system to store energy as lipids than as glycogen or as glucose, because of the significant differences in the energy densities of these substances. The energy density is the grams of tissue or fluid per kilocalorie available. If an isotonic solution of glucose is the energy source, it takes 5 grams of this solution to provide 1 kcal of energy. Its energy density is 5 g/kcal. However, if the glucose is changed into glycogen *and no longer is in solution*, there is little associated water. The energy density of wet glycogen is about 0.6 g/kcal. However, the energy density for triacylglycerol is only 0.13 g/kcal. One factor that accounts for this is the greater number of C—H bonds per molecule in triacylglycerols compared to glucose. The electrons of this bond enter the respiratory chain. As we will see, we can make roughly 120 ATP molecules by the catabolism of one molecule of a fatty acid, but we get only 38 at the most from each glucose molecule.

Because lipids are water-insoluble, they attract the least amount of associated water in storage.

Adipose Tissue as the Lipid Storage Depot.   The chief depot for the storage of fatty acids is adipose tissue, a very metabolically active tissue. There are two kinds, brown and white. Both types are associated with internal organs, and they cushion the organs against mechanical bumps and shocks, and they insulate them from swings in temperature. White adipose tissue stores energy as triacylglycerols chiefly on behalf of the energy budgets of other tissues. For a discussion of the possible relationship between brown adipose tissue and a severely overweight condition — obesity — see Special Topic 28.1. The discussion that continues concerns the metabolic activities of white adipose tissue.

These data are for information; they're certainly not recommendations!

A 70-kg adult male has about 12 kg of triacylglycerol in storage. If he had to exist on no food, just water and a vitamin-mineral supplement, and if he needed 2500 kcal/day, this fat would supply his caloric needs for 43 days. Of course, during this time the body proteins would also be wasting away, and metabolic acidosis would be a problem of growing urgency.

Mobilization of Lipid Reserves for Energy.   Lipid material comes and goes constantly from adipose tissue, and the balance between its receiving lipids or releasing them is struck by the energy requirements elsewhere. In a nutritional state that is abundant in glucose, little if any fatty acids are needed for energy. In fact, considerable glucose is converted into fat (to the delight of weight-reducing salons). However, sometimes glucose is either in very low supply (as in starvation) or what is available can't be used (as in diabetes), and the body now turns to its fatty acids for energy.

Figure 28.2 outlines the many steps involved in tapping the lipid reserves for their energy. Triacylglycerol molecules in adipose tissue are first hydrolyzed to free fatty acids and glycerol. The lipase needed for this is activated by a process involving cyclic AMP and such hormones as epinephrine and glucagon. Insulin suppresses this enzyme by suppressing the formation of cyclic AMP at adipose cells. Thus when insulin is in circulation—and it's there only because there is a good supply of glucose in the blood—it means that the fatty acids aren't as much needed for energy.

The glycerol that is produced is changed to dihydroxyacetone phosphate, and it enters the pathway of glycolysis.

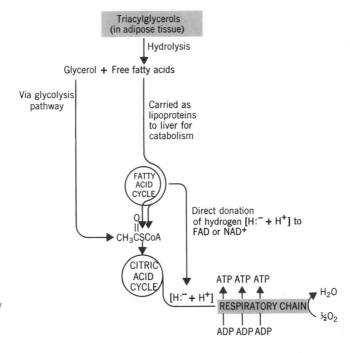

FIGURE 28.2
Pathways for the mobilization of energy reserves in the triacylglycerols of adipose tissue.

The fatty acids are carried as lipoprotein complexes to the liver, the chief site for their catabolism. Both fatty acids and short molecules made from them can also be used for energy in the heart and by skeletal muscles. In fact, most of the energy needs of resting muscle tissue are met by intermediates of fatty acid catabolism, not from glucose. Given enough time for adjustment, even brain cells can obtain some energy from fatty acid breakdown products.

28.3 THE CATABOLISM OF FATTY ACIDS

Acetyl groups are produced by the beta-oxidation of fatty acids and are fed into the citric acid cycle and respiratory chain.

The degradation of fatty acids takes place inside mitochondria by a repeating series of steps known as the **fatty acid cycle,** or as **beta-oxidation.** Figure 28.3 outlines its chief steps.

To enter the cycle, a fatty acid has to be joined to coenzyme A. It costs one ATP to do this, but now the fatty acyl unit is activated. (The ATP itself breaks down to AMP and PP_i, so the actual cost in high-energy phosphate is *two* high-energy phosphate bonds. Breaking ATP to AMP is the equivalent of breaking two molecules of ATP to two of ADP.)

The repeating sequence of the fatty acid cycle consists of four steps. Each turn of the cycle degrades the fatty acyl unit by two carbons, and produces one molecule of $FADH_2$, one of NADH, and one of acetyl coenzyme A. The now shortened fatty acyl unit is carried again through the four steps, and the process is repeated until no more two-carbon acetyl units can be made. The $FADH_2$ and the NADH fuel the respiratory chain. The acetyl groups pass into the citric acid cycle, or they enter the general pool of acetyl coenzyme A that the body draws from to make other substances (e.g., cholesterol). Let's now look at the four steps in greater detail. The numbers that follow refer to Figure 28.3.

Although the whole series isn't exactly a cycle of a true variety, like the citric acid cycle, it's still referred to as the fatty acid cycle.

1. The first step is dehydrogenation. FAD accepts ($H:^- + H^+$) from the α- and the β-carbons of the fatty acyl unit.

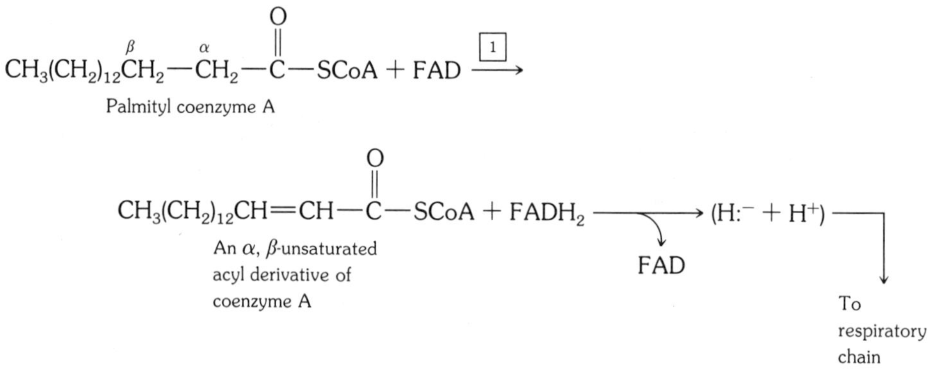

$$CH_3(CH_2)_{12}\overset{\beta}{C}H_2-\overset{\alpha}{C}H_2-\overset{\overset{\textstyle O}{\|}}{C}-SCoA + FAD \xrightarrow{\boxed{1}}$$

Palmityl coenzyme A

$$CH_3(CH_2)_{12}CH{=}CH-\overset{\overset{\textstyle O}{\|}}{C}-SCoA + FADH_2 \longrightarrow (H:^- + H^+) \longrightarrow$$

An α, β-unsaturated acyl derivative of coenzyme A

FAD

To respiratory chain

2. The second step is hydration. Water adds to the alkene double bond and a secondary alcohol group forms.

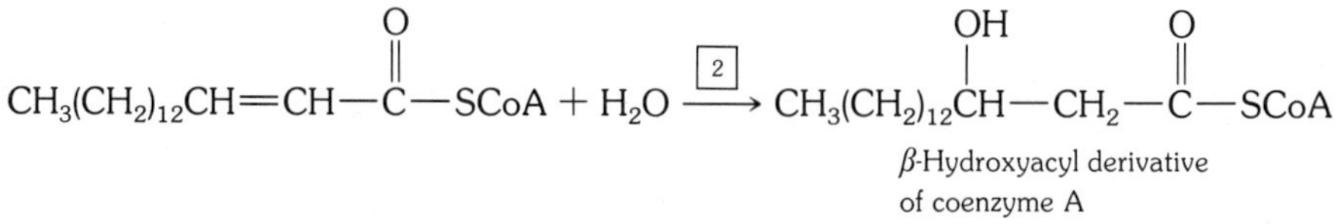

$$CH_3(CH_2)_{12}CH{=}CH-\overset{\overset{\textstyle O}{\|}}{C}-SCoA + H_2O \xrightarrow{\boxed{2}} CH_3(CH_2)_{12}\overset{\overset{\textstyle OH}{|}}{C}H-CH_2-\overset{\overset{\textstyle O}{\|}}{C}-SCoA$$

β-Hydroxyacyl derivative of coenzyme A

3. The third step is another dehydrogenation—a loss of ($H:^- + H^+$). This oxidizes the secondary alcohol to a keto group. Notice that these steps end in the oxidation of the

FIGURE 28.3
The fatty acid cycle (β-oxidation). The numbers refer to the numbered steps discussed in the text.

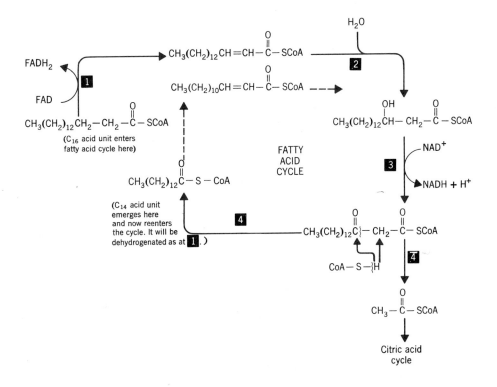

β-position of the original fatty acyl group to a keto group. This is why the fatty acid cycle is sometimes called *beta oxidation*.

$$CH_3(CH_2)_{12}\overset{\underset{|}{OH}}{CH}-CH_2-\overset{\underset{\|}{O}}{C}-SCoA + NAD^+ \xrightarrow{\boxed{3}}$$

$$CH_3(CH_2)_{12}\overset{\underset{\|}{O}}{C}-CH_2-\overset{\underset{\|}{O}}{C}-SCoA + NADH + H^+$$

β-Keto acyl coenzyme A

$\longrightarrow (H:^- + H^+)$

$\searrow$ NAD$^+$ $\downarrow$

To respiratory chain

Franz Knoop directed much of the research on the fatty acid cycle, so this pathway is sometimes called *Knoop oxidation*.

4. The fourth step breaks the bond between the α and the β carbons. This has been weakened by the stepwise oxidation of the β-carbon, and now this bond breaks to release one unit of acetyl coenzyme A.

$$CH_3(CH_2)_{12}\overset{\underset{\|}{O}}{C}\left(-CH_2-\overset{\underset{\|}{O}}{C}-SCoA \xrightarrow{\boxed{4}} CH_3(CH_2)_{12}\overset{\underset{\|}{O}}{C}-SCoA + CH_3-\overset{\underset{\|}{O}}{C}-SCoA\right.$$

CoAS$-$) H

Myristyl coenzyme A

Acetyl coenzyme A

To citric acid cycle

12 ATP $\longleftarrow$ via respiratory chain

TABLE 28.1
Maximum Yield of ATP from Palmityl CoA

Seven Turns of the Cycle Produce:	ATP from Each Energy-Rich Intermediate	Total ATP Produced
7FADH$_2$	2	14
7NADH	3	21
8CH$_3$C̈—SCoA (O)	12	96
		131 ATP
Deduct two high-energy phosphate bonds for activating the acyl unit		−2
Net ATP yield per palmityl unit		129 ATP

The remaining acyl unit, the original now shortened by two carbons, now goes through the cycle of steps again — dehydrogenation, hydration, dehydrogenation, and cleavage. After seven such cycles, one molecule of palmityl coenzyme A is broken into eight molecules of acetyl coenzyme A.

Yield of ATP via the Fatty Acid Cycle.   Table 28.1 shows how the maximum yield of ATP from the oxidation of one unit of palmityl CoA adds up to 131 ATPs. The net from palmitic acid is two ATP fewer, or 129 ATP, because the activation of the palmityl unit — joining it to CoA — requires this initial investment, as we mentioned earlier.

28.4 BIOSYNTHESIS OF FATTY ACIDS

Acetyl CoA molecules that are not needed to make ATP can be made into fatty acids.

Acetyl CoA stands at a major metabolic crossroads. It can be made from any monosaccharide in the diet, from virtually all amino acids, and from fatty acids. Once made, it can be shunted into the citric acid cycle where its chemical energy can be used to make ATP; or its acetyl group can be made into other compounds that the body needs. In this section we'll see how acetyl CoA can be made into long-chain fatty acids by a series of steps called **lipigenesis.**

The cytosol is the fluid outside of cellular organelles such as mitochondria and nuclei.

Lipigenesis.   Whenever acetyl coenzyme A molecules are made within mitochondria but aren't needed for the citric acid cycle and respiratory chain, they are exported to the cytosol. The enzymes for lipigenesis are found in the cytosol, not within the mitochondria, which illustrates the general rule that the body segregates its sequences of catabolism from those of anabolism.

As might be expected, because lipigenesis is in the direction of climbing an energy hill, the cell has to invest some energy of ATP to make fatty acids from smaller molecules. The first payment occurs in the first step in which the bicarbonate ion reacts with acetyl CoA.

The enzyme for this step, acetyl CoA carboxylase, requires the vitamin biotin.

$$CH_3—\overset{O}{\overset{\|}{C}}—SCoA + HCO_3^- + ATP \longrightarrow$$

Acetyl CoA

$$^-O—\overset{O}{\overset{\|}{C}}—CH_2—\overset{O}{\overset{\|}{C}}—SCoA + 2H^+ + ADP + P_i$$

Malonyl CoA

This activates the acetyl system for lipigenesis.

Enzymes

1. Malonyl transferase
2. 3-Ketoacyl-ACP synthase
3. 3-Ketoacyl-ACP reductase
4. 3-Hydroxyacyl-ACP dehydratase
5. Enoyl-ACP reductase
6. Acetyl transferase
7. Acyl carrier protein

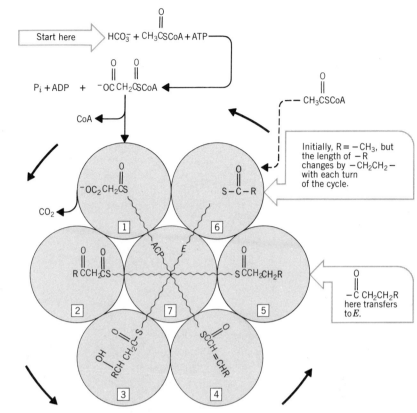

FIGURE 28.4

The lipigenesis cycle. At the top, an acetyl group is activated and joined as a malonyl unit to an arm of the acyl carrier protein, ACP. Another acetyl group transfers from acetyl CoA to site E. In a second transfer, this acetyl group and is then joined to the malonyl unit as CO_2 splits back out. This gives a β-ketoacyl system whose keto group is reduced to CH_2 by the next series of steps. One turn of the cycle adds a CH_2CH_2 unit to the growing acyl chain.

The enzyme that now takes over is actually a huge complex of seven enzymes called *fatty acid synthase*. (See Figure 28.4.) In the center of this complex is a molecular unit long enough to serve as a swinging arm carrier. It's called the *acyl carrier protein*, or ACP, and like the boom of a construction crane, this arm swings from site to site in the synthase. Thus the arm brings what it carries over first one enzyme and then another, and at each stop a reaction is catalyzed that contributes to chain lengthening. Let's see how it works.

The malonyl unit in malonyl CoA, which we just made, transfers to the swinging arm of ACP:

$$\overset{O}{\underset{||}{^-OCCH_2}}\overset{O}{\underset{||}{C}}-S-CoA + ACP \longrightarrow \overset{O}{\underset{||}{^-OCCH_2}}\overset{O}{\underset{||}{C}}-S-ACP + CoA$$

Malonyl ACP

In the meantime, a similar reaction occurs to another molecule of acetyl CoA at a different unit of the synthase, a unit that we'll call simply E:

$$\overset{O}{\underset{||}{CH_3C}}-S-CoA + E \longrightarrow \overset{O}{\underset{||}{CH_3C}}-S-E + CoA$$

Acetyl E

Next, the acetyl group of acetyl E is transferred to the malonyl group of malonyl ACP. Carbon dioxide, the initial activator, is ejected — it has served its activating purpose — and a four-carbon derivative of ACP, acetoacetyl ACP, forms. The E unit is vacated.

$$CH_3\overset{O}{\overset{\|}{C}}-S-E + {}^-O\overset{O}{\overset{\|}{C}}CH_2\overset{O}{\overset{\|}{C}}-S-ACP \longrightarrow CH_3\overset{O}{\overset{\|}{C}}CH_2\overset{O}{\overset{\|}{C}}-S-ACP + CO_2 + E$$
Acetoacetyl ACP

The ketone group in acetoacetyl ACP is next reduced to a secondary alcohol. Then this alcohol is dehydrated to introduce a double bond. And the double bond is next reduced to give butyryl ACP. The overall effect of these steps is to reduce the keto group to $-CH_2-$. Notice that NADPH, the reducing agent manufactured by the pentose phosphate pathway of glucose catabolism, is used here, not NADH.

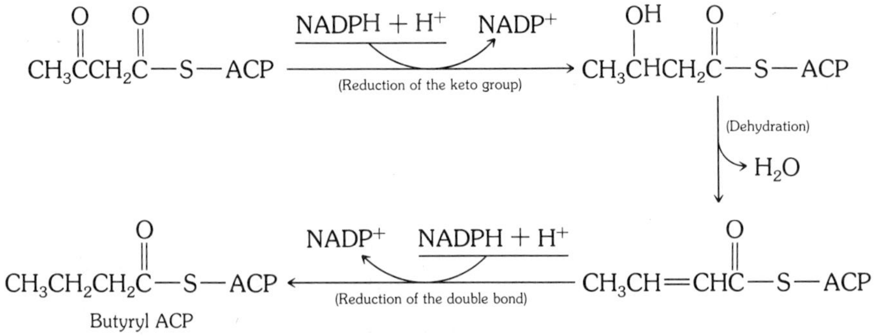

$$CH_3\overset{O}{\overset{\|}{C}}CH_2\overset{O}{\overset{\|}{C}}-S-ACP \xrightarrow[\text{(Reduction of the keto group)}]{NADPH + H^+ \quad NADP^+} CH_3\overset{OH}{\overset{|}{C}}HCH_2\overset{O}{\overset{\|}{C}}-S-ACP$$

(Dehydration) $\longrightarrow H_2O$

$$CH_3CH_2CH_2\overset{O}{\overset{\|}{C}}-S-ACP \xleftarrow[\text{(Reduction of the double bond)}]{NADP^+ \quad NADPH + H^+} CH_3CH=CH\overset{O}{\overset{\|}{C}}-S-ACP$$
Butyryl ACP

The butyryl group is now transferred to the *vacant E* unit of the synthase, the unit that initially held an acetyl group. This ends one complete turn of the cycle. To recapitulate, we have gone from two two-carbon acetyl units to one four-carbon acyl unit.

The steps now repeat as shown in Figure 28.4. A new malonyl unit is joined to the ACP. Then the newly made *butyryl* group is made to transfer to the malonyl unit as CO_2 is again ejected. This elongates the fatty acyl chain to six carbons in length, positions it on the swinging arm, and gets it ready for the several-step reduction of the keto group to $-CH_2-$. The swinging arm mechanism and the enzymes of the synthase complex go to work until the chain is that of the hexanoyl group, $CH_3CH_2CH_2CH_2CH_2CO-$.

In the next turn, this six-carbon acyl group will be elongated to an eight-carbon group. And the process will repeat until the chain is as long as the system requires. Overall, the net equation for the synthesis of the palmitate ion from acetyl CoA is:

$$8CH_3\overset{O}{\overset{\|}{C}}SCoA + 7ATP + 14NADPH \longrightarrow$$
$$CH_3(CH_2)_{14}CO_2^- + 7ADP + 7P_i + 8CoA + 14NADP^+ + 6H_2O$$

Glucagon, epinephrine, and cyclic AMP — all stimulators of the use of glucose to make ATP — depress the synthesis of fatty acids in the liver. Insulin, however, promotes it.

Because the symbols ATP, ADP, and P_i are not given with their electrical charges, we can't provide an electrical balance to equations such as this.

28.5 BIOSYNTHESIS OF CHOLESTEROL

Excessive cholesterol can inhibit the formation of a key enzyme required in the multistep synthesis of cholesterol.

In addition to serving as a raw material for making fatty acids, acetyl CoA can be used to make the steroid nucleus. Cholesterol, an alcohol with this nucleus, is the end product of a long,

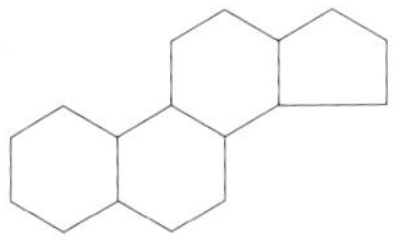

Steroid nucleus

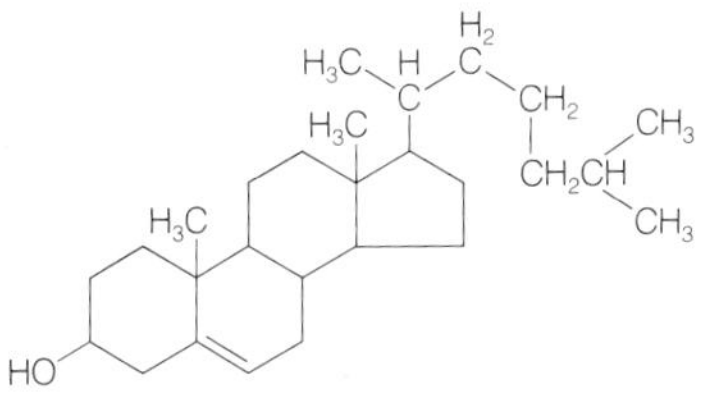

Cholesterol

multistep process, and once it is made the body makes various bile salts and sex hormones.

In mammals, about 80 to 95% of all cholesterol synthesis takes place in cells of the liver and the intestines. We won't go into all of the details, but we will go far enough to learn more about how the body normally controls the process. If sufficient cholesterol is provided by the diet for use in making cell membranes or to make other steroids, then the body's synthesis should be shut down. Let's see how this is done.

When the level of acetyl CoA builds up in, say, the liver, the following equilibrium shifts to the right:

$$2CH_3\overset{O}{\overset{||}{C}}-SCoA \rightleftharpoons CH_3\overset{O}{\overset{||}{C}}CH_2\overset{O}{\overset{||}{C}}-SCoA + CoA-SH$$

Acetyl CoA   Acetoacetyl CoA

When cholesterol synthesis is switched on, then acetoacetyl CoA combines with another acetyl CoA:

$$CH_3\overset{O}{\overset{||}{C}}CH_2\overset{O}{\overset{||}{C}}-SCoA + CH_3\overset{O}{\overset{||}{C}}-SCoA \xrightleftharpoons{\text{HMG—CoA synthase}}$$

$$^-O\overset{O}{\overset{||}{C}}CH_2\overset{OH}{\underset{|}{\underset{CH_3}{C}}}CH_2\overset{O}{\overset{||}{C}}-SCoA + CoA-SH$$

HMG—CoA
(β-Hydroxy-β-methyl-
glutaryl CoA)

Both a reduction and a hydrolysis occur in the next step, which is a complex change catalyzed by the enzyme HMG—CoA reductase:

$$HMG-CoA + 2NADPH + 2H^+ \xrightarrow[\text{reductase}]{\text{HMG—CoA}}$$

$$HOCH_2CH_2\overset{OH}{\underset{|}{\underset{CH_3}{C}}}CH_2CO_2^- + 2NADP^+ + CoA-SH$$

Mevalonate

The control of HMG—CoA reductase is the major factor in the overall control of the biosynthesis of cholesterol. Cholesterol itself is one inhibitor, and it works by inhibiting the *synthesis* of the enzyme. In the presence of cholesterol the enzyme isn't deactivated. There is just *less* of it. Thus if the diet is relatively rich in cholesterol, the body tends to make less of it. If the diet is very low in cholesterol, the body makes more of it. Of course, another control mechanism over the cholesterol level in the blood is the efficiency with which excess cholesterol can be exported via the lower intestinal tract, as we discussed in Section 28.1.

Figure 28.5 provides a summary of much of what we have covered in this and the previous chapter. One point emphasized by this figure is that triacylglycerols can be made from any of the three dietary components—carbohydrates, lipids, and proteins. Notice the central position occupied by acetyl CoA.

FIGURE 28.5
Principal sources of triacylglycerols for
adipose tissue and the chief uses of
acetyl CoA.

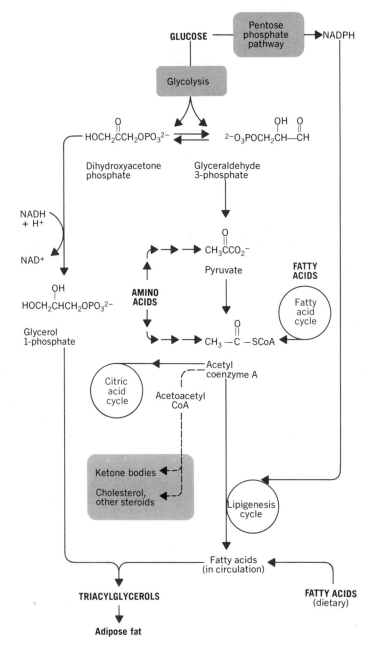

28.6   KETOACIDOSIS

**An acceleration of the fatty acid cycle tips some equilibria in a direction that
leads to ketoacidosis.**

Cells of certain tissues have to engage in gluconeogenesis in two serious conditions — starvation and uncontrolled diabetes mellitus. In starvation, the blood sugar level drops because of nutritional deficiencies, so the body (principally the liver) tries to compensate by making glucose.

In type II diabetes, despite a surplus of glucose in circulation, insulin-dependent cells such as those of adipose tissue and skeletal muscles can't absorb enough from the blood. The body therefore has to mobilize alternative methods for manufacturing ATP, and the principal route taken is to draw chemical energy at an accelerated rate from the catabolism of fatty acids. The liver is the principal organ to run the fatty acid cycle, and some of the intermediate, shorter-chain products made by this cycle can get into circulation and be further catabolized to supply energy to muscle and adipose tissues. However, the shorter chain intermediates of the fatty acid cycle as well as acetyl CoA, the principal end product of this cycle, accumulate in the liver more rapidly than they are removed by subsequent processes. These compounds activate a key enzyme that regulates the rate of gluconeogenesis in the liver. The result is that the liver in the untreated diabetic individual manufactures glucose at an abnormally high rate, and much of this new glucose goes out into circulation to drive the blood sugar level higher. Moreover, the gradual rise in the level of acetyl CoA opens the way to a form of metabolic acidosis known as ketoacidosis.

Ketone Bodies. As the supply of acetyl CoA rises in the liver, more and more of this compound is converted into a group of three compounds called the *ketone bodies*. In an abundance of acetyl CoA, the following equilibrium shifts to the right to make acetoacetyl CoA:

Le Chatelier's principle is again at work. The stress in this equilibrium is an increase in the concentration of acetyl CoA, so the shift is in the direction that uses it up.

$$2CH_3\overset{O}{\overset{\|}{C}}-SCoA \rightleftharpoons CH_3\overset{O}{\overset{\|}{C}}CH_2\overset{O}{\overset{\|}{C}}-SCoA + CoA-SH$$

Acetyl CoA   Acetoacetyl CoA

The level of acetoacetyl CoA builds up more rapidly than it can be used by normal processes, and some of it eventually hydrolyzes:

$$CH_3\overset{O}{\overset{\|}{C}}CH_2\overset{O}{\overset{\|}{C}}-SCoA + H_2O \longrightarrow CH_3\overset{O}{\overset{\|}{C}}CH_2\overset{O}{\overset{\|}{C}}OH + HS-CoA$$

Acetoacetic acid

Each molecule of acetoacetic acid has to be neutralized by the buffer:

$$CH_3\overset{O}{\overset{\|}{C}}CH_2\overset{O}{\overset{\|}{C}}OH + HCO_3^- \longrightarrow CH_3\overset{O}{\overset{\|}{C}}CH_2CO_2^- + H_2O + CO_2$$

Acetoacetate ion

In other words, the increasingly rapid production of acetoacetic acid depletes the chief base in the buffer system of the blood, the bicarbonate ion. The result is a slow but steady drop in the pH of the blood or acidosis — *metabolic* acidosis, because the cause lies in a disorder of metabolism. Because the chief species responsible for this acidosis has a keto group, the condition is often called **ketoacidosis,** and the *acetoacetate ion* is called one of the **ketone bodies.** The other two ketone bodies are *acetone* and the *β-hydroxybutyrate ion*. Both are produced from the acetoacetate ion. Acetone arises from this ion by the loss of the carboxyl group:

β-Hydroxybutyrate is called a *ketone* body not because it has a keto group but because it is made from and is found together with one that does.

$$H_2O + CH_3\overset{O}{\overset{\|}{C}}CH_2\overset{O}{\overset{\|}{C}}O^- \longrightarrow CH_3\overset{O}{\overset{\|}{C}}CH_3 + HCO_3^-$$

Acetoacetate   Acetone

β-Hydroxybutyrate is produced when the keto group of acetoacetate is reduced by NADH:

$$CH_3\overset{O}{\overset{\|}{C}}CH_2CO_2^- + NADH + H^+ \longrightarrow CH_3\overset{OH}{\overset{|}{C}}HCH_2CO_2^- + NAD^+$$

Acetoacetate   *β*-Hydroxybutyrate

The vapor pressure of acetone at body temperature is nearly 400 mm Hg, so it readily evaporates from the blood in the lungs.

The ketone bodies enter general circulation. Because acetone is volatile, most of it leaves the body via the lungs, and individuals with severe ketoacidosis have "acetone breath"—the noticeable odor of acetone on the breath.

Acetoacetate and β-hydroxybutyrate can be used in skeletal muscles to make ATP. In fact, some tissues such as heart muscle use these two for energy in preference to glucose. Even the brain, given time, can adapt to using these ions for energy when the blood sugar level drops in starvation or prolonged fasting. The ketone bodies are not in themselves abnormal constituents of blood. There is always a low but finite level of these substances in the blood. In fact, there are normal uses of them, which we just described, that help to keep their levels low. Only when they are produced at a rate faster than the blood buffer can handle them are they a problem.

Ketosis and Ketoacidosis.   Normally, the levels of acetoacetate and β-hydroxybutyrate in the blood are, respectively, 2 μmol/dL and 4 μmol/dL. In prolonged, undetected, and untreated diabetes, these values can increase as much as 200-fold. The condition of excessive levels of ketone bodies in the blood is called **ketonemia.**

1 μmol = 1 micromole = 10^{-6} mol

As ketonemia becomes more and more advanced, the ketone bodies begin to appear in the urine—a condition called **ketonuria.** When there is a combination of ketonemia, ketonuria, and acetone breath, the overall state is called **ketosis.** The individual will be described as *ketotic.* As unchecked ketosis becomes more severe, the associated ketoacidosis worsens, and the pH of the blood continues its fatal descent.

To leave the anions of the ketone bodies in the urine, the kidneys have to leave positive ions in the urine with them to keep everything electrically neutral. Na^+ ions, the most abundant cations, are used. It is common among specialists in these matters to describe this loss of Na^+ as the loss of *base* from the blood even though Na^+ is not itself a base. What they mean is that each Na^+ that leaves the body in the urine corresponds to one H^+ ion from acetoacetic acid (or other organic acid) that the true base, HCO_3^-, has to neutralize. Hence, the loss of Na^+ is taken as an indicator of the loss of this true base. Another way to understand the urinary loss of Na^+ as the loss of base from the blood is that a Na^+ ion has to accompany a bicarbonate ion when it goes from the kidneys into the blood. The kidneys manufacture HCO_3^- ions normally in order to replenish the blood buffer system. The more that Na^+ ions have to leave in the urine in order to clear ketone bodies from the blood, the less can the true base, HCO_3^-, be put into the blood.

Condition	$[HCO_3^-]_{blood}$ in mmol/L
Normal	22–30
Mild acidosis	16–20
Moderate acidosis	10–16
Severe acidosis	<10

The solutes that are leaving the body in the urine cannot, of course, be allowed to make the urine too concentrated. Otherwise, osmotic pressure balances are upset. Therefore increasing quantities of water must be excreted. To satisfy this need, the individual has a powerful thirst. Other wastes, such as urea, are also being produced at higher than normal rates, because amino acids are being sacrificed in gluconeogenesis. These wastes add to the demand for water to make urine.

Polyuria is the technical name for the overproduction of urine.

If, during a state of ketosis, insufficient water is drunk, then water is simply taken from extracellular fluids. The blood volume therefore tends to drop, and the blood becomes more concentrated. It also thickens and becomes more viscous, which makes the delivery of blood more difficult. Because the brain has the highest priority for blood flow, some of this flow is diverted from the kidneys to try to ensure that the brain gets what it needs. This only worsens the situation in the kidneys, and they have an increasingly difficult time in clearing wastes. As the water shortage worsens, some water is borrowed from the intracellular supply. This, in addition to a combination of other developments, leads to coma and eventually death. Figure 28.6 outlines the succession of events in untreated, type I diabetes. It is nothing short of remarkable how the absence of one chemical, insulin, can release such a vast train of biochemical events. But at the molecular level of life, this is the kind of story that occurs very often.

FIGURE 28.6
The principal sequence of events in
untreated diabetes.

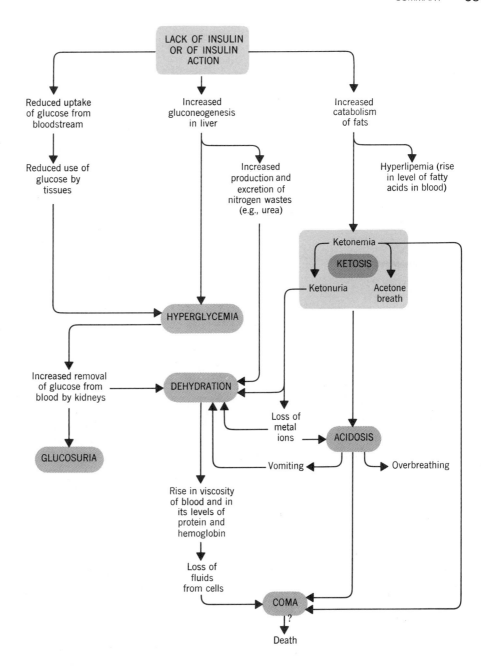

SUMMARY

Lipid absorption and distribution   As fatty acids and glycerol migrate out of the digestive tract they become reconstituted as triacylglycerols. These, in addition to cholesterol and other lipids, are picked up by chylomicrons. As they migrate through the vascular compartment, they unload some of their triacylglycerols and finally are taken up by the liver. The liver organizes the remaining lipids together with those that the liver makes itself, which include cholesterol and triacylglycerols, and then sends very low density lipoprotein complexes, VLDL, into circulation. Triacylglycerols are again unloaded where they are needed, and the VLDL become slightly more dense and change into intermediate density complexes, IDL. Much of these are reabsorbed by the liver. Those that aren't become more dense and change over to low density lipoprotein complexes, LDL. Some of the LDL finds its way to extrahepatic tissues, including endocrine glands that need cholesterol to make steroid hormones, and much of the LDL is reabsorbed by the liver to recycle their lipids.

Any extra cholesterol not needed in extrahepatic tissue is carried back to the liver by high density lipoprotein complexes, HDL. The liver excretes excess cholesterol via the bile, or it makes bile salts. Severe hypercholesterolemia is experienced by individuals who have inherited an inability to make enough of the receptor proteins for IDL and LDL.

Storage and mobilization of lipids   The favorable energy density of triacylglycerol means that more energy is stored per gram of this material than can be stored by any other chemical system. The adipose tissue is the principal storage site, and fatty material comes and goes from this tissue according to the energy budget of the body. When the energy of fatty acids is needed, they are liberated from triacylglycerols, and carried to the liver, the chief site of fatty acid catabolism.

The catabolism of fatty acids   Fatty acyl groups, after being pinned to coenzyme A, are catabolized by the fatty acid cycle. By a succession of four steps — dehydrogenation, hydration of a double bond, oxidation of the resulting alcohol, and cleavage of the bond from the α to the β carbon — one turn of the cycle removes one two-carbon acetyl group. The cycle then repeats as the shortened fatty acyl group continues to be degraded. Each turn of the cycle produces one $FADH_2$ and one NADH, which pass $H:^-$ to the respiratory chain for the synthesis of ATP. Each turn also sends one acetyl group into the citric acid cycle which, via the respiratory chain, leads to several more ATPs. The net ATP production is 129 ATPs per palmityl residue.

Biosynthesis of fatty acids   Fatty acids can be made by a repetitive cycle of steps called the lipigenesis cycle. It begins by building one butyryl group from two acetyl groups. The four-carbon butyryl group is attached to an acyl carrier protein that acts as a swinging arm on the enzyme complex. This arm moves the growing fatty acyl unit first over one enzyme and then another as additional two carbon units are added. The process consumes ATP and NADPH.

Biosynthesis of cholesterol   Cholesterol is made from acetyl groups by a long series of reactions. The synthesis of one of the enzymes is inhibited by excess cholesterol, which gives the system a mechanism for keeping its own cholesterol synthesis under control.

Ketoacidosis   Acetoacetate, β-hydroxybutyrate, and acetone build up in the blood — ketonemia — in starvation or in diabetes. The first two are normal sources of energy in some tissues. When they are made faster than they can be metabolized, however, there is an accompanying increased loss of bicarbonate ion from the blood's carbonate buffer. This leads to a form of metabolic acidosis called ketoacidosis. The kidneys try to leave the anionic ketone bodies in the urine, but this takes Na^+, too (for electrical neutrality), and water (for osmotic pressure balances). The excessive loss of Na^+ in the urine is interpreted as a loss of "base." With Na^+ leaving the body in the urine, less is available to accompany replacement HCO_3^-, made by the kidneys, when this base should be going into the blood. Under developing ketoacidosis, the kidneys have extra nitrogen wastes and, in diabetes, extra glucose to leave in the urine being made. This also demands an increased volume of water. Unless this is brought in by the thirst mechanism, it has to be sought from within. However, the brain has first call on blood flowage, so the kidneys suffer more. Eventually, if these events continue unchecked, the victim goes into a coma and dies.

KEY TERMS

The following terms were emphasized in this chapter and should be mastered.

beta oxidation	ketoacidosis	ketonuria	lipoprotein complex
chylomicron	ketone bodies	ketosis	
fatty acid cycle	ketonemia	lipigenesis	

SELECTED REFERENCES

1 Thomas Skillman. "Diabetes Mellitus." Chapter 30 in *Clinical Chemistry, Theory, Analysis, and Correlation,* by L. A. Kaplan and A. J. Pesce. The C. V. Mosby Company, St. Louis, MO, 1984.

2 M. Davidson. *Diabetes Mellitus: Diagnosis and Treatment.* John Wiley and Sons, Inc., New York, 1981.

3 J. L. Goldstein, T. Kita, and M. S. Brown. "Defective Lipoprotein Receptors and Atherosclerosis." *The New England Journal of Medicine,* August 4, 1983, page 288.

4 M. S. Brown and J. L. Goldstein. "How LDL Receptors Influence Cholesterol and Atherosclerosis." *Scientific American,* November 1984, page 58.

5 Jean Himms-Hagen. "Thermogenesis in Brown Adipose Tissue as an Energy Buffer. Implications for Obesity." *The New England Journal of Medicine,* December 13, 1984, page 1549.

6 Jeanette K. Chambers. "Save Your Diabetic Patient from Early Kidney Damage." *Nursing83,* May 1983, page 58.

7 Patricia Stock-Barkman. "Confusing Concepts. Is It Diabetic Shock or Diabetic Coma?" *Nursing83,* June 1983, page 33.

REVIEW EXERCISES

The answers to these Review Exercises are in the *Study Guide* that accompanies this book.

Absorption and Distribution of Lipids

28.1   What are the end products of the digestion of triacylglycerols?

28.2   What happens to the products of the digestion of triacylglycerols as they migrate out of the intestinal tract?

28.3   What are chylomicrons and what is their function?

28.4   What happens to chylomicrons as they move through capillaries of, say, adipose tissue?

28.5   What happens to chylomicrons when they reach the liver?

28.6   What are the two chief sources of cholesterol that the liver exports?

28.7   What do the following symbols stand for?
(a) VLDL   (b) IDL
(c) LDL   (d) HDL

28.8   The loss of what kind of substance from the VLDL converts them into IDL?

28.9   What do IDL release as they change over to LDL?

28.10   What tissue can reabsorb IDL complexes?

28.11   What is the chief constitutent of LDL?

28.12   In extrahepatic tissue, what two general uses await delivered cholesterol?

28.13   If the liver lacks the key receptor proteins, which specific lipoprotein complexes can't be reabsorbed?

28.14   Explain the relationship between the liver's receptor proteins for lipoprotein complexes and the control of the cholesterol level of the blood.

28.15   What is the chief job of the HDL?

28.16   Some scientists believe that a relatively high level of HDL provides protection against atherosclerosis, and that one's risk of having heart disease declines when the level of HDL is raised by exercise and losing weight. Why would a low level of HDL tend to promote heart disease?

Storage and Mobilization of Lipids

28.17   With reference to the storage of chemical energy in the body, what is meant by *energy density?*

28.18   Arrange the following in their order of increasing quantity of energy that they store per gram.

Wet glycogen   Adipose lipids   Isotonic glucose
          1   2   3

28.19   Briefly describe two conditions in which the body would have to turn to fatty acids for energy.

28.20   How does insulin suppress the mobilization of fatty acids from adipose tissue?

28.21   Arrange the following processes in the order in which they occur when the energy in storage in triacylglycerols is mobilized.

Fatty acid   Oxidative   Citric acid
cycle   phosphorylation   cycle
  1   2   3

Lipoprotein   Lipolysis in
formation   adipose
           tissue
    4   5

28.22   What specific function does the fatty acid cycle have in obtaining energy from fatty acids?

28.23   What specific function does the citric acid cycle have in the use of fatty acids for energy?

28.24   Name two hormones that activate the lipase in adipose tissue. Referring to the previous chapter, what does the presence of these hormones do for the blood sugar level?

28.25   Explain how a rise in the blood sugar level indirectly inhibits the mobilization of energy from adipose tissue.

28.26   When lipolysis occurs in adipose tissue, what happens to the glycerol?

Catabolism of Fatty Acids

28.27   How are long chain fatty acids activated for entry into the fatty acid cycle?

28.28   Complete the following equations for one turn of the fatty acid cycle by which a six-carbon fatty acyl group is catabolized.

(a) $CH_3CH_2CH_2CH_2CH_2\overset{\displaystyle O}{\overset{\displaystyle \|}{C}}$—SCoA + FAD ⟶
_____ + _____

(b) _____ + H_2O ⟶ _____

(c) _____ + NAD^+ ⟶
_____ + NADH + H^+

(d) _____ + CoA—SH ⟶
_____ + _____

28.29   Write the equations for the four steps in the fatty acid cycle as it operates on butyryl CoA. How many more turns of the cycle are possible after this one?

28.30   How is the FAD-enzyme recovered from its reduced form, $FADH_2$, when the fatty acid cycle operates?

28.31   How is the reduced form of the NAD^+-enzyme used in the fatty acid cycle restored to its oxidized form?

28.32   Why is the fatty acid cycle sometimes called beta oxidation?

28.33   Myristic acid, $CH_3(CH_2)_{12}CO_2H$, can be catabolized by the fatty acid cycle just like palmitic acid.
(a) How many units of acetyl CoA can be made from it?
(b) In producing this much acetyl CoA, how many times does $FADH_2$ form and then deliver its hydrogen to the respiratory chain?

(c) Referring again to part (a), how many times does NADH form as acetyl CoA is produced and then deliver its hydrogen to the respiratory chain?

(d) Complete the following table by supplying the missing numbers of molecules that are involved in the catabolism of myristic acid to acetyl CoA.

Intermediate	Maximum Number of ATP from Each	Total Number of ATP Possible from Each as Acetyl CoA Forms
_____ FADH$_2$	_____	_____
_____ NADH	_____	_____
_____ CH$_3$C—SCoA (with O double bonded to C)	_____	_____
	Sum =	_____

Deduct _____ high-energy phosphate bonds for activating the myristyl group   − _____

Net ATP produced for each myristyl group as it changes to acetyl CoA   _____

Biosynthesis of Fatty Acids

28.34 Where are the principal sites for each activity in a liver cell?
(a) Fatty acid catabolism
(b) Lipigenesis

28.35 Outline the steps that make butyryl ACP out of acetyl CoA.

28.36 What metabolic pathway in the body is the chief supplier of NADPH for lipigenesis?

Biosynthesis of Cholesterol

28.37 The enzyme for the formation of which intermediate in cholesterol synthesis is the major control point in this pathway?

28.38 How does cholesterol itself work to inhibit the activity of the enzyme referred to in Review Exercise 28.37?

Ketoacidosis

28.39 Which nutrient becomes increasingly important as a source of energy (ATP) if the net effect of either starvation or diabetes is the reduced availability of glucose as a source of ATP?

28.40 The catabolism of the nutrient of Review Exercise 28.39 produces what intermediate that is further catabolized by the citric acid cycle?

28.41 Two molecules of acetyl CoA can combine to give the coenzyme A derivative of what keto acid? Give the structure of this keto acid.

28.42 Give the names and structures of the ketone bodies. (For those that occur as anions at body pH, give the structures as anions.)

28.43 What is ketonemia?

28.44 What is ketonuria?

28.45 What is meant by acetone breath?

28.46 Ketosis consists of what collection of conditions?

28.47 What is ketoacidosis? What form of acidosis is it, metabolic or respiratory?

28.48 The formation of which particular compound most lowers the supply of HCO$_3^-$ in ketoacidosis?

28.49 What are the reasons for the increase in the volume of urine that is excreted in someone with untreated, type I diabetes?

28.50 If the ketone bodies (other than acetone) can normally be used by heart and skeletal muscle, what makes them dangerous in starvation or in diabetes?

28.51 Why does the rate of urea production increase in untreated, type I diabetes?

28.52 When a physician refers to the loss of Na$^+$ as the loss of *base*, what is actually meant?

Chapter 29
Metabolism of Nitrogen Compounds

Cats need meat and big cats need a lot of it. Getting all the essential amino acids from grass, leaves, and grain doesn't work for this lion. We study how amino acids are made and catabolized in this chapter.

29.1   THE SYNTHESIS OF AMINO ACIDS IN THE BODY

The body can manufacture a number of amino acids from intermediates that appear in the catabolism of nonprotein substances.

The Nitrogen Pool.   Amino acids, the end products of protein digestion, are rapidly transported across the walls of the small intestine. Some very small, simple peptides can also be absorbed. Once amino acids enter circulation, they become part of what is called the **nitrogen pool,** the name given to the whole collection of nitrogen compounds found anywhere in the body in any state of chemical combination.

Amino acids enter the nitrogen pool not only from the intestinal tract but also by means of the breakdown of proteins in body tissues and fluids, which are always undergoing degradation and resynthesis. Most of the proteins in the body are in a dynamic state. They experience a constant turnover, which is fairly rapid among the proteins of the liver and the blood but is quite slow among muscle proteins. Figure 29.1 illustrates the various compartments of the nitrogen pool and how they are interrelated.

As indicated in Figure 29.1, individual amino acids can be used in any one of the following ways, depending on the body's needs of the moment.

1. The synthesis of new or replacement proteins.
2. The synthesis of such nonprotein nitrogen compounds as heme, creatine, nucleic acids, and certain hormones and neurotransmitters.
3. The production of ATP or of glycogen and fatty acids, substances with the potential for making ATP.
4. The synthesis of any needed nonessential amino acids.

In Chapter 23 we learned that we do not need all of the 20 amino acids in the diet, that we can make roughly half of them. We labeled those that we *must* obtain via the diet as the *essential amino acids.* The others, the *nonessential amino acids,* can be synthesized in the body. Be sure to remember that *nonessential* in this context refers *only* to a dietary need. In a larger context, the body must have all of the amino acids to make polypeptides.

The pathways for the synthesis of nonessential amino acids have several steps, and we won't examine any in detail. However, Figure 29.2 gives an overview that illustrates in general terms how some of the nonessential amino acids can be made from the intermediates of glycolysis and the citric acid cycle.

The Synthesis of Amino Acids.   Many of the syntheses outlined in Figure 29.2 depend on the availability of glutamic acid (actually, at body pH, the glutamate ion), and it is made from α-ketoglutarate by a reaction called **reductive amination.** This reaction involves the use of the ammonium ion as a source of the amino group and NADPH as a reducing agent. (In some cells, NADH-enzymes can also work.) The margin describes *how* it happens. The overall result is:

$$^-O_2CCH_2CH_2\overset{\overset{\displaystyle O}{\|}}{C}CO_2^- + NH_4^+ + NADPH + H^+ \rightleftharpoons$$

α-Ketoglutarate

$$^-O_2CCH_2CH_2\overset{\overset{\displaystyle NH_3^+}{|}}{C}HCO_2^- + NADP^+ + H_2O$$

Glutamate

The polypeptides in proteins that serve as enzymes have a particularly rapid turnover.

Both in starvation and diabetes the body draws down its amino acid pool to make glucose.

Steps in reductive amination

$$\downarrow$$

O   (Keto group)

$$\|$$

$$-C-$$

$NH_3 \rightharpoondown \rightharpoonup NH_3$

OH
|
$-C-$
|
NH_2

$H_2O \leftharpoondown \rightharpoonup H_2O$

$-C-$ (Imine group)

$\|$
NH

$H^+ + NADPH \rightharpoondown \rightharpoonup NADPH + H^+$

$NADP^+ \leftharpoondown \rightharpoonup NADP^+$

$-CH-$
|
NH_2 (Amino group)

$\uparrow$

Steps in oxidative deamination

FIGURE 29.1
The nitrogen pool.

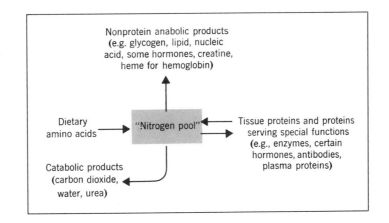

FIGURE 29.2
The biosynthesis of some
nonessential amino acids.

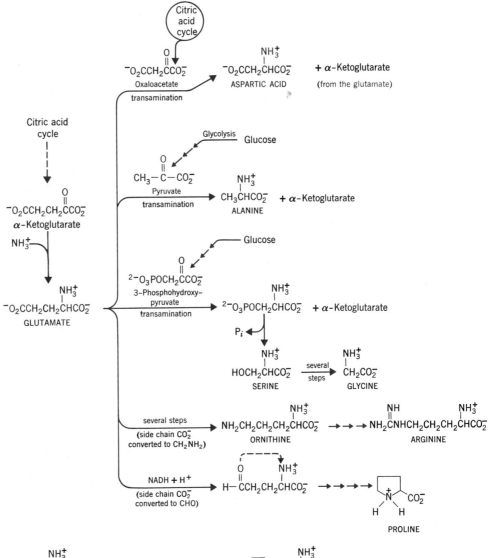

Glutamate now becomes the source of amino groups for still other amino acids that can be made by a reaction called **transamination**—the transfer of an amino group. We'll illustrate the general case; Figure 29.2 has several examples.

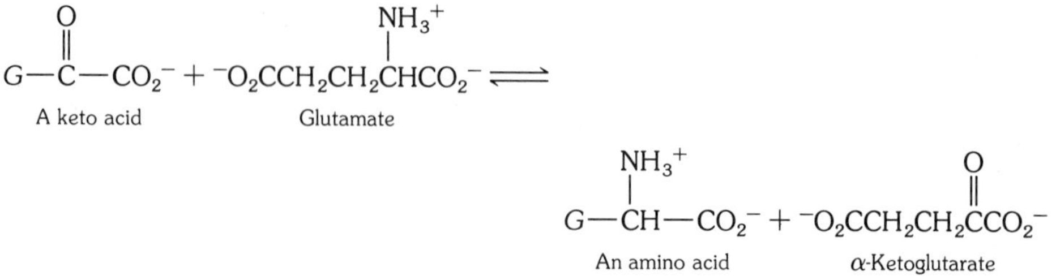

$$G-\overset{\overset{\displaystyle O}{\|}}{C}-CO_2^- + {}^-O_2CCH_2CH_2\overset{\overset{\displaystyle NH_3^+}{|}}{C}HCO_2^- \rightleftharpoons$$

A keto acid   Glutamate

$$G-\overset{\overset{\displaystyle NH_3^+}{|}}{C}H-CO_2^- + {}^-O_2CCH_2CH_2\overset{\overset{\displaystyle O}{\|}}{C}CO_2^-$$

An amino acid   α-Ketoglutarate

The enzymes for transaminations, the *transaminases,* use a B-vitamin, pyridoxal, to make their cofactors.

29.2 THE CATABOLISM OF AMINO ACIDS

The breakdown products of amino acid catabolism eventually enter the pathways of catabolism of either carbohydrates or lipids.

Amino acids in the nitrogen pool that aren't needed to make other amino acids or other nitrogen compounds are catabolized. There is no special storage system for them analogous to glycogen or to the fat in adipose tissue.

The end products of the complete catabolism of amino acids are urea, carbon dioxide, and water. On the way to these end products, several intermediates that belong to other pathways form. Thus as Figure 29.3 shows, all of the pathways for the use of carbohydrates, lipids, and proteins are interconnected in one way or another. This figure provides a broad, overall summary of what we have been studying in these latter chapters.

We won't study in detail how each amino acid is catabolized, because each requires its own particular scheme, usually quite complicated. However, there are three kinds of reactions, besides transamination, that occur often: oxidative deamination, direct deamination, and decarboxylation. We'll study these and how they apply to certain selected amino acids.

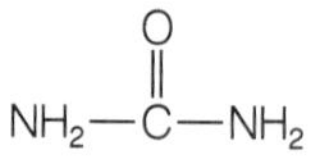

Oxidative Deamination.   Reductive amination, which we studied on page 664, involves a series of equilibria, all of which can be shifted into reverse. The reverse of reductive amination is called **oxidative deamination.** It occurs chiefly as a step in one of the processes that shuttles amino groups to urea. In the display that follows, the step on the left is a transamination. The next step is oxidative deamination, and the arrowheads are in the direction of catabolism.

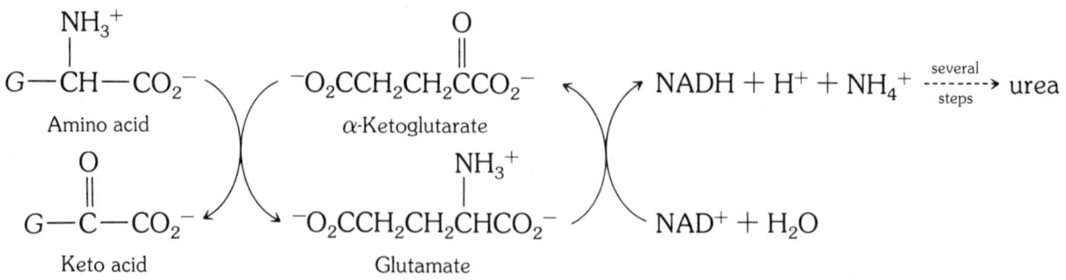

$$NH_2-\overset{\overset{\displaystyle O}{\|}}{C}-NH_2$$
Urea

$$G-\overset{\overset{\displaystyle NH_3^+}{|}}{C}H-CO_2^-$$
Amino acid

$$G-\overset{\overset{\displaystyle O}{\|}}{C}-CO_2^-$$
Keto acid

$${}^-O_2CCH_2CH_2\overset{\overset{\displaystyle O}{\|}}{C}CO_2^-$$
α-Ketoglutarate

$${}^-O_2CCH_2CH_2\overset{\overset{\displaystyle NH_3^+}{|}}{C}HCO_2^-$$
Glutamate

$$NADH + H^+ + NH_4^+ \xrightarrow[\text{steps}]{\text{several}} urea$$

$$NAD^+ + H_2O$$

Notice that the nitrogen of the amino acid on the upper left ends up in urea, and that the α-ketoglutarate–glutamate pair provides a switching mechanism to convey this nitrogen in the right direction for catabolism.

FIGURE 29.3
Interrelationships of major metabolic pathways.

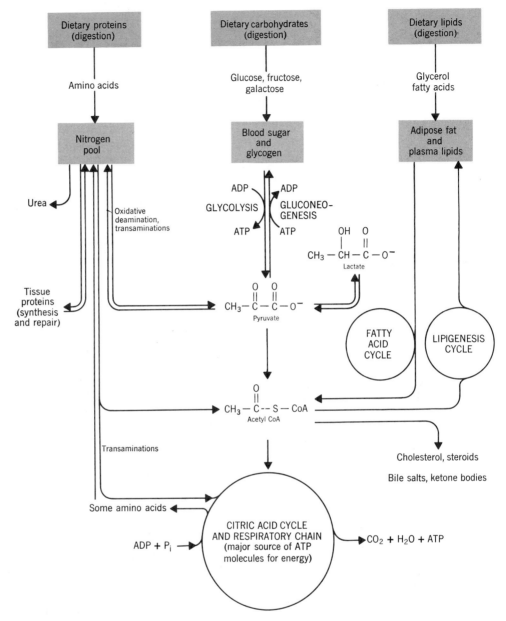

Direct Deamination.   Two amino acids, one of them serine, undergo a simultaneous loss both of water and ammonia in a reaction that removes an amino group without consuming an oxidizing agent (such as NAD^+). This can take place in serine because the carbon adjacent to the amino group's carbon atom holds an —OH group. Here's how **direct deamination** happens:

Imine groups easily hydrolyze because they can add water and then split out ammonia.

The first step is the dehydration of the alcohol system of serine to give an unsaturated amine. This spontaneously rearranges into an imine, a compound with a carbon-nitrogen double bond. Water can add to this double bond, but the product spontaneously breaks up, so the net effect is the hydrolysis of the imine group to a keto group and ammonia. Thus serine breaks down to pyruvate, which, as we now well know, can send an acetyl group into the citric acid cycle, or can contribute an acetyl group to lipigenesis, or can be used to make glucose.

Decarboxylation.   Some special enzymes can split out just the —CO_2^- groups from amino acids and make amines. This reaction is called **decarboxylation,** and it is used to make some neurotransmitters and hormones. Dopamine, norepinephrine, and epinephrine all are made by steps that begin with the decarboxylation of dihydroxyphenylalanine, which the body makes from the amino acid tyrosine.

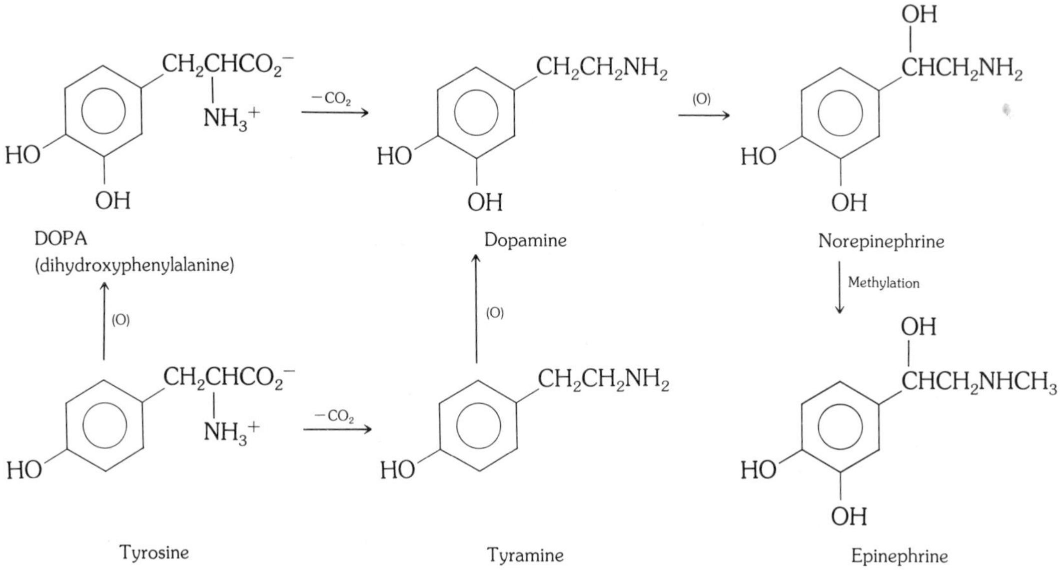

The Catabolism of Some Specific Amino Acids.   To illustrate the reactions we have just introduced, we'll look in more detail at the catabolism of two amino acids, alanine and aspartic acid. Figure 29.4 puts these and others into the contexts of the major metabolic pathways of the body, so this figure fills in some of the details of Figure 29.3.

Notice in Figure 29.4 that oxaloacetate occurs in two places, as an intermediate in the citric acid cycle and as the product of the oxidative deamination of aspartate. For a *net gain* of glucose molecules via gluconeogenesis, the oxaloacetate that is involved in the citric acid cycle can't be counted as available. Only oxaloacetate made from amino acids can give a net gain of glucose this way. This is why gluconeogenesis under conditions of starvation necessarily breaks down body proteins. It needs some of their amino acids to make oxaloacetate.

The transamination of alanine gives pyruvate, which can go into the citric acid cycle, into gluconeogenesis, or it can be used for the biosynthesis of fatty acids.

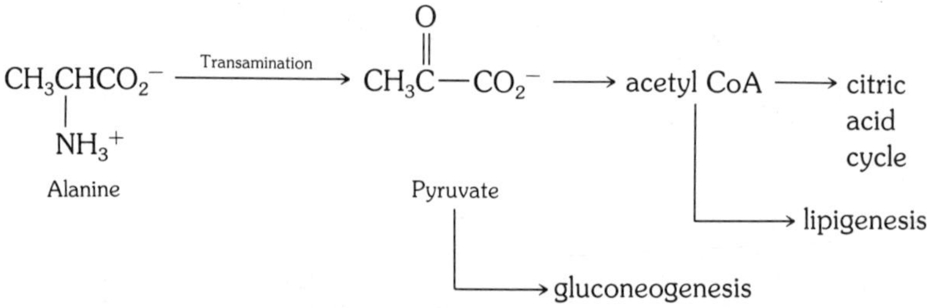

FIGURE 29.4
The catabolism of some amino acids.

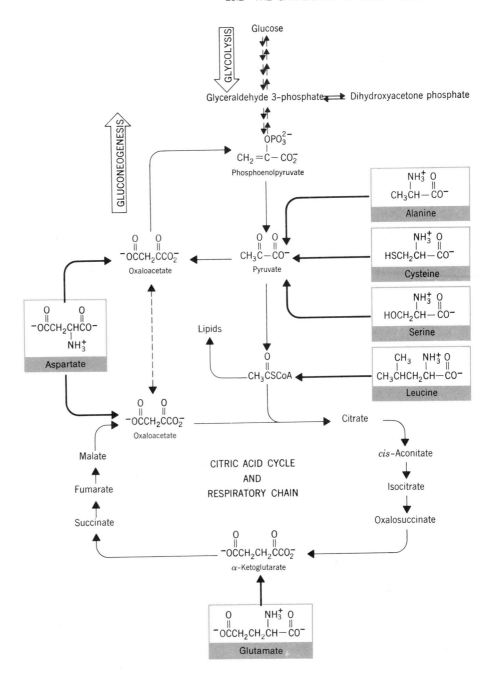

The transamination of aspartic acid gives oxaloacetate, an intermediate in both gluconeogenesis and the citric acid cycle.

29.3 THE FORMATION OF UREA

Ammonia and amino groups are converted into urea by a complex cycle of reactions called the urea cycle.

Urea, as we have learned, is the chief nitrogen waste made by the body. Most of its nitrogen indirectly comes from amino acids, but some comes (also indirectly) from two of the side-chain bases of nucleic acids.

There are two direct sources for the nitrogen atoms in urea. One is the ammonium ion that is produced by the oxidative deamination of glutamate that we studied on page 666. This glutamate nitrogen therefore stems from the shuttle mechanism that can remove the amino group nitrogens from any amino acid. The other nitrogen atom in urea comes from a specific amino acid, aspartate. But because asparate is made by a transamination that takes its nitrogen from glutamate, you can see that glutamate is close to being the direct source of both nitrogens in urea.

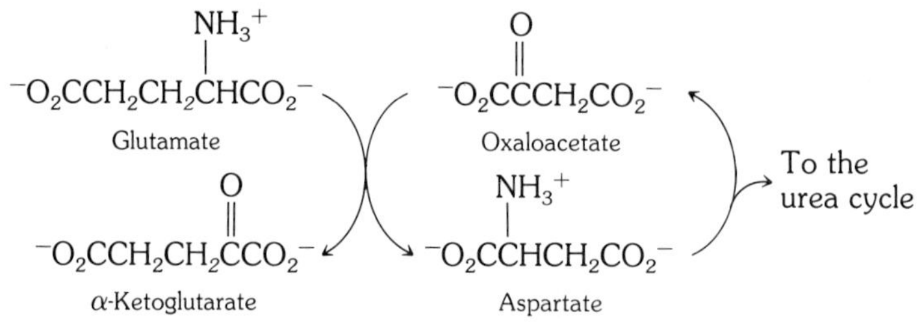

The urea cycle is sometimes called the Kreb's ornithine cycle.

Urea Cycle. A series of reactions called the **urea cycle** manufactures urea from ammonium ion, carbon dioxide, and asparate. Figure 29.5 displays the steps in the urea cycle, and the boxed numbers in this figure refer to the following steps:

1. Ammonia, with the help of ATP, reacts with CO_2 to form carbamoyl phosphate, a high-energy phosphate. In a sense, this is an activation of ammonia that launches it into the next step that takes it into the cycle.

2. The carbamoyl group transfers to the carrier unit, ornithine, as P_i is ejected. This consumes high-energy phosphate energy. Citrulline forms.

3. Citrulline condenses with the alpha amino group of aspartate to give argininosuccinate.

4. Fumarate forms from the original aspartate as the amino group stays with the arginine that emerges. Fumarate is an intermediate in the citric acid cycle. By a transamination that involves glutamate, it is reconverted to aspartate.

5. Arginine is hydrolyzed. Urea forms and ornithine is regenerated to start another turn of the cycle.

The overall result of the urea cycle is given by the following equation (which, as Figure 29.5 makes clear, is extremely simplified):

$$2NH_3 + H_2CO_3 \longrightarrow NH_2 - \overset{\displaystyle O}{\overset{\displaystyle \|}{C}} - NH_2 + 2H_2O$$

Urea

Some inherited genetic defects produce enzymes for this cycle that have reduced activity. In such individuals, the level of ammonium ion in blood rises, a condition called **hyperammonemia.** The ammonium ion is toxic, and hyperammonemia means mental retardation for some. Infants that have this genetic defect improve on low-protein diets.

FIGURE 29.5
The urea cycle. The boxed numbers refer to the text discussion. The dashed-line circle is the aspartate-oxaloacetate shuttle also discussed in the text.

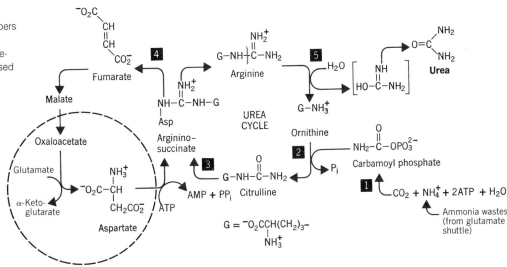

29.4 THE CATABOLISM OF OTHER NITROGEN COMPOUNDS

Uric acid and the bile pigments are other end products of the catabolism of nitrogen compounds.

The nitrogen of the purine bases of nucleic acids, adenine (A) and guanine (G), is excreted as uric acid, which also has the purine nucleus. After studying how uric acid forms, we'll see how defects in this pathway can lead to gout or to a particularly difficult disease of children, the Lesch-Nyhan syndrome.

Formation of Uric Acid. The numbered steps in Figure 29.6 are discussed below to show how adenine can be used to make uric acid.

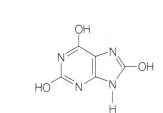

Uric acid

Purine

1. A transamination removes the amino group of the adenine side chain in adenosine monophosphate, AMP.

2. Ribose phosphate is removed and will enter the pentose phosphate pathway of carbohydrate catabolism. The product is hypoxanthine.

3. An oxidation produces xanthine. (The steps from guanine lead to xanthine, too.)

4. Another oxidation produces the keto form of uric acid, which exists partly in the form of a phenol. Actually, it's the salt of uric acid that forms, sodium urate, because the acid is neutralized by base in the buffer system.

In the disease known as **gout,** the rate of formation of sodium urate increases more rapidly than its rate of elimination. Crystals of this salt precipitate in joints where they cause painful inflammations and lead to a form of arthritis. Kidney stones may form as this salt comes out of solution in this organ.

Just why the formation of sodium urate accelerates isn't well known, but genetic factors are involved. Normally, some of the hypoxanthine made in step 2 is recycled back to nucleotide bases that are needed to make nucleic acids or high-energy phosphates. Some individuals with gout are known to have a partial deficiency of the enzyme system required for this recycling of hypoxanthine. Hence, most if not all of their hypoxanthine ends up as more sodium urate than normal.

The lack of one enzyme usually means great personal and family suffering.

In the Lesch-Nyhan syndrome, the enzyme for recycling hypoxanthine is totally lacking. The result is both bizarre and traumatic. Infants with this syndrome develop compulsive,

FIGURE 29.6
The catabolism of adenosine monophosphate, AMP. The boxed numbers refer to the discussion in the text.

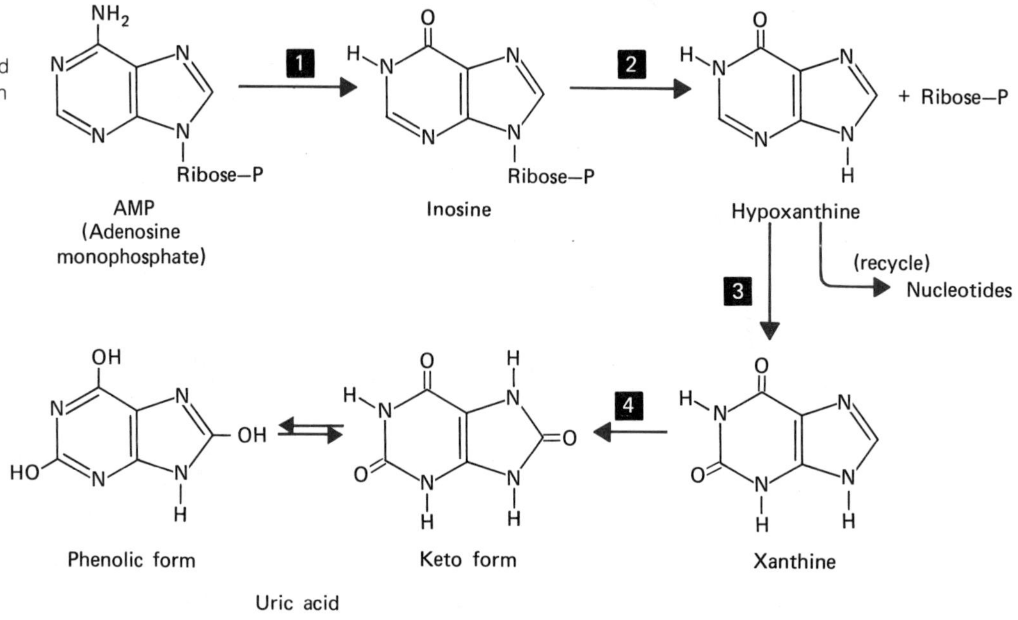

AMP
(Adenosine monophosphate)

Inosine

Hypoxanthine

+ Ribose—P

(recycle)
Nucleotides

Phenolic form

Keto form

Xanthine

Uric acid

self-destructive behavior at age 2 or 3. Unless their hands are wrapped in cloth, they will bite themselves to the point of mutilation. They act with dangerous aggression toward others. Some become spastic and mentally retarded. Kidney stones develop early, and gout comes later.

The Catabolism of Heme.   Erythrocytes have life spans of only about 120 days. Eventually they split open. Their hemoglobin spills out and then is degraded. Its breakdown products are eliminated via the feces and, to some extent, in the urine. In fact, the characteristic colors of feces and urine are caused by partially degraded heme molecules called the **bile pigments.**

The degradation of heme begins before the globin portion breaks away. The heme molecule partly opens up to give a system that has a chain of four small rings called pyrrole rings. (This is why the bile pigments are sometimes called the *tetrapyrrole pigments*.)

Pyrrole skeleton

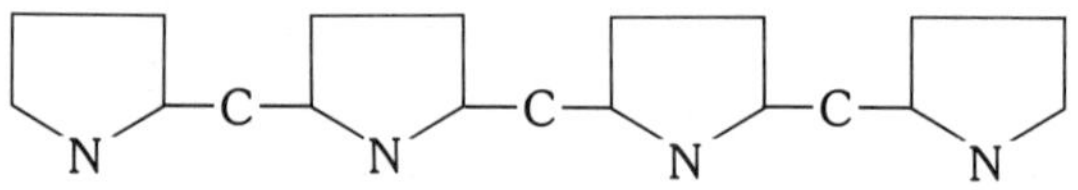

Carbon skeleton of the bile pigments

The rings have varying numbers of double bonds according to the state of oxidation of the pigment.

The slightly broken hemoglobin molecule, now called verdohemoglobin, then splits into globin, iron(II) ion, and a greenish pigment called **biliverdin** (Latin *bilis*, bile, + *virdus*, green). Globin enters the nitrogen pool. Iron is conserved in a storage protein called ferritin and is reused. Biliverdin is changed in the liver to a reddish-orange pigment called **bilirubin** (Latin *bilis*, bile + *rubin*, red). Bilirubin is not only made by the liver but is also removed from circulation by the liver, which transfers it to the bile. In this fluid it finally enters the intestinal tract.

Bile pigments are also responsible for the color of bile.

The pathway from hemoglobin to bilirubin after the rupture of an erythrocyte as well as the fate of bilirubin are shown in Figure 29.7.

FIGURE 29.7
The formation and the
elimination of the products of
the catabolism of hemoglobin.

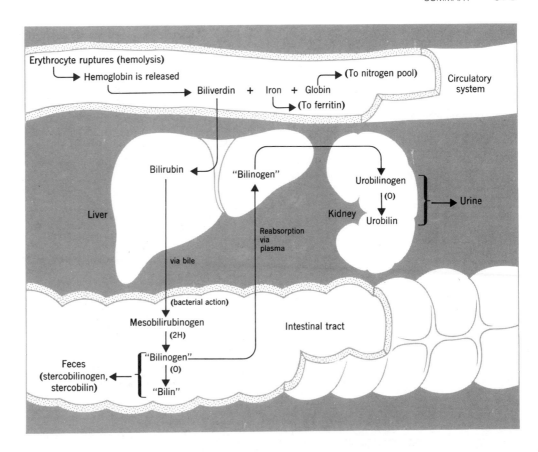

Bilirubin is the principal bile pigment in humans. In the intestinal tract, bacterial enzymes convert bilirubin to a colorless substance called mesobilirubinogen. This is further processed to form a substance known as **bilinogen,** which usually goes by other names that describe differences in destination rather than structure. Thus bilinogen that leaves the body in the feces is called *stercobilinogen* (Latin, *stercus,* dung). Some bilinogen is reabsorbed via the bloodstream, comes to the liver, and finally leaves the body in the urine. Now it is called *urobilinogen.* Some bilinogen is reoxidized to give **bilin,** a brownish pigment. Depending on its destination, bilin is called *stercobilin* or *urobilin.*

Special Topic 29.1 describes how the bile pigments are involved in jaundice.

SUMMARY

Amino acid distribution The nitrogen pool receives amino acids from the diet, from the breakdown of proteins in body fluids or tissues, and from any synthesis of nonessential amino acids that occurs. Amino acids are used to build and repair tissue, replace proteins of body fluids, make nonprotein nitrogen compounds, provide chemical energy if needed, and supply molecular parts for gluconeogenesis or lipigenesis.

Amino acid metabolism By reactions of transamination, oxidative deamination, direct deamination, and decarboxylation, the α-amino acids shuffle amino groups between themselves and intermediates of the citric acid cycle, or the synthesis of urea and nonprotein nitrogen compounds.

Deaminated amino acids eventually become acetyl coenzyme A,

acetoacetyl CoA, pyruvate, or an intermediate in the citric acid cycle. The skeletons of most amino acids can be used to make glucose, or fatty acids, or the ketone bodies. Their nitrogen atoms become part of urea.

Metabolism of other nitrogen compounds The nitrogen atoms in some of the side-chain bases of nucleic acids end up in urea and those of the others are excreted as sodium urate. Urea is made by a complex cycle of reactions—the urea cycle.

Heme is catabolized to bile pigments and its iron is reused. The pigments—first, biliverdin (green), then bilirubin (red), then mesobilirubinogen (colorless), and finally bilinogen and bilin (brown)—become stercobilin or stercobilinogen, urobilin or urobilinogen, depending on the route of elimination.

SPECIAL TOPIC 29.1 JAUNDICE AND THE TETRAPYRROLE PIGMENTS

Jaundice (French, *jaune,* yellow) is a condition that is symptomatic of a malfunction somewhere along the pathway of heme metabolism. If bile pigments accumulate in the plasma in concentrations high enough to impart a yellowish coloration to the skin, the condition of *jaundice* is said to exist. Jaundice can result from one of three kinds of malfunctions.

Hemolytic jaundice results when hemolysis takes place at an abnormally fast rate. Bile pigments, particularly bilirubin, form faster than the liver can clear them. Hepatic diseases such as infectious hepatitis and cirrhosis sometimes prevent the liver from removing bilirubin from circulation. The stools are usually clay-colored, because the pyrrole pigments do not reach the intestinal tract.

Obstructions of bile ducts can prevent release of bile into the intestinal tract, and the tetrapyrrole pigments in bile cannot be eliminated. Under these circumstances, they tend to reenter general circulation. The kidneys remove large amounts of bilirubin, but the stools are usually clay-colored. As the liver works harder and harder to handle its task of removing excess bilirubin, it can weaken and become permanently damaged.

KEY TERMS

The following terms were emphasized in this chapter.

bile pigment
bilin
bilinogen
bilirubin
biliverdin
deamination, direct
decarboxylation
gout
hyperammonemia
nitrogen pool
oxidative deamination
reductive amination
transamination
urea cycle

SELECTED REFERENCES

1 A. L. Lehninger. *Principles of Biochemistry.* Worth Publishers, Inc., New York, 1982

2 L. Stryer. *Biochemistry,* 2d ed. W. H. Freeman and Company, San Francisco, CA. 1981.

3 W. L. Nyhan (ed.). *Heritable Disorders of Amino Acid Metabolism.* John Wiley and Sons, Inc., New York, 1974.

4 M. Msall, M. L. Batshaw, R. Suss, S. W. Brusilow, and E. D. Mellits. "Neurologic Outcome in Children with Inborn Errors of Urea Synthesis." *The New England Journal of Medicine,* June 7, 1984, page 1500.

REVIEW EXERCISES

The answers to these Review Exercises are in the *Study Guide* that accompanies this book.

Nitrogen Pool

29.1 What is the nitrogen pool?

29.2 What are four ways in which amino acids are used in the body?

29.3 When the body retains more nitrogen than it excretes in all forms, the system is said to be on a *positive nitrogen balance.* Would this state characterize infancy or old age?

29.4 What happens to amino acids that are obtained in the diet but aren't needed to make any nitrogeneous compounds?

Biosynthesis of Amino Acids

29.5 Write the equation for the reductive amination that produces glutamate. (Use NADPH as the reducing agent.)

29.6 Write the structure of the keto acid that forms when phenylalanine undergoes transamination with α-ketoglutarate.

29.7 When valine and α-ketoglutarate undergo transamination, what new keto acid forms? Write its structure.

The Catabolism of Amino Acids

29.8 By means of two successive equations, one a transamination and the other an oxidative deamination, write the reactions that illustrate how the amino group of alanine can be removed as NH_4^+.

29.9 Arrange the following compounds in the order in which they would be produced if the carbon skeleton of alanine were to appear in one of the ketone bodies.

Pyruvate   Acetoacetyl CoA   Acetoacetate
1   **2**   **3**

Alanine   Acetyl CoA
4   **5**

29.10 In what order would the following compounds appear if some of the carbon atoms in glutamate were to become part of glycogen?

Oxaloacetate   α-Ketoglutarate   Glucose
1   **2**   **3**

Glycogen   Glutamate
4   **5**

29.11 From a study of the figures in this chapter, can the carbon atoms of serine become a part of a molecule of palmitic acid? If so, write the names of the compounds, beginning with serine, in the sequence to the start of the lipigenesis cycle.

29.12 Can any of the carbon atoms of glucose become part of alanine? If so, explain (in general terms).

29.13 Write the structure of the keto acid that forms by the direct deamination of threonine.

29.14 When tyrosine undergoes decarboxylation, what forms? Write its structure.

29.15 Write the structure of the product of the decarboxylation of tryptophan.

29.16 In the conditions of starvation or diabetes, what can the amino acids be used for?

The Formation of Urea

29.17 In the biosynthesis of urea, what are the sources of (a) the two —NH_2 groups and (b) the C$=$O group?

29.18 What is hyperammonemia and, in general terms, how does it arise and how can it be handled in infants?

29.19 What is the overall equation for the synthesis of urea?

The Catabolism of Nonprotein Nitrogen Compounds

29.20 What compounds are catabolized to make uric acid?

29.21 What product of catabolism accumulates in the joints in gout?

29.22 Arrange the names of the following substances in the order in which they appear during the catabolism of heme.

Biliverdin   Heme   Hemoglobin   Bilirubin
1   **2**   **3**   **4**

Mesobilirubinogen   Bilin   Bilinogen
5   **6**   **7**

Appendix A
Mathematical Concepts

A.1 EXPONENTIALS

When numbers are either very large or very small, it's often more convenient to express them in what is called *exponential notation*. Several examples are given in Table A.1, which shows how multiples of 10, such as 10,000 and submultiples of 10, such as 0.0001 can be expressed in exponential notation.

Exponential notation expresses a number as the product of two numbers. The first is a digit between 1 and 10, and this is multiplied by the second, 10 raised to some whole-number power or exponent. For example, 55,000,000 is expressed in exponential notation as 5.5×10^7, in which 7 (meaning +7) is the exponent.

Exponents can be negative numbers, too. For example, 3.4×10^{-3} is a number with a negative exponent. Now let's learn how to move back and forth between the exponential and the expanded expressions.

Positive Exponents. A positive exponent is a number that tells how many times the number standing before the 10 has to be multiplied by 10 to give the same number in its expanded form. For example:

$$5.5 \times 10^7 = 5.5 \times \underbrace{10 \times 10 \times 10 \times 10 \times 10 \times 10 \times 10}_{10^7}$$

$$6 \times 10^3 = 6 \times 10 \times 10 \times 10 = 6000$$

$$8.576 \times 10^2 = 8.576 \times 10 \times 10 = 857.6$$

TABLE A.1

Number	Exponential Form
1	1×10^0
10	1×10^1
100	1×10^2
1,000	1×10^3
10,000	1×10^4
100,000	1×10^5
1,000,000	1×10^6
0.1	1×10^{-1}
0.01	1×10^{-2}
0.001	1×10^{-3}
0.0001	1×10^{-4}
0.00001	1×10^{-5}
0.000001	1×10^{-6}

The number before the 10 doesn't always have to be a number between 1 and 10. This is just a convention, which we sometimes might find useful to ignore. However, we can't ignore the rules of arithmetic in conversions from one form to another. For example,

$$0.00045 \times 10^5 = 0.00045 \times 10 \times 10 \times 10 \times 10 \times 10$$
$$= 45$$
$$87.5 \times 10^3 = 87.5 \times 10 \times 10 \times 10$$
$$= 87,500$$

In most problem-solving situations you find that the problem is given the other way around. You encounter a large number, and (after you've learned the usefulness of exponential notation) you know that the next few minutes of your life could actually be easier if you could quickly restate the number in exponential form. This is very easy to do. Just count the number of places that you have to move the decimal point to the *left* to put it just after the first digit of the given number. For example, you might have to work with a number such as 1500 (as in 1500 mL). You'd have to move the decimal point three places leftward from where it's understood to be (1500 = 1500.) to put it just after the first digit.

$$1\,5\,0\,0$$
$$3\ 2\ 1$$

Each of these leftward moves counts as one unit for the exponent. Three leftward moves means an exponent of $+3$. Therefore 1500 can be rewritten as 1.500×10^3. A really huge number and one that you'll certainly meet somewhere during the course is 602,000,000,000,000,000,000,000. It's called Avogadro's number, and you can see that manipulating it would be awkward. (How in the world does one even pronounce it?) In exponential notation, it's written simply as 6.02×10^{23}. Check it out. Do you have to move the decimal point leftward 23 places? (And now you could pronounce it: "six point oh two times ten to the twenty-third" — but saying "Avogadro's number" is easier.) Work these exercises for practice.

EXERCISE A.1   Expand each of these exponential numbers.

(a) 5.050×10^6 (b) 0.0000344×10^8 (c) 324.4×10^3

EXERCISE A.2   Write each of these numbers in exponential form.

(a) 422,045
(b) 24,000,000,000,000,000,000
(c) 24.32

Answers to Exercises A.1 and A.2

A.1   (a) 5,050,000   (b) 3,440   (c) 324,400
A.2   (a) 4.22045×10^5 (b) 2.4×10^{19} (c) 2.432×10^1

Negative Exponents.   A negative exponent is a number that tells how many times the number standing before the 10 has to be *divided* by 10 to give the number in its expanded form. For example:

$$1 \times 10^{-4} = 1 \div 10 \div 10 \div 10 \div 10$$
$$= \frac{1}{10 \times 10 \times 10 \times 10} = \frac{1}{10,000} = \frac{1}{10^4}$$
$$= 0.0001$$

$$6 \times 10^{-3} = 6 \div 10 \div 10 \div 10$$
$$= \frac{6}{10 \times 10 \times 10} = \frac{6}{1000}$$
$$= 0.006$$
$$8.576 \times 10^{-2} = \frac{8.576}{10 \times 10} = \frac{8.576}{100} = 0.08576$$

You'll see negative exponents often when you study aqueous solutions that have very low concentrations.

Sometimes, you'll want to convert a very small number into its equivalent in exponential notation. This is also easy. This time we count *rightward* the number of times that you have to move the decimal point, one digit at a time, to place the decimal just to the right of the first nonzero digit in the number. For example, if the number is 0.00045, you have to move the decimal four times to the right to place it after the 4.

$$0.\underset{1}{\underset{\frown}{0}}\,\underset{2}{\underset{\frown}{0}}\,\underset{3}{\underset{\frown}{0}}\,\underset{4}{\underset{\frown}{4}}\,5$$

Therefore we can write: $0.00045 = 4.5 \times 10^{-4}$. Similarly, we can write: $0.0012 = 1.2 \times 10^{-3}$. And $0.00000000000000011 = 1.1 \times 10^{-16}$. Now try these exercises.

EXERCISE A.3   Write each number in expanded form.

(a) 4.3×10^{-2} (b) 5.6×10^{-10}

(c) 0.00034×10^{-2} (d) 4523.34×10^{-4}

EXERCISE A.4   Write the following numbers in exponential forms.

(a) 0.115   (b) 0.00005000041   (c) 0.000000000000345

Answers to Exercises A.3 and A.4

A.3   (a) 0.043   (b) 0.00000000056   (c) 0.0000034   (d) 0.452334

A.4   (a) 1.15×10^{-1} (b) 5.000041×10^{-5} (c) 3.45×10^{-13}

Now that we can write numbers in exponential notation, let's learn how to manipulate them.

How to Add and Subtract Numbers in Exponential Notation.   We'll not spend too much time on this, because it doesn't come up very often. The only rule is that when you add or subtract exponentials, all of the numbers must have the same exponents of 10. If they don't, we have to reexpress them to achieve this condition. Suppose you have to add 4.41×10^3 and 2.20×10^3. The result is simply 6.61×10^3.

$$(4.41 \times 10^3) + (2.20 \times 10^3) = [4.41 + 2.20] \times 10^3$$
$$= 6.61 \times 10^3$$

However, we can't add 4.41×10^3 to 2.20×10^4 without first making the exponents equal. We can do this in either of the following ways. In one, we notice that $2.20 \times 10^4 = 2.20 \times 10 \times 10^3 = 22.0 \times 10^3$, so we have:

$$(4.41 \times 10^3) + (22.0 \times 10^3) = 26.41 \times 10^3 = 2.641 \times 10^4$$

Alternatively, we could notice that $4.41 \times 10^3 = 4.41 \times 10^{-1} \times 10^4 = 0.441 \times 10^4$, so we can do the addition as follows:

$$(0.441 \times 10^4) + (2.20 \times 10^4) = 2.641 \times 10^4$$

The result is the same both ways. The extension of this to subtraction should be obvious.[1]

How to Multiply Numbers Written In Exponential Form.   Use the following two steps to multiply numbers that are expressed in exponential forms.

Step 1.   Multiply the numbers in front of the 10s.

Step 2.   *Add* the exponents of the 10s algebraically.

EXAMPLE A.1

$$(2 \times 10^4) \times (3 \times 10^5) = 2 \times 3 \times 10^{(4+5)}$$
$$= 6 \times 10^9$$

Usually, the problem you want to solve involves very large or very small numbers that aren't yet stated in exponential form. When this happens, convert the given numbers into their exponential forms first, and then carry out the operation. The next example illustrates this and shows how exponentials can make a calculation easier.

EXAMPLE A.2

$$6576 \times 2000 = (6.576 \times 10^3) \times (2 \times 10^3)$$
$$= 13.152 \times 10^6$$
$$= 1.3152 \times 10^7$$

Work the following exercise to practice.

EXERCISE A.5   Calculate the following products after you have converted large or small numbers to exponential forms.
(a) $6,000,000 \times 0.0000002$ (b) $10^6 \times 10^{-7} \times 10^8 \times 10^{-7}$
(c) $0.003 \times 0.002 \times 0.000001$ (d) $1,500 \times 3,000,000,000,000$
Answers: (a) 1.2   (b) 1   (c) 6×10^{-12} (d) 4.5×10^{15}

How to Divide Numbers Written in Exponential Form.   To divide numbers expressed in exponential forms, use the following two steps.

Step 1.   Divide the numbers that stand in front of the 10s.

Step 2.   *Subtract* the exponents of the 10s algebraically.

EXAMPLE A.3

$$(8 \times 10^4) \div (2 \times 10^3) = (8 \div 2) \times 10^{(4-3)}$$
$$= 4 \times 10^1$$

EXAMPLE A.4

$$(8 \times 10^4) \div (2 \times 10^{-3}) = (8 \div 2) \times 10^{[4-(-3)]}$$
$$= 4 \times 10^7$$

[1] Whenever these operations are with pure numbers and not with physical quantities obtained by measurements, we are not concerned about significant figures.

For practice, try the following exercise.

EXERCISE A.6   Do the following calculations using exponential forms of the numbers.

(a) $6,000,000 \div 1500$ (b) $7460 \div 0.0005$

(c) $\dfrac{3,000,000 \times 6,000,000,000}{20,000}$ (d) $\dfrac{0.016 \times 0.0006}{0.000008}$

(e) $\dfrac{400 \times 500 \times 0.002 \times 500}{2,500,000}$

Answers: (a) 4×10^3 (b) 1.492×10^7 (c) 9×10^{11} (d) 1.2 (e) 8×10^{-2}

The Pocket Calculator and Exponentials.   The foregoing was meant to refresh your memory about exponentials, because you almost certainly have studied them in any course in algebra or some earlier course. You probably own a good pocket calculator, at least one that can take numbers in exponential form. Go ahead and use it, but be sure that you understand exponentials well, first. Otherwise, there are a lot of pitfalls.

Most pocket calculators have a key marked *EE* or *EXP*. This is used to enter exponentials. Here is where an ability to *read* exponentials comes in handy. For example, the number 2.1×10^4 reads "two point one times ten to the fourth." The *EE* or *EXP* key on most calculators stands for ". . . times ten to the . . ." Therefore to enter 2.1×10^4, punch the following keys:

$$\boxed{2}\;\boxed{\cdot}\;\boxed{1}\;\boxed{EE}\;\boxed{4}$$

Try this on your own calculator, and be sure to see that the display is correct. If it isn't, you may have a calculator that works differently than most, so recheck your operations of entering and then check your owner's manual.

To enter an exponential with a negative exponent, you have to use one more key, the $\boxed{+/-}$ key. This switches a positive number to its negative, and you *must* use it rather than the $\boxed{-}$ key in this situation. Thus the number 2.1×10^{-5} enters as follows:

$$\boxed{2}\;\boxed{\cdot}\;\boxed{1}\;\boxed{EE}\;\boxed{+/-}\;\boxed{5}$$

Try it and check the display. To see what happens if you use the $\boxed{-}$ key instead of the $\boxed{+/-}$ key, clear the display and enter this number only using the $\boxed{-}$ key instead of the $\boxed{+/-}$ key.

A.2   CROSS-MULTIPLICATION

In this section we will learn how to solve for x in such expressions as

$$\frac{12}{x} = \frac{16}{25} \quad \text{or} \quad \frac{32.0}{11.2} = \frac{6.15x}{13.1}$$

The operation is called *cross-multiplication,* and its object is to get x to stand alone, all by itself, on one side of the $=$ sign, and above any real or understood divisor line.

> To cross-multiply, move a number or a symbol both across the $=$ sign and across a divisor line, and then multiply.

EXAMPLE A.5

Problem: Solve for x in $\dfrac{25}{x} = 5$

Solution: Notice first the divisor lines; one is understood.

divisor line $\longrightarrow \dfrac{25}{x} = 5 \longleftarrow$ divisor line understood because $5 = \dfrac{5}{1}$

Remember, we want x to stand alone, on top of a divisor line (even if this line is understood). To make this happen, we carry out cross-multiplication as indicated:

$$\frac{25}{x} = \boxed{5}$$

Notice that the arrows show moves that carry the quantities not only across the $=$ sign but also across their respective divisor lines. *It is essential that both crossing-overs be done.* Now we have x standing alone above its (understood) divisor line.

$$\frac{25}{5} = x$$

Now we can do the arithmetic. $x = 5$.

EXAMPLE A.6

Problem: Solve for x in: $\dfrac{40}{4} = 5x$

Solution: To get x to stand alone, only the 5 has to be moved:

The result is:

$$\frac{40}{4} = \boxed{5}\,x$$

$$\frac{40}{4 \times 5} = x \quad \text{or} \quad x = 2$$

EXAMPLE A.7

Problem: Solve for x in: $\dfrac{25 \times 60}{12} = \dfrac{625}{x}$

Solution: To get x to stand alone, we carry out the following cross-multiplication.

$$\frac{\boxed{25 \times 60}}{\boxed{12}} = \frac{625}{\boxed{x}}$$

The result is:

$$x = \frac{625 \times 12}{25 \times 60}$$
$$= 5$$

Now try the following exercise for practice.

EXERCISE A.7   Solve for x in the following.

(a) $\dfrac{12}{x} = \dfrac{16}{25}$ (b) $\dfrac{32.0}{11.2} = \dfrac{6.15x}{13.1}$

Answers: (a) $x = 18.75$ (b) $x = 6.085946574$

In Example 2.3 on page 39, the problem was to solve for Δt in the equation:

$$\frac{0.106 \text{ cal}}{g \, °C} = \frac{115 \text{ cal}}{25.4 \text{ g} \times \Delta t}$$

Here is a problem with both units and numbers, so now we have to add one more and very important principle. *We cross-multiply units as well as numbers.*

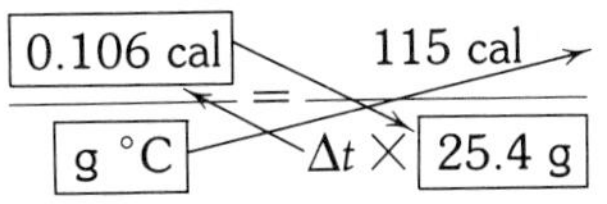

The result is the following, in which the cancel lines show how the units cancel.

$$\Delta t = \frac{(115 \text{ cal}) \times (g \, °C)}{(25.4 \text{ g}) \times (0.106 \text{ cal})} = 42.7 \, °C \qquad \text{(correctly rounded)}$$

How to Do Chain Calculations with the Pocket Calculator.   Sometimes the steps in solving a problem lead to something like the following:

$$x = \frac{24.2 \times 30.2 \times 55.6}{2.30 \times 18.2 \times 4.44}$$

Many people will first calculate the value of the numerator and write it down. Then they'll compute the denominator and write it down. Finally, they'll divide the two results to get the final answer. There's no need to do this much work. All you have to do is enter the first number you see in the numerator, 24.2 in our example. Then use the $\boxed{\times}$ key for any number in the numerator and the $\boxed{\div}$ key for any number in the denominator. *Each number in the denominator is entered with the $\boxed{\div}$ key.* Any of the following sequences work. Try them.

$$24.2 \times 30.2 \times 55.6 \div 2.30 \div 18.2 \div 4.44 = 218.632 \ldots$$

Or

$$24.2 \div 2.30 \times 30.2 \div 18.2 \times 55.6 \div 4.44 = 218.632 \ldots$$

Appendix B
Electron Configurations of the Elements

Atomic Number	Element	1s	2s	2p	3s	3p	3d	4s	4p	4d	4f	5s	5p	5d	5f	5g
1	H	1														
2	He	2														
3	Li	2	1													
4	Be	2	2													
5	B	2	2	1												
6	C	2	2	2												
7	N	2	2	3												
8	O	2	2	4												
9	F	2	2	5												
10	Ne	2	2	6												
11	Na	2	2	6	1											
12	Mg	2	2	6	2											
13	Al	2	2	6	2	1										
14	Si	2	2	6	2	2										
15	P	2	2	6	2	3										
16	S	2	2	6	2	4										
17	Cl	2	2	6	2	5										
18	Ar	2	2	6	2	6										
19	K	2	2	6	2	6		1								
20	Ca	2	2	6	2	6		2								
21	Sc	2	2	6	2	6	1	2								
22	Ti	2	2	6	2	6	2	2								
23	V	2	2	6	2	6	3	2								
24	Cr	2	2	6	2	6	5	1								
25	Mn	2	2	6	2	6	5	2								
26	Fe	2	2	6	2	6	6	2								
27	Co	2	2	6	2	6	7	2								
28	Ni	2	2	6	2	6	8	2								
29	Cu	2	2	6	2	6	10	1								
30	Zn	2	2	6	2	6	10	2								
31	Ga	2	2	6	2	6	10	2	1							
32	Ge	2	2	6	2	6	10	2	2							
33	As	2	2	6	2	6	10	2	3							
34	Se	2	2	6	2	6	10	2	4							
35	Br	2	2	6	2	6	10	2	5							
36	Kr	2	2	6	2	6	10	2	6							
37	Rb	2	2	6	2	6	10	2	6			1				
38	Sr	2	2	6	2	6	10	2	6			2				
39	Y	2	2	6	2	6	10	2	6	1		2				
40	Zr	2	2	6	2	6	10	2	6	2		2				
41	Nb	2	2	6	2	6	10	2	6	4		1				
42	Mo	2	2	6	2	6	10	2	6	5		1				
43	Tc	2	2	6	2	6	10	2	6	6		1				
44	Ru	2	2	6	2	6	10	2	6	7		1				
45	Rh	2	2	6	2	6	10	2	6	8		1				
46	Pd	2	2	6	2	6	10	2	6	10						
47	Ag	2	2	6	2	6	10	2	6	10		1				
48	Cd	2	2	6	2	6	10	2	6	10		2				
49	In	2	2	6	2	6	10	2	6	10		2	1			
50	Sn	2	2	6	2	6	10	2	6	10		2	2			
51	Sb	2	2	6	2	6	10	2	6	10		2	3			
52	Te	2	2	6	2	6	10	2	6	10		2	4			
53	I	2	2	6	2	6	10	2	6	10		2	5			
54	Xe	2	2	6	2	6	10	2	6	10		2	6			

Electron Configurations of the Elements *(continued)*

Atomic Number	Element	K	L	M	4s	4p	4d	4f	5s	5p	5d	5f	5g	6s	6p	6d	7s
55	Cs	2	8	18	2	6	10		2	6				1			
56	Ba	2	8	18	2	6	10		2	6				2			
57	La	2	8	18	2	6	10		2	6	1			2			
58	Ce	2	8	18	2	6	10	2	2	6				2			
59	Pr	2	8	18	2	6	10	3	2	6				2			
60	Nd	2	8	18	2	6	10	4	2	6				2			
61	Pm	2	8	18	2	6	10	5	2	6				2			
62	Sm	2	8	18	2	6	10	6	2	6				2			
63	Eu	2	8	18	2	6	10	7	2	6				2			
64	Gd	2	8	18	2	6	10	7	2	6	1			2			
65	Tb	2	8	18	2	6	10	9	2	6				2			
66	Dy	2	8	18	2	6	10	10	2	6				2			
67	Ho	2	8	18	2	6	10	11	2	6				2			
68	Er	2	8	18	2	6	10	12	2	6				2			
69	Tm	2	8	18	2	6	10	13	2	6				2			
70	Yb	2	8	18	2	6	10	14	2	6				2			
71	Lu	2	8	18	2	6	10	14	2	6	1			2			
72	Hf	2	8	18	2	6	10	14	2	6	2			2			
73	Ta	2	8	18	2	6	10	14	2	6	3			2			
74	W	2	8	18	2	6	10	14	2	6	4			2			
75	Re	2	8	18	2	6	10	14	2	6	5			2			
76	Os	2	8	18	2	6	10	14	2	6	6			2			
77	Ir	2	8	18	2	6	10	14	2	6	7			2			
78	Pt	2	8	18	2	6	10	14	2	6	9			1			
79	Au	2	8	18	2	6	10	14	2	6	10			1			
80	Hg	2	8	18	2	6	10	14	2	6	10			2			
81	Tl	2	8	18	2	6	10	14	2	6	10			2	1		
82	Pb	2	8	18	2	6	10	14	2	6	10			2	2		
83	Bi	2	8	18	2	6	10	14	2	6	10			2	3		
84	Po	2	8	18	2	6	10	14	2	6	10			2	4		
85	At	2	8	18	2	6	10	14	2	6	10			2	5		
86	Rn	2	8	18	2	6	10	14	2	6	10			2	6		
87	Fr	2	8	18	2	6	10	14	2	6	10			2	6		1
88	Ra	2	8	18	2	6	10	14	2	6	10			2	6		2
89	Ac	2	8	18	2	6	10	14	2	6	10			2	6	1	2
90	Th	2	8	18	2	6	10	14	2	6	10			2	6	2	2
91	Pa	2	8	18	2	6	10	14	2	6	10	2		2	6	1	2
92	U	2	8	18	2	6	10	14	2	6	10	3		2	6	1	2
93	Np	2	8	18	2	6	10	14	2	6	10	5		2	6		2
94	Pu	2	8	18	2	6	10	14	2	6	10	6		2	6		2
95	Am	2	8	18	2	6	10	14	2	6	10	7		2	6		2
96	Cm	2	8	18	2	6	10	14	2	6	10	7		2	6	1	2
97	Bk	2	8	18	2	6	10	14	2	6	10	8		2	6	1	2
98	Cf	2	8	18	2	6	10	14	2	6	10	10		2	6		2
99	Es	2	8	18	2	6	10	14	2	6	10	11		2	6		2
100	Fm	2	8	18	2	6	10	14	2	6	10	12		2	6		2
101	Md	2	8	18	2	6	10	14	2	6	10	13		2	6		2
102	No	2	8	18	2	6	10	14	2	6	10	14		2	6		2
103	Lw	2	8	18	2	6	10	14	2	6	10	14		2	6	1	2

Appendix C
Some Rules for Naming Inorganic Compounds

Only those rules considered sufficient to meet most of the needs of the users of this text are in this Appendix. The latest edition of the *Handbook of Chemistry and Physics,* published annually by the CRC Publishing Company, Cleveland, OH, under the general editorship of R. C. Weast, has a section on all of the rules. Virtually all college libraries have this reference.

I. **Binary Compounds**—those made from only two elements
    A. One element is a metal and the other is a nonmetal
        1. The name of the metal is written first in the name of the compound, and its symbol is placed first in the formula.
        2. The name ending of the nonmetal is changed to *-ide.* Thus the names of the simple ions of groups VIA and VIIA of the periodic table are:

Group VIIA	Group VIA
Fluoride	Oxide
Chloride	Sulfide
Bromide	Selenide
Iodide	Telluride

        3. If the metal and the nonmetal each have just one oxidation number, a binary compound of the two is named simply by writing the name of the metal and then that of the nonmetal with its ending modified by *-ide,* as shown above. Greek prefixes such as mono-, di-, tri-, etc., are not necessary. Examples are:

Some Compounds Between Elements of Groups IA and VIIA		Some Compounds Between Elements of Groups IA and VIA	
NaF	Sodium fluoride	Na_2O	Sodium oxide (not disodium oxide)
KCl	Potassium chloride	K_2S	Potassium sulfide
$LiBr$	Lithium bromide	Li_2O	Lithium oxide
RbI	Rubidium iodide	Cs_2S	Cesium sulfide
$CsCl$	Cesium chloride	Rb_2O	Rubidium oxide
Some Compounds Between Elements of Groups IIA and VIIA		**Some Compounds Between Elements of Groups IIA and VIA**	
$BeCl_2$	Beryllium chloride	BeO	Beryllium oxide
$MgBr_2$	Magnesium bromide	MgS	Magnesium sulfide
CaF_2	Calcium fluoride	CaO	Calcium oxide
SrI_2	Strontium iodide	SrS	Strontium sulfide
$BaCl_2$	Barium chloride	BaO	Barium oxide
Some Compounds Between Elements of Groups IIIA and VIIA		**Some Compounds Between Elements of Groups IIIA and VIA**	
$AlCl_3$	Aluminum chloride	Al_2O_3	Aluminum oxide
AlF_3	Aluminum fluoride	Al_2S_3	Aluminum sulfide

4. If the metal has more than one oxidation number, but the nonmetal has just one, the formal name of the compound includes a roman numeral in parentheses following the name of the metal. This numeral stands for the oxidation number of the metal. Greek prefixes such as mono-, di-, etc., are not needed.

EXAMPLE 1   Compounds of Iron in Oxidation States of 2+ or 3+

	Formal Name	Common Name
$FeCl_2$	Iron(II) chloride[a]	ferrous chloride
FeO	Iron(II) oxide	ferrous oxide
Fe_2O_3	Iron(III) oxide	ferric oxide
$FeCl_3$	Iron(III) chloride	ferric chloride

[a] Pronounced "iron two chloride."

EXAMPLE 2   Compounds of Copper in Oxidation States of 1+ and 2+

	Formal Name	Common Name
Cu_2O	Copper(I) oxide	cuprous oxide
$CuBr$	Copper(I) bromide	cuprous bromide
$CuCl_2$	Copper(II) chloride	cupric chloride
CuS	Copper(II) sulfide	cupric sulfide

---

5. Molecular compounds of two elements. Greek prefixes such as mono-, di-, etc., are used, sometimes for *both* elements.
   (a) Oxides of Nonmetals
       (1) Oxides of Carbon   (2) Oxides of Sulfur

CO	Carbon monoxide	SO_2	Sulfur dioxide
CO_2	Carbon dioxide	SO_3	Sulfur trioxide

   (3) Oxides of Nitrogen (Older Names in Parentheses)

N_2O	Dinitrogen monoxide (nitrous oxide)
NO	Nitrogen oxide (nitric oxide)
N_2O_3	Dinitrogen trioxide
NO_2	Nitrogen dioxide
N_2O_4	Dinitrogen tetroxide
N_2O_5	Dinitrogen pentoxide

   (4) Oxides of Some Halogens

F_2O	Difluorine monoxide
Cl_2O	Dichlorine monoxide
Cl_2O_7	Dichlorine heptoxide

   (b) Some Halides of Carbon

CCl_4	Carbon tetrachloride
CBr_4	Carbon tetrabromide

   (c) Some Exceptions

H_2O Water	NH_3 Ammonia	CH_4 Methane

II. **Compounds of Three or More Elements**

A. A positive and a negative ion are combined

1. The name of the positive ion is first followed by the name of the negative ion, just as with binary compounds between metals and nonmetals. Greek prefixes are not needed except where they occur in the name of an ion. (Older names are shown in parentheses.)

Li_2SO_4	Lithium sulfate	$MgSO_4$	Magnesium sulfate
Na_2SO_4	Sodium sulfate	$CaSO_4$	Calcium sulfate
K_2SO_4	Potassium sulfate	$Al_2(SO_4)_3$	Aluminum sulfate
$LiHCO_3$	Lithium hydrogen carbonate (lithium bicarbonate)[a]		
$NaHCO_3$	Sodium hydrogen carbonate (sodium bicarbonate)		
Li_2CO_3	Lithium carbonate	$KMnO_4$	Potassium permanganate
Na_2CO_3	Sodium carbonate	Na_2CrO_4	Sodium chromate
$CaCO_3$	Calcium carbonate	$Mg(NO_3)_2$	Magnesium nitrate
$Al_2(CO_3)_2$	Aluminum carbonate	$NaNO_2$	Sodium nitrite
$NaHSO_4$	Sodium hydrogen sulfate (sodium bisulfate)		
NaH_2PO_4	Sodium dihydrogen phosphate		
K_2HPO_4	Potassium monohydrogen phosphate		
$MgHPO_4$	Magnesium monohydrogen phosphate		
$(NH_4)_2HPO_4$	Ammonium monohydrogen phosphate		
Na_3PO_4	Sodium phosphate		
$Ca_3(PO_4)_2$	Calcium phosphate		

[a] *Bicarbonate* instead of *hydrogen carbonate* is used in this text because it is judged to be the more commonly used name for this ion, particularly among health scientists.

B. Molecular compounds of two or more elements. Most are organic compounds, so their rules of nomenclature are given in the chapters on organic compounds.

III. **Important Inorganic Acids and Their Anions**

Formula	Name	Formula	Name
H_2CO_3	Carbonic acid	HCO_3^-	Hydrogen carbonate ion (bicarbonate ion)
		CO_3^{2-}	Carbonate ion
HNO_3	Nitric acid	NO_3^-	Nitrate ion
HNO_2	Nitrous acid	NO_2^-	Nitrite ion
H_2SO_4	Sulfuric acid	HSO_4^-	Hydrogen sulfate ion (bisulfate ion)
		SO_4^{2-}	Sulfate ion
H_2SO_3	Sulfurous acid	HSO_3^-	Hydrogen sulfite ion (bisulfite ion)
		SO_3^{2-}	Sulfite ion
H_3PO_4	Phosphoric acid (orthophosphoric acid)	$H_2PO_4^-$	Dihydrogen phosphate ion
		HPO_4^{2-}	Monohydrogen phosphate ion
		PO_4^{3-}	Phosphate ion
$HClO_4$	Perchloric acid	ClO_4^-	Perchlorate ion
$HClO_3$	Chloric acid	ClO_3^-	Chlorate ion
$HClO_2$	Chlorous acid	ClO_2^-	Chlorite ion
$HClO$	Hypochlorous acid	ClO^-	Hypochlorite ion
HCl	Hydrochloric acid[a]	Cl^-	Chloride ion

[a] The name of the aqueous solution of gaseous HCl.

Some generalizations about names of acids and their anions

1. Names of ions from acids whose names end in *-ic* all end in *-ate*.

2. When a nonmetal forms an oxyacid whose name ends in *-ic* also forms an acid with one fewer oxygen atoms, the name of the latter acid ends in *-ous*. (Compare nitric acid, HNO_3, and nitrous acid, HNO_2.)

3. When a nonmetal forms an oxyacid with one fewer oxygen atoms than are in an *-ous* acid, then the prefix *hypo-* is used. (Compare chlorous acid, $HClO_2$, and hypochlorous acid, $HClO$.)

4. The binary hydrohalogen acids are called hydrogen halides when they occur as pure gases but are called hydrohalic acids when they occur as aqueous solutions. Thus hydrogen fluoride in water becomes hydrofluoric acid; hydrogen chloride in water becomes hydrochloric acid; and so on.

Appendix D
Answers to Practice Exercises and Selected Review Exercises

PRACTICE EXERCISES, CHAPTER 1

1. 310 K
2. (a) 5.45×10^8 (b) 5.67×10^{12} (c) 6.454×10^3 (d) 2.5×10^1 (e) 3.98×10^{-5} (f) 4.26×10^{-3} (g) 1.68×10^{-1} (h) 9.87×10^{-12}
3. (a) 10^{-6} (b) 10^{-9} (c) 10^{-6} (d) 10^3
4. (a) mL (b) μL (c) dL (d) mm (e) cm (f) kg (g) μg (h) mg
5. (a) kilogram (b) centimeter (c) deciliter (d) microgram (e) milliliter (f) milligram (g) millimeter (h) microliter
6. (a) 1.5 Mg (b) 3.45 μL (c) 3.6 mg (d) 6.2 mL (e) 1.68 kg (f) 5.4 dm
7. (a) 275 kg (b) 62.5 μL (c) 82 nm or 0.082 μm
8. (a) 95 (b) 11.36 (c) 0.0263 (d) 1.3000 (e) 16.1 (f) 3.8×10^2 (g) 9.31 (h) 9.1×10^2
9. (a) $\dfrac{1 \text{ g}}{1000 \text{ mg}}$ or $\dfrac{1000 \text{ mg}}{1 \text{ g}}$
   (b) $\dfrac{1 \text{ kg}}{2.205 \text{ lb}}$ or $\dfrac{2.205 \text{ lb}}{1 \text{ kg}}$
10. 0.324 g of aspirin
11. (a) 324 mg of aspirin
    (b) 3.28×10^4 ft
    (c) 18.5 mL
    (d) 38.87 g
    (e) 4.78×10^3 μL
12. 40.0 °C
13. 59 °F (Quite cool)
14. 20.7 mL
15. 32.1 g

REVIEW EXERCISES, CHAPTER 1

1.31 273 K and 373 K

1.33 410 °F

1.35 88 °F

1.37 42.8 °C. No, regardless of the temperature *scale,* it's just as hot.

1.55 (a) 192 cm (b) 111 lb

1.57 16.9 liq oz

1.59 Do not cross; 2.1×10^3 kg > 1.5×10^3 kg (the limit)

1.61 3.5 g/pat

1.63 8.8477×10^3 m; 8.8477 km

1.71 23.7 lb of lead

1.73 28.3 mL of acetic acid

PRACTICE EXERCISES, CHAPTER 2

1. Na_2S
2. Potassium, carbon, and oxygen in an atom ratio of 2:1:3.
3. Final temperature = 24.5 °C. ($\Delta t = 4.54$ °C)
4. 5.8×10^2 kcal
5. 2.99×10^4 cal
   29.9 kcal
6. 8.85 kcal
7. 4.0×10^3 kcal/day

REVIEW EXERCISES, CHAPTER 2

2.21 12.2/6.10 = 2:1, a ratio of small whole numbers.

2.44 (a) $KE = (1/2)mv^2 = 5.86 \times 10^5$ J
(b) $KE = 1.40 \times 10^5$ cal
$= 1.40 \times 10^2$ kcal

2.51 For 1.00 g of water, heat capacity = 1.00 cal/°C.
For 10.00 g of water, heat capacity = 10.0 cal/°C.

2.53 When ice melts at 0 °C and when liquid water boils at 100 °C.

2.65 $t = 282$ °C (2.8×10^2 °C when properly rounded)

2.67 265 g of ethyl alcohol

2.81 3.3×10^3 kcal

2.83 1.7×10^2 kcal/cup milk (rounded from 168 kcal)

PRACTICE EXERCISES, CHAPTER 3

1. $^{16}_{8}O$

2. (a) 9 n   (b) 7 n   (c) 20 n   (d) 18 n
         8 p   7 p   17 p   17 p

3. The atomic number of carbon is 6, not 7.

4. (a) $1s^2 2s^2 2p_x^2 2p_y^2 2p_z^2 3s^2$
    (b) $1s^2 2s^2 2p_x^2 2p_y^2 2p_z^2 3s^2 3p_x^2 3p_y^2 3p_z^1$
    (c) $1s^2 2s^2 2p_x^2 2p_y^2 2p_z^2 3s^2 3p_x^2 3p_y^2 3p_z^2$
    (d) $1s^2 2s^2 2p_x^2 2p_y^2 2p_z^2 3s^2 3p_x^2 3p_y^2 3p_z^2 4s^2$

5. (a) Sn   (b) Cl   (c) Rb   (d) Mg   (e) Ar

6. (a) 1   (b) 6   (c) 5   (d) 7

REVIEW EXERCISES, CHAPTER 3

3.9 (a) 4.031884 amu/He nucleus (calculated)
     (b) 4.001507 amu/He nucleus (observed)
     (c) 0.030377 amu/He nucleus (mass loss)
     (d) 4.5400×10^{-12} J/He nucleus
     (e) $2.7340 \times ^{12}$ J per 4 g of helium that forms by fusion

3.41 12 times heavier

3.42 (a) 1.33 times as heavy
     (b) 16.0 g of oxygen atoms

PRACTICE EXERCISES, CHAPTER 4

1. (a) $1s^2 2s^2 2p_x^2 2p_y^2 2p_z^2 3s^2 3p_x^2 3p_y^2 3p_z^2 4s^1$ Ion's charge = 1+
    (b) $1s^2 2s^2 2p_x^2 2p_y^2 2p_z^2 3s^2 3p_x^2 3p_y^1 3p_z^1$ Ion's charge = 2−
    (c) $1s^2 2s^2 2p_x^2 2p_y^2 2p_z^2 3s^2 3p_x^1 3p_y^1$ No ion exists

2. (a) $1s^2 2s^2 2p_x^2 2p_y^2 2p_z^2 3s^2 3p_x^2 3p_y^2 3p_z^2$
    (b) $1s^2 2s^2 2p_x^2 2p_y^2 2p_z^2 3s^2 3p_x^2 3p_y^2 3p_z^2$
    (c) No ion exists

3. (a) Cs^+ (b) F^- (c) no ion   (d) Sr^{2+}

4. (a) AgBr   (b) Na_2O (c) Fe_2O_3 (d) $CuCl_2$

5. (a) Copper(II) sulfide; cupric sulfide
    (b) Sodium fluoride
    (c) Iron(II) iodide; ferrous iodide
    (d) Zinc bromide
    (e) Copper(I) oxide; cuprous oxide

6. (a) 3+   (b) 2+   (c) 2+

7. (a) Mg is oxidized; S is reduced
    Mg is the reducing agent; S is the oxidizing agent
    (b) Zn is oxidized; Cu^{2+} is reduced
    Zn is the reducing agent; Cu^{2+} is the oxidizing agent

8. ·Na ·Mg· ·Al· ·Si· ·P· ·S· :Cl· :Ar:

9. ·S· H:S:H H—S—H

10. 4

11. H—Si—H (with H above and H below Si)

12. H—C—C—C—H (with H H H above and H H H below)

13. Structure (d) is not of a possible compound.

14. (a) $KHCO_3$ (b) Na_2HPO_4 (c) $(NH_4)_3PO_4$

15. (a) Sodium cyanide
    (b) Potassium nitrate
    (c) Sodium hydrogen sulfite
    (d) Ammonium carbonate
    (e) Sodium acetate

16. The molecule is polar. (structure: C bonded to H with δ+, and three Cl with δ−)

PRACTICE EXERCISES, CHAPTER 5

1. $3O_2 \rightarrow 2O_3$

2. $4Al + 3O_2 \rightarrow 2Al_2O_3$

3. (a) $2Ca + O_2 \rightarrow 2CaO$
    (b) $2KOH + H_2SO_4 \rightarrow 2H_2O + K_2SO_4$
    (c) $Cu(NO_3)_2 + Na_2S \rightarrow CuS + 2NaNO_3$
    (d) $2AgNO_3 + CaCl_2 \rightarrow 2AgCl + Ca(NO_3)_2$
    (e) $2Al + 3H_2SO_4 \rightarrow Al_2(SO_4)_3 + 3H_2$
    (f) $CH_4 + 2O_2 \rightarrow 2H_2O + CO_2$

4. 8.68×10^{22} atoms of gold per ounce

5. (a) 180   (b) 58.3   (c) 859

6. 0.500 mol of H_2O

7. 4.20 mol of N_2 and 4.20 mol of O_2

8. 450 mol of H_2 and 150 mol of N_2

9. 408 g of NH_3

10. 0.0380 mol of aspirin

11. 11.5 g of O_2

12. 18.4 g of Na

13. (a) 2.45 g of H_2SO_4
    (b) 9.00 g of $C_6H_{12}O_6$

14. 156 mL of 0.800 M Na_2CO_3 solution

15. 9.82 mL of 0.112 M H_2SO_4, when calculated step-by-step with rounding after each step.
    9.84 mL of 0.112 m H_2SO_4, when found by a chain calculation.

16. Dilute 14 mL of 18 M H_2SO_4 to a final volume of 250 mL.

REVIEW EXERCISES, CHAPTER 5

5.8 2.01×10^{23} formula units of H_2O

5.10 0.150 g of medication

5.13 (a) 40.0 (b) 100 (c) 98.1
      (d) 106 (e) 158 (f) 128

5.20 (a) 100 g (b) 250 g (c) 245 g
      (d) 265 g (e) 395 g (f) 320 g

5.22 (a) 1.88 mol (b) 0.750 mol (c) 0.765 mol
      (d) 0.708 mol (e) 0.475 mol (f) 0.586 mol

5.24 (a) $\dfrac{4\ \text{mol Fe}}{3\ \text{mol}\ O_2}$ and $\dfrac{3\ \text{mol}\ O_2}{4\ \text{mol Fe}}$

      (b) $\dfrac{4\ \text{mol Fe}}{2\ \text{mol}\ Fe_2O_3}$ and $\dfrac{2\ \text{mol}\ Fe_2O_3}{4\ \text{mol Fe}}$

      (c) $\dfrac{3\ \text{mol}\ O_2}{2\ \text{mol}\ Fe_2O_3}$ and $\dfrac{2\ \text{mol}\ Fe_2O_3}{3\ \text{mol}\ O_2}$

5.26 (a) 26 mol of O_2
      (b) 50 mol of H_2O
      (c) 26 mol of O_2

5.28 (a) 75.0 mol of H_2
      (b) 43 mol of CH_4

5.30 (a) 1.28×10^3 g of NaCl
      (b) 872 g of NaOH (stepwise calculation with rounding after each step)
          873 g of NaOH (when found by a chain calculation)
      (c) 21.8 g of H_2 (using 2.00 g/mol for H_2)

5.32 48.7 g of Na_2CO_3 (stepwise calculation with rounding after each step)
      48.6 g of Na_2CO_3 (when found by a chain calculation)

5.40 (a) 5.85 g of NaCl
      (b) 5.63 g of $C_6H_{12}O_6$
      (c) 0.981 g of H_2SO_4
      (d) 11.2 g of KOH

5.42 2.5×10^2 mL of 0.10 M HCl solution

5.44 1.0×10^2 mL of 0.010 M $NaHCO_3$ solution

5.46 250 mL of 1.00 M NaOH solution

5.48 167 mL of 0.150 M Na_2SO_4 solution

5.50 540 mL of 0.100 M HCl solution

5.52 Dissolve 12 mL of 17 M $HC_2H_3O_2$ in water and make the final volume equal to 100 mL.

PRACTICE EXERCISES, CHAPTER 6

1. 1.31×10^3 mL of helium

2. 2.6 L of anesthetic

3. 52 mm Hg

4. Partial pressure of $N_2 = 720$ mm Hg
   Volume of $N_2 = 308$ mL

5. 547 mL of cyclopropane

6. 1.29×10^{-2} mol of O_2

7. Initially, $R = P_1V_1/nT_1$, and
   finally, $R = P_2V_2/nT_2$.
   Because $R = R$,
         $P_1V_1/nT_1 = P_2V_2/nT_2$
   The n's cancel, so
         $P_1V_1/T_1 = P_2V_2/T_2$

8. 2.63×10^4 mL of oxygen

REVIEW EXERCISES, CHAPTER 6

6.10 2.08 lb/in.^{2}

6.12 350 mm Hg or 350 torr

6.14 735 mm Hg

6.28 829 mm Hg

6.30 73 mm Hg

6.34 257 mL of wet O_2

6.36 1.84 L of O_2

6.44 (a) 3.54×10^{-2} mol of gas
      (b) 32.1
      (c) oxygen

6.46 (a) 2.00 mol of H_2
      (b) 200 L of H_2
      (c) 1.55×10^5 mL of H_2

PRACTICE EXERCISES, CHAPTER 7

1. 0.00147 g N_2/100 g H_2O

2. 10.2 g of 96% H_2SO_4

3. 1.25 g of glucose and 499 g of water (rounded from 498.75, and assuming that the density of water is 1.00 g/mL)

4. 1.67×10^4 mm Hg

5. (a) 0.020 Osm (b) 0.015 Osm
   (c) 0.100 Osm (d) 0.150 Osm

REVIEW EXERCISES, CHAPTER 7

7.46 $Z \cdot 10H_2O$

7.48 0.0207 g/L

7.59 (a) 0.500 g of NaI
      (b) 1.25 g of NaBr
      (c) 6.25 g of $C_6H_{12}O_6$
      (d) 15.0 g of H_2SO_4

7.61 (a) 12.5 g of $Mg(NO_3)_2$
      (b) 5.00 g of NaBr
      (c) 2.50 g of KI
      (d) 1.68 g of $Ca(NO_3)_2$

7.63 25.0 mL of ethyl alcohol

7.65 (a) 11.0 mL of KOH solution
      (b) 30.0 mL of HCl solution
      (c) 650 mL of NaCl solution
      (d) 281 mL of KOH solution

7.67 24.7 g of $Na_2SO_4 \cdot 10H_2O$

7.69 12.5 g of stock solution, or 11.9 mL of stock solution

7.71 (a) 71.0% (w/w) HNO_3
(b) 24.8 mL of concentrated solution

7.80 10% NaCl, which has 1.7 mmol/100 g solution versus only 0.67 mmol/100 g solution for the NaI

7.83 372 mm Hg

PRACTICE EXERCISES, CHAPTER 8

1. $:\!\ddot{O}\!: + H\!-\!\ddot{B}\ddot{r}\!: \rightleftharpoons \overset{+}{\ddot{O}}\!-\!H + :\!\ddot{B}\ddot{r}\!:^-$ (with H's on oxygen)

$:\!\ddot{O}\!: + H\!-\!\ddot{I}\!: \rightleftharpoons \overset{+}{\ddot{O}}\!-\!H + :\!\ddot{I}\!:^-$

2. $:\!\ddot{O}\!: + H\!-\!\ddot{F}\!: \rightarrow \overset{+}{\ddot{O}}\!-\!H + :\!\ddot{F}\!:^-$

3. $H_2O + H_2SO_3(aq) \rightleftharpoons H_3O^+ + HSO_3^-(aq)$
$H_2O + HSO_3^-(aq) \rightleftharpoons H_3O^+ + SO_3^{2-}(aq)$

4. $HNO_3(aq) + KOH(aq) \rightarrow H_2O + KNO_3(aq)$
$H^+(aq) + NO_3^-(aq) + K^+(aq) + OH^-(aq) \rightarrow$
$\qquad\qquad\qquad\qquad H_2O + K^+(aq) + NO_3^-(aq)$
$H^+(aq) + OH^-(aq) \rightarrow H_2O$

5. K_2SO_4

6. $2NaHCO_3(aq) + H_2SO_4(aq) \rightarrow$
$\qquad\qquad\qquad 2CO_2(g) + 2H_2O + Na_2SO_4(aq)$
$2Na^+(aq) + 2HCO_3^-(aq) + 2H^+(aq) + SO_4^{2-}(aq) \rightarrow$
$\qquad\qquad 2CO_2(g) + 2H_2O + 2Na^+(aq) + SO_4^{2-}(aq)$
$HCO_3^-(aq) + H^+(aq) \rightarrow CO_2(g) + H_2O$

7. $K_2CO_3(aq) + H_2SO_4(aq) \rightarrow CO_2(g) + H_2O + K_2SO_4(aq)$
$2K^+(aq) + CO_3^{2-}(aq) + 2H^+(aq) + SO_4^{2-}(aq) \rightarrow$
$\qquad\qquad CO_2(g) + H_2O + 2K^+(aq) + SO_4^{2-}(aq)$
$CO_3^{2-}(aq) + 2H^+(aq) \rightarrow CO_2(g) + H_2O$

8. $MgCO_3(s) + 2HNO_3(aq) \rightarrow CO_2(g) + H_2O + Mg(NO_3)_2(aq)$
$MgCO_3(s) + 2H^+(aq) + 2NO_3^-(aq) \rightarrow$
$\qquad\qquad CO_2(g) + H_2O + Mg^{2+}(aq) + 2NO_3^-(aq)$
$MgCO_3(s) + 2H^+(aq) \rightarrow CO_2(g) + H_2O + Mg^{2+}(aq)$

9. $Mg(OH)_2(s) + 2HCl(aq) \rightarrow 2H_2O + MgCl_2(aq)$
$Mg(OH)_2(s) + 2H^+(aq) \rightarrow 2H_2O + Mg^{2+}(aq)$

10. (a) $NH_3(aq) + HBr(aq) \rightarrow NH_4Br(aq)$
$NH_3(aq) + H^+(aq) \rightarrow NH_4^+(aq)$
(b) $2NH_3(aq) + H_2SO_4(aq) \rightarrow (NH_4)_2SO_4(aq)$
$NH_3(aq) + H^+(aq) \rightarrow NH_4^+(aq)$

11. $Mg(s) + 2HCl(aq) \rightarrow H_2(g) + MgCl_2(aq)$
$Mg(s) + 2H^+(aq) \rightarrow H_2(g) + Mg^{2+}(aq)$

12. (a) HNO_3 (b) HSO_3^- (c) HCO_3^-
(d) HSO_4^- (e) HCl (f) H_3O^+
(g) H_2O

13. (a) CO_3^{2-} (b) PO_4^{3-} (c) HSO_4^-
(d) SO_4^{2-} (e) Br^- (f) H_2O
(g) OH^-

14. All are weak Brønsted bases.

15. (a) Weak (b) Weak (c) Strong (d) Strong

16. Yes. $H^+(aq) + NO_2^-(aq) \rightleftharpoons HNO_2(aq)$

17. $Cu(NO_3)_2(aq) + Na_2S(aq) \rightarrow CuS(s) + 2NaNO_3(aq)$
$Cu^{2+}(aq) + S^{2-}(aq) \rightarrow CuS(s)$

18. The acetate ion, $C_2H_3O_2^-(aq)$, binds H^+ ions from $HCl(aq)$ because $C_2H_3O_2^-$ is a relatively strong Brønsted base:
$C_2H_3O_2^-(aq) + H^+(aq) \rightleftharpoons HC_2H_3O_2(aq)$

19. (a) $Ag^+(aq) + Cl^-(aq) \rightarrow AgCl(s)$
(b) $CaCO_3(s) + 2H^+(aq) \rightarrow Ca^{2+}(aq) + H_2O + CO_2(g)$
(c) No reaction

REVIEW EXERCISES, CHAPTER 8

8.45 0.250 mol of $NaHCO_3$

8.47 6.68 g of Na_2CO_3

8.49 4.91 g of $NaHCO_3$

8.51 29.5 mL of NaOH solution

8.53 $CaCO_3(s) + 2HCl(aq) \rightarrow CO_2(g) + H_2O + CaCl_2(aq)$
$CaCO_3(s) + 2H^+(aq) \rightarrow CO_2(g) + H_2O + Ca^{2+}(aq)$
47.8 g of $CaCO_3$
191 mL of 5.00 M HCl

PRACTICE EXERCISES, CHAPTER 9

1. (a) 2.5×10^{-6} mol OH^-/L. Basic
(b) 9.1×10^{-8} mol OH^-/L. Acidic
(c) 1.1×10^{-7} mol OH^-/L. Basic

2. (a) 1×10^{-7} to 1×10^{-8} mol H^+/L
(b) Slightly basic
(c) Acidosis
(d) 6.90

3. $NaHCO_3(aq)$
$HCO_3^-(aq) + H^+(aq) \rightarrow CO_2(g) + H_2O$

4. (a) $CO_3^{2-}(aq) + H_2O \rightleftharpoons HCO_3^-(aq) + OH^-(aq)$. Basic
(b) $HPO_4^{2-}(aq) + H_2O \rightleftharpoons H_2PO_4^-(aq) + OH^-(aq)$. Basic
(c) $S^{2-}(aq) + H_2O \rightleftharpoons HS^-(aq) + OH^-(aq)$. Basic
(d) Neutral
(e) $NO_2^-(aq) + H_2O \rightleftharpoons HNO_2(aq) + OH^-(aq)$. Basic
(f) $F^-(aq) + H_2O \rightleftharpoons HF(aq) + OH^-(aq)$. Basic

5. The solution is basic.

6. No

7. Tends to lower the pH

8. 1.250 g of NaOH

9. 0.105 N NaOH

REVIEW EXERCISES, CHAPTER 9

9.4 2.43×10^{-14}

9.33 3.76 g or 3.76×10^3 mg of Cl^-

9.35 5.01 meq of K^+

9.37 Anion gap = 10 meq/L. This is in the normal range of 5–14 meq/L, so no serious disturbance in metabolism is indicated.

9.49 (a) 100.46 g/eq (b) 46.03 g/eq
(c) 26.00 g/eq (d) 64.04 g/eq

9.51 (a) 1.000 eq (b) 0.5579 eq
(c) 1.090 eq (d) 0.007808 eq or 7.808 meq

9.53 (a) 150 meq (b) 5.50 meq
(c) 4.56×10^3 meq (d) 450 meq

9.55 (a) 0.4000 M HCl (b) 0.4000 M HBr

9.57 400.0

9.59 (a) 9.115 g of HCl (b) 5.435 g of HNO_3
(c) 0.3678 g of H_2SO_4

9.61 (a) 0.08913 N Na_2CO_3 (b) 4.723 g of Na_2CO_3/L

PRACTICE EXERCISES, CHAPTER 10

1. $^{131}_{53}I \rightarrow {}^{131}_{54}Xe + {}^{0}_{-1}e + {}^{0}_{0}\gamma$

2. $^{239}_{94}Pu \rightarrow {}^{235}_{92}U + {}^{4}_{2}He + {}^{0}_{0}\gamma$

3. 10,000 units

4. 8.5 m

REVIEW EXERCISES, CHAPTER 10

10.11 (a) $^{241}_{94}Pu$ (b) $^{22}_{10}Ne$

10.13 (a) $^{252}_{99}Es \rightarrow {}^{4}_{2}He + {}^{248}_{97}Bk$
(b) $^{28}_{12}Mg \rightarrow {}^{0}_{-1}e + {}^{28}_{13}Al$
(c) $^{20}_{8}O \rightarrow {}^{0}_{-1}e + {}^{20}_{9}F$
(d) $^{251}_{98}Cf \rightarrow {}^{4}_{2}He + {}^{247}_{96}Cm + {}^{0}_{0}\gamma$

10.17 0.750 ng

10.34 15.1 m

10.50 Iron-55. $^{55}_{25}Mn + {}^{1}_{1}H \rightarrow {}^{1}_{0}n + {}^{55}_{26}Fe$

10.52 $^{113}_{49}In \rightarrow {}^{111}_{49}In + 2\,{}^{1}_{0}n$

10.54 $^{67}_{31}Ga + {}^{0}_{-1}e \xrightarrow{\text{electron capture}} {}^{67}_{30}Zn$

10.56 $^{10}_{5}B + {}^{4}_{2}He \rightarrow {}^{13}_{7}N + {}^{1}_{0}n$

10.58 $^{27}_{13}Al + {}^{6}_{3}Li \rightarrow {}^{32}_{15}P + {}^{1}_{1}H$

PRACTICE EXERCISES, CHAPTER 11

1. (a) $CH_3-CH_2-CH_3$ (b) 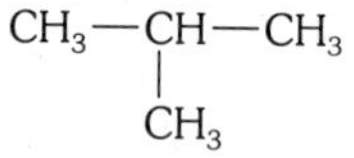 $CH_3-CH-CH_3$ with CH_3 below

(c)

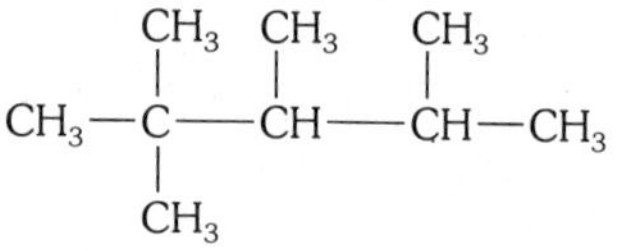

2. (a) $CH_3CH_2CH_3$ (b) 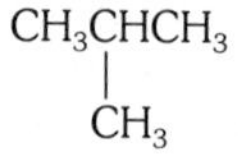 CH_3CHCH_3 with CH_3 below

(c)

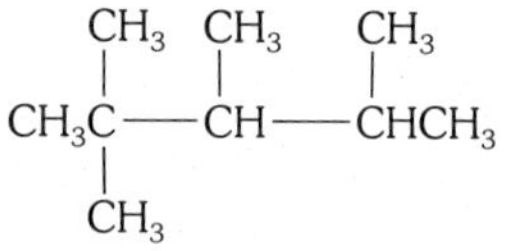

Note that *vertical* bonds are always shown and that sometimes horizontal bonds are left in so that there will be room for other groups.

3. (a)

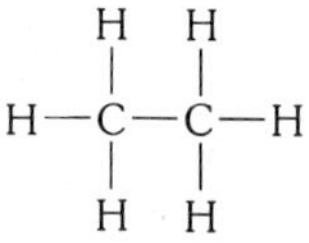

(b)

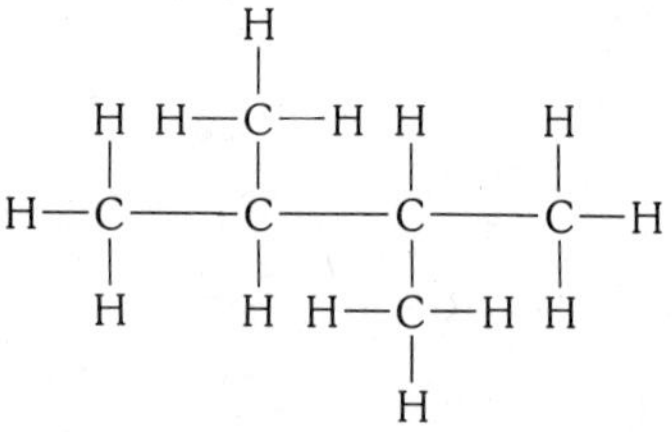

(c)

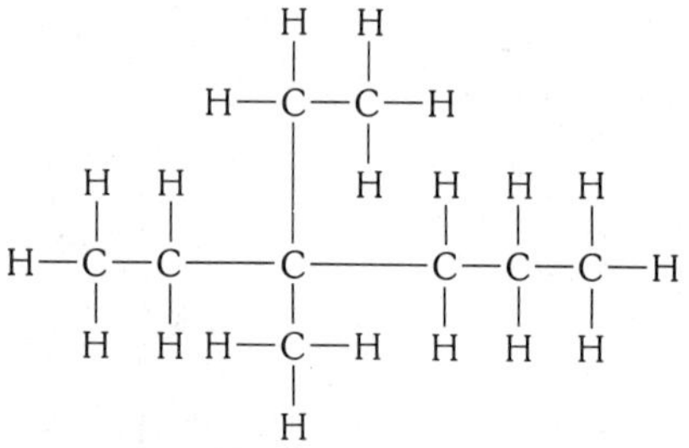

4. Structures (b) and (c) cannot represent real compounds.

5. (a) Identical
(b) Isomers
(c) Identical
(d) Isomers
(e) Different in another way.

PRACTICE EXERCISES, CHAPTER 12

1. (a)

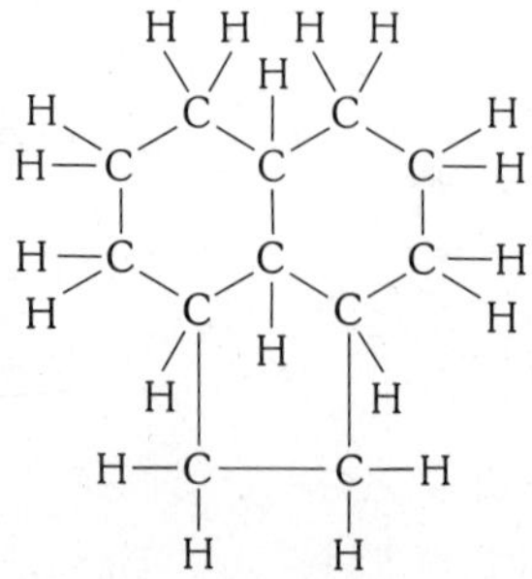

(b)

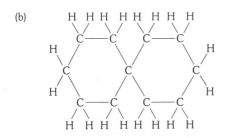

2. CH$_3$CH$_2$— (cyclohexane with HO substituent)—(cyclopentane)

3. The second structure

4. (a) 3-Methylhexane
   (b) 2,3-Dimethyl-4-t-butylheptane
   (c) 2,4-Dimethyl-5-sec-butylnonane
   (d) 1-Chloro-2-bromo-3-iodopropane
   (e) 4-Isopropyl-5-t-butyloctane or 5-isopropyl-4-t-butyloctane
   (f) 1,3-Dinitro-2,2-dimethylpropane
   (g) 1-Chloro-2,3-dimethylpentane
   (h) 2,6-Dimethyl-3-ethylheptane

5. (a) BrCH$_2$CHCH$_2$CH$_2$CH$_3$
            |
            NO$_2$

   (b)   CH$_3$
                          |
         CH$_3$ CH$_3$ CH$_3$ CHCH$_3$
          | | | |
      CH$_3$C — C — C — CHCH$_2$CH$_2$CH$_3$
          | | |
         CH$_3$ CH$_3$ CH$_3$

   (c)   I   CH$_3$ CH$_3$CHCH$_2$CH$_3$
              | | |
      CH$_3$C — CH — CH — CH — CHCH$_2$CH$_2$CH$_3$
              | | |
              I   CHCH$_3$ CH$_3$CCH$_3$
                          | |
                        CH$_3$ CH$_3$

   (d)   CH$_3$
              |
      ClCHCHCH$_3$
         |
         Br

   (e)   CH$_3$CHCH$_2$CH$_3$
                      |
      CH$_3$CH$_2$CH$_2$CH$_2$CCH$_2$CH$_2$CH$_2$CH$_2$CH$_3$
                      |
                  CH$_3$CHCH$_2$CH$_3$

6.
                         ↓
         CH$_3$—(CH)—CH$_2$—CH$_3$

   CH$_3$—CH$_2$—CH$_2$—CH$_2$—C—CH$_2$—CH$_2$—CH$_2$—CH$_2$—CH$_3$
        ↑   ↑   ↑   ↑   ↑   ↑   ↑
                     CH$_3$—(CH)—CH$_2$—CH$_3$
                              ↑

7. (a) Ethyl chloride
   (b) Butyl bromide
   (c) Isobutyl chloride
   (d) t-Butyl bromide

8. (a) Two. Butyl chloride or 1-chlorobutane and sec-butyl chloride or 2-chlorobutane
   (b) Two. Isobutyl chloride or 1-chloro-2-methylpropane and t-butyl chloride or 2-chloro-2-methylpropane

PRACTICE EXERCISES, CHAPTER 13

1. (a)   CH$_3$CH$_2$ CH$_3$ CH$_3$CH$_2$ H
                  \   /   \   /
                   C==C   and   C==C
                  /   \   /   \
            H$_3$C   H   H$_3$C   CH$_3$

   (b)   Cl   Cl   H   Cl
          \   /   \   /
           C==C   and   C==C
          /   \   /   \
         H   H   Cl   H

   (c) None
   (d) None

2. (a) 2-Methylpropene   (b) 3,6-Dimethyl-4-isobutyl-3-heptene
   (c) 1-Chloropropene   (d) 3-Bromopropene
   (e) 4-Methyl-1-hexene   (f) 4-Methylcyclohexene

3. (a)   CH$_3$
                    |
       CH$_3$CH==CHCHCH$_3$

   (b) CH$_2$==CHCHCH$_2$CH$_2$CH$_2$CH$_3$
                 |
              CH$_2$CH$_2$CH$_3$

   (c)   CH$_3$
                    |
       CH$_2$==CHCCH$_2$Cl
                    |
                 CH$_3$

   (d)   CH$_3$
              |
       CH$_3$C==CCH$_3$
                |
              CH$_3$

4. (a) CH$_3$CH$_2$CH$_3$
   (b) No reaction
   (c) (cyclohexane)—CH$_3$
   (d) CH$_3$(CH$_2$)$_{16}$CO$_2$H

5. (a)   CH$_3$
              |
       CH$_3$CCH$_2$Br
              |
              Br
   (b) No reaction
   (c) ClCH$_2$CHCH$_2$CH$_3$
                |
                Cl
   (d) CH$_3$CH$_2$CH$_2$CH$_3$

6. (a) 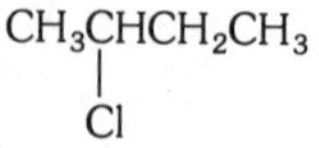 $CH_3CHCH_2CH_3$
         |
         Cl

   (b) 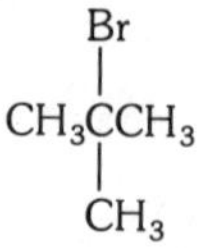 CH_3CCH_3 (with Br above and CH_3 below the central C)

   (c) CH_3CH_2C— (cyclohexyl), with CH_3 above and OH below

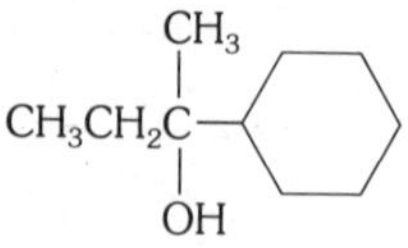

   (d) 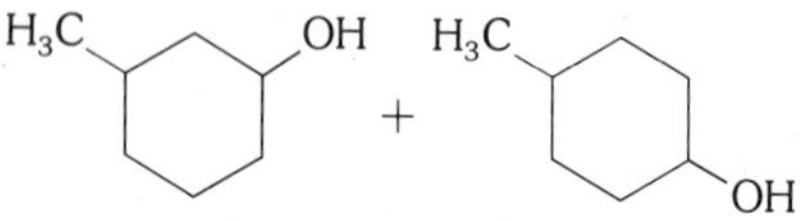 +

   (e) No reaction

7. (a) $CH_3CH_2\overset{+}{C}HCH_3$, not $CH_3CH_2CH_2\overset{+}{C}H_2$
      (Preferred)

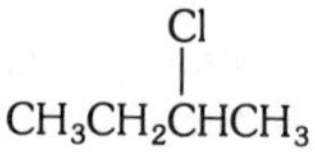

      Cl
      |
      $CH_3CH_2CHCH_3$
      (Product)

   (b) 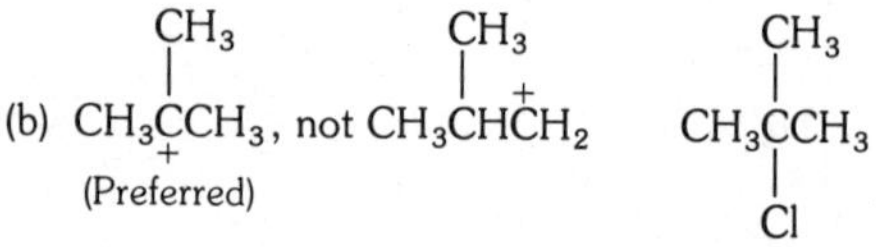 $CH_3\overset{CH_3}{\underset{+}{C}}CH_3$, not $CH_3CH\overset{+}{C}H_2$ (with CH₃) $CH_3\overset{CH_3}{\underset{Cl}{C}}CH_3$
      (Preferred)   (Product)

   (c) (cyclohexyl cation with CH₃, Preferred)   not   (cyclohexyl with CH₃ and +)   (cyclohexyl with Cl and CH₃, Product)

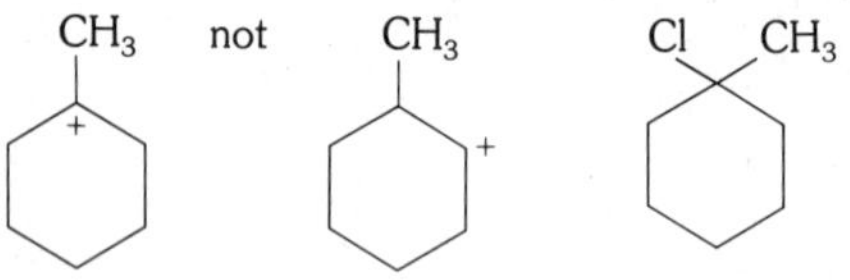

      (Preferred)   (Product)

   (d) $CH_3\overset{+}{C}HCH_2CH_3$ (the only cation) $CH_3CHCH_2CH_3$
                                                                     |
                                                                     Cl

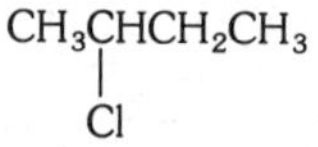

   (e) $CH_3\overset{+}{C}HCH_2CH_2CH_3$ and $CH_3CH_2\overset{+}{C}HCH_2CH_3$ 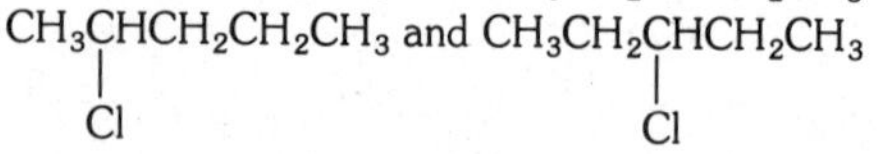
      $CH_3CHCH_2CH_2CH_3$ and $CH_3CH_2CHCH_2CH_3$
           | |
           Cl   Cl
      (Both are equally stable.)

   (f) 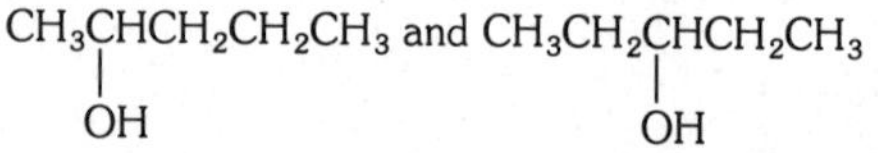 $CH_3CHCH_2CH_2CH_3$ and $CH_3CH_2CHCH_2CH_3$
           | |
           OH   OH

      The two possible cations, $CH_3\overset{+}{C}HCH_2CH_2CH_3$ and
      $CH_3CH_2\overset{+}{C}HCH_2CH_3$ are about equally stable, so both can
      and do form. This leads to the mixture of alcohols.

PRACTICE EXERCISES, CHAPTER 14

1. (a) Alcohol   (b) Phenol
   (c) Carboxylic acid   (d) Alcohol
   (e) Alcohol   (f) Alcohol

2. (a) Monohydric, secondary
   (b) Monohydric, secondary
   (c) Dihydric, unstable
   (d) Dihydric
   (e) Monohydric, primary
   (f) Monohydric, primary
   (g) Monohydric, tertiary
   (h) Monohydric, secondary
   (i) Trihydric, unstable

3. (a) 4-Methyl-1-pentanol
   (b) 2-Methyl-2-propanol
   (c) 2-Methyl-2-ethyl-1-pentanol
   (d) 2-Methyl-1,3-propanediol

4. In 1,2-propanediol. Its boiling point is 189 °C, much higher than
   that of 1-butanol (b.p. 117 °C).

5. (a) $CH_3CH=CH_2$ (b) $CH_3CH=CH_2$

   (c) 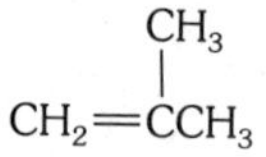 $CH_2=\overset{CH_3}{\underset{}{C}}CH_3$ (d)

6. (a) 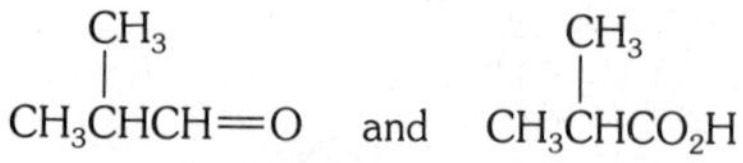 $CH_3\overset{CH_3}{\underset{}{C}}HCH=O$ and $CH_3\overset{CH_3}{\underset{}{C}}HCO_2H$

   (b) 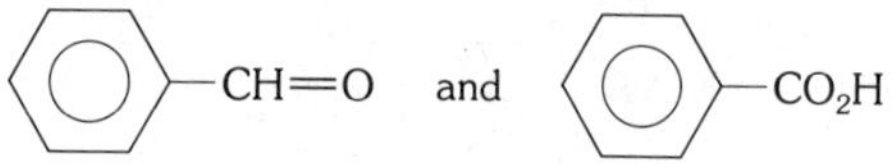 (phenyl)—CH=O and (phenyl)—CO₂H

   (c) $CH_2=O$ and HCO_2H

7. (a) 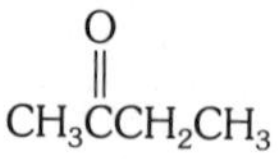 $CH_3\overset{O}{\overset{\|}{C}}CH_2CH_3$ (b) (phenyl)—$\overset{O}{\overset{\|}{C}}CH_3$

   (c) (cyclopentanone)=O

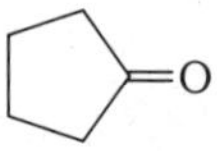

8. (a) 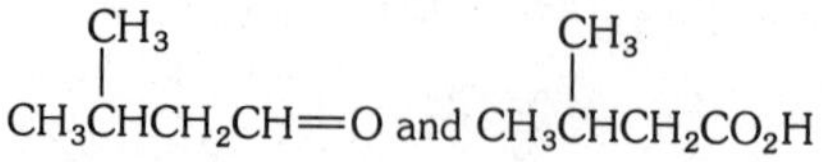 $CH_3\overset{CH_3}{\underset{}{C}}HCH_2CH=O$ and $CH_3\overset{CH_3}{\underset{}{C}}HCH_2CO_2H$
   (b) No reaction
   (c) 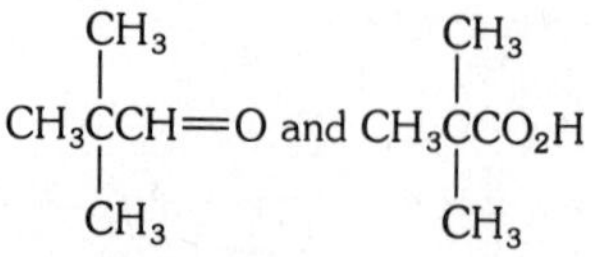 $CH_3\overset{CH_3}{\underset{CH_3}{C}}CH=O$ and $CH_3\overset{CH_3}{\underset{CH_3}{C}}CO_2H$

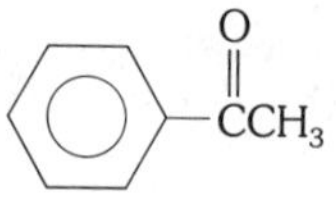

   (d) 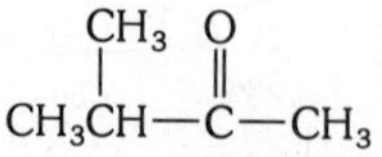 $CH_3\overset{CH_3}{\underset{}{C}}H—\overset{O}{\overset{\|}{C}}—CH_3$

9. (a) $2CH_3SH$
   (b) $(CH_3)_2CH—S—S—CH(CH_3)_2$
   (c) $HSCH_2CH_2CH_2CH_2SH$
   (d) 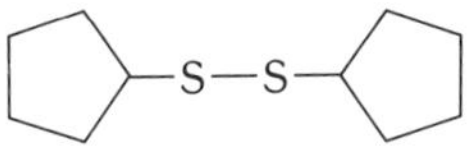

10. (a) $CH_3—O—CH_3$
    (b) $CH_3CH_2CH_2—O—CH_2CH_2CH_3$
    (c) 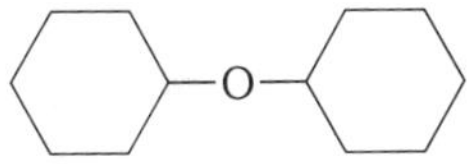

PRACTICE EXERCISES, CHAPTER 15

1. (a) 2-Methylpropanal
   (b) 3-Bromobutanal
   (c) 2,4,6-Trimethyl-4-ethylheptanal
2. 2-Isopropylpropanal would have the structure:

   $CH_3CHCH{=}O$
   |
   CH_3CHCH_3

   and it should be named 2,3-dimethylbutanal.
3. (a) 2-Butanone
   (b) 6-Methyl-2-heptanone
   (c) 2-Methylcyclohexanone
4. (a)
   O
   ‖
   $CH_3CH_2CCHCH_3$
   |
   CH_3

   (b) 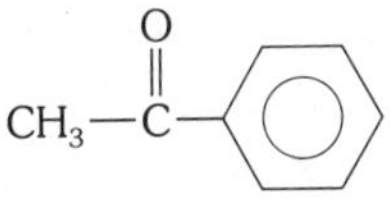

   (c)
   O
   ‖
   $CH_3CH_2CH_2CCH_2CH_2CH_3$

   (d)
   CH_3 O CH_3
   | ‖ |
   $CH_3C—C—CCH_3$
   | |
   CH_3 CH_3
5. (a)
   OH
   |
   $CH_3CHCH_2CH_3$

   (b)
   CH_3
   |
   $CH_3CHCH_2CH_2OH$

   (c) 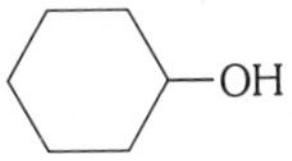
6. (a) Not a hemiacetal (b) Not a hemiacetal
   (c) $HO—CH_2—O—CH_2CH_3$
   ↑
   (d) 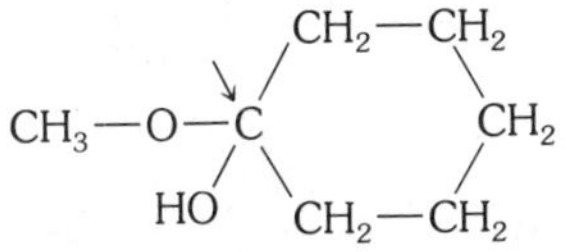

7. (a)
   CH_3CHOCH_3
   |
   OH

   (b)
   $CH_3CH_2CH_2CH—O—CH_2CH_3$
   |
   OH

   (c) 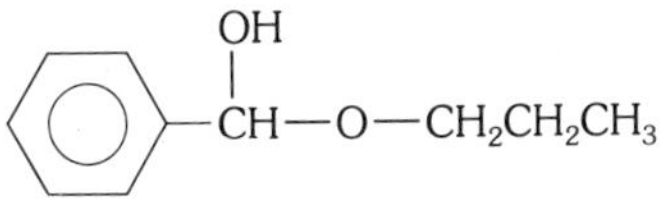

   (d) $HO—CH_2—O—CH_3$
8. (a) $CH_3CH_2CH{=}O + HOCH_3$
   (b) $CH_3CH_2OH + O{=}CHCH_2CH_3$
9. (a) Not a hemiketal
   (b)
   CH_3
   | ↙
   $HO—C—O—CH_2CH_2CH_3$
   |
   CH_3
10.
    OCH_3
    |
    CH_3CCH_3
    |
    OH
11.
    O
    ‖
    $CH_3CH_2CH_2CCH_2CH_3 + HOCH_2CH_3$
12. (a) Neither
    (b) A ketal,

    CH_3
    | ↙
    $CH_3CH_2—O—C—O—CH_2CH_3$
    |
    CH_3
13. (a) $2CH_3OH + CH_2{=}O$
    (b) No reaction
    (c)
    CH_3
    |
    $CH_3CHC{=}O + 2\ HOCH_3$
    |
    CH_3

PRACTICE EXERCISES, CHAPTER 16

1. (a) 2,2-Dimethylpropanoic acid
   (b) 3-Methyl-5-ethyl-5-isopropyloctanoic acid
   (c) Sodium ethanoate
   (d) Sodium 5-chloro-3-methylheptanoate
2. Pentanedioic acid
3. 9-Octadecenoic acid
4. (a) $CH_3CH_2CO_2^-$
   (b) 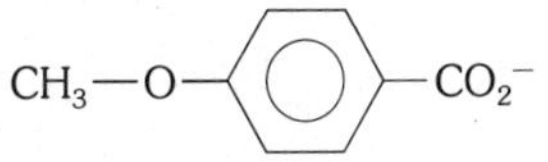
   (c) $CH_3CH{=}CHCO_2^-$

5. (a) 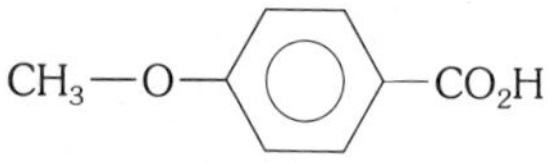 CH_3-O-⟨benzene ring⟩$-CO_2H$

(b) $CH_3CH_2CO_2H$
(c) $CH_3CH=CHCO_2H$

6. (a) $CH_3CO_2CH_3$
(b) $CH_3CO_2CH_2CH_2CH_3$
(c) 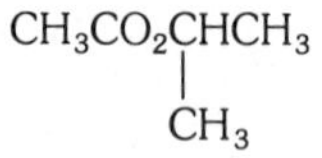 $CH_3CO_2CHCH_3$
   $\quad\quad\quad\quad |$
   $\quad\quad\quad CH_3$

7. (a) $HCO_2CH_2CH_3$
(b) $CH_3CH_2CO_2CH_2CH_3$
(c) ⟨benzene ring⟩$-CO_2CH_2CH_3$

8. (a) Methyl propanoate
(b) Propyl 3-methylpentanoate

9. (a) t-Butyl acetate
(b) Ethyl butyrate

10. (a) $CH_3CO_2H + CH_3OH$
(b) 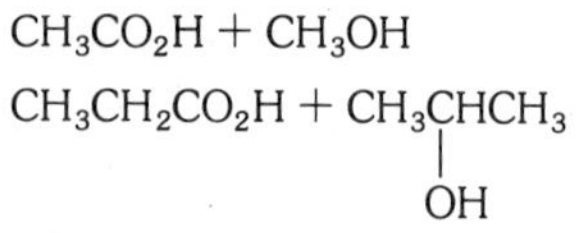 $CH_3CH_2CO_2H + CH_3CHCH_3$
    $\quad\quad\quad\quad\quad\quad\quad\quad\quad |$
    $\quad\quad\quad\quad\quad\quad\quad\quad\quad OH$

   $\quad\quad\quad\quad CH_3$
   $\quad\quad\quad\quad |$
(c) 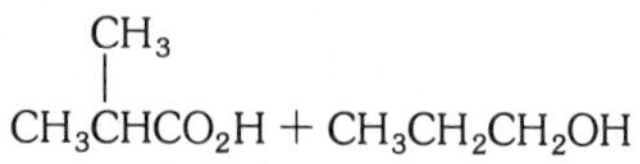 $CH_3CHCO_2H + CH_3CH_2CH_2OH$

11. (a) 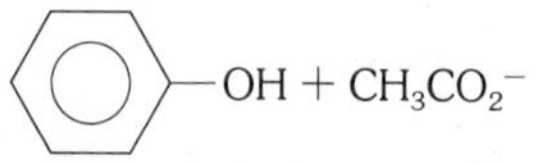 ⟨benzene ring⟩$-OH + CH_3CO_2^-$

(b) 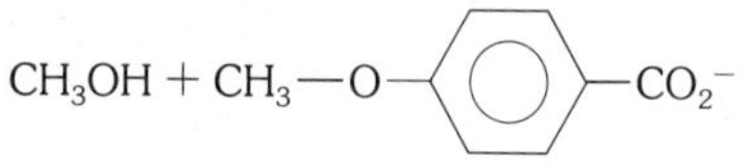 $CH_3OH + CH_3-O-$⟨benzene ring⟩$-CO_2^-$

PRACTICE EXERCISES, CHAPTER 17

1. (a) Dimethylisopropylamine
(b) Cyclohexylamine
(c) Isobutyl-t-butylamine

2. (a)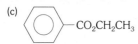

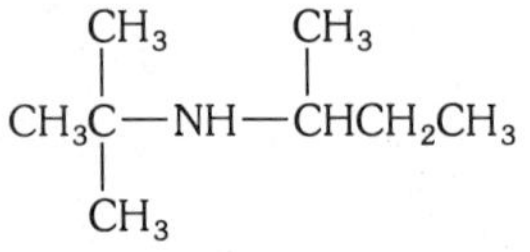

(b) 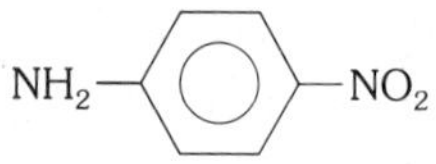 NH_2-⟨benzene ring⟩$-NO_2$

(c) 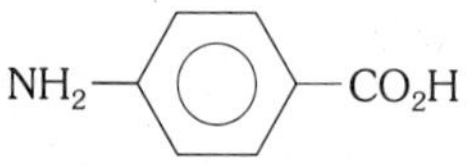 NH_2-⟨benzene ring⟩$-CO_2H$

3. (a) 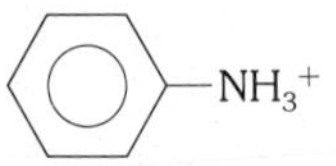 ⟨benzene ring⟩$-NH_3^+$

(b) 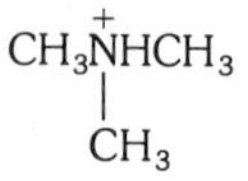 $CH_3\overset{+}{N}HCH_3$
    $\quad\quad |$
    $\quad\quad CH_3$

(c) $^+NH_3CH_2CH_2NH_3^+$

4. (a) 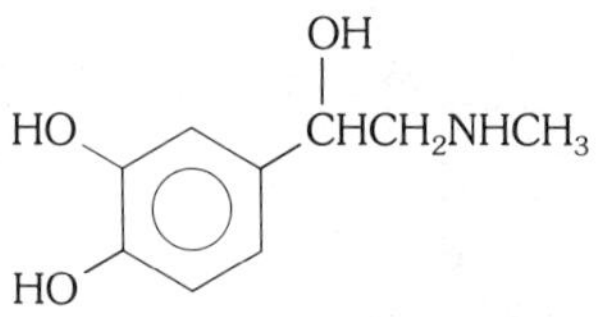
$HO-$⟨benzene ring, HO⟩$-\overset{\displaystyle OH}{\overset{|}{C}}HCH_2NHCH_3$

(b) 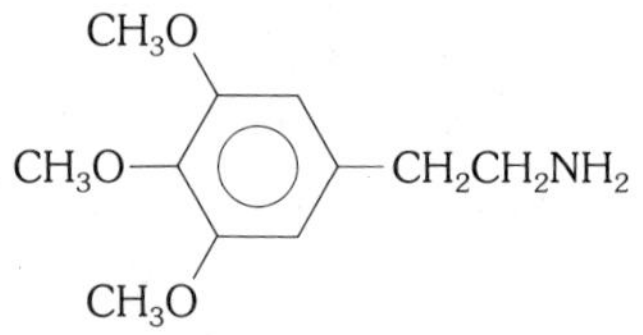
CH_3O-⟨benzene ring, CH_3O, CH_3O⟩$-CH_2CH_2NH_2$

5. (a) 4-Methylhexanamide
(b) 2-Ethylbutanamide

6. (a)  CH_3
      $\quad |$
   $CH_3CHCONHCH_3$

(b) 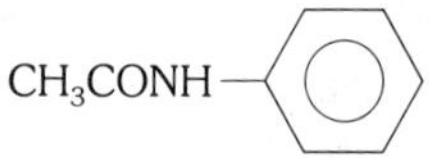 CH_3CONH-⟨benzene ring⟩

(c) No amide can form.   (d) No amide can form.

7. (a) 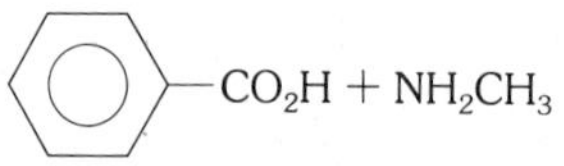 ⟨benzene ring⟩$-CO_2H + NH_2CH_3$

(b) No hydrolysis can occur.

(c) 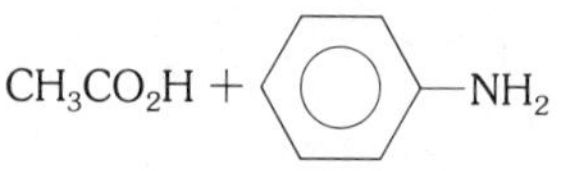 $CH_3CO_2H + $⟨benzene ring⟩$-NH_2$

(d) $2CH_3CO_2H + NH_2CH_2CH_2NH_2$

8. 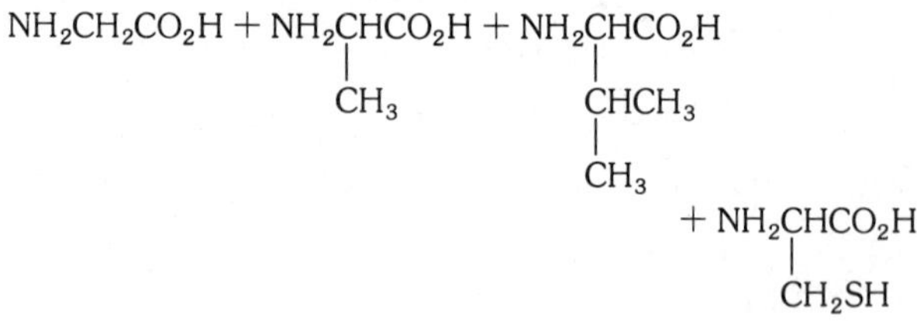 $NH_2CH_2CO_2H + NH_2CHCO_2H + NH_2CHCO_2H$
   $\quad\quad\quad\quad\quad\quad\quad\quad\quad\quad\quad |\quad\quad\quad\quad\quad\quad\quad\quad |$
   $\quad\quad\quad\quad\quad\quad\quad\quad\quad\quad\quad CH_3\quad\quad\quad\quad\quad CHCH_3$
   $\quad\quad\quad\quad\quad\quad\quad\quad\quad\quad\quad\quad\quad\quad\quad\quad\quad\quad\quad\quad |$
   $\quad\quad\quad\quad\quad\quad\quad\quad\quad\quad\quad\quad\quad\quad\quad\quad\quad\quad\quad\quad CH_3$
   $\quad\quad\quad\quad\quad\quad\quad\quad\quad\quad\quad\quad\quad\quad\quad + NH_2CHCO_2H$
   $\quad\quad\quad\quad\quad\quad\quad\quad\quad\quad\quad\quad\quad\quad\quad\quad\quad\quad |$
   $\quad\quad\quad\quad\quad\quad\quad\quad\quad\quad\quad\quad\quad\quad\quad\quad\quad\quad CH_2SH$

PRACTICE EXERCISES, CHAPTER 18

1. (a) 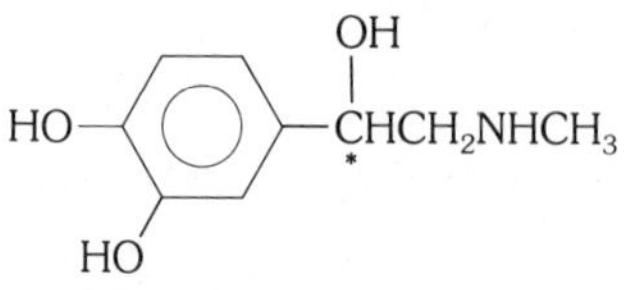
$HO-$⟨benzene ring, HO⟩$-\overset{\displaystyle OH}{\overset{|}{\underset{*}{C}}}HCH_2NHCH_3$

(b) $CH_3\overset{*}{C}HCO_2H$
    $\quad\quad\quad |$
    $\quad\quad\quad OH$

(c) $CH_3\overset{*}{C}H\overset{*}{C}HCO_2^-$
    $\quad\quad\quad |\quad\quad |$
    $\quad\quad\quad HO\quad NH_3^+$

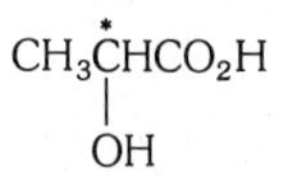

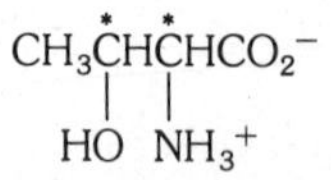

(d) HOCH$_2$ĊH—ĊH—ĊH—CH=O
           | | |
           OH OH OH

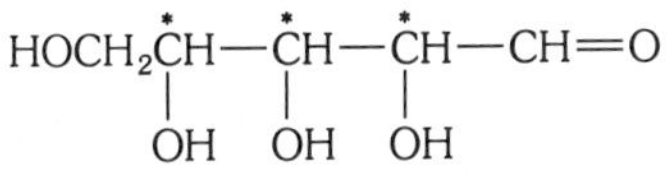

2. (a) Three
   (b) $2^3 = 8$ optical isomers
   (c) Four pairs of enantiomers

3. CH$_3$—ĊH—ĊH—CH$_3$
        | |
       OH OH

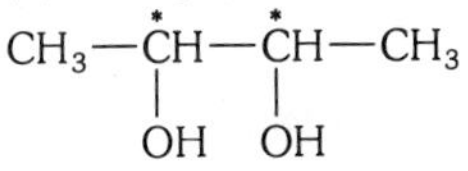

No, they are not different. Each chiral carbon holds the identical set of four different groups.

REVIEW EXERCISES, CHAPTER 18

18.19 3.01 g/dL

18.21 $[\alpha] = -139°$. The unknown is strychnine.

PRACTICE EXERCISES, CHAPTER 19

1. (a) = (b)
   (c) = (e) = a meso compound.
   (a) and (d) are enantiomers.

2.
    CO$_2$H   CO$_2$H

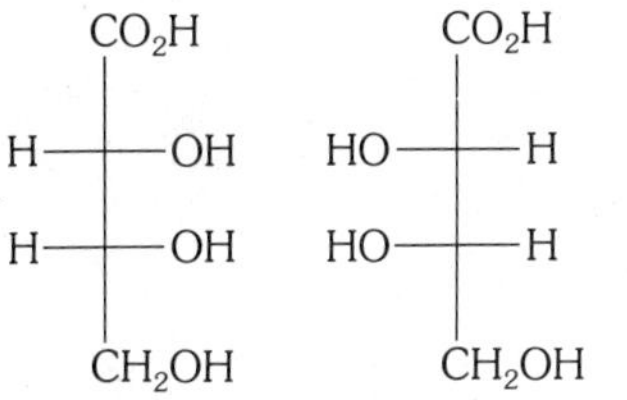

  H——OH   HO——H

  H——OH   HO——H

    CH$_2$OH   CH$_2$OH
    One pair of enantiomers

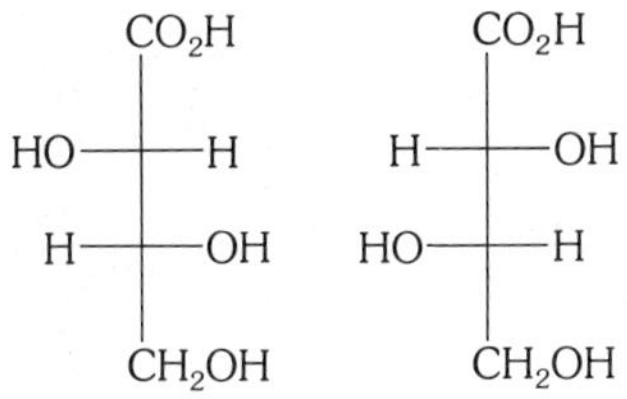

    CO$_2$H   CO$_2$H

  HO——H   H——OH

  H——OH   HO——H

    CH$_2$OH   CH$_2$OH
    Another pair of enantiomers

PRACTICE EXERCISES, CHAPTER 20

1. CH$_3$(CH$_2$)$_{26}$ C̈O(CH$_2$)$_{25}$CH$_3$
               (O, double bond above CO)

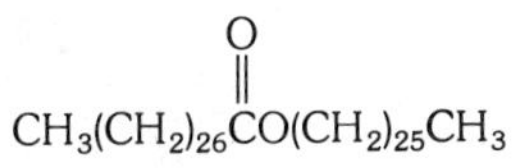

2. CH$_3$(CH$_2$)$_7$ (CH$_2$)$_7$CO$_2$H
        \C=C/
      /   \
     H   H

3. **1** + 3NaOH →
   HOCH$_2$CHCH$_2$OH + NaO$_2$C(CH$_2$)$_7$CH=CH(CH$_2$)$_7$CH$_3$
       | + NaO$_2$C(CH$_2$)$_{14}$CH$_3$
      OH
   + NaO$_2$C(CH$_2$)$_7$CH=CHCH$_2$CH=CH(CH$_2$)$_4$CH$_3$

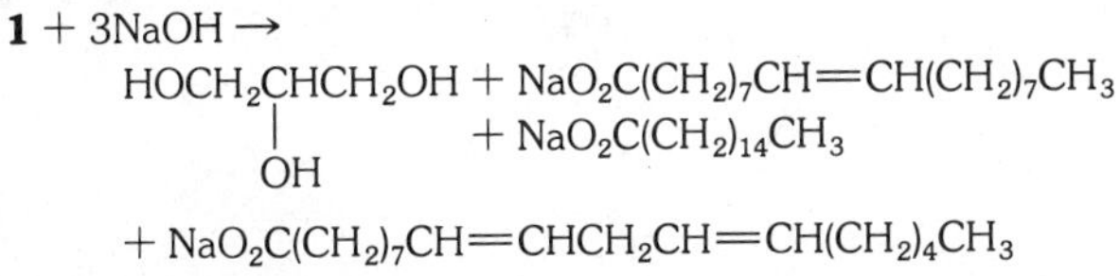

4. **1** + 3H$_2$ $\xrightarrow[\text{Heat, pressure}]{\text{Catalyst}}$ CH$_2$—O—C(CH$_2$)$_{16}$CH$_3$ (=O above C)
                   |
              CH—O—C(CH$_2$)$_{14}$CH$_3$ (=O above C)
               |
             CH$_2$—O—C(CH$_2$)$_{16}$CH$_3$ (=O above C)

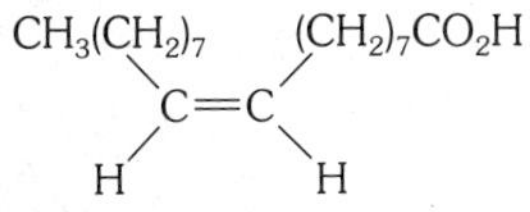

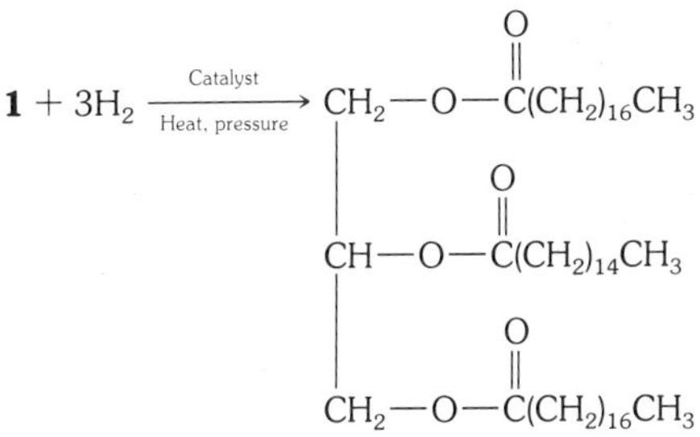

PRACTICE EXERCISES, CHAPTER 21

1. Glycine: $^+$NH$_3$CH$_2$CO$_2$$^-$
   Alanine: $^+$NH$_3$CHCO$_2$$^-$
             |
            CH$_3$
   Lysine: $^+$NH$_3$CHCO$_2$$^-$
            |
           (CH$_2$)$_4$
            |
           NH$_2$
   Glutamic acid: $^+$NH$_3$CHCO$_2$$^-$
                  |
                (CH$_2$)$_2$
                |
                CO$_2$H

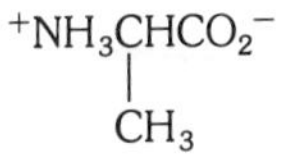

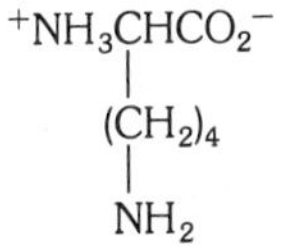

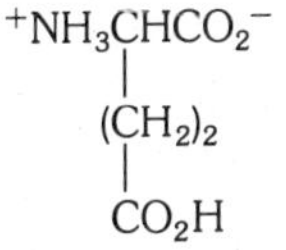

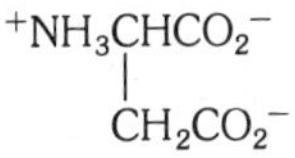

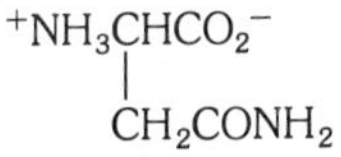

2. (a) $^+$NH$_3$CHCO$_2$$^-$
            |
          CH$_2$CO$_2$$^-$
   (b) $^+$NH$_3$CHCO$_2$$^-$
            |
          CH$_2$CONH$_2$

3. $^+$NH$_3$CHCO$_2$$^-$ NH$_2$$^+$
     | ‖
   CH$_2$CH$_2$CH$_2$NHCNH$_2$

4. Hydrophilic; neutral side chain.

5. $^+$NH$_3$CHCONHCHCO$_2$$^-$ $^+$NH$_3$CHCONHCHCO$_2$$^-$
     | | | |
    CH$_3$ CH$_2$CH$_2$CO$_2$H   CH$_2$ CH$_3$
                         |
                       CH$_2$CO$_2$H

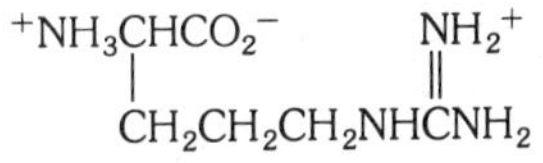

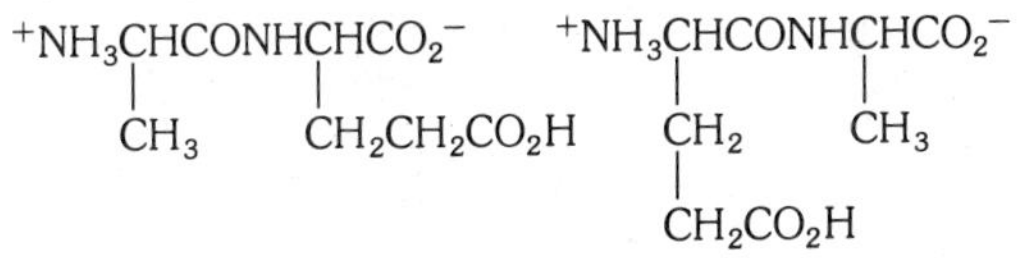

PRACTICE EXERCISES, CHAPTER 22

1. (a) Proline   (b) Arginine
   (c) Glutamic acid (d) Lysine

2. (a) Serine   (b) CT (chain termination)
   (c) Glutamic acid (d) Isoleucine

PRACTICE EXERCISE, CHAPTER 26

1. (a) Yes   (b) Yes   (c) No

Glossary

Absolute Configuration The actual arrangement in space about each chiral center in a molecule.

Acetyl Coenzyme A The molecule from which acetyl groups are transferred into the citric acid cycle or into the lipigenesis cycle.

$$
\underset{\text{CH}_3\text{C}-\text{S}-\text{CoA}}{\overset{\overset{\text{O}}{\|}}{}}
$$

Accuracy In science, the degree of conformity to some accepted standard or reference; freedom from error or mistake; correctness.

Acetal Any organic compound in which two ether linkages extend from one CH unit, as in:

$$
\underset{\text{R}-\text{O}-\text{CH}-\text{O}-\text{R}'}{\overset{|}{}}
$$

Achiral Not possessing chirality; that quality of a molecule (or other object) that allows it to be superimposed on its mirror image.

Acid *Arrhenius theory:* Any substance that makes H$^+$ available in water. *Brønsted theory:* Any substance that can donate a proton (H$^+$).

Acid-Base Neutralization The reaction of an acid with a base.

Acid Anhydride In organic chemistry, a compound formed by splitting water out between two —OH groups of the acid function of an organic acid. The structural features are

$$
\underset{\text{Carboxylic acid}}{\overset{\overset{\text{O} \quad\quad \text{O}}{\| \quad\quad \|}}{-\text{C}-\text{O}-\text{C}-}} \qquad \underset{\substack{\text{Phosphoric acid} \\ \text{anhydride system}}}{\overset{\overset{\text{O} \quad\quad\ \text{O}}{\| \quad\quad\ \|}}{-\text{P}-\text{O}-\text{P}-}}
$$

Carboxylic acid anhydride system   Phosphoric acid anhydride system

Acid Chloride A derivative of an acid in which the —OH group of the acid has been replaced by —Cl.

$$
\underset{\text{R}-\text{C}-\text{Cl}}{\overset{\overset{\text{O}}{\|}}{}}
$$

Acid Derivative Any organic compound that can be made from an organic acid or that can be changed back to the acid by hydrolysis. (Examples are acid chlorides, acid anhydrides, esters, and amides.)

Acid Ionization Constant (K_a) A modified equilibrium constant for the following equilibrium: $HA + H_2O \rightleftharpoons H_3O^+ + A^-$

$$
K_a = \frac{[H_3O^+][A^-]}{[HA]}
$$

Acid-Base Indicator (See *Indicator.*)

Acidic Solution A solution in which the molar concentration of hydronium ions is greater than that of hydroxide ions.

Acidosis A condition in which the pH of the blood is below normal. *Metabolic acidosis* is brought on by a defect in some metabolic pathway. *Respiratory acidosis* is caused by a defect in the respiratory centers or in the mechanisms of breathing.

Active Site That region of an enzyme molecule most directly responsible for the catalytic effect of the enzyme.

Active Transport The movement of a substance through a biological membrane against a concentration gradient and caused by energy-consuming chemical changes that involve parts of the membrane.

Activity Series A list of elements (or other substances) in the order of the ease with which they release electrons under standard conditions and become oxidized.

Acyl Group

$$
\underset{\text{R}-\text{C}-}{\overset{\overset{\text{O}}{\|}}{}}
$$

Acyl Group Transfer Reaction Any reaction in which an acyl group transfers from a donor to an acceptor.

Addition Reaction Any reaction in which two parts of a reactant molecule add to a double or a triple bond.

Adenosine Diphosphate (ADP) A high-energy diphosphate ester obtained from adenosine triphosphate (ATP) when part of the chemical energy in ATP is tapped for some purpose in a cell.

Adenosine Monophosphate (AMP) A low-energy phosphate ester that can be obtained by the hydrolysis of ATP or ADP; a monomer for the biosynthesis of nucleic acids.

Adenosine Triphosphate (ATP) A high-energy triphosphate ester used in living systems to provide chemical energy for metabolic needs.

Adequate Protein A protein that, when digested, makes available all of the essential amino acids in suitable proportions to satisfy both the amino acid and total nitrogen requirements of good nutrition without providing excessive calories.

ADP (See *Adenosine Diphosphate*)

Aerobic Sequence An oxygen-consuming sequence of catabolism that starts with glucose or with glucose units in glycogen, and proceeds through glycolysis, the citric acid cycle, and the respiratory chain.

Agonist A compound whose molecules can bind to a receptor on a cell membrane and cause a response by the cell.

Albumin One of a family of globular proteins that tend to dissolve in water, and that in blood contribute to the blood's colloidal osmotic pressure and aid in the transport of metal ions, fatty acids, cholesterol, triacylglycerols, and other water-insoluble substances.

Alcohol Any organic compound whose molecules have the —OH group attached to a saturated carbon; R—OH.

Alcohol Group The —OH group when it is joined to a saturated carbon.

Aldehyde An organic compound that has a carbonyl group joined to H on one side and C on the other; R—CH=O.

Aldehyde Group —CH=O

Aldohexose A monosaccharide whose molecules have six carbon atoms and an aldehyde group.

Aldose A monosaccharide whose molecules have an aldehyde group.

Aldosterone A steroid hormone, made in the adrenal cortex, secreted into the bloodstream when the sodium ion level is low, and that signals the kidneys to leave sodium ions in the bloodstream.

Aliphatic Compound Any organic compound whose molecules lack a benzene ring or a similar structural feature.

Alkali A strongly basic substance such as sodium hydroxide or potassium hydroxide.

Alkali Metal The elements of Group IA of the periodic table — lithium, sodium, potassium, rubidium, cesium, and francium.

Alkaline Earth Metals The elements of Group IIA of the periodic table — beryllium, magnesium, calcium, strontium, barium, and radium.

Alkaloid A physiologically active, heterocyclic amine isolated from plants.

Alkalosis A condition in which the pH of the blood is above normal. *Metabolic alkalosis* is caused by a defect in metabolism. *Respiratory alkalosis* is caused by a defect in the respiratory centers of the brain or in the apparatus of breathing.

Alkane Any saturated hydrocarbon, one that has only single bonds. A *normal* alkane is any whose molecules have straight chains.

Alkene Any hydrocarbon whose molecules have one or more double bonds.

Alkyl Group A substituent group that is an alkane minus one H atom.

Allosteric Activation The activation of an enzyme's catalytic site by the binding of some molecule at a position elsewhere on the enzyme.

Allosteric Inhibition The inhibition of the activity of an enzyme caused by the binding of an inhibitor molecule at some site other than the enzyme's catalytic site.

Alloy A mixture of two or more metals made by mixing them in their molten states.

Alpha (α) Particle The nucleus of a helium atom; $^4_2$He.

Alpha (α) Radiation A stream of high-energy alpha particles.

Amide Any organic compound whose molecules have a carbonyl-nitrogen unit,

$$\underset{\underset{|}{C}-\underset{|}{N}-}{\overset{\overset{O}{\|}}{}}$$

Amide Bond The single bond that holds the carbonyl group to the nitrogen atom in an amide.

Amine Any organic compound whose molecules have a trivalent nitrogen atom, as in R—NH$_2$, R—NH—R, or R$_3$N.

Amine Salt Any organic compound whose molecules have a positively charged, tetravalent, protonated nitrogen atom, as in RNH$_3^+$, R$_2$NH$_2^+$, or R$_3$NH$^+$.

Amino Acid Any organic compound whose molecules have both an amino group and a carboxyl group.

Amino Acid Residue A structural unit in a polypeptide, —NH—CH—CO—, furnished by an amino acid, where G is a side-chain group.

Aminopeptidase An enzyme that catalyzes the hydrolysis of N-terminal amino acid residues from small polypeptides.

AMP (See *Adenosine Monophosphate*.)

Amphipathic Compound A substance whose molecules have both hydrophilic and hydrophobic groups.

α-Amylase An enzyme that catalyzes the hydrolysis of amylose.

Amylopectin A polymer found in starch in which linear amylose chains have joined to each other by $\alpha(1 \rightarrow 6)$ glycosidic bonds to give a branched polymer of α-glucose.

Amylose A linear polymer of glucose in which glucose units are joined by $\alpha(1 \rightarrow 4)$ glycosidic bonds.

Anaerobic Sequence The oxygen-independent catabolism of glucose or of glucose units in glycogen to lactate ion.

Anhydrous Without water.

Anion A negatively charged ion.

Anion Gap

$$\text{anion gap} = \frac{\text{meq of Na}^+}{L} - \left(\frac{\text{meq of Cl}^-}{L} + \frac{\text{meq of HCO}_3^-}{L} \right)$$

Anode The positive electrode to which negatively charged ions — anions — are attracted during electrolysis.

Anoxia A condition of a tissue in which it receives no oxygen.

Antagonist A compound that can bind to a membrane receptor but not cause any response by the cell.

Anticodon A sequence of three adjacent side-chain bases on a molecule of *t*RNA that is complementary to a codon and that fits to its codon on an *m*RNA chain during polypeptide synthesis.

Antimetabolite A substance that inhibits the growth of bacteria.

Apoenzyme The wholly polypeptide part of an enzyme.

Aromatic Compound Any organic compound whose molecules have a benzene ring (or a feature very similar to this).

Arrhenius Theory Acids in water release hydrogen ions. Bases in water release hydroxide ions.

Atmosphere, Standard (See *Standard Atmosphere*.)

Atom A small particle with one nucleus and zero charge; the smallest particle of a given element that bears the chemical properties of the element.

Atomic Mass Number (See *Mass Number*.)

Atomic Mass Unit (amu) 1.66606×10^{-24} g. A mass very close to that of a proton or a neutron.

Atomic Number The positive charge on an atom's nucleus; the number of protons in an atom's nucleus.

Atomic Orbital A region in space close to an atom's nucleus in which one or two electrons can reside.

Atomic Weight The average mass, in amu, of the atoms of the isotopes of a given element as they occur naturally.

ATP (See *Adenosine Triphosphate*.)

Aufbau Rules The rules for constructing electron configurations.

Avogadro's Law Equal volumes of gases contain equal numbers of moles when they are compared at identical temperatures and pressures.

Avogadro's Number 6.023×10^{23}. The number of formula units in one mole of any element or compound.

Background Radiation Cosmic rays plus the natural atomic radiation emitted by the traces of radioactive isotopes in soils and rocks.

Balanced Equation (See *Equation, Balanced*.)

Barometer An instrument for measuring atmospheric pressure.

Basal Activities The minimum activities of the body needed to maintain muscle tone, control body temperature, circulate the blood, handle wastes, breathe, and carry out other essential activities.

Basal Metabolic Rate The rate at which energy is expended to maintain basal activities.

Basal Metabolism The total of all of the chemical reactions that support basal activities.

Base *Arrhenius Theory:* A compound that provides the hydroxide ion. *Brønsted theory:* A proton-acceptor; a compound that neutralizes hydrogen ions.

Base, Heterocyclic A heterocyclic amine obtained from the hydrolysis of nucleic acids: adenine, thymine, guanine, cytosine, or uracil.

Base Ionization Constant (K_b) For the equilibrium (where B is some base) $B + H_2O \rightleftharpoons BH^+ + OH^-$

$$K_b = \frac{[BH^+][OH^-]}{[B]}$$

Base Pairing In nucleic acid chemistry, the association by means of hydrogen bonds of two heterocyclic, side-chain bases — adenine with thymine (or uracil) and guanine with cytosine.

Base Quantity A fundamental quantity of physical measurement such as mass, length, and time; a quantity used to define derived quantities such as mass/volume for density.

Base Unit A fundamental unit of measurement for a base quantity — such as the kilogram for mass, the meter for length, the second for time, the kelvin for temperature degree, and the mole for quantity of chemical substance; a unit to which derived units of measurement are related.

Basic Solution A solution in which the molar concentration of hydroxide ions is greater than that of hydronium ions.

Becquerel (Bq) The SI unit for the activity of a radioactive source; one nuclear disintegration (or other transformation) per second.

$$1 \text{ curie} = 3.7 \times 10^{10} \text{ Bq}$$

Benedict's Reagent A solution of copper(II) sulfate, sodium citrate, and sodium carbonate that is used in the Benedict's test.

Benedict's Test The use of Benedict's reagent to detect the presence of any compound whose molecules have easily oxidized functional groups — α-hydroxyaldehydes and α-hydroxyketones — such as those present in monosaccharides. In a positive test the intensely blue color of the reagent disappears and a reddish precipitate of copper(I) oxide separates.

Beta Oxidation The fatty acid cycle of catabolism.

Beta (β) Particle A high-energy electron; $_{-1}^{0}e$.

Beta (β) Radiation A stream of high-energy electrons.

Bile A secretion of the gall bladder that empties into the upper intestine and furnishes bile salts; a route of excretion for cholesterol and bile pigments.

Bile Pigment Colored products of the partial catabolism of heme that are transferred from the liver to the gall bladder for secretion via the bile.

Bile Salts Steroid-base detergents in bile that emulsify fats and oils during digestion.

Bilin The brownish pigment that is the end product of the catabolism of heme and that contributes to the characteristic colors of feces and urine.

Bilinogen A product of the catabolism of heme that contributes to the characteristic colors of feces and urine and some of which is oxidized to bilin.

Bilirubin An reddish-orange substance that forms from biliverdin during the catabolism of heme and which enters the intestinal tract via the bile and is eventually changed into bilinogen and bilin.

Biliverdin A greenish pigment that forms when partly catabolized hemoglobin (as verdohemoglobin) is further broken down, and which is changed in the liver to bilirubin.

Binding Site That part of an enzyme molecule that holds the substrate molecule and positions it over the active site.

Biochemistry The study of the structures and properties of substances found in living systems.

Biological Value In nutrition, the percentage of the nitrogen of ingested protein that is absorbed from the digestive tract and retained by the body when the total protein intake is less than normally required.

Biotin A water-soluble vitamin needed to make enzymes used in fatty acid synthesis.

Blood Sugar The carbohydrates — mostly glucose — that are present in blood.

Blood Sugar Level The concentration of carbohydrate — mostly glucose — in the blood; usually stated in units of mg/dL.

Binary Compound A compound made from just two elements.

Bohr Model of the Atom The solar system model of the structure of an atom, proposed by Niels Bohr, that pictures the electrons circling the nucleus in discrete energy states called orbits.

Boiling The turbulent behavior in a liquid when its vapor pressure equals the atmospheric pressure and when the liquid absorbs heat while experiencing no rise in temperature.

Boiling Point, Normal The temperature at which a substance boils when the atmospheric pressure is 760 mm Hg (1 atm).

Bond, Chemical A net electrical force of attraction that holds atomic nuclei near each other within compounds.

Boyle's Law (See *Pressure-Volume Law*.)

Branched Chain A sequence of atoms to which additional atoms are attached at points other than the ends.

Brønsted Theory An acid is a proton-donor and a base is a proton-acceptor.

Brownian Movement The random, chaotic movements of particles in a colloidal dispersion that can be seen with a microscope.

Buffer A combination of solutes that holds the pH of a solution relatively constant even if small amounts of acids or bases are added.

Calorie The amount of heat that raises the temperature of 1 g of water by 1 degree Celsius from 14.5 °C to 15.5 °C.

Carbaminohemoglobin Hemoglobin that carries chemically bound carbon dioxide.

Carbohydrate Any naturally-occurring substance whose molecules are polyhydroxyaldehydes or polyhydroxyketones or can be hydrolyzed to such compounds.

Carbon Family The group IVA elements in the periodic table—carbon, silicon, germanium, tin, and lead.

Carbonate Buffer A mixture or a solution that includes bicarbonate ions and carbonic acid (dissolved carbon dioxide) in which the bicarbonate ion can neutralize added acid and carbonic acid can neutralize added base.

Carbocation Any cation in which a carbon atom has just six outer level electrons; a carbonium ion.

Carboxylic Acid A compound whose molecules have the carboxyl group, $-CO_2H$.

Carboxypeptidase A digestive enzyme that catalyzes the hydrolysis of C-terminal amino acid residues from small polypeptides.

Carcinogen A chemical or physical agent that induces the onset of cancer or the formation of a tumor that might or might not become cancerous.

Cardiovascular Compartment The entire network of blood vessels and the heart.

Catabolism The reactions of metabolism that break molecules down.

Catalysis The phenomenon of an increase in the rate of a chemical reaction brought about by a relatively small amount of a chemical—the catalyst—that is not permanently changed by the reaction.

Catalyst A substance that is able, in relatively low concentrations, to accelerate the rate of a chemical reaction without itself being permanently changed. (In living systems, the catalysts are called enzymes.)

Cathode The negative electrode to which positively charged ions—cations—are attracted during electrolysis.

Cation A positively charged ion.

Cellulose A linear polymer of β-glucose in which glucose units are joined by $\beta(1 \rightarrow 4)$ glycosidic linkages. The chief constituent of cotton.

Centimeter (cm) A length equal to one-hundreth of the meter. 1 cm = 0.01 m = 0.394 in.

Charles' Law (See *Temperature-Volume Law.*)

Chemical Bond (See *Bond, Chemical.*)

Chemical Energy The potential energy that substances have because their arrangements of electrons and atomic nuclei are not as stable as are alternative arrangements that become possible in chemical reactions.

Chemical Equation A shorthand representation of a chemical reaction that uses formulas instead of names for reactants and products; that separates reactant formulas from product formulas by an arrow; that separates formulas on either side of the arrow by plus signs; and that expresses the mole proportions of the chemicals by simple numbers (coefficients) placed before the formulas.

Chemical Property Any chemical reaction that a substance can undergo and the ability to undergo such a reaction.

Chemical Reaction Any event in which substances change into different chemical substances.

Chemiosmotic Theory An explanation of how oxidative phosphorylation is related to a flow of protons in a proton gradient that is established by the respiratory chain, and that extends across the inner membrane of a mitochondrion.

Chiral Having handedness in a molecular structure. (See also *Chirality.*)

Chiral Carbon A carbon that holds four different atoms or groups.

Chirality The quality of handedness that a molecular structure has that prevents this structure from being superimposible on its mirror image.

Chloride Shift An interchange of chloride ions and bicarbonate ions between a red blood cell and the surrounding blood serum.

Choline A compound needed to make complex lipids and acetylcholine; classified as a vitamin.

Chromatin Filaments that consist of DNA and polypeptides in a cell nucleus.

Chromosome Small thread-like bodies in a cell nucleus that carry genes in a linear array and that are microscopically visible during cell division.

Chylomicron A microdroplet of lipid material with a trace amount of protein that carries lipids picked up from the digestive tract to adipose tissue and to the liver.

Chyme The mixture of partially digested food that forms in the stomach and that is released through the pyloric valve into the duodenum.

Chymotrypsin A digestive enzyme that catalyzes the hydrolysis of peptide bonds in large polypeptides.

Citric Acid Cycle A series of reactions that dismantle acetyl units and send electrons (and protons) into the respiratory chain; a major source of metabolites for the respiratory chain.

Codon A sequence of three adjacent side-chain bases in a molecule of *m*RNA that codes for a specific amino acid residue when the *m*RNA participates in polypeptide synthesis.

Coefficient of Digestibility The proportion of an ingested protein's nitrogen that enters circulation rather than elimination (in feces); the difference between the nitrogen ingested and the nitrogen in the feces divided by the nitrogen ingested.

Coefficients Numbers placed before formulas in chemical equations to indicate the mole proportions of reactants and products.

Coenzyme An organic compound needed to make a complete enzyme from an apoenzyme.

Cofactor A nonprotein compound or ion that is an essential part of an enzyme.

Collagen The fibrous protein of connective tissue that changes to gelatin in boiling water.

Colligative Property A property of a solution that depends only on the concentrations of the solute and the solvent and not on their chemical identities (e.g., osmotic pressure).

Colloidal Dispersion A relatively stable, uniform distribution in some dispersing medium of colloidal particles — those with at least one dimension between 1 and 1000 nm.

Colloidal Osmotic Pressure The contribution made to the osmotic pressure of a solution by substances colloidally dispersed in it.

Combustion A rapid chemical reaction with oxygen that produces heat, light, a flame, and mostly gaseous products.

Common Ion Effect The reduction in the solubility of a salt in some solution by the addition of another solute that furnishes one of the ions of this salt.

Competitive Inhibition The inhibition of an enzyme by the binding of a molecule that can compete with the substrate for the occupation of the catalytic site.

Compound A substance made from the atoms of two or more elements that are present in a definite proportion by mass and by atoms.

Concentration The quantity of some component of a mixture in a unit of volume or a unit of mass of the mixture.

Condensation The physical change of a substance from its gaseous state to its liquid state.

Condensed Structure (See *Structural Formula*.)

Conduction In the science of heat energy, the transfer of heat from a region of higher temperature to a region of lower temperature by means of the transfer of the kinetic energy of atoms, ions, or molecules to their neighbors. In the science of electrical energy, the movement of electrons in a conductor.

Configuration Any one of the many geometric forms that molecules can have by means of cis-trans or optical isomerism. *Absolute configuration* is the actual configuration at a chiral center.

Configurational Isomer An optical isomer. (See *Optical Isomer*.)

Conformation One of the infinite number of contortions of a molecule that is permitted by free rotations around single bonds.

Conjugate Acid-Base Pair Two particles whose formulas differ by only one H^+, such as NH_4^+ and NH_3, or HCl and Cl^-.

Convection In the science of heat energy, the transfer of heat by the circulation of a warmer fluid throughout the remainder of the fluid.

Conversion Factor A fraction that expresses a relationship between quantities that have different units, such as 2.54 cm/in.

Coordinate Covalent Bond A covalent bond in which both of the electrons of the shared pair originated from one of the atoms involved in the bond.

Cori Cycle The sequence of chemical events and transfers of substances in the body that describes the distribution, storage, and mobilization of blood sugar, including the reconversion of lactate to glycogen.

Cosmic Radiation A stream of ionizing radiations, from the sun and outer space, that consists mostly of protons but also includes alpha particles, electrons, and the nuclei of atoms up to atomic number 28.

Covalence Number The number of electron-pair (covalent) bonds that an atom can have in a molecule.

Covalent Bond The net force of attraction that arises as two atomic nuclei share a pair of electrons. One pair is shared in a single bond, two pairs in a double bond, and three electrons pairs are shared in a triple bond.

Crenation The shrinkage of red blood cells when they are in contact with a hypertonic solution.

Crick-Watson Theory A theory that uses the double-helix structure of DNA to explain how genes store genetic information and how this information is translated into sequences of amino acids in proteins.

Curie (Ci) A unit of activity of a radioactive source. 1 Ci = 3.70 × 10^{10} disintegrations/s

Dalton's Law (See *Law of Partial Pressures*.)

Dalton's Theory A theory that accounts for the laws of chemical combination by postulating that matter consists of indestructible atoms; that all atoms of the same element are identical in mass and other properties; that the atoms of different elements are different in mass and other properties; and that in the formation of a compound, atoms join together in definite, whole-number ratios.

Deamination The removal of an amino group from an amino acid.

Decarboxylation The removal of a carboxyl group.

Degree Celsius One-one hundredth (1/100) of the interval on a thermometer between the freezing point and the boiling point of water.

Degree Fahrenheit One-one hundred and eightieth (1/180) of the interval on a thermometer between the freezing point and the boiling point of water.

Deliquescence The ability of a substance to attract water vapor to itself to form a concentrated solution.

Denaturation The loss of the natural shape and form of a protein molecule together with its ability to function biologically, but not necessarily accompanied by the rupture of any of its peptide bonds.

Density The ratio of the mass of an object to its volume; the mass per unit volume. Density = mass/volume (usually expressed in g/mL).

Deoxyribonucleic Acid (DNA) The chemical of a gene; one of a large number of polymers of deoxyribonucleotides and whose sequences of side-chain bases constitute the genetic messages of genes.

Deoxyribonucleotides The monomers of deoxyribonucleic acids (DNA) that consist of deoxyribose-phosphate esters with each deoxyribose unit carrying a side-chain base (one of four heterocyclic amines, adenine, thymine, guanine, or cytosine).

Derived Quantity A quantity based on a relationship that involves one or more base quantities of measurement such as volume (length³) or density (mass/volume).

Derived Unit A unit of a derived quantity such as g/mL (density).

Desiccant A substance that combines with water vapor to form a hydrate and thereby reduces the concentration of water vapor in the air space around the substance.

Dextrin A polymer of α-glucose that forms from the incomplete hydrolysis of starch.

Dextrorotatory That property of an optically active substance by which it can cause the plane of plane-polarized light to rotate clockwise.

D-Family; L-Family The names of the two optically active families to which substances can belong when they are considered solely according to one kind of molecular chirality (molecular handedness) or the other.

Diabetes Mellitus A disease in which there is an insufficiency of effective insulin and an impairment of glucose tolerance.

Dialysis The passage through a dialyzing membrane of water and particles in solution, but not of particles that have colloidal size.

Diastereomer. One of a pair of stereoisomers not related as an object is to its mirror image.

Diatomic Molecule A molecule made of two atoms.

Dietetics The application of the findings of the science of nutrition to the feeding of individual humans, whether well or ill.

Diffusion A physical process whereby particles, by random motions, intermingle and spread out so as to erase concentration gradients.

Digestive Juice A secretion into the digestive tract that consists of a dilute aqueous solution of digestive enzymes (or their zymogens) and inorganic ions.

Dihydric Alcohol An alcohol with two —OH groups, a glycol.

Dipeptide A compound whose molecules have two α-amino acid residues joined by a peptide (amide) bond.

2,3-Diphosphoglycerate (DPG) An organic ion that nestles within the hemoglobin molecule in deoxygenated blood but is expelled from the hemoglobin molecule during oxygenation.

Dipolar Ion A molecule that carries one plus charge and one minus charge, such as an α-amino acid.

Dipole, Electrical A pair of equal but opposite (and usually partial) electrical charges separated by a small distance in a molecule.

Diprotic Acid An acid with two protons available per molecule to neutralize a base (e.g., H_2SO_4).

Disaccharide A carbohydrate that can be hydrolyzed into two monosaccharides.

Disulfide Link The sulfur-sulfur covalent bond in polypeptides.

Disulfide System —S—S— as in R—S—S—R.

DNA (See *Deoxyribonucleic Acid*.)

Double Bond A covalent bond in which two pairs of electrons are shared.

Double Decomposition A reaction in which a compound is made by the exchange of partner-ions between two salts.

Double Helix DNA Model A spiral arrangement of two intertwining DNA molecules held together by hydrogen bonds between side-chain bases.

DPG (See *2,3-Diphosphoglycerate*.)

Duodenum The upper 12 in. of the intestinal tract immediately below the stomach.

Dynamic Equilibrium (See *Equilibrium*.)

Dyspnea Air hunger.

Edema The swelling of tissue caused by the retention of water.

Effector A chemical other than a substrate that can allosterically activate an enzyme.

Elastin The fibrous protein of tendons and arteries.

Electrical Balance The condition of a net ionic equation wherein the algebraic sum of the positive and negative charges of the reactants equals that of the products.

Electrode A metal object, usually a wire, suspended in an electrically conducting medium through which electricity passes to or from an external circuit.

Electrolysis A procedure in which an electrical current is passed through a solution that contains ions, or through a molten salt, for the purpose of bringing about a chemical change.

Electrolyte Any substance whose solution in water conducts electricity; or the solution itself of such a substance.

Electrolytes, Blood The ionic substances dissolved in the blood.

Electron A subatomic particle that bears one unit of negative charge and has a mass that is 1/1836 the mass of a proton.

Electron Cloud A mental model that views the one or two rapidly moving electrons of an orbital as creating a cloud-like distribution of negative charge.

Electron Configuration The most stable arrangement (i.e., the arrangement of lowest energy) of the electrons of an atom, ion, or molecule.

Electron Shell An alternative name for *principal energy level*.

Electron Volt (eV) A very small unit of energy used to describe the energy of a radiation.

$$1 \text{ eV} = 1.6 \times 10^{-19} \text{ joule}$$
$$1 \text{ eV} = 3.8 \times 10^{-20} \text{ calorie}$$
$$1000 \text{ eV} = 1 \text{ KeV} \qquad (1 \text{ kiloelectron volt})$$
$$1000 \text{ keV} = 1 \text{ MeV} \qquad (1 \text{ megaelectron volt})$$

Electronegativity The ability of an atom joined to another by a covalent bond to attract the electrons of the bond toward itself.

Electron Sharing The joint attraction of two atomic nuclei toward a pair of electrons situated between the nuclei and between which, therefore, a covalent bond exists.

Element A substance that cannot be broken down into anything that is both stable and more simple; a substance in which all of the atoms have the same atomic number and the same electron configuration; one of the three broad kinds of matter, the others being compounds and mixtures.

Emulsion A colloidal dispersion of tiny microdroplets of one liquid in another liquid.

Enantiomer One of a pair of stereoisomers that are related as an object is related to its mirror image but that cannot be superimposed one on the other.

End Point The stage in a titration when the operation is stopped.

Endergonic Describing a change that needs a constant supply of energy to happen.

Endocrine Gland An organ that makes one or more hormones.

Endothermic Describing a change that needs a constant supply of heat energy to happen.

Energy A capacity to cause a change that can, in principle, be harnessed for useful work.

Energy Density The energy per gram of stored glycogen or fat.

Energy Level A principal energy state in which electrons of an atom can be.

Energy of Activation The minimum energy that must be provided by the collision between reactant particles to initiate the rearrangement of electrons relative to nuclei that must happen if the reaction is to occur.

Enterokinase An enzyme of intestinal juice that changes trypsinogen into trypsin.

Enzyme A catalyst in a living system.

Enzyme-Substrate Complex The temporary combination that an enzyme must form with its substrate before catalysis can occur.

Epinephrine A hormone of the adrenal medulla that activates the enzymes needed to release glucose from glycogen.

Equation, Balanced A chemical equation in which all of the atoms represented in the formulas of the reactants are present in identical numbers among the products, and in which any net electrical charge provided by the reactants equals the same charge indicated by the products. (See also *Chemical Equation.*)

Equilibrium A situation in which two opposing events occur at identical rates so that no net change happens.

Equilibrium Constant The value that the mass action expression has when a chemical system is at equilibrium.

Equivalence Point The stage in a titration when the reactants have been mixed in the exact molar proportions represented by the balanced equation; in an acid-base titration, the stage when the moles of hydrogen ions furnished by the acid match the moles of hydroxide ions (or other proton acceptor) supplied by the base.

Equivalent For an acid, its mass in grams that can neutralize one mole of hydroxide ion; for a base, its mass in grams that can neutralize one mole of hydrogen ion; for an ion, usually its mass in grams divided by the amount of its electrical charge.

Equivalent Weight A synonym for *Equivalent.*

Erythrocyte A red blood cell.

Essential Amino Acid An α-amino acid that the body cannot make from other amino acids and that must be supplied by the diet.

Essential Fatty Acid A fatty acid that must be supplied by the diet.

Ester A derivative of an acid and an alcohol that can be hydrolyzed to these parent compounds. Esters of carboxylic acids and phosphoric acid occur in living systems.

$$
\begin{array}{cc}
\overset{\displaystyle O}{\underset{|}{\overset{||}{-C-O-C-}}} & \overset{\displaystyle O}{\underset{|}{\overset{||}{-C-O-P-OH}}} \\
& OH
\end{array}
$$

System in an ester of   System in an ester of
a carboxylic acid   phosphoric acid

Esterase An enzyme that catalyzes the hydrolysis of an ester.

Esterification The formation of an ester.

Ether An organic compound whose molecules have an oxygen attached by single bonds to separate carbon atoms neither of which is a carbonyl carbon atom: R—O—R′.

Evaporation The conversion of a substance from its liquid to its vapor state.

Exergonic Describing a change by which energy of any form is released from the system.

Exon A segment of a DNA strand that eventually becomes expressed as a corresponding sequence of aminoacyl residues in a polypeptide.

Exothermic Describing a change by which heat energy is released from the system.

Extensive Property Any property whose value is directly proportional to the size of the sample, such as volume or mass.

Extracellular Fluids Body fluids that are outside of cells.

Fact In science, something that has physical existence, that can be experienced or observed, and that can be measured by independent observers.

Factor-Label Method A strategy for solving computational problems that uses conversion factors and the cancellation of the units of physical quantities as an aid in working toward the solution.

Fatty Acid Any carboxylic acid that can be obtained by the hydrolysis of animal fats or vegetable oils.

Fatty Acid Cycle The catabolism of a fatty acid by a series of repeating steps that produce acetyl units (in acetyl CoA).

Feedback Inhibition The competitive inhibition of an enzyme by a product of its own action.

Fibrin The fibrous protein of a blood clot that forms from fibrinogen during clotting.

Fibrinogen A protein in blood that is changed to fibrin during clotting.

Fibrous Proteins Water-insoluble proteins found in fibrous tissues.

Fission The splitting of the nucleus of a heavy atom approximately in half and that is accompanied by the release of one or a few neutrons and energy.

Folacin A vitamin supplied by folic acid or pteroylglutamic acid and that is needed to prevent megaloblastic anemia.

Food A material that supplies one or more nutrients without contributing materials that, either in kind or quantity, would be harmful to most healthy people.

Formula, Empirical A chemical symbol for a compound that gives just the ratios of the atoms and not necessarily the composition of a complete molecule.

Formula, Molecular A chemical symbol for a substance that gives the composition of a complete molecule.

Formula, Structural A chemical symbol for a substance that uses atomic symbols and lines to describe the pattern in which the atoms are joined together in a molecule.

Formula Unit A small particle — an atom, a molecule, or a set of ions — that has the composition given by the chemical formula of the substance.

Formula Weight The sum of the atomic weights of the atoms represented in a chemical formula.

Free Rotation The absence of a barrier to the rotation of two groups with respect to each other when they are joined by a single, covalent bond.

Fructose A ketohexose present in honey and one product of the hydrolysis of sucrose; levulose.

Functional Group An atom or a group of atoms in a molecule that is responsible for the particular set of reactions that all compounds with this group have.

Galactose An aldohexose that forms, together with glucose, when lactose (milk sugar) is hydrolyzed.

Gamma (γ) One microgram; 1×10^{-6} g.

Gamma Radiation A natural radiation similar to but more powerful than X rays.

Gas Any substance that must be contained in a wholly closed space and whose shape and volume is determined entirely by the shape and volume of its container; a state of matter.

Gas Constant, Universal (R) The ratio of PV to nT for a gas, where $P =$ the gas pressure, $V =$ volume, $n =$ number of moles, and $T =$ the Kelvin temperature. When P is in mm Hg and V is in mL,

$$R = 6.23 \times 10^4 \text{ mm Hg mL/mol K}$$

Gas Tension The partial pressure of a gas over its solution in some liquid when the system is in equilibrium.

Gastric Juice The digestive juice secreted into the stomach and that contains pepsinogen, hydrochloric acid, and gastric lipase.

Gay-Lussac's Law (See *Pressure-Temperature Law*.)

Gel A colloidal dispersion of a solid in a liquid that has adopted a semisolid form.

Gene A unit of heredity carried on a cell's chromosomes and consisting of DNA.

Genetic Code The set of correlations that specifies which codons on *m*RNA chains are responsible for which amino acyl residues when the latter are steered into place during the *m*RNA-directed synthesis of polypeptides.

Geometric Isomerism Stereoisomerism caused by restricted rotation that gives different geometries to the same structural organization; cis-trans isomerism.

Geometric Isomers Stereoisomers whose molecules have identical atomic organizations but different geometries; cis-trans isomers.

Globular Proteins Proteins that are soluble in water or in water that contains certain dissolved salts.

Globulins Globular proteins in the blood that include γ-globulin, an agent in the body's defense against infectious diseases.

Glucagon A hormone secreted by the α-cells of the pancreas in response to a decrease in the blood sugar level, that stimulates the liver to release glucose from its glycogen stores.

Gluconeogenesis The synthesis of glucose from compounds with smaller molecules or ions.

Glucose An aldohexose whose molecules serve as building blocks for glycogen, starch, cellulose, dextrin, maltose, sucrose, and lactose; the chief carbohydrate in blood; *blood sugar; dextrose.*

Glucose Tolerance The ability of the body to manage the intake of dietary glucose while keeping the blood sugar level from fluctuating widely.

Glucose Tolerance Test A series of measurements of the blood sugar level after the ingestion of a considerable amount of glucose; used to obtain information about an individual's glucose tolerance.

Glucoside An acetal formed from glucose (in its cyclic, hemiacetal form) and an alcohol.

Glucosuria The presence of glucose in urine.

Glycogen The starch-like polymer of α-glucose that serves as an animal's means of storing glucose units.

Glycogenesis The synthesis of glycogen.

Glycogenolysis The breakdown of glycogen to glucose.

Glycol A dihydric alcohol.

Glycolipid A lipid whose molecules include a glucose unit, a galactose unit, or some other carbohydrate unit.

Glycolysis A series of chemical reactions that break down glucose or glucose units in glycogen until pyruvate remains (when the series is operated aerobically) or lactate forms (when the conditions are anaerobic).

Glycoside An acetal or a ketal formed from the cyclic form of a monosaccharide and an alcohol.

Glycosidic Link The oxygen bridge between one monosaccharide unit and another in a disaccharide or a polysaccharide.

Gout A disease of the joints in which deposits of salts of uric acid accumulate and cause inflammation, swelling, and pain.

Gradient The presence of a change in value of some physical quantity with distance, as in a *concentration* gradient in which the concentration of a solute is different in different parts of the system.

Gram (g) A mass equal to one-thousandth of the kilogram mass, the SI standard mass.

$$1 \text{ g} = 0.001 \text{ kg} = 1000 \text{ mg}; \ 1 \text{ lb} = 454 \text{ g}$$

Gray (Gy) The SI unit of absorbed dose of radiation equal to one joule of energy absorbed per kilogram of tissue.

Group A vertical column in the periodic table; a family of elements.

Half-Life The time needed for half of the atoms in a sample of a particular radioactive isotope to undergo radioactive decay.

Halogens The elements of group VIIA of the periodic table — fluorine, chlorine, bromine, iodine, and astatine.

Hard Water Water that contains one or more of the metallic ions Mg^{2+}, Ca^{2+}, Fe^{2+} or Fe^{3+}. The negative ions present are usually Cl^- and $SO_4{}^{2-}$. If $HCO_3{}^-$ is the chief negative ion, the water is said to be *temporary hard water;* otherwise it is *permanent hard water*.

Heat The form of energy that transfers between two objects in contact that have initially different temperatures.

Heat Capacity The quantity of heat that a given object can absorb (or release) per degree Celsius change in temperature.

$$\text{heat capacity} = \text{heat}/\Delta t$$

where Δt is the change in temperature.

Heat of Fusion The quantity of heat that one gram of a substance absorbs when it changes from its solid to its liquid state at its melting point.

Heat of Reaction The net energy difference between the reactants and the products of a reaction.

Heat of Vaporization The quantity of heat that one gram of a substance absorbs when it changes from its liquid to its gaseous state.

Heisenberg Uncertainty Principle It is impossible simultaneously to determine with precision and accuracy both the position and the velocity of an electron.

α-Helix One kind of secondary structure of a polypeptide in which its molecules are coiled.

Heme The deep-red, iron-containing prosthetic group in hemoglobin and myoglobin.

Hemiacetal Any compound whose molecules have both an —OH group and an ether linkage coming to a —CH— unit:

$$\begin{array}{ccc} & \text{OH} & \\ & | & \\ -\text{CH}-\text{O}-\text{C}- & \quad \text{as in} \quad & \text{R}-\text{CH}-\text{O}-\text{R}' \\ \end{array}$$

Hemiketal Any compound whose molecules have both an —OH group and an ether linkage coming to a carbon that otherwise bears no H atoms:

$$\begin{array}{ccc} & \text{OH} & \\ & | & \\ -\text{C}-\text{O}-\text{C}- & \quad \text{as in} \quad & \text{R}-\text{C}-\text{O}-\text{R}' \\ & | & \quad\quad\quad | \\ & & \quad\quad\quad \text{R} \end{array}$$

Hemoglobin (HHb) The oxygen-carrying protein in red blood cells.

Hemolysis The bursting of a red blood cell.

Henry's Law See *Pressure-Solubility Law.*

Heterogeneous Nuclear RNA (*hn*RNA) RNA made directly at the guidance of DNA and from which messenger RNA (*m*RNA) is made; primary transcript RNA.

High-Energy Phosphate An organophosphate with a phosphate group transfer potential equal to or higher than that of ADP or ATP.

Homeostasis The response of an organism to a stimulus such that the organism is restored to its prestimulated state.

Homogeneous Mixture A mixture in which the composition and properties are uniform throughout.

Homolog Any member of a homologous series of organic compounds.

Homologous Series A series of organic compounds in the same family whose successive members differ by individual CH_2 units.

Hormone A primary chemical messenger made by an endocrine gland and carried by the bloodstream to a target organ where a particular chemical response is initiated.

Human Growth Hormone One of the hormones that affects the blood sugar level; a stimulator of the release of the hormone glucagon.

Hund's Rule Electrons become distributed among *different* orbitals of the same general energy level insofar as there is room.

Hybrid Orbital An atomic orbital obtained by mixing two or more pure orbitals (those of the *s, p, d,* or *f* types).

Hydrate A compound in which intact molecules of water are held in a definite molar proportion to the other components.

Hydration The association of water molecules with dissolved ions or polar molecules.

Hydrocarbon Any organic compound that consists entirely of carbon and hydrogen.

Hydrogen Bond The force of attraction between a $\delta+$ on a hydrogen held by a covalent bond to oxygen or nitrogen (or fluorine) and a $\delta-$ charge on a nearby atom of oxygen or nitrogen (or fluorine).

Hydrolysis of Salts Any reaction in which a cation (other than H^+) or an anion (other than OH^-) changes the ratio of the molar concentrations of hydrogen and hydroxide ions in an aqueous solution.

Hydronium Ion H_3O^+

Hydrophilic Group Any part of a molecular structure that attracts water molecules; a polar or ionic group such as —OH, $-CO_2^-$, $-NH_3^+$, or $-NH_2$.

Hydrophobic Group Any part of a molecular structure that has no attraction for water molecules; a nonpolar group such as any alkyl group.

Hydroxide Ion OH^-

Hygroscopic Describing a substance that can reduce the concentration of water vapor in the surrounding air by forming a hydrate.

Hyperammonemia An elevated level of ammonium ion in the blood.

Hypercalcemia An elevated level of calcium ion in blood—above 5.2 meq/L.

Hyperglycemia An elevated level of sugar in the blood—above 95 mg/dL in whole blood.

Hyperkalemia An elevated level of potassium ion in blood—above 5.0 meq/L.

Hypermagnesemia An elevated level of magnesium ion in blood—above 2.0 meq/L.

Hypernatremia An elevated level of sodium ion in blood—above 145 meq/L.

Hyperthermia An elevated body temperature.

Hypertonic Having an osmotic pressure greater than some reference; having a total concentration of all solute particles higher than that of some reference.

Hyperventilation Breathing considerably faster and deeper than normal.

Hypocalcemia A low level of calcium ion in blood—below 4.2 meq/L.

Hypoglycemia A low level of glucose in blood—below 65 mg/dL of whole blood.

Hypokalemia A low level of potassium ion in blood—below 3.5 meq/L.

Hypomagnesemia A low level of magnesium ion in blood— below 1.5 meq/L.

Hyponatremia A low level of sodium ion in blood—below 135 meq/L.

Hypothermia A low body temperature.

Hypothesis A conjecture, subject to being disproved, that explains a set of facts in terms of a common cause and that serves as the basis for the design of additional tests or experiments.

Hypotonic Having an osmotic pressure less than some reference; having a total concentration of dissolved solute particles less than that of some reference.

Hypoventilation Breathing more slowly and less deeply than normal; shallow breathing.

Hypoxia A condition of a low supply of oxygen.

Ideal Gas A hypothetical gas that obeys the gas laws exactly.

Indicator A dye that, in solution, has one color below a measured pH range and a different color above this range.

Induced Fit Theory Certain enzymes are induced by their substrate molecules to modify their shapes to accommodate the substrate.

Inducer A substance whose molecules remove repressor molecules from operator genes and so open the way for structural genes to direct the overall syntheses of particular polypeptides.

Inertia The resistance of an object to a change in its position or its motion.

Inhibitor A substance that interacts with an enzyme to prevent its acting as a catalyst.

Inner Transition Elements The elements of the lanthanide and actinide series of the periodic table.

Inorganic Compound Any compound that is not an organic compound.

Insensible Perspiration The loss of water from the body with no visible sweating; evaporative losses from the skin and the lungs.

Insulin A protein hormone made by the pancreas, released in response to a rise in the blood sugar level, and used by certain tissues to help them take up glucose from circulation.

Insulin Shock Shock brought on by a drastic reduction in the blood sugar level, usually after an overdose of insulin.

Intensive Property Any property whose value is independent of the size of the sample, such as temperature and density.

Internal Environment Everything enclosed within an organism.

International System of Units (SI) The successor to the metric system with new reference standards for the base units but with the same names for the units and the same decimal relationships.

International Union of Pure and Applied Chemistry System (IUPAC System) A set of systematic rules for naming compounds and designed to give each compound one unique name and for which only one structure can be drawn; the Geneva system of nomenclature.

Interstitial Fluids Fluids in tissues but not inside cells.

Intestinal Juice The digestive juice that empties into the duodenum from the intestinal mucosa and whose enzymes also work within the intestinal mucosa as molecules migrate through.

Intron A segment of a DNA strand that separates exons and that does not become expressed as a segment of a polypeptide.

Inverse Square Law The intensity of radiation varies inversely with the square of the distance from its source.

Invert Sugar A 1 : 1 mixture of glucose and fructose.

Iodine Test A test for starch by which a drop of iodine produces an intensely purple color if starch is present.

Ion An electrically charged, atomic or molecular-sized particle; a particle that has one or a few atomic nuclei and either one or two (seldom, three) too many or too few electrons to render the particle electrically neutral.

Ionic Bond The force of attraction between oppositely charged ions in an ionic compound.

Ionic Compound A compound that consists of an orderly aggregation of oppositely charged ions that assemble in whatever ratio ensures overall electrical neutrality.

Ionic Equation A chemical equation that explicitly shows all of the particles — ions, atoms, or molecules — that are involved in a reaction even if some are only spectator particles. (See also *Net Ionic Equation; Equation, Balanced.*)

Ionization A change, usually involving solvent molecules, whereby molecules change into ions.

Ionizing Radiation Any radiation that can create ions from molecules within the medium that it enters, such as alpha, beta, gamma, X, and cosmic radiation.

Ion Product Constant of Water (K_w) The product of the molar concentrations of hydrogen ions and hydroxide ions in water at a given temperature.

$$K_w = [H^+][OH^-]$$
$$= 1.0 \times 10^{-14} \quad \text{(at 25 °C)}$$

Isoelectric A condition of a molecule in which it has an equal number of positive and negative sites.

Isoelectric Point (pI) The pH of a solution in which a specified amino acid or a protein is in an isoelectric condition; the pH at which there is no net migration of the amino acid or protein in an electric field.

Isoenzymes Enzymes that have identical catalytic functions but which are made of slightly different polypeptides.

Isohydric Shift In actively metabolizing tissue, the use of a hydrogen ion released from newly formed carbonic acid to react with and liberate oxygen from oxyhemoglobin; in the lungs, the use of hydrogen ion released when hemoglobin oxygenates to combine with bicarbonate ion and liberate carbon dioxide for exhaling.

Isomerism The phenomenon of the existence of two or more compounds with identical molecular formulas but different structures.

Isomers Compounds with identical molecular formulas but different structures.

Isotonic Having an osmotic pressure identical to that of a reference; having a concentration equivalent to the reference with respect to the ability to undergo osmosis.

Isotope A substance in which all of the atoms are identical in atomic number, mass number, and electron configuration.

IUPAC System (See *International Union of Pure and Applied Chemistry System.*)

K_a (See *Acid Dissociation Constant.*)
K_b (See *Base Dissociation Constant.*)
K_w (See *Ion Product Constant of Water.*)

Kelvin The SI unit of temperature degree and equal to 1/100th of the interval between the freezing point and the boiling point of water when measured under standard conditions.

Keratin The fibrous protein of hair, fur, fingernails, and hooves.

Ketal A substance whose molecules have two ether linkages joined to a carbon that also holds two hydrocarbon groups as in

$$R_2C(OR')_2$$

Ketoacidosis The acidosis caused by untreated ketonemia.

Keto Group The carbonyl group when it is joined on each side to carbon atoms.

Ketohexose A monosaccharide whose molecules contain six carbon atoms and have a keto group.

Ketone Any compound with a carbonyl group attached to two carbon atoms, as in $R_2C{=}O$.

Ketone Bodies Acetoacetate, β-hydroxybutyrate — or their parent acids — and acetone.

Ketonemia An elevated concentration of ketone bodies in the blood.

Ketonuria An elevated concentration of ketone bodies in the urine.

Ketose A monosaccharide whose molecules have a ketone group.

Ketosis The combination of ketonemia, ketonuria, and acetone breath.

Kilocalorie (kcal) The quantity of heat equal to 1000 calories.

Kilogram (kg) The SI base unit of mass; 1000 g; 2.205 lb.

Kilometer (km) A length equal to 1000 meters or 0.621 miles.

Kinase An enzyme that catalyzes the transfer of a phosphate group.

Kindling Temperature The temperature at which a substance spontaneously bursts into flame in air.

Kinetic Energy The energy of an object by virtue of its motion.

$$\text{Kinetic Energy} = \tfrac{1}{2}(\text{mass})(\text{velocity})^2$$

Kinetics The field of chemistry that deals with the rates of chemical reactions.

Kinetic Theory of Gases A set of postulates about the nature of an ideal gas: that it consists of a large number of very small particles in constant, random motion; that in the collisions the particles lose no frictional energy; that between collisions the particles neither attract nor repel each other; and that the motions and collisions of the particles obey all the laws of physics.

Kreb's Cycle The citric acid cycle.

Lactase A digestive enzyme that catalyzes the hydrolysis of lactose.

Lactose A disaccharide that can be hydrolyzed to glucose and galactose; milk sugar.

Lambda (λ) One microliter; 1×10^{-6} L.

Law of Conservation of Energy Energy can be neither created nor destroyed but only transformed from one form to another.

Law of Conservation of Mass Matter is neither created nor destroyed in chemical reactions; the masses of all products equal the masses of all reactants.

Law of Definite Proportions The elements in a compound occur in definite proportions by mass.

Law of Mass Action (Law of Guldberg and Waage) The molar proportions of the interacting substances in a chemical equilibrium are related by the following equation (in which the ratio on the left is called the *mass action expression* for the system).

$$\frac{[C]^c[D]^d}{[A]^a[B]^b} = K_{eq}$$

The symbols refer to the following generalized equilibrium:

$$aA + bB \rightleftharpoons cC + dD$$

and the brackets, [ ], denote molar concentrations. (When an equilibrium involves additional substances, the equation for the equilibrium constant is adjusted accordingly.)

Law of Multiple Proportions When two elements can combine to form more than one compound, the different masses of the first that can combine with the same mass of the second are in the ratio of small whole numbers.

Law of Partial Pressures (Dalton's Law) The total pressure of a mixture of gases is the sum of their individual partial pressures.

Le Chatelier's Principle If a system is in equilibrium and a change is made in its conditions, the system will change in whichever way most directly restores equilibrium.

Length The base quantity for expressing distances or how long something is.

Levorotatory The property of an optically active substance that causes a counterclockwise rotation of the plane of plane-polarized light.

Like-Dissolves-Like Rule Polar solvents dissolve polar or ionic solutes and nonpolar solvents dissolve nonpolar or weakly polar solutes.

Limiting Amino Acid The essential amino acid most poorly provided by a dietary protein.

Lipase An enzyme that catalyzes the hydrolysis of lipids.

Lipid A plant or animal product that tends to dissolve in such nonpolar solvents as ether, carbon tetrachloride, and benzene.

Lipid Bilayer The sheet-like array of two layers of lipid molecules, interspersed with molecules of cholesterol and proteins, that make up the membranes of cells in animals.

Lipigenesis The synthesis of fatty acids from two-carbon acetyl units.

Lipoprotein Complex A combination of a lipid molecule with a protein molecule that serves as the vehicle for carrying the lipid in the bloodstream.

Liquid A state of matter in which a substance's volume but not its shape is independent of the shape of its container.

Liter (L) A volume equal to 1000 cm^3 or 1000 mL or 1.057 liquid quart.

Lock-and-Key Theory The specificity of an enzyme for its substrate is caused by the need for the substrate molecule to fit to the enzyme's surface much as a key fits to and turns only one tumbler lock.

London Force A net force between molecules that arises from temporary polarities induced in the molecules by collisions or near-collisions with neighboring molecules.

Macromolecule Any molecule with a very high formula weight—generally several thousand or more.

Maltase A digestive enzyme that catalyzes the hydrolysis of maltose.

Maltose A disaccharide that can be hydrolyzed to two glucose molecules; malt sugar.

Manometer A device for measuring gas pressure.

Markovnikov's Rule In the addition of an unsymmetrical reactant to an unsymmetrical double bond of a simple alkene, the positive part of the reactant molecule (usually H^+) goes to the carbon with the greater number of hydrogen atoms and the negative part goes to the other carbon of the double bond.

Mass A quantitative measure of inertia based on an artifact at Sèvres, France, called the standard kilogram mass; a measure of the quantity of matter in an object relative to this reference standard.

Mass Number The sum of the numbers of protons and neutrons in one atom of an isotope.

Material Balance The condition of a chemical equation in which all of the atoms present among the reactants are also found in the products.

Matter Anything that occupies space and has mass.

Melting Point The temperature at which a solid changes into its liquid form; the temperature at which equilibrium exists between the solid and liquid forms of a substance.

Mercaptan A thioalcohol; R—S—H.

Meso Compound One of a set of optical isomers whose own molecules are not chiral and which, therefore, is optically inactive.

Messenger RNA (mRNA) RNA that carries the genetic code as a specific series of codons for a specific polypeptide from the cell's nucleus to the cytoplasm.

Metabolism The sum total of all of the chemical reactions that occur in an organism.

Metal Any element that is shiny, conducts electricity well, and (if a solid) can be hammered into sheets and drawn into wires.

Metalloids Elements that have some metallic and some nonmetallic properties.

Meter (m) The base unit of length in the International System of Measurements (SI).

$$1 \text{ m} = 100 \text{ cm} = 39.37 \text{ in.} = 3.280 \text{ ft} = 1.093 \text{ yd}$$

Metric System A decimal system of weights and measures in which the conversion of a base unit of measurement into a multiple or a submultiple is done by moving the decimal point; the predecessor to the International System of Measurements (SI).

Microangiopathy A change in the thickness, composition, and metabolism of the basement membrane of blood capillaries.

Microgram (μg) A mass equal to one-thousandth of a milligram.

$$1 \mu g = 0.001 \text{ mg} = 1 \times 10^{-6} \text{ g}$$

(Its symbol is sometimes given as mcg or as γ in pharmaceutical work.)

Microliter (μL) A volume equal to one-thousandth of a milliliter.

$$1 \mu L = 0.001 \text{ mL} = 1 \times 10^{-6} \text{ L}$$

(Its symbol is sometimes given as λ in pharmaceutical work.)

Milliequivalent (meq) A quantity of substance equal to one-thousandth of an equivalent.

Milligram (mg) A mass equal to one-thousandth of a gram.

$$1 \text{ mg} = 0.001 \text{ g} \quad 1000 \text{ mg} = 1 \text{ g} \quad 1 \text{ grain} = 64.8 \text{ mg}$$

Milliliter (mL) A volume equal to one-thousandth of a liter.

$$1 \text{ mL} = 0.001 \text{ L} = 16.23 \text{ minim} = 1 \text{ cm}^3$$

$$1 \text{ liquid ounce} = 29.57 \text{ mL} \quad 1 \text{ liquid quart} = 946.4 \text{ mL}$$

Millimeter (mm) A length equal to one-thousandth of a meter.

$$1 \text{ mm} = 0.001 \text{ m} = 0.0394 \text{ in.}$$

Millimeter of Mercury (mm Hg) A unit of pressure equal to 1/760 atm.

Millimole (mmol) One-thousandth of a mole. 1000 mmol = 1 mol

Minerals Ions that must be provided in the diet at levels of 100 mg/day or more; Ca^{2+}, Mg^{2+}, Na^+, K^+, Cl^-, and phosphate.

Mixture One of the three kinds of matter (together with elements and compounds); any substance made up of two or more elements or compounds combined physically in no particular proportion by mass and separable into its component parts by physical means.

Model, Scientific A mental construction, often involving pictures or diagrams, that is used to explain a number of facts.

Molar Concentration (M) A solution's concentration in units of moles of solute per liter of solution; molarity.

Molar Mass The number of grams per mole of a substance.

Molar Volume The volume occupied by one mole of a gas under standard conditions of temperature and pressure; 22.4 L at 273 K and 1 atm.

Molarity (See *Molar Concentration.*)

Mole (mol) A mass of a compound or of an element that equals its formula weight in grams; Avogadro's number of a substance's formula units.

Molecular Compound A compound whose smallest representative particle is a molecule; a covalent compound.

Molecular Equation An equation that shows the complete formulas of all of the substances present in a mixture undergoing a reaction. (See also *Net Ionic Equation; Equation, Balanced.*)

Molecular Orbital A region in the space that envelopes two (or sometimes more) atomic nuclei where a shared pair of electrons of a covalent bond resides.

Molecular Weight The formula weight of a substance.

Molecule An electrically neutral (but often polar) particle made up of the nuclei and electrons of two or more atoms and held together by covalent bonds; the smallest representative sample of a molecular compound.

Monoamine Oxidase An enzyme that catalyzes the inactivation of neurotransmitters or other amino compounds of the nervous system.

Monohydric Alcohol An alcohol whose molecules have one —OH group.

Monomer Any compound that can be used to make a polymer.

Monoprotic Acid An acid with one proton per molecule that can neutralize a base.

Monosaccharide A carbohydrate that cannot be hydrolyzed.

Mucin A viscous glycoprotein released in the mouth and the stomach that coats and lubricates food particles and protects the stomach from the acid and pepsin of gastric juice.

Mutagen Any chemical or physical agent that can induce the mutation of a gene without preventing the gene from replicating.

Mutarotation The gradual change in the specific rotation of a substance in solution but without a permanent, irreversible chemical change occurring.

Myosins Proteins in contractile muscle.

Net Ionic Equation A chemical equation in which all spectator particles are omitted so that only the particles that participate directly are represented.

Neurotransmitter A substance released by one nerve cell to carry a signal to the next nerve cell.

Neutral Solution A solution in which the molar concentration of hydronium ions exactly equals the molar concentration of hydroxide ions.

Neutralization, Acid-Base A reaction between an acid and a base.

Neutralizing Capacity The capacity of a solution or a substance to neutralize an acid or a base — expressed as a molar concentration.

Neutron An electrically neutral subatomic particle with a mass of 1 amu.

Niacin A water-soluble vitamin needed to prevent pellagra and essential to the coenzymes in NAD^+ and $NADP^+$; nicotinic acid or nicotinamide.

Nitrogen Balance A condition of the body in which it excretes as much nitrogen as it receives in the diet.

Nitrogen Family The elements of group VA of the periodic table: nitrogen, phosphorus, arsenic, antimony, and bismuth.

Nitrogen Pool The sum total of all nitrogen compounds in the body.

Noble Gases The elements of group 0 of the periodic table: helium, neon, argon, krypton, xenon, and radon.

Nomenclature The system of names and the rules for devising such names, given structures, or for writing structures, given names.

Nonelectrolyte Any substance that cannot furnish ions when dissolved in water or when melted.

Nonfunctional Group A section of an organic molecule that remains unchanged during a chemical reaction at a functional group.

Nonmetal Any element that is not a metal. (See *Metal.*)

Nonsaponifiable Lipid Any lipid, such as the steroids, that cannot be hydrolyzed or similarly broken down by aqueous alkali.

Nonvolatile Liquid Any liquid with a very low vapor pressure at room temperature and that does not readily evaporate.

Normal Fasting Level The normal concentration of something in the blood, such as blood sugar, after about 4 hours without food.

Normality The concentration of a solution in units of equivalents per liter.

Nuclear Chain Reaction The mechanism of nuclear fission by which one fission event makes enough fission initiators (neutrons) to cause more than one additional fission event.

Nuclear Equation A representation of a nuclear transformation in which the chemical symbols of the reactants and products include mass numbers and atomic numbers.

Nuclease An enzyme that catalyzes the hydrolysis of nucleic acids.

Nucleic Acid A polymer of nucleotides in which the repeating units are pentose phosphate esters, each pentose unit bearing a side-chain base (one of four heterocyclic amines); polymeric compounds that are involved in the storage, transmission, and expression of genetic messages.

Nucleotide A monomer of a nucleic acid that consists of a pentose phosphate ester in which the pentose unit carries one of five heterocyclic amines as a side-chain base.

Nucleus In chemistry and physics, the subatomic particle that serves as the core of an atom and that is made up of protons and neutrons. In biology, the organelle in a cell that houses DNA.

Nutrient Any one of a large number of substances in food and drink that is needed to sustain growth and health.

Nutrition The science of the substances of the diet that are necessary for growth, operation, energy, and repair of bodily tissues.

Octet, Outer A condition of an atom or ion in which its highest occupied energy level has eight electrons — a condition of stability.

Octet Rule The atoms of a reactive element tend to undergo those chemical reactions that most directly give them the electron configuration of the noble gas that stands nearest the element in the periodic table (all but one of which have outer octets).

Olefin An alkene.

One-Substance–One-Structure Rule If two samples of matter have identical physical and chemical properties, they have identical molecules.

Optical Activity The ability of a substance to rotate the plane of polarization of plane-polarized light.

Optical Isomer One of a set of compounds whose molecules differ only in their chiralities.

Optical Rotation The degrees of rotation of the plane of plane-polarized light caused by an optically active solution; the observed rotation of such a solution.

Orbital (See *Atomic Orbital.*)

Orbital Hybridization The mixing of two or more ordinary atomic orbitals to give an equal number of modified atomic orbitals, called *hybrid orbitals,* each of which possesses some of the characteristics of the originals.

Orbital Overlap The interpenetration of one atomic orbital by another from an adjacent atom to form a molecular orbital.

Organic Compounds Compounds of carbon other than those related to carbonic acid and its salts, or to the oxides of carbon, or to the cyanides.

Osmolarity The molar concentration of all osmotically active solute particles in a solution.

Osmosis The passage of water only, without any solute, from a less concentrated solution (or pure water) to a more concentrated solution when the two solutions are separated by a semipermeable membrane.

Osmotic Membrane A semipermeable membrane that permits only osmosis, not dialysis.

Osmotic Pressure The pressure that would have to be applied to a solution to prevent osmosis if the solution were separated from water by an osmotic membrane.

Outer Octet (See *Octet, Outer.*)

Outside Level In an atom the highest principal energy level that holds at least one electron.

Oxidase An enzyme that catalyzes an oxidation.

Oxidation The loss of one or more electrons from an atom, molecule, or ion; in organic chemistry, the loss of hydrogen or the gain of oxygen.

Oxidation Number For simple monoatomic ions, the quantity and sign of the electrical charge on the ion.

Oxidative Deamination The change of an amino group to a keto group with loss of nitrogen.

Oxidative Phosphorylation The synthesis of high-energy phosphates such as ATP from lower-energy phosphates and inorganic phosphate by the reactions that involve the respiratory chain.

Oxidizing Agent A substance that can cause an oxidation.

Oxidoreductase An enzyme that catalyzes the formation of an oxidation-reduction equilibrium.

Oxygen Affinity The percentage to which all of the hemoglobin molecules in the blood are saturated with oxygen molecules.

Oxygen Debt The condition in a tissue when anaerobic glycolysis has operated and lactate has been excessively produced.

Oxygen Family The elements in group VIA of the periodic table: oxygen, sulfur, selenium, tellurium, and polonium.

P_i Inorganic phosphate ion(s) of whatever mix of PO_4^{3-}, HPO_4^{2-}, $H_2PO_4^-$, and possibly even traces of H_3PO_4 that is possible at the particular pH of the system, but almost entirely $HPO_4^{2-} + H_2PO_4^-$.

Pancreatic Juice The digestive juice that empties into the duodenum from the pancreas.

Pantothenic Acid A water-soluble vitamin needed to make coenzyme A.

Partial Pressure The pressure contributed by an individual gas in a mixture of gases.

Parts per Billion (ppb) The number of parts in a billion parts. (Two drops of water in a railway tank car that holds 34,000 gallons of water correspond roughly to 1 ppb.)

Parts per Million (ppm) The number of parts in a million parts. (Two drops of water in a large 32-gallon trash can correspond roughly to 1 ppm.)

Pauli Exclusion Principle No more than two electrons can occupy the same orbital at the same time, and two can be present only if they have opposite spin.

Pentose Phosphate Pathway The synthesis of NADPH that uses chemical energy in glucose 6-phosphate and that involves pentoses as intermediates.

Pepsin A digestive, proteolytic enzyme in gastric juice that forms by the action of acid on pepsinogen.

Pepsinogen The zymogen of pepsin.

Peptidase A digestive enzyme that catalyzes the hydrolysis of a peptide.

Peptide Bond The amide linkage in a protein; a carbonyl-to-nitrogen bond.

Percent (%) A measure of concentration. *Vol/vol percent:* The number of volumes of solute in 100 volumes of solution. *Wt/wt percent:* The number of grams of solute in 100 g of the solution. *Wt/vol percent:* The number of grams of solute in 100 mL of the solution. *Milligram percent:* The number of milligrams of the solute in 100 mL of the solution.

Period A horizontal row in the periodic table.

Periodic Law Many properties of the elements are periodic functions of their atomic numbers.

Periodic Table A display of the elements that emphasizes the family relationships.

pH The negative power to which the base 10 must be raised to express the molar concentration of hydrogen ions in an aqueous solution.

$$[H^+] = 1 \times 10^{-pH}$$

or

$$-\log [H^+] = pH$$

Phenol Any organic compound whose molecules have an —OH group attached to a benzene ring.

Phenyl Group The benzene ring minus one H atom; C_6H_5—.

Phosphate Buffer Usually a mixture or a solution that contains dihydrogen phosphate ions ($H_2PO_4^-$) to neutralize OH^- and monohydrogen phosphate ions (HPO_4^{2-}) to neutralize H^+.

Phosphate Group Transfer Potential The relative ability of an organophosphate to transfer a phosphate group to some acceptor.

Phosphoglyceride A phospholipid such as a plasmalogen or a lecithin whose molecules include a glycerol unit.

Phospholipid Lipids such as the phosphoglycerides, the plasmalogens, and the sphingomyelins whose molecules include phosphate ester units.

Photon A package of energy released when an electron in an atom moves from a higher to a lower energy state; a unit of light energy.

Photosynthesis The synthesis in plants of complex compounds from carbon dioxide, water, and minerals with the aid of sunlight captured by the plant's green pigment, chlorophyll.

Physical Property Any observable characteristic of a substance other than a chemical property, such as color, density, melting point, boiling point, temperature, and quantity.

Physical Quantity A property of something to which we assign both a numerical value and a unit, such as mass, volume, or temperature; physical quantity = number × unit.

Physiological Saline Solution A solution of sodium chloride with an osmotic pressure equal to that of blood.

pI (See *Isoelectric Point.*)

Pi Bond (π Bond) A covalent bond formed when two electrons fill a molecular orbital created by the side-to-side overlap of two *p* orbitals.

Pi (π) Electrons The pair of electrons in a pi bond.

Plane-Polarized Light Light whose electrical field vibrations are all in the same plane.

Plasmalogens Glycerol-based phospholipids whose molecules also include an unsaturated fatty alcohol unit.

Plasmid A circular molecule of supercoiled DNA in a bacterial cell.

β-Pleated Sheet A secondary structure for a polypeptide in which the molecules are aligned side by side in a sheet-like array with the sheet partially pleated.

pOH The negative power to which the base 10 must be raised to express the concentration of hydroxide ions in an aqueous solution in mol/L.

$$[OH^-] = 1 \times 10^{-pOH}$$

At 25 °C,

$$pH + pOH = 14.00$$

Vascular Compartment The entire network of blood vessels and their contents.

Vasopressin A hypophysis hormone that acts at the kidneys to help regulate the concentrations of solutes in the blood by instructing the kidneys to retain water (if the blood is too concentrated) or to excrete water (if the blood is too dilute).

Virus One of a large number of substances that consist of nucleic acid (usually RNA) surrounded (usually) by a protein overcoat and that can enter host cells, multiply, and destroy the host.

Vital Force Theory A discarded theory that organic compounds could be made in the laboratory only if the chemicals possessed a vital force contributed by some living thing.

Vitamin An organic substance that must be in the diet; whose absence causes a deficiency disease; which is present in foods in trace concentrations; and that isn't a carbohydrate, lipid, protein, or amino acid.

Vitamin A Retinol; a fat-soluble vitamin in yellow-colored foods and needed to prevent night blindness and certain conditions of the mucous membranes.

Vitamin B_6 Pyridoxine, pyridoxal, or pyridoxamine; a vitamin needed to prevent hypochromic microcytic anemia and used in enzymes of amino acid catabolism.

Vitamin B_{12} Cobalamin; A vitamin needed to prevent pernicious anemia.

Vitamin C Ascorbic acid; a vitamin needed to prevent scurvy.

Vitamin D Cholecalciferol (D_3) or ergocalciferol (D_2); a fat-soluble vitamin needed to prevent rickets and to ensure the formation of healthy bones and teeth.

Vitamin Deficiency Diseases Diseases caused not by bacteria or viruses but by the absence of specific vitamins, such as pernicious anemia (B_{12}), hypochromic microcytic anemia (B_6), pellagra (niacin), the breakdown of certain tissues (riboflavin), megaloblastic anemia (folacin), beriberi (thiamin), scruvy (C), hemorrhagic disease (K), rickets (D), and night blindness (A).

Vitamin E A mixture of tocopherols; a fat-soluble vitamin apparently needed for protection against edema and anemia (in infants) and possibly against dystrophy, paralysis, and heart attacks.

Vitamin K The antihemorrhagic vitamin that serves as a cofactor in the formation of a blood clot.

Volatile Liquid A liquid that has a high vapor pressure and readily evaporates at room temperature.

Volume The capacity of an object to occupy space.

Water of Hydration Water molecules held in a hydrate in some definite mole ratio to the rest of the compound.

Wax A lipid whose molecules are esters of long-chain monohydric alcohols and long-chain fatty acids.

Weak Acid An acid with a low percentage ionization in solution and with a low value of acid ionization constant, K_a.

Weak Base A base with a low percentage ionization in solution.

Weak Brønsted Acid Any species, molecule or ion, that has a weak tendency to donate a proton and poorly serves as a proton donor.

Weak Brønsted Base Any species, molecule or ion, that weakly holds an accepted proton and poorly serves as a proton-acceptor.

Weak Electrolyte Any electrolyte that has a low percentage ionization in solution.

Weight The gravitational force of attraction on an object as compared to that of some reference.

Zymogen A polypeptide that is changed into an enzyme by the loss of a few amino acid residues or by some other change in its structure; a proenzyme.

Photo Credits

Chapter 1
Opener: U.S. Dept. of Education.
Page 4: © Joel Gordon 1979.
Page 5: (left) The Granger Collection; (center) Ann L. Stewart/Black Star; (right) Mettler Instrument Corp.
Figure 1.1: Courtesy National Bureau of Standards, Washington, D.C.
Page 8 (bottom): Peter Lerman.
Page 9 (top): Peter Lerman.
Figure 1.3: Peter Lerman.

Chapter 2
Opener: George Bellerose/Stock, Boston.
Page 30: Ben Asen.
Figure 2.1: Mimi Forsyth/Monkmeyer.
Page 32: Smith, A.L. *John Dalton 1766-1844; A Bibliography of Works by and About Him.* Manchester: The University Press, 1966. Courtesy AIP Niels Bohr Library.
Figure 2.4: Courtesy A.V. Crewe. From A.V. Crewe, T.B. Park, and J. Biggins. *Science.* © 1970 by the American Association for the Advancement of Science.
Page 37: Robert A. Lisak.
Page 38: Robert A. Lisak.

Chapter 3
Opener: Helmut Wimmer, The American Museum of Natural History—Hayden Planetarium.
Page 52: Peter Lerman.
Page 55: drawing by William Numeroff.
Page 56: Courtesy Nobel Foundation, Stockholm.
Page 64: New York Public Library Picture Collection.

Chapter 4
Opener: Russ Kinne/Photo Researchers.
Page 77: Smithsonian Institution.

Chapter 5
Opener: Mettler Instrument Corp.
Page 107: Courtesy AIP Niels Bohr Library.
Figure 5.2: J. Brady & K. Bendo, from J.E. Brady and J.R. Holum, *Fundamentals of Chemistry,* John Wiley & Sons, New York, 1981. Used by permission.
Figure 5.5: Peter Lerman.

Chapter 6
Opener: Grant Heilman.
Page 128: Courtesy AIP Niels Bohr Library.
Page 146: (top) J. Brady & K. Bendo, from J.E. Brady and G.E. Humiston, *General Chemistry,* John Wiley & Sons, New York, 1982.
Page 146: (bottom) U.S.D.A. photo by Fred Faurot.

Chapter 7
Opener: Georg/New York State Department of Environmental Conservation.
Figure 7.9: Photo by David Crouch. From T.R. Dickson, *Introduction to Chemistry,* 1971. John Wiley & Sons, Inc.

Chapter 8
Opener: © The Denver Post, 1985.
Page 188: American Institute of Physics.
Page 195: Peter Lerman.
Figure 8.6: K. Bendo.
Figure 8.7: Peter Lerman.
Page 215: The Permutit Company, Divison Sybron Corporation.
Figure 8.8: K. Bendo.

Chapter 9
Opener: Joe Munroe/Photo Researchers.
Figure 9.3: Courtesy Whatman Lab Sales, Inc. & the Orgeon Graduate Center.
Figure 9.5: Peter Lerman.

Chapter 10
Opener: Courtesy Brookhaven National Laboratory
Page 262: (left) Courtesy General Electric Company.
Page 262: (right) General Electric, Medical Systems Division, Milwaukee.

Chapter 11
Opener: Mark Antman/Stock, Boston.
Page 270: Peter Lerman.
Figure 11.1: Peter Lerman.
Page 274: Peter Lerman.

Chapter 12
Opener: Frederic Lewis/Lambert.
Page 288: K. Bendo & J. Brady.
Page 294: Peter Lerman.
Page 295: Peter Lerman.

Chapter 13
Opener: Courtesy AG-BAG, Astoria, Oregon.
Page 307: Peter Lerman.
Page 308: Peter Lerman.
Figure 13.2: K. Bendo.

Chapter 14
Opener: Laurence Pringle/From National Audubon Society/Photo Researchers.

Chapter 15
Opener: John Schultz/PAR, New York City.

Chapter 16
Opener: Christopher Brown/Stock, Boston.
Figure 16.2: Courtesy of the Department of Surgery, Baylor University College of Medicine.

Chapter 17
Opener: Craig Aurness/Woodfin Camp.

Chapter 18
Opener: Robert A. Lisak.
Page 437: K. Bendo.
Figure 18.7: Courtesy Polaroid Corporation.

Chapter 19
Opener: U.S.D.A.

Chapter 20
Opener: Townsend Godsey/Monkmeyer.

Chapter 21
Opener: J.C. Kendrew & H.C. Wilson, Medical Research Council, Cambridge.
Page 504: (left) Courtesy of Francois Morel; from *J. Cell Biology,* 48, 91–100, 1971; (right)
Courtesy Springer-Verlag Publishers, from *Corpuscels,* by Marcel Bessis.

Chapter 22
Opener: Doris Pinney/Monkmeyer.
Figure 22.8: Reprinted from *Psychology Today Magazine,* May 1967. Copyright © Ziff-Davis
Publishing Company. Photo by John Oldenkamp.

Chapter 23
Opener: Courtesy Monsanto.

Chapter 24
Opener: Courtesy Zeta-Meter, Inc.

Chapter 25
Opener: Bruce Roberts/Photo Researchers.

Chapter 26
Opener: Gerry Cranham/Photo Researchers.

Chapter 27
Opener: Janice Fullman/The Picture Cube.

Chapter 28
Opener: U.S. Dept. of Housing and Urban Development.

Chapter 29
Opener: Eric Hosking, F.R.P.S./Photo Researchers.

Index